Analytical Methods for Kolmogorov Equations

Second Edition

MONOGRAPHS AND RESEARCH NOTES IN MATHEMATICS

Published Titles

Analytical Methods for Kolmogorov Equations, Second Edition, Luca Lorenzi

Application of Fuzzy Logic to Social Choice Theory, John N. Mordeson, Davender S. Malik and Terry D. Clark

Blow-up Patterns for Higher-Order: Nonlinear Parabolic, Hyperbolic Dispersion and Schrödinger Equations, Victor A. Galaktionov, Enzo L. Mitidieri, and Stanislav Pohozaev

Complex Analysis: Conformal Inequalities and the Bieberbach Conjecture, Prem K. Kythe

Computational Aspects of Polynomial Identities: Volume I, Kemer's Theorems, 2nd Edition Alexei Kanel-Belov, Yakov Karasik, and Louis Halle Rowen

A Concise Introduction to Geometric Numerical Integration, Fernando Casas and Sergio Blanes

Cremona Groups and Icosahedron, Ivan Cheltsov and Constantin Shramov

Delay Differential Evolutions Subjected to Nonlocal Initial Conditions Monica-Dana Burlică, Mihai Necula, Daniela Roşu, and Ioan I. Vrabie

Diagram Genus, Generators, and Applications, Alexander Stoimenow

Difference Equations: Theory, Applications and Advanced Topics, Third Edition Ronald E. Mickens

Dictionary of Inequalities, Second Edition, Peter Bullen

Finite Element Methods for Eigenvalue Problems, Jiguang Sun and Aihui Zhou

Introduction to Abelian Model Structures and Gorenstein Homological Dimensions Marco A. Pérez

Iterative Optimization in Inverse Problems, Charles L. Byrne

Line Integral Methods for Conservative Problems, Luigi Brugnano and Felice Iavernaro

Lineability: The Search for Linearity in Mathematics, Richard M. Aron, Luis Bernal González, Daniel M. Pellegrino, and Juan B. Seoane Sepúlveda

Modeling and Inverse Problems in the Presence of Uncertainty, H. T. Banks, Shuhua Hu, and W. Clayton Thompson

Monomial Algebras, Second Edition, Rafael H. Villarreal

Nonlinear Functional Analysis in Banach Spaces and Banach Algebras: Fixed Point Theory Under Weak Topology for Nonlinear Operators and Block Operator Matrices with Applications, Aref Jeribi and Bilel Krichen

Partial Differential Equations with Variable Exponents: Variational Methods and Qualitative Analysis, Vicenţiu D. Rădulescu and Dušan D. Repovš

Published Titles Continued

A Practical Guide to Geometric Regulation for Distributed Parameter Systems
Eugenio Aulisa and David Gilliam

Reconstruction from Integral Data, Victor Palamodov

Signal Processing: A Mathematical Approach, Second Edition, Charles L. Byrne

Sinusoids: Theory and Technological Applications, Prem K. Kythe

Special Integrals of Gradshteyn and Ryzhik: the Proofs – Volume I, Victor H. Moll

Special Integrals of Gradshteyn and Ryzhik: the Proofs – Volume II, Victor H. Moll

Stochastic Cauchy Problems in Infinite Dimensions: Generalized and Regularized Solutions, Irina V. Melnikova

Submanifolds and Holonomy, Second Edition, Jürgen Berndt, Sergio Console, and Carlos Enrique Olmos

The Truth Value Algebra of Type-2 Fuzzy Sets: Order Convolutions of Functions on the Unit Interval, John Harding, Carol Walker, and Elbert Walker

Forthcoming Titles

Actions and Invariants of Algebraic Groups, Second Edition, Walter Ferrer Santos and Alvaro Rittatore

Geometric Modeling and Mesh Generation from Scanned Images, Yongjie Zhang

Groups, Designs, and Linear Algebra, Donald L. Kreher

Handbook of the Tutte Polynomial, Joanna Anthony Ellis-Monaghan and Iain Moffat

Microlocal Analysis on R^n and on NonCompact Manifolds, Sandro Coriasco

Practical Guide to Geometric Regulation for Distributed Parameter Systems, Eugenio Aulisa and David S. Gilliam

Symmetry and Quantum Mechanics, Scott Corry

Published Titles (Continued)

MONOGRAPHS AND RESEARCH NOTES IN MATHEMATICS

Analytical Methods for Kolmogorov Equations

Second Edition

Luca Lorenzi

University of Parma

Italy

CRC Press
Taylor & Francis Group
Boca Raton London New York

CRC Press is an imprint of the
Taylor & Francis Group, an **informa** business
A CHAPMAN & HALL BOOK

CRC Press
Taylor & Francis Group
6000 Broken Sound Parkway NW, Suite 300
Boca Raton, FL 33487-2742

Printed on acid-free paper
Version Date: 20160607

International Standard Book Number-13: 978-1-4822-4332-1 (Hardback)

Visit the Taylor & Francis Web site at
http://www.taylorandfrancis.com

and the CRC Press Web site at
http://www.crcpress.com

Printed and bound in the United States of America by Publishers Graphics,
LLC on sustainably sourced paper.

To my parents

Contents

Preface to the second edition

This is the second edition of the book "*Analytical Methods for Markov Semigroups*" which Marcello Bertoldi and I wrote in 2006.

When the staff at Taylor and Francis asked me to write a second edition of the monograph, I accepted with a great enthusiasm. I had many ideas of new arguments to add, due to large number of papers on Kolmogorov equations which have appeared since 2006. What I actually underestimated at the very beginning was the difficulty of combining and homogenizing the old material with the new to avoid useless repetitions and to simplify the presentation of the arguments as much as possible. This work took me long time to get to a satisfactory (at least from my point of view) result. I hope that the reader will appreciate the book and find it useful.

All the material in the first edition is still contained in this book and new chapters have been added. The already existing material has been re-arranged in the book which now is split into just three parts. The first one deals with the autonomous case when the coefficients of the elliptic operators depend only on the spatial variables. Two new chapters have been written (Chapters 7 and 8) which deal, respectively, with Markov semigroups in L^p-spaces of the Lebesgue measure and with the Green functions associated with such semigroups. Particular attention is paid to the relevant case of Schrödinger type operators. Chapter 9 and Chapter 11 (Chapters 7 and 13 in the first edition) have been considerably updated. More precisely, two new sections have been added to Chapter 9: the first one deals with the summability improving properties of Markov semigroups and the other one contains a class of elliptic operators with unbounded coefficients (more general than the one in the first edition of the book) which are analytic in the L^p-spaces of the invariant measure. On the other hand, in Chapter 11 we have considered a slightly more general class of degenerate elliptic operators than the ones in the first edition of the book.

The second part of the book (consisting of five chapters) is completely new, and deals with nonautonomous Kolmogorov equations in $\mathbb{R}^N$. Particular attention has been paid to the analysis of the associated evolution operator $\{G(t,s)\}$ in L^p-spaces with respect to the so-called evolution systems of measures (the natural counterpart of the invariant measure of the autonomous case) and to the asymptotic behaviour of the function $G(t,s)f$ (when f is bounded and continuous or when it belongs to the above L^p-spaces) as t tends to $+\infty$ (and in the periodic setting even as s tends to $-\infty$).

Finally, the last part of the book (the Appendices) contains (almost) all the classical tools that are used in the book. New material has been added with the aim of keeping the book as self-contained as possible.

Acknowledgments

I would like to express my gratitude to:

☞ my parents: if I have achieved my dream of being a mathematician, this is thanks to you!

- ☞ Marcello Bertoldi: he has not taken part in this new edition of the book but he was a precious co-author of the first edition;
- ☞ all my co-authors and friends: besides those already mentioned in the preface to the first edition of the book, I would like to mention Angela Albanese, Luciana Angiuli, Balint Farkas, Matthias Geissert, Matthias Hieber, Markus Kunze, Elisabetta Mangino, Jan Prüss, Abdelaziz Rhandi, Roland Schnaubelt, and Alessandro Zamboni. They shared their expertise with me, and from each of them I learned something new;
- ☞ Chiara Spina, who read a preliminary version of Chapters 7 and 8 and gave me useful comments and suggestions;
- ☞ the staff at Taylor and Francis who followed and helped me during the preparation of this book. In particular, I would like to thank Sarfraz Khan, for his huge patience with me during all these months when it seemed that this project would never come to an end(!), and Shashi Kumar, who solved all my (many) problems with LaTeX.

Parma, February 2016

Luca Lorenzi

Preface to the first edition

The main aim of this book is to study Markov semigroups, starting from the partial differential equations associated (parabolic and elliptic equations involving a second-order elliptic operator which has unbounded coefficients).

The theory of Markov processes and of related transition semigroups has numerous applications in many fields of science and economics. Historically, in the mathematical literature the subject is studied using several approaches, with ideas and methods from partial differential equations, Dirichlet forms, stochastic processes, stochastic differential equations, and Martingale theory.

Somehow the classical semigroup theory is unfit to study this particular class of equations, as well as the classical theory of elliptic differential operators. Indeed, the fact that the coefficients of the operator are unbounded is not merely a technical difficulty, but has significant consequences for the solutions (non-uniqueness of continuous bounded solutions, semigroup not strongly continuous, failure of regularity properties).

These facts lead us to look for specific techniques and results, and here we describe some of them.

The semigroup is studied in spaces of continuous functions and in L^p-spaces of the invariant measure, which is the stationary distribution of the Markov process and it exists under suitable assumptions.

In the first part of the book we study the general properties of the semigroup in spaces of continuous functions: the existence of solutions to the elliptic and to the parabolic equation, the uniqueness properties (and counterexamples to uniqueness), the definition and the properties of the weak generator, which is a specific notion that substitutes the infinitesimal generator of strongly continuous semigroups.

We see also some properties of the associated Markov process and the connection with the uniqueness of the solutions.

Then we focus on the proof of regularity results: global and pointwise estimates of the space derivatives of the semigroup. In particular, first we prove global estimates for the space derivatives and Schauder estimates, similar to those for operators with bounded coefficients. Then, we prove a number of pointwise estimates, which relate the derivatives of the semigroup at a point with the semigroup applied to the derivatives of the function, or to the function itself, at the same point. These estimates are truly characteristic of this class of operators and have interesting consequences (for instance in terms of Liouville theorems).

We make the same analysis for boundary value problems in unbounded domains, both with Dirichlet and Neumann boundary conditions, and for problems involving degenerate operators.

In the part about the invariant measure, we study some different approaches to the problem of the existence of the invariant measure, and we study the properties of the semigroup in L^p-spaces, including the asymptotic behaviour, the Poincaré inequality and the log-Sobolev inequality.

Also, we devote a chapter to the Ornstein-Uhlenbeck semigroup, the most studied example of an operator with unbounded coefficients.

Acknowledgments

The authors acknowledge the financial support of the research projects "Equazioni di evoluzione deterministiche e stocastiche" and "Equazioni di Kolmogorov" of the Italian Ministero dell'Istruzione, dell'Università e della Ricerca (M.I.U.R.), and the financial support of the European Community's Human Potential Programme under contract HPRN-CT-2002-00281 "Evolution Equations". They also wish to thank S. Fornaro, A. Lunardi, G. Metafune, D. Pallara, G. Tessitore and L. Tubaro for many useful discussions and suggestions during the preparation of this book.

The authors express their gratitude also to K. Payne for many useful suggestions concerning English language.

M. Bertoldi wishes to thank Ph. Clément and the members of the Department of Applied Mathematics of Delft University of Technology for the warm hospitality.

L. Lorenzi wishes to thank the Department of Mathematics of the University of Milan who allowed him usage of its rich library during the preparation of this book.

Finally, the authors thank the staff of Taylor and Francis for their help and useful suggestions they supplied during the preparation of this book.

Parma, February 2006

Marcello Bertoldi
Luca Lorenzi

Symbol Description

Sets

$\mathbb{N}$ set of all positive natural numbers
$\mathbb{Z}$ set of all relative integers
$\mathbb{R}$ set of all real numbers
$\mathbb{C}$ set of all complex numbers
$\mathbb{R}^N$ set of all real N-tuples
$\mathbb{C}^N$ set of all complex N-tuples
$A \Subset B$ given two subsets $A, B \subset \mathbb{R}^N$ with B open, it means that $\overline{A}$ is contained in B
$A \hookrightarrow B$ given two subsets A, B of a Banach space X, it means that A is continuously embedded in B
$\overline{A}$ the closure of A
∂A the boundary of A
$\mathcal{B}(\Omega)$ the σ algebra of of all the Borel subsets of Ω
$B_R(x)$ open ball in $\mathbb{R}^N$ with centre at x and radius $R > 0$
$\overline{B}_R(x)$ the closure of $B_R(x)$
B_R $:= B_R(0)$
$\overline{B}_R$ $:= \overline{B}_R(0)$
$E + F$ the algebraic sum of E and F
$E \oplus F$ the set $E + F$ when E and F are two subspaces of a vector space X and $E \cap F = \{0\}$
Δ_J $:= \{(t, s) \in J \times J : t \geq s\}$ for any $J \subset \mathbb{R}$
$L(X, Y)$ the set of all the bounded linear operators from X to Y
$L(X)$ $:= L(X, X)$
X' $:= L(X, \mathbb{K})$, where $\mathbb{K} = \mathbb{R}, \mathbb{C}$

Functions

δ_x the delta function centered at x
$\operatorname{div} f$ the divergence of $f : \Omega \subset \mathbb{R}^N \to \mathbb{R}^N$, i.e., $\operatorname{div} f = \sum_{i=1}^N D_i f$
$\|f\|_\infty$ the sup-norm of $f : \Omega \subset \mathbb{R}^N \to \mathbb{R}$
f' (or f_x) the first-order derivative of a function f of one variable. We use a similar notation for higher order derivatives
$f^{(k)}$ the k-th-order derivative of a function f of one variable
$D_{i_1,\ldots,i_r} f$ the derivative $\frac{\partial^r f}{\partial x_{i_1},\ldots,\partial x_{i_r}}$, where $i_1, \ldots, i_r \in \{1, \ldots, N\}$
$D^\alpha f$ the derivative $\frac{\partial^{|\alpha|} f}{\partial x^{\alpha_1},\ldots,\partial x^{\alpha_N}}$, where $\alpha = (\alpha_1, \ldots, \alpha_N)$
$D^k f$ (resp. $D_x^k f$) the vector consisting of all the k-th order (resp. spatial) derivatives of f
∇f (resp. $\nabla_x f$) $:= Df$ (resp. $D_x f$)
$|D^k f(x)|^2 := \sum_{|\alpha|=k} |D^\alpha f(x)|^2$ for $k \in \mathbb{N}$
$\|D^k f\|_{L^p(\Omega,\mu)}$ the L^p-norm of the function $|D^k f|$ with respect to the measure μ
$\|D^k f\|_\infty$ the sup norm of the function $|D^k f|$
$\operatorname{Jac} g$ the Jacobian matrix of the vector-valued function $g : \Omega \subset \mathbb{R}^N \to \mathbb{R}^m$
Γ the Euler-Gamma function
χ_A the characteristic function of the set A
$\mathbb{1}$ $:= \chi_{\mathbb{R}^N}$

Operators

$D(A)$ the domain of the operator A
$R(A)$ the range of the operator A
$r(A)$ the numerical range of the operator A
$\rho(A)$ resolvent set of the operator A
$\sigma(A)$ spectrum of the operator A
I (in a Banach space X) the identity operator
$[A, B]$ the commutator between the operators A and B, i.e., $[A, B] = AB - BA$ defined on $D(AB) \cap D(BA)$

Matrix and linear algebra

$\det B$ the determinant of the matrix B
$\operatorname{diag}(\lambda_1, \ldots, \lambda_N)$ the diagonal matrix whose entries on the main diagonal are $\lambda_1, \ldots, \lambda_N$
$e_j = e_j^{(N)}$ the j-th vector of the canonical basis of $\mathbb{R}^N$
I_r the identity matrix with r rows and r columns. When there is no damage of confusion we simply write I
$\lambda_{\min}(A)$ the minimum eigenvalue of the matrix A

$\lambda_{\max}(A)$	the maximum eigenvalue of the matrix A
$\operatorname{rank} A$	the maximum number of linearly independent columns of the matrix A
$\operatorname{span}(E)$	the set of finite linear combinations of elements of E
$\operatorname{Tr} Q$	the trace of the matrix Q
q_{ij}	the entries of the matrix Q
b_j	the entries of the vector b
$x \otimes x$	the $N \times N$-matrix whose entries are $(x \otimes x)_{ij} = x_i x_j$ for any $i, j = 1, \ldots, N$
$\langle x, y \rangle$	the Euclidean inner product between the vectors $x, y \in \mathbb{R}^N$
$x \cdot y$	$:= \langle x, y \rangle$
$\|Q\|_\infty^2$	$:= \sum_{i,j=1}^N q_{ij}^2$. If Q depends on $x \in \Omega$, $\|Q\|_\infty^2 = \sum_{i,j=1}^N \|q_{ij}\|_\infty^2$
$\|DQ\|_\infty^2$	$:= \sup_\Omega \sum_{i,j,h=1}^N \|D_h q_{ij}\|^2$

Miscellanea

$a \vee b$	the maximum between a and b
$a \wedge b$	the minimum between a and b
a^+	$:= a \vee 0$
a^-	$:= a \wedge 0$
$\|\alpha\|$	the length of the multi-index α
$\alpha!$	the factorial of the multi-index α, i.e., $\alpha! = \prod_{i=1}^N \alpha_i!$
$\deg p$	the degree of the polynomial $p : \mathbb{R}^N \to \mathbb{R}$
δ_{ij}	the Kronecker delta, i.e., $\delta_{ij} = 1$ if $i = j$ and $\delta_{ij} = 0$ otherwise
$\operatorname{dist}(x, F)$	the distance of the point x from F, i.e., $\operatorname{dist}(x, F) = \inf_{y \in F} \|x - y\|$
$\operatorname{dist}(F, G)$	the distance of the set F from the set G
dx	the Lebesgue measure in $\mathbb{R}^N$
$\operatorname{Re} \lambda$	the real part of the complex number λ
$\operatorname{Im} \lambda$	the imaginary part of the complex number λ
$\nu(x)$	the unit outward vector normal to $\partial\Omega$ at x
ω_N	the Lebesgue measure of the unit ball in $\mathbb{R}^N$

About the author

Luca Lorenzi obtained his Ph.D. degree in Mathematics from the University of Pisa (Italy) in 2001. In 2000 he got a permanent position as assistant professor in Mathematical Analysis at the University of Parma (Italy) and he was promoted to associate professor in 2006 still at the University of Parma. In 2013 he got the Italian national habilitation as full professor in Mathematical analysis. His mathematical interests include the analysis of partial differential equations of parabolic and elliptic type, mainly with unbounded coefficients, identification and ill-posed problems and the analysis of mathematical models in Combustion Theory. He has authored or co-authored over 50 papers published on international journals. He is very active as a reviewer for several mathematical journals.

Introduction

Elliptic operators with bounded coefficients have been widely studied in the literature both in $\mathbb{R}^N$ and in open subsets of $\mathbb{R}^N$, starting from the 1950s, and nowadays they are well understood.

The study of operators with unbounded coefficients is a generalization of this case and the interest in this class of operators has grown considerably due to their numerous applications in many fields of science and economics. The most famous example of an autonomous elliptic operator with unbounded coefficients in $\mathbb{R}^N$ ($N \geq 1$) is the Ornstein-Uhlenbeck operator, defined on smooth functions ψ by

$$\mathcal{A}\psi(x) = \frac{1}{2}\sum_{i,j=1}^{N} q_{ij} D_{ij}\psi(x) + \sum_{i,j=1}^{N} b_{ij} x_j D_i\psi(x), \qquad x \in \mathbb{R}^N, \tag{0.1}$$

where (q_{ij}) is a constant strictly positive definite matrix and (b_{ij}) is a constant real matrix. Such an operator displays all the main peculiarities of the autonomous elliptic operators with unbounded coefficients. For instance, the associated semigroup $\{T(t)\}$ in $BUC(\mathbb{R}^N)$ is neither strongly continuous nor analytic, whereas the semigroups associated with operators with bounded coefficients are typically analytic in $C_b(\mathbb{R}^N)$. Nevertheless, it has smoothing effects similar to those enjoyed by analytic semigroups associated with uniformly elliptic operators with bounded coefficients.

Starting from the pioneering papers [25, 130, 150, 151, 152, 153] the literature on elliptic operators with unbounded coefficients has spread out considerably and now we are able to treat uniformly elliptic operators of the type

$$\mathcal{A}(t)\psi(x) = \sum_{i,j=1}^{N} q_{ij}(t,x) D_{ij}\psi(x) + \sum_{i=1}^{N} b_i(t,x) D_i\psi(x) + c(t,x)\psi(t,x), \tag{0.2}$$

under rather weak assumptions on the coefficients, both with analytic and probabilistic methods.

The aim of this book is to present most of the old and recent results on the semigroups and evolution operators associated with Kolmogorov equations, using analytic methods. We mainly consider the case when the coefficients of the operator $\mathcal{A}$ are defined in $I \times \mathbb{R}^N$, I being a right halfline (possibly $I = \mathbb{R}$), but we also provide some results in the case when $\mathbb{R}^N$ is replaced by an open and unbounded domain.

The book is split into three parts.

Part I: the autonomous case

In the first part of the book we consider the autonomous case and we are concerned with the following topics:

$\ast$ Existence of the Markov semigroup associated in $C_b(\mathbb{R}^N)$ with the nondegenerate operator $\mathcal{A}$ in (0.2) with time independent coefficients. In the case when the coefficients of the operator $\mathcal{A}$ are bounded, the natural way to construct analytically such a semigroup consists of defining, for any $f \in C_b(\mathbb{R}^N)$ (the space of all bounded and continuous functions defined in $\mathbb{R}^N$) and $t > 0$, $T(t)f$ as the value at t of the classical solution to the Cauchy problem

$$\begin{cases} D_t u(t,x) = \mathcal{A}u(t,x), & t > 0, \quad x \in \mathbb{R}^N, \\ u(0,x) = f(x), & \qquad\quad x \in \mathbb{R}^N. \end{cases} \tag{0.3}$$

To generalize this procedure to the case when the coefficients are unbounded, we are led to prove existence results for the classical solution of the Cauchy problem (0.3) which is bounded in any strip $[0,T] \times \mathbb{R}^N$ $(T > 0)$, when $\mathcal{A}$ is given by (0.2). Here, by classical solution of (0.3), we mean a function u which is continuous in $[0,+\infty) \times \mathbb{R}^N$ and admits first-order time derivative and first- and second-order spatial derivatives, which are continuous in $(0,+\infty) \times \mathbb{R}^N$. This has been done, first, by S. Itô in [130] and, more recently, by R. Azencott in [25], under quite minimal regularity assumptions on the coefficients of the operator $\mathcal{A}$. The arguments used to prove the existence of a classical solution (which we denote by u_f) are very simple and elegant. They are based both on an approximation argument with Cauchy-Dirichlet problems in bounded and smooth domains and on classical Schauder estimates. Without additional assumptions, problem (0.3) may admit more than a unique classical solution which is bounded in $[0,T] \times \mathbb{R}^N$ for any $T > 0$. This is a typical feature of elliptic operators with unbounded coefficients. Nevertheless, u_f enjoys a nice property: when $f \geq 0$, u_f is the minimal positive solution to problem (0.3). This minimality property allows us to define the semigroup $\{T(t)\}$ by setting $T(t)f = u_f(t,\cdot)$ for any $t \geq 0$.

In general, such a semigroup is neither strongly continuous nor analytic in $C_b(\mathbb{R}^N)$. In fact, $T(t)f$ converges to f as t tends to 0, uniformly on compact subsets, but, in general, not uniformly in $\mathbb{R}^N$, and this happens even if f is uniformly continuous. Sufficient conditions on the coefficients of the operator $\mathcal{A}$ which imply that the semigroup $\{T(t)\}$ is (resp. is not) analytic in $C_b(\mathbb{R}^N)$ are known in the literature and here we describe some of them.

There is also a natural, purely probabilistic way to introduce the semigroup $\{T(t)\}$. It is known that, under suitable assumptions on the coefficients of the operator $\mathcal{A}$, there exists a Markov process X in $\mathbb{R}^N$ such that $\{T(t)\}$ is the transition semigroup of X, i.e.

$$(T(t)f)(x) = \mathbb{E}_x[\chi_{t<\tau} f(X_t)], \qquad t > 0, \; x \in \mathbb{R}^N, \; f \in C_b(\mathbb{R}^N), \tag{0.4}$$

where $\tau \in (0,+\infty]$ is the lifetime of X (X_t is defined for $t \in [0,\tau)$), and $\mathbb{E}_x$ is the expectation under the probability measure $\mathbb{P}_x$. Note that $X_0 = x$ $\mathbb{P}_x$-almost surely. The process X associated with $\{T(t)\}$ is equivalent, under the probability $\mathbb{P}_x$, to the solution ξ_t^x of the stochastic differential equation

$$d\xi_t^x = b(\xi_t^x)dt + \sigma(\xi_t^x)dW_t, \qquad \xi_0^x = x, \tag{0.5}$$

where W_t is a N-dimensional Brownian motion, $b : \mathbb{R}^N \to \mathbb{R}^N$ is the coefficient of $\mathcal{A}$ and $\sigma : \mathbb{R}^N \to L(\mathbb{R}^N)$ is such that $Q(x) = \frac{1}{2}\sigma(x)\sigma^*(x)$, where $Q(x)$ is the matrix with elements $q_{ij}(x)$ appearing in the formula (0.1).

Conversely, the stochastic equation (0.5) can be the starting point for the construction of $\{T(t)\}$: if ξ_t^x is a solution of equation (0.5) and $\{T(t)\}$ is defined by (0.4) with $X_t = \xi_t^x$, then the function $u = T(\cdot)f$ is a solution of the parabolic problem (0.3). This follows using results from the stochastic calculus (the Itô formula is the main tool).

The equation (0.5) can be considered a random perturbation of the ordinary differential equation

$$\frac{d}{dt}\xi_t^x = b(\xi_t^x).$$

Of course, here it is not natural to assume that the function b is bounded; in almost all the significant cases b is unbounded. This is an important reason for studying operators with unbounded coefficients.

Note that, according to the formula (0.4), the properties of $\{T(t)\}$ may be deduced from the study of the process X. This probabilistic approach has been widely used in the literature (see e.g. [55, 92, 97, 115, 129, 138, 145, 245]).

Under the same smoothness assumptions on the coefficients as above and using similar approximation arguments by Dirichlet problems in bounded domains, the existence of a solution $v_f \in D_{\max}(\mathcal{A})$ to the elliptic equation

$$\lambda v - \mathcal{A}v = f, \qquad \lambda > c_0 := \sup_{x\in\mathbb{R}^N} c(x), \tag{0.6}$$

can be proved for any $f \in C_b(\mathbb{R}^N)$. Here,

$$D_{\max}(\mathcal{A}) = \Big\{u \in C_b(\mathbb{R}^N) \cap \bigcap_{1\le p<+\infty} W^{2,p}_{\rm loc}(\mathbb{R}^N) : \mathcal{A}u \in C_b(\mathbb{R}^N)\Big\}.$$

As in the parabolic case, problem (0.6) may admit several solutions in $D_{\max}(\mathcal{A})$. In any case v_f is the minimal positive solution when $f \ge 0$.

As in the case when the coefficients of the operator $\mathcal{A}$ are bounded, there is a connection between the functions u_f and v_f. In fact, for any $\lambda > c_0$, v_f is the Laplace transform of u_f, in the sense that

$$v_f(x) = \int_0^{+\infty} e^{-\lambda t} u_f(t,x)dt, \qquad x \in \mathbb{R}^N. \tag{0.7}$$

The main difference with the classical case is that now, in general, the integral term in the right-hand side of (0.7) does not converge in $C_b(\mathbb{R}^N)$.

Since, in general, the semigroup $\{T(t)\}$ is neither strongly continuous nor analytic, then the infinitesimal generator does not exist in the classical sense. This gap is filled introducing the concept of a "weak generator". In fact, we can define a family $\{R(\lambda),\ \lambda > c_0\}$ of bounded operators in $C_b(\mathbb{R}^N)$ by setting $R(\lambda)f = v_f$. Such a family is the resolvent family associated with some (closed) operator $\widehat{A}$. We call $\widehat{A}$ the "weak generator" of the semigroup. The reason for this name is due to the fact that $\widehat{A}$ is a generalization of the classical concept of the infinitesimal generator of a strongly continuous semigroup. Indeed, the weak generator $\widehat{A}$ can be defined, in a equivalent way, using the bounded pointwise convergence: a function f belongs to $D(\widehat{A})$ if and only if

$$\frac{|(T(t)f)(x) - f(x)|}{t} \le M, \qquad t \in (0,1),\ x \in \mathbb{R}^N,$$

for some positive constant M, and $(T(t)f - f)/t$ converges pointwise to some bounded and continuous function g as t tends to 0^+. In such a case $\widehat{A}f = g$.

In general, $D(\widehat{A})$ is properly contained in $D_{\max}(\mathcal{A})$ and $\widehat{A}$ coincides with $\mathcal{A}$ on $D(\widehat{A})$. The previous set inclusion is actually a set equality if and only if the elliptic equation (0.6) is uniquely solvable in $D_{\max}(\mathcal{A})$.

As in the classical case of bounded coefficients, it is possible to associate a transition family $\{p(t,x;dy) : t > 0,\ x \in \mathbb{R}^N\}$ (or equivalently a Green function g) with the semigroup, for which the following representations hold:

$$(T(t)f)(x) = \int_{\mathbb{R}^N} f(y)p(t,x;dy) = \int_{\mathbb{R}^N} g(t,x,y)f(y)dy, \qquad t > 0,\ x \in \mathbb{R}^N. \tag{0.8}$$

Formula (0.8) is a keystone for proving some interesting and very useful continuity properties of the semigroup, as well as for showing that the semigroup can be extended to the space $B_b(\mathbb{R}^N)$ of bounded Borel functions. Such a semigroup, still denoted by $\{T(t)\}$, is both irreducible and strong Feller.

All the previous results can be obtained without assuming the uniqueness of the solutions to problems (0.3) and (0.6) with the claimed regularity properties. Of course, it is natural to investigate what conditions imply the uniqueness of the classical solution to the Cauchy problem (0.3), which is bounded in each strip $[0,T]\times\mathbb{R}^N$, and of the solution to the elliptic equation (0.6) in $D_{\max}(\mathcal{A})$. The two problems are not independent of one another. In fact, there exists a unique solution to the elliptic equation in $D_{\max}(\mathcal{A})$ if and only if there exists a unique classical solution to problem (0.3), which is bounded in $[0,T]\times\mathbb{R}^N$ for any $T>0$.

In the case when the coefficients are bounded, the uniqueness results are straightforward consequences of the classical maximum principles. If the coefficients of the operator $\mathcal{A}$ are unbounded, then the classical maximum principles may fail. This is the reason why, in general, the elliptic equation and the parabolic Cauchy-Dirichlet problem admit more than one solution. Hence, to prove uniqueness results some additional assumptions on the operator $\mathcal{A}$ need to be imposed. The typical assumptions which we assume are those that allow us to prove a generalized maximum principle: the existence of a so-called Lyapunov function φ, i.e. a sufficiently smooth function φ such that

$$(i)\ \lim_{|x|\to+\infty}\varphi(x)=+\infty \quad\text{and}\quad (ii)\ \sup_{x\in\mathbb{R}^N}\mathcal{A}\varphi(x)-\lambda\varphi(x)<+\infty \tag{0.9}$$

for some $\lambda > c_0$. At a first glance, the condition (0.9)(ii) may seem a condition related merely to the growth rate of the coefficients at infinity, but actually this is not the case. Indeed, the differential operators $\mathcal{A}_-$ and $\mathcal{A}_+$, defined on smooth functions $\psi:\mathbb{R}^N\to\mathbb{R}$ by

$$\mathcal{A}_-\psi(x)=\Delta\psi(x)-|x|^2\langle x,\nabla\psi(x)\rangle, \qquad x\in\mathbb{R}^N$$

and

$$\mathcal{A}_+\psi(x)=\Delta\psi(x)+|x|^2\langle x,\nabla\psi(x)\rangle, \qquad x\in\mathbb{R}^N,$$

differ just in the sign of the drift term, but this difference is essential. Indeed, the Cauchy problem associated with the first operator admits a unique classical solution, while the Cauchy problem associated with the second operator admits several classical solutions.

In the case when $c\equiv 0$, both the functions $\mathbb{1}$ (i.e., the constant function identically equal to 1) and $T(\cdot)\mathbb{1}$ solve the Cauchy problem (0.3). Hence, $T(t)\mathbb{1}\equiv\mathbb{1}$ for any $t>0$ is a necessary condition for problem (0.3) to admit a unique classical solution. Actually, such a condition is also sufficient. When this condition is satisfied we say that the semigroup is conservative or that the conservation of the probability holds. The reason for this nomenclature is based on the fact that, in this case, the family of measures $\{p(t,x;dy):\ t>0,\ x\in\mathbb{R}^N\}$ consists of probability measures.

There are also some conditions under which problem (0.3) admits several classical solutions. For instance, this is the case when there exists a nonnegative bounded and smooth function ψ such that $\lambda\psi-\mathcal{A}\psi\le 0$ for a suitable $\lambda\in\mathbb{R}$.

The one-dimensional case is easier and has been studied by W. Feller (see [102]). In such a setting there are necessary and sufficient conditions which give uniqueness of the solution to the elliptic problem (and, consequently, to the parabolic problem) and can be written in terms of the integrability at infinity of some functions which are strictly related to the coefficients of the operator $\mathcal{A}$.

The arguments that we have briefly described here are discussed in detail in Chapters 1, 2 and 3.

$*$ Study of the main properties of the semigroup $\{T(t)\}$ in the space $C_b(\mathbb{R}^N)$. We discuss two main topics in this book. The first one is the compactness of the semigroup in $C_b(\mathbb{R}^N)$ which has been studied mainly by E.B. Davies in [82] and by G. Metafune, D. Pallara, and M. Wacker in [196].

In the conservative case the semigroup is compact if and only if for any $t > 0$ the family of measures $\{p(t,x;dy) : x \in \mathbb{R}^N\}$ is tight. This means that for any $t > 0$ the measures $p(t,x;dy)$ are ε-concentrated in a compact set, uniformly with respect to $x \in \mathbb{R}^N$. This, for instance, is the case when there exist a nonnegative function $\varphi \in C^2(\mathbb{R}^N)$ and a convex function $g \in C^1([0,+\infty))$ such that

$$\lim_{|x|\to\infty} \varphi(x) = +\infty, \quad 1/g \text{ is integrable at } +\infty, \quad \mathcal{A}\varphi(x) \le -g(\varphi(x)), \quad x \in \mathbb{R}^N.$$

In the case when the family of measures $\{p(t,x;dy) : x \in \mathbb{R}^N\}$ is tight for any $t > 0$, the semigroup, besides being compact, enjoys another important property: it is norm-continuous in $(0,+\infty)$. This means that the function $t \mapsto T(t)$ is continuous with respect to the operator topology at any point $t > 0$.

In the nonconservative case the compactness of the semigroup can be read, roughly speaking, in terms of some regularity properties of the function $T(t)\mathbb{1}$. Indeed, $T(t)\mathbb{1}$ belongs to $C_0(\mathbb{R}^N)$ (the space of all continuous functions defined in $\mathbb{R}^N$ which vanish at infinity), for any $t > 0$, if and only if the semigroup is compact and maps $C_0(\mathbb{R}^N)$ into itself. In such a case, as in the conservative one, the semigroup is norm-continuous and $R(\lambda, \widehat{A})$ is compact as well.

Even though the semigroup $\{T(t)\}$ may fail to be strongly continuous in $C_b(\mathbb{R}^N)$, the function $t \mapsto T(t)f$ is continuous up to $t = 0$, with respect to the sup-norm, for any $f \in C_0(\mathbb{R}^N)$. Therefore, if the semigroup maps $C_0(\mathbb{R}^N)$ into itself, then its restriction to $C_0(\mathbb{R}^N)$ gives rise to a strongly continuous semigroup. So, it is important to determine suitable conditions on the coefficients of the operator $\mathcal{A}$ which imply that $C_0(\mathbb{R}^N)$ is invariant under the action of the semigroup. Note that, in general, $\{T(t)\}$ does not map $C_0(\mathbb{R}^N)$ into itself. This is the case, for instance, of the semigroup associated with the operator $\mathcal{A}_-$ defined above.

All the arguments that we have described here are discussed in Chapter 4.

$*$ Study of the smoothing effects of $\{T(t)\}$. We pay attention to the estimates of the spatial derivatives of $T(t)f$ when $f \in C_b(\mathbb{R}^N)$ or it is smooth. These estimates can be split into two main families: uniform (in the spatial variables) and pointwise estimates. Starting from the pioneering papers on the Ornstein-Uhlenbeck semigroup it has become clear that, under suitable assumptions on the growth rate at infinity of the coefficients of the operator $\mathcal{A}$, the spatial derivatives of $T(t)f$ should behave as in the case of the bounded coefficients. Therefore, under reasonable assumptions on the coefficients, one can expect that

$$\|D_x^k T(t)f\|_\infty \le C t^{-\frac{k-h}{2}} e^{\omega t} \|f\|_{C_b^h(\mathbb{R}^N)}, \qquad t > 0, \tag{0.10}$$

for any $0 \le h \le k$, any $f \in C_b^h(\mathbb{R}^N)$ (the space of all continuously differentiable up to the $[h]$-order functions $f : \mathbb{R}^N \to \mathbb{R}$, which are bounded together with their derivatives and have $[h]$-order derivatives which are $(h\text{-}[h])$-Hölder continuous in $\mathbb{R}^N$) and some positive constants C and ω.

We focus our attention on the case when $k \le 3$. Indeed, in this case, estimates (0.10) are a fundamental tool in order to prove existence results and Schauder estimates for the nonhomogeneous Cauchy problem and the nonhomogeneous elliptic equation associated with the operator $\mathcal{A}$.

Estimate (0.10) has been proved in [34, 182] with analytic methods and in [54, 55] with probabilistic methods. Here, we describe the results of [34]. The choice of focusing attention on the case when $c \equiv 0$ is due to the fact that such an assumption is necessary for the validity of some of the pointwise estimates that we present.

The method used to prove estimate (0.10) is very easy and at the same time very elegant. It consists of adapting to our situation the classical Bernstein method. To make such a method work, one needs to assume stronger assumptions than those needed to construct the semigroup. In particular, some dissipativity conditions, i.e., some bound from above on the inner product $\langle (\mathrm{Jac}_x b)\xi, \xi\rangle$ is assumed to prove (0.10). Note that the assumptions that we assume are not merely technical. Indeed, without any dissipativity assumption, estimates (0.10) may fail to hold even in the one-dimensional case.

As it has been claimed, the previous estimates allow us to prove optimal Schauder estimates for both the nonhomogeneous Cauchy problem and the nonhomogeneous elliptic equation associated with the operator $\mathcal{A}$. This is obtained by an interpolation argument under the same assumptions on f and g as in the case of bounded coefficients. As in this latter case, the solution to the nonhomogeneous Cauchy problem

$$\begin{cases} D_t u(t,x) = \mathcal{A}u(t,x) + g(t,x), & t>0, \quad x \in \mathbb{R}^N, \\ u(0,x) = f(x), & \qquad\quad x \in \mathbb{R}^N \end{cases}$$

can be represented by means of the usual variation-of-constants formula

$$u(t,x) = (T(t)f)(x) + \int_0^t (T(t-s)g(t,\cdot))(x)ds, \qquad t \geq 0, \ x \in \mathbb{R}^N, \tag{0.11}$$

where now the integral term, in general, does not converge in $C_b(\mathbb{R}^N)$.

The optimal Schauder estimates may also be used to prove that $D(\widehat{A})$ is continuously embedded in $C_b^{1+\beta}(\mathbb{R}^N)$ for any $\beta \in (0,1)$, thus providing a partial characterization of $D(\widehat{A})$.

The latter type of estimates that we discuss in this book are pointwise estimates. There are two types of pointwise estimates. The first ones are of the following type:

$$|(D_x^k T(t)f)(x)|^p \leq M_{k,p}\big(T(t)(|f|^2 + \ldots + |D^k f|^2)^{\frac{p}{2}}\big)(x), \tag{0.12}$$

for any $t>0$, $x \in \mathbb{R}^N$, $f \in C_b^k(\mathbb{R}^N)$ $(k = 1,2,3)$, $p \in (1,+\infty)$ and some positive constant $M_{k,p}$. They can be proved under the same hypotheses used to get the uniform estimates. Under somewhat heavier assumptions on the coefficients of the operator $\mathcal{A}$, estimate (0.12) can be improved, eliminating the dependence on f^2 from its right-hand side. In particular, it can be shown that

$$|\nabla_x T(t)f|^p \leq e^{\sigma_p t} T(t)|\nabla f|^p, \qquad t>0, \tag{0.13}$$

for any $p>1$. Note that for (0.13) to hold it is necessary that $T(t)\mathbb{1} \equiv \mathbb{1}$ for any $t>0$ and, hence, $c \equiv 0$. Such more restrictive assumptions cover, for instance, some cases when the coefficients have polynomial growth rate at infinity.

The second type of estimate allows us to give an upper bound of the left-hand side of (0.12) for any $t>0$ and $x \in \mathbb{R}^N$ in terms of $\zeta(t)\big(T(t)(|f|^2+\ldots+|D^{k-1}f|^2)^{\frac{p}{2}}\big)(x)$, ζ being a positive function which behaves like $t^{-kp/2}$ near $t=0$ and it is bounded or decreases exponentially at infinity. Iterating the arguments and taking the semigroup property into account, we can derive the estimates

$$|(D_x^k T(t)f)(x)|^p \leq C_{k,p} t^{-\frac{pk}{2}} \zeta_{k,p}(t)(T(t)|f|^p)(x), \qquad t>0, \ x \in \mathbb{R}^N, \tag{0.14}$$

for any $k = 1,2,3$, where the function $\zeta_{k,p}$ is bounded at 0 and it behaves like $t^{\frac{pk}{2}}$ or it decreases exponentially to 0 at $+\infty$. This latter estimate improves the uniform estimates (0.10), since now $\omega = \omega_{k,p}$ may be a negative constant. Consequently, estimates (0.14) allow for a better asymptotic analysis of the semigroup at infinity.

In general, estimates (0.12) can not be extended to the case $p = 1$, as a well-known counterexample by F.-Y. Wang shows. However, in the particular case when $\mathcal{A} = \Delta + \sum_{i=1}^N b_i(\cdot)D_i$ and the b_i's satisfy suitable growth conditions at infinity, they can be extended also to the case $p = 1$. The importance of obtaining such estimates with $p = 1$ will be clarified later on in this introduction. On the other hand, we stress that estimates (0.14) with $p = 1$ may fail also in the case when the coefficients of the operator $\mathcal{A}$ are bounded. A very easy counterexample is given by the Gauss-Weierstrass semigroup.

Liouville type theorems are an important consequence of the previous gradient estimates. Indeed, in the case when $\omega_{1,p} < 0$ in (0.14), one can show that if $u \in D_{\max}(\mathcal{A})$ satisfies $\mathcal{A}u = 0$, then u is constant. But this is not the only application of these estimates. Other important applications will be seen later on in this introduction.

All the arguments discussed here are contained in Chapters 5 and 6.

∗ **Study of the semigroup in L^p-spaces related to the Lebesgue measure.** As a matter of fact, in general elliptic operators with unbounded coefficients do not exhibit a nice behaviour in the L^p-spaces related to the Lebesgue measure. This is for instance the case of the one-dimensional elliptic operator $\mathcal{A}$, defined on smooth functions $\psi : \mathbb{R}^N \to \mathbb{R}$ by $\mathcal{A}\psi = \psi'' + b\psi'$, where $b(x) = -x|x|^\varepsilon$ for any $x \subset \mathbb{R}$ and ε is any positive real number. For any $\lambda > 0$, any $p \in [1,+\infty)$ and any not identically vanishing function $f \in C_c^\infty(\mathbb{R})$, the equation $\lambda u - \mathcal{A}u = f$ admits no solution $u \in L^p(\mathbb{R}^N)$. This shows that the L^p-spaces related to the Lebesgue measure are not the L^p-spaces which fit best the properties of elliptic operators with unbounded coefficients.

Anyway, sufficient conditions (which are both algebraic and growth assumptions on the coefficients of $\mathcal{A}$) can be provided which guarantee that the restriction to $C_c(\mathbb{R}^N)$ of the minimal semigroup $\{T(t)\}$ associated with the operator $\mathcal{A} = \mathrm{Tr}(QD^2) + \langle b, \nabla\rangle + c$ can be extended to $L^p(\mathbb{R}^N)$ with a strongly continuous semigroup $\{T_p(t)\}$ for $p \in (1,+\infty)$ and, in some cases, also for $p = 1$. In general, the characterization of the domain of the infinitesimal generator of the semigroup $\{T_p(t)\}$ is not known. There are some situations in which the domain can be characterized as the set of all functions $u \in W^{2,p}(\mathbb{R}^N)$ such that $cu \in L^p(\mathbb{R}^N)$. For instance this is the case when $\mathcal{A}$ is in divergence form, i.e., $\mathcal{A} = \mathrm{div}(Q\nabla) + \langle b, \nabla\rangle + c$ and suitable conditions on its coefficients are prescribed.

Schrödinger type operators have a particular relevance among the elliptic operators with unbounded coefficients. They are elliptic operators $\mathcal{A}$, defined on smooth functions $\psi : \mathbb{R}^N \to \mathbb{R}$ by $\mathcal{A}\psi(x) = q(x)\Delta\psi(x) + c(x)\psi(x)$ for any $x \in \mathbb{R}^N$, where $q(x) = 1 + |x|^\alpha$, $c(x) = -|x|^\beta$ for any $x \in \mathbb{R}^N$ and $\alpha, \beta \geq 0$ with $\alpha^2 + \beta^2 \neq 0$. When $\alpha \in [0,2]$, $\{T_p(t)\}$ is a strongly continuous analytic semigroup in $L^p(\mathbb{R}^N)$ and the domain of its infinitesimal generator A_p is the set $D(A_p) = \{u \in W^{2,p}(\mathbb{R}^N) : q|D^2u|, \sqrt{q}|\nabla u|, cu \in L^p(\mathbb{R}^N)\}$.

The picture changes drastically when $\alpha > 2$ and $\beta = 0$, since now the generation results depend upon N. More specifically, if $N = 1,2$ no realization of the operator $\mathcal{A}$ in $L^p(\mathbb{R}^N)$ generates a strongly continuous (resp. analytic) semigroup. The same happens if $N \geq 3$ and $p \leq N/(N-2)$. On the other hand, if $N \geq 3$, $p > N/(N-2)$, then the maximal realization A_p of the operator $\mathcal{A}$ in $L^p(\mathbb{R}^N)$ generates a positive strongly continuous analytic semigroup of contractions.

∗ **Estimates of the Green function of the semigroup $\{T(t)\}$.** Formula (0.8) shows that it is possible to associate a Green function g with the semigroup $\{T(t)\}$ as in the classical case when the coefficients are (sufficiently smooth and) bounded. In this latter case it is known that the Green function g satisfies the estimate

$$g(t,x,y) \le C_1 t^{-\frac{N}{2}} e^{\delta t} e^{-C_2 \frac{|x-y|^2}{t}}, \qquad t>0,\ x,y \in \mathbb{R}^N,$$

for some positive constants C_1, C_2 and $\delta \in \mathbb{R}$. In this book we present some of the results available in the literature in the case of nondegenerate elliptic operators $\mathcal{A}$ with unbounded coefficients, paying particular attention to the case of Schrödinger type operators. We illustrate two techniques used in the literature to prove estimates for the Green function g. The first one is inspired by the arguments in the book by E.B. Davies [84] and relies on the use of form method and the well-known Nash inequality. On the other hand, the second technique is a more typical PDE's technique and it is based on sharp estimates in suitable Sobolev spaces (the so-called Krylov spaces) of solutions to parabolic equations associated with the adjoint operator $\mathcal{A}^*$. This latter technique makes strong use of Lyapunov functions and provides an upper bound for the function ωg, ω being a suitable smooth functions. Proper choices of the function ω allow to prove precise estimates of the Green function g.

We consider both the general case when the operator $\mathcal{A}$ is in divergence form i.e., $\mathcal{A} = \mathrm{div}(Q\nabla) + \langle b, \nabla \rangle + c$ and the case when $\mathcal{A} = (1+|x|^\alpha)\Delta + c$, assuming that at least $\alpha > 0$ or c is negative and unbounded.

∗ Study of both the invariant measures associated with $\{T(t)\}$ and of its regularity properties in L^p-spaces related to this measure. By definition, an invariant measure is a probability measure μ such that

$$\int_{\mathbb{R}^N} T(t)f\,d\mu = \int_{\mathbb{R}^N} f\,d\mu, \qquad t>0,\ f \in B_b(\mathbb{R}^N).$$

The spaces $L^p(\mathbb{R}^N,\mu)$ are the "right" L^p-spaces for $\{T(t)\}$. Indeed, if an invariant measure exists, then $\{T(t)\}$ maps $L^p(\mathbb{R}^N,\mu)$ into itself. As it has already been noticed, this property fails for the L^p-spaces related to other measures, even in the case of the Lebesgue measure. In $L^p(\mathbb{R}^N,\mu)$, $\{T(t)\}$ is a strongly continuous semigroup for any $p \in [1,+\infty)$. Note that, if $c \equiv 0$ and an invariant measure exists, then the semigroup is conservative. So, from the probabilistic point of view, an invariant measure is a stationary distribution for the Markov process X.

When it exists, the invariant measure of $\{T(t)\}$ is unique and is absolutely continuous with respect to the Lebesgue measure. Moreover, its density is a positive function (not necessarily bounded). In general, an explicit expression of the density of μ is not available. In any case, in some situations (e.g. in the case when $\mathcal{A}$ is in divergence form), under suitable integrability and smoothness assumptions on the coefficients, one can prove global L^p and Sobolev regularity properties of the density, which is locally Hölder continuous.

So, the main problem consists of determining suitable conditions ensuring the existence of the invariant measure. The main result is the Khas'minskii theorem. It states that if there exists a regular function φ such that

$$\varphi \ge 0, \qquad \lim_{|x|\to+\infty} \mathcal{A}\varphi(x) = -\infty,$$

then the invariant measure of $\{T(t)\}$ exists.

Let L_p be the infinitesimal generator of $\{T(t)\}$ in $L^p(\mathbb{R}^N,\mu)$. There are only a few situations in which a complete characterization of $D(L_p)$ is available. This is, for instance, the case of the Ornstein-Uhlenbeck semigroup. In any case, one can prove that $D_{\max}(\mathcal{A})$ is always a core of $D(L_p)$. This means that $D_{\max}(\mathcal{A})$ is dense in $D(L_p)$ which is endowed with the graph norm. Moreover, if the pointwise estimate (0.14) holds, then one can partially characterize the domain of L_p, showing that it is continuously embedded in $W^{1,p}(\mathbb{R}^N,\mu)$ (the set of all the functions whose first-order distributional derivatives are in $L^p(\mathbb{R}^N,\mu)$).

A description of the asymptotic behaviour of the semigroup is available. In fact, for any $f \in L^p(\mathbb{R}^N, \mu)$ ($p \in (1, +\infty)$) it holds that

$$\lim_{t \to +\infty} \int_{\mathbb{R}^N} |T(t)f - \overline{f}|^p d\mu = 0, \tag{0.15}$$

where $\overline{f}$ is the projection of f on the subspace of the constant functions. This subspace is the eigenspace corresponding to the eigenvalue $\lambda = 0$ of the operator $\mathcal{A}$.

If there exists a *spectral gap*, i.e., if $\sigma(L_2) \setminus \{0\} \subset \{\mathrm{Re}\,\lambda \le -\delta\}$ for some $\delta > 0$, then the convergence in (0.15) is of exponential type. This is the case, for instance, when the Poincaré inequality

$$\int_{\mathbb{R}^N} |f - \overline{f}|^2 d\mu \le C \int_{\mathbb{R}^N} |\nabla f|^2 d\mu, \qquad f \in W^{1,2}(\mathbb{R}^N, \mu),$$

is satisfied. A sufficient condition for the validity of the Poincaré inequality is the compactness of the embedding $W^{1,2}(\mathbb{R}^N, \mu) \subset L^2(\mathbb{R}^N, \mu)$. In turn, this is implied by the validity of the log-Sobolev inequality:

$$\int_{\mathbb{R}^N} f^2 \log(|f|) d\mu \le \|f\|^2_{L^2(\mathbb{R}^N,\mu)} \log(\|f\|_{L^2(\mathbb{R}^N,\mu)}) + C \int_{\mathbb{R}^N} |\nabla f|^2 d\mu, \qquad f \in W^{1,2}(\mathbb{R}^N, \mu), \tag{0.16}$$

for some positive constant C. For instance, this happens whenever the gradient estimate (0.13), with $p = 1$ and $\sigma_1 = -\alpha < 0$, hold. Here, we see the importance of the pointwise gradient estimates with $p = 1$ and exponentials of negative type.

The logarithmic-Sobolev inequality (0.16) implies that $f^2 \log|f|$ is integrable whenever $f \in W^{1,2}(\mathbb{R}^N, \mu)$. This Sobolev-type result is very precise, as an example of L. Gross shows: if μ is the Gaussian measure, then there exists a function $f \in W^{1,2}(\mathbb{R}^N, \mu)$ such that $f^2(\log f)(\log\log f)$ is not integrable. In particular, one can not expect any Sobolev embedding, such as $W^{1,2}(\mathbb{R}^N, \mu) \subset L^p(\mathbb{R}^N, \mu)$, when $p > 2$.

The arguments described here are discussed in detail in Chapter 9.

* **The Ornstein-Uhlenbeck operator.** Due to its importance, we devote Chapter 10 to describing most of the significant results on the Ornstein-Uhlenbeck semigroup both in $C_b(\mathbb{R}^N)$ and in the L^p-spaces related to its invariant measure and to the Lebesgue measure. We mainly focus on the nondegenerate Ornstein-Uhlenbeck operator, but some of the results that we describe hold also in the degenerate case.

* **Degenerate autonomous elliptic operators.** We consider degenerate elliptic operators $\mathcal{A}$, defined on smooth functions $\psi : \mathbb{R}^N \to \mathbb{R}$ by

$$\mathcal{A}\psi(x) = \sum_{i,j=1}^{p_0} q_{ij}(x) D_{ij}\psi(x) + \sum_{i,j=1}^{N} b_{ij} x_j D_i\psi(x) + \sum_{j=1}^{p_0} F_j(x) D_j\psi(x) \quad x \in \mathbb{R}^N, \tag{0.17}$$

for some $p_0 < N$, where the matrix function $B = (b_{ij})$ can be split into blocks as follows

$$B = \begin{pmatrix} \star & \star & \cdots & \cdots & \star \\ B_1 & \star & \cdots & \cdots & \star \\ 0 & B_2 & \star & \cdots & \star \\ \vdots & \vdots & \ddots & \ddots & \vdots \\ 0 & 0 & 0 & B_r & \star \end{pmatrix},$$

B_h being a $p_h \times p_{h-1}$ matrix with full rank ($h = 1, \ldots, r$), $r \in \mathbb{N}$ and the integers $p_0, \ldots, p_r$

satisfy the condition $p_0 \geq p_1 > \ldots \geq p_r$. Finally, the symbol "$\star$" denotes suitable matrices, the matrix $Q(x) = (q_{ij}(x))$ is strictly positive definite at any $x \in \mathbb{R}^N$, with coefficients which may grow at most quadratically at infinity, and F is a sufficiently smooth vector valued function.

It is possible to associate a semigroup $\{T(t)\}$ with the operator $\mathcal{A}$ and to prove uniform estimates for the spatial derivatives of the function $T(t)f$ up to the third-order, when $f \in C_b(\mathbb{R}^N)$. Such estimates can be used to prove Schauder estimates for the distributional solutions to both the elliptic equation and the nonhomogeneous Cauchy problem associated with the operator $\mathcal{A}$.

Unfortunately, the techniques used in the nondegenerate case can not be easily adapted to this situation. The construction of the semigroup associated with problem (0.3) and the determination of uniform estimates are two topics to be treated simultaneously: we need *a priori* estimates on the behaviour of the spatial derivatives of the function $T(t)f$ in order to guarantee its existence. This forces us to assume stronger assumptions on the coefficients than the ones of the nondegenerate case.

The underlying idea used to solve the problem is simple: we approximate the operator $\mathcal{A}$ by a sequence of uniformly elliptic operators $\mathcal{A}_\varepsilon$ ($\varepsilon > 0$) converging to the operator $\mathcal{A}$ in a sense to be made precise. With each of such operators we can associate a semigroup of bounded linear operators $\{T_\varepsilon(t)\}$. We prove uniform estimates for $\{T_\varepsilon(t)\}$, with constants being independent of ε, and, then, letting ε tend to 0^+, we can show that $\{T_\varepsilon(t)\}$ converges to a semigroup of bounded operators. Such a semigroup is associated with the operator $\mathcal{A}$, in the sense that $T(\cdot)f$ turns out to be the (unique) smooth function, satisfying $D_t u = \mathcal{A}u$ in $(0, +\infty) \times \mathbb{R}^N$ and the initial condition $u(0, \cdot) = f$, which is bounded in $[0, T] \times \mathbb{R}^N$ for any $T > 0$.

A fundamental tool for such a machinery work is given by the maximum principle which must hold for both of the operators $\mathcal{A}$ and $\mathcal{A}_\varepsilon$.

The behaviour of the derivatives of $T(t)f$ as t tends to 0 is worse than in the nondegenerate case, since it depends on the variable we are differentiating with respect to. To illustrate this difference, we consider the simplest case when $r = 1$, so that

$$B = \begin{pmatrix} \star & \star \\ B_1 & \star \end{pmatrix}.$$

In this situation one can show that there exists a positive constant C such that

$$\|D_i T(t)f\|_\infty \leq C\|f\|_\infty \begin{cases} t^{-\frac{1}{2}}, & i \leq p_0, \\ t^{-\frac{3}{2}}, & i > p_0, \end{cases} \tag{0.18}$$

for any $t > 0$ and $f \in C_b(\mathbb{R}^N)$, i.e., the behaviour, as t tends to 0^+, of the partial derivatives computed with respect to the first p_0 variables is the same as in the nondegenerate case. On the other hand, the behaviour with respect to the variables along which the operator degenerates is worse. This situation is not surprising at all, since this is just what happens for the degenerate Ornstein-Uhlenbeck operator defined by (0.17) with constant diffusion coefficients. In such a case, estimates (0.18) are known to be optimal near $t = 0$.

Many results holding in the nondegenerate case can be recovered also in this situation. In particular, it is still possible to associate a weak generator with the semigroup and, taking advantage of the uniform estimates, one can (at least partially) characterize its domain. Further, one can deal with the elliptic equation $\lambda u - \mathcal{A}u = f \in C_b(\mathbb{R}^N)$ whose continuous distributional solution is still formally given by the Laplace transform of the semigroup, where the integral, as in the nondegenerate case, is meant in pointwise sense.

Similarly, the continuous distributional solution to the nonhomogeneous Cauchy problem

$$\begin{cases} D_t u(t,x) = \mathcal{A}u(t,x) + g(t,x), & t > 0, \quad x \in \mathbb{R}^N, \\ u(0,x) = f(x), & \qquad\quad\ x \in \mathbb{R}^N, \end{cases}$$

when f and g are sufficiently smooth, is still given by the variation-of-constants formula (0.11). Such a formula allows us, also in this situation, to determine Schauder estimates for the solution u to the Cauchy problem via an interpolation argument.

* **The analysis of autonomous Kolmogorov equations in domains.** We consider the case when $\mathbb{R}^N$ is replaced with an open and unbounded domain of $\mathbb{R}^N$ and we associate homogeneous Dirichlet or Neumann boundary conditions with the operator $\mathcal{A}$. In these two situations the geometry of the open set Ω plays a crucial role in proving the existence of a semigroup associated with the operator $\mathcal{A}$. Besides the existence of a semigroup associated with the operator $\mathcal{A}$, we prove some uniform gradient estimates (and pointwise gradient estimates in the case of Neumann boundary conditions).

As in the case of $\mathbb{R}^N$, the semigroup $\{T(t)\}$ (in $C_b(\Omega)$ if Dirichlet boundary conditions are prescribed and in $C_b(\overline{\Omega})$ otherwise) is defined through the bounded classical solution to the Cauchy problem

$$\begin{cases} D_t u(t,x) = \mathcal{A}u(t,x), & t > 0, \quad x \in \Omega, \\ \mathcal{B}u(t,x) = 0, & t > 0, \quad x \in \partial\Omega, \\ u(0,x) = f(x), & \qquad\quad\ x \in \Omega, \end{cases} \tag{0.19}$$

where $\mathcal{B}u = u$ or $\mathcal{B}u = \partial u/\partial \nu$ on $\partial\Omega$, ν denoting the outward unit normal vector to $\partial\Omega$. In general, such a semigroup is neither strongly continuous nor analytic in the space of bounded and continuous functions. We confine ourselves to the case when the potential is nonpositive (or even identically vanishes in Ω, in view of some applications) but most of the results can be extended to potential terms which are just bounded from above in Ω.

Both in the cases of Dirichlet and Neumann boundary conditions, the uniqueness of the classical solution to problem (0.19) follows from a variant of the classical maximum principle.

In the case of Dirichlet boundary conditions, the existence of the bounded solution $u \in C^{1,2}((0,+\infty) \times \Omega) \cap C(([0,+\infty) \times \overline{\Omega}) \setminus (\{0\} \times \overline{\Omega}))$ to the Cauchy problem (0.19), for any $f \in C_b(\Omega)$, is obtained by an approximation argument, where, differently from the case of $\mathbb{R}^N$, we approximate the coefficients of the operator $\mathcal{A}$ rather than the domain Ω. The gradient estimate is obtained by employing the Bernstein method, which requires more regularity on the solution u and an apriori knowledge of $|\nabla_x u|$ on $(0,+\infty) \times \partial\Omega$. This latter information is obtained by a comparison argument with the solution to a one-dimensional parabolic problem. The existence of the solution to problem (0.19) with the additional degree of smoothness is obtained proving generation results in L^p-spaces and this technique to work requires the boundedness of the diffusion coefficients of the operator $\mathcal{A}$.

Generation results in L^p-spaces are also used to prove the existence of the solution $u \in C^{1,2}((0,+\infty) \times \Omega) \cap C_b([0,+\infty) \times \overline{\Omega})$ to the homogeneous Cauchy-Neumann problem (0.19) when Ω is a sufficiently smooth unbounded domain and $f \in C_b(\overline{\Omega})$. Again the Bernstein method allows to prove the uniform gradient estimate. As a matter of fact, some technical difficulties arise to make this method work. They are overcome by taking advantage of the geometry of Ω. In the particular case when Ω is an exterior domain, things are easier and the existence of the solution $u \in C^{1,2}((0,+\infty) \times \Omega) \cap C_b([0,+\infty) \times \overline{\Omega})$ to the Cauchy-Neumann problem (0.19) is proved using an approximation argument similar to that used in the case of homogeneous Dirichlet boundary conditions. In particular, no boundedness conditions on the diffusion coefficients are prescribed. Finally, when Ω is a convex set, the existence of the semigroup is proved by approximating the Cauchy-Neumann problem (0.19) with a sequence of Cauchy-Neumann problems in suitable bounded and convex domains Ω_n. Using the same arguments as in the case when $\Omega = \mathbb{R}^N$ and the gradient estimate satisfied by the solutions u_n to such approximating problems, one can show that u_n converges to the solution to the Cauchy-Neumamm problem (0.19). Also in this situation, we are not forced to assume the boundedness of the diffusion coefficients.

Unfortunately, the Bernstein method can not be adapted to prove estimates for the higher order derivatives and, to the best of our knowledge, such estimates are available only in some particular situations.

Part II: the non-autonomous case

As in the autonomous case, non-autonomous elliptic operators with unbounded coefficients occur naturally in the study of stochastic differential equations. Indeed, consider the stochastic differential equation

$$\begin{cases} dX_t = \mu(t, X_t)dt + \sigma(t, X_t)dW_t, & t > s, \\ X_s = x, \end{cases} \tag{0.20}$$

where W_t is a standard N-dimensional Brownian motion and μ (resp. σ) are smooth $\mathbb{R}^N$- (resp. $\mathbb{R}^{N\times N}$-) valued coefficients. If problem (0.20) admits a solution $X_t = X(t, s, x)$ for any $x \in \mathbb{R}^N$, then Itô's formula reveals that, for $f \in C_b^2(\mathbb{R}^N)$ and $t \in \mathbb{R}$, the function $u(s, x) := \mathbb{E}\left(f(X(t, s, x))\right)$, where $\mathbb{E}$ denotes expectation, solves the Cauchy problem

$$\begin{cases} D_s u(s, x) = -\mathcal{A}(-s)u(s, x), & (s, x) \in (-\infty, t) \times \mathbb{R}^N, \\ u(t, x) = f, & x \in \mathbb{R}^N, \end{cases}$$

where $\mathcal{A}$ is given by (0.2).

In the second part of this book, we consider non-autonomous elliptic operators and, to be consistent with the first part of the book, forward Cauchy problems. Hence, we assume that the coefficients of the operator $\mathcal{A}$ are defined in $I \times \mathbb{R}^N$, where I is a right-halfline (possibly $I = \mathbb{R}$), even if some results hold true also when I is a bounded interval. We are concerned with the following topics.

∗ Existence of an evolution operator associated with the operator $\mathcal{A}$ and analysis of its main properties. Under conditions, which are the natural counterpart of the ones of the autonomous case, one can prove that an evolution operator $\{G(t, s) : t \geq s \in I\}$ can be associated with the operator $\mathcal{A}$. As in the autonomous case, one can associate a Green function g with the evolution operator (which we simply denote by $\{G(t, s)\}$ from now). More precisely, there exists a function $g : \{(t, s) \in \mathbb{R}^2 : t > s \in I\} \times \mathbb{R}^N \times \mathbb{R}^N \to \mathbb{R}$ such that

$$(G(t, s)f)(x) = \int_{\mathbb{R}^N} g(t, s, x, y)f(y)dy, \qquad t > s \in I, \;\; x, y \in \mathbb{R}^N.$$

Some estimates of the Green function g are provided in Chapter 15.

It is interesting and important for many applications to study the behaviour of the function $s \mapsto (G(t, s)f)(x)$ when t and x are fixed and $f \in C_b(\mathbb{R}^N)$. If $f \in C_c^2(\mathbb{R}^N)$, then this function is differentiable and $(D_s G(t, s)f)(x) = -(G(t, s)\mathcal{A}(s)f)(x)$. By a straightforward density argument, the continuity of the function $s \mapsto (G(t, s)f)(x)$ can be guaranteed for any $f \in C_0(\mathbb{R}^N)$. Such a result can be extended to any $f \in C_b(\mathbb{R}^N)$ assuming the existence of a suitable family of Lyapunov functions, which allow to prove that the family of measures

$\{p(t,s,x,dy) : (t,s,x) \in \{(t,s) \in I \times I : t \geq s\} \times \overline{B}_r\}$ is tight for any $J \subset I$ and $r > 0$, where $p(t,s,x,dy) = g(t,s,x,y)dy$. This property yields the continuity of the function $(t,s,x) \mapsto (G(t,s)f)(x)$ in $\{(t,s) \in \mathbb{R}^2 : t \geq s \in I\} \times \mathbb{R}^N$.

Other properties discussed in this part of the book are the compactness of each operator $G(t,s)$ in $C_b(\mathbb{R}^N)$, the invariance of $C_0(\mathbb{R}^N)$ under the action of the evolution operator, and the gradient estimates (uniform and pointwise) satisfied by the function $G(t,s)f$. These latter estimates will play a crucial role in the study of the evolution operator $\{G(t,s)\}$ in suitable L^p-spaces and in the asymptotic analysis of the function $G(t,s)f$ as t tends to $+\infty$.

As in the classical case of bounded coefficients, when $I = \mathbb{R}$ one can associate an evolution semigroup $\{\mathscr{T}(t)\}$ in $C_b(\mathbb{R}^{N+1})$ with the evolution operator $\{G(t,s)\}$. Besides the interest in itself, this evolution semigroup turns out to play a relevant role in the study of the asymptotic behaviour of the function $G(t,s)f$ as t tends to $+\infty$. In general, the semigroup $\{\mathscr{T}(t)\}$, which is defined by $(\mathscr{T}(t)f)(s,x) = (G(s,s-t)f(s-t,\cdot)(x)$ for any $(s,x) \in \mathbb{R}^{N+1}$ and $t > 0$, fails to be strongly continuous.

We briefly discuss the main properties of the semigroup $\{\mathscr{T}(t)\}$ in $C_b(\mathbb{R}^{N+1})$ characterizing the domain of its weak generator. We also consider the case when the coefficients of the operator $\mathcal{A}$ are T-time periodic. In such a situation, the semigroup $\{\mathscr{T}(t)\}$ leaves invariant the set $C_\sharp(\mathbb{R}^{N+1})$ of all the functions $f : \mathbb{R}^{N+1} \to \mathbb{R}$ such that $f(t+T,x) = f(t,x)$ for any $(t,x) \in \mathbb{R}^{N+1}$.

$*$ **Analysis of the evolution operator $\{G(t,s)\}$ in L^p-spaces.** As the autonomous case shows, the L^p-spaces related to the Lebesgue measure are not the L^p-spaces which fit best the properties of elliptic operators with unbounded coefficients.

The invariant measure of the autonomous case is now replaced by a family of probability measures $\{\mu_t : t \in I\}$ characterized by the property

$$\int_{\mathbb{R}^N} G(t,s)f d\mu_t = \int_{\mathbb{R}^N} f d\mu_s, \qquad t > s \in I, \ f \in C_b(\mathbb{R}^N). \tag{0.21}$$

Such a family, called *evolution system of measures* in [77] and *entrance laws at* $-\infty$ in [93], consists of measures which are equivalent to the Lebesgue measure. Motivated by the autonomous case, we assume that the potential c of the operator $\mathcal{A}$ identically vanishes in $I \times \mathbb{R}^N$.

A sufficient condition for the existence of a family of probability measures $\{\mu_t : t \in I\}$ satisfying condition (0.21) is given in terms of the existence of a Lyapunov function, i.e. a positive function $\varphi : \mathbb{R}^N \to \mathbb{R}$, blowing up as $|x|$ tends to $+\infty$, such that $\mathcal{A}(t)\varphi(x) \leq a - d\varphi(x)$ for any $t \in J$, any $x \in \mathbb{R}^N$ and some positive constants a and d, where $J \subset I$ is a right-halfline.

Through the formula (0.21) one can extend each operator $G(t,s)$ to a contraction from $L^p(\mathbb{R}^N, \mu_s)$ to $L^p(\mathbb{R}^N, \mu_t)$. As the non-autonomous Ornstein-Uhlenbeck operator shows, in general the measures μ_t and μ_s differ if $t \neq s$ and, consequently, the corresponding spaces $L^p(\mathbb{R}^N, \mu_t)$ and $L^p(\mathbb{R}^N, \mu_s)$ differ as well for any $p \in [1, +\infty)$. Hence, the classical theory of evolution semigroups in L^p-spaces such as the one developed in [56] can not be applied. Moreover, still the non-autonomous Ornstein-Uhlenbeck operator reveals that evolution systems of measures are infinitely many in general. Among all these systems there exists one system with a relevant role: the eventually tight one, where eventually tight means that there exists $t_0 \in I$ such that the family $\{\mu_t : t \geq t_0\}$ is tight, i.e. for any $\varepsilon > 0$ there exists $R > 0$ such that $\mu_t(B_R) \geq 1 - \varepsilon$ for any $t \geq t_0$. Under suitable assumptions (see e.g. Theorem 16.2.11), the eventually tight evolution system of measures is unique.

We concentrate on this particular evolution system of measure since it is related to the asymptotic behaviour of the function $G(t,s)f$ as t tends to $+\infty$. We study the summability properties of the evolution operator $\{G(t,s)\}$ in the L^p-spaces related to this evolution system of measures. More precisely, we study the hypercontractivity, supercontractivity, ultraboundedness and ultracontractivity of the operators $G(t,s)$. Roughly speaking the evolution operator is

- hypercontractive if for any $p \in (1,+\infty)$ and $t > s \in I$ then the operator $G(t,s)$ is bounded from $L^p(\mathbb{R}^N, \mu_s)$ into $L^q(\mathbb{R}^N, \mu_t)$ for *some* $q \in (p,+\infty)$ (which depends on $t-s$);

- supercontractive if for any $p \in (1,+\infty)$ and $t > s \in I$ then the operator $G(t,s)$ is bounded from $L^p(\mathbb{R}^N, \mu_s)$ into $L^q(\mathbb{R}^N, \mu_t)$ for *any* $q \in (p,+\infty)$;

- ultrabounded if for any $p \in (1,+\infty)$ and $t > s \in I$ then the operator $G(t,s)$ is bounded from $L^p(\mathbb{R}^N, \mu_s)$ into $L^\infty(\mathbb{R}^N)$;

- ultracontractive if for any $t > s \in I$ then the operator $G(t,s)$ is bounded from $L^1(\mathbb{R}^N, \mu_s)$ into $L^\infty(\mathbb{R}^N)$.

The previous properties are not straightforward since the Sobolev embedding theorems fail to hold in general when the underlying measure is not the Lebesgue one. Hence, additional assumptions should be assumed to guarantee the previous property and, under such assumptions, it can be proved that

$$G(t,s)[|f|^p\log(|f|^p)] \le C_p(1-e^{2r_0(t-s)})G(t,s)(|f|^{p-2}|\nabla f|^2) + (G(t,s)|f|^p)\log(G(t,s)|f|^p),$$

in $\mathbb{R}^N$ for any $I \ni s < t$, $p \in [1,+\infty)$, $f \in C_b(\mathbb{R}^N)$ (with positive infimum, if $p < 2$) and some constants $C_p > 0$ and $r_0 < 0$. As a straightforward consequence of this estimate one can prove that each measure μ_t satisfies the Logarithmic Sobolev inequality (0.16) for any $f \in C^1_b(\mathbb{R}^N)$ and $p \in (1,+\infty)$ and from this estimate one can deduce the hypercontractivity property of the evolution operator $\{G(t,s)\}$.

The supercontractivity of the evolution operator is characterized by the validity of a refined version of the Logarithmic-Sobolev inequality. More precisely, it turns out that the evolution operator $\{G(t,s)\}$ is supercontractive if and only if the following family of logarithmic Sobolev inequalities

$$\int_{\mathbb{R}^N} f^2\log\left(\frac{|f|}{\|f\|_{L^2(\mathbb{R}^N,\mu_t)}}\right)d\mu_t \le \varepsilon\|\nabla f\|^2_{L^2(\mathbb{R}^N,\mu_t)} + \beta(\varepsilon)\|f\|^2_{L^2(\mathbb{R}^N,\mu_t)}$$

is satisfied by any $f \in W^{1,2}(\mathbb{R}^N, \mu_s)$, $t \in I$, $\varepsilon > 0$ and some decreasing function $\beta : \mathbb{R}_+ \to \mathbb{R}$, blowing up as ε tends to 0^+. Equivalently, the supercontractivity can be characterized in terms of the integrability, for any $\lambda > 0$, of the functions φ_λ, defined by $\varphi_\lambda(x) := e^{\lambda|x|^2}$ for any $x \in \mathbb{R}^N$.

Ultraboundedness can also be characterized in terms of the functions φ_λ. More precisely, the evolution operator $\{G(t,s)\}$ is ultrabounded if and only if, for any $\lambda > 0$ and $t > s$, the function $G(t,s)\varphi_\lambda$ belongs to $L^\infty(\mathbb{R}^N)$ and $\sup\{\|G(t,s)\varphi_\lambda\|_\infty : t,s \in I,\ t \ge s+\delta\} < +\infty$ for any $\delta, \lambda > 0$.

As far as ultracontractivity is concerned, the main step consists of determining suitable conditions which imply that each operator $G(t,s)$ maps $L^1(\mathbb{R}^N, \mu_s)$ into $L^2(\mathbb{R}^N, \mu_t)$, since then the evolution property and the ultraboundedness of the evolution operator imply that, actually, each operator $G(t,s)$ maps $L^p(\mathbb{R}^N, \mu_s)$ into $L^\infty(\mathbb{R}^N)$.

$\ast$ **Study of the evolution semigroup in L^p-spaces.** As it has been already remarked, the evolution semigroup $\{\mathscr{T}(t)\}$ plays a crucial role in the study of the long-term behaviour of the function $G(t,s)f$. In the case when the coefficients of the operator $\mathcal{A}$ are defined in

$\mathbb{R}^{N+1}$ and $\{\mu_t : t \in \mathbb{R}\}$ is an evolution system of measures for $\{G(t,s)\}$ there exists a unique measure μ defined on the σ-algebra of all the Borel subsets of $\mathbb{R}^{N+1}$ such that

$$\mu(A \times B) = \int_A \mu_s(B)ds$$

for any pair of Borel sets $A \subset \mathbb{R}$ and $B \subset \mathbb{R}^N$. Since each measure μ_t is a probability measure, μ is not a finite measure. Anyway, to some extent it can still be called the invariant measure of the semigroup $\{\mathscr{T}(t)\}$ since it is characterized by the invariance property

$$\int_{\mathbb{R}^{N+1}} \mathscr{T}(t) f d\mu = \int_{\mathbb{R}^{N+1}} f d\mu, \qquad t > 0, \tag{0.22}$$

for any $f \in C_b(\mathbb{R}^{N+1})$ with support contained in some strip $[a,b] \times \mathbb{R}^N$. The measure μ is absolutely continuous with respect to the Lebesgue measure and, if the coefficients of the operator $\mathcal{A}$ are smooth enough, then the density is a locally Hölder continuous function.

Through the formula (0.22) the evolution semigroup $\{\mathscr{T}(t)\}$ can be extended to a strongly continuous semigroup (denoted by $\{\mathscr{T}_p(t)\}$) in $L^p(\mathbb{R}^{N+1},\mu)$ for any $p \in [1,+\infty)$. The characterization of its infinitesimal generator G_p is known in the particular case when $\mathcal{A}$ is a non-autonomous Ornstein-Uhlenbeck operator. For more general operators only a partial characterization is available. So, it is extremely important to determine subspaces of "nice" functions which are cores of G_p. Under rather mild assumptions on the coefficients of the operator $\mathcal{A}$, the set

$$D_{\rm comp}(\mathcal{G}) = \Big\{ \psi \in C_b(\mathbb{R}^{N+1}) \cap \bigcap_{p<+\infty} W^{1,2}_{p,\rm loc}(\mathbb{R}^{N+1}) : \\ \mathcal{G}\psi \in C_b(\mathbb{R}^{N+1}),\ \mathrm{supp}(\psi) \subset [-M,M] \times \mathbb{R}^N \text{ for some } M > 0 \Big\}, \tag{0.23}$$

where $\mathcal{G} = \mathcal{A} - D_s$, and, under stronger assumptions, also the space $C^\infty(\mathbb{R}^{N+1})$ of smooth and compactly supported functions defined in $\mathbb{R}^{N+1}$ is a core of G_p for any $p \in [1,+\infty)$.

Using the core in (0.23), one can prove the inequality

$$\int_{\mathbb{R}^{N+1}} |u|^{p-2} |\sqrt{Q}\nabla_x u|^2 \chi_{\{u \neq 0\}} d\mu \le \frac{1}{1-p} \int_{\mathbb{R}^{N+1}} u|u|^{p-2} G_p u d\mu, \tag{0.24}$$

satisfied by any $u \in D(G_p)$ and $p \in (1,+\infty)$. Under stronger assumptions on the inner product $\langle b(t,x), x\rangle$ or assuming that the diffusion coefficients are bounded in $\mathbb{R}^{N+1}$, the previous inequality is, in fact an equality (the so called time dependent *identité de carré du champs*) satisfied by any $u \in D(G_p)$ for any $p \ge 2$. Formula (0.24) will play a crucial role in the study of the asymptotic behaviour of the evolution operator $\{G(t,s)\}$.

In the case when the coefficients of the operator $\mathcal{A}$ are T-time periodic for some $T > 0$, one can extend the semigroup $\{\mathscr{T}(t)\}$ with a strongly continuous semigroup (denoted by $\{\mathscr{T}_p^\sharp(t)\}$) in $L^p_\sharp(\mathbb{R}^{N+1},\mu^\sharp)$, where $L^p_\sharp(\mathbb{R}^{N+1},\mu^\sharp)$ denotes the set of all (equivalence classes of) functions $f : \mathbb{R}^{N+1} \to \mathbb{R}$ such that $f(\cdot + T, \cdot) = f$, $\mu^\sharp$-almost everywhere in $(0,T) \times \mathbb{R}^N$. Here, $\mu^\sharp$ is the probability measure on the σ-algebra of all the Borel subsets of $(0,T) \times \mathbb{R}^N$, defined by

$$\mu^\sharp(A \times B) = \frac{1}{T} \int_A \mu_s^\sharp(B)ds$$

for any pair of Borel subsets $A \subset (0,T)$ and $B \subset \mathbb{R}^N$, $\{\mu_t^\sharp : t \in \mathbb{R}\}$ being a T-periodic evolution system of measures for $\{G(t,s)\}$, where T-periodic means that $\mu^\sharp_{t+T} = \mu^\sharp_t$ for any $t \in \mathbb{R}$.

Clearly, the measure $\mu^\sharp$ satisfies the invariance property

$$\int_{(0,T)\times\mathbb{R}^N} \mathscr{T}^\sharp(t) f d\mu^\sharp = \int_{(0,T)\times\mathbb{R}^N} f d\mu^\sharp, \qquad t > 0,$$

for any bounded and continuous function $f : \mathbb{R}^{N+1} \to \mathbb{R}$ which is T-time periodic. Differently from the nonperiodic setting, where in general there exist several measures μ which satisfy the formula (0.22), in this periodic setting the measure μ is unique and it is induced by a T-periodic evolution system of measures for $\{G(t,s)\}$. The proof of the uniqueness is much more tricky than in the autonomous case due to the fact the semigroup $\{\mathscr{T}^\sharp(t)\}$ fails to be strong Feller and irreducible.

Also for the infinitesimal generator $G^\sharp_p$ of the semigroup $\{\mathscr{T}^\sharp_p(t)\}$ some suitable cores can be provided and the counterpart of inequality (0.24) in the periodic setting and a logarithmic Sobolev type inequality satisfied by all the functions $f \in D_\sharp(\mathcal{G})$ can be proved.

∗ **Asymptotic behaviour of the evolution operator $\{G(t,s)\}$.** The long-term behaviour of the function $G(t,s)f$ as t tends to $+\infty$ is of particular interest. Under suitable assumptions on the coefficients of the operator $\mathcal{A}$ it turns out that the function $\|G(t,s)f - m_s(f)\|_{L^p(\mathbb{R}^N,\mu_t)}$ vanishes as t tends to $+\infty$ for any $f \in L^p(\mathbb{R}^N,\mu_s)$, any $p \in [1,+\infty)$ and $s \in I$. Moreover, if $f \in C_b(\mathbb{R}^N)$ then $G(t,s)f - m_s(f)$ converges to 0 locally uniformly in $\mathbb{R}^N$ as t tends to $+\infty$. Assuming that the pointwise gradient estimate

$$|(\nabla_x G(t,s)f)(x)| \le e^{\ell_p(t-s)}(G(t,s)|\nabla_x f|^p)(x), \qquad \tau > s \in I,\ x \in \mathbb{R}^N,\ f \in C_b(\mathbb{R}^N), \tag{0.25}$$

for some negative constant ℓ_p, the decay rate of $\|G(t,s)f - m_s(f)\|_{L^p(\mathbb{R}^N,\mu_t)}$ to zero, as t tends to $+\infty$, is of exponential type.

In the particular case when the diffusion coefficients are independent of x and (0.25) holds true with $p = 1$ and some negative constant ℓ_1, the long-term behaviour of the functions $\|G(t,s)f - m_s(f)\|_{L^p(\mathbb{R}^N,\mu_t)}$ and $\|\nabla_x G(t,s)f\|_{L^p(\mathbb{R}^N,\mu_t)}$ as t tends to $+\infty$ can be compared. More precisely, the sets

$$\begin{aligned}\mathfrak{C}_p =& \big\{\omega \in \mathbb{R} : \|G(t,s)f - m_s(f)\|_{L^p(\mathbb{R}^N,\mu_t)} \le M_{p,\omega} e^{\omega(t-s)}\|f\|_{L^p(\mathbb{R}^N,\mu_s)} \\ &\quad \text{for any } I \ni s < t, \text{ any } f \in L^p(\mathbb{R}^N,\mu_s) \text{ and some } M_{p,\omega} > 0\big\},\\ \mathfrak{D}_p =& \big\{\omega \in \mathbb{R} : \|\nabla_x G(t,s)f\|_{L^p(\mathbb{R}^N,\mu_t)} \le N_{p,\omega} e^{\omega(t-s)}\|f\|_{L^p(\mathbb{R}^N,\mu_s)} \\ &\quad \text{for any } s,t \in I,\ t-s \ge 1, \text{ any } f \in L^p(\mathbb{R}^N,\mu_s) \text{ and some } N_{p,\omega} > 0\big\}\end{aligned}$$

coincide for any $p \in (1,+\infty)$ and are independent of p. As one can expect, the case $p = 1$ is particular. Indeed, the previous equality does not hold as the autonomous Ornstein-Uhlenbeck operator shows.

When the coefficients of the operator $\mathcal{A}$ are T-time periodic the long-term behaviours of $\|G(t,s)f - m_s(f)\|_{L^p(\mathbb{R}^N,\mu_s)}$, as t tends to $+\infty$ and as s tends to $-\infty$, and the behaviour of the function $t \mapsto \|\mathscr{T}^\sharp_p(t)f - \Pi f\|_{L^p(\mathbb{R}^{N+1},\mu^\sharp)}$ at infinity are strictly related each other. Here, Π is the projection of $L^p(\mathbb{R}^{N+1},\mu^\sharp)$ on the space of x-independent functions. More precisely, under suitable (not too restrictive) assumptions on the coefficients of the operator $\mathcal{A}$, $\|\mathscr{T}^\sharp_p(t)f - \Pi f\|_{L^p(\mathbb{R}^{N+1},\mu^\sharp)}$ vanishes as t tends to $+\infty$ if and only if $\|G(t,s)f - m_s(f)\|_{L^p(\mathbb{R}^N,\mu_s)}$ vanishes both as t tends to $+\infty$ and as s tends to $-\infty$.

Part III: appendices

With the aim of making the book as self-contained as possible, we collect in four appendices (from B to D) all the classical results of functional analysis and of the theory of partial differential equations, as well as some properties of the distance function. Moreover, in Appendix A we collect the function spaces and some of their relevant properties that we use in this book.

Part I

Autonomous Kolmogorov equations

Chapter 1

The elliptic equation and the Cauchy problem in $C_b(\mathbb{R}^N)$: the uniformly elliptic case

1.0 Introduction

In this chapter we begin the analysis of linear elliptic and parabolic problems in $C_b(\mathbb{R}^N)$ associated with the differential operator $\mathcal{A}$ defined by

$$\mathcal{A}u(x) = \sum_{i,j=1}^{N} q_{ij}(x)D_{ij}u(x) + \sum_{i=1}^{N} b_i(x)D_iu(x) + c(x)u(x), \qquad x \in \mathbb{R}^N,$$

on smooth functions. We assume the following hypotheses on the coefficients of the operator $\mathcal{A}$.

Hypotheses 1.0.1 (i) $q_{ij} \equiv q_{ji}$ for any $i,j = 1,\dots,N$ and there exists a function $\kappa : \mathbb{R}^N \to (0,+\infty)$ such that $\langle Q(x)\xi,\xi\rangle \geq \kappa(x)|\xi|^2$ for any $\xi, x \in \mathbb{R}^N$;

(ii) q_{ij}, b_i $(i,j = 1,\dots,N)$ and c belong to $C^{\alpha}_{\rm loc}(\mathbb{R}^N)$ for some $\alpha \in (0,1)$;

(iii) there exists $c_0 \in \mathbb{R}$ such that $c(x) \leq c_0$ for any $x \in \mathbb{R}^N$.

Besides, we introduce the realization A of $\mathcal{A}$ in $C_b(\mathbb{R}^N)$, with domain $D_{\max}(\mathcal{A})$, defined as follows:

$$D_{\max}(\mathcal{A}) = \left\{u \in C_b(\mathbb{R}^N) \cap \bigcap_{p<+\infty} W^{2,p}_{\rm loc}(\mathbb{R}^N) : \ \mathcal{A}u \in C_b(\mathbb{R}^N)\right\}, \qquad Au = \mathcal{A}u. \qquad (1.0.1)$$

In Section 1.1 we prove that, for any $\lambda > c_0$ and $f \in C_b(\mathbb{R}^N)$, the elliptic equation

$$\lambda u(x) - \mathcal{A}u(x) = f(x), \qquad x \in \mathbb{R}^N, \qquad (1.0.2)$$

admits a solution $u \in D_{\max}(\mathcal{A})$. The idea of the proof is the following. For each $n \in \mathbb{N}$, we consider the Dirichlet problem

$$\begin{cases} \lambda u_n(x) - \mathcal{A}u_n(x) = f(x), & x \in B_n, \\ u_n(x) = 0, & x \in \partial B_n. \end{cases} \qquad (1.0.3)$$

This problem has a unique solution $u_n \in C(\overline{B}_n)$ (in Appendix D we recall the results about elliptic and parabolic problems in bounded domains that we need throughout this chapter). Using an interior estimate (see Theorem D.1.1), we prove that we can define a function $u : \mathbb{R}^N \to \mathbb{R}$ by setting $u(x) := \lim_{n\to+\infty} u_n(x)$ for any $x \in \mathbb{R}^N$. The function u belongs to $D_{\max}(\mathcal{A})$, is a solution of equation (1.0.2) and $\|u\|_\infty \leq (\lambda - c_0)^{-1}\|f\|_\infty$.

We stress that, in general, u is not the unique solution of equation (1.0.2) in $D_{\max}(\mathcal{A})$. It is the unique solution provided further conditions on the coefficients are satisfied. The problem of the uniqueness of this solution is treated in Chapter 3. At any rate, if $f \geq 0$ then u is the minimal positive solution of (1.0.2) in $D_{\max}(\mathcal{A})$.

Next, we prove that u is given by the formula

$$u(x) = \int_{\mathbb{R}^N} K_\lambda(x,y) f(y)dy, \qquad x \in \mathbb{R}^N, \tag{1.0.4}$$

where K_λ is a positive function; K_λ is the so-called Green function associated with problem (1.0.2). To prove (1.0.4) we recall that the solution u_n of (1.0.3) is given by the formula $u_n(x) = \int_{B_n} K_\lambda^n(x,y)f(y)dy$ for any $x \in B_n$, where K_λ^n is the Green function associated with problem (1.0.3). Using the classical maximum principle we prove that the sequence $\{K_\lambda^n\}$ is increasing (with respect to $n \in \mathbb{N}$). This leads to the formula (1.0.4) with K_λ being the pointwise limit in $\mathbb{R}^{2N}$ of the sequence of functions $\{K_\lambda^n\}$. Thus, for any $\lambda > c_0$ we can define the linear operator $R(\lambda)$ in $C_b(\mathbb{R}^N)$ by setting

$$(R(\lambda)f)(x) = \int_{\mathbb{R}^N} K_\lambda(x,y)f(y)dy, \qquad x \in \mathbb{R}^N.$$

$R(\lambda)$ is a bounded operator, with $\|R(\lambda)\|_{L(C_b(\mathbb{R}^N))} \leq (\lambda - c_0)^{-1}$. In Section 1.3 we prove that $\{R(\lambda) : \lambda > c_0\}$ is the resolvent family of a linear operator $\widehat{A}$ in $C_b(\mathbb{R}^N)$. The operator $\widehat{A}$ is called weak generator. This terminology is justified by the theory of parabolic problems and it will be much clearer later on in this chapter.

The parabolic problem that we consider is the following one:

$$\begin{cases} D_t u(t,x) = \mathcal{A}u(t,x), & t > 0, \quad x \in \mathbb{R}^N, \\ u(0,x) = f(x), & \qquad\quad\; x \in \mathbb{R}^N, \end{cases} \tag{1.0.5}$$

with $f \in C_b(\mathbb{R}^N)$. In Section 1.2 we prove the existence of a classical solution of such a problem (i.e., a function $u \in C([0,+\infty) \times \mathbb{R}^N) \cap C^{1,2}((0,+\infty) \times \mathbb{R}^N))$, which is bounded in $[0,T] \times \mathbb{R}^N$, for any $T > 0$, and such that $D_t u$, $D_x^2 u$ belong to $C_{\rm loc}^{\alpha/2,\alpha}((0,+\infty) \times \mathbb{R}^N)$. The idea of the proof is similar to the one used in the elliptic case. More precisely, for each $n \in \mathbb{N}$ we consider the Cauchy-Dirichlet problem

$$\begin{cases} D_t u_n(t,x) = \mathcal{A}u_n(t,x), & t > 0, \quad x \in B_n, \\ u_n(t,x) = 0, & t > 0, \quad x \in \partial B_n, \\ u_n(0,x) = f(x), & \qquad\quad\; x \in B_n, \end{cases} \tag{1.0.6}$$

in the ball B_n, which, by classical results for parabolic Cauchy problems in bounded domains, admits a unique solution $u_n \in C([0,+\infty) \times \overline{B}_n \setminus (\{0\} \times \partial B_n)) \cap C_{\rm loc}^{1+\alpha/2,2+\alpha}((0,+\infty) \times B_n)$. Using Schauder interior estimates (see Theorem D.1.5) and a compactness argument, we show that u_n pointwise converges in $[0,+\infty) \times \mathbb{R}^N$ to some function u, which is a classical solution to problem (1.0.5), belongs to $C_{\rm loc}^{1+\alpha/2,2+\alpha}((0,+\infty) \times \mathbb{R}^N)$ and satisfies the estimate $|u(t,x)| \leq e^{c_0 t}\|f\|_\infty$ for any $t > 0$ and $x \in \mathbb{R}^N$.

In general, as in the elliptic case, u is not the unique classical solution of problem (1.0.5), which is bounded in $[0,T] \times \mathbb{R}^N$ for any $T > 0$. It turns out to be the unique solution under further assumptions on the coefficients (see Chapter 3). Even though u is not the unique solution to problem (1.0.5), at any rate it has a relevant property: in fact, for any $f \geq 0$, u is the minimal solution to problem (1.0.5) in the sense that, if v is another positive solution to the same Cauchy problem, then $v \geq u$. This is shown in Remark 1.2.3.

Then, we prove that u can be represented by the formula

$$u(t,x) = \int_{\mathbb{R}^N} g(t,x,y)f(y)dy, \qquad t > 0,\ x \in \mathbb{R}^N, \tag{1.0.7}$$

where g is a positive function, called the Green function. To prove (1.0.7) we argue as in the elliptic case, recalling that the solution u_n of problem (1.0.6) is given by the formula $u_n(t,x) = \int_{B_n} g_n(t,x,y)f(y)dy$ for any $t > 0$ and $x \in B_n$, where g_n is the Green function of (1.0.6). Using the classical maximum principle we show that the sequence $\{g_n\}$ is increasing with respect to $n \in \mathbb{N}$. This gives the formula (1.0.7), with g being the pointwise limit in $(0,+\infty) \times \mathbb{R}^{2N}$ of the sequence $\{g_n\}$, and it allows us to define the linear operator $T(t)$ in $C_b(\mathbb{R}^N)$, for any $t > 0$, by setting

$$(T(t)f)(x) = \int_{\mathbb{R}^N} g(t,x,y)f(y)dy, \qquad x \in \mathbb{R}^N. \tag{1.0.8}$$

We prove that the family $\{T(t)\}$ is a semigroup of linear operators in $C_b(\mathbb{R}^N)$. In general, $\{T(t)\}$ is a strongly continuous semigroup neither in $C_b(\mathbb{R}^N)$ nor in $BUC(\mathbb{R}^N)$ (see Proposition 10.2.6 and [261, Theorem 4.2]). Nevertheless, $T(t)f$ converges to f as t tends to 0, uniformly on compact sets. If f vanishes at infinity, then, actually, $T(t)f$ converges to f as t tends to 0, uniformly in $\mathbb{R}^N$. But this does not mean that the restriction of $\{T(t)\}$ to $C_0(\mathbb{R}^N)$ is a strongly continuous semigroup, because, in general, $C_0(\mathbb{R}^N)$ is not invariant for $\{T(t)\}$ (see Section 4.3).

In Section 1.3 we study the relation between the semigroup $\{T(t)\}$ and the weak generator $\widehat{A}$ introduced above. In fact, $\widehat{A}$ is the generator of $\{T(t)\}$ with respect to the bounded pointwise convergence. This means that $f \in D(\widehat{A})$ if and only if the function $(t,x) \mapsto t^{-1}[(T(t)f)(x) - f(x)]$ is bounded in $(0,1) \times \mathbb{R}^N$ and $t^{-1}(T(t)f - f)$ converges to $\widehat{A}f$ pointwise in $\mathbb{R}^N$, as t tends to 0^+. Further, $\widehat{A}$ is also the generator of the semigroup with respect to the mixed topology, which is the finest locally convex topology which agrees on every norm-bounded subsets of $C_b(\mathbb{R}^N)$ with the topology of the uniform convergence on compact sets. A sequence of functions converges in the mixed topology if and only if it is bounded and it converges locally uniformly. Thus, the convergence in the mixed topology is very similar to the bounded pointwise convergence.

In general, $D(\widehat{A})$ is a proper subset of $D_{\max}(\mathcal{A})$. It turns out to coincide with $D_{\max}(\mathcal{A})$ whenever the elliptic equation (1.0.2) has a unique solution in $D_{\max}(\mathcal{A})$ for any $f \in C_b(\mathbb{R}^N)$.

Formula (1.0.8) allows to extend the semigroup $\{T(t)\}$ to the space $B_b(\mathbb{R}^N)$ of Borel measurable and bounded functions. We prove that $\{T(t)\}$ is irreducible and it has the strong Feller property in $B_b(\mathbb{R}^N)$ (see Definition 1.2.12). These properties are useful for the theory of the invariant measures treated in Chapter 9.

In, general $\{T(t)\}$ is not analytic in $C_b(\mathbb{R}^N)$, as is shown in Section 1.4. In that section we also provide sufficient conditions on the coefficients of the operator $\mathcal{A}$, which yield the analyticity of the semigroup. The results of this section are taken from [198].

In the case when $c \equiv 0$, the semigroup $\{T(t)\}$ is associated with a transition function. This leads to the existence of a Markov process associated with $\{T(t)\}$. We briefly deal with the probabilistic approach in Section 1.5. Of course, there is a huge literature on the subject. Here, we just recall the definitions and the main properties of the Markov processes associated with the semigroup. Among them, we introduce the Dynkin formula and the link with the theory of differential stochastic equations.

Most of the results of this chapter are taken from [197].

1.1 The elliptic equation and the resolvent $R(\lambda)$

In this section we consider the elliptic equation (1.0.2) with $f \in C_b(\mathbb{R}^N)$ and $\lambda > c_0$. We prove that such an equation admits *at least* a solution u belonging to the domain $D_{\max}(\mathcal{A})$ defined in (1.0.1).

Theorem 1.1.1 *For any $f \in C_b(\mathbb{R}^N)$ there exists a function $u \in D_{\max}(\mathcal{A})$ which solves (1.0.2). Moreover, the following estimate holds true:*

$$\|u\|_\infty \le (\lambda - c_0)^{-1} \|f\|_\infty. \tag{1.1.1}$$

Finally, if $f \ge 0$, then $u \ge 0$.

Proof For any $n \in \mathbb{N}$, we denote by A_n the realization of the operator $\mathcal{A}$ with homogeneous Dirichlet boundary conditions in $C(\overline{B}_n)$. By Proposition D.3.3, the elliptic problem (1.0.3) admits a unique solution $u_n = R(\lambda, A_n)f$ in $\bigcap_{1\le p<+\infty} W^{2,p}(B_n)$. Moreover, by (D.2.3), u_n satisfies the estimate

$$\|u_n\|_\infty \le (\lambda - c_0)^{-1} \|f\|_\infty. \tag{1.1.2}$$

Let us prove that the sequence $\{u_n\}$ converges, uniformly on compact sets and in $W^{2,p}_{\mathrm{loc}}(\mathbb{R}^N)$, to a function $u \in D_{\max}(\mathcal{A})$ which satisfies the statement. For this purpose, consider first the case when $f \ge 0$. By the maximum principle, the functions u_n and $u_{n+1} - u_n$ are nonnegative in B_n for any $n \in \mathbb{N}$, i.e., the sequence $\{u_n\}$ is nonnegative and increasing. Therefore, it converges pointwise to some nonnegative function $u : \mathbb{R}^N \to \mathbb{R}$. By (1.1.2), u satisfies (1.1.1). Moreover, according to Theorem D.1.1 and (1.1.2), the sequence $\{u_n\}$ is bounded in $W^{2,p}(B_R)$ for any $p \in [1, +\infty)$ and any fixed $R > 0$. Then, by the Sobolev embedding theorems (see [3, Theorem 5.4]), it is bounded in $C^1(\overline{B}_R)$ too, and the Arzelà-Ascoli Theorem implies that it converges to u in $C(\overline{B}_R)$. Besides, applying again Theorem D.1.1 to the function $u_n - u_m$, we deduce that u belongs to $W^{2,p}(B_R)$ and that u_n converges to u in $W^{2,p}(B_R)$, for any $p \in [1, +\infty)$. Since $\mathcal{A}u_n = \lambda u_n - f$ in B_n, it follows that $u \in D_{\max}(\mathcal{A})$ and $\mathcal{A}u = \lambda u - f$. This concludes the proof in the case when $f \ge 0$. For an arbitrary $f \in C_b(\mathbb{R}^N)$, it suffices to split $f = f^+ - f^-$, $u_n = R(\lambda, A_n)(f^+) - R(\lambda, A_n)(f^-) =: u_{n,1} + u_{n,2}$ and to apply the previous arguments separately to the sequences $\{u_{n,1}\}$ and $\{u_{n,2}\}$. □

Remark 1.1.2 In general, equation (1.0.2) admits more than one solution in $D_{\max}(\mathcal{A})$. In Chapters 2 and 4 we consider some situations in which the elliptic equation $\lambda u - \mathcal{A}u = 0$ admits a nontrivial solution in $D_{\max}(\mathcal{A})$ (see Examples 2.2.5 and 4.2.5). Nevertheless, in the case when the datum f is nonnegative the solution u provided by Theorem 1.1.1 can be characterized as the minimal nonnegative solution. Indeed, if v is another nonnegative solution, then by the maximum principle it follows that $v(x) \ge u_n(x)$ for any $x \in B_n$ and $n \in \mathbb{N}$. Letting n tend to $+\infty$ gives $v \ge u$.

Now we prove that we can associate a positive Green function with equation (1.0.2) and define the resolvent operator $R(\lambda)$ for any $\lambda > c_0$.

Theorem 1.1.3 *For any $\lambda > c_0$ there exists a linear operator $R(\lambda)$ in $C_b(\mathbb{R}^N)$ such that for any $f \in C_b(\mathbb{R}^N)$ the solution of equation (1.0.2), provided by Theorem 1.1.1, is represented by $u = R(\lambda)f$. The family of operators $\{R(\lambda) :\ \lambda > c_0\}$ satisfies the estimate*

$$\|R(\lambda)f\|_\infty \le (\lambda - c_0)^{-1}\|f\|_\infty, \qquad f \in C_b(\mathbb{R}^N), \tag{1.1.3}$$

and the resolvent identity

$$R(\lambda)f - R(\mu)f = (\mu - \lambda)R(\mu)R(\lambda)f, \qquad c_0 < \lambda < \mu. \tag{1.1.4}$$

Moreover, $R(\lambda)$ is injective for any $\lambda > c_0$. Finally, there exists a positive function $K_\lambda : \mathbb{R}^N \times \mathbb{R}^N \to \mathbb{R}$ such that

$$(R(\lambda)f)(x) = \int_{\mathbb{R}^N} K_\lambda(x,y)f(y)dy, \qquad x \in \mathbb{R}^N,\ f \in C_b(\mathbb{R}^N). \tag{1.1.5}$$

Proof Let K_λ^n be the Green function associated with the realization A_n of the operator $\mathcal{A}$ in $C(\overline{B}_n)$ with homogeneous Dirichlet boundary conditions (see Proposition D.3.3). We extend it by zero to the whole of $\mathbb{R}^N \times \mathbb{R}^N$ and we still denote by K_λ^n this extension. With any nonnegative function $f \in C_0(B_n)$, let

$$v_n(x) = \int_{B_n} (K_\lambda^{n+1}(x,y) - K_\lambda^n(x,y))f(y)dy, \qquad x \in B_n.$$

Since $f \geq 0$ and $v_n = u_{n+1} - u_n$ by Proposition D.2.2, v_n is nonnegative in B_n. The arbitrariness of $f \geq 0$ and the continuity in $B_n \times B_n$ of the function K_λ^n imply that $K_\lambda^{n+1}(x,y) - K_\lambda^n(x,y) \geq 0$ for any $x, y \in B_n$. Hence, $\{K_\lambda^n(x,y)\}$ is an increasing sequence, for any $x, y \in \mathbb{R}^N$, and we can define the function $K_\lambda : \mathbb{R}^N \times \mathbb{R}^N \to \mathbb{R}$ by setting

$$K_\lambda(x,y) := \lim_{n\to+\infty} K_\lambda^n(x,y), \qquad x, y \in \mathbb{R}^N. \tag{1.1.6}$$

Note that this limit is finite for any $x \in \mathbb{R}^N$ and almost any $y \in \mathbb{R}^N$.[1] Indeed, estimate (D.2.3) with $f \equiv \mathbf{1}$ yields $\|K_\lambda^n(x,\cdot)\|_{L^1(B_n)} \leq (\lambda - c_0)^{-1}$ for any $x \in B_n$ and $n \in \mathbb{N}$. By monotone convergence, we can infer that $\|K_\lambda(x,\cdot)\|_{L^1(\mathbb{R}^N)} \leq (\lambda - c_0)^{-1}$ for any $x \in \mathbb{R}^N$. Hence, for any $x \in \mathbb{R}^N$, $K(x,y)$ is finite for almost any $y \in \mathbb{R}^N$. Moreover, since K_λ^n is strictly positive in $B_n \times B_n$ for any $n \in \mathbb{N}$, also K_λ is.

Now, we observe that the solution of equation (1.0.2) given by Theorem 1.1.1 can be represented by

$$u(x) = \lim_{n\to+\infty} \left(\int_{\mathbb{R}^N} K_\lambda^n(x,y)f^+(y)dy - \int_{\mathbb{R}^N} K_\lambda^n(x,y)f^-(y)dy \right)$$

for any $x \in \mathbb{R}^N$. Since both f^+ and f^- are nonnegative, the monotone convergence theorem implies that $u(x) = \int_{\mathbb{R}^N} K_\lambda(x,y)f(y)dy$ for any $x \in \mathbb{R}^N$. Thus, the operator $R(\lambda)$ in (1.1.5) is well defined and $u = R(\lambda)f$. Clearly, $R(\lambda)$ is linear, and, by virtue of (1.1.1), it satisfies (1.1.3). Moreover, $R(\lambda)$ is injective. Indeed, if $u \equiv R(\lambda)f \equiv 0$, then $f \equiv 0$ since $R(\lambda)f$ solves, by construction, the elliptic equation $\lambda u - \mathcal{A}u = f$.

To conclude the proof, it remains to prove the resolvent identity (1.1.4). For this purpose, we fix $f \in C_b(\mathbb{R}^N)$, $n \in \mathbb{N}$ and observe that $R(\lambda, A_n)f - R(\mu, A_n)f = (\mu - \lambda)R(\lambda, A_n)R(\mu, A_n)f$ for any $\mu > \lambda > c_0$ (see Appendices B.1 and D). By virtue of (D.3.5), this formula can be rewritten as

$$\int_{B_n} K_\lambda^n(x,y)f(y)dy - \int_{B_n} K_\mu^n(x,y)f(y)dy = (\mu - \lambda)\int_{B_n} dy \int_{B_n} K_\mu^n(x,y)K_\lambda^n(y,z)f(z)dz.$$

Letting n tend to $+\infty$, the dominated convergence theorem gives

$$\int_{\mathbb{R}^N} K_\lambda(x,y)f(y)dy - \int_{\mathbb{R}^N} K_\mu(x,y)f(y)dy = (\mu - \lambda)\int_{\mathbb{R}^N} dy \int_{\mathbb{R}^N} K_\mu(x,y)K_\lambda(y,z)f(z)dz,$$

i.e., the formula (1.1.4). □

[1] Theorem 1.2.7 shows that, in fact, $K(x,y) \in \mathbb{R}$ for any $x, y \in \mathbb{R}^N$.

1.2 The Cauchy problem and the semigroup

Now we prove that for any $f \in C_b(\mathbb{R}^N)$ the Cauchy problem (1.0.5) admits a classical solution.

Theorem 1.2.1 *For any $f \in C_b(\mathbb{R}^N)$, problem* (1.0.5) *admits a solution $u \in C([0,+\infty) \times \mathbb{R}^N) \cap C^{1+\alpha/2,2+\alpha}_{\rm loc}((0,+\infty) \times \mathbb{R}^N)$ which satisfies the estimate*

$$|u(t,x)| \le e^{c_0 t}\|f\|_\infty, \qquad t>0, \;\; x \in \mathbb{R}^N. \tag{1.2.1}$$

Proof We split the proof into two steps. First, we show that there exists a solution $u \in C^{1+\alpha/2,2+\alpha}_{\rm loc}((0,+\infty)\times\mathbb{R}^N)$ to the differential equation in (1.0.5) and it satisfies (1.2.1). Then, in Step 2, we show that u can be extended by continuity up to $t=0$ and $u(0,\cdot)=f$.

Step 1. For any $n \in \mathbb{N}$, let $u_n \in C([0,+\infty)\times \overline{B}_n \setminus (\{0\}\times\partial B_n)) \cap C^{1,2}((0,+\infty)\times B_n)$ be the solution of the Cauchy-Dirichlet problem (1.0.6) (see Proposition D.3.1). Estimate (D.3.2) shows that, for any $n \in \mathbb{N}$,

$$|u_n(t,x)| \le e^{c_0 t}\|f\|_\infty, \qquad t>0, \;\; x \in B_n. \tag{1.2.2}$$

Now, fix $M \in \mathbb{N}$ and set $D(M) = (0,M)\times B_M$ and $D'(M) = [1/M, M]\times B_{M-1}$. From the interior Schauder estimate (D.1.11) we deduce that

$$\|u_n\|_{C^{1+\alpha/2,2+\alpha}(D'(M))} \le C_M \|u_n\|_{L^\infty(D(M))} \le C_M e^{c_0^+ M}\|f\|_\infty \tag{1.2.3}$$

for any $n \ge M$, where C_M is a positive constant independent of $n \in \mathbb{N}$. Fix $\beta \in (0,\alpha)$. By (1.2.3) there exists a subsequence $\{u_n^{(M)}\}$ of $\{u_n\}$ converging in $C^{1+\beta/2,2+\beta}(D'(M))$ to some function $u_\infty^{(M)} \in C^{1+\alpha/2,2+\alpha}(D'(M))$. Without loss of generality we can assume that $\{u_n^{(M+1)}\}$ is a subsequence of $\{u_n^{(M)}\}$. Thus, $u_\infty^{(M)}$ and $u_\infty^{(M+1)}$ coincide in the domain $D'(M)$ and we can define the function $u \in C^{1+\alpha/2,2+\alpha}_{\rm loc}((0,+\infty)\times\mathbb{R}^N)$ by setting $u = u_\infty^{(M)}$ in $D'(M)$. Moreover, the diagonal subsequence $\{\widetilde{u}_n\}$, defined by $\widetilde{u}_n = u_n^{(n)}$ for any $n\in\mathbb{N}$, converges to u in $C^{1+\beta/2,2+\beta}([T_1,T_2]\times K)$ for any compact set $K \subset \mathbb{R}^N$ and any $0<T_1<T_2$. Hence, letting n tend to $+\infty$ in the differential equation satisfied by $\widetilde{u}_n$, it follows that $D_t u = \mathcal{A}u$ in $(0,+\infty)\times\mathbb{R}^N$. Besides, (1.2.1) follows from (1.2.2).

Step 2. To complete the proof we show that $u \in C([0,+\infty)\times\mathbb{R}^N)$ and $u(0,\cdot) = f$. For this purpose, we take advantage of the semigroup theory. In particular, we will use the representation formula of solutions to Cauchy-Dirichlet problems in bounded domains through semigroups.

Fix $M \in \mathbb{N}$ and let ϑ be any smooth function such that $\chi_{B_{M-1}} \le \vartheta \le \chi_{B_M}$. For any $n > M$, let $v_n = \vartheta \widetilde{u}_n$. The function v_n belongs to $C([0,+\infty)\times\overline{B}_M)$ and solves the Cauchy-Dirichlet problem

$$\begin{cases} D_t v_n(t,x) = \mathcal{A}v_n(t,x) + \psi_n(t,x), & t>0, \quad x\in B_M,\\ v_n(t,x) = 0, & t>0, \quad x \in \partial B_M,\\ v_n(0,x) = \vartheta(x) f(x), & \qquad\quad x \in B_M,\end{cases}$$

where $\psi_n = -2\langle Q\nabla_x \widetilde{u}_n, \nabla\vartheta\rangle - \widetilde{u}_n {\rm Tr}(QD_{ij}\vartheta) - \widetilde{u}_n\langle b,\nabla\vartheta\rangle$. For any $t>0$ and $x\in B_M$ we can estimate

$$|\psi_n(t,x)| \le K_M(e^{c_0 t}\|f\|_\infty + \|\nabla_x \widetilde{u}_n(t,\cdot)\|_{L^\infty(B_M)}), \tag{1.2.4}$$

where $K_M := \|{\rm Tr}(QD^2\vartheta)\|_{L^\infty(B_M)} + \|\langle b,\nabla\vartheta\rangle\|_{L^\infty(B_M)} + 2\|Q\nabla\vartheta\|_{L^\infty(B_M)}$.

We use again the interior estimate of Theorem D.1.5. By (D.1.12) it follows that

$$|\sqrt{t}\nabla_x \widetilde{u}_n(t,x)| \le C\|\widetilde{u}_n\|_{L^\infty(D(M+1))} \le Ce^{c_0^+}\|f\|_\infty =: C'\|f\|_\infty$$

for any $x \in B_M$, $t \le 1 = \text{dist}\,(B_M, \partial B_{M+1})$ and some positive constant C, independent of n. This yields $\|\nabla_x \widetilde{u}_n(t,\cdot)\|_{L^\infty(B_M)} \le t^{-1/2}C'\|f\|_\infty$ for any $t \le 1$. Then, by (1.2.4) we deduce that

$$|\psi_n(t,x)| \le K'_M(1+t^{-1/2})\|f\|_\infty, \qquad t \in (0,1],\ x \in B_M, \tag{1.2.5}$$

for any $n > M$, where $K'_M > 0$ is a constant independent of n. Therefore, ψ_n belongs to $L^1(0,T,C_0(B_M))$ and v_n can be represented by means of the variation-of-constants formula

$$v_n(t,\cdot) = T_M(t)(\vartheta f) + \int_0^t T_M(t-s)\psi_n(s,\cdot)ds, \qquad t > 0,$$

where $\{T_M(t)\}$ is the semigroup in $C(\overline{B}_M)$ associated with the operator $\mathcal{A}$ with homogeneous Dirichlet conditions on ∂B_M. Since $v_n \equiv \widetilde{u}_n$ and $\vartheta \equiv \mathbb{1}$ in B_{M-1}, by (1.2.5) and (D.3.2) it follows

$$|\widetilde{u}_n(t,x) - f(x)| \le \|T_M(t)(\vartheta f) - \vartheta f\|_\infty + K'_M\|f\|_\infty \int_0^t e^{c_0(t-s)}(1+s^{-\frac{1}{2}})ds$$

for any $t > 0$ and $x \in B_{M-1}$. Letting n tend to $+\infty$ (and taking (D.3.2) into account) we get

$$|u(t,x) - f(x)| \le \|T_M(t)(\vartheta f) - \vartheta f\|_\infty + K'_M\|f\|_\infty \int_0^t e^{c_0(t-s)}(1+s^{-\frac{1}{2}})ds,$$

which shows that u can be extended by continuity at $(0,x)$ for any $x \in B_{M-1}$. Since $M \in \mathbb{N}$ is arbitrary, u belongs to $C([0,+\infty) \times \mathbb{R}^N)$ and $u(0,\cdot) \equiv f$. □

Remark 1.2.2 We note that the maximum principle implies that, if $f \ge 0$ does not identically vanish, then the sequence $\{u_n\}$ is positive and increasing. Therefore, u is positive and the whole sequence $\{u_n\}$ converges to u. In the general case when $f \in C_b(\mathbb{R}^N)$, we write $f = f^+ - f^-$ and conclude again that the whole sequence $\{u_n\}$ converges to u in $C^{1+\beta/2,2+\beta}([T_1,T_2] \times K)$ for any $\beta \in (0,\alpha)$, any compact set $K \subset \mathbb{R}^N$ and any $0 < T_1 < T_2$, applying the previous argument to the functions u_{f^+} and u_{f^-} which are the solutions to problem (1.0.5) provided by Theorem 1.2.1, corresponding, respectively, to the data f^+ and f^-.

Remark 1.2.3 If $f \ge 0$ does not identically vanish, then the solution u given by Theorem 1.2.1 is the minimal positive solution of problem (1.0.5). Indeed, if v is another positive solution, then the maximum principle yields $v(t,x) \ge u_n(t,x)$ for any $t > 0$, $x \in B_n$, $n \in \mathbb{N}$ and, eventually, $v \ge u$.

We stress that, in general, problem (1.0.5) admits more than a unique classical solution which satisfies (1.2.1). In Example 2.2.5 (see also Proposition 3.1.1) it is shown that the Cauchy problem (1.0.5), associated with the one-dimensional operator $\mathcal{A}u = u'' + x^3u'$, admits a nontrivial bounded solution u satisfying $u(0,\cdot) = 0$.

We now recall the definition of transition function as is given in [92]. Here, $\mathcal{B}(E)$ denotes the σ-algebra of Borel sets of a topological space E.

Definition 1.2.4 *A family of Borel measures $\{p(t,x;\cdot) : t \ge 0, x \in E\}$ is a transition function if the function $p(t,\cdot;B) : E \to \mathbb{R}$ is Borel measurable for any $t \ge 0$ and any $B \in \mathcal{B}(E)$, and* *(i)* $p(t,x;E) \le 1$ *for any* $t \ge 0$ *and* $x \in E$;

(ii) $p(0,x;E \setminus \{x\}) = 0$ *for any* $x \in E$;

(iii) $p(t+s,x;B) = \int_E p(s,y;B)p(t,x;dy)$ *for any* $s,t \ge 0$, $x \in E$ *and* $B \in \mathcal{B}(E)$.

A transition function is normal if $\lim_{t\to 0^+} p(t,x;E) = 1$ *for any* $x \in E$; *it is stochastically continuous if, for any open set* $U \subset E$ *and any* $x \in U$, *then* $p(t,x;U)$ *converges to* 1 *as* t *tends to* 0^+.

Theorem 1.2.5 *There exists a semigroup of linear operators* $\{T(t)\}$ *defined in* $C_b(\mathbb{R}^N)$ *such that, for any* $f \in C_b(\mathbb{R}^N)$, *the solution of problem* (1.0.5), *given by Theorem* 1.2.1, *is represented by* $u(t,x) = (T(t)f)(x)$ *for any* $t \ge 0$ *and* $x \in \mathbb{R}^N$. *For any* $t > 0$, $T(t)$ *satisfies the estimate*

$$\|T(t)f\|_\infty \le e^{c_0 t}\|f\|_\infty, \qquad f \in C_b(\mathbb{R}^N). \tag{1.2.6}$$

Moreover, there exist a family of Borel measures $p(t,x;dy)$ *in* $\mathbb{R}^N$, *such that*

$$(T(t)f)(x) = \int_{\mathbb{R}^N} f(y)p(t,x;dy), \qquad t \ge 0,\ x \in \mathbb{R}^N, \tag{1.2.7}$$

and a function $g : (0,+\infty) \times \mathbb{R}^N \times \mathbb{R}^N \to \mathbb{R}$ *such that*

$$p(t,x;dy) = g(t,x,y)dy, \qquad t > 0,\ x,y \in \mathbb{R}^N. \tag{1.2.8}$$

The function g *is strictly positive and the functions* $g(t,\cdot,\cdot)$ *and* $g(t,x,\cdot)$ *are measurable for any* $t > 0$ *and* $x \in \mathbb{R}^N$. *Further, for almost every fixed* $y \in \mathbb{R}^N$, *the function* $g(\cdot,\cdot,y)$ *belongs to* $C^{1+\alpha/2,2+\alpha}_{\rm loc}((0,+\infty) \times \mathbb{R}^N)$, *and it is a solution of the equation* $D_t u = \mathcal{A}u$. *Finally, if* $c_0 \le 0$ *then* $p(t,x;dy)$ *is a stochastically continuous transition function.*

Proof We split the proof into two steps. In the first one we define the function g and prove some of its properties. Finally, in Step 2, we complete the proof.

Step 1. For any $k \in \mathbb{N}$, let $g_k \in C((0,+\infty) \times B_k \times B_k)$ be the Green function of the Cauchy Dirichlet problem, associated with the operator $\mathcal{A}$ in B_k, given by Proposition D.3.1. We extend the function g_k to $(0,+\infty) \times \mathbb{R}^N \times \mathbb{R}^N$ with value zero for $x, y \notin B_k$ and still denote by g_k the so obtained function. A straightforward computation shows that, for any fixed $t \in (0,+\infty)$ and $x,y \in \mathbb{R}^N$, the sequence $\{g_k(t,x,y)\}$ is increasing. Indeed, for any positive $f \in C_b(B_k)$, the function

$$w(t,x) = \int_{B_k} f(y)(g_{k+1}(t,x,y) - g_k(t,x,y))dy, \qquad t \ge 0,\ x \in \mathbb{R}^N,$$

is positive as well by virtue of Remark 1.2.2. Recalling that, for any $t > 0$ and $x \in B_k$, the function $g_{k+1}(t,x,\cdot) - g_k(t,x,\cdot)$ is continuous in B_k, we conclude that $g_{k+1} \ge g_k$ in $(0,+\infty) \times B_k \times B_k$, implying that the sequence $\{g_k\}$ is pointwise increasing in $(0,+\infty) \times \mathbb{R}^N \times \mathbb{R}^N$. Hence, we can define the function $g : (0,+\infty) \times \mathbb{R}^N \times \mathbb{R}^N \to \mathbb{R}$ as the pointwise limit in $(0,+\infty) \times \mathbb{R}^N \times \mathbb{R}^N$ of the sequence $\{g_k\}$. Estimate (D.3.2) with $f \equiv \mathbb{1}$ yields $\|g_k(t,x,\cdot)\|_{L^1(B_k)} \le e^{c_0 t}$ for any $t > 0$, $x \in B_k$, $k \in \mathbb{N}$, and then, by monotone convergence,

$$\int_{\mathbb{R}^N} g(t,x,y)dy \le e^{c_0 t}, \qquad t > 0,\ x \in \mathbb{R}^N, \tag{1.2.9}$$

so that $g(t,x,y)$ is finite for any $(t,x) \times (0,+\infty) \times \mathbb{R}^N$ and $y \in E(t,x)$, where $\mathbb{R}^N \setminus E(t,x)$

is a negligible set. Since $g_k(t,x,\cdot) > 0$ almost everywhere in B_k for any $t > 0$ and $x \in B_k$, then g is strictly positive.

Clearly, $g(t,\cdot,\cdot)$, $g(t,x,\cdot)$ and $g(t,\cdot,y)$ are measurable functions for any $t > 0$ and $x, y \in \mathbb{R}^N$ since they are the pointwise limit of measurable functions.

We now prove the regularity properties of g. Fix $R, T > 0$, $x_0 \in B_R$ and let $y_0 \in \mathbb{R}^N$ be such that $g(T, x_0, y_0) < +\infty$; actually we have seen that this holds true for almost any $y_0 \in \mathbb{R}^N$. If $h, k \in \mathbb{N}$ satisfy $R + 1 < h < k$, the functions $g_h(\cdot,\cdot,y_0)$ and $g_k(\cdot,\cdot,y_0)$ are solutions of the equation $D_t u - \mathcal{A}u = 0$ in $(0,+\infty) \times B_{R+1}$ (see Proposition D.3.1), and hence $g_k(\cdot,\cdot,y_0) - g_h(\cdot,\cdot,y_0)$ is as well. Moreover, $g_k(\cdot,\cdot,y_0) - g_h(\cdot,\cdot,y_0)$ is positive and, for any fixed $0 < t_0 < t_1 < T$, it satisfies the following Harnack inequality (see [161]):

$$\begin{aligned}\sup_{[t_0,t_1]\times \overline{B}_R} (g_k(\cdot,\cdot,y_0) - g_h(\cdot,\cdot,y_0)) &\le C \inf_{x\in \overline{B}_R} (g_k(T,\cdot,y_0) - g_h(T,\cdot,y_0))\\ &\le C\big(g_k(T,x_0,y_0) - g_h(T,x_0,y_0)\big),\end{aligned}$$

where $C > 0$ is a constant, independent of h and k. Since $g(T,x_0,y_0) < +\infty$, $\{g_n(\cdot,\cdot,y_0)\}$ turns out to be a Cauchy sequence in $C([t_0,t_1] \times \overline{B}_R)$. Recalling that it converges pointwise to the function $g(\cdot,\cdot,y_0)$, we conclude that $g(\cdot,\cdot,y_0) \in C([t_0,t_1] \times \overline{B}_R)$. Moreover, from Theorem D.1.5 it follows that, for any $t_0' > t_0$, $t_1' < t_1$ and $R' < R$, the sequence $\{g_n(\cdot,\cdot,y_0)\}$ converges also in $C^{1+\beta/2,2+\beta}([t_0',t_1'] \times \overline{B}_{R'})$ for any $\beta \in (0,\alpha)$. Hence, $g(\cdot,\cdot,y_0)$ belongs to $C^{1+\alpha/2,2+\alpha}([t_0',t_1'] \times \overline{B}_{R'})$. The arbitrariness of T, R, R', t_0, t_0', t_1, t_1' implies that $g(\cdot,\cdot,y_0) \in C^{1+\alpha/2,2+\alpha}_{\rm loc}((0,+\infty) \times \mathbb{R}^N)$.

Finally, since $D_t g_n = \mathcal{A} g_n$ for any $n \in \mathbb{N}$, letting n tend to $+\infty$ it follows that $D_t g = \mathcal{A} g$.

Step 2. Now, for any $t > 0$ and $x \in \mathbb{R}^N$ we define the measure $p(t,x;dy)$ by (1.2.8), while for $t = 0$ we set $p(t,x;dy) = \delta_x$. Moreover, for any $t > 0$, we define the operator $T(t)$ by (1.2.7). Let us prove that, for any $f \in C_b(\mathbb{R}^N)$, the solution u of problem (1.0.5), found in Theorem 1.2.1, is $T(\cdot)f$. For this purpose, we recall that

$$u(t,x) = \lim_{k\to+\infty} \int_{\mathbb{R}^N} f(y) g_k(t,x,y)dy, \qquad t > 0,\ \ x \in \mathbb{R}^N,$$

and we can split it as

$$u(t,x) = \lim_{k\to+\infty} \int_{\mathbb{R}^N} f^+(y) g_k(t,x,y)dy - \lim_{k\to+\infty} \int_{\mathbb{R}^N} f^-(y) g_k(t,x,y)dy.$$

By the monotone convergence theorem we deduce that

$$u(t,x) = \int_{\mathbb{R}^N} f^+(y) g(t,x,y)dy - \int_{\mathbb{R}^N} f^-(y) g(t,x,y)dy = \int_{\mathbb{R}^N} f(y) g(t,x,y)dy,$$

that is $u \equiv T(\cdot)f$.

To show that $\{T(t)\}$ is a semigroup it suffices to observe that from (D.3.3), using the monotone convergence theorem, it follows that

$$g(t+s,x,y) = \int_{\mathbb{R}^N} g(s,x,z) g(t,z,y)dz, \qquad t,s > 0,\ \ x,y \in \mathbb{R}^N. \tag{1.2.10}$$

Moreover, (1.2.6) is an immediate consequence of (1.2.1).

It remains to prove that, if $c_0 \le 0$, then $\{p(t,x;dy) : t \ge 0,\ x \in \mathbb{R}^N\}$ is a stochastically continuous transition function. The condition (i) in Definition 1.2.4 follows immediately from (1.2.9), and the condition (ii) is obvious. The condition (iii) will follow from (1.2.10) once we have proved that the function $p(t,\cdot;B)$ is Borel measurable in $\mathbb{R}^N$, for any fixed

$t \geq 0$ and any $B \in \mathcal{B}(\mathbb{R}^N)$. So, let us prove this property. If $t = 0$ then the property is clear. If $t > 0$, then consider a sequence of functions $\{f_n\} \subset C_b(\mathbb{R}^N)$ converging almost everywhere to χ_B and such that $0 \leq f_n \leq 1$ for any $n \in \mathbb{N}$.[2] The dominated convergence theorem yields

$$p(t,x;B) = \lim_{n\to+\infty} \int_{\mathbb{R}^N} f_n(y)p(t,x;dy) = \lim_{n\to+\infty} (T(t)f_n)(x), \qquad x \in \mathbb{R}^N.$$

Hence, $p(t,\cdot;B)$ is Borel measurable, being the pointwise limit of a sequence of continuous functions.

Finally, we show that $p(t,x;dy)$ is stochastically continuous; clearly this implies that it is also normal. For this purpose, for any $x_0 \in \mathbb{R}^N$ and $R > 0$, we introduce the function $f_{x_0,r} : \mathbb{R}^N \to \mathbb{R}$ defined by $f_{x_0,r}(x) = (1 - r^{-1}|x - x_0|)\chi_{B_r(x_0)}(x)$ for any $x \in \mathbb{R}^N$. As is easily seen,

$$f_{x_0,r}(x_0) - (T(t)f_{x_0,r})(x_0) = 1 - \int_{B_r(x_0)} f_{x_0,r}(y)p(t,x_0;dy) \geq 1 - p(t,x_0;B_r(x_0)).$$

Letting t tend to 0^+, Theorem 1.2.1 yields $\lim_{t\to0^+} p(t,x_0;B_r(x_0)) = 1$.

Now, we fix an arbitrary open set $U \subset \mathbb{R}^N$. Then, with any $x \in U$, we associate an open ball $B_r(x)$ contained in U. Since $1 \geq p(t,x;U) \geq p(t,x;B_r(x))$, we deduce that $p(t,x;U)$ tends to 1 as t tends to 0^+ and we are done. □

Remark 1.2.6 For estimates on the behaviour of the function g as t tends to 0^+ and as $|x|$, $|y|$ tend to $+\infty$, we refer the reader to Chapter 8.

Theorem 1.2.7 *For any $\lambda > c_0$ we have*

$$K_\lambda(x,y) = \int_0^{+\infty} e^{-\lambda t} g(t,x,y)dt, \qquad x,y \in \mathbb{R}^N, \tag{1.2.11}$$

and, for any $f \in C_b(\mathbb{R}^N)$,

$$(R(\lambda)f)(x) = \int_0^{+\infty} e^{-\lambda t}(T(t)f)(x)dt, \qquad x \in \mathbb{R}^N. \tag{1.2.12}$$

Proof The equality (1.2.11) is a consequence of (D.3.6) with $\Omega = B_n$, (1.1.6) and the monotone convergence theorem. Then, (1.2.12) follows from (1.2.7), (1.2.11) and the Fubini theorem. □

In general, $\{T(t)\}$ is a strongly continuous semigroup neither in $C_b(\mathbb{R}^N)$ nor in $BUC(\mathbb{R}^N)$ (see [261, Theorem 4.2] and Proposition 10.2.6). Nevertheless, as a straightforward consequence of Theorems 1.2.1 and 1.2.5, we deduce that, for a general $f \in C_b(\mathbb{R}^N)$, $T(t)f$ converges to f as t tends to 0, locally uniformly in $\mathbb{R}^N$. Actually, as the next proposition shows, if f vanishes at infinity, then $T(t)f$ converges to f in $C_b(\mathbb{R}^N)$, as t tends to 0.

Proposition 1.2.8 ([197], Prop. 4.3) *For any function $f \in C_0(\mathbb{R}^N)$, $T(t)f$ tends to f in $C_b(\mathbb{R}^N)$, as t tends to 0^+.*

---

[2] If B has finite positive Lebesgue measure, then one can take, up to a subsequence, $f_n = \rho_n \star \chi_B$ ($n \in \mathbb{N}$), where ρ is a standard mollifier and "$\star$" denotes convolution. If B has infinite Lebesgue measure, set $B^M = B \cap B_M$ for any $M \in \mathbb{N}$. By the above construction, one can determine a sequence $\{f_n^M\}$ converging almost everywhere in $\mathbb{R}^N$ to χ_{B^M} as n tends to $+\infty$, such that $\{f_n^{M+1}\} \subset \{f_n^M\}$ for any $M \in \mathbb{N}$. Then, it suffices to set $f_n = f_n^n$ for any $n \in \mathbb{N}$.

Proof We prove the statement assuming that $f \in C_c^\infty(\mathbb{R}^N)$. The general case then will follow by density. So, let us fix $f \in C_c^\infty(\mathbb{R}^N)$ and $x \in \mathbb{R}^N$. Moreover, let $k \in \mathbb{N}$ be such that B_k contains both x and $\mathrm{supp}(f)$. Then, $A_k f(x) = \mathcal{A}f(x)$, where A_k denotes the realization of the operator $\mathcal{A}$ in $C(\overline{B}_k)$ with homogeneous Dirichlet boundary conditions. Let $u_k = T_k(\cdot)f$, where $\{T_k(t)\}$ is the analytic semigroup generated by A_k. For any $t > 0$ we can write

$$\begin{aligned}u_k(t,x) - f(x) &= \int_0^t \frac{\partial}{\partial s} u_k(s,x)ds = \int_0^t (A_k T_k(s)f)(x)ds\\ &= \int_0^t (T_k(s)\mathcal{A}f)(x)ds = \int_0^t ds \int_{\mathbb{R}^N} g_k(s,x,y)\mathcal{A}f(y)dy,\end{aligned}$$

where we have extended $g_k(t,x,\cdot)$ to the whole of $\mathbb{R}^N$ by setting $g_k(t,x,y) = 0$ for any $y \notin B_k$. Letting k tend to $+\infty$, from the dominated convergence theorem it follows that

$$|(T(t)f)(x) - f(x)| = \left|\int_0^t (T(s)\mathcal{A}f)(x)ds\right| \le \|\mathcal{A}f\|_\infty \int_0^t e^{c_0 s}ds.$$

The assertion follows taking the supremum over $\mathbb{R}^N$ and letting t tend to 0^+. □

Remark 1.2.9 The results of the previous proposition do not imply that the restriction of the semigroup to $C_0(\mathbb{R}^N)$ gives rise to a strongly continuous semigroup. Indeed, as is shown in Section 4.3, in general, $\{T(t)\}$ does not map $C_0(\mathbb{R}^N)$ into itself.

Taking advantage of Theorem 1.2.5 we can prove some interesting properties of the semigroup $\{T(t)\}$.

Proposition 1.2.10 *Let $\{f_n\} \subset C_b(\mathbb{R}^N)$ be a bounded sequence of continuous functions converging pointwise to a function $f \in C_b(\mathbb{R}^N)$ as n tends to $+\infty$. Then, $T(\cdot)f_n$ tends to $T(\cdot)f$ in $C^{1,2}(K)$ for any compact set $K \subset (0,+\infty) \times \mathbb{R}^N$.*

Further, if f_n tends to f uniformly on compact subsets of $\mathbb{R}^N$, then $T(t)f_n$ converges to $T(t)f$ locally uniformly in $[0,+\infty) \times \mathbb{R}^N$ as n tends to $+\infty$.

Proof To prove the first part of the proof, we fix $0 < T_1 < T_2$, $R > 0$, a sequence $\{f_n\} \subset C_b(\mathbb{R}^N)$, converging pointwise to $f \in C_b(\mathbb{R}^N)$ and such that $\sup_{n\in\mathbb{N}} \|f_n\|_\infty \le K$ for some $K > 0$, and we prove that $T(\cdot)f_n$ converges to $T(\cdot)f$ in $C^{1,2}([T_1,T_2] \times \overline{B}_R)$ as n tends to $+\infty$. Formula (1.2.7) and the dominated convergence theorem show that $T(\cdot)f_n$ converges pointwise to $T(\cdot)f$ in $(0,+\infty) \times \mathbb{R}^N$ as n tends to $+\infty$. Moreover, taking (1.2.6) into account, we conclude that $\sup_{n\in\mathbb{N}} \|T(t)f_n\|_\infty \le K(e^{c_0 T} \vee 1)$ for any $t \in [0,T]$. The interior Schauder estimates in Theorem D.1.5 then imply that the sequence $\{T(\cdot)f_n\}$ is bounded in $C^{1+\alpha/2,2+\alpha}([T_1,T_2] \times \overline{B}_R)$. Hence, by the Arzelà-Ascoli Theorem, there exists a subsequence $\{T_{n_k}(\cdot)\}$ converging in $C^{1,2}([T_1,T_2] \times \overline{B}_R)$ to a function $v \in C^{1+\alpha/2,2+\alpha}([T_1,T_2] \times \overline{B}_R)$. Since, $T(\cdot)f_n$ converges pointwise to $T(\cdot)f$ in $(0,+\infty) \times \mathbb{R}^N$, we deduce that $v = T(\cdot)f$ and the whole sequence $\{T(\cdot)f_n\}$ converges to $T(\cdot)f$ in $C^{1,2}([T_1,T_2] \times \overline{B}_R)$.

Now, we suppose that the sequence $\{f_n\} \subset C_b(\mathbb{R}^N)$ converges uniformly to f on compact subsets of $\mathbb{R}^N$ and we show that, for any $R, T > 0$, $T(\cdot)f_n$ tends to $T(\cdot)f$ uniformly in $[0,T] \times \overline{B}_R$. Possibly replacing f_n with $f_n - f$, we can suppose that $f \equiv 0$.

For any $n \in \mathbb{N}$, let $\varphi_n \in C_0(\mathbb{R}^N)$ be a nonnegative function such that $\chi_{B_{n-1}} \le \varphi_n \le \chi_{B_n}$. Moreover, for any $\varepsilon, R > 0$, let

$$C_{\varepsilon,R} = \left\{s \ge 0 : \ \exists n \in \mathbb{N} \ \text{ s.t. } \inf_{(t,x)\in[0,s]\times B_R} (T(t)(\varphi_n - \mathbf{1}))(x) \ge -\varepsilon\right\}.$$

Let us prove that $C_{\varepsilon,R}=[0,+\infty)$ for any $\varepsilon, R>0$. For this purpose, it suffices to prove that $C_{\varepsilon,R}$ contains some $\widetilde{s}>0$, i.e., that there exists $n_0\in\mathbb{N}$ such that $T(\cdot)(\varphi_{n_0}-\mathbb{1})\geq -\varepsilon$ in $[0,\widetilde{s}]\times B_R$. Indeed, if this is the case, then $C_{\varepsilon,R}$ clearly contains the interval $[0,\widetilde{s}]$. Moreover, by the first part of the proof, we know that $T(\cdot)(\varphi_n-\mathbb{1})$ converges to 0 uniformly to zero in $[\widetilde{s},M]\times B_R$ for any $M>\widetilde{s}$. Therefore, there exists $n_1\in\mathbb{N}$ such that $T(\cdot)(\varphi_{n_1}-\mathbb{1})\geq -\varepsilon$ in $[s,M]\times B_R$. Observing that the sequence $\{T(\cdot)(\varphi_n-\mathbb{1})\}$ is increasing pointwise in $[0,+\infty)\times\mathbb{R}^N$, we conclude that $T(\cdot)(\varphi_{n_0\vee n_1}-\mathbb{1})\geq-\varepsilon$ for any $t\in[0,M]$ and $M\in C_{\varepsilon,R}$ or, equivalently, $[0,M]\subset C_{\varepsilon,R}$ for any $M>0$, i.e., $C_{\varepsilon,R}=[0,+\infty)$. To prove that there exists $0<\widetilde{s}\in C_{\varepsilon,R}$, we fix $n_0\in\mathbb{N}$ larger than R. Since $\varphi_{n_0}-\mathbb{1}=0$ in B_R and the function $T(\cdot)(\varphi_{n_0}-\mathbb{1})$ is continuous in $[0,+\infty)\times B_R$, $T(\cdot)(\varphi_{n-0}-\mathbb{1})$ vanishes, uniformly in B_R, as t tends to 0^+. Hence, there exists $\widetilde{s}>0$ such that $T(t)(\varphi_{n_0}-\mathbb{1})\geq-\varepsilon$ in B_R for any $t\in[0,\widetilde{s}]$, and we are done.

Now, since $p(t,x;B_m)\geq (T(t)\varphi_m)(x)$ for any $t>0$, $x\in\mathbb{R}^N$ and $m\in\mathbb{N}$, and $C_{\varepsilon,R}=[0,+\infty)$, we easily deduce that, for any arbitrarily fixed $T>0$ and $R>0$, there exists $m\in\mathbb{N}$ such that $p(t,x;B_m)\geq (T(t)\mathbb{1})(x)-\varepsilon=p(t,x;\mathbb{R}^N)-\varepsilon$ for any $t\in[0,T]$ and $x\in B_R$. Therefore, we can estimate

$$
\begin{aligned}
|(T(t)f_n)(x)|\leq&\int_{B_m}|f_n(y)|p(t,x;dy)+\int_{\mathbb{R}^N\setminus B_m}|f_n(y)|p(t,x;dy)dy\\
\leq& e^{c_0t}\sup_{y\in B_m}|f_n(y)|+Kp(t,x;\mathbb{R}^N\setminus B_m)\leq e^{c_0t}\sup_{y\in B_m}|f_n(y)|+K\varepsilon
\end{aligned}
$$

for any $t\in[0,T]$ and $x\in B_R$. Now, the assertion follows. □

Remark 1.2.11 Using the formula (1.2.7), the semigroup $\{T(t)\}$ can be extended to a semigroup (which we still denote by $\{T(t)\}$) in the space $B_b(\mathbb{R}^N)$ of all the bounded Borel measurable functions, and, for any $f\in B_b(\mathbb{R}^N)$ and any bounded sequence $\{f_n\}\in C_b(\mathbb{R}^N)$ converging pointwise to f, $T(\cdot)f_n$ converges to $T(\cdot)f$ pointwise in $(0,+\infty)\times\mathbb{R}^N$. Clearly, estimate (1.2.6) holds also for $f\in B_b(\mathbb{R}^N)$. Similarly, if $\{f_n\}\subset B_b(\mathbb{R}^N)$ is a bounded sequence converging pointwise to $f\in B_b(\mathbb{R}^N)$, then $T(\cdot)f_n$ converges to $T(\cdot)f$ pointwise in $[0,+\infty)\times\mathbb{R}^N$. Actually, taking the forthcoming Proposition 1.2.13, into account, we can show that $T(\cdot)f_n$ converges locally uniformly in $(0,+\infty)\times\mathbb{R}^N$. Indeed, if $[a,b]\times K$ is a compact set in $(0,+\infty)\times\mathbb{R}^N$, then we can split $T(t)f_n=T(t-a/2)T(a/2)f_n$ for any $t\geq a/2$. By the above result and Proposition 1.2.13, the sequence $\{T(a/2)f_n\}$ is contained in $C_b(\mathbb{R}^N)$, is bounded and it converges pointwise to $T(a/2)f$, as n tends to $+\infty$. Proposition 1.2.10 now implies that $T(\cdot)f_n$ converges to $T(\cdot)f$ in $[a,b]\times K$, as n tends to $+\infty$.

Let us now prove that $\{T(t)\}$ is irreducible and has the strong Feller property. For this purpose, we recall the following definition.

Definition 1.2.12 *A semigroup $\{S(t)\}$ in $B_b(\mathbb{R}^N)$ is irreducible if for any nonempty open set $U\subset\mathbb{R}^N$ it holds that $(S(t)\chi_U)(x)>0$ for any $t>0$ and $x\in\mathbb{R}^N$. It has the strong Feller property if $S(t)f\in C_b(\mathbb{R}^N)$ for any $f\in B_b(\mathbb{R}^N)$.*

Proposition 1.2.13 *$\{T(t)\}$ is irreducible and has the strong Feller property.*

Proof Showing that the semigroup is irreducible is easy due to the fact that g is strictly positive (see Theorem 1.2.5).

To prove that $\{T(t)\}$ is strong Feller, fix $f\in B_b(\mathbb{R}^N)$ and let $\{f_n\}\subset C_b(\mathbb{R}^N)$ be a bounded sequence converging almost everywhere in $\mathbb{R}^N$ to f as n tends to $+\infty$. Applying the interior Schauder estimates in Theorem D.1.5 we deduce that for any compact set $F\subset(0,+\infty)\times\mathbb{R}^N$ there exists a positive constant $C=C(F)$ such that $\|T(\cdot)f_n\|_{C^{1+\alpha/2,2+\alpha}(F)}\leq$

$C\|f_n\|_\infty$ for any $n \in \mathbb{N}$. Since $\sup_{n\in\mathbb{N}} \|f_n\|_\infty$ is finite and $T(\cdot)f_n$ converges to $T(\cdot)f$ pointwise in $[0,+\infty) \times \mathbb{R}^N$ (see Remark 1.2.11), we deduce that $T(t)f$ is continuous in $\mathbb{R}^N$ for any $t > 0$. □

Remark 1.2.14 Since g is strictly positive, $T(t)\chi_E$ is strictly positive as well in $\mathbb{R}^N$ for any $t > 0$ and any Borel set $E \subset \mathbb{R}^N$ with a positive Lebesgue measure.

1.3 The weak generator of $\{T(t)\}$

Since $\{T(t)\}$ is not strongly continuous in $C_b(\mathbb{R}^N)$ and in general is not strongly continuous in either $C_0(\mathbb{R}^N)$ or $BUC(\mathbb{R}^N)$, we can not define the infinitesimal generator in the usual sense. Nevertheless, we can still associate a "generator" with $\{T(t)\}$, the so-called weak generator, which has properties similar to those of the infinitesimal generator.

We will provide three equivalent definitions of the weak generator. The first definition that we introduce was considered in [53, 131]: the resolvent operators $\{R(\lambda) : \lambda > c_0\}$ given by Theorem 1.1.3 satisfy the resolvent identity (1.1.4) and $R(\lambda)$ is injective in $C_b(\mathbb{R}^N)$ for any $\lambda > c_0$. Hence, by a classical result of functional analysis (see Proposition B.1.3), we can define the weak generator as the unique linear operator A_1 in $C_b(\mathbb{R}^N)$ such that

$$R(\lambda) = R(\lambda, A_1) \quad \text{and} \quad \operatorname{Im}(R(\lambda)) = D(A_1), \quad \lambda > c_0. \tag{1.3.1}$$

The second definition is based on the bounded pointwise convergence: a sequence $\{f_n\} \subset C_b(\mathbb{R}^N)$ is said to be boundedly and pointwise convergent to $f \in C_b(\mathbb{R}^N)$ if there exists a positive constant C such that $\|f_n\|_\infty \le C$ for any $n \in \mathbb{N}$ and if f_n converges to f pointwise in $\mathbb{R}^N$ (see [97]). This notion of convergence leads to the following definition of the weak generator which was introduced in [215, 216]:

$$\begin{cases} D(A_2) = \Big\{f \in C_b(\mathbb{R}^N) : \displaystyle\sup_{t\in(0,1)} \frac{\|T(t)f - f\|_\infty}{t} < +\infty \text{ and } \exists g \in C_b(\mathbb{R}^N) : \\ \qquad\qquad \displaystyle\lim_{t\to 0^+} \frac{(T(t)f)(x) - f(x)}{t} = g(x) \ \ \forall x \in \mathbb{R}^N\Big\}, \\ (A_2 f)(x) = \displaystyle\lim_{t\to 0^+} \frac{(T(t)f)(x) - f(x)}{t}, \qquad x \in \mathbb{R}^N, \ f \in D(A_2). \end{cases} \tag{1.3.2}$$

The third definition is based on the notion of mixed topology introduced in [258]. The mixed topology τ^M is the finest locally convex topology which agrees on every norm-bounded subsets of $C_b(\mathbb{R}^N)$ with the topology of the uniform convergence on compact sets. Equivalently, it can be defined by means of the family of seminorms

$$p_{\{a_n\},\{K_n\}}(f) = \sup_{n\in\mathbb{N}} \Big\{a_n \sup_{x\in K_n} |f(x)|\Big\}, \qquad f \in C_b(\mathbb{R}^N),$$

where $\{a_n\}$ is any sequence of positive numbers converging to zero and $\{K_n\}$ is any sequence of compact subsets of $\mathbb{R}^N$. Given a sequence $\{f_n\} \subset C_b(\mathbb{R}^N)$ and a function $f \in C_b(\mathbb{R}^N)$ it holds that

$$\tau^M\text{-}\lim_{n\to+\infty} f_n = f \iff \begin{cases} \|f_n\|_\infty \le C, & n \in \mathbb{N}, \\ f_n \to f & \text{locally uniformly}. \end{cases} \tag{1.3.3}$$

For results on transition semigroups and mixed topology we refer the reader to [118].

We define the generator of the semigroup in the mixed topology, i.e., the operator $A_3 : D(A_3) \to C_b(\mathbb{R}^N)$, as follows:

$$\begin{cases} D(A_3) = \left\{ f \in C_b(\mathbb{R}^N) : \; \exists g \in C_b(\mathbb{R}^N) : \; \tau^M\text{-}\lim_{t\to 0^+} \dfrac{T(t)f - f}{t} = g \right\}, \\[2ex] A_3 f = \tau^M\text{-}\lim_{t\to 0^+} \dfrac{T(t)f - f}{t}, \qquad f \in D(A_3). \end{cases} \tag{1.3.4}$$

Proposition 1.3.1 *The three operators A_1, A_2 and A_3 coincide.*

To prove Proposition 1.3.1 we need the following lemma.

Lemma 1.3.2 *For any $f \in D(A_2)$, $t > 0$ and $\lambda > c_0$ the functions $T(t)f$ and $R(\lambda)f$ belong to $D(A_2)$. Moreover, $A_2T(t)f = T(t)A_2 f$ and $A_2R(\lambda)f = R(\lambda)A_2 f$.*

Proof As a first step, we prove that

$$(T(t)R(\lambda)f)(x) = \int_0^{+\infty} e^{-\lambda s}(T(t+s)f)(x)ds \tag{1.3.5}$$

for any $f \in C_b(\mathbb{R}^N)$, $t > 0$, $x \in \mathbb{R}^N$ and $\lambda > 0$. We show first that, for any $\delta > 0$,

$$T(t)\Big(\int_0^\delta e^{-\lambda s}T(s)f ds\Big)(x) = \int_0^\delta e^{-\lambda s}(T(t+s)f)(x)ds, \qquad t > 0, \;\; x \in \mathbb{R}^N. \tag{1.3.6}$$

For this purpose, we observe that, for any $x \in \mathbb{R}^N$, it holds that

$$\int_0^\delta e^{-\lambda s}(T(s)f)(x)ds = \lim_{k\to+\infty} \frac{1}{k}\sum_{j=0}^{k-1} e^{-\lambda\delta j/k}(T(\delta j/k)f)(x) =: \lim_{k\to+\infty} \sigma_k(f)(x).$$

Clearly, $(T(t)\sigma_k(f))(x)$ converges to $\int_0^\delta e^{-\lambda s}(T(s+t)f)(x)ds$ for any $t \ge 0$ and $x \in \mathbb{R}^N$, as k tends to $+\infty$. On the other hand, since $\{\sigma_k(f)\}$ is a bounded sequence in $C_b(\mathbb{R}^N)$, $(T(t)\sigma_k(f))(x)$ converges to $T(t)\big(\int_0^\delta e^{-\lambda s}T(s)f\big)(x)ds$ as k tends to $+\infty$, by Proposition 1.2.10. Thus, (1.3.6) follows.

Letting δ tend to $+\infty$ and using again Proposition 1.2.10, from (1.3.6) we get (1.3.5).

Now, we fix $f \in D(A_2)$, $t > 0$ and $K > 0$ such that

$$\sup_{h\in(0,1)} \frac{\|T(h)f - f\|_\infty}{h} = K. \tag{1.3.7}$$

Then, taking (1.2.6) into account, we get

$$\frac{\|T(h)T(t)f - T(t)f\|_\infty}{h} = \left\|T(t)\left(\frac{T(h)f - f}{h}\right)\right\|_\infty \le e^{c_0 t}K, \qquad h \in (0,1).$$

Moreover, since $(T(h)f - f)/t$ converges to $A_2 f$ pointwise as h tends to 0^+, from (1.3.7) and Proposition 1.2.10 we deduce that $(T(h)T(t)f - T(t)f)/h$ converges pointwise to $T(t)A_2 f$ as h tends to 0^+, i.e., $T(t)f \in D(A_2)$ and $A_2T(t)f = T(t)A_2 f$.

Next, from (1.3.5) we deduce that

$$\frac{(T(h)R(\lambda)f)(x) - (R(\lambda)f)(x)}{h} = \int_0^{+\infty} e^{-\lambda s}T(s)\left(\frac{T(h)f - f}{h}\right)(x)ds$$

for any $\lambda > c_0$ and $x \in \mathbb{R}^N$. Thus, using (1.3.7) we get

$$\sup_{h\in(0,1)} \frac{\|T(h)R(\lambda)f - R(\lambda)f\|_\infty}{h} \le \frac{K}{\lambda - c_0}. \tag{1.3.8}$$

Finally, by Proposition 1.2.10 and the dominated convergence theorem it follows that $(T(h)R(\lambda)f - R(\lambda)f)/h$ converges pointwise to $R(\lambda)A_2 f$ as h tends to 0^+. This implies that $R(\lambda)f \in D(A_2)$ and $A_2 R(\lambda)f = R(\lambda)A_2 f$. □

Proof of Proposition 1.3.1 Let us prove the inclusion $A_1 \subset A_3$. For this purpose, fix $\lambda > c_0$ and $f \in C_b(\mathbb{R}^N)$. Taking (1.3.5) into account, we get

$$\begin{aligned}
&\frac{(T(t)R(\lambda)f)(x) - (R(\lambda)f)(x)}{t}\\
=&\frac{1}{t}\left(\int_t^{+\infty} e^{-\lambda(s-t)}(T(s)f)(x)ds - \int_0^{+\infty} e^{-\lambda s}(T(s)f)(x)ds\right)\\
=&\frac{e^{\lambda t}-1}{t}(R(\lambda)f)(x) - \frac{e^{\lambda t}}{t}\int_0^t e^{-\lambda s}(T(s)f)(x)ds
\end{aligned} \tag{1.3.9}$$

for any $t > 0$ and $x \in \mathbb{R}^N$. From estimate (1.3.8) we deduce that the function $(t,x) \mapsto t^{-1}(T(t)R(\lambda)f - R(\lambda)f)(x)$ is bounded in $(0,1] \times \mathbb{R}^N$. Moreover, the last side of (1.3.9) tends to $\lambda R(\lambda)f - f$ locally uniformly in $\mathbb{R}^N$ as t tends to 0^+. Indeed, for any $R > 0$ and $x \in \overline{B}_R$ it holds that

$$\begin{aligned}
&\left|\frac{e^{\lambda t}}{t}\int_0^t e^{-\lambda s}(T(s)f)(x)ds - f(x)\right|\\
\le&\left|\frac{e^{\lambda t}}{t}\int_0^t e^{-\lambda s}\left((T(s)f)(x) - f(x)\right)ds\right| + \frac{1}{t}\left|\int_0^t (e^{-\lambda(s-t)} - 1)f(x)ds\right|\\
\le&e^{(\lambda\vee 0)t}\sup_{s\in[0,t]}\|T(s)f - f\|_{C(\overline{B}_R)} + \|f\|_\infty\frac{1}{t}\int_0^t (e^{-\lambda(s-t)} - 1)ds
\end{aligned} \tag{1.3.10}$$

and the last side of (1.3.10) converges to 0 as t tends to 0^+. By the characterization (1.3.3), it follows that $(T(t)R(\lambda)f - R(\lambda)f)/t$ converges to $\lambda R(\lambda)f - f$ in the mixed topology. This implies that $R(\lambda)f \in D(A_3)$ and

$$(A_3 R(\lambda)f)(x) = \lambda(R(\lambda)f)(x) - f(x) = (A_1 R(\lambda)f)(x) \tag{1.3.11}$$

for any $x \in \mathbb{R}^N$. Therefore, we conclude that $A_1 \subset A_3$.

The inclusion $A_3 \subset A_2$ is clear from the characterization (1.3.3).

We complete the proof by proving that $A_2 \subset A_1$. For this purpose, fix $f \in D(A_2)$ and $\lambda > c_0$. From Lemma 1.3.2 and the formula (1.3.11) we deduce that $f = (\lambda - A_1)R(\lambda)f = (\lambda - A_2)R(\lambda)f = R(\lambda)(\lambda f - A_2 f)$, implying that $f \in D(A_1)$ and $A_1 f = A_2 f$. □

Definition 1.3.3 *The operator* $\widehat{A} := A_1 = A_2 = A_3$, *as defined in* (1.3.1), (1.3.2) *and* (1.3.4), *is called the weak generator of* $\{T(t)\}$.

The weak generator fulfils the following further properties.

Proposition 1.3.4 *For any* $f \in D(\widehat{A})$ *and any fixed* $x \in \mathbb{R}^N$, *the function* $(T(\cdot)f)(x)$ *is continuously differentiable in* $[0,+\infty)$ *and*

$$\frac{d}{dt}(T(t)f)(x) = (T(t)\widehat{A}f)(x), \qquad t \ge 0. \tag{1.3.12}$$

For any sequence $\{f_n\} \subset D(\widehat{A})$ such that f_n and $\widehat{A}f_n$ converge boundedly and pointwise to some functions $f, g \in C_b(\mathbb{R}^N)$, respectively, it holds that $f \in D(\widehat{A})$ and $\widehat{A}f = g$.

Finally, $D(\widehat{A})$ is dense in $C_b(\mathbb{R}^N)$ in the mixed topology.

Proof Fix $f \in D(\widehat{A})$ and $x \in \mathbb{R}^N$. By Lemma 1.3.2 the right derivative

$$\frac{d^+}{dt}(T(t)f)(x) := \lim_{h\to 0^+}\frac{(T(t+h)f)(x)-(T(t)f)(x)}{h}$$

exists at any $t \geq 0$ and its value is $(T(t)\widehat{A}f)(x)$. Moreover, by Theorems 1.2.1 and 1.2.5 the function $t \mapsto (T(t)\widehat{A}f)(x)$ is continuous in $[0,+\infty)$. Hence, $(T(\cdot)f)(x)$ is differentiable in $[0,+\infty)$ and (1.3.12) holds.

Next, let $\{f_n\} \subset D(\widehat{A})$ be as in the statement. By the previous step, for any $x \in \mathbb{R}^N$ and $n \in \mathbb{N}$, the function $(T(\cdot)f_n)(x)$ is differentiable in $[0,+\infty)$ and its derivative is the function $(T(\cdot)\widehat{A}f_n)(x)$. Hence,

$$\frac{(T(t)f_n)(x)-f_n(x)}{t} = \frac{1}{t}\int_0^t (T(s)\widehat{A}f_n)(x)ds, \qquad t \geq 0.$$

Letting n tend to $+\infty$ and taking Proposition 1.2.10 into account, from the dominated convergence theorem we get

$$\frac{(T(t)f)(x)-f(x)}{t} = \frac{1}{t}\int_0^t (T(s)g)(x)ds.$$

Letting t tend to 0 from the right, we conclude that $f \in D(\widehat{A})$ and $\widehat{A}f = g$.

Finally, to prove that $D(\widehat{A})$ is dense in $C_b(\mathbb{R}^N)$ in the mixed topology it suffices to observe that any bounded and continuous function can be approximated by a sequence of functions $f_n \in C_c^\infty(\mathbb{R}^N)$ ($n \in \mathbb{N}$), bounded in the sup-norm and converging locally uniformly to f as n tends to $+\infty$. □

The next proposition shows the connections between $D_{\max}(\mathcal{A})$ (see (1.0.1)) and $D(\widehat{A})$. We recall that A denotes the realization of the operator $\mathcal{A}$, with $D_{\max}(\mathcal{A})$ as domain.

Proposition 1.3.5 *We have*

$$D_{\max}(\mathcal{A}) \cap C_0(\mathbb{R}^N) \subseteq D(\widehat{A}) \subseteq D_{\max}(\mathcal{A}). \tag{1.3.13}$$

Moreover, the following conditions are equivalent:

(i) $\lambda \in \rho(A)$ *for some* $\lambda > c_0$*, i.e., equation* (1.0.2) *has a unique bounded solution* $u \in D_{\max}(\mathcal{A})$*;*

(ii) $(c_0, +\infty) \subset \rho(A)$*;*

(iii) $(A, D_{\max}(\mathcal{A})) = (\widehat{A}, D(\widehat{A}))$.

Proof To prove the inclusion $D_{\max}(\mathcal{A}) \cap C_0(\mathbb{R}^N) \subseteq D(\widehat{A})$ we fix a function $u \in D_{\max}(\mathcal{A}) \cap C_0(\mathbb{R}^N)$, $\lambda > c_0$ and set $f = \lambda u - \mathcal{A}u$. Clearly, $f \in C_b(\mathbb{R}^N)$. For any $n \in \mathbb{N}$, let u_n be, as in the proof of Theorem 1.1.1, the solution of the elliptic problem

$$\begin{cases} \lambda u_n - \mathcal{A}u_n = f, & \text{in } B_n,\\ u_n = 0, & \text{on } \partial B_n. \end{cases}$$

The function $u_n - u$ solves the equation $\lambda v - \mathcal{A}v = 0$ and $u_n - u = -u$ on ∂B_n. By the classical maximum principle, it follows that $\|u_n - u\|_{C(\overline{B}_n)} = \sup_{|x|=n}|u(x)|$ and, consequently, u_n converges to u as n tends to $+\infty$, locally uniformly in $\mathbb{R}^N$. On the other hand, from the proof of Theorem 1.1.1 we know that u_n converges to the function $R(\lambda)f$ as n tends to $+\infty$, locally uniformly in $\mathbb{R}^N$. It follows that $R(\lambda)f = u$ and, consequently, $u \in D(\widehat{A})$.

The inclusion $D(\widehat{A}) \subseteq D_{\max}(\mathcal{A})$ follows from Theorems 1.1.1 and 1.1.3.

Now, we prove the second part of the proposition. It suffices to prove that "(i)$\Rightarrow$(iii)" and "(iii)$\Rightarrow$(ii)", since "(ii)$\Rightarrow$(i)" is trivial.

"(i)$\Rightarrow$(iii)". By Theorems 1.1.1 and 1.1.3 it follows immediately that $\widehat{A} \subset A$. Hence, we only need to prove that $A \subset \widehat{A}$. For this purpose, fix $u \in D_{\max}(\mathcal{A})$ and set $f = \lambda u - Au$ and $v = R(\lambda, \widehat{A})f$. Since $\widehat{A} \subset A$, it follows that $\lambda v - Av = f$. From property (i) we conclude that $u = v \in D(\widehat{A})$ and, therefore, property (iii) holds true.

"(iii)$\Rightarrow$(ii)". We observe that property (iii) implies that $\rho(A) = \rho(\widehat{A})$, which yields property (ii) by virtue of Theorems 1.1.1 and 1.1.3. □

We will provide some examples of situations in which $D_{\max}(\mathcal{A}) \neq D(\widehat{A})$, in Chapter 3, showing that, in general, the problem (1.0.5) is not uniquely solvable in $C_b(\mathbb{R}^N)$.

1.4 Analytic and non-analytic semigroups in $C_b(\mathbb{R}^N)$

In this section we confine ourselves to the case when the operator $\mathcal{A}$ is defined on smooth functions $u : \mathbb{R}^N \to \mathbb{R}$ by

$$\mathcal{A}u(x) = \Delta u(x) + \langle b(x), \nabla u(x)\rangle, \qquad x \in \mathbb{R}^N,$$

and the components b_j $(j = 1,\ldots,N)$ of the vector valued function $b \in C^1(\mathbb{R}^N)$ are unbounded functions. We begin by providing sufficient conditions on b which imply that $\widehat{A}$ (see Section 1.3) is not sectorial in $C_b(\mathbb{R}^N)$.

Theorem 1.4.1 *Suppose that there exist sequences $\{r_n\}$, $\{\lambda_n\} \subset (0,+\infty)$ and $\{\sigma_n\} \subset \mathbb{R}^N$ such that $r_n \leq M$, for any $n \in \mathbb{N}$ and some positive constant M, and*

$$(i)\ \lim_{n\to+\infty} \lambda_n^{-2} r_n = 0, \qquad (ii)\ \lim_{n\to+\infty} r_n b(\lambda_n x + \sigma_n)\lambda_n^{-1} = h \in \mathbb{R}^N \setminus \{0\}, \qquad (1.4.1)$$

uniformly with respect to x on compact subsets of $\mathbb{R}^N$. Then, $(\widehat{A}, D(\widehat{A}))$ is not sectorial in $C_b(\mathbb{R}^N)$. In particular, if $r_n = 1$ for any $n \in \mathbb{N}$, then the spectrum of $(\widehat{A}, D(\widehat{A}))$ contains the imaginary axis.

Proof We argue by contradiction and show that, if $(\widehat{A}, D(\widehat{A}))$ were sectorial, then the semigroup $\{S(t)\} \subset L(C_0(\mathbb{R}^N))$, defined by $(S(t)f)(x) = f(x+th)$ for any $t > 0$, $x \in \mathbb{R}^N$ and $f \in C_0(\mathbb{R}^N)$, should be analytic. But this is not the case, since the infinitesimal generator of $\{S(t)\}$ is the operator $B : \{u \in C_0(\mathbb{R}^N) : \frac{\partial u}{\partial h} \in C_0(\mathbb{R}^N)\} \to C_0(\mathbb{R}^N)$ defined by $Bu = \frac{\partial u}{\partial h}$ for any $u \in D(B)$, whose spectrum contains the imaginary axis.

So, let us assume that $(\widehat{A}, D(\widehat{A}))$ is sectorial. For any $n \in \mathbb{N}$ we introduce the isometries $I_n : C_b(\mathbb{R}^N) \to C_b(\mathbb{R}^N)$, defined by $I_n u = u(\lambda_n^{-1}(\cdot - \sigma_n))$ for any $u \in C_b(\mathbb{R}^N)$, and denote by $\{T_n(t)\}$ the minimal positive semigroup of contractions in $C_b(\mathbb{R}^N)$ associated with the operator $\mathcal{A}_n$ defined by

$$\mathcal{A}_n u = r_n\lambda_n^{-2}\Delta u + r_n\lambda_n^{-1}\langle b(\lambda_n \cdot + \sigma_n), \nabla u\rangle$$

on smooth functions $u:\mathbb{R}^N\to\mathbb{R}$ (see Theorem 1.2.5). Since $\mathcal{A}_n=r_nI_n^{-1}\mathcal{A}I_n$, the weak generator $\widehat{A}_n$ of the semigroup $\{T_n(t)\}$ is the operator $\widehat{A}_n:D(\widehat{A}_n):=I_n^{-1}(D(\widehat{A}))\to C_b(\mathbb{R}^N)$ defined by $\widehat{A}_n=r_n(I_n^{-1}\widehat{A}I_n)$ for any $n\in\mathbb{N}$. Therefore, $\rho(\widehat{A}_n)=r_n\rho(\widehat{A})$ and $R(\lambda,\widehat{A}_n)=r_n^{-1}I_n^{-1}R(r_n^{-1}\lambda,\widehat{A})I_n$ for any $n\in\mathbb{N}$. Since we are assuming that $(\widehat{A},D(\widehat{A}))$ is sectorial, there exist two positive constants C and K such that $\|R(\lambda,\widehat{A})\|_{L(C_b(\mathbb{R}^N))}\le C|\lambda|^{-1}$ for any $\lambda\in\mathbb{C}$ with real part greater than K. From this estimate, the previous formula and the assumptions on the sequence $\{r_n\}$, we conclude that

$$\|R(\lambda,\widehat{A}_n)\|_{L(C_b(\mathbb{R}^N))}\le C|\lambda|^{-1},\qquad \operatorname{Re}\lambda>MK. \tag{1.4.2}$$

Now, we fix $f\in C_c^\infty(\mathbb{R}^N)$ and check that for any $\lambda\in\mathbb{C}$, with $\operatorname{Re}\lambda>MK$, and $x\in\mathbb{R}^N$ it holds that

$$\lim_{n\to+\infty}(R(\lambda,\widehat{A}_n)f)(x)=(R(\lambda,B)f)(x). \tag{1.4.3}$$

Of course, once (1.4.3) is proved, letting n tend to $+\infty$ in (1.4.2) we will get $\|R(\lambda,B)f\|_\infty\le C|\lambda|^{-1}\|f\|_\infty$ for any $\operatorname{Re}\lambda>MK$, first for any $f\in C_c^\infty(\mathbb{R}^N)$ and then, by density, for any $f\in C_0(\mathbb{R}^N)$. Theorem C.2.7 then will imply that B is sectorial: a contradiction.

To show (1.4.3) we modify the proof of the classical Trotter-Kato theorem (see Theorem C.1.15). As a first step we prove that, for any $f\in C_c^\infty(\mathbb{R}^N)$, the function $T_n(\cdot)f$ converges to $S(\cdot)f$ in $[0,T]\times\mathbb{R}^N$ as n tends to $+\infty$, for any $T>0$. For this purpose, we observe that $S(t)$ maps $C_c^\infty(\mathbb{R}^N)\subset D(B)$ into itself for any $t\ge0$ and $C_c^\infty(\mathbb{R}^N)$ is contained in $D(\widehat{A}_n)$ (see Proposition 1.3.5). By Propositions 1.3.4 and since $D_tS(t)\psi=S(t)B\psi$ for any $t>0$ and any $\psi\in D(B)\subset C_c^\infty(\mathbb{R}^N)$, it follows that

$$\begin{aligned}(T_n(t)f)(x)-(S(t)f)(x)&=-\int_0^t\frac{d}{ds}[(T_n(t-s)S(s)f)(x)]ds\\&=\int_0^t(T_n(t-s)(\widehat{A}_n-B)S(s)f)(x)ds\end{aligned}$$

for any $t>0$ and $x\in\mathbb{R}^N$, which implies that

$$\|T_n(t)f-S(t)f\|_\infty\le T\sup_{s\in[0,T]}\|(\widehat{A}_n-B)S(s)f\|_\infty \tag{1.4.4}$$

for any $t\in[0,T]$ and any $T>0$. Note that the right-hand side of (1.4.4) converges to 0 as n tends to $+\infty$. Indeed, recalling that $\widehat{A}_nS(s)f=\mathcal{A}_nS(s)f$, we can estimate

$$|\widehat{A}_nS(\cdot)f-BS(\cdot)f|\le r_n\lambda_n^{-2}\|\Delta S(\cdot)f\|_\infty+|r_n\lambda_n^{-1}b(\lambda_n\cdot+\sigma_n)-h|\|\nabla_xS(\cdot)f\|_\infty\chi_{\operatorname{supp}(S(\cdot)f)}$$

in $[0,T]\times\mathbb{R}^N$. Since $\operatorname{supp}(S(s)f)\subset\overline{B}_{|h|T}(\operatorname{supp}(f))$ for any $s\in[0,T]$, using conditions (1.4.1), we conclude that $\|(\widehat{A}_n-B)S(s)f\|_\infty$ converges to 0 as n tends to $+\infty$. Therefore, from (1.4.4) it follows that $T_n(\cdot)$ converges to $S(\cdot)f$ uniformly in $[0,T]\times\mathbb{R}^N$.

Now, by (1.2.12) and (C.1.2) we can write

$$(R(\lambda,\widehat{A}_n)f)(x)-(R(\lambda,B)f)(x)=\int_0^{+\infty}e^{-\lambda t}(T_n(t)-S(t)f)(x)dt$$

for any $x\in\mathbb{R}^N$. Consequently,

$$|R(\lambda,\widehat{A}_n)f-R(\lambda,B)f|\le\int_0^{+\infty}e^{-\operatorname{Re}\lambda t}|(T_n(t)f)(\cdot)-(S(t)f)(\cdot)|dt \tag{1.4.5}$$

in $\mathbb{R}^N$, and the right-hand side of (1.4.5) converges to 0 as n tends to $+\infty$, by the dominated convergence theorem. Therefore, (1.4.3) follows.

To conclude the proof we assume that $r_n = 1$ for any $n \in \mathbb{N}$. Then, the resolvent sets of the operators $\widehat{A}$ and A_n coincide and $\|R(\lambda, \widehat{A})\|_{L(C_b(\mathbb{R}^N))} = \|R(\lambda, \widehat{A}_n)\|_{L(C_b(\mathbb{R}^N))}$ for any $\lambda \in \rho(\widehat{A})$ and $n \in \mathbb{N}$. Since $\rho(\widehat{A})$ contains the halfplane $\{\lambda \in \mathbb{C} : \mathrm{Re}\,\lambda > 0\}$, by (1.4.3) we can estimate $\|R(\lambda, B)\|_{L(C_0(\mathbb{R}^N))} \leq \|R(\lambda, \widehat{A})\|_{L(C_b(\mathbb{R}^N))}$. But, since $i\mathbb{R} \subset \sigma(B)$, $\|R(\lambda, B)\|_{L(C_0(\mathbb{R}^N))}$ (and consequently $\|R(\lambda, A)\|_{L(C_b(\mathbb{R}^N))}$) diverges to $+\infty$ as λ, with positive real part, tends to some point which lies on the imaginary axis, so that $i\mathbb{R} \subset \sigma(\widehat{A})$ (see Proposition B.1.4). This shows that the spectrum of $(\widehat{A}, D(\widehat{A}))$ contains the imaginary axis. □

Now we look for sufficient conditions under which the assumptions of Theorem 1.4.1 are satisfied.

Corollary 1.4.2 *Assume that there exist two sequences $\{\tau_n\} \subset \mathbb{R}^N$ and $\{\widetilde{\lambda}_n\} \subset (0, +\infty)$ such that $b(\tau_n) \neq 0$ for any $n \in \mathbb{N}$, $\{\widetilde{\lambda}_n\}$ is bounded, $\widetilde{\lambda}_n|b(\tau_n)|$ tends to $+\infty$ as n tends to $+\infty$ and*

$$\lim_{n\to+\infty} |b(\tau_n)|^{-1}|b(\tau_n + \widetilde{\lambda}_n x) - b(\tau_n)| = 0, \tag{1.4.6}$$

uniformly on compact subsets of $\mathbb{R}^N$. Then, $(\widehat{A}, D(\widehat{A}))$ is not sectorial. In particular, the previous conditions are satisfied in the following cases:

(i) *there exists a sequence $\{\tau_n\} \subset \mathbb{R}^N$, diverging to infinity, such that $|b(\tau_n)| \geq K|\tau_n|^\alpha$ for some $K, \alpha > 0$ and any $n \in \mathbb{N}$; moreover, $|\mathrm{Jac}\, b(x)| \leq K(1 + |x|^\beta)$ for any $x \in \mathbb{R}^N$ and some $\beta < \alpha$;*

(ii) *there exist a sequence $\{\tau_n\} \subset \mathbb{R}^N$, such that $|b(\tau_n)|$ diverges to $+\infty$ as n tends to $+\infty$, and positive numbers γ, δ, C such that $|\mathrm{Jac}\, b(x)| \leq \gamma|b(x)|^{\frac{3}{2}} + C$ for any $x \in \mathbb{R}^N$ with $|x - \tau_n| \leq \delta|b(\tau_n)|^{-1/2}$;*

(iii) *there exists a sequence $\{\tau_n\} \subset \mathbb{R}^N$, diverging to infinity, such that $|b(\tau_n)|$ tends to $+\infty$ as n tends to $+\infty$; moreover, $|\mathrm{Jac}\, b(x)|^2$ and $|b(x)|/|x|$ converge to 0 as $|x|$ tends to $+\infty$. In this case, $\sigma(\hat{A})$ contains the imaginary axis.*

Proof To prove the first part of the statement, let $\{\tau_{k_n}\}$ be a subsequence of $\{\tau_n\}$ such that $b(\tau_{k_n})/|b(\tau_{k_n})|$ converges to some $h \in \mathbb{R}^N$ as n tends to $+\infty$ If we set $\sigma_n = \tau_{k_n}$, $\lambda_n = \widetilde{\lambda}_{k_n}$, $r_n = \lambda_n|b(\sigma_n)|^{-1}$ for any $n \in \mathbb{N}$, then the three sequences $\{r_n\}$, $\{\lambda_n\}$ and $\{\sigma_n\}$ satisfy the conditions in Theorem 1.4.1.

To complete the proof, we show that, under one of the conditions in (i), (ii) and (iii), the operator $(\widehat{A}, D(\widehat{A}))$ is not sectorial. In the rest of the proof, we arbitrarily fix $R > 0$ and prove that the limit in (1.4.6) is uniform with respect to $x \in B_R$.

(i) The mean value theorem yields

$$|b(\tau_n + x) - b(\tau_n)| \leq R \sup_{0\leq\theta\leq1} |\mathrm{Jac}\, b(\tau_n + \theta x)| \leq KR(1 + (|\tau_n| + MR)^\beta)$$

for any $x \in B_R$ and some $M = M(\beta)$. Dividing both sides of the previous estimate by $|b(\tau_n)|$, condition (1.4.6) follows (with $\widetilde{\lambda}_n = 1$ for any $n \in \mathbb{N}$) since, by assumption, $|b(\tau_n)|$ tend to $+\infty$ faster than $|\tau_n|^\beta$.

(ii) Without loss of generality, we can assume that $\delta < (4\gamma)^{-1}$ and that $b(\tau_n) \neq 0$ for any $n \in \mathbb{N}$. Next, we fix an arbitrary $n \in \mathbb{N}$ and observe that, for any $s \in (0, \delta)$ and $|x| \leq s|b(\tau_n)|^{-1/2}$, we can estimate

$$
\begin{aligned}
|b(\tau_n)|^{-1}|b(\tau_n+x)-b(\tau_n)| \leq& |x||b(\tau_n)|^{-1}\sup_{0\leq\theta\leq 1}|\mathrm{Jac}\, b(\tau_n+\theta x)|\\
\leq& \gamma s\sup_{0\leq\theta\leq 1}|b(\tau_n)|^{-\frac{3}{2}}|b(\tau_n+\theta x)|^{\frac{3}{2}}+Cs|b(\tau_n)|^{-\frac{3}{2}}. \qquad (1.4.7)
\end{aligned}
$$

Taking the supremum with respect to $x \in \overline{B}_{s|b(\tau_n)|^{-1/2}}$, we conclude that the function $F_n : [0,\delta) \to \mathbb{R}$, defined by $F_n(s) = \sup\{|b(\tau_n+x)||b(\tau_n)|^{-1} : |x| \leq s|b(\tau_n)|^{-\frac{1}{2}}\}$ for any $s \in [0,\delta)$, satisfies the estimate $F_n \leq 1 + \frac{1}{4}F_n^{\frac{3}{2}} + C\delta|b(\tau_n)|^{-\frac{3}{2}}$. We claim that $F_n(s) < 4$ for any $s \in [0,\delta)$ and n sufficiently large. To prove the claim, we begin by noting that, for n large enough, it holds that $1 + C\delta|b(\tau_n)|^{-\frac{3}{2}} \leq 2$. It follows that $F_n(s) < 2 + \frac{1}{4}(F_n(s))^{\frac{3}{2}}$ for any $s \in [0,\delta)$. Since F_n is continuous in $[0,+\infty)$ with $F_n(0) = 1$ for any $n \in \mathbb{N}$ and $[0,4)$ is the biggest interval containing 1 in which the inequality $x < 2+\frac{1}{4}x^{3/2}$ holds, we deduce that $F_n(s) \in [0,4)$ for any $s \in [0,\delta)$, and the claim follows. Thus, for n large enough, equation (1.4.7) yields, for $|x| \leq s|b(\tau_n)|^{-1/2}$,

$$
|b(\tau_n)|^{-1}|b(\tau_n+x)-b(\tau_n)| \leq 8\gamma s + Cs|b(\tau_n)|^{-\frac{3}{2}}. \qquad (1.4.8)
$$

Now, let $\widetilde{\lambda}_n = |b(\tau_n)|^{-\frac{3}{4}}$. For any $x \in B_R$ and n sufficiently large that $R|b(\tau_n)|^{-1/4} < \delta$, we can estimate $|\widetilde{\lambda}_n x| \leq s|b(\tau_n)|^{-1/2}$ with $s = R|b(\tau_n)|^{-1/4}$. Applying (1.4.8), we get $|b(\tau_n)|^{-1}|b(\tau_n+\widetilde{\lambda}_n x)-b(\tau_n)| \leq 8R\gamma|b(\tau_n)|^{-\frac{1}{4}} + CR|b(\tau_n)|^{-\frac{7}{4}}$ and (1.4.6) follows.

(iii) We take $\widetilde{\lambda}_n = |b(\tau_n)|$, $r_n = 1$ for any $n \in \mathbb{N}$ and show that $\sup_{0\leq\theta\leq 1}|\mathrm{Jac}\, b(\tau_n+\theta\widetilde{\lambda}_n x)|$ vanishes as n tends to $+\infty$, uniformly with respect to $x \in \overline{B}_R$. For this purpose, it suffices to show that $|\tau_n + \theta\widetilde{\lambda}_n x|$ diverges to $+\infty$ as n tends to $+\infty$, uniformly with respect to $\theta \in [0,1]$ and $x \in \overline{B}_R$. This follows observing that, since b is sublinear, we have $|\tau_n+\theta\lambda_n x| \geq |\tau_n| - R|b(\tau_n)| \geq \frac{1}{2}|\tau_n|$ for any $\theta \in [0,1]$, $x \in \overline{B}_R$ and n sufficiently large. Using the mean value theorem, we can estimate $|b(\tau_n+\widetilde{\lambda}_n x) - b(\tau_n)| \leq R\widetilde{\lambda}_n \sup_{0\leq\theta\leq 1}|\mathrm{Jac}\, b(\tau_n+\theta\widetilde{\lambda}_n x)|$. Hence, we conclude that condition (1.4.6) is satisfied. Finally, since $r_n = 1$ for any $n \in \mathbb{N}$, Theorem 1.4.1 shows that $\sigma(\widehat{A})$ contains the imaginary axis. □

Remark 1.4.3 The conditions in Corollary 1.4.2(i) are always satisfied in the case when b_j are polynomials for any $j = 1,\ldots,N$. This is clear if $\deg(b_i) \leq 1$ for any $i = 1,\ldots,N$. Suppose that b_i is the polynomial of maximal degree, greater than one, and set $\tau_n = nx_0$ for any $n \in \mathbb{N}$, where $x_0 \in \mathbb{R}^N$ is such that the homogeneous part of b_i of maximum degree does not vanish at x_0. It is immediate to check that $|b(\tau_n)| \geq |b_i(\tau_n)| \geq Kn^{\deg(b_i)}$ for some positive constant $K > 0$ and n sufficiently large. Moreover, since $|\mathrm{Jac}\, b|^2$ is a polynomial of degree $2(\deg(b_i)-1)$, the condition on the Jacobian matrix of b in Corollary 1.4.2(i) is satisfied with $\beta = \deg(b_i) - 1 < \alpha := \deg(b_i)$.

Up to now we have shown sufficient conditions implying that $(\widehat{A}, D(\widehat{A}))$ is not sectorial. In the next theorem we provide suitable conditions on b which guarantee that $(\widehat{A}, D(\widehat{A}))$ is sectorial.

Theorem 1.4.4 *Let $b_j \in C^1(\mathbb{R}^N)\cap L^p(\mathbb{R}^N)$, for some $p \in (N,+\infty)$ and any $j = 1,\ldots,N$, be such that* $\mathrm{div}\, b$ *is bounded from below. Then, $(\widehat{A}, D(\widehat{A}))$ is sectorial in $C_b(\mathbb{R}^N)$.*

Proof As a first step we introduce the formal adjoint of $\mathcal{A}$, i.e., the operator $\mathcal{A}^* = \Delta - \langle b, \nabla\rangle - \mathrm{div}\, b$, and prove that its realization A^* in $L^1(\mathbb{R}^N)$ with domain $D(A^*) = \{u \in L^1(\mathbb{R}^N) : \Delta u, (\mathrm{div}\, b)u \in L^1(\mathbb{R}^N)\}$ generates an analytic semigroup in $L^1(\mathbb{R}^N)$ (here, Δu is meant in the sense of distributions). Then, we use this result to show that $(\widehat{A}, D(\widehat{A}))$ is sectorial.

Since $\operatorname{div} b$ is bounded from below and it is locally bounded, according to [136, Theorem I, Lemmata 6 and 8] the operator $B_0 = \Delta - \operatorname{div} b + \inf_{\mathbb{R}^N}(\operatorname{div} b) =: B + \inf_{\mathbb{R}^N}(\operatorname{div} b)$, with domain $D(B_0) = D(A^*)$, generates an analytic semigroup in $L^1(\mathbb{R}^N)$. Moreover,

$$\|\Delta u\|_{L^1(\mathbb{R}^N)} + \|(\operatorname{div} b)u\|_{L^1(\mathbb{R})} \le C(\|u\|_{L^1(\mathbb{R}^N)} + \|Bu\|_{L^1(\mathbb{R}^N)}) \tag{1.4.9}$$

for any $u \in D(A^*)$.

We now recall that, if $u \in L^1(\mathbb{R}^N)$ is such that $\Delta u \in L^1(\mathbb{R}^N)$, then $u \in W^{1,q}(\mathbb{R}^N)$ for any $q \in [1, N/(N-1))$ and there exists a positive constant $C = C(N,q)$, independent of u, such that

$$\|\nabla u\|_{L^q(\mathbb{R}^N)} \le C(N,q)\left(\|u\|_{L^1(\mathbb{R}^N)} + \|\Delta u\|_{L^1(\mathbb{R}^N)}\right), \tag{1.4.10}$$

see [248, Theorem 5.8]. Applying estimate (1.4.10), with $q = p/(p-1)$, to the function $v_\lambda : \mathbb{R}^N \to \mathbb{R}$ ($\lambda > 0$), defined by $v_\lambda(x) = u(\lambda x)$ for any $x \in \mathbb{R}^N$, and minimizing with respect to $\lambda > 0$, yields that

$$\|\nabla u\|_{L^{\frac{p}{p-1}}(\mathbb{R}^N)} \le C'(N,p)\|u\|_{L^1(\mathbb{R}^N)}^{\frac{p-N}{2p}} \|\Delta u\|_{L^1(\mathbb{R}^N)}^{\frac{N+p}{2p}}$$

for some positive constant $C'(N,p)$. Hölder inequality then shows that

$$\begin{aligned}\|\langle b, \nabla u\rangle\|_{L^1(\mathbb{R}^N)} \le& \|b\|_{L^p(\mathbb{R}^N)}\|\nabla u\|_{L^{\frac{p}{p-1}}(\mathbb{R}^N)}\\ \le& C'(N,p)\|b\|_{L^p(\mathbb{R}^N)}\|u\|_{L^1(\mathbb{R}^N)}^{\frac{p-N}{2p}}\|\Delta u\|_{L^1(\mathbb{R}^N)}^{\frac{N+p}{2p}}\end{aligned} \tag{1.4.11}$$

for any $u \in L^1(\mathbb{R}^N)$ such that $\Delta u \in L^1(\mathbb{R}^N)$.

Now, from (1.4.9) and (1.4.11) it follows that the term $u \mapsto \langle b, \nabla u\rangle$ is a small perturbation of $(B, D(B))$. Therefore, according to Theorem C.2.14, $(A^*, D(A^*))$ generates an analytic semigroup in $L^1(\mathbb{R}^N)$.

We can now prove that $(\widehat{A}, D(\widehat{A}))$ is sectorial in $C_b(\mathbb{R}^N)$. Possibly replacing A^* with $A^* - \omega$, with ω large enough, we can assume that $\|R(\lambda, A^*)\|_{L(L^1(\mathbb{R}^N))} \le M/|\lambda|$ for any $\lambda \in \mathbb{C}$ with positive real part. Let us set $D_\lambda = (\lambda - A^*)(C_c^\infty(\mathbb{R}^N))$ for such λ's. Since $C_c^\infty(\mathbb{R}^N)$ is a core of A^* (see [136, Theorem IV]), D_λ is dense in $L^1(\mathbb{R}^N)$ for any $\operatorname{Re}\lambda > 0$. Therefore, if $u \in D(\widehat{A}) \subset C_b(\mathbb{R}^N)$ and λ is as above, then

$$\begin{aligned}\|u\|_\infty =& \sup\left\{\int_{\mathbb{R}^N} u\varphi\, dx : \varphi \in D_\lambda,\ \|\varphi\|_{L^1(\mathbb{R}^N)} \le 1\right\}\\ \le& \sup\left\{\int_{\mathbb{R}^N} u(\lambda - A^*)v\, dx : v \in C_c^\infty(\mathbb{R}^N),\ \|v\|_{L^1(\mathbb{R}^N)} \le \frac{M}{|\lambda|}\right\}\\ =& \sup\left\{\int_{\mathbb{R}^N} v(\lambda - \widehat{A})u\, dx : v \in C_c^\infty(\mathbb{R}^N),\ \|v\|_{L^1(\mathbb{R}^N)} \le \frac{M}{|\lambda|}\right\}\\ \le& M|\lambda|^{-1}\|(\lambda - \widehat{A})u\|_\infty.\end{aligned}$$

We conclude that $\|R(\lambda, \widehat{A})\|_{L(C_b(\mathbb{R}^N))} \le M/|\lambda|$ and the assertion follows from Theorem C.2.7. □

Example 1.4.5 Let $\mathcal{A}$ be the operator defined on smooth functions u by $\mathcal{A}u(x,y) = D_x^2 u(x,y) + D_y^2 u(x,y) + b_0(\sqrt{x^2+y^2})(xu_y - yu_x)$ for any $(x,y) \in \mathbb{R}^2$, where $b_0 \in C^1(\mathbb{R})$. Note that the drift $b(x,y) = b_0(\sqrt{x^2+y^2})(-y,x)$ has null divergence. The assumptions of Theorem 1.4.4 hold if the function $r \mapsto r^{p+1}|b_0(r)|$ belongs to $L^1((0,+\infty))$ for some $p > 2$. In such a case, the semigroup associated with the operator $\mathcal{A}$ is analytic in $C_b(\mathbb{R}^N)$.

1.5 The Markov process

In this section we briefly consider the Markov process associated with the semigroup $\{T(t)\}$ and show the Dynkin formula. In the whole section we assume that c is nonpositive in $\mathbb{R}^N$.

We introduce a few notation. Let E be a topological space and let $\mathcal{B}$ be the σ-algebra of Borel subsets of E. Moreover, let Ω be an arbitrary set, $\mathcal{F}$ be a σ-algebra on it and $\tau : \Omega \to [0, +\infty]$ be a $\mathcal{F}$-measurable function. For any $t \geq 0$ we denote by $\mathcal{F}_t$ a σ-algebra on the set $\Omega_t = \{\omega : \ t < \tau(\omega)\}$, such that $(\mathcal{F}_s)_{|\Omega_t} \subset \mathcal{F}_t \subset \mathcal{F}$ for any $0 < s < t$.

Next, we denote by $X = \{X_t : \Omega_t \to E, \ t \geq 0\}$ a family of functions defined in Ω such that, for any $\omega \in \Omega$, $X_t(\omega) \in E$ is a trajectory defined for $t \in [0, \tau(\omega))$ and such that X_t is $\mathcal{F}_t$-measurable on Ω_t. Finally, let $\{\mathbb{P}_x : \ x \in E\}$ be a family of probability measures on $(\Omega, \mathcal{F}_0)$ such that the function

$$x \mapsto p(t, x; B) = \mathbb{P}_x(X_t \in B) \tag{1.5.1}$$

is Borel measurable for any fixed $t > 0$ and $B \in \mathcal{B}$, and such that $\mathbb{P}_x(X_0 = x) = 1$, and let $\mathbb{E}_x$ denote the expectation under the probability measure $\mathbb{P}_x$ $(x \in E)$.

The following definition of Markov process is taken from [92].

Definition 1.5.1 *X is a Markov process if, for any $x \in E$, any $s, t \geq 0$ and any $B \in \mathcal{B}$, we have*

$$\mathbb{P}_x(X_{t+s} \in B|\mathcal{F}_t) = \mathbb{P}_{X_t}(X_s \in B), \qquad \mathbb{P}_x\text{-a.s.} \tag{1.5.2}$$

and, for any $\omega \in \Omega_t$, there exists $\omega' \in \Omega_t$ such that $\tau(\omega') = \tau(\omega) - t$ and $X_s(\omega') = X_{t+s}(\omega)$ for any $s \in [0, \tau(\omega'))$.

We say that X is continuous if all the trajectories are continuous. Moreover, we say that two Markov processes are equivalent if they have same transition probabilities $\{p(t, x; dy)\}$.

Definition 1.5.2 *A random variable τ' with values in $[0, +\infty]$ is a Markov time of a Markov process X if $\tau' \leq \tau$ and $\{t < \tau'\} \in \mathcal{F}_t$ for any $t > 0$.*

The family of measures $\{p(t, x; dy)\}$ defined in (1.5.1) is a transition function (see Definition 1.2.4). Indeed, conditions (i) and (ii) in Definition 1.2.4 are straightforward, while the condition (iii) follows from (1.5.2). Indeed, for any $B \in \mathcal{B}$ we have

$$\begin{aligned} p(t + s, x; B) =& \mathbb{P}_x(X_{t+s} \in B) = \mathbb{E}_x \mathbb{P}_x(X_{t+s} \in B|\mathcal{F}_t) \\ =& \mathbb{E}_x p(s, X_t; B) = \int_E p(s, y; B) p(t, x; dy). \end{aligned}$$

Conversely, given a transition function $\{p(t, x; dy)\}$, we say that a Markov process X is associated with it if (1.5.1) holds. It is known that for any normal transition function there exists an associated Markov process, see [92, Theorem 3.2]. In particular, as far as the semigroup $\{T(t)\}$ is concerned, we have the following result.

Theorem 1.5.3 *There exists a continuous Markov process X associated with the semigroup $\{T(t)\}$. Moreover, for any $f \in B_b(\mathbb{R}^N)$, we have*

$$(T(t)f)(x) = \mathbb{E}_x[\chi_{t<\tau} f(X_t)], \qquad t > 0, \ x \in \mathbb{R}^N, \tag{1.5.3}$$

$$(R(\lambda)f)(x) = \mathbb{E}_x \int_0^\tau e^{-\lambda s} f(X_s) ds, \qquad \lambda > 0, \ x \in \mathbb{R}^N. \tag{1.5.4}$$

Proof The existence of X is proved in [92, Theorem 3.2], using the fact that the transition function of $\{T(t)\}$ is normal (see Theorem 1.2.5). See also [115, Theorem 1.6.3]. The continuity of X is proved in [25].

Then, (1.5.3) is straightforward: if $f \equiv \chi_B$ for some Borel set $B \subset \mathbb{R}^N$, then (1.5.3) is just (1.5.1). By linearity, the formula (1.5.3) can be extended, first, to any simple function f and, then, to any $f \in B_b(\mathbb{R}^N)$, by approximating with simple functions.

Finally, (1.5.4) follows from the resolvent formula (1.2.12) and (1.5.3), applying the Fubini theorem. □

For any set $U \subset \mathbb{R}^N$, we define the first exit time of X from U by

$$\tau_U = \inf\{t : \ X_t \notin U\}, \qquad x \in U, \tag{1.5.5}$$

and we denote by X^U the process induced by X in U, which is defined by $X_t^U = X_t$, if $t < \tau_U$, and $X_t^U = \infty$, if $t \geq \tau_U$.

Theorem 1.5.4 ([25]) *Let $U \subset \mathbb{R}^N$ be a regular bounded domain. Then, X^U is the Markov process associated with the semigroup $\{T^U(t)\}$.*

In the next theorem we state the Dynkin formula, in a slightly different form from the one in [92, Theorem 5.1].

Theorem 1.5.5 *Let $U \subset \mathbb{R}^N$ be a regular bounded domain, let $\tau' \leq \tau_U$ be a Markov time, and fix $\lambda \geq 0$. If $u \in W^{2,p}_{\rm loc}(\mathbb{R}^N)$, for any $p \in [1, +\infty)$, is such that $\mathcal{A}u \in C(\mathbb{R}^N)$, then*

$$u(x) = \mathbb{E}_x e^{-\lambda\tau'} u(X_{\tau'}) + \mathbb{E}_x \int_0^{\tau'} e^{-\lambda s}(\lambda u - \mathcal{A}u)(X_s)ds, \qquad x \in U. \tag{1.5.6}$$

Proof Let $U_1 \subset \mathbb{R}^N$ be a regular bounded domain such that $\overline{U} \subset U_1$, and let $\vartheta \in C_c^\infty(U_1)$ be a function such that $\vartheta \equiv 1$ in U. Define $u' = \vartheta u$ and $f' = \lambda u' - \mathcal{A}u'$. Then, we have $f' \in C_c(U_1)$ and $u' = R(\lambda, \mathcal{A}_{U_1})f'$, where $R(\lambda, \mathcal{A}_{U_1})$ is the resolvent operator of $\{T^{U_1}(t)\}$.

By [25, Theorem 1.6] (see also [115, Theorem 1.6.3]) X is a strong Markov process. Then, also X^{U_1} is strong Markov. So, we can apply [92, Theorem 5.1] to X^{U_1} and we get the formula

$$u'(x) = \mathbb{E}_x e^{-\lambda\tau'} u'(X_{\tau'}) + \mathbb{E}_x \int_0^{\tau'} e^{-\lambda s} f'(X_s)ds, \qquad x \in U_1.$$

Since $\tau' \leq \tau_U$ and since $u = u'$ and $f' = \lambda u - \mathcal{A}u$ in U, the restriction of this formula to $x \in U$ gives (1.5.6). □

Notice that, taking $\tau' = \tau_U$ in (1.5.6), it follows that the solution of the boundary value problem

$$\begin{cases} \lambda u(x) - \mathcal{A}u(x) = f(x), & x \in U, \\ u(x) = h(x), & x \in \partial U, \end{cases}$$

with $h \in C(\partial U)$, $f \in C(\overline{U})$ and $\lambda \geq 0$, can be represented by the formula

$$u(x) = \mathbb{E}_x e^{-\lambda\tau_U} h(X_{\tau_U}) - \mathbb{E}_x \int_0^{\tau_U} e^{-\lambda s} f(X_s)ds, \qquad x \in U.$$

1.6 The associated stochastic differential equation

In this section we consider the stochastic differential equation associated with the differential operator $\mathcal{A}$.

Let $\{W_t : t \geq 0\}$ be a N-dimensional Wiener process and let $\{\mathcal{F}_t^W : t \geq 0\}$ be the filtration generated by W_t. For any $x \in \mathbb{R}^N$, let $\sigma(x) \in \mathcal{L}(\mathbb{R}^N)$ be the unique positive definite matrix such that $Q(x) = \frac{1}{2}\sigma(x)\sigma(x)^*$. We consider the stochastic differential equation

$$\begin{cases} d\xi_t^x = b(\xi_t^x)dt + \sigma(\xi_t^x)dW_t, & t > 0, \\ \xi_0^x \equiv x, \end{cases} \tag{1.6.1}$$

where $x \in \mathbb{R}^N$ is fixed. This equation is a short writing of the integral equation

$$\xi_t^x = x + \int_0^t b(\xi_s^x)ds + \int_0^t \sigma(\xi_s^x)dW_s, \qquad t > 0. \tag{1.6.2}$$

We say that ξ_t^x is a solution of problem (1.6.1) if it is a continuous real process, defined for any $t \geq 0$, adapted to the filtration $\{\mathcal{F}_t^W\}$ and such that (1.6.2) holds almost surely. Note that the continuity of ξ_t^x ensures the existence of the stochastic integral in (1.6.2).

Let us consider the following hypothesis.

Hypothesis 1.6.1 The functions b and σ are continuous and satisfy

$$\|\sigma(x) - \sigma(y)\|_2^2 + 2\langle b(x) - b(y), x - y\rangle \leq K_R|x-y|^2, \qquad x, y \in B_R, \tag{1.6.3}$$

$$\mathcal{A}(1+|x|^2) = \|\sigma(x)\|_2^2 + 2\langle b(x), x\rangle \leq K(1+|x|^2), \qquad x \in \mathbb{R}^N, \tag{1.6.4}$$

where $K, K_R > 0$ are constants and $\|\sigma\|_2^2 = \mathrm{Tr}\,(\sigma\sigma^*)$.

Theorem 1.6.2 *There exists a unique (up to equivalence) solution ξ_t^x of problem* (1.6.1), *which is equivalent to the Markov process X.*

Proof For the proof of the existence and uniqueness of the solution ξ_t^x of equation (1.6.1) we refer the reader to [97, Theorems 3.7 and 3.11] or [145, Theorem V.1.1]. Then, using the Itô formula we can show that ξ_t^x is equivalent to X_t. Indeed, fix $t_0 > 0$ and define $v(t,x) = (T(t_0-t)f)(x)$ for $t \in [0,t_0]$ and $x \in \mathbb{R}^N$. For any $n \in \mathbb{N}$, let $\tau_n(x)$ be the first exit time of ξ_t^x from the ball B_n. Since ξ_t^x is continuous and defined for any $t > 0$, τ_n diverges to $+\infty$ as n tends to $+\infty$, almost surely. Since $D_t v = -\mathcal{A}v$ in $(0,t_0) \times \mathbb{R}^N$, the Itô formula gives

$$v(t \wedge \tau_n(x), \xi_{t\wedge\tau_n(x)}^x) = v(0,x) + \int_0^{t\wedge\tau_n(x)} \sigma(\xi_s^x)Dv(s,\xi_s^x)dW_s, \tag{1.6.5}$$

for any $t \in [0,t_0]$ and $x \in B_n$. If $t < t_0$, then, for any $s \in (0, t\wedge\tau_n(x))$, we have $|\sigma(\xi_s^x)Dv(s,\xi_s^x)| \leq \sup_{B_n}|\sigma| \sup_{[0,t]\times B_n}|Dv| < +\infty$ and, therefore,

$$\mathbb{E}\int_0^{t\wedge\tau_n(x)} |\sigma(\xi_s^x)Dv(s,\xi_s^x)|^2 ds < +\infty.$$

Then, the expectation of the stochastic integral in (1.6.5) is zero, and we get

$$\mathbb{E}\, v(t\wedge\tau_n(x), \xi_{t\wedge\tau_n(x)}(x)) = v(0,x) = (T(t_0)f)(x), \qquad t < t_0, \; x \in B_n, \; n \in \mathbb{N}.$$

Letting n tend to $+\infty$, from dominated convergence it follows that $\mathbb{E}\, v(t, \xi_t^x) = (T(t_0)f)(x)$ for any $t < t_0$ and, finally, letting t tend to t_0, we get $\mathbb{E}\, f(\xi_{t_0}^x) = (T(t_0)f)(x)$. This is equivalent to say that $\mathbb{P}(\xi_t^x \in B) = p(t, x; B)$ for any $t \geq 0$, $x \in \mathbb{R}^N$ and any Borel set B, where $\{p(t, x; dy)\}$ is the transition function of X. Thus, ξ and X are equivalent. □

Chapter 2

One-dimensional theory

2.0 Introduction

The one-dimensional case is particular and easier. It has been studied by W. Feller ([102]). Here, we consider the case when the second-order differential operator $\mathcal{A}$ is defined by

$$\mathcal{A}\psi(x) = q(x)\psi''(x) + b(x)\psi'(x), \qquad x \in \mathbb{R},$$

on smooth functions $\psi : \mathbb{R} \to \mathbb{R}$ under the following assumptions on q and b.

Hypothesis 2.0.1 *The coefficients q and b are continuous in $\mathbb{R}$. Moreover $q(x) > 0$ for any $x \in \mathbb{R}$.*

Under such a hypothesis, the existence of a solution of the elliptic equation $\lambda u - \mathcal{A}u = f$ in $D_{\max}(\mathcal{A}) = C_b(\mathbb{R}) \cap C_b^2(\mathbb{R})$ can be proved. Note that Hypothesis 2.0.1 is weaker than the ones in Chapter 1. This is not surprising at all, since already in the case of bounded coefficients, it is well known that the one-dimensional theory of elliptic equations requires less smoothness on the coefficients of the operator $\mathcal{A}$.

A deep analysis of the solutions to the homogeneous equation $\lambda u - \mathcal{A}u = 0$ allows us to prove that, for any $f \in C_b(\mathbb{R})$, there exists at least a solution $u = R_\lambda f$ to the equation $\lambda u - \mathcal{A}u = f$ which is in $C_b(\mathbb{R}) \cap C^2(\mathbb{R})$. We see that the boundedness at infinity of the solutions to the homogeneous equation is strictly connected to the integrability at infinity of certain functions $\mathscr{P}$ and $\mathscr{R}$ which depend on the coefficients q and b. This analysis will allow us to determine (integral) conditions on the coefficients which guarantee the uniqueness of the solution $u \in C_b(\mathbb{R}) \cap C^2(\mathbb{R})$ to the equation $\lambda u - \mathcal{A}u = f$. Then, we show that the arguments used in Chapter 1 to find a solution $u = R(\lambda)f \in D_{\max}(\mathcal{A})$ to the equation $\lambda u - \mathcal{A}u = f$ apply also in this situation in which the coefficients are less regular. Moreover, we prove that $R_\lambda f$ and $R(\lambda)f$ actually coincide. Finally, under suitable additional assumptions on q and b, we show that $D_{\max}(\mathcal{A}) = \{u \in C_b^2(\mathbb{R}) : \; qu'', bu' \in C_b(\mathbb{R})\}$.

2.1 The homogeneous equation

In this section, for any fixed $\lambda > 0$, we study the homogeneous equation

$$\lambda u - (qu'' + bu') = 0. \tag{2.1.1}$$

Remark 2.1.1 Let $u \in C^2(\mathbb{R})$ be a solution to equation (2.1.1). Then, u can attain neither a positive maximum nor a negative minimum. Indeed, suppose for instance that $x_0 \in \mathbb{R}$ is a positive maximum of the function u. Then, $u'(x_0) = 0$ and $u''(x_0) \le 0$. From (2.1.1) we get $\lambda u(x_0) = q(x_0)u''(x_0) \le 0$, which leads us to a contradiction. If u has a negative minimum, then it suffices to apply the previous argument to the function $-u$. As a consequence, if u solves (2.1.1) and vanishes at two different points x_0 and x_1, then u identically vanishes in $\mathbb{R}$.

The following lemma is a crucial step in order to understand the behaviour of the solutions of (2.1.1) in a neighborhood of $-\infty$ and $+\infty$.

Lemma 2.1.2 *There exist a positive, strictly decreasing function $\overline{u}_1$ and a positive, strictly increasing function $\overline{u}_2$, which solve equation* (2.1.1).

Proof To prove the existence of a positive, strictly decreasing solution $\overline{u}_1$ to equation (2.1.1) we introduce the set

$$\begin{aligned} B =&\{b \in \mathbb{R} : \exists u_b \text{ solution to (2.1.1) with } u_b(0) = 1,\ u_b'(0) = b \\ &\text{and } u_b(x) = 0 \text{ for some } x > 0\}. \end{aligned} \tag{2.1.2}$$

Let us prove that B is an interval. First of all, we observe that B is not empty. Indeed, denote by v_1 and v_2 two linearly independent solutions to equation (2.1.1). Then, the more general solution to (2.1.1) is given by $v = c_1v_1 + c_2v_2$. By the last part of Remark 2.1.1 the matrix whose rows are $(v_1(0), v_2(0))$ and $(v_1(x_0), v_2(x_0))$ is invertible for any $x_0 > 0$, since $v \equiv 0$ is the unique solution to equation (2.1.1) such that $v(0) = v(x_0) = 0$. This implies that for any $x_0 > 0$ there exists a unique solution v to (2.1.1) such that $v(0) = 1$, $v(x_0) = 0$. Hence, $b = v'(0)$ belongs to B.

To prove that B is an interval, we show that, if $b \in B$, then $(-\infty, b] \subset B$. For this purpose we observe that if $c < b$ then $u_c < u_b$ in $(0, +\infty)$. Indeed, since $u_c'(0) < u_b'(0)$ there exists at least an interval $(0, x_1)$ in which $u_c < u_b$. Suppose that $x_1 < +\infty$. Then, $u_b(x_1) = u_c(x_1)$. This would imply that the function $v = u_b - u_c$, which solves equation (2.1.1), should have two zeroes. Hence, it should be constant in $\mathbb{R}$, which is a contradiction. Therefore, $u_c < u_b$ in $(0, +\infty)$ and, consequently, u_c vanishes at some point $x \in (0, +\infty)$, so that $c \in B$.

Now we show that, if $b \in B$, then u_b is decreasing in $\mathbb{R}$. As a consequence we deduce that $B \subset (-\infty, 0]$. Fix $b \in B$ and let $x > 0$ be the unique zero of the function u_b. Then, u_b is decreasing in $(-\infty, x)$, otherwise it should have a positive maximum. Similarly, u_b is decreasing in $[x, +\infty)$. To check it, we observe that $u_b'(x) < 0$. Indeed, if $u_b'(x) = 0$, then u should coincide with the null solution. Therefore, u_b is strictly decreasing in a neighborhood of x. If u_b were not decreasing in $(x, +\infty)$ it should have a negative minimum, which is a contradiction.

Set $\overline{b} = \sup B$ and $\overline{u}_1 = u_{\overline{b}}$. We claim that $\overline{u}_1$ is strictly decreasing and positive in $\mathbb{R}$ and u_b converges to $\overline{u}_1$ in $C^2([-M, M])$ for any $M > 0$, as b tends to $\overline{b}$ from the left. Since $u_b' \le 0$ for any $b < \overline{b}$, it follows that $\overline{u}_1' \le 0$ as well. Actually, $\overline{u}_1$ is strictly decreasing, otherwise it would be constant on some interval, violating Remark 2.1.1. To prove that $\overline{u}_1$ is positive in $\mathbb{R}$, we equivalently prove that $\overline{b} \notin B$. By contradiction, suppose that $\overline{b} \in B$ and denote by $x > 0$ a positive zero of $\overline{u}_1$. Further, let u be the solution to (2.1.1) such that $u(0) = 1$ and $u(2x) = 0$ (whose existence has been shown in the first part of the proof). Since the function $v = u - \overline{u}_1$ vanishes at zero and is positive at x, it should be increasing in $[0, x]$, otherwise it would have a negative minimum, which could not be the case. Hence, $u'(0) \ge \overline{u}_1 = \overline{b}$. Actually, the strict inequality holds, otherwise, since $u(0) = \overline{u}_1(0)$, u and $\overline{u}_1$ should coincide, which, of course, is not the case. Hence, $u'(0) \in B$, but this is a contradiction since $\overline{b} = \sup B$.

Finally, to prove the existence of a positive increasing solution to (2.1.1) it suffices to set $\overline{u}_2(x) = v(-x)$ for any $x \in \mathbb{R}$, where v is the positive solution to the equation $\lambda v(x) - q(-x)v''(x) + b(-x)v'(x) = 0$, provided by the previous arguments. $\square$

In order to describe the behaviour of the solutions to equation (2.1.1), we introduce the functions

$$\mathscr{W}(x) = \exp\left(-\int_0^x \frac{b(s)}{q(s)}ds\right), \qquad x \in \mathbb{R}, \tag{2.1.3}$$

$$\mathscr{P}(x) = \frac{1}{q(x)\mathscr{W}(x)}\int_0^x \mathscr{W}(s)ds, \qquad x \in \mathbb{R}, \tag{2.1.4}$$

$$\mathscr{R}(x) = \mathscr{W}(x)\int_0^x \frac{1}{q(s)\mathscr{W}(s)}ds, \qquad x \in \mathbb{R}. \tag{2.1.5}$$

Moreover, we observe that $u \in C^2(\mathbb{R})$ is a solution to equation (2.1.1) if and only if u solves the differential equation

$$\left(\frac{u'}{\mathscr{W}}\right)' = \lambda\frac{u}{q\mathscr{W}}. \tag{2.1.6}$$

Therefore, any solution u of (2.1.1) satisfies

$$u'(x) = \mathscr{W}(x)\left(u'(0) + \lambda\int_0^x \frac{u(s)}{q(s)\mathscr{W}(s)}ds\right), \qquad x \in \mathbb{R}. \tag{2.1.7}$$

Remark 2.1.3 Notice that

$$\overline{u}_1(x)\overline{u}_2'(x) - \overline{u}_1'(x)\overline{u}_2(x) = w_0\mathscr{W}(x), \qquad x \in \mathbb{R}, \tag{2.1.8}$$

for some positive constant w_0. Formula (2.1.8) is immediately checked since the function $\mathscr{W}^{-1}(\overline{u}_1\overline{u}_2' - \overline{u}_1'\overline{u}_2)$ is positive[1] and its first-order derivative identically vanishes in $\mathbb{R}$.

Proposition 2.1.4 *The following properties are satisfied:*

(i) *all the solutions to equation* (2.1.1) *admit finite limit at* $+\infty$ *if and only if the function* $\mathscr{R}$ *belongs to* $L^1((0,+\infty))$*;*

(ii) *if* $\mathscr{P} \in L^1((0,+\infty))$ *and* $\mathscr{R} \notin L^1((0,+\infty))$ *then, for any positive decreasing solution* u *to equation* (2.1.1), *the function* $u'/\mathscr{W}$ *converges to* 0 *as* x *tends to* $+\infty$*;*

(iii) *if* $\mathscr{P}$, $\mathscr{R}$ *belong to* $L^1((0,+\infty))$ *then, for any solution* u *to equation* (2.1.1), *the functions* u *and* $u'/\mathscr{W}$ *admit finite limits at* $+\infty$. *Moreover, there exist two decreasing solutions* u_1 *and* u_2 *of* (2.1.1) *such that*

$$\lim_{x\to+\infty} u_j(x) = j-1, \qquad \lim_{x\to+\infty}\frac{u_j'(x)}{\mathscr{W}(x)} = -2+j, \qquad j=1,2; \tag{2.1.9}$$

(iv) *equation* (2.1.1) *admits a decreasing solution, with strictly positive limit at* $+\infty$, *if and only if* $\mathscr{P} \in L^1((0,+\infty))$.

[1] by virtue of Lemma 2.1.2, noting that $\overline{u}_j'$ ($j = 1,2$) could not vanish at any points $x \in \mathbb{R}$ since, in this case, from the differential equation (2.1.1), we would conclude that $u_j''(x) > 0$ and u_j' would be positive in a right-neighborhood of x.

Proof (i) Since any solution to equation (2.1.1) is given by a linear combination of the functions $\overline{u}_1$, $\overline{u}_2$ and $\overline{u}_1$ is decreasing (see Lemma 2.1.2), it suffices to show that $\lim_{x\to+\infty}\overline{u}_2(x)\in\mathbb{R}$ if and only if $\mathscr{R}\in L^1((0,+\infty))$. Recalling that $\overline{u}_2$ is increasing and $\overline{u}_2(0)=1$, we can estimate

$$\mathscr{R}(x)\le \mathscr{W}(x)\int_0^x \frac{\overline{u}_2(s)}{q(s)\mathscr{W}(s)}ds\le \overline{u}_2(x)\mathscr{R}(x),\qquad x>0. \tag{2.1.10}$$

Moreover, by (2.1.7) we can write

$$\overline{u}_2'(x)=\mathscr{W}(x)\left(\overline{u}_2'(0)+\lambda\int_0^x\frac{\overline{u}_2(s)}{q(s)\mathscr{W}(s)}ds\right),\qquad x>0. \tag{2.1.11}$$

Suppose that $\overline{u}_2$ is bounded in a neighborhood of $+\infty$. Then, the two terms in the right-hand side of (2.1.11) are in $L^1((0,+\infty))$, since they are both positive. Therefore, from (2.1.10) it follows that $\mathscr{R}\in L^1((0,+\infty))$.

Conversely, suppose that $\mathscr{R}\in L^1((0,+\infty))$. Plugging (2.1.10) into (2.1.11) we deduce that $\overline{u}_2'\le\overline{u}_2'(0)W+\lambda R\overline{u}_2$ in $(0,+\infty)$. Therefore,

$$\overline{u}_2(x)\le\exp\left(\lambda\int_0^x\mathscr{R}(t)dt\right)\left[1+\overline{u}_2'(0)\int_0^x\mathscr{W}(t)\exp\left(-\lambda\int_0^t\mathscr{R}(s)ds\right)dt\right] \tag{2.1.12}$$

for any $x>0$. Since $\mathscr{R}\in L^1((0,+\infty))$ and

$$\mathscr{W}(x)\le\left(\int_0^1\frac{1}{q(s)\mathscr{W}(s)}ds\right)^{-1}\mathscr{R}(x),\qquad x\ge 1,$$

$\mathscr{W}$ belongs to $L^1((0,+\infty))$ as well. Thus, from (2.1.12) it follows that $\overline{u}_2$ is bounded in $(0,+\infty)$.

(ii) Let u be a positive decreasing solution to (2.1.1). Then, equation (2.1.6) implies that the function $u'/\mathscr{W}$ is negative and increasing in $\mathbb{R}$. Therefore, it converges to a nonpositive number, which we denote by k, as x tends to $+\infty$. Let us prove that $k=0$. Integrating (2.1.6) from x to c and, then, letting c tend to $+\infty$, gives

$$u'(x)=\mathscr{W}(x)\left(k-\lambda\int_x^{+\infty}\frac{u(s)}{q(s)\mathscr{W}(s)}ds\right),\qquad x\in\mathbb{R}. \tag{2.1.13}$$

Since $\mathscr{P}$ belongs to $L^1((0,+\infty))$, the function $x\mapsto\mathscr{W}(x)\int_x^{+\infty}(q(s)\mathscr{W}(s))^{-1}ds$ belongs to $L^1((0,+\infty))$ as well, by the Fubini theorem. The boundedness of u in $(0,+\infty)$ yields the summability in $(0,+\infty)$ of the integral function in the right-hand side of (2.1.13). Therefore, from this quoted formula it follows that, if $k\neq 0$, then the function $\mathscr{W}$ is integrable in $(0,+\infty)$. Since

$$\frac{1}{q(x)\mathscr{W}(x)}\le\left(\int_0^1\mathscr{W}(s)ds\right)^{-1}\mathscr{P}(x),\qquad x\ge 1, \tag{2.1.14}$$

the function $1/(q\mathscr{W})$ is integrable in $(0,+\infty)$. But this implies that $\mathscr{R}$ is integrable in $(0,+\infty)$ as well. Hence, we get a contradiction.

(iii) Let us prove that, if $\mathscr{P},\mathscr{R}\in L^1((0,+\infty))$, then, for any solution u to (2.1.1), the functions $u'/\mathscr{W}$ and u itself admit finite limits at $+\infty$. By property (i) we can limit ourselves to showing that the limit $\lim_{x\to+\infty}u'/\mathscr{W}$ is finite. For this purpose we observe that from (2.1.14) and the boundedness of u at $+\infty$ it follows that $u/(q\mathscr{W})$ is integrable in $(0,+\infty)$. Therefore, dividing both sides of (2.1.7) by $\mathscr{W}(x)$ and letting x tend to $+\infty$, we conclude that $\lim_{x\to+\infty}u'(x)/\mathscr{W}(x)$ is finite.

Let us now set $\overline{u} = \overline{u}_1 - c\overline{u}_2$ where c is a constant such that $\lim_{x\to+\infty}\overline{u}(x) = 0$. Let us observe that $c \in [0,1)$. Of course, $c > 0$ since $\overline{u}_1$ and $\overline{u}_2$ are positive in $\mathbb{R}$. Hence, $\overline{u}$ is strictly decreasing in $\mathbb{R}$ and, since $\overline{u}(0) = 1 - c$, we conclude that $1 - c > 0$, i.e., $c < 1$. Arguing as in the proof of property (ii) we see that $\overline{u}'(x)/\mathscr{W}(x)$ tends to a nonpositive limit k as x tends to $+\infty$. If $k = 0$, then from (2.1.13) we would get

$$\frac{\overline{u}'(x)}{\overline{u}(x)} = -\lambda\frac{\mathscr{W}(x)}{\overline{u}(x)}\int_x^{+\infty}\frac{\overline{u}(s)}{q(s)\mathscr{W}(s)}ds \geq -\lambda\mathscr{W}(x)\int_x^{+\infty}\frac{1}{q(s)\mathscr{W}(s)}ds \tag{2.1.15}$$

for any $x \in \mathbb{R}$. As it has been shown in the proof of property (ii) the last side of (2.1.15) belongs to $L^1((0,+\infty))$ since $\mathscr{P} \in L^1((0,+\infty))$. It follows that the function $\log(\overline{u})$ is bounded from below in $(0,+\infty)$ and, consequently, $\lim_{x\to+\infty}\overline{u}(x) = l > 0$. Since this is a contradiction, k should be strictly negative. Therefore, setting $u_1 = -\tilde{u}/k$, we obtain a solution to (2.1.1) satisfying (2.1.9), with $j = 1$.

Now, let us show that problem (2.1.1) admits a positive decreasing solution u_2 satisfying (2.1.9), with $j = 2$. For this purpose, let w be the solution to (2.1.1) satisfying $w(0) = 0$ and $w'(0) = 1$. According to Remark 2.1.1, w is increasing in $\mathbb{R}$. Therefore, from (2.1.6) it follows that $w'/\mathscr{W}$ is increasing as well and, consequently, $\lim_{x\to+\infty} w'(x)/\mathscr{W}(x) \in (1,+\infty)$.

Let us set $v = u_1 + dw$, where $d > 0$ is a constant such that $v'/\mathscr{W}$ converges to zero as x tends to $+\infty$. Since v is nonnegative, by (2.1.6) we conclude that v is decreasing in $\mathbb{R}$. Moreover, since $\lim_{x\to+\infty} v'(x)/\mathscr{W}(x) = 0$, the previous arguments show that $\ell := \lim_{x\to+\infty} v(x)$ is positive. Thus, the function $u_2 = v/\ell$ is a solution to (2.1.1) satisfying (2.1.9) with $j = 2$.

(iv) Let u be a decreasing solution to equation (2.1.1) with positive limit ℓ at $+\infty$. Arguing as in the proof of property (ii), it can be shown that the function $u'/\mathscr{W}$ has nonpositive limit k at $+\infty$ and u' is given by (2.1.13). It follows that the function $u/(q\mathscr{W})$ is in $L^1((0,+\infty))$. Since all the terms in the right-hand side of (2.1.13) are nonpositive and the left-hand side is summable in $(0,+\infty)$, the function $x \mapsto \mathscr{W}(x)\int_x^{+\infty} u(s)/(q(s)\mathscr{W}(s))ds$ is in $L^1((0,+\infty))$ as well. From the condition $u \geq \ell$ in $(0,+\infty)$, it follows that also the function $x \mapsto \mathscr{W}(x)\int_x^{+\infty}(q(s)\mathscr{W}(s))^{-1}ds$ is summable in $(0,+\infty)$. Using the Fubini Theorem, this implies that $\mathscr{P} \in L^1((0,+\infty))$.

Conversely, let us suppose that $\mathscr{P} \in L^1((0,+\infty))$ and let us prove that equation (2.1.1) admits a positive decreasing solution u with $\lim_{x\to+\infty} u(x) = \ell > 0$. If $\mathscr{R} \in L^1((0,+\infty))$, then property (iii) gives us the wished function u. If $\mathscr{R} \notin L^1((0,+\infty))$, then any positive and decreasing solution u to (2.1.1) is given by (2.1.13) and $k = 0$ by property (ii). Hence u'/u satisfies (2.1.15) and, consequently, arguing as in the proof of property (iii), we conclude that $\lim_{x\to+\infty} u(x) = \ell > 0$. □

The behaviour of the solutions to equation (2.1.1) in a neighborhood of $-\infty$, is illustrated in the following proposition.

Proposition 2.1.5 *The following properties are satisfied:*

(i) *all the solutions to equation* (2.1.1) *admit finite limit at* $-\infty$ *if and only if the function* $\mathscr{R}$ *belongs to* $L^1((-\infty,0))$;

(ii) *if* $\mathscr{P} \in L^1((-\infty,0))$ *and* $\mathscr{R} \notin L^1((-\infty,0))$ *then, for any positive increasing solution* u *to equation* (2.1.1), *it holds that* $u'/\mathscr{W}$ *converges to* 0 *as* x *tends to* $-\infty$;

(iii) *if* $\mathscr{P}$, $\mathscr{R}$ *belong to* $L^1((-\infty,0))$ *then any solution to equation* (2.1.1) *is such that* u *and* $u'/\mathscr{W}$ *admit finite limits at* $-\infty$. *Moreover, there exist two increasing solutions* u

and v *of equation* (2.1.1) *such that*

$$(i)\quad \lim_{x\to-\infty} u(x)=0,\qquad \lim_{x\to-\infty}\frac{u'(x)}{\mathscr{W}(x)}=1,$$
$$(ii)\quad \lim_{x\to-\infty} v(x)=1,\qquad \lim_{x\to-\infty}\frac{v'(x)}{\mathscr{W}(x)}=0;$$

(iv) *equation* (2.1.1) *admits an increasing solution, with strictly positive limit at* $-\infty$*, if and only if* $\mathscr{P}\in L^1((-\infty,0))$.

Proof Let $\widetilde{\mathcal{A}}$ be the second-order differential operator defined by $\widetilde{\mathcal{A}}u=\widetilde{q}u''+\widetilde{b}u'$, where $\widetilde{q}(x)=q(-x)$, and $\widetilde{b}(x)=-b(-x)$ for any $x\in\mathbb{R}$. Further, let $\widetilde{\mathscr{W}}$, $\widetilde{\mathscr{P}}$ and $\widetilde{\mathscr{R}}$ be defined according to (2.1.3)-(2.1.5) with q, b being replaced with $\widetilde{q}$ and $\widetilde{b}$. Since $\widetilde{\mathscr{W}}(x)=\mathscr{W}(-x)$, $\widetilde{\mathscr{P}}(x)=-\mathscr{P}(-x)$, $\widetilde{\mathscr{R}}(x)=-\mathscr{P}(-x)$, the functions $\widetilde{\mathscr{W}}$, $\widetilde{\mathscr{P}}$ and $\widetilde{\mathscr{R}}$ are integrable in a neighborhood of $-\infty$ if and only if $\mathscr{W}$, $\mathscr{P}$ and $\mathscr{R}$ are integrable in a neighborhood of $+\infty$. Moreover, u solves $\lambda u-\mathcal{A}u=0$ if and only if the function $x\mapsto v(x):=u(-x)$ solves the differential equation $\lambda v-\widetilde{\mathcal{A}}v=0$. Now the assertion follows from Proposition 2.1.4. □

Now we introduce the following definitions.

Definition 2.1.6 *The point* $+\infty$ *is said*

$$\text{accessible if}\quad\begin{cases}\text{regular, i.e.,} & \mathscr{P}\in L^1((0,+\infty)),\ \mathscr{R}\in L^1((0,+\infty)),\\ \text{exit, i.e.,} & \mathscr{P}\notin L^1((0,+\infty)),\ \mathscr{R}\in L^1((0,+\infty)),\end{cases}$$

$$\text{unaccessible if}\quad\begin{cases}\text{entrance, i.e.,} & \mathscr{P}\in L^1((0,+\infty)),\ \mathscr{R}\notin L^1((0,+\infty)),\\ \text{natural, i.e.,} & \mathscr{P}\notin L^1((0,+\infty)),\ \mathscr{R}\notin L^1((0,+\infty)).\end{cases}$$

Similarly, the point $-\infty$ *is said*

$$\text{accessible if}\quad\begin{cases}\text{regular, i.e.,} & \mathscr{P}\in L^1((-\infty,0)),\ \mathscr{R}\in L^1((-\infty,0)),\\ \text{exit, i.e.,} & \mathscr{P}\notin L^1((-\infty,0)),\ \mathscr{R}\in L^1((-\infty,0)),\end{cases}$$

$$\text{unaccessible if}\quad\begin{cases}\text{entrance, i.e.,} & \mathscr{P}\in L^1((-\infty,0)),\ \mathscr{R}\notin L^1((-\infty,0)),\\ \text{natural, i.e.,} & \mathscr{P}\notin L^1((-\infty,0)),\ \mathscr{R}\notin L^1((-\infty,0)).\end{cases}$$

The results in Propositions 2.1.4 and 2.1.5 can be rephrased as follows.

Proposition 2.1.7 *The following properties are satisfied:*

(i) $+\infty$ (*resp.* $-\infty$) *is regular if and only if the differential equation* (2.1.1) *admits two positive decreasing* (*resp. increasing*) *solutions* u_1 *and* u_2 *such that*

$$\lim_{x\to+\infty} u_j(x)=j-1,\qquad \lim_{x\to+\infty}\frac{u_j'(x)}{\mathscr{W}(x)}=-2+j,\qquad j=1,2,$$

$$\left(\text{resp. }\lim_{x\to-\infty} u_j(x)=j-1,\qquad \lim_{x\to-\infty}\frac{u_j'(x)}{\mathscr{W}(x)}=2-j,\qquad j=1,2\right).$$

In this case all the solutions to equation (2.1.1) *are bounded in* $(0,+\infty)$ (*resp. in* $(-\infty,0)$);

(ii) $+\infty$ (*resp.* $-\infty$) *is an exit if and only if all the solutions of equation* (2.1.1) *are bounded in* $(0,+\infty)$ (*resp. in* $(-\infty,0)$) *and any positive decreasing* (*resp. increasing*) *solution* u *vanishes at* $+\infty$ (*resp. at* $-\infty$)*;*

(iii) $+\infty$ (*resp.* $-\infty$) *is an entrance if and only if the differential equation* (2.1.1) *admits a positive decreasing* (*resp. increasing*) *solution* u *such that the functions* u *and* $u'/\mathscr{W}$ *converge, respectively, to* 1 *and* 0 *as* x *tends to* $+\infty$ (*resp. as* x *tends to* $-\infty$)*, and any other solution of equation* (2.1.1)*, which is independent of* u*, is unbounded in* $(0,+\infty)$ (*resp. in* $(-\infty,0)$)*;*

(iv) $+\infty$ (*resp.* $-\infty$) *is natural if and only the differential equation* (2.1.1) *admits a positive decreasing* (*resp. increasing*) *solution* u *such that the functions* u *and* $u'/\mathscr{W}$ *vanish as* x *tends to* $+\infty$ (*resp. as* x *tends to* $-\infty$)*, and any other solution of equation* (2.1.1)*, which is independent of* u*, is unbounded in* $(0,+\infty)$ (*resp. in* $(-\infty,0)$).

Proof We just prove property (iv) since the other properties follow easily from Propositions 2.1.4 and 2.1.5. Moreover, we limit ourselves to dealing with the point $+\infty$ since the other case can be deduced from this one using Proposition 2.1.5 instead of Proposition 2.1.4.

Let us suppose that $+\infty$ is natural. Then, according to Proposition 2.1.4(iv), $\overline{u}_1$ vanishes at $+\infty$. To prove that $\overline{u}_1'/\mathscr{W}$ vanishes at $+\infty$ as well, we begin by observing that from (2.1.6), written with $\overline{u}_1$ instead of u, we deduce that $\overline{u}_1'/\mathscr{W}$ admits finite (and nonpositive) limit as x tends to $+\infty$ since it is negative and increasing. By contradiction, we assume that the previous limit is negative and we denote it by k. Since any solution of (2.1.1) is a linear combination of $\overline{u}_1$ and $\overline{u}_2$ and, according to Proposition 2.1.4(i), equation (2.1.1) admits solutions which are unbounded in a neighborhood of $+\infty$, it follows that $\overline{u}_2(x)$ diverges $+\infty$ as x tends to $+\infty$. Taking Remark 2.1.3 into account, we can write

$$1=\frac{\mathscr{W}(x)}{\mathscr{W}(x)}=\frac{1}{w_0}\left(\frac{\overline{u}_2'(x)}{\mathscr{W}(x)}\overline{u}_1(x)-\frac{\overline{u}_1'(x)}{\mathscr{W}(x)}\overline{u}_2(x)\right),\qquad x>0.$$

Recalling that $\overline{u}_2'$ and $\overline{u}_1$ are positive in $\mathbb{R}$ and taking the limit as x tends to $+\infty$, we are led to a contradiction.

Conversely, let us assume that there exists a positive decreasing solution $\widetilde{u}$ to equation (2.1.1) vanishing at $+\infty$ together with the function $\widetilde{u}'/\mathscr{W}$ and that any other solution to (2.1.1), independent of $\widetilde{u}$, is unbounded at $+\infty$. According to Proposition 2.1.4(i), it is clear that $\mathscr{R}\notin L^1((0,+\infty))$. To show that $\mathscr{P}\notin L^1((0,+\infty))$, we denote by v a solution to (2.1.1) linearly independent of $\widetilde{u}$. Then, any solution u to problem (2.1.1) is given by $u=c_1\widetilde{u}+c_2v$, for some $c_1,c_2\in\mathbb{R}$. Therefore, u is bounded at $+\infty$ if and only if $c_2=0$. But in such a case, u vanishes at $+\infty$. Therefore, Proposition 2.1.4(iv) implies that $\mathscr{P}\notin L^1((0,+\infty))$ and we are done. □

2.2 The nonhomogeneous equation

In this section we study the solutions $u\in C^2(\mathbb{R})$ of the nonhomogeneous equation

$$\lambda u-\mathcal{A}u=f, \tag{2.2.1}$$

when $f \in C_b(\mathbb{R})$. The classical theory of ordinary differential equations shows that the more general solution to the differential equation (2.2.1) is given by

$$u(x) = -\int_0^x \langle W(x)W^{-1}(t)e_2, e_1\rangle \frac{f(t)}{q(t)}dt + c_1\overline{u}_1(x) + c_2\overline{u}_2(x), \qquad x \in \mathbb{R}, \tag{2.2.2}$$

where c_1, c_2 are arbitrary real constants, $e_1 = (1,0)$, $e_2 = (0,1)$, $\overline{u}_1$, $\overline{u}_2$ are as in Lemma 2.1.2 and W is the Wronskian matrix

$$W(x) = \begin{pmatrix} \overline{u}_1(x) & \overline{u}_2(x) \\ \overline{u}_1'(x) & \overline{u}_2'(x) \end{pmatrix}, \qquad x \in \mathbb{R}.$$

Taking (2.1.8) into account, from (2.2.2) we can write, for any $x \in \mathbb{R}$,

$$u(x) = \left(c_1 + \frac{1}{w_0}\int_0^x \frac{f(t)\overline{u}_2(t)}{q(t)\mathscr{W}(t)}dt\right)\overline{u}_1(x) + \left(c_2 - \frac{1}{w_0}\int_0^x \frac{f(t)\overline{u}_1(t)}{q(t)\mathscr{W}(t)}dt\right)\overline{u}_2(x). \tag{2.2.3}$$

We can now prove the following result.

Proposition 2.2.1 *For any $f \in C_b(\mathbb{R})$ and $\lambda > 0$, the function $u = R_\lambda f$ defined by*

$$(R_\lambda f)(x) = \int_{-\infty}^{+\infty} G_\lambda(x,s)f(s)ds, \qquad x \in \mathbb{R}, \tag{2.2.4}$$

where $G_\lambda : \mathbb{R}^2 \to \mathbb{R}$ is given by

$$G_\lambda(x,s) = \begin{cases} \dfrac{\overline{u}_1(s)\overline{u}_2(x)}{w_0 q(s)\mathscr{W}(s)}, & x < s, \\[2ex] \dfrac{\overline{u}_1(x)\overline{u}_2(s)}{w_0 q(s)\mathscr{W}(s)}, & x \geq s, \end{cases}$$

belongs to $C_b(\mathbb{R})\cap C^2(\mathbb{R})$ and solves the differential equation (2.2.1). *Moreover, the operator $R_\lambda : C_b(\mathbb{R}) \to C_b(\mathbb{R})$ is bounded and $\|R_\lambda\|_{L(C_b(\mathbb{R}^N))} \leq 1/\lambda$.*

Proof A formal computation reveals that

$$(R_\lambda f)(x) = \frac{\overline{u}_1(x)}{w_0}\int_{-\infty}^x \frac{\overline{u}_2(s)}{q(s)\mathscr{W}(s)}f(s)ds + \frac{\overline{u}_2(x)}{w_0}\int_x^{+\infty} \frac{\overline{u}_1(s)}{q(s)\mathscr{W}(s)}f(s)ds \tag{2.2.5}$$

for any $x \in \mathbb{R}$. Let us prove that the integral terms in the right-hand side of (2.2.5) are well defined for any $x \in \mathbb{R}$. Since the function $\mathscr{W}^{-1}\overline{u}_2'$ is positive and increasing in $\mathbb{R}$ (as it easily follows from (2.1.6)), for any $c < x$ we can estimate

$$\begin{aligned}
\int_c^x \frac{\overline{u}_2(s)}{q(s)\mathscr{W}(s)}|f(s)|ds \leq& \|f\|_\infty \int_c^x \frac{\overline{u}_2(s)}{q(s)\mathscr{W}(s)}ds = \frac{1}{\lambda}\|f\|_\infty \int_c^x \left(\frac{\overline{u}_2'}{\mathscr{W}}\right)'(s)ds \\
=& \frac{1}{\lambda}\|f\|_\infty \left(\frac{\overline{u}_2'(x)}{\mathscr{W}(x)} - \frac{\overline{u}_2'(c)}{\mathscr{W}(c)}\right) \\
\leq& \frac{1}{\lambda}\|f\|_\infty \left(\frac{\overline{u}_2'(x)}{\mathscr{W}(x)} - \lim_{c\to-\infty}\frac{\overline{u}_2'(c)}{\mathscr{W}(c)}\right) \leq \frac{1}{\lambda}\|f\|_\infty \frac{\overline{u}_2'(x)}{\mathscr{W}(x)}.
\end{aligned} \tag{2.2.6}$$

Hence, the first integral in the right-hand side of (2.2.5) is well defined. A similar argument shows that, for any $c > x$,

$$\int_x^c \frac{\overline{u}_1(s)}{q(s)\mathscr{W}(s)}|f(s)|ds \leq -\frac{1}{\lambda}\|f\|_\infty \frac{\overline{u}_1'(x)}{\mathscr{W}(x)}. \tag{2.2.7}$$

Hence, also the other integral is well defined. Of course, they both define continuous functions in $\mathbb{R}$. Therefore, $R_\lambda f$ is a continuous function in $\mathbb{R}$ and it is bounded since (2.1.8), (2.2.6) and (2.2.7) imply that

$$|(R_\lambda f)(x)| \le \|f\|_\infty \left(\frac{\overline{u}_1(x)}{w_0\lambda}\frac{\overline{u}_2'(x)}{\mathscr{W}(x)} - \frac{\overline{u}_1'(x)}{w_0\lambda}\frac{\overline{u}_2(x)}{\mathscr{W}(x)}\right) = \frac{1}{\lambda}\|f\|_\infty$$

for any $x \in \mathbb{R}$. Finally, a straightforward computation shows that $R_\lambda f$ solves the differential equation (2.2.1). This finishes the proof. □

Thanks to Propositions 2.1.7 and 2.2.1, it is now easy to solve the problem of the uniqueness of the solution $u \in C_b(\mathbb{R}) \cap C^2(\mathbb{R})$ to equation (2.2.1).

Theorem 2.2.2 *The elliptic equation* (2.2.1) *is uniquely solvable in* $C_b(\mathbb{R}) \cap C^2(\mathbb{R})$ *for any* $\lambda > 0$ *and any* $f \in C_b(\mathbb{R})$ *if and only if* $-\infty$ *and* $+\infty$ *are unaccessible. If both* $-\infty$ *and* $+\infty$ *are accessible, then any solution* $u \in C^2(\mathbb{R})$ *to* (2.2.1) *is bounded in* $\mathbb{R}$.

In Chapter 1 we have seen that, if the coefficients q and b belong to $C^\alpha_{\rm loc}(\mathbb{R})$, then a solution $u \in C^{2+\alpha}_{\rm loc}(\mathbb{R})$ can be obtained by approximating equation (2.1.1) with the Dirichlet problems

$$\begin{cases} \lambda u(x) - q(x)u''(x) - b(x)u'(x) = f(x), & x \in (-n, n), \\ u(-n) = u(n) = 0 \end{cases} \tag{2.2.8}$$

and, then, letting n tend to $+\infty$. Now two questions arise naturally. Does this approximation argument work also in the case when the coefficients are merely continuous in $\mathbb{R}$ and, if this is the case, what are the relations between the solution provided by this method and the solution given by (2.2.4)? The answers to the previous two questions are contained in the following proposition.

Proposition 2.2.3 *For any* $f \in C_b(\mathbb{R})$ *and* $n \in \mathbb{N}$, *let* u_n *be the solution to problem* (2.2.8). *Then,* u_n *converges to the function* $R_\lambda f$ *(see* (2.2.4)*) in* $C^2([-a, a])$ *for any* $a > 0$, *as* n *tends to* $+\infty$.

Proof Throughout the proof, $\overline{u}_1$ and $\overline{u}_2$ denote the functions in Lemma 2.1.2. Moreover, to simplify the notation, we write $u(+\infty)$ (resp. $u(-\infty)$) to denote the limit of u at $+\infty$ (resp. at $-\infty$), whenever these limits exist. Finally, for any fixed function $f \in C_b(\mathbb{R})$, we denote by g_f and h_f the functions defined by

$$g_f(x) = \frac{1}{w_0}\int_0^x \frac{f(t)\overline{u}_1(t)}{q(t)\mathscr{W}(t)}dt, \qquad h_f(x) = \frac{1}{w_0}\int_0^x \frac{f(t)\overline{u}_2(t)}{q(t)\mathscr{W}(t)}dt$$

for any $x \in \mathbb{R}$. First of all, we prove that the sequence $\{u_n\}$ converges in $C^2([-a, a])$, for any $a > 0$, to a function $u \in C^2(\mathbb{R}) \cap C_b(\mathbb{R})$. For this purpose, we observe that a simple computation shows that u_n is given by (2.2.3), with $c = c_1(n)$ and $c_2 = c_2(n)$ satisfying

$$\begin{cases} (c_1(n) + h_f(n))\overline{u}_1(n) + (c_2(n) - g_f(n))\overline{u}_2(n) = 0, \\ (c_1(n) + h_f(-n))\overline{u}_1(-n) + (c_2(n) - g_f(-n))\overline{u}_2(-n) = 0. \end{cases} \tag{2.2.9}$$

Moreover, by the classical maximum principle (see Theorem D.2.1) it follows that

$$\sup_{x \in [-n,n]} |u_n(x)| \le \frac{1}{\lambda}\|f\|_\infty, \qquad n \in \mathbb{N}. \tag{2.2.10}$$

Thus, writing (2.2.3) at $x = \pm 1$, we deduce that $|c_1(n)\overline{u}_1(1)+c_1(n)\overline{u}_2(1)|$ and $|c_1(n)\overline{u}_1(-1)+$

$c_2(n)\overline{u}_2(-1)|$ are bounded by a positive constant, independent of n. Since the matrix A, whose rows are $(\overline{u}_1(-1), \overline{u}_2(-1))$ and $(\overline{u}_1(1), \overline{u}_2(1))$, is invertible, $|c_1(n)|$ and $|c_2(n)|$ are bounded uniformly with respect to $n \in \mathbb{N}$. Therefore, up to a subsequence, we can assume that $c_1(n)$ and $c_2(n)$ converge to some real numbers $c_1(+\infty)$ and $c_2(+\infty)$. Now, from (2.2.3) it is immediate to check that u_n converges in $C^2([-a,a])$ to some function $u \in C^2(\mathbb{R})$ which, of course, satisfies the differential equation (2.2.1). Moreover, estimate (2.2.10) shows that u is bounded in $\mathbb{R}$. Therefore, the function $R(\lambda)f$ is well defined for any $f \in C_b(\mathbb{R})$.

Now, let us prove that $R(\lambda)f = R_\lambda f$, where $R_\lambda f$ is given by (2.2.4). For this purpose, we just need to show that $c_1(n)$ and $c_2(n)$ converge as n tends to $+\infty$ and $c_1(+\infty) = -h_f(-\infty)$, $c_2(+\infty) = g_f(+\infty)$. Of course, according to Theorem 2.2.2 it suffices to consider the case when at least one between $-\infty$ and $+\infty$ is accessible. To begin with we observe that (2.2.6) and (2.2.7) show that the functions $\overline{u}_1/(q\mathscr{W})$ and $\overline{u}_2/(q\mathscr{W})$ are summable in $(0,+\infty)$ and in $(-\infty,0)$, respectively. Therefore, $g_f(+\infty)$ and $h_f(-\infty)$, as well as $g_1(+\infty)$ and $h_1(-\infty)$, exist and are finite. Moreover, if $\overline{u}_1(+\infty) = 0$ and $\overline{u}_2(+\infty)$ is finite, then $(h_f\overline{u}_1)(+\infty) = 0$. Indeed,

$$
\begin{aligned}
|h_f(n)\overline{u}_1(n)| =& \frac{\overline{u}_1(n)}{w_0}\left|\int_0^n \frac{f(t)\overline{u}_2(t)}{q(t)\mathscr{W}(t)}dt\right| \le \|f\|_\infty \frac{\overline{u}_1(n)}{w_0}\int_0^n \frac{\overline{u}_2(t)}{q(t)\mathscr{W}(t)}dt \\
\le& \|f\|_\infty \frac{\overline{u}_1(n)\overline{u}_2(n)}{w_0}\left(\int_0^x \frac{1}{q(t)\mathscr{W}(t)}dt + \int_x^n \frac{1}{q(t)\mathscr{W}(t)}dt\right) \\
\le& \|f\|_\infty \left\{\frac{\overline{u}_1(n)\overline{u}_2(n)}{w_0}g_1(x) + \overline{u}_2(n)(g_1(n) - g_1(x))\right\}
\end{aligned}
$$

for any $x > 0$. Thus, letting n tend to $+\infty$ we get

$$
\limsup_{n\to+\infty} |h_f(n)\overline{u}_1(n)| \le \|f\|_\infty u_2(+\infty)(g_1(+\infty) - g_1(x)).
$$

Then, letting x tend to $+\infty$, yields $(h_f\overline{u}_1)(+\infty) = 0$. Similarly, if $\overline{u}_1(-\infty)$ is real and $\overline{u}_2(-\infty) = 0$, then $\overline{u}_2(-n)g_f(-n)$ vanishes as n tends to $+\infty$.

To complete the proof we need a deeper analysis of the functions $\overline{u}_1$ and $\overline{u}_2$. For this purpose we split the remainder of the proof into three steps.

Step 1: the case when $+\infty$ and $-\infty$ are both accessible. In such a situation, the functions $\overline{u}_1$ and $\overline{u}_2$ are bounded in $\mathbb{R}$ and $\overline{u}_1(+\infty) = \overline{u}_2(-\infty) = 0$. The boundedness of $\overline{u}_1$ and $\overline{u}_2$ follows from properties (i) and (ii) in Proposition 2.1.7. Let us now show that $\overline{u}_1(+\infty) = 0$. This is clear by Proposition 2.1.7(ii) if $+\infty$ is an exit point. So, let us assume that $+\infty$ is regular. A similar argument then can be used to show that $\overline{u}_2(-\infty) = 0$ when $-\infty$ is regular. For this purpose, let u be a decreasing solution of (2.1.1) with $u(+\infty) = 0$, provided by Proposition 2.1.7. As is immediately seen, the function $v = u/u(0)$ vanishes at $+\infty$ and $v(0) = 1$. We claim that $v \ge \overline{u}_1$ in $(0,+\infty)$. Since $\overline{u}_1 > 0$ in $\mathbb{R}$, it then follows that $\overline{u}_1(+\infty) = 0$. By contradiction suppose that there exists $x > 0$ such that $\overline{u}_1(x) > v(x)$. Then, $\overline{u}_1 > v$ in $(0,+\infty)$. Indeed if this were not the case, then there should exist a point $x_1 > 0$ such that $\overline{u}_1(x_1) = v(x_1)$. By the last part of Remark 2.1.1 it would follow that $\overline{u}_1 \equiv v$: a contradiction. Therefore, if $\overline{u}_1(x) > v(x)$ for some $x > 0$, then $\overline{u}_1 > v$ in $(0,+\infty)$. But this implies that $\overline{u}_1'(0) > v'(0)$ ($\overline{u}'(0) = v(0)$ is not allowed, otherwise $\overline{u}_1$ and v would coincide). As a byproduct, $v'(0) \in B$ (see (2.1.2)) since B is an interval, as the proof of Lemma 2.1.2 reveals, and $\overline{u}_1(0)$ is its supremum. It thus follows that v vanishes somewhere in $(0,+\infty)$, which clearly is a contradiction. Therefore, $v \ge \overline{u}_1$ in $(0,+\infty)$ and, by comparison, $\overline{u}_1(+\infty) = 0$.

Now, from (2.2.9) we deduce that $c_1(+\infty)$ and $c_2(+\infty)$ exist and $c_1(+\infty) = -h_f(-\infty)$, $c_2(+\infty) = g_f(+\infty)$.

Step 2: the case when $+\infty$ is accessible and $-\infty$ is unaccessible. According to properties

(i) and (ii) in Proposition 2.1.7 and arguing as in the previous step, it follows that $\overline{u}_1(+\infty) = 0$, $\overline{u}_1(-\infty) = +\infty$, $\overline{u}_2(+\infty) \in (0, +\infty)$. Hence, letting n tend to $+\infty$ in the first equation in (2.2.9) we deduce that $c_2(+\infty) = g_f(+\infty)$. To prove that $c_1(+\infty) = -h_f(-\infty)$, we begin by studying the behaviour of $\overline{u}_2$ at $-\infty$. For this purpose, we observe that we can write $\overline{u}_2 = d_1\overline{u}_1 + d_2 u$, where u is an increasing solution to (2.1.1) satisfying $\lim_{x\to-\infty} u'(x)/\mathscr{W}(x) = 0$ and $u(-\infty) = 0$, if $-\infty$ is natural, and $u(-\infty) = 1$, if $-\infty$ is an entrance. Since $\overline{u}_2(-\infty)$ is real, we conclude that $d_1 = 0$ and, consequently, $\overline{u}_2 = u/u(0)$. Therefore, $\overline{u}_2(-\infty) = 0$, if $-\infty$ is natural, whereas $\overline{u}_2(-\infty) > 0$, if $-\infty$ is an entrance. In any case, $\overline{u}_2'/\mathscr{W}$ converges to 0 as x tends to $-\infty$. Therefore, using (2.1.6) and (2.1.8), we get

$$\begin{aligned}
|g_f(-n)|\overline{u}_2(-n) \leq & \frac{\|f\|_\infty}{w_0}\overline{u}_2(-n)\int_{-n}^0 \frac{\overline{u}_1(t)}{q(t)\mathscr{W}(t)}dt \\
= & \frac{\|f\|_\infty}{w_0}\frac{\overline{u}_2(-n)}{\lambda}\left(\overline{u}_1'(0) - \frac{\overline{u}_1'(-n)}{\mathscr{W}(-n)}\right) \\
= & \frac{\|f\|_\infty}{w_0}\frac{\overline{u}_2(-n)}{\lambda}\overline{u}_1'(0) - \frac{\|f\|_\infty}{w_0}\frac{\overline{u}_2'(-n)}{\lambda\mathscr{W}(-n)}\overline{u}_1(-n) + \frac{1}{\lambda}\|f\|_\infty,
\end{aligned}$$

which shows that $g_f(-n)\overline{u}_2(-n)(\overline{u}_1(-n))^{-1}$ converges to 0 as n tends to $+\infty$. Thus, letting n tend to $+\infty$ in the second equation in (2.2.9), we easily conclude that $c_1(+\infty) = -h_f(-\infty)$.

Step 3: the case when $+\infty$ is unaccessible and $-\infty$ is accessible. By changing x to $-x$, we go back again to the previous case (see the proof of Proposition 2.1.5) and we still conclude that $c_1(+\infty) = -h_f(-\infty)$ and $c_2(+\infty) = g_f(+\infty)$. □

Now, we consider some examples.

Example 2.2.4 Let $\mathcal{A}$ be the operator defined on smooth functions ψ by $\mathcal{A}\psi(x) = \psi''(x) - x^3\psi(x)$ for any $x \in \mathbb{R}$, Let us show that $+\infty$ and $-\infty$ are both unaccessible. By Theorem 2.2.2 this will imply that the equation $\lambda u - \mathcal{A}u = f$ is uniquely solvable in $C_b(\mathbb{R}) \cap C^2(\mathbb{R})$ for any $f \in C_b(\mathbb{R})$. A straightforward computation shows that

$$\mathscr{W}(x) = e^{\frac{x^4}{4}}, \qquad \mathscr{P}(x) = e^{-\frac{x^4}{4}}\int_0^x e^{\frac{s^4}{4}}ds, \qquad \mathscr{R}(x) = e^{\frac{x^4}{4}}\int_0^x e^{-\frac{s^4}{4}}ds$$

for any $x \in \mathbb{R}$. It is readily seen that $\mathscr{R}$ belongs neither to $L^1((-\infty, 0))$ nor to $L^1((0, +\infty))$. Moreover, $\lim_{x\to\pm\infty} x^3\mathscr{P}(x) = 1$. Hence, $\mathscr{P} \in L^1((-\infty, 0)) \cap L^1((0, +\infty))$. We conclude that $+\infty$ and $-\infty$ are entrance points.

We will generalize this example to the N-dimensional setting in Chapter 3 (see Example 3.1.13).

Example 2.2.5 Let $\mathcal{A}$ be the one-dimensional differential operator defined on smooth functions ψ by $\mathcal{A}\psi(x) = \psi''(x) + x^3\psi'(x)$ for any $x \in \mathbb{R}$. In this case

$$\mathscr{W}(x) = e^{-\frac{x^4}{4}}, \qquad \mathscr{P}(x) = e^{\frac{x^4}{4}}\int_0^x e^{-\frac{s^4}{4}}ds, \qquad \mathscr{R}(x) = e^{-\frac{x^4}{4}}\int_0^x e^{\frac{s^4}{4}}ds.$$

By Example 2.2.4, $\mathscr{P} \notin L^1((-\infty, 0)) \cup L^1((0, +\infty))$ and $\mathscr{R} \in L^1((-\infty, 0)) \cap L^1((0, +\infty))$. We conclude that $+\infty$ and $-\infty$ are both exit points, i.e., they are accessible. Therefore, according to Theorem 2.2.2, for any $f \in C_b(\mathbb{R})$ the equation $\lambda u - \mathcal{A}u = f$ admits more than one solution belonging to $C_b(\mathbb{R}) \cap C^2(\mathbb{R})$.

We will generalize also this example to the N-dimensional setting in Chapter 4 (see Example 4.2.5).

Remark 2.2.6 The previous two examples show us that the uniqueness of the solution $u \in D_{\max}(\mathcal{A})$ to the elliptic equation $\lambda u - \mathcal{A}u = f$ does not depend merely on the growth at infinity of the coefficients of the operator $\mathcal{A}$. In fact, the operators defined in Examples 2.2.4 and 2.2.5 differ only in the sign of the drift term, but this difference is crucial. Indeed, using the notation of Chapter 1, we have $D(\widehat{A}) = D_{\max}(\mathcal{A})$, if $\mathcal{A}$ is the operator in Example 2.2.4, whereas $D(\widehat{A})$ is properly contained in $D_{\max}(\mathcal{A})$ if $\mathcal{A}$ is the operator in Example 2.2.5. (See also Example 4.2.5 and [145, Section 5.2] for a discussion about these two previous examples in the probabilistic framework.)

To conclude this section we characterize $D_{\max}(\mathcal{A})$ under some additional assumptions of the coefficients q and b. The following theorem has been proved in [105].

Theorem 2.2.7 *Suppose that $q \in C(\mathbb{R})$ and $q(x) \geq \kappa_0$ for any $x \in \mathbb{R}$ and some positive constant κ_0. Further assume that $b \in C^1(\mathbb{R})$ satisfies*

$$q(x)b'(x) \leq c_1 + c_2(b(x))^2, \qquad x \in \mathbb{R}, \tag{2.2.11}$$

for some constants $c_1 \in \mathbb{R}$ and $c_2 < 1$. Finally, assume that $+\infty$ and $-\infty$ are both unaccessible. Then,

$$D_{\max}(\mathcal{A}) = \{u \in C_b^2(\mathbb{R}) : \ qu'', bu' \in C_b(\mathbb{R})\}. \tag{2.2.12}$$

Moreover, there exist two positive constants C_1 and C_2 such that

$$C_1(\|u\|_\infty + \|\mathcal{A}u\|_\infty) \leq \|u\|_\infty + \|qu''\|_\infty + \|bu'\|_\infty \leq C_2(\|u\|_\infty + \|\mathcal{A}u\|_\infty) \tag{2.2.13}$$

for any $u \in D_{\max}(\mathcal{A})$.

Proof Clearly, the inclusion "$\supset$" and the first inequality in (2.2.13) are trivially satisfied. Let us prove the other inclusion. Throughout the proof, C denotes a positive constant, which is independent of n and may vary from line to line. Similarly, C_ε denotes a positive constant which depends only on the parameter ε and may vary from line to line. Fix $u \in D_{\max}(\mathcal{A})$ and set $u - \mathcal{A}u =: f \in C_b(\mathbb{R})$. Since the points $+\infty$ and $-\infty$ are both unaccessible, u is the unique solution in $C_b(\mathbb{R}) \cap C^2(\mathbb{R})$ of the previous equation. We will see that u is the limit in $C^2([-a,a])$ (for any $a > 0$) of the solution $u_n \in C^2([-n,n])$ to the Cauchy-Neumann problem

$$\begin{cases} u(x) - \mathcal{A}u(x) = f(x), & x \in (-n,n), \\ u'(-n) = u'(n) = 0. \end{cases}$$

According to Theorem D.2.1, we have

$$\|u_n\|_{C([-n,n])} \leq \|f\|_{C([-n,n])}, \qquad \|\mathcal{A}u_n\|_{C([-n,n])} \leq 2\|f\|_{C([-n,n])} \tag{2.2.14}$$

for any $n \in \mathbb{N}$. Using these estimates we now show that

$$\|qu_n''\|_{C([-n,n])} + \|bu_n'\|_{C([-n,n])} \leq C\|f\|_{C([-n,n])} \leq C\|f\|_\infty \tag{2.2.15}$$

for any $n \in \mathbb{N}$. This is the crucial step of the proof.

Fix $n \in \mathbb{N}$ and denote by $x_0 = x_0(n)$ a point in $[-n,n]$ such that $|b(x_0)u_n'(x_0)| = \|bu_n'\|_{C([-n,n])}$. Up to replacing u_n with $-u_n$ and f with $-f$, we can assume that x_0 is a maximum of the function bu_n'. Moreover, we can also assume that $x_0 \in (-n,n)$ and $b(x_0) \neq 0$, otherwise $b(x_0)u_n'(x_0) = 0$ and (2.2.15) would follow immediately. Since $(bu_n')'(x_0) = 0$, we have $u_n''(x_0) = -b'(x_0)(b(x_0))^{-1}u_n'(x_0)$. Hence, (2.2.11) implies that

$$q(x_0)u_n''(x_0) = -q(x_0)b'(x_0)\frac{u_n'(x_0)}{b(x_0)} \geq -c_1\frac{u_n'(x_0)}{b(x_0)} - c_2 b(x_0)u_n'(x_0).$$

Taking (2.2.14) into account, we deduce that

$$\begin{aligned}2\|f\|_{C([-n,n])} \geq & - f(x_0) + u_n(x_0) = q(x_0)u_n''(x_0) + b(x_0)u_n'(x_0) \\ \geq & (1 - c_2)b(x_0)u_n'(x_0) - c_1\frac{u_n'(x_0)}{b(x_0)}.\end{aligned} \tag{2.2.16}$$

Multiplying the first and the last sides of (2.2.16) by $b(x_0)u_n'(x_0)$ we get

$$\begin{aligned}2\|bu_n'\|_{C([-n,n])}\|f\|_{C([-n,n])} \geq & (1 - c_2)\|bu_n'\|_{C([-n,n])}^2 - c_1(u_n'(x_0))^2 \\ \geq & (1 - c_2)\|bu_n'\|_{C([-n,n])}^2 - c_1^+\|u_n'\|_{C([-n,n])}^2.\end{aligned}$$

Setting $\alpha = 2(1 - c_2)^{-1}\|f\|_{C([-n,n])}$, $\beta = c_1^+(1 - c_2)^{-1}\|u_n'\|_{C([-n,n])}^2$ and $x = \|bu_n'\|_{C([-n,n])}$, we conclude that x satisfies the inequality $x^2 \leq \alpha x + \beta$, which implies that $x \leq \alpha + \sqrt{\beta}$ or, equivalently,

$$\|bu_n'\|_{C([-n,n])} \leq \frac{2}{1 - c_2}\|f\|_{C([-n,n])} + \left(\frac{c_1^+}{1 - c_2}\right)^{\frac{1}{2}}\|u_n'\|_{C([-n,n])}. \tag{2.2.17}$$

Using the well known estimate[2]

$$\|v'\|_{C([-n,n])} \leq C\|v\|_{C([-n,n])}^{\frac{1}{2}}\|v''\|_{C([-n,n])}^{\frac{1}{2}}, \qquad v \in C^2([-n, n]), \ n \in \mathbb{N}, \tag{2.2.18}$$

and (2.2.14), we get

$$\|u_n'\|_{C([-n,n])} \leq \varepsilon\|u_n''\|_{C([-n,n])} + C_\varepsilon\|u_n\|_{C([-n,n])} \leq \varepsilon\|u_n''\|_{C([-n,n])} + C_\varepsilon\|f\|_{C([-n,n])}$$

for any $n \in \mathbb{N}$ and $\varepsilon > 0$. Replacing this inequality in (2.2.17) we obtain $\|bu_n'\|_{C([-n,n])} \leq C_\varepsilon\|f\|_{C([-n,n])} + \varepsilon\|u_n''\|_{C([-n,n])}$ so that, taking (2.2.14) into account, we can estimate

$$\begin{aligned}\|u_n''\|_{C([-n,n])} \leq \kappa_0^{-1}\|qu_n''\|_{C([-n,n])} \leq & \kappa_0^{-1}(\|\mathcal{A}u_n\|_{C([-n,n])} + \|bu_n'\|_{C([-n,n])}) \\ \leq & \kappa_0^{-1}(\|bu_n'\|_{C([-n,n])} + 2\|f\|_{C([-n,n])}) \\ \leq & \kappa_0^{-1}[\varepsilon\|u_n''\|_{C([-n,n])} + (C_\varepsilon' + 2)\|f\|_{C([-n,n])}].\end{aligned}$$

Taking $\varepsilon = \kappa_0/2$ in the previous estimate yields, first, that $\|u_n''\|_{C([-n,n])} \leq C\|f\|_\infty$ and, then, that $\|bu_n'\|_{C([-n,n])} \leq C\|f\|_\infty$ for any $n \in \mathbb{N}$. By difference, using again (2.2.14) we estimate $\|qu_n''\|_{C([-n,n])}$ and (2.2.15) follows.

Now, we can complete the proof. Indeed, the previous estimates show that the sequence $\{u_n\}$ is bounded in $C^2([-R, R])$ for any $R > 0$. Hence, using the Arzelà-Ascoli theorem, a compactness argument, and taking into account that $u_n'' = (u_n - f - bu_n')/q$ for any $n \in \mathbb{N}$, we conclude that there exists a subsequence $\{u_{n_k}\}$ which converges in $C^2([-R, R])$, for any $R > 0$, to a function $v \in C^2(\mathbb{R})$, which solves the equation $v - \mathcal{A}v = f$. Further, writing estimates (2.2.14) and (2.2.15) with u_n being replaced by u_{n_k} and letting k tend to $+\infty$, we deduce that v is bounded in $\mathbb{R}$ and $\|qv''\|_\infty + \|bv'\|_\infty \leq C\|f\|_\infty$. Hence, v belongs to $D_{\max}(\mathcal{A})$. Since the equation $w - \mathcal{A}w = f$ is uniquely solvable in $D_{\max}(\mathcal{A})$, $u \equiv v$ and, as a consequence, u belongs to the set in the right-hand side of (2.2.12) and the second part of estimate (2.2.13) follows.

As a final remark, we observe that, actually, all the sequence $\{u_n\}$ converges to u in $C^2([-R, R])$ for any $R > 0$. Indeed, the previous arguments show that any subsequence of $\{u_n\}$ admits a subsequence which, together with its first- and second-order derivatives, converges to u locally uniformly in $\mathbb{R}$. □

[2]If $n = 1$, then estimate (2.2.18) follows from the Landau inequality applied to any extension $w \in C_b^2(\mathbb{R})$ of v. To get (2.2.18) for a general n, with a constant being independent of n, it suffices to apply the Landau inequality with $n = 1$ to the function $w : [-1, 1] \to \mathbb{R}$ defined by $w(x) = v(nx)$ for any $x \in [-1, 1]$.

Chapter 3

Uniqueness results, conservation of probability and maximum principles

3.0 Introduction

In this chapter we deal with the problem of the uniqueness in $D_{\max}(\mathcal{A})$ (see (1.0.1)) of the solution of the elliptic equation

$$\lambda v(x) - \mathcal{A}v(x) = f(x), \qquad x \in \mathbb{R}^N, \tag{3.0.1}$$

(with $f \in C_b(\mathbb{R}^N)$) and of the classical solution (i.e., it belongs to $C([0,+\infty) \times \mathbb{R}^N) \cap C^{1,2}((0,+\infty) \times \mathbb{R}^N)$) to the parabolic Cauchy problem

$$\begin{cases} D_t u(t,x) = \mathcal{A}u(t,x), & t > 0, \quad x \in \mathbb{R}^N, \\ u(0,x) = f(x), & x \in \mathbb{R}^N, \end{cases} \tag{3.0.2}$$

which is bounded in $[0,T] \times \mathbb{R}^N$ for any $T > 0$.

Hypotheses 1.0.1 are assumed throughout the chapter. Under such hypotheses, in Chapter 1 we have proved the existence of solutions to problems (3.0.1) and (3.0.2), with the regularity properties claimed above, for any $f \in C_b(\mathbb{R}^N)$ and $\lambda > c_0 = \sup_{\mathbb{R}^N} c$ (see Theorems 1.1.1 and 1.2.1).

The two problems of uniqueness are strictly connected. We exploit this connection in Proposition 3.1.1 (in the general case) and in Proposition 3.1.10 (in the case when $c \equiv 0$).

The following assumption is often considered to get uniqueness results (see e.g. [138, 178, 245]).

Hypothesis 3.0.1 There exist a positive function $\varphi \in C^2(\mathbb{R}^N)$ and $\lambda_0 > c_0$ such that

$$\lim_{|x|\to+\infty} \varphi(x) = +\infty, \qquad \mathcal{A}\varphi(x) - \lambda_0\varphi(x) \le 0, \qquad x \in \mathbb{R}^N. \tag{3.0.3}$$

Remark 3.0.2 Equivalently one can assume that there exists $\varphi \in C^2(\mathbb{R}^N)$ such that

$$\lim_{|x|\to+\infty} \varphi(x) = +\infty, \qquad \mathcal{A}\varphi(x) - \lambda_0\varphi(x) \le C, \qquad x \in \mathbb{R}^N, \tag{3.0.4}$$

for some $C \in \mathbb{R}$. Indeed, if φ satisfies (3.0.4), then the function $\varphi + M$ satisfies Hypothesis 3.0.1 provided the constant M is sufficiently large.

A function $\varphi \in C^2(\mathbb{R}^N)$ satisfying condition (3.0.3) or condition (3.0.4) is usually called a Lyapunov function for the operator $\mathcal{A}$.

The condition (3.0.3) can be made clearer with a particular choice of φ. For instance, if we take $\varphi(x) = \log(m + |x|^2)$, where m is a positive constant, then (3.0.4) reads as follows:

$$(m + |x|^2)\mathrm{Tr}(Q(x)) - 2\langle Q(x)x, x\rangle + (m + |x|^2)\langle b(x), x\rangle$$

$$+\frac{c(x)}{2}(m+|x|^2)^2\log(m+|x|^2)\le\frac{\lambda_0}{2}(m+|x|^2)^2\log(m+|x|^2) \qquad (3.0.5)$$

for any $x\in\mathbb{R}^N$. Assuming Hypotheses 1.0.1 and 3.0.1 we prove some maximum principles for the elliptic equation and for the parabolic Cauchy problem. The maximum principles yield the uniqueness of both the bounded and continuous solution of the elliptic equation (3.0.1) and the classical solution of the parabolic Cauchy problem (3.0.2), which is bounded in $[0,T]\times\mathbb{R}^N$ for any $T>0$.

Afterwards, we focus our attention on the case when $c\equiv 0$. In such a situation, the function $u_1\equiv\mathbb{1}$ is clearly a bounded solution of problem (3.0.2) with initial value $f\equiv\mathbb{1}$. Also the function $u_2=T(\cdot)\mathbb{1}$ is a bounded solution of the same problem. Therefore, if problem (3.0.2) admits a unique bounded classical solution for any $f\in C_b(\mathbb{R}^N)$, then, necessarily, $T(\cdot)\mathbb{1}\equiv\mathbb{1}$. In Proposition 3.1.10 we prove that, actually, the previous one is also a sufficient condition guaranteeing the uniqueness of the classical solution to problem (3.0.2), which is bounded in $[0,T]\times\mathbb{R}^N$ for any $T>0$. In this case also equation (3.0.1) has a unique solution in $D_{\max}(\mathcal{A})$ for any $f\in C_b(\mathbb{R}^N)$.

When the condition $T(\cdot)\mathbb{1}\equiv\mathbb{1}$ holds we say that $\{T(t)\}$ is conservative or that conservation of probability holds.

Finally, in Section 3.2 we provide a non-uniqueness result. In the case when $c\equiv 0$, some other sufficient conditions implying that problem (3.0.2) is not uniquely solvable are provided in Section 4.2.

3.1 Conservation of probability and uniqueness

The following proposition describes the relation between the uniqueness of the solution to the elliptic equation (3.0.1) and the uniqueness of the solution to the parabolic Cauchy problem (3.0.2).

Proposition 3.1.1 *Consider the following conditions:*

(i) *for any $\lambda>c_0$ and $f\in C_b(\mathbb{R}^N)$, the function $u=R(\lambda)f$ is the unique solution of the elliptic equation* (3.0.1) *in $D_{\max}(\mathcal{A})$;*

(ii) *for any $f\in C_b(\mathbb{R}^N)$, the function $u=T(\cdot)f$ is the unique solution of the parabolic Cauchy problem* (3.0.2) *which belongs to $C_b([0,T]\times\mathbb{R}^N)\cap C^{1,2}((0,T)\times\mathbb{R}^N)$ for any $T>0$;*

(iii) *for any $f\in C_b(\mathbb{R}^N)$, the function $u=T(\cdot)f$ is the unique solution of the parabolic Cauchy problem* (3.0.2) *in $C([0,+\infty)\times\mathbb{R}^N)\cap C^{1,2}((0,+\infty)\times\mathbb{R}^N)$ satisfying $|u(t,x)|\le M\exp(c_0t)$ for any $t>0$, $x\in\mathbb{R}^N$ and some $M>0$.*

Then, (i) *implies* (iii) *and* (ii) *implies* (i).

Proof "(i)$\Rightarrow$(iii)". Let $u\in C([0,+\infty)\times\mathbb{R}^N)\cap C^{1,2}((0,+\infty)\times\mathbb{R}^N)$ be a solution of the parabolic Cauchy problem (3.0.2), with $u(0,\cdot)=0$, satisfying

$$|u(t,x)|\le M\exp(c_0t),\qquad t>0,\ \ x\in\mathbb{R}^N, \qquad (3.1.1)$$

for some $M>0$. To prove that $u\equiv 0$, we fix $\lambda>c_0$, and consider the functions

$$v_\infty=\int_0^{+\infty}e^{-\lambda t}u(t,\cdot)dt,\qquad v_n=\int_{1/n}^{n}e^{-\lambda t}u(t,\cdot)dt,\quad n\in\mathbb{N}.$$

Estimate (3.1.1) shows that $v_n \in C_b(\mathbb{R}^N)$ for any $n \in \mathbb{N} \cup \{\infty\}$ and

$$\|v_n\|_\infty \le M(\lambda - c_0)^{-1}, \qquad n \in \mathbb{N} \cup \{\infty\}. \tag{3.1.2}$$

Further, for any $x \in \mathbb{R}^N$ and $n \in \mathbb{N}$, we can write

$$\begin{aligned}\mathcal{A}v_n(x) =& \int_{1/n}^n e^{-\lambda t}\mathcal{A}u(t,x)dt = \int_{1/n}^n e^{-\lambda t}D_t u(t,x)dt\\ =& e^{-\lambda n}u(n,x) - e^{-\lambda/n}u(1/n,x) + \lambda v_n(x),\end{aligned} \tag{3.1.3}$$

which implies that $\mathcal{A}v_n \in C_b(\mathbb{R}^N)$, so that $v_n \in D_{\max}(\mathcal{A})$. Moreover, by (3.1.1)-(3.1.3) it follows that there exists a constant $C > 0$ such that $\|\mathcal{A}v_n\|_\infty \le C$ for any $n \in \mathbb{N}$ and $\mathcal{A}v_n$ converges to λv_∞, pointwise in $\mathbb{R}^N$, as n tends to $+\infty$.

For any $n \in \mathbb{N}$ let us introduce the function $\psi_n = \lambda v_n - \mathcal{A}v_n$, which belongs to $C_b(\mathbb{R}^N)$ and vanishes pointwise as n tends to $+\infty$. By (i) it follows that

$$v_n(x) = (R(\lambda)\psi_n)(x) = \int_{\mathbb{R}^N} K_\lambda(x,y)\psi_n(y)dy, \qquad x \in \mathbb{R}^N,$$

where $R(\lambda)$ and K_λ are given by Theorem 1.1.3. Since the sequence $\{\psi_n\}$ is bounded in $C_b(\mathbb{R}^N)$, from the dominated convergence theorem we obtain that v_n vanishes pointwise in $\mathbb{R}^N$, as n tends to $+\infty$, i.e., $v_\infty \equiv 0$. Since $\lambda > c_0$ is arbitrary, by the uniqueness of the Laplace transform we conclude that $u \equiv 0$ in $(0,+\infty) \times \mathbb{R}^N$.

"(ii) ⇒ (i)". Let $v \in D_{\max}(\mathcal{A})$ be a solution of the equation $\lambda v - \mathcal{A}v = 0$; let us prove that $v \equiv 0$. By local regularity results for elliptic equations in bounded domains, v belongs to $C^{2+\alpha}_{\rm loc}(\mathbb{R}^N)$. Then, the function u, defined by $u(t,x) = e^{\lambda t}v(x)$ for any $t \ge 0$ and $x \in \mathbb{R}^N$, belongs to $C([0,+\infty) \times \mathbb{R}^N) \cap C^{1,2}((0,+\infty) \times \mathbb{R}^N)$ and it is a solution of problem (3.0.2), with $u(0,\cdot) = v$; besides u is bounded in $[0,T] \times \mathbb{R}^N$ for any $T > 0$. By (ii) it follows that $u = T(\cdot)v$ in $(0,+\infty) \times \mathbb{R}^N$ and, then, by Theorem 1.2.5, $|u(t,x)| \le \exp(c_0 t)\|v\|_\infty$ for any $(t,x) \in (0,+\infty) \times \mathbb{R}^N$. Since $\sup_{x \in \mathbb{R}^N}|u(t,x)| = \exp(\lambda t)\|v\|_\infty$ and $\lambda > c_0$, we conclude that $\|v\|_\infty = 0$. □

3.1.1 Maximum principles

In this subsection, we prove two maximum principles which provide us with uniqueness results for the solution to equation (3.0.1) and the Cauchy problem (3.0.2). First we prove the following lemma which is a local maximum principle for functions in $W^{2,p}_{\rm loc}(\mathbb{R}^N)$.

Lemma 3.1.2 *Suppose that u belongs to $W^{2,p}_{\rm loc}(\mathbb{R}^N)$, for any $p \in [1,+\infty)$, and that $\mathcal{A}u \in C(\mathbb{R}^N)$. If x_0 is a local maximum (resp. minimum) of u, then $\mathcal{A}u(x_0) - c(x_0)u(x_0) \le 0$ (resp. $\mathcal{A}u(x_0) - c(x_0)u(x_0) \ge 0$).*

Proof We limit ourselves to considering the case when x_0 is a local maximum point for u, since the case when x_0 is a local minimum follows from this one replacing u with $-u$. Clearly, we can suppose that $c \equiv 0$. Possibly replacing u with $u(\cdot - x_0) + C$ for a suitable constant $C > 0$, we can assume that $x_0 = 0$ and $u(0) > 0$. Let $r > 0$ be such that the restriction of u to $\overline{B}_r$ attains its maximum value at 0. Moreover, let $\psi \in C^\infty_c(\mathbb{R}^N)$ satisfy $\chi_{B_{r/2}} \le \psi \le \chi_{B_r}$ and set $v = \psi u$. Then, 0 is a global maximum point of the function v. Moreover, the function $\mathcal{A}v = \psi\mathcal{A}u + u\mathcal{A}\psi + 2\langle Q\nabla u, \nabla\psi\rangle$ belongs to $C_b(\mathbb{R}^N)$ since, by the Sobolev embedding theorems (see [3, Theorem 5.4]), $u \in C^1(\mathbb{R}^N)$. Therefore, $v \in D_{\max}(\mathcal{A}) \cap C_0(\mathbb{R}^N)$ and Proposition 1.3.5 implies that $v \in D(\widehat{A})$. Now, since $c \equiv 0$, the family of measures $\{p(t,x;dy): \ t>0, \ x\in\mathbb{R}^N\}$, introduced in Theorem 1.2.5, defines a transition function and, by (1.2.7), we have

$$\frac{(T(t)v)(0)-v(0)}{t}\le\frac{1}{t}\int_{B_r}(v(y)-v(0))p(t,0;dy)\le 0$$

for any $t>0$. Letting t tend to 0^+ and taking Proposition 1.3.5 into account, we conclude that $\mathcal{A}u(0)=\mathcal{A}v(0)=\widehat{A}v(0)\le 0$. □

The proof of Lemma 3.1.2 strongly relies on the representation formula (1.2.7) and on the characterization of $D(\widehat{A})$. Now we provide an alternative proof which can be easier extended to the case when $\mathbb{R}^N$ is replaced by an open set Ω and $u\in\bigcap_{1\le p<+\infty}W^{2,p}_{\rm loc}(\Omega)$ satisfies $\lambda u-\mathcal{A}u\le 0$ in Ω.

A second proof of Lemma 3.1.2 As in the previous proof, we deal only with the case when x_0 is a relative maximum. Moreover, we assume that $c\equiv 0$, $x_0=0$ and $u(x)\ge 0$ for any $x\in B_r$ and some $r>0$.

Let us fix a function $\psi\in C^\infty_c(\mathbb{R}^N)$ with support in B_r and such that $\psi(0)=1$, $\nabla\psi(0)=0$, $D^2\psi(0)=0$ and $0\le\psi(x)<1$ for any $x\in B_r\setminus\{0\}$. As is easily seen, $x=0$ is the unique maximum point of the function $v:=\psi u$, and $\mathcal{A}v\in C(\mathbb{R}^N)$. Moreover, since $\nabla\psi(0)=0$ and $D^2\psi(0)=0$, we deduce that $\mathcal{A}u(0)=\mathcal{A}v(0)$.

Now let $\varrho\in C^\infty_c(\mathbb{R}^N)$ be a smooth function such that $0\le\varrho(x)\le 1$ for any $x\in\mathbb{R}^N$, $\mathrm{supp}(\varrho)\subset B_1$ and $\|\varrho\|_{L^1(\mathbb{R}^N)}=1$. For any $n\in\mathbb{N}$ and $x\in\mathbb{R}^N$, we set $\varrho_n(x)=n^N\varrho(nx)$ and $v_n=\varrho_n\star v\in C^\infty_c(\mathbb{R}^N)$, where "$\star$" denotes the convolution operator. Since $v\in C_c(\mathbb{R}^N)$, v_n converges to v, uniformly in $\mathbb{R}^N$, as n tends to $+\infty$. Moreover, for any $n\in\mathbb{N}$, v_n has an absolute maximum at some point $x_n\in\overline{B}_{r+1/n}$ so that $\mathcal{A}v_n(x_n)\le 0$. A straightforward computation reveals that

$$\begin{aligned}\mathcal{A}v_n(x)&=(\varrho_n\star\mathcal{A}v)(x)+(\varrho_n\star\mathrm{Tr}((Q(x)-Q)D^2v))(x)+(\varrho_n\star\langle b(x)-b,\nabla v\rangle)(x)\\&=:(\varrho_n\star\mathcal{A}v)(x)+G_n(x)\end{aligned}$$

for any $x\in\mathbb{R}^N$. Since $\mathcal{A}v\in C_0(\mathbb{R}^N)$, $\varrho_n\star\mathcal{A}v$ tends to $\mathcal{A}v$ uniformly in $\mathbb{R}^N$ as n tends to $+\infty$. Moreover, G_n vanishes as n tends to $+\infty$, uniformly with respect to $x\in\mathbb{R}^N$. Indeed, if $p>N/\alpha$ (α being as in Hypothesis 1.0.1(ii)), then

$$\begin{aligned}\|G_n\|_\infty&\le C_1n^{-\alpha}\|\varrho_n\|_{L^{p'}(\mathbb{R}^N)}(\|\nabla v\|_{L^p(\mathbb{R}^N)}+\|D^2v\|_{L^p(\mathbb{R}^N)})\\&=C_1n^{-\alpha+N/p}\|\varrho\|_{L^{p'}(\mathbb{R}^N)}(\|\nabla v\|_{L^p(\mathbb{R}^N)}+\|D^2v\|_{L^p(\mathbb{R}^N)})\end{aligned}$$

for some positive constant C_1, independent of n, where $1/p+1/p'=1$. We have so proved that v_n and $\mathcal{A}v_n$ converge, respectively, to v and $\mathcal{A}v$ uniformly in $\mathbb{R}^N$ as n tends to $+\infty$.

Since $\{x_n\}$ is a bounded sequence, up to a subsequence $\{x_n\}$ converges to some point $\widehat{x}\in\overline{B}_r$ as n tends to $+\infty$. By continuity, $v(x_n)$ tends to $v(\widehat{x})$ as n tends to $+\infty$. Hence, $v(\widehat{x})=\lim_{n\to+\infty}v(x_n)=\lim_{n\to+\infty}v_n(x_n)=\lim_{n\to+\infty}\|v_n\|_\infty=\|v\|_\infty=v(0)$. Recalling that 0 is the unique point where v attains its maximum value, it follows that $\widehat{x}=0$. Since $\mathcal{A}v_n(x_n)\le 0$ for any $n\in\mathbb{N}$ and it converges uniformly to $\mathcal{A}v$, we conclude that $\mathcal{A}u(0)=\mathcal{A}v(0)\le 0$. □

Theorem 3.1.3 *Let Hypothesis* 3.0.1 *be satisfied and fix* $T>0$, $f\in C(\mathbb{R}^N)$, $g:(0,T]\times\mathbb{R}^N\to\mathbb{R}$. *Further suppose that the function* $u\in C([0,T]\times\mathbb{R}^N)$ *is such that* $D_tu\in C((0,T]\times\mathbb{R}^N)$, $u(t,\cdot)\in W^{2,p}_{\rm loc}(\mathbb{R}^N)$, $\mathcal{A}u(t,\cdot)\in C(\mathbb{R}^N)$, *for any* $t\in(0,T]$ *and* $p\in[1,+\infty)$, *and it solves the Cauchy problem*

$$\begin{cases}D_tu(t,x)=\mathcal{A}u(t,x)+g(t,x), & t\in(0,T], \quad x\in\mathbb{R}^N,\\ u(0,x)=f(x), & x\in\mathbb{R}^N.\end{cases}\tag{3.1.4}$$

If f is bounded from above (resp. from below) in $\mathbb{R}^N$, g is nonpositive (resp. nonnegative) in $(0,T]\times\mathbb{R}^N$ and

$$\limsup_{|x|\to+\infty}\sup_{t\in[0,T]}\frac{u(t,x)}{\varphi(x)}\le 0,\qquad\left(\textit{resp.}\ \liminf_{|x|\to+\infty}\inf_{t\in[0,T]}\frac{u(t,x)}{\varphi(x)}\ge 0,\right)\tag{3.1.5}$$

then

$$u(t,x)\le e^{c_0t}\Big(0\vee\sup_{\mathbb{R}^N}f\Big)\qquad\left(\textit{resp.}\ u(t,x)\ge e^{c_0t}\Big(0\wedge\inf_{\mathbb{R}^N}f\Big)\right)\tag{3.1.6}$$

for any $t\in[0,T]$ and $x\in\mathbb{R}^N$. In particular, for any $f\in C(\mathbb{R}^N)$ such that the function f/φ vanishes as $|x|$ tends to $+\infty$, there exists at most one solution $u\in C([0,+\infty)\times\mathbb{R}^N)$ of problem (3.1.4) such that $D_tu\in C((0,T]\times\mathbb{R}^N)$, $u(t,\cdot)\in W^{2,p}_{\rm loc}(\mathbb{R}^N)$, $\mathcal{A}u(t,\cdot)\in C(\mathbb{R}^N)$ for any $p\in[1,+\infty)$ and $t\in(0,T]$, and satisfying both the conditions in (3.1.5) for any $T>0$.

Proof It suffices to consider the case when f is bounded from above and g is nonpositive, since the case when f is bounded from below and g is nonnegative follows applying the previous case to the function $-u$. Finally, the last part of the statement is a straightforward consequence of both the estimates in (3.1.6).

We make some reductions. First, we assume that $c_0\le 0$. Indeed, in the case when $c_0>0$ it suffices to consider the function $\widetilde{u}:[0,T]\times\mathbb{R}^N\to\mathbb{R}$, defined by $\widetilde{u}(t,x)=e^{-c_0t}u(t,x)$ for any $t\in[0,T]$ and $x\in\mathbb{R}^N$, which satisfies (3.1.5) as well as the equation $D_t\widetilde{u}(t,x)=\mathcal{A}_0\widetilde{u}(t,x)+e^{-c_0t}g(t,x)$, for any $t\in(0,T]$ and $x\in\mathbb{R}^N$, where $\mathcal{A}_0=\mathcal{A}-c_0$ has nonpositive zero-order coefficient and satisfies Hypothesis 3.0.1 with the same Lyapunov function φ and the same λ_0. Thus, from the case $c_0=0$ it follows that $\widetilde{u}(t,x)\le 0\vee\sup_{\mathbb{R}^N}f$ for any $t\in[0,T]$ and $x\in\mathbb{R}^N$, which yields (3.1.6). Besides, we can also assume that $\sup_{\mathbb{R}^N}f\le 0$. Indeed, in the case when $\sup_{\mathbb{R}^N}f>0$, it suffices to consider the function $\overline{u}=u-\sup_{\mathbb{R}^N}f$ that has a nonpositive initial value and satisfies the equation $D_t\overline{u}=\mathcal{A}\overline{u}+c\sup_{\mathbb{R}^N}f+g$ in $(0,T]\times\mathbb{R}^N$, where the right-hand side is nonpositive. Then, from the case $\sup_{\mathbb{R}^N}f\le 0$ we get $\overline{u}\le 0$, that is (3.1.6).

Taking all these reductions into account, we have to prove that $u\le 0$ in $[0,T]\times\mathbb{R}^N$. For this purpose, we introduce the function v, defined by $v(t,x)=e^{-\lambda_0t}u(t,x)$ for any $(t,x)\in[0,T]\times\mathbb{R}^N$, where λ_0 is as in Hypothesis 3.0.1, and, for any $k\in\mathbb{N}$, the functions $v_k=v-k^{-1}\varphi$. We claim that, for any $k\in\mathbb{N}$, v_k is nonpositive in $[0,T]\times\mathbb{R}^N$. Note that, by Hypothesis 3.0.1,

$$D_tv_k(t,\cdot)-(\mathcal{A}-\lambda_0)v_k(t,\cdot)=e^{-\lambda_0t}g(t,\cdot)+k^{-1}(\mathcal{A}\varphi-\lambda_0\varphi)\le 0\tag{3.1.7}$$

in $\mathbb{R}^N$, for any $t\in(0,T]$. Moreover, by Hypothesis 3.0.1 and assumption (3.1.5), v_k has a maximum point (t_k,x_k) in $[0,T]\times\mathbb{R}^N$ for any $k\in\mathbb{N}$. If $t_k=0$, then $v_k(t_k,x_k)\le 0$ since $f\le 0$ and $\varphi\ge 0$. On the other hand, if $t_k\in(0,T]$, then by Lemma 3.1.2 it holds that $D_tv_k(t_k,x_k)\ge 0$ and $(\mathcal{A}-c(x_k))v_k(t_k,x_k)\le 0$, which, combined with (3.1.7), yields $(\mathcal{A}-c(x_k))v_k(t_k,x_k)\le(\mathcal{A}-\lambda_0)v_k(t_k,x_k)$. Since $\lambda_0>c(x_k)$, we conclude that $v_k(t_k,x_k)\le 0$. We have so proved that $v_k\le 0$ in $[0,T]\times\mathbb{R}^N$ or, equivalently, that $v\le k^{-1}\varphi$ in $[0,T]\times\mathbb{R}^N$ for any $k\in\mathbb{N}$. Letting k tend to $+\infty$, we get $v\le 0$ and, then, also $u\le 0$. □

Remark 3.1.4 The maximum principle in Theorem 3.1.3 implies in particular that, under Hypotheses 1.0.1 and 3.0.1, the function $u=T(\cdot)f$ is the unique classical solution of the Cauchy problem (3.0.2), which is bounded in $[0,T]\times\mathbb{R}^N$ for any $T>0$.

Next, we prove two maximum principles for the elliptic equation (3.0.1). The former holds for any $\lambda>c_0$ and gives uniqueness of the bounded solution, while the latter holds only in the case when $\lambda>\lambda_0$ (recall that $\lambda_0>c_0$, see Hypothesis 3.0.1) but it ensures uniqueness also for unbounded solutions.

Theorem 3.1.5 *Let Hypothesis 3.0.1 be satisfied and $u \in C(\mathbb{R}^N) \cap W^{2,p}_{\rm loc}(\mathbb{R}^N)$, for any $p \in [1,+\infty)$, be such that $\mathcal{A}u \in C(\mathbb{R}^N)$. Further, fix $\lambda > c_0$. If u is bounded from above (resp. from below) in $\mathbb{R}^N$ and $\lambda u - \mathcal{A}u \le 0$ (resp. ≥ 0) in $\mathbb{R}^N$, then $u \ge 0$ (resp. $u \le 0$) in $\mathbb{R}^N$. In particular, for any $f \in C_b(\mathbb{R}^N)$ and $\lambda > c_0$, the function $R(\lambda)f$ (see Theorem 1.1.3) is the unique bounded solution to the elliptic equation (3.0.1) in $D_{\max}(\mathcal{A})$.*

Proof We limit ourselves to proving that if u is bounded from above and $\lambda u - \mathcal{A}u \le 0$ in $\mathbb{R}^N$, then $u \ge 0$ in $\mathbb{R}^N$, since the second statement follows applying this one to the function $-u$ and the last statement is a straightforward consequence of the first two.

Consider the function $\widetilde{u} : [0,+\infty) \times \mathbb{R}^N \to \mathbb{R}$ defined by $\widetilde{u}(t,x) = e^{\lambda t}u(x)$ for any $(t,x) \in [0,+\infty) \times \mathbb{R}^N$. As is easily seen, $\widetilde{u}$ satisfies the regularity assumptions in Theorem 3.1.3, condition (3.1.5) and $D_t\widetilde{u}(t,x) - \mathcal{A}\widetilde{u}(t,x) = e^{\lambda t}(\lambda u(x) - \mathcal{A}u(x)) \le 0$ for any $(t,x) \in (0,+\infty) \times \mathbb{R}^N$. Therefore, $u(x) \le e^{(c_0-\lambda)t}(0 \vee \sup_{\mathbb{R}^N} u)$ for any $t > 0$ and $x \in \mathbb{R}^N$. Letting t tend to $+\infty$, we conclude that $u \le 0$. □

Theorem 3.1.6 *Assume Hypothesis 3.0.1 and let u belong to $C(\mathbb{R}^N) \cap W^{2,p}_{\rm loc}(\mathbb{R}^N)$, for any $p \in [1,+\infty)$, with $\mathcal{A}u$ in $C(\mathbb{R}^N)$. Further, assume that $\limsup_{|x|\to+\infty} u(x)(\varphi(x))^{-1} \le 0$ (resp. $\liminf_{|x|\to+\infty} u(x)(\varphi(x))^{-1} \ge 0$) and $\lambda u - \mathcal{A}u \le 0$ in $\mathbb{R}^N$ (resp. $\lambda u - \mathcal{A}u \ge 0$ in $\mathbb{R}^N$) for some $\lambda \ge \lambda_0$. Then, u is nonpositive (nonnegative) in $\mathbb{R}^N$.*

Proof We assume that $\limsup_{|x|\to+\infty} u(x)(\varphi(x))^{-1} \le 0$, $\lambda u - \mathcal{A}u \le 0$ in $\mathbb{R}^N$ and prove that $u \le 0$ in $\mathbb{R}^N$. The other statement follows applying this one to the function $-u$.

Without loss of generality we can assume that $c_0 \le 0$. Indeed, in the case when $c_0 > 0$, it suffices to replace the operator $\mathcal{A}$ with the operator $\mathcal{A}_0 = \mathcal{A} - c_0$ which has a nonpositive zero-order coefficient. Then, the Lyapunov function φ and the function u satisfy $\mathcal{A}_0\varphi - (\lambda_0 - c_0)\varphi \le 0$ and $(\lambda - c_0)u - \mathcal{A}_0 u \le 0$ in $\mathbb{R}^N$. Therefore, from the case $c_0 \le 0$ it follows that $u \le 0$ in $\mathbb{R}^N$.

So, let us assume that $c_0 \le 0$. Then, Hypothesis 3.0.1 implies that $\mathcal{A}\varphi - \lambda\varphi \le 0$ in $\mathbb{R}^N$ for any $\lambda \ge \lambda_0$. Now, for any $k \in \mathbb{N}$, we consider the function $u_k = u - k^{-1}\varphi$. Since $\lambda u - \mathcal{A}u \le 0$ in $\mathbb{R}^N$, it follows that $\lambda u_k - \mathcal{A}u_k \le 0$ in $\mathbb{R}^N$ for any $k \in \mathbb{N}$. Moreover, the condition $\limsup_{|x|\to+\infty} u(x)(\varphi(x))^{-1} \le 0$ and Hypothesis 3.0.1 show that the function u_k has a maximum point x_k in $\mathbb{R}^N$. By Lemma 3.1.2, $(\mathcal{A} - c(x_k))u_k(x_k) \le 0$ and, thus, $\lambda u_k(x_k) \le c(x_k)u_k(x_k)$. Since $\lambda > c(x_k)$, $u_k(x_k)$ is nonpositive. This means that $u_k \le 0$ in $\mathbb{R}^N$ for any $k \in \mathbb{N}$ and, therefore, $u \le 0$. □

3.1.2 The case when $c \equiv 0$

In this situation both the functions $T(\cdot)\mathbb{1}$ and $\mathbb{1}$ are bounded solutions of problem (3.0.2) with initial value $f \equiv \mathbb{1}$. In general, they can differ at some point $x \in \mathbb{R}^N$. In any case, we have $0 \le T(t)\mathbb{1} \le \mathbb{1}$ for any $t > 0$, as it follows from Remark 1.2.3. With this remark in mind, we introduce the following definition.

Definition 3.1.7 *We say that $\{T(t)\}$ is conservative, or that the conservation of probability holds, if $T(t)\mathbb{1} \equiv \mathbb{1}$ for any $t > 0$.*

Let us prove the following preliminary result.

Lemma 3.1.8 *For any fixed $x \in \mathbb{R}^N$ the functions $t \mapsto (T(t)\mathbb{1})(x)$ and $\lambda \mapsto \lambda(R(\lambda)\mathbb{1})(x)$ are, respectively, decreasing and increasing in $(0,+\infty)$.*

Proof To prove the first part of the lemma, we observe that, for any $t, s > 0$ and $x \in \mathbb{R}^N$, it holds that $(T(t+s)\mathbb{1})(x) = (T(t)T(s)\mathbb{1})(x) \le (T(t)\mathbb{1})(x)$. The inequality follows easily observing that the semigroup $\{T(t)\}$ is order preserving since it maps nonnegative functions into nonnegative functions.

To prove the last assertion, we begin by observing that, according to Theorem 1.1.3, $0 \le \lambda R(\lambda)\mathbb{1} \le \mathbb{1}$ for any $\lambda > 0$. Let us now fix $0 < \lambda < \mu$. From the resolvent identity (1.1.4) we deduce that

$$\begin{aligned}\lambda R(\lambda)\mathbb{1} - \mu R(\mu)\mathbb{1} =& \lambda(R(\mu)\mathbb{1} + (\mu - \lambda)R(\mu)R(\lambda)\mathbb{1}) - \mu R(\mu)\mathbb{1}\\ =&(\lambda-\mu)R(\mu)\mathbb{1} + \lambda(\mu-\lambda)R(\mu)R(\lambda)\mathbb{1}\\ \le&(\lambda-\mu)R(\mu)\mathbb{1} + (\mu-\lambda)R(\mu)\mathbb{1} = 0\end{aligned}$$

and this completes the proof. □

Remark 3.1.9 As the above proof shows, the lemma also holds in the case when $c \le 0$. We will use this generalization of Lemma 3.1.8 in Chapter 4.

Proposition 3.1.10 *The following conditions are equivalent:*

(i) $T(t)\mathbb{1} \equiv \mathbb{1}$ *for some* $t > 0$;

(ii) $\{T(t)\}$ *is conservative;*

(iii) $R(\lambda)\mathbb{1} \equiv \mathbb{1}/\lambda$ *for some* $\lambda > 0$;

(iv) $R(\lambda)\mathbb{1} \equiv \mathbb{1}/\lambda$ *for any* $\lambda > 0$;

(v) *for any* $f \in C_b(\mathbb{R}^N)$, *the function* $u \equiv T(\cdot)f$ *is the unique solution of problem* (3.0.2) *which belongs to* $C_b([0,T] \times \mathbb{R}^N) \cap C^{1,2}((0,T) \times \mathbb{R}^N)$ *for any* $T > 0$;

(vi) *for any* $\lambda > 0$ *and* $f \in C_b(\mathbb{R}^N)$, *the function* $u \equiv R(\lambda)f$ *is the unique solution of the elliptic equation* (3.0.1) *in* $D_{\max}(\mathcal{A})$.

Proof Showing that "(ii) ⇒ (i)", "(iv) ⇒ (iii)" and "(vi) ⇒ (iv)" is trivial, while the implication "(v) ⇒ (vi)" has been shown in Proposition 3.1.1. Hence, we just need to prove that "(i) ⇒ (ii)", "(iii) ⇒ (ii)" and "(ii) ⇒ (v)".

"(i) ⇒ (ii)". Fix $x \in \mathbb{R}^N$. If $T(t_0)\mathbb{1} \equiv \mathbb{1}$ for some $t_0 > 0$, then, according to the semigroup law, the function $t \mapsto (T(t)\mathbb{1})(x)$ is periodic, with period t_0. By Lemma 3.1.8 it is also decreasing. Hence, it is constant.

"(iii) ⇒ (ii)" follows from the formula (1.2.12). Indeed, if $(T(t)\mathbb{1})(x)$ were less than 1 for some $t > 0$ and $x \in \mathbb{R}^N$, then from (1.2.12) it would follow that $(R(\lambda)\mathbb{1})(x) < 1/\lambda$.

"(ii) ⇒ (v)". Let u be a solution of the parabolic Cauchy problem (3.0.2), with $f \equiv 0$, which belongs to $C_b([0,T] \times \mathbb{R}^N) \cap C^{1,2}((0,T) \times \mathbb{R}^N)$ for any $T > 0$. We prove that $u \equiv 0$. On the contrary, suppose that $u \not\equiv 0$. By changing sign and multiplying u by a suitable constant if needed, we can assume that $\sup_{(0,T)\times\mathbb{R}^N} u > 0$ and $\sup_{(0,T)\times\mathbb{R}^N} |u| < 1$. Set $v = \mathbb{1} - u$. Then $v(t,x) > 0$ for any $(t,x) \in (0,T) \times \mathbb{R}^N$ and $\inf_{(0,T)\times\mathbb{R}^N} v < 1$. Moreover, v is a bounded solution of problem (3.0.2) with initial value $v(0,\cdot) = \mathbb{1}$. Then, by Remark 1.2.3, it follows that $T(\cdot)\mathbb{1} \le v$. This implies that $\inf_{\mathbb{R}^N} T(t)\mathbb{1} < \mathbb{1}$ for some $t \in (0,T)$, which contradicts statement (ii). □

The following propositions are straightforward consequences of the formula (1.2.7) and Theorem 3.1.3, respectively.

Proposition 3.1.11 *Let* $\{p(t,x;dy) : t > 0, x \in \mathbb{R}^N\}$ *be the transition function associated with the semigroup* $\{T(t)\}$ *(see Theorem* 1.2.5*). Then,* $\{T(t)\}$ *is conservative if and only if* $p(t,x;\mathbb{R}^N) = 1$ *for any* $t > 0$ *and* $x \in \mathbb{R}^N$.

Proposition 3.1.12 *Assume Hypothesis* 3.0.1 *with* $c \equiv 0$. *Then, the conservation of probability holds.*

Example 3.1.13 Consider the operator $\mathcal{A}$ defined on smooth functions $\zeta : \mathbb{R}^N \to \mathbb{R}$ by $\mathcal{A}\zeta = \Delta\zeta + \langle b, \nabla\zeta\rangle$, where $b \in C^\alpha_{\rm loc}(\mathbb{R}^N)$ for some $\alpha \in (0,1)$. Choosing $\varphi(x) = \log(m + |x|^2)$ for some $m > 1$, condition (3.0.4) becomes $\langle b(x), x\rangle \le M(m + |x|^2)\log(m + |x|^2)$ for any $x \in \mathbb{R}^N$ and some $M > 0$ (see (3.0.5)). If this condition holds, then the corresponding semigroup $\{T(t)\}$ is conservative.

For instance, this is the case when $b(x) = -x|x|^2$ for any $x \in \mathbb{R}^N$. Note that the one-dimensional version of the operator $\mathcal{A}$, with $b(x) = -x^3$ for any $x \in \mathbb{R}$, has been already considered in Example 2.2.4, where, in fact, we proved, with a different argument, that the associated semigroup is conservative.

Remark 3.1.14 The papers [23, 35, 94] contain some uniqueness results for the classical solution to the Cauchy problem (3.0.2), with a prescribed growth rate at infinity. In particular such results apply to prove the uniqueness of the bounded solution to the Cauchy problem (3.0.2). We stress that the results in [23, 35, 94] apply to operators with diffusion coefficients with at most quadratic growth at infinity.

3.2 Non-uniqueness

By the results in Chapter 2 (see Theorem 2.2.2), in the one-dimensional setting necessary and sufficiently conditions for the elliptic equation (3.0.1) to be not uniquely solvable in $D_{\max}(\mathcal{A})$ can be given in terms of integrability properties at $\pm\infty$ of the functions $\mathscr{P}$ and $\mathscr{R}$ in (2.1.4) and (2.1.5).

In this section we prove a sufficient condition for the non-uniqueness of the solution $v \in D_{\max}(\mathcal{A})$ of the elliptic equation (3.0.1) in the N-dimensional setting (see [138, Section 3.4], or [197, Remark 3.16]). According to Proposition 3.1.1, this result will imply that the parabolic Cauchy problem (3.0.2) admits several solutions belonging to $C([0,+\infty) \times \mathbb{R}^N) \cap C^{1,2}((0,+\infty) \times \mathbb{R}^N)$ which are bounded in $[0,T] \times \mathbb{R}^N$ for any $T > 0$. Here, we do not assume that $c \equiv 0$.

Proposition 3.2.1 *Suppose that there exists a nontrivial nonnegative function* ψ *which belongs to* $C^2(\mathbb{R}^N) \cap C_b(\mathbb{R}^N)$ *and satisfies* $\lambda\psi - \mathcal{A}\psi \le 0$ *in* $\mathbb{R}^N$ *for some* $\lambda > c_0$. *Then, there exists a bounded, continuous and not identically vanishing solution of the equation* $\lambda u - \mathcal{A}u = 0$.

Proof For any $n \in \mathbb{N}$, let $u_n \in C(\overline{B}_n)$ be the solution of the Dirichlet problem

$$\begin{cases} \lambda u_n(x) - \mathcal{A}u_n(x) = 0, & x \in B_n, \\ u_n(x) = \|\psi\|_\infty, & x \in \partial B_n. \end{cases}$$

The assumptions on ψ and the classical maximum principle show that $\psi \le u_n \le \|\psi\|_\infty$. Moreover, $\{u_n\}$ is a decreasing sequence. Indeed, the function $v = u_{n+1} - u_n$ satisfies the equation $\lambda v - \mathcal{A}v = 0$ in B_n and $v_n = u_{n+1} - \|\psi\|_\infty \le 0$ on ∂B_n. Then, it is nonpositive, still by virtue of the maximum principle. Thus, we can define the function $u : \mathbb{R}^N \to \mathbb{R}$ as the pointwise limit of the bounded sequence $\{u_n\}$. Taking the *a priori* interior estimates of Theorem D.1.1 into account and arguing as in the proof of Theorem 1.1.1, it follows that u is a continuous and nontrivial solution of the equation $\lambda u - \mathcal{A}u = 0$ in $\mathbb{R}^N$. □

Remark 3.2.2 In Section 4.2 we will consider other conditions implying non-uniqueness of the bounded solutions to the elliptic equation (3.0.1) and the Cauchy problem (3.0.2). Actually, we will provide sufficient conditions ensuring that $T(t)(C_b(\mathbb{R}^N)) \subset C_0(\mathbb{R}^N)$ for any $t > 0$. It is clear that, in such a situation uniqueness does not hold, if $c \equiv 0$, since the previous condition implies that $T(\cdot)\mathbf{1} \not\equiv \mathbf{1}$.

Chapter 4

Properties of $T(t)$ in spaces of continuous functions

4.0 Introduction

In this chapter we study several properties of the semigroup $\{T(t)\}$. More precisely, we deal with: the compactness of the semigroup in $C_b(\mathbb{R}^N)$, the inclusion of $T(t)(C_b(\mathbb{R}^N))$ in $C_0(\mathbb{R}^N)$ and the invariance of $C_0(\mathbb{R}^N)$ under the action of the semigroup.

∗ **Compactness.** In Section 4.1 we study the compactness of $\{T(t)\}$ in $C_b(\mathbb{R}^N)$. First, we consider the conservative case and, then, the non-conservative one. In the conservative case $T(t)$ is a compact operator, for any $t > 0$, provided that the family $\{p(t, x; dy) : x \in \mathbb{R}^N\}$ is tight for any $t > 0$ (see Definition 4.1.1). Here, $\{p(t, x; dy) : t > 0,\ x \in \mathbb{R}^N\}$ is the transition family defined in Theorem 1.2.5. In such a case the function $t \mapsto T(t)$ is continuous in $(0, +\infty)$ with respect to the operator topology.

We show that the family $\{p(t, x; dy) : x \in \mathbb{R}^N\}$ is tight for any $t > 0$ if there exist a strictly positive function $\varphi \in C^2(\mathbb{R}^N)$ and a convex function $g : [0, +\infty) \to \mathbb{R}$ such that φ blows up as $|x|$ tends to $+\infty$, $1/g$ is integrable at $+\infty$ and $\mathcal{A}\varphi \le -g \circ \varphi$ in $\mathbb{R}^N$. It is not difficult to give sufficient conditions on the coefficients of the operator $\mathcal{A}$ guaranteeing that the previous conditions are satisfied. For instance, this is the case of the semigroup generated by the operator $\mathcal{A}$, defined on smooth functions φ by $\mathcal{A}\varphi(x) = \Delta\varphi(x) - |x|^2\langle x, \nabla\varphi(x)\rangle$ for any $x \in \mathbb{R}^N$.

In the non-conservative case, there are some connections between the compactness of $T(t)$ $(t > 0)$ and the fact that $T(t)$ maps $C_b(\mathbb{R}^N)$ into $C_0(\mathbb{R}^N)$. More precisely, $T(t)$ is compact and maps $C_0(\mathbb{R}^N)$ into itself, for some $t > 0$, if and only if $T(t)\mathbb{1}$ belongs to $C_0(\mathbb{R}^N)$. In such a case (see Proposition 4.1.11), the resolvent operator[1] $R(\lambda, \widehat{A})$ is compact for any $\lambda > 0$ and the semigroup is norm-continuous.

∗ **The inclusion $T(t)(C_b(\mathbb{R}^N)) \subset C_0(\mathbb{R}^N)$.** It is not always true that $T(t)$ maps $C_b(\mathbb{R}^N)$ into $C_0(\mathbb{R}^N)$ for any $t > 0$, since it is not always true that $C_0(\mathbb{R}^N)$ is invariant under $\{T(t)\}$. For instance, as has been remarked above, if $T(t)$ is conservative and compact for some $t > 0$, then $C_0(\mathbb{R}^N)$ is not invariant under $T(t)$. In fact, $T(t)$ maps $C_b(\mathbb{R}^N)$ into $C_0(\mathbb{R}^N)$, for some $t > 0$, if and only if $T(t)\mathbb{1} \in C_0(\mathbb{R}^N)$.

As is shown in Proposition 4.2.2, $T(t)$ maps $C_b(\mathbb{R}^N)$ into $C_0(\mathbb{R}^N)$ for any $t > 0$ if and only if the resolvent operator $R(\lambda) = R(\lambda, \widehat{A})$ does, for some (and, hence, for any) $\lambda > 0$ or, equivalently, if and only if $C_0(\mathbb{R}^N) \cap D_{\max}(\mathcal{A})$ is the domain of the weak generator $\widehat{A}$ of the semigroup $\{T(t)\}$.

A sufficient condition implying that $T(t)\mathbb{1} \in C_0(\mathbb{R}^N)$ is provided in Proposition 4.2.4 by assuming the existence of a function $\varphi \in C^2(\mathbb{R}^N) \cap C_0(\mathbb{R}^N)$, positive outside of a compact

[1] See Sections 1.1 and 1.3 for the definitions of $\widehat{A}$ and the resolvent operator $R(\lambda, \widehat{A})$.

subset K and such that the function $\lambda\varphi - \mathcal{A}\varphi$ has positive infimum in $\mathbb{R}^N \setminus K$ for some $\lambda > c_0$, where, as usual, $c_0 = \sup_{x\in\mathbb{R}^N} c(x)$.

$*$ Invariance of $C_0(\mathbb{R}^N)$. As has already been remarked above, it is not always true that $T(t)(C_0(\mathbb{R}^N))$ is contained in $C_0(\mathbb{R}^N)$ for any $t > 0$. The previous inclusion holds true, for instance, when there exists a strictly positive function $\varphi \in C^2(\mathbb{R}^N) \cap C_0(\mathbb{R}^N)$ such that $\lambda\varphi - \mathcal{A}\varphi \geq 0$ for some $\lambda > 0$. In such a case the operator A_0, defined by $A_0 u = \mathcal{A}u$ for any $u \in D(A_0) = \{u \in D_{\max}(\mathcal{A}) \cap C_0(\mathbb{R}^N) : \mathcal{A}u \in C_0(\mathbb{R}^N)\}$ (i.e., the part of $\widehat{A}$ in $C_0(\mathbb{R}^N)$), is the generator of a strongly continuous semigroup.

4.1 Compactness of $T(t)$

In this section we study the compactness of the semigroup $\{T(t)\}$ in $C_b(\mathbb{R}^N)$. The results that we present here are taken from [82] and [196]. In this and in the subsequent sections we assume that the coefficients of the operator $\mathcal{A}$, defined on smooth functions by $\mathcal{A}u = \mathrm{Tr}(QD^2u) + \langle b, \nabla u\rangle + cu$, satisfy the same hypotheses as in Chapter 1. We consider both the cases when the semigroup is conservative and non-conservative.

4.1.1 The conservative case

We begin this subsection recalling that $\{T(t)\}$ is conservative whenever $T(t)\mathbb{1} = \mathbb{1}$ for any $t > 0$ or, equivalently, whenever the elliptic equation $\lambda u - \mathcal{A}u = f$ and the Cauchy problem

$$\begin{cases} D_t u(t,x) = \mathcal{A}u(t,x), & t > 0, \quad x \in \mathbb{R}^N, \\ u(0,x) = f(x), & \qquad\quad x \in \mathbb{R}^N, \end{cases}$$

are uniquely solvable, respectively, in $D_{\max}(\mathcal{A})$ (see (1.0.1)) and in $C_b([0,T] \times \mathbb{R}^N) \cap C^{1+\alpha/2,2+\alpha}_{\mathrm{loc}}((0,+\infty) \times \mathbb{R}^N)$ for any $T > 0$ (see Proposition 3.1.10). Of course, if $\{T(t)\}$ is conservative then, necessarily, $c \equiv 0$.

We also recall that we can associate a transition family $\{p(t,x;dy) : t > 0, \ x \in \mathbb{R}^N\}$ with the semigroup (see Definition 1.2.4 and Theorem 1.2.5). The connection between the semigroup and the transition family is explained by the formula (1.2.7) which allows us to write $T(t)f$ for any $t > 0$ and $f \in C_b(\mathbb{R}^N)$ in terms of the Borel measures $p(t,x;dy)$.

With all these remarks in mind, now we can show that the compactness of the semigroup $\{T(t)\}$ is equivalent to the tightness of the family of measures $\{p(t,x;dy) : x \in \mathbb{R}^N\}$ for any $t > 0$ and it implies that the semigroup is norm-continuous in $(0,+\infty)$ (see Definition C.2.4).

Definition 4.1.1 *A family of Borel probability measures $\{\mu_\alpha\}_{\alpha\in\mathcal{F}}$ is said to be tight if for any $\varepsilon > 0$ there exists $\rho > 0$ such that $\mu_\alpha(B_\rho) \geq 1 - \varepsilon$ for any $\alpha \in \mathcal{F}$.*

Proposition 4.1.2 *Assume that the semigroup $\{T(t)\}$ is conservative. Then, $T(t)$ is compact in $C_b(\mathbb{R}^N)$, for any $t > 0$, if and only if the family of measures $\{p(t,x;dy) : x \in \mathbb{R}^N\}$ is tight for any $t > 0$. In this case the semigroup is norm-continuous in $(0,+\infty)$ and the resolvent $R(\lambda,\widehat{A})$ is compact for any $\lambda > 0$.*

Proof Fix $t, \varepsilon > 0$ and suppose that the operator $T(t)$ is compact. Let $\{f_n\}$ be a sequence of bounded and continuous functions in $\mathbb{R}^N$ such that $\chi_{B_{n-1}} \le f_n \le \chi_{B_n}$ for any $n \in \mathbb{N}$. By Proposition 1.2.10, $T(t)f_n$ tends to $T(t)\mathbb{1} \equiv \mathbb{1}$ as n tends to $+\infty$, locally uniformly in $\mathbb{R}^N$. Since $T(t)$ is compact, $T(t)f_n$ tends to $\mathbb{1}$ uniformly in $\mathbb{R}^N$. Indeed, if this were not the case one could find a positive number ε_0 and a subsequence $\{f_{n_k}\}$ such that $\|T(t)f_{n_k} - \mathbb{1}\|_\infty \ge \varepsilon_0$. Hence, no subsequences uniformly converging in $\mathbb{R}^N$ could be extracted from the sequence $\{T(t)f_{n_k}\}$, contradicting the compactness of $T(t)$. It follows that the family of measures $\{p(t,x;dy) : x \in \mathbb{R}^N\}$ is tight for any $t > 0$, since (see (1.2.7)) $p(t,x;B_n) \ge (T(t)f_n)(x)$ for any $x \in \mathbb{R}^N$.

Conversely, assume that, for any $t > 0$, the family $\{p(t,x;dy) : x \in \mathbb{R}^N\}$ is tight. Fix $t > 0$ and let $\rho_n > 0$ be such that

$$p(t/2, x; B_{\rho_n}) \ge 1 - n^{-1}, \qquad x \in \mathbb{R}^N. \tag{4.1.1}$$

Define the operator $S_{t,n}$ in $C_b(\mathbb{R}^N)$ by

$$(S_{t,n}f)(x) = \int_{B_{\rho n}} (T(t/2)f)(y)p(t/2,x;dy)$$

for any $x \in \mathbb{R}^N$ and any $f \in C_b(\mathbb{R}^N)$. By (4.1.1) it follows that

$$|(T(t)f)(x) - (S_{t,n}f)(x)| \le \int_{\mathbb{R}^N \setminus B_{\rho n}} |(T(t/2)f)(y)|p(t/2,x;dy) \le n^{-1}\|f\|_\infty \tag{4.1.2}$$

for any $f \in C_b(\mathbb{R}^N)$ and $x \in \mathbb{R}^N$, since $\{T(t)\}$ is a semigroup of contractions (see (1.2.6)). Moreover, $S_{t,n}$ is compact for any $n \in \mathbb{N}$. Indeed, $S_{t,n} = G_n \circ R_n \circ T(t/2)$ for any $n \in \mathbb{N}$, where $R_n : C_b(\mathbb{R}^N) \to C(\overline{B}_{\rho_n})$ is the restriction operator (i.e., $R_n f(x) = f(x)$ for any $x \in \overline{B}_{\rho_n}$ and any $f \in C_b(\mathbb{R}^N)$) and $G_n : C(\overline{B}_{\rho_n}) \to C_b(\mathbb{R}^N)$ is the bounded operator defined by

$$(G_n h)(x) = \int_{B_{\rho_n}} h(y)p(t/2,x;dy) = T(t/2)(h\chi_{B_n}), \qquad x \in \mathbb{R}^N,$$

for any $h \in C(\overline{B}_{\rho_n})$. We observe that $G_n h$ belongs to $C_b(\mathbb{R}^N)$ for any $h \in C(\overline{B}_{\rho_n})$, since $\{T(t)\}$ has the strong Feller property (see Proposition 1.2.13). By the interior Schauder estimates in Theorem D.1.5, we deduce that $R_n \circ T(t/2)$ is bounded from $C_b(\mathbb{R}^N)$ into $C^{2+\alpha}(\overline{B}_{\rho_n})$ and the Arzelà-Ascoli theorem implies that $R_n \circ T(t/2)$ is compact. It follows that $S_{t,n}$ is compact as well. Since $\{S_{t,n}\}$ is a sequence of compact operators converging to $T(t)$ in the operator topology (see (4.1.2)), $T(t)$ is, itself, a compact operator. Due to the arbitrariness of $t > 0$, $T(t)$ is compact for any $t > 0$.

Let us now prove that $T(t)$ is norm-continuous for any $t > 0$. We begin by showing that $T(\cdot)f$ is continuous in $(0,+\infty)$ for any $f \in C_b(\mathbb{R}^N)$ with $\|f\|_\infty \le 1$. For this purpose, we fix $\varepsilon > 0$, $t_0 > 0$ and let ρ be sufficiently large that $p(t_0,x;\mathbb{R}^N \setminus B_\rho) \le \varepsilon$ for any $x \in \mathbb{R}^N$. Then, for any $h > 0$ it holds that

$$\begin{aligned}(T(t_0+h)f)(x) - (T(t_0)f)(x) &= \int_{B_\rho} ((T(h)f)(y) - f(y))p(t_0,x;dy)\\ &\quad + \int_{\mathbb{R}^N \setminus B_\rho} ((T(h)f)(y) - f(y))p(t_0,x;dy)\\ &\le \|T(h)f - f\|_{C(\overline{B}_\rho)} + 2\varepsilon \end{aligned} \tag{4.1.3}$$

and the right-hand side of (4.1.3) is less than 3ε provided h is sufficiently small since

$T(\cdot)f \in C([0,+\infty)\times\mathbb{R}^N)$ and $T(0)f \equiv f$ (see Theorems 1.2.1, 1.2.5). Hence, $\|T(t_0+h)f - T(t_0)f\|_\infty$ vanishes as h tends to 0^+. For negative values of h we fix $\delta \in (0,t_0)$. Then, for any $h \in (\delta - t_0, 0)$ we split $T(t_0+h)f - T(t_0)f = T(t_0+h-\delta)(T(\delta)f - T(\delta-h)f)$ and estimate

$$\begin{aligned}\|T(t_0+h)f - T(t_0)f\|_\infty =&\|T(t_0+h-\delta)\left(T(\delta)f - T(\delta-h)f\right)\|_\infty\\ \leq&\|T(\delta)f - T(\delta-h)f\|_\infty.\end{aligned}$$

Letting h tend to 0^-, by the previous step we get

$$\lim_{h\to 0^-} T(t_0+h)f = T(t_0)f. \tag{4.1.4}$$

Hence,

$$\lim_{h\to 0}\|T(t_0+h)f - T(t_0)f\|_\infty = 0, \qquad t>0. \tag{4.1.5}$$

Now, we can show that (4.1.5) and the compactness of the semigroup imply that the function $t \mapsto T(t)$ is norm-continuous in $(0,+\infty)$. Suppose by contradiction, that $T(\cdot)$ is not norm-continuous in $(0,+\infty)$. Then, we can determine $t_0, \varepsilon > 0$, two sequences $\{f_n\} \subset C_b(\mathbb{R}^N)$, with $\|f_n\|_\infty = 1$ for any $n \in \mathbb{N}$, and $\{t_n\} \subset (0,+\infty)$, converging to t_0 as n tends to $+\infty$, such that

$$\|(T(t_n) - T(t_0))f_n\|_\infty \geq \varepsilon, \qquad n \in \mathbb{N}. \tag{4.1.6}$$

Without loss of generality, we can assume that $\{t_n\}$ is a monotone sequence. We first assume that it is decreasing and we set $a_n = t_n - t_0$ for any $n \in \mathbb{N}$. Since $T(t_0)$ is a compact operator, up to a subsequence we can assume that $T(t_0)f_n$ converges uniformly to a function $g \in C_b(\mathbb{R}^N)$ as n tends to $+\infty$. Hence, from (4.1.6) we get

$$\begin{aligned}\|(T(a_n)-I)g\|_\infty \geq&\|(T(a_n)-I)T(t_0)f_n\|_\infty - \|(T(a_n)-I)(g - T(t_0)f_n)\|_\infty\\ \geq&\varepsilon - 2\|g - T(t_0)f_n\|_\infty,\end{aligned}$$

which implies that $\limsup_{n\to+\infty}\|(T(a_n)-I)g\|_\infty \geq \varepsilon$, contradicting (4.1.5). Therefore, $\lim_{t\to t_0^+} T(t) = T(t_0)$ in $L(C_b(\mathbb{R}^N))$. Now to prove that $\{T(t)\}$ is norm-continuous from the left in $(0,+\infty)$, it suffices to argue as in the proof of (4.1.4).

To conclude the proof, we observe that, since $\{T(\cdot)\}$ is norm-continuous and bounded in $(0,+\infty)$, the integral $\int_0^{+\infty} e^{-\lambda t}T(t)dt$ converges in the operator topology to $R(\lambda)$, for any $\lambda > 0$. Since $T(t)$ is compact for any $t > 0$, the operator $R(\lambda)$ is compact as well. □

Now, we prove a lemma.

Lemma 4.1.3 *Suppose that $\{T(t)\}$ is conservative. Let $\varphi \in C^2(\mathbb{R}^N)$ be any function, which diverges to $+\infty$ as $|x|$ tends to $+\infty$ and satisfies the inequality $\mathcal{A}\varphi \leq K$ in $\mathbb{R}^N$ for some positive constant K. Then, for any $t > 0$ and $x \in \mathbb{R}^N$ it holds that*

$$(T(t)\varphi)(x) := \int_{\mathbb{R}^N}\varphi(y)p(t,x;dy) \in \mathbb{R}, \tag{4.1.7}$$

$$(T(t)\mathcal{A}\varphi)(x) := \int_{\mathbb{R}^N}\mathcal{A}\varphi(y)p(t,x;dy) \in \mathbb{R}. \tag{4.1.8}$$

Moreover, $T(\cdot)\varphi$ belongs to $C^{1+\alpha/2,2+\alpha}_{\rm loc}((0,+\infty)\times\mathbb{R}^N)\cap C([0,+\infty)\times\mathbb{R}^N)$ and

$$\lim_{t\to 0^+}(T(t)\varphi)(x) = \varphi(x), \qquad (D_tT(t)\varphi)(x) \leq (T(t)\mathcal{A}\varphi)(x) \tag{4.1.9}$$

for any $t > 0$ and $x \in \mathbb{R}^N$.

Proof Without loss of generality we can suppose that $\varphi \geq 0$. Indeed, if this is not the case, we replace φ with $\varphi - \inf_{x\in\mathbb{R}^N}\varphi(x)$. Observe that $\inf_{x\in\mathbb{R}^N}\varphi(x)$ is a real number since φ diverges to $+\infty$ as $|x|$ tends to $+\infty$. Moreover, $\mathcal{A}\varphi = \mathcal{A}(\varphi - \inf_{x\in\mathbb{R}^N}\varphi(x))$ since the zero-order coefficient of $\mathcal{A}$ identically vanishes in $\mathbb{R}^N$.

For any $n \in \mathbb{N}$, let $\psi_n \in C_b^2([0,+\infty))$ be any function such that $\psi_n(s) = s$ for any $s \in [0,n]$, $\psi_n(s) = n+1$ for any $s \geq n+2$, $0 \leq \psi_n' \leq 1$ and $\psi_n'' \leq 0$ in $[0,+\infty)$. Then, $\psi_n \circ \varphi$ is bounded and, in particular, it is constant outside some compact set of $\mathbb{R}^N$. Consequently, the function $\mathcal{A}(\psi_n \circ \varphi)$ has compact support. This implies that $\psi_n \circ \varphi$ belongs to $D_{\max}(\mathcal{A})$. By Propositions 1.3.5 and 3.1.10, we know that $D_{\max}(\mathcal{A}) = D(\widehat{A})$. Hence, by Lemma 1.3.2 we deduce that

$$T(t)\mathcal{A}(\psi_n \circ \varphi) = T(t)\widehat{A}(\psi_n \circ \varphi) = \widehat{A}T(t)(\psi_n \circ \varphi) = \mathcal{A}T(t)(\psi_n \circ \varphi) \tag{4.1.10}$$

for any $t \geq 0$. Now, set

$$u_n(t,x) = (T(t)(\psi_n \circ \varphi))(x) = \int_{\mathbb{R}^N} \psi_n(\varphi(y))p(t,x;dy), \qquad t>0,\ x\in\mathbb{R}^N.$$

Thanks to our assumptions on the functions ψ_n ($n \in \mathbb{N}$), the sequence $\{\psi_n(s)\}$ is increasing to $\varphi(s)$ for any $s \in \mathbb{R}$. Hence, the monotone convergence theorem yields

$$\lim_{n\to+\infty} u_n(t,x) = \lim_{n\to+\infty}\int_{\mathbb{R}^N} \psi_n(\varphi(y))p(t,x;dy) = u(t,x) := (T(t)\varphi)(x) \tag{4.1.11}$$

for any $t>0$ and $x \in \mathbb{R}^N$. From (4.1.10) we deduce that, for any $s>0$ and $x\in\mathbb{R}^N$,

$$\begin{aligned} D_t u_n(s,x) =& \mathcal{A}u_n(s,x) = (T(s)\mathcal{A}(\psi_n\circ\varphi))(x) \\ =& \int_{\mathbb{R}^N} \{(\psi_n'\circ\varphi)\mathcal{A}\varphi + (\psi_n''\circ\varphi)\langle Q\nabla\varphi,\nabla\varphi\rangle\} p(s,x;dy) \\ \leq& \int_{\mathbb{R}^N} (\psi_n'\circ\varphi)\mathcal{A}\varphi\, p(s,x;dy) \leq K. \end{aligned} \tag{4.1.12}$$

Integrating (4.1.12) with respect to $s \in [0,t]$ and recalling that the semigroup $\{T(t)\}$ preserves positivity, gives

$$0 \leq u_n(t,x) \leq \psi_n(\varphi(x)) + Kt \leq \varphi(x) + Kt, \qquad t>0,\ x\in\mathbb{R}^N,\ n\in\mathbb{N}. \tag{4.1.13}$$

Then, letting n tend to $+\infty$, we get $0 \leq u(t,x) \leq \varphi(x) + Kt < +\infty$ for any $(t,x) \in (0,+\infty)\times\mathbb{R}^N$ and (4.1.7) follows.

Now we prove (4.1.8) and (4.1.9) and, at the same time, we show that u belongs to $C^{1+\alpha/2,2+\alpha}_{\rm loc}((0,+\infty)\times\mathbb{R}^N)$. By (4.1.13) and Theorem D.1.5, the sequence $\{u_n\}$ is bounded in $C^{1+\alpha/2,2+\alpha}([\varepsilon,T]\times F)$ for any $\varepsilon, T>0$ and any compact set $F\subset\mathbb{R}^N$. From (4.1.11) and the Arzelà-Ascoli theorem, we deduce that $u \in C^{1+\alpha/2,2+\alpha}([\varepsilon,T]\times F)$ and u_n converges to u in $C^{1+\beta/2,2+\beta}([\varepsilon,T]\times F)$ for any $\beta<\alpha$ and any ε, T and F as above. In particular,

$$\lim_{n\to+\infty} D_t u_n(t,x) = D_t u(t,x) \tag{4.1.14}$$

for any $t>0$ and $x\in\mathbb{R}^N$. From (4.1.12) it follows that

$$D_t u_n(t,x) \leq \int_{\{\mathcal{A}\varphi\geq 0\}} \psi_n'(\varphi(y))\mathcal{A}\varphi(y)p(t,x;dy) + \int_{\{\mathcal{A}\varphi<0\}} \psi_n'(\varphi(y))\mathcal{A}\varphi(y)p(t,x;dy). \tag{4.1.15}$$

As n tends to $+\infty$, we get (4.1.8) and (4.1.9) by (4.1.14) and the monotone convergence theorem. Note that the first integral in the right-hand side of (4.1.15) converges for any $x \in \mathbb{R}$, since $0 \le \psi_n' \le 1$, for any $n \in \mathbb{N}$, and $\mathcal{A}\varphi$ is bounded from above. Therefore, the second integral in the right-hand side of (4.1.15) converges as well for any $x \in \mathbb{R}^N$ and the second formula in (4.1.9) follows.

To show that u is continuous up to $t = 0$ we argue as in the proof of Theorem 1.2.1. We fix $m \in \mathbb{N}$ and introduce a cut-off function $\vartheta = \vartheta_m \in C_c^\infty(\mathbb{R}^N)$ such that $\chi_{B_{m-1}} \le \vartheta \le \chi_{B_m}$. Then, we consider the function $v_n = u_n\vartheta_m$ and take $n \ge \sup_{x\in B_m} |\varphi(x)|$. This choice of n implies that $\vartheta\psi_n(\varphi) = \vartheta\varphi$. As is easily seen, v_n solves the Cauchy-Dirichlet problem

$$\begin{cases} D_t v_n(t,x) = \mathcal{A}v_n(t,x) + g_n(t,x), & t>0, \quad x \in B_m, \\ v_n(t,x) = 0, & t > 0, \quad x \in \partial B_m, \\ v_n(0,x) = \vartheta(x)\varphi(x), & x \in \overline{B}_m, \end{cases}$$

for any $n \ge m$, where $g_n = -2\langle Q\nabla_x u, \nabla\vartheta\rangle - u_n \mathrm{Tr}(QD^2\vartheta) - u_n\langle b, \nabla\vartheta\rangle$ satisfies $|g_n(t,x)| \le C_1(\|\varphi\|_{L^\infty(B_m)} + \|u_n\|_{L^\infty(B_m)} + \|\nabla_x u_n(t,\cdot)\|_{L^\infty(B_m)})$ for some positive constant C_1, independent of n. Estimate (4.1.13) and the interior estimates in Theorem D.1.5 (see (D.1.12)) yield $\|\sqrt{t}D_i u_n(t,\cdot)\|_{L^\infty(B_m)} \le C\|u_n\|_{L^\infty((0,T)\times B_{m+1})} \le C_2(\|\varphi\|_{L^\infty(B_{m+1})} + K)$ for any $t \in (0,1)$, $x \in B_m$, $i = 1,\ldots,N$ and some positive constant C_2, independent of n. Repeating step by step the last part of the proof of Theorem 1.2.1, we deduce that the function $u(t,\cdot)$ converges to φ uniformly in B_{m-1}, as t tends to 0^+. Due to the arbitrariness of m, it follows that $u \in C([0,+\infty) \times \mathbb{R}^N)$. □

Thanks to Lemma 4.1.3, we can now prove the following theorem, which gives a sufficient condition ensuring the compactness of $T(t)$ for any $t > 0$.

Theorem 4.1.4 *Assume that the semigroup $\{T(t)\}$ is conservative and suppose that there exist a strictly positive function $\varphi \in C^2(\mathbb{R}^N)$ and a convex function $g \in C^1([0,+\infty))$ such that*

$$\lim_{|x|\to+\infty} \varphi(x) = +\infty, \quad \frac{1}{g} \text{ is integrable at } +\infty, \quad \mathcal{A}\varphi(x) \le -g(\varphi(x)), \quad x \in \mathbb{R}^N. \tag{4.1.16}$$

Then, $T(t)$ is compact in $C_b(\mathbb{R}^N)$ for any $t > 0$.

Proof Without loss of generality we can suppose that $g(0) < 0$. Indeed, if this is not the case we replace g with $g - C$ for some large enough positive constant C. Let us set

$$u(t,x) = (T(t)\varphi)(x) = \int_{\mathbb{R}^N} \varphi(y)p(t,x;dy), \qquad t > 0, \ x \in \mathbb{R}^N.$$

Observe that, since g is convex and $1/g$ is integrable in a neighborhood of $+\infty$, $g(x)$ diverges to $+\infty$ as x tends to $+\infty$. In particular, g is bounded from below in $(0,+\infty)$. Hence, by (4.1.16), $\mathcal{A}\varphi$ is bounded from above. Consequently, by Lemma 4.1.3, it follows that $u \in C^{1,2}((0,+\infty)\times\mathbb{R}^N) \cap C([0,+\infty)\times\mathbb{R}^N)$. Moreover,

$$D_t u(t,x) \le \int_{\mathbb{R}^N} \mathcal{A}\varphi(y)p(t,x;dy) \le -\int_{\mathbb{R}^N} g(\varphi(y))p(t,x;dy) \le -g(u(t,x))$$

for any $t > 0$ and $x \in \mathbb{R}^N$, where the last inequality follows from Jensen's inequality.

For any fixed $x \in \mathbb{R}^N$, let $z(\cdot,x)$ be the solution of the Cauchy problem

$$\begin{cases} z'(t,x) = -g(z(t,x)), & t > 0, \quad x \in \mathbb{R}^N, \\ z(0,x) = \varphi(x), & x \in \mathbb{R}^N. \end{cases} \tag{4.1.17}$$

Note that $z(\cdot, x)$ is defined in $(0, +\infty)$. Indeed, if $\varphi(x) = x_g$, where x_g denotes the zero of g, then $z(\cdot, x) \equiv x_g$, whereas $z(\cdot, x)$ is increasing and greater than x_g if $\varphi(x) > x_g$. Similarly, $z(\cdot, x)$ is decreasing and less than x_g if $\varphi(x) < x_g$. In any case, $z(\cdot, x)$ exists in $[0, +\infty)$.

Let us now prove that $z(\cdot, x)$ is bounded from above in $[\tau, +\infty)$ for any $\tau > 0$, uniformly with respect to $x \in \mathbb{R}^N$. First, we assume that $\varphi(x) > x_g$. From (4.1.17) it follows that

$$\int_{z(t,x)}^{+\infty} \frac{dz}{g(z)} = \int_{\varphi(x)}^{+\infty} \frac{dz}{g(z)} - \int_{\varphi(x)}^{z(t,x)} \frac{dz}{g(z)} \geq - \int_{\varphi(x)}^{z(t,x)} \frac{dz}{g(z)} = t > \tau. \tag{4.1.18}$$

Note that all the integral terms are finite since the function $1/g$ is integrable in a neighborhood of $+\infty$. Since $1/g$ is not integrable in a neighborhood of x_g, there exists a unique value $\bar{z} = \bar{z}(\tau) > x_g$ such that

$$\int_{\bar{z}}^{+\infty} \frac{dz}{g(z)} = \tau. \tag{4.1.19}$$

The positivity of g in $(x_g, +\infty)$, (4.1.18) and (4.1.19) imply that $z(t, x) < \bar{z}$ for any $t > \tau$. On the other hand, if $\varphi(x) \leq x_g$ then $z(t, x) \leq x_g$ for any $t \in (\tau, +\infty)$. Summing up, we have proved that $z(t, x) \leq \max(\bar{z}, x_g) = \bar{z}$ for any $t > \tau$ and $x \in \mathbb{R}^N$.

Now, we fix $\varepsilon > 0$ and let $\rho > 0$ be such that $\varphi(x) \geq \bar{z}/\varepsilon$ for any $x \in \mathbb{R}^N \setminus B_\rho$. Then, we have

$$p(t, x; \mathbb{R}^N \setminus B_\rho) \leq \varepsilon \bar{z}^{-1} \int_{\mathbb{R}^N \setminus B_\rho} \varphi(y) p(t, x; dy) \leq \varepsilon \bar{z}^{-1} u(t, x) \leq \varepsilon$$

for any $x \in \mathbb{R}^N$ and $t > \tau$. The compactness of $T(t)$, for any $t > 0$, now follows from Proposition 4.1.2 and the arbitrariness of $\tau > 0$. Note that the proof of the quoted proposition shows that $T(t_0)$ is compact whenever the family $\{p(t_0, x; dy) : x \in \mathbb{R}^N\}$ is tight. □

Example 4.1.5 Consider the operator $\mathcal{A}$ defined by

$$\mathcal{A}\zeta(x) = \Delta\zeta(x) + \langle b(x), \nabla\zeta(x) \rangle, \qquad x \in \mathbb{R},$$

on smooth functions $\zeta : \mathbb{R}^N \to \mathbb{R}$, with the drift term b satisfying $\langle b(x), x \rangle \leq C - M|x|^{2+\varepsilon}$ for any $x \in \mathbb{R}^N$ and some constants $C \in \mathbb{R}$, $M, \varepsilon > 0$. Then, the associated semigroup $\{T(t)\}$ is compact. To see it, it suffices to apply Theorem 4.1.4 with the functions φ and g defined by $\varphi(x) = |x|^2$ for any $x \in \mathbb{R}^N$ and $g(s) = -(2N + C) + Ms^{1+\varepsilon/2}$ for any $s > 0$.

Finally, we give another condition implying that $T(t)$ is a compact operator for any $t > 0$. For this purpose, we introduce the one-dimensional differential operator $\mathscr{C}$, defined on smooth functions u by $\mathscr{C}u(r) = q(r)\left[u''(r) + r^{-1}b(r)u'(r)\right]$ for any $r > 0$, where

$$q(r) = \min_{x \in \partial B_r} \left(\frac{1}{|x|^2} \sum_{i,j=1}^{N} q_{ij}(x) x_i x_j \right) := \min_{x \in \partial B_r} \mathcal{Q}(x), \tag{4.1.20}$$

$$b(r) = \max_{x \in \partial B_r} \left(\frac{1}{\mathcal{Q}(x)} \sum_{i=1}^{N} (q_{ii}(x) + b_i(x) x_i) - 1 \right) := \max_{x \in \partial B_r} \mathcal{B}(x). \tag{4.1.21}$$

We introduce a slight modification of the functions $\mathscr{P}$ and $\mathscr{R}$ considered in Chapter 2, i.e., which we still denote by the same letters and are defined by

$$\mathscr{P}(x) = \frac{1}{q(x)\mathscr{W}(x)} \int_1^x \mathscr{W}(s) ds, \qquad \mathscr{R}(x) = \mathscr{W}(x) \int_1^x \frac{1}{q(s)\mathscr{W}(s)} ds, \qquad x \geq 1,$$

where

$$\mathscr{W}(x) = \exp\left(-\int_1^x \frac{b(s)}{q(s)} ds \right), \qquad x \geq 1.$$

Proposition 4.1.6 *If $+\infty$ is an entrance point for the operator $\mathscr{C}$ (i.e., $\mathscr{P} \in L^1((1,+\infty))$ and $\mathscr{R} \notin L^1((1,+\infty))$), then $T(t)$ is a compact operator for any $t>0$.*

To prove the proposition we need two preliminary lemmata.

Lemma 4.1.7 *Suppose that $+\infty$ is an entrance point for the operator $\mathscr{C}$. Then, for any $R>0$, there exists a positive function $\varphi \in C^2(\mathbb{R}^N \setminus B_R)$ diverging to $+\infty$ as $|x|$ tends to $+\infty$ and such that $\varphi - \mathcal{A}\varphi \ge 0$ in $\mathbb{R}^N \setminus B_R$. As a consequence, if $z \in C_b([0,T] \times \mathbb{R}^N \setminus B_R) \cap C^{1,2}((0,T) \times \mathbb{R}^N \setminus \overline{B}_R)$ satisfies*

$$\begin{cases} D_t z(t,x) - \mathcal{A}z(t,x) \ge 0, & t \in (0,T), \quad x \in \mathbb{R}^N \setminus \overline{B}_R, \\ z(t,x) \ge 0, & t \in (0,T), \quad x \in \partial B_R, \\ z(0,x) \ge 0, & \qquad\qquad\quad\; x \in \mathbb{R}^N \setminus B_R, \end{cases}$$

for some $R>0$, then $z \ge 0$.

Proof We limit ourselves to considering the case when $R=1$, the general case being completely similar.

To determine a function φ satisfying the assertion, we observe that, since $+\infty$ is an entrance, adapting the proofs of Propositions 2.1.4 and 2.1.7(iii) we can determine an increasing positive solution $\psi : [1,+\infty) \to \mathbb{R}$ of the equation $\psi - \mathscr{C}\psi = 0$ such that $\lim_{r\to+\infty} \psi(r) = +\infty$. The function φ can now be defined by setting $\varphi(x) = \psi(|x|)$ for any $x \in \mathbb{R}^N \setminus B_1$. Indeed, a straightforward computation shows that

$$\varphi(x) - \mathcal{A}\varphi(x) = \psi(|x|) - \mathcal{Q}(x)\left(\psi''(|x|) + \frac{\mathcal{B}(x)}{|x|}\psi'(|x|)\right), \qquad x \in \mathbb{R}^N \setminus B_1,$$

where $\mathcal{Q}$ and $\mathcal{B}$ are given by (4.1.20) and (4.1.21). Since ψ and ψ' are both positive, it is now easy to check that

$$\varphi(x) - \mathcal{A}\varphi(x) \ge \mathcal{Q}(x)\left(\frac{\psi(|x|)}{q(|x|)} - \psi''(|x|) - \frac{b(|x|)}{|x|}\psi'(|x|)\right) = 0, \qquad x \in \mathbb{R}^N \setminus B_1.$$

To prove the last part of the lemma, for any $\varepsilon > 0$ we introduce the function z_ε defined by $z_\varepsilon(t,x) = z(t,x) + \varepsilon e^t \varphi(x) + \varepsilon t$ for any $t \in [0,T]$ and $x \in \mathbb{R}^N \setminus B_1$. As is immediately seen, $D_t z_\varepsilon - \mathcal{A} z_\varepsilon \ge \varepsilon$ in $(0,T) \times (\mathbb{R}^N \setminus \overline{B}_1)$ and z_ε diverges to $+\infty$ as $|x|$ tends to $+\infty$, uniformly with respect to $t \in [0,T]$. This implies that z_ε has an absolute minimum at some point $(t_0, x_0) \in [0,T] \times (\mathbb{R}^N \setminus B_1)$. The case $t_0 > 0$ and $|x_0| > 1$ can not occur, otherwise $D_t z_\varepsilon(t_0,x_0) - \mathcal{A} z_\varepsilon(t_0,x_0) \le 0$ should be negative, which is a contradiction. Therefore, $t_0 = 0$ or $|x_0| = 1$. In any case $z_\varepsilon(t_0,x_0)$ is nonnegative. Letting ε tend to 0^+ the assertion follows. □

Lemma 4.1.8 *Suppose that $+\infty$ is an entrance point for the operator $\mathscr{C}$ and fix $\delta > 0$. Then, for any $f \in C_b([\delta,+\infty))$ the Cauchy-Dirichlet problem*

$$\begin{cases} D_t u(t,x) = \mathscr{C}u(t,x), & t>0, \quad x>\delta, \\ u(t,\delta) = 0, & t>0, \\ u(0,x) = f(x), & \qquad\quad\; x>\delta, \end{cases}$$

admits a solution $u \in C_b(([0,+\infty) \times [\delta,+\infty)) \setminus \{(0,\delta)\}) \cap C^{1,2}((0,+\infty) \times (\delta,+\infty))$. Moreover, $\|u\|_\infty \le \|f\|_\infty$. Further, the family of linear operators $\{S(t)\}$ defined by $S(t)f = u(t,\cdot)$ for any $t>0$, where u is as above, is a positive semigroup of contractions. Finally, the function $v = S(\cdot)\mathbf{1}$ satisfies $D_t v \le 0$ and $D_r v \ge 0$ in $(0,+\infty) \times (\delta,+\infty)$.

Proof By a straightforward computation we can check that the coefficients of the operator $\mathscr{C}$ belong to $C^{\alpha}_{\rm loc}([\delta,+\infty))$. Moreover, the proof of Lemma 4.1.7 shows that there exists a positive function $\psi \in C^2([\delta,+\infty))$, blowing up as x tends to $+\infty$, such that $\psi - \mathscr{C}\psi = 0$ in $(\delta,+\infty)$. Therefore, to prove the assertion, it suffices to apply the same arguments as in the proof of the forthcoming Theorem 12.1.2.

Let us just give the proof of the last part of the lemma. We set $u = S(\cdot)\mathbb{1}$ and prove that $D_t u$ and $-D_r u$ are nonpositive in $(0,+\infty)\times(\delta,+\infty)$. As a first step we observe that from the estimate $\|u\| \le \|f\|_\infty$ it follows that $u \le \mathbb{1}$. Using the positivity of the semigroup, we get

$$u(t+s,\cdot) = S(t+s)\mathbb{1} = S(s)S(t)\mathbb{1} \le S(t)\mathbb{1} = u(t,\cdot), \qquad s,t>0,$$

so that the function $u(\cdot,r)$ is nonincreasing in $(0,+\infty)$ for any $r>\delta$ and, consequently, $D_t u \le 0$ in $(0,+\infty)\times(\delta,+\infty)$.

Now we show that, for any $t>0$, the function $u(t,\cdot)$ is nondecreasing in $(0,+\infty)$. Noting that

$$\mathscr{C}u(t,r) = q(r)\mathscr{W}(r)D_r\left(\frac{D_r u(t,\cdot)}{\mathscr{W}}\right)(r), \qquad t>0,\ r>\delta,$$

where $\mathscr{W}$ is given by (2.1.3), we deduce that, for any $0<r<s$ and $t>0$,

$$\frac{D_r u(t,s)}{\mathscr{W}(s)} - \frac{D_r u(t,r)}{\mathscr{W}(r)} = \int_r^s \frac{D_t u(t,\xi)}{q(\xi)\mathscr{W}(\xi)}d\xi.$$

Since $D_t u \le 0$ in $(0,+\infty)\times(\delta,+\infty)$, the function $w(t,\cdot) := D_r u(t,\cdot)/\mathscr{W}$ is nonincreasing in $(\delta,+\infty)$ for any $t>0$. Set $\ell_t = \lim_{r\to+\infty} w(t,r)$. We claim that $\ell_t \ge 0$ for any $t>0$. Of course, once the claim is proved, we will immediately obtain that $w\ge 0$ and, consequently, $D_r u \ge 0$ in $(0,+\infty)\times(\delta,+\infty)$ since $\mathscr{W}$ is a strictly positive function. Suppose by contradiction that $\ell_t<0$ for some $t>0$. Then, there exist $r^*>0$ and a negative constant c such that $D_t u(t,r) \le c\mathscr{W}(r)$ for any $r\ge r^*$. Integrating the previous inequality in $[r^*,r]$, we get

$$v(t,r) = v(t,r^*) + \int_{r^*}^r D_t u(t,s)ds \le u(t,\delta) + c\int_{r^*}^r \mathscr{W}(s)ds.$$

The boundedness of v in $(0,+\infty)\times(\delta,+\infty)$ implies that $\mathscr{W}$ is integrable in $(r^*,+\infty)$ and, consequently, in $(\delta,+\infty)$. But this is a contradiction. Indeed, since $\mathscr{P}$ is integrable in $(\delta,+\infty)$, the function $1/(q\mathscr{W})$ is integrable in $(\delta,+\infty)$ as well. The integrability of $\mathscr{W}$ in $(\delta,+\infty)$ should imply the integrability of $\mathscr{R}$ in $(\delta,+\infty)$, which contradicts our assumptions. □

Now, we can prove Proposition 4.1.6.

Proof of Proposition 4.1.6 According to Proposition 4.1.2, to prove that $T(t)$ is compact for any $t>0$ we show that the family of measures $\{p(t,x;dy) : x\in\mathbb{R}^N\}$ is tight for any $t>0$, i.e., we fix $t_0>0$ and prove that for any $\varepsilon>0$ there exists $\rho>0$ such that $p(t_0,x;B_\rho) \ge 1-\varepsilon$ for any $x\in\mathbb{R}^N$.

Let $\{f_n\}\subset C_b(\mathbb{R}^N)$ be a sequence of continuous functions in $\mathbb{R}^N$ such that $\chi_{B_{n-1}} \le f_n \le \chi_{B_n}$ for any $n\in\mathbb{N}$. Observe that

$$p(t_0,x;B_n) = \int_{B_n} p(t_0,x;dy) \ge \int_{\mathbb{R}^N} f_n(y)p(t_0,x;dy) = (T(t_0)f_n)(x)$$

for any $x\in\mathbb{R}^N$ (see Theorem 1.2.5). Since f_n converges to f as n tends to $+\infty$, locally uniformly in $\mathbb{R}^N$, by Proposition 1.2.10 it follows that the function $T(t_0)f_n$ converges locally uniformly in $\mathbb{R}^N$ to the function $\mathbb{1}$. Hence, it suffices to show that there exist $\rho > 0$ and $n_1 \in \mathbb{N}$ such that

$$(T(t_0)f_n)(x) \geq 1 - \varepsilon, \qquad x \in \mathbb{R}^N \setminus \overline{B}_\rho, \ n \geq n_1. \tag{4.1.22}$$

To prove (4.1.22) we use a comparison argument with radial functions. Being rather long, we split the rest of the proof into two steps. First, in Step 1, we study the main properties of the function $v_\delta = S(\cdot)\mathbb{1}$ that we need in the sequel. Here, $\{S(t)\}$ is the semigroup in $C_b([\delta, +\infty))$ defined in Lemma 4.1.8. Then, in Step 2, by means of the function v_δ, we construct a radial function w such that $w(t_0, \cdot) \geq 1 - \varepsilon$ and such that, for $|x|$ and n large enough, it satisfies the inequality $(T(t_0)f_n)(x) \geq w(t_0, x)$.

Step 1. We claim that, for any δ sufficiently large,

$$0 \leq v_\delta(t_0, r) \leq \eta := 1 - (1 - \varepsilon)^{1/2}, \qquad r \geq \delta. \tag{4.1.23}$$

For this purpose, for any $\delta > 0$ we set

$$\psi_\delta(r) = 1 - \int_0^{+\infty} e^{-t} v(t, r)dt, \qquad r \geq \delta.$$

The function ψ_δ is well defined since v is bounded in $(0, +\infty) \times (\delta, +\infty)$. Moreover, it solves the Cauchy problem

$$\begin{cases} \psi_\delta(r) - \mathscr{C}\psi_\delta(r) = 0, & r > \delta, \\ \psi_\delta(\delta) = 1. \end{cases} \tag{4.1.24}$$

Of course, $\psi_\delta(\delta) = 1$ since $v(t, 0) = 0$ for any $t > 0$. To show that ψ_δ satisfies the differential equation in (4.1.24), for any $n \in \mathbb{N}$ we introduce the function $\psi_{\delta,n} : [\delta, +\infty) \to \mathbb{R}$ defined by

$$\psi_{\delta,n}(r) = \int_0^n e^{-t} v(t, r)dt, \qquad r \geq \delta.$$

It is immediate to check that $\psi_{\delta,n} \in C^2((\delta, +\infty))$ and it solves the differential equation

$$\psi_{\delta,n}(r) - \mathscr{C}\psi_{\delta,n}(r) = 0, \qquad r > \delta. \tag{4.1.25}$$

Moreover, $\psi_{\delta,n}$ converges uniformly to ψ_δ in $[\delta, +\infty)$. Let us now show that $\psi_\delta \in C^2((\delta, +\infty))$ and it solves the differential equation in (4.1.24) pointwise. For this purpose, we observe that, since the sequence $\{\psi_{\delta,n}\}$ is bounded in $C_b([\delta, +\infty))$, using the same compactness arguments as in the proof of Theorem 1.1.1, we can easily show that, up to a subsequence, $\psi_{\delta,n}$ converges to ψ_δ in $W^{2,p}(r_1, r_2)$ for any $\delta < r_1 < r_2$ and $p \in (1, +\infty)$. Using the Sobolev embedding theorems (see [5, Theorem 5.4]), we deduce that $\psi_\delta \in C^1((\delta, +\infty))$ and $\psi'_{\delta,n}$ converges to ψ'_δ locally uniformly in $(\delta, +\infty)$. Since $q(r)\psi''_{\delta,n}(r) = \psi_{\delta,n}(r) - r^{-1}q(r)b(r)\psi'_{\delta,n}(r)$ for any $r \in (\delta, +\infty)$ and $n \in \mathbb{N}$, it follows that $q\psi''_{\delta,n}$ (and consequently $\psi''_{\delta,n}$ since q does not vanish in $[\delta, +\infty)$) converges locally uniformly in $(\delta, +\infty)$. Therefore, $\psi_\delta \in C^2((\delta, +\infty))$. Letting n tend to $+\infty$ in (4.1.25), we conclude that ψ_δ solves the differential equation in (4.1.24).

Let us now prove that δ can be chosen sufficiently large that

$$\psi_\delta(r) \geq 1 - \eta t_0 e^{-t_0}, \qquad r \geq \delta. \tag{4.1.26}$$

For this purpose, denote by φ the unique positive decreasing solution of the equation $\varphi -$

$\mathscr{C}\varphi = 0$ in $(1, +\infty)$, such that $\lim_{r\to+\infty} \varphi(r) = 1$ (the existence of such a solution can be established arguing as in the proof of properties (ii) and (iv) in Proposition 2.1.4). Then, $\psi_\delta = \varphi/\varphi(\delta)$. Indeed, the same arguments as in the proof of Proposition 2.1.4 show that, up to a multiplicative constant, ψ_δ is the unique bounded solution to problem (4.1.24). It is clear that we can fix $\widehat{\delta}$ such that $\psi_{\widehat{\delta}}$ satisfies (4.1.26).

Now, for any $m \in \mathbb{N}$, let $E_m = \{t \geq 0 : e^{-t} \lim_{r\to+\infty} v_{\widehat{\delta}}(t, r) \geq \lambda_m\}$, where $\lambda_m = e^{-t_0}\eta + m^{-1}$. Since v is bounded, the set E_m is bounded for any $m \in \mathbb{N}$ and it is not empty since it contains $t = 0$. Moreover, using the Chebyshev inequality, we get

$$\begin{aligned}\int_{E_m} dr =& \frac{1}{\lambda_m}\int_{E_m} \lambda_m dr \leq \frac{1}{\lambda_m}\int_{E_m} e^{-t} \lim_{r\to+\infty} v_{\widehat{\delta}}(t,r)dt\\ =& \frac{1}{\lambda_m}\lim_{r\to+\infty}\int_{E_m} e^{-t} v_{\widehat{\delta}}(t,r)dt \leq \frac{1}{\lambda_m}\limsup_{r\to+\infty}(1-\psi_{\widehat{\delta}}(r)) \leq \frac{\eta}{\lambda_m} t_0 e^{-t_0}.\end{aligned}$$

Therefore, due to the choice of λ_m, the Lebesgue measure of E_m is strictly less than t_0. From this fact and recalling that $v_{\widehat{\delta}}(t, \cdot)$ is increasing in $[\widehat{\delta}, +\infty)$ for any $t > \widehat{\delta}$, it follows that there exists $t_{1,m} < t_0$ such that $e^{-t_{1,m}} v_{\widehat{\delta}}(t_{1,m}, \cdot) < \lambda_m$ in $(\widehat{\delta}, +\infty)$. Up to a subsequence we can assume that $\{t_{1,m}\}$ converges to a point $t_1 \in (0, t_0]$ as m tends to $+\infty$. Moreover, by continuity, $e^{-t_1} v_{\widehat{\delta}}(t_1, r) \leq e^{-t_0}$ for any $r \geq \widehat{\delta}$ and, consequently, $v_{\widehat{\delta}}(t_1, \cdot) \leq \eta$ in $[\widehat{\delta}, +\infty)$. Since $v_{\widehat{\delta}}(\cdot, r)$ is nonincreasing in $(0, +\infty)$ for any $r > \widehat{\delta}$, estimate (4.1.23) follows.

Step 2. Now, we prove (4.1.22) with $\rho = \widehat{\delta} + 1$. Since, as has already pointed out, $T(t)f_n$ converges locally uniformly in $[0, +\infty) \times \mathbb{R}^N$ to $\mathbb{1}$, we can fix $n_1 \in \mathbb{N}$ such that $(T(t)f_n)(x) \geq 1 - \eta$ for any $t \in (0, t_0)$, $x \in \overline{B}_{\widehat{\delta}+1}$ and $n \geq n_1$. Let us now introduce the function $w : [0, +\infty) \times (\mathbb{R}^N \setminus B_{\widehat{\delta}+1}) \to \mathbb{R}$ defined by $w(t, x) = (1-\eta)(1-v_{\widehat{\delta}}(t, |x|))$ for any $t \geq 0$ and $x \in \mathbb{R}^N \setminus B_{\widehat{\delta}+1}$. Clearly, $w \in C_b([0, +\infty) \times (\mathbb{R}^N \setminus B_{\widehat{\delta}+1})) \cap C^{1,2}((0, +\infty) \times (\mathbb{R}^N \setminus \overline{B}_{\widehat{\delta}+1}))$, it satisfies the conditions $w \leq 1 - \eta$ in $[0, +\infty) \times (\mathbb{R}^N \setminus B_{\widehat{\delta}+1})$, $w(0, \cdot) = 0$ and, by (4.1.23), $w(t_0, \cdot) \geq (1 - \eta)^2 = 1 - \varepsilon$ in $\mathbb{R}^N \setminus B_{\widehat{\delta}+1}$. Moreover, it satisfies the differential equation $D_t w - \mathcal{A}w = g$, where

$$g(t,x) = (\eta-1)\mathcal{Q}(x)\bigg(\frac{D_t v(t,|x|)}{\mathcal{Q}(x)} - D_{rr}v(t,|x|) - \frac{\mathcal{B}(x)}{|x|}D_r v(t,|x|)\bigg)$$

for any $t > 0$ and $x \in \mathbb{R}^N \setminus B_{\widehat{\delta}+1}$. Here, $\mathcal{Q}$ and $\mathcal{B}$ are defined by (4.1.20) and (4.1.21). Since $D_t v$ and $-D_r v$ are nonpositive in $(0, +\infty) \times (\widehat{\delta}, +\infty)$, we deduce that

$$g(t,x) \leq (\eta-1)(q(|x|))^{-1}\mathcal{Q}(x)(D_t v(t,|x|) - \mathscr{C}v(t,|x|)) = 0.$$

From all the results above, it follows that the function $z = T(\cdot)f_n - w$ solves the problem

$$\begin{cases} D_t z(t,x) - \mathcal{A}z(t,x) \geq 0, & t \in (0,t_0), \quad x \in \mathbb{R}^N \setminus \overline{B}_{\widehat{\delta}+1},\\ z(t,x) \geq 0, & t \in (0,t_0), \quad x \in \mathbb{R}^N \setminus \partial B_{\widehat{\delta}+1},\\ z(0,x) = f_n(x), & \qquad\qquad\quad x \in \mathbb{R}^N \setminus B_{\widehat{\delta}+1},\end{cases}$$

for any $n \geq n_1$. The maximum principle in Lemma 4.1.7 implies that $z \geq 0$ in $(0, t_0) \times (\mathbb{R}^N \setminus B_{\widehat{\delta}+1})$ and (4.1.22) follows. □

4.1.2 The non-conservative case

We now consider the non-conservative case. In particular, here we do not assume that the coefficient c identically vanishes in $\mathbb{R}^N$.

As we are going to show, $T(t)$ is compact and maps $C_0(\mathbb{R}^N)$ into itself, for some $t > 0$, if and only if it maps the function $\mathbb{1}$ into a function vanishing at infinity. To prove such a result, we need the following lemma.

Lemma 4.1.9 *Suppose that $\mathscr{F}$ is a bounded subset of $C_0(\mathbb{R}^N)$ with the following property: for any $\varepsilon > 0$ there exists $\delta > 0$ such that $\sup_{f\in\mathscr{F}}\sup_{\mathbb{R}^N\setminus B_\delta}|f| < \varepsilon$. Further, suppose that, for any compact set $H \subset \mathbb{R}^N$ and $\varepsilon > 0$, there exists $\sigma > 0$ such that $\sup\{|f(x) - f(y)| : x, y \in H,\ |x - y| \le \sigma,\ f \in \mathscr{F}\} \le \varepsilon$. Then, $\mathscr{F}$ is totally bounded in $C_0(\mathbb{R}^N)$.*

Proof Fix $\varepsilon > 0$ and let $\delta > 0$ be such that $|f(x)| \le \varepsilon$ for any $|x| \ge \delta$ and $f \in \mathscr{F}$. According to our assumptions and the Arzelà-Ascoli theorem, the set $\mathscr{F}_{|B_{2\delta}}$, of all the restrictions to $B_{2\delta}$ of the functions in $\mathscr{F}$, is totally bounded. Hence, we can find $k \in \mathbb{N}$ and $g_1, \dots, g_k \in C(\overline{B}_{2\delta})$ such that $\mathscr{F}_{|\overline{B}_{2\delta}} \subset \bigcup_{j=1}^k \overline{B}_\varepsilon(g_j)$, where $\overline{B}_\varepsilon(g_j)$ $(j = 1, \dots, k)$ denotes the closed ball in $C(\overline{B}_{2\delta})$ with centre at g_j and radius ε. Now, let φ be any smooth function such that $\chi_{B_\delta} \le \varphi \le \chi_{B_{2\delta}}$. Then,

$$\mathscr{F} \subset \bigcup_{j=1}^{k} \overline{B}_{3\varepsilon}(\varphi g_j), \tag{4.1.27}$$

where the closed balls $\overline{B}_{3\varepsilon}(\varphi g_j)$ are now meant in $C_b(\mathbb{R}^N)$. To check (4.1.27), fix $f \in \mathscr{F}$ and let $j \in \{1, \dots, k\}$ be such that $\|f - g_j\|_{C(\overline{B}_{2\delta})} \le \varepsilon$. Then, in particular, $\|f - \varphi g_j\|_{C(\overline{B}_\delta)} \le \varepsilon$. Moreover,

$$\begin{aligned}\|f - \varphi g_j\|_{C_b(\mathbb{R}^N\setminus B_\delta)} \le& \|f\|_{C_b(\mathbb{R}^N\setminus B_\delta)} + \|g_j\|_{C(\overline{B}_{2\delta}\setminus B_\delta)}\\ \le& 2\|f\|_{C_b(\mathbb{R}^N\setminus B_\delta)} + \|f - g_j\|_{C(\overline{B}_{2\delta})} \le 3\varepsilon.\end{aligned}$$

Summing up, $\|f - \varphi g_j\|_{C_b(\mathbb{R}^N)} \le 3\varepsilon$ and (4.1.27) follows. □

Now, we can prove the following result.

Theorem 4.1.10 *Fix $t > 0$. Then $T(t)\mathbb{1}$ belongs to $C_0(\mathbb{R}^N)$ if and only if $T(t)$ is compact and $C_0(\mathbb{R}^N)$ is invariant for $T(t)$.*

Proof Suppose that $T(t)\mathbb{1} \in C_0(\mathbb{R}^N)$. Then, using (1.2.7), we get

$$|(T(t)f)(x)| \le \|f\|_\infty \int_{\mathbb{R}^N} p(t, x; dy) = \|f\|_\infty (T(t)\mathbb{1})(x) \tag{4.1.28}$$

for any $f \in C_b(\mathbb{R}^N)$ and $x \in \mathbb{R}^N$. Therefore, $T(t)$ maps $C_b(\mathbb{R}^N)$ into $C_0(\mathbb{R}^N)$. In particular, $C_0(\mathbb{R}^N)$ is invariant.

Let us now prove that $T(t)$ is a compact operator. For this purpose, let $B_\infty(1) = \{f \in C_b(\mathbb{R}^N) : \|f\|_\infty \le 1\}$ be the unit ball in $C_b(\mathbb{R}^N)$. According to (4.1.28), $T(t)(B_\infty(1))$ is a set of functions vanishing uniformly at infinity. This implies that, for any $\varepsilon > 0$, there exists $\delta > 0$ such that

$$|(T(t)f)(x)| \le \varepsilon, \qquad |x| \ge \delta, \tag{4.1.29}$$

for any $f \in B_\infty(1)$. Moreover, from the interior Schauder estimates in Theorem D.1.5, it follows that, for any $\rho > 0$, there exists a positive constant $C = C(\rho, t)$ such that

$$\|T(t)f\|_{C^{2+\alpha}(\overline{B}_\rho)} \le C\|f\|_{C(\overline{B}_{2\rho})} \le C \tag{4.1.30}$$

for any $f \in B_\infty(1)$, where α is as in Hypothesis 1.0.1(ii). Therefore, the set of all the restrictions to $\overline{B}_\rho$ of the functions in $T(t)(B_\infty(1))$ is a bounded set of equicontinuous functions. From (4.1.29), (4.1.30) and Lemma 4.1.9 it follows that $T(t)(B_\infty(1))$ is compact in $C_b(\mathbb{R}^N)$.

Conversely, fix $t > 0$ and suppose that $T(t)$ is compact and $C_0(\mathbb{R}^N)$ is invariant for $T(t)$. Let $\{f_n\}$ be a sequence of continuous functions such that $\chi_{B_{n-1}} \le f_n \le \chi_{B_n}$ for any $n \in \mathbb{N}$. By Proposition 1.2.10, $(T(t)f_n)(x)$ converges to $(T(t)\mathbb{1})(x)$ for any $x \in \mathbb{R}^N$, as n tends to $+\infty$. Since $T(t)$ is compact, $T(t)f_n$ tends to $T(t)\mathbb{1}$ uniformly in $\mathbb{R}^N$. Moreover, the invariance of $C_0(\mathbb{R}^N)$ under the action of $T(t)$ shows that $T(t)f_n \in C_0(\mathbb{R}^N)$ for any $n \in \mathbb{N}$ and, therefore, $T(t)\mathbb{1} \in C_0(\mathbb{R}^N)$ as well. □

In Section 4.2 we will determine some sufficient conditions implying that $T(t)\mathbb{1} \in C_0(\mathbb{R}^N)$ for any $t > 0$. By virtue of Theorem 4.1.10, such conditions turn out to be sufficient conditions for the compactness of $T(t)$ for any $t > 0$.

We conclude this section with the following proposition.

Proposition 4.1.11 *Suppose that, for any $t > 0$, $T(t)$ is a compact operator and $C_0(\mathbb{R}^N)$ is invariant under $T(t)$. Then, the semigroup is norm-continuous in $(0, +\infty)$ and the resolvent $R(\lambda, \widehat{A})$ is compact for any $\lambda > c_0$.*

Proof It suffices to show that the map $t \mapsto T(t)f$ is continuous from the right in $(0, +\infty)$ for any $f \in B_\infty(1) = \{f \in C_b(\mathbb{R}^N) : \|f\|_\infty \le 1\}$. Indeed, once this property is proved, repeating the proof of Proposition 4.1.2, we can easily show that the map $t \mapsto T(t)$ is norm-continuous in $(0, +\infty)$ and $R(\lambda, \widehat{A})$ is compact for any $\lambda > c_0$. Observe that, by Theorem 4.1.10, $T(t)\mathbb{1} \in C_0(\mathbb{R}^N)$ for any $t > 0$. Hence, applying (4.1.28), written for $f \in C_b(\mathbb{R}^N)$, we conclude that $T(t)$ maps $C_b(\mathbb{R}^N)$ into $C_0(\mathbb{R}^N)$ for any $t > 0$. By Proposition 1.2.8, $T(t)g$ converges to g uniformly in $\mathbb{R}^N$, as t tends to 0^+, for any $g \in C_0(\mathbb{R}^N)$. Hence, $\|(T(h) - I)T(t)f\|_\infty$ converges to 0 as h tends to 0^+ for any $t > 0$ and $f \in C_b(\mathbb{R}^N)$, and this implies that the function $t \mapsto T(t)f$ is right-continuous in $(0, +\infty)$. □

4.2 On the inclusion $T(t)(C_b(\mathbb{R}^N)) \subset C_0(\mathbb{R}^N)$

In this section we determine sufficient conditions implying that $T(t)$ maps $C_b(\mathbb{R}^N)$ into $C_0(\mathbb{R}^N)$. If not explicitly specified, we do not assume that $c \equiv 0$. Without loss of generality, we can limit ourselves to determining sufficient conditions which ensure that $T(t)\mathbb{1}$ belongs to $C_0(\mathbb{R}^N)$. Indeed, as has already been pointed out in the proof of Proposition 4.1.11, from (4.1.28) it follows that

$$|(T(t)f)(x)| \le \|f\|_\infty (T(t)\mathbb{1})(x), \qquad t > 0, \ x \in \mathbb{R}^N, \ f \in C_b(\mathbb{R}^N). \tag{4.2.1}$$

Hence, if $T(t)\mathbb{1} \in C_0(\mathbb{R}^N)$, then $T(t)f \in C_0(\mathbb{R}^N)$ as well, for any $f \in C_b(\mathbb{R}^N)$. Using (4.2.1), we can show that

$$|(R(\lambda, \widehat{A})f)(x)| \le \|f\|_\infty |(R(\lambda, \widehat{A})\mathbb{1})(x)|, \qquad x \in \mathbb{R}^N, \ \lambda > c_0, \tag{4.2.2}$$

which implies that $R(\lambda, \widehat{A})$ maps $C_b(\mathbb{R}^N)$ into $C_0(\mathbb{R}^N)$ if and only if $R(\lambda, \widehat{A})\mathbb{1} \in C_0(\mathbb{R}^N)$. We recall that, for any $f \in C_b(\mathbb{R}^N)$,

$$(R(\lambda, \widehat{A})f)(x) = \int_0^{+\infty} e^{-\lambda t}(T(t)f)(x)dt, \qquad x \in \mathbb{R}^N, \tag{4.2.3}$$

where $\widehat{A}$ is the weak generator of the semigroup. See Sections 1.1 and 1.3.

To prove that $T(t)$ maps $C_b(\mathbb{R}^N)$ into $C_0(\mathbb{R}^N)$ if and only if $R(\lambda, \widehat{A})$ does for some (and hence all) $\lambda > c_0$, we need the following lemma.

Lemma 4.2.1 *The function $t \mapsto (S(t)\mathbb{1})(x) := e^{-c_0 t}(T(t)\mathbb{1})(x)$ is not increasing in $[0, +\infty)$ for any $x \in \mathbb{R}^N$. Here, as usual, $c_0 = \sup_{x\in\mathbb{R}^N} c(x)$.*

Proof It can be obtained applying the proof of Lemma 3.1.8. See Remark 3.1.9. □

We can now prove the following proposition.

Proposition 4.2.2 *The following properties are equivalent:*

(*i*) *$T(t)\mathbb{1} \in C_0(\mathbb{R}^N)$ for any $t > 0$;*

(*ii*) *$R(\lambda, \widehat{A})\mathbb{1} \in C_0(\mathbb{R}^N)$ for any $\lambda > c_0$;*

(*iii*) *$R(\lambda, \widehat{A})\mathbb{1} \in C_0(\mathbb{R}^N)$ for some $\lambda > c_0$;*

(*iv*) *$D(\widehat{A}) = D_{\max}(\mathcal{A}) \cap C_0(\mathbb{R}^N)$, where $D(\widehat{A})$ is defined in Section 1.3.*

Proof It suffices to prove that "(i) ⇒ (ii)" and "(iii) ⇒ (i)", since the implication "(ii) ⇒ (iii)" is trivial and "(ii) ⇔ (iv)" follows immediately from Proposition 1.3.5 observing that (ii) and (4.2.2) imply that $R(\lambda, \widehat{A})f \in C_0(\mathbb{R}^N)$ for any $f \in C_b(\mathbb{R}^N)$.

"(i) ⇒ (ii)". Fix $\lambda > c_0$. Taking (4.2.3) and Lemma 4.2.1 into account, we deduce that

$$\begin{aligned}(R(\lambda, \widehat{A})\mathbb{1})(x) =& \int_0^{+\infty} e^{-\lambda t}(T(t)\mathbb{1})(x)dt = \int_0^{+\infty} e^{-(\lambda-c_0)t}(S(t)\mathbb{1})(x)dt \\ =& \int_0^{\varepsilon} e^{-(\lambda-c_0)t}(S(t)\mathbb{1})(x)dt + \int_\varepsilon^{+\infty} e^{-(\lambda-c_0)t}(S(t)\mathbb{1})(x)dt \\ \le& \varepsilon + (S(\varepsilon)\mathbb{1})(x)\int_0^{+\infty} e^{-(\lambda-c_0)t}dt\end{aligned}$$

for any $\varepsilon > 0$ and $x \in \mathbb{R}^N$, where we used the fact that $\{S(t)\}$ is a semigroup of contractions. Since $T(\varepsilon)\mathbb{1} \in C_0(\mathbb{R}^N)$, $S(\varepsilon)\mathbb{1}$ belongs to $C_0(\mathbb{R}^N)$ as well. This implies that $\limsup_{|x|\to+\infty}(R(\lambda, \widehat{A})\mathbb{1})(x) \le \varepsilon$. From the arbitrariness of $\varepsilon > 0$ and the nonnegativity of the function $R(\lambda, \widehat{A})\mathbb{1}$, we deduce that $R(\lambda, \widehat{A})\mathbb{1} \in C_0(\mathbb{R}^N)$.

"(iii) ⇒ (i)". Fix $t > 0$. Again by the formula (4.2.3) and Lemma 4.2.1 we get

$$\begin{aligned}(R(\lambda, \widehat{A})\mathbb{1})(x) =& \int_0^{+\infty} e^{-(\lambda-c_0)s}(S(s)\mathbb{1})(x)ds \\ \ge& \int_0^{t} e^{-(\lambda-c_0)s}(S(s)\mathbb{1})(x)ds \ge (S(t)\mathbb{1})(x)\int_0^{t} e^{-(\lambda-c_0)s}ds\end{aligned}$$

for any $x \in \mathbb{R}^N$, i.e.,

$$0 \le S(t)\mathbb{1} \le \left(\frac{\lambda - c_0}{1 - e^{-(\lambda-c_0)t}}\right) R(\lambda, \widehat{A})\mathbb{1}.$$

Thus, we conclude that $S(t)\mathbb{1}$ (and hence $T(t)\mathbb{1}$) belongs to $C_0(\mathbb{R}^N)$. □

An interesting consequence of the fact that the semigroup maps $C_b(\mathbb{R}^N)$ into $C_0(\mathbb{R}^N)$ is contained in the next proposition.

Proposition 4.2.3 *Suppose that $c \equiv 0$ and $T(t_0)$ maps $C_b(\mathbb{R}^N)$ into $C_0(\mathbb{R}^N)$ for some $t_0 > 0$. Then, $\|T(t)\|_{L(C_b(\mathbb{R}^N))}$ decreases exponentially to 0 as t tends to $+\infty$.*

Proof We begin the proof observing that, since $T(t)\mathbb{1} = T(t_0)T(t-t_0)\mathbb{1}$ for any $t \ge t_0$, $T(t)$ maps $C_b(\mathbb{R}^N)$ into $C_0(\mathbb{R}^N)$ for any $t \ge t_0$. Next we prove that there exists a positive constant $k_0 \in (0,1)$ such that $(T(t_0)\mathbb{1})(x) \le k_0$ for any $x \in \mathbb{R}^N$. For this purpose, we notice that, since $T(t_0)\mathbb{1} \in C_0(\mathbb{R}^N)$, there exists $\delta > 0$ such that $(T(t_0)\mathbb{1})(x) \le 1/2$ for any $|x| \ge \delta$. Moreover, since $T(t)\mathbb{1} \le \mathbb{1}$ for any $t > 0$, $(T(t_0)\mathbb{1})(x) < 1$ for any $x \in B_\delta$ otherwise by the classical maximum principle (see Proposition D.2.2(ii)), $T(t_0)\mathbb{1}$ should be constant in B_δ and equal to $\mathbb{1}$, which, of course, can not be the case.

Now, the semigroup property allows us to show that $\|T(t)\|_{L(C_b(\mathbb{R}^N))}$ decreases to 0 exponentially as t tends to $+\infty$. For this purpose, we split any $t > t_0$ as $t = nt_0 + r$ with $n \in \mathbb{N}$, $r \in [0, t_0)$. Recalling that $\|T(r)\|_{L(C_b(\mathbb{R}^N))} = \|T(r)\mathbb{1}\|_\infty \le 1$ for any $r > 0$, we get

$$\begin{aligned}\|T(t)\|_{L(C_b(\mathbb{R}^N))} =& \|(T(t_0))^n T(r)\mathbb{1}\|_{L(C_b(\mathbb{R}^N))}\\ \le& \|T(t_0)\|^n_{L(C_b(\mathbb{R}^N))}\|T(r)\mathbb{1}\|_\infty \le \exp(n\log(k_0))\end{aligned}$$

for any $t \ge t_0$. Observing that $n \ge t/t_0 - 1$, we deduce that $\exp(n\log(k_0)) \le k_0^{-1}e^{\omega t}$, where $\omega = t_0^{-1}\log(k_0) < 0$. □

The following result gives a sufficient condition which guarantees that $T(t)\mathbb{1}$ belongs to $C_0(\mathbb{R}^N)$, in terms of a suitable Lyapunov function φ.

Proposition 4.2.4 *Suppose that there exist $\lambda_0 > c_0$, a compact set K and a strictly positive function $\varphi \in C^2(\mathbb{R}^N \setminus K) \cap C_b(\mathbb{R}^N \setminus K)$, which vanishes at infinity, such that*

$$\inf_{x \in \mathbb{R}^N \setminus K} \big(\lambda_0\varphi(x) - \mathcal{A}\varphi(x)\big) =: a > 0. \tag{4.2.4}$$

Then, $T(t)\mathbb{1}$ belongs to $C_0(\mathbb{R}^N)$ for any $t > 0$.

Proof By virtue of Proposition 4.2.2, it suffices to prove that $R(\lambda_0, \widehat{A})\mathbb{1}$ belongs to $C_0(\mathbb{R}^N)$. Let n_0 be the smallest integer such that $K \subset B_{n_0}$. Moreover, let $\gamma = \inf_{\partial B_{n_0}} \varphi > 0$ and set $\delta = \min\{a, (\lambda_0 - c_0)\gamma\}$. We will prove that $R(\lambda_0, \widehat{A})\delta \in C_0(\mathbb{R}^N)$. Of course, this will imply that $R(\lambda_0, \widehat{A})\mathbb{1} = \delta^{-1}R(\lambda_0, \widehat{A})\delta$ belongs to $C_0(\mathbb{R}^N)$. For this purpose, we recall that $R(\lambda_0, \widehat{A})\delta$ is the pointwise limit of the sequence $\{u_n\}$, where, for any $n \in \mathbb{N}$, the function u_n is the unique solution in $\bigcap_{1 \le p < +\infty} W^{2,p}(B_n)$ to the problem

$$\begin{cases} \lambda_0 u_n(x) - \mathcal{A}u_n(x) = \delta, & x \in B_n,\\ u_n(x) = 0, & x \in \partial B_n.\end{cases}$$

Moreover, $0 \le u_n \le (\lambda_0 - c_0)^{-1}\delta$, see the proof of Theorem 1.1.1. Next we observe that, for any $n > n_0$, the function φ satisfies

$$\begin{cases} \lambda_0\varphi(x) - \mathcal{A}\varphi(x) \ge a, & x \in B_n \setminus \overline{B}_{n_0},\\ \varphi(x) \ge \gamma, & x \in \partial B_{n_0},\\ \varphi(x) > 0, & x \in \partial B_n.\end{cases}$$

Therefore, the classical maximum principle (see Theorem D.2.1) implies that $\varphi \ge u_n$ in $B_n \setminus \overline{B}_{n_0}$. Letting n tend to $+\infty$, we get $0 \le R(\lambda_0, \widehat{A})\delta \le \varphi$ in $\mathbb{R}^N \setminus \overline{B}_{n_0}$. Hence, $R(\lambda_0, \widehat{A})\delta$ belongs to $C_0(\mathbb{R}^N)$. □

Example 4.2.5 Consider the operator $\mathcal{A}$ defined on smooth functions $\zeta : \mathbb{R}^N \to \mathbb{R}$ by

$$\mathcal{A}\zeta(x) = \Delta\zeta(x) + \langle b(x), \nabla\zeta(x)\rangle, \qquad x \in \mathbb{R}^N,$$

where $b \in C^\alpha_{\rm loc}(\mathbb{R}^N)$ ($\alpha \in (0,1)$). Moreover, let φ be the function defined by $\varphi(x) = (1+|x|^2)^{-1}$ for any $x \in \mathbb{R}^N$. As is easily seen,

$$\mathcal{A}\varphi(x) = 8|x|^2(\varphi(x))^3 - 2N(\varphi(x))^2 - 2(\varphi(x))^2\langle b(x), x\rangle, \qquad x \in \mathbb{R}^N. \tag{4.2.5}$$

Condition (4.2.4) reads as

$$\langle b(x), x\rangle \geq \frac{a}{2}(1+|x|^2)^2 - \frac{\lambda_0}{2}(1+|x|^2) - N + 4\frac{|x|^2}{1+|x|^2}, \qquad x \in \mathbb{R}^N \setminus K,$$

for some compact set $K \subset \mathbb{R}^N$ and some positive constants λ_0 and a. Such a condition is satisfied, for instance, if

$$\langle b(x), x\rangle \geq C + M|x|^4 \tag{4.2.6}$$

for some constants $C \in \mathbb{R}$ and $M > a/2$. In such a case we can take as λ_0 any positive constant. According to Proposition 4.2.4, the semigroup $\{T(t)\}$, associated with the operator $\mathcal{A}$, is such that $T(t)\mathbb{1} \in C_0(\mathbb{R}^N)$ for any $t > 0$.

Condition (4.2.6) is satisfied, for instance, when the operator $\mathcal{A}$ is defined on smooth functions $\zeta : \mathbb{R}^N \to \mathbb{R}$ by $\mathcal{A}_+\zeta(x) = \Delta\zeta(x) + |x|^2\langle x, \nabla\zeta(x)\rangle$ for any $x \in \mathbb{R}^N$.

This example should be compared with Example 3.1.13, where we considered the operator $\mathcal{A}_-$, defined on smooth functions ζ by $\mathcal{A}_-\zeta(x) = \Delta\zeta(x) - |x|^2\langle x, \nabla\zeta(x)\rangle$ for any $x \in \mathbb{R}^N$, and showed that the associated semigroup is conservative. The operators $\mathcal{A}_+$ and $\mathcal{A}_-$ differ only in the sign of the drift term. The reason for the difference in the behaviour of the two semigroups can be understood looking at the stochastic equations associated with these two operators. The operator $\mathcal{A}_-$ satisfies Hypotheses 1.6.3 and 1.6.4. Therefore, all the results in Section 1.6 hold. The stochastic equation associated with it is $d\xi^x_t = -|\xi^x_t|^2\xi^x_t dt + dW_t$ for any $t > 0$ with the initial condition $\xi^x_0 = x \in \mathbb{R}^N$. This equation has a solution defined for any $t > 0$. Heuristically, one can think to this equation as a perturbation of the ordinary differential equation

$$\frac{d}{dt}\xi^x_t = -|\xi^x_t|^2\xi^x_t, \qquad t > 0,$$

whose solutions exist for any $t > 0$.

Repeating the same argument in the case of the operator $\mathcal{A}_+$ leads to the differential equation

$$\frac{d}{dt}\xi^x_t = |\xi^x_t|^2\xi^x_t, \qquad t > 0,$$

whose not vanishing solutions explode in a finite time. And also the solutions of the stochastic equation $d\xi^x_t = |\xi^x_t|^2\xi^x_t dt + dW_t$ for any $t > 0$, with the initial condition $\xi^x_0 = x \in \mathbb{R}^N$, do not exist for all $t > 0$. Indeed, the life time of the Markov process associated with $\{T(t)\}$ is finite since $\mathbb{P}_x\{t < \tau\} = (T(t)\mathbb{1})(x) < 1$.

Example 4.2.6 Let $\mathcal{A}$ be the one-dimensional operator defined by

$$\mathcal{A}\zeta(x) = \zeta''(x) + |x|^3\zeta'(x), \qquad x \in \mathbb{R},$$

on smooth functions $\zeta : \mathbb{R} \to \mathbb{R}$. Then, the associated semigroup $\{T(t)\}$ is not conservative. Indeed, using the notation of Chapter 2 (see the formulae (2.1.3), (2.1.4) and (2.1.5)), it is easy to check that $\mathscr{P} \in L^1((-\infty,0))$, $\mathscr{P} \notin L^1((0,+\infty))$, $\mathscr{R} \notin L^1((-\infty,0))$ and $\mathscr{R} \in L^1((0,+\infty))$. Therefore, the point $-\infty$ is an entrance and, hence, unaccessible while the point $+\infty$ is an exit and, hence, accessible. According to Theorem 2.2.2, we conclude that the semigroup $\{T(t)\}$ is not conservative.

We now claim that $R(\lambda,\widehat{A})\mathbb{1}$ and $T(t)\mathbb{1}$ do not belong to $C_0(\mathbb{R}^N)$, for any $\lambda > 0$ and $t > 0$, respectively. To show that $u = R(\lambda,\widehat{A})\mathbb{1}$ is not in $C_0(\mathbb{R})$, we observe that $\lambda u(x) - u''(x) - |x|^3 u'(x) = 1$ for any $x \in \mathbb{R}$ and $0 \le u \le 1/\lambda$ in $\mathbb{R}^N$. Moreover, since the coefficients of the operator $\mathcal{A}$ belong to $C^{2+\theta}_{\rm loc}(\mathbb{R})$ for any $\theta \in (0,1)$, it is readily seen that $u \in C^{4+\theta}_{\rm loc}(\mathbb{R})$. Let us prove that $u'(x) \le 0$ for any $x \in \mathbb{R}$. This, of course, will imply that $u \notin C_0(\mathbb{R})$. By contradiction, suppose that $u'(x) > 0$ for any $x \in (a,b)$ and some $a,b \in \mathbb{R}$. Let $a_0 = \inf\{\alpha : u' > 0 \text{ in } (\alpha,b)\}$. If $a_0 > -\infty$ then it holds that $u'(a_0) = 0$ and $u'(x) > 0$ for any $x \in (a_0,b)$. Therefore, for such values of x, $u''(x) = \lambda u(x) - 1 - |x|^3 u'(x)$ is nonpositive. But this is a contradiction. Similarly, if $a_0 = -\infty$ then it holds that $u'(x) > 0$ and $u''(x) \le 0$ for any $x < b$, so that $u'(x)$ admits a positive limit as x tends to $-\infty$, which is a contradiction since u is bounded. We conclude that $u' \le 0$. Since u does not identically vanish in $\mathbb{R}$, u does not belong to $C_0(\mathbb{R})$.

To prove that for any $t > 0$ $T(t)\mathbb{1}$ does not belong to $C_0(\mathbb{R}^N)$, we show that $D_x T(t)\mathbb{1} \le 0$. For this purpose, set $u = T(\cdot)\mathbb{1}$. According to Lemma 3.1.8, we know that $D_t u(t,x) \le 0$ for any $(t,x) \in \mathbb{R}_+ \times \mathbb{R}$. Suppose that $D_x u(t,x) > 0$ at some point $(t,x) \in \mathbb{R}_+ \times \mathbb{R}$. Then, it holds that $D_{xx} u(t,x) = D_t u(t,x) - |x|^3 D_x u(t,x) \le 0$. Repeating the same arguments as above, we see that the assumption on the positiveness of $D_x u(t,x)$ leads us to a contradiction. Therefore, $D_x u(t,x) \le 0$ for any $(t,x) \in \mathbb{R}_+ \times \mathbb{R}$. Since $T(t)\mathbb{1}$ does not identically vanish in $\mathbb{R}^N$, we conclude that $T(t)\mathbb{1} \notin C_0(\mathbb{R}^N)$.

4.3 Invariance of $C_0(\mathbb{R}^N)$

In this section we determine sufficient conditions implying that $C_0(\mathbb{R}^N)$ is invariant under $T(t)$ for $t > 0$. By the results in Section 4.2 we know that this is the case if $T(t)\mathbb{1}$ vanishes at infinity. But this is the case also under weaker assumptions as the next propositions show.

In Proposition 4.2.2 we have seen that $T(t)$ maps $C_b(\mathbb{R}^N)$ into $C_0(\mathbb{R}^N)$ for any $t > 0$ if and only if $R(\lambda,\widehat{A})$ does for some (and hence all) $\lambda > c_0 := \sup_{x\in\mathbb{R}^N} c(x)$. A similar property holds true as far as the invariance of $C_0(\mathbb{R}^N)$ is concerned. Moreover, if $C_0(\mathbb{R}^N)$ is invariant under the semigroup, then the restriction of the semigroup to $C_0(\mathbb{R}^N)$ defines a strongly continuous semigroup.

Proposition 4.3.1 *$C_0(\mathbb{R}^N)$ is invariant under the semigroup $\{T(t)\}$ if and only if it is invariant under $R(\lambda,\widehat{A})$ for any (equivalently, some) $\lambda > c_0$. In such a case, the restriction of $\{T(t)\}$ to $C_0(\mathbb{R}^N)$ defines a strongly continuous semigroup. Its infinitesimal generator A_0 is the part of $\widehat{A}$ in $C_0(\mathbb{R}^N)$, i.e.,*

$$\begin{cases} D(A_0) = \{u \in D_{\max}(\mathcal{A}) \cap C_0(\mathbb{R}^N) : \mathcal{A}u \in C_0(\mathbb{R}^N)\}, \\ A_0 u = \mathcal{A}u \qquad \text{for any } u \in D(A_0). \end{cases} \tag{4.3.1}$$

Proof Suppose that $T(t)$ maps $C_0(\mathbb{R}^N)$ into itself for any $t > 0$. Then, by Proposition 1.2.8 we deduce that the restriction of $\{T(t)\}$ to $C_0(\mathbb{R}^N)$ is a strongly continuous semigroup in $C_0(\mathbb{R}^N)$. Since the resolvent operator associated with the restriction to $C_0(\mathbb{R}^N)$ of $\{T(t)\}$ is the restriction of $R(\lambda,\widehat{A})$ to $C_0(\mathbb{R}^N)$, $R(\lambda,\widehat{A})$ maps $C_0(\mathbb{R}^N)$ into itself for any $\lambda > c_0$.

Conversely, let us assume that $R(\lambda, \widehat{A})$ maps $C_0(\mathbb{R}^N)$ into itself for any $\lambda > c_0$. It is easy to show that, for any $\lambda > c_0$, the operator $\lambda I - A_0$, with domain given by (4.3.1), is bijective from $D(A_0)$ into $C_0(\mathbb{R}^N)$. Moreover,

$$R(\lambda, A_0) = R(\lambda, \widehat{A})_{|C_0(\mathbb{R}^N)}, \qquad \lambda > c_0, \tag{4.3.2}$$

and, consequently,

$$\|R(\lambda, A_0)\|_{L(C_0(\mathbb{R}^N))} \leq \frac{1}{\lambda - c_0}, \qquad \lambda > c_0. \tag{4.3.3}$$

Indeed, if $u \in D(A_0)$ is such that $\lambda u - A_0 u = 0$, then u belongs to $D(\widehat{A})$ (see (1.3.13)) and consequently $\lambda u - \widehat{A}u = 0$, implying that $u = 0$. Now take $f \in C_0(\mathbb{R}^N)$. The function $u = R(\lambda, \widehat{A})f$ is the unique solution in $D(\widehat{A})$ of the equation $\lambda u - \mathcal{A}u = f$. Since $R(\lambda, \widehat{A})$ maps $C_0(\mathbb{R}^N)$ into itself, u belongs to $D(\widehat{A}) \cap C_0(\mathbb{R}^N)$ and $\mathcal{A}u \in C_0(\mathbb{R}^N)$. Still from (1.3.13) we deduce that $u \in D(A_0)$ and it solves the equation $\lambda u - A_0 u = f$. Formula (4.3.2) easily follows. Similarly, inequality (4.3.3) follows recalling that $\|R(\lambda, \widehat{A})\|_{L(C_b(\mathbb{R}^N))} \leq (\lambda - c_0)^{-1}$ for any $\lambda > c_0$ (see (1.1.3) and Section 1.3).

Now, the Hille-Yosida theorem (see Theorem C.1.6) implies that the operator A_0 with domain (4.3.1) is the generator of a strongly continuous semigroup $\{T_0(t)\}$ in $C_0(\mathbb{R}^N)$. By general results for strongly continuous semigroups we know that

$$R(\lambda, A_0)f = \int_0^{+\infty} e^{-\lambda t} T_0(t) f dt$$

for any $\lambda > c_0$, where the integral converges in $C_0(\mathbb{R}^N)$. Since $R(\lambda, A_0)$ is the restriction of $R(\lambda, \widehat{A})$ to $C_0(\mathbb{R}^N)$ and

$$(R(\lambda, \widehat{A})g)(x) = \int_0^{+\infty} e^{-\lambda t} (T(t)g)(x) dt, \qquad x \in \mathbb{R}^N,$$

for any $g \in C_b(\mathbb{R}^N)$ (see (1.2.12), (1.3.1)), we deduce that

$$\int_0^{+\infty} e^{-\lambda t}((T(t) - T_0(t))f)(x) dt = 0$$

for any $f \in C_0(\mathbb{R}^N)$, $x \in \mathbb{R}^N$ and $\lambda > c_0$. The uniqueness of the Laplace transform, implies that $T(t)f = T_0(t)f$ for any $t > 0$. Hence, $T(t)$ maps $C_0(\mathbb{R}^N)$ into itself for any $t > 0$.

Finally, we show that, if $C_0(\mathbb{R}^N)$ is invariant under $R(\lambda_0, \widehat{A})$ for some $\lambda_0 > c_0$, then it is invariant under $R(\lambda, \widehat{A})$ for any $\lambda > c_0$. For this purpose, we set $\mathcal{F} = \{\lambda \in (c_0, +\infty) : C_0(\mathbb{R}^N)$ is invariant under $R(\lambda, \widehat{A})\}$ and prove that $\mathcal{F}$ is open and closed in $(c_0, +\infty)$. This will easily imply that $\mathcal{F} = (c_0, +\infty)$, since, by assumptions $\mathcal{F}$ is not empty. We begin by proving that $\mathcal{F}$ is an open set. We fix $\lambda_1 \in \mathcal{F}$, $f \in C_0(\mathbb{R}^N)$ and prove that, if λ is sufficiently close to λ_1, then the equation $\lambda u - \mathcal{A}u = f$ admits a solution $u \in D(A_0)$. As is immediately seen, $u \in D(A_0)$ is a solution to the previous equation if and only if

$$(\lambda - \lambda_1) R(\lambda_1, \widehat{A}) u + u = R(\lambda_1, \widehat{A}) f. \tag{4.3.4}$$

Since $C_0(\mathbb{R}^N)$ is invariant under $R(\lambda_1, \widehat{A})$, the operator $G = (\lambda - \lambda_1) R(\lambda_1, \widehat{A}) + I$ is bounded from $C_0(\mathbb{R}^N)$ into itself. Moreover, if $|\lambda - \lambda_1| < |\lambda_1 - c_0|$, then $\|(\lambda - \lambda_1) R(\lambda_1, \widehat{A})\|_{C_0(\mathbb{R}^N)} < 1$ (see again Sections 1.1 and 1.3). Therefore, G is bijective in $C_0(\mathbb{R}^N)$ and equation (4.3.4) is uniquely solvable in $C_0(\mathbb{R}^N)$. This implies that $R(\lambda, \widehat{A})f$ belongs to $C_0(\mathbb{R}^N)$.

Let us now suppose that $\lambda \in (c_0, +\infty)$ is the limit of a sequence $\{\lambda_n\} \subset \mathcal{F}$. Then, for any $n \in \mathbb{N}$, $R(\lambda_n, \widehat{A})$ is a bounded operator in $C_0(\mathbb{R}^N)$. Since $R(\lambda_n, \widehat{A})$ converges to $R(\lambda, \widehat{A})$ in $L(C_b(\mathbb{R}^N))$, $R(\lambda, \widehat{A})$ belongs to $L(C_0(\mathbb{R}^N))$ as well and $\mathcal{F}$ is closed. □

The next theorem gives a sufficient condition for the invariance of $C_0(\mathbb{R}^N)$. It should be compared with Proposition 4.2.4.

Theorem 4.3.2 *Suppose that there exist $\lambda_0 > 0$, a compact set F and a strictly positive function $\varphi \in C^2(\mathbb{R}^N \setminus F) \cap C_b(\mathbb{R}^N \setminus F)$, vanishing at infinity, such that*

$$\lambda_0\varphi(x) - \mathcal{A}\varphi(x) \geq 0, \qquad x \in \mathbb{R}^N \setminus F. \tag{4.3.5}$$

Then, $C_0(\mathbb{R}^N)$ is invariant under $T(t)$ for any $t > 0$.

Proof We prove that $C_0(\mathbb{R}^N)$ is invariant under $R(\lambda_0 + 1, \widehat{A})$. Then, the conclusion will follow from Proposition 4.3.1.

We take a compactly supported function $f \in C_0(\mathbb{R}^N)$ and prove that $u = R(\lambda_0, \widehat{A})f$ belongs to $C_0(\mathbb{R}^N)$. The general case will follow by density since any function $f \in C_0(\mathbb{R}^N)$ can be approximated uniformly in $\mathbb{R}^N$ by a sequence of compactly supported functions.

We fix $n_0 \in \mathbb{N}$ such that B_{n_0} contains both F and the support of f. Replacing φ with $\delta\varphi$ for a sufficiently large positive constant δ, we can assume that $\varphi \geq (\lambda_0 - c_0)^{-1}\|f\|_\infty$ on ∂B_{n_0}. For any $n \in \mathbb{N}$, we denote by u_n the (unique) solution in $\bigcap_{1 \leq p < +\infty} W^{2,p}(B_n)$ of the equation

$$\begin{cases} \lambda_0 u_n(x) - \mathcal{A}u_n(x) = f(x), & x \in B_n, \\ u_n(x) = 0, & x \in \partial B_n. \end{cases}$$

According to the proof of Theorem 1.1.1, u is the pointwise limit of the sequence $\{u_n\}$ and $\|u_n\|_\infty \leq (\lambda_0 - c_0)^{-1}\|f\|_\infty$ for any $n \in \mathbb{N}$. Since f vanishes outside B_{n_0}, for any $n > n_0$ the function $w_n = \varphi - u_n$ satisfies the inequality $\lambda_0 w - \mathcal{A}w \geq 0$ in $B_n \setminus \overline{B}_{n_0}$ as well as the boundary condition $w \geq 0$ on $\partial B_{n_0} \cup \partial B_n$. The classical maximum principle (see Theorem D.2.1) implies that $w \geq 0$ in $B_n \setminus \overline{B}_{n_0}$. Thus, letting n tend to $+\infty$, we obtain that $0 \leq R(\lambda_0, \widehat{A})f \leq \varphi$ in $\mathbb{R}^N \setminus \overline{B}_{n_0}$. This completes the proof. □

Example 4.3.3 Consider again the operator $\mathcal{A}$ and the function φ defined in Example 4.2.5. Taking (4.2.5) into account, it is immediate to see that the condition (4.3.5) reads as

$$\langle b(x), x\rangle \geq -\frac{\lambda_0}{2}(1 + |x|^2) - N + 4\frac{|x|^2}{1 + |x|^2}, \qquad x \in \mathbb{R}^N,$$

for some $\lambda_0 > 0$. If it is satisfied then $C_0(\mathbb{R}^N)$ is invariant. This holds, for instance, if

$$\langle b(x), x\rangle \geq C - M|x|^2, \qquad x \in \mathbb{R}^N, \tag{4.3.6}$$

for some $C \in \mathbb{R}$ and $M > 0$. Indeed, it suffices to take $\lambda_0 > 2M$.

Condition (4.3.6) should be compared with condition (4.2.6), which guarantees that the semigroup associated with the operator $\mathcal{A}$ maps $C_b(\mathbb{R}^N)$ into $C_0(\mathbb{R}^N)$ and hence, in particular, leaves $C_0(\mathbb{R}^N)$ invariant.

We conclude this section with the following proposition which gives a sufficient condition ensuring that $C_0(\mathbb{R}^N)$ is not preserved by the action of the semigroup. It is an immediate corollary of Theorem 4.1.10.

Proposition 4.3.4 *Fix $t > 0$. If $T(t)\mathbf{1} = \mathbf{1}$ and $T(t)$ is compact, then $C_0(\mathbb{R}^N)$ is not invariant for $T(t)$.*

Chapter 5

Uniform estimates for the derivatives of the function $T(t)f$

5.0 Introduction

In this chapter we prove some uniform estimates for the derivatives of the function $T(t)f$, when $f \in C_b(\mathbb{R}^N)$ and $\{T(t)\}$ is the semigroup associated with the uniformly elliptic operator

$$\mathcal{A} = \sum_{i,j=1}^{N} q_{ij}(x)D_{ij} + \sum_{j=1}^{N} b_j(x)D_j = \mathrm{Tr}(QD^2) + \langle b, \nabla\rangle,$$

with possibly unbounded coefficients in $\mathbb{R}^N$.

The problem of estimating the derivatives of $T(t)f$ has been studied in literature with both analytic ([29, 31, 182]) and probabilistic methods ([54, 55, 254]).

Here, we present the results of [34]. More precisely, we prove uniform estimates for the first-, second- and third-order derivatives of $T(t)f$. First, we show that, for any $\omega > 0$ and $k, l \in \mathbb{N}$, with $0 \le k \le l \le 3$, there exists a positive constant $C_{k,l} = C_{k,l}(\omega)$ such that

$$\|T(t)f\|_{C_b^l(\mathbb{R}^N)} \le C_{k,l}t^{-\frac{l-k}{2}}e^{\omega t}\|f\|_{C_b^k(\mathbb{R}^N)}, \qquad f \in C_b^k(\mathbb{R}^N),\ t > 0. \tag{5.0.1}$$

Although we limit ourselves to the case when $l \le 3$, the techniques that we present work as well for $l > 3$ under suitable additional assumptions on the coefficients.

To prove (5.0.1) we use the Bernstein method (see [30]) and approximate $T(t)f$ by solutions of Cauchy problems in bounded domains. We assume dissipativity-type and growth conditions on the coefficients of $\mathcal{A}$. We notice that some dissipativity condition is necessary, because in general estimate (5.0.1) fails, see Example 5.1.12.

By interpolation, we can then extend estimate (5.0.1) to the case when $k, l \in \mathbb{R}_+$, $0 \le k \le l \le 3$. This allows us to prove optimal Schauder estimates for the solution of the nonhomogeneous Cauchy problem

$$\begin{cases} D_t u(t,x) = \mathcal{A}u(t,x) + g(t,x), & t > 0, \quad x \in \mathbb{R}^N, \\ u(0,x) = f(x), & x \in \mathbb{R}^N, \end{cases} \tag{5.0.2}$$

as well as for the elliptic equation $\lambda u - \mathcal{A}u = f$ for any $\lambda > 0$.

To conclude this introduction, we stress that estimate (5.0.1) for elliptic operators $\mathcal{A}$, with a potential term which does not identically vanish, has been proved in [182] and in [172], which deals with the non-autonomous case[1]. In such papers also optimal Schauder estimates for the solution to the non-autonomous version of the Cauchy problem (5.0.2) have been proved. For further details, we refer the reader to Subsection 14.4.1.

[1] but, of course, they hold true also in the autonomous case.

5.1 Uniform estimates

Throughout this section we assume the following hypotheses:

Hypotheses 5.1.1 (i) $q_{ij}(x) = q_{ji}(x)$ for any $i,j = 1,\dots,N$ and $x \in \mathbb{R}^N$; moreover, $\langle Q(x)\xi,\xi\rangle \ge \kappa(x)|\xi|^2$ for any $\xi, x \in \mathbb{R}^N$ and some function $\kappa : \mathbb{R}^N \to \mathbb{R}$ with positive infimum κ_0;

(ii) there exist a function $\varphi \in C^2(\mathbb{R}^N)$, which blows up at $+\infty$ as $|x|$ tends to $+\infty$, and two positive constants λ_0 and C such that $\mathcal{A}\varphi - \lambda_0\varphi$ is bounded from above in $\mathbb{R}^N$;

(iii) there exists a constant $C > 0$ such that

$$|(Q(x)x)_i| \le C(1+|x|^2)\kappa(x), \qquad \mathrm{Tr}(Q(x)) \le C(1+|x|^2)\kappa(x), \tag{5.1.1}$$

$$\langle b(x), x\rangle \le C(1+|x|^2)\kappa(x), \tag{5.1.2}$$

for any $x \in \mathbb{R}^N$, $i = 1,\dots,N$.

Moreover, in the next theorems we always assume that one of the following hypotheses holds true:

(iv-1) $q_{ij}, b_j \in C^{1+\delta}_{\mathrm{loc}}(\mathbb{R}^N)$ for some $\delta \in (0,1)$ and any $i,j = 1,\dots,N$, and there exist a positive constant C and a function $d : \mathbb{R}^N \to \mathbb{R}$, with $L_1 := \sup_{x\in\mathbb{R}^N}\{d(x)/\kappa(x)\} < +\infty$, such that $|D_k q_{ij}(x)| \le C\kappa(x)$ for any $i,j,k = 1,\dots,N$ and

$$\langle \mathrm{Jac}\, b(x)\xi,\xi\rangle \le d(x)|\xi|^2, \qquad x,\xi \in \mathbb{R}^N; \tag{5.1.3}$$

(iv-2) $q_{ij}, b_j \in C^{2+\delta}_{\mathrm{loc}}(\mathbb{R}^N)$ for some $\delta \in (0,1)$ and any $i,j = 1,\dots,N$; Hypothesis 5.1.1(iv-1) holds true and there exist a positive function $r : \mathbb{R}^N \to \mathbb{R}$ and three constants $K_1 \in \mathbb{R}$, $L_2, L_3 > 0$ such that $|D^\beta b_j| \le r$ and $d + L_2 r \le L_3\kappa$ in $\mathbb{R}^N$ for any $j = 1,\dots,N$ and $|\beta| = 2$, and

$$\sum_{i,j,h,k=1}^N D_{hk}q_{ij}(x)m_{ij}m_{hk} \le K_1\kappa(x)\sum_{h,k=1}^N m_{hk}^2, \qquad x \in \mathbb{R}^N,$$

for any symmetric matrix $M = (m_{hk})$;

(iv-3) $q_{ij}, b_j \in C^{3+\delta}_{\mathrm{loc}}(\mathbb{R}^N)$ for some $\delta \in (0,1)$ and any $i,j = 1,\dots,N$; Hypothesis 5.1.1(iv-2) holds true and there exists a constant $C > 0$ such that $|D^\beta b_j(x)| \le r(x)$ and $|D^\beta q_{ij}(x)| \le C\kappa(x)$ for any $i,j = 1,\dots,N$, $|\beta| = 3$ and $x \in \mathbb{R}^N$.

Remark 5.1.2 Differently from what we did in the previous chapters, we assume that the function κ is far from zero.

Remark 5.1.3 We remark that in Hypothesis 5.1.1(iv-l) ($l = 2,3$) we can take $r = L_4(1+|d|)$ for any $L_4 > 0$. It suffices to take $L_2 < L_4^{-1}$ and $L_3 = (1+L_2L_4)(\sup_{\mathbb{R}^N} d\kappa^{-1})^+ + L_2L_4\kappa_0^{-1}$ in Hypothesis 5.1.1(iv-2).

Remark 5.1.4 In some situation Hypothesis 5.1.1(iii) is easily implied by Hypothesis 5.1.1(iv-1). This is the case, for instance, when there exists a positive constant K such that

$$\int_0^1 \kappa(tx)dt \le K\kappa(x), \qquad x \in \mathbb{R}^N. \tag{5.1.4}$$

It suffices to prove that in such a situation condition (5.1.2) follows from Hypothesis 5.1.1(iv-1). The same argument can also be used to prove that (5.1.1) is a consequence of Hypothesis 5.1.1(iv-1).

To show (5.1.2) we observe that, taking $(x,\xi)=(tz,z)$ in (5.1.3), gives

$$\frac{d}{dt}\langle b(tz),z\rangle \le L_1\kappa(tz)|z|^2, \qquad z\in\mathbb{R}^N. \tag{5.1.5}$$

Integrating (5.1.5) with respect to $t\in[0,1]$ and taking (5.1.4) into account, we get

$$\langle b(z),z\rangle \le \langle b(0),z\rangle + L_1|z|^2\int_0^1\kappa(tz)dt \le |z||b(0)| + KL_1|z|^2\kappa(z) \le C'|z|(1+|z|)\kappa(z) \tag{5.1.6}$$

for any $z\in\mathbb{R}^N$, where $b=(b_1,\dots,b_N)$, $C'=\max\{|b(0)|\kappa_0^{-1},KL_1\}$.

Let us observe that (5.1.4) is satisfied, for instance, in the case when $\kappa(x)=\overline{\kappa}(|x|)$ for any $x\in\mathbb{R}^N$ and some nondecreasing function $\overline{\kappa}:\mathbb{R}_+\to\mathbb{R}_+$, and in the case when $\kappa(x)=c+\widetilde{\kappa}(x)$, c and $\widetilde{\kappa}$ being, respectively, a positive constant and a nonnegative homogeneous function of degree $\alpha>0$.

In the sequel, for any $n\in\mathbb{N}$, we denote by u_n the (unique) classical solution of the Cauchy-Dirichlet problem

$$\begin{cases} D_tu_n(t,x)=\mathcal{A}u_n(t,x), & t>0, \quad x\in B_n,\\ u_n(t,x)=0, & t>0, \quad x\in\partial B_n,\\ u_n(0,x)=\vartheta_n(x)f(x), & \qquad\quad x\in B_n, \end{cases} \tag{5.1.7}$$

corresponding to $f\in C_b(\mathbb{R}^N)$ (i.e., the unique solution in $C_b([0,+\infty)\times\overline{B}_n)\cap C^{1,2}((0,+\infty)\times\overline{B}_n))$. Here, $\vartheta_n:\mathbb{R}^N\to\mathbb{R}$ is defined by $\vartheta_n(x)=\vartheta(|x|/n)$ for any $x\in\mathbb{R}^N$ and $n\in\mathbb{N}$, $\vartheta\in C_c^\infty(\mathbb{R})$ being any nonincreasing function such that $\chi_{(-1/2,1/2)}\le\vartheta\le\chi_{(-1,1)}$.

Remark 5.1.5 Repeating the same arguments as in Theorems 1.2.1 and in Remark 1.2.3 and taking Remark 3.1.4 and Theorem 3.1.3 into account, it is easy to check that u_n converges to $u:=T(\cdot)f$ locally uniformly in $(0,+\infty)\times\mathbb{R}^N$ as n tends to $+\infty$. Moreover, from the interior Schauder estimate in Theorem D.1.5(ii), we deduce that $\sup_{\varepsilon\le t\le T}\|u_n(t,\cdot)-u(t,\cdot)\|_{C^3(B_k)}$ vanishes as n tends to $+\infty$ for any $0<\varepsilon<T$ and $k\in\mathbb{N}$.

To prove the main result of this section we need the following lemma.

Lemma 5.1.6 *Let $\Omega\subset\mathbb{R}^N$ be a smooth bounded open set. Assume that Hypothesis* 5.1.1(*i*) *is satisfied in $\overline{\Omega}$ and that the coefficients of the operator $\mathcal{A}$ belong to $C^{k+\alpha}(\overline{\Omega})$ for some $\alpha\in(0,1)$ and $k\in\{1,2,3\}$. Then, for any $f\in C_c^j(\overline{\Omega})$ $(j=0,\dots,k)$ the function $(t,x)\mapsto t^{(k-j)/2}(D_x^kT_n(t)f)(x)$ is continuous in $[0,T]\times\overline{\Omega}_\delta$ for any $T>0$ and $\delta>0$, where $\Omega_\delta=\{x\in\Omega:\delta\le\inf_{y\in\partial\Omega}|x-y|\}$.*

Proof The proof follows from a density argument. We limit ourselves to proving it in the case when $j=k=3$, the other cases being completely similar, and even simpler.

It is well known that for any $f \in C_c^\infty(\Omega)$ the function $D_x^k T_n(\cdot)f$ ($k = 1,2,3$) is continuous in $[0,T] \times \overline{\Omega}$ (see Proposition D.3.1). Using a method similar to the one used in the proof of the following Theorem 5.1.7, we can easily show that there exists a positive constant $C = C_{T,\delta}$, independent of f, such that $\|T(t)f\|_{C^3(\overline{\Omega}_\delta)} \le C\|f\|_{C^3(\overline{\Omega})}$ for any $t \in [0,T]$. Now with any $f \in C_c^3(\overline{\Omega})$ we associate a sequence of smooth functions $\{f_n\} \subset C_c^\infty(\Omega)$ converging to f in $C^k(\overline{\Omega})$. Since, for any $n \in \mathbb{N}$ and $k \in \{1,2,3\}$, the function $D^k T(\cdot)f_n$ is continuous in $[0,T] \times \overline{\Omega}$ for any $T > 0$, taking the previous estimate into account we easily get the assertion. □

Now, we prove the main result of this section.

Theorem 5.1.7 *Let Hypotheses 5.1.1(i)-5.1.1(iii), 5.1.1(iv-l) be satisfied for some $l \in \{1,2,3\}$. Then, for any $\omega > 0$ and $k = 0,\dots,l$, there exist constants $C_{k,l} = C_{k,l}(\omega) > 0$ such that*

$$\|T(t)f\|_{C_b^l(\mathbb{R}^N)} \le C_{k,l}t^{-\frac{l-k}{2}}e^{\omega t}\|f\|_{C_b^k(\mathbb{R}^N)}, \qquad f \in C_b^k(\mathbb{R}^N),\ t > 0. \tag{5.1.8}$$

In particular, if $k = l$ we can take $\omega = 0$ in (5.1.8).

Proof To ease the notation, throughout the proof, we set

$$\mathfrak{q}_0(v) = |\sqrt{Q}\nabla v|^2, \qquad \mathfrak{q}_1(v) = \sum_{i,j,h=1}^N D_h q_{ij} D_h v D_{ij} v, \tag{5.1.9}$$

$$\mathfrak{q}_2(v) = \sum_{i,j,h,k=1}^N D_{hk} q_{ij} D_{ij} v D_{hk} v, \qquad \mathfrak{b}_1(v) = \langle \mathrm{Jac}\, b\, \nabla_x v, \nabla_x v\rangle, \tag{5.1.10}$$

$$\mathfrak{b}_2(v) = \sum_{j,h,k=1}^N D_{hk} b_j D_j v D_{hk} v, \qquad \mathfrak{q}_3(v) = \sum_{i,j,h,k,l=1}^N D_{hkl} q_{ij} D_{ij} v D_{hkl} v, \tag{5.1.11}$$

$$\mathfrak{b}_3(v) = \sum_{j,h,k,l=1}^N D_{hkl} b_j D_j u D_{hkl} v, \tag{5.1.12}$$

for any smooth function $v : \mathbb{R}^N \to \mathbb{R}$.

We begin the proof considering the case when $k = 0$ and $l = 3$. For any $n \in \mathbb{N}$, let ϑ_n be the smooth function defined by $\vartheta_n(x) = \vartheta(|x|/n)$ for any $x \in \mathbb{R}^N$, where $\vartheta \in C_c^\infty(\mathbb{R})$ is a nonincreasing function such that $\chi_{(-1/2,1/2)} \le \vartheta \le \chi_{(-1,1)}$. We fix $\omega > 0$ and $t_0 > 0$ such that $e^{\omega t_0}t^{-\frac{3}{2}} \ge 1$ for any $t \ge t_0$ and define the function $v_{0,3,n}$ by setting

$$v_{0,3,n}(t,x) = |u_n(t,x)|^2 + at\vartheta_n^2|\nabla_x u_n(t,x)|^2 + a^2t^2\vartheta_n^4|D_x^2 u_n(t,x)|^2 + a^3t^3\vartheta_n^6|D_x^3 u_n(t,x)|^2$$

for any $t \in [0,t_0]$ and $x \in B_n$ (see (5.1.7)). For notational convenience, throughout the rest of the proof, we drop out the dependence on n and t, when no confusion may arise. Observing that u_n is in $C([0,+\infty);C(\overline{B}_n))$ (see Proposition C.2.5(i) and Theorem D.3.5(iv)) and taking Lemma 5.1.6 into account, we can infer that $v_{0,3}$ is continuous in $[0,t_0] \times \overline{B}_n$ and, with some computations, we can show that $v_{0,3}$ solves the Cauchy problem

$$\begin{cases} D_t v_{0,3}(t,x) = \mathcal{A}v_{0,3}(t,x) + g(t,x), & t \in (0,t_0],\quad x \in B_n, \\ v_{0,3}(t,x) = 0, & t \in [0,t_0],\quad x \in \partial B_n, \\ v_{0,3}(0,x) = (\vartheta_n(x)f(x))^2, & x \in \overline{B}_n, \end{cases} \tag{5.1.13}$$

where $g = \sum_{j=1}^{8} g_j$ with

$$g_1 = -2\mathfrak{q}_0(u) - 2at\vartheta^2 \sum_{i=1}^{N} \mathfrak{q}_0(D_i u) - 2a^2t^2\vartheta^4 \sum_{i,j=1}^{N} \mathfrak{q}_0(D_{ij}u) - 2a^3t^3\vartheta^6 \sum_{i,j,h=1}^{N} \mathfrak{q}_0(D_{ijh}u), \tag{5.1.14}$$

$$g_2 = -2at\mathfrak{q}_0(\vartheta)(|\nabla_x u|^2 + 6at\vartheta^2|D_x^2 u|^2 + 15a^2t^2\vartheta^4|D_x^3 u|^2), \tag{5.1.15}$$

$$\begin{aligned} g_3 =& -2at\vartheta(\mathcal{A}\vartheta)(|\nabla_x u|^2 + 2at\vartheta^2|D_x^2 u|^2 + 3a^2t^2\vartheta^4|D_x^3 u|^2) \\ &- 8at\vartheta\langle Q\nabla\vartheta, D_x^2 u\nabla_x u\rangle - 16a^2t^2\vartheta^3 \sum_{i=1}^{N}\langle Q\nabla\vartheta, D^2 D_i u\nabla_x D_i u\rangle \\ &- 24a^3t^3\vartheta^5 \sum_{i,j=1}^{N} \langle Q\nabla\vartheta, D_x^2 D_{ij}u\nabla_x D_{ij}u\rangle, \end{aligned} \tag{5.1.16}$$

$$g_4 = 2at\vartheta^2\mathfrak{b}_1(u) + 4a^2t^2\vartheta^4 \sum_{i=1}^{N} \mathfrak{b}_1(D_i u) + 6a^3t^3\vartheta^6 \sum_{i,j=1}^{N} \mathfrak{b}_1(D_{ij}u), \tag{5.1.17}$$

$$g_5 = 2at\vartheta^2\mathfrak{q}_1(u) + 4a^2t^2\vartheta^4 \sum_{i=1}^{N} \mathfrak{q}_1(D_i u) + 6a^3t^3\vartheta^6 \sum_{i,j=1}^{N} \mathfrak{q}_1(D_{ij}u), \tag{5.1.18}$$

$$g_6 = 2a^2t^2\vartheta^4(\mathfrak{q}_2(u) + \mathfrak{b}_2(u)) + 6a^3t^3\vartheta^6 \sum_{i=1}^{N}(\mathfrak{q}_2(D_i u) + \mathfrak{b}_2(D_i u)), \tag{5.1.19}$$

$$g_7 = 2a^3t^3\vartheta^6(\mathfrak{q}_3(u) + \mathfrak{b}_3(u)), \tag{5.1.20}$$

$$g_8 = a\vartheta^2|\nabla_x u|^2 + 2a^2t\vartheta^4|D_x^2 u|^2 + 3a^3t^2\vartheta^6|D_x^3 u|^2. \tag{5.1.21}$$

Taking the ellipticity condition in Hypothesis 5.1.1(i) into account, we easily deduce that

$$g_1 \le -2\kappa|\nabla_x u|^2 - 2at\vartheta^2\kappa|D_x^2 u|^2 - 2a^2t^2\vartheta^4\kappa|D_x^3 u|^2 - 2a^3t^3\vartheta^6\kappa|D_x^4 u|^2 \tag{5.1.22}$$

and that $g_2 \le 0$.

To estimate the function g_3 we observe that, by virtue of (5.1.1), (5.1.2), and recalling that ϑ is nonincreasing in $\mathbb{R}$ and $\vartheta' = 0$ in $(-1/2, 1/2)$, it can be shown that (i) $|\mathrm{Tr}(QD^2\vartheta)| \le C'\kappa$, (ii) $|(Q\nabla\vartheta)_i| \le C'\kappa$, (iii) $\langle b, \nabla\vartheta\rangle \ge -C'\kappa$ in $\mathbb{R}^N$ for any $n \ge 1$, any $i = 1, \ldots, N$ and some positive constant C'. Taking these estimates into account and recalling that for any $a, b, \varepsilon > 0$, it holds that $ab \le (4\varepsilon)^{-1}a^2 + \varepsilon^2 b^2$, we get

$$\begin{aligned} g_3 \le& 2atC'\kappa\vartheta|\nabla_x u|^2 + 4a^2t^2C'\kappa\vartheta^3|D_x^2 u|^2 + 6a^3t^3C'\kappa\vartheta^5|D_x^3 u|^2 \\ &+ 2atC'\kappa\vartheta|\nabla_x u|^2 + 4a^2C't^2\kappa\vartheta^3|D_x^2 u|^2 + 6a^3t^3C'\kappa\vartheta^5|D_x^3 u|^2 \\ &+ 8atC'\kappa\left(\frac{N}{4\varepsilon}|\nabla_x u|^2 + \varepsilon\vartheta^2|D_x^2 u|^2\right) + 16a^2t^2C'\kappa\left(\frac{N}{4\varepsilon}\vartheta^2|D_x^2 u|^2 + \varepsilon\vartheta^4|D_x^3 u|^2\right) \\ &+ 24a^3t^3C'\kappa\left(\frac{N}{4\varepsilon}\vartheta^4|D_x^3 u|^2 + \varepsilon\vartheta^6|D_x^4 u|^2\right) \end{aligned}$$

$$
\begin{aligned}
&\le 2aC'\left(2+\frac{N}{\varepsilon}\right)t\kappa|\nabla_x u|^2+4aC'\left(2at+2\varepsilon+at\frac{N}{\varepsilon}\right)t\kappa\vartheta^2|D^2_x u|^2\\
&+2a^2C'\left(6at+8\varepsilon+3at\frac{N}{\varepsilon}\right)t^2\kappa\vartheta^4|D^3_x u|^2+24a^3t^3C'\varepsilon\kappa\vartheta^6|D^4_x u|^2. \qquad (5.1.23)
\end{aligned}
$$

Taking advantage of Hypothesis 5.1.1(iv-1), we deduce that

$$
g_4\le 2atd\,\vartheta^2|\nabla_x u|^2+4a^2t^2d\,\vartheta^4|D^2_x u|^2+6a^3t^3d\,\vartheta^6|D^3_x u|^2. \qquad (5.1.24)
$$

Arguing as in (5.1.23) and taking Hypothesis 5.1.1(iv-3) into account, we can estimate

$$
\mathfrak{q}_{2j+1}(v)\le C\kappa\left(\frac{N^{2(1+j)}}{4\varepsilon}|D^{j+1}v|^2+\varepsilon N|D^{2+j}v|^2\right), \qquad (5.1.25)
$$

$$
\mathfrak{q}_2(v)\le K_1\kappa|D^2v|^2,\qquad \mathfrak{b}_{2+j}(v)\le r\left(\frac{N^{2+j}}{4\varepsilon}|\nabla v|^2+\varepsilon N|D^{2+j}v|^2\right) \qquad (5.1.26)
$$

in $\mathbb{R}^N$, for any smooth function $v:\mathbb{R}^N\to\mathbb{R}$ and $j=0,1$. Using the previous estimates we conclude that

$$
\begin{aligned}
g_5\le\, &atC\frac{N^2}{2\varepsilon}\kappa|\nabla_x u|^2+aCN\left(2\varepsilon+at\frac{N}{\varepsilon}\right)t\kappa\,\vartheta^2|D^2_x u|^2\\
&+a^2CN\left(4\varepsilon+3at\frac{N}{2\varepsilon}\right)t^2\kappa\,\vartheta^4|D^3_x u|^2+6a^3t^3\varepsilon CN\kappa\,\vartheta^4|D^4_x u|^2, \qquad (5.1.27)
\end{aligned}
$$

$$
\begin{aligned}
g_6\le\, &a^2t^2\vartheta^2r\frac{N^2}{2\varepsilon}|\nabla_x u|^2+a^2t^2\vartheta^4\left[2K_1\kappa+N\left(2\varepsilon+3at\frac{N}{2\varepsilon}\right)r\right]|D^2_x u|^2\\
&+6a^3t^3\vartheta^6(K_1\kappa+\varepsilon Nr)|D^3_x u|^2, \qquad (5.1.28)
\end{aligned}
$$

$$
g_7\le a^3t^3\vartheta^2\left[r\frac{N^3}{2\varepsilon}|\nabla_x u|^2+\vartheta^2C\kappa\frac{N^4}{2\varepsilon}|D^2_x u|^2+2\vartheta^4\varepsilon N(C\kappa+r)|D^3_x u|^2\right]. \qquad (5.1.29)
$$

From (5.1.22)-(5.1.24) and (5.1.27)-(5.1.29) we deduce that

$$
\begin{aligned}
g\le&\bigg\{-\kappa_0+a+\kappa\left[-1+2at\left(2C'+C'\frac{N}{\varepsilon}+C\frac{N^2}{4\varepsilon}\right)+a^2t^2C\frac{N^2}{2\varepsilon}\right]\\
&+at\left[2d+at(1+atN)\frac{N^2}{2\varepsilon}r\right]\bigg\}|\nabla_x u|^2\\
&+a\bigg\{-\kappa_0+2a+\kappa\bigg[-1+2(4C'+CN)\varepsilon+a^2t^2C\frac{N^3}{2\varepsilon}\\
&\qquad+4at\left(2C'+C'\frac{N}{\varepsilon}+C\frac{N^2}{4\varepsilon}+\frac{1}{2}K_1^+\right)\bigg]\\
&+at\left[4d+\left(2\varepsilon N+3at\frac{N^2}{2\varepsilon}\right)r\right]\bigg\}t\vartheta^2|D^2_x u|^2\\
&+a^2\bigg\{-\kappa_0+3a+2at(3d+4\varepsilon Nr)\\
&\qquad+\kappa\bigg[-1+4(4C'+CN)\varepsilon+2at\varepsilon CN
\end{aligned}
$$

$$+6at\left(2C'+C'\frac{N}{\varepsilon}+C\frac{N^2}{4\varepsilon}+K_1^+\right)\Bigg]\Bigg\}t^2\vartheta^4|D_x^3u|^2$$
$$+2a^3(-1+12C'\varepsilon+3\varepsilon CN)t^3\kappa\vartheta^6|D_x^4u|^2 \qquad (5.1.30)$$

for any $t\in(0,t_0]$. Choosing (a,ε) properly and independently of n, we can make g non positive in $(0,t_0]\times B_n$. It suffices to fix $0<\varepsilon\le\min\{(32C'+8CN)^{-1},3L_2(4N)^{-1}\}$ and choose a small enough. From the classical maximum principle we conclude that $|v_{0,3,n}(t,x)|\le\|\vartheta_n f\|_\infty^2\le\|f\|_\infty^2$, for any $t\in[0,t_0]$ and $x\in B_n$. Taking the limit as n tends to $+\infty$, by Remark 5.1.5 estimate (5.0.1) follows for any $t\in[0,t_0]$, some constant $C_{0,3}=C_{0,3}(t_0)>0$ and with $\omega=0$. Using the semigroup property we can then extend the estimate to any $t>0$. Indeed, taking the choice of t_0 into account, we get, for any $t>t_0$,

$$\begin{aligned}\|T(t)f\|_{C_b^3(\mathbb{R}^N)}=&\|T(t_0)T(t-t_0)f\|_{C_b^3(\mathbb{R}^N)}\le C_{0,3}t_0^{-\frac{3}{2}}\|T(t-t_0)f\|_\infty\\ \le& C_{0,3}t_0^{-\frac{3}{2}}\|f\|_\infty\le C_{0,3}t_0^{-\frac{3}{2}}e^{\omega t}t^{-\frac{3}{2}}\|f\|_\infty\end{aligned}$$

and (5.0.1) follows with $C_{0,3}(\omega)=\max\{C_{0,3}(t_0),C_{0,3}(t_0)t_0^{-3/2}\}$.

In the other cases the proof is very similar. It suffices to apply the quoted arguments to the function $v_{k,l,n}$, defined by

$$v_{k,l,n}(t,x)=\sum_{j=0}^{l}a^jt^{(j-k)^+}(\vartheta_n(x))^{2j}|D_x^ju_n(t,x)|^2,\qquad t\in(0,t_0],\ x\in B_n.$$

Let us just show that, if $k=l=3$, then we can take $\omega=0$ in (5.0.1). A straightforward computation shows that $v_{3,3,n}$ is a classical solution to the Cauchy-Dirichlet problem (5.1.13) with $v_{3,3,n}(0,\cdot)=\sum_{j=0}^3|D^j(\vartheta_n f)|^2$ and g_n being replaced with $\widetilde{g}_n=\sum_{j=1}^7\widetilde{g}_{j,n}$, where $\widetilde{g}_{j,n}$ $(j=1,\dots,7)$ are defined by the right-hand sides of (5.1.14)-(5.1.20) after replacing each t, therein appearing as coefficients of u_n or of its derivatives, with $t=1$. Arguing as above, we can easily show that $\widetilde{g}_n$ can be estimated, for any $t>0$, by the last side of (5.1.30), where we set $t=1$ and replace the terms $-\kappa_0+a$, $-\kappa_0+2a$ and $-\kappa_0+3a$ simply with $-\kappa$. It is now clear that we can take (a,ε) such that $\widetilde{g}_n\le 0$ in $(0,+\infty)\times\mathbb{R}^N$ for any $n\in\mathbb{N}$ and, consequently,

$$|v_{3,3,n}|\le\|\vartheta f\|_\infty^2+\|\nabla_x(\vartheta f)\|_\infty^2+\|D_x^2(\vartheta f)\|_\infty^2+\|D_x^3(\vartheta f)\|_\infty^2 \qquad (5.1.31)$$

in $[0,t_0]\times\overline{B}_n$ for any $n\in\mathbb{N}$. Estimate (5.1.31) yields (5.0.1) taking the limit as n tends to $+\infty$. □

Remark 5.1.8 In the case when $l=1$ and $k=0,1$, the gradient estimate (5.1.8) can be proved under weaker assumptions on the diffusion coefficients of the operator $\mathcal{A}$. This can be done, approximating the function $T(\cdot)f$ by solutions u_n of Cauchy-Neumann problems in B_n. We refer the reader to Proposition 13.1.4 and Theorem 14.4.2. Unfortunately, this approximation argument can not be used to prove estimates for higher order spatial derivatives of the function $T(\cdot)f$ since no information on the sign of the normal derivative of the functions $|D_x^ju_n|^2$ on ∂B_n is available when $j>1$.

Now, by interpolation, we can extend (5.0.1) to the case when at least one between k and l is not integer.

Theorem 5.1.9 *Let Hypotheses* 5.1.1(*i*)-5.1.1(*iii*), 5.1.1(*iv-l*) *be satisfied for some* $l\in$

$\{1,2,3\}$. *Then, for any* $\omega > 0$, $k, m = 0, \ldots, l-1$ *and* $\alpha, \theta \in [0,1]$ *such that* $k+\alpha \leq m+\theta$, *there exist constants* $C_{k+\alpha,m+\theta} = C_{k+\alpha,m+\theta}(\omega) > 0$ *such that*

$$\|T(t)f\|_{C_b^{m+\theta}(\mathbb{R}^N)} \leq C_{k+\alpha,m+\theta} t^{-\frac{m+\theta-k-\alpha}{2}} e^{\omega t} \|f\|_{C_b^{k+\alpha}(\mathbb{R}^N)} \tag{5.1.32}$$

for any $f \in C_b^{k+\alpha}(\mathbb{R}^N)$ *and any* $k = 0, \ldots, m$. *In particular, if* $k+\alpha = m+\theta$ *then we can take* $\omega = 0$ *in* (5.1.32).

Proof The proof follows from an interpolation argument. We limit ourselves to sketching it in a particular case, since the same techniques can also be applied to all the other cases. So, let us assume that $k = m = 2$ and $0 < \alpha \leq \theta < 1$. Moreover, fix $\omega > 0$ and $t > 0$. Setting $\omega' = \omega/(\theta - \alpha)$, from (5.1.8) with $(k,l) = (2,2)$ and $(k,l) = (2,3)$, we deduce that

$$\|T(t)\|_{L(C_b^2(\mathbb{R}^N),C_b^2(\mathbb{R}^N))} \leq C_{2,2}, \qquad \|T(t)\|_{L(C_b^2(\mathbb{R}^N),C_b^3(\mathbb{R}^N))} \leq C_{2,3} t^{-\frac{1}{2}} e^{\omega' t} \tag{5.1.33}$$

for any $t > 0$. Recalling that $(C_b^2(\mathbb{R}^N); C_b^3(\mathbb{R}^N))_{\beta,\infty} = C_b^{2+\beta}(\mathbb{R}^N)$, for any $\beta \in (0,1)$ (see Theorem B.3.8), and applying Proposition B.3.2, we conclude from (5.1.33) that $T(t)$ is bounded from $C_b^2(\mathbb{R}^N)$ into $C_b^{2+\beta}(\mathbb{R}^N)$ for any $\beta \in (0,1)$ and

$$\|T(t)\|_{L(C_b^2(\mathbb{R}^N),C_b^{2+\beta}(\mathbb{R}^N))} \leq C_{2,2+\beta} t^{-\frac{\beta}{2}} e^{\beta\omega' t}, \qquad t > 0, \tag{5.1.34}$$

where $C_{2,2+\beta} = C_{2,2}^{1-\beta} C_{2,3}^{\beta}$.

Applying the same argument to (5.1.8), with $(k,l) = (3,3)$, and to (5.1.34), we deduce that $T(t)$ is bounded from $C_b^{2+\alpha}(\mathbb{R}^N)$ into $C_b^{2+\beta+(1-\beta)\alpha}(\mathbb{R}^N)$ and

$$\|T(t)\|_{L(C_b^{2+\alpha}(\mathbb{R}^N),C_b^{2+\beta+(1-\beta)\alpha}(\mathbb{R}^N))} \leq C_{2+\alpha,2+\beta+(1-\beta)\alpha} t^{-(1-\alpha)\frac{\beta}{2}} e^{(1-\alpha)\beta\omega' t},$$

where $C_{2+\alpha,2+\beta+(1-\beta)\alpha} = C_{2,2+\beta}^{1-\alpha} C_{3,3}^{\alpha}$. Now the assertion follows, taking $\beta = (\theta-\alpha)/(1-\alpha)$. □

In some situations we can extend Theorems 5.1.7 and 5.1.9 to the case when in Hypothesis 5.1.1(iv-j), the j-th-order derivatives of the coefficients are merely continuous in $\mathbb{R}^N$. As the following theorem shows, we can do this when condition (5.1.4) is satisfied and there exist $m, M > 0$ such that

$$\int_0^1 d(tx+y)dt \leq M\kappa(x+y), \qquad x \in \mathbb{R}^N, \ y \in B_m. \tag{5.1.35}$$

As is immediately seen, (5.1.35) is satisfied, for instance, when d is bounded from above.

Theorem 5.1.10 *Suppose that Hypotheses* 5.1.1(*i*)-5.1.1(*iii*), 5.1.1(*iv-j*) ($j = 1,2,3$) (*with the j-th-order derivatives of the coefficients merely continuous in* $\mathbb{R}^N$) *and conditions* (5.1.4) *and* (5.1.35) *are satisfied. Then, estimates* (5.1.8) *and* (5.1.32) *hold true for any* $k, l \in \mathbb{N}$ *such that* $k \leq l \leq j$.

Proof The proof follows from a density argument. Of course, we can limit ourselves to dealing with estimate (5.1.8), since, as Theorem 5.1.9 shows, (5.1.32) follows easily from (5.0.1).

For any $\varepsilon > 0$ let $\varphi^\varepsilon(x) = \varepsilon^{-N}\varphi(x/\varepsilon)$, where $\varphi \in C_c^\infty(\mathbb{R}^N)$ is any nonnegative even function compactly supported in B_1 and with integral 1. We denote by f^ε the convolution between f and φ^ε.

Let $\mathcal{A}^\varepsilon$ be defined as $\mathcal{A}$ with q_{ij} and b_j being replaced, respectively, with q_{ij}^ε and b_j^ε for any $i,j=1,\dots,N$. As is immediately seen, q_{ij}^ε and b_j^ε $(i,j=1,\dots,N)$ satisfy Hypotheses 5.1.1(i) and 5.1.1(iv-j), with κ, d, r being replaced with $\kappa^\varepsilon, d^\varepsilon$ and r^ε, and with $C^\varepsilon = C$, $L_i^\varepsilon = L_i$ for $i=1,2,3$, $K_1^\varepsilon = K_1$ (if $j>1$) and $\kappa_0^\varepsilon \ge \kappa_0$.

Let us check that the coefficients q_{ij}^ε and b_j^ε $(i,j=1,\dots,N)$ satisfy Hypothesis 5.1.1(iii) for some positive constant, independent of ε. For this purpose, we observe that (5.1.4) implies that $|q_{ij}(x)| \le |q_{ij}(0)| + CK\sqrt{N}|x|\kappa(x)$ for any $x \in \mathbb{R}^N$ and $i,j=1,\dots,N$, which yields (5.1.1).

A straightforward computation shows that $|q_{ij}^\varepsilon(x)| \le |q_{ij}(0)| + CK\sqrt{N}(|x|+\varepsilon)\kappa^\varepsilon(x)$ for any $x\in\mathbb{R}^N$ and $i,j=1,\dots,N$, so that the coefficients q_{ij}^ε satisfy condition (5.1.1) with a constant independent of $\varepsilon \in (0,1]$.

Similarly, we can show that the coefficients b_j^ε satisfy (5.1.2) with a positive constant independent of $\varepsilon \le m$. Indeed, combining (5.1.6) and (5.1.35), we deduce that

$$\int_0^1 d^\varepsilon(tx)dt \le M\int_{B_\varepsilon} dy \int_0^1 \kappa(x-y)\varphi^\varepsilon(y)dt = M\kappa^\varepsilon(x)$$

for any $x\in\mathbb{R}^N$ and $\varepsilon \le m$. Now, arguing as in the proof of (5.1.6), we get

$$\langle b^\varepsilon(x), x\rangle \le \langle b^\varepsilon(0), x\rangle + |x|^2\int_0^1 d^\varepsilon(tx)dt \le \sup_{y\in B_\varepsilon}|b_j(y)||x| + M|x|^2\kappa^\varepsilon(x)$$

for any $x\in\mathbb{R}^N$, which yields (5.1.2) with a constant being independent of $\varepsilon\le m$.

Now, applying the same arguments as in the proof of Theorem 5.1.7, we can show that, for any $\omega>0$ and $k,l\in\mathbb{N}$, such that $0\le k\le l\le j$, there exists a positive constant $C_{k,l}=C_{k,l}(\omega)$, independent of ε and n, such that

$$\|T_n^\varepsilon(t)f\|_{C^l(\overline{B}_n)} \le C_{k,l}t^{-\frac{l-k}{2}}e^{\omega t}\|f\|_{C^k(\overline{B}_n)}, \qquad t>0,\ \ f\in C_b^k(\overline{B}_n),$$

where $\{T_n^\varepsilon(t)\}$ is the semigroup associated with the realization of the operator $\mathcal{A}^\varepsilon$ in $C_b(B_n)$ with homogeneous Dirichlet boundary conditions. As n tends to $+\infty$, $T_n^\varepsilon(t)f$ tends to a solution $u^\varepsilon =: T^\varepsilon(t)f$ of the Cauchy problem

$$\begin{cases} D_t u(t,x) = \mathcal{A}^\varepsilon u(t,x), & t>0, \quad x\in\mathbb{R}^N,\\ u(0,x)=f(x), & \qquad\quad\ x\in\mathbb{R}^N,\end{cases}$$

which satisfies

$$\|T^\varepsilon(t)f\|_{C^l(\mathbb{R}^N)} \le C_{k,l}t^{-\frac{l-k}{2}}e^{\omega t}\|f\|_{C^k(\mathbb{R}^N)}, \qquad t>0. \tag{5.1.36}$$

Theorem D.1.5 and estimate (5.1.36) imply that there exists an infinitesimal sequence $\{\varepsilon_n\}$ such that $T^{\varepsilon_n}(t)f$ and its spatial derivatives up to the $(j-1)$-th-order converge in $C_{\rm loc}^{1+\delta/2,2+\delta}((0,+\infty)\times\mathbb{R}^N)$ to a function $u =: S(\cdot)f$ satisfying

$$\|S(t)f\|_{C^l(\mathbb{R}^N)} \le C_{k,l}t^{-\frac{l-k}{2}}e^{\omega t}\|f\|_{C^k(\mathbb{R}^N)}, \qquad t>0. \tag{5.1.37}$$

Since the coefficients q_{ij} and b_j $(i,j=1,\dots,N)$ are locally Lipschitz continuous, $q_{ij}^{\varepsilon_n}$ and $b_j^{\varepsilon_n}$ converge locally uniformly in $\mathbb{R}^N$, as n tends to $+\infty$, respectively to q_{ij} and b_j, so that $S(t)f$ satisfies the differential equation $D_t w = \mathcal{A}w$ in $(0,+\infty)\times\mathbb{R}^N$. Moreover, for any $f\in C_b^2(\mathbb{R}^N)$, $S(t)f$ converges to f as t tends to 0, locally uniformly in x. This can be seen by a localization argument similar to the one used in the proof of Theorem 1.2.1. For this purpose, we fix $k \in \mathbb{N}$ and let $\vartheta = \vartheta_k$ be as in the proof of Theorem 5.1.7. Moreover, for any $n \in \mathbb{N}$, we set $v_n = \vartheta T^{\varepsilon_n}(\cdot)f$ and observe that v_n is a solution to the Cauchy problem

$$\begin{cases} D_t v_n(t,x) = \mathcal{A}^{\varepsilon_n} v_n(t,x) + \psi_n(t,x), & t \in [0,T], \quad x \in B_k, \\ v_n(t,x) = 0, & t \in [0,T], \quad x \in \partial B_k, \\ v_n(0,x) = \vartheta(x)f(x), & \qquad\qquad x \in \overline{B}_k, \end{cases}$$

where $\psi_n = -2(T^{\varepsilon_n}(\cdot)f)\mathcal{A}^{\varepsilon_n}(\vartheta) - 2\langle Q_{ij}^{\varepsilon_n}\nabla_x T^{\varepsilon_n}(\cdot)f, \nabla\vartheta\rangle$.

Since the coefficients q_{ij}^{ε} and b_j^{ε} $(i,j = 1,\ldots,N)$ converge locally uniformly in $\mathbb{R}^N$, they are equibounded (with respect to $n \in \mathbb{N}$) in B_k and there exists a positive constant $\widetilde{C}$ such that $\|\mathcal{A}^{\varepsilon_n} g\|_{C(\overline{B}_k)} \le \widetilde{C}\|g\|_{C^2(\overline{B}_k)}$ for any $g \in C^2(\overline{B}_k)$. Therefore, from (5.1.36) we deduce that there exists a constant $\overline{C}$, independent of n, such that

$$|\psi_n(t,x)| \le \overline{C}t^{-\frac{1}{2}}\|f\|_\infty, \qquad t \in (0,T], \quad x \in B_k, \quad n \in \mathbb{N}. \tag{5.1.38}$$

Estimate (5.1.38) implies that v_n can be written by the usual variation-of-constants formula as

$$v_n(t,\cdot) = T_{k,n}(t)(\vartheta f) + \int_0^t T_{k,n}(t-s)\psi_n(s,\cdot)ds, \qquad t \in (0,T],$$

where $\{T_{k,n}(t)\}$ is the semigroup associated with the realization $A_{k,n}$ of $\mathcal{A}^{\varepsilon_n}$ in $C(\overline{B}_k)$ with homogeneous Dirichlet boundary conditions (see Theorem D.3.5(iv)).

Since $\vartheta f \in D(A_{k,n}) = \{u \in \bigcap_{1\le p<+\infty} W^{2,p}(B_k) : \mathcal{A}^{\varepsilon_n} u \in C(\overline{B}_k)\}$, the domain of the realization of $\mathcal{A}^{\varepsilon_n}$ in $C(\overline{B}_k)$, (see Theorem D.3.5), and $\{T_{k,n}(t)\}$ is a semigroup of contractions in $C(\overline{B}_k)$ for any $n \in \mathbb{N}$, we can estimate

$$\begin{aligned} \|T_{k,n}(t)(\vartheta f) - \vartheta f\|_{C(\overline{B}_k)} &= \left\| \int_0^t T_{k,n}(s)\mathcal{A}^{\varepsilon_n}(\vartheta f)ds \right\|_{C(\overline{B}_k)} \\ &\le t\|\mathcal{A}^{\varepsilon_n}(\vartheta f)\|_{C(\overline{B}_k)} \le \overline{C}t\|\vartheta f\|_{C^2(\overline{B}_k)} \end{aligned}$$

for any $t \in (0,T)$, which readily shows that $\sup_{n\in\mathbb{N}} \|T_{k,n}(t)(\vartheta f) - \vartheta f\|_{C(\overline{B}_k)}$ vanishes as t tends to 0^+. From this result and (5.1.38) it follows that $\sup_{n\in\mathbb{N}} \|v_n(t,\cdot) - \vartheta f\|_{C(\overline{B}_k)}$ vanishes as t tends to 0^+. Recalling that $\vartheta \equiv 1$ in $B_{k/2}$, we deduce that

$$\begin{aligned} |(S(t)f)(x) - f(x)| &\le \sup_{n\in\mathbb{N}} |(T^{\varepsilon_n}(t)f)(x) - \vartheta(x)f(x)| + \limsup_{n\to+\infty} |(T^{\varepsilon_n}(t)f)(x) - (S(t)f)(x)| \\ &\le \sup_{n\in\mathbb{N}} \|v_k(t,\cdot) - \vartheta f\|_{C(\overline{B}_k)} \end{aligned} \tag{5.1.39}$$

for any $t \in (0,T)$ and $x \in B_{k/2}$. Taking the limit as t tends to 0 in both the first and the last side of (5.1.39) we conclude that $\sup_{x\in B_{k/2}} |u(t,x) - f(x)|$ converges to 0 as t tends to 0^+. The arbitrariness of $k \in \mathbb{N}$ shows that $S(t)f$ converges locally uniformly (with respect to x) to f as t tends to 0. Hence, $S(\cdot)f$ is a classical solution to the Cauchy problem

$$\begin{cases} D_t w(t,x) = \mathcal{A}w(t,x), & t > 0, \quad x \in \mathbb{R}^N, \\ w(0,x) = f(x), & \qquad\quad x \in \mathbb{R}^N, \end{cases}$$

and Hypothesis 5.1.1(ii) and the maximum principle in Theorem 3.1.3 imply that $T(\cdot)f = S(\cdot)f$. Since $C_b^2(\mathbb{R}^N)$ is dense in $C_b^1(\mathbb{R}^N)$ (which is endowed with the sup-norm) we can extend the previous equality to all the functions $f \in C_b^1(\mathbb{R}^N)$, obtaining (5.0.1) in the case when $k \geq 1$.

Finally, with any $f \in C(\mathbb{R}^N)$ and any $t > 0$ we split $T(t)f = T(t/2)T(t/2)f$ and observe that since $T(t/2)f \in C_b^1(\mathbb{R}^N)$, then $T(t)f = S(t/2)T(t/2)f$. Applying (5.1.37) with (t, f) being replaced with $(t/2, T(t/2)f)$, we easily get (5.0.1) also in the case when $k = 0$. □

Remark 5.1.11 In Theorem 5.1.7 we have shown that, when $k < l$, estimate (5.1.8) holds true for some $\omega \geq 0$. In Chapter 6 we will show that, actually, we can take $\omega = 0$ in (5.1.8) (and, consequently, in (5.1.32)) and that, under somewhat heavier assumptions on the coefficients, we can also take $\omega < 0$.

To conclude this section, we show that estimates (5.0.1) may fail to hold without any dissipativity assumption.

Example 5.1.12 Let $\mathcal{A}$ be the one-dimensional operator, defined on smooth functions $\zeta : \mathbb{R} \to \mathbb{R}$ by $\mathcal{A}\zeta = \zeta'' + p'\zeta' = e^{-p}(e^p\zeta')'$, where $p(x) = -x^4 + \log(h(x))$ for any $x \in \mathbb{R}$, h being a regular function such that $\varepsilon_n = h(n - \delta_n/2) \leq h(x) \leq 1$ for any $x \in (n - \delta_n, n)$ and $n \in \mathbb{N}$, $h(x) = 1$ otherwise in $\mathbb{R}$ and $\varepsilon_n = n^{-1}e^{-4n^3-n}$, $\delta_n = n^{-2}e^{-n^4}\varepsilon_n$.

Let $\psi : \mathbb{R} \to \mathbb{R}$ be the function defined by

$$\psi(s) = e^{p(s)} \int_0^s e^{-p(t)} dt, \qquad s \in \mathbb{R}.$$

A straightforward computation shows that $x^3\psi(x)$ converges to $1/4$ as x tends to $-\infty$ and thus $\psi \in L^1((-\infty, 0))$. On the other hand,

$$\begin{aligned}\psi(x) \leq & e^{-x^4} \int_0^x \frac{e^{t^4}}{h(t)} dt \leq e^{-x^4} \int_0^x e^{t^4} dt + e^{-x^4} \sum_{n=1}^{[x]+1} \int_{n-\delta_n}^n \frac{e^{n^4}}{\varepsilon_n} dt \\ \leq & e^{-x^4} \int_0^x e^{t^4} dt + \frac{\pi^2}{6} e^{-x^4}\end{aligned}$$

for any $x > 0$, which shows that $\psi \in L^1((0, +\infty))$. We thus conclude that $\psi \in L^1(\mathbb{R})$. As a byproduct, the function e^p belongs to $L^1(\mathbb{R})$ as well.

Fix a function $f \in C_b(\mathbb{R})$ such that $f \equiv \mathbb{1}$ in $\mathbb{R}_+$ and $\int_{\mathbb{R}} f(t)e^{p(t)} dt = 0$. Then, the function $u : \mathbb{R} \to \mathbb{R}$, defined by

$$u(x) = C_1 - \int_0^x e^{-p(t)} dt \int_t^{+\infty} f(s)e^{p(s)} ds, \qquad x \in \mathbb{R},$$

belongs to $C^2(\mathbb{R})$ and $\mathcal{A}u = f$. Moreover, $|u(x)| \leq |C_1| + \|f\|_\infty \|\psi\|_{L^1(\mathbb{R})}$ for any $x \in \mathbb{R}$ and this implies that $u \in C_b(\mathbb{R})$. Therefore, u belongs to $D_{\max}(\mathcal{A})$. Since

$$|u'(n - \delta_n/2)| = \frac{e^{(n-\delta_n/2)^4}}{\varepsilon_n} \int_{n-\frac{\delta_n}{2}}^{+\infty} h(t)e^{-t^4} dt \geq \frac{e^{(n-1/2)^4}}{\varepsilon_n} \int_n^{n+\frac{1}{2}} e^{-t^4} dt \geq \frac{n}{2},$$

u' is unbounded. It turns out that the function u belongs to $D_{\max}(\mathcal{A})$ but not to $C_b^1(\mathbb{R})$. This means that (5.0.1) with $k = 0$, $l = 1$ fails to hold.

Note that in this case the dissipativity assumption (5.1.3) fails since p'' is unbounded from above. Indeed, let $g = \log(h)$. Since $g(n - \delta_n) = g(n) = 0$ and $g(n - \delta_n/2) = \log(\varepsilon_n)$ for any $n \in \mathbb{N}$, by the mean value theorem, there exist two points $y_n \in (n - \delta_n, n - \delta_n/2)$

and $z_n \in (n-\delta_n/2, n)$ such that $g'(y_n) = -g'(z_n) = 2\delta_n^{-1}\log(\varepsilon_n)$ for any $n \in \mathbb{N}$. Applying again the mean value theorem, it follows that there exists $x_n \in (y_n, z_n)$ such that

$$g''(x_n) = -\frac{4\log(\varepsilon_n)}{\delta_n(z_n - y_n)} \geq -\frac{4\log(\varepsilon_n)}{\delta_n^2} = -\frac{4\log(\varepsilon_n)}{\varepsilon_n^2} n^4 e^{2n^4}, \qquad n \in \mathbb{N}.$$

Since ε_n vanishes as n tends to $+\infty$, for n large enough it holds that $p''(x_n) = -12x_n^2 + g''(x_n) \geq -12n^2 + n^4e^{2n^4}$, which shows that p'' is unbounded from above.

Remark 5.1.13 In [54, 55], the author, starting from the stochastic equation

$$\begin{cases} d\xi_t^x = b(\xi_t^x)dt + \sigma(\xi_t^x)dW_t, & t > 0, \\ \xi_0 \equiv x, \end{cases}$$

and using probabilistic methods, proves the uniform estimates (5.0.1) with $l = 1, 2, 3$ and $f \in B_b(\mathbb{R}^N)$, when $\{T(t)\}$ is the semigroup associated with the stochastic differential equation. More precisely, she assumes the following set of assumptions:

(i) σ_{ij} belongs to $C^3(\mathbb{R}^N)$ and there exists $k \geq 0$ such that for any $j = 0, \ldots, 3$ it holds that

$$\sup_{x\in\mathbb{R}^N} \frac{\|D^\beta\sigma(x)\|}{1+|x|^{k-j}} < +\infty, \qquad |\beta| = j;$$

(ii) b_i belongs to $C^3(\mathbb{R}^N)$ and there exists $m \geq k$ such that for any $j = 0, \ldots, 3$ it holds that

$$\sup_{x\in\mathbb{R}^N} \frac{|D^\beta b(x)|}{1+|x|^{2m+1-j}} < +\infty, \qquad |\beta| = j;$$

there exist $a, \gamma > 0$ and $c \in \mathbb{R}$ such that for any $x, h \in \mathbb{R}^N$ it holds that

$$\langle b(x+h) - b(x), h\rangle \leq -a|h|^{2m+2} + c(1+|x|^\gamma);$$

(iii) for any $p > 0$ there exists a real constant c_p such that

$$\langle b(x) - b(y), x-y\rangle + p\|\sigma(x) - \sigma(y)\|^2 \leq c_p|x-y|^2.$$

Note that also the estimates in Theorem 5.1.7 can be extended to any $f \in B_b(\mathbb{R}^N)$. Indeed, since $\{T(t)\}$ is a semigroup of contractions in $B_b(\mathbb{R}^N)$ and it is also strong Feller (see Remark 1.2.11 and Proposition 1.2.13), we can split $T(t)f = T(t/2)T(t/2)f$ for any $f \in B_b(\mathbb{R}^N)$ and write, for any $\omega > 0$ and some positive constant $C = C(\omega)$,

$$\|D_x^kT(t)f\|_\infty \leq \|D_x^kT(t/2)\|_{L(C_b(\mathbb{R}^N);C_b^k(\mathbb{R}^N))}\|T(t/2)f\|_{C_b(\mathbb{R}^N)} \leq Ce^{\frac{\omega}{2}t}t^{-\frac{k}{2}}\|f\|_\infty \qquad (5.1.40)$$

for any $t > 0$ and $k = 1, 2, 3$. Actually, in view of the results in Section 6.2 (see Remark 6.2.4), we can get rid of the exponential term in (5.1.40).

Remark 5.1.14 Estimates (5.0.1) have been proved also in [182] under assumptions on the coefficients of the operator $\mathcal{A}$ which are comparable to Hypotheses 5.1.1. The main differences are in the method used in [182] to prove such uniform estimates. Indeed, A. Lunardi approximates the coefficients of the operator $\mathcal{A}$ rather than the whole space $\mathbb{R}^N$ by balls, centered at the origin, as in Theorem 5.1.7. Moreover, she does not assume that the potential c identically vanishes in $\mathbb{R}^N$. Here, we assume such an assumption in view of the pointwise estimates of Chapter 6, but the estimates in Theorem 5.1.7 can be extended to the case of not identically vanishing potentials.

5.2 Some consequences

Estimates (5.1.32) can be used to prove optimal Schauder estimates for the elliptic equation

$$\lambda u - \mathcal{A}u = f, \tag{5.2.1}$$

as well as for the Cauchy problem

$$\begin{cases} D_t u(t,x) = \mathcal{A}u(t,x) + g(t,x), & t > 0, \quad x \in \mathbb{R}^N, \\ u(0,x) = u_0(x), & \quad\quad\quad x \in \mathbb{R}^N. \end{cases} \tag{5.2.2}$$

We begin with the elliptic equation and prove the following lemma.

Lemma 5.2.1 *Let $\theta \in (0,3)$ not be an integer, let $I \subset \mathbb{R}$ be an interval, and let $\psi : I \to C_b^\theta(\mathbb{R}^N)$ be such that, for any $x \in \mathbb{R}^N$, the real function $t \mapsto \psi(t)(x)$ is continuous in I and $\|\psi(t)\|_{C_b^\theta(\mathbb{R}^N)} \le c(t)$ for any $t \in I$ and some function $c \in L^1(I)$. Then, the function $f : \mathbb{R}^N \to \mathbb{R}$, defined by*

$$f(x) = \int_I \psi(t)(x)dt, \qquad x \in \mathbb{R}^N, \tag{5.2.3}$$

belongs to $C_b^\theta(\mathbb{R}^N)$ and $\|f\|_{C_b^\theta(\mathbb{R}^N)} \le K\|c\|_{L^1(I)}$ for some positive constant K.

Proof We begin the proof observing that, according to [250, Section 2.7.2], $C_b^\theta(\mathbb{R}^N)$ can be characterized, for any $\theta \in (0,3)$, as the space of functions $g \in C_b(\mathbb{R}^N)$ such that

$$[[g]]_\theta = \sup_{\substack{x,h \in \mathbb{R}^N \\ h \neq 0}} |h|^{-\theta} \left| \sum_{l=0}^{3} (-1)^l f(x+lh) \right| < +\infty$$

and the norm $g \mapsto \|g\|_\infty + [[g]]_\theta$ is equivalent to the classical norm of $C_b^\theta(\mathbb{R}^N)$. Hence, if ψ is as in the statement of the lemma, then for any $x, h \in \mathbb{R}^N$, with $h \neq 0$, we can estimate

$$\left| \sum_{l=0}^{3} (-1)^l \int_I \psi(t)(x+lh)dt \right| \le \int_I \left| \sum_{l=0}^{3} (-1)^l \psi(t)(x+lh) \right| dt \le K|h|^\theta \int_I c(t)dt \tag{5.2.4}$$

for some positive constant K. Estimate (5.2.4) implies that the function f in (5.2.3) belongs to $C_b^\theta(\mathbb{R}^N)$ and the statement follows. □

Now, we observe that, by virtue of Hypothesis 5.1.1(ii), Theorem 3.1.5 implies that, for any $f \in C_b(\mathbb{R}^N)$ and any $\lambda > c_0$, the function $R(\lambda)f$, defined by

$$(R(\lambda)f)(x) = \int_0^{+\infty} e^{-\lambda t}(T(t)f)(x)dt, \qquad x \in \mathbb{R}^N,$$

is the unique solution to the elliptic equation (5.2.1) in $D_{\max}(\mathcal{A})$ (see (1.0.1)). As a consequence, Proposition 1.3.5 implies that $(A, D_{\max}(\mathcal{A})) = (\widehat{A}, D(\widehat{A}))$ with equivalence of the corresponding norms.

With this remark in mind, we can now prove the following theorem, which provides us a partial characterization of $D_{\max}(\mathcal{A})$ and optimal Schauder estimates for the solution of the elliptic equation (5.2.1).

Theorem 5.2.2 *Suppose that Hypotheses 5.1.1(i)-5.1.1(iii), 5.1.1(iv-j) are satisfied (for some $j=1,2$). Then, $D_{\max}(\mathcal{A})$ is contained in $C_b^\theta(\mathbb{R}^N)$ for any $\theta\in(0,1]$, if $j=1$, and for any $\theta\in(0,2)$, if $j=2$. Moreover, for any $\omega>0$ and $\theta\in(0,j)$, and also for $\theta=1$, if $j=1$, there exists a positive constant $C=C(\theta,\omega)$ such that*

$$\|u\|_{C_b^\theta(\mathbb{R}^N)}\le C\|u\|_\infty^{1-\frac{\theta}{2}}\|(\omega-\mathcal{A})u\|_\infty^{\frac{\theta}{2}},\qquad u\in D_{\max}(\mathcal{A}). \tag{5.2.5}$$

Finally, let the previous assumptions be satisfied with $j=3$. Then, for any $f\in C_b^\theta(\mathbb{R}^N)$ ($\theta\in(0,1)$) and any $\lambda>0$, there exist a unique solution $u\in C_b^{2+\theta}(\mathbb{R}^N)$ of the elliptic equation $\lambda u-\mathcal{A}u=f$ and a positive constant $C=C(\theta,\lambda)$ such that

$$\|u\|_{C_b^{2+\theta}(\mathbb{R}^N)}\le C\|f\|_{C_b^\theta(\mathbb{R}^N)}. \tag{5.2.6}$$

Proof To prove the first part we assume that $j=3$, the other case being completely similar and even simpler. Let $u\in D_{\max}(\mathcal{A})$ be a not identically vanishing function and fix $\omega>0$. For any $\lambda>\omega$ set $\varphi=\lambda u-\mathcal{A}u$. Then,

$$u(x)=\int_0^{+\infty}e^{-\lambda t}(T(t)\varphi)(x)dt,\qquad x\in\mathbb{R}^N,$$

so that, according to Lemma 5.2.1 and estimate (5.1.32), f belongs to $C_b^\theta(\mathbb{R}^N)$ and, if $\theta\neq1$, then

$$\|u\|_{C_b^\theta(\mathbb{R}^N)}\le C\frac{\Gamma(1-\theta/2)}{(\lambda-\omega)^{1-\theta/2}}\|\varphi\|_\infty\le C\frac{\Gamma(1-\theta/2)}{(\lambda-\omega)^{1-\theta/2}}[(\lambda-\omega)\|u\|_\infty+\|(\omega-\mathcal{A})u\|_\infty] \tag{5.2.7}$$

for some positive constant C, independent of λ, where Γ denotes the Gamma function. Taking the minimum with respect to $\lambda\in(\omega,+\infty)$ in (5.2.7), estimate (5.2.5) follows.

Now, we prove that, for any $f\in C_b^\theta(\mathbb{R}^N)$ ($\theta\in(0,1)$) and any $\lambda>0$, there exists a (unique) solution $u\in C_b^{2+\theta}(\mathbb{R}^N)$ of equation (5.2.1). So, we assume that Hypothesis 5.1.1(iv-3) is satisfied. By virtue of Theorem 3.1.5, to prove the assertion it suffices to show that the function $u=R(\lambda)f$ has the claimed regularity properties.

We set

$$u(x)=\int_0^{+\infty}e^{-\lambda t}(T(t)\varphi)(x)dt,\qquad x\in\mathbb{R}^N$$

and prove that, for any $\alpha\in(\theta,1)$, u belongs to $(C_b^\alpha(\mathbb{R}^N),C_b^{2+\alpha}(\mathbb{R}^N))_{1-(\alpha-\theta)/2,+\infty}$. This is enough for our aims. Indeed, a well known result in interpolation theory (see Theorem B.3.8) shows that $(C_b^\alpha(\mathbb{R}^N),C_b^{2+\alpha}(\mathbb{R}^N))_{1-(\alpha-\theta)/2,+\infty}=C_b^{2+\theta}(\mathbb{R}^N)$.

In order to prove that $u\in(C_b^\alpha(\mathbb{R}^N),C_b^{2+\alpha}(\mathbb{R}^N))_{1-(\alpha-\theta)/2,+\infty}$, we apply the same arguments as in [179]. For this purpose, we split, for any $\xi>0$, $u=a_\xi+b_\xi$, where

$$a_\xi(x)=\int_0^\xi e^{-\lambda t}(T(t)f)(x)dt,\qquad b_\xi(x)=\int_\xi^{+\infty}e^{-\lambda t}(T(t)f)(x)dt$$

for any $x\in\mathbb{R}^N$. From Lemma 5.2.1 and estimate (5.1.32), where we take $\omega=\lambda/2$, we deduce that

$$\|a_\xi\|_{C_b^\alpha(\mathbb{R}^N)}\le C\|f\|_{C_b^\theta(\mathbb{R}^N)}\int_0^\xi t^{-\frac{\alpha-\theta}{2}}dt=C'\xi^{1-\frac{\alpha-\theta}{2}}\|f\|_{C_b^\theta(\mathbb{R}^N)},$$

$$\|b_\xi\|_{C_b^{2+\alpha}(\mathbb{R}^N)}\le C\|f\|_{C_b^\theta(\mathbb{R}^N)}\int_\xi^{+\infty}t^{-1-\frac{\alpha-\theta}{2}}dt=C''\xi^{-\frac{\alpha-\theta}{2}}\|f\|_{C_b^\theta(\mathbb{R}^N)}$$

for some positive constants C, C', C'', independent of $\xi > 0$. Hence,

$$\xi^{-1+\frac{\alpha-\theta}{2}}(\|a_\xi\|_{C_b^\alpha(\mathbb{R}^N)} + \xi\|b_\xi\|_{C_b^{2+\alpha}(\mathbb{R}^N)}) \le C'\|f\|_{C_b^\theta(\mathbb{R}^N)} + C''\|f\|_{C_b^\theta(\mathbb{R}^N)}, \qquad \xi > 0.$$

This implies that (see Definition B.3.1)

$$\sup_{\xi\in(0,1)} \xi^{-1+\frac{\alpha-\theta}{2}} K(\xi, u) \le (C' + C'')\|\varphi\|_{C_b^\theta(\mathbb{R}^N)},$$

so that $u \in (C_b^\alpha(\mathbb{R}^N), C_b^{2+\alpha}(\mathbb{R}^N))_{1-(\alpha-\theta)/2,\infty}$ and, consequently, (5.2.6) holds true. □

Remark 5.2.3 Actually, the results in Chapter 6 will show that, in some situations, estimate (5.1.8) and, consequently, estimate (5.1.32) holds with an exponential term of negative type (see Remark 6.2.4). In such a situation, the same arguments as in the proof of Theorem 5.2.2 show that we can take $\omega = 0$ in (5.2.5).

As far as the Cauchy problem (5.2.2) is concerned, we can prove two results which provide us with both a (unique) solution u to such a problem and sharp estimates for its spatial derivatives. As in the classical case, we give the following definition.

Definition 5.2.4 *The function $u : [0, +\infty) \times \mathbb{R}^N \to \mathbb{R}$, given by*

$$u(t, x) = (T(t)u_0)(x) + \int_0^t (T(t-s)g(s, \cdot))(x)ds, \qquad t > 0,\ x \in \mathbb{R}^N, \tag{5.2.8}$$

is called mild solution to problem (5.2.2).

Throughout the rest of this section we always assume that Hypotheses 5.1.1(i)-5.1.1(iii) and 5.1.1(iv-3) are satisfied.

Theorem 5.2.5 *Fix $\alpha, \theta \in (0, 1)$, $\theta < \beta \le 2 + \theta$. Moreover, let $u_0 \in C_b(\mathbb{R}^N)$ and let g be a continuous function in $(0, T] \times \mathbb{R}^N$ such that $g(t, \cdot) \in C_b^\beta(\mathbb{R}^N)$ for any $t \in (0, T]$ and $\sup_{0<t\le T} t^\alpha \|g(t, \cdot)\|_{C_b^\beta(\mathbb{R}^N)} < +\infty$. Then, the function u defined in* (5.2.8) *belongs to $C([0, T] \times \mathbb{R}^N) \cap C^{1,2}((0, T] \times \mathbb{R}^N)$. Moreover, u is the unique bounded classical solution to problem* (5.2.2). *Finally, $u(t, \cdot) \in C_b^{2+\theta}(\mathbb{R}^N)$ for any $t \in (0, T]$ and there exists a positive constant $C > 0$, independent of u, such that*

$$\|u(t, \cdot)\|_\infty + t^{1+\frac{\theta}{2}}\|u(t, \cdot)\|_{C_b^{2+\theta}(\mathbb{R}^N)} \le C\bigg(\|u_0\|_\infty + \sup_{0<t\le T} t^\alpha\|g(t, \cdot)\|_{C_b^\beta(\mathbb{R}^N)}\bigg), \qquad t \in [0, T].$$

Proof The uniqueness of the bounded classical solution to problem (5.2.8) is a straightforward consequence of Hypothesis 5.1.1(ii) and Theorem 3.1.3. So, let us prove that the function u in (5.2.8) is actually a bounded classical solution to problem (5.2.2).

Throughout the proof, we denote by C positive constants, independent of s and t, which may vary from line to line.

To begin with, we will deal with the function defined by the integral term in (5.2.8), which we denote by v. Let us show that v is well defined and continuous in $[0, T] \times \mathbb{R}^N$. For this purpose, we observe that the function $(r, s, x) \mapsto (T(r)g(s, \cdot))(x)$ is continuous in $[0, T] \times (0, T] \times \mathbb{R}^N$. Indeed, for any $s \in [0, T]$, the function $(r, x) \mapsto (T(r)g(s, \cdot))(x)$ is continuous in $[0, T] \times \mathbb{R}^N$ and, by virtue of Proposition 1.2.10, for any $r \in [0, T]$ and $x \in \mathbb{R}^N$, the function $s \mapsto (T(r)g(s, \cdot))(x)$ is continuous in $(0, T]$, uniformly with respect to r and x on compact subsets of $\mathbb{R}^N$. Moreover, $|(T(r)g(s, \cdot))(x)| \le \|g(s, \cdot)\|_\infty \le Cs^{-\alpha}$ for any $s \in (0,T]$, any $x \in \mathbb{R}^N$ and some positive constant C. This is enough to conclude that v is continuous in $[0,T] \times \mathbb{R}^N$.

Let us now prove that $v(t,\cdot) \in C_b^{2+\theta}(\mathbb{R}^N)$ for any $t \in (0,T]$. Using estimate (5.1.32), with $k+\alpha = \beta$ and $m = 2$, we get

$$\|T(t-s)g(s)\|_{C_b^{2+\theta}(\mathbb{R}^N)} \leq \frac{C}{s^\alpha (t-s)^{1-(\beta-\theta)/2}} \sup_{0<t\leq T} t^\alpha \|g(t,\cdot)\|_{C_b^\beta(\mathbb{R}^N)}.$$

Since the function $s \mapsto (s^\alpha (t-s)^{1-(\beta-\alpha)/2})^{-1}$ is in $L^1((0,t))$, by Lemma 5.2.1, $v(t,\cdot)$ belongs to $C_b^{2+\theta}(\mathbb{R}^N)$ for any $t \in (0,T]$ and

$$\begin{aligned}\|v(t,\cdot)\|_{C_b^{2+\theta}(\mathbb{R}^N)} \leq & C\left(\int_0^t \frac{1}{s^\alpha(t-s)^{1-(\beta-\theta)/2}}ds\right) \sup_{0<t\leq T} t^\alpha\|g(t,\cdot)\|_{C_b^\beta(\mathbb{R}^N)} \\ = & C' t^{-\alpha+\frac{\beta-\theta}{2}} \sup_{0<t\leq T} t^\alpha \|g(t,\cdot)\|_{C_b^\beta(\mathbb{R}^N)}.\end{aligned}$$

Therefore, the function $t \mapsto v(t,\cdot)$ is bounded in $[\varepsilon,T]$ with values in $C^{2+\theta}(K)$ for any compact set $K \subset \mathbb{R}^N$ and any $\varepsilon \in (0,T)$. Since v is continuous, it belongs to $C([0,T];C(K))$. By Propositions B.3.4 and B.3.6(i), the map v belongs to $C([\varepsilon,T] \times C^2(K))$. Therefore, the first- and second-order derivatives of v are continuous in $[\varepsilon,T]\times K$ and, hence, in $(0,T]\times\mathbb{R}^N$.

Let us now consider the regularity of v with respect to t. For this purpose we observe that, for any $t>0$, any $x \in \mathbb{R}^N$ and $h>0$, sufficiently close to 0, it holds that

$$\begin{aligned}\frac{v(t+h,x)-v(t,x)}{h} = & \frac{1}{h}\int_t^{t+h} (T(t+h-s)g(s,\cdot))(x)ds \\ & + \frac{1}{h}\int_0^t ((T(t+h-s)-T(t-s))g(s,\cdot))(x)ds. \end{aligned} \tag{5.2.9}$$

As is easily seen, the first integral term in the right-hand side of (5.2.9) tends to $g(t,x)$ as h tends to 0^+. As far as the second term is concerned, we observe that, taking Theorem 1.2.1 into account, we can show that, for any $s \in (0,T]$ and $x \in \mathbb{R}^N$, the function $t \mapsto (T(t-s)g(s,\cdot))(x)$ is continuously differentiable in $(s,T]$ and $(D_tT(t-s)g(s,\cdot))(x) = (\mathcal{A}T(t-s)g(s,\cdot))(x)$. Therefore,

$$\begin{aligned}((T(t+h-s)-T(t-s))g(s,\cdot))(x) = & \int_0^h (D_tT(t-s+r)g(s,\cdot))(x)dr \\ = & \int_0^h (\mathcal{A}T(t-s+r)g(s,\cdot))(x)dr.\end{aligned}$$

Now, using (5.1.32), we can show that

$$|(\mathcal{A}T(t+r-s)g(s,\cdot))(x)| \leq \frac{C}{s^\alpha}(t-s)^{(\frac{\beta}{2}-1)\wedge 0} \sup_{0<t\leq T} t^\alpha \|g(t,\cdot)\|_{C_b^\beta(\mathbb{R}^N)}$$

for any $t \in (s,T]$, $x \in B_n$, $r \in (0,h)$ and some positive constant $C = C(n,T)$. Hence,

$$\begin{aligned}&|((T(t+h-s)-T(t-s))g(s,\cdot))(x)| \\ \leq & \frac{C}{s^\alpha}(t-s)^{\min(\frac{\beta}{2}-1,0)}|h| \sup_{0<t\leq T} t^\alpha \|g(t,\cdot)\|_{C_b^\beta(\mathbb{R}^N)}\end{aligned} \tag{5.2.10}$$

for any s,t,h as above. From (5.2.10) and the dominated convergence theorem, we obtain that the second term in the right-hand side of (5.2.9) converges to $\mathcal{A}v(t,x)$ as h tends to 0^+, for any $t>0$ and $x\in\mathbb{R}^N$. Therefore, $v(\cdot,x)$ is differentiable from the right in $(0,T)$ for any $x\in\mathbb{R}^N$ and

$$D_t^+v(t,x)=\int_0^t(\mathcal{A}T(t-s)g(s,\cdot))(x)ds+g(t,x),\qquad t\in(0,T)\times\mathbb{R}^N.$$

Let us now prove that

$$\int_0^t(\mathcal{A}T(t-s)g(s,\cdot))(x)ds=\mathcal{A}\int_0^t(T(t-s)g(s,\cdot))(x)ds \tag{5.2.11}$$

for any $t\in(0,T]$ and $x\in\mathbb{R}^N$. This will imply that $D_t^+v=\mathcal{A}v+g$. But, since we have already shown that the functions $D^\gamma v$ ($|\gamma|=1,2$) are continuous in $(0,T]\times\mathbb{R}^N$, it follows that $\mathcal{A}v$ is continuous in $(0,T]\times\mathbb{R}^N$ as well. Therefore, we will conclude that D_tv exists in $(0,T]\times\mathbb{R}^N$ and it is therein continuous, so that v is differentiable with respect to the time variable in $(0,T]\times\mathbb{R}^N$ and $D_tv=\mathcal{A}v+g$.

We begin by proving (5.2.11). Observe that the function $s\mapsto D_x^\gamma T(t-s)g(s,\cdot)$ belongs to $C((0,T);C(K))\cap L^1((0,T);C(K))$ for any compact set $K\subset\mathbb{R}^N$, any multi-index γ, with $|\gamma|=1,2$, and $t\in(0,T]$. Since the realization of the derivative D_x^γ in $C(K)$ is a closed operator, it follows that

$$D_x^\gamma\bigg(\int_0^t(T(t-s)g(s,\cdot))(\cdot)ds\bigg)(x)=\int_0^t(D_x^\gamma T(t-s)g(s,\cdot))(x)ds$$

for any $t\in(0,T]$ and $x\in K$. Hence, the formula (5.2.11) immediately follows.

To conclude the proof, let us deal with the function $T(\cdot)u_0$. According to Theorem 1.2.1 we already know that such a function belongs to $C^{1,2}((0,+\infty)\times\mathbb{R}^N)\cap C_b([0,+\infty)\times\mathbb{R}^N)$ and $D_tT(t)u_0=\mathcal{A}T(t)u_0$ for any $t\in(0,+\infty)$. We now observe that, by virtue of (5.1.32), the function $t\mapsto t^{1+\theta/2}T(t)u_0$ is bounded with values in $C_b^{2+\theta}(\mathbb{R}^N)$. Hence, the function $u=T(\cdot)u_0+v$ belongs to $C^{1,2}((0,T]\times\mathbb{R}^N)$, it satisfies (5.2.2), and

$$\sup_{0<t\le T}t^{1+\frac{\theta}{2}}\|u(t,\cdot)\|_{C_b^{2+\theta}(\mathbb{R}^N)}\le C\bigg(\|u_0\|_\infty+\sup_{0<t\le T}t^\alpha\|g(t,\cdot)\|_{C_b^\beta(\mathbb{R}^N)}\bigg).$$

If, in addition, g is bounded with values in $C_b^{2+\theta}(\mathbb{R}^N)$, then $v(t,\cdot)$ is bounded up to $t=0$ with values in $C_b^{2+\theta}(\mathbb{R}^N)$ since, by (5.1.32),

$$\|T(t-s)\|_{L(C_b^\beta(\mathbb{R}^N);C_b^{2+\theta}(\mathbb{R}^N))}\le C(t-s)^{-1+\frac{\beta-\theta}{2}},\qquad 0<s<t\le T,$$

and the right-hand side of the previous inequality defines a function (of the variable s) which belongs to $L^1((0,t))$. Once more taking advantage of (5.1.32), we obtain that also the function $T(\cdot)u_0$ is bounded with values in $C_b^{2+\theta}(\mathbb{R}^N)$. Arguing as above, we conclude that $u\in C^{1,2}([0,T]\times\mathbb{R}^N)$. □

Theorem 5.2.6 *Let $T>0$, $\theta\in(0,1)$ and let $g:[0,T]\times\mathbb{R}^N\to\mathbb{R}$ be a bounded and continuous function such that $g(t,\cdot)\in C_b^\theta(\mathbb{R}^N)$ for any t and $\sup_{t\in[0,T]}\|g(t,\cdot)\|_{C_b^\theta(\mathbb{R}^N)}<+\infty$. Moreover, let u_0 belong to $C_b^{2+\theta}(\mathbb{R}^N)$. Then, the function u in (5.2.8) is the unique strict solution to problem (5.2.2) (i.e., D_tu, D_xu, $D_{xx}u\in C_b([0,T]\times\mathbb{R}^N)$ and u solves (5.2.2)). Moreover, there exists a positive constant $C=C_T$ such that*

$$\sup_{t\in[0,T]}\|u(t,\cdot)\|_{C_b^{2+\theta}(\mathbb{R}^N)}\le C\bigg(\|u_0\|_{C_b^{2+\theta}(\mathbb{R}^N)}+\sup_{t\in[0,T]}\|g(t,\cdot)\|_{C_b^\theta(\mathbb{R}^N)}\bigg). \tag{5.2.12}$$

Proof From Theorem 5.2.5 we know that, under our assumptions, the function u defined by (5.2.8) is the unique classical solution to problem (5.2.2) belonging to $C^{1,2}([0,T]\times\mathbb{R}^N)$. For notational convenience, throughout the rest of the proof, we denote by v the function defined by the integral term in (5.2.8), so that we can write $u(t,\cdot)=T(t)u_0+v(t,\cdot)$ for any $t\in[0,T]$.

Let us prove that u is bounded with values in $C^{2+\theta}_b(\mathbb{R}^N)$ and it satisfies (5.2.12). For this purpose, we begin by observing that, due to (5.1.32), we already know that $T(t)u_0\in C^{2+\theta}_b(\mathbb{R}^N)$ for any $t\geq 0$ and

$$\|T(t)\|_{C^{2+\theta}_b(\mathbb{R}^N)}\leq C_1\|u_0\|_{C^{2+\theta}_b(\mathbb{R}^N)} \tag{5.2.13}$$

for some positive constant C_1 and any $t\in[0,T]$. To show that v is bounded with values in $C^{2+\theta}_b(\mathbb{R}^N)$ we adapt the techniques used in the proof of Theorem 5.2.2, i.e., we show that $v(t,\cdot)$ belongs to $(C^\alpha_b(\mathbb{R}^N),C^{2+\alpha}_b(\mathbb{R}^N))_{1-(\alpha-\theta)/2,\infty}$ for any $t\in[0,T]$ and $\alpha\in(\theta,1)$, recalling that $(C^\alpha_b(\mathbb{R}^N),C^{2+\alpha}_b(\mathbb{R}^N))_{1-(\alpha-\theta)/2,+\infty}=C^{2+\theta}_b(\mathbb{R}^N)$ with equivalence of the corresponding norms (see Theorem B.3.8). For this purpose, we split $v(t,\cdot)=a_\xi(t,\cdot)+b_\xi(t,\cdot)$, where, for any $\xi\in(0,1)$, we set

$$a_\xi(t,x)=\begin{cases}\displaystyle\int_{t-\xi}^t (T(t-s)g(s,\cdot))(x)ds, & \text{if } \xi\leq t,\\[2ex] \displaystyle\int_0^t (T(t-s)g(s,\cdot))(x)ds, & \text{if } \xi>t,\end{cases}$$

for any $x\in\mathbb{R}^N$. Taking Lemma 5.2.1 and estimate (5.1.32) into account, we deduce that, for any $\alpha\in(\theta,1)$ and $t\in[0,T]$, $a_\xi(t,\cdot)$ and $b_\xi(t,\cdot)$ belong, respectively to $\in C^\alpha_b(\mathbb{R}^N)$ and $C^{2+\alpha}_b(\mathbb{R}^N)$. Moreover,

$$\begin{aligned}\|a_\xi(t,\cdot)\|_{C^\alpha_b(\mathbb{R}^N)}\leq& C_2\sup_{0\leq s\leq t}\|g(s,\cdot)\|_{C^\theta_b(\mathbb{R}^N)}\sup_{0\leq s\leq t}\int_{(t-\xi)^+}^t (t-s)^{-\frac{\alpha-\theta}{2}}ds\\ \leq& C_3\xi^{1-\frac{\alpha-\theta}{2}}\sup_{0\leq s\leq t}\|g(s,\cdot)\|_{C^\theta_b(\mathbb{R}^N)}\end{aligned}$$

and, similarly,

$$\|b_\xi(t,\cdot)\|_{C^{2+\alpha}_b(\mathbb{R}^N)}\leq C_4\xi^{-\frac{\alpha-\theta}{2}}\sup_{0\leq s\leq t}\|g(s,\cdot)\|_{C^\theta_b(\mathbb{R}^N)}$$

for some positive constants C_2, C_3 and C_4, independent of ε, so that

$$\begin{aligned}\xi^{-1+\frac{\alpha-\theta}{2}}K(\xi,v(t,\cdot))\leq&\xi^{-1+\frac{\alpha-\theta}{2}}\|a_\xi(t,\cdot)\|_{C^\alpha_b(\mathbb{R}^N)}+\xi^{\frac{\alpha-\theta}{2}}\|b_\xi(t,\cdot)\|_{C^\alpha_b(\mathbb{R}^N)}\\ \leq&(C_3+C_4)\sup_{0\leq s\leq T}\|g(s,\cdot)\|_{C^\theta_b(\mathbb{R}^N)}\end{aligned}$$

for any $\xi\in(0,1)$, implying that $v(t,\cdot)$ belongs to $(C^\alpha_b(\mathbb{R}^N),C^{2+\alpha}_b(\mathbb{R}^N))_{1-(\alpha-\theta)/2,\infty}$ for any $t\in(0,T)$ and that

$$\sup_{0\leq t\leq T}\|v(t,\cdot)\|_{C^{2+\theta}_b(\mathbb{R}^N)}\leq C_5\sup_{0\leq t\leq T}\|g(t,\cdot)\|_{C^\theta_b(\mathbb{R}^N)} \tag{5.2.14}$$

for some positive constant C_5. Summing up, we have proved that $u(t,\cdot)\in C^{2+\theta}_b(\mathbb{R}^N)$ for any $t\in[0,T]$. Moreover, combining (5.2.13) and (5.2.14), estimate (5.2.12) follows at once. □

Chapter 6

Pointwise estimates for the derivatives of the function $T(t)f$

6.0 Introduction

This chapter is devoted to proving some pointwise estimates for the first-, second- and third-order derivatives of the function $T(t)f$, when $f \in C_b^k(\mathbb{R}^N)$ ($k = 0, 1, 2, 3$). Under the same assumptions as in Chapter 5, we prove that, for any $k = 1, 2, 3$ and $p \in (1, +\infty)$, there exists a constant $M_{k,p} > 0$ such that

$$\left(\sum_{|\alpha|\le k} |D_x^\alpha T(t)f|^2\right)^{\frac{p}{2}} \le M_{k,p} T(t)\left(\sum_{|\alpha|\le k} |D^\alpha f|^2\right)^{\frac{p}{2}} \tag{6.0.1}$$

in $\mathbb{R}^N$, for any $t > 0$ and $f \in C_b^k(\mathbb{R}^N)$. Under somewhat heavier assumptions, we show that

$$\left(\sum_{|\alpha|\le k} |D_x^\alpha T(t)f|^2\right)^{\frac{p}{2}} \le \widehat{M}_{k,p} e^{\omega_{k,p} t} T(t)\left(\sum_{|\alpha|\le k} |D^\alpha f|^2\right)^{\frac{p}{2}} \tag{6.0.2}$$

in $\mathbb{R}^N$, for any $t > 0$, $p > p_0$, where p_0 is given by Hypothesis 6.1.3(*ii*-1) and $\widehat{M}_{k,p}$ and $\omega_{k,p}$ are, respectively, a positive and a negative constant.

Estimates (6.0.1) and (6.0.2) are then used to prove the much more precise estimate

$$|D_x^k T(t)f|^p \le \left(\frac{\psi_{k,p\wedge 2}(t)}{1 - e^{-\widehat{\omega}_{k,p\wedge 2} t}}\right)^{1\vee\frac{p}{2}} T(t)\left(\sum_{|\alpha|\le k-1} |D^\alpha f|^2\right)^{\frac{p}{2}} \tag{6.0.3}$$

in $\mathbb{R}^N$ for any $t > 0$, $f \in C_b^{k-1}(\mathbb{R}^N)$ and $p > \widehat{p}$ (which, depending on the assumptions, is equal or greater than 1), where $\widehat{\omega}_{k,r}$ is a nonpositive constant[1], $\psi_{k,r} \in C([0, +\infty))$ is a suitable function which behaves as $t^{1-r/2}$ near 0 and the term in the first brackets in the right-hand side of (6.0.3) stays bounded at infinity, or it decreases to 0 exponentially. Taking the semigroup property into account, from (6.0.3) we get

$$|D_x^k T(t)f|^p \le C_{k,p} t^{-pk/2} e^{\overline{\omega}_{k,p} t} T(t)|f|^p \tag{6.0.4}$$

in $\mathbb{R}^N$, for any $t > 0$, $f \in C_b(\mathbb{R}^N)$, $p > 1$ and some constant $C_{k,p} > 0$, blowing up as p tends to 1^+, and $\overline{\omega}_{k,p} \le 0$. This is an improvement of estimate (5.1.8) since the right-hand side of (6.0.4) stays bounded or decreases exponentially as t tends to $+\infty$.

In the particular case when $\mathcal{A} = \Delta + \langle b, \nabla\rangle$, we prove estimate (6.0.1) also for $p = 1$.

[1] When $\widehat{\omega}_{k,p\wedge 2} = 0$, we agree that $(1 - e^{-\widehat{\omega}_{k,p\wedge 2} t})/\widehat{\omega}_{k,p\wedge 2} = t$.

Such pointwise estimates are typical for transition semigroups of Markov processes and they have been studied in [26, 29, 31, 34]. Here, we present the results of [34].

We stress that estimate (6.0.4) can not be extended, in general, to the case when $p = 1$. Counterexamples are easily obtained in the simple case $\mathcal{A} = \Delta$ (see Example 6.3.3).

In the case when $\omega_{1,p} \le 0$, estimate (6.0.4) with $k = 1$ allows us to obtain a Liouville type theorem, namely, in such a situation it can be proved that, if $\mathcal{A}u = 0$, then u is constant. When $\omega_{1,p} > 0$, in general, such a result fails. Counterexamples are given in [220] also in the one-dimensional case.

6.1 The first type of pointwise gradient estimates

We begin by considering the following lemma which will be essential to prove the first type of pointwise estimates.

Proposition 6.1.1 *Let $k \in \{1, 2, 3\}$ and let Hypotheses 5.1.1(i)-5.1.1(iii) and 5.1.1(iv-k) be satisfied. Then, for any $f \in C_b^k(\mathbb{R}^N)$, the function $D_x^k T(\cdot)f$ is continuous in $[0, +\infty) \times \mathbb{R}^N$.*

Proof Fix $f \in C_b^k(\mathbb{R}^N)$ and set $u = T(\cdot)f$. The regularity of u for $t > 0$ is a classical result, recalled in Theorem D.1.5. Thus, we have only to prove the regularity at $t = 0$. We will do it using a localization argument.

Fix $x_0 \in \mathbb{R}^N$ and let Ω be a smooth bounded neighborhood of x_0. Moreover, let $\vartheta \in C_c^\infty(\Omega)$ be such that $\vartheta \equiv 1$ in a smaller neighborhood $\Omega_0 \subset \Omega$ of x_0. The function $v = u\vartheta$ vanishes on $(0, +\infty) \times \partial\Omega$ and satisfies the equation $D_t v - \mathcal{A}v = \psi$ in $(0, +\infty) \times \Omega$, where $\psi = -u\mathcal{A}\vartheta - 2\langle Q\nabla_x u, \nabla\vartheta\rangle$. Moreover, it is readily seen that there exists a constant $C > 0$ such that $\|\psi(t, \cdot)\|_\infty \le C\|u(t, \cdot)\|_{C^1(\overline{\Omega})} \le CC_{0,1}e^T t^{-1/2}\|f\|_\infty$ for any $t \in (0, T]$, where we have used estimate (5.0.1) with $(k, l) = (0, 1)$. In particular, the function $t \mapsto \|\psi(t, \cdot)\|_\infty$ belongs to $L^1((0, T))$ for any $T > 0$ and, therefore, we can write

$$v(t, x) = (T_\Omega(t)(\vartheta f))(x) + \int_0^t (T_\Omega(t - s)\psi(s, \cdot))(x)ds, \qquad t > 0, \;\; x \in \Omega,$$

where $\{T_\Omega(t)\}$ is the semigroup associated with the realization of the operator $\mathcal{A}$ in $C(\overline{\Omega})$ with homogeneous Dirichlet boundary conditions on $\partial\Omega$.

Using the classical gradient estimates for $\{T_\Omega(t)\}$ (see Proposition D.3.1) and estimate (5.0.1), we get

$$\begin{aligned}\|D_x^k T_\Omega(t - s)\psi(s, \cdot)\|_\infty \le& \frac{C_1}{\sqrt{t - s}}\|\psi(s, \cdot)\|_{C^{k-1}(\overline{\Omega})} \le \frac{C_2}{\sqrt{t - s}}\|u(s, \cdot)\|_{C^k(\overline{\Omega})}\\ \le& C_3\frac{C_T}{\sqrt{t - s}}\|f\|_{C_b^k(\mathbb{R}^N)}\end{aligned}$$

for any $0 < s < t \le T$, where C_1, C_2 and C_3 are positive constants, independent of t. We thus deduce that

$$|D_x^k v(t, x) - (D_x^k T_\Omega(t)(\vartheta f))(x)| = \left|\int_0^t (D_x^k T_\Omega(t - s)\psi(s, \cdot))(x)ds\right| \le 2C_3 t\|f\|_{C_b^k(\mathbb{R}^N)}$$

for any $t \in (0, T]$ and $x \in \Omega$. This implies that the function $D_x^k v$ is continuous in $[0, T] \times \Omega_0$ since, by virtue of Lemma 5.1.6, $D_x^k T_\Omega(t)(\vartheta f)$ converges to $D^k(\vartheta f)$ uniformly in $\Omega_\delta := \{x \in$

$\Omega : \mathrm{dist}(x, \partial\Omega) \geq \delta\}$, for any $\delta > 0$, as t tends to 0^+. Since $v \equiv u$ in a neighborhood of x_0, $D^k_x u$ is continuous at $(0, x_0)$. □

Now, we can prove the following theorem.

Theorem 6.1.2 *Fix $k \in \{1,2,3\}$ and let Hypotheses* 5.1.1(*i*)-5.1.1(*iii*) *and* 5.1.1(*iv-k*) *be satisfied. Then, for any $f \in C^k_b(\mathbb{R}^N)$ and $p \in (1,+\infty)$, there exists a positive constant $M_{k,p}$ such that, for any $t > 0$,*

$$\left(\sum_{|\alpha|\leq k} |D^\alpha_x T(t)f|^2\right)^{\frac{p}{2}} \leq M_{k,p} T(t)\left(\sum_{|\alpha|\leq k} |D^\alpha f|^2\right)^{\frac{p}{2}} \qquad \text{in } \mathbb{R}^N. \tag{6.1.1}$$

Proof We begin with the case when $k = 3$ and $p \in (1,2]$. For any $\delta > 0$ we introduce the function $v_\delta : [0,+\infty) \times \mathbb{R}^N \to \mathbb{R}$, defined by

$$v_\delta = (\alpha|u|^2 + \beta|\nabla_x u|^2 + \gamma|D^2_x u|^2 + |D^3_x u|^2 + \delta)^{\frac{p}{2}}, \tag{6.1.2}$$

where α, β, γ are positive constants to be fixed later on, and $u = T(\cdot)f$. By Theorems 1.2.1, D.1.5(ii) and Proposition 6.1.1, v_δ belongs to $C_b([0,+\infty) \times \mathbb{R}^N) \cap C^{1,2}((0,+\infty) \times \mathbb{R}^N)$ and it solves the Cauchy problem

$$\begin{cases} D_t v_\delta(t,\cdot) = \mathcal{A} v_\delta(t,\cdot) + g_\delta(t,\cdot), & t > 0, \\ v_\delta(0,\cdot) = \left(\alpha|f|^2 + \beta|\nabla f|^2 + \gamma|D^2 f|^2 + |D^3 f|^2 + \delta\right)^{\frac{p}{2}}, \end{cases} \tag{6.1.3}$$

where $g_\delta = p v_\delta^{1-2/p} g_{1,\delta} + p(2-p) v_\delta^{1-4/p} g_{2,\delta}$,

$$\begin{aligned} g_{1,\delta} =& -\alpha\mathfrak{q}_0(u) - \beta\sum_{i=1}^N \mathfrak{q}_0(D_i u) - \gamma\sum_{i,j=1}^N \mathfrak{q}_0(D_{ij}u) - \sum_{i,j,h=1}^N \mathfrak{q}_0(D_{ijh}u) + \beta\mathfrak{q}_1(u) \\ &+ 2\gamma\sum_{i=1}^N(\mathfrak{q}_1(D_i u) + \mathfrak{b}_1(D_i u)) + 3\sum_{i,j=1}^N(\mathfrak{q}_1(D_{ij}u) + \mathfrak{b}_1(D_{ij}u)) + \gamma\mathfrak{q}_2(u) \\ &+ 3\sum_{i=1}^N(\mathfrak{q}_2(D_i u) + \mathfrak{b}_2(D_i u)) + \beta\mathfrak{b}_1(u) + \mathfrak{b}_2(u) + \mathfrak{q}_3(u) + \mathfrak{b}_3(u), \end{aligned}$$

with $\mathfrak{q}_i$ $(i = 0,\dots,3)$ and $\mathfrak{b}_j$ $(j = 1,2,3)$ being defined in (5.1.9)-(5.1.12), and $g_{2,\delta} = |\sqrt{Q}\mathfrak{h}(u)|^2$ with $(\mathfrak{h}(u))_i = \alpha u D_i u + \beta(D^2_x u \nabla_x u)_i + \gamma \mathrm{Tr}(D^2_x u D^2_x D_i u) + \sum_{j=1}^N \mathrm{Tr}(D^2_x D_h u D^2_x D_{ij} u)$ for any $i = 1,\dots,N$.

Applying the Cauchy-Schwarz inequality twice (first to the inner product $(\xi,\eta) \mapsto \langle Q(x)\xi,\eta\rangle$ and then to the Euclidean one) we deduce that

$$\begin{aligned} &\sum_{i,j=1}^N q_{ij} \sum_{|\alpha|=h} D^\alpha_x u D_i D^\alpha_x u \sum_{|\beta|=k} D^\beta_x u D_j D^\beta_x u \\ \leq& \sum_{|\alpha|=h} |D^\alpha_x u||\sqrt{Q}\nabla_x D^\alpha_x u| \sum_{|\beta|=k} |D^\beta_x u||\sqrt{Q}\nabla_x D^\beta_x u| \\ \leq& |D^h_x u||D^k_x u|\sqrt{\mathfrak{q}_0(D^\alpha_x u)}\sqrt{\mathfrak{q}_0(D^\beta_x u)} \end{aligned} \tag{6.1.4}$$

for any $h, k \in \{0,1,2,3\}$. This estimate can be used in order to get

$$|\sqrt{Q}\mathfrak{h}(u)|^2 \leq \Bigg[\alpha|u|\sqrt{\mathfrak{q}_0(u)} + \beta|\nabla_x u|\left(\sum_{i=1}^N \mathfrak{q}_0(D_i u)\right)^{\frac{1}{2}}$$

$$+\gamma|D^2_x u|\bigg(\sum_{i,j=1}^N \mathfrak{q}_0(D_{ij}u)\bigg)^{\frac{1}{2}}+|D^3_x u|\bigg(\sum_{i,j,h=1}^N \mathfrak{q}_0(D_{ijh}u)\bigg)^{\frac{1}{2}}\Bigg]^2$$

$$\le w^{\frac{2}{p}}\bigg(\alpha\mathfrak{q}_0(u)+\beta\sum_{i=1}^N \mathfrak{q}_0(D_iu)+\gamma\sum_{i,j=1}^N \mathfrak{q}_0(D_{ij}u)+\sum_{i,j,h=1}^N \mathfrak{q}_0(D_{ijh}u)\bigg).$$

It thus follows that

$$\begin{aligned}g_\delta \le p\bigg\{&(1-p)\bigg(\alpha\mathfrak{q}_0(u)+\beta\sum_{i=1}^N \mathfrak{q}_0(D_iu)+\gamma\sum_{i,j=1}^N \mathfrak{q}_0(D_{ij}u)+\sum_{i,j,h=1}^N \mathfrak{q}_0(D_{ijh}u)\bigg)+\beta\mathfrak{q}_1(u)\\&+2\gamma\sum_{i=1}^N \mathfrak{q}_1(D_iu)+3\sum_{i,j=1}^N \mathfrak{q}_1(D_{ij}u)+\beta\mathfrak{b}_1(u)+2\gamma\sum_{i=1}^N \mathfrak{b}_1(D_iu)+3\sum_{i,j=1}^N \mathfrak{b}_1(D_{ij}u)\\&+\gamma\mathfrak{q}_2(u)+3\sum_{i=1}^N \mathfrak{q}_2(D_iu)+\gamma\mathfrak{b}_2(u)+3\sum_{i=1}^N \mathfrak{b}_2(D_iu)+\mathfrak{q}_3(u)+\mathfrak{b}_3(u)\bigg\}v_\delta^{1-\frac{2}{p}}.\end{aligned} \tag{6.1.5}$$

We now use (5.1.25), with $\varepsilon=(p-1)(3CN)^{-1}$, and the first part of (5.1.26), to estimate the terms in (6.1.5) which contain $\mathfrak{q}_j(\cdot)$ $(j=1,2,3)$. Then, we use the last part of (5.1.26) to estimate $\mathfrak{b}_2(u)$ (taking $\varepsilon=\gamma^{-1}$) and the terms $\mathfrak{b}_2(D_iu)$, $\mathfrak{b}_3(u)$ (taking $\varepsilon=3L_2(4N)^{-1}$, where L_2 is defined in Hypothesis 5.1.1(iv-3)). Finally, we use Hypothesis 5.1.1(i) to estimate from below the terms depending on $\mathfrak{q}_0(\cdot)$. Putting everything together, and choosing β, γ such that $N^4/(3L_2)+N^2\gamma^2/4<L_2\beta$ and $N(1+N^2/L_2)\le 2L_2\gamma$, from (6.1.5) we get

$$\begin{aligned}g_\delta \le\, & p\bigg\{\bigg((1-p)\alpha\kappa+\frac{3C^2N^3}{4(p-1)}\beta\kappa+\beta(d+L_2r)\bigg)|\nabla_x u|^2\\&+\bigg[\bigg(\frac{2-2p}{3}\beta+\frac{3C^2N^3}{4(p-1)}(2\gamma+N^2)+\gamma K_1\bigg)\kappa+2\gamma(d+L_2r)\bigg]|D^2_xu|^2\\&+\bigg[\bigg(\frac{p-1}{3}(3-\gamma)+\frac{9C^2N^3}{4(p-1)}+3K_1\bigg)\kappa+3(d+L_2r)\bigg]|D^3_xu|^2\bigg\}v_\delta^{1-\frac{2}{p}}\\ \le\, & p\bigg\{\bigg((1-p)\alpha+\frac{3C^2N^3}{4(p-1)}\beta+\beta L_3\bigg)\kappa|\nabla_x u|^2\\&+\bigg(\frac{2-2p}{3}\beta+\frac{3C^2N^3}{4(p-1)}(2\gamma+N)+\gamma K_1+2\gamma L_3\bigg)\kappa|D^2_xu|^2\\&+\bigg(\frac{p-1}{3}(3-\gamma)+\frac{9C^2N^3}{4(p-1)}+3(K_1+L_3)\bigg)\kappa|D^3_xu|^2\bigg\}v_\delta^{1-\frac{2}{p}},\end{aligned} \tag{6.1.6}$$

where we used Hypothesis 5.1.1(iv-3), which shows that $d+L_2r\le L_3\kappa$, to obtain the last side of (6.1.6). Hence, up to taking larger β and γ and fixing α suitably large, we can make the right-hand side of (6.1.6) non positive in $(0,+\infty)\times\mathbb{R}^N$. Theorem 3.1.3 now implies that $v_\delta\le T(\cdot)(\alpha|f|^2+\beta|\nabla f|^2+\gamma|D^2f|^2+|D^3f|^2+\delta)$ in $(0,+\infty)\times\mathbb{R}^N$ for any $\delta>0$. Taking the limit as δ tends to 0^+, from Proposition 1.2.10 estimate (6.1.1) follows.

Finally, we consider the case when $p>2$. Since

$$\bigg(\sum_{|\alpha|\le 3}|D^\alpha_x T(t)f|^2\bigg)^{\frac{p}{2}}\le M_{3,2}^{\frac{p}{2}}\bigg(T(t)\bigg(\sum_{|\alpha|\le 3}|D^\alpha_x f|^2\bigg)\bigg)^{\frac{p}{2}},$$

we get (6.1.1) observing that, since $T(t)\psi$ is given by (1.2.7) and the measures $p(t,x;dy)$ in (1.2.8) are probability measures for any $t>0$ and $x\in\mathbb{R}^N$ (since $c\equiv 0$), the Jensen inequality implies that $(T(t)\psi)^{p/2}\le T(t)(\psi^{p/2})$ for any $t>0$ and any nonnegative function $\psi\in C_b(\mathbb{R}^N)$.

To prove (6.1.1) in the case when $p\in(1,2)$ and $k=1,2$, one can apply the previous arguments to the function $v_\delta=(\alpha u^2+|\nabla_x u|^2+\delta)^{p/2}$, if $k=1$, and to the function $v_\delta=(\alpha u^2+\beta|\nabla_x u|^2+|D_x^2 u|^2+\delta)^{p/2}$, if $k=2$, where α and β are positive constants to be properly fixed. Actually, one gets

$$\left(\sum_{|\alpha|\le k}|D_x^\alpha T(t)f|^2+\delta\right)^{\frac{p}{2}}\le M_{k,p}T(t)\left(\sum_{|\alpha|\le k}|D^\alpha f|^2+\delta\right)^{\frac{p}{2}} \tag{6.1.7}$$

in $\mathbb{R}^N$, for any $t>0$ and $k=1,2$.

In the case when $p>2$, estimate (6.1.1) then follows from the case $p=2$, applying the Jensen inequality. □

Now, we show that, under more restrictive assumptions on the coefficients of the operator $\mathcal{A}$, the estimates in Theorem 6.1.2 can be improved.

Hypotheses 6.1.3 (i) Hypotheses 5.1.1(i)-(iii) are satisfied.

Moreover, we always assume that one of the following hypotheses holds true:

(ii-1) $q_{ij},b_j\in C^{1+\delta}_{\rm loc}(\mathbb{R}^N)$ for some $\delta\in(0,1)$ and any $i,j=1,\dots,N$, and there exist $\sigma\in(0,1)$ and a constant $C>0$ such that $|D^\beta q_{ij}|\le C\kappa^\sigma$ in $\mathbb{R}^N$ for any $|\beta|=1$ and $i,j=1,\dots,N$. Moreover, there exist two constants $L>0$, $p_0\in(1,2]$ and a function $d:\mathbb{R}^N\to\mathbb{R}$ such that

$$\langle {\rm Jac}\, b(x)\xi,\xi\rangle\le d(x)|\xi|^2,\qquad x,\xi\in\mathbb{R}^N, \tag{6.1.8}$$

and

$$0>C_1(p_0):=\sup_{\mathbb{R}^N}\left(\frac{C^2N^3}{4(p_0-1)\kappa_0^{1-\sigma}}\kappa^\sigma+d\right);$$

(ii-2) $q_{ij},b_j\in C^{2+\delta}_{\rm loc}(\mathbb{R}^N)$ for some $\delta\in(0,1)$, $i,j=1,\dots,N$ and there exist $\sigma\in(0,1)$, a constant $C>0$ and a positive function $r:\mathbb{R}^N\to\mathbb{R}$ such that $|D^2b_i|\le r$ and $|D^\beta q_{ij}|\le C\kappa^\sigma$ in $\mathbb{R}^N$, for any $|\beta|=1,2$ and $i,j=1,\dots,N$. Moreover, the condition (6.1.8) is satisfied and there exist two constants $L>0$ and $p_0\in(1,2]$ such that

$$0>C_2(p_0):=\sup_{\mathbb{R}^N}\left(d+Lr+\frac{C^2N^3}{4(p_0-1)}\kappa_0^{\sigma-1}\kappa^\sigma\right). \tag{6.1.9}$$

Finally, there exists a constant $K_1\in\mathbb{R}$ such that

$$\sum_{i,j,h,k=1}^N D_{hk}q_{ij}(x)m_{ij}m_{hk}\le K_1(\kappa(x))^\sigma\sum_{h,k=1}^N m_{hk}^2$$

for any symmetric matrix $M=(m_{hk})_{h,k=1}^N$ and any $x\in\mathbb{R}^N$;

(ii-3) $q_{ij},b_j\in C^{3+\delta}_{\rm loc}(\mathbb{R}^N)$ for some $\delta\in(0,1)$ and any $i,j=1,\dots,N$; Hypothesis 6.1.3(ii-2) holds true. Moreover, $|D^\beta b_i|\le r$ and $|D^\beta q_{ij}|\le C\kappa^\sigma$ in $\mathbb{R}^N$, for any $i,j=1,\dots,N$ and $|\beta|=3$.

Remark 6.1.4 Hypotheses 6.1.3(ii-k) ($k = 1, 2, 3$) allow us to consider the case when the coefficients are of polynomial type.

We are now able to improve the result in Theorem 6.1.2.

Theorem 6.1.5 *Let $k = 1, 2, 3$ and let Hypotheses 6.1.3(i), 6.1.3(ii-k) be satisfied. Then, there exist positive constants $\widehat{M}_{k,p}$ and $\omega_{k,p}$ such that*

$$\left(\sum_{1\le|\alpha|\le k}|D^\alpha_x T(t)f|^2\right)^{\frac{p}{2}} \le \widehat{M}_{k,p}e^{\omega_{k,p}t}T(t)\left(\sum_{1\le|\alpha|\le k}|D^\alpha f|^2\right)^{\frac{p}{2}} \tag{6.1.10}$$

in $\mathbb{R}^N$, for any $t > 0$, $f \in C^k_b(\mathbb{R}^N)$ and $p \in (p_0, +\infty)$.

Proof The proof is close to that of Theorem 6.1.2. Hence, we just point out the main differences. To get (6.1.10), in the case when $p \in (p_0, 2]$ and $k = 3$, one has to deal with the function v_δ defined in (6.1.2) with $\alpha = 0$ and β, γ to be determined. This function solves the Cauchy problem (6.1.3) with g_δ satisfying (6.1.5) (with $\alpha = 0$ everywhere).

Taking Hypothesis 6.1.3(ii-3) into account and using the same techniques as in the proof of Theorem 6.1.2 to estimate the right-hand side of (6.1.5), we get

$$\begin{aligned} g_\delta \le p\Bigg\{&\left[\left(C\frac{N^2}{4\varepsilon_2}\kappa^\sigma + d\right)\beta + r\frac{N^2}{4\varepsilon_1}(N+\gamma^2)\right]|\nabla_x u|^2\\ &+\left[((1-p)\kappa_0^{1-\sigma} + C\varepsilon_2 N)\beta\kappa^\sigma + \left(\frac{CN^2(2\gamma+N)}{4\varepsilon_3} + K_1\gamma\right)\kappa^\sigma\right.\\ &\qquad\left.+2d\gamma + rN\left(\varepsilon_1 + \frac{3N}{4\varepsilon_1}\right)\right]|D^2_x u|^2\\ &+\left[((1-p)\kappa_0^{1-\sigma} + 2C\varepsilon_3 N)\gamma\kappa^\sigma + \left(C\frac{3N^2}{4\varepsilon_3} + \varepsilon_3 CN^2 + 3K_1\right)\kappa^\sigma\right.\\ &\qquad\left.+3d + 4r\varepsilon_1 N\right]|D^3_x u|^2 + [(1-p)\kappa_0^{1-\sigma} + 3C\varepsilon_3 N]\kappa^\sigma|D^4_x u|^2\Bigg\}v_\delta^{1-\frac{2}{p}}. \end{aligned} \tag{6.1.11}$$

We now fix $\tau = (p - p_0)/(2p - 2)$ and choose $\varepsilon_2 = (p-1)(1-\tau)\kappa_0^{1-\sigma}/(CN)$, $\varepsilon_3 = (p-1)\kappa_0^{1-\sigma}/(3CN)$ and ε_1 in order to minimize the function $h : (0, +\infty) \to \mathbb{R}$ defined by $h(x) = N\max\{4x/3, (4x^2 + 3N)/(8\gamma x)\}$. We obtain $\varepsilon_1 = 3\sqrt{N}/(2\sqrt{8\gamma-3})$, if $\gamma \ge 1/2$. With these choices of ε_1, ε_2 and ε_3, from (6.1.11) we conclude that

$$\begin{aligned} g_\delta \le p\Bigg\{&\beta\left(\frac{C^2N^3}{4(p-1)(1-\tau)}\kappa_0^{\sigma-1}\kappa^\sigma + d + r\frac{\sqrt{(8\gamma-3)N^3}}{6\beta}(N+\gamma^2)\right)|\nabla_x u|^2\\ &+\left[\left(\tau(1-p)\kappa_0^{1-\sigma}\beta + \frac{3C^2N^3(2\gamma+N)}{4(p-1)\kappa_0^{1-\sigma}} + K_1\gamma\right)\kappa^\sigma + 2\gamma\left(d + \frac{2N^{\frac{3}{2}}}{\sqrt{8\gamma-3}}r\right)\right]|D^2_x u|^2\\ &+\left[\left(\frac{1}{3}(1-p)\kappa_0^{1-\sigma}\gamma + \frac{9C^2N^3}{4(p-1)\kappa_0^{1-\sigma}} + \frac{1}{3}(p-1)N\kappa_0^{1-\sigma} + 3K_1\right)\kappa^\sigma\right.\\ &\left.+3\left(d + \frac{2N^{\frac{3}{2}}}{\sqrt{8\gamma-3}}r\right)\right]|D^3_x u|^2\Bigg\}v_\delta^{1-\frac{2}{p}}. \end{aligned} \tag{6.1.12}$$

Due to estimate (6.1.9), it holds that

$$\sup_{x\in\mathbb{R}^N}\left(\frac{d(x) + Lr(x)}{(\kappa(x))^\sigma} + \frac{C^2N^3}{4q}\kappa_0^{\sigma-1}\right), \qquad q \ge p_0 - 1.$$

Using this inequality we can make negative all the coefficients in front of the derivatives of u in the right-hand side of (6.1.12), by properly choosing the parameters β, γ and, consequently, we can estimate

$$g_\delta \leq p\omega_{3,p}\big(\beta|\nabla_x u|^2+\gamma|D_x^2u|^2+|D_x^3u|^2\big)v_\delta^{1-\frac{2}{p}} \leq p\omega_{3,p}v_\delta - p\omega_{3,p}\delta^{\frac{p}{2}}$$

in $(0,+\infty)\times\mathbb{R}^N$ for some negative constant $\omega_{3,p}$. As a byproduct, we can infer that $D_t v_\delta \leq \mathcal{A}v_\delta + p\omega_{3,p}v_\delta - p\omega_{3,p}\delta^{p/2}$ in $(0,+\infty)\times\mathbb{R}^N$. Set $z_\delta = e^{-p\omega_{3,p}t}(v_\delta - \delta^{p/2})$ and observe that, since $D_t z_\delta \leq \mathcal{A}z_\delta$ in $(0,+\infty)\times\mathbb{R}^N$, Theorem 3.1.3 implies that,

$$z_\delta \leq T(\cdot)(\beta|\nabla f|^2+\gamma|D^2 f|^2+|D^3 f|^2+\delta)^{\frac{p}{2}}$$

in $(0,+\infty)\times\mathbb{R}^N$, for any $\delta>0$. Taking the limit as δ tends to 0^+, from Proposition 1.2.10 we get (6.1.10) with $\widehat{M}_{3,p}=\beta^{p/2}\vee\gamma^{p/2}$.

Next, to prove (6.1.10) in the case when $p>2$, it suffices to repeat the same arguments as in the proof of Theorem 6.1.2.

To obtain (6.1.1) in the case when $p\in(1,2)$ and $k=2$, we can apply the above arguments to the function $v_\delta=(\beta|\nabla_x T(\cdot)f|^2+|D_x^2T(\cdot)f|^2+\delta)^{\frac{p}{2}}$ ($\delta>0$), where β is a positive constant to be determined. From (6.1.11), where we disregard the terms whose coefficients do not depend on β,γ, take $\gamma=1$, $\varepsilon_1=2L/N$, $\varepsilon_3=(p-1)\kappa_0^{1-\sigma}/(2CN)$ and make the same choices of ε_2 as above, we get

$$\begin{aligned} g_\delta \leq\, & p\bigg\{\beta\left(\frac{C^2N^3}{4(p-1)(1-\tau)}\kappa_0^{\sigma-1}\kappa^\sigma+d+\frac{N^3}{8\beta L}r\right)|\nabla_x u|^2 \\ & +\left(\tau(1-p)\beta\kappa_0^{1-\sigma}+\frac{C^2N^3}{(p-1)\kappa_0^{1-\sigma}}+K_1\right)\kappa^\sigma|D_x^2u|^2\bigg\}v_\delta^{1-\frac{2}{p}}. \end{aligned}$$

As above, choosing β properly we get

$$(\beta|\nabla_x u(t,\cdot)|^2+|D_x^2u(t,\cdot)|^2+\delta)^{\frac{p}{2}} \leq e^{p\omega_{2,p}t}\big(T(t)(\beta|\nabla f|^2+|D_x^2u|^2+\delta)^{\frac{p}{2}}\big)+\delta^{\frac{p}{2}} \qquad (6.1.13)$$

for any $t>0$ and some negative constant $\omega_{2,p}$.

Finally, in the case when $k=1$, taking $\beta=1$, $\varepsilon_2=(p-1)\kappa_0^{1-\sigma}/(CN)$ in (6.1.11) and disregarding the terms whose coefficients are independent of β, we get

$$g_\delta \leq p(C^2N^3\kappa_0^{\sigma-1}(4(p-1))^{-1}\kappa^\sigma+d)|\nabla_x u|^2$$

and this inequality allows to conclude the proof. □

6.2 The second type of pointwise gradient estimates

In this section we introduce a second type of pointwise estimates. For this purpose, we first prove a lemma.

Lemma 6.2.1 *The following properties are satisfied.*

(i) *If $\psi_n,\psi:[0,T]\times\mathbb{R}^N\to\mathbb{R}$ ($n\in\mathbb{N}$), are continuous functions such that* $\sup_{n\in\mathbb{N}}\|\psi_n\|_\infty<+\infty$ *and ψ_n tends to ψ, locally uniformly in $(0,T)\times\mathbb{R}^N$, then $T_n(\cdot)\psi_n$ converges to $T(\cdot)\psi$ locally uniformly in $(0,T)\times\mathbb{R}^N$, as n tends to $+\infty$.*

(ii) *If $f_\varepsilon, f_0 \in C_b(\mathbb{R}^N)$ ($\varepsilon > 0$) are such that $\sup_{\varepsilon>0} \|f_\varepsilon\|_\infty < +\infty$ and f_ε converges to f_0, locally uniformly in $\mathbb{R}^N$ as ε tends to 0^+, then, for any $t > 0$, $T(t-\varepsilon)f_\varepsilon$ converges to $T(t)f_0$ locally uniformly in $\mathbb{R}^N$, as ε tends to 0^+.*

Proof Throughout the proof, in order to simplify the notation, sometimes we do not explicitly write the dependence on x of the functions that we will consider. All the estimates that we write are meant pointwise in x.

(i) To begin with observe that $|T(t)f - T_n(t)f| \le T(t)|f| - T_n(t)|f|$ in B_n, for any $t > 0$, $n \in \mathbb{N}$ and $f \in C_b(\mathbb{R}^N)$. Indeed, we recall that (see Theorem 1.2.5 and Proposition D.3.1)

$$(T_n(t)f)(x) = \int_{B_n} g_n(t,x,y)f(y)dy, \qquad (T(t)f)(x) = \int_{\mathbb{R}^N} g(t,x,y)f(y)dy$$

for any $t > 0$, $x \in B_n$ and $f \in C_b(\mathbb{R}^N)$ and the functions g_n and g are positive, respectively, in $(0,+\infty) \times B_n \times B_n$ and in $(0,+\infty) \times \mathbb{R}^N \times \mathbb{R}^N$. Moreover $g_n \le g$ in $(0,+\infty) \times B_n \times B_n$. Hence,

$$\begin{aligned}|(T(t)f)(x) - (T_n(t)f)(x)| =& \left|\int_{\mathbb{R}^N} (g(t,x,y) - g_n(t,x,y)\chi_{B_n})f(y)dy\right| \\ \le& \int_{\mathbb{R}^N} (g(t,x,y) - g_n(t,x,y)\chi_{B_n})|f(y)|dy \\ =& (T(t)|f|)(x) - (T_n(t)|f|)(x)\end{aligned}$$

for any $t > 0$ and $x \in B_n$. Hence, if ψ_n and ψ are as in the statement, then we can estimate

$$\begin{aligned}|T_n(t)\psi_n(t,\cdot) - T(t)\psi(t,\cdot)| \le& |T_n(t)(\psi_n(t,\cdot) - \psi(t,\cdot))| + |(T(t) - T_n(t))\psi(t,\cdot)| \\ \le& T_n(t)|\psi_n(t,\cdot) - \psi(t,\cdot)| + (T(t) - T_n(t))|\psi(t,\cdot)| \\ \le& T(t)|\psi_n(t,\cdot) - \psi(t,\cdot)| + M(T(t) - T_n(t))\mathbb{1}\end{aligned} \tag{6.2.1}$$

in $\mathbb{R}^N$, for any $n \in \mathbb{N}$, where $M = \sup_{n\in\mathbb{N}} \|\psi_n\|_\infty$.

Let us observe that the second term in the last side of (6.2.1) converges to zero, locally uniformly in $(0,T) \times \mathbb{R}^N$, by Theorem 1.2.1. To estimate the first term we observe that

$$T(t)|\psi_n(t,\cdot) - \psi(t,\cdot)| \le T(t) \sup_{t\in[\varepsilon,T-\varepsilon]} |\psi_n(t,\cdot) - \psi(t,\cdot)| \tag{6.2.2}$$

in $\mathbb{R}^N$, for any $t \in [\varepsilon, T-\varepsilon]$ and $\varepsilon \in (0,T)$. Since $\sup_{t\in[\varepsilon,T-\varepsilon]} |\psi_n(t,\cdot) - \psi(t,\cdot)|$ is a bounded and continuous function in $\mathbb{R}^N$, converging to 0 locally uniformly in $\mathbb{R}^N$, from Proposition 1.2.10 we deduce that the right-hand side of (6.2.2) converges to 0, locally uniformly in $\mathbb{R}^N$, as n tends to $+\infty$, and the conclusion follows.

(ii) Let f_ε, f_0 be as in the statement; we can estimate

$$|T(t-\varepsilon)f_\varepsilon - T(t)f_0| \le |T(t-\varepsilon)f_\varepsilon - T(t-\varepsilon)f_0| + |T(t-\varepsilon)f_0 - T(t)f_0| \tag{6.2.3}$$

for any $\varepsilon > 0$ and $t > \varepsilon$. By Proposition 1.2.10, for any fixed $t > 0$ the first term in the right-hand side of (6.2.3) converges to 0 as ε tends to 0^+, locally uniformly in $\mathbb{R}^N$. The second term in the right-hand side of (6.2.3) converges to zero, locally uniformly in $\mathbb{R}^N$, as well, since the function $T(\cdot)f_0$ in continuous in $[0,+\infty) \times \mathbb{R}^N$ and, consequently, uniformly continuous in $[0,T] \times B_k$ for any $T, k > 0$. □

We can now prove our estimates. For notational convenience, we set

$$\widehat{\omega}_k = \widehat{\omega}_{k,p} = \begin{cases} 0, & \text{if Hypotheses 5.1.1 hold} \\ \omega_{k,p}, & \text{if Hypotheses 6.1.3 hold} \end{cases}, \qquad k = 1,2,3,$$

where $\omega_{k,p}$ are defined in the statement of Theorem 6.1.5. Moreover, we denote by $M_k = M_{k,p}$ both the constants in (6.1.1) and in (6.1.10). Finally, we set $\widetilde{L} = L_2$, if Hypotheses 5.1.1 are satisfied, and $\widetilde{L} = L$, if Hypotheses 6.1.3 are satisfied.

Theorem 6.2.2 *Let $k \in \{1,2,3\}$ and suppose that either of Hypotheses 5.1.1(i)-5.1.1(iii) and 5.1.1(iv-k) or Hypotheses 6.1.3(i) and 6.1.3(ii-k) hold. Then, for any $f \in C_b^{k-1}(\mathbb{R}^N)$ the estimate*

$$|D_x^k T(t) f|^p \le \left(\frac{\psi_{k,p\wedge 2}(t)}{1-e^{-\widehat{\omega}_{k,p\wedge 2}t}}\right)^{1\vee \frac{p}{2}} T(t)\left(\sum_{|\alpha|\le k-1} |D^\alpha f|^2\right)^{\frac{p}{2}} \tag{6.2.4}$$

holds true in $\mathbb{R}^N$, for any $t > 0$ and $p > \widehat{p}$, where $\widehat{p} = 1$, under Hypotheses 5.1.1, and $\widehat{p} = p_0$, under Hypothesis 6.1.3, whereas $\psi_{1,r}(t) = \widehat{\omega}_{1,r} b_{1,r} t^{1-p/2}(1+t)^{p/2}$ for any $t > 0$, $r \in (1,2]$, and[2]

$$\psi_{k,r}(t) = \widehat{\omega}_{k,r} b_{k,r}\left(1+\frac{e^{\widehat{\omega}_{k-1}t}-1}{\widehat{\omega}_{k-1}}\right)^{\frac{r}{2}}\left(at+\widehat{M}_{k,r}\frac{e^{\widehat{\omega}_{k-1}t}-1}{\widehat{\omega}_{k-1}}\right)^{1-\frac{r}{2}} \tag{6.2.5}$$

for any $t > 0$, $r \in (1,2]$ and $k = 2,3$, $b_{k,r}$ ($r \in (1,2]$, $k = 1,2,3$) being positive constants that can be explicitly determined from the data (see the proof). Finally, $\widehat{M}_{k,r}$ is the constant in (6.1.7) and $a = 1$, under Hypotheses 5.1.1, whereas $\widehat{M}_{k,r}$ is the constant in (6.1.13) and $a = 0$, under Hypotheses 6.1.3.

Proof We first consider the case when $k = 3$ and $p \in (\widehat{p}, 2)$. We fix $\delta, t > 0$, $n \in \mathbb{N}$ and let $\vartheta_n : \mathbb{R}^N \to \mathbb{R}$ be the cut-off function defined by $\vartheta_n(x) = \varrho(|x|/n)$, where $\varrho \in C^\infty([0,+\infty))$ is any nonincreasing function such that $\chi_{(0,1/2)} \le \varrho \le \chi_{(0,1)}$. For any $\alpha > 0$ and $f \in C_b^2(\mathbb{R}^N)$ we consider the function

$$g_{n,\delta} = T_n(\cdot)\{(\alpha^2|T_n(t-\cdot)f|^2 + \alpha\vartheta_n^2|\nabla_x T_n(t-\cdot)f|^2 + \vartheta_n^4|D_x^2 T_n(t-\cdot)f|^2+\delta)^{\frac{p}{2}} - \delta^{\frac{p}{2}}\}$$

for any $0 \le s \le t$, where $\{T_n(\cdot)\}$ is the semigroup generated in $C_b(\overline{B}_n)$ by the realization A_n of the operator $\mathcal{A}$ with homogeneous Dirichlet boundary conditions (see Appendix D). To simplify the notation, we set $\zeta_n(r,\cdot) := T_n(t-r)f$ for any $r \in [0,t]$ and $v_{n,\delta} = (\alpha^2|\zeta_n(r,\cdot)|^2 + \alpha\vartheta_n^2|\nabla_x\zeta_n(r,\cdot)|^2 + \vartheta_n^4|D_x^2\zeta_n(r,\cdot)|^2+\delta)^{p/2}$. Finally, $\widetilde{\sigma} = 1$, under Hypothesis 5.1.1(iv-3) and $\widetilde{\sigma} = \sigma$, under Hypotheses 6.1.3.

As is easily seen, the function $v_{n,\delta}(r,\cdot) - \delta^{p/2}$ belongs to $D(A_n)$ for any $r \in [0,t)$. Recalling that $\mathcal{A}(\delta^{p/2}) = 0$, we get $g_{n,\delta}' = pT_n(\cdot)\chi_{n,\delta}$, where

$$\begin{aligned}
\chi_{n,\delta} =& \bigg(\alpha^2\mathfrak{q}_0(\zeta_n) + \alpha\vartheta_n^2\sum_{i=1}^N \mathfrak{q}_0(D_i\zeta_n) + \vartheta_n^4\sum_{i,j=1}^N \mathfrak{q}_0(D_{ij}\zeta_n) + \frac{\alpha}{2}\mathcal{A}(\vartheta_n^2)|\nabla_x\zeta_n|^2\\
&+\frac{1}{2}\mathcal{A}(\vartheta_n^4)|D_x^2\zeta_n|^2 + 4\alpha\vartheta_n\langle Q\nabla\vartheta_n, D_x^2\zeta_n\nabla_x\zeta_n\rangle - \alpha\vartheta_n^2(\mathfrak{q}_1(\zeta_n)+\mathfrak{b}_1(\zeta_n))\\
&-2\vartheta_n^4\sum_{i=1}^N(\mathfrak{q}_1(D_i\zeta_n)+\mathfrak{b}_1(D_i\zeta_n)) + 8\vartheta_n^3\sum_{i=1}^N\langle Q\nabla\vartheta_n, (D_x^2D_i\zeta_n)\nabla_x D_i\zeta_n\rangle\\
&-\vartheta_n^4(\mathfrak{q}_2(\zeta_n)+\mathfrak{b}_2(\zeta_n))\bigg)(\alpha^2|\zeta_n|^2 + \alpha\vartheta_n^2|\nabla_x\zeta_n|^2 + \vartheta_n^4|D_x^2\zeta_n|^2+\delta)^{\frac{p}{2}-1}\\
&-\frac{2-p}{4}\mathfrak{q}_0(\alpha^2\zeta_n^2 + \alpha\vartheta_n^2|\nabla_x\zeta_n|^2 + \vartheta_n^4|D_x^2\zeta_n|^2)v_{n,\delta}^{1-\frac{4}{p}},
\end{aligned} \tag{6.2.6}$$

[2]When $\widehat{\omega}_{k-1} = 0$ we agree that $(1-e^{-\widehat{\omega}_{k-1}t})/\widehat{\omega}_{k-1} = t$.

where $\mathfrak{q}_i$ ($i=0,1,2$) and $\mathfrak{b}_j$ ($j=1,2$) are defined in (5.1.9)-(5.1.12). Arguing as in the proof of (6.1.4) and observing that $(a+b)^2 \le (3-p)[(4-2p)^{-1}a^2+(p-1)^{-1}b^2]$ for any $a,b>0$, we deduce that

$$\begin{aligned}
&\frac{1}{4}\mathfrak{q}_0(\alpha^2\zeta_n^2+\alpha\vartheta_n^2|\nabla_x\zeta_n|^2+\vartheta_n^4|D_x^2\zeta_n|^2)\\
\le&\frac{3-p}{4-2p}\left[\alpha^2|\zeta_n|\sqrt{\mathfrak{q}_0(\zeta_n)}+\alpha\vartheta_n^2|\nabla_x\zeta_n|\left(\sum_{i=1}^N\mathfrak{q}_0(D_i\zeta_n)\right)^{\frac{1}{2}}+\vartheta_n^4|D_x^2\zeta_n|\left(\sum_{i,j=1}^N\mathfrak{q}_0(D_{ij}\zeta_n)\right)^{\frac{1}{2}}\right]^2\\
&+\frac{3-p}{p-1}(\alpha\vartheta_n|\nabla_x\zeta_n|^2+2\vartheta_n^3|D_x^2\zeta_n|^2)^2\mathfrak{q}_0(\vartheta_n)\\
\le&\frac{3-p}{4-2p}\left[\alpha^2\mathfrak{q}_0(\zeta_n)+\alpha\vartheta_n^2\sum_{i=1}^N\mathfrak{q}_0(D_i\zeta_n)+\vartheta_n^4\sum_{i,j=1}^N\mathfrak{q}_0(D_{ij}\zeta_n)\right]\\
&\qquad\times(\alpha^2|\zeta_n|^2+\alpha\vartheta_n^2|\nabla_x\zeta_n|^2+\vartheta_n^4|D_x^2\zeta_n|^2)\\
&+\frac{3-p}{p-1}(\alpha|\nabla_x\zeta_n|^2+4\vartheta_n^2|D_x^2\zeta_n|^2)(\alpha\vartheta_n^2|\nabla_x\zeta_n|^2+\vartheta_n^4|D_x^2\zeta_n|^2)\mathfrak{q}_0(\vartheta_n).
\end{aligned}\tag{6.2.7}$$

Taking Hypothesis 5.1.1(iii) into account, it is easy to check that $|\mathfrak{q}_0(\vartheta_n)|\le C_1\kappa$, $\mathcal{A}(\vartheta_n^2)\ge -C_1\kappa$, $\mathcal{A}(\vartheta_n^4)\ge 2\vartheta_n^2\mathcal{A}(\vartheta_n^2)\ge -2C_1\vartheta_n^2\kappa$ in B_n for some positive constant C_1, independent of n. Moreover, by using the Young inequality, it can be proved that

$$|\vartheta_n\langle Q\nabla\vartheta_n,D_x^2\zeta_n\nabla_x\zeta_n\rangle|\le\frac{C_2}{p-1}\kappa|\nabla_x\zeta_n|^2+\frac{p-1}{16}\vartheta_n^2\kappa|D_x^2\zeta_n|^2,\tag{6.2.8}$$

$$\left|\vartheta_n^3\sum_{i=1}^N\langle Q\nabla\vartheta_n,(D_x^2D_i\zeta_n)\nabla_xD_i\zeta_n\rangle\right|\le\frac{C_2}{p-1}\vartheta_n^2\kappa|D_x^2\zeta_n|^2+\frac{p-1}{32}\vartheta_n^4\kappa|D_x^3\zeta_n|^2\tag{6.2.9}$$

in B_n for some positive constant C_2, independent of n. Finally, we can estimate $\mathfrak{q}_0(D_x^\alpha\zeta_n)\ge\kappa|\nabla_xD_x^\alpha\zeta_n|^2$, $\mathfrak{q}_2(\zeta_n)\le K_1^+\kappa^{\sigma_a}|D^2\zeta_n|^2$, $\mathfrak{b}_1(D_x^\alpha\zeta_n)\le d|\nabla_xD_x^\alpha\zeta_n|^2$ and, using again the Young inequality,

$$\alpha\mathfrak{q}_1(\zeta_n)\le\kappa^{\widetilde{\sigma}}\left(\frac{C^2N^3}{8\widetilde{L}}\alpha^{3/2}|\nabla_x\zeta_n|^2+2\sqrt{\alpha}\widetilde{L}|D_x^2\zeta_n|^2\right),\tag{6.2.10}$$

$$\begin{aligned}
\sum_{i=1}^N\mathfrak{q}_1(D_i\zeta_n)\le&\kappa^{\widetilde{\sigma}}\left(\frac{4C^2N^3}{p-1}\kappa_0^{\widetilde{\sigma}-1}|D_x^2\zeta_n|^2+\frac{p-1}{16}\kappa_0^{1-\widetilde{\sigma}}\right)\\
\le&\frac{4C^2N^3}{p-1}\kappa_0^{\widetilde{\sigma}-1}\kappa^{\widetilde{\sigma}}|D_x^2\zeta_n|^2+\frac{p-1}{16}\kappa,
\end{aligned}\tag{6.2.11}$$

$$\mathfrak{b}_2(\zeta_n)\le r\left(\frac{N^3}{8\widetilde{L}}|\nabla_x\zeta_n|^2+2\widetilde{L}|D_x^2\zeta_n|^2\right),\tag{6.2.12}$$

where $\widetilde{L}=L_2$, if Hypothesis 5.1.1(*iv*-3) is satisfied, and $\widetilde{L}=L$, if Hypotheses 6.1.3 are satisfied. From (6.2.7)-(6.2.12) and taking $\alpha>N^3/(8\widetilde{L}^2)$, we get

$$g'_\delta\ge pT_n(\cdot)\left\{\kappa\left(H_1(\alpha,p)|\nabla_x\zeta_n|^2+H_2(\alpha,p)\vartheta_n^2|D_x^2\zeta_n|^2+\frac{p-1}{8}\vartheta_n^4|D_x^3\zeta_n|^2\right)v_{n,\delta}^{1-\frac{2}{p}}\right\}$$

with

$$H_1(\alpha,p):=\frac{p-1}{2}\alpha^2-\left(\frac{C_1}{2}+4\frac{C_2}{p-1}+\frac{p-3}{p-1}(2-p)C_1\right)\alpha-\frac{CN^3}{8\widetilde{L}}\kappa_0^{\widetilde{\sigma}}\alpha^{\frac{3}{2}}-A\alpha,$$

$$H_2(\alpha,p) := \frac{p-1}{4}\alpha - C_1 - 8\frac{C_2}{p-1} + 4\frac{3-p}{p-1}(p-2)C_1 - \left(\frac{8C^2N^3}{p-1}\kappa_0^{a-1} + K_1^+ + 2\sqrt{\alpha}\widetilde{L}\right)\kappa_0^{a-1} - 2B,$$

where $A = (\beta - N^3/(8\widetilde{L}^2))L_1 + N^3/(8\widetilde{L}^2)L_3$ and $B = L_3^+$, under Hypotheses 5.1.1, whereas $A = B = 0$ otherwise.

Both in the cases when $a = 1$ and $a = \sigma$, we can fix α sufficiently large so that $H_1(\alpha,p)$ and $H_2(\alpha,p)$ are positive and $\alpha \geq N^3/(8\widetilde{L}^2)$. For such α's we get

$$g'_\delta \geq c_p T_n(\cdot)\left(\vartheta_n^4(|\nabla_x\zeta_n|^2 + |D_x^2\zeta_n|^2 + |D_x^3\zeta_n|^2)v_{n,\delta}^{1-\frac{2}{p}}\right), \tag{6.2.13}$$

where $c_p = c_p(\alpha) := p\kappa_0\min\{H_1(\alpha,p), H_2(\alpha,p), (p-1)\kappa_0^{1-a}/8\}$. Integrating (6.2.13) in $[\varepsilon, t-\varepsilon]$ ($\varepsilon \in (0,t)$), and recalling that $\{T_n(t)\}$ is a positive semigroup, yields

$$\begin{aligned}
&c_p\int_\varepsilon^{t-\varepsilon}\left(T_n(s)\left(\vartheta_n^4(|\nabla_x\zeta_n|^2 + |D_x^2\zeta_n|^2 + |D_x^3\zeta_n|^2)v_{n,\delta}^{1-\frac{2}{p}}\right)(x)\,ds\right.\\
&\leq\big(T_n(t-\varepsilon)\big((\alpha^2|T_n(\varepsilon)f|^2 + \alpha\vartheta_n^2|\nabla_x T_n(\varepsilon)f|^2 + \vartheta_n^4|D_x^2T_n(\varepsilon)f|^2 + \delta)^{\frac{p}{2}} - \delta^{\frac{p}{2}}\big)\big)(x).
\end{aligned} \tag{6.2.14}$$

Now, we observe that the same arguments as in Remark 5.1.5 show that $D_x^\gamma\zeta_n$ converges to $D_x^\gamma T(t-\cdot)f$ uniformly in $[\varepsilon, t-\varepsilon]\times B_k$, as n tends to $+\infty$, for any $\varepsilon \in (0,t/2)$, $k \in \mathbb{N}$ and $|\gamma| \leq 3$. Therefore, applying Lemma 6.2.1 and the dominated convergence theorem, we can let n tend to $+\infty$ in (6.2.14) and get

$$\begin{aligned}
&c_p\int_\varepsilon^{t-\varepsilon}\Big(T(s)\Big((|\nabla_x T(t-s)f|^2 + |D_x^2T(t-s)f|^2 + |D_x^3T(t-s)f|^2)\\
&\qquad\qquad\times(\alpha^2|T(t-s)f|^2 + \alpha|\nabla_x T(t-s)f|^2 + |D_x^2T(t-s)f|^2 + \delta)^{\frac{p}{2}-1}\Big)\Big)(x)\,ds\\
&\leq\Big(T(t-\varepsilon)\Big(\alpha^2|T(\varepsilon)f|^2 + \alpha|\nabla_x T(\varepsilon)f|^2 + |D^2T(\varepsilon)f|^2)^{\frac{p}{2}}\Big)\Big)(x)
\end{aligned} \tag{6.2.15}$$

for any $x \in \mathbb{R}^N$. Now, from Proposition 6.1.1 we can infer that $D_x^jT(\varepsilon)f$ converges to D^jf, locally uniformly in $\mathbb{R}^N$, as ε tends to 0^+, for any $j = 0,1,2$. Therefore, from Lemma 6.2.1 it follows that the right-hand side of (6.2.15) converges to $(T(t)(\alpha^2|f|^2 + \alpha|\nabla f|^2 + |D^2f|^2)^{p/2})(x)$ for any $x \in \mathbb{R}^N$, as ε tends to 0^+.

As far as the first integral term in (6.2.15) is concerned, we observe that, by virtue of Theorem 6.1.2, for any $x \in \mathbb{R}^N$ the function under the integral sign can be estimated from above by a constant (independent of x) times the function $s \mapsto (t-s)^{-p/2}$. Therefore, using the dominated convergence theorem we can take the limit as ε tends to 0^+ in (6.2.15), getting

$$\begin{aligned}
&c_p\int_0^t\Big(T(s)\Big((|\nabla_x T(t-s)f|^2 + |D_x^2T(t-s)f|^2 + |D_x^3T(t-s)f|^2)\\
&\qquad\qquad\times(\alpha^2|T(t-s)f|^2 + \alpha|\nabla_x T(t-s)f|^2 + |D_x^2T(t-s)f|^2 + \delta)^{\frac{p}{2}-1}\Big)\Big)(x)ds\\
&\leq\Big(T(t)(\alpha^2f^2 + \alpha|\nabla f|^2 + |D^2f|^2)^{\frac{p}{2}}\Big)(x)
\end{aligned} \tag{6.2.16}$$

for any $x \in \mathbb{R}^N$.

From now on, for notational convenience we set $\zeta(s,\cdot) = T(t-s)f$ and $v_\delta(s,\cdot) = \alpha^2|\zeta(s,\cdot)|^2 + \alpha|\nabla_x\zeta(s,\cdot)|^2 + |D_x^2\zeta(s,\cdot)|^2 + \delta$ for any $s \in (0,t)$. Theorem 1.2.5 and the Young inequality show that

$$|T(t)(g_1g_2)| \leq |T(t)g_1^{2/p}|^{p/2}|T(t)g_2^{2/(2-p)}|^{1-p/2}$$

$$\leq \frac{p}{2}\varepsilon^{2/p}|T(t)g_1^{2/p}| + \frac{2-p}{2}\varepsilon^{2/(p-2)}|T(t)g_2^{2/(2-p)}|$$

for any $g_1, g_2 \in C_b(\mathbb{R}^N)$ and $t > 0$, Thus, from this estimate and (6.1.1) or (6.1.10) (with $k = 3$), we get

$$\begin{aligned}
&\left(|\nabla_x T(t)f|^2 + |D_x^2 T(t)f|^2 + |D_x^3 T(t)f|^2\right)^{\frac{p}{2}} \\
=&\left(|\nabla_x T(s)\zeta(s,\cdot)|^2 + |D_x^2 T(s)\zeta(s,\cdot)|^2 + |D_x^3 T(s)\zeta(s,\cdot)|^2\right)^{\frac{p}{2}} \\
\leq& M_3 e^{\widehat{\omega}_3 s} T(s)\Big[(a|\zeta(s,\cdot)|^2 + |\nabla_x\zeta(s,\cdot)|^2 + |D_x^2\zeta(s,\cdot)|^2 + |D_x^3\zeta(s,\cdot)|^2)^{\frac{p}{2}} \\
&\qquad\qquad \times (\upsilon_\delta(s,\cdot))^{-\frac{p(2-p)}{4}}(\upsilon_\delta(s,\cdot))^{\frac{p(2-p)}{4}}\Big] \\
\leq& M_3 e^{\widehat{\omega}_3 s}\Big\{\frac{p}{2}\varepsilon^{\frac{2}{p}} T(s)\Big[(a|\zeta(s,\cdot)|^2 + |\nabla_x\zeta(s,\cdot)|^2 + |D_x^2\zeta(s,\cdot)|^2 + |D_x^3\zeta(s,\cdot)|^2)(\upsilon_\delta(s,\cdot))^{\frac{p}{2}-1}\Big] \\
&\qquad + \frac{2-p}{2}\varepsilon^{\frac{2}{p-2}} T(s)(\upsilon_\delta(s,\cdot))^{\frac{p}{2}}\Big\} \\
\leq& M_3 e^{\widehat{\omega}_3 s}\Big\{\frac{p}{2}\varepsilon^{\frac{2}{p}} T(s)\Big[\left(|\nabla_x\zeta(s,\cdot)|^2 + |D_x^2\zeta(s,\cdot)|^2 + |D_x^3\zeta(s,\cdot)|^2\right)(\upsilon_\delta(s,\cdot))^{\frac{p}{2}-1}\Big] \\
&\qquad + \frac{p}{2}\varepsilon^{\frac{2}{p}}\frac{a}{\alpha} T(s)[(\upsilon_\delta(s,\cdot))^{\frac{p}{2}}] + \frac{2-p}{2}\varepsilon^{\frac{2}{p-2}} T(s)[(\upsilon_\delta(s,\cdot))^{\frac{p}{2}}]\Big\}
\end{aligned} \tag{6.2.17}$$

for any $s \in (0,t)$ and some positive constant M_3, where $a = 1$, under Hypotheses 5.1.1(*i*)-(*iii*), 5.1.1(*iv-k*), and $a = 0$, under Hypotheses 6.1.3(*i*) and 6.1.3(ii-k). Since

$$\begin{aligned}
\upsilon_\delta^{\frac{p}{2}} \leq& (1 \vee \alpha^p)(|\zeta(s,\cdot)|^2 + |\nabla_x\zeta(s,\cdot)|^2 + |D_x^2\zeta(s,\cdot)|^2 + \delta)^{\frac{p}{2}} \\
\leq& (1 \vee \alpha^p) M_2 e^{\widehat{\omega}_2(t-s)} T(t-s)(f^2 + |\nabla f|^2 + |D^2 f|^2 + \delta)^{\frac{p}{2}} \\
&+ (1 \vee \alpha^p)(1-a)(\delta^{\frac{p}{2}} + T(t-s)|f|^p)
\end{aligned}$$

for any $s \in (0,t)$ (see (6.1.7) and (6.1.13)), from (6.2.17) we get, for any $\varepsilon > 0$,

$$\begin{aligned}
&\left(|\nabla_x T(t)f|^2 + |D_x^2 T(t)f|^2 + |D_x^3 T(t)f|^2\right)^{\frac{p}{2}} \\
\leq& M_3 e^{\widehat{\omega}_3 s}\Big\{\frac{p}{2}\varepsilon^{\frac{2}{p}} T(s)[\left(|\nabla_x\zeta(s,\cdot)|^2 + |D_x^2\zeta(s,\cdot)|^2 + |D_x^3\zeta(s,\cdot)|^2\right)(\upsilon_\delta(s,\cdot))^{\frac{p}{2}-1}] \\
&\qquad + \left(\frac{2-p}{2}\varepsilon^{\frac{2}{p-2}} + \frac{p}{2}\varepsilon^{\frac{2}{p}}\frac{a}{\alpha}\right)(1 \vee \alpha^p) M_2 e^{\widehat{\omega}_2(t-s)} \\
&\qquad\qquad \times T(t)\left(|f|^2 + |\nabla f|^2 + |D^2 f|^2 + \delta\right)^{\frac{p}{2}} \\
&\qquad + \frac{2-p}{2}(1 \vee \alpha^p)\varepsilon^{\frac{2}{p-2}}(1-a)(\delta^{\frac{p}{2}} + T(t)|f|^p)
\end{aligned} \tag{6.2.18}$$

for any $s \in (0,t)$. Now, we multiply the first and the last side of (6.2.18) by $e^{-\widehat{\omega}_3 s}$, integrate in $(0,t)$, pointwise with respect to $x \in \mathbb{R}^N$, taking (6.2.16) into account. Finally, we let δ tend to 0^+ and minimize with respect to $\varepsilon > 0$. We get

$$\begin{aligned}
&\frac{1-e^{-\widehat{\omega}_3 t}}{\widehat{\omega}_3}\left(|\nabla_x T(t)f|^2 + |D_x^2 T(t)f|^2 + |D_x^3 T(t)f|^2\right)^{\frac{p}{2}} \\
\leq& M_3(\alpha \vee \alpha^2)^{\frac{p}{2}}\left(\frac{1}{c_p} + \frac{a}{\alpha}M_2\frac{e^{\widehat{\omega}_2 t}-1}{\widehat{\omega}_2}\right)^{\frac{p}{2}}\left(M_2\frac{e^{\widehat{\omega}_2 t}-1}{\widehat{\omega}_2} + (1-a)t\right)^{1-\frac{p}{2}} \\
&\qquad \times T(t)(f^2 + |\nabla f|^2 + |D^2 f|^2)^{\frac{p}{2}},
\end{aligned}$$

where, as usual, $\widehat{\omega}_2^{-1}(e^{\widehat{\omega}_2 t}-1)=t$ if $\widehat{\omega}_2=0$. Now, (6.2.4) with $p\in(1,2)$ follows taking as $b_{3,p}$ the infimum of $M_3\{(\alpha\vee\alpha^2)[(c_p(\alpha))^{-1}\vee(a\alpha^{-2}M_2)]\}^{p/2}$ over the set $\{\alpha>N^3/(8\widetilde{L}): H_1(\alpha,p)>0,\ H_2(\alpha,p)>0\}$.

The case when $p=2$ follows by letting p tend to 2^- in (6.2.4). For $p>2$, we observe that

$$|D_x^3T(t)f|^p=(|D_x^3T(t)f|^2)^{\frac{p}{2}}\le\left(\frac{\widehat{\omega}_{3,2}\psi_{3,2}(t)}{1-e^{-\widehat{\omega}_{3,2}t}}\right)^{\frac{p}{2}}(T(t)(f^2+|\nabla f|^2+|D^2f|^2))^{\frac{p}{2}}$$

in $\mathbb{R}^N$, for any $t>0$. Applying the Hölder inequality, estimate (6.2.4) follows also in this case.

The proof of (6.2.4) in the case when $k=1,2$ is completely similar. In the case when $k=2$ one can apply the previous arguments to the function $g_{n,\delta}=T_n(\cdot)(\alpha\zeta_n^2+\vartheta_n^2|\nabla_x\zeta_n|^2+\delta)^{p/2}-\delta^{p/2}$. After some computations one finds that $g_\delta'=pT_n(\cdot)\chi_{n,\delta}$, where $\chi_{n,\delta}$ is still given by (6.2.6) with the obvious changes induced by the changes in the definition of the function g_δ. Therefore, taking $\varepsilon=(p-1)\kappa_0^{1-a}/(8CN)$ and $\varepsilon_1=(p-1)/(16C_2)$, one gets

$$g_\delta'\ge pT_n(\cdot)\left\{\left(H_3(\alpha,p)\kappa|\nabla_x\zeta|^2+\frac{p-1}{8}\kappa_0^{1-a}\kappa^a\vartheta^2|D_x^2\zeta|^2\right)v_\delta^{\frac{p}{2}-1}\right\},$$

where

$$H_3(\alpha,p)=\frac{p-1}{2}\alpha-\frac{C_1}{2}-\frac{16C_2^2N}{p-1}+(p-2)C_1\frac{3-p}{p-1}-\frac{2C^2N^3}{p-1}\kappa_0^{2a-2}-A,$$

again with $A=L_3^+$, under Hypotheses 5.1.1, and $A=0$ otherwise. Repeating the same arguments as above, one can now prove that

$$\left(\sum_{1\le|\alpha|\le 2}^{2}|D_x^\alpha T(t)f|^2\right)^{\frac{p}{2}}\le\left(\frac{\omega_{2,p\wedge 2}\psi_{2,p\wedge 2}(t)}{1-e^{-\omega_{2,p\wedge 2}t}}\right)^{1\vee\frac{p}{2}}T(t)\left(\sum_{|\alpha|\le 1}|D^\alpha f|^2\right)^{\frac{p}{2}}\tag{6.2.19}$$

in $\mathbb{R}^N$, for any $t>0$ and $p\in(1,+\infty)$, where $\psi_{2,p\wedge 2}$ is as in the statement with

$$b_{2,p}=M_2\left[M_1\vee(1-a)\right]^{1-\frac{p}{2}}\min\left\{\alpha^{\frac{p}{2}},[(c_p'(\alpha))^{-1}\vee(aM_1)]^{\frac{p}{2}}:\alpha\ge 1,H_3(\alpha,p)>0\right\}.$$

Here, $c_p'(\alpha)=c_p'(\alpha,p):=p\kappa_0\min\{H_3(\alpha,p),(p-1)/8\}$.

In the case when $k=1$ the previous arguments actually show that

$$b_{1,p}=M_1((p(p-1))^{-1}\vee(a-\sigma)(1-\sigma)^{-1})^{\frac{p}{2}}$$

and

$$|\nabla_xT(t)f|^p\le\left(\frac{\omega_{1,p\wedge 2}\psi_{1,p\wedge 2}(t)}{1-e^{-\omega_{1,p\wedge 2}t}}\right)^{1\vee\frac{p}{2}}T(t)|f|^p,\qquad t>0.$$

This completes the proof. □

Corollary 6.2.3 *Let $k=2,3$. Moreover, let Hypotheses 5.1.1(i)-5.1.1(iii) and 5.1.1(iv-k), or Hypotheses 6.1.3(i) and 6.1.3(ii-k) hold. Then, for any $f\in C_b(\mathbb{R}^N)$ and $p\in(\widehat{p},+\infty)$ (see Theorem 6.2.2), the estimate*

$$|D_x^kT(t)f|^p\le\left(\sum_{j=0}^{k-1}k^j\prod_{h=0}^{j}\frac{\widetilde{\omega}_{k-h,p\wedge 2}\psi_{k-h,p\wedge 2}(t/k)}{1-e^{-\widetilde{\omega}_{k-h,p\wedge 2}t}}\right)^{1\vee\frac{p}{2}}T(t)|f|^p\tag{6.2.20}$$

holds true in $\mathbb{R}^N$, for any $t>0$, where $\widetilde{\omega}_{k,r}=\widehat{\omega}_{j,r}/k$, and $\psi_{j,r}$ and $\widehat{\omega}_{j,r}$ are as in Theorem 6.2.2.

Proof We begin by proving (6.2.20) in the case when $k = 3$ and $p \le 2$. We fix $f \in C_b(\mathbb{R}^N)$ and $t > 0$. Applying (6.2.4) to the function $T(2t/3)f$ we get

$$|D^3_xT(t)f|^p \le \frac{\widehat{\omega}_{3,p}\psi_{3,p}(t/3)}{1-e^{-\widehat{\omega}_{3,p}t/3}}T(t/3)\Big\{\big(|T(2t/3)f|^2+|\nabla_xT(2t/3)f|^2+|D^2_xT(2t/3)f|^2\big)^{\frac{p}{2}}\Big\} \tag{6.2.21}$$

in $\mathbb{R}^N$. Now, taking (6.2.19) into account, with $T(t)f$ replaced with $T(2t/3)f$, we get

$$\begin{aligned}&\big(|T(2t/3)f|^2+|\nabla_xT(2t/3)f|^2+|D^2_xT(2t/3)f|^2\big)^{\frac{p}{2}}\\ &\le \frac{\widehat{\omega}_{2,p}\psi_{2,p}(t/3)}{1-e^{-\widehat{\omega}_{2,p}t/3}}T(t/3)\Big\{\big(|T(t/3)f|^2+|\nabla_xT(t/3)f|^2\big)^{\frac{p}{2}}\Big\}+T(2t/3)|f|^p\\ &\le \frac{\widehat{\omega}_{2,p}\psi_{2,p}(t/3)}{1-e^{-\widehat{\omega}_{2,p}t/3}}\bigg(\frac{\widehat{\omega}_{1,p}\psi_{1,p}(t/3)}{1-e^{-\widehat{\omega}_{1,p}t/3}}+1\bigg)T(2t/3)|f|^p+T(2t/3)|f|^p.\end{aligned} \tag{6.2.22}$$

Combining (6.2.21) and (6.2.22) gives (6.2.20) for $p \le 2$. If $p > 2$ and $k = 3$, then estimate (6.2.20) follows from the case when $p = 2$ applying the Hölder inequality. The proof of (6.2.20) (with $k = 2$) is similar: we write $|D^2_xT(t)f|^2 = |D^2_xT(t/2)T(t/2)f|^2$ and apply (6.2.4) to $T(t/2)f$ with, respectively, $k = 1$ and $k = 2$. We omit the details. □

Remark 6.2.4 The estimates in (6.2.4) (with $k = 1$) and in Corollary 6.2.3 allow us to improve the uniform estimates for the derivatives of $T(t)$ in Theorem 5.1.7. Indeed, they show that under Hypotheses 5.1.1(i)-5.1.1(iii) and 5.1.1(ii-k) ($k = 1, 2, 3$), the sup-norm of $D^k_xT(t)f$ stays bounded as t tends to $+\infty$, whereas it decreases exponentially to 0 as t tends to $+\infty$ if Hypotheses 6.1.3(i) and 6.1.3(ii-k) are satisfied.

Estimate (6.2.4) (with $k = 1$) can also be used to prove a Liouville type theorem. As it has already been stressed in the introduction, such a result fails in general, without any additional assumptions besides the basic Hypotheses 1.0.1. We refer the reader to [220] for further details.

Theorem 6.2.5 *Suppose that Hypotheses* 6.1.3(*i*) *and* 6.1.3(*ii*-1) *hold. If $f \in C_b(\mathbb{R}^N)$ satisfies $\mathcal{A}f = 0$, then f is constant.*

Proof We observe that, if $f \in C_b(\mathbb{R}^N)$ is such that $\mathcal{A}f = 0$, then, by local elliptic regularity, $u \in D_{\max}(\mathcal{A})$ (see (1.0.1)). Moreover, for any $g \in D_{\max}(\mathcal{A})$ and any $x \in \mathbb{R}^N$, the function $t \mapsto (T(t)g)(x)$ is continuously differentiable in $[0, +\infty)$ and $(D_tT(t)g)(x) = (\mathcal{A}T(t)g)(x) = (T(t)\mathcal{A}g)(x)$ (see Lemma 1.3.2, Propositions 1.3.4, 1.3.5 and 3.1.10). Hence, $D_tT(t)f = 0$ for any $t > 0$, so that $T(t)f \equiv f$ for any $t > 0$. From (6.2.4) we deduce that

$$|\nabla_xT(t)f| \le \frac{\widetilde{C}e^{-\omega t}}{\sqrt{t}}\sqrt{T(t)f^2} \le \frac{\widetilde{C}e^{-\omega t}}{\sqrt{t}}\|f\|_\infty, \qquad t > 0, \tag{6.2.23}$$

in $\mathbb{R}^N$, for some constants $\widetilde{C}, \omega > 0$. Letting t tend to $+\infty$ in (6.2.23), we get $\nabla f = \lim_{t\to+\infty}\nabla_xT(t)f = 0$ pointwise in $\mathbb{R}^N$, so that f is constant. □

6.3 Further estimates when $\mathcal{A} = \Delta + \langle b, \nabla\rangle$

Now, we consider the particular case when $q_{ij} \equiv \delta_{ij}$ ($i, j = 1, \ldots, N$). In this case, under Hypotheses 6.1.3, the gradient estimate (6.1.1) can be proved also for $p = 1$.

Theorem 6.3.1 *Fix $k = 1$ and suppose that $q_{ij} \equiv \delta_{ij}$ $(i,j = 1,\dots,N)$. Moreover, let Hypotheses 6.1.3(i), 6.1.3(ii-k) hold with (6.1.9) being replaced with the condition $d + Lr \le K$ in $\mathbb{R}^N$, to be satisfied for some constants $L > 0$ and $K \in \mathbb{R}$. Then,*

$$|D_x^k T(t)f| \le M_k e^{\omega t} T(t)\left(\sum_{j=1}^{k} |D^j f|^2\right)^{\frac{1}{2}}, \qquad t > 0, \tag{6.3.1}$$

in $\mathbb{R}^N$, where $\omega = K \vee (2K)$ and $M_k > 0$ are constants which can be explicitly determined (see the proof).

Finally, if $k = 1$, $b_j \in C^{1+\delta}_{\rm loc}(\mathbb{R}^N)$ for some $\delta \in (0,1)$ and any $j = 1,\dots,N$, and (6.1.8) is satisfied with the function d being replaced by a real constant d_0, then

$$|(\nabla_x T(t)f)(x)| \le e^{d_0 t}(T(t)|\nabla f|)(x). \qquad t > 0,\ x \in \mathbb{R}^N. \tag{6.3.2}$$

Proof We limit ourselves to sketching the proof of (6.3.1) in the case when $k = 3$, which is similar to that of Theorem 6.1.2. The proofs of (6.3.1) (with $k = 2$) and (6.3.2) are similar and even simpler.

Fix $\delta > 0$, $f \in C^1_b(\mathbb{R}^N)$ and let $v_\delta = \left(\alpha^2|\nabla_x u|^2 + \alpha|D^2_x u|^2 + |D^3_x u|^2 + \delta\right)^{1/2}$, where, as usual, $u = T(\cdot)f$ and α is a positive constant to be fixed later on. Arguing as in the proof of Theorem 6.1.2 and taking the Cauchy-Schwarz inequality into account, we can show that the function v_δ satisfies $D_t v_\delta = \mathcal{A}v_\delta + g_\delta$ in $(0,+\infty)\times\mathbb{R}^N$, where (see (6.1.5))

$$g_\delta \le v_\delta^{-1}\left(\alpha^2\mathfrak{b}_1(u) + 2\alpha\sum_{i=1}^N \mathfrak{b}_1(D_i u) + 3\sum_{i,j=1}^N \mathfrak{b}_1(D_{ij}u) + \alpha\mathfrak{b}_2(u) + \mathfrak{b}_3(u) + 3\sum_{i=1}^N \mathfrak{b}_2(D_i u)\right)$$

and $\mathfrak{b}_j$ $(j = 1,2,3)$ are defined in (5.1.9)-(5.1.12). From Hypotheses 6.1.3(ii-3) and (5.1.26) we can estimate

$$\mathfrak{b}_2(u) \le r\left(\frac{N^3}{4L}|\nabla_x u|^2 + L|D^2_x u|^2\right),$$

$$\mathfrak{b}_2(D_i u) \le r\left(\frac{N^3}{3L}|\nabla_x D_i u|^2 + \frac{3}{4}L|D^2_x D_i u|^2\right),$$

$$\mathfrak{b}_3(u) \le r\left(\frac{N^4}{3L}|\nabla_x u|^2 + \frac{3}{4}L|D^3_x u|^2\right).$$

Taking $\alpha > \frac{N^3}{L^2}$ such that $\alpha^2 > \frac{\alpha N^3}{4L^2} + \frac{N^4}{3L^2}$ and using the assumption $d + Lr \le K$ in $\mathbb{R}^N$, we get

$$g_\delta \le v_\delta^{-1}(\alpha^2 K|\nabla_x u|^2 + 2\alpha K|D^2_x u|^2 + K|D^3_x u|^2) \le (K \vee (2K))v_\delta - K^-\delta^{\frac{1}{2}}.$$

Repeating the same arguments in the last part of the proofs of Theorems 6.1.2 and 6.1.5, we obtain (6.3.1). □

Remark 6.3.2 Using the Jensen inequality as in the proof of Theorem 6.1.2, from (6.3.2) we get

$$|(\nabla_x T(t)f)(x)|^p \le e^{d_0 p t}(T(t)|\nabla f|^p)(x), \qquad t > 0,\ x \in \mathbb{R}^N, \tag{6.3.3}$$

for any $p \in [1,+\infty)$. Moreover, adapting the proof of Theorem 6.2.2 to this case, we can show that, for any $f \in C_b(\mathbb{R}^N)$, $t > 0$ and $x \in \mathbb{R}^N$, it holds:

$$|(\nabla_x T(t)f)(x)|^p \le \frac{2^p p^{1-p/2} d_0}{(p-1)^{p/2}(1 - e^{-p d_0 t})} t^{1-\frac{p}{2}}(T(t)|f|^p)(x) \tag{6.3.4}$$

for any $p \in (1, 2]$ and

$$|(\nabla_x T(t)f)(x)|^p \le \left(\frac{d_0}{1 - e^{-2d_0 t}}\right)^{\frac{p}{2}} (T(t)|f|^p)(x)$$

for any $p \in (2, +\infty)$.

We will use these estimates in Section 9.3.

In general, estimates (6.2.4) and (6.2.20) fail for $p = 1$ also in the case when the coefficients of the operator $\mathcal{A}$ are bounded. Here, we provide a simple situation in which this happens, taking $\mathcal{A} = \Delta$.

Example 6.3.3 Let $\{T(t)\}$ be the heat semigroup in $\mathbb{R}$, i.e.,

$$(T(t)f)(x) = \frac{1}{\sqrt{4\pi t}} \int_{\mathbb{R}} e^{-\frac{(x-y)^2}{4t}} f(y)dy, \qquad t > 0, \ x \in \mathbb{R}.$$

Fix $R > 0$ and let $f \in C_b(\mathbb{R})$ be such that $0 \le f \le 1$, $f(x) = 0$ for any $x < R - R^{-1}$ and $f(x) = 1$ for any $x > R$. Then,

$$(T(t)f)(0) \le \frac{1}{\sqrt{4\pi t}} \int_{R-R^{-1}}^{+\infty} e^{-\frac{|y|^2}{4t}} dy,$$

$$(D_x T(t)f)(0) \ge \frac{1}{2t\sqrt{4\pi t}} \int_{R}^{+\infty} y e^{-\frac{|y|^2}{4t}} dy.$$

Therefore, $(D_x T(t)f)(0) \ge c_R (T(t)f)(0)$, where

$$c_R = \frac{1}{2t} \int_R^{+\infty} y e^{-\frac{|y|^2}{4t}} dy \left(\int_{R-R^{-1}}^{+\infty} e^{-\frac{|y|^2}{4t}} dy\right)^{-1}.$$

It is readily seen that c_R diverges to $+\infty$ as R tends to $+\infty$. This means that no pointwise estimates similar to (6.2.4) can hold for $p = 1$.

Chapter 7

Markov semigroups in L^p-spaces

7.0 Introduction

In this chapter we consider some classes of elliptic operators with unbounded coefficients whose realizations in L^p-spaces related to the Lebesgue measure generate strongly continuous (and in some cases also analytic) semigroups. As is shown in Example 7.1.1, the L^p-spaces related to the Lebesgue measure are not the L^p-spaces which fit best the properties of elliptic operators with unbounded coefficients: for any choice of $\varepsilon > 0$ the operator $\mathcal{A}$, defined on smooth functions $\psi : \mathbb{R}^N \to \mathbb{R}$ by $\mathcal{A}\psi = \psi'' - x|x|^{\varepsilon}\psi'$, does not admit any realization in $L^p(\mathbb{R})$ generating a strongly continuous semigroup.

In Theorem 7.1.2 sufficient conditions are provided which guarantee that the restriction to $C_c(\mathbb{R}^N)$ of the minimal semigroup $\{T(t)\}$, associated with the operator $\mathcal{A} = \mathrm{Tr}(QD^2) + \langle b, \nabla\rangle + c$, can be extended to $L^p(\mathbb{R}^N)$ with a strongly continuous semigroup for $p \in (1, +\infty)$ and, in some cases, also for $p = 1$. If $\mathcal{A}$ is in divergence form and suitable conditions on its coefficients are assumed, then for any $p \in (1, +\infty)$ the realization of the operator $\mathcal{A}$ in $L^p(\mathbb{R}^N)$, with domain $D(A_p) = \{u \in W^{2,p}(\mathbb{R}^N) : cu \in L^p(\mathbb{R}^N)\}$, generates a strongly continuous analytic semigroup $\{T_p(t)\}$, which agrees with $\{T(t)\}$ on $C_c(\mathbb{R}^N)$.

In Sections 7.2 and 7.3 we confine ourselves to the case when $\mathcal{A}$ is a Schrödinger type operator. More specifically, we deal with the elliptic operator $\mathcal{A}$, defined by

$$\mathcal{A}\psi(x) = q(x)\Delta\psi(x) + c(x)\psi(x), \qquad x \in \mathbb{R}^N, \tag{7.0.1}$$

on smooth functions $\psi : \mathbb{R}^N \to \mathbb{R}$, where $q(x) = 1 + |x|^{\alpha}$ for any $x \in \mathbb{R}^N$ and some nonnegative constant α, c is a radial decreasing function (typically, $c(x) = -|x|^{\beta}$ for some $\beta \geq 0$) and at least one between q and c is unbounded.

In the case when $\alpha \in [0, 2]$, which we deal with in Section 7.2, generation results of analytic semigroups for suitable realizations A_p of the operator $\mathcal{A}$ in $L^p(\mathbb{R}^N)$ have been proved in [104, 176, 203]. The results in [176] cover the case when $\alpha \in [0, 2]$, $c(x) = -|x|^{\beta}$ for any $x \in \mathbb{R}^N$ and some $\beta \geq 0$. They show that the realization A_p in $L^p(\mathbb{R}^N)$, with domain

$$D(A_p) = \{u \in W^{2,p}(\mathbb{R}^N) : q|D^2u|, \sqrt{q}|\nabla u|, cu \in L^p(\mathbb{R}^N)\}, \tag{7.0.2}$$

generates a strongly continuous analytic semigroup $\{T_p(t)\}$ for any $p \in (1, +\infty)$. Such a semigroup agrees with $\{T(t)\}$ on $C_c(\mathbb{R}^N)$. In Subsection 7.2.1 we prove some remarkable properties of $\{T_p(t)\}$. In particular, we show that, for any $p \in (1, +\infty)$, $D(A_p)$ coincides with the domain of the maximal realization of the operator $\mathcal{A}$ in $L^p(\mathbb{R}^N)$. When $\beta > 0$, we prove that the spectrum of the operator A_p consists of a sequence of eigenvalues which accumulate at $-\infty$. Moreover, the spectrum is independent of p.

The picture changes drastically in the case when $\alpha > 2$. As it has been proved by G. Metafune and C. Spina in [203], the generation results depend upon N. More specifically, if $N = 1, 2$ no realization of $\mathcal{A}$ in $L^p(\mathbb{R}^N)$ generates a strongly continuous (resp. analytic) semigroup. The same happens if $N \geq 3$ and $p \leq N/(N-2)$. On the other hand, if $N \geq 3$, $p >$

$N/(N-2)$, then the maximal realization A_p of the operator $\mathcal{A}$ in $L^p(\mathbb{R}^N)$ generates a positive strongly continuous analytic semigroup, which is also contractive if $p > (N+\alpha-2)(N-2)^{-1}$. The case $\alpha > 2$ is treated in Section 7.3.

Finally, in Section 7.4 we consider the slightly more general class of operators $\mathcal{A}$, defined on smooth functions $\psi : \mathbb{R}^N \to \mathbb{R}$ by

$$\mathcal{A}\psi(x) = (1 + |x|^2)^{\frac{\alpha}{2}} \operatorname{Tr}(Q(x)D^2\psi(x)), \qquad x \in \mathbb{R}^N, \tag{7.0.3}$$

where the coefficients of the matrix valued function Q converge as $|x|$ tends to $+\infty$. This class of operators has been studied in [206, 238]. In the case $\alpha \le 2$, the authors of [206] have shown that the realization A_p of the operator $\mathcal{A}$ in (7.0.3), with domain $D(A_p)$ given by (7.0.2), generates a strongly continuous analytic semigroup in $L^p(\mathbb{R}^N)$ for any $p \in (1, +\infty)$. The case $\alpha > 2$ has been considered in [238] where the authors have proved generation results for $p \in (N(N-2)^{-1}, +\infty)$.

7.1 The general case

To begin with, we show by an example (due to J. Prüss, A. Rhandi, and R. Schnaubelt, see [223, Section 2]) that, in general, the L^p-spaces related to the Lebesgue measure are not the L^p-spaces which fit best the properties of elliptic operators with unbounded coefficients.

Example 7.1.1 Let $\mathcal{A}$ be the one-dimensional elliptic operator, defined on smooth functions $\psi : \mathbb{R} \to \mathbb{R}$ by $\mathcal{A}\psi = \psi'' + b\psi'$, where $b(x) = -x|x|^\varepsilon$ for any $x \in \mathbb{R}$ and ε is any positive real number. Fix $p \in [1, +\infty)$, $\lambda > 0$, a nonnegative and not identically vanishing function $f \in C_c^\infty(\mathbb{R})$ and suppose that the equation $\lambda u - \mathcal{A}u = f$ admits a solution $u \in L^p(\mathbb{R})$. By local elliptic regularity, u belongs to $C^2(\mathbb{R})$ and the classical maximum principle shows that u is positive in $\mathbb{R}$. We claim that u is bounded in $\mathbb{R}$. For this purpose, we take advantage of the Feller theory developed in Chapter 2. In this situation the functions $\mathscr{P}$ and $\mathscr{R}$ are given, respectively, by

$$\mathscr{P}(x) = e^{-\frac{|x|^{\varepsilon+2}}{\varepsilon+2}} \int_0^x e^{\frac{|t|^{\varepsilon+2}}{\varepsilon+2}} dt, \qquad \mathscr{R}(x) = e^{\frac{|x|^{\varepsilon+2}}{\varepsilon+2}} \int_0^x e^{-\frac{|t|^{\varepsilon+2}}{\varepsilon+2}} dt, \qquad x \in \mathbb{R}^N.$$

A straightforward computation reveals that $\lim_{x\to\pm\infty} x^{1+\varepsilon/2}\mathscr{P}(x) = 0$. Hence, $\mathscr{P} \in L^1(\mathbb{R})$. On the other hand, the function $\mathscr{R}$ does not belong to $L^1(\mathbb{R})$. This shows that $+\infty$ and $-\infty$ are both entrance points and the equation $\lambda u - \mathcal{A}u = 0$ admits a decreasing solution u_1 which tends to 1 at $+\infty$ and an increasing solution u_2 which tends to 1 at $-\infty$ (see Proposition 2.1.7). Clearly, u_1 and u_2 are linearly independent. Moreover u_1 and u_2 diverge to $+\infty$ at $-\infty$ and at $+\infty$, respectively. By Proposition 2.2.1 the equation $\lambda u - \mathcal{A}u = f$ admits a solution $R_\lambda f$ which is bounded. All the other solutions are given by $R_\lambda f + c_1u_1 + c_2u_2$ for some real constants c_1 and c_2. In view of the previous remarks, if $(c_1, c_2) \ne (0, 0)$ then the function $R_\lambda f + c_1u_1 + c_2u_2$ does not belong to $L^p(\mathbb{R})$. This shows that, if u is a solution in $L^p(\mathbb{R})$ to the equation $\lambda u - \mathcal{A}u = f$, then $u = R_\lambda f$. In particular, u is bounded.

We now set $V(x) = x^2$ for any $x \in \mathbb{R}$ and $W(x) = \varepsilon(2\lambda)^{-1} + |x|^{-\varepsilon}$ for $x \ne 0$. Then, there exists $r > 0$ such that $\lambda V - \mathcal{A}V \ge 0$ and $\lambda W - \mathcal{A}W \le 0$ in $\mathbb{R} \setminus (-r, r)$. Let $\beta > 0$ be sufficiently small that $u(\pm r) \ge \beta W(\pm r)$. For any $\delta > 0$ the function $w = u - \beta W + \delta V$ satisfies the inequality $\lambda w - \mathcal{A}w \ge 0$ in $\mathbb{R} \setminus (-r, r)$ and diverges to $+\infty$ as x tends to $\pm\infty$. Hence, it admits a minimum in $\mathbb{R} \setminus (-r, r)$. Such a minimum should be nonnegative, otherwise the inequality $\lambda w - \mathcal{A}w \ge 0$ would lead us to a contradiction. It thus follows that $u - \beta W + \delta V \geq 0$ in $\mathbb{R} \setminus (-r, r)$ for any $\delta > 0$. Letting δ tend to 0^+ we conclude that $u \geq \beta W$. Since $W \notin L^p(\mathbb{R})$ the function u does not belong to $L^p(\mathbb{R})$ as well and we obtain a contradiction.

In view of the previous example, it turns out that, besides Hypotheses 1.0.1, some additional conditions on the coefficients of the operator $\mathcal{A}$ should be assumed to guarantee that the semigroup $\{T(t)\}$, associated in $C_b(\mathbb{R}^N)$ with the operator $\mathcal{A} = \mathrm{Tr}(QD^2) + \langle b, \nabla\rangle + c$, can be extended to $L^p(\mathbb{R}^N)$.

Theorem 7.1.2 *Suppose that Hypotheses* 1.0.1 *hold true and the diffusion coefficients* q_{ij} $(i, j = 1, \ldots, N)$ *are continuously differentiable in* $\mathbb{R}^N$. *Further, let the function* $\beta : \mathbb{R}^N \to \mathbb{R}^N$ *be defined by* $\beta_i = b_i - \sum_{j=1}^N D_j q_{ij}$ *for any* $i = 1, \ldots, N$ *and assume that one of the following additional conditions is satisfied:*

(i) *the drift coefficients* b_j $(i, j = 1, \ldots, N)$ *are continuously differentiable in* $\mathbb{R}^N$, *the derivatives* $D_{ij} q_{ij}$ $(i, j = 1, \ldots, N)$ *exist in* $\mathbb{R}^N$ *and there exists a positive constant* K *such that* $\mathrm{div}_x \beta(x) - c(x) \geq -K$ *for any* $x \in \mathbb{R}^N$;

(ii) *the function* $\kappa : \mathbb{R}^N \to \mathbb{R}$ *in Hypothesis* 1.0.1(i) *is measurable and there exists a positive constant* K_p' *such that*

$$\frac{|\beta(x)|^2}{4(p-1)\kappa(x)} + c(x) \leq K_p', \qquad x \in \mathbb{R}^N.$$

Then, the minimal semigroup $\{T(t)\}$ *in* $C_b(\mathbb{R}^N)$ *associated with the operator* $\mathcal{A}$ *extends with a strongly continuous semigroup in* $L^p(\mathbb{R}^N)$ *for any* $p \in (1, +\infty)$ *(and even for* $p = 1$ *under condition* (i)*) and, for such values of* p,

$$\|T(t)f\|_{L^p(\mathbb{R}^N)} \leq e^{K_p t}\|f\|_{L^p(\mathbb{R}^N)}, \qquad t > 0, \tag{7.1.1}$$

where $K_p = p^{-1}(K - (p-1)c_0)$, *under condition* (i), *and* $K_p = K_p'$ *under condition* (ii).

Proof We prove the assertions for nonnegative functions $f \in C_c^\infty(\mathbb{R}^N)$. The density of $C_c^\infty(\mathbb{R}^N)$ in $L^p(\mathbb{R}^N)$ $(p \in [1, +\infty))$ and the estimate $|T(t)f| \leq T(t)|f|$ (which follows from the positivity of the semigroup $\{T(t)\}$, see Theorem 1.2.5), then allow us to extend the result to any $f \in L^p(\mathbb{R}^N)$.

For any $n \in \mathbb{N}$, let u_n be the classical solution of the Cauchy-Dirichlet problem

$$\begin{cases} D_t u_n(t, x) = \mathcal{A} u_n(t, x), & t > 0, \quad x \in B_n, \\ u_n(t, x) = 0, & t > 0, \quad x \in \partial B_n, \\ u_n(0, x) = f(x), & \qquad\quad\; x \in B_n. \end{cases}$$

By Theorem 1.2.1, u_n is nonnegative in $[0, +\infty) \times \mathbb{R}^N$ and therein converges to u pointwise as n tends to $+\infty$.

The main step of the proof consists of showing that

$$\|u_n(t, \cdot)\|_{L^p(B_n)} \leq e^{K_p t}\|u_n(s, \cdot)\|_{L^p(B_n)} = e^{K_p t}\|f\|_{L^p(B_n)} \tag{7.1.2}$$

for any $t > 0$ and $n \in \mathbb{N}$ such that $\mathrm{supp}(f) \subset B_n$. Letting n tend to $+\infty$ in (7.1.2), by Fatou's lemma one then completes the proof. Indeed, as the proof of Theorem 1.2.1 shows, the sequence $\{u_n\}$ converges pointwise in $(0, +\infty) \times \mathbb{R}^N$ to $T(\cdot)f$, as n tends to $+\infty$.

Fix $p \in (1, +\infty)$, $\varepsilon > 0$ and observe that

$$\frac{d}{dt}\|u_n(t, \cdot) + \varepsilon\|_{L^p(B_n)}^p = p\int_{B_n} (u_n(t, \cdot) + \varepsilon)^{p-1} \mathcal{A} u_n(t, \cdot)\, dx.$$

We first assume that the additional condition (i) is satisfied. Integrating by parts and observing that u_n vanishes on ∂B_n, we get

$$\begin{aligned}
&\int_{B_n}(u_n(t,\cdot)+\varepsilon)^{p-1}\mathcal{A}u_n(t,\cdot)\,dx\\
=&\varepsilon^{p-1}\int_{\partial B_n}\langle Q\nabla_x u_n(t,\cdot),\nu\rangle d\sigma-\varepsilon\int_{B_n}c(u_n(t,\cdot)+\varepsilon)^{p-1}\,dx\\
&-(p-1)\int_{B_n}(u_n(t,\cdot)+\varepsilon)^{p-2}\langle Q\nabla_x u_n(t,\cdot),\nabla_x u_n(t,\cdot)\rangle dx\\
&+\frac{1}{p}\int_{B_n}\langle\beta(t,\cdot),\nabla_x(u_n(t,\cdot)+\varepsilon)^p\rangle\,dx+\int_{B_n}c(u_n(t,\cdot)+\varepsilon)^p dx,
\end{aligned}\tag{7.1.3}$$

where, as usual, $\nu=\nu(x)$ is the outward unit normal at $x\in\partial B_n$ and $d\sigma$ denotes the surface measure on ∂B_n. Integrating by parts the term containing β, we conclude that

$$\begin{aligned}
\frac{d}{dt}\|u_n(t,\cdot)+\varepsilon\|_{L^p(B_n)}^p\leq&\varepsilon^{p-1}p\int_{\partial B_n}\langle Q\nabla_x u_n(t,\cdot),\nu\rangle d\sigma-\varepsilon p\int_{B_n}c(u_n(t,\cdot)+\varepsilon)^{p-1}\,dx\\
&+\int_{B_n}(u_n(t,\cdot)+\varepsilon)^p\big(pc-\operatorname{div}_x\beta\big)dx+\varepsilon^p\int_{\partial B_n}\langle\beta,\nu\rangle\,d\sigma\\
=:&\,g_n^{\varepsilon,p}(t)+\int_{B_n}(u_n(t,\cdot)+\varepsilon)^p(pc-\operatorname{div}_x\beta)dx.
\end{aligned}\tag{7.1.4}$$

Since $pc-\operatorname{div}_x\beta=p(c-c_0)-\operatorname{div}_x\beta+pc_0\leq c-c_0-\operatorname{div}_x\beta+pc_0\leq pK_p$, from (7.1.4) we get

$$\frac{d}{dt}\|u_n(t,\cdot)+\varepsilon\|_{L^p(B_n)}^p\leq g_n^{\varepsilon,p}(t)+pK_p\|u_n^\varepsilon(t,\cdot)\|_{L^p(B_n)}^p,\qquad t>0.$$

As is easily seen, the function $g_n^{\varepsilon,p}$ vanishes as ε tends to 0 from the right. Hence, estimate (7.1.2) with $p>1$ follows from applying the Gronwall lemma and the dominated convergence theorem to let ε tend to 0^+. The case $p=1$ follows letting p tend to 1^+ in (7.1.2).

Let us now suppose that the condition (ii) is satisfied. Using the Hölder and Young inequalities we can estimate the integral term containing β in the right-hand side of (7.1.3) as follows

$$\begin{aligned}
&\int_{B_n}\langle\beta,\nabla_x(u_n(t,\cdot)+\varepsilon)^p\rangle dx\\
\leq&p\bigg(\int_{B_n}\kappa(u_n(t,\cdot)+\varepsilon)^{p-2}|\nabla_x u_n(t,\cdot)|^2\,dx\bigg)^{\frac{1}{2}}\bigg(\int_{B_n}\frac{|\beta|^2}{\kappa}(u_n(t,\cdot)+\varepsilon)^p dx\bigg)^{\frac{1}{2}}\\
\leq&\delta p\int_{B_n}\kappa(u_n(t,\cdot)+\varepsilon)^{p-2}|\nabla_x u_n(t,\cdot)|^2dx+\frac{p}{4\delta}\int_{B_n}\frac{|\beta|^2}{\kappa}(u_n(t,\cdot)+\varepsilon)^p dx
\end{aligned}$$

for any $\delta>0$. Choosing $\delta=p-1$ we get

$$\begin{aligned}
\frac{d}{dt}\|u_n(t,\cdot)+\varepsilon\|_{L^p(B_n)}^p\leq&K_p'\int_{B_n}(u_n(t,\cdot)+\varepsilon)^p\,dx-\varepsilon\int_{B_n}c(u_n(t,\cdot)+\varepsilon)^{p-1}dx\\
&+\varepsilon^{p-1}\int_{\partial B_n}\langle Q\nabla_x u_n(t,\cdot),\nu\rangle d\sigma.
\end{aligned}$$

Again, using Gronwall lemma and letting ε tend to 0^+, estimate (7.1.2) follows also in this case.

To conclude the proof, let us prove that $T(t)f$ converges to f in $L^p(\mathbb{R}^N)$, as t tends to 0^+. It suffices to prove the assertion in the case when $f \in C_c^2(\mathbb{R}^N)$. Indeed, the general case then follows from a standard density argument through estimate (7.1.1). So, let us fix $f \in C_c^2(\mathbb{R}^N)$ and $t > 0$. Using Proposition 1.3.4 we can write

$$(T(t)f)(x) - f(x) = \int_0^t (T(s)\mathcal{A}f)(x)ds, \qquad x \in \mathbb{R}^N,$$

and, using inequality (7.1.1), we deduce that $\|T(t)f - f\|_{L^p(\mathbb{R}^N)} \le C_p(e^{K_p t} - 1)$, where $C_p = K_p^{-1}\|\mathcal{A}f\|_{L^p(\mathbb{R}^N)}$. Clearly, this estimate implies that $T(t)f$ converges to f in $L^p(\mathbb{R}^N)$ as t tends to 0^+. □

Analyticity of the semigroup $\{T(t)\}$ in $L^p(\mathbb{R}^N)$ has been studied by N. Okazawa in [211] (for Schrödinger operators), by P. Cannarsa and V. Vespri in [50, 51], by G. Da Prato and V. Vespri in [78] and by G. Metafune, J. Prüss, A. Rhandi, and R. Schnaubelt in [200]. Here, we present the results in [200] which deal with the class of elliptic operators in divergence form $\mathcal{A} = \mathrm{div}(Q\nabla) + \langle b, \nabla\rangle + c$ under the following set of assumptions.

Hypotheses 7.1.3 (i) *The diffusion coefficients belong to $C_b^1(\mathbb{R}^N)$ and there exists a positive constant κ_0 such that $\langle Q(x)\xi, \xi\rangle \ge \kappa_0|\xi|^2$ for any $x, \xi \in \mathbb{R}^N$;*

(ii) *$b_j \in C^1(\mathbb{R}^N)$ $(j = 1, \ldots, N)$ and there exist positive constants θ, γ, K_0, ξ_0, K_0, C_γ and a function $\zeta \in C^1(\mathbb{R}^N)$ such that*

$$\frac{\theta}{p} + (p-1)\gamma\left(\frac{K_0}{p} + \frac{\Lambda\gamma}{4}\right) < 1, \tag{7.1.5}$$

$\zeta \ge \xi_0$, $|\nabla\zeta| \le \gamma\zeta^{3/2} + C_\gamma$ and $|b| \le K_0\sqrt{\zeta}$ in $\mathbb{R}^N$, where Λ denotes the supremum over $\mathbb{R}^N$ of the maximum eigenvalue of the matrix $Q(x)$. Moreover, $\theta\zeta + \mathrm{div}\, b \ge 0$ in $\mathbb{R}^N$ for some $\theta < p$;

(iii) *the function c belongs to $C^\alpha_{\mathrm{loc}}(\mathbb{R}^N)$ and $-K_1\zeta \le c \le -\zeta$ for some constant $K_1 \ge 1$.*

Remark 7.1.4 Actually in [200] the assumption on the smoothness of c is relaxed since only its measurability is assumed. Here, to avoid some technical difficulties and to use the results of Chapter 1, we assume that c is locally Hölder continuous in $\mathbb{R}^N$.

Under the previous set of hypotheses, slightly modifying the arguments in [200], we can prove the following result.

Theorem 7.1.5 *The restriction of the semigroup $\{T(t)\}$ to $C_c(\mathbb{R}^N)$ extends to $L^p(\mathbb{R}^N)$ $(p \in (1, +\infty))$ with a strongly continuous analytic semigroup of contractions. Moreover the domain of its infinitesimal generator is $D(A_p) = \{u \in W^{2,p}(\mathbb{R}^N) : cu \in L^p(\mathbb{R}^N)\}$ and the graph norm of $D(A_p)$ is equivalent to the norm $\|\cdot\|$ defined by $\|u\|_{W^{2,p}(\mathbb{R}^N)} + \|cu\|_{L^p(\mathbb{R}^N)}$ for any $u \in D(A_p)$.*

In the proof of Theorem 7.1.5 we take advantage of the following proposition.

Proposition 7.1.6 ([64], Prop. 6.1) *Let $\{B_{\rho(x)}(x) : x \in \mathbb{R}^N\}$ be a covering of $\mathbb{R}^N$, where $\rho : \mathbb{R}^N \to \mathbb{R}_+$ is a Lipschitz continuous function, with Lipschitz constant L strictly less than $1/2$. Then, there exist a countable subcovering $\{B_{\rho(x_n)}(x_n) : n \in \mathbb{N}\}$ and a natural number $\varsigma = \varsigma(N, L)$ such that at most ς among the doubled balls $\{B_{2\rho(x_n)}(x_n) : n \in \mathbb{N}\}$ overlap.*

Proof We set $\rho_n = \rho(0)(1+L)^n$ for any $n \in \mathbb{N}$ and $\rho_0 = \rho(0)$. As is easily seen, the family $\{\Omega_n : n \in \mathbb{N} \cup \{0\}\}$, where $\Omega_n = B_{2\rho_n} \setminus B_{2\rho_{n-1}}$, if $n \geq 1$, and $\Omega_0 = B_{2\rho_0}$, is a covering of $\mathbb{R}^N$. By the Besicovitch covering theorem (see, e.g. [12, Theorem 2.18]), there exists $\xi(N) \in \mathbb{N}$, and for any $n \in \mathbb{N} \cup \{0\}$ an at most countable set $S_n \subset \Omega_n$, such that $\Omega_n \subset C_n := \bigcup_{j \in S_n} B_{\rho(x_j)}(x_j)$ for any $n \in \mathbb{N} \cup \{0\}$. Since $\rho(x) \leq \rho(0) + L|x|$ for any $x \in \mathbb{R}^N$, it follows that $C_n \subset B_{2\rho_{n+1}+\rho_0} \setminus B_{2\rho_{n-1}(1-L)-\rho_0}$ for any $n \in \mathbb{N} \cup \{0\}$.

We claim that there exists $h = h(L)$ such that $C_m \cap C_n = \varnothing$ if $n \geq m + h$. For this purpose, it suffices to show that there exists $h \in \mathbb{N}$ such that $B_{2\rho_{m+1}+\rho_0} \subset B_{2\rho_{n-1}(1-L)-\rho_0}$ if $n - m \geq h$. This is the case, if $1 < (1+L)^{m+1}((1-L)(1+L)^{n-m-2} - 1)$. Since $L < 1$, we can fix $h = h(L) \in \mathbb{N}$ such that $(1-L)(1+L)^{h-2} - 1 > 1$. With this choice of h, the claim follows at once. We have so proved that each set C_n overlaps at most $2h$ sets from the family $\{C_m : m \in \mathbb{N} \cup \{0\}\}$.

Let us order the elements of the set $S = \bigcup_{n \in \mathbb{N} \cup \{0\}} S_n$ into a sequence $\{x_n\}$. We claim that the family $\mathcal{F} = \{B_{\rho(x_n)}(x_n) : n \in \mathbb{N}\}$ is a subcovering of $\mathbb{R}^N$ with the properties in the statement. This family is clearly a covering of $\mathbb{R}^N$, since the union of the sets C_n covers the union of the sets Ω_n ($n \in \mathbb{N} \cup \{0\}$) which is $\mathbb{R}^N$, as has been observed. To control the overlaps of the double balls $B_{2\rho(x_n)}(x_n)$, we begin by observing that the intersection of more than $\xi(N, L) := h(L)\xi(N)$ balls of the family $\mathcal{F}$ is empty. Indeed, let $J \subset \mathbb{N}$ be a finite set such that $\bigcap_{j \in J} B_{\rho(x_j)}(x_j) \neq \varnothing$. Then, there exist $m \in \mathbb{N}$ and $k_1, \ldots, k_m \in \mathbb{N}$ such that $J_i = J \cap S_{k_i} \neq \varnothing$ and $\bigcup_{i=1} J_i = J$. Therefore, we can split $\bigcap_{j \in J} B_{\rho(x_j)}(x_j) = \bigcap_{i=1}^m \bigcap_{j \in J_i} B_{\rho(x_j)}(x_j)$. Since $\bigcap_{j \in J_i} B_{\rho(x_j)}(x_j) \subset C_{k_i}$, for any $i = 1, \ldots, m$, from the above results it follows that $m \leq h$. Moreover, since no more than $\xi(N)$ from the balls which constitute each set C_{k_i} may intersect, the cardinality of J_i does not exceed $\xi(N)$ for any $i = 1, \ldots, m$. Summing up, we conclude that the cardinality of J does not exceed $\xi(L, N)$. Let us fix $i, j \in \mathbb{N}$ such that $B_{2\rho(x_i)}(x_i) \cap B_{2\rho(x_j)}(x_j) \neq \varnothing$. Then, $|x_i - x_j| \leq 2(\rho(x_i) + \rho(x_j))$. Since ρ is L-Lipschitz continuous, we deduce that $|\rho(x_i) - \rho(x_j)| \leq 2L(\rho(x_i) + \rho(x_j))$ or, equivalently,

$$\frac{1-2L}{1+2L}\rho_i \leq \rho_j \leq \frac{1+2L}{1-2L}\rho_i. \tag{7.1.6}$$

Thus, $B_{\rho(x_j)}(x_j) \subset B_{\frac{5+2L}{1-2L}\rho_i}(x_i)$. Set $K_i = \{j \in \mathbb{N} : B_{2\rho(x_i)}(x_i) \cap B_{2\rho(x_j)}(x_j) \neq \varnothing\}$. Then, $\bigcup_{j \in K_i} B_{2\rho(x_j)}(x_j) \subset B_{\frac{5+2L}{1-2L}\rho_i}(x_i)$, which implies that

$$\sum_{j \in K_i} \chi_{B_{\rho(x_j)}(x_j)} \leq \xi(N, L)\chi_{B_{\frac{5+2L}{1-2L}\rho_i}(x_i)}.$$

Integrating over $\mathbb{R}^N$ and using the first inequality in (7.1.6), we obtain that

$$\left(\frac{1-2L}{1+2L}\right)^N \rho_i^N \mathrm{card}(K_i) \leq \sum_{j \in K_i} \rho(x_j)^N \leq \xi(N, L)\left(\frac{5+2L}{1-2L}\right)^N \rho_i^N,$$

i.e.,

$$\mathrm{card}(K_i) \leq \zeta(N, L)\left(\frac{4L^2 + 12L + 5}{(1-2L)^2}\right)^N =: \varsigma(N, L) - 1.$$

This completes the proof. □

Remark 7.1.7 In the proof of Theorem 7.1.5 as well as in the proofs of Proposition 7.2.1 and Theorem 7.2.5 we should perform some integration by parts, which are straightforward if $p \geq 2$ and demand some approximation arguments to be justified when $p \in (1, 2)$. For further details, we refer the reader to [202].

Proof of Theorem 7.1.5 The proof is rather long. Hence, we split it into steps. Throughout the proof, we assume that $|\nabla\zeta| \le \gamma\zeta^{3/2}$. We can reduce to this situation by replacing ζ and c with $\zeta+k$ and $c-k$ for a suitable positive constant k. Moreover, we denote by C a positive constant, independent of u, which may vary from line to line, and we assume that p is arbitrarily fixed in $(1,+\infty)$.

Step 1. Here, we prove that there exist two positive constants $\varepsilon_{0,p}$ and C such that

$$\|\sqrt{|c|}\nabla u\|_{L^p(\mathbb{R}^N)} \le \varepsilon\|\Delta u\|_{L^p(\mathbb{R}^N)} + C\varepsilon^{-1}\|cu\|_{L^p(\mathbb{R}^N)} \tag{7.1.7}$$

for any $u \in D(A_p)$ and $\varepsilon \in (0,\varepsilon_{0,p}]$. Observe that it suffices to prove the previous estimate for functions in $C_c^\infty(\mathbb{R}^N)$ and with c being replaced by ζ. Indeed, $C_c^\infty(\mathbb{R}^N)$ is dense in $D(A_p)$ endowed with the norm $\|u\|_{D(A_p)} = \|u\|_{W^{2,p}(\mathbb{R}^N)} + \|cu\|_{L^p(\mathbb{R}^N)}$ for any $u \in D(A_p)$, as it can be easily deduced from the proof of the forthcoming Lemma 7.2.2 and, by Hypothesis 7.1.3(iii), $\zeta \le |c| \le K_1\zeta$.

So, let us fix $u \in C_c^\infty(\mathbb{R}^N)$. The starting point of the proof of (7.1.7) is the well-known interpolation inequality (see e.g. [250])

$$\|\nabla u\|_{L^p(\mathbb{R}^N)} \le C\|u\|_{L^p(\mathbb{R}^N)}^{\frac12}\|D^2u\|_{L^p(\mathbb{R}^N)}^{\frac12}, \qquad u \in W^{2,p}(\mathbb{R}^N), \tag{7.1.8}$$

which, together with Calderón-Zygmund inequality, allows us to estimate

$$\|\nabla u\|_{L^p(\mathbb{R}^N)} \le C\|\Delta u\|_{L^p(\mathbb{R}^N)}^{\frac12}\|u\|_{L^p(\mathbb{R}^N)}^{\frac12}, \qquad u \in W^{2,p}(\mathbb{R}^N). \tag{7.1.9}$$

The estimate $|\nabla\zeta| \ge \gamma\zeta^{3/2}$ implies that $|\nabla\sqrt{\zeta}| \le 2^{-1}\gamma|\zeta|$, which reveals that for any $x_0 \in \mathbb{R}^N$

$$\frac12\sqrt{\zeta(x)} \le \sqrt{\tilde{\zeta}(x_0)} \le \frac32\sqrt{\zeta(x)}, \qquad x \in B_{2r_{x_0}}(x_0), \tag{7.1.10}$$

where $r_{x_0} = (2\gamma\sqrt{\zeta(x_0)})^{-1}$. Now, for any $x_0 \in \mathbb{R}^N$ we set $\vartheta_{x_0}(x) = \vartheta(r_{x_0}^{-1}(x-x_0))$, where $\vartheta \in C_c^\infty(\mathbb{R}^N)$ satisfies the condition $\chi_{B_1} \le \vartheta \le \chi_{B_2}$. Applying estimate (7.1.9) to the function $\vartheta_{x_0}u$ and using the Young inequality, we get

$$\begin{aligned}
&\|\sqrt{\zeta(x_0)}\nabla u\|_{L^p(B_{r_{x_0}}(x_0))} \le \|\sqrt{\zeta(x_0)}\nabla(\vartheta_{x_0}u)\|_{L^p(\mathbb{R}^N)}\\
\le& C\|\zeta(x_0)\vartheta_{x_0}u\|_{L^p(\mathbb{R}^N)}^{\frac12}\|\Delta(\vartheta_{x_0}u)\|_{L^p(\mathbb{R}^N)}^{\frac12}\\
\le& C\left(\varepsilon\|\Delta(\vartheta_{x_0}u)\|_{L^p(\mathbb{R}^N)} + \frac{1}{4\varepsilon}\|\zeta(x_0)\vartheta_{x_0}u\|_{L^p(\mathbb{R}^N)}\right)\\
\le& C\left(\varepsilon\|\Delta u\|_{L^p(B_{2r_{x_0}}(x_0))} + \varepsilon\|\sqrt{\zeta(x_0)}\nabla u\|_{L^p(B_{2r_{x_0}}(x_0))} + \varepsilon^{-1}\|\zeta(x_0)u\|_{L^p(B_{2r_{x_0}}(x_0))}\right)
\end{aligned}$$

for any $\varepsilon \in (0,1]$. Using (7.1.10) we can estimate

$$\begin{aligned}
&\|\sqrt{\zeta}\nabla u\|_{L^p(B_{r_{x_0}}(x_0))} \le \frac32\|\sqrt{\zeta(x_0)}\nabla u\|_{L^p(B_{r_{x_0}}(x_0))}\\
\le& C\big(\varepsilon\|\Delta u\|_{L^p(B_{2r_{x_0}}(x_0))} + \varepsilon\|\sqrt{\zeta}\nabla u\|_{L^p(B_{2r_{x_0}}(x_0))} + \varepsilon^{-1}\|\zeta u\|_{L^p(B_{2r_{x_0}}(x_0))}\big).
\end{aligned}$$

By Proposition 7.1.6 there exist a sequence $\{x_n\}$ and a positive number ς such that $\mathcal{F} = \{B_{r_{x_n}}(x_n) : n \in \mathbb{N}\}$ is a covering of $\mathbb{R}^N$ and the intersection of more than ς balls from $\mathcal{F}$ is empty. Proposition 7.1.6 can be rephrased by claiming that

$$1 \le \sum_{n=1}^{+\infty}\chi_{B_{r_{x_n}}(x_n)} \le \sum_{n=1}^{+\infty}\chi_{B_{2r_{x_n}}(x_n)} \le \varsigma \ \text{ in } \mathbb{R}^N.$$

Hence, we can write

$$\begin{aligned}\|\sqrt{\zeta}\nabla u\|^p_{L^p(\mathbb{R}^N)} \le & C\varepsilon^p \int_{\mathbb{R}^N} |\Delta u|^p \sum_{n=1}^{+\infty} \chi_{B_{2r_{x_n}}(x_n)} dx + C\varepsilon^p \int_{\mathbb{R}^N} \zeta^{\frac{p}{2}} |\nabla u|^p \sum_{n=1}^{+\infty} \chi_{B_{2r_{x_n}}(x_n)} dx \\ & + C\varepsilon^{-p} \int_{\mathbb{R}^N} |u|^p \sum_{n=1}^{+\infty} \chi_{B_{2r_{x_n}}(x_n)} dx \\ \le & C\varepsilon^p \|\Delta u\|^p_{L^p(\mathbb{R}^N)} + C\varepsilon^p \|\sqrt{\zeta}\nabla u\|^p_{L^p(\mathbb{R}^N)} + C\varepsilon^{-p} \|\zeta u\|^p_{L^p(\mathbb{R}^N)}.\end{aligned}$$

Due to the arbitrariness of $\varepsilon \in (0,1]$, from the above estimate we deduce that there exist $\varepsilon_{0,p} \in (0,1]$ and C such that $\|\sqrt{\zeta}\nabla u\|_{L^p(\mathbb{R}^N)} \le \varepsilon \|\Delta u\|_{L^p(\mathbb{R}^N)} + C\varepsilon^{-1}\|\zeta u\|_{L^p(\mathbb{R}^N)}$ for any $\varepsilon \in (0, \varepsilon]$.

Step 2. Here, we prove that for any $p \in (1,+\infty)$ there exist two positive constants $C_{1,p}$ and $C_{2,p}$ such that

$$\|u\|_{W^{2,p}(\mathbb{R}^N)} + \|cu\|_{L^p(\mathbb{R}^N)} \le C_{1,p}\|u - \mathcal{A}u\|_{L^p(\mathbb{R}^N)} \le C_{2,p}(\|u\|_{W^{2,p}(\mathbb{R}^N)} + \|cu\|_{L^p(\mathbb{R}^N)}) \quad (7.1.11)$$

for any $u \in D(A_p)$.

In view of estimate (7.1.7), the realization A_p of the operator $\mathcal{A}$ with domain $D(A_p)$ is well defined and $\|u - \mathcal{A}u\|_{L^p(\mathbb{R}^N)} \le C(\|u\|_{W^{2,p}(\mathbb{R}^N)} + \|cu\|_{L^p(\mathbb{R}^N)})$ for any $u \in D(A_p)$. To prove the other estimate in (7.1.11), we fix a real-valued function $u \in C^\infty_c(\mathbb{R}^N)$ and set $f = -\mathcal{A}u$. An integration by parts shows that

$$\begin{aligned}\int_{\mathbb{R}^N} \zeta^{p-1}\langle b, \nabla u\rangle u|u|^{p-2}dx =& \frac{1}{p}\int_{\mathbb{R}^N} \zeta^{p-1}\langle b, \nabla(|u|^p)\rangle dx \\ =& -\frac{1}{p}\int_{\mathbb{R}^N} (\operatorname{div} b)\zeta^{p-1}|u|^p dx - \frac{p-1}{p}\int_{\mathbb{R}^N} \zeta^{p-2}|u|^p\langle b, \nabla\zeta\rangle dx,\end{aligned}$$

$$\begin{aligned}\int_{\mathbb{R}^N} \operatorname{div}(Q\nabla u)\zeta^{p-1}u|u|^{p-2}dx =& -(p-1)\int_{\mathbb{R}^N} \zeta^{p-1}|u|^{p-2}|\sqrt{Q}\nabla u|^2 dx \\ & -(p-1)\int_{\mathbb{R}^N} \zeta^{p-1}u|u|^{p-2}\zeta^{p-2}\langle Q\nabla u, \nabla\zeta\rangle dx.\end{aligned}$$

Hence, multiplying both sides of the equation $\mathcal{A}u = -f$ by $\zeta^{p-1}u|u|^{p-2}$, we get

$$\begin{aligned}\int_{\mathbb{R}^N} c\zeta^{p-1}|u|^p dx =& (p-1)\int_{\mathbb{R}^N} \zeta^{p-1}|u|^{p-2}|\sqrt{Q}\nabla u|^2 dx + \frac{1}{p}\int_{\mathbb{R}^N} (\operatorname{div} b)\zeta^{p-1}|u|^p dx \\ & + (p-1)\int_{\mathbb{R}^N} u|u|^{p-2}\zeta^{p-2}\langle Q\nabla u, \nabla\zeta\rangle dx + \frac{p-1}{p}\int_{\mathbb{R}^N} \zeta^{p-2}|u|\langle b, \nabla\zeta\rangle dx \\ & - \int_{\mathbb{R}^N} f\zeta^{p-1}u|u|^{p-2}dx.\end{aligned}$$

By Hypotheses 7.1.3, $p^{-1}\operatorname{div} b - c \ge (p-\theta)p^{-1}\zeta$ and, consequently, from the previous estimate, the Hölder and Young inequalities, and recalling that $|c| \ge \zeta$, we conclude that

$$\begin{aligned}&\left(1 - \frac{\theta}{p}\right)\int_{\mathbb{R}^N} \zeta^p|u|^p dx + (p-1)\int_{\mathbb{R}^N} |u|^{p-2}\zeta^{p-1}|\sqrt{Q}\nabla u|^2 dx \\ \le & (p-1)\int_{\mathbb{R}^N} \zeta^{p-2}|u|^{p-1}|\langle Q\nabla u, \nabla\zeta\rangle| dx + \frac{p-1}{p}\int_{\mathbb{R}^N} \zeta^{p-2}|u|^p|b||\nabla\zeta| dx \\ & + \varepsilon\|\zeta u\|^p_{L^p(\mathbb{R}^N)} + C_\varepsilon\|f\|^p_{L^p(\mathbb{R}^N)} \qquad (7.1.12)\end{aligned}$$

for any $\varepsilon > 0$. Using Hypotheses 7.1.3(ii), (iii) we get

$$\int_{\mathbb{R}^N} \zeta^{p-2}|u|^p|b||\nabla\zeta|dx \le K_0\gamma\|\zeta u\|^p_{L^p(\mathbb{R}^N)} \tag{7.1.13}$$

and

$$\begin{aligned}\int_{\mathbb{R}^N} \zeta^{p-2}|u|^{p-1}|\langle Q\nabla u, \nabla\zeta\rangle|dx \le& \sqrt{M}\gamma \int_{\mathbb{R}^N} \zeta^{p-\frac{1}{2}}|u|^{p-1}|\sqrt{Q}\nabla u|dx\\ \le& \sqrt{M}\gamma\bigg(\int_{\mathbb{R}^N} |u|^{p-2}\zeta^{p-1}|\sqrt{Q}\nabla u|^2dx\bigg)^{\frac{1}{2}}\|\zeta u\|^{\frac{p}{2}}_{L^p(\mathbb{R}^N)}.\end{aligned} \tag{7.1.14}$$

From (7.1.12)-(7.1.14) it follows that

$$\begin{aligned}&\bigg(1-\frac{\theta}{p}-K_0\gamma\frac{p-1}{p}-\varepsilon\bigg)\|\zeta u\|^p_{L^p(\mathbb{R}^N)} + (p-1)\int_{\mathbb{R}^N} |u|^{p-2}\zeta^{p-1}|\sqrt{Q}\nabla u|^2dx\\ \le&(p-1)\sqrt{M}\gamma\bigg(\int_{\mathbb{R}^N} |u|^{p-2}\zeta^{p-1}|\sqrt{Q}\nabla u|^2dx\bigg)^{\frac{1}{2}}\|u\zeta\|^{\frac{p}{2}}_{L^p(\mathbb{R}^N)} + C_\varepsilon\|f\|^p_{L^p(\mathbb{R}^N)}\\ \le&\frac{M}{4}(p-1)\gamma^2\|\zeta u\|^p_{L^p(\mathbb{R}^N)} + (p-1)\int_{\mathbb{R}^N} |u|^{p-2}\zeta^{p-1}|\sqrt{Q}\nabla u|^2dx + C_\varepsilon\|f\|^p_{L^p(\mathbb{R}^N)}\end{aligned}$$

so that

$$\bigg(1-\frac{\theta}{p}-K_0\gamma\frac{p-1}{p}-\frac{M}{4}(p-1)\gamma^2-\varepsilon\bigg)\|\zeta u\|^p_{L^p(\mathbb{R}^N)} \le C_\varepsilon\|f\|^p_{L^p(\mathbb{R}^N)}.$$

Due to condition (7.1.5), we can fix ε sufficiently small that the term in brackets in the previous inequality is positive and we thus conclude that $\|\zeta u\|_{L^p(\mathbb{R}^N)} \le C\|f\|_{L^p(\mathbb{R}^N)}$ so that $\|cu\|_{L^p(\mathbb{R}^N)} \le CK_1\|f\|_{L^p(\mathbb{R}^N)}$. This estimate combined with (7.1.7) and Hypotheses 7.1.3, yields $\|\langle b, \nabla u\rangle\|_{L^p(\mathbb{R}^N)} \le C\|f\|_{L^p(\mathbb{R}^N)}$. By the well-known estimate $\|D^2u\|_{L^p(\mathbb{R}^N)} \le C(\|\text{div}(Q\nabla u)\|_{L^p(\mathbb{R}^N)} + \|u\|_{L^p(\mathbb{R}^N)})$ and all the previous estimates we conclude that

$$\|D^2u\|_{L^p(\mathbb{R}^N)} \le C(\|f\|_{L^p(\mathbb{R}^N)} + \|\langle b, \nabla u\rangle\|_{L^p(\mathbb{R}^N)} + \|cu\|_{L^p(\mathbb{R}^N)}) \le C\|f\|_{L^p(\mathbb{R}^N)}.$$

Finally, the interpolation formula (7.1.8) allows us to bound $\|\nabla u\|_{L^p(\mathbb{R}^N)}$ from above in terms of $\|f\|_{L^p(\mathbb{R}^N)}$. The first inequality in (7.1.11) follows for functions in $C^\infty_c(\mathbb{R}^N)$. The density of this space in $D(A_p)$ can be used to extend (7.1.11) to any $u \in D(A_p)$.

Step 3. Here, we prove that

$$\bigg|\text{Im}\bigg(\int_{\mathbb{R}^N} |u|^{p-2}\overline{u}Audx\bigg)\bigg| \le -\delta_p\text{Re}\bigg(\int_{\mathbb{R}^N} |u|^{p-2}\overline{u}Audx\bigg), \qquad u \in D(A_p), \tag{7.1.15}$$

with $\delta_p = \frac{|p-2|^2}{4(p-1)} + \frac{K_0^2p}{4\kappa_0(p-\theta)} + 2$.

As in the previous steps, it suffices to prove (7.1.15) for functions in $C^\infty_c(\mathbb{R}^N)$. We fix $u \in C^\infty_c(\mathbb{R}^N)$. An integration by parts and Hypothesis 7.1.3(ii) reveal that

$$\begin{aligned}\text{Re}\bigg(\int_{\mathbb{R}^N} |u|^{p-2}\overline{u}Audx\bigg) =& -(p-1)\int_{\mathbb{R}^N} |u|^{p-4}|\sqrt{Q}\text{Re}(\overline{u}\nabla u)|^2dx\\ &-\int_{\mathbb{R}^N} |u|^{p-4}|\sqrt{Q}\text{Im}(\overline{u}\nabla u)|^2dx + \int_{\mathbb{R}^N}(c-p^{-1}\text{div}\, b)|u|^pdx\\ \le& -(p-1)\int_{\mathbb{R}^N} |u|^{p-4}|\sqrt{Q}\text{Re}(\overline{u}\nabla u)|^2dx\end{aligned}$$

$$
-\int_{\mathbb{R}^N}|u|^{p-4}|\sqrt{Q}\mathrm{Im}(\overline{u}\nabla u)|^2dx-\left(1-\frac{\theta}{p}\right)\int_{\mathbb{R}^N}|c||u|^pdx
\tag{7.1.16}
$$

and we conclude that the operator $\mathcal{A}$ is dissipative in $C_c^\infty(\mathbb{R}^N)$. Arguing similarly, and using Hypotheses 7.1.3 and the Hölder inequality, we can show that

$$
\begin{aligned}
&\left|\mathrm{Im}\left(\int_{\mathbb{R}^N}|u|^{p-2}\overline{u}\mathcal{A}udx\right)\right|\\
\le&|p-2|\int_{\mathbb{R}^N}|u|^{p-4}|\langle Q\mathrm{Re}(\overline{u}\nabla u),\mathrm{Im}(\overline{u}\nabla u)\rangle|dx+\int_{\mathbb{R}^N}|b||u|^{p-2}|\mathrm{Im}(\overline{u}\nabla u)|^2dx\\
\le&|p-2|\int_{\mathbb{R}^N}|u|^{p-4}|\sqrt{Q}\mathrm{Re}(\overline{u}\nabla u)||\sqrt{Q}\mathrm{Im}(\overline{u}\nabla u)|^2dx\\
&+K_0\int_{\mathbb{R}^N}\sqrt{\zeta}|\mathrm{Im}(\overline{u}\nabla u)||u|^{\frac{p}{2}}|u|^{\frac{p-4}{2}}dx\\
\le&|p-2|\left(\int_{\mathbb{R}^N}|u|^{p-4}|\sqrt{Q}\mathrm{Re}(\overline{u}\nabla u)|^2dx\right)^{\frac12}\left(\int_{\mathbb{R}^N}|u|^{p-4}|\sqrt{Q}\mathrm{Im}(\overline{u}\nabla u)|^2dx\right)^{\frac12}\\
&+\frac{K_0}{\sqrt{\kappa_0}}\left(\int_{\mathbb{R}^N}|c||u|^pdx\right)^{\frac12}\left(\int_{\mathbb{R}^N}|u|^{p-4}|\sqrt{Q}\mathrm{Im}(\overline{u}\nabla u)|^2dx\right)^{\frac12}\\
\le&\frac{(p-2)^2}{4}\int_{\mathbb{R}^N}|u|^{p-4}|\sqrt{Q}\mathrm{Re}(\overline{u}\nabla u)|^2dx+2\int_{\mathbb{R}^N}|u|^{p-4}|\sqrt{Q}\mathrm{Im}(\overline{u}\nabla u)|^2dx\\
&+\frac{K_0^2}{4\kappa_0}\int_{\mathbb{R}^N}|c||u|^pdx.
\end{aligned}
\tag{7.1.17}
$$

Now, from (7.1.16) and (7.1.17) estimate (7.1.15) follows at once.

Step 4. Here we show that $(A_p,D(A_p))$ is a sectorial operator. From the formula (7.1.15) it follows that the numerical range of the operator A_p is contained in the sector $\{z\in\mathbb{C}: |\mathrm{Im}z|\le-\delta_p\mathrm{Re}z\}$. Hence, if we prove that $1\in\rho(A_p)$, then we can apply Theorem C.2.9 to get the sectoriality of the operator $(A_p,D(A_p))$.

Fix $f\in L^p(\mathbb{R}^N)$, $n\in\mathbb{N}$ and denote by u_n the unique solution to the equation $u_n-\mathcal{A}u_n=f$ in $W^{2,p}(B_n)\cap W_0^{1,p}(B_n)$. The same arguments as in the previous steps can be applied to prove that $\|u_n\|_{W^{2,p}(B_n)}+\|cu_n\|_{L^p(B_n)}\le C\|f\|_{L^p(B_n)}$ for some positive constant C, independent of f and n. By compactness, up to a subsequence, u_n converges weakly in $W^{2,p}(\mathbb{R}^N)$ and strongly in $W^{1,p}_{\mathrm{loc}}(\mathbb{R}^N)$ to a solution $u\in W^{2,p}(\mathbb{R}^N)$ of the equation $u-\mathcal{A}u=f$. Moreover, Fatou's lemma shows that $\|cu\|_{L^p(\mathbb{R}^N)}\le C\|f\|_{L^p(\mathbb{R}^N)}$. Hence, u belongs to $D(A_p)$. By estimate (7.1.11) this is the unique solution in $D(A_p)$ of the equation $v-\mathcal{A}v=f$.

The above arguments can be used to show that, for any $\lambda>0$ and $f\in C_c(\mathbb{R}^N)$, the solution $v_n\in W^{2,p}(B_n)\cap W_0^{1,p}(B_n)$ of the equation $\lambda v_n-\mathcal{A}v_n=f$ converges to $R(\lambda,A_p)f$ in $L^p_{\mathrm{loc}}(\mathbb{R}^N)$ as n tends to $+\infty$. Moreover, it converges to $R(\lambda,\widehat{A})f$ locally uniformly in $\mathbb{R}^N$, as n tends to $+\infty$. It follows that $R(\lambda,A_p)f=R(\lambda,\widehat{A})$ almost everywhere in $\mathbb{R}^N$. If we denote by $\{T_p(t)\}$ the analytic semigroup generated by the operator $(A_p,D(A_p))$, then from the previous equation we conclude that, for any $t>0$, $T(t)f=T_p(t)f$ almost everywhere in $\mathbb{R}^N$. This completes the proof. □

Remark 7.1.8 In the particular case when $\mathcal{A}=\Delta+\langle b,\nabla\rangle+c$, the authors of [200] proved also that for any $p\in(1,+\infty)$ the operator A_p has maximal L^p-regularity, i.e., for any $f\in L^p((0,T)\times\mathbb{R}^N)$ the mild solution to the Cauchy problem

$$
\begin{cases}
D_tu=\mathcal{A}u+f, & \text{in } (0,T)\times\mathbb{R}^N,\\
u(0,\cdot)=0, & \text{in } \mathbb{R}^N,
\end{cases}
$$

is actually a strong solution and it belongs to $W^{1,p}((0,T);L^p(\mathbb{R}^N)) \cap L^p((0,T);D(A_p))$. We refer the reader to [200, Proposition 6.5] for a proof of this result that we will use in Chapter 8.

7.2 Schrödinger type operators: the case $\alpha \le 2$

For any $\alpha \in [0,2]$ and $\beta \ge 0$, we denote by A_p the realization in $L^p(\mathbb{R}^N)$ $(p \in (1,+\infty))$ of the operator $\mathcal{A}$, defined in (7.0.1), with domain given by (7.0.2). We endow $D(A_p)$ with the norm $\|\cdot\|_{D(A_p)}$, defined for any $u \in D(A_p)$ by

$$\|u\|_{D(A_p)} = \|u\|_{L^p(\mathbb{R}^N)} + \|cu\|_{L^p(\mathbb{R}^N)} + \|\sqrt{q}\nabla u\|_{L^p(\mathbb{R}^N)} + \|qD^2u\|_{L^p(\mathbb{R}^N)}. \tag{7.2.1}$$

We begin this section with some preliminary results that we need to prove Theorem 7.2.5.

Proposition 7.2.1 *Fix $p \in (1,+\infty)$ and let $a,\sigma : \mathbb{R}^N \to \mathbb{R}$ be two functions with the following properties:*

(i) *$a \in C(\mathbb{R}^N) \cap C^1(\mathbb{R}^N \setminus \{0\})$ has positive infimum a_0 and there exist two positive constants r and K such that $|\nabla a| \le K\sqrt{a}$ in $\mathbb{R}^N \setminus B_r$;*

(ii) *$\sigma \in C(\mathbb{R}^N) \cap C^1(\mathbb{R}^N \setminus \{0\})$ has negative supremum σ_0 and there exist two constants $C_{1,p} > 0$ and $C_{2,p} \in (0, 4/(p-1))$ such that*

$$|\nabla\zeta(x)|^2 \le C_{1,p}|\zeta(x)|^2 + C_{2,p}|\zeta(x)|^3, \qquad x \in \mathbb{R}^N \setminus B_r, \tag{7.2.2}$$

where $\zeta = a^{-1}\sigma$. Further $0 > \xi_0 := \sup_{x\in\mathbb{R}^N} \zeta(x)$.

Then, there exist three positive constants ε_0, C (depending on κ, $C_{1,p}$, $C_{2,p}$, ξ_0, as well as on $\|a\|_{C^1(B_{2r}\setminus B_r)}$, $\|\sigma\|_{C^1(B_{2r}\setminus B_r)}$) and C_ε (depending also on ε and blowing up as ε tend to 0^+) such that

$$\|\sqrt{a}\nabla u\|_{L^p(\mathbb{R}^N)} \le \varepsilon\|a\Delta u + \sigma u\|_{L^p(\mathbb{R}^N)} + C_\varepsilon\|u\|_{L^p(\mathbb{R}^N)} \tag{7.2.3}$$

and

$$\|aD^2u\|_{L^p(\mathbb{R}^N)} \le C(\|u\|_{L^p(\mathbb{R}^N)} + \|a\Delta u + \sigma u\|_{L^p(\mathbb{R}^N)}) \tag{7.2.4}$$

for any $\varepsilon \in (0,\varepsilon_0]$ and $u \in W^{2,p}(\mathbb{R}^N)$ with $\sqrt{a}|\nabla u|$, $a|D^2u|$, $\sigma u \in L^p(\mathbb{R}^N)$.

To prove Proposition 7.2.1 we need two preliminary results. The first one is a density result.

Lemma 7.2.2 *Let $a : \mathbb{R}^N \to \mathbb{R}$ be a positive continuous function such that $a(x) \le C|x|^2$ for any $x \in \mathbb{R}^N$ and some positive constant C. Further, let $\sigma \in C(\mathbb{R}^N)$ be a nonnegative function. Then, for any $p \in (1,+\infty)$, $C_c^\infty(\mathbb{R}^N)$ is dense in the space $\mathscr{X} := \{u \in W^{2,p}(\mathbb{R}^N) : \sqrt{a}|\nabla u|, a|D^2u|, \sigma u \in L^p(\mathbb{R}^N)\}$, endowed with the norm $\|u\|_{\mathscr{X}} = \|u\|_{L^p(\mathbb{R}^N)} + \|\sigma u\|_{L^p(\mathbb{R}^N)} + \|\sqrt{a}\nabla u\|_{L^p(\mathbb{R}^N)} + \|aD^2u\|_{L^p(\mathbb{R}^N)}$.*

Proof Even if the proof can be obtained employing a standard argument, for the reader's convenience we enter into detail. If $u \in W^{2,p}(\mathbb{R}^N)$ has compact support, then we regularize it by convolution with standard mollifiers, obtaining a sequence $\{u_n\} \subset C_c^\infty(\mathbb{R}^N)$ converging to u in $W^{2,p}(\mathbb{R}^N)$. Since $\text{supp}(u_n) \subset \text{supp}(u) + \overline{B}_1$, for any $n \in \mathbb{N}$, $\sqrt{a}D_iu_n$,

$aD_{ij}u_n$ and σu_n converge, respectively, to $\sqrt{a}D_iu$, $aD_{ij}u$ and σu in $L^p(\mathbb{R}^N)$, as n tends to $+\infty$, for any $i,j=1,\ldots,N$. Hence, u_n tends to u in $\mathscr{X}$. In the general case when $u\in\mathscr{X}$ is not compactly supported, for any $n\in\mathbb{N}$ we consider the function $u_n=u\vartheta_n$, where $\vartheta_n(x)=\vartheta(n^{-1}x)$ for any $x\in\mathbb{R}^N$ and ϑ is a smooth function such that $\chi_{B_1}\le\vartheta\le\chi_{B_2}$. By dominated convergence, u_n and σu_n tend to u and cu in $L^p(\mathbb{R}^N)$, respectively. Similarly, one has

$$\begin{aligned}\|\sqrt{a}\nabla(u_n-u)\|_{L^p(\mathbb{R}^N)}^p \le& 2^{p-1}\|\sqrt{a}(\vartheta_n-1)\nabla u\|_{L^p(\mathbb{R}^N)}^p\\ &+\frac{2^{p-1}}{n^p}\sup_{n\le|x|\le 2n}(a(x))^{\frac{p}{2}}\|u\nabla\vartheta(n^{-1}\cdot)\|_{L^p(\mathbb{R}^N)}^p\\ \le& 2^{p-1}(\|\sqrt{a}(\vartheta_n-1)\nabla u\|_{L^p(\mathbb{R}^N)}^p+2^pC^{\frac{p}{2}}\|u\nabla\vartheta(n^{-1}\cdot)\|_{L^p(\mathbb{R}^N)}^p)\end{aligned}$$

and, by dominated convergence, the last side of the previous chain of inequalities vanishes as n tends to $+\infty$.

A completely similar computation shows that $aD_{ij}u_n$ tends to $aD_{ij}u$ in $L^p(\mathbb{R}^N)$ as n tends to $+\infty$, for any $i,j=1,\ldots,N$. This completes the proof. □

The second result that we need is an *a priori* estimate due to N. Okazawa.

Theorem 7.2.3 ([211], Lemma 1.4 and Theorem 2.1) *Let $V:\mathbb{R}^N\to\mathbb{R}$ be a nonpositive C^1-function such that*

$$|\nabla V(x)|^2\le K_{1,p}|V(x)|^2+K_{2,p}|V(x)|^3,\qquad x\in\mathbb{R}^N,\tag{7.2.5}$$

for some constants $K_{1,p}>0$ and $K_{2,p}\in(0,4/(p-1))$. Then,

$$\|\Delta u\|_{L^p(\mathbb{R}^N)}\le\frac{4}{4-(p-1)K_{2,p}}\|\Delta u+Vu\|_{L^p(\mathbb{R}^N)}+\left(\frac{(p-1)K_{1,p}}{4-(p-1)K_{2,p}}\right)\|u\|_{L^p(\mathbb{R}^N)}$$

for any $u\in C_c^\infty(\mathbb{R}^N)$.

Proof of Proposition 7.2.1 In view of Lemma 7.2.2 it suffices to prove (7.2.3) and (7.2.4) when $u\in C_c^\infty(\mathbb{R}^N)$. Being rather long, we split the proof into steps. Throughout the proof p is arbitrarily fixed in $(1,+\infty)$.

Step 1. Let us prove that

$$\|D^2u\|_{L^p(\mathbb{R}^N)}\le C_p\|\Delta u+\widetilde{\zeta}u\|_{L^p(\mathbb{R}^N)}\tag{7.2.6}$$

for any $u\in C_c^\infty(\mathbb{R}^N)$ and some positive constant C_p. Here, $\widetilde{\zeta}=-\varphi+(1-\varphi)\zeta$, where φ is any smooth function such that $\chi_{B_{\widetilde{r}}}\le\varphi\le\chi_{B_{2\widetilde{r}}}$, and $\widetilde{r}=r\vee 1$. Note that $\widetilde{\zeta}(x)\le\widetilde{\xi_0}:=\xi_0\vee(-1)$ for any $x\in\mathbb{R}^N$.

To prove (7.2.6) we begin by observing that the function $\widetilde{\zeta}$ satisfies estimate (7.2.5). Clearly, it satisfies this estimate in $\mathbb{R}^N\setminus B_{2\widetilde{r}}$, by virtue of (7.2.2), and in $B_{\widetilde{r}}$, since $\nabla\zeta$ therein identically vanishes. For any $x\in B_{2\widetilde{r}}\setminus B_{\widetilde{r}}$ we can estimate

$$|\nabla\widetilde{\zeta}(x)|^2\le 2(\|\nabla\varphi\|_\infty^2\|\zeta-1\|_{L^\infty(B_{2\widetilde{r}}\setminus B_{\widetilde{r}})}^2+\|\nabla\zeta\|_{L^\infty(B_{2\widetilde{r}}\setminus B_{\widetilde{r}})}^2)\widetilde{\xi_0}^{-2}|\widetilde{\zeta}(x)|^2.$$

Hence, inequality (7.2.5) follows in the whole of $\mathbb{R}^N$ with $K_{2,p}=C_{2,p}$ and

$$K_{1,p}=\max\{C_{1,p},2(\|\nabla\varphi\|_\infty^2\|\zeta-1\|_{L^\infty(B_{2\widetilde{r}}\setminus B_{\widetilde{r}})}^2+\|\nabla\zeta\|_{L^\infty(B_{2\widetilde{r}}\setminus B_{\widetilde{r}})}^2)\widetilde{\xi_0}^{-2}\}.$$

Theorem 7.2.3 then yields

$$\|\Delta u\|_{L^p(\mathbb{R}^N)} \le \frac{4}{4-(p-1)C_{2,p}}\|\Delta u + \widetilde{\zeta} u\|_{L^p(\mathbb{R}^N)} + \left(\frac{(p-1)K_{1,p}}{4-(p-1)C_{2,p}}\right)\|u\|_{L^p(\mathbb{R}^N)}. \quad (7.2.7)$$

To complete the proof of estimate (7.2.6) we observe that, in view of (7.2.7) and the well-known Calderon-Zygmund inequality (see (D.1.1)), it suffices to show that, for any $p \in (1,+\infty)$, there exists a positive constant C_p, independent of u, such that

$$\|u\|_{L^p(\mathbb{R}^N)} \le C_p\|\Delta u + \widetilde{\zeta} u\|_{L^p(\mathbb{R}^N)}. \quad (7.2.8)$$

Set $f := \Delta u + \widetilde{\zeta} u$ and multiply both sides of the equation $\Delta u + \widetilde{\zeta} u = f$ by $u|u|^{p-2}$. Without loss of generality, we assume that u is real. Indeed, if u is complex-valued, then (7.2.8) will follow, arguing on its real and imaginary parts, and using the estimate

$$(\|f_1\|_{L^p(\mathbb{R}^N)} + \|f_2\|_{L^p(\mathbb{R}^N)})^p \le (2^{p/2} \vee 2^{p-1}) \int_{\mathbb{R}^N} (|f_1|^2 + |f_2|^2)^{\frac{p}{2}}\, dx,$$

satisfied by any $f_1, f_2 \in L^p(\mathbb{R}^N)$.

An integration by parts shows that

$$-(p-1)\int_{\mathbb{R}^N} |u|^{p-2}|\nabla u|^2 \chi_{\{u\neq 0\}} dx + \widetilde{\xi}_0 \int_{\mathbb{R}^N} |u|^p dx \ge \int_{\mathbb{R}^N} f u|u|^{p-2} dx$$

and from this estimate we get (7.2.8) with $C_p = \widetilde{\xi}_0^{-1}$.

Step 2. Let us now prove estimate (7.2.3) by a covering argument. The starting point is inequality (7.1.8) which, in view of (7.2.6), allows us to estimate

$$\|\nabla u\|_{L^p(\mathbb{R}^N)} \le C'\|\Delta u + \widetilde{\zeta} u\|_{L^p(\mathbb{R}^N)}^{\frac{1}{2}} \|u\|_{L^p(\mathbb{R}^N)}^{\frac{1}{2}} \quad (7.2.9)$$

for some positive constant C', depending on p but being independent of $u \in C_c^\infty(\mathbb{R}^N)$.

To apply the covering argument, we set $\widetilde{a} := 2^{-4}\varphi + (1-\varphi)a$, where the function φ is as above. Arguing as in Step 1 we can show that $|\nabla \widetilde{a}| \le \widetilde{K}\sqrt{\widetilde{a}(x)}$ in $\mathbb{R}^N$, where $\widetilde{K} = \max\{1, K, a_0^{-1/2}(\|\nabla\varphi\|_\infty \|a - 2^{-4}\|_{L^\infty(B_{2\widetilde{r}}\setminus B_{\widetilde{r}})} + \|\nabla a\|_{L^\infty(B_{2\widetilde{r}}\setminus B_{\widetilde{r}})})\}$. Further, we introduce the function $\rho : \mathbb{R}^N \to \mathbb{R}$ defined by $\rho(x) = (2\widetilde{K})^{-1}\sqrt{\widetilde{a}(x)}$ for any $x \in \mathbb{R}^N$. Clearly, ρ is a Lipschitz continuous function, with Lipschitz constant not greater than $1/4$. Moreover,

$$\frac{1}{2}\sqrt{\widetilde{a}(x_0)} \le \sqrt{\widetilde{a}(x)} \le \frac{3}{2}\sqrt{\widetilde{a}(x_0)}, \qquad x \in B_{2\rho(x_0)}(x_0), \quad (7.2.10)$$

and $\rho(x) \le |x|/4 + 1/8$ for any $x \in \mathbb{R}^N$. This latter inequality implies that

$$B_{\rho(x_0)}(x_0) \subset B_{(1+10|x_0|)/8}, \qquad B_{2\rho(x_0)}(x_0) \subset \mathbb{R}^N \setminus B_{-1/4+|x_0|/2}, \quad (7.2.11)$$

for any $|x_0| \ge 1/2$.

Now, for any $x_0 \in \mathbb{R}^N$ we set $\vartheta_{x_0}(x) = \vartheta((\rho(x_0))^{-1}(x-x_0))$, where $\vartheta \in C_c^\infty(\mathbb{R}^N)$ satisfies the condition $\chi_{B_1} \le \vartheta \le \chi_{B_2}$. Moreover, to fix the notation, we set $L := \|\nabla\vartheta\|_\infty + \|\Delta\vartheta\|_\infty$. Applying estimate (7.2.9) to the function $\vartheta_{x_0} u$ and using the Young inequality, we get

$$\begin{aligned}
&\|\sqrt{\widetilde{a}(x_0)}\nabla u\|_{L^p(B_{\rho(x_0)}(x_0))} \le \|\sqrt{\widetilde{a}(x_0)}\nabla(\vartheta_{x_0} u)\|_{L^p(\mathbb{R}^N)}\\
\le& C'\|\vartheta_{x_0} u\|_{L^p(\mathbb{R}^N)}^{\frac{1}{2}} \|\widetilde{a}(x_0)\Delta(\vartheta_{x_0} u) + \widetilde{a}(x_0)\widetilde{\zeta}\vartheta_{x_0} u\|_{L^p(\mathbb{R}^N)}^{\frac{1}{2}}\\
\le& C'\left(\varepsilon\|\widetilde{a}(x_0)\Delta(\vartheta_{x_0} u) + \widetilde{a}(x_0)\widetilde{\zeta}\vartheta_{x_0} u\|_{L^p(\mathbb{R}^N)} + \frac{1}{4\varepsilon}\|\vartheta_{x_0} u\|_{L^p(\mathbb{R}^N)}\right)
\end{aligned}$$

$$\leq C'\bigg(\varepsilon\|\widetilde{a}(x_0)\Delta u+\widetilde{a}(x_0)\widetilde{\zeta}u\|_{L^p(B_{2\rho(x_0)}(x_0))}+\frac{2L}{\rho(x_0)}\varepsilon\|\widetilde{a}(x_0)\nabla u\|_{L^p(B_{2\rho(x_0)}(x_0))}$$
$$+\frac{L}{\rho(x_0)^2}\varepsilon\|\widetilde{a}(x_0)\,u\|_{L^p(B_{2\rho(x_0)}(x_0))}+\frac{1}{4\varepsilon}\|u\|_{L^p(B_{2\rho(x_0)}(x_0))}\bigg)$$

for any $\varepsilon>0$. Now, from (7.2.10) and the definition of ρ we can estimate

$$\begin{aligned}\|\sqrt{\widetilde{a}}\nabla u\|_{L^p(B_{\rho(x_0)}(x_0))}\leq&\frac{3}{2}\|\sqrt{\widetilde{a}(x_0)}\nabla u\|_{L^p(B_{\rho(x_0)}(x_0))}\\ \leq&6C'\bigg\{\varepsilon\|\widetilde{a}\Delta u+\widetilde{a}\widetilde{\zeta}u\|_{L^p(B_{2\rho(x_0)}(x_0))}+2\varepsilon L\widetilde{K}\|\sqrt{\widetilde{a}}\nabla u\|_{L^p(B_{2\rho(x_0)}(x_0))}\\ &+\left(\varepsilon L\widetilde{K}^2+\frac{1}{16\varepsilon}\right)\|u\|_{L^p(B_{2\rho(x_0)}(x_0))}\bigg\}.\end{aligned}$$

By Proposition 7.1.6 there exist a sequence $\{x_n\}$ and a positive number ς such that $\mathcal{F}=\{B_{\rho(x_n)}(x_n):n\in\mathbb{N}\}$ is a covering of $\mathbb{R}^N$ and the intersection of more than ς balls from $\mathcal{F}$ is empty. From (7.2.11) it is immediate to conclude that $\{B_{\rho(x_n)}(x_n):|x_n|\geq 6\widetilde{r}\}$ is a covering of $\mathbb{R}^N\setminus B_{8\widetilde{r}}$ and $B_{2\rho(x_n)}(x_n)\subset\mathbb{R}^N\setminus B_{2\widetilde{r}}$ for $|x_n|\geq 6\widetilde{r}$ and any $n\in\mathbb{N}$. Hence, recalling that $\widetilde{a}=a$ and $\widetilde{a}\widetilde{\zeta}=\sigma$ in $\mathbb{R}^N\setminus B_{2\widetilde{r}}$, we can write

$$\begin{aligned}\|\sqrt{a}\nabla u\|^p_{L^p(\mathbb{R}^N\setminus B_{8\widetilde{r}})}\leq&C\varepsilon^p\int_{\mathbb{R}^N}|a\Delta u+\sigma u|^p\sum_{|x_n|\geq 6\widetilde{r}}\chi_{B_{2\rho(x_n)}(x_n)}dx\\ &+C(\varepsilon L\widetilde{K})^p\int_{\mathbb{R}^N}a^{\frac{p}{2}}|\nabla u|^p\sum_{|x_n|\geq 6\widetilde{r}}\chi_{B_{2\rho(x_n)}(x_n)}dx\\ &+C\left(\varepsilon L\widetilde{K}^2+\frac{1}{16\varepsilon}\right)^p\int_{\mathbb{R}^N}|u|^p\sum_{|x_n|\geq 6\widetilde{r}}\chi_{B_{2\rho(x_n)}(x_n)}dx\\ \leq&C\varepsilon^p\|a\Delta u+\sigma u\|^p_{L^p(\mathbb{R}^N\setminus B_{2\widetilde{r}})}+C(\varepsilon L\widetilde{K})^p\|\sqrt{a}\nabla u\|^p_{L^p(\mathbb{R}^N\setminus B_{2\widetilde{r}})}\\ &+C\left(\varepsilon L\widetilde{K}^2+\frac{1}{16\varepsilon}\right)^p\|u\|^p_{L^p(\mathbb{R}^N\setminus B_{2\widetilde{r}})}\end{aligned}$$

for some positive constant C, independent of ε and u. Due to the arbitrariness of $\varepsilon>0$, from the above estimate we get

$$\|\sqrt{a}\nabla u\|_{L^p(\mathbb{R}^N\setminus B_{8\widetilde{r}})}\leq\varepsilon\|a\Delta u+\sigma u\|_{L^p(\mathbb{R}^N)}+\varepsilon\|\sqrt{a}\nabla u\|_{L^p(\mathbb{R}^N)}+C_\varepsilon\|u\|_{L^p(\mathbb{R}^N)}\tag{7.2.12}$$

for any $\varepsilon>0$ and some positive constant C_ε, possibly blowing up as ε tends to 0^+.

To extend the previous inequality to $\mathbb{R}^N$ we use the classical interior L^p-estimate (see Theorem D.1.1)

$$\|u\|_{W^{2,p}(B_{8\widetilde{r}})}\leq M_1(\|a\Delta u+\sigma u\|_{L^p(\mathbb{R}^N)}+\|u\|_{L^p(\mathbb{R}^N)})\tag{7.2.13}$$

and the interpolation estimate $\|\nabla v\|_{L^p(B_{8\widetilde{r}})}\leq M_2\|v\|^{1/2}_{L^p(B_{8\widetilde{r}})}\|v\|^{1/2}_{W^{2,p}(B_{8\widetilde{r}})}$, which hold true for some positive constants M_1, M_2 independent of u and $v\in W^{2,p}(B_{8\widetilde{r}})$, to infer that

$$\begin{aligned}\|\sqrt{a}\nabla u\|_{L^p(B_{8\widetilde{r}})}\leq&\|\sqrt{a}\|_{L^\infty(B_{8\widetilde{r}})}\|\nabla u\|_{L^p(B_{8\widetilde{r}})}\\ \leq&M_3\|u\|^{\frac{1}{2}}_{L^p(B_{8\widetilde{r}})}(\|a\Delta u+\sigma u\|_{L^p(\mathbb{R}^N)}+\|u\|_{L^p(\mathbb{R}^N)})^{\frac{1}{2}}\\ \leq&M_3\|u\|^{\frac{1}{2}}_{L^p(B_{8\widetilde{r}})}\|a\Delta u+\sigma u\|^{\frac{1}{2}}_{L^p(\mathbb{R}^N)}+M_3\|u\|_{L^p(\mathbb{R}^N)}\end{aligned}$$

$$\le \varepsilon \|a\Delta u + \sigma u\|_{L^p(\mathbb{R}^N)} + C'_\varepsilon \|u\|_{L^p(\mathbb{R}^N)} \tag{7.2.14}$$

for any $\varepsilon > 0$ and some positive constants M_3 and C'_ε, this latter one possibly blowing up as ε tends to 0^+. From estimates (7.2.12) and (7.2.14) we deduce that

$$\begin{aligned}\|\sqrt{a}\nabla u\|_{L^p(\mathbb{R}^N)} \le& \|\sqrt{a}\nabla u\|_{L^p(B_{8\tilde{r}})} + \|\sqrt{a}\nabla u\|_{L^p(\mathbb{R}^N\setminus B_{8\tilde{r}})}\\ \le& 2\varepsilon\|a\Delta u + \sigma u\|_{L^p(\mathbb{R}^N)} + C''_\varepsilon\|u\|_{L^p(\mathbb{R}^N)} + \varepsilon\|\sqrt{a}\nabla u\|_{L^p(\mathbb{R}^N)}\end{aligned}$$

for any $\varepsilon > 0$ and some positive constant C''_ε possibly blowing up as ε tends to 0^+. Hence, taking $\varepsilon < 1$, we get (7.2.3).

Step 3. Here we conclude the proof, checking estimate (7.2.4). From (7.2.6), applied to the function $u\vartheta_{x_0}$, we deduce that

$$\|a(x_0)D^2(\vartheta_{x_0}u)\|_{L^p(\mathbb{R}^N)} \le C_p\|a(x_0)\Delta(\vartheta_{x_0}u) + a(x_0)\widetilde{\zeta}\vartheta_{x_0}u\|_{L^p(\mathbb{R}^N)}.$$

Therefore, taking (7.2.10) into account and arguing as in the proof of Step 2, we first get

$$\begin{aligned}\|\widetilde{a}D^2u\|_{L^p(B_{\rho(x_0)}(x_0))} \le& \frac{9}{4}\|\widetilde{a}(x_0)D^2u\|_{L^p(B_{\rho(x_0)}(x_0))}\\ \le& M_4\big(\|\widetilde{a}(x_0)\Delta u + \widetilde{a}(x_0)\widetilde{\zeta}u\|_{L^p(B_{2\rho(x_0)}(x_0))}\\ &+ \|\langle\widetilde{a}(x_0)\nabla u, \vartheta_{x_0}\rangle\|_{L^p(\mathbb{R}^N)} + \|\widetilde{a}(x_0)u\Delta\vartheta_{x_0}\|_{L^p(\mathbb{R}^N)}\big)\\ \le& M_5\big(\|\widetilde{a}\Delta u + \widetilde{a}\widetilde{\zeta}u\|_{L^p(B_{2\rho(x_0)}(x_0))} + \|\sqrt{\widetilde{a}}\nabla u\|_{L^p(B_{2\rho(x_0)}(x_0))}\\ &+ \|u\|_{L^p(B_{2\rho(x_0)}(x_0))}\big)\end{aligned}$$

and, then, applying the same covering argument as above, we conclude that

$$\|aD^2u\|_{L^p(\mathbb{R}^N\setminus B_{8\tilde{r}})} \le M_6(\|a\Delta u + \sigma u\|_{L^p(\mathbb{R}^N)} + \|\sqrt{a}\nabla u\|_{L^p(\mathbb{R}^N)} + \|u\|_{L^p(\mathbb{R}^N)}). \tag{7.2.15}$$

Here, M_4, M_5 and M_6 are positive constants, independent of u. Combining (7.2.3) and (7.2.15) we get

$$\|aD^2u\|_{L^p(\mathbb{R}^N\setminus B_{8\tilde{r}})} \le M_7(\|a\Delta u + \sigma u\|_{L^p(\mathbb{R}^N)} + \|u\|_{L^p(\mathbb{R}^N)}) \tag{7.2.16}$$

for some positive constant M_7, independent of u. Estimate (7.2.4) now follows from (7.2.16) and (7.2.13) (with u replacing v). □

Taking $a = q$, $\sigma = c$ and recalling that $\alpha \le 2$, we obtain the following result.

Corollary 7.2.4 *The norm of $D(A_p)$ defined in (7.2.1) is equivalent to the graph norm of $D(A_p)$, i.e., to the norm defined by $\|u\|_{D(A_p)} = \|u\|_{L^p(\mathbb{R}^N)} + \|A_pu\|_{L^p(\mathbb{R}^N)}$ for any $u \in D(A_p)$.*

We can now prove that the operator $(A_p, D(A_p))$ is sectorial in $L^p(\mathbb{R}^N)$.

Theorem 7.2.5 *For any $p \in (1, +\infty)$ the operator A_p generates a strongly continuous analytic semigroup $\{T_p(t)\}$ in $L^p(\mathbb{R}^N)$, which is also positive and consistent.*

Proof Being rather long, we split the proof into several steps.

Step 1. For any $n \in \mathbb{N}$ we consider the functions q_n and c_n defined by

$$q_n(x) = \frac{n(1 + |x|^\alpha)}{n + 1 + |x|^\alpha}, \qquad c_n(x) = -\frac{n(|x|^\beta + 1)}{n + 1 + |x|^\beta}, \qquad x \in \mathbb{R}^N.$$

These functions are bounded in $\mathbb{R}^N$, $q_n \ge 1/2$ and $c_n \le -1/2$ in $\mathbb{R}^N$ for any $n \in \mathbb{N}$. By

Theorem D.3.5, for any $n \in \mathbb{N}$ and $p \in (1,+\infty)$ the realization $L_{n,p}$ of operator $\mathcal{L}_n = q_n\Delta + c_n$ in $L^p(\mathbb{R}^N)$, with $W^{2,p}(\mathbb{R}^N)$ as a domain, is the generator of a strongly continuous, analytic semigroup.

Let us prove that

$$\|\sqrt{q_n}\nabla u\|_{L^p(\mathbb{R}^N)} \leq \varepsilon\|L_{n,p}u\|_{L^p(\mathbb{R}^N)} + C_\varepsilon\|u\|_{L^p(\mathbb{R}^N)}, \tag{7.2.17}$$

$$\|q_n D^2 u\|_{L^p(\mathbb{R}^N)} \leq C(\|u\|_{L^p(\mathbb{R}^N)} + \|L_{n,p}u\|_{L^p(\mathbb{R}^N)}) \tag{7.2.18}$$

for any $\varepsilon > 0$, any $u \in W^{2,p}(\mathbb{R}^N)$ and some positive constants C and C_ε, independent of $n \in \mathbb{N}$ and u, the latter constant possibly blowing up as ε tends to 0^+. For this purpose, we introduce the function $\zeta_{1,n} = c_n/q_n$. Observing that $\zeta_{1,n} \leq -1/2$ in B_1, a straightforward computation reveals that $|\nabla\zeta_{1,n}| \leq 4|\beta-\alpha| + 6\alpha + 8\beta \leq 4(2|\beta-\alpha| + 3\beta + 4\alpha)|\zeta_{1,n}|$ in $B_1 \setminus B_{1/2}$ and

$$\begin{aligned}
|\nabla\zeta_{1,n}(x)| \leq & \bigg(|\beta-\alpha|\frac{|x|^{\beta-1}}{1+|x|^\alpha} + \beta\frac{|x|^{\beta-1}}{1+|x|^\alpha} + \frac{\alpha}{|x|(1+|x|^\alpha)}\bigg)\frac{n+1+|x|^\alpha}{n+1+|x|^\beta} \\
& + \frac{1+|x|^\beta}{|x|(1+|x|^\alpha)}\bigg(\alpha\frac{|x|^\alpha}{n+1+|x|^\beta} + \beta\frac{n+1+|x|^\alpha}{n+1+|x|^\beta}\bigg) \\
\leq & (|\beta-\alpha| + 2\alpha + 2\beta)|\zeta_{1,n}(x)||x|^{-1} \\
\leq & (|\beta-\alpha| + 2\alpha + 2\beta)|\zeta_{1,n}(x)|,
\end{aligned}$$

if $x \in \mathbb{R}^N \setminus B_1$. Hence, $|\nabla\zeta_{1,n}| \leq 4(2|\beta-\alpha| + 3\beta + 4\alpha)|\zeta_{1,n}|$ in $\mathbb{R}^N \setminus B_{1/2}$. As a byproduct, we can infer that $|\nabla\zeta_{1,n}|^2 \leq C_1\zeta_{1,n}^2$ in $\mathbb{R}^N \setminus B_{1/2}$, where $C_1 = 16(2|\beta-\alpha| + 3\beta + 4\alpha)^2$. Since, in addition, $|\nabla q_n| \leq 2^{1-\frac{\alpha}{2}}\alpha\sqrt{q_n}$ in $\mathbb{R}^N \setminus B_{1/2}$, estimates (7.2.17) and (7.2.18) follow at once from Proposition 7.2.1.

Step 2. Here we prove that, for any $p \in (1,+\infty)$, there exist $\omega_0 \in \mathbb{R}$ and $M_p > 0$ such that

$$|\lambda|\|u\|_{L^p(\mathbb{R}^N)} \leq M_p\|\lambda u - L_{n,p}u\|_{L^p(\mathbb{R}^N)} \tag{7.2.19}$$

for any $u \in D(A_p)$, any $\lambda \in \mathbb{C}$ with $\mathrm{Re}\lambda \geq \omega_0$ and any $n \in \mathbb{N}$.

We begin by considering the case when $\alpha \in [1,2]$. We fix $p \in (1,+\infty)$, $\lambda \in \mathbb{C}$, $u \in W^{2,p}(\mathbb{R}^N)$ and set $f := \lambda u - L_{n,p}u$. We multiply both sides of this equation by $\overline{u}|u|^{p-2}$ and integrate by parts. We get

$$\begin{aligned}
\int_{\mathbb{R}^N} f\overline{u}|u|^{p-2}dx = & \lambda\int_{\mathbb{R}^N}|u|^p dx - \int_{\mathbb{R}^N} c_n|u|^p dx - \int_{\mathbb{R}^N} q_n|u|^{p-2}\overline{u}\Delta u dx \\
= & \lambda\int_{\mathbb{R}^N}|u|^p dx - \int_{\mathbb{R}^N} c_n|u|^p dx + (p-1)\int_{\mathbb{R}^N} q_n|u|^{p-4}|\mathrm{Re}(\overline{u}\nabla u)|^2\chi_{\{u\neq 0\}}dx \\
& + \int_{\mathbb{R}^N} q_n|u|^{p-4}|\mathrm{Im}(\overline{u}\nabla u)|^2\chi_{\{u\neq 0\}}dx - \int_{\mathbb{R}^N}\langle\nabla q_n, \nabla u\rangle|u|^{p-2}\overline{u}dx \\
& + i(p-2)\int_{\mathbb{R}^N} q_n|u|^{p-4}\langle\mathrm{Re}(\overline{u}\nabla u), \mathrm{Im}(\overline{u}\nabla u)\rangle\chi_{\{u\neq 0\}}dx.
\end{aligned} \tag{7.2.20}$$

Taking the real part of the first and last side of (7.2.20) gives

$$\begin{aligned}
\int_{\mathbb{R}^N}\mathrm{Re}(f\overline{u})|u|^{p-2}dx = & \mathrm{Re}\lambda\int_{\mathbb{R}^N}|u|^p dx + (p-1)\int_{\mathbb{R}^N} q_n|u|^{p-4}|\mathrm{Re}(\overline{u}\nabla u)|^2\chi_{\{u\neq 0\}}dx \\
& - \int_{\mathbb{R}^N} c_n|u|^p dx + \int_{\mathbb{R}^N} q_n|u|^{p-4}|\mathrm{Im}(\overline{u}\nabla u)|^2\chi_{\{u\neq 0\}}dx \\
& - \int_{\mathbb{R}^N}\langle\nabla q_n, \mathrm{Re}(\overline{u}\nabla u)\rangle|u|^{p-2}\chi_{\{u\neq 0\}}dx.
\end{aligned}$$

For notational convenience we set

$$A := \int_{\mathbb{R}^N} q_n |u|^{p-4} |\mathrm{Re}(\overline{u}\nabla u)|^2 \chi_{\{u\neq 0\}} dx,$$

$$B := \int_{\mathbb{R}^N} q_n |u|^{p-4} |\mathrm{Im}(\overline{u}\nabla u)|^2 \chi_{\{u\neq 0\}} dx.$$

Recalling that $|\nabla q_n| \le 2\sqrt{q_n}$ and using the Hölder and Young inequalities we can estimate

$$\begin{aligned}\|f\|_{L^p(\mathbb{R}^N)}\|u\|_{L^p(\mathbb{R}^N)}^{p-1} \ge& (\mathrm{Re}\lambda)\|u\|_{L^p(\mathbb{R}^N)}^p + (p-1)A + B - 2\int_{\mathbb{R}^N} \sqrt{q_n}|\mathrm{Re}(\overline{u}\nabla u)||u|^{p-2}dx\\ \ge& (\mathrm{Re}\lambda)\|u\|_{L^p(\mathbb{R}^N)}^p + (p-1)A + B - 2\sqrt{A}\|u\|_{L^p(\mathbb{R}^N)}^{\frac{p}{2}}\\ \ge& \left(\mathrm{Re}\lambda - \frac{2}{p-1}\right)\|u\|_{L^p(\mathbb{R}^N)}^p + \frac{p-1}{2}A + B.\end{aligned}$$

We thus deduce that

$$\frac{1}{2}(\mathrm{Re}\lambda)\|u\|_{L^p(\mathbb{R}^N)} \le \|f\|_{L^p(\mathbb{R}^N)}, \tag{7.2.21}$$

$$A \le \left(\frac{p-1}{2}\right)^{p-2}\|f\|_{L^p(\mathbb{R}^N)}^p, \qquad B \le \left(\frac{p-1}{2}\right)^{p-1}\|f\|_{L^p(\mathbb{R}^N)}^p \tag{7.2.22}$$

for any $\lambda \in \mathbb{C}$ with $\mathrm{Re}\lambda \ge 4(p-1)^{-1}$. Now, taking the imaginary part of (7.2.20), we get

$$\begin{aligned}\int_{\mathbb{R}^N} \mathrm{Im}(f\overline{u})|u|^{p-2}dx =& (\mathrm{Im}\lambda)\int_{\mathbb{R}^N}|u|^p dx - \int_{\mathbb{R}^N}\langle\nabla q_n, \mathrm{Im}(\overline{u}\nabla u)\rangle|u|^{p-2}\chi_{\{u\neq 0\}}dx\\ &+ (p-2)\int_{\mathbb{R}^N} q_n|u|^{p-4}\langle\mathrm{Re}(\overline{u}\nabla u), \mathrm{Im}(\overline{u}\nabla u)\rangle\chi_{\{u\neq 0\}}dx.\end{aligned}$$

Hence, $|\mathrm{Im}\lambda|\|u\|_{L^p(\mathbb{R}^N)}^p \le \|f\|_{L^p(\mathbb{R}^N)}\|u\|_{L^p(\mathbb{R}^N)}^{p-1} + |p-2|\sqrt{A}\sqrt{B} + 2\sqrt{B}\|u\|_{L^p(\mathbb{R}^N)}^{p/2}$. Using (7.2.21) and (7.2.22) we obtain

$$|\mathrm{Im}\lambda|\|u\|_{L^p(\mathbb{R}^N)} \le C_p\|f\|_{L^p(\mathbb{R}^N)}, \qquad \lambda \in \mathbb{C},\ \mathrm{Re}\lambda \ge \frac{4}{p-1}, \tag{7.2.23}$$

where the constant C_p is independent of λ, u and f. From (7.2.21), (7.2.23), estimate (7.2.19) follows at once with $\omega_0 = 4(p-1)^{-1}$.

If $\alpha \in [0,1)$ then the function q_n does not belong to $W^{1,\infty}(\mathbb{R}^N)$ and, consequently, we can not control ∇q_n by $\sqrt{q_n}$. To overcome this difficulty, we regularize the function q_n in a neighborhood of the origin by introducing the function $\widehat{q}_n := \varphi + (1-\varphi)q_n$, where φ is a smooth function such that $\chi_{B_1} \le \varphi \le \chi_{B_2}$. Since $|\nabla\widehat{q}_n| \le \widehat{K}\sqrt{\widehat{q}_n}$ in $\mathbb{R}^N$ for some positive constant $\widehat{K}$ independent of n, by the above results we deduce that there exist two positive constants $\widehat{\omega}_0$ and $\widehat{M_p}$ such that $|\lambda|\|u\|_{L^p(\mathbb{R}^N)} \le \widehat{M_p}\|\lambda u - \widehat{L}_{n,p}u\|_{L^p(\mathbb{R}^N)}$ for any $u \in W^{2,p}(\mathbb{R}^N)$, $\lambda \in \mathbb{C}$, with $\mathrm{Re}\lambda \ge \widehat{\omega}_0$, and $n \in \mathbb{N}$, where $\widehat{L}_{n,p}$ denotes the realization of the operator $\widehat{\mathcal{A}} = \widehat{q}_n\Delta + c$ in $L^p(\mathbb{R}^N)$, with $W^{2,p}(\mathbb{R}^N)$ as a domain. Taking into account that $\widehat{q}_n = q_n$ in $\mathbb{R}^N \setminus B_2$, we can estimate

$$\begin{aligned}|\lambda|\|u\|_{L^p(\mathbb{R}^N)} \le& \|\lambda u - \widehat{L}_{n,p}u\|_{L^p(B_2)} + \|\lambda u - L_{n,p}u\|_{L^p(\mathbb{R}^N\setminus B_2)}\\ \le& \|(\widehat{q}_n - q_n)\Delta u\|_{L^p(B_2)} + \|\lambda u - L_{n,p}u\|_{L^p(B_2)} + \|\lambda u - L_{n,p}u\|_{L^p(\mathbb{R}^N\setminus B_2)}\\ \le& \|\widehat{q}_n - q_n\|_\infty\|\Delta u\|_{L^p(B_2)} + 2\|\lambda u - L_{n,p}u\|_{L^p(\mathbb{R}^N)}\\ \le& 4\|\Delta u\|_{L^p(B_2)} + 2\|\lambda u - L_{n,p}u\|_{L^p(\mathbb{R}^N)}.\end{aligned} \tag{7.2.24}$$

We now apply estimate (D.3.7) with $r = 2$ to the operator $L_n = \widetilde{q}_n\Delta + \widetilde{c}$, where $\widetilde{q}_n = q_n\psi + 1 - \psi$, $\widetilde{c} = \psi c$ and ψ is any smooth function such that $\chi_{B_4} \le \psi \le \chi_{B_8}$. Note that the sup-norm and the modulus of continuity of the function $\widetilde{q}_n$ can be estimated independently of n. So, we can determine two positive constants $\widetilde{\omega}_{0,p}$ and K_p, independent of n and u, such that $\|\Delta u\|_{L^p(B_2)} \le K_p(\|\lambda u - L_{n,p}u\|_{L^p(\mathbb{R}^N)} + \|u\|_{L^p(\mathbb{R}^N)})$ for any $\lambda \in \mathbb{C}$ with $\mathrm{Re}\lambda \ge \widetilde{\omega}_{0,p}$, which replaced in (7.2.24) yields (7.2.19).

Step 3. Here we fix $p \in (1, +\infty)$ and prove that the equation $\lambda u - A_p u = f$ admits a unique solution $u_\lambda \in D(A_p)$ for any $f \in L^p(\mathbb{R}^N)$ and any $\lambda \in \mathbb{C}$ with real part not less than $\omega_0 + 1$. As a first step, we observe that $\rho(L_{n,p}) \supset \Sigma := \{\lambda \in \mathbb{C} : \mathrm{Re}\lambda > \omega_0\}$ for any $n \in \mathbb{N}$. Indeed, as we have already remarked, $L_{n,p}$ is a sectorial operator; hence, its resolvent set contains a right-halfplane. Such a right-halfplane contains Σ. Indeed, it is well-known that the function $\lambda \mapsto \|R(\lambda, L_{n,p})\|_{L(L^p(\mathbb{R}^N))}$ blows up as λ tends to the boundary of $\rho(L_{n,p})$ and (7.2.19) shows that this can not be case at any point of Σ.

Now, for any $n \in \mathbb{N}$ and $\lambda \in \Sigma$, we denote by u_n the unique solution of the equation $\lambda u_n - L_{n,p}u_n = f$ in $W^{2,p}(\mathbb{R}^N)$. By (7.2.19) the sequence $\{u_n\}$ is bounded in $L^p(\mathbb{R}^N)$. Since $L_{n,p}u_n = \lambda u_n - f$, also the sequence $\{L_{n,p}u_n\}$ is bounded in $L^p(\mathbb{R}^N)$. Hence, by (7.2.17) and (7.2.18) we can infer that

$$\sup_{n\in\mathbb{N}} \|\sqrt{q_n}\nabla u_n\|_{L^p(\mathbb{R}^N)} + \sup_{n\in\mathbb{N}} \|q_n D^2 u_n\|_{L^p(\mathbb{R}^N)} \le C\|f\|_{L^p(\mathbb{R}^N)} \tag{7.2.25}$$

for some positive constant C, independent of f. Recalling that q_n is bounded from below by $1/2$, we deduce that the sequence $\{u_n\}$ is bounded in $W^{2,p}(\mathbb{R}^N)$. By a classical compactness argument, u_n converges (up to a subsequence) to some function $u \in W^{2,p}(\mathbb{R}^N)$, weakly in $W^{2,p}(B_R)$ and strongly in $W^{1,p}(B_R)$, for any $R > 0$. Again, up to a subsequence, we can assume that u_n and ∇u_n converge, respectively, to u and ∇u, pointwise in $\mathbb{R}^N$. Since $\Delta u_n = q_n^{-1}(\lambda u_n - f - c_n u_n)$ and q_n and c_n converge, respectively, to q and $c-1$ locally uniformly in $\mathbb{R}^N$, Δu_n converges in $L^p_{\mathrm{loc}}(\mathbb{R}^N)$ to the function $q^{-1}((\lambda+1)u - f - cu)$. But we already know that, for any $R > 0$, Δu_n converges weakly in $L^p(B_R)$ to Δu. Hence, the function u solves the equation $(\lambda+1)u - \mathcal{A}u = f$. Finally, from (7.2.25) we get $\|\sqrt{q}\nabla u\|_{L^p(\mathbb{R}^N)} + \|qD^2u\|_{L^p(\mathbb{R}^N)} \le C\|f\|_{L^p(\mathbb{R}^N)}$, the constant C being independent of u and f. By difference, $cu = f + q\Delta u - \lambda u$ belongs to $L^p(\mathbb{R}^N)$. Hence, u belongs to $D(A_p)$. We have so proved that the equation $\lambda u - A_p u = f$ admits a solution $u_\lambda \in D(A_p)$ for any $\lambda \in \mathbb{C}$ with $\mathrm{Re}\lambda > \omega_0 + 1$. This is the unique solution to the previous equation in $D(A_p)$. Indeed, if $v \in D(A_p)$ solves the equation $\lambda v - A_p v = 0$, then $v \in W^{2,p}(\mathbb{R}^N)$ and $\lambda q^{-1}v - \Delta v - q^{-1}cv = 0$. Multiplying both sides of this equality by $\overline{v}|v|^{p-2}$ and integrating by parts gives

$$\begin{aligned}
0 =& \lambda \int_{\mathbb{R}^N} q^{-1}|v|^p dx + (p-1)\int_{\mathbb{R}^N} |\mathrm{Re}(\overline{v}\nabla v)|^2|v|^{p-4}\chi_{\{v\neq 0\}}dx \\
&+ \int_{\mathbb{R}^N} |\mathrm{Im}(\overline{v}\nabla v)|^2|v|^{p-4}\chi_{\{v\neq 0\}}dx \\
&+ i(p-2)\int_{\mathbb{R}^N} \langle \mathrm{Re}(\overline{v}\nabla v), \mathrm{Im}(\overline{v}\nabla v)\rangle |v|^{p-4}\chi_{\{v\neq 0\}}dx - \int_{\mathbb{R}^N} q^{-1}c|v|^p dx.
\end{aligned}$$

Taking the real part and recalling that $\mathrm{Re}\lambda > 0$, we conclude that $v \equiv 0$. Hence $\{\lambda \in \mathbb{C} : \mathrm{Re}\lambda \ge \omega_0 + 1\}$ is contained in $\rho(A_p)$.

Finally, letting n tend to $+\infty$ in (7.2.19) gives $|\lambda|\|u\|_{L^p(\mathbb{R}^N)} \le M_p\|\lambda u - A_p u\|_{L^p(\mathbb{R}^N)}$ for any $\alpha \in [0, 2]$, $u \in D(A_p)$ and $\lambda \in \mathbb{C}$ with $\mathrm{Re}\lambda \ge \omega_0 + 1$. By Theorem C.2.7, we conclude that A_p is a sectorial operator and, therefore, it generates an analytic semigroup $\{T_p(t)\}$ in $L^p(\mathbb{R}^N)$. Such a semigroup is strongly continuous. Indeed, $D(A_p)$ is dense in $L^p(\mathbb{R}^N)$ since it contains $C_c^\infty(\mathbb{R}^N)$.

Step 4. To complete the proof we check that the semigroups $\{T_p(t)\}$ preserve positivity and are all consistent. In view of the exponential formula (see Theorem C.1.7)

$$T_p(t)f = \lim_{n\to+\infty}\left[\frac{n}{t}R\left(\frac{n}{t},A_p\right)\right]^n f, \qquad t>0,\ f\in L^p(\mathbb{R}^N),$$

where the limit is meant in the norm topology of $L^p(\mathbb{R}^N)$, it suffices to prove that the resolvent families $\mathcal{R}_p := \{R(\lambda, A_p) : \lambda > 0\}$ ($p \in (1,+\infty)$) are consistent and preserve positivity. The positivity of the resolvent family $\mathcal{R}_p$ for any $p \in (1,+\infty)$ follows immediately by recalling that, for any $f \in L^p(\mathbb{R}^N)$, $R(\lambda, A_p)f$ is the limit in $L^p_{\rm loc}(\mathbb{R}^N)$ of the sequence of functions $\{R(\lambda, L_{n,p})f\}$ and, by classical results, each operator $R(\lambda, L_{n,p})$ preserves positivity.

Similarly, since for any $n \in \mathbb{N}$ the resolvent families $\{R(\lambda, L_{n,p}) : \lambda > 0\}$ ($p \in (1,+\infty)$) are consistent, for any $p, q \in (1,+\infty)$, $n \in \mathbb{N}$ and $f \in L^p(\mathbb{R}^N) \cap L^q(\mathbb{R}^N)$ the function $R(\lambda, L_{n,p})f$ belongs to $L^p(\mathbb{R}^N) \cap L^q(\mathbb{R}^N)$ and $R(\lambda, A_{n,p})f = R(\lambda, A_{n,q})f$. Letting n tend to $+\infty$ we conclude that $R(\lambda, A_p)f = R(\lambda, A_q)f$. □

7.2.1 The semigroup $\{T_p(t)\}$ and the spectrum of operator A_p

Proposition 7.2.6 *For any $p \in (1,+\infty)$, $f \in L^p(\mathbb{R}^N)$, $\gamma \in (0,1)$ and $t > 0$, the function $T_p(t)f$ belongs to $C^{1+\gamma}_b(\mathbb{R}^N) \cap C_0(\mathbb{R}^N)$. In particular, the semigroup $\{T_p(t)\}$ is ultracontractive.*

Proof Fix $f \in L^p(\mathbb{R}^N)$, $\gamma \in (0,1)$ and $t > 0$. Since $\{T_p(t)\}$ is an analytic semigroup in $L^p(\mathbb{R}^N)$, $T_p(t/2)f$ belongs to $D(A_p) \subset W^{2,p}(\mathbb{R}^N)$. If $p \ge N/2$ then $T_p(t/2)f$ belongs to $L^q(\mathbb{R}^N)$ for any $q \in [p,+\infty)$. Since the semigroups $\{T_r(t)\}$ ($r \in (1,+\infty)$) are consistent, $T_p(t)f = T_q(t/2)T_p(t/2)f$ belongs to $D(A_q)$. Choosing q such that $W^{2,q}(\mathbb{R}^N) \hookrightarrow C^{1+\gamma}_b(\mathbb{R}^N)$, we conclude that $T_p(t)f \in C^{1+\gamma}_b(\mathbb{R}^N) \cap C_0(\mathbb{R}^N)$.

Let us now suppose that $p < N/2$. Consider the sequence $\{q_n\}$ defined by $q_n = r_n^{-1}$, where $r_n = 1/p - 2n/N$ for any $n \in \mathbb{N}$. Let n_0 be the smallest integer such that $r_{n_0} \le 2/N$; note that $r_{n_0} > 0$. Then, $T_p(t/(n_0+2))f$ belongs to $D(A_p) \subset L^{q_1}(\mathbb{R}^N) \cap L^p(\mathbb{R}^N)$, by the Sobolev embedding theorem. Hence, $T_p(2t/(n_0+2))f = T_{q_1}(t/(n_0+2))T_p(t/(n_0+2))f \in D(A_{q_1}) \subset L^{q_2}(\mathbb{R}^N)$. Iterating this argument, we obtain that $T_p((n_0+1)t/(n_0+2))f \in D(A_{q_{n_0}})$ and we can conclude that $T_p(t)f$ belongs to $C^{1+\gamma}_b(\mathbb{R}^N) \cap C_0(\mathbb{R}^N)$ arguing as in the previous case. The last statement of the proposition is now immediate. □

By Theorems 1.2.5 and 3.1.3 (taking, as a Lyapunov function, the function $\varphi : \mathbb{R}^N \to \mathbb{R}$ defined by $\varphi(x) = 1+|x|^2$ for any $x \in \mathbb{R}^N$), we can associate a semigroup $\{T(t)\}$ of bounded operators in $C_b(\mathbb{R}^N)$ with operator $\mathcal{A}$. For any $f \in C_b(\mathbb{R}^N)$, $T(t)f$ is the value at $t > 0$ of the unique solution $u \in C([0,+\infty)\times\mathbb{R}^N) \cap C^{1,2}((0,+\infty)\times\mathbb{R}^N)$ of the Cauchy problem

$$\begin{cases} D_t u(t,x) = \mathcal{A}u(t,x), & t>0, \quad x\in\mathbb{R}^N,\\ u(0,x) = f(x), & \qquad\quad x\in\mathbb{R}^N, \end{cases} \tag{7.2.26}$$

which is bounded in $[0,+\infty)\times\mathbb{R}^N$.

Proposition 7.2.7 *The semigroups $\{T_p(t)\}$ and $\{T(t)\}$ agree on $B_b(\mathbb{R}^N) \cap L^p(\mathbb{R}^N)$ for any $p \in (1,+\infty)$. Moreover, for any $f \in L^p(\mathbb{R}^N)$ and $t > 0$, the function $T_p(t)f$ belongs to $C^{2+\gamma}_{\rm loc}(\mathbb{R}^N)$, where $\gamma = \alpha \wedge \beta$, if $\alpha, \beta > 0$, and $\gamma = \alpha \vee \beta$, if $\alpha\beta = 0$.*

Proof Fix $f \in C^2_c(\mathbb{R}^N) \subset D(A_q)$ for any $q \in (1,+\infty)$. Since $\{T_p(t)\}$ is a strongly continuous analytic semigroup in $L^p(\mathbb{R}^N)$ and $D(A_p) \hookrightarrow W^{2,p}(\mathbb{R}^N)$ for any $p \in (1,+\infty)$, the function $v = T_p(\cdot)f$ is in $C^\infty((0,+\infty); W^{2,p}(\mathbb{R}^N)) \cap C([0,+\infty); W^{2,p}(\mathbb{R}^N))$ and solves the Cauchy problem (7.2.26). Taking Proposition 7.2.6 and the Sobolev embedding theorem into account, we can infer that v belongs to $C^\infty((0,+\infty);C_b^{1+\theta}(\mathbb{R}^N))\cap C([0,+\infty)\times\mathbb{R}^N)$ for any $\theta\in(0,1)$. In particular, v belongs to $C^{1+\theta}_{\rm loc}((0,+\infty)\times\mathbb{R}^N)$ for any θ as above and, by difference, $q\Delta v+cv\in C^\theta_{\rm loc}((0,+\infty)\times\mathbb{R}^N)$. As a byproduct, Δv belongs to $C^\theta_{\rm loc}((0,+\infty)\times\mathbb{R}^N)$ and, by elliptic regularity, $D_{ij}v\in C((0,+\infty)\times\mathbb{R}^N)$ for any $i,j=1,\ldots,N$. Hence, $v\in C([0,+\infty)\times\mathbb{R}^N)\cap C^{1,2}((0,+\infty)\times\mathbb{R}^N)$ is a classical solution to problem (7.2.26) and is bounded in each strip $[0,T]\times\mathbb{R}^N$. By uniqueness, $T_p(\cdot)f=T(\cdot)f$.

Let us now assume that $f\in L^p(\mathbb{R}^N)\cap B_b(\mathbb{R}^N)$. For any $n\in\mathbb{N}$, let us consider the function $f_n=\vartheta_n(\varrho_n\star f)$, where $\{\varrho_n\}$ is a standard sequence of mollifiers, $\{\vartheta_n\}$ is a (standard) sequence of cut-off functions such that $\chi_{B_n}\le\vartheta_n\le\chi_{B_{2n}}$ for any $n\in\mathbb{N}$. The sequence $\{f_n\}$ converges to f in $L^p(\mathbb{R}^N)$ and pointwise almost everywhere in $\mathbb{R}^N$, as n tends to $+\infty$. Hence, $T_p(t)f_n$ converges to $T_p(t)f$ in $L^p(\mathbb{R}^N)$, as n tends to $+\infty$, for any $t>0$. Moreover, since $\|f_n\|_\infty\le\|f\|_\infty$ for any $n\in\mathbb{N}$, $T(t)f_n$ converges to $T(t)f$ as n tends to $+\infty$, pointwise in $\mathbb{R}^N$, for any $t>0$ (as it easily follows from the formula (1.2.7)). Hence, recalling that $\{T(t)\}$ is strong Feller (see Proposition 1.2.13), splitting $T(t)f_n=T(t/2)T(t/2)f_n$ and using Proposition 1.2.10, we can infer that, for any $t>0$, $T(t)f_n$ converges to $T(t)f$, locally uniformly in $\mathbb{R}^N$, as n tends to $+\infty$. We thus conclude that $T_p(t)f\equiv T(t)f$ for any $t>0$ and we are done.

To complete the proof we observe that, for any $g\in C_b(\mathbb{R}^N)$ and $t>0$, the function $T(t)g$ belongs to $C^{2+\gamma}_{\rm loc}(\mathbb{R}^N)$. Since $T_p(t/2)f\in C_b(\mathbb{R}^N)$ for any $f\in L^p(\mathbb{R}^N)$, by Proposition 7.2.6, we obtain that $T_p(t)f=T(t/2)T_p(t/2)f\in C^{2+\gamma}_{\rm loc}(\mathbb{R}^N)$ as has been claimed. □

Corollary 7.2.8 *For any $p\in(1,+\infty)$ the semigroup $\{T_p(t)\}$ is irreducible, i.e., for any nonnegative and not identically vanishing function $f\in L^p(\mathbb{R}^N)$ and any $t>0$, $T_p(t)f$ is everywhere positive in $\mathbb{R}^N$.*

Proof Since the semigroup $\{T(t)\}$ is irreducible, from Proposition 7.2.7 we deduce that $T_p(t)\chi_E>0$ in $\mathbb{R}^N$ for any measurable set $E\subset\mathbb{R}^N$ with positive and finite Lebesgue measure and any $t>0$.

Let us now fix a nonnegative and not identically vanishing function $f\in L^p(\mathbb{R}^N)$. Then, there exists $m\in\mathbb{N}$ such that the set $E_m=\{x\in\mathbb{R}^N:\ f(x)>1/m\}$ has a positive Lebesgue measure. Up to intersecting E_m with a sufficiently large ball, we can assume that the set E_m is bounded and non-empty. Since $f\ge m^{-1}\chi_{E_m}$, from the positivity of the semigroup $\{T_p(t)\}$ we conclude that $T_p(t)f\ge T_p(t)\chi_{E_m}>0$ everywhere in $\mathbb{R}^N$. □

For any $p\in(1,+\infty)$, $(A_p,D(A_p))$ is the minimal realization of the operator $\mathcal{A}$ in $L^p(\mathbb{R}^N)$. The following proposition shows that $(A_p,D(A_p))$ actually coincides with the maximal realization of the operator $\mathcal{A}$ in $L^p(\mathbb{R}^N)$.

Proposition 7.2.9 *For any $p\in(1,+\infty)$ it holds that*

$$D(A_p)=D_{\max,p}(\mathcal{A}):=\{u\in L^p(\mathbb{R}^N)\cap W^{2,p}_{\rm loc}(\mathbb{R}^N):\ \mathcal{A}u\in L^p(\mathbb{R}^N)\}.$$

Proof Clearly, we have only to prove the inclusion "$\supset$". Fix $p\in(1,+\infty)$, $u\in D_{\max,p}(\mathcal{A})$, $\lambda\in\rho(A_p)\cap\mathbb{R}$ and set $f:=\lambda u-\mathcal{A}u$. Without loss of generality, we can assume that u is a real-valued function. The function $v:=u-R(\lambda,A_p)f$ satisfies the equation $\lambda v-\mathcal{A}v=0$. We shall show that $v\equiv 0$, provided λ is large enough.

We first consider the case when α belongs to $[1,2]$. Let $\{\vartheta_n\}$ be a standard sequence of cut-off functions. Integrating the identity $(\lambda v-\mathcal{A}v)v|v|^{p-2}\vartheta_n^2=0$ by parts on $\mathbb{R}^N$, we get

$$\begin{aligned}0=&\lambda\int_{\mathbb{R}^N}|v|^p\vartheta_n^2dx-\int_{\mathbb{R}^N}c|v|^p\vartheta_n^2dx+(p-1)\int_{\mathbb{R}^N}q|\nabla v|^2|v|^{p-2}\vartheta_n^2\chi_{\{v\neq 0\}}dx\\&+2\int_{\mathbb{R}^N}q\vartheta_n|v|^{p-2}v\langle\nabla v,\nabla\vartheta_n\rangle dx+\int_{\mathbb{R}^N}\vartheta_n^2|v|^{p-2}v\langle\nabla q,\nabla v\rangle dx.\end{aligned}\tag{7.2.27}$$

Note that $C_1 := \sup_{n\in\mathbb{N}} \|\sqrt{q}\nabla\vartheta_n\|_\infty < +\infty$. Hence,

$$
\begin{aligned}
\left|\int_{\mathbb{R}^N} q\vartheta_n|v|^{p-2}v\langle\nabla v, \nabla\vartheta_n\rangle dx\right| &\le C_1\int_{\mathbb{R}^N\setminus B_n} \sqrt{a}|v|^{p-1}|\nabla v|\vartheta_n dx\\
\le& C_1\left(\int_{\mathbb{R}^N} q|\nabla v|^2|v|^{p-2}\vartheta_n^2\chi_{\{v\neq 0\}}dx\right)^{\frac{1}{2}}\left(\int_{\mathbb{R}^N\setminus B_n}|v|^p dx\right)^{\frac{1}{2}}\\
\le& \varepsilon\int_{\mathbb{R}^N} q|\nabla v|^2|v|^{p-2}\vartheta_n^2\chi_{\{v\neq 0\}}dx + \frac{C_1^2}{4\varepsilon}\int_{\mathbb{R}^N\setminus B_n}|v|^p dx
\end{aligned}
\tag{7.2.28}
$$

and, since $|\nabla q| \le \alpha\sqrt{q}$, we can estimate (in a completely similar way)

$$
\left|\int_{\mathbb{R}^N}\vartheta_n^2|v|^{p-2}v\langle\nabla q, \nabla v\rangle dx\right| \le \varepsilon\int_{\mathbb{R}^N} q|\nabla v|^2|v|^{p-2}\vartheta_n^2\chi_{\{v\neq 0\}}dx + \frac{\alpha^2}{4\varepsilon}\int_{\mathbb{R}^N}|v|^p\vartheta_n^2 dx \tag{7.2.29}
$$

for any $\varepsilon > 0$. Replacing (7.2.28) and (7.2.29) in (7.2.27) and taking $\varepsilon = (p-1)/3$ gives

$$
\left(\lambda - \frac{3\alpha^2}{4(p-1)}\right)\int_{\mathbb{R}^N}|v|^p\vartheta_n^2 dx - \frac{3C_1^2}{2(p-1)}\int_{\mathbb{R}^N\setminus B_n}|v|^p dx \le 0. \tag{7.2.30}
$$

Letting n tend to $+\infty$ yields $v \equiv 0$, if we take λ large enough.

Let us now assume that $\alpha \in [0,1)$ and consider the function $\widehat{q}$ defined by $\widehat{q} = \varphi\widetilde{q}+(1-\varphi)q$, where $\widetilde{q}$ is any smooth function such that $\widetilde{q} \ge 1$, $\|q-\widetilde{q}\|_{L^\infty(B_2)} \le 1/2$ and $\varphi \in C_c^\infty(\mathbb{R}^N)$ satisfies the condition $\chi_{B_1} \le \varphi \le \chi_{B_2}$. Function $\widetilde{q}$ can be obtained, for instance, regularizing by convolution q. Let $\widehat{\mathcal{A}}$ be the operator defined as $\mathcal{A}$ with q being replaced by $\widehat{q}$. As in the case when $\alpha \ge 1$, we multiply the equation $\lambda v - \mathcal{A}v = 0$ by $v|v|^{p-2}\vartheta_n^2$ and integrate in $\mathbb{R}^N$, getting

$$
\begin{aligned}
0 =& \int_{\mathbb{R}^N}(\lambda v - \mathcal{A}v)v|v|^{p-2}\vartheta_n^2 dx\\
=& \int_{\mathbb{R}^N}(\lambda v - \widehat{\mathcal{A}}v)v|v|^{p-2}\vartheta_n^2 dx + \int_{B_2}(\widehat{q}-q)v|v|^{p-2}\vartheta_n^2\Delta v dx.
\end{aligned}
\tag{7.2.31}
$$

Since $\widehat{q}$ is continuously differentiable in $\mathbb{R}^N$ and $|\nabla\widehat{q}| \le \widehat{K}\sqrt{\widehat{q}}$ for some positive constant $\widehat{K}$, we can integrate by parts the first term in the last side of (7.2.31). Arguing as in the proof of (7.2.30) we can estimate

$$
\int_{\mathbb{R}^N}(\lambda v - \widehat{\mathcal{A}}v)v|v|^{p-2}\vartheta_n^2 dx \ge \left(\lambda - \frac{3\widehat{K}^2}{4(p-1)}\right)\int_{\mathbb{R}^N}|v|^p\vartheta_n^2 dx - \frac{3\widehat{C}_1^2}{2(p-1)}\int_{\mathbb{R}^N\setminus B_n}|v|^p dx, \tag{7.2.32}
$$

where $\widehat{C}_1 = \sup_{n\in\mathbb{N}}\|\sqrt{\widehat{q}}\nabla\vartheta_n\|_\infty$. As far as the other term in the last side of (7.2.31) is concerned, we observe that, since $q \ge 1$ in $\mathbb{R}^N$, it holds that

$$
\begin{aligned}
\left|\int_{B_2}(\widehat{q}-q)v|v|^{p-2}\vartheta_n^2\Delta v dx\right| \le& \|q-\widetilde{q}\|_{L^\infty(B_2)}\|v\|_{L^p(B_2)}^{p-1}\|q\Delta v\|_{L^p(B_2)}\\
\le& \frac{1}{2}\|v\|_{L^p(B_2)}^{p-1}\|(c+\lambda)v\|_{L^p(B_2)}\\
\le& \frac{1}{2}(2^\beta+\lambda)\|v\|_{L^p(\mathbb{R}^N)}^p.
\end{aligned}
\tag{7.2.33}
$$

From (7.2.31)-(7.2.33) we obtain

$$0 \geq \left(\lambda - \frac{3\widehat{K}^2}{4(p-1)}\right)\int_{\mathbb{R}^N} |v|^p \vartheta_n^2 dx - \frac{3\widehat{C}_1^2}{2(p-1)}\int_{\mathbb{R}^N\setminus B_n} |v|^p dx - \frac{1}{2}(2^\beta + \lambda)\int_{\mathbb{R}^N} |v|^p dx.$$

Again, letting n tend to $+\infty$ and, then, choosing λ large enough, we conclude that $v \equiv 0$.
□

Theorem 7.2.10 *For any $\alpha \in [0,2]$, $\beta > 0$ and $p \in (1,+\infty)$ the spectrum of A_p consists of a sequence of negative eigenvalues which accumulates at $-\infty$. Moreover, $\sigma(A_p)$ is independent of p.*

Proof The proof is split into three steps. In the first one we prove that $\sigma(A_p)$ consists of isolated eigenvalues for any $p \in (1,+\infty)$. Then, we prove that $\sigma(A_p)$ is independent of p and, finally, we show that the eigenvalues of A_p are real and negative.

Step 1. Fix $p \in (1,+\infty)$. Let us show that $D(A_p)$ is compactly embedded into $L^p(\mathbb{R}^N)$ for any $p \in (1,+\infty)$. This will yield immediately that the resolvent operator $R(\lambda, A_p)$ is compact in $L^p(\mathbb{R}^N)$ for any $\lambda \in \rho(A_p)$. Hence, the spectrum of A_p consists of eigenvalues.

Since

$$\int_{\mathbb{R}^N} |x|^{\beta p}|u(x)|^p dx \leq C_1\left(\int_{\mathbb{R}^N} |u(x)|^p dx + \int_{\mathbb{R}^N} |A_p u(x)|^p dx\right)$$

for some positive constant C_1, independent of u, taking Corollary 7.2.4 into account we easily conclude that there exists a positive constant C_2, independent of u as well, such that

$$\int_{\mathbb{R}^N} |x|^{\beta p}|u(x)|^p dx \leq C_2 \tag{7.2.34}$$

for any $u \in \mathcal{B} := \{v \in D(A_p) : \|v\|_{D(A_p)} \leq 1\}$. This estimate yields the compactness of $\mathcal{B}$ in $L^p(\mathbb{R}^N)$. Indeed, estimate (7.2.34) shows that

$$\int_{\mathbb{R}^N\setminus B_M} |u(x)|^p dx \leq M^{-\beta p}\int_{\mathbb{R}^N\setminus B_M} |x|^{\beta p}|u(x)|^p dx \leq C_2 M^{-\beta p}, \qquad u \in \mathcal{B}. \tag{7.2.35}$$

Let us fix $\varepsilon > 0$ and let M_ε be large enough that $\|u\|_{L^p(\mathbb{R}^N\setminus B_{M_\varepsilon})} \leq 2^{-1/p}\varepsilon$ for any $u \in \mathcal{B}$. Since the set $\mathcal{B}_{|B_{M_\varepsilon}}$ of the restrictions to B_{M_ε} of all the functions in $\mathcal{B}$ is continuously embedded in $W^{2,p}(B_{M_\varepsilon})$, which is compactly embedded in $L^p(B_{M_\varepsilon})$, there exist $n_\varepsilon \in \mathbb{N}$ and functions $f_1, \ldots, f_{n_\varepsilon}$ in $L^p(B_{M_\varepsilon})$ such that, for any $u \in \mathcal{B}$ and some $j = j(u) \in \{1, \ldots, n_\varepsilon\}$, $\|u - f_j\|_{L^p(B_{M_\varepsilon})} \leq 2^{-1/p}\varepsilon$. Let us now denote by $\widetilde{f}_j$ ($j = 1, \ldots, n_\varepsilon$) the function which equals f_j in B_{M_ε} and identically vanishes elsewhere in $\mathbb{R}^N$. Using (7.2.35) and this last estimate we obtain that $\|u - \widetilde{f}_j\|_{L^p(\mathbb{R}^N)} \leq \varepsilon$ and this shows that $\mathcal{B}$ is totally bounded (i.e., is compact) in $L^p(\mathbb{R}^N)$.

Step 2. To show that the spectrum of A_p is independent of $p \in (1,+\infty)$, we adapt the arguments in the proof of [84, Corollary 1.6.2]. Fix $p, q \in (1,+\infty)$ and $f \in C_c^\infty(\mathbb{R}^N)$. By the proof of Theorem 7.2.5 we know that A_p and A_q are sectorial operators. Hence, we can determine $\omega > 0$ such that the interval $(\omega, +\infty)$ is contained in both the resolvent sets of the operators A_p and A_q. The proof of the quoted theorem also shows that $R(\lambda, A_p) \equiv R(\lambda, A_q)$ on $L^p(\mathbb{R}^N) \cap L^p(\mathbb{R}^N)$ for any $\lambda > \omega$. Since $\sigma(A_p)$ and $\sigma(A_q)$ consist of isolated eigenvalues, $\mathbb{C} \setminus (\sigma(A_p) \cup \sigma(A_q))$ is a connected open set in $\mathbb{C}$. Hence, $R(\lambda, A_p)$ and $R(\lambda, A_q)$ coincide in $L^p(\mathbb{R}^N) \cap L^p(\mathbb{R}^N)$ for any $\lambda \in \mathbb{C} \setminus (\sigma(A_p) \cup \sigma(A_q))$.

Let us now fix $\lambda_0 \in \sigma(A_p)$ and observe that λ_0 is isolated in $\sigma(A_p) \cup \sigma(A_q)$ since neither

$\sigma(A_p)$ nor $\sigma(A_q)$ admit accumulation points in $\mathbb{C}$. Hence, we can determine $\varepsilon > 0$ small enough such that $B_\varepsilon(\lambda_0) \setminus \{\lambda_0\} \subset \mathbb{C} \setminus (\sigma(A_p) \cup \sigma(A_q))$. Let P be the spectral projection associated with the eigenvalue $\lambda_0 \in \sigma(A_p)$, which is defined by

$$Ph = \frac{1}{2\pi i} \int_{\partial B_\varepsilon(\lambda_0)} R(\lambda, A_p) h d\lambda, \qquad h \in L^p(\mathbb{R}^N),$$

where $\partial B_\varepsilon(\lambda_0)$ is oriented counterclockwise. If $\lambda_0 \notin \sigma(A_q)$, then from the above arguments we obtain that $Pf = 0$, which implies that $P \equiv 0$ by density: a contradiction. Hence, $\sigma(A_p) \subset \sigma(A_q)$. The arbitrariness of p and q yields $\sigma(A_q) = \sigma(A_p)$.

Step 3. Now, we prove that the spectrum of A_p consists of negative eigenvalues for any $p \in (1, +\infty)$. In view of Step 2, we can limit ourselves to dealing with the case $p = 2$. Let $\lambda \in \sigma(A_2)$ and $u \in D(A_2)$ be such that $\lambda u - A_2 u = 0$. Multiplying both sides of this equality by $q^{-1}\overline{u}$ and integrating by parts we get

$$\begin{aligned} 0 =& \lambda \int_{\mathbb{R}^N} q^{-1}|u|^2 dx - \int_{\mathbb{R}^N} \overline{u} \Delta u dx - \int_{\mathbb{R}^N} q^{-1} c |u|^2 dx \\ =& \lambda \int_{\mathbb{R}^N} q^{-1}|u|^2 dx + \int_{\mathbb{R}^N} |\nabla u|^2 dx - \int_{\mathbb{R}^N} q^{-1} c |u|^2 dx \end{aligned}$$

and conclude that λ is real and negative.

Finally, we observe that, since A_p is an unbounded operator, the eigenvalues of operator A_p can be ordered into a sequence diverging to $-\infty$. This concludes the proof. □

We conclude this subsection with the following result which will be used in Chapter 8.

Proposition 7.2.11 *Suppose that $\alpha \in [0, 2]$ and $\beta > 0$. Then, for any $p \in (1, +\infty)$, the eigenspace corresponding to the largest eigenvalue λ_0 of A_p is one-dimensional and is spanned by a strictly positive function ψ_0, which is radial, belongs to $C_b^{1+\gamma}(\mathbb{R}^N) \cap C_0(\mathbb{R}^N) \cap C^2(\mathbb{R}^N)$ for any $\gamma \in (0, 1)$ and converges to 0 as $|x|$ tends to $+\infty$. As a byproduct, for any $p \in (1, +\infty)$ there exists a positive constant M_p such that*

$$\|T_p(t)\|_{L(L^p(\mathbb{R}^N))} \le M_p e^{\lambda_0 t}, \qquad t \ge 0. \tag{7.2.36}$$

Proof Fix $p \in (1, +\infty)$. By Theorem 7.2.10, the spectrum of A_p consists of a sequence of negative isolated eigenvalues with finite geometric multiplicities. From Proposition B.1.6 and the Riesz-Schauder theory for compact operators it follows that λ_0 is a pole of $R(\cdot, A_p)$ with finite algebraic multiplicity.

Corollary 7.2.8 and Theorem C.1.16 then show that the eigenspace corresponding to λ_0 is one-dimensional and is spanned by a strictly positive function ψ_0. Since $e^{\lambda_0}\psi_0 = T_p(1)\psi_0$ Propositions 7.2.6 and 7.2.7 imply that ψ_0 belongs to $C_b^{1+\gamma}(\mathbb{R}^N) \cap C^2(\mathbb{R}^N) \cap C_0(\mathbb{R}^N)$ for any $\gamma \in (0, 1)$.

To prove that ψ_0 is radial, we observe that, for any orthogonal matrix R, the function $x \mapsto \psi_R(x) := \psi_0(Rx)$ is an eigenfunction of A_p associated with the eigenvalue λ_0. Hence, ψ_R and ψ_0 should be proportional. Since they coincide at $x = 0$, they coincide everywhere in $\mathbb{R}^N$ so that ψ_0 is radial.

Finally, estimate (7.2.36) follows from Theorem C.2.13 since λ_0 is a simple eigenvalue of the operator A_p. □

7.2.2 Concluding remarks

In [104] the operator $\mathcal{A} = q(x)\Delta + \langle b(x), \nabla \rangle$ is considered under the following set of assumptions.

Hypotheses 7.2.12 (i) The positive function a belongs to $C^1(\mathbb{R}^N)$ and $|\nabla q(x)| \le C\sqrt{q(x)}$ for any $x \in \mathbb{R}^N$ and some positive constant $C > 0$;

(ii) $b : \mathbb{R}^N \to \mathbb{R}^N$ is a measurable function such that $|b(x)| \le C\sqrt{a(x)}$ for almost every $x \in \mathbb{R}^N$.

The previous assumptions are satisfied, in particular, when $\mathcal{A} = (1 + |x|^\alpha)\Delta + |x|^{\alpha-2}\langle x, \nabla\rangle$ and $\alpha \in (1, 2]$. Actually, using the same arguments used in the proof of Proposition 7.2.1 also the case when $\alpha \in (0, 1]$ can be considered.

The results in [104] show that for any $p \in (1, +\infty)$ the realization A_p in $L^p(\mathbb{R}^N)$ of the operator $\mathcal{A}$, with domain $D(A_p) = \{u \in L^p(\mathbb{R}^N) \cap W^{2,p}_{\rm loc}(\mathbb{R}^N) : \sqrt{q}|\nabla u|, q\Delta u \in L^p(\mathbb{R}^N)\}$, generates a strongly continuous analytic semigroup and A_p coincides with the maximal realization of the operator $\mathcal{A}$ in $L^p(\mathbb{R}^N)$. If, in addition to the previous assumptions, $q \in C^2(\mathbb{R}^N)$ has positive infimum and bounded Laplacian, then the realization of the operator $\mathcal{A}$ in $L^1(\mathbb{R}^N)$ with domain $\{u \in L^1(\mathbb{R}^N) : \sqrt{q}|\nabla u|,\ q\Delta u \in L^p(\mathbb{R}^N)\}$ is sectorial.

7.3 Schrödinger type operators: the case $\alpha > 2$

In this section we consider the operator $\mathcal{A} = (1 + |x|^\alpha)\Delta$ with $\alpha > 2$. First of all, we show that, if a resolvent operator can be associated with the maximal realization A_p of the operator $\mathcal{A}$ in $L^p(\mathbb{R}^N)$ (with domain $D(A_p) = \{u \in L^p(\mathbb{R}^N) \cap W^{2,p}_{\rm loc}(\mathbb{R}^N) : \mathcal{A}u \in L^p(\mathbb{R}^N)\}$), then this operator is positive.

Lemma 7.3.1 *If $\lambda \in \rho(A_p) \cap (0, +\infty) \neq \varnothing$, then, for any $f \in L^p(\mathbb{R}^N)$, $R(\lambda, A_p)$ belongs to $W^{2,p}(\mathbb{R}^N)$. Moreover, $R(\lambda, A_p)f \ge R(\lambda, \widehat{A})f$ for any nonnegative function $f \in C_c(\mathbb{R}^N)$, where $R(\lambda, \widehat{A})$ is the resolvent operator associated with the realization of the operator $\mathcal{A}$ in $C_b(\mathbb{R}^N)$.*

Proof As a first step, we prove that $D_{\max,p}(\mathcal{A}) = \{u \in W^{2,p}(\mathbb{R}^N) : q\Delta u \in L^p(\mathbb{R}^N)\}$. This is enough to conclude that any solution of the equation $\lambda u - A_p u = f$ in $D(A_p)$ is actually a $W^{2,p}(\mathbb{R}^N)$-function. The inclusion "$\supset$" clearly holds. On the other hand, if $u \in D(A_p)$, then $q\Delta u$ belongs to $L^p(\mathbb{R}^N)$. Hence, Δu is in $L^p(\mathbb{R}^N)$. Since u and Δu are in $L^p(\mathbb{R}^N)$, Theorem D.1.4(iii) shows that $u \in W^{2,p}(\mathbb{R}^N)$ and the inclusion "$\subset$" follows.

To prove that $R(\lambda, \widehat{A})f \le R(\lambda, A_p)f$ when $C_c(\mathbb{R}^N) \ni f \ge 0$, it suffices to show that $R(\lambda, A_p)f$ actually belongs to $D_{\max}(\mathcal{A})$ (see (1.0.1)) and is nonnegative. The minimality of $R(\lambda, \widehat{A})f$ among all the nonnegative solutions in $D_{\max}(\mathcal{A})$ of the equation $\lambda v - \mathcal{A}v = f$ (see Remark 1.1.2), will allow us to conclude the proof. Suppose first that $f \in C^\infty_c(\mathbb{R}^N)$ is nonnegative (the general case then will follow by a straightforward density argument). By local elliptic regularity, $R(\lambda, A_p)f$ belongs to $C^2(\mathbb{R}^N)$. Let us prove that u belongs to $C_0(\mathbb{R}^N)$; this will allow us to conclude that $u \in D_{\max}(\mathcal{A})$. If $p > N/2$ then this is clear, due to the Sobolev embedding theorem. When $p < N/2$ the claim follows from a bootstrap argument. To make this argument work, we fix $R > 0$ large enough that ${\rm supp}(f) \subset B_R$ and observe that u solves the equation $\Delta u = u/q$ in $\mathbb{R}^N \setminus B_R$. Since $u \in W^{2,p}(\mathbb{R}^N)$, u belongs to $L^{p_1}(\mathbb{R}^N)$, where $1/p_1 = 1/p - 2/N$. Hence $\Delta u \in L^{p_1}(\mathbb{R}^N \setminus B_R)$ and, again by Theorem D.1.4(iii), u belongs to $W^{2,p_1}(\mathbb{R}^N \setminus B_R)$. If $p_1 > N/2$, then we conclude that $u \in C_0(\mathbb{R}^N)$, otherwise we repeat this argument and, in a finite number of steps, we obtain an exponent p_n greater than $N/2$. Indeed, the sequence $\{p_k^{-1}\}$, defined recursively by $p_{k+1}^{-1} = p_k^{-1} - 2/N$

is definitively negative. Finally, the maximum principle shows that $R(\lambda, A_p)$ is nonnegative. □

The following proposition is crucial for the rest of this section.

Proposition 7.3.2 *If $N = 1, 2$ and $p \in [1, +\infty)$ or $N \geq 3$ and $p \leq N(N-2)^{-1}$, then the set $\rho(A_p) \cap [0, +\infty)$ is empty. Hence, A_p generates neither a strongly continuous nor an analytic semigroup in $L^p(\mathbb{R}^N)$.*

Proof In the case when $N = 1$ the proof follows arguing as in Example 7.1.1 since both $+\infty$ and $-\infty$ are entrance points for the operator $\mathcal{A}$.

The case when $N = 2$ is similar. Fix a radial positive function $f \in L^p(\mathbb{R}^2) \cap C_0(\mathbb{R}^2)$ and $\lambda > 0$. Since $R(\lambda, \widehat{A})f$ is the pointwise limit in $\mathbb{R}^2$ of the sequence of solutions $\{u_n\}$ to equation $\lambda u - \mathcal{A}u = f$ in B_n, with homogeneous Dirichlet boundary conditions on ∂B_n (see Theorem 1.1.1) and each function u_n is radial (since f is a radial function), we conclude that u is radial as well, i.e., $u(x) = v(|x|)$ for any $x \in \mathbb{R}^2$. We are thus led to the study of the equation

$$\lambda v(r) - \left((1 + r^\alpha)w''(r) + \frac{N-1}{r}v'(r)\right) = g(r), \qquad r \in (0, +\infty),$$

where $f(x) = g(|x|)$ for any $x \in \mathbb{R}^N$.

The Feller theory, developed in Chapter 2 in $\mathbb{R}$, works as well in the halfine $(0, +\infty)$, with obvious changes. Since the operator in brackets in the above equation has $+\infty$ as an entrance point, we can still apply the arguments in Example 7.1.1 and conclude that the nonnegative function $R(\lambda, \widehat{A})f$ does not belong to $L^p(\mathbb{R}^2)$. As a consequence of Lemma 7.3.1, we conclude that $\rho(A_p) \cap [0, +\infty) = \varnothing$.

Now, we suppose that $N \geq 3$, $p \leq N(N-2)^{-1}$ and slightly modify the above arguments. Fix $\lambda > 0$ and a smooth and nonnegative function $\psi : [0, +\infty) \to \mathbb{R}$, with support contained in $[0, 1]$ and such that $\psi(0) = 1$. Finally, set $f(x) = \psi(|x|)$ for any $x \in \mathbb{R}^N$. Then, Proposition 4.3.1 and Theorem 4.3.2, with $\varphi(x) = (1 + |x|^2)^{-1}$ for any $x \in \mathbb{R}^N$, show $R(\lambda, \widehat{A})f$ belongs to $C_0(\mathbb{R}^N)$ and is strictly positive. As in the case when $N = 2$, u is radial, i.e., $u(x) = v(|x|)$ for any $x \in \mathbb{R}^N$. If we set $v(r) = w(r)r^{2-N}$ for any $r \geq 0$, we easily realize that

$$\lambda w(r) - \left((1 + r^\alpha)w''(r) + \frac{3-N}{r}w'(r)\right) = 0, \qquad r \geq 1. \tag{7.3.1}$$

Straightforward computations reveal that the functions $\mathscr{P}$ and $\mathscr{R}$, defined accordingly to the formulae (2.1.4), (2.1.5) (both the integrals being taken in the interval $[1, r]$), are given by $\mathscr{P}(r) = (N-2)^{-1}(1 + r^\alpha)^{-1}(r - r^{3-N})$ for any $r \geq 1$ and

$$\mathscr{R}(r) = r^{N-3}\int_1^r \frac{1}{(1 + s^\alpha)s^{N-3}}ds \geq \frac{r^{N-3}(r^{4-\alpha-N} - 1)}{4 - \alpha - N}, \qquad r \geq 1.$$

The choice of α shows that $\mathscr{P} \in L^1((1, +\infty))$ and $\mathscr{R} \notin L^1((1, +\infty))$. By Proposition 2.1.7(iii), equation (7.3.1) admits a positive decreasing solution w_1, which converges to 1 as r tends to $+\infty$, and all the other solutions, which are linearly independent from w_1, are unbounded in $(1, +\infty)$. It thus follows that $v(r) \geq r^{2-N}$ in a neighborhood of $+\infty$. The condition on p reveals that $R(\lambda, \widehat{A})f$ does not belong to $L^p(\mathbb{R}^N)$. *Ad fortiori*, $u \notin L^p(\mathbb{R}^N)$ since $u \geq R(\lambda, \widehat{A})f$, and this completes the proof. □

In view of Proposition 7.3.2 in the rest of this section we assume that $p > N(N-2)^{-1}$. The following lemma plays a crucial role in all this section.

Lemma 7.3.3 *The operator A_p is invertible (i.e., $0 \in \rho(A_p)$) and*

$$(R(0, A_p)f)(x) = \frac{1}{N(N-2)\omega_N} \int_{\mathbb{R}^N} \frac{f(y)}{|x-y|^{N-2}(1+|y|^\alpha)} dy \tag{7.3.2}$$

for almost every $x \in \mathbb{R}^N$ and any $f \in L^p(\mathbb{R}^N)$.

Proof The operator A_p is injective since the null function is the only harmonic function which belongs to $L^p(\mathbb{R}^N)$. On the other hand, if $f \in C_c^\infty(\mathbb{R}^N)$, then the equation $-A_p u = f$ is equivalent to the equation $\Delta u = -f/q$. By classical results, this equation is solvable with solution given by the right-hand side of the previous formula. If $f \in L^p(\mathbb{R}^N)$, then we consider a sequence $\{f_n\} \in C_c^\infty(\mathbb{R}^N)$ converging to $f \in L^p(\mathbb{R}^N)$. A straightforward computation (see also the proof of the forthcoming Theorem 7.3.4) shows that the right-hand side of (7.3.2) defines a bounded operator in $L^p(\mathbb{R}^N)$. Hence, if u_n denotes the right-hand side of this formula, with f being replaced by f_n, then u_n and Δu_n converge in $L^p(\mathbb{R}^N)$ as n tends to $+\infty$. By the Calderón-Zygmund inequality (D.1.1), we can infer that u_n converges to the right-hand side of the formula (7.3.2) in $W^{2,p}(\mathbb{R}^N)$ and, hence, $A_p u = f$. □

We now provide a different characterization of the domain $D(A_p)$ of the operator A_p in the case when $\alpha \in (2, N/p')$ and an embedding result for any value of $\alpha > 2$.

Theorem 7.3.4 *Let $\alpha > 2$ and fix $f \in L^p(\mathbb{R}^N)$. Then, for any $\beta, \gamma \geq 0$, which satisfy the inequalities $\beta \leq \alpha - 2$, $\beta < N/p' - 2$, $\gamma \leq \alpha - 1$ and $\gamma < N/p' - 1$, there exists a positive constant C such that*

$$\|q^{\beta/\alpha} u\|_{L^p(\mathbb{R}^N)} + \|q^{\gamma/\alpha} \nabla u\|_{L^p(\mathbb{R}^N)} \leq C \|Au\|_{L^p(\mathbb{R}^N)}, \qquad u \in D(A_p).$$

In particular, if $\alpha \in (2, N/p')$, then $D(A_p) = \mathscr{D}_p$ with equivalence of the corresponding norms, where

$$\mathscr{D}_p = \{u \in W^{2,p}(\mathbb{R}^N) : q^{1-2/\alpha} u,\ q^{1-1/\alpha} |\nabla u|,\ q|D^2 u| \in L^p(\mathbb{R}^N)\} \tag{7.3.3}$$

and $\|u\|_{\mathscr{D}_p} = \|q^{1-2/\alpha} u\|_{L^p(\mathbb{R}^N)} + \|q^{1-1/\alpha} \nabla u\|_{L^p(\mathbb{R}^N)} + \|q D^2 u\|_{L^p(\mathbb{R}^N)}$ for any $u \in \mathscr{D}_p$.

Proof The core of the proof consists of showing that

$$\|q^{\beta/\alpha} R(0, A_p) f\|_{L^p(\mathbb{R}^N)} + \|q^{\gamma/\alpha} \nabla R(0, A_p) f\|_{L^p(\mathbb{R}^N)} \leq C \|f\|_{L^p(\mathbb{R}^N)}, \qquad f \in L^p(\mathbb{R}^N), \tag{7.3.4}$$

for α, β and γ as in the statement. Here and throughout the proof, we denote by C a positive constant, independent of f, which may vary from line to line.

From estimate (7.3.4) the assertion follows at once since $u = -R(0, A_p) A_p u$ for any $u \in D(A_p)$, due to Lemma 7.3.3. Note that, if $\alpha < N/p'$, then $\beta = \alpha - 2$ and $\gamma = \alpha - 1$ are admissible choices in (7.3.4).

As a first step we prove that

$$\frac{1}{N(2-N)\omega_N} \int_{\mathbb{R}^N} \frac{1}{|x-y|^{N-2} |y|^\beta} dy = \frac{|x|^{2-\beta}}{(2-\beta)(N-\beta)}, \qquad x \in \mathbb{R}^N, \tag{7.3.5}$$

for any $\beta \in (2, N)$. For this purpose, we begin by observing that the right-hand side of (7.3.5) is invariant under rotations. Therefore, we can replace x by $|x|e_1$, where e_1 denotes the first vector of the Euclidean basis of $\mathbb{R}^N$. As is immediately seen,

$$\int_{\mathbb{R}^N} \frac{1}{||x|e_1 - y|^{N-2} |y|^\beta} dy = \int_0^{+\infty} \rho^{N-1-\beta} d\rho \int_{\partial B_1} \frac{1}{||x|e_1 - \rho\sigma|^{N-2}} d\sigma$$

$$=|x|^{2-\beta}\int_0^{+\infty} s^{N-1-\beta}ds\int_{\partial B_1}\frac{1}{|e_1-s\sigma|^{N-2}}d\sigma = C|x|^{2-\beta},$$

where $d\sigma$ denotes the surface measure on ∂B_1 and in the last integral we have performed the change of variable $\rho = |x|s$. To identify the constant C, it suffices to observe that the left-hand side of (7.3.5), which we denote by u, satisfies the equation $\Delta u(x) = |x|^{-\beta}$ for any $x \in \mathbb{R}^N \setminus \{0\}$.

Now, we can prove estimate (7.3.4). We limit ourselves to proving that

$$\|q^{\beta/\alpha}R(0,A_p)f\|_{L^p(\mathbb{R}^N)} \le C\|f\|_{L^p(\mathbb{R}^N)}, \qquad f \in L^p(\mathbb{R}^N), \tag{7.3.6}$$

since the other part of (7.3.4) can be proved likewise. Let us split $x \in \mathbb{R}^N \setminus \{0\}$ in the pair (ρ, ω) where $\rho = |x|$ and $\omega \in \partial B_1$. We set $v(x) = |x|^\beta|(R(0,A_p)f)(x)|$ and observe that

$$\begin{aligned}v(x) =&v(\rho\omega) = C\rho^\beta\int_0^{+\infty} s^{N-1}ds\int_{\partial B_1}\frac{f(s\sigma)}{(1+s^\alpha)|\rho\omega - s\sigma|^{N-2}}d\sigma\\ =&C\rho^{2+\beta}\int_0^{+\infty} r^{N-1}dr\int_{\partial B_1}\frac{f(r\rho\sigma)}{(1+(r\rho)^\alpha)|\omega - r\sigma|^{N-2}}d\sigma.\end{aligned}$$

The Minkowski inequality shows that

$$\begin{aligned}\left(\int_0^{+\infty}\rho^{N-1}(v(\rho\omega))^p d\rho\right)^{\frac1p} \le&C\int_0^{+\infty} r^{N-1}dr\int_{\partial B_1}\frac{d\sigma}{|\omega - r\sigma|^{N-2}}\\ &\times\left(\int_0^{+\infty}\rho^{N-1+\beta p+2p}\frac{|f(r\rho\sigma)|^p}{(1+(r\rho)^\alpha)^p}d\rho\right)^{\frac1p}.\end{aligned}$$

We consider the last integral term, perform the change of unknowns $r\rho = \tau$ and use the estimate $\sup_{z>0}|z|^{2+\beta}(1+|z|^\alpha)^{-1} < +\infty$, which holds true since $\beta \le \alpha - 2$, to conclude that

$$\left(\int_0^{+\infty}\rho^{N-1+(\beta+2)p}\frac{|f(r\rho\sigma)|^p}{(1+(r\sigma)^\alpha)^p}d\rho\right)^{\frac1p} \le Cr^{-(\frac{N}{p}+\beta+2)}\left(\int_0^{+\infty}\tau^{N-1}|f(\tau\sigma)|^p d\tau\right)^{\frac1p}.$$

Hence, setting $\xi = r^{-1}$, we get

$$\begin{aligned}&\left(\int_0^{+\infty}\rho^{N-1}(v(\rho\omega))^p d\rho\right)^{\frac1p}\\ =&C\int_0^{+\infty} r^{\frac{N}{p}-\beta-1}dr\int_{\partial B_1}\frac{d\sigma}{|r^{-1}\omega-\sigma|^{N-2}}\left(\int_0^{+\infty}\tau^{N-1}|f(\tau\sigma)|^p d\tau\right)^{\frac1p}\\ =&C\int_0^{+\infty}\xi^{-\frac{N}{p}+\beta-1}d\xi\int_{\partial B_1}\frac{d\sigma}{|\xi\omega-\sigma|^{N-2}}\left(\int_0^{+\infty}\tau^{N-1}|f(\tau\sigma)|^p d\tau\right)^{\frac1p}.\end{aligned}\tag{7.3.7}$$

Integrating the first- and last-side of (7.3.7) with respect to $\omega \in \partial B_1$, taking (7.3.5) into account, we get

$$\begin{aligned}&\left(\int_{\mathbb{R}^N}|x|^\beta|R(0,A_p)f)(x)|^p dx\right)^{\frac1p}\\ \le&C\int_{\partial B_1}d\sigma\int_{\mathbb{R}^N}\frac{1}{|\sigma-y|^{N-2}|y|^{N/p'-\beta}}dy\left(\int_0^{+\infty}\tau^{N-1}|f(\tau\sigma)|^p d\rho\right)^{\frac1p}\end{aligned}$$

$$=C\int_{\partial B_1} d\sigma\left(\int_0^{+\infty}\tau^{N-1}|f(\tau\sigma)|^p d\rho\right)^{\frac{1}{p}}. \tag{7.3.8}$$

Finally, raising the first- and last-side of (7.3.8) to the power p and using the Hölder inequality, we obtain (7.3.6) if we take into account that $R(0,A_p)f \in W^{2,p}(\mathbb{R}^N)$ and $\|\Delta R(0,A_p)f\|_{L^p(\mathbb{R}^N)} \le C\|R(0,A_p)f\|_{L^p(\mathbb{R}^N)}$, by Lemma 7.3.3, and $\|\nabla R(0,A_p)f\|_{L^p(\mathbb{R}^N)} \le C(\|R(0,A_p)u\|_{L^p(\mathbb{R}^N)} + \|\Delta R(0,A_p)f\|_{L^p(\mathbb{R}^N)})$, by a well-known interpolation estimate. □

Remark 7.3.5 In the case when $\alpha \ge N/p'$, $\mathscr{D}_p$ is a proper subspace of $D(A_p)$. Indeed, take a smooth radial function f such that $\chi_{B_1} \le f \le \chi_{B_2}$. Clearly, f belongs to $L^p(\mathbb{R}^N)$. On the other hand, the solution $u \in D(A_p)$ of the equation $\mathcal{A}u = f$ is radial as well. Hence, $u(x) = \psi(|x|)$ for any $x \in \mathbb{R}^N$. In particular, for $\rho \ge 2$, the function ψ satisfies the equation $\psi''(\rho) + (N-1)\rho^{-1}\psi'(\rho) = 0$ and, therefore, $\psi(\rho) = C\rho^{2-N}$ for some positive constant C and any $\rho \ge 2$. It thus follows that

$$\int_{\mathbb{R}^N} q^p|u|^p dx \ge \int_{\mathbb{R}^N\setminus B_2} q^p|u|^p dx \ge \widetilde{C}\int_2^{+\infty}\rho^{p\left(\alpha-\frac{N}{p'}\right)-1}d\rho = +\infty$$

for some positive constant $\widetilde{C}$. Hence, u does not belong to $\mathscr{D}_p$. In a completely similar way, one can show that neither $q^{1-1/\alpha}|\nabla u|$ nor $q^{1-2/\alpha}|D^2u|$ belong to $L^p(\mathbb{R}^N)$.

As a consequence of Theorem 7.3.4 we can prove the following characterization of the spectrum of the operator A_p.

Theorem 7.3.6 *The spectrum of the operator A_p consists of a sequence of negative eigenvalues which accumulate at $-\infty$. Moreover, $\sigma(A_p)$ is independent of $p > N/(N-2)$.*

Proof The assertion can be checked by following the same lines as in the proof of Theorem 7.2.10. The only difference is the proof of the equality $R(\lambda,A_p)f = R(\lambda,A_q)f$ for any $\lambda \in \rho(A_p)\cap\rho(A_q)$, $f \in L^p(\mathbb{R}^N)\cap L^q(\mathbb{R}^N)$ and $p,q > N/(N-2)$, which is used in Step 2 of the proof of the quoted theorem. By Lemma 7.3.3, the previous formula is satisfied with $\lambda = 0$. Moreover, if $|\lambda| \le 2^{-1}(\|R(0,A_p)\|_{L(L^p(\mathbb{R}^N))}\wedge\|R(0,A_q)\|^{-1}_{L(L^q(\mathbb{R}^N))})$, then the operators $I+\lambda R(0,A_p)$ and $I+\lambda R(0,A_q)$ are invertible in $L^p(\mathbb{R}^N)$ and in $L^q(\mathbb{R}^N)$ respectively. For such values of λ and for $f \in L^p(\mathbb{R}^N)\cap L^q(\mathbb{R}^N)$, $R(\lambda,A_p)f$ is the fixed point of the operator $\lambda R(0,A_p) + R(0,A_p)f$ in $L^p(\mathbb{R}^N)$ and $R(\lambda,A_q)f$ is the fixed point of the operator $\lambda R(0,A_q) + R(0,A_q)f$ in $L^q(\mathbb{R}^N)$. These fixed points can be obtained by Picard iteration starting from a function $u_0 \in L^p(\mathbb{R}^N)\cap L^q(\mathbb{R}^N)$. Since the operators $\lambda R(0,A_p) + R(0,A_p)f$ and $\lambda R(0,A_q) + R(0,A_q)f$ coincide on $L^p(\mathbb{R}^N)\cap L^q(\mathbb{R}^N)$, we conclude that $R(\lambda,A_p)f = R(\lambda,A_q)$. To extend this equality to any $\lambda \in \rho(A_p)\cap\rho(A_q)$ it suffices to observe that the spectra of A_p and A_q are discrete, so that $\rho(A_p)\cap\rho(A_q)$ is connected, and use the analyticity of the resolvent operator in its domain. □

Now, we can state and prove the main result of this section.

Theorem 7.3.7 *The operator A_p generates a strongly continuous analytic semigroup, which is dissipative if and only if $p > (N+\alpha-2)(N-2)^{-1}$.*

Two different techniques are used to prove this result, depending on the value of p. If $p > (N+\alpha-2)(N-2)^{-1}$ then we deduce the assertion from the dissipativity of the operator A_p. More sophisticated arguments are used for the other values of p.

Remark 7.3.8 We stress that, for $p \in (N(N-2)^{-1}, (N+\alpha-2)(N-2)^{-1}]$, the operator A_p is not dissipative. Suppose by contradiction that A_p is dissipative for some p in such an interval. Then,

$$\int_{\mathbb{R}^N} qu|u|^{p-2}\Delta u dx \le 0$$

for any real-valued function $u \in C_c^\infty(\mathbb{R}^N)$. Fix $u \in C_c^\infty(\mathbb{R}^N)$ and apply the previous inequality to the function u_λ defined by $u_\lambda(x) = u(\lambda x)$ for any $x \in \mathbb{R}^N$. Integrating by parts and applying the change of variable $y = \lambda x$, we get

$$\begin{aligned}
0 \ge &-\int_{\mathbb{R}^N} \langle \nabla(q(x)u_\lambda(x)|u_\lambda(x)|^{p-2}), \nabla u_\lambda(x)\rangle dx\\
= &-(p-1)\int_{\mathbb{R}^N} q(x)|\nabla u_\lambda(x)|^2|u_\lambda(x)|^{p-2}dx - \frac{1}{p}\int_{\mathbb{R}^N}\langle \nabla q(x), \nabla(|u_\lambda(x)|^p)\rangle dx\\
= &-(p-1)\int_{\mathbb{R}^N} q(x)|\nabla u_\lambda(x)|^2|u_\lambda(x)|^{p-2}dx + \frac{1}{p}|u_\lambda(x)|^p\int_{\mathbb{R}^N}\Delta q(x)dx\\
= &-(p-1)\lambda^2\int_{\mathbb{R}^N} q(x)|\nabla u(\lambda x)|^2|u_\lambda(x)|^{p-2}dx + \frac{\alpha(\alpha+N-2)}{p}\int_{\mathbb{R}^N}|x|^{\alpha-2}|u(\lambda x)|^p dx\\
= &-(p-1)\lambda^{2-N}\int_{\mathbb{R}^N} q(\lambda^{-1}y)|\nabla u(y)|^2|u(y)|^{p-2}dy\\
&+\frac{\alpha(\alpha+N-2)}{p}\lambda^{2-\alpha-N}\int_{\mathbb{R}^N}|y|^{\alpha-2}|u(y)|^p dx,
\end{aligned}$$

so that

$$\int_{\mathbb{R}^N}|x|^{\alpha-2}|u(x)|^p dx \le \frac{p(p-1)}{\alpha(\alpha+N-2)}\int_{\mathbb{R}^N}(\lambda^\alpha + |x|^\alpha)|u(x)|^{p-2}|\nabla u(x)|^2 dx.$$

Letting λ tend to 0^+ we conclude that

$$\int_{\mathbb{R}^N}|x|^{\alpha-2}|u(x)|^p dx \le \frac{p(p-1)}{\alpha(\alpha+N-2)}\int_{\mathbb{R}^N}|x|^\alpha|u(x)|^{p-2}|\nabla u(x)|^2 dx, \tag{7.3.9}$$

which can be extended to any $u \in W^{1,p}(\mathbb{R}^N)$ with compact support.

Now, we fix $\delta > 0$ and consider the function $u_\delta : \mathbb{R}^N \to \mathbb{R}$ defined by $u_\delta(x) = |x|^{-(N+\alpha-2-\delta)/p}$, if $x \in B_1$, and $u_\delta(x) = |x|^{-(N+\alpha-2+\delta)/p}$, if $x \in \mathbb{R}^N \setminus B_1$. Further, for any $n \in \mathbb{N}$, we introduce the function $u_{n,\delta} = u_\delta\vartheta_n$, where $\vartheta_n \in C_c^\infty(\mathbb{R}^N)$ satisfies the condition $\chi_{B_n} \le \vartheta_n \le \chi_{B_{2n}}$. Clearly, each function $u_{n,\delta}$ belongs to $W^{1,p}(\mathbb{R}^N)$, has compact support and

$$\begin{aligned}
\int_{\mathbb{R}^N}|x|^{\alpha-2}|u_{n,\delta}(x)|^2dx \ge &\int_{B_n}|x|^{\alpha-2}|u_\delta(x)|^2dx\\
= &\int_{B_1}|x|^{-N+\delta}dx + \int_{B_n\setminus B_1}|x|^{-(N+\delta)}dx\\
\ge &\frac{p^2}{(N+\alpha-2+\delta)^2}\int_{B_n}|u_\delta(x)|^{p-2}|\nabla u_\delta(x)|^2dx\\
= &\frac{p^2}{(N+\alpha-2+\delta)^2}\int_{\mathbb{R}^N}|u_{n,\delta}(x)|^{p-2}|\nabla u_{n,\delta}(x)|^2dx\\
&-\frac{p^2}{(N+\alpha-2+\delta)^2}\int_{B_{2n}\setminus B_n}|u_{n,\delta}(x)|^{p-2}|\nabla u_{n,\delta}(x)|^2dx
\end{aligned}$$

$$=: \frac{p^2}{(N+\alpha-2+\delta)^2}\int_{\mathbb{R}^N}|u_{n,\delta}(x)|^{p-2}|\nabla u_{n,\delta}(x)|^2dx + a_n, \quad (7.3.10)$$

where the sequence $\{a_n\}$ vanishes as n tends to $+\infty$. Comparing (7.3.9) (with $u = u_{n,\delta}$) and (7.3.10) gives

$$\left(\frac{p(p-1)}{\alpha(\alpha+N-2)} - \frac{p^2}{(N+\alpha-2+\delta)^2}\right)\int_{\mathbb{R}^N}|x|^\alpha|u_{n,\delta}(x)|^{p-2}|\nabla u_{n,\delta}(x)|^2dx \geq a_n.$$

Hence,

$$\left(\frac{p(p-1)}{\alpha(\alpha+N-2)} - \frac{p^2}{(N+\alpha-2+\delta)^2}\right)\liminf_{n\to+\infty}\int_{\mathbb{R}^N}|x|^\alpha|u_{n,\delta}(x)|^{p-2}|\nabla u_{n,\delta}(x)|^2dx \geq 0$$

and

$$\liminf_{n\to+\infty}\int_{\mathbb{R}^N}|x|^\alpha|u_{n,\delta}(x)|^{p-2}|\nabla u_{n,\delta}(x)|^2dx \geq \int_{B_1}|x|^\alpha|u_\delta(x)|^{p-2}|\nabla u_\delta(x)|^2dx > 0,$$

so that

$$\frac{p(p-1)}{\alpha(\alpha+N-2)} - \frac{p^2}{(N+\alpha-2+\delta)^2} \geq 0$$

and, letting δ tend to 0^+, we conclude that

$$\frac{p(p-1)}{\alpha(N+\alpha-2)} - \frac{p^2}{(N+\alpha-2)^2} \geq 0.$$

But this estimate to hold requires that $p \geq (N+\alpha-2)(N-2)^{-1}$. The contradiction follows.

Proof of Theorem 7.3.7 We split the proof into two parts. In the first one we prove that the operator A_p is sectorial if $p > (N+\alpha-2)(N-2)^{-1}$ and dissipative if $p \geq (N+\alpha-2)(N-2)^{-1}$. In the second part, we address the case when $p \in (N(N-2)^{-1}, (N+\alpha-2)(N-2)^{-1}]$.

Part I. The uniqueness part follows from Lemma 7.3.1. To prove the existence of a solution $u \in D(A_p)$ of the equation $\lambda u - \mathcal{A}u = f$ for any $f \in L^p(\mathbb{R}^N)$, $p > (N+\alpha-2)(N-2)^{-1}$ and λ in a right-halfplane, we use an approximation argument similar to that used in Chapter 1, i.e., for any $n \in \mathbb{N}$, we introduce the Cauchy-Dirichlet problem

$$\begin{cases} \lambda u - \mathcal{A}u = f, & \text{in } B_n, \\ u = 0, & \text{on } \partial B_n. \end{cases} \quad (7.3.11)$$

By Theorem D.3.5 this problem admits a unique solution $u_n \in W^{2,p}(B_n) \cap W_0^{1,p}(B_n)$ for any $n \in \mathbb{N}$ and λ in a right-halfplane. In principle, this halfplane might depend on n and the intersection of all these halfplanes might contain no right-halfplanes. To prove that this is not the case, we follow the same argument as in the proof of Theorem 7.2.5, i.e., we fix $\lambda \in \mathbb{C}$ with positive real part, multiply both sides of the equation $\lambda u_n - \mathcal{A}u_n = f$ by $\overline{u_n}|u_n|^{p-2}$ and integrate by parts. Since u_n vanishes on ∂B_n, we obtain the following two equations

$$\begin{aligned}\int_{B_n}\text{Re}(f\overline{u_n})|u_n|^{p-2}dx =& \text{Re}\lambda\int_{B_n}|u_n|^pdx + \int_{B_n}q|u_n|^{p-4}|\text{Im}(\overline{u_n}\nabla u_n)|^2dx \\ &+ (p-1)\int_{B_n}q|u_n|^{p-4}|\text{Re}(\overline{u_n}\nabla u_n)|^2dx\end{aligned}$$

$$+\alpha\int_{B_n}|u_n(x)|^{p-2}|x|^{\alpha-2}\langle x, \mathrm{Re}(\overline{u_n(x)}\nabla u_n(x))\rangle dx, \tag{7.3.12}$$

$$\int_{B_n}\mathrm{Im}(f\overline{u_n})|u_n|^{p-2}dx = \mathrm{Im}\lambda\int_{B_n}|u_n|^p dx + (p-2)\int_{B_n}q|u_n|^{p-4}\langle \mathrm{Re}(\overline{u_n}\nabla u_n), \mathrm{Im}(\overline{u_n}\nabla u_n)\rangle dx$$
$$+\alpha\int_{B_n}|u_n(x)|^{p-2}|x|^{\alpha-2}\langle x, \mathrm{Im}(\overline{u_n(x)}\nabla u_n(x))\rangle dx. \tag{7.3.13}$$

To go further, we need to prove the Hardy-type inequality

$$\int_{B_n}|u_n(x)|^p|x|^\gamma dx \le \left(\frac{p}{\gamma+N}\right)^2\int_{B_n}|u_n(x)|^{p-4}|\mathrm{Re}(\overline{u_n(x)}\nabla u_n(x))|^2|x|^{\gamma+2}dx. \tag{7.3.14}$$

Clearly, it suffices to prove it with u_n being replaced by a function $v \in W^{1,p}_0(B_n)$. In such a case, we can write

$$|v(x)|^p = -\int_1^{+\infty}\frac{\partial}{\partial t}|v(tx)|^p dx = -p\int_1^{+\infty}|v(tx)|^{p-2}\langle x, \mathrm{Re}(\overline{v(tx)}\nabla v(tx))\rangle dt.$$

Therefore,

$$\begin{aligned}
&\int_{B_n}|v(x)|^p|x|^\gamma dx\\
\le& p\int_1^{+\infty}dt\int_{B_n}|v(tx)|^{p-2}|\mathrm{Re}(\overline{v(tx)}\nabla v(tx))||x|^{\gamma+1}dx\\
=& p\int_1^{+\infty}dt\int_{\partial B_1}d\sigma\int_0^n|v(\rho\sigma)|^{p-2}|\mathrm{Re}(\overline{v(t\rho\sigma)}\nabla v(t\rho\sigma))|\rho^{\gamma+N}d\rho\\
=& p\int_1^{+\infty}t^{-\gamma-N-1}dt\int_{\partial B_1}d\sigma\int_0^n|v(r\sigma)|^{p-2}|\mathrm{Re}(\overline{v(r\sigma)}\nabla v(r\sigma))|r^{\gamma+N}dr\\
=& \frac{p}{\gamma+N}\int_{B_n}|v(x)|^{p-2}|\mathrm{Re}(\overline{v(x)}\nabla v(x))||x|^{\gamma+1}dx\\
\le& \frac{p}{\gamma+N}\left(\int_{B_n}|v(x)|^p|x|^\gamma dx\right)^{\frac12}\left(\int_{B_n}|v(x)|^{p-4}|\mathrm{Re}(\overline{v(x)}\nabla v(x))|^2|x|^{\gamma+2}dx\right)^{\frac12},
\end{aligned}$$

where $d\sigma$ denotes the surface measure on ∂B_1. Estimate (7.3.14) follows.

Using (7.3.14), we can estimate

$$\begin{aligned}
&\alpha\left|\int_{B_n}|u_n(x)|^{p-2}|x|^{\alpha-2}\langle x, \mathrm{Re}(\overline{u_n(x)}\nabla u_n(x))\rangle dx\right|\\
\le& \alpha\int_{B_n}|u_n(x)|^{p-2}|x|^{\alpha-1}|\mathrm{Re}(\overline{u_n(x)}\nabla u_n(x))|dx\\
\le& \left(\int_{B_n}|u_n(x)|^{p-4}|x|^\alpha|\mathrm{Re}(\overline{u_n(x)}\nabla u_n(x))|^2dx\right)^{\frac12}\left(\int_{B_n}|u_n(x)|^p|x|^{\alpha-2}dx\right)^{\frac12}\\
\le& \frac{p\alpha}{\alpha-2+N}\int_{B_n}|u_n(x)|^{p-4}q(x)|\mathrm{Re}(\overline{u_n(x)}\nabla u_n(x))|^2dx
\end{aligned}$$

and a similar estimate holds true with the real part being replaced everywhere by the imaginary part. Hence, from (7.3.12) and (7.3.13) it follows that

$$\|f\|_{L^p(\mathbb{R}^N)}\|u_n\|^{p-1}_{L^p(B_n)} \ge \mathrm{Re}\lambda\int_{B_n}|u|^p dx + \int_{B_n}q|u_n|^{p-4}|\mathrm{Im}(\overline{u_n}\nabla u_n)|^2dx$$

$$+\left[\left(\frac{N-2}{N-2+\alpha}\right)p-1\right]\int_{B_n} q|u_n|^{p-4}|\mathrm{Re}(\overline{u_n}\nabla u_n)|^2dx \qquad (7.3.15)$$

and

$$\begin{aligned}|\mathrm{Im}\lambda|\|u_n\|_{L^p(B_n)}^p \leq &\|f\|_{L^p(\mathbb{R}^N)}\|u_n\|_{L^p(B_n)}^{p-1}\\ &+\left(p-2+\frac{p\alpha}{\alpha-2+N}\right)\left(\int_{B_n}|x|^\alpha|u_n(x)|^{p-4}|\mathrm{Re}(\overline{u_n(x)}\nabla u_n(x))|^2dx\right)^{\frac{1}{2}}\\ &\times\left(\alpha\int_{B_n}|u_n(x)|^{p-4}|x|^\alpha|\mathrm{Im}(\overline{u_n(x)}\nabla u_n(x))|^2)dx\right)^{\frac{1}{2}}.\end{aligned}$$

Since $p > (N+\alpha-2)(N-2)^{-1}$ all the coefficients in front of the integral terms in the right-hand side of (7.3.15) are positive. Hence, arguing as in the last part of the proof of (7.2.19), we conclude that there exist two positive constants C_p and M_p, independent of $n \in \mathbb{N}$, such that

$$|\lambda|\|u_n\|_{L^p(B_n)} \leq C_p\|f\|_{L^p(\mathbb{R}^N)} \qquad (7.3.16)$$

for any $\lambda \in \mathbb{C}$ with $\mathrm{Re}\lambda \geq M_p$. This shows that the resolvent set of the realization of the operator $\mathcal{A}$ in $L^p(B_n)$, with homogeneous Dirichlet boundary conditions on ∂B_n, contains the halfplane $\{\lambda \in \mathbb{C} : \mathrm{Re}\lambda \geq M_p\}$ for any $n \in \mathbb{N}$. In particular, taking $\lambda > 0$, from (7.3.15) it follows that $\lambda\|u_n\|_{L^p(B_n)} \leq \|\lambda u_n - \mathcal{A}u_n\|_{L^p(B_n)}$.

Now, to make the sequence $\{u_n\}$ (of solutions to (7.3.11) with $\mathrm{Re}\lambda \geq M_p$) converge to a solution $u \in D(A_p)$ of the equation $\lambda u - \mathcal{A}u = f$, we apply a compactness argument. From estimate (7.3.16) it follows that $\|\Delta u_n\|_{L^p(B_n)} \leq (C_p+1)\|f\|_{L^p(B_n)}$. By the Calderon-Zygmund inequality (D.1.1), we can infer that for any $r > 0$ the $W^{2,p}(B_r)$-norm of the sequence $\{u_n\}$ is bounded by a constant independent of n. Hence, u_n weakly converges in $W^{2,p}(B_r)$, for any $r > 0$, to a function which belongs to $D(A_p)$, solves the equation $\lambda u - \mathcal{A}u = f$ and satisfies the estimate $|\lambda|\|u\|_{L^p(\mathbb{R}^N)} \leq C_p\|f\|_{L^p(\mathbb{R}^N)}$. In particular, if $\lambda > 0$, then we can take $C_p = 1$ so that A_p is a dissipative operator. Theorem C.2.7 allows us to conclude that A_p is a sectorial operator. Hence, it generates an analytic semigroup which is also strongly continuous since, clearly, $D(A_p)$ is dense in $L^p(\mathbb{R}^N)$. The contractivity of the semigroup follows from the dissipativity of A_p.

Finally, we observe that estimate (7.3.15) implies that $\lambda\|u_n\|_{L^p(B_n)} \leq \|\lambda u_n - \mathcal{A}u_n\|_{L^p(B_n)}$ for any $\lambda > 0$ and $n \in \mathbb{N}$, also in the case when $p = (N+\alpha-2)(N-2)^{-1}$. The above arguments show that u_n converges to $R(\lambda, A_{(N+\alpha-2)(N-2)^{-1}})u$ and the dissipativity of the operator $A_{(N+\alpha-2)(N-2)^{-1}}$ follows.

Part II. For any $s \in (0,1]$, let $q_{2,s} : \mathbb{R}^N \to \mathbb{R}$ be the function defined by $q_{2,s}(x) = s^\alpha + |x|^\alpha$ for any $x \in \mathbb{R}^N$ and let $A_{p,s}$ be the realization of the operator $\mathcal{A}_s = q_s\Delta$ in $L^p(\mathbb{R}^N)$ with $D(A_{p,s}) = \{u \in L^p(\mathbb{R}^N)\cap W^{2,p}_{\mathrm{loc}}(\mathbb{R}^N) : \mathcal{A}_s u \in L^p(\mathbb{R}^N)\}$. Further, for $j = 1, 2, 3$ let $A_{j,p,s}$ be the realization in $L^p(\mathbb{R}^N)$ of the operator $\mathcal{A}_{j,s} = q_{j,s}\Delta$, with domain $D(A_{p,1,s}) = D(A_{p,3,s}) = W^{2,p}(\mathbb{R}^N)$ and $D(A_{p,2,s}) = \{u \in L^p(\mathbb{R}^N) \cap W^{2,p}_{\mathrm{loc}}(\mathbb{R}^N) : \mathcal{A}u \in L^p(\mathbb{R}^N)\}$, where $q_{1,s} = \psi_1 q_s + 1 - \psi_1$, $q_{2,s} = \psi_2 + (1-\psi_2)q_s$, $q_{3,s} = \psi_1 q_{2,s} + 1 - \psi_1$ and the radial functions $\psi_1, \psi_2 \in C^\infty_c(\mathbb{R}^N)$ satisfy $\chi_{B_2} \leq \psi_1 \leq \chi_{B_4}$ and $\chi_{B_{1/2}} \leq \psi_2 \leq \chi_{B_1}$ in $\mathbb{R}^N$.

We split the proof into steps. For $r \in \mathbb{R}$, $\rho > 0$ and $\theta \in (\pi/2, \pi)$, we find it convenient to introduce the sets $\Sigma_{\theta,\rho} := \{\lambda \in \mathbb{C} : |\lambda| \geq \rho, |\arg\lambda| \leq \theta\}$ and $\Pi_r = \{\lambda \in \mathbb{C} : \mathrm{Re}\lambda \geq r\}$.

Step 1. Here we prove that for any $p > N/(N-2)$ there exist $\rho_1 > 0$ and $\theta_1 \in (\pi/2, \pi)$ such that $\rho(A_{p,1,s}) \supset \Sigma_{\theta_1,\rho_1}$ and

$$|\lambda|\|R(\lambda, A_{p,k,s})\|_{L(L^p(\mathbb{R}^N))} + \sqrt{|\lambda|}\|\nabla R(\lambda, A_{p,k,s})\|_{L(L^p(\mathbb{R}^N);(L^p(\mathbb{R}^N))^N)} \leq C_1, \qquad \lambda \in \Sigma_{\theta_1,\rho_1}, \qquad (7.3.17)$$

for $k = 1, 3$, the constant C_1 being independent of $s \in (0, 1]$.

We also prove that, for any $p > (N+\alpha-2)/(N-2)$, there exist $\rho_2 > 0$ and $\theta_2 \in (\pi/2, \pi)$ such that $\rho(A_{p,2,s}) \supset \Sigma_{\rho_2,\theta_2}$ and

$$|\lambda|\|R(\lambda, A_{p,2,s})\|_{L(L^p(\mathbb{R}^N))} + \sqrt{|\lambda|}\|\nabla R(\lambda, A_{p,2,s})\|_{L(L^p(\mathbb{R}^N);(L^p(\mathbb{R}^N))^N)} \le C_2, \qquad \lambda \in \Sigma_{\theta_2,\rho_2}, \tag{7.3.18}$$

also the constant C_2 being independent of $s \in (0, 1]$. From now on, we denote by C any positive constant which is independent of s and may vary from line to line.

We begin by checking the properties of the operator $A_{p,1,s}$. For this purpose, we fix $p > N/(N-2)$. A straightforward computation shows that

$$\begin{aligned}|\nabla q_{1,s}(x)| \le & |\nabla\psi_1(x)|(s^\alpha + 1 + |x|^\alpha) + \alpha\psi_1(x)|x|^{\alpha-1} \\ \le & (2 + 4^\alpha)\|\nabla\psi_1\|_\infty \chi_{B_4\setminus B_2}(x) + 4^{\frac{\alpha}{2}-1}\alpha\sqrt{\psi_1(x)}|x|^{\frac{\alpha}{2}}\end{aligned}$$

for any $x \in \mathbb{R}^N$. Since $\psi_1 \ge 1$ in $B_4 \setminus B_2$, from the previous estimate we conclude that $|\nabla_x q_{1,s}| \le C\sqrt{q_{1,s}}$ in $\mathbb{R}^N$, where $C = 4^{\alpha/2-1}\alpha + 2 + 4^\alpha\|\nabla\psi_1\|_\infty$.

Now, we can repeat the arguments in the proof of Theorem 7.2.5 (where we take $\beta = 0$) with slight changes and show that there exist constants $\rho_1 > 0$ and $\theta_1 \in (\pi/2, \pi)$, independent of $s \in (0, 1]$, such that $|\lambda|\|R(\lambda, A_{p,1,s})\|_{L(L^p(\mathbb{R}^N))} \le C$ for any $\lambda \in \Sigma_{\theta_1,\rho_1}$. Moreover, an inspection of the proof of (7.2.3) shows that $C_\varepsilon = C\varepsilon^{-1}$. Hence, up to replacing ρ_1 with a larger value (still independent of s) if needed, we can take $\varepsilon = |\lambda|^{-1}$, with $\lambda \in \Sigma_{\theta,\rho}$, in (7.2.3) and get

$$\begin{aligned}\|\nabla R(\lambda, A_{p,1,s})f\|_{L^p(\mathbb{R}^N)} \le & |\lambda|^{-\frac{1}{2}}\|A_{p,1,s}R(\lambda, A_{p,1,s})f\|_{L^p(\mathbb{R}^N)} + C|\lambda|^{\frac{1}{2}}\|R(\lambda, A_{p,1,s})f\|_{L^p(\mathbb{R}^N)} \\ \le & C|\lambda|^{-\frac{1}{2}}\|f\|_{L^p(\mathbb{R}^N)} + C|\lambda|^{\frac{1}{2}}\|R(\lambda, A_{p,1,s})f\|_{L^p(\mathbb{R}^N)} \\ \le & C|\lambda|^{-\frac{1}{2}}\|f\|_{L^p(\mathbb{R}^N)}\end{aligned}$$

for any $f \in L^p(\mathbb{R}^N)$. Thus, estimate (7.3.17) follows.

Estimate (7.3.17), with $k = 3$, can be proved in the same way. Indeed, arguing as above and taking also the forthcoming estimates on $q_{2,s}$ into account, it can be shown that $|\nabla q_{3,s}| \le C\sqrt{q_{3,s}}$ in $\mathbb{R}^N$.

Finally, we prove estimate (7.3.18). We fix $p > (N + \alpha - 2)/(N - 2)$ and observe that

$$|\nabla q_{2,s}(x)| \le 3\|\nabla\psi_2\|_\infty\chi_{B_1\setminus B_{1/2}}(x) + \alpha|x|^{\alpha-1} \le (2^{\alpha-1}3\|\nabla\psi_2\|_\infty + \alpha)|x|^{\alpha-1}, \qquad x \in \mathbb{R}^N.$$

Moreover, $q_{2,s}(x) \ge 2^{-\alpha}|x|^\alpha$ for any $x \in \mathbb{R}^N$. Hence, we can apply the arguments in the first part of the proof with minor changes and conclude that $\rho(A_{p,2,s}) \supset \Sigma_{\theta_2,\rho_2}$ and $|\lambda|\|R(\lambda, A_{p,2,s})\|_{L(L^p(\mathbb{R}^N))} \le C$ for any $\lambda \in \Sigma_{\theta_2,\rho_2}$, $s \in (0, 1]$ and some constants $\rho_2 > 0$ and $\theta_2 \in (\pi/2, \pi)$. From this estimate and observing that $q_{2,s} \ge 2^{-\alpha}$ in $\mathbb{R}^N$, it follows that $\|\Delta R(\lambda, A_{p,2,s})\|_{L(L^p(\mathbb{R}^N))} \le C\|A_{p,2,s}R(\lambda, A_{p,2,s})\|_{L(L^p(\mathbb{R}^N))} \le C$ for any $\lambda \in \Sigma_{\theta_2,\rho_2}$. Using the well-known interpolation estimate

$$\|\nabla u\|_{L^p(\mathbb{R}^N)} \le C\|u\|^{\frac{1}{2}}_{L^p(\mathbb{R}^N)}\|\Delta u\|^{\frac{1}{2}}_{L^p(\mathbb{R}^N)}, \qquad u \in W^{2,p}(\mathbb{R}^N), \tag{7.3.19}$$

we can complete the proof of (7.3.18).

Step 2. Here we consider the case when $p \in (N/(N-2), (N+\alpha-2)/(N-2)]$ and prove that $\rho(A_{p,2,s}) \supset \Pi_{\lambda_0}$ and

$$\|R(\lambda, A_{p,2,s})\|_{L(L^p(\mathbb{R}^N))} \le C, \qquad \lambda \in \Pi_{\lambda_0}, \ s \in (0, 1], \tag{7.3.20}$$

for some constant $\lambda_0 < 0$ independent of s.

Arguing as in Lemma 7.3.3 we can show that $0 \in \rho(A_{p,2,s})$ and

$$(R(0, A_{p,2,s})f)(x) = \frac{1}{N(2-N)\omega_N}\int_{\mathbb{R}^N}\frac{f(y)}{|x-y|^{N-2}q_{2,s}(y)}dy \tag{7.3.21}$$

for almost any $x \in \mathbb{R}^N$. Since $q_{2,s}(x) = s^\alpha + |x|^\alpha \geq 2^{-1}(|x|^\alpha + 1)$ for any $x \in \mathbb{R}^N \setminus B_1$ and $q_{2,s} \geq 2^{-\alpha}$ in B_1, it follows that $q_{2,s}(x) \geq 2^{-(\alpha+1)}(1 + |x|^\alpha)$ for any $x \in \mathbb{R}^N$. Hence, from (7.3.21) it follows that $\|R(0, A_{p,2,s})\|_{L(L^p(\mathbb{R}^N))} \leq C$ for any $s \in (0,1]$. From this estimate and the resolvent identity it follows that there exists $\rho_0 > 0$ such that $\overline{B}_{\rho_0} \subset \rho(A_{p,2,s})$ and $\|R(\lambda, A_{p,2,s})\|_{L(L^p(\mathbb{R}^N))} \leq C$ for any $\lambda \in \overline{B}_{\rho_0}$ and $s \in (0,1]$. Arguing as in the proof of Theorem 7.3.6 it can be shown that the spectrum of the operator $A_{p,2,s}$ consists of a sequence of negative eigenvalues and we thus deduce that $\rho(A_{p,2,s}) \supset \Pi_{\lambda_0}$, where $\lambda_0 = -\rho_0$.

To complete this step of the proof, let us check estimate (7.3.20). For this purpose, let us consider the minimal semigroup $\{T_{2,s}(t)\}$ associated in $C_b(\mathbb{R}^N)$ with the operator $\mathcal{A}_{2,s}$ (see Theorem 1.2.1). Its restriction to $C_0(\mathbb{R}^N)$ is a strongly continuous semigroup. Indeed, the one-dimensional Feller theory[1] shows that $+\infty$ is an accessible point of the operator L defined on functions $\zeta \in C^2([1,+\infty))$ by $(L\zeta)(r) = (s^\alpha + r^\alpha)\zeta''(r) + (N-1)r^{-1}(s^\alpha + r^\alpha)\zeta'(r)$ for any $r \geq 1$. Hence, for any $\lambda > 0$ there exists a positive solution $\zeta \in C^2([1,+\infty))$ of the equation $\lambda\zeta - L\zeta = 0$. The function $u : \mathbb{R}^N \setminus B_1 \to \mathbb{R}$, defined by $u(x) = \zeta(|x|)$ for any $x \in \mathbb{R}^N \setminus B_1$, belongs to $C^2(\mathbb{R}^N \setminus B_1) \cap C_b(\mathbb{R}^N \setminus B_1)$, vanishes at infinity and solves the equation $\lambda u - \mathcal{A}u = 0$. Hence, we can invoke Proposition 4.2.4 and conclude that $T_{2,s}(t)$ maps $C_b(\mathbb{R}^N)$ in $C_0(\mathbb{R}^N)$ for any $t > 0$ and $s \in (0,1]$. Then, by Proposition 4.3.1, the restriction of the semigroup $\{T_{2,s}(t)\}$ to $C_0(\mathbb{R}^N)$ is strongly continuous and it is generated by the part $\widehat{A}_{s,0}$ in $C_0(\mathbb{R}^N)$ of the weak generator $\widehat{A}_{2,s}$ of the semigroup $\{T_{2,s}(t)\}$ (see Section 1.3). Moreover, by Theorem 4.1.10 and Proposition 4.1.11, the operators $R(\lambda, \widehat{A}_{2,s})$ and $T_{2,s}(t)$ are compact in $C_0(\mathbb{R}^N)$ for any $t > 0$ and $\lambda \in \rho(\widehat{A}_{2,s})$.

Using the same arguments as in the proof of Theorems 7.2.10 and 7.3.6, it can be proved that $A_{p,2,s}$ and $\widehat{A}_{2,s}$ have the same spectrum and that the resolvent operators $R(\lambda, \widehat{A}_{2,s})$ and $R(\lambda, A_{p,2,s})$ agree on $L^p(\mathbb{R}^N) \cap C_0(\mathbb{R}^N)$. The main step to make the arguments in the proofs of the quoted theorems work, consists of showing that $0 \in \rho(\widehat{A}_{2,s})$ and

$$(R(0, \widehat{A}_{2,s})f)(x) = \frac{1}{N(N-2)\omega_N}\int_{\mathbb{R}^N}\frac{f(y)}{|x-y|^{N-2}q_{2,s}(y)}dy, \qquad x \in \mathbb{R}^N. \tag{7.3.22}$$

For this purpose, we adapt the arguments in Lemma 7.3.3. Since the trivial function is the only harmonic function which vanishes at infinity, there exists at most one solution $u \in D(\widehat{A}_{2,s})$ to the equation $\widehat{A}_{2,s}u = f$. On the other hand, writing this equation in the form $\Delta u(x) = -f(x)(s^\alpha + |x|^\alpha)^{-1}$ for any $x \in \mathbb{R}^N$ and $f \in C_c^\infty(\mathbb{R}^N)$, by classical results we know that the right-hand side of (7.3.22) is the (unique) solution to the previous equation which belongs to $C_b^2(\mathbb{R}^N)$ and vanishes at infinity. If $f \in C_0(\mathbb{R}^N)$, then there exists a sequence $\{f_n\} \in C_c^\infty(\mathbb{R}^N)$ converging to f uniformly in $\mathbb{R}^N$. Denote by u_n the solution to the equation $\mathcal{A}_{2,s}u_n = f_n$ given by the right-hand side of (7.3.22) (with f_n replacing f). Clearly, u_n converges uniformly in $\mathbb{R}^N$ to the function

$$u(x) = \frac{1}{N(N-2)\omega_N}\int_{\mathbb{R}^N}\frac{f(y)}{|x-y|^{N-2}q_{2,s}(y)}dy, \qquad x \in \mathbb{R}^N,$$

and $\mathcal{A}_{2,s}u_n$ converges to f uniformly in $\mathbb{R}^N$. Using the interior estimate in Theorem D.1.1(i), we conclude that $\{u_n\}$ is a Cauchy sequence in $W^{2,q}(B_R)$ for any $q \in [1,+\infty)$ and $R > 0$.

[1]As it has been already remarked in the proof of Proposition 7.3.2, such a theory, which has been developed in Chapter 2 in $\mathbb{R}$, works as well in the right-halfline $[1,+\infty)$, with the integrals in the definition of the functions $\mathscr{W}$, $\mathscr{P}$ and $\mathscr{R}$ being taken in the interval $[1,x]$.

Hence, u belongs to $D(\widehat{A}_{2,s})$. Formula (7.3.22) is proved. As a byproduct, the operators $R(0, A_{p,2,s})$ and $R(0, \widehat{A}_{2,s})$ agree on $C_0(\mathbb{R}^N) \cap L^p(\mathbb{R}^N)$.

By the above results and the spectral mapping theorem for compact semigroups (see Proposition C.1.12) it follows that

$$\begin{aligned}|(R(\lambda, \widehat{A}_{s,0})f)(x)| &= \left|\int_0^{+\infty} e^{\lambda t}(T_{2,s}(t)f)(x)dt\right| \\ &\le \int_0^{+\infty} e^{\mathrm{Re}\lambda t}(T_{2,s}(t)|f|)(x)dt = (R(\mathrm{Re}\lambda, \widehat{A}_{2,s})|f|)(x)\end{aligned}$$

for any $\lambda \in \Pi_{\lambda_0}$ and $f \in C_c^\infty(\mathbb{R}^N)$, where we have taken into account that the semigroup $\{T_{2,s}(t)\}$ preserves positivity (see Remark 1.2.3). Raising the first and last side of the previous chain of inequalities to the power p and integrating in $\mathbb{R}^N$, we conclude that

$$\begin{aligned}\|R(\lambda, A_{p,2,s})f\|_{L^p(\mathbb{R}^N)} &= \|R(\lambda, \widehat{A}_{2,s})f\|_{L^p(\mathbb{R}^N)} \le \|R(\mathrm{Re}\lambda, \widehat{A}_{2,s})|f|\|_{L^p(\mathbb{R}^N)} \\ &= \|R(\mathrm{Re}\lambda, A_{p,2,s})|f|\|_{L^p(\mathbb{R}^N)} \le C\|f\|_{L^p(\mathbb{R}^N)}.\end{aligned}$$

Finally, by the density of $C_c^\infty(\mathbb{R}^N)$ in $L^p(\mathbb{R}^N)$, we deduce that $\|R(\lambda, A_{p,2,s})f\|_{L^p(\mathbb{R}^N)} \le C$ for any $\lambda \in \Pi_{\lambda_0}$.

Step 3. Suppose that

$$|\lambda|\|R(\lambda, A_{p_1,2,s})\|_{L(L^p(\mathbb{R}^N))} \le C, \qquad \lambda \in \Sigma_{\overline{\rho},\overline{\theta}}, \tag{7.3.23}$$

for some constants $\overline{\rho} > 0$, $\overline{\theta} \in (\pi/2, \pi)$, independent of $s \in (0, 1]$, and $p_1 > N/(N-2)$. Fix $q \in (N/(N-2), p)$ and let p be defined by the formula

$$\frac{1}{p} = \frac{3}{4p_1} + \frac{1}{4q}.$$

Here we prove that there exist constants $\rho_* > 0$ and $\theta_* \in (\pi/2, \pi)$, independent of s, such that

$$(i)\ |\lambda|\|R(\lambda, A_{p,s})\|_{L(L^p(\mathbb{R}^N))} \le C, \qquad (ii)\ |\lambda|\|R(\lambda, A_{p,2,s})\|_{L(L^p(\mathbb{R}^N))} \le C, \qquad \lambda \in \Sigma_{\rho_*,\theta_*}. \tag{7.3.24}$$

From these estimates we deduce, in particular, that the operators $A_{p,s}$ and $A_{p,2,s}$ are sectorial in $L^p(\mathbb{R}^N)$.

To begin with, we observe that, arguing as in the previous step, starting from the formula (7.3.2), we can easily show that $R(\lambda, A_p) = R(\lambda, A_{p_1})$ on $L^p(\mathbb{R}^N) \cap L^{p_1}(\mathbb{R}^N)$. Hence, from (7.3.20) and (7.3.23) and the Riesz-Thorin interpolation theorem, we deduce that

$$\|R(\lambda, A_{p,2,s})\|_{L(L^p(\mathbb{R}^N))} \le C|\lambda|^{-\frac{3}{4}}, \qquad \lambda \in \Sigma_{\overline{\rho},\pi/2}. \tag{7.3.25}$$

Starting from this estimate we get

$$\begin{aligned}\|\Delta R(\lambda, A_{p,2,s})\| &\le \|A_{p,2,s}R(\lambda, A_{p,2,s})\|_{L(L^p(\mathbb{R}^N))} \\ &= \|I - \lambda R(\lambda, A_{p,2,s})\|_{L(L^p(\mathbb{R}^N))} \le C(1 + |\lambda|^{\frac{1}{4}})\end{aligned}$$

for any $\lambda \in \Sigma_{\overline{\rho},\pi/2}$. Using again estimate (7.3.19), we obtain

$$\|\nabla R(\lambda, A_{p,2,s})\|_{L(L^p(\mathbb{R}^N);(L^p(\mathbb{R}^N))^N)} \le C|\lambda|^{-\frac{1}{4}}, \qquad \lambda \in \Sigma_{\widehat{\rho},\pi/2}, \tag{7.3.26}$$

for some positive constant $\widehat{\rho} \ge \overline{\rho}$.

Now, we can prove estimate (7.3.24), adapting to our situation a classical argument and using estimates (7.3.25) and (7.3.26). We fix two functions $\eta_1, \eta_2 \in C_c^\infty(\mathbb{R}^N)$ such that $\mathrm{supp}(\eta_1) \subset B_2$, $\mathrm{supp}(\eta_2) \subset \mathbb{R}^N \setminus B_1$ and $\eta_1^2 + \eta_2^2 = 1$ in $\mathbb{R}^N$.

For any $\lambda \in \Sigma_{\widehat{\rho},\pi/2}$ let $R_\lambda \in L(L^p(\mathbb{R}^N))$ be the operator defined by

$$R_\lambda f = \eta_1 R(\lambda, A_{p,1,s})(\eta_1 f) + \eta_2 R(\lambda, A_{p,2,s})(\eta_2 f), \qquad f \in L^p(\mathbb{R}^N).$$

Since $q_s = q_{1,s}$ in B_2 and $q_s = q_{2,s}$ in $\mathbb{R}^N \setminus B_1$, it follows that $A_{p,s}\psi = A_{p,1,s}\psi$ for any $\psi \in W^{2,p}(\mathbb{R}^N)$, with $\mathrm{supp}(\psi) \subset B_2$, and $A_{p,s}\psi = A_{p,2,s}\psi$ for any $\psi \in L^p(\mathbb{R}^N) \cap W^{2,p}_{\mathrm{loc}}(\mathbb{R}^N)$, with $\mathrm{supp}(\psi) \subset \mathbb{R}^N \setminus B_1$ and such that $\mathcal{A}_s\psi \in L^p(\mathbb{R}^N)$. Taking these formulae into account we can write

$$(\lambda - A_{p,s})R_\lambda f = f + \sum_{i=1}^{2}[\eta_i, A_{p,s}]R(\lambda, A_{p,i,s})(\eta_i f) =: f + S_\lambda f, \qquad \lambda \in \Sigma_{\widehat{\rho},\pi/2}.$$

By the gradient estimates in (7.3.17) and in (7.3.26), there exists $\rho_* \geq \widehat{\rho}$ such that $\|S_\lambda f\|_{L(L^p(\mathbb{R}^N))} \leq 1/2$ for any $\lambda \in \Sigma_{\rho_*,\pi/2}$. Hence, the operator $I + S_\lambda$ is invertible in $L^p(\mathbb{R}^N)$ and the function $u = R_\lambda(I + S_\lambda)^{-1}f$ is a solution to the equation $\lambda u - A_{p,s}u = f$ which satisfies the estimate $\|u\|_{L^p(\mathbb{R}^N)} \leq C\lambda^{-3/4}\|f\|_{L^p(\mathbb{R}^N)}$. Actually, this is the unique solution to the previous equation in $D(A_{p,s})$ since the spectrum of the operator $A_{p,s}$ is a sequence of negative eigenvalues. It thus follows that $\|R(\lambda, A_{p,s})\|_{L(L^p(\mathbb{R}^N))} \leq C|\lambda|^{-3/4}$ for any $\lambda \in \Sigma_{\rho_*,\pi/2}$.

Since the points $\pm i\rho_*$ belong to $\rho(A_{p,s})$ and $\|R(\pm i\rho_*, A_{p,s})\|_{L(L^p(\mathbb{R}^N))} \leq C$, writing the equation $\lambda u - A_{p,2,s}u = f$ in the form $u + (\lambda \mp i\rho_*)R(\pm i\rho_*, A_{p,s})u = R(\pm i\rho_*, A_{p,s})f$, we easily see that, if $\lambda \in B(i\rho_*, M) \cup B(-i\rho_*, M)$, where $M = (2\|R(\pm i\rho_*, A_{p,2,s})\|_{L(L^p(\mathbb{R}^N))})^{-1}$, then $\lambda \in \rho(A_{p,s})$ and $\|R(\lambda, A_{p,s})\|_{L(L^p(\mathbb{R}^N))} \leq C$. Thus, there exists $\theta_* \in (\pi/2, \pi)$ such that $\|R(\lambda, A_{p,2,s})\|_{L(L^p(\mathbb{R}^N))} \leq C$ for any $\lambda \in \Theta := \{\rho_* e^{i\theta} : |\theta| \leq \theta_*\}$.

We now observe that $\Sigma_{\rho_*,\theta_*} = \{rz \in \mathbb{C} : r \geq 1, z \in \Theta\}$. Fix $\lambda \in \Sigma_{\rho_*,\theta_*}$. Set $r = |\lambda|\rho_*^{-1}$ and $z = r^{-1}\lambda$. Further, consider the operator $\mathscr{I}_r$ in $L^p(\mathbb{R}^N)$ defined by $\mathscr{I}_r u(x) = u(r^{1/(2-\alpha)}x)$ for almost every $x \in \mathbb{R}^N$. A straightforward computation shows that $\lambda I - A_p = r\mathscr{I}_r(z - A_{p,r})\mathscr{I}_r^{-1}$. Hence, $\|\lambda I - A_p\|_{L(L^p(\mathbb{R}^N))} \leq r^{-1}\|\omega - A_{p,s}\|_{L(L^p(\mathbb{R}^N))} \leq C|\lambda|^{-1}$ and (7.3.24)(i) follows with $s = 1$. To get this estimate with $s \in (0, 1)$, it suffices to observe that $\lambda I - A_{s,p} = s^{\alpha-2}\mathscr{I}_s(s^{2-\alpha}\lambda I - L_1)\mathscr{I}_s^{-1}$.

Finally, to prove (7.3.24)(ii) it suffices to apply the above arguments to the operators $A_{p,s}$ and $A_{p,3,s}$, taking Step 1 into account.

Step 4. We fix $p \in (N/(N-2), (N + \alpha - 2)/(N-2)]$, $p_1 > (N + \alpha - 2)/(N-2)$ and $q \in (N/(N-2), p)$. Applying recursively the arguments in Step 3, we conclude that the operator A_{p_n} is sectorial in $L^{p_n}(\mathbb{R}^N)$, for any $n \in \mathbb{N}$, where the sequence $\{p_n\}$ is defined recursively by the formula

$$\frac{1}{p_{n+1}} = \frac{3}{4p_n} + \frac{1}{4q}, \qquad n \in \mathbb{N}.$$

The sequence $\{p_n\}$ is decreasing and it converges to q as n tends to $+\infty$. Hence, we can fix $n_0 \in \mathbb{N}$ such that $p_{n_0} < p$. Since the operators A_{p_1} and $A_{p_{n_0}}$ are sectorial, applying the Riesz-Thorin theorem, we conclude that the operator A_p is sectorial as well. This completes the proof. □

Now, we can prove that the semigroups generated by the operators A_p are consistent.

Proposition 7.3.9 *Denote by $\{T_p(t)\}$ the semigroup in $L^p(\mathbb{R}^N)$ associated with the operator A_p. Finally, let $\{T(t)\}$ be the minimal semigroup associated with the operator $\mathcal{A}$ in $C_b(\mathbb{R}^N)$. Then, $T_p(\cdot) = T(\cdot)$ on $L^p(\mathbb{R}^N) \cap C_0(\mathbb{R}^N)$. In particular, $T_p(\cdot) \equiv T_q(\cdot)$ on $L^p(\mathbb{R}^N) \cap L^q(\mathbb{R}^N)$ for any $p, q > N(N-2)^{-1}$.*

Proof The arguments in the second part of the proof of Theorem 7.3.7 show that $R(\lambda, A_p)f = R(\lambda, \widehat{A})f$ almost everywhere in $\mathbb{R}^N$ for any $p > N(N-2)^{-1}$, $f \in L^p(\mathbb{R}^N) \cap C_0(\mathbb{R}^N)$ and $\lambda \in \sigma(A_p)$. Fix $t > 0$ and $f \in L^p(\mathbb{R}^N) \cap C_0(\mathbb{R}^N)$. By Theorem C.1.7, we can write

$$T_p(t)f = \lim_{n\to+\infty}\left[\frac{n}{t}R\left(\frac{n}{t}, A_p\right)\right]^n f, \qquad T(t)f = \lim_{n\to+\infty}\left[\frac{n}{t}R\left(\frac{n}{t}, \widehat{A}\right)\right]^n f,$$

where the first limit is meant in $L^p(\mathbb{R}^N)$ and the second one in $C_0(\mathbb{R}^N)$. Since both $C_0(\mathbb{R}^N)$ and $L^p(\mathbb{R}^N)$ are continuously embedded in $L^p(B_r)$ for any $r > 0$, from the previous two formulae it follows that $T_p(t)f = T(t)f$ almost everywhere in $\mathbb{R}^N$.

To prove the last part of the statement, we fix $q > N(N-2)^{-1}$ and observe that, if $f \in C_c(\mathbb{R}^N)$, then $T_p(t)f = T(t)f = T_q(t)f$ almost everywhere in $\mathbb{R}^N$, for any $t > 0$. Hence, by density, the operators $T_p(t)$ and $T_q(t)$ coincide in $L^p(\mathbb{R}^N) \cap L^q(\mathbb{R}^N)$ for any $t > 0$. □

7.3.1 Concluding remarks

In [205] the more general operator $\mathcal{A}_\gamma = q\Delta + \gamma|x|^{\alpha-2}\langle x, \nabla\rangle$ is considered in the case when $\alpha > 2$, assuming that γ is a constant greater than $2-N$. In that paper, G. Metafune, C. Spina, and C. Tacelli prove that, if $p \in (N(N-2+\gamma)^{-1}, +\infty)$, then the realization $A_{p,\gamma}$ of the operator $\mathcal{A}_\gamma$ in $L^p(\mathbb{R}^N)$ with $D(A_{p,\gamma}) = \{u \in L^p(\mathbb{R}^N) \cap W^{2,p}_{\rm loc}(\mathbb{R}^N) : \sqrt{q}|\nabla u|, q\Delta u \in L^p(\mathbb{R}^N)\}$ generates an analytic semigroup, which is contractive if and only if $p \geq (N+\alpha-2)(N+\gamma-2)^{-1}$. The condition $\gamma > 2-N$ is necessary: as the authors show (using the same techniques of this section), if $\gamma \leq 2-N$, then there exist no realizations of the operator $\mathcal{A}_\gamma$ in $L^p(\mathbb{R}^N)$ that generate a semigroup in $L^p(\mathbb{R}^N)$. Also the condition $p > N(N+\gamma-2)^{-1}$ is necessary for the generation result to hold.

Also in this situation, the starting point of all the analysis is the operator $R(0, A_{p,\gamma})$, which for any $f \in L^p(\mathbb{R}^N)$ is given by

$$(R(0, A_{p,\gamma})f)(x) = \int_{\mathbb{R}^N} g(x,y)f(y)dy$$

for almost every $x \in \mathbb{R}^N$, where now g is not anymore explicit. Providing suitable upper bounds of the function g, the authors of [205] prove that $D(A_{p,\gamma}) = \mathscr{D}_p$ (see (7.3.3)) if and only if $p > N(N+\gamma-\alpha)^{-1}$ with equivalence of the corresponding norms. Further, they prove that, for any $p > N(N-2+\gamma)^{-1}$, $\beta \in [0, \alpha-2] \cap [0, N/p' + \gamma - 2)$, there exists a positive constant C such that $\|q^{\beta/\alpha}u\|_{L^p(\mathbb{R}^N)} \leq C\|A_{p,\gamma}u\|_{L^p(\mathbb{R}^N)}$ for any $u \in D(A_{p,\gamma})$. This estimate is used to show that the spectrum of the operator $A_{p,\gamma}$ consists of a sequence of negative eigenvalues which accumulate at $-\infty$.

7.4 Some slightly more general operators

In this section we consider the operator $\mathcal{A} = q{\rm Tr}(QD^2)$, where $q(x) = (1+|x|^2)^{\alpha/2}$ for any $x \in \mathbb{R}^N$ and assume the following conditions on the matrix-valued function Q.

Hypothesis 7.4.1 *The diffusion coefficients q_{ij} belong to $C^\gamma_{\rm loc}(\mathbb{R}^N)$ for some $\gamma > 0$. Moreover, there exists a positive constant κ_0 such that $\langle Q(x)\xi, \xi\rangle \geq \kappa_0|\xi|^2$ for any $x, \xi \in \mathbb{R}^N$. Finally, for any $i, j \in \{1, \ldots, N\}$ the coefficient $q_{ij}(x)$ converges in $\mathbb{R}$ as $|x|$ tends to $+\infty$.*

Remark 7.4.2 In view of the previous hypotheses, we introduce the constant matrix $Q^{(0)} = (q_{ij}^{(0)})$, where $q_{ij}^{(0)} = \lim_{|x|\to+\infty} q_{ij}(x)$ for any $i,j = 1,\dots,N$. As is immediately seen, $\langle Q^{(0)}\xi,\xi\rangle \ge \kappa_0|\xi|^2$ for any $\xi\in\mathbb{R}^N$.

We first consider the case $\alpha\in(0,2]$. The following theorem has been proved by G. Metafune, C. Spina and C. Tacelli.

Theorem 7.4.3 ([206]) *Let $p\in(1,+\infty)$, $\alpha\in(0,2]$ and denote by A_p the realization of the operator $\mathcal{A}$ in $L^p(\mathbb{R}^N)$, with domain $D(A_p) = \{u\in W^{2,p}(\mathbb{R}^N) : q|D^2u|, \sqrt{q}|\nabla u|\in L^p(\mathbb{R}^N)\}$. Then, there exist two positive constants λ_0 and C such that $\{\lambda\in\mathbb{C} : \mathrm{Re}\lambda > \lambda_0\}\subset\rho(A_p)$ and*

$$\|qD^2u\|_{L^p(\mathbb{R}^N)} + \sqrt{|\lambda|}\|\sqrt{q}\nabla u\|_{L^p(\mathbb{R}^N)} + |\lambda|\|u\|_{L^p(\mathbb{R}^N)} \le C\|\lambda u - \mathcal{A}u\|_{L^p(\mathbb{R}^N)} \tag{7.4.1}$$

for any $u\in D(A_p)$. As a byproduct, $(A_p, D(A_p))$ generates an analytic semigroup.

Proof Throughout the proof, we denote by C a positive constant, independent of n, s, θ, ε, u, v and w, which may vary from line to line. Being rather long, we split the proof into steps.

Step 1. Here we prove that there exists $\varepsilon_0 > 0$ such that

$$\|\sqrt{q}\nabla_x v\|_{L^p(\mathbb{R}^{N+1})} \le \varepsilon\|qD_x^2 v\|_{L^p(\mathbb{R}^{N+1})} + C\varepsilon^{-1}\|v\|_{L^p(\mathbb{R}^{N+1})}. \tag{7.4.2}$$

$$\|D_t v\|_{L^p(\mathbb{R}^{N+1})} \le \varepsilon\|D_{tt}v\|_{L^p(\mathbb{R}^{N+1})} + C\varepsilon^{-1}\|v\|_{L^p(\mathbb{R}^{N+1})} \tag{7.4.3}$$

for any $\varepsilon\in(0,\varepsilon_0]$ and $v\in C^\infty(\mathbb{R}^{N+1})$. We limit ourselves to proving estimate (7.4.2) since the proof of the other estimate is completely similar.

Set $\rho_1 = \sqrt{q}/(2\alpha)$. Since $|\nabla q|\le\alpha\sqrt{q}$, the function ρ_1 is Lipschitz continuous and its Lipschitz constant does not exceed $1/4$. Choose $\vartheta\in C_c^\infty(\mathbb{R})$ such that $\chi_{(-1,1)}\le\vartheta\le\chi_{(-2,2)}$ and set $\vartheta_1(x) = \vartheta(r_{x_0}^{-1}|x-x_0|)$, $\vartheta_2(t) = \vartheta(t-t_0)$ for any $x\in\mathbb{R}^N$ and $t\in\mathbb{R}$, where $r_{x_0} = \rho_1(x_0)$. Clearly, $\chi_{B_{\rho_1}(x_0)\times(t_0-1,t_0+1)}\le\vartheta_1\vartheta_2\le\chi_{B_{2r_{x_0}}(x_0)\times(t_0-2,t_0+2)}$. Fix $\varepsilon>0$. By applying the well known interpolation inequality $\|\nabla w\|_{L^p(\mathbb{R}^{N+1})}\le C(N,p)\|w\|_{L^p(\mathbb{R}^{N+1})}^{1/2}\|D^2w\|_{L^p(\mathbb{R}^{N+1})}^{1/2}$ to the function $\vartheta_1\vartheta_2 v$ we obtain

$$\begin{aligned}
&\|\sqrt{q}\nabla_x v\|_{L^p(B_{r_{x_0}}(x_0)\times(t_0-1,t_0+1))}\\
\le& C\|\sqrt{q(x_0)}\nabla_x(\vartheta_1\vartheta_2 v)\|_{L^p(\mathbb{R}^{N+1})}\\
\le& C(\varepsilon\|q(x_0)D_x^2(\vartheta_1\vartheta_2 v)\|_{L^p(\mathbb{R}^{N+1})} + \varepsilon^{-1}\|\vartheta_1\vartheta_2 v\|_{L^p(\mathbb{R}^{N+1})})\\
\le& C\Big(\varepsilon\|q(x_0)D_x^2 v\|_{L^p(B_{2r_{x_0}}(x_0)\times(t_0-2,t_0+2))} + r_{x_0}^{-1}\varepsilon\|q(x_0)\nabla_x v\|_{L^p(B_{2r_{x_0}}(x_0)\times(t_0-2,t_0+2))}\\
&\qquad + r_{x_0}^{-2}\|q(x_0)v\|_{L^p(B_{2r_{x_0}}(x_0)\times(t_0-2,t_0+2))} + \varepsilon^{-1}\|v\|_{L^p(B_{2r_{x_0}}(x_0)\times(t_0-2,t_0+2))}\Big)\\
\le& C\Big(\varepsilon\|q(x_0)D_x^2 v\|_{L^p(B_{2r_{x_0}}(x_0)\times(t_0-2,t_0+2))} + \varepsilon\|\sqrt{q(x_0)}\nabla_x v\|_{L^p(B_{2r_{x_0}}(x_0)\times(t_0-2,t_0+2))}\\
&\qquad + (1+\varepsilon^{-1})\|v\|_{L^p(B_{2r_{x_0}}(x_0)\times(t_0-2,t_0+2))} + \varepsilon^{-1}\|v\|_{L^p(B_{2r_{x_0}}(x_0)\times(t_0-2,t_0+2))}\Big).
\end{aligned}$$

Since $\frac{1}{2}\sqrt{q(x_0)}\le\sqrt{q(x)}\le\frac{3}{2}\sqrt{q(x_0)}$ for any $x\in\overline{B}_{2\rho_1}(x_0)$ we can estimate

$$\begin{aligned}
&\|\sqrt{q}\nabla_x v\|_{L^p(B_{r_{x_0}}(x_0)\times(t_0-1,t_0+1))}\\
\le& C\Big(\varepsilon\|qD_x^2 v\|_{L^p(B_{2r_{x_0}}(x_0)\times(t_0-2,t_0+2))} + \varepsilon\|\sqrt{q}\nabla_x v\|_{L^p(B_{2r_{x_0}}(x_0)\times(t_0-2,t_0+2))}\\
&\qquad + (1+\varepsilon^{-1})\|v\|_{L^p(B_{2r_{x_0}}(x_0)\times(t_0-2,t_0+2))}\Big).
\end{aligned} \tag{7.4.4}$$

Let $\{B_{r_{x_n}}(x_n) : n \in \mathbb{N}\}$ be a covering of $\mathbb{R}^N$ such that at most ς_1 among the double balls $B_{2r_{x_n}(x_n)}(x_n)$ $(n \in \mathbb{N})$ overlap and let $\{(t_n - 1, t_n + 1) : n \in \mathbb{N}\}$ be a covering of $\mathbb{R}$ such that at most ς_2 among the intervals $(t_n - 2, t_n + 2)$ $(n \in \mathbb{N})$ overlap (see Proposition 7.1.6). Writing (7.4.4) with x_n instead of x_0 and summing over n, it follows that

$$\|\sqrt{q}\nabla_x v\|_{L^p(\mathbb{R}^{N+1})} \le C(\varepsilon\|qD_x^2 v\|_{L^p(\mathbb{R}^{N+1})} + \varepsilon\|\sqrt{q}\nabla_x v\|_{L^p(\mathbb{R}^{N+1})} + (1+\varepsilon^{-1})\|v\|_{L^p(\mathbb{R}^{N+1})}).$$

Choosing ε_0 small enough we obtain estimate (7.4.2) for any $\varepsilon \in (0, \varepsilon_0]$.

Step 2. Here, we prove that

$$\begin{aligned}&\|qD_x^2 v\|_{L^p(\mathbb{R}^{N+1})} + \|D_{tt}v\|_{L^p(\mathbb{R}^{N+1})} + \|\sqrt{q}D_t\nabla_x v\|_{L^p(\mathbb{R}^{N+1})}\\ \le&C(\|\mathcal{A}_s v + e^{i\theta}D_{tt}v\|_{L^p(\mathbb{R}^{N+1})} + \|v\|_{L^p(\mathbb{R}^{N+1})})\end{aligned} \tag{7.4.5}$$

for any $v \in C_c^\infty(\mathbb{R}^{N+1})$, $s \in [0,1]$, $\theta \in (-\pi/2, \pi/2)$, where $\mathcal{A}_s = \mathrm{Tr}(Q_s D_x^2)$ and $Q_s = sQ + (1-s)I$. For this purpose, we fix $\varepsilon \in (0,1)$ and set $\widetilde{r}_{x_0} = \varepsilon r_{x_0}$. Since

$$\begin{aligned}|\langle Q_s(x)\xi,\xi\rangle + e^{i\theta}\xi_{N+1}^2|^2 =&|\langle Q_s(x)\xi,\xi\rangle + \cos(\theta)\xi_{N+1}^2|^2 + |\sin(\theta)|^2|\xi_{N+1}|^4\\ =&(\langle Q_s(x)\xi,\xi\rangle + |\cos(\theta)|^2\xi_{N+1}^2)^2 + |\sin(\theta)|^2\xi_{N+1}^4\\ \ge&(\kappa_0^2\wedge 1)(|\xi|^4 + \xi_{N+1}^4) \ge \frac{1}{2}(\kappa_0^2\wedge 1)(|\xi|^2 + |\xi_{N+1}|^2)^2\end{aligned}$$

for any $x \in \mathbb{R}^N$ and $(\xi, \xi_{N+1}) \in \mathbb{R}^{N+1}$, the same estimate is satisfied if we replace $Q(x)$ with the constant matrix $Q^{(0)}$. Hence, by a well-known estimate for elliptic operators with constant coefficients, it follows that

$$\|D_t\nabla_x w\|_{L^p(\mathbb{R}^{N+1})} + \|D_x^2 w\|_{L^p(\mathbb{R}^{N+1})} + \|D_{tt}w\|_{L^p(\mathbb{R}^{N+1})} \le C(\|\mathcal{A}_s^{(0)}D_x^2 w) + e^{i\theta}D_{tt}w\|_{L^p(\mathbb{R}^N)}).$$

for any $w \in W^{2,p}(\mathbb{R}^N)$. Let us apply this estimate to the function $v : \mathbb{R}^{N+1} \to \mathbb{R}$, defined by $v(x,t) = \widetilde{\vartheta}_1(x)\widetilde{\vartheta}_2(t)u(\gamma x, t)$ for any $(x,t) \in \mathbb{R}^{N+1}$, where $\widetilde{\vartheta}_1$ is defined as ϑ_1, with r_{x_0} being replaced with $\widetilde{r}_{x_0}$, $\widetilde{\vartheta}_2(t) = \vartheta(\varepsilon^{-1}(t - t_0))$ for any $t \in \mathbb{R}$, and $\gamma = \sqrt{q(x_0)}$. Observing that $q(x_0)\mathrm{Tr}(Q_s^{(0)}D_x^2 v) = \mathcal{A}_s v + (q(x_0) - q)\mathrm{Tr}(Q_s^{(0)}D_x^2 v) + q\mathrm{Tr}((Q^{(0)} - Q)D_x^2 v)$, we obtain that

$$\begin{aligned}&\|\sqrt{q(x_0)}D_t\nabla_x v\|_{L^p(B_{\widetilde{r}_{x_0}}(x_0)\times(t_0-\varepsilon,t_0+\varepsilon))} + \|q(x_0)D_x^2 v\|_{L^p(B_{\widetilde{r}_{x_0}}(x_0)\times(t_0-\varepsilon,t_0+\varepsilon))}\\ &+ \|D_{tt}v\|_{L^p(B_{\widetilde{r}_{x_0}}(x_0)\times(t_0-\varepsilon,t_0+\varepsilon))}\\ \le&C\Big(\|\mathcal{A}_{\theta,s}v\|_{L^p(B_{2\widetilde{r}_{x_0}}(x_0)\times(t_0-2\varepsilon,t_0+2\varepsilon))} + \varepsilon^{-1}\|D_t v\|_{L^p(B_{2\widetilde{r}_{x_0}}(x_0)\times(t_0-2\varepsilon,t_0+2\varepsilon))}\\ &\quad+ \varepsilon^{-1}\|\sqrt{q(x_0)}\nabla_x v\|_{L^p(B_{2\widetilde{r}_{x_0}}(x_0)\times(t_0-2\varepsilon,t_0+2\varepsilon))} + \varepsilon^{-2}\|v\|_{L^p(B_{2\widetilde{r}_{x_0}}(x_0)\times(t_0-2\varepsilon,t_0+2\varepsilon))}\\ &\quad+ \|(q(x_0) - q)\mathrm{Tr}(Q_s^{(0)}D_x^2 v)\|_{L^p(B_{2\widetilde{r}_{x_0}}(x_0)\times(t_0-2\varepsilon,t_0+2\varepsilon))}\\ &\quad+ \|q\mathrm{Tr}((Q^{(0)} - Q)D_x^2 v)\|_{L^p(B_{2\widetilde{r}_{x_0}}(x_0)\times(t_0-2\varepsilon,t_0+2\varepsilon))}\Big).\end{aligned} \tag{7.4.6}$$

Now, we fix $R > 0$ such that $|q_{ij} - q_{ij}^{(0)}| < \varepsilon$ in $\mathbb{R}^N \setminus B_R$ and take advantage of the interior L^p-estimate

$$\begin{aligned}&\|D_x^2 v\|_{L^p(B_{2\widetilde{r}_{x_0}}\times(t_0-2\varepsilon,t_0+2\varepsilon))}\\ \le&C(\|v\|_{L^p(B_{4\widetilde{r}_{x_0}}\times(t_0-4\varepsilon,t_0+4\varepsilon))} + \|\mathcal{A}_s v + e^{i\theta}D_{tt}v\|_{L^p(B_{4\widetilde{r}_{x_0}}\times(t_0-4\varepsilon,t_0+4\varepsilon))}),\end{aligned}$$

to conclude that

$$\|q\mathrm{Tr}((Q^{(0)} - Q)D_x^2 v)\|_{L^p(B_{2\widetilde{r}_{x_0}}(x_0)\times(t_0-2\varepsilon,\tau_0+2\varepsilon))}$$

$$
\begin{aligned}
&\le \|q\mathrm{Tr}((Q^{(0)}-Q)D_x^2 v)\|_{L^p(((\mathbb{R}^N\setminus B_R)\cap B_{2\tilde r_{x_0}}(x_0))\times(t_0-2\varepsilon,\tau_0+2\varepsilon))}\\
&\quad+\|q\mathrm{Tr}((Q^{(0)}-Q)D_x^2 v)\|_{L^p((B_R\cap B_{2\tilde r_{x_0}}(x_0))\times(t_0-2\varepsilon,\tau_0+2\varepsilon))}\\
&\le \varepsilon N\|qD_x^2 v\|_{L^p(B_{2\tilde r_{x_0}}(x_0)\times(t_0-2\varepsilon,t_0+2\varepsilon))}+C_R\|v\|_{L^p(B_{4\tilde r_{x_0}}(x_0)\times(t_0-4\varepsilon,\tau_0+4\varepsilon))}\\
&\quad+C_R\|\mathcal{A}_s v+e^{i\theta}D_{tt}v\|_{L^p(B_{4\tilde r_{x_0}}(x_0)\times(t_0-4\varepsilon,t_0+4\varepsilon))}.
\end{aligned}
\tag{7.4.7}
$$

Replacing (7.4.6) in (7.4.7), gives

$$
\begin{aligned}
&\|\sqrt{q(x_0)}D_t\nabla_x v\|_{L^p(B_{\tilde r_{x_0}}(x_0)\times(t_0-\varepsilon,t_0+\varepsilon))}+\|q(x_0)D_x^2 v\|_{L^p(B_{\tilde r_{x_0}}(x_0)\times(t_0-\varepsilon,t_0+\varepsilon))}\\
&+\|D_{tt}v\|_{L^p(B_{\tilde r_{x_0}}(x_0)\times(t_0-\varepsilon,t_0+\varepsilon))}\\
\le& C_R\Big(\|\mathcal{A}_s v+e^{i\theta}D_{tt}v\|_{L^p(B_{\tilde r_{x_0}}(x_0)\times(t_0-\varepsilon,t_0+\varepsilon))}+\varepsilon^{-2}\|v\|_{L^p(B_{2\tilde r_{x_0}}(x_0)\times(t_0-2\varepsilon,t_0+2\varepsilon))}\\
&\quad+\|\mathcal{A}_s v+e^{i\theta}D_{tt}v\|_{L^p(B_{4\tilde r_{x_0}}(x_0)\times(t_0-4\varepsilon,t_0+4\varepsilon))}\\
&\quad+\|v\|_{L^p(B_{4\tilde r_{x_0}}(x_0)\times(t_0-4\varepsilon,t_0+4\varepsilon))}+\|(q(x_0)-q)D_x^2 v\|_{L^p(B_{2\tilde r_{x_0}}(x_0)\times(t_0-2\varepsilon,t_0+2\varepsilon))}\\
&\quad+\varepsilon\|qD_x^2 v\|_{L^p(B_{2\tilde r_{x_0}}(x_0)\times(t_0-2\varepsilon,t_0+2\varepsilon))}+\varepsilon\|D_{tt}v\|_{L^p(B_{2\tilde r_{x_0}}(x_0)\times(t_0-2\varepsilon,t_0+2\varepsilon))}\\
&\quad+\varepsilon^{-1}\|\sqrt{q(x_0)}\nabla_x v\|_{L^p(B_{2\tilde r_{x_0}}(x_0)\times(t_0-2\varepsilon,t_0+2\varepsilon))}+\varepsilon^{-1}\|D_t v\|_{L^p(B_{2\tilde r_{x_0}}(x_0)\times(t_0-2\varepsilon,t_0+2\varepsilon))}\Big).
\end{aligned}
$$

Since $\|\nabla_x\sqrt{q}\|_\infty\le\alpha/2$, for any $(x,t)\in B_{\tilde r_{x_0}}(x_0)\times(t_0-\varepsilon,t_0+\varepsilon)$ it holds that

$$
\frac{2-\varepsilon}{2}\sqrt{q(x_0)}\le\sqrt{q(x)}\le\frac{2+\varepsilon}{2}\sqrt{q(x_0)},
$$

which can be used to prove that $|q(x)-q(x_0)|\le C\varepsilon q(x)$ for any $x\subset B_{\tilde r_{x_0}}$. Then,

$$
\begin{aligned}
&\|\sqrt{q}D_t\nabla_x v\|_{L^p(B_{\tilde r_{x_0}}(x_0)\times(t_0-\varepsilon,t_0+\varepsilon))}+\|qD_x^2 v\|_{L^p(B_{\tilde r_{x_0}}(x_0)\times(t_0-\varepsilon,t_0+\varepsilon))}\\
&+\|D_{tt}v\|_{L^p(B_{\tilde r_{x_0}}(x_0)\times(t_0-\varepsilon,t_0+\varepsilon))}\\
\le& C_R\Big(\varepsilon^{-1}\|D_t v\|_{L^p(B_{2\tilde r_{x_0}}(x_0)\times(t_0-2\varepsilon,t_0+2\varepsilon))}+(1+\varepsilon^{-2})\|v\|_{L^p(B_{4\tilde r_{x_0}}(x_0)\times(t_0-4\varepsilon,t_0+4\varepsilon))}\\
&\quad+\|\mathcal{A}_s v+e^{i\theta}D_{tt}v\|_{L^p(B_{4\tilde r_{x_0}}(x_0)\times(t_0-4\varepsilon,t_0+4\varepsilon))}\\
&\quad+\varepsilon\|qD_x^2 v\|_{L^p(B_{2\tilde r_{x_0}}(x_0)\times(t_0-2\varepsilon,t_0+2\varepsilon))}+\varepsilon\|D_{tt}v\|_{L^p(B_{2\tilde r_{x_0}}(x_0)\times(t_0-2\varepsilon,t_0+2\varepsilon))}\\
&\quad+\varepsilon^{-1}\|\sqrt{q(x_0)}\nabla_x v\|_{L^p(B_{2\tilde r_{x_0}}(x_0)\times(t_0-2\varepsilon,t_0+2\varepsilon))}\Big).
\end{aligned}
$$

The same covering argument as in Step 1 reveals that

$$
\begin{aligned}
&\|\sqrt{q}D_t\nabla_x v\|_{L^p(\mathbb{R}^{N+1})}+\|qD_x^2 v\|_{L^p(\mathbb{R}^{N+1})}+\|D_{tt}v\|_{L^p(\mathbb{R}^{N+1})}\\
\le& C_R\Big(\|\mathcal{A}_s v+e^{i\theta}D_{tt}v\|_{L^p(\mathbb{R}^{N+1})}+(1+\varepsilon^{-2})\|v\|_{L^p(\mathbb{R}^{N+1})}+\varepsilon\|qD_x^2 v\|_{L^p(\mathbb{R}^{N+1})}\\
&\quad+\varepsilon\|D_{tt}v\|_{L^p(\mathbb{R}^{N+1})}+\varepsilon^{-1}\|\sqrt{q}\nabla_x v\|_{L^p(\mathbb{R}^{N+1})}+\varepsilon^{-1}\|D_t v\|_{L^p(\mathbb{R}^{N+1})}\Big).
\end{aligned}
$$

Finally, using (7.4.2), (7.4.3) and taking ε sufficiently small, estimate (7.4.5) follows at once.

Step 3. Now, we can complete the proof. We fix a cut off function $\eta\in C_c^\infty(\mathbb{R})$, such that $\chi_{(-1/2,1/2)}\le\eta\le\chi_{(-1,1)}$, a function u in $C_c^\infty(\mathbb{R})$ and $r>0$. Applying estimate (7.4.5) to the function v, defined by $v(t,x):=\eta(t)e^{irt}u(x)$ for any $t\in\mathbb{R}$ and $x\in\mathbb{R}^N$, we get

$$
\|qD_x^2 v\|_{L^p(\mathbb{R}^{N+1})}+\|D_{tt}v\|_{L^p(\mathbb{R}^{N+1})}\le C(\|\mathcal{A}_s u-r^2e^{i\theta}u\|_{L^p(\mathbb{R}^N)}+(1+r)\|u\|_{L^p(\mathbb{R}^N)}).
$$

Since $\eta \equiv 1$ in $[-1/2, 1/2]$, from the previous estimate we can infer that

$$\|qD^2u\|_{L^p(\mathbb{R}^N)} + r^2\|u\|_{L^p(\mathbb{R}^N)} \le C(\|(\mathcal{A}_s - r^2e^{i\theta})u\|_{L^p(\mathbb{R}^N)} + (1+r)\|u\|_{L^p(\mathbb{R}^N)}).$$

By choosing r_0 large enough and setting $\lambda = r^2e^{i\theta}$ for any $r \ge r_0$, we deduce that

$$\|\sqrt{q}D^2u\|^p_{L^p(\mathbb{R}^N)} + |\lambda|\|u\|^p_{L^p(\mathbb{R}^N)} \le C\|\mathcal{A}_su - \lambda u\|_{L^p(\mathbb{R}^N)}. \tag{7.4.8}$$

Finally, using estimate (7.4.2) and taking $\varepsilon = |\lambda|^{-1/2}$ for Reλ large enough, we get

$$\sqrt{|\lambda|}\|\sqrt{q}\nabla u\|_{L^p(\mathbb{R}^N)} \le C(\|\mathcal{A}_su - \lambda u\|_{L^p(\mathbb{R}^N)} + \|\lambda\|\|u\|_{L^p(\mathbb{R}^N)}). \tag{7.4.9}$$

Thanks to Lemma 7.2.2, we can extend estimates (7.4.8) and (7.4.9) to any function $u \in D(A_p)$ The method of continuity allows us to prove that the halfplane $\{\lambda \in \mathbb{C} : \mathrm{Re}\lambda \ge \lambda_0\}$ is contained in $\rho(A_p)$ if λ_0 is sufficiently large.

Finally, from (7.4.8) and (7.4.9), with $s = 0$, we deduce that

$$\begin{aligned}\|\sqrt{q}D^2u\|^p_{L^p(\mathbb{R}^N)} + \sqrt{|\lambda|}\|\sqrt{q}\nabla u\|_{L^p(\mathbb{R}^N)} + |\lambda|\|u\|^p_{L^p(\mathbb{R}^N)} \le& C(\|A_pu - \lambda u\|_{L^p(\mathbb{R}^N)} + \|u\|_{L^p(\mathbb{R}^N)})\\ \le& C\|A_pu - \lambda u\|_{L^p(\mathbb{R}^N)}\end{aligned}$$

for any $u \in D(A_p)$, and (7.4.1) follows. □

Remark 7.4.4 We stress that the generation results in [206] can also be applied to some more general operators than the ones we consider in Theorem 7.4.3. Moreover, arguing as in Section 7.2, we can prove a similar result when q is replaced by the function $q : \mathbb{R}^N \to \mathbb{R}$, defined by $q(x) = 1 + |x|^\alpha$ for any $x \in \mathbb{R}^N$, thus providing a generalization of the generation result in Theorem 7.2.5.

We now consider the case when $\alpha > 2$ and state the following generation results proved by M. Sobajima and C. Spina.

Theorem 7.4.5 ([238]) *Suppose that $\alpha > 2$ and $p \in (N(N-2)^{-1}, +\infty)$. Then, the following properties are satisfied.*

(i) *If $\alpha < N/p'$, then the minimal realization of the operator $\mathcal{A}$ in $L^p(\mathbb{R}^N)$, i.e., the closure in $L^p(\mathbb{R}^N)$ of the restriction of $\mathcal{A}$ to $C^\infty_c(\mathbb{R}^N)$, generates an analytic semigroup.*

(ii) *If $\alpha \ge N/p'$, then the (maximal) realization A_p of the operator $\mathcal{A}$ in $L^p(\mathbb{R}^N)$, with domain $D(A_p) = \{u \in W^{2,p}_{\mathrm{loc}}(\mathbb{R}^N) \cap L^p(\mathbb{R}^N) : \mathcal{A}u \in L^p(\mathbb{R}^N)\}$, generates an analytic semigroup.*

Chapter 8

Estimates on the Green function

8.0 Introduction

In Chapter 1, under weak assumptions on the coefficients of the elliptic operator $\mathcal{A} = \mathrm{Tr}(QD^2) + \langle b, \nabla \rangle + c$, we have shown that the semigroup $\{T(t)\}$, associated in $C_b(\mathbb{R}^N)$ with the operator $\mathcal{A}$, admits a Green function, i.e., there exists a function $g : (0, +\infty) \times \mathbb{R}^N \times \mathbb{R}^N \to \mathbb{R}$ such that

$$T(t)f(x) = \int_{\mathbb{R}^N} g(t, x, y) f(y) dy, \qquad t > 0, \;\; x \in \mathbb{R}^N, \;\; f \in C_b(\mathbb{R}^N).$$

Green functions have been widely studied in the literature starting from the case when the operator $\mathcal{A}$ has bounded and smooth enough coefficients. The first estimates on Green functions (also called Gaussian kernels in the literature) are due to D.G. Aronson (see [22]). The more recent results by E.B. Davies (see [83, 84]) have given new impetus to the analysis of Green functions g associated with elliptic operators (with bounded coefficients) and to the determination of optimal constants in the estimates of g. In this direction, form methods were extensively used to prove estimates on the kernel g associated with elliptic operators in divergence form and provided very elegant proofs. In the case of elliptic operators $\mathcal{A}$ with bounded coefficients defined in $\mathbb{R}^N$ it is known that the Green functions g satisfy the estimate

$$g(t, x, y) \le C_1 t^{-\frac{N}{2}} e^{\delta t} e^{-C_2 \frac{|x-y|^2}{t}}, \qquad t > 0, \;\; x, y \in \mathbb{R}^N,$$

for some positive constants C_1, C_2 and $\delta \in \mathbb{R}$. This estimate states that, even if the operator $\mathcal{A}$ is "far" from being the Laplacian in $\mathbb{R}^N$, its Green function g is "close" to the Green function of the heat semigroup, which is given by

$$g(t, x, y) = (4\pi t)^{-\frac{N}{2}} e^{-\frac{|x-y|^2}{4t}}, \qquad t > 0, \;\; x, y \in \mathbb{R}^N. \tag{8.0.1}$$

Gaussian estimates led to very important consequences for the spectral theory and allowed to prove regularity properties for solutions to parabolic equations. A few of these consequences are illustrated in this chapter (see Proposition 8.3.17 and Corollary 8.4.9)

For an overview on the results available in the literature (mainly in the case of bounded coefficients in $\mathbb{R}^N$ and in domains of $\mathbb{R}^N$) we refer the reader e.g. to the monographs [18, 47, 84, 107, 159, 212, 256] and the papers [21, 81, 249].

In this chapter we illustrate some of the results which are available in the literature in the case of nondegenerate elliptic operators $\mathcal{A}$ with unbounded coefficients, having a particular attention for Schrödinger type operators. Due to the number of results available in the literature, providing an exhaustive survey is out of the scope of this chapter. For a more detailed overview we refer the reader to the Ph.D. thesis [240]. Here, we rather prefer to present two main different techniques which have been used to prove such estimates: the first one is inspired by the arguments in [84] and relies on the use of form methods and the well-known Nash inequality. The second technique is a more typical PDE's technique and it is based on sharp estimates in Krylov spaces of solutions to parabolic equations associated with the adjoint operator $\mathcal{A}^*$ (note that the function $g(\cdot,x,\cdot)$ is actually a solution to the adjoint equation $D_t g + \mathcal{A}^* g = 0$ subject to the initial condition $g(0,x,\cdot) = \delta_x$). This latter technique makes strong use of Lyapunov functions and provides upper bounds for the function ωg, ω being a suitable smooth function. Proper choices of ω allow us to prove sharp estimates of the function g.

Typically when dealing with estimates for Green functions one is interested in proving estimates which are "good" with respect to t or with the spatial variables, and to see how these estimates are far from the behaviour of the Green function of the heat semigroup.

In Section 8.1 we consider the case when the operator $\mathcal{A}$ is in divergence form i.e., $\mathcal{A} = \operatorname{div}(Q\nabla) + \langle b, \nabla\rangle + c$ and discuss the results in [20], when c identically vanishes in $\mathbb{R}^N$, and in [191] in the general case. In both situations the diffusion coefficients are assumed to be bounded in $\mathbb{R}^N$ together with their gradients. In particular, in the case when $c \equiv 0$, the only algebraic condition assumed on b is related to its divergence, which is assumed to be bounded from below in $\mathbb{R}^N$. On the contrary, in [191] an additional bound on $|b|$ in terms of a suitable power of $|c|$ is prescribed. Hence, in the case when $c \equiv 0$, the assumptions in [191] do not include the ones in [20]. In particular, the main results of [191] aim at studying the behaviour of g with respect to t and show that, under Hypotheses 8.1.1, for fixed $x, y \in \mathbb{R}^N$, the function $g(\cdot,x,y)$ behaves like the Green function of the heat semigroup (see Theorem 8.1.2). On the other hand, the main result of [20] is a *pseudo-Gaussian estimate* for g. More precisely, the authors show that

$$g(t,x,y) \le C_1 e^{\omega t} t^{-\frac{N}{2}} \exp\bigg(-\frac{C_2|x-y|^{\frac{1}{1-\alpha}}}{t}\bigg)^{\frac{\alpha}{1-\alpha}}$$

for any $t > 0$ and $x, y \in \mathbb{R}^N$, where ω, C_1 and C_2 are positive constants and α is such that, up to an additive constant, $|c|^\alpha$ bounds from above the modulus of the drift term b. This estimate is of Gaussian type[1] when $\alpha = 1/2$.

For further estimates on the Green function g for a rather general class of nondegenerate elliptic operators with unbounded coefficients, we refer the reader to Subsection 8.1.1 and to Chapter 15.

In the rest of the chapter, we focus our attention on the case when $\mathcal{A} = (1+|x|^\alpha)\Delta + c$, assuming that at least $\alpha > 0$ or c is negative and unbounded. First, in Section 8.2 we consider the case when $\alpha > 0$ and $c \equiv 0$. In this situation estimates for the Green function of the semigroup associated with $\mathcal{A}$ are obtained by using the well known equivalence between the Nash inequality and ultracontractivity in the case of symmetric Markov semigroups (see e.g. [84, 212]).

Next, in Section 8.3, we consider the case when $\mathcal{A} = \Delta + c$. In such a situation the semigroup $\{T(t)\}$ associated with the operator $\mathcal{A}$ in $C_b(\mathbb{R}^N)$ is dominated from above by the heat semigroup, as it follows from applying the maximum principle. Therefore, the Green function g of the semigroup $\{T(t)\}$ is dominated from above by the function in the right-hand side of (8.0.1). It is important to understand how the presence of a negative and unbounded potential may provide us with sharper upper bounds on g. Also lower bounds of the function g are of particular interest. These questions have been investigated in the literature starting from the paper [85] by E.B. Davies and B. Simon in the case when $c(x) = -|x|^\beta$ for any $x \in \mathbb{R}^N$ and some $\beta > 2$. Here, we present the results proved by G. Metafune, D. Pallara and A. Rhandi in [194] and (part of the results proved) by G. Metafune and C. Spina in [201]. All these results imply, in particular, that the function $x \mapsto g(t,x,x)$

[1] i.e., g is pointwise estimated by a function which has the form of the right-hand side of (8.0.1).

decreases to zero with an exponential rate as $|x|$ tends to $+\infty$, which, of course, is not the case of the Green function of the heat semigroup. This shows explicitly how the presence of a negative unbounded potential term improves the decay rate of g with respect to the spatial variables. The results in [194] apply in the case when, roughly speaking, the oscillations of the function c at infinity are controlled in terms of $|c|^{3/2}$. For related results on the Schrödinger operator $\Delta + c$ in $\mathbb{R}^N$ (or on manifold), c being a nonpositive and unbounded function, we refer the reader to [84, Section 4.5], [169, 211, 236].

Finally, in the last section of this chapter, we consider the case when $\mathcal{A} = (1+|x|^\alpha)\Delta - |x|^\beta$ with α and β satisfying the condition $0 < \alpha \le 2 < \beta$. Here, the arguments used to prove the estimates for g are inspired by the ones in [84] but the non symmetry in $L^2(\mathbb{R}^N)$ of the associated semigroup $\{T(t)\}$ causes additional technical difficulties.

8.1 The general case

The results that we present here have been proved by W. Arendt, G. Metafune, and D. Pallara in [20] and by G. Metafune, E.M. Ouhabaz, and D. Pallara in [191].

We first consider the case when the operator $\mathcal{A}$ has null potential term, i.e., $\mathcal{A}\psi = \mathrm{Tr}(QD^2\psi) + \langle b, \nabla\psi\rangle$ on smooth functions $\psi : \mathbb{R}^N \to \mathbb{R}$, and we assume the following conditions on its coefficients.

Hypotheses 8.1.1 (i) *The diffusion coefficients $q_{ij} = q_{ji}$ and the drift coefficients b_j $(i,j = 1,\ldots,N)$ belong to $C_b(\mathbb{R}^N) \cap C^1(\mathbb{R}^N)$ and $C^1(\mathbb{R}^N)$, respectively;*

(ii) *there exists a positive constant κ_0 such that $\langle Q(x)\xi,\xi\rangle \ge \kappa_0|\xi|^2$ for any $x,\xi \in \mathbb{R}^N$;*

(iii) *the divergence of b is bounded from below by a real constant K.*

Under the previous assumptions we can prove the following estimates for the kernel g.

Theorem 8.1.2 ([191], Theorems 1 and 2) *The following properties hold true.*

(i) *There exists a positive constant C_0, depending only on N and κ_0, such that*

$$g(t,x,y) \le \left(\frac{NC_0}{4\kappa_0}\right)^{\frac{N}{2}} \begin{cases} t^{-\frac{N}{2}}, & \text{if } K = 0,\\[2mm] e^{-\frac{K}{2}t}\left(\dfrac{N}{|K|}\right)^{-\frac{N}{2}} (\cosh(N^{-1}Kt)-1)^{-\frac{N}{4}}, & \text{if } K \neq 0, \end{cases} \tag{8.1.1}$$

for any $t > 0$ and $x, y \in \mathbb{R}^N$.

(ii) *If $K \neq 0$ then there exists a constant C_1, depending only on N and κ_0, such that*

$$g(t,x,y) \le C_1 e^{-t(K\vee 0)}, \qquad t > 2|K|^{-1}. \tag{8.1.2}$$

(iii) *For any compact set $\mathscr{K} \subset \mathbb{R}^N$, there exists a positive constant C_2 such that $g(t,x,x) \ge C_2 t^{-N/2}$ for any $x \in \mathscr{K}$.*

Remark 8.1.3 Note that, if $K \neq 0$, then the right-hand side of the formula (8.1.1) behaves (up to a multiplicative constant) like $t^{-\frac{3}{4}Kt}$ as t tends to $+\infty$. Hence, the formula (8.1.2) provides us with much more precise information about the longtime behaviour of the function g.

Proof of Theorem 8.1.2 (i) For any $n \in \mathbb{N}$, let g_n be the Green function of the semigroup $\{T_n(t)\}$ associated in $C_b(B_n)$ with the operator $\mathcal{A}$, with homogeneous Dirichlet boundary conditions on ∂B_n. Note that it suffices to prove estimate (8.1.1) with g being replaced by g_n and with and a constant M_0, independent of $n \in \mathbb{N}$. Indeed, the proof of Theorem 1.2.5 shows that the sequence $\{g_n(t,x,y)\}$ increases to $g(t,x,y)$ a n tends to $+\infty$ for any $t>0$ and $x,y \in \mathbb{R}^N$.

The main ingredient of the proof is the Nash inequality (see (D.1.2)) in the ball B_n ($n \in \mathbb{N}$). If u belongs to $W^{2,2}(B_n)\cap W^{1,2}_0(B_n)$, then we can integrate by parts and write

$$\int_{B_n} u\mathcal{A}u dx = -\int_{B_n} |\sqrt{Q}\nabla u|^2 dx - \frac{1}{2}\int_{B_n}(\operatorname{div} b)|u|^2 dx \le -\kappa_0\int_{B_n}|\nabla u|^2 dx - \frac{K}{2}\int_{B_n}|u|^2 dx.$$

Replacing this estimate in the Nash inequality gives

$$\|u\|^{2+4/N}_{L^2(B_n)} \le -\frac{C_0}{\kappa_0}\|u\|^{4/N}_{L^1(B_n)}\left(\int_{B_n} u\mathcal{A}u dx + \frac{K}{2}\int_{B_n}|u|^2 dx\right) \tag{8.1.3}$$

for any $u \in W^{2,2}(B_n)\cap W^{1,2}_0(B_n)$ and $n \in \mathbb{N}$.

We use (8.1.3) to show that, for any $n \in \mathbb{N}$ and $t>0$, the operator $T_n(t)$ is bounded from $L^1(B_n)$ into $L^2(B_n)$ and

$$\|T_n(t)\|_{L(L^1(\mathbb{R}^N),L^2(\mathbb{R}^N))} \le \left(\frac{NC_0}{4\kappa_0}\right)^{\frac{N}{4}}\begin{cases} t^{-\frac{N}{4}}, & K=0,\\[1ex] e^{-\frac{K}{2}t}\left(\dfrac{N}{2K}\right)^{-\frac{N}{4}}(e^{\frac{2Kt}{N}}-1)^{-\frac{N}{4}}, & K\neq 0.\end{cases} \tag{8.1.4}$$

For this purpose, we fix $n \in \mathbb{N}$, $f \in C^\infty_c(B_n)$ and consider the function $\psi_n : [0,+\infty) \to \mathbb{R}$ defined by $\psi_n(s) := e^{Ks}\|T_n(s)f\|^2_{L^2(B_n)}$ for any $s>0$. Since $T_n(s)f \in W^{1,2}_0(B_n)\cap L^1(B_n)$ (see Theorem D.3.5) and $\|T_n(s)\|_{L(L^1(B_n))} \le e^{-Ks}$ for any $s>0$, $p \in [1,+\infty]$ and $n \in \mathbb{N}$ (which follows from the proof of Theorem 7.1.2), from (8.1.3) with $u = T_n(s)f$ we get

$$\begin{aligned}-\psi_n'(s) =& 2\left(-\int_{B_n} T_n(s)f\mathcal{A}T_n(s)f dx - \frac{K}{2}\int_{B_n}|T_n(s)f|^2 dx\right)e^{sK}\\ \ge& \frac{2\kappa_0}{C_0}(\psi_n(s))^{1+\frac{2}{N}}\|e^{sK/2}T_n(s)f\|^{-\frac{4}{N}}_{L^1(B_n)}\\ \ge& \frac{2\kappa_0}{C_0}(\psi_n(s))^{1+\frac{2}{N}}\|f\|^{-\frac{4}{N}}_{L^1(B_n)}e^{\frac{2K}{N}s}\end{aligned}$$

or, equivalently,

$$\frac{d}{ds}[(\psi_n(s))^{-\frac{2}{N}}] \ge \frac{4\kappa_0}{NC_0}\|f\|^{-\frac{4}{N}}_{L^1(B_n)}e^{\frac{2Ks}{N}}, \qquad s>0.$$

Integrating this inequality from 0 to $t>0$, we get (8.1.4).

Next, we consider the adjoint semigroup $\{T_n(t)^*\}$, which is generated by the realization in $C_b(B_n)$ of the adjoint of the operator $\mathcal{A}$ with Dirichlet boundary conditions. Arguing as above and taking into account that $\|T_n(t)^*\|_{L(L^1(B_n))} \le 1$ for any $t>0$ (see the proof of Theorem 7.1.2), we can show that

$$\|T_n(t)^*\|_{L(L^1(B_n),L^2(B_n))} \le \left(\frac{NC_0}{4\kappa_0}\right)^{\frac{N}{4}}\begin{cases} t^{-\frac{N}{4}}, & \text{if } K=0,\\[1ex] e^{-\frac{K}{2}t}\left(\dfrac{N}{2K}\right)^{-\frac{N}{4}}(1-e^{-\frac{2Kt}{N}})^{-\frac{N}{4}}, & \text{if } K\neq 0.\end{cases}$$

By duality we thus conclude that $T_n(t) \in L(L^2(B_n), L^\infty(B_n))$ and

$$\|T_n(t)\|_{L(L^2(B_n),L^\infty(B_n))} \le \left(\frac{NC_0}{4\kappa_0}\right)^{\frac{N}{4}} \begin{cases} t^{-\frac{N}{4}}, & \text{if } K=0,\\ e^{-\frac{K}{2}t}\left(\frac{N}{2K}\right)^{-\frac{N}{4}}\left(1-e^{-\frac{2Kt}{N}}\right)^{-\frac{N}{4}}, & \text{if } K\neq 0, \end{cases} \tag{8.1.5}$$

for any $t > 0$.

Splitting $\|T_n(t)\|_{L(L^1(B_n),L^\infty(B_n))} \le \|T_n(t/2)\|_{L(L^1(B_n),L^2(B_n))}\|T_n(t/2)\|_{L(L^2(B_n),L^\infty(B_n))}$ for any $t > 0$ and using (8.1.4) and (8.1.5), we get

$$\|T_n(t)\|_{L(L^1(B_n),L^\infty(B_n))} \le \left(\frac{NC_0}{2\kappa_0}\right)^{\frac{N}{2}} \begin{cases} t^{-\frac{N}{2}}, & \text{if } K=0,\\ e^{-\frac{K}{2}t}\left(\frac{|K|}{N}\right)^{\frac{N}{2}}\left(\cosh(N^{-1}Kt)-1\right)^{-\frac{N}{4}}, & \text{if } K\neq 0. \end{cases} \tag{8.1.6}$$

Since

$$(T(t)f)(x) = \int_{B_n} g_n(t,x,y)f(y)dy, \qquad t>0,\ \ x\in\mathbb{R}^N,\ \ f\in C_b(B_n),$$

from (8.1.6) it follows that g_n satisfies estimate (8.1.1).

(ii) Let us check the formula (8.1.2). For this purpose we consider the cases $K < 0$ and $K > 0$ separately. In the first case, using the semigroup property, we estimate

$$\|T_n(t)\|_{L(L^1(B_n),L^\infty(B_n))} \le \|T_n(1/|K|)\|_{L(L^1(B_n),L^2(B_n))}\|T_n(t-1/|K|)\|_{L(L^2(B_n),L^\infty(B_n))}$$

for any $t > 2/|K|$. On the other hand, if $K > 0$, then we estimate

$$\|T_n(t)\|_{L(L^1(B_n),L^\infty(B_n))} \le \|T_n(t-1/K)\|_{L(L^1(B_n),L^2(B_n))}\|T_n(1/K)\|_{L(L^2(\mathbb{R}^N),L^\infty(\mathbb{R}^N))},$$

still for any $t > 2/K$. Combining these estimates with (8.1.4) and (8.1.5) and then letting n tend to $+\infty$, the formula (8.1.2) follows at once.

(iii) Fix $x_0 \in \mathbb{R}^N$. Throughout this step by C we denote a positive constant, independent of $t \in (0,1]$ and x_0, which may vary from line to line.

Let $\{S_{x_0}(t)\}$ be the semigroup associated with the realization of the operator $\mathcal{A}$ in $C_b(B_1(x_0))$ with homogeneous Dirichlet boundary conditions on $\partial B_1(x_0)$ and denote by $\widetilde{g}_{x_0}$ its Green function. As is easily seen, $g \ge \widetilde{g}_{x_0}$ in $(0,+\infty)\times B_1(x_0)\times B_1(x_0)$. Indeed, for any nonnegative function $f \in C_c(B_1(x_0))$, $T(\cdot)f$ is strictly positive in $(0,+\infty)\times\mathbb{R}^N$. Hence, the function $w = T(\cdot)f - S_{x_0}f$, which satisfies the equation $D_tw - \mathcal{A}w = 0$ in $(0,+\infty)\times B_1(x_0)$, is nonnegative on $(0,+\infty)\times\partial B_1(x_0)$ and vanishes on $\{0\}\times B_1(x_0)$. By the maximum principle $T(\cdot)f \ge S_{x_0}(\cdot)f$, i.e.,

$$\int_{\mathbb{R}^N} g(t,x,y)f(y)dy \ge \int_{B_1(x_0)} \widetilde{g}_{x_0}(t,x,y)f(y)dy, \qquad t>0,\ \ x\in B_1(x_0),$$

and, as a byproduct, $g \ge \widetilde{g}_{x_0}$ in $(0,+\infty)\times B_1(x_0)\times B_1(x_0)$.

Let $\{R_{x_0}(t)\}$ be the semigroup in $C_b(B_1(x_0))$ associated with the principal part of the operator $\mathcal{A}$ in $C_b(B_n)$, with homogeneous Dirichlet boundary conditions. Since the infinitesimal generators of the semigroups $\{R_{x_0}(t)\}$ and $\{S_{x_0}(t)\}$ have the same domain (see Theorem D.3.5), we can apply the variation-of-constants formula to obtain

$$S_{x_0}(t)f - R_{x_0}(t)f = \int_0^t \frac{d}{ds}(S_{x_0}(s)R_{x_0}(t-s)f)ds$$

$$
\begin{aligned}
&= \int_0^t S_{x_0}(t-s)\langle b, \nabla R_{x_0}(s)f\rangle ds \\
&= \int_0^{t/2} S_{x_0}(t-s)\langle b, \nabla R_{x_0}(s)f\rangle ds + \int_{t/2}^t S_{x_0}(t-s)\langle b, \nabla R_{x_0}(s)f\rangle ds \\
&=: I_1(t,\cdot) + I_2(t,\cdot)
\end{aligned} \tag{8.1.7}
$$

for any $f \in C_b(B_1(x_0))$.

Fix $p \in (1,+\infty)$ such that $N(p-1) < p$. The proof of property (i) shows that $S_{x_0}(t) \in L(L^1(B_1(x_0)), L^\infty(B_1(x_0)))$ and $\|S_{x_0}(t)\|_{L(L^1(B_1(x_0)),L^\infty(B_1(x_0)))} \le Ct^{-N/2}$ for any $t \in (0,1]$. Since the operator $S_{x_0}(t)$ is a contraction in $L^\infty(B_1(x_0))$ for any $t > 0$, from applying the Riesz-Thorin theorem we deduce that $\|S_{x_0}(s)\|_{L(L^q(B_1(x_0)),L^\infty(B_1(x_0)))} \le C_q t^{-N/(2q)}$ for any $t \in (0,1]$ and $q \in (1,+\infty)$. Similarly,

$$
\sup_{t\in(0,1]} \|R_{x_0}(t)\|_{L(L^1(B_1(x_0)))} < +\infty, \qquad \sup_{t\in(0,1]} t^{\frac{N}{2}} \|R_{x_0}(t)\|_{L(L^1(B_1(x_0)),L^\infty(B_1(x_0)))} < +\infty
$$

and the Riesz-Thorin theorem implies that

$$
\|R_{x_0}(t)\|_{L(L^1(B_1(x_0)),L^p(B_1(x_0)))} \le Ct^{-\frac{N}{2}(1-\frac{1}{p})}, \qquad t \in (0,1]. \tag{8.1.8}
$$

Finally, the general theory of analytic semigroups (see Theorem C.2.2(iii)) shows that $\|\mathrm{div}(Q\nabla R_{x_0}(t))\|_{L(L^p(B_1(x_0)))} \le Ct^{-1}$ for any $t \in (0,1]$. This and the well-known estimate $\|\nabla u\|_{L^p(B_1(x_0))} \le C(\|u\|_{L^p(B_1(x_0))} + \|\mathrm{div}(Q\nabla u)\|_{L^p(B_1(x_0))})$, which is satisfied by any function $u \in W^{2,p}(B_1(x_0)) \cap W_0^{1,p}(B_1(x_0))$, yield

$$
\|\nabla R_{x_0}(t)f\|_{L^p(B_1(x_0))} \le Ct^{-\frac{1}{2}}\|f\|_{L^p(B_1(x_0))}, \qquad t \in (0,1]. \tag{8.1.9}
$$

We can thus estimate

$$
\begin{aligned}
&\|S_{x_0}(t-s)\langle b, \nabla R_{x_0}(s)f\rangle\|_{L^\infty(B_1(x_0))} \\
\le&\|S_{x_0}(t-s)\|_{L(L^p(B_1(x_0)),L^\infty(B_1(x_0)))}\|b\|_{L^\infty(B_1(x_0))}\|\nabla R_{x_0}(s)f\|_{L^p(B_1(x_0))} \\
\le&C(t-s)^{-\frac{N}{2p}}\|b\|_{L^\infty(B_1(x_0))}\|\nabla R_{x_0}(s/2)\|_{L(L^p(B_1(x_0)))} \\
&\qquad \times \|R_{x_0}(s/2)\|_{L(L^1(B_1(x_0)),L^p(B_1(x_0)))}\|f\|_{L^1(B_1(x_0))} \\
\le&C\|b\|_{L^\infty(B_1(x_0))}(t-s)^{-\frac{N}{2p}}s^{-\frac{1}{2}-\frac{N}{2}(1-\frac{1}{p})}\|f\|_{L^1(B_1(x_0))}
\end{aligned}
$$

and, by the choice of p, we conclude that

$$
\|I_1(t)\|_{L^\infty(B_1(x_0))} \le C\|b\|_{L^\infty(B_1(x_0))}\|f\|_{L^1(B_1(x_0))}t^{\frac{1-N}{2}}, \qquad t \in (0,1]. \tag{8.1.10}
$$

Now, let us estimate the term $I_2(t)$ in (8.1.7). As a first step, we observe that, by duality and the estimate $\|S_{x_0}(r)\|_{L(W^{1,p}(B_1(x_0)),L^p(B_1(x_0)))} \le Cr^{-1/2}$ for any $r \in (0,1]$, we get

$$
\begin{aligned}
\|S_{x_0}(r)\langle b, \nabla\rangle\|_{L(L^{p'}(B_1(x_0)))} =&\|\mathrm{div}(bS_{x_0}(r))\|_{L(L^p(B_1(x_0)))} \\
\le&\|b\|_{W^{1,\infty}(B_1(x_0))}\|S_{x_0}(r)\|_{L(W^{1,p}(B_1(x_0)),L^p(B_1(x_0)))} \\
\le&C\|b\|_{W^{1,\infty}(B_1(x_0))}r^{-\frac{1}{2}}
\end{aligned}
$$

for any $r \in (0,1]$. Hence, splitting $S_{x_0}(t-s)\langle b, \nabla\rangle = S_{x_0}((t-s)/2)S_{x_0}((t-s)/2)\langle b, \nabla\rangle$ for any $0 < s < t < 1$, from (8.1.8), (8.1.9) and the previous estimate we conclude that

$$
\|S_{x_0}(t-s)\langle b, \nabla R_{x_0}(s)f\rangle\|_{L^\infty(B_1(x_0))}
$$

$$
\begin{aligned}
\leq&\|S_{x_0}((t-s)/2)\|_{L(L^{p'}(B_1(x_0)),L^\infty(B_1(x_0)))}\|S_{x_0}((t-s)/2)\langle b,\nabla\rangle\|_{L(L^{p'}(B_1(x_0)))}\\
&\times\|R_{x_0}f\|_{L(L^1(B_1(x_0)),L^{p'}(B_1(x_0)))}\\
\leq&C(t-s)^{-\frac{N}{2q}-\frac{1}{2}}s^{-\frac{N}{2}(1-\frac{1}{q})}\|b\|_{W^{1,\infty}(B_1(x_0))}\|f\|_{L^1(B_1(x_0))}.
\end{aligned}
$$

This gives for any $t\in(0,1]$

$$
\|I_2(t)\|_{L^\infty(B_1(x_0))}\leq Ct^{\frac{1-N}{2}}\|b\|_{W^{1,\infty}(B_1(x_0))}\|f\|_{L^1(B_1(x_0))}. \tag{8.1.11}
$$

Combining (8.1.10) and (8.1.11) we deduce that

$$
\|S_{x_0}(t)-R_{x_0}(t)\|_{L(L^1(B_1(x_0)),L^\infty(B_1(x_0)))}\leq Ct^{\frac{1-N}{2}}\|b\|_{W^{1,\infty}(B_1(x_0))},\qquad t\in(0,1].
$$

It thus follows that $|\widetilde{g}_{x_0}(t,x,x)-\widetilde{g}_{x_0,0}(t,x,x)|\leq Ct^{\frac{1-N}{2}}$ for $x\in B_1(x_0)$. Now, we are almost done. Indeed, by classical results (see e.g. [84, Lemma 3.3.3]) there exists a positive constant C, independent of $x\in B_1(x_0)$, such that $\widetilde{g}_{x_0,0}(t,x,x)\geq Ct^{-N/2}$ for $t\in(0,1]$. These last two estimates yield $g(t,x,x)\geq\widetilde{g}_{x_0}(t,x,x)\geq C_{x_0}t^{-N/2}$ for $t\in(0,1]$ and $x\in B_1(x_0)$. A covering argument allows us to complete the proof. □

Now, we consider operators $\mathcal{A}=\mathrm{Tr}(QD^2)+\langle b,\nabla\rangle+c$ with a not identically vanishing potential c under the following conditions on their coefficients.

Hypotheses 8.1.4 (i) *The diffusion coefficients $q_{ij}=q_{ji}$ and the drift coefficients b_j $(i,j=1,\ldots,N)$ belong to $C^1_b(\mathbb{R}^N)$ and $C^1(\mathbb{R}^N)$, respectively. Moreover, $c\in C(\mathbb{R}^N)$ is nonpositive in $\mathbb{R}^N$;*

(ii) *there exists a positive constant κ_0 such that $\langle Q(x)\xi,\xi\rangle\geq\kappa_0|\xi|^2$ for any $x,\xi\in\mathbb{R}^N$;*

(iii) *there exist constants $k_1,k_2\geq 0$, $\alpha\in[1/2,1)$ and $\beta\in(0,1)$ such that $|b|\leq k_1|c|^\alpha+k_2$ and $\mathrm{div}\,b\geq\beta c$ in $\mathbb{R}^N$.*

Under this set of assumptions we can prove the following so called *pseudo-Gaussian estimates.*

Theorem 8.1.5 ([20]) *The Green function of the minimal semigroup associated in $C_b(\mathbb{R}^N)$ with the operator $\mathcal{A}$ satisfies the estimate*

$$
g(t,x,y)\leq\widehat{C}_1e^{\omega t}t^{-\frac{N}{2}}\exp\left(-\frac{\widehat{C}_2|x-y|^{\frac{1}{\alpha}}}{t^{\frac{1-\alpha}{\alpha}}}\right) \tag{8.1.12}
$$

for $t>0$, $x,y\in\mathbb{R}^N$ and some positive constants ω, $\widehat{C}_1$ and $\widehat{C}_2$.

Proof The proof of (8.1.12) is based on the so-called *Davies' trick.* For this purpose, we introduce the set $\mathscr{E}$ of all functions $\zeta\in C_b(\mathbb{R}^N)\cap C^\infty(\mathbb{R}^N)$ such that $\|D_j\zeta\|_\infty\leq 1$ and $\|D_{ij}\zeta\|_\infty\leq 1$ for any $i,j=1,\ldots,N$. Moreover, for any $a\in\mathbb{R}$ and $\zeta\in\mathscr{E}$, we introduce the semigroup $\{T_a(t)\}$ on $L^2(\mathbb{R}^N)$, defined by $T_a(t)f=e^{-a\zeta}T(t)e^{a\zeta}f$ for any $f\in L^2(\mathbb{R}^N)$ and $t>0$. Clearly, $\{T_a(t)\}$ admits the function $g_{a,\zeta}$, defined by $e^{-a\zeta(x)}g(t,x,y)e^{a\zeta(y)}$ for any $t>0$ and $x,y\in\mathbb{R}^N$, as Green function. A straightforward computation reveals that the semigroup $\{T_{a,\zeta}(t)\}$ is associated in $C_b(\mathbb{R}^N)$ with the operator $\mathcal{A}_{a,\zeta}$ defined by $\mathcal{A}_{a,\zeta}\psi=\mathcal{A}\psi+a\langle Q\nabla\zeta,\nabla\psi\rangle+a(\mathcal{A}\zeta-c\zeta)\psi=:\mathrm{div}(Q\nabla\psi)+\langle b_{a,\zeta},\nabla\psi\rangle+c_{a,\zeta}\psi$ for any smooth function $\psi:\mathbb{R}^N\to\mathbb{R}$. Note that

$$
\mathrm{div}(b_{a,\zeta})=\mathrm{div}\,b+a\mathrm{Tr}(QD_{ij}\zeta)+a\sum_{i,j=1}^N D_iq_{ij}D_j\zeta
$$

$$\geq \beta c - |a| \sum_{i,j=1}^{N} \|q_{ij}\|_\infty - |a| \sum_{i,j=1}^{N} \|D_i q_{ij}\|_\infty. \tag{8.1.13}$$

Due to Hypothesis 8.1.4(iii) we can estimate

$$\begin{aligned} c_{a,\zeta} \leq& c + |a| \sum_{i,j=1}^{N} \|q_{ij}\|_\infty + |a| \sum_{j=1}^{N} \|b_j\|_\infty \\ \leq& c + |a| \sum_{i,j=1}^{N} \|q_{ij}\|_\infty + |a|\sqrt{N}(k_1|c|^\alpha + k_2). \end{aligned} \tag{8.1.14}$$

Applying the well known estimate $ab \leq a^p/p + b^q/q$ (which holds true for any $a, b \geq 0$ and $p, q \geq 1$ such that $1/p + 1/q = 1$) with $a = \varepsilon^\alpha |c|^\alpha$, $b = \varepsilon^{-\alpha} k_1 \sqrt{N} a$, $p = 1/\alpha$, $q = 1/(1-\alpha)$, we can estimate

$$k_1\sqrt{N}|a||c|^\alpha \leq \varepsilon|c| + \varepsilon^{\frac{1}{\alpha(1-\alpha)}} k_1^{\frac{1}{\alpha-1}} N^{\frac{1}{2(\alpha-1)}}$$

for any $\varepsilon > 0$. Replacing this estimate in (8.1.14) we conclude that for any $\varepsilon > 0$ there exists a positive constant $K_\varepsilon > 0$ such that

$$c_{a,\zeta} \leq (1-\varepsilon)c + (|a| + |a|^{\frac{1}{1-\alpha}})K_\varepsilon \tag{8.1.15}$$

for any $\zeta \in \mathscr{E}$. From (8.1.13) and (8.1.15), where we take $\varepsilon = 1 - \beta$, it follows that

$$\operatorname{div} b_{a,\zeta} \geq c_{a,\zeta} - \widehat{K}(1 + |a|^{\frac{1}{1-\alpha}}) \tag{8.1.16}$$

in $\mathbb{R}^N$, for some positive constant $\widehat{K}$ independent of $\zeta \in \mathscr{E}$.

We now consider the operator $\widetilde{\mathcal{A}}_{a,\zeta} = \mathcal{A}_{a,\zeta} - \widehat{K}(1 + |a|^{\frac{1}{1-\alpha}})$ and the semigroup $\{\widetilde{T}_{a,\zeta}(t)\}$ associated with it in $C_b(\mathbb{R}^N)$. We denote by $\widetilde{c}_{a,\zeta}$ the potential term of the operator $\widetilde{\mathcal{A}}_{a,\zeta}$. Arguing as in the proof of Theorem 8.1.2(i), we can show that the semigroup $\{\widetilde{T}_{a,\zeta}(t)\}$ maps $L^1(\mathbb{R}^N)$ into $L^\infty(\mathbb{R}^N)$. We do not enter too deeply into detail, referring the readers to the proof of the quoted theorem for all the missing details. Here, we just notice that the condition (8.1.16) and an integration by parts reveal that

$$\int_{B_n} u\widetilde{\mathcal{A}}_{a,\zeta} u dx = -\int_{B_n} |\sqrt{Q}\nabla u_n|^2 dx + \int_{B_n} \left(c - \frac{1}{2}\operatorname{div} b\right)|u|^2 dx \leq -\int_{B_n} |\sqrt{Q}\nabla u|^2 dx,$$

i.e.,

$$\|\nabla u\|_{L^2(B_n)}^2 \leq -\int_{B_n} u\widetilde{\mathcal{A}}_{a,\zeta} u dx, \qquad u \in W^{2,2}(B_n) \cap W_0^{1,2}(B_n).$$

This estimate and the Nash inequality show that the semigroup $\{\widetilde{T}_{n,a,\zeta}(t)\}$, associated in $C_b(B_n)$ with the operator $\mathcal{A}_{a,\zeta}$ with homogeneous Dirichlet boundary conditions, maps $L^1(B_n)$ into $L^2(B_n)$ and $\|\widetilde{T}_{n,a,\zeta}(t)\|_{L(L^1(\mathbb{R}^N), L^2(\mathbb{R}^N))} \leq C_N t^{-N/4}$ for any $t > 0$ and some positive constant C_N, independent of t and n. The same arguments can be applied to the adjoint semigroup $\{\widetilde{T}_{n,a,\zeta}(t)\}$ so that, by duality, we can infer that $\|\widetilde{T}_{n,a,\zeta}(t)\|_{L(L^2(\mathbb{R}^N), L^\infty(\mathbb{R}^N))} \leq C_N t^{-N/4}$ for any $t > 0$. Splitting $\widetilde{T}_{n,a,\zeta}(t) = \widetilde{T}_{n,a,\zeta}(t/2)\widetilde{T}_{n,a,\zeta}(t/2)$ and using the previous two estimates, we deduce that each operator $\widetilde{T}_{n,a,\zeta}(t)$ maps $L^1(B_n)$ into $L^\infty(B_n)$ and

$\|\widetilde{T}_{n,a,\zeta}(t)\|_{L(L^1(B_n),L^\infty(B_n))} \le 2^{N/2}C_N^2 t^{-N/2}$ for any $t>0$. Letting n tend to $+\infty$ we conclude that $\widetilde{T}_{a,\zeta}(t) \in L(L^1(\mathbb{R}^N), L^\infty(\mathbb{R}^N))$ and $\|\widetilde{T}_{a,\zeta}(t)\|_{L(L^1(\mathbb{R}^N),L^\infty(\mathbb{R}^N))} \le 2^{N/2}C_N^2 t^{-N/2}$ for any $t>0$. As a byproduct, it follows that $T_{a,\zeta}(t) \in L(L^1(\mathbb{R}^N), L^\infty(\mathbb{R}^N))$ and

$$\|T_{a,\zeta}(t)\|_{L(L^1(\mathbb{R}^N),L^\infty(\mathbb{R}^N))} \le 2^{\frac{N}{2}} C_N^2 t^{-\frac{N}{2}} e^{\widehat{K}(1+|a|^{\frac{1}{1-\alpha}})t}, \qquad t>0.$$

Hence,

$$e^{-a\zeta(x)} g(t,x,y) e^{a\zeta(y)} \le 2^{\frac{N}{2}} C_N^2 t^{-\frac{N}{2}} e^{\widehat{K}(1+|a|^{\frac{1}{1-\alpha}})t}, \qquad t>0,\ x,y\in\mathbb{R}^N,$$

i.e.,

$$g(t,x,y) \le 2^{\frac{N}{2}} C_N^2 t^{-\frac{N}{2}} e^{\widehat{K}(1+|a|^{\frac{1}{1-\alpha}})t} \inf_{\zeta\in\mathscr{E}} e^{-a|\zeta(y)-\zeta(x)|} = C_N' t^{-\frac{N}{2}} e^{\widehat{K}(1+|a|^{\frac{1}{1-\alpha}})t} e^{-a\widehat{d}(x,y)} \tag{8.1.17}$$

for any $t>0$ and $x,y\in\mathbb{R}^N$, where $\widehat{d}(x,y) = \sup\{|\zeta(x)-\zeta(y)| : \zeta\in\mathscr{E}\}$ for any $x,y\in\mathbb{R}^N$.

Now, let us prove that there exists a positive constant γ_1 such that

$$\widehat{d}(x,y) \ge \gamma_1|x-y|, \qquad x,y\in\mathbb{R}^N. \tag{8.1.18}$$

From this estimate and (8.1.17) it will follow that

$$g(t,x,y) \le C_N' t^{-\frac{N}{2}} e^{\widehat{K}(1+|a|^{\frac{1}{1-\alpha}})t} e^{-a\gamma_1|x-y|}, \qquad t>0,\ x,y\in\mathbb{R}^N. \tag{8.1.19}$$

Since $a\in\mathbb{R}$ is arbitrary fixed, minimizing the right-hand side of (8.1.19) with respect to $a\in\mathbb{R}$, i.e., taking $a = (\gamma_1|x-y|(1-\alpha)/(\widehat{K}t))^{(1-\alpha)/\alpha}$, we conclude the proof.

So, let us prove (8.1.18). Clearly it suffices to prove that

$$\widehat{d}(x,0) = \sup\{|\zeta(x)-\zeta(0)| : \zeta\in\mathscr{E}\} \ge \gamma_1|x|, \qquad x\in\mathbb{R}^N. \tag{8.1.20}$$

We follow the same procedure as in [228, pp. 200-202]. We fix a nonnegative smooth function $\vartheta\in C_c^\infty(B_1)$ such that $\|\vartheta\|_{L^1(\mathbb{R}^N)} = 1$. Further, we set $\lambda = 6\int_{\mathbb{R}^N}|y|\vartheta(y)dy$ and introduce the function $\chi:\mathbb{R}^N\to\mathbb{R}$ defined by

$$\chi(x) = \int_{\mathbb{R}^N} (|y-x|+|y|)\vartheta(y)dy, \qquad x\in\mathbb{R}^N.$$

The function χ is smooth in $\mathbb{R}^N$ and the triangle inequality shows that $|x|\le\chi(x)\le|x|+\lambda/3$ for any $x\in\mathbb{R}^N$. Next, for any $\varepsilon>0$ we introduce the function $\zeta_\varepsilon = \chi/(1+\varepsilon\chi)$, which satisfies the condition $\zeta_\varepsilon(0) = \frac{\lambda}{3+\varepsilon\lambda}$. Moreover, the above estimates on χ reveal that

$$\frac{|x|}{1+\varepsilon|x|} \le \zeta_\varepsilon(x) \le \frac{3|x|+\lambda}{3+\varepsilon(3|x|+\lambda)}, \qquad x\in\mathbb{R}^N.$$

Hence,

$$\begin{aligned}\zeta_\varepsilon(x)-\zeta_\varepsilon(0) \ge& \frac{|x|}{1+\varepsilon|x|} - \frac{\lambda}{3+\varepsilon\lambda} = \frac{3|x|-\lambda}{(1+\varepsilon|x|)(3+\lambda\varepsilon)}\\ \ge& \frac{|x|}{(1+\varepsilon|x|)(3+2\varepsilon|x|)} \ge \frac{|x|}{3(1+\varepsilon|x|)^2}\end{aligned}$$

for any $x\notin B_{\lambda/2}$. A straightforward computation shows that the first- and second-order derivatives of ζ_ε are bounded in $\mathbb{R}^N$, uniformly with respect to $\varepsilon>0$. Let M_1 be any positive constant such that $\|D_j\zeta_\varepsilon\|_\infty+\|D_{ij}\zeta_\varepsilon\|_\infty \le M_1$ for any $\varepsilon \in (0,1]$ and $i,j=1,\ldots,N$. If $x \notin B_{\lambda/2}$ then we can estimate

$$\sup\{|\zeta(x)-\zeta(0)| : \zeta \in \mathscr{E}\} \ge \sup_{\varepsilon\in(0,1]} \frac{|\zeta_\varepsilon(x)-\zeta_\varepsilon(0)|}{M_1} \ge \sup_{\varepsilon\in(0,1]} \frac{|x|}{3M_1(1+\varepsilon|x|)^2} = \frac{|x|}{3M_1}.$$

Finally, let $h_\lambda : \mathbb{R}^N \to \mathbb{R}$ be the function defined by $h_\lambda(x) = x$, if $x \in B_\lambda$, and $h_\lambda \equiv 0$ otherwise in $\mathbb{R}^N$. Denote by $\widetilde{\zeta}$ the convolution of h_λ with a function $\vartheta \in C^\infty_c(\mathbb{R}^N)$ such that $\chi_{B_{n_0}} \le \vartheta \le \chi_{B_{2n_0}}$ where n_0 is any integer greater than $2/\lambda$. Of course, $\widetilde{\zeta}(x) = x$ for any $x \in B_{\lambda/2}$. Hence, if we set $\zeta = \widetilde{\zeta}/K_0$, where K_0 is any positive constant such that $\|D_j\widetilde{\zeta}\|_\infty + \|D_{ij}\widetilde{\zeta}\|_\infty \le M_2$ for any $\varepsilon \in (0,1]$ and $i,j=1,\ldots,N$, then we can infer that $\widehat{d}(x,0) \ge |x|/M_2$. Summing up, (8.1.20) follows with $\gamma_1 = (3M_1)^{-1} \wedge M_2^{-1}$. □

8.1.1 Bibliographic remarks

For other estimates from above of the Green function g, with a different set of assumptions, we refer the reader to the papers [195], by G. Metafune. D. Pallara and A. Rhandi, and [239], by C. Spina, which deal with operators in divergence form with no potential term, assuming that the diffusion coefficients are bounded in $\mathbb{R}^N$. The arguments used in these papers are similar to that in Section 8.3 and in Chapter 15 (which deals with the non-autonomous case[2]) but the hypotheses assumed by the authors are slightly different from the ones that we assume in this book. Both of the papers mentioned use Lyapunov functions, which are independent of t in the first paper and depend on t in the second paper. Moreover, the estimates of the function g are written in terms of the Lyapunov functions and are obtained proving sharp estimates in the Krylov spaces $\mathscr{H}^{p,1}((a,b)\times\mathbb{R}^N)$ for weak solutions to parabolic equations. These techniques have been used previously in [41, 42] to prove regularity results for solutions (in distributional sense) of parabolic equations in $(a,b)\times\mathbb{R}^N$, subject to suitable initial conditions. In the particular case when $b(x) = -|x|^{r-1}x$ for any $x \in \mathbb{R}^N$ and some $r > 0$, the use of time dependent Lyapunov functions allows one to prove a better estimate for the behaviour as t tends to 0^+ of the associated Green function g. More precisely, the results in [195] show that for any $\gamma < (r+1)^{-1}$ there exist two positive constants C_1 and C_2 such that

$$g(t,x,y) \le C_1 \exp\left(C_2 t^{-\frac{r+1}{r-1}}\right) e^{-\gamma|y|^{r+1}}, \qquad t \in (0,1],\ \ x,y, \in \mathbb{R}^N. \tag{8.1.21}$$

On the other hand, the estimate proved in [239] shows that

$$g(t,x,y) \le C_1 t^{-\delta} e^{-\gamma t^\alpha |y|^{r+1}}, \qquad t \in (0,1],\ \ x,y, \in \mathbb{R}^N, \tag{8.1.22}$$

for γ, C_1 and C_2 as above and some constants $\alpha, \delta > 0$. Clearly, the behaviour with respect to y in (8.1.21) is much better than the behaviour in (8.1.22). On the other hand, for any fixed $x, y \in \mathbb{R}^N$, estimate (8.1.22) shows that $g(t,x,y)$ blows up at most polynomially, whereas (8.1.21) provides us with an at most exponential blow up of the function $g(\cdot,x,y)$ at the origin.

We stress that in [195] also pointwise estimates for the first-order time derivative and first- and second-order spatial derivatives of the function g are provided (see Section 5 of the quoted paper).

We also quote [8] where, as in [239], time dependent Lyapunov functions have been used

[2]But, of course, Theorem 15.2.2, which also covers some situations in which the diffusion coefficients are unbounded, can also be applied in the autonomous case.

to prove estimates for Green functions associated with elliptic operators with unbounded coefficients.

Finally, for short time estimates on g we also refer the reader to [169, 262] which deal with elliptic operators in open sets.

8.2 The operator $\mathcal{A} = (1 + |x|^\alpha)\Delta$

Here, we present the results in [204]. Throughout this section, we set $q(x) = 1 + |x|^\alpha$ for any $x \in \mathbb{R}^N$ and some $\alpha > 0$ and assume that $N \geq 3$.

Under the previous conditions, the maximal realization A_p of the operator $\mathcal{A}$ in $L^p(\mathbb{R}^N)$, with domain $D(A_p) = \{u \in L^p(\mathbb{R}^N) \cap W^{2,p}_{\rm loc}(\mathbb{R}^N) : \ \mathcal{A}u \in L^p(\mathbb{R}^N)\}$, generates a strongly continuous analytic semigroup $\{T_p(t)\}$ (see Theorem 7.3.7).

To begin with, we introduce first the positive measure μ, whose density with respect to the Lebesgue measure is the function $1/q$, and, then, the corresponding L^2-space on $\mathbb{R}^N$, which we denote by $L^2(\mathbb{R}^N, \mu)$. Clearly, μ is a finite measure if and only if $\alpha > N$. We also introduce the Sobolev space $\mathscr{H}$ consisting of all complex-valued functions[3] $u \in L^2(\mathbb{R}^N, \mu)$ whose distributional gradient belongs to $(L^2(\mathbb{R}^N))^N$. We endow $\mathscr{H}$ with the inner product

$$\langle u, v\rangle_{\mathscr{H}} = \int_{\mathbb{R}^N} u\overline{v}d\mu + \int_{\mathbb{R}^N} \langle \nabla u, \nabla \overline{v}\rangle dx, \qquad u, v \in \mathscr{H}.$$

We finally denote by $\mathscr{V}$ the closure of $C^1_c(\mathbb{R}^N)$ in $\mathscr{H}$ with respect to the norm induced by the above inner product.

Let $\mathfrak{a} : \mathscr{V} \times \mathscr{V} \to \mathbb{C}$ be the form defined by

$$\mathfrak{a}(u, v) = \int_{\mathbb{R}^N} \langle \nabla u, \nabla \overline{v}\rangle dx, \qquad u, v \in \mathscr{V},$$

and let A be the operator associated with the form $\mathfrak{a}$, i.e., $u \in D(A)$ if and only if there exists $f \in L^2(\mathbb{R}^N, \mu)$ such that $\mathfrak{a}(u, v) = -\int_{\mathbb{R}^N} f\overline{v}d\mu$ for any $v \in \mathscr{V}$ and $Au = f$.

The following lemma summarizes the main properties of the operator A that we need in what follows.

Lemma 8.2.1 *The operator A is selfadjoint in $L^2(\mathbb{R}^N, \mu)$. Hence, it generates a strongly continuous analytic semigroup $\{\widetilde{T}(t)\}$ in $L^2(\mathbb{R}^N, \mu)$. Such a semigroup is positive and L^∞-contractive and $D(A)$ is contained in the set $\{u \in \mathscr{V} \cap W^{2,2}(\mathbb{R}^N) : q\Delta u \in L^2(\mathbb{R}^N, \mu)\}$. Finally, $\widetilde{T}(t)f = T_p(t)f$ for any $t > 0$, $p \in [1, +\infty]$ and $f \in L^2(\mathbb{R}^N, \mu) \cap L^p(\mathbb{R}^N)$, where $T_\infty(\cdot) = T(\cdot)$.*

Proof As is immediately seen, $\mathfrak{a}(u, u) \geq 0$ and $\mathfrak{a}(u, v) = \overline{\mathfrak{a}(v, u)}$ for any $u, v \in \mathscr{V}$. Hence, it follows immediately that the operator A is selfadjoint. It is also dissipative. Indeed,

$$\int_{\mathbb{R}^N} (\lambda u - Au)\overline{u}dx = \lambda\|u\|^2_{L^2(\mathbb{R}^N,\mu)} + \mathfrak{a}(u, u) \geq \lambda\|u\|^2_{L^2(\mathbb{R}^N,\mu)}, \qquad \lambda > 0,$$

which shows that $\|\lambda u - Au\|_{L^2(\mathbb{R}^N,\mu)} \geq \lambda\|u\|_{L^2(\mathbb{R}^N,\mu)}$. By Theorem (C.2.10), A generates an

---

[3] To avoid cumbersome notation, we still denote by $L^2(\mathbb{R}^N, \mu)$ the space of all complex-valued functions u such that $|u| \in L^2(\mathbb{R}^N, \mu)$.

analytic semigroup in $L^2(\mathbb{R}^N,\mu)$, which clearly is also strongly continuous since $C_c^\infty(\mathbb{R}^N)\subset D(A)$. It is easy to check that $D(A)\subset W^{2,2}(\mathbb{R}^N)$ and an integration by parts reveals that

$$\int_{\mathbb{R}^N}(q\Delta u)\overline{v}d\mu=\int_{\mathbb{R}^N}\overline{v}\Delta u dx=\int_{\mathbb{R}^N}f\overline{v}d\mu,\qquad v\in C_c^\infty(\mathbb{R}^N),$$

for any $u\in D(A)$. Hence, $q\Delta u$ belongs to $L^2(\mathbb{R}^N,\mu)$. It is also easy to check that $D(A_2)\subset D(A)$ and $Au=A_2u$ for any $u\in D(A_2)$. Hence, the functions $R(\lambda,A)f$ and $R(\lambda,A_2)f$ coincide for any $f\in L^2(\mathbb{R}^N)$ and $\lambda>0$. In view of the exponential formula (see Theorem C.1.7), it thus follows that

$$\widetilde{T}(t)f=\lim_{n\to+\infty}\left[\frac{n}{t}R\left(\frac{n}{t},A\right)\right]^n f=\lim_{n\to+\infty}\left[\frac{n}{t}R\left(\frac{n}{t},A_2\right)\right]^n f,$$
$$T_2(t)f=\lim_{n\to+\infty}\left[\frac{n}{t}R\left(\frac{n}{t},A_2\right)\right]^n f$$

for any $t>0$ and $f\in L^2(\mathbb{R}^N)$, where the limits are taken in $L^2(\mathbb{R}^N,\mu)$ and in $L^2(\mathbb{R}^N)$, respectively, and we conclude that $\widetilde{T}(t)f=T_2(t)f$. Next, recalling that the semigroups $\{T_p(t)\}$ and $\{T_2(t)\}$ coincide in $L^p(\mathbb{R}^N)\cap L^2(\mathbb{R}^N)$ (as it follows from Proposition 7.3.9), we deduce that $\widetilde{T}(t)f=T_p(t)f$ for any $t>0$, $f\in L^p(\mathbb{R}^N)\cap L^2(\mathbb{R}^N)$ and $p\in(1,+\infty)$. In particular, since $C_c^\infty(\mathbb{R}^N)$ is dense in $L^2(\mathbb{R}^N,\mu)\cap L^p(\mathbb{R}^N)$, as can be easily seen using the construction in the proof of Lemma 7.2.2, the semigroups $\{\widetilde{T}(t)\}$ and $\{T_p(t)\}$ coincide in $L^2(\mathbb{R}^N,\mu)\cap L^p(\mathbb{R}^N)$.

Proposition 7.3.9 shows also that $T_p(t)f=T(t)f$ for any $t>0$ and $f\in C_c^\infty(\mathbb{R}^N)$. Recalling that the semigroup $\{T(t)\}$ is order preserving (see Theorem 1.2.5), we conclude that, for any $t>0$, $\widetilde{T}(t)f$ is almost everywhere positive in $\mathbb{R}^N$ if $f\in C_c^\infty(\mathbb{R}^N)$ is nonnegative and it does not identically vanish. Since any nonnegative function $f\in L^2(\mathbb{R}^N,\mu)$ is the $L^2(\mathbb{R}^N,\mu)$-limit of a sequence of nonnegative functions which belong to $C_c^\infty(\mathbb{R}^N)$,[4] it follows that $\widetilde{T}(t)f\geq 0$ almost everywhere in $\mathbb{R}^N$, for any $t>0$. Finally, the above results also show that the semigroup $\{\widetilde{T}(t)\}$ maps $C_c^\infty(\mathbb{R}^N)$ into $L^\infty(\mathbb{R}^N)\cap L^2(\mathbb{R}^N,\mu)$. If $f\in L^2(\mathbb{R}^N,\mu)\cap L^\infty(\mathbb{R}^N)$ then we can take a sequence $\{f_n\}\subset C_c^\infty(\mathbb{R}^N)$ which converges to f in $L^2(\mathbb{R}^N,\mu)$, almost everywhere in $\mathbb{R}^N$ and such that $\sup_{n\in\mathbb{N}}\|f_n\|_\infty<+\infty$. Since $\|\widetilde{T}(t)f_n\|_\infty=\|T(t)f_n\|_\infty\leq\|f_n\|_\infty$ for any $n\in\mathbb{N}$, up to a subsequence we can assume that $\widetilde{T}(t)f_n$ converges almost everywhere in $\mathbb{R}^N$ to $\widetilde{T}(t)f$, as n tends to $+\infty$. Similarly, by Theorem 1.2.5, $T(t)f_n$ tends to $T(t)f$ pointwise in $\mathbb{R}^N$ for any $t>0$. We thus conclude that $\widetilde{T}(t)f=T(t)f$ almost everywhere in $\mathbb{R}^N$. Since the semigroup $\{T(t)\}$ is strong Feller and irreducible (see Proposition 1.2.13), $T(t)f$ is positive in $\mathbb{R}^N$ if $0\not\equiv f\in L^\infty(\mathbb{R}^N)$ is nonnegative almost everywhere in $\mathbb{R}^N$. We have so proved that the semigroup $\{\widetilde{T}(t)\}$ is L^∞-contractive and positive. This completes the proof. □

In view of Lemma 8.2.1 the semigroup $\{\widetilde{T}(t)\}$ is symmetric. Moreover, since $\widetilde{T}(\cdot)\equiv T(\cdot)$ on $C_c^\infty(\mathbb{R}^N)$, as has been shown in the proof of the previous lemma, taking Theorem 1.2.5 into account we can associate a Green function g with this semigroup. Clearly, we can also write

$$(\widetilde{T}(t)f)(x)=\int_{\mathbb{R}^N}g_\mu(t,x,y)f(y)d\mu$$

for any $f\in L^2(\mathbb{R}^N,\mu)$, any $t>0$ and almost every $x\in\mathbb{R}^N$, where $g_\mu(t,x,y)=q(y)g(t,x,y)$ for any $t>0$, $x,y\in\mathbb{R}^N$.

[4] It is enough to approximate first f by the sequence $\{f\chi_{B_n}\}$ and, then, each function $f\chi_{B_n}$ by convolution with a standard sequence of mollifiers.

Lemma 8.2.2 *Let $p \in (1, N)$, $q \in [p, p^*]$ satisfy*[5] *the conditions $N(q-p) = pq(1-\gamma+\beta)$ and $N + p(\gamma-1) \neq 0$ for some real constants γ and $\beta \in [\gamma-1, \gamma]$. Then, for any $\sigma \in \mathbb{R}$ there exists a positive constant C such that*

$$\|q^{\beta/\alpha}u\|_{L^q(\mathbb{R}^N)} \le C(\|q^{\gamma/\alpha}\nabla u\|_{L^p(\mathbb{R}^N)} + \|q^{\sigma/\alpha}u\|_{L^p(\mathbb{R}^N)}), \qquad u \in C_c^\infty(\mathbb{R}^N). \tag{8.2.1}$$

Proof We split $u = \eta u + (1-\eta)u$, where η is a smooth function such that $\chi_{B_1} \le \eta \le \chi_{B_2}$. Clearly, ηu satisfies estimate (8.2.1) as it can be checked applying the Sobolev embedding theorem. To show that $(1-\eta)u$ satisfies (8.2.1) as well, we prove that

$$\|\zeta_\beta v\|_{L^q(\mathbb{R}^N)} \le C\|\zeta_\gamma \nabla v\|_{L^p(\mathbb{R}^N)} \tag{8.2.2}$$

for any $v \in C_c^\infty(\mathbb{R}^N \setminus \{0\})$ and some positive constant C, where $\zeta_\delta(x) = |x|^\delta$ for any $\delta > 0$. Throughout the proof, C denotes a positive constant which may vary from line to line, but is independent of u and v.

Writing (8.2.2) with $v = (1-\eta)u$ and observing that $\nabla\eta$ is supported in B_2 we get

$$\begin{aligned}\|\zeta_\beta(1-\eta)u\|_{L^q(\mathbb{R}^N)} \le& C(\|\zeta_\gamma u\nabla\eta\|_{L^p(\mathbb{R}^N)} + \|\zeta_\gamma\nabla u\|_{L^p(\mathbb{R}^N)})\\ \le&(1+2^{\gamma-\sigma})C(\|\zeta_\sigma u\|_{L^p(\mathbb{R}^N)} + \|\zeta_\gamma\nabla u\|_{L^p(\mathbb{R}^N)}).\end{aligned}$$

We begin by proving (8.2.2) in the particular cases $(q,\beta) = (p, \gamma-1)$ and $(q,\beta) = (p^*, \gamma)$. The general case then will follow from these cases using the Hölder inequality.

To prove (8.2.2) with $(q,\beta) = (p,\gamma-1)$, we fix $v \in C_c^\infty(\mathbb{R}^N \setminus \{0\})$ and observe that, since $\mathrm{div}(x|x|^{(\gamma-1)p}) = (N + (\gamma-1)p)|x|^{(\gamma-1)p}$ for any $x \in \mathbb{R}^N \setminus \{0\}$, using the divergence theorem we can estimate

$$\begin{aligned}\|\zeta_{\gamma-1}v\|_{L^p(\mathbb{R}^N)}^p =& \int_{\mathbb{R}^N} \mathrm{div}\left(\frac{x|x|^{(\gamma-1)p}}{N+p(\gamma-1)}\right)|v(x)|^p dx\\ =& -\int_{\mathbb{R}^N} \frac{p|x|^{(\gamma-1)p}}{N+p(\gamma-1)} v(x)|v(x)|^{p-2}\langle x, \nabla v(x)\rangle dx\\ \le& \frac{p}{|N+p(\gamma-1)|}\int_{\mathbb{R}^N} (|x|^{(\gamma-1)}|v(x)|)^{p-1}|x|^\gamma|\nabla v(x)|dx\\ \le& \frac{p}{|N+p(\gamma-1)|}\|\zeta_{\gamma-1}v\|_{L^p(\mathbb{R}^N)}^{p-1}\|\zeta_\gamma\nabla v\|_{L^p(\mathbb{R}^N)}\end{aligned}$$

and (8.2.2) follows in this case.

To prove (8.2.2) with $(q,\beta) = (p^*,\gamma)$ we apply the Sobolev embedding theorem to the function $\zeta_\gamma v$ and use estimate (8.2.2) with $(q,\beta) = (p,\gamma-1)$. We deduce that

$$\|\zeta_\gamma v\|_{L^{p^*}(\mathbb{R}^N)}^{p^*} \le C\|\nabla(\zeta_\gamma v)\|_{L^p(\mathbb{R}^N)}^{p^*} \le C(\|\zeta_{\gamma-1}v\|_{L^p(\mathbb{R}^N)}^{p^*} + \|\zeta_\gamma\nabla v\|_{L^p(\mathbb{R}^N)}^{p^*}) \le C\|\zeta_\gamma\nabla v\|_{L^p(\mathbb{R}^N)}^{p^*}.$$

Now, we consider the general case and fix $\theta_1 \in [0,1]$ such that $\beta = (1-\theta_1)(\gamma-1)+\theta_1\gamma = \gamma - 1 + \theta_1$. The condition on p and q in the statement implies that $q = (1-\theta_2)p + \theta_2 p^*$ where $\theta_2 = \theta_1(N-p)(N-\theta_1 p)^{-1}$. Moreover,

$$\begin{aligned}\beta q =& [(1-\theta_1)(\gamma-1)+\theta_1\gamma][(1-\theta_2)p + \theta_2 p^*]\\ =&(1-\theta_2)(\gamma-1)p - \theta_1(1-\theta_2)(\gamma-1)p + (1-\theta_1)\theta_2(\gamma-1)p^*\\ &+ \theta_1(1-\theta_2)\gamma p + \theta_2\gamma p^* - (1-\theta_1)\theta_2\gamma p^*\\ =&(1-\theta_2)(\gamma-1)p + \theta_2\gamma p^* + \theta_1(1-\theta_2)p - (1-\theta_1)\theta_2 p^*\end{aligned}$$

[5] p^* is the Sobolev exponent defined by $1/p^* = 1/p - 1/N$.

$$=(1-\theta_2)(\gamma-1)p+\theta_2\gamma p^*.$$

Hence, the Hölder inequality shows that

$$\|\zeta_\beta v\|^q_{L^q(\mathbb{R}^N)}=\int_{\mathbb{R}^N}|x|^{(1-\theta_2)(\gamma-1)p+\gamma\theta_2 p^*}|v(x)|^{(1-\theta_2)p+\theta_2 p^*}dx\le\|\zeta_{\gamma-1}v\|^{(1-\theta_2)p}_{L^p(\mathbb{R}^N)}\|\zeta_\gamma v\|^{\theta_2 p}_{L^{p^*}(\mathbb{R}^N)}.$$

Using the already established estimates we complete the proof of (8.2.2). □

To prove the forthcoming Theorem 8.2.5, we also need some information on the spectrum of the operator A_p (when $\alpha>2$), which is contained in Proposition 8.2.4. To prove it, we first establish some estimate of the function

$$\mathscr{I}_\beta(x)=\int_{\mathbb{R}^N}\frac{dy}{|x-y|^{N-2}(1+|y|)^\beta},\qquad x\in\mathbb{R}^N,$$

which is one of the Riesz potentials applied to the function $y\mapsto(1+|y|)^{-\beta}$.

Lemma 8.2.3 *For any $\beta>2$ the function $\mathscr{I}_\beta$ is bounded in $\mathbb{R}^N$ and*

$$\mathscr{I}_\beta(x)\le C\begin{cases}|x|^{2-(\beta\wedge N)}, & \text{if } \beta\ne N,\\ |x|^{2-N}\log(|x|), & \text{if } \beta=N,\end{cases}$$

for any $x\in\mathbb{R}^N\setminus B_1$ and some positive constant C.

Proof As is immediately seen, $\mathscr{I}_\beta$ is a radial function. Hence, $\mathscr{I}_\beta(x)=\mathscr{I}_\beta(|x|e_1)$ for any $x\in\mathbb{R}^N$. Moreover, it is clearly locally bounded in $\mathbb{R}^N$. Using spherical coordinates and the change of variable $\rho=|x|\xi$, we can write

$$\begin{aligned}\mathscr{I}_\beta(x)=&\int_{\partial B_1}d\sigma\int_0^{+\infty}\frac{\rho^{N-1}}{(1+\rho)^\beta||x|e_1-\rho\sigma|^{N-2}}d\rho\\
=&|x|^2\int_{\partial B_1}d\sigma\int_0^{+\infty}\frac{\xi^{N-1}}{(1+|x|\xi)^\beta|e_1-\xi\sigma|^{N-2}}d\xi\\
=&|x|^2\int_{\partial B_1}d\sigma\int_0^{1/2}\frac{\xi^{N-1}}{(1+|x|\xi)^\beta|e_1-\xi\sigma|^{N-2}}d\xi\\
&+|x|^2\int_{\partial B_1}d\sigma\int_{1/2}^{+\infty}\frac{\xi^{N-1}}{(1+|x|\xi)^\beta|e_1-\xi\sigma|^{N-2}}d\xi=:\mathscr{I}_{1,\beta}(x)+\mathscr{I}_{2,\beta}(x),\end{aligned}$$

where $d\sigma$ denotes the surface measure on ∂B_1. Using the change of variable $z=|x|\xi$, we can estimate $\mathscr{I}_{1,\beta}(x)$ as follows:

$$\mathscr{I}_{1,\beta}(x)=|x|^{2-N}\int_{\partial B_1}d\sigma\int_0^{|x|/2}\frac{z^{N-1}}{(1+z)^\beta|e_1-|x|^{-1}\sigma z|^{N-2}}dz.$$

Since $1\le 2|e_1-y|$ for any $y\in B_{1/2}$, we deduce that

$$\mathscr{I}_{1,\beta}(x)\le 2^{N-2}N\omega_N|x|^{2-N}\int_0^{|x|/2}\frac{z^{N-1}}{(1+z)^\beta}dz=:2^{N-2}N\omega_N|x|^{2-N}G_\beta(|x|)$$

and, as t tends to $+\infty$, the function G_β behaves as $t^{N-\beta}$, if $\beta<N$, as $\log(t)$, if $\beta=N$, and as a positive constant, if $\beta>N$. It thus follows that

$$\mathscr{I}_{1,\beta}(x)\le C\begin{cases}|x|^{2-(\beta\wedge N)}, & \text{if } \beta\ne N,\\ |x|^{2-N}\log(|x|), & \text{if } \beta=N,\end{cases}\tag{8.2.3}$$

for any $|x| \ge 1$. As far as $\mathscr{I}_{2,\beta}$ is concerned, we observe that

$$|x|^{\beta-2}\mathscr{I}_{2,\beta}(x) = \int_{\partial B_1} d\sigma \int_{1/2}^{+\infty} \frac{\xi^{N-1}}{(|x|^{-\beta} + \xi^\beta)|e_1 - \xi\sigma|^{N-2}} d\xi.$$

By dominated convergence we conclude that

$$\begin{aligned}\lim_{x\to+\infty} |x|^{\beta-2}\mathscr{I}_{2,\beta}(x) &= \int_{\partial B_1} d\sigma \int_{1/2}^{+\infty} \frac{\xi^{N-1-\beta}}{|e_1 - \xi\sigma|^{N-2}} d\xi \\ &= \int_{\mathbb{R}^N\setminus B_{1/2}} \frac{1}{|y|^\beta |e_1 - y|^{N-2}} dy =: C_{1,\beta}. \end{aligned} \tag{8.2.4}$$

Now, the assertion follows from (8.2.3) and (8.2.4). □

Proposition 8.2.4 *If $\alpha > 2$ then the spectrum of the operator A_p consists of a sequence of negative eigenvalues which are simple poles of the resolvent operator and diverge to $-\infty$ as n tends to $+\infty$. Each eigenspace is finite dimensional and independent of p. In particular, the eigenspace corresponding to the largest eigenvalue λ_0 is spanned by a positive function ψ_0, which satisfies the estimate*

$$C_1(1 + |x|)^{2-N} \le \psi_0(x) \le C_2(1 + |x|)^{2-N}, \qquad x \in \mathbb{R}^N, \tag{8.2.5}$$

for some positive constants C_1 and C_2. Moreover, any eigenfunction ψ of the operator A_p satisfies the upper bound in (8.2.5), with a possibly different constant.

Proof The first part of the statement has been proved in Theorem 7.3.6. Moreover, the arguments in the proof of Proposition 7.2.11 can be used to show that the eigenspace corresponding to the eigenvalue λ_0 of the operator A_2 is one-dimensional and spanned by a positive radial function.

To complete the proof, let us check estimate (8.2.5). Since the operator A_p is invertible (see Lemma 7.3.3), we can write

$$\psi_0(x) = \frac{1}{N(2-N)\omega_N} \int_{\mathbb{R}^N} \frac{\lambda_0\psi_0(y)}{|x-y|^{N-2}(1+|y|^\alpha)} dy, \qquad x \in \mathbb{R}^N. \tag{8.2.6}$$

Denote by C_* the minimum of the function ψ_0 in the ball B_1. Then,

$$\begin{aligned}\psi_0(x) \ge& \frac{|\lambda_0|C_*}{2^\alpha N(N-2)\omega_N} \int_{B_1} \frac{dy}{|x-y|^{N-2}} \\ \ge& \frac{|\lambda_0|C_*}{2^\alpha N(N-2)\omega_N} \int_{B_1} \frac{dy}{(|x|+|y|)^{N-2}} \\ \ge& \frac{|\lambda_0|C_*}{2^\alpha N(N-2)\omega_N} \int_{B_1} \frac{dy}{(|x|+1)^{N-2}} = C(1+|x|)^{2-N}\end{aligned}$$

for any $x \in \mathbb{R}^N$, and the first estimate in (8.2.5) follows. Here and in the rest of the proof, we denote by C a positive constant, independent of x, which may vary from line to line.

To prove the upper bound in (8.2.5), we begin by observing that, since the spectrum of A_p is independent of p, ψ_0 belongs to $W^{2,p}(\mathbb{R}^N)$ for any $p \in (1, +\infty)$. In particular, ψ_0 is bounded in $\mathbb{R}^N$. Hence, using again the formula (8.2.6) we can estimate

$$\psi_0(x) \le \frac{2^{\alpha-1}}{N(2-N)\omega_N} \int_{\mathbb{R}^N} \frac{\lambda_0\psi_0(y)}{|x-y|^{N-2}(1+|y|)^\alpha} dy, \qquad x \in \mathbb{R}^N. \tag{8.2.7}$$

Taking Lemma 8.2.3 into account we thus deduce that

$$\psi_0(x) \le \frac{2^{\alpha-1}|\lambda_0|\|\psi_0\|_\infty}{N(N-2)\omega_N}\mathscr{I}_\alpha(x) \le C\begin{cases} |x|^{2-(\alpha\wedge N)}, & \alpha \neq N, \\ |x|^{2-N}\log(|x|), & \alpha = N, \end{cases}$$

for $x \in \mathbb{R}^N \setminus B_1$. Hence, if $\alpha > N$, then the estimate from above of ψ_0 follows at once. On the other hand, if $\alpha = N$, then we get

$$\psi_0(x) \le C|x|^{2-N}\log(|x|) \le C|x|^{1-\frac{N}{2}}, \qquad x \in \mathbb{R}^N \setminus B_1,$$

and we can extend this estimate to any $x \in \mathbb{R}^N$ to get $\psi_0(x) \le C(1+|x|)^{1-\frac{N}{2}}$ for any $x \in \mathbb{R}^N$, which replaced in (8.2.7) yields

$$\psi_0(x) = C\int_{\mathbb{R}^N}\frac{dy}{|x-y|^{N-2}(1+|y|)^{\frac{3}{2}N-1}}, \qquad x \in \mathbb{R}^N.$$

Since $N \ge 3$, $3N/2 - 1$ is greater than N and we conclude that $\psi_0(x) \le C(1+|x|)^{2-N}$ for any $x \in \mathbb{R}^N$. If $\alpha < N$, then we get $\psi_0(x) \le C(1+|x|)^{2-\alpha}$ for any $x \in \mathbb{R}^N$. Hence, again from (8.2.7), we deduce that

$$\psi_0(x) \le C\begin{cases} |x|^{2-2(\alpha-1)\wedge N}, & \alpha \neq \frac{N}{2}+1, \\ |x|^{2-N}\log(|x|), & \alpha = \frac{N}{2}+1, \end{cases}$$

for any $x \in \mathbb{R}^N$. If $2(\alpha-1) > N$ then we are done. If $2(\alpha-1) = N$, then, arguing as above, in one additional step, we conclude that $\psi_0(x) \le C(1+|x|)^{2-N}$ for any $x \in \mathbb{R}^N$. Otherwise we iterate this procedure, and in a finite number of steps we get the wished for estimate. Indeed, the sequence $\{n(\alpha-2)+2\}$ diverges to $+\infty$; hence, there exists $n \in \mathbb{N}$ such that $n(\alpha-2)+2 = N$ or $n(\alpha-2)+2 > N$.

Clearly, the same arguments can be applied to any eigenfunction of the operator A_p starting from the formula

$$\psi(x) = \frac{1}{N(2-N)\omega_N}\int_{\mathbb{R}^N}\frac{\lambda\psi(y)}{|x-y|^{N-2}(1+|y|^\alpha)}dy, \qquad x \in \mathbb{R}^N,$$

where λ is the eigenvalue of A_p corresponding to ψ. □

Now we can state the first main result of this section.

Theorem 8.2.5 *The kernel g satisfies the estimate*

$$g(t,x,y) \le C\Big(1 + t^{-\frac{N-(4-\alpha)^+}{(\alpha-2)\wedge 2}}\Big)e^{\lambda_0 t}(1+|x|)^{2-N}(1+|y|)^{2-N-\alpha} \tag{8.2.8}$$

for any $t > 0$, $x, y \in \mathbb{R}^N$ and some positive constant C.

Proof Note that it suffices to prove estimate (8.2.8) for $t \in (0,1]$, since using the symmetry of the Green function g_μ of the semigroup $\{\widetilde{T}(t)\}$ with respect to the spatial variables it can be then extended to any positive time. Indeed, by the semigroup rule we can write

$$g_\mu(t,x,y) = \int_{\mathbb{R}^N} g_\mu(t-1,x,z)g_\mu(1,y,z)\mu(dz), \qquad s,t > 0, \ \ x,y \in \mathbb{R}^N. \tag{8.2.9}$$

Since $g_\mu = qg$, if estimate (8.2.8) holds true for any $t \in (0,1]$, then the function $g_\mu(1,y,\cdot)$

belongs to $L^2(\mathbb{R}^N,\mu)$ and, thanks to (8.2.5), $g_\mu(1,y,z) \le \widetilde{C}\psi_0(y)\psi_0(z)$ for some positive constant $\widetilde{C}$ and any $y,z\in\mathbb{R}^N$. Hence, from (8.2.9) and observing that the semigroups $\{\widetilde{T}(t)\}$ and $\{T_2(t)\}$ agree on $L^2(\mathbb{R}^N)$, we get

$$g_\mu(t,x,y) = (\widetilde{T}(t-1)g_\mu(1,y,\cdot))(x) \le \widetilde{C}\psi_0(y)(\widetilde{T}(t-1)\psi_0)(x) = \widetilde{C}e^{\lambda_0(t-1)}\psi_0(x)\psi_0(y)$$

for any $t>1$ and $x,y\in\mathbb{R}^N$. Using (8.2.5) again, estimate (8.2.8) follows for $t>1$.

We split the rest of the proof into two steps. In the first one we consider the case when $\alpha\in(2,4]$ and in the latter one the case when $\alpha>4$.

Step 1. We fix $u\in\mathscr{V}$ such that $u\psi_0\in L^1(\mathbb{R}^N,\mu)$ and set $\theta = 2(N+\alpha-4)(\alpha-2)^{-1}>2$. Applying the Hölder inequality, we can estimate

$$\|u\|^2_{L^2(\mathbb{R}^N,\mu)} = \int_{\mathbb{R}^N}\psi_0^{-\frac{4}{\theta+4}}|u|^{\frac{2\theta}{\theta+2}}(|u|\psi_0)^{\frac{4}{\theta+2}}d\mu \le \left(\int_{\mathbb{R}^N}|u|^{\frac{2\theta}{\theta-2}}\psi_0^{\frac{4}{2-\theta}}d\mu\right)^{\frac{\theta-2}{\theta+2}}\|u\psi_0\|^{\frac{4}{\theta-2}}_{L^1(\mathbb{R}^N,\mu)}.$$

Taking (8.2.5) into account we deduce that

$$\|u\|^{2+\frac{4}{\theta}}_{L^2(\mathbb{R}^N,\mu)} \le C_1\bigg(\int_{\mathbb{R}^N}|u(x)|^{\frac{2\theta}{\theta-2}}(1+|x|)^{\frac{4(N-2)}{\theta-2}-\alpha}dx\bigg)\|u\psi_0\|^{\frac{4}{\theta}}_{L^1(\mathbb{R}^N,\mu)}$$

for some positive constant C_1, independent of u. From estimate (8.2.1) with $q=2\theta(\theta-2)^{-1}$, $\beta=(\alpha-4)(N-2)/(2(N+\alpha-4))$, $\gamma=0$, $p=2$ and $\sigma=-\alpha$, the following weighted Nash inequality

$$\|u\|^{2+\frac{4}{\theta}}_{L^2(\mathbb{R}^N,\mu)} \le C_2\big(\|u\|^2_{L^2(\mathbb{R}^N,\mu)} + \|\nabla u\|^2_{L^2(\mathbb{R}^N)}\big)\|u\psi_0\|^{\frac{4}{\theta}}_{L^1(\mathbb{R}^N,\mu)} \tag{8.2.10}$$

follows for some positive constant C_2, independent of u

Now, the proof can be completed using the arguments in [27, Theorems 3.1 and 3.5], which we describe here below. Since the semigroups $\{\widetilde{T}(t)\}$ and $\{T_2(t)\}$ agree on $L^2(\mathbb{R}^N)$, it follows that $\widetilde{T}(t)\psi_0 = e^{\lambda_0 t}\psi_0$ for any $t>0$. Hence, using the symmetry of the semigroup $\{\widetilde{T}(t)\}$, we conclude that

$$\int_{\mathbb{R}^N}\psi_0\widetilde{T}(t)f\,d\mu = \int_{\mathbb{R}^N}f\widetilde{T}(t)\psi_0 d\mu = e^{\lambda_0 t}\int_{\mathbb{R}^N}\psi_0 f\,d\mu, \qquad f\in C_c(\mathbb{R}^N),\ \ t>0. \tag{8.2.11}$$

Fix $t\in(0,1]$, a nonnegative function $f\in C_c^\infty(\mathbb{R}^N)$, which does not identically vanish in $\mathbb{R}^N$, and introduce the function $F:[0,t]\to\mathbb{R}$ defined by

$$F(s) = e^{-2(s+(\lambda_0-1)t)}\|f\psi_0\|^{-2}_{L^1(\mathbb{R}^N,\mu)}\|\widetilde{T}(s)f\|^2_{L^2(\mathbb{R}^N,\mu)}, \qquad s\in[0,t].$$

This function is decreasing, since

$$F'(s) = -2e^{-2(s+(\lambda-1)t)}\|f\psi_0\|^{-2}_{L^1(\mathbb{R}^N,\mu)}(\|\widetilde{T}(s)f\|^2_{L^2(\mathbb{R}^N,\mu)} + \|\nabla_x\widetilde{T}(s)f\|^2_{L^2(\mathbb{R}^N)}), \qquad s\in[0,t].$$

Using estimate (8.2.10) with $u=\widetilde{T}(s)f$, yields

$$-\frac{1}{2}F'(s) \ge \frac{1}{C_2}(F(s))^{1+\frac{2}{\theta}}e^{\frac{4}{\theta}(\lambda_0-1)t}\bigg(e^s\frac{\|\psi_0 f\|_{L^1(\mathbb{R}^N,\mu)}}{\|\psi_0\widetilde{T}(s)f\|_{L^1(\mathbb{R}^N,\mu)}}\bigg)^{\frac{4}{\theta}} \ge C_3(F(s))^{1+\frac{2}{\theta}}$$

for any $s\in[0,t]$, where we have used (8.2.11), and $C_3 = C_2^{-1}e^{\frac{4}{\theta}(\lambda_0-1)}$. Note that $\widetilde{T}(s)f$ is positive almost everywhere in $\mathbb{R}^N$ (with respect to the measure μ), due to Lemma 8.2.1. It thus follows that the function $s\mapsto 2Cs-\frac{\theta}{2}(F(s))^{-2/\theta}$ is not increasing and, consequently,

$F(t) \le (C_\theta t)^{-\frac{\theta}{2}}$ for any $t \in (0,1]$ and some positive constant C_θ. The arbitrariness of $T > 0$ yields

$$\|\widetilde{T}(t)f\|_{L^2(\mathbb{R}^N,\mu)} \le (C'_\theta t)^{-\frac{\theta}{4}}\|f\psi_0\|_{L^1(\mathbb{R}^N,\mu)} = \zeta(t)\|f\psi_0\|_{L^1(\mathbb{R}^N,\mu)}, \qquad t \in (0,1], \quad (8.2.12)$$

where $C'_\theta = e^{-4\lambda_0/\theta}C_\theta$. Since $|T(t)f| \le T(t)|f|$ (see Theorem 1.2.5), estimate (8.2.12) is satisfied by any function $f \in C_c(\mathbb{R}^N)$. Then, by a density argument, we extend it to any $f \in L^2(\mathbb{R}^N,\mu)$.

For any $t \in (0,1]$ let $S(t)$ be the operator defined by $S(t)f = (\zeta(t/2))^{-1}\widetilde{T}(t/2)(\psi_0 f)/\psi_0$ for any $f \in L^1(\mathbb{R}^N,\psi_0^2 d\mu)$. By estimate (8.2.12), $S(t)$ is a contraction from $L^1(\mathbb{R}^N,\mu_1)$ into $L^2(\mathbb{R}^N,\mu_1)$, where $d\mu_1 = \psi_0^2 d\mu$. Since it is symmetric with respect to the measure μ_1, by duality $S(t)$ is also bounded from $L^2(\mathbb{R}^N,\mu_1)$ into $L^\infty(\mathbb{R}^N)$. Hence, the operator $S(t/2)\circ S(t/2) = (\xi(t/2))^{-2}\widetilde{T}(t)(\psi_0\cdot)/\psi_0$ is a contraction from $L^1(\mathbb{R}^N,\mu_1)$ into $L^\infty(\mathbb{R}^N)$ and

$$\frac{(C_\theta t)^{\frac{\theta}{2}}}{2^{\frac{\theta}{2}}e^{(\lambda_0+1)^+}\psi_0(x)}\left|\int_{\mathbb{R}^N}\frac{g_\mu(t,x,y)}{\psi_0(y)}f(y)\mu_1(dy)\right| \le \|f\|_{L^1(\mathbb{R}^N,\mu_1)}, \qquad f \in L^1(\mathbb{R}^N,\mu_1).$$

This inequality shows that the function $g_\mu(t,x,\cdot)/\psi_0$ belongs to $L^\infty(\mathbb{R}^N)$,

$$g_\mu(t,x,y) \le 2^{\frac{\theta}{2}}(C_\theta t)^{-\frac{\theta}{2}}e^{(\lambda_0+1)t}\psi_0(x)\psi_0(y), \qquad x,y \in \mathbb{R}^N,$$

and (8.2.8) follows in this case.

Step 2. Let us now assume that $\alpha > 4$. Applying the Hölder inequality and using (8.2.5) we can estimate

$$\begin{aligned}
\|u\|_{L^2(\mathbb{R}^N,\mu)}^{2+\frac{4}{N}} \le & \left(\int_{\mathbb{R}^N}|u|^{\frac{2N}{N-2}}\psi_0^{\frac{4}{2-N}}d\mu\right)^{1-\frac{2}{N}}\|u\psi_0\|_{L^1(\mathbb{R}^N,\mu)}^{\frac{4}{N}}\\
\le & C_1^{-\frac{4}{N}}C_2^{\frac{4}{N}}\left(\int_{\mathbb{R}^N}|u(x)|^{\frac{2N}{N-2}}(1+|x|)^{4-\alpha}dx\right)^{\frac{N-2}{N}}\left(\int_{\mathbb{R}^N}|u(x)|(1+|x|)^{2-N}\mu(dx)\right)^{\frac{4}{N}}\\
\le & C_1^{-\frac{4}{N}}C_2^{\frac{4}{N}}\left(\int_{\mathbb{R}^N}|u(x)|^{\frac{2N}{N-2}}dx\right)^{\frac{N-2}{N}}\left(\int_{\mathbb{R}^N}|u(x)|(1+|x|)^{2-N}\mu(dx)\right)^{\frac{4}{N}}.
\end{aligned}$$

Hence, the Sobolev embedding theorem yields

$$\|u\|_{L^2(\mathbb{R}^N,\mu)}^{2+\frac{4}{N}} \le C\|\nabla u\|_{L^2(\mathbb{R}^N)}^2\|u\psi_0\|_{L^1(\mathbb{R}^N,\mu)}^{\frac{4}{N}}. \qquad (8.2.13)$$

Replacing the weighted Nash inequality (8.2.10) with this estimate and arguing as in the proof of Step 1, with the function F therein defined being now replaced by the function F, defined by $F(s) = e^{-2\lambda_0 t}\|f\psi_0\|_{L^1(\mathbb{R}^N,\mu)}^{-2}\|\widetilde{T}(s)f\|_{L^2(\mathbb{R}^N,\mu)}^2$ for any $s \in [0,t]$, estimate (8.2.8) can be proved also in this case. □

As far as the behaviour with respect to the spatial variables is concerned, estimate (8.2.8) is rather optimal. More precisely, if one looks for an upper bound of the type $g_\mu(t,x,y) \le C(t)h(x)h(y)$ for some function h, one can easily realize that the choice $h(x) = (1+|x|)^{2-N}$ is the best one. Indeed, $\widetilde{T}(t)\psi_0 = e^{\lambda_0 t}\psi_0$ for any $t > 0$ and $x \in \mathbb{R}^N$. Hence,

$$\begin{aligned}
e^{\lambda_0}\psi_0(x) =&(\widetilde{T}(1)\psi_0)(x) = \int_{\mathbb{R}^N}g_\mu(1,x,y)\psi_0(y)\mu(dy)\\
=&C(1)h(x)\int_{\mathbb{R}^N}h(y)\psi_0(y)dy = \widetilde{C}(1)h(x)
\end{aligned}$$

for any $x \in \mathbb{R}^N$. It thus follows that h coincides, up to a multiplicative constant, with the function ψ_0 which behaves as the function $x \mapsto (1+|x|)^{2-N}$ (see Proposition 8.2.4). On the other hand, when $\alpha \in (2,4]$, the behaviour of the kernel g for large t can be improved as is shown in the next theorem. We also stress that, if $\alpha \in (2,4)$, then the behaviour with respect to the spatial variable, as $|x|$ tends to $+\infty$, is worse than that in estimate (8.2.8). Note that the next theorem also covers the case when $\alpha \in (0,2]$.

Theorem 8.2.6 *Let $\alpha \in (0,4]$. Then, there exists a positive constant C such that*

$$g(t,x,y) \le C(1+t^{-\frac{N}{2}})e^{\lambda_0 t}(1+|x|^\alpha)^{\frac{2-N}{4}}(1+|y|^\alpha)^{-\frac{2+N}{4}}, \qquad t>0,\ x,y \in \mathbb{R}^N. \quad (8.2.14)$$

Proof The proof is similar to that of Theorem 8.2.5, with two main differences: the starting point is the weighted Nash inequality

$$\|u\|^2_{L^2(\mathbb{R}^N,\mu)} \le C\|\nabla u\|^{\frac{2N}{N+2}}_{L^2(\mathbb{R}^N)}\bigg(\int_{\mathbb{R}^N}|u|\psi d\mu\bigg)^{\frac{4}{N+2}}, \quad (8.2.15)$$

with $\psi = q^{(2-N)/4}$, which holds true for any $u \in \mathscr{V}$ and some positive constant C. This estimate replaces (8.2.10) and (8.2.13). Then, we need to show that

$$\int_{\mathbb{R}^N}\psi\widetilde{T}(t)f d\mu = e^{\lambda_0 t}\int_{\mathbb{R}^N}\psi f d\mu, \qquad f \in C_c(\mathbb{R}^N),\ t>0. \quad (8.2.16)$$

Once these two estimates are checked, the proof can be completed following the same lines as in the proof of Step 1 of the quoted theorem. So, we do not elaborate all the details.

To prove (8.2.15), it suffices to apply the Hölder inequality to estimate

$$\|u\|^2_{L^2(\mathbb{R}^N,\mu)} \le \|u\|^{\frac{2N}{N+2}}_{L^{2^*}(\mathbb{R}^N)}\|u\psi\|^{\frac{4}{N+2}}_{L^1(\mathbb{R}^N,\mu)}, \qquad u \in \mathscr{V},$$

where $2^* = 2N/(N-2)$ and, then, the Sobolev embedding theorem to bound from above the $L^{2^*}(\mathbb{R}^N)$-norm of u in terms of the $L^2(\mathbb{R}^N)$-norm of its gradient.

Let us now prove estimate (8.2.16). For this purpose we observe that

$$(\mathcal{A}\psi)(x) = \frac{2-N}{4}\alpha|x|^{\alpha-2}\psi(x)\bigg(-\alpha\frac{N+2}{4}\frac{|x|^\alpha}{1+|x|^\alpha}+\alpha-2+N\bigg)$$

for any $x \in \mathbb{R}^N$, and the right-hand side of the previous formula is nonpositive since $\alpha \le 4$. By Theorem 4.3.2 (with $\varphi = \psi$), the minimal semigroup $\{T(t)\}$ associated with the operator $\mathcal{A}$ in $C_b(\mathbb{R}^N)$ preserves $C_0(\mathbb{R}^N)$. This and Proposition 1.2.8 show that the restriction of $\{T(t)\}$ to $C_0(\mathbb{R}^N)$ is a strongly continuous semigroup. Moreover, $\lambda R(\lambda,\widehat{A})\psi \le \psi$ for any $\lambda > 0$, where $\widehat{A}$ is the weak generator of the semigroup $\{T(t)\}$. Indeed, the function $w = \lambda^{-1}\psi - R(\lambda,\widehat{A})\psi$ satisfies the differential inequality $\lambda w - \mathcal{A}w \ge 0$ in $\mathbb{R}^N\setminus\{0\}$ and belongs to $C_0(\mathbb{R}^N)$. Suppose that w is not everywhere nonnegative in $\mathbb{R}^N$. Hence, it admits a negative minimum at some point $x_0 \in \mathbb{R}^N\setminus\{0\}$ (note that, by Theorem 1.1.3, $w(0) = \lambda^{-1} - (R(\lambda,\widehat{A})\psi)(0) \ge \lambda^{-1} - \|R(\lambda,\widehat{A})\|_{L(C_b(\mathbb{R}^N))} \ge 0$). The second proof of Lemma 3.1.2 shows that $\mathcal{A}w(x_0) \ge 0$ and this clearly contradicts the condition $\mathcal{A}w(x_0) \le \lambda w(x_0)$. We thus conclude that $w \ge 0$ in $\mathbb{R}^N$ and, consequently, $\psi \ge \lambda R(\lambda,\widehat{A})\psi$.

By iteration we then get $\psi \le \lambda^n(R(\lambda,\widehat{A})\psi)^n$ and, consequently, using the exponential formula (see Theorem C.1.7), we get

$$\psi \ge \lim_{n\to+\infty}\bigg[\frac{n}{t}R\Big(\frac{n}{t},A_2\Big)\psi\bigg]^n = T(t)\psi, \qquad t \ge 0.$$

We now fix $f \in C_c(\mathbb{R}^N)$. Lemma 8.2.1 and Proposition 7.3.9 show that $\widetilde{T}(t)g = T(t)g \in L^p(\mathbb{R}^N)$ for any $g \in C_c(\mathbb{R}^N)$, $t > 0$ and $p \in [1, +\infty)$, if $\alpha \le 2$, and $p > N/(N-2)$ otherwise. Since $\psi \in L^p(\mathbb{R}^N)$ if p is sufficiently large, the left-hand side of (8.2.16) is well defined. Moreover, if $\{\vartheta_n\} \subset C_c(\mathbb{R}^N)$ is a sequence of functions satisfying $\chi_{B_n} \le \vartheta_n \le \chi_{B_{2n}}$ for any $n \in \mathbb{N}$, then, using the symmetry of the semigroup $\{\widetilde{T}(t)\}$ in $L^2(\mathbb{R}^N, \mu)$, we can write

$$\begin{aligned}\int_{\mathbb{R}^N} \psi \widetilde{T}(t) f d\mu &= \lim_{n\to+\infty} \int_{\mathbb{R}^N} \psi\vartheta_n \widetilde{T}(t) f d\mu = \lim_{n\to+\infty} \int_{\mathbb{R}^N} \widetilde{T}(t)(\psi\vartheta_n) f d\mu \\ &= \lim_{n\to+\infty} \int_{\mathbb{R}^N} T(t)(\psi\vartheta_n) f d\mu = \int_{\mathbb{R}^N} f T(t)\psi d\mu = e^{\lambda_0 t} \int_{\mathbb{R}^N} \psi f d\mu\end{aligned}$$

for any $t > 0$ and (8.2.16) follows. □

To conclude this section we use the above kernel estimates to provide some growth bound of the eigenfunctions of the operator A_2. The proof is similar to that of Corollary 8.4.9 if one takes estimates (8.2.8) and (8.2.14) into account. Hence, the details are omitted.

Theorem 8.2.7 *Suppose that $\alpha > 2$ and let $\{\lambda_n\}$ be the decreasing sequence of all the eigenvalues of the operator A_p, each of them repeated according to its geometric multiplicity. Further, let $\{\psi_n\}$ be the sequence of the corresponding linearly independent eigenfunctions. Then, for any $n \in \mathbb{N}$, there exists a positive constant C_n such that $|\psi_n| \le C_n|\lambda_n|^{N/4}\psi_0$ in $\mathbb{R}^N$. Moreover,*

$$|\psi_n(x)| \le C_n \begin{cases} |\lambda_n|^{\frac{N}{4}}(1+|x|^\alpha)^{\frac{2-N}{4}}, & \alpha \in [2,4], \\ |\lambda_n|^{\frac{N+\alpha-4}{2(\alpha-2)}}, & \alpha > 4, \end{cases}$$

for any $x \in \mathbb{R}^N$.

8.3 The Schrödinger operator $\mathcal{A} = \Delta + c$

The results of this section are taken from [194, 201]. We begin by presenting the results in [194]. We assume the following assumptions.

Hypotheses 8.3.1 (i) *The function c belongs to $C^1(\mathbb{R}^N)$, is nonpositive and for any $\gamma > 0$ there exists a positive constant C_γ such that $|\nabla c| \le \gamma |c|^{3/2} + C_\gamma$;*

(ii) *there exists a positive function $\omega \in C^2(\mathbb{R}^N)$ which satisfies one of the following additional conditions:*

(a) *ω is a Lyapunov function for the operator $\mathcal{A} = \Delta + c$, i.e., $\mathcal{A}\omega \le \lambda\omega$ in $\mathbb{R}^N$ for some positive constant λ; moreover, for any $\gamma > 0$,*

$$\frac{|\nabla \omega|^2}{\omega^2} + \frac{|\Delta\omega|}{\omega} \le \gamma|c| + C_\gamma; \tag{8.3.1}$$

(b) *estimate (8.3.1) holds true for some value of $\gamma > 0$. Moreover, there exist a Lyapunov function $\widetilde{\omega}$ of the operator $\mathcal{A}$, $k > N/2+1$ and a positive constant C such that $(1+|c|^k)\omega \le C\widetilde{\omega}$.*

To begin with, we observe that, by Theorems 1.2.5 and 3.1.3, with $\varphi(x) = 1 + |x|^2$ for any $x \in \mathbb{R}^N$, one can associate a unique semigroup in $C_b(\mathbb{R}^N)$, which admits a kernel g, i.e.,

$$(T(t)f)(x) = \int_{\mathbb{R}^N} g(t,x,y)f(y)dy, \qquad t > 0, \; x \in \mathbb{R}^N.$$

Next, for any positive function W we set

$$\zeta_W(t,x) = \int_{\mathbb{R}^N} g(t,x,y)W(y)dy, \qquad t > 0, \; x \in \mathbb{R}^N.$$

The following proposition extends Lemma 4.1.3 to this particular nonconservative situation.

Proposition 8.3.2 *Let W be a positive function which belongs to $C^2(\mathbb{R}^N)$ and satisfies the inequality $\mathcal{A}W \le \lambda W$ in $\mathbb{R}^N$. Then, $\xi_W(t,x) \le e^{\lambda t}W(x)$ for any $(t,x) \in (0,+\infty) \times \mathbb{R}^N$. Moreover, $\mathcal{A}\zeta_W \in L^1(\mathbb{R}^N, p(t,x;dy))$ for any $(t,x) \in (0,+\infty) \times \mathbb{R}^N$, where $p(t,x;dy) = g(t,x,y)dy$, the function ζ_W belongs to $C([0,+\infty) \times \mathbb{R}^N) \cap C^{1,2}((0,+\infty) \times \mathbb{R}^N)$ and*

$$D_t\zeta_W(t,x) \le \int_{\mathbb{R}^N} \mathcal{A}W(y)g(t,x,y)dy, \qquad t > 0, \; x \in \mathbb{R}^N. \tag{8.3.2}$$

Proof We can not apply the same arguments as in the proof of Lemma 4.1.3. Indeed, the function $\psi_n \circ \varphi$ (where we use the same notation as in the quoted lemma) does not belong to $D_{\max}(\mathcal{A})$ since $\mathcal{A}(\psi_n \circ \varphi)$ is not bounded. To overcome this difficulty, we approximate the potential c by a sequence $\{c_m\}$ of bounded potentials, by setting $c_m = c \vee (-m)$ for any $m \in \mathbb{N}$, and denote by $\{T_m(t)\}$ the analytic semigroup associated in $C_b(\mathbb{R}^N)$ with the operator $\mathcal{A}_m = \Delta + c_m$. Moreover, for any $\gamma > 0$ and $n \in \mathbb{N}$, we introduce a smooth nondecreasing function $\psi_{n,\gamma}$ such that $\psi_{n,\gamma}(t) = t$ for any $t \in [0,\gamma]$ and $\psi_{n,\gamma}(t) = \gamma + n^{-1}$ for any $t \ge \gamma + 2/n$. We also assume that $\psi'_{n,\gamma} \ge 0$ and $\psi''_{n,\gamma} \le 0$ in $[0,+\infty)$. Finally, for any $n \in \mathbb{N}$ we set $W_{n,\gamma} = \psi_n \circ W$.

Fix $n \in \mathbb{N}$ and $m \ge n+1$. Applying the arguments in the proof of Lemma 4.1.3 to the semigroup $\{T_m(t)\}$ (and with the function ψ_n therein defined being replaced by the function $\psi_{n,\gamma}$), we conclude that

$$(D_tT_m(t)W_{n,\gamma})(x) \le (T_m(t)[(\psi'_{n,\gamma} \circ W)\mathcal{A}_mW])(x) = (T_m(t)[(\psi'_{n,\gamma} \circ W)\mathcal{A}W])(x) \tag{8.3.3}$$

for any $t > 0$ and $x \in \mathbb{R}^N$, where g_m is the Green function associated with the semigroup $\{T_m(t)\}$ and the equality follows from observing that the function $\psi'_{n,\gamma} \circ W$ vanishes in the set $\{x \in \mathbb{R}^N : W(x) \ge n+1\}$.

Now, we observe that, for any $f \in C_b(\mathbb{R}^N)$, $T_m(\cdot)f$ converges to $T(\cdot)f$ in $C^{1,2}(K)$ for any compact set $K \subset (0,+\infty) \times \mathbb{R}^N$. Indeed, Theorem 3.1.3 shows that $\|T_m(t)f\|_\infty \le \|f\|_\infty$ for any $m \in \mathbb{N}$ and $t > 0$. Therefore, using the classical interior Schauder estimates in Theorem D.1.5, we conclude that, for any $n \in \mathbb{N}$, the sequence $\{T_m(\cdot)W_{n,\gamma}\}$ is bounded in $C^{1+\alpha/2,2+\alpha}(K)$ for any K as above (we refer the reader to Theorem 1.2.1 for further details). By compactness and using a diagonal argument, we conclude that, as m tends to $+\infty$, $T_m(t)W_{n,\gamma}$ converges to a classical solution to the equation $D_tu = \mathcal{A}u$, which is also bounded. Moreover, u is continuous in $[0,+\infty) \times \mathbb{R}^N$. This can be checked, for instance by using the localization argument in the proof of Theorem 1.2.1. Again, Theorem 3.1.3 shows that $u = T(\cdot)W_{n,\gamma}$. Hence, we can let m tend to $+\infty$ in the first- and last-side of (8.3.3) and conclude that

$$(D_tT(t)W_{n,\gamma})(x) \le \int_{\mathbb{R}^N} \psi'_{n,\gamma}(W(y))\mathcal{A}W(y)g(t,x,y)dy, \qquad t > 0, \; x \in \mathbb{R}^N. \tag{8.3.4}$$

Since the sequence $\{\psi_{n,\gamma}\}$ is positive, bounded from above by $\gamma+1$, the sequence $\{W_{n,\gamma}\}$ is bounded as well and converges to the function $W \wedge \gamma$ pointwise in $\mathbb{R}^N$. Proposition 1.2.10 shows that $T(\cdot)W_{n,\gamma}$ converges to $T(\cdot)(W \wedge \gamma)$ in $C^{1,2}(K)$ for any compact set $K \subset (0,+\infty)\times\mathbb{R}^N$. Moreover, $\psi'_{n,\gamma}$ converges to $\chi_{[0,\gamma]}$ as n tends to $+\infty$. Therefore, we can take the limit in both sides of (8.3.4) and, using the condition $\mathcal{A}W \le \lambda W$, we conclude that

$$\begin{aligned}(D_tT(t)(W\wedge\gamma))(x) \le &\int_{\{W\le\gamma\}} \mathcal{A}W(y)g(t,x,y)dy \le \lambda \int_{\{W\le\gamma\}} W(y)g(t,x,y)dy\\ \le&\lambda(T(t)(W\wedge\gamma))(x)\end{aligned}$$

for any $t>0$ and $x\in\mathbb{R}^N$. The Gronwall lemma shows that $(T(t)(W\wedge\gamma))(x) \le e^{\lambda t}W(x)$ for any t and x as above. By monotone convergence, we can let γ tend to $+\infty$ and this gives $\zeta_W(t,x) \le e^{\lambda t}W(x)$. The local boundedness in $(0,+\infty)\times\mathbb{R}^N$ of the family $\{T(\cdot)(W\wedge\gamma) : \gamma > 0\}$, together with the interior Schauder estimates in Theorem D.1.5 show that the function ζ_W belongs to $C^{1+\alpha/2,2+\alpha}_{\rm loc}((0,+\infty)\times\mathbb{R}^N)$ and $D_t\zeta = \mathcal{A}\zeta$ in $(0,+\infty)\times\mathbb{R}^N$. Moreover, since $(T(t)(W\wedge\gamma))(x) \le \zeta_W(t,x) \le e^{\lambda t}W(x)$ for any $(t,x)\in(0,+\infty)\times\mathbb{R}^N$, the function ζ_W is continuous at $t=0$.

To prove that the function $\mathcal{A}W$ is in $L^1(\mathbb{R}^N, p(t,x;dy))$ for any $(t,x)\in(0,+\infty)\times\mathbb{R}^N$, we fix such a pair (t,x) and observe that

$$\begin{aligned}(D_tT(t)(W\wedge n))(x) \le &\int_{\{W\le n,\ \mathcal{A}W\ge 0\}} \mathcal{A}W(y)p(t,x;dy)\\ &-\int_{\{W\le n,\ \mathcal{A}W<0\}} |\mathcal{A}W(y)|p(t,x;dy).\end{aligned}\tag{8.3.5}$$

Since the first integral term in the right-hand side of (8.3.5) is positive and can be bounded from above by $\zeta_W(t,x)$, by the monotone convergence theorem it converges to $\int_{\{\mathcal{A}W\ge 0\}} \mathcal{A}W(y)p(t,x;dy)$ as n tends to $+\infty$. Finally,

$$\int_{\{W\le n,\ \mathcal{A}W<0\}} |\mathcal{A}W(y)|p(t,x;dy) \le \lambda\zeta_W(t,x) + (D_tT(t)(W\wedge n))(x)$$

and, consequently,

$$\begin{aligned}\int_{\{\mathcal{A}W<0\}} |\mathcal{A}W(y)|p(t,x;dy) \le &\liminf_{n\to+\infty}\int_{\{W\le n,\ \mathcal{A}W<0\}} |\mathcal{A}W(y)|p(t,x;dy)\\ \le&\int_{\{\mathcal{A}W\ge 0\}} \mathcal{A}W(y)p(t,x;dy) + (D_tT(t)W)(x).\end{aligned}$$

We thus conclude that $\mathcal{A}W$ belongs to $L^1(\mathbb{R}^N, p(t,x;dy))$ and (8.3.2) follows. □

Corollary 8.3.3 *Suppose that $W \in C^2(\mathbb{R}^N)$ is a positive function which satisfies the estimate $\mathcal{A}W \le -h\circ W$ for some convex function $h\in C^1([0,+\infty))$, which is nonpositive at zero and diverges to $+\infty$ as s tends to $+\infty$. Further, suppose that $1/h$ is integrable in a neighborhood of $+\infty$. Then, the function ζ_W is bounded in $[a,+\infty)\times\mathbb{R}^N$ for any $a>0$.*

Proof Since h is convex and diverges to $+\infty$ as s tends to $+\infty$, there exists $s_0>0$ such that $h'(s_0)>0$. Hence, $h(s)\ge h(s_0)+h'(s_0)(s-s_0)$ for any $s>0$. It thus follows that $\mathcal{A}W \le -h(s_0) - h'(s_0)W + h'(s_0)s_0$ in $(0,+\infty)\times\mathbb{R}^N$. Denoting by ω_0 the positive minimum of the function W and observing that $h'(s_0)s_0 - h(s_0)\ge 0$ since h is convex, we get $\mathcal{A}W \le (g'(s_0) + (g'(s_0)s_0 - g(s_0))\omega_0^{-1})W$. Hence, by Proposition 8.3.2 the function ζ_W is well defined.

Now, we fix $(t,x) \in (0,+\infty) \times \mathbb{R}^N$. Using again the convexity of the function h, we can estimate $h(W(y)) \geq h(\zeta_W(t,x)) + h'(\zeta_W(t,x))(W(y) - s_1)$ for any $y \in \mathbb{R}^N$. Integrating both sides of the previous inequality with respect to the measure $p(t,x;dy)$ gives

$$\begin{aligned}
&\int_{\mathbb{R}^N} h(W(y))p(t,x;dy)\\
\geq& h(\zeta_W(t,x))p(t,x;\mathbb{R}^N) + h'(\zeta_W(t,x))\zeta_W(t,x)(1 - p(t,x;\mathbb{R}^N))\\
=& h'(\zeta_W(t,x))\zeta_W(t,x) + [h(\zeta_W(t,x)) - h'(\zeta_W(t,x))\zeta_W(t,x)]p(t,x;\mathbb{R}^N)\\
\geq& h(\zeta_W(t,x)).
\end{aligned}$$

The previous estimate together with (8.3.2) give $D_t\zeta_W(t,x) \leq -h(\zeta_W(t,x))$. Now, by a comparison argument with the solution to the Cauchy problem

$$\begin{cases} z'(t) = -h(z(t)), & t > 0,\\ z(0) = W(x) \end{cases}$$

we can complete the proof. We refer the reader to the last part of the proof of Theorem 4.1.4 for further details. □

The results in the following proposition will be used in the proof of the main (abstract) Theorem 8.3.5.

Lemma 8.3.4 *The following properties are satisfied.*

(i) *Let $k > (1+\beta)(2+N)/2$ ($\beta = 0,1$) and $T > 0$. Then, there exists a positive constant C such that*

$$\|u\|_{L^\infty((0,T)\times\mathbb{R}^N)} \leq C\|u\|_{L^1((0,T)\times\mathbb{R}^N)}^{1-\theta}\|u\|_{W_k^{1,2}((0,T)\times\mathbb{R}^N)}^{\theta} \tag{8.3.6}$$

for any $u \in W_k^{1,2}((0,T)\times\mathbb{R}^N)$, where $\theta = k(N+2+\beta)[(N+2)(k-1)+2k]^{-1}$.

(ii) *For any $\sigma > 0$ and $k \in (1,+\infty)$ there exists a positive constant C such that*

$$\|\nabla_x u\|_{L^k((0,T)\times\mathbb{R}^N)} \leq \sigma\|u\|_{W_k^{1,2}((0,T)\times\mathbb{R}^N)} + C\sigma^{-1}\|u\|_{L^k((0,T)\times\mathbb{R}^N)} \tag{8.3.7}$$

for any $u \in W_k^{1,2}((0,T)\times\mathbb{R}^N)$. Moreover, under Hypothesis 8.3.1(ii)-(a), for any $k \in (1,+\infty)$ there exist two positive constants C and σ_0 such that

$$\|\sqrt{|c|}\nabla_x u\|_{L^k((0,T)\times\mathbb{R}^N)} \leq \sigma\|u\|_{W_k^{1,2}((0,T)\times\mathbb{R}^N)} + C\sigma^{-1}\|cu\|_{L^k((0,T)\times\mathbb{R}^N)} \tag{8.3.8}$$

for any $u \in W^{1,2}((0,T)\times\mathbb{R}^N)$, such that $cu \in L^k((0,T)\times\mathbb{R}^N)$, and any $\sigma \in (0,\sigma_0]$.

(iii) *Under Hypothesis 8.3.1(ii)-(a), let f belong to $L^k((0,T)\times\mathbb{R}^N)$ for some $k \in (1,+\infty)$. Further, assume that u belongs to $L^k((0,T)\times\mathbb{R}^N)\cap W^{1,2}((0,T)\times B_R)$ for any $R > 0$ and solves the Cauchy problem*

$$\begin{cases} D_t u = \mathcal{A}u + f, & \text{in } (0,T)\times\mathbb{R}^N,\\ u(0,\cdot) = 0, & \text{in } \mathbb{R}^N. \end{cases} \tag{8.3.9}$$

Then, $u \in W_k^{1,2}((0,T)\times\mathbb{R}^N)$ and there exists a constant C, independent of f and u, such that

$$\|u\|_{W_k^{1,2}((0,T)\times\mathbb{R}^N)} + \|cu\|_{L^k((0,T)\times\mathbb{R}^N)} \leq C\|f\|_{L^k((0,T)\times\mathbb{R}^N)}. \tag{8.3.10}$$

Proof (i) We first assume that $\beta = 0$. Since any function $u \in W_k^{1,2}((0,T)\times\mathbb{R}^N)$ can be extended to $\mathbb{R}^{N+1}$ with a function in $W_k^{1,2}(\mathbb{R}^{N+1})$, by means of a bounded and linear extension operator, which is also bounded from $L^q((0,T)\times\mathbb{R}^N)$ into $L^q(\mathbb{R}^{N+1})$ for any $1\le q\le+\infty$, it suffices to prove (8.3.6) with $(0,T)\times\mathbb{R}^N$ being replaced by $\mathbb{R}^{N+1}$.

Let $Q=(0,1)^{N+1}$. By classical results, $W_k^{1,2}(Q)$ is compactly embedded in $C(\overline{Q})$. From this property it follows easily that there exists a positive constant C such that

$$\|u\|_{L^\infty(Q)} \le C(\|u\|_{L^1(Q)}+\|D_tu\|_{L^k(Q)}+\|D_x^2u\|_{L^k(Q)}),\qquad u\in W_k^{1,2}(Q). \tag{8.3.11}$$

Indeed, suppose that this is not the case. Then, there exists a sequence $\{u_n\}\subset W_k^{1,2}(Q)$ such that $\|u_n\|_{W_k^{1,2}(Q)}=1$ and

$$\|u_n\|_{L^\infty(Q)} \ge n(\|u_n\|_{L^1(Q)}+\|D_tu_n\|_{L^k(Q)}+\|D_x^2u_n\|_{L^k(Q)}),\qquad n\in\mathbb{N}. \tag{8.3.12}$$

Note that the right-hand side of the previous estimate is bounded from below by a positive constant. Indeed, if this were not the case, a subsequence of $\{u_n\}$ would vanish in $L^1(Q)$ as n tends to $+\infty$. But clearly this can not be the case since $\|u_n\|_{W_k^{1,2}(Q)}=1$ for any $n\in\mathbb{N}$. From (8.3.12) it thus follows that $\|u_n\|_{L^\infty(Q)}$ diverges to $+\infty$ as n tends to $+\infty$, which contradicts the compactness of the embedding of $W_k^{1,2}(Q)$ in $C(\overline{Q})$. Estimate (8.3.11) is thus proved.

Since the Lebesgue measure is invariant for translations, estimate (8.3.11) is satisfied by any cube x_0+Q. Covering $\mathbb{R}^{N+1}$ with the sequence of cubes $\{x_n+Q\}$ where $\mathbb{Z}^{N+1}=\{x_1,\ldots,x_n,\ldots\}$, we obtain that

$$\|u\|_{L^\infty(\mathbb{R}^{N+1})} \le C(\|u\|_{L^1(\mathbb{R}^{N+1})}+\|D_tu\|_{L^k(\mathbb{R}^{N+1})}+\|D_x^2u\|_{L^k(\mathbb{R}^{N+1})}),\qquad u\in W_k^{1,2}(\mathbb{R}^{N+1}). \tag{8.3.13}$$

To get (8.3.6) it now suffices to fix $0\not\equiv u\in W_k^{1,2}(\mathbb{R}^{N+1})$, apply (8.3.13) to the function $(t,x)\mapsto u(\lambda t,\lambda^2x)$ and minimize with respect to $\lambda>0$.

The proof of (8.3.6), with $\beta=1$, is completely similar. Hence, the details are skipped.

(ii) Estimates (8.3.7) and (8.3.8) are straightforward consequences of (7.1.7) and (7.1.8). Hence, the details are skipped.

(iii) By Remark 7.1.8, the Cauchy problem (8.3.9) admits a unique solution $v\in W_k^{1,2}((0,T)\times\mathbb{R}^N)$ which satisfies estimate (8.3.10). To complete the proof we just need to show that $v=u$. For this purpose, we observe that the function $w=u-v$ belongs to $L^k((0,T)\times\mathbb{R}^N)\cap W_k^{1,2}((0,T)\times\mathbb{R}^N)$ and solves the variational equation

$$\int_{(0,T)\times\mathbb{R}^N} w(-D_t\psi-\Delta\psi-c\psi)dtdx=0 \tag{8.3.14}$$

for any $\psi\in C^{1,2}([0,T]\times\mathbb{R}^N)$ with support contained in $[0,T]\times B_R$ for some $R>0$ and such that $\psi(T,\cdot)=0$. By density, we can extend (8.3.14) to any $\psi\in W_{k'}^{1,2}((0,T)\times\mathbb{R}^N)$. Again Remark 7.1.8 shows that the Cauchy problem

$$\begin{cases} D_t\widetilde{\psi}=\Delta\widetilde{\psi}+c\widetilde{\psi}+\widetilde{|w|^{k-2}w}, & \text{in } (0,T)\times\mathbb{R}^N,\\ \widetilde{\psi}(0,\cdot)=0, & \text{in } \mathbb{R}^N,\end{cases}$$

where $\widetilde{w}(t,x)=w(-t,x)|w(-t,x)|^{k-2}$ for $(t,x)\in(0,T)\times\mathbb{R}^N$, admits a unique solution $\widetilde{\psi}\in W_{k'}^{1,2}((0,T)\times\mathbb{R}^N)$ such that $c\widetilde{\psi}\in L^{k'}((0,T)\times\mathbb{R}^N)$. The function ψ, defined by $\psi(t,x)=-\widetilde{\psi}(-t,x)$ for $(t,x)\in(0,T)\times\mathbb{R}^N$, belongs to $W_{k'}^{1,2}((0,T)\times\mathbb{R}^N)$, vanishes at $t=T$ and solves the equation $-D_t\psi-\Delta\psi-c\psi=w|w|^{k-2}$. Hence, from (8.3.14) it follows that $w\equiv0$, so that $u\equiv v$. □

Theorem 8.3.5 *Let Hypothesis* 8.3.1(*i*) *be satisfied. Further, fix* $T > 0$ *and* $0 < a_0 < a < b < b_0 < T$ *such that* $b_0 - b \geq a - a_0$. *Then, the following properties are satisfied.*

(i) *If Hypothesis* 8.3.1(*ii*)-(*a*) *is satisfied, then there exists a positive constant* C *such that*

$$\omega(y)g(t,x,y) \leq C(a-a_0)^{-\frac{N+2}{2}} \int_{a_0}^{b_0} \zeta_\omega(t,x)dt, \qquad t \in [a,b],\ x,y \in \mathbb{R}^N. \quad (8.3.15)$$

Moreover,

$$\begin{aligned} &\|\omega(y)g(\cdot,x,\cdot)\|_{W^{1,2}_k((a,b)\times\mathbb{R}^N)} + \|\omega c g(\cdot,x,\cdot)\|_{L^k((a,b)\times\mathbb{R}^N)} \\ \leq &C_k(a-a_0)^{-\frac{N(k-1)}{2k}-2+\frac{1}{k}} \int_{a_0}^{b_0} \zeta_\omega(t,x)dt \end{aligned} \quad (8.3.16)$$

for any $k > 1 + N/2$, *any* $x \in \mathbb{R}^N$ *and some positive constant* C_k.

(ii) *If Hypothesis* 8.3.1(*ii*)-(*b*) *is satisfied, then estimate* (8.3.15) *holds true with* ω *being replaced by* $\widetilde{\omega}$. *Moreover, estimate* (8.3.16) *is satisfied as well, with* k *being given by Hypothesis* 8.3.1(*ii*)-(*b*).

Proof We begin the proof, by observing that, for any nonnegative function $f \in C_b(\mathbb{R}^N)$, the function $T(\cdot)f$ is positive as well (see Theorem 1.2.5), so that $D_tT(\cdot)f \leq \Delta T(\cdot)f$. Hence, a comparison argument, based on the maximum principle in Theorem 3.1.3, shows that $T(\cdot)f$ is bounded from above by the solution of the heat equation in the whole space (with f as initial datum), i.e.,

$$\int_{\mathbb{R}^N} f(y)g(t,x,y)dy \leq (4\pi t)^{-\frac{N}{2}} \int_{\mathbb{R}^N} e^{-\frac{|x-y|^2}{4t}} f(y)dy, \qquad t > 0,\ x \in \mathbb{R}^N.$$

We thus deduce that

$$g(t,x,y) \leq (4\pi t)^{-\frac{N}{2}} e^{-\frac{|x-y|^2}{4t}}, \qquad t > 0,\ x,y \in \mathbb{R}^N.$$

As a byproduct, $g(\cdot,x,\cdot)$ belongs to $L^k((a_0,b_0)\times\mathbb{R}^N)\cap C_b([a_0,b_0]\times\mathbb{R}^N)$ for any $k \in (1,+\infty)$.

(i) We fix a function $\vartheta \in C^\infty_c(\mathbb{R})$ such that $\chi_{[a,b]} \leq \vartheta \leq \chi_{[a_0,b_0]}$ and $|\vartheta'| \leq 2(a-a_0)^{-1}$. Then, we set $v = \vartheta^k g(\cdot,x,\cdot)$. Note that, for any $n \in \mathbb{N}$, the semigroup $\{T_n(t)\}$ is symmetric in $L^2(B_n)$. Hence, the restriction of $\{T(t)\}$ to $C_c(\mathbb{R}^N)$ extends with a symmetric semigroup in $L^2(\mathbb{R}^N)$. This implies that $g(t,x,y) = g(t,y,x)$ for $t > 0$, $x,y \in \mathbb{R}^N$ and, by Theorem 1.2.5, that $g(\cdot,x,\cdot)$ belongs to $C^{1+\alpha/2,2+\alpha}_{\rm loc}(\mathbb{R}^N)$. As a byproduct, the function v belongs to $L^k((0,T)\times\mathbb{R}^N) \cap W^{1,2}_k((0,T)\times B_R)$, for any $R > 0$, and it solves the parabolic Cauchy problem

$$\begin{cases} D_tv = \mathcal{A}v + k\vartheta^{k-1}\vartheta' g(\cdot,x,\cdot), & \text{in } (0,T)\times\mathbb{R}^N, \\ v(0,\cdot) = 0, & \text{in } \mathbb{R}^N. \end{cases}$$

Since $\vartheta' g(\cdot,x,\cdot) \in L^k((0,T)\times\mathbb{R}^N)$, the function v actually belongs to $W^{1,2}_k((0,T)\times\mathbb{R}^N)$ and $cv \in L^q((0,T)\times\mathbb{R}^N)$ by Lemma 8.3.4.

For any $\varepsilon > 0$ we now introduce the function $\omega_\varepsilon = \omega(1+\varepsilon\omega)^{-1}$, which is clearly bounded and, up to replacing C_γ with a larger constant C'_γ, satisfies estimate (8.3.1). In the rest of the proof, we denote by K a positive constant, which is independent of a_0, b_0, $a - a_0$ ω_ε, c, g and v, which may vary from line to line. We write K_{γ_1} (resp. K_{γ_1,γ_2}) to denote constants which diverge to $+\infty$ as γ_1 (resp $\gamma_1 + \gamma_2$) vanishes.

A straightforward computation reveals that the function $w_\varepsilon = \omega_\varepsilon v$ solves the Cauchy problem

$$\begin{cases} D_t w_\varepsilon = \mathcal{A} w_\varepsilon + g_\varepsilon, & \text{in } (0,T)\times\mathbb{R}^N,\\ v(0,\cdot)=0, & \text{in } \mathbb{R}^N,\end{cases}$$

where $g_\varepsilon = k\vartheta^{k-1}\vartheta' g(\cdot,x,\cdot)\omega_\varepsilon - v\Delta\omega_\varepsilon - 2\langle\nabla\omega,\nabla_x v\rangle$. We claim that g_ε belongs to $L^k((0,T)\times\mathbb{R}^N)$. For this purpose, we begin by observing that

$$\begin{aligned}\|\vartheta^{k-1}\vartheta' g(\cdot,x,\cdot)\omega_\varepsilon\|_{L^k((0,T)\times\mathbb{R}^N)} \le& \frac{K}{a-a_0}\|\vartheta^{k-1}g(\cdot,x,\cdot)\omega_\varepsilon\|_{L^k((0,T)\times\mathbb{R}^N)}\\ \le& \frac{K}{a-a_0}\|\omega_\varepsilon v\|_{L^\infty((0,T)\times\mathbb{R}^N)}^{\frac{k-1}{k}}\left(\int_{a_0}^{b_0}\zeta_\omega(t,x)dt\right)^{\frac1k};\end{aligned}$$

moreover,

$$v\Delta\omega_\varepsilon + 2\langle\nabla\omega,\nabla_x v\rangle = \left(\Delta\omega_\varepsilon - 2\frac{|\nabla\omega_\varepsilon|^2}{\omega_\varepsilon}\right)v + \frac{2}{\omega_\varepsilon}\langle\nabla\omega_\varepsilon,\nabla_x w_\varepsilon\rangle$$

and

$$\frac{|\nabla\omega_\varepsilon|^2}{\omega_\varepsilon^2} \le \frac{|\nabla\omega|^2}{\omega^2} + \frac{|\Delta\omega|}{\omega} \le \delta|c| + C_\delta, \qquad \frac{|\Delta\omega_\varepsilon|}{\omega_\varepsilon} \le 2\frac{|\nabla\omega|^2}{\omega^2} + \frac{|\Delta\omega|}{\omega} \le 2\delta|c| + 2C_\delta$$

for any $\varepsilon,\delta>0$. Hence, using twice the interpolation estimates (8.3.7) and (8.3.8), we obtain

$$\begin{aligned}&\|v\Delta\omega_\varepsilon + 2\langle\nabla\omega,\nabla_x v\rangle\|_{L^k((0,T)\times\mathbb{R}^N)}\\ \le&\left\|\left(\Delta\omega_\varepsilon - 2\frac{|\nabla\omega_\varepsilon|^2}{\omega_\varepsilon}\right)v\right\|_{L^k((0,T)\times\mathbb{R}^N)} + \left\|\frac{2}{\omega_\varepsilon}\langle\nabla\omega_\varepsilon,\nabla_x w_\varepsilon\rangle\right\|_{L^k((0,T)\times\mathbb{R}^N)}\\ \le&\gamma\|c\omega_\varepsilon v\|_{L^k((0,T)\times\mathbb{R}^N)} + K_\gamma\|\omega_\varepsilon v\|_{L^k((0,T)\times\mathbb{R}^N)}\\ &+\gamma\|\sqrt{|c|}\nabla_x(\omega_\varepsilon v)\|_{L^k((0,T)\times\mathbb{R}^N)} + K_\gamma\|\nabla_x(\omega_\varepsilon v)\|_{L^k((0,T)\times\mathbb{R}^N)}\\ \le&\gamma\|c\omega_\varepsilon v\|_{L^k((0,T)\times\mathbb{R}^N)} + K_{\delta,\gamma}\|\omega_\varepsilon v\|_{L^k((0,T)\times\mathbb{R}^N)}\\ &+K\gamma(\|\omega_\varepsilon v\|_{W^{1,2}_k((0,T)\times\mathbb{R}^N)} + \|c\omega_\varepsilon v\|_{L^k((0,T)\times\mathbb{R}^N)}) + K_\lambda\delta\|\omega_\varepsilon v\|_{W^{1,2}_k((0,T)\times\mathbb{R}^N)}\\ \le&K\gamma\|c\omega_\varepsilon v\|_{L^k((0,T)\times\mathbb{R}^N)} + (K\gamma + K_\lambda\delta)\|\omega_\varepsilon v\|_{W^{1,2}_k((0,T)\times\mathbb{R}^N)}\\ &+K_{\delta,\gamma}\|\omega_\varepsilon v\|_{L^\infty((0,T)\times\mathbb{R}^N)}^{1-\frac1k}\left(\int_{a_0}^{b_0}\zeta_\omega(t,x)dt\right)^{\frac1k}\end{aligned}\tag{8.3.17}$$

for any $\delta,\gamma>0$. Lemma 8.3.4 shows that $\omega_\varepsilon v$ belongs to $W^{1,2}_k((0,T)\times\mathbb{R}^N)$ and, putting everything together, we deduce that

$$\begin{aligned}&\|\omega_\varepsilon v\|_{W^{1,2}_k((0,T)\times\mathbb{R}^N)} + \|c\omega_\varepsilon v\|_{L^k((0,T)\times\mathbb{R}^N)}\\ \le&K\Bigg[K_{\gamma,\delta}\|\omega_\varepsilon v\|_\infty^{1-\frac1k}\frac{1}{a-a_0}\left(\int_{a_0}^{b_0}\zeta_\omega(t,x)dt\right)^{\frac1k} + \gamma\|c\omega_\varepsilon v\|_{L^k((0,T)\times\mathbb{R}^N)}\\ &+(\gamma + K_\gamma\delta)\|\omega_\varepsilon v\|_{W^{1,2}_k((0,T)\times\mathbb{R}^N)}\Bigg].\end{aligned}\tag{8.3.18}$$

Fix γ and, then, δ sufficiently small such that $K\gamma\le 1/2$, $K(\gamma+K_\gamma\delta)<1/2$. Then, estimate (8.3.18) reduces to

$$\|\omega_\varepsilon v\|_{W^{1,2}_k((0,T)\times\mathbb{R}^N)} + \|c\omega_\varepsilon v\|_{L^k((0,T)\times\mathbb{R}^N)} \le K\|\omega_\varepsilon v\|_\infty^{1-\frac1k}\frac{1}{a-a_0}\left(\int_{a_0}^{b_0}\zeta_\omega(t,x)dt\right)^{\frac1k}.\tag{8.3.19}$$

Now, we apply (8.3.6) with $\theta = k(N+2)[(N+2)(k-1)+2k]^{-1}$ to estimate the L^∞-norm in the right-hand side of (8.3.19) and get

$$\|\omega_\varepsilon v\|_{W_k^{1,2}((0,T)\times\mathbb{R}^N)} \le K\|\omega_\varepsilon v\|_{L^1((0,T)\times\mathbb{R}^N)}^{\frac{(k-1)(1-\theta)}{k-\theta(k-1)}}\left(\frac{1}{a-a_0}\int_{a_0}^{b_0}\zeta_\omega(t,x)dt\right)^{\frac{k}{k-\theta(k-1)}}. \tag{8.3.20}$$

Replacing (8.3.20) in (8.3.6) gives

$$\|\omega_\varepsilon v\|_\infty \le K\|\omega_\varepsilon v\|_{L^1((0,T)\times\mathbb{R}^N)}^{\frac{k(1-\theta)}{k-\theta(k-1)}}\left(\frac{1}{a-a_0}\int_{a_0}^{b_0}\zeta_\omega(t,x)dt\right)^{\frac{\theta k}{k-\theta(k-1)}}$$

for any $\varepsilon > 0$. Letting ε tend to 0^+ and observing that

$$\|\omega v\|_{L^1((0,T)\times\mathbb{R}^N)} \le K\int_{a_0}^{b_0}\zeta_\omega(t,x)dt,$$

estimate (8.3.15) follows at once. Finally, (8.3.16) follows from (8.3.15) and the above estimates.

(ii) The proof is similar to that of property (i). The main difference is that now γ is no more an arbitrary positive constant. Hence, we can not make the coefficients in front of the terms $\|c\omega_\varepsilon v\|_{L^k((0,T)\times\mathbb{R}^N)}$ and $\|\omega_\varepsilon v\|_{W_k^{1,2}((0,T)\times\mathbb{R}^N)}$ as small as we like. To overcome this difficulty, we use (8.3.8) with a small enough value of σ to estimate

$$\gamma\|\sqrt{V}\nabla_x(\omega_\varepsilon v)\|_{L^k((0,T)\times\mathbb{R}^N)} \le K\gamma(\sigma\|\omega_\varepsilon v\|_{W_k^{1,2}((0,T)\times\mathbb{R}^N)} + C\sigma^{-1}\|c\omega_\varepsilon v\|_{L^k((0,T)\times\mathbb{R}^N)})$$

in (8.3.17). We thus get

$$\begin{aligned}
&\|\omega_\varepsilon v\|_{W_k^{1,2}((0,T)\times\mathbb{R}^N)} + \|c\omega_\varepsilon v\|_{L^k((0,T)\times\mathbb{R}^N)}\\
&\le K\bigg[K_{\gamma,\delta}\|\omega_\varepsilon v\|_\infty^{1-\frac{1}{k}}\frac{1}{a-a_0}\left(\int_{a_0}^{b_0}\zeta_\omega(t,x)dt\right)^{\frac{1}{k}} + \gamma\sigma^{-1}\|c\omega_\varepsilon v\|_{L^k((0,T)\times\mathbb{R}^N)}\\
&\qquad + (\gamma\sigma + K_\gamma\delta)\|\omega_\varepsilon v\|_{W_k^{1,2}((0,T)\times\mathbb{R}^N)}\bigg]
\end{aligned} \tag{8.3.21}$$

and, choosing σ and δ small enough, we can make the term $\|\omega_\varepsilon v\|_{W_k^{1,2}((0,T)\times\mathbb{R}^N)}$ disappear from the right-hand side of (8.3.21). Finally, since $(1+|c|^k)\omega \le \widetilde{\omega}$, we can estimate

$$\begin{aligned}
\|c\omega_\varepsilon v\|_{L^k((0,T)\times\mathbb{R}^N)} &= \left(\int_0^T dt\int_{\mathbb{R}^N}|c|^k\omega_\varepsilon^k v^k dx\right)^{\frac{1}{k}}\\
&\le K\|\omega_\varepsilon v\|_\infty^{\frac{k-1}{k}}\left(\int_0^T dt\int_{\mathbb{R}^N}\widetilde{\omega}v dx\right)^{\frac{1}{k}}\\
&\le K_T\|\omega_\varepsilon v\|_\infty^{\frac{k-1}{k}}\frac{1}{a-a_0}\left(\int_{a_0}^{b_0}\zeta_{\widetilde{\omega}}(t,x)dt\right)^{\frac{1}{k}}
\end{aligned}$$

and $\zeta_\omega \le \zeta_{\widetilde{\omega}}$. Thus,

$$\|\omega_\varepsilon v\|_{W_k^{1,2}((0,T)\times\mathbb{R}^N)} + \|c\omega_\varepsilon v\|_{L^k((0,T)\times\mathbb{R}^N)} \le K_T\|\omega_\varepsilon v\|_\infty^{1-\frac{1}{k}}\frac{1}{a-a_0}\left(\int_{a_0}^{b_0}\zeta_{\widetilde{\omega}}(t,x)dt\right)^{\frac{1}{k}}$$

and we can complete the proof arguing as in the previous case. □

Remark 8.3.6 In [194] the authors also proved some pointwise estimates for the gradient of the kernel g. More precisely, in Theorem 4.14 they proved that, if c satisfies Hypotheses 8.3.1 and the kernel g satisfies an estimate of the form $g(t,x,y) \le a(t)\psi(x)\psi(y)$ for any $t \in (0,T]$, any $x,y \in \mathbb{R}^N$, some $T>0$, some nonincreasing function a and a positive function ψ such that $\omega = 1/\psi$ satisfies Hypothesis 8.3.1(ii)-(a), then

$$|D_x g(t,x,y)| + |D_y g(t,x,y)| \le K_1 t^{-\frac{1}{2}} a(t/4)\psi(x)\psi(y),$$
$$|\Delta_x g(t,x,y)| + |\Delta_y g(t,x,y)| + |D_t g(t,x,y)| \le K_1 t^{-1} a(t/4)\psi(x)\psi(y)$$

for any $t \in (0,T]$, any $x,y \in \mathbb{R}^N$ and some positive constant K_1.

In the following two propositions we apply Theorem 8.3.5 to the particular case when $c(x) = -|x|^\beta$ for any $x \in \mathbb{R}^N$ and some $\beta > 0$. We will also consider this particular potential term in Subsection 8.3.1.

Proposition 8.3.7 *Suppose that $c(x) = -|x|^\beta$ for any $x \in \mathbb{R}^N$ and some $\beta > 0$. Then, for any $\alpha > 0$, the Green function g of the semigroup $\{T(t)\}$ associated with the operator $\mathcal{A} = \Delta + c$ satisfies the following estimate*

$$g(t,x,y) \le Ct^{-\frac{N}{2}-\frac{2\alpha}{\beta}}(1+|x|^2)^{-\frac{\alpha}{2}}(1+|y|^2)^{-\frac{\alpha}{2}}, \qquad t \in (0,1],\ x,y \in \mathbb{R}^N, \tag{8.3.22}$$

for some positive constant $C = C(\alpha)$. In particular, the behaviour as t tends to 0^+ in the previous estimate is optimal, where optimal means that, if $g(t,x,y) \le C(t)(1+|x|^2)^{-\frac{\alpha}{2}}(1+|y|^2)^{-\frac{\alpha}{2}}$ for any $t \in (0,1]$ and $x,y \in \mathbb{R}^N$, then $C(t) \ge Kt^{-\frac{N}{2}-\frac{2\alpha}{\beta}}$ for any $t \in (0,1]$ and a suitable positive constant K.

Proof Fix $\alpha > 0$ and let $\omega : \mathbb{R}^N \to \mathbb{R}$ be the function defined by $\omega(x) = (1+|x|^2)^\alpha$ for any $x \in \mathbb{R}^N$. A straightforward computation reveals that

$$(\mathcal{A}\omega)(x) = \omega(x)\left(\frac{2\alpha N}{1+|x|^2} + \frac{4\alpha(\alpha-1)|x|^2}{(1+|x|^2)^2} - |x|^\beta\right) \le C_1 - C_2(1+|x|^2)^{\frac{2\alpha+\beta}{2}} = -h(\omega(x))$$

for any $x \in \mathbb{R}^N$ and some positive constants C_1 and C_2, where $h(s) = C_2 s^{1+\beta/(2\alpha)} - C_1$ for any $s \ge 0$. Since h is convex and integrable in a neighborhood of $+\infty$, Corollary 8.3.3 shows that the function ζ_ω is bounded in $[a,+\infty) \times \mathbb{R}^N$ for any $a > 0$. In particular, for any $x \in \mathbb{R}^N$ and $t > 0$, $\zeta_\omega(t,x) \le z_x(t)$, where z_x is the solution of the Cauchy problem

$$\begin{cases} z_x'(t) = C_1 - C_2(z_x(t))^{1+\frac{\beta}{2\alpha}}, & t > 0, \\ z_x(0) = \omega(x). \end{cases}$$

Since $z = (C_1/C_2)^{2\alpha/(2\alpha+\beta)}$ is a stationary solution to the above differential equation, it follows that $z_x(t) \le (C_1/C_2)^{2\alpha/(2\alpha+\beta)}$ for any $t \in (0,1]$, if $\omega(x) \le (C_1/C_2)^{2\alpha/(2\alpha+\beta)}$. Let us now suppose that $\omega(x) > (C_1/C_2)^{2\alpha/(2\alpha+\beta)}$. Since $h(s) \ge 2^{-1}C_2 s^{1-\beta/(2\alpha)}$, if $s \ge 2(C_1/C_2)^{2\alpha/(2\alpha+\beta)}$, and z_x is implicitly defined by the equation

$$\int_{z_x(t)}^{\omega(x)} \frac{1}{C_2 s^{1+(2\alpha)^{-1}\beta} - C_1} ds = t, \qquad t \in (0,1],$$

if $z_x(t) \ge 2(C_1/C_2)^{2\alpha/(2\alpha+\beta)}$ for some $t \in (0,1)$, then

$$t \le \frac{2}{C_2}\int_{z_x(t)}^{+\infty} s^{-1-\frac{\beta}{2\alpha}} ds =: C_3(z_x(t))^{-\frac{\beta}{2\alpha}}$$

and this implies that $z_x(t) \le C_4 t^{-2\alpha/\beta}$ for some positive constant C_4. We have so proved that

$$\zeta_\omega(t,x) \le z_x(t) \le \max\left\{2(C_1/C_2)^{2\alpha/(2\alpha+\beta)}, C_4 t^{-2\alpha/\beta}\right\} \le C_5 t^{-2\alpha/\beta}, \qquad t \in (0,1],$$

for some positive constant C_5.

Clearly, the function c satisfies Hypothesis 8.3.1(i). Moreover, the function ω satisfies Hypothesis 8.3.1(ii)-(a). Indeed,

$$\frac{|\nabla\omega|^2}{\omega^2} + \frac{|\Delta\omega|}{\omega} \le \alpha^2 + 2\alpha N + 4\alpha|\alpha - 1|$$

and, clearly, we can bound the right-hand side of the previous inequality from above by $\gamma|c| + \widehat{C}_\gamma$ for any $\gamma > 0$ and some positive constant $\widehat{C}_\gamma$. Now, applying estimate (8.3.15) with $a_0 = t/2$, $a = t$, $b_0 = 2t$ (and e.g. $b = 4$) we conclude that

$$(1+|y|^2)^\alpha g(t,x,y) \le C_5 t^{-\frac{N+2}{2}} \int_{t/2}^{2t} s^{-\frac{2\alpha}{\beta}}\, ds = C_6 t^{-\frac{N}{2}-\frac{2\alpha}{\beta}}.$$

Finally, using the symmetry of the semigroup $\{T(t)\}$, the proof of (8.3.22) can be easily completed.

To prove the optimality of (8.3.22) with respect to t, we fix $t \in (0,1/2]$ and $x_0 \in \mathbb{R}^N$ such that $|x_0|^\beta = t^{-1}$. As is immediately seen, $|c(x)| \ge 2^{-\beta}|x_0|^\beta$ for any $x \in B_1(x_0)$. Denote by $\{S_1(t)\}$ the semigroup associated with the realization of the operator $\mathcal{A}$ in $C_b(B_1(x_0))$ endowed with homogeneous Dirichlet boundary conditions. The maximum principle shows that

$$T(s)f \ge S_1(s)f, \qquad s > 0,\ f \ge 0. \tag{8.3.23}$$

Note that $c \le -2^{-\beta}|x_0|^\beta$ in $B_1(x_0)$ and, as a consequence, $S_1(s)f \ge e^{-2^{-\beta}t^{-1}s}S_2(s)f$ for any $s \ge 0$ and any nonnegative function f, where $\{S_2(s)\}$ denotes the heat semigroup in $C_b(B_1(x_0))$ with homogeneous Dirichlet boundary conditions, whose Green function g_2 satisfies the estimate $g_2(s,x_0,x_0) = s^{-N/2}$ for any $s \in (0,1]$ (see [84, Theorem 3.3.4]). Hence, if $g(s,x,x) \le C(s)(1+|x|^2)^{-\alpha}$ for any $s \in (0,1]$ and some function C, then from (8.3.23) we conclude that $c(t)(1+t^{-\beta})^{-\alpha} = g(t,x_0,x_0) \ge g_2(t,x_0,x_0) = t^{-N/2}$, so that $c(t) \ge \widetilde{C}t^{-N/2-2\alpha/\beta}$ for some positive constant $\widetilde{C}$, independent of t. The optimality of the behaviour in t of estimate (8.3.22) follows. □

In the next proposition, we improve the behaviour of the function g with respect to the spatial variables. Estimate (8.3.24) should be compared with (8.3.38) and (8.3.39).

Proposition 8.3.8 *Let $\beta > 2$ and $\delta < 2(\beta+2)^{-1}$. Then, the estimate*

$$g(t,x,y) \le C\exp\left(t^{-\frac{\beta+2}{\beta-2}}\right)\exp\left(-\delta|x|^{1+\frac{\beta}{2}} - \delta|y|^{1+\frac{\beta}{2}}\right) \tag{8.3.24}$$

holds true for any $t \in (0,1]$, $x, y \in \mathbb{R}^N$ and some positive constant C. The behaviour in t is optimal in the sense that if $g(t,x,y) \le C(t)\exp(-\delta(|x|^{1+\beta/2} + |y|^{1+\beta/2}))$ for any $t \in (0,1]$ and $x, y \in \mathbb{R}^N$, then $C(t) \ge \exp(t^{-(\beta+2)/(\beta-2)})$ for any $t \in (0,1]$.

Proof We fix δ as in the statement and $\widetilde{\delta} \in (\delta, 2(\beta+2)^{-1})$. Moreover, we set $\omega(y) = \exp(\delta|y|^{1+\beta/2})$ and $\widetilde{\omega}(y) = \exp(\widetilde{\delta}|y|^{1+\beta/2})$ for any $y \in \mathbb{R}^N$.

We claim that both ω and $\widetilde{\omega}$ are Lyapunov functions for the operator $\mathcal{A}$. Since the proofs are just the same, we limit ourselves to considering the function ω and observe that

$$(\mathcal{A}\omega)(x) = \frac{1}{2}\delta(\beta+2)|c(x)|^{\frac{1}{\beta}}\omega(x)\left(\frac{2N+\beta-2}{2|x|} + \frac{\delta^2(\beta+2)^2-4}{2\delta(\beta+2)}|x|^{\frac{\beta}{2}}\right).$$

The coefficient in front of $|x|^{\beta/2}$ is negative due to our choice of δ. Hence, we can determine two positive constants C_1 and C_2 such that $(\mathcal{A}\omega)(x) \le C_1 - C_2|x|^\beta\omega(x) = -h(\omega(x))$ for any $x \in \mathbb{R}^N$, where $h(s) = C_2 s(0 \vee \log(s))^{2\beta/(\beta+2)} - C_1$ for any $s > 0$. Using the same arguments as in the proof of Proposition 8.3.7, we can prove that

$$\zeta_\omega(t,x) \vee \zeta_{\widetilde{\omega}}(t,x) \le C_3 \exp\left(C_4 t^{-\frac{\beta+2}{\beta-2}}\right), \qquad t \in (0,1]. \tag{8.3.25}$$

Indeed, if $W(x)$ is less than the last positive zero s_0 of the function h, then $z_x(t) \le s_0$. On the other hand, $g(s) \ge 2^{-1}C_2 s(0 \vee \log(s))^{2\beta/(\beta+2)}$ if $s \ge 2s_0$. Hence, if $z_x(t) > 2s_0$, then

$$t \le \frac{2}{C_2}\int_{z_x(t)}^{+\infty} \frac{1}{s|\log(s)|^{2\beta/(\beta+2)}}ds$$

and (8.3.25) follows.

Next, we claim that the functions ω and $\widetilde{\omega}$ satisfy Hypothesis 8.3.1(ii). Clearly, $(1 + |c|^k)\omega \le C\widetilde{\omega}$ in $\mathbb{R}^N$ for some constant $C > 0$. Moreover, the above computations show that $(|\nabla\omega|/\omega)^2 + |\Delta\omega|/\omega \le \gamma(|c| \vee 1)$, where $\gamma = (1+\beta/2)^2\delta^2 + \delta(1+\beta/2)[2N+\beta-2+\delta(\beta+2)]/2$ and Hypothesis 8.3.1(ii)-(a) follows.

Now, we fix $t \in (0,1]$. Applying Theorem 8.3.5 with $a_0 = t/2$, $a = t$, $b_0 = 2t$ and $b = 4t$, we obtain that

$$g(t,x,y) \le \exp\left(t^{-\frac{\beta+2}{\beta-2}}\right)\exp\left(-\delta|y|^{1+\frac{\beta}{2}}\right), \qquad t \in (0,1],\ x,y \in \mathbb{R}^N. \tag{8.3.26}$$

Since $g(t,x,y) = g(t,y,x)$ for $t > 0$ and $x,y \in \mathbb{R}^N$, as has been shown in the proof of Theorem 8.3.5, we can estimate

$$g(t,x,y) \le \exp\left(t^{-\frac{\beta+2}{\beta-2}}\right)\exp\left[-\frac{1}{2}\delta\left(|y|^{1+\frac{\beta}{2}} + |x|^{1+\frac{\beta}{2}}\right)\right], \qquad t \in (0,1],\ x,y \in \mathbb{R}^N,$$

which is worse than (8.3.24) since δ is subject to the constraint $\delta < 2(\beta+2)^{-1}$. We thus need to use a different argument to deduce (8.3.24) from (8.3.26). For this purpose, we use the symmetry of $g(t,\cdot,\cdot)$ to infer that

$$g(t,x,y) \le \exp\left(t^{-\frac{\beta+2}{\beta-2}}\right)\exp\left(-\delta|x|^{1+\frac{\beta}{2}}\right), \qquad t \in (0,1],\ x,y \in \mathbb{R}^N. \tag{8.3.27}$$

Combining (8.3.26) and (8.3.27) gives

$$g(t,x,y) \le \exp\left(t^{-\frac{\beta+2}{\beta-2}}\right)\exp\left(-\alpha\delta|x|^{1+\frac{\beta}{2}}\right)\exp\left((\alpha-1)\delta|y|^{1+\frac{\beta}{2}}\right)$$

for any $t \in (0,1]$, $x,y \in \mathbb{R}^N$ and $\alpha \in (0,1)$. The Chapman-Kolmogorov equation implies that

$$\begin{aligned}
g(t,x,y) =& \int_{\mathbb{R}^N} g(t/2,x,z)g(t/2,z,y)dz\\
\le& \exp\left(t^{-\frac{\beta+2}{\beta-2}}\right)\exp\left(-\delta|x|^{1+\frac{\beta}{2}}\right)\exp\left((\alpha-1)\delta|y|^{1+\frac{\beta}{2}}\right)\int_{\mathbb{R}^N}\exp\left(-\alpha\delta|z|^{1+\frac{\beta}{2}}\right)dz\\
=& C\exp\left(t^{-\frac{\beta+2}{\beta-2}}\right)\exp\left(-\delta|x|^{1+\frac{\beta}{2}}\right)\exp\left((\alpha-1)\delta|y|^{1+\frac{\beta}{2}}\right)
\end{aligned}$$

and, again by symmetry,

$$g(t,z,y) \le C\exp\left(t^{-\frac{\beta+2}{\beta-2}}\right)\exp\left(-\delta|y|^{1+\frac{\beta}{2}}\right)\exp\left((\alpha-1)\delta|z|^{1+\frac{\beta}{2}}\right).$$

Using the Chapman-Kolmogorov equation once more, we conclude that

$$\begin{aligned} g(t,x,y) =& \int_{\mathbb{R}^N} g(t/2,x,z)g(t/2,z,y)dz \\ \leq& \exp\Big(t^{-\frac{\beta+2}{\beta-2}}\Big)\exp\Big(-\delta|x|^{1+\frac{\beta}{2}}\Big)\exp\Big(\delta|y|^{1+\frac{\beta}{2}}\Big)\int_{\mathbb{R}^N}\exp\Big((\alpha-1)\delta|z|^{1+\frac{\beta}{2}}\Big)dz \\ =& C\exp\Big(t^{-\frac{\beta+2}{\beta-2}}\Big)\exp\Big(-\delta|x|^{1+\frac{\beta}{2}}\Big)\exp\Big(\delta|y|^{1+\frac{\beta}{2}}\Big) \end{aligned}$$

and we are done.

To prove the optimality of estimate (8.3.24) with respect to t, we argue as in the last part of the proof of Proposition 8.3.7. We assume that $g(t,x,x) \leq C(t)\exp(-2\delta|x|^{1+\beta/2})$ for any $t \in (0,1]$ and $x \in \mathbb{R}^N$. From (8.3.23) we infer that

$$C(t)\exp(-2\delta|x|^{1+\beta/2}) \geq t^{-\frac{N}{2}}e^{-2^{-\beta}|x|^\beta t}, \qquad t \in (0,1],\ x \in \mathbb{R}^N \setminus B_2.$$

We fix $t \in (0,1]$ and take $x \in \mathbb{R}^N$ such that $|x| = Mt^{-2(\beta-2)^{-1}}$, where M is a sufficiently large constant. With this choice of x, from the previous estimate we deduce that

$$C(t) \geq t^{-\frac{N}{2}}\exp\Big[\Big(-2^{-\beta}M^\beta + 2\delta M^{1+\frac{\beta}{2}}\Big)t^{-\frac{\beta+2}{\beta-2}}\Big].$$

Since $\beta > 2$, $-2^{-\beta}M^\beta + 2\delta M^{1+\beta/2}$ diverges to $-\infty$ as M tends to $+\infty$. Hence, choosing M large enough, we deduce that $C(t) \geq \exp(t^{-(\beta+2)/(\beta-2)})$. ☐

Remark 8.3.9 In view of Remark 8.3.6 we can infer that

$$\begin{aligned} &|D_xg(t,x,y)| + |D_yg(t,x,y)| \leq K_1t^{-\frac{1}{2}}c(t/4)\psi(x)\psi(y), \\ &|\Delta_xg(t,x,y)|+|\Delta_yg(t,x,y)|+(|x|^\beta+|y|^\beta)g(t,x,y) \leq K_1t^{-1}c(t/4)\psi(x)\psi(y), \end{aligned}$$

where

$$(C(t/4),\psi(x)) = \Big(t^{-\frac{N}{2}-\frac{2\alpha}{\beta}}, (1+|x|^2)^{-\frac{\alpha}{2}}\Big)$$

or

$$(C(t/4),\psi(x)) = \Big(\exp\Big(\frac{1}{4}t^{-\frac{\beta+2}{\beta-2}}\Big), \exp(-\delta|x|^{1+\frac{\beta}{2}})\Big).$$

8.3.1 The case when c is a decreasing radial potential

In this subsection, we present the results proved by G. Metafune and C. Spina in [201] and apply them in the particular case when $c(x) = -|x|^\beta$ for any $x \in \mathbb{R}^N$ and some $\beta > 0$.

Throughout the subsection we prescribe the following condition on c.

Hypothesis 8.3.10 The potential c is radial, nonpositive and belongs to $C^\beta_{\rm loc}(\mathbb{R}^N)$ for some $\beta \in (0,1)$ and $c(x) < c(y)$ for any $x, y \in \mathbb{R}^N$ such that $|x| < |y|$.

As a first step, for any $s > 0$ we introduce the potential $c_s = c \vee (-s)$. By domination, we can infer that the Green function g_s of the semigroup $\{T_s(t)\}$ associated with the operator $\Delta + c_s$ can be bounded from above by the Green function of the semigroup $\{S_s(t)\}$ associated with the operator $\Delta - s$, i.e.,

$$0 \leq g_s(t,x,y) \leq (4\pi t)^{-\frac{N}{2}}\exp\Big(-\frac{|x-y|^2}{4t}\Big)e^{-ts}, \qquad t > 0,\ x,y \in \mathbb{R}^N. \tag{8.3.28}$$

Indeed, if $f \in C_b(\mathbb{R}^N)$ is a nonnegative function, then the function $u = T_s(\cdot)f - S_s(\cdot)f$ belongs to $C_b([0,+\infty)\times\mathbb{R}^N)\cap C^{1,2}((0,+\infty)\times\mathbb{R}^N)$, vanishes on $\{0\}\times\mathbb{R}^N$ and $D_tu-\Delta u+su \le 0$. The classical maximum principle implies that $u \le 0$ in $[0,+\infty)\times\mathbb{R}^N$, i.e.,

$$\int_{\mathbb{R}^N} g_s(t,x,y)f(y)dy \le (4\pi t)^{-\frac{N}{2}}e^{-st}\int_{\mathbb{R}^N} e^{-\frac{|x-y|^2}{4t}}f(y)dy, \qquad t>0,\ x,y\in\mathbb{R}^N.$$

The arbitrariness of f yields (8.3.28). In a completely similar way, we can show that g is dominated from above by the kernel of the heat semigroup.

Theorem 8.3.11 *There exists a positive constant C_N such that*

$$g(t,x,x) \le (4\pi t)^{-\frac{N}{2}}e^{tc(\varepsilon|x|)} + C_Nt^{-\frac{N}{2}}\left(\frac{\varepsilon}{1-\varepsilon}\right)^N \exp\left(-\frac{(1-\varepsilon)^2|x|^2}{4t}\right) \tag{8.3.29}$$

for any $t>0$, $x\in\mathbb{R}^N$ and $\varepsilon>0$.

Proof We split the proof into two steps.
Step 1. Here, we prove that there exists a positive constant $\widetilde{C}_N$ such that

$$g(t,x,x) \le (4\pi t)^{-\frac{N}{2}}e^{-ts}+\widetilde{C}_Nt^{-\frac{N}{2}}\int_{E_s}|x-y|^{-N}\exp\left(-\frac{|x-y|^2}{4t}\right)dy, \qquad t>0,\ x,y\in\mathbb{R}^N, \tag{8.3.30}$$

for any $s>0$, where $E_s=\{x\in\mathbb{R}^N : c(x)\ge -s\}$.
To prove (8.3.30) it suffices to show that

$$|g_s(t,x,x)-g(s,x,x)| \le Ct^{-\frac{N}{2}}\int_{E_s}|x-y|^{-N}\exp\left(-\frac{|x-y|^2}{4t}\right)dy \tag{8.3.31}$$

for some positive constant C and any $t>0$, $x,y\in\mathbb{R}^N$, and take estimate (8.3.28) into account. Here and below by C we denote a positive constant which depends only on N and may vary from line to line.

We fix $f\in C_b(\mathbb{R}^N)$ and, using the notation introduced above, we consider the function $u=T(\cdot)f-T_s(\cdot)f$, which solves the Cauchy problem

$$\begin{cases} D_tu = \Delta u + c_su + (c-c_s)T(\cdot)f, & \text{in } (0,+\infty)\times\mathbb{R}^N,\\ u(0,\cdot)=0, & \text{in } \mathbb{R}^N.\end{cases}$$

The variation-of-constants formula shows that

$$u(t,\cdot) = \int_{\mathbb{R}^N} T_s(t-r)((c-c_s)T(r)f)dr, \qquad t>0.$$

Hence, using (8.3.28), estimating g from above by the kernel of the heat semigroup and observing that $|c-c_s|\le s\chi_{B_{s^{1/\beta}}}$, we can estimate

$$\begin{aligned}|u(t,x)| \le & \int_0^t dr\int_{\mathbb{R}^N} g_s(t-r,x,y)|c(y)-c_s(y)|dy\int_{\mathbb{R}^N} g(r,y,z)f(z)dz\\ \le & \frac{s}{(4\pi)^N}\int_0^t (r(t-r))^{-\frac{N}{2}}e^{(r-t)s}dr\int_{E_s}\exp\left(-\frac{|x-y|^2}{4(t-r)}\right)dy\\ & \times\int_{\mathbb{R}^N}\exp\left(-\frac{|z-y|^2}{4r}\right)f(z)dz.\end{aligned}$$

On the other hand,

$$u(t,x) = \int_{\mathbb{R}^N} (g(t,x,z) - g_s(t,x,z))f(z)dz, \qquad t > 0,\ x \in \mathbb{R}^N.$$

Comparing these two formulae we conclude that

$$\begin{aligned}
|g(t,x,x) - g_s(t,x,x)| \leq & \frac{s}{(4\pi)^N} \int_0^t (r(t-r))^{-\frac{N}{2}} e^{(r-t)s} dr \int_{E_s} \exp\left(-\frac{t|x-y|^2}{4r(t-r)}\right) dy \\
= & \frac{s}{(4\pi)^N} \int_0^{t/2} (r(t-r))^{-\frac{N}{2}} e^{(r-t)s} dr \int_{E_s} \exp\left(-\frac{t|x-y|^2}{4r(t-r)}\right) dy \\
& + \frac{s}{(4\pi)^N} \int_{t/2}^t (r(t-r))^{-\frac{N}{2}} e^{(r-t)s} dr \int_{E_s} \exp\left(-\frac{t|x-y|^2}{4r(t-r)}\right) dy
\end{aligned} \tag{8.3.32}$$

for $t > 0$ and $x \in \mathbb{R}^N$. Denote, respectively, by $\mathscr{I}_1$ and $\mathscr{I}_2$ the two integral terms in the last side of (8.3.32). As far as $\mathscr{I}_1$ is concerned, we observe that

$$\mathscr{I}_1(t,x) \leq \frac{2^N s}{(4\pi\sqrt{t})^N} \int_0^{t/2} r^{-\frac{N}{2}} e^{(r-t)s} dr \int_{E_s} \exp\left(-\frac{(t+r)|x-y|^2}{4rt}\right) dy, \qquad t > 0,\ x \in \mathbb{R}^N.$$

Since

$$r^{-\frac{N}{2}} \exp\left(-\frac{|x-y|^2}{4r}\right) \leq \left(\frac{2N}{e}\right)^{\frac{N}{2}} |x-y|^{-N}, \qquad r > 0,\ x, y \in \mathbb{R}^N,$$

we can go further in the previous estimate and write

$$\begin{aligned}
\mathscr{I}_1(t,x) \leq & Ct^{-\frac{N}{2}} \int_{E_s} \exp\left(-\frac{|x-y|^2}{4t}\right) |x-y|^{-N} dy \int_0^{t/2} s e^{(r-t)s} dr \\
\leq & Ct^{-\frac{N}{2}} \int_{E_s} \exp\left(-\frac{|x-y|^2}{4t}\right) |x-y|^{-N} dy
\end{aligned} \tag{8.3.33}$$

for any $t > 0$ and $x \in \mathbb{R}^N$. Arguing similarly, we can estimate

$$\mathscr{I}_2(t,x) \leq Ct^{-\frac{N}{2}} \int_{E_s} \exp\left(-\frac{|x-y|^2}{4t}\right) |x-y|^{-N} dy, \qquad t > 0,\ x \in \mathbb{R}^N. \tag{8.3.34}$$

From (8.3.32)-(8.3.34), estimate (8.3.31) follows at once.

Step 2. Here, we complete the proof. From (8.3.30) we obtain that

$$g(t,x,x) \leq (4\pi t)^{-N/2} e^{-ts} + Ct^{-\frac{N}{2}} m(E_s) |d(x, E_s)|^{-N} \exp\left(-\frac{|d(x,E_s)|^2}{4t}\right)$$

for any $t > 0$ and $x \in \mathbb{R}^N$, where $d(x, E_s)$ and $m(E_s)$ denote, respectively, the distance of x from E_s and the Lebesgue measure of the set E_s.

Fix $\varepsilon \in (0,1)$ and $x \in \mathbb{R}^N \setminus \{0\}$. Taking $s = -c(\varepsilon x)$ in the previous formula, estimate (8.3.29) follows immediately if we observe that $E_s = \overline{B}_{\varepsilon|x|}$. □

In the particular case when $\lim_{|x|\to+\infty} |x|^{-\beta} c(x) = -1$ for some $\beta \in (0,1)$ (which clearly contains the case when $c(x) = -|x|^\beta$ for any $x \in \mathbb{R}^N$), from Theorem 8.3.11 we get the following result.

Corollary 8.3.12 *Suppose that $\lim_{|x|\to+\infty}|x|^{-\beta}c(x)=-1$ for some $\beta\in(0,1)$. Then, for any $0<\varepsilon_1<\varepsilon_2<1$, there exists a positive constant $K=K(\varepsilon_1,\varepsilon_2)$ such that the Green function g of the semigroup $\{T(t)\}$, associated in $C_b(\mathbb{R}^N)$ with the operator $\Delta+c$, satisfies the estimate*

$$g(t,x,x)\le e^{-Kt}t^{-\frac{N}{2}}\left\{(4\pi)^{-\frac{N}{2}}e^{-t\varepsilon_1^\beta|x|^\beta}+C_N\left(\frac{\varepsilon_2}{1-\varepsilon_2}\right)^N\exp\left(-\frac{(1-\varepsilon_2)^2|x|^2}{4t}\right)\right\} \tag{8.3.35}$$

for any $t>0$ and $x\in\mathbb{R}^N$. In the particular case when $c(x)=-|x|^\beta$ for any $x\in\mathbb{R}^N$, we can take $\varepsilon_1=\varepsilon_2$ and $K=1$.

Proof In the case when $c(x)=-|x|^\beta$, the assertion follows straightforwardly from the formula (8.3.29). In the general case, fix $\delta\in(0,1)$ and let $R=R(\delta)>0$ be such that $-1-\delta\le|x|^{-\beta}c(x)\le-1+\delta$ for any $x\in\mathbb{R}^N\setminus B_R$. From the continuity of the function c and the previous estimate, we can determine two real constants $C_{1,\delta}$ and $C_{2,\delta}$ such that

$$-(1+\delta)|x|^{-\beta}+C_{1,\delta}\le c(x)\le(\delta-1)|x|^{-\beta}+C_{2,\delta},\qquad x\in\mathbb{R}^N.$$

As at the beginning of this section, by domination we can show that g is bounded from above by $e^{C_{2,\delta}\cdot}g_\delta$, where g_δ is the Green function of the semigroup associated in $C_b(\mathbb{R}^N)$ with the operator $\Delta+(\delta-1)|x|^{-\beta}$. This function is immediately estimated using (8.3.29). Hence,

$$g(t,x,x)\le e^{-tC_{2,\delta}}\left\{(4\pi t)^{-\frac{N}{2}}e^{-t(1-\delta)\varepsilon^\beta|x|^\beta}+C_Nt^{-\frac{N}{2}}\left(\frac{\varepsilon}{1-\varepsilon}\right)^N\exp\left(-\frac{(1-\varepsilon)^2|x|^2}{4t}\right)\right\}$$

for $t>0$, $x\in\mathbb{R}^N$ and $\delta,\varepsilon\in(0,1)$. To conclude the proof, it suffices to fix $0<\varepsilon_1<\varepsilon_2$ and choose $\delta\in(0,1)$ such that $(1-\delta)\varepsilon_1^\beta=\varepsilon_2^\beta$. □

From now on, we confine ourselves to the case when $c(x)=-|x|^\beta$ for any $x\in\mathbb{R}^N$ and some $\beta>0$, and estimate the function g in space-time regions.

Theorem 8.3.13 *There exist four positive constants $C_1,\ldots,C_4$ such that*

$$g(t,x,x)\le C_2t^{-\frac{N}{2}}e^{-C_1t|x|^\beta} \tag{8.3.36}$$

for any $(t,x)\in(0,+\infty)\times\mathbb{R}^N$, such that $t\le(1+|x|)^{1-\beta/2}$, and

$$g(t,x,x)\le C_4t^{-\frac{N}{2}}e^{\lambda_0t}e^{-C_3t|x|^{1+\beta/2}} \tag{8.3.37}$$

for any $(t,x)\in(0,+\infty)\times\mathbb{R}^N$, such that $t>(1+|x|)^{1-\beta/2}$.

Moreover, if $\beta\in(0,2)$, then for any $\varepsilon\in(0,1)$ there exist two positive constants $C_{1,\varepsilon}$ and $C_{2,\varepsilon}$ such that

$$C_{1,\varepsilon}t^{-\frac{N}{2}}e^{-(1+\varepsilon)t|x|^\beta}\le g(t,x,x)\le C_{2,\varepsilon}t^{-\frac{N}{2}}e^{-(1-\varepsilon)t|x|^\beta},\qquad t\in(0,1],\ x\in\mathbb{R}^N. \tag{8.3.38}$$

Finally, if $\beta>2$ then there exist two positive constants C_5 and C_6 such that

$$g(t,x,x)\le C_6t^{-\frac{N}{2}}e^{-C_5t|x|^{1+\beta/2}},\qquad t\in(0,1],\ x\in\mathbb{R}^N. \tag{8.3.39}$$

Proof We first prove estimate (8.3.36). For this purpose, we assume that $(t,x)\in(0,+\infty)\times\mathbb{R}^N$ satisfies the condition $t\le(1+|x|)^{1-\beta/2}$ and estimate

$$t|x|^\beta\le(1+|x|)^{1-\frac{\beta}{2}}|x|^\beta\le\frac{(1+|x|)^2}{(1+|x|)^{1-\frac{\beta}{2}}}\le\frac{2}{(1+|x|)^{1-\frac{\beta}{2}}}+\frac{2|x|^2}{(1+|x|)^{1-\frac{\beta}{2}}}\le2+2\frac{|x|^2}{t},$$

if $0 < \beta \leq 2$, and

$$t|x|^\beta = t|x|^2|x|^{\beta-2} \leq t|x|^2(1+|x|)^{\beta-2} \leq t|x|^2t^{-2} = t^{-1}|x|^2,$$

if $\beta > 2$. Replacing these inequalities in (8.3.35), estimate (8.3.36) follows.

Let us now prove estimate (8.3.37). For this purpose we observe that the semigroup rule shows that

$$g(r+s,x,y) = \int_{\mathbb{R}^N} g(r,x,z)g(s,z,y)dz \tag{8.3.40}$$

for $r,s > 0$ and $x,y \in \mathbb{R}^N$. In particular, taking $r = s$, $x = y$ and observing that the function $g(t,\cdot,\cdot)$ is symmetric since the semigroup $\{T_2(t)\}$, associated with the realization in $L^2(\mathbb{R}^N)$ of the operator $\mathcal{A}$, is symmetric, we conclude that $g(r,x,\cdot)$ belongs to $L^2(\mathbb{R}^N)$ and $g(2r,x,x) = \|g(r,x,\cdot)\|_{L^2(\mathbb{R}^N)}$ for $r > 0$ and $x \in \mathbb{R}^N$. Hence, from (8.3.40) we can write $g(r+s,x,y) = (T_2(r)g(s,\cdot,y))(x)$ and taking Proposition 7.2.11 into account we can estimate

$$\begin{aligned}\|g(r+s,\cdot,y)\|_{L^2(\mathbb{R}^N)} =&\|T_2(r)g(s,\cdot,y)\|_{L^2(\mathbb{R}^N)}\\ \leq& M_2e^{\lambda_0 r}\|g(s,\cdot,y)\|_{L^2(\mathbb{R}^N)} = M_2e^{\lambda_0 r}\sqrt{g(2s,y,y)}\end{aligned} \tag{8.3.41}$$

for $r,s > 0$ and $y \in \mathbb{R}^N$

Now, we fix $(t,x) \in (0,+\infty)\times\mathbb{R}^N$ such that $t > (1+|x|)^{1-\beta/2}$ and set $t_0 = (1+|x|)^{1-\beta/2}$. Taking $r = t/2 - t_0/2$, $s = t_0/2$ and $y = x$ in (8.3.41) to infer that

$$g(t,x,x) = \|g(t/2,\cdot,y)\|^2_{L^2(\mathbb{R}^N)} \leq M_2^2e^{\lambda_0(t-t_0)}g(t_0,x,x)$$

and using (8.3.36), we conclude that

$$g(t,x,x) \leq M_2^2e^{\lambda_0(1+|x|)^{1-\beta/2}}e^{\lambda_0 t}(1+|x|)^{-N\left(\frac{1}{2}-\frac{\beta}{4}\right)}e^{-C_1|x|^{1+\beta/2}}.$$

Since $e^{\lambda_0(1+|x|)^{1-\beta/2}}(1+|x|)^{-N(1/2-\beta/4)}e^{-\varepsilon|x|^{1+\beta/2}}$ vanishes as $|x|$ tends to $+\infty$ for any $\varepsilon > 0$, estimate (8.3.37) follows, C_3 being any positive constant smaller than C_1.

To conclude the proof, let us check estimates (8.3.38) and (8.3.39). We begin by proving estimate (8.3.38). For this purpose, we fix $z \in \mathbb{R}^N$ and denote by $\{\widetilde{T}_1(t)\}$ the semigroup in $C_b(B_1(z))$ associated with the operator $\Delta + c$ with homogeneous Dirichlet boundary conditions. By the classical maximum principle, $T(t)f \geq T_1(t)f$ in $B_1(z)$ for any nonnegative function $f \in C_b(\mathbb{R}^N)$. Moreover, since $c \geq -(1+|z|)^\beta$ in $B_1(z)$, again from the maximum principle we can infer that $T(t)f \geq T_1(t)f \geq e^{-t(1+|z|)^\beta}S_\Delta(t)f$ in $B_1(z)$, where $\{S_\Delta(t)\}$ denotes the semigroup in $C_b(B_1(z))$ associated with the Laplacian with homogeneous Dirichlet boundary conditions. Since the Green function associated with this latter semigroup can be bounded from below by $Ct^{-N/2}$ (see [84, Lemma 3.13]), with C being independent of z, we conclude that

$$g(t,z,z) \geq Ce^{-t(1+|z|)^\beta}t^{-\frac{N}{2}}, \qquad t > 0,\ z \in \mathbb{R}^N. \tag{8.3.42}$$

Fix $\varepsilon > 0$. Since $t \in (0,1]$, we can estimate

$$(1+|z|)^\beta \leq (1+\varepsilon)|z|^\beta + C_\varepsilon \leq (1+\varepsilon)|z|^\beta + t^{-1}C_\varepsilon$$

for any $z \in \mathbb{R}^N$, where the constant C_ε blows up as ε tends to 0^+. Replacing this estimate in (8.3.42), the lower bound in (8.3.38) follows. To prove the upper bound in (8.3.38), we apply estimate (8.3.35) with $\varepsilon_1 = \varepsilon_2$ being replaced by $(1-\varepsilon)^{1/\beta}$, which yields

$$g(t,x,x) \leq C_N't^{-\frac{N}{2}}\left\{e^{-t(1-\varepsilon)|x|^\beta} + \left(\frac{(1-\varepsilon)^{1/\beta}}{1-(1-\varepsilon)^{1/\beta}}\right)^N \exp\left(-\frac{(1-(1-\varepsilon)^{1/\beta})^2|x|^2}{4t}\right)\right\}. \tag{8.3.43}$$

The upper bound in (8.3.38) follows from this estimate, observing that

$$\frac{(1-(1-\varepsilon)^{1/\beta})^2|x|^2}{4t} \geq (1-\varepsilon)|x|^\beta \geq (1-\varepsilon)t|x|^\beta$$

for any $x \in \mathbb{R}^N \setminus B_{R_\varepsilon}$ and some $R_\varepsilon > 0$, which implies that

$$\exp\left(-\frac{(1-(1-\varepsilon)^{1/\beta})^2|x|^2}{4t}\right) \leq e^{-(1-\varepsilon)t|x|^\beta}, \qquad x \in \mathbb{R}^N \setminus B_{R_\varepsilon}.$$

On the other hand, if $x \in B_{R_\varepsilon}$, then

$$\exp\left(-\frac{(1-(1-\varepsilon)^{1/\beta})^2|x|^2}{4t}\right) \leq 1 = e^{(1-\varepsilon)t|x|^\beta}e^{-(1-\varepsilon)t|x|^\beta} \leq e^{(1-\varepsilon)R_\varepsilon^\beta}e^{-(1-\varepsilon)t|x|^\beta}.$$

Replacing these last two estimates in (8.3.43), the upper bound in (8.3.38) follows.

As far as estimate (8.3.39) is concerned, we observe that, since $\alpha \geq 2$, $(1+|x|)^{1-\beta/2} \leq 1$ for any $x \in \mathbb{R}^N$. Fix $(t,x) \in (0,1] \times \mathbb{R}^N$ and note that $|x|^\beta \geq |x|^{1+\beta/2}$ if $|x| \geq 1$, whereas $|x|^\beta \geq |x|^{1+\beta/2} - K_\beta$, where $K_\beta = [(\beta+2)(2\beta)^{-1}]^{(\beta+2)/(\beta-2)}(2-\beta)(2\beta)^{-1}$, for any $x \in B_1$. Using this estimate from (8.3.36) it follows that

$$g(t,x,x) \leq C_2 t^{-\frac{N}{2}} e^{-C_1 t|x|^{1+\beta/2}} e^{C_1 K_\beta t} \leq C_2 e^{C_1 K_\beta} t^{-\frac{N}{2}} e^{-C_1 t|x|^{1+\beta/2}}, \tag{8.3.44}$$

if $t \leq (1+|x|)^{1-\beta/2}$. On the other hand if $t > (1+|x|)^{1-\beta/2}$, using (8.3.37) it follows immediately that

$$y(l,x,x) \leq C_2 t^{-\frac{N}{2}} e^{\lambda_0} e^{-C_1 t|x|^{1+\beta/2}}. \tag{8.3.45}$$

From (8.3.44) and (8.3.45), we get estimate (8.3.39) at once. □

Remark 8.3.14 Estimates (8.3.36) and (8.3.37) was proved for the first time by A. Sikora in [236], using some tools from functional calculus. Actually, A. Sikora also proved lower bound estimates for the function g similar to the upper bounds. More specifically, he proved that there exist positive constants C_j' ($j = 1, \ldots, 4$) such that

$$g(t,x,x) \geq C_2' t^{-\frac{N}{2}} e^{-C_1' t|x|^\beta}$$

for any $(t,x) \in (0,+\infty) \times \mathbb{R}^N$, such that $t \leq (1+|x|)^{1-\beta/2}$, and

$$g(t,x,x) \geq C_4' e^{\lambda_0 t} e^{-C_3'|x|^{1+\frac{\beta}{2}}}$$

for any $(t,x) \in (0,+\infty) \times \mathbb{R}^N$, such that $t \geq (1+|x|)^{1-\beta/2}$. In view of Sikora's estimates, Theorem 8.3.13 shows that the result of Corollary 8.3.12 is sharp.

From Theorem 8.3.13 we immediately get the following result.

Corollary 8.3.15 *The following properties hold true.*

(i) *Fix $\beta \in (0,2)$. Then, for any $\varepsilon \in (0,1)$*

$$g(t,x,y) \leq C_{2,\varepsilon} t^{-\frac{N}{2}} e^{-(1-\varepsilon)t|x|^\beta} e^{-(1-\varepsilon)t|y|^\beta}$$

for $t \in (0,1]$ and $x, y \in \mathbb{R}^N$.

(ii) *Fix* $\beta \geq 2$. *Then,*

$$g(t,x,y) \leq C_6 t^{-\frac{N}{2}} \exp\Big(- C_5 t(1+|x|^2)^{\frac{\beta}{2}}\Big) \exp\Big(- C_5 t(1+|y|^2)^{\frac{\beta}{2}}\Big)$$

for $t \in (0,1]$ *and* $x,y \in \mathbb{R}^N$.

Here, $C_{2,\varepsilon}$, C_5 *and* C_6 *are the same constants as in Theorem* 8.3.13.

Proof The Chapman-Kolmogorov equation shows that

$$g(t,x,y) = \int_{\mathbb{R}^N} g(t/2,x,z)g(t/2,z,y)dz, \qquad t>0, \ x,y\in\mathbb{R}^N. \tag{8.3.46}$$

In particular, taking $x = y$ and using the symmetry of g we conclude that $g(t,x,x) = \|g(t/2,x,\cdot)\|^2_{L^2(\mathbb{R}^N)}$. Therefore, from (8.3.46) and the Hölder inequality it follows that

$$g(t,x,y) \leq \sqrt{g(t,x,x)}\sqrt{g(t,y,y)}, \qquad t>0, \ x,y\in\mathbb{R}^N.$$

From this estimate and (8.3.39) the assertion follows immediately. □

We recall that, by Theorem 7.2.10 and its proof, the spectrum of the realization A_2 of the operator $\mathcal{A}$ in $L^2(\mathbb{R}^N)$ consists of a sequence of negative eigenvalues, which accumulate at $-\infty$. Moreover, $L^2(\mathbb{R}^N)$ admits an orthonormal basis consisting of eigenfunctions of the operator A_2. From now on, we order the eigenvalues in a nonincreasing sequence, where each eigenvalue is repeated according to its geometric multiplicity and λ_0 denotes the largest eigenvalue.

As a second application of Theorem 8.3.11 we prove a rather classical result: an estimate of the distribution of the eigenvalues of the operator A_2 when $\beta > 2$. More precisely, the following proposition provides us with the asymptotic behaviour of the function $\lambda \mapsto \mathscr{N}(\lambda)$ (where $\mathscr{N}(\lambda)$ is the number of the eigenvalues (counted with their multiplicity) of the operator A_2 which are greater than $-\lambda$). To prove it we shall use the following Tauberian theorem (for a proof we refer the reader to e.g. [237]).

Theorem 8.3.16 *Let* μ *be a positive measure on* $[0,+\infty)$ *such that the function* $x \mapsto e^{-\lambda x}$ *belongs to* $L^1([0,+\infty),\mu)$ *for any* $\lambda > 0$. *Suppose that there exist two nonnegative constants* C *and* γ *such that*

$$\lim_{t\to 0} t^\gamma \int_0^{+\infty} e^{-tx} d\mu = C.$$

Then,

$$\lim_{\lambda\to+\infty} \lambda^{-\gamma}\mu([0,\lambda)) = \frac{C}{\Gamma(\gamma+1)}.$$

Proposition 8.3.17 *It holds that*

$$\lim_{\lambda\to-\infty} \frac{\mathscr{N}(\lambda)}{|\lambda|^{N(1/2+1/\beta)}} = \frac{N\omega_N \Gamma(\beta^{-1}N)}{(4\pi)^{N/2}\beta\Gamma(N(1/\beta+1/2)+1)}. \tag{8.3.47}$$

Proof The main step of the proof consists of proving that the series $\sum_{n=0}^{+\infty} e^{\lambda_n t}$ converges and

$$\lim_{t\to 0^+} t^{N(\frac{1}{2}+\frac{1}{\beta})} \sum_{n=0}^{+\infty} e^{\lambda_n t} = (4\pi)^{-\frac{N}{2}}\beta^{-1}\Gamma(\beta^{-1}N). \tag{8.3.48}$$

Indeed, suppose that the formula (8.3.48) holds true and introduce the measure μ on the σ algebra of all the subsets of $[0,+\infty)$ defined as follows: for any $\lambda>0$, $\mu\{\lambda\}$ is the cardinality of the set $\{n\in\mathbb{N}\cup\{0\}:\lambda_n=-\lambda\}$. As is easily seen $\mu([0,\lambda))=\mathscr{N}(\lambda)$ and

$$\int_0^{+\infty} e^{-tx}d\mu=\sum_{n=0}^{+\infty}e^{\lambda_n t}$$

for any $t>0$. Hence, applying Theorem 8.3.16 the assertion follows at once.

So, let us prove that the series $\sum_{n=0}^{+\infty}e^{\lambda_n t}$ converges for any $t>0$ and the formula (8.3.48) holds true. By the estimate in Corollary 8.3.12, the function $x\mapsto g(t,x,x)$ belongs to $L^1(\mathbb{R}^N)$ for any $t>0$. Moreover, the Chapman-Kolmogorov equation and the symmetry of the function $g(t,\cdot,\cdot)$ (which follows from the symmetry of the semigroup $\{T(t)\}$) imply that

$$g(2t,x,x)=\int_{\mathbb{R}^N}g(t,x,y)g(t,y,x)dy=\int_{\mathbb{R}^N}|g(t,x,y)|^2dy,\qquad t>0,\ \ x\in\mathbb{R}^N.$$

Integrating in $\mathbb{R}^N$ and using the Tonelli theorem, we conclude that the function $g(t/2,\cdot,\cdot)$ belongs to $L^2(\mathbb{R}^{2N})$. This is enough to conclude that[6] $T(t)$ is a Hilbert-Schmidt operator in $L^2(\mathbb{R}^N)$ for any $t>0$. Since $L^2(\mathbb{R}^N)$ admits an orthonormal basis consisting of the eigenfunctions of the realization of the operator $\Delta+|x|^\beta$ in $L^2(\mathbb{R}^N)$, it follows that

$$\int_{\mathbb{R}^N}g(t,x,x)dx=\sum_{n=0}^{+\infty}e^{\lambda_n t},\qquad t>0,\tag{8.3.49}$$

so that the series converges for any $t>0$.

Let us now prove (8.3.48). Using Corollary 8.3.12 we can estimate

$$\begin{aligned}&\int_{\mathbb{R}^N}g(t,x,x)dx\\ \le&(4\pi t)^{-\frac{N}{2}}\int_{\mathbb{R}^N}\exp(-t\varepsilon^\beta|x|^\beta)dx+t^{-\frac{N}{2}}C_N\left(\frac{\varepsilon}{1-\varepsilon}\right)^N\int_{\mathbb{R}^N}\exp\left(-\frac{(1-\varepsilon)^2|x|^2}{4t}\right)dx\\ =&(4\pi)^{-\frac{N}{2}}t^{-\frac{N}{2}-\frac{N}{\beta}}\int_{\mathbb{R}^N}\exp(-\varepsilon^\beta|x|^\beta)dx+2^NC_N\left(\frac{\varepsilon}{1-\varepsilon}\right)^N\int_{\mathbb{R}^N}\exp(-(1-\varepsilon)^2|x|^2)dx.\end{aligned}\tag{8.3.50}$$

Multiplying the first and last side of (8.3.50) by $t^{N(1/2+1/\beta)}$, taking the formula (8.3.49) into account, and letting, first, t tend to 0^+ and, then, ε tend to 1^-, we deduce that

$$\limsup_{t\to0^+}t^{N\left(\frac12+\frac1\beta\right)}\sum_{n=0}^{+\infty}e^{\lambda_n t}\le(4\pi)^{-\frac{N}{2}}\int_{\mathbb{R}^N}e^{-|x|^\beta}dx.\tag{8.3.51}$$

To prove that

$$\liminf_{t\to0^+}t^{N\left(\frac12+\frac1\beta\right)}\sum_{n=0}^{+\infty}e^{\lambda_n t}\ge(4\pi)^{-\frac{N}{2}}\int_{\mathbb{R}^N}e^{-|x|^\beta}dx\tag{8.3.52}$$

we first show that, for any $r>0$,

$$g(t,x,x)\ge(4\pi t)^{-\frac{N}{2}}\left(1-e^{\frac{r^2}{4t}}\right)\exp(-t(|x|+r)^\beta),\qquad 0<t\le\frac{r^2}{2N},\ x\in\mathbb{R}^N.\tag{8.3.53}$$

[6] See e.g. [226].

For this purpose, we fix r, t and x as in (8.3.53). Since $c(z) \geq -(|x|+r)^\beta$ for any $z \in B_r(x)$, by the maximum principle it follows that $g(t,\cdot,\cdot) \geq \exp(-t(|x|+r)^\beta)g_*(t,\cdot,\cdot)$ in $B_r(x) \times B_r(x)$, where g_* denotes the Green function of the semigroup in $C_b(B_r(x))$ associated with the Laplacian with homogeneous Dirichlet boundary conditions on $\partial B_r(x)$. To complete the proof of estimate (8.3.53) it suffices to show that

$$g_*(t,x,x) \geq (4\pi t)^{-\frac{N}{2}}\left(1 - e^{\frac{r^2}{4t}}\right), \qquad 0 < t \leq \frac{r^2}{2N}. \tag{8.3.54}$$

For this purpose, we consider the function $h : [0,+\infty) \times \overline{B}_r(x) \to \mathbb{R}$ defined by

$$h(t,z) = (4\pi t)^{-\frac{N}{2}} \exp\left(-\frac{|z-x|^2}{4t}\right) - g_*(t,z,x), \qquad t \geq 0,\ \ z \in \overline{B}_r(x).$$

It belongs to $C^{1,2}((0,+\infty) \times B_r(x))$ and $D_t h = \Delta h$ in $(0,+\infty) \times B_r(x)$. Moreover, it converges pointwise to 0 as t tends to 0^+ and $h(t,z) = (4\pi t)^{-N/2}e^{-r^2/4t}$ for any $(t,z) \in (0,+\infty) \times \partial B_r(x)$. By the maximum principle and observing that the function $s \mapsto (4\pi s)^{-N/2}e^{-r^2/4s}$ attains its maximum at $s = r^2/(2N)$, we can infer that

$$h(t,z) \leq \sup_{0<s\leq t} (4\pi s)^{-\frac{N}{2}} e^{-\frac{r^2}{4s}} = (4\pi t)^{-\frac{N}{2}} e^{-\frac{r^2}{4t}}.$$

The previous estimate with $z = x$ leads immediately to (8.3.54).

Using (8.3.54) we can complete the proof of (8.3.52). Indeed, it allows us to estimate

$$\begin{aligned}
\int_{\mathbb{R}^N} g(t,x,x)dx \geq &(4\pi t)^{-\frac{N}{2}}\left(1 - e^{\frac{r^2}{4t}}\right)\int_{\mathbb{R}^N} e^{-t(|x|+r)^\beta}dx \\
=&(4\pi t)^{-\frac{N}{2}}N\omega_N\left(1 - e^{\frac{r^2}{4t}}\right)\int_0^{+\infty} e^{-t(\rho+r)^\beta}\rho^{N-1}d\rho \\
=&(4\pi)^{-\frac{N}{2}}N\omega_N t^{-N\left(\frac{1}{2}+\frac{1}{\beta}\right)}\left(1 - e^{\frac{r^2}{4t}}\right)\int_{rt^{1/\beta}}^{+\infty} e^{-s^\beta}(s - t^{1/\beta}r)^{N-1}ds.
\end{aligned}$$

Multiplying the first and last side of the previous formula by $t^{N(1/2+1/\beta)}$ and letting t tend to 0^+, we obtain estimate (8.3.52).

Now, the formula (8.3.48) follows immediately from (8.3.49), (8.3.51) and (8.3.52). □

Remark 8.3.18 In Theorem 8.3.13 we proved that

$$g(t,x,x) \leq C_4 t^{-\frac{N}{2}} e^{-C_3 t|x|^{1+\beta/2}}, \qquad t \in (0,1],\ \ x \in \mathbb{R}^N,$$

for some positive constants C_3 and C_4. Note that an estimate of the type

$$g(t,x,x) \leq C_4' t^{-\frac{N}{2}} e^{-C_3' t|x|^{1+\beta/2}}, \qquad t \in (0,1],\ \ x \in \mathbb{R}^N, \tag{8.3.55}$$

can not hold. Indeed, suppose that estimate (8.3.55) holds true. Then, adapting the arguments in the proof of Proposition 8.3.17 we can show that

$$\begin{aligned}
\sum_{n=0}^{+\infty} e^{\lambda_n t} =& \int_{\mathbb{R}^N} g(t,x,x)dx \\
\geq& C_4' t^{-\frac{N}{2}} \int_{\mathbb{R}^N} e^{-C_3' t|x|^{1+\beta/2}}dx
\end{aligned}$$

$$
\begin{aligned}
&= C_4' t^{-\frac{N}{2}} N\omega_N \int_0^{+\infty} e^{-C_3 t\rho^{1+\beta/2}} \rho^{N-1} d\rho \\
&= C_4' N\omega_N t^{-N\left(\frac{1}{2}+\frac{2}{\beta+2}\right)} \int_0^{+\infty} e^{-C_3' s^{1+\beta/2}} s^{N-1} ds.
\end{aligned}
$$

Hence,

$$
\liminf_{t\to 0^+} t^{N\left(\frac{1}{2}+\frac{2}{\beta+2}\right)} \sum_{n=0}^{+\infty} e^{\lambda_n t} \geq C_4' N\omega_N \int_0^{+\infty} e^{-C_3' s^{1+\beta/2}} s^{N-1} ds =: M. \tag{8.3.56}
$$

We use this formula to get to a contradiction. Fix $\lambda > 0$ and split $\mathbb{N}$ into the disjoint union of the sets $A_k = \{n \in \mathbb{N} : -(k+1)\lambda < \lambda_n \leq -k\lambda\}$ ($k \in \mathbb{N}\cup\{0\}$). Since the cardinality of the set A_k does not exceed $\mathscr{N}((k+1)\lambda)$ and $e^{\lambda_n t} \leq e^{-\lambda k t}$ for any $n \in A_k$ and $k \in \mathbb{N}\cup\{0\}$, we can estimate

$$
\sum_{n=0}^{+\infty} e^{\lambda_n t} = \sum_{k=0}^{+\infty} \sum_{n\in A_k} e^{\lambda_n t} \leq \sum_{k=0}^{+\infty} e^{-\lambda k t} \mathscr{N}((k+1)\lambda).
$$

From (8.3.56) it follows that there exists $t_0 > 0$ such that

$$
t^{N\left(\frac{1}{2}+\frac{2}{\beta+2}\right)} \sum_{n=0}^{+\infty} e^{\lambda_n t} \geq \frac{1}{2} M, \qquad t \in (0, t_0).
$$

For such values of t and any $k_0 \in \mathbb{N}$, taking the formula (8.3.47) into account we can thus estimate

$$
\begin{aligned}
\frac{1}{2} M t^{-N\left(\frac{1}{2}+\frac{2}{\beta+2}\right)} &\leq \sum_{k=0}^{k_0} e^{-\lambda k t} \mathscr{N}((k+1)\lambda) + \sum_{k=k_0+1}^{+\infty} e^{-\lambda k t} \mathscr{N}((k+1)\lambda) \\
&\leq (k_0+1)\mathscr{N}((k_0+1)\lambda) + \sum_{k_0+1}^{+\infty} e^{-\lambda k t} \mathscr{N}((k+1)\lambda) \\
&\leq (k_0+1)\mathscr{N}((k_0+1)\lambda) + \lambda^{N\left(\frac{1}{2}+\frac{1}{\beta}\right)} \sum_{k=k_0+1}^{+\infty} (k+1)^{N\left(\frac{1}{2}+\frac{1}{\beta}\right)} e^{-\lambda k t}.
\end{aligned}
$$

Setting $t = 1/\lambda$ in the first and last side of the previous chain of inequalities (with $\lambda > 1$ sufficiently large), we get

$$
\begin{aligned}
(k_0+1)\mathscr{N}((k_0+1)\lambda) &\geq \frac{1}{2} M \lambda^{N\left(\frac{1}{2}+\frac{2}{\beta+2}\right)} - \lambda^{N\left(\frac{1}{2}+\frac{1}{\beta}\right)} \sum_{k=k_0+1}^{+\infty} (k+1)^{N\left(\frac{1}{2}+\frac{1}{\beta}\right)} e^{-\lambda k} \\
&\geq \frac{1}{2} M \lambda^{N\left(\frac{1}{2}+\frac{2}{\beta+2}\right)} - \lambda^{N\left(\frac{1}{2}+\frac{2}{\beta+2}\right)} \sum_{k=k_0+1}^{+\infty} (k+1)^{N\left(\frac{1}{2}+\frac{1}{\beta}\right)} e^{-k}.
\end{aligned}
$$

Now, we fix k_0 sufficiently large that $\sum_{k=k_0+1}^{+\infty} (k+1)^{N\left(\frac{1}{2}+\frac{1}{\beta}\right)} e^{-k} \leq M/4$ and conclude that

$$
\mathscr{N}((k_0+1)\lambda) \geq \frac{M}{4(k_0+1)} \lambda^{N\left(\frac{1}{2}+\frac{2}{\beta+2}\right)}
$$

for λ sufficiently large. But this contradicts (8.3.47) since $\lambda^{N\left(\frac{1}{2}+\frac{2}{\beta+2}\right)} > \lambda^{N\left(\frac{1}{2}+\frac{1}{\beta}\right)}$ for large values of λ.

Remark 8.3.19 In [201] the authors prove that estimate (8.3.47) also holds true in the slightly more general case when $\lim_{|x|\to+\infty} |x|^{-\beta} c(x) = -1$.

8.4 The Schrödinger operator $\mathcal{A} = (1+|x|^\alpha)\Delta + c$ when $\alpha \in (0,2)$

Throughout this section we assume the following assumptions.

Hypothesis 8.4.1 *$N \ge 3$ and $c(x) = -|x|^\beta$ for any $x \in \mathbb{R}^N$ and some $\beta > 2$.*

Moreover, we use the Landau symbol $o(|x|^\gamma)$ ($\gamma \in \mathbb{R}$) to denote any function f such that $\lim_{|x|\to+\infty} |x|^{-\gamma} f(x) = 0$.

To prove an upper bound for the Green function g, we follow the ideas in [84]. For this purpose, we begin by estimating the eigenfunction ψ_0 corresponding to the largest eigenvalue λ_0 of the operator $\mathcal{A}$.

Proposition 8.4.2 *Assume that $\beta > 0$. Then, there exist two positive constants C_1 and C_2 such that $C_1 f_0(x) \le \psi_0(x) \le C_2 f_{2\lambda_0}(x)$ for any $x \in \mathbb{R}^N \setminus B_1$, where*

$$f_\lambda(x) := |x|^{\frac{\alpha-\beta}{4} - \frac{N-1}{2}} \exp\left(-\int_1^{|x|} \frac{s^{\beta/2}}{(1+s^\alpha)^{1/2}} ds - \frac{\lambda}{2}\int_1^{|x|} \frac{1}{s^{\beta/2}(1+s^\alpha)^{1/2}} ds \right). \tag{8.4.1}$$

Proof Let us set $f_\lambda(x) = \zeta(|x|)e^{-h_\lambda(|x|)}$, where λ is a real constant,

$$h_\lambda'(r) = \frac{r^{\beta/2}}{(1+r^\alpha)^{1/2}} + \frac{\lambda}{2r^{\beta/2}(1+r^\alpha)^{1/2}}, \qquad r > 0,$$

and $\zeta(r) = r^{-\frac{N-1}{2}}(1+r^\alpha)^{1/4} r^{-\beta/4}$ for any $r > 0$. A straightforward computation reveals that

$$\begin{aligned}\mathcal{A}f_\lambda(x) - \lambda f_\lambda(x) = &\bigg[(1+|x|^\alpha)\frac{\zeta''(|x|)}{\zeta(|x|)} - \lambda\frac{(1+|x|^\alpha)^{1/2}\zeta'(|x|)}{|x|^{\beta/2}\zeta(|x|)} + \frac{N-1}{|x|}(1+|x|^\alpha)\frac{\zeta'(|x|)}{\zeta(|x|)} \\ &+ \frac{\lambda^2}{4|x|^\beta} - \frac{\lambda}{2}(1+|x|^\alpha)\left(\frac{d}{dr}\frac{1}{r^{\beta/2}(1+r^\alpha)^{1/2}}\right)_{|r=|x|} \\ &- \lambda\frac{(N-1)(1+|x|^\alpha)^{1/2}}{2|x|^{1+\beta/2}}\bigg] f_\lambda(x)\end{aligned}$$

for any $x \in \mathbb{R}^N \setminus \{0\}$, so that

$$\mathcal{A}f_\lambda(x) - \lambda f_\lambda(x) = o(1) f_\lambda(x), \tag{8.4.2}$$

as $|x|$ tends to $+\infty$.

Denote by A_2 the realization of the operator $\mathcal{A}$ in $L^2(\mathbb{R}^N)$ with domain given by (7.0.2) and let $\psi_0 \in C_0(\mathbb{R}^N) \cap C^2(\mathbb{R}^N)$ be a positive eigenfunction of the operator A_2 associated with the largest eigenvalue $\lambda_0 < 0$ (see Proposition 7.2.11). Since $f_0 > 0$ it follows that $0 = \mathcal{A}f_0 - o(1)f_0 \le \mathcal{A}f_0 - \lambda_0 f_0$ in $\mathbb{R}^N \setminus B_R$, if R is large enough. This estimate yields $\mathcal{A}(f_0 - \psi_0) - \lambda_0(f_0 - \psi_0) \ge 0$ in $\mathbb{R}^N \setminus B_R$. Up to replacing R with a larger value if needed, we can assume also that $|x|^\beta + \lambda_0 > 0$ for any $x \in \mathbb{R}^N \setminus B_R$. Since both f_0 and ψ_0 vanish as $|x|$ tends to $+\infty$ and $f_0^{-1}\varphi \ge C_1$ on ∂B_R, the maximum principle shows that $C_1 f_0 - \psi_0 \le 0$ in $\mathbb{R}^N \setminus B_R$ for some constant $C_1 > 0$.

Analogously, by (8.4.2) it follows that $\mathcal{A}f_{2\lambda_0} - (2\lambda_0 + o(1))f_{2\lambda_0} = 0$ in $\mathbb{R}^N \setminus \{0\}$. Now, up to replacing R with a larger value, we can assume that $0 < |x|^\beta + 2\lambda_0 + o(1) < |x|^\beta + \lambda_0$ for any $x \in \mathbb{R}^N \setminus B_R$. Hence, $\mathcal{A}f_{2\lambda_0} - \lambda_0 f_{2\lambda_0} \le 0$ in $\mathbb{R}^N \setminus B_R$. This implies that

$$\mathcal{A}(f_{2\lambda_0} - \psi_0) - \lambda_0(f_{2\lambda_0} - \psi_0) \le 0$$

in $\mathbb{R}^N \setminus B_R$ and as above we can conclude that $C_2 f_{2\lambda_0} - \psi_0 \geq 0$ in $\mathbb{R}^N \setminus B_R$ for some constant $C_2 > 0$. □

Note that the function $f_{2\lambda_0}/f_0$ is bounded in a neighborhood of ∞ if and only if $\beta+\alpha > 2$. In such a situation the following result holds true.

Corollary 8.4.3 *If $2 < \alpha + \beta < 2 + \beta$ then there exist constants C_1, $C_2 > 0$ such that $C_1 f_0 \leq \psi_0 \leq C_2 f_0$ in $\mathbb{R}^N \setminus B_1$.*

Now, let us introduce on $L^2(\mathbb{R}^N, \mu)$ the bilinear form $\mathfrak{a}_\mu$, defined by

$$\mathfrak{a}_\mu(u,v) = \int_{\mathbb{R}^N} \langle \nabla u, \nabla \overline{v}\rangle dx - \int_{\mathbb{R}^N} cu\overline{v}\, d\mu, \qquad u, v \in D(\mathfrak{a}_\mu),$$

where $\mu(dx) = (1+|x|^\alpha)^{-1}dx$ and $D(\mathfrak{a}_\mu) = \{u \in L^2(\mathbb{R}^N, \mu) : \sqrt{|c|}u, |\nabla u| \in L^2(\mathbb{R}^N)\}$. $D(\mathfrak{a}_\mu)$ is a Hilbert space when endowed with the inner product

$$\langle u, v\rangle_{D(\mathfrak{a}_\mu)} = \int_{\mathbb{R}^N} (1-c)u\overline{v}\, d\mu + \int_{\mathbb{R}^N} \langle \nabla u, \nabla \overline{v}\rangle dx.$$

Since $\mathfrak{a}_\mu$ is a closed, symmetric, and accretive form, with $\mathfrak{a}_\mu$ we associate the self-adjoint operator A_μ defined by

$$\begin{cases} D(A_\mu) = \left\{ u \in L^2(\mathbb{R}^N, \mu) : \exists g \in L^2(\mathbb{R}^N, \mu) \text{ s.t. } \mathfrak{a}_\mu(u,v) = -\displaystyle\int_{\mathbb{R}^N} g\overline{v}\, d\mu,\ \forall v \in D(\mathfrak{a}_\mu)\right\}, \\ A_\mu u = g, \end{cases}$$

see e.g. [212, Proposition 1.24]. By Theorem C.2.11, A_μ generates a positive analytic semigroup $\{S_\mu\}$ in $L^2(\mathbb{R}^N, \mu)$.

Lemma 8.4.4 *The following properties are satisfied.*

(i) *$D(A_\mu) = \{u \in D(\mathfrak{a}_\mu) \cap W^{2,2}_{\rm loc}(\mathbb{R}^N) : Au \in L^2(\mathbb{R}^N, \mu)\}$ and $A_\mu u = Au$ for any $u \in D(A_\mu)$;*

(ii) *$S_\mu(t)f = T_p(t)f$ for any $t \geq 0$, $p \in (1, +\infty)$ and $f \in L^p(\mathbb{R}^N) \cap L^2(\mathbb{R}^N, \mu)$;*

(iii) *there exists a function $g_\mu : (0, +\infty) \times \mathbb{R}^N \times \mathbb{R}^N \to \mathbb{R}$ such that*

$$S_\mu(t)f = \int_{\mathbb{R}^N} g_\mu(t, \cdot, y) f(y)\mu(dy), \qquad t > 0,\ f \in L^2(\mathbb{R}^N, \mu).$$

As a byproduct, it follows that

$$g_\mu(t,x,y) = (1+|y|^\alpha)g(t,x,y), \qquad t > 0,\ x, y \in \mathbb{R}^N. \tag{8.4.3}$$

Proof (i) To prove the inclusion "$\subset$" we fix $u \in D(A_\mu) \subset W^{1,2}_{\rm loc}(\mathbb{R}^N)$ and observe that, since $A_\mu u - cu \in L^2_{\rm loc}(\mathbb{R}^N)$, we can estimate

$$\left|\int_{\mathbb{R}^N} \langle \nabla u, \nabla \overline{v}\rangle dx\right| = \left|\int_{\mathbb{R}^N} (A_\mu u - cu)\overline{v} d\mu\right| \leq \|A_\mu u - cu\|_{L^2(B_R)} \|v\|_{L^2(\mathbb{R}^N)}$$

for any $v \in C^\infty_c(B_R)$ and $R > 0$. By density this estimate can be extended to any $v \in W^{1,2}_0(B_R)$ and $R > 0$ and, using a standard argument, it is immediate to check that $\nabla u \in (W^{1,2}_{\rm loc}(\mathbb{R}^N))^N$. Hence, $u \in W^{2,2}_{\rm loc}(\mathbb{R}^N)$.

Finally, integrating by parts we conclude that $A_\mu u = \mathcal{A}u$. The inclusion "$\subset$" follows at once.

Let us now prove the inclusion "$\supset$". We fix $u \in D(\mathfrak{a}_\mu) \cap W^{2,2}_{\rm loc}(\mathbb{R}^N)$ such that $f := \mathcal{A}u \in L^2(\mathbb{R}^N, \mu)$ and $\psi \in C^1_c(\mathbb{R}^N)$. Integrating by parts we get

$$\int_{\mathbb{R}^N} f\overline{\psi}d\mu = -\int_{\mathbb{R}^N} \langle \nabla u, \nabla\overline{\psi}\rangle dx + \int_{\mathbb{R}^N} cu\overline{\psi}d\mu = -\mathfrak{a}_\mu(u, \psi).$$

To conclude that $u \in D(A_\mu)$ we need to extend the previous formula to any $\psi \in D(\mathfrak{a}_\mu)$. But this follows immediately from the density of $C^1_c(\mathbb{R}^N)$ in $D(\mathfrak{a}_\mu)$, which can be checked arguing as in the proof of Lemma 7.2.2.

(ii) Since $L^2(\mathbb{R}^N) \hookrightarrow L^2(\mathbb{R}^N, \mu)$, by Proposition 7.2.9 it follows that $D(A_2) \hookrightarrow D(A_\mu)$ (when the two previous spaces are endowed with the graph norms). Hence, for any $f \in L^2(\mathbb{R}^N)$, both the functions $t \mapsto T_2(t)f$ and $t \mapsto S_\mu(t)f$ belong to $C^1((0, +\infty); L^2(\mathbb{R}^N, \mu)) \cap C([0, +\infty); L^2(\mathbb{R}^N, \mu)) \cap C((0, +\infty); D(A_\mu))$. Moreover, they solve the Cauchy problem

$$\begin{cases} v' = \mathcal{A}v, & \text{in } [0, +\infty) \times \mathbb{R}^N, \\ v(0, \cdot) = f, & \text{in } \mathbb{R}^N. \end{cases}$$

Since the previous problem admits a unique solution with the claimed regularity properties, it holds that $T_2(t)f = S_\mu(t)f$ for any $t > 0$. Recalling that $T_2(t)$ and $T_p(t)$ agree on $L^p(\mathbb{R}^N) \cap L^2(\mathbb{R}^N)$ for any $t > 0$, $S_\mu(t)$ and $T_p(t)$ coincide in $L^p(\mathbb{R}^N) \cap L^2(\mathbb{R}^N, \mu)$ for any $t > 0$. Formula (8.4.3) now follows immediately.

(iii) Let $\{\widetilde{T}(t)\}$ be the analytic semigroup in Lemma 8.2.1. Such a semigroup is L^∞-contractive and Theorem 8.2.6 shows that $\widetilde{T}(t) \in L(L^1(\mathbb{R}^N, \mu), L^\infty(\mathbb{R}^N))$. Stein interpolation theorem implies that $\{\widetilde{T}(t)\}$ is ultracontractive, i.e., each operator $\widetilde{T}(t)$ is bounded from $L^2(\mathbb{R}^N, \mu)$ into $L^\infty(\mathbb{R}^N)$.

To complete the proof of (iii) it now suffices to check that $S_\mu(t)f \leq \widetilde{T}(t)f$ for any $t > 0$ and any $0 \leq f \in L^2(\mathbb{R}^N, \mu)$. Indeed, this will show that the semigroup $\{S_\mu(t)\}$ is ultracontractive and, hence, it admits a kernel g_μ. In fact, we prove such a property for any nonnegative function $f \in C^\infty_c(\mathbb{R}^N)$. By Proposition 7.2.7, both the functions $t \mapsto S_\mu(t)f = T_2(t)f$ and $\widetilde{T}(\cdot)f$ belong to $C_b([0, +\infty) \times \mathbb{R}^N) \cap C^{1,2}((0, +\infty) \times \mathbb{R}^N)$ and their difference v satisfies the differential inequality $D_t v - a\Delta v \leq 0$ and vanishes at $t = 0$. By Theorem 3.1.3, we can infer that $v \leq 0$. Hence, $S_\mu(t)f \leq \widetilde{T}(t)f$ for any $t \geq 0$. □

Let us now give the first application of Proposition 8.4.2.

Proposition 8.4.5 *If $\beta > 0$ then the estimate $g(t, x, x) \geq Me^{\lambda_0 t}(f_0(x))^2(1 + |x|^\alpha)^{-1}$ holds true for any $t > 0$, any $x \in \mathbb{R}^N \setminus B_1$ and some constant $M > 0$. Here, f_0 is given by (8.4.1).*

Proof From the semigroup law and the symmetry of $g_\mu(t, \cdot, \cdot)$ for any $t > 0$, we deduce that

$$g_\mu(t + s, x, y) = \int_{\mathbb{R}^N} g_\mu(t, x, z)g_\mu(s, y, z)\mu(dz), \qquad t, s > 0, \;\; x, y \in \mathbb{R}^N. \tag{8.4.4}$$

Hence,

$$g_\mu(t, x, x) = \int_{\mathbb{R}^N} (g_\mu(t/2, x, \cdot))^2 d\mu, \qquad t > 0, \;\; x \in \mathbb{R}^N. \tag{8.4.5}$$

Let us denote by $\widetilde{\psi}_0$ the normalized (in $L^2(\mathbb{R}^N)$) eigenfunction of $\mathcal{A}$ corresponding to the eigenvalue λ_0. Using the Hölder inequality we get

$$e^{\lambda_0 \frac{t}{2}}\widetilde{\psi}_0(x) = T_2(t/2)\widetilde{\psi}_0(x) = \int_{\mathbb{R}^N} g_\mu(t/2, x, y)\widetilde{\psi}_0(y)\mu(dy)$$

$$\leq\left(\int_{\mathbb{R}^N}(g_\mu(t/2,x,y))^2\mu(dy)\right)^{\frac{1}{2}}=g_\mu(t,x,x)^{\frac{1}{2}}$$

for $t>0$ and $x\in\mathbb{R}^N$. The assertion now follows from Proposition 8.4.2. □

We now state the main result of this section.

Theorem 8.4.6 *Let $b>\frac{\beta+2}{\beta-2}$ and f_0 be given by (8.4.1). Then, there exist positive constants K and C such that*

$$0<g(t,x,y)\leq\frac{Ke^{\lambda_0 t}e^{Ct^{-b}}f_0(x)f_0(y)}{1+|y|^\alpha},\qquad t>0,\ x,y\in\mathbb{R}^N\setminus B_1. \tag{8.4.6}$$

Remark 8.4.7 Theorem 8.4.6 applies also in the case when $\alpha=0$ and $\beta>2$. From estimate (8.4.6) we deduce that for any $b>(\beta+2)(\beta-2)^{-1}$ there exist positive constants C and K such that

$$g(t,x,y)\leq Ke^{\lambda_0 t}e^{Ct^{-b}}|x|^{-\frac{\beta}{4}-\frac{N-1}{2}}|y|^{-\frac{\beta}{4}-\frac{N-1}{2}}\exp\left(-\frac{2}{2+\beta}|x|^{\frac{\beta}{2}+1}\right)\exp\left(-\frac{2}{2+\beta}|x|^{\frac{\beta}{2}+1}\right) \tag{8.4.7}$$

for any $t>0$ and $x,y\in\mathbb{R}^N\setminus B_1$. This is a well-known result (see e.g. [84, Corollary 4.5.5]) and shows that the estimate in Theorem 8.4.6 is sharp. Actually, in [201] G. Metafune and C. Spina have shown that also the value $b=(\beta+2)/(\beta-2)$ can be taken in (8.4.7). This estimate should be compared with inequality (8.3.39).

The proof of Theorem 8.4.6 is based on the following result whose proof (in a more general context) can be found in [84, Lemma 2.1.2, Theorem 2.4.2, Example 2.3.4].

Proposition 8.4.8 *Suppose that μ is a positive Borel measure on $\mathbb{R}^N$, absolutely continuous with respect to the Lebesgue measure. Further, let $V\hookrightarrow L^2(\mathbb{R}^N,\mu)$ be a Hilbert space and let H be the nonnegative self-adjoint operator associated with the sesquilinear continuous and coercive form $\mathfrak{a}:V\times V\to\mathbb{C}$. Further, suppose that the semigroup $\{S(t)\}$, associated in $L^2(\mathbb{R}^N,\mu)$ with $\mathfrak{a}$, is positivity preserving. Then, the following properties are satisfied.*

(i) *The estimate $\|S(t)\|_{L(L^2(\mathbb{R}^N,\mu),L^\infty(\mathbb{R}^N))}\leq C_1t^{-\frac{N}{4}}$ holds true for any $t>0$ if and only if $L^{\frac{2N}{N-2}}(\mathbb{R}^N,\mu)\hookrightarrow V$ and $\|u\|^2_{L^{\frac{2N}{N-2}}(\mathbb{R}^N,\mu)}\leq b_0\mathfrak{a}(u,u)$ for any $u\in V$ and some positive constant b_0.*

(ii) *Suppose that the logarithmic Sobolev inequality*

$$\int_{\mathbb{R}^N}f^2\log(f)d\mu\leq\varepsilon\mathfrak{a}(f,f)+\beta(\varepsilon)\|f\|^2_{L^2(\mathbb{R}^N,\mu)}+\|f\|^2_{L^2(\mathbb{R}^N,\mu)}\log(\|f\|^2_{L^2(\mathbb{R}^N,\mu)})$$

holds true for any $f\in L^1(\mathbb{R}^N,\mu)\cap L^\infty(\mathbb{R}^N)\cap D(\mathfrak{a})$ and $\varepsilon>0$, with $\beta(\varepsilon)=C(1+\varepsilon^{-\delta})$ for some $C,\delta>0$. Then, the semigroup $\{S(t)\}$ maps $L^1(\mathbb{R}^N,\mu)$ into $L^\infty(\mathbb{R}^N)$ and there exist two positive constants K_1 and K_2 such that

$$\|S(t)\|_{L(L^1(\mathbb{R}^N,\mu),L^\infty(\mathbb{R}^N))}\leq K_1e^{K_2t^{-\delta}},\qquad t\in(0,1].$$

Proof of Theorem 8.4.6 The proof is an adaption of the arguments in [84, Sections 4.4 and 4.5], which cover the case $\alpha=0$. We need to prove the existence of positive constants K_1, K_2 and K_3 such that

$$g_\mu(t,x,y)\leq K_1e^{K_2t^{-b}}\widetilde{\psi}_0(x)\widetilde{\psi}_0(y),\qquad 0<t\leq 1,\ x,y\in\mathbb{R}^N \tag{8.4.8}$$

and

$$g_\mu(t,x,y) \le K_3 e^{\lambda_0 t}\widetilde{\psi}_0(x)\widetilde{\psi}_0(y), \qquad t > 1,\ x,y \in \mathbb{R}^N. \tag{8.4.9}$$

Indeed, combining (8.4.8) and (8.4.9) and taking (8.4.3) and Corollary 8.4.3 into account, estimate (8.4.6) follows at once.

Estimates (8.4.8) and (8.4.9) are proved in Steps 1 and 2, respectively.

Step 1. Let $\{\widetilde{T}(t)\}$ be the semigroup defined in Lemma 8.2.1. By Theorem 8.2.5 each operator $\widetilde{T}(t)$ is bounded from $L^1(\mathbb{R}^N, \mu)$ into $L^\infty(\mathbb{R}^N)$ and $\|\widetilde{T}(t)\|_{L(L^1(\mathbb{R}^N,\mu),L^\infty(\mathbb{R}^N))} \le Ct^{-N/2}$ for any $t > 0$. Here and below we denote by C a positive constant, independent of t and the functions that we consider, that may vary from line to line. Since the semigroup $\{\widetilde{T}(t)\}$ is bounded in $L^\infty(\mathbb{R}^N)$, applying the Stein interpolation theorem we conclude that $\|\widetilde{T}(t)f\|_{L(L^2(\mathbb{R}^N,\mu),L^\infty(\mathbb{R}^N))} \le Ct^{-\frac{N}{4}}\|f\|_{L^2(\mathbb{R}^N,\mu)}$ for any $t > 0$. A similar estimate is satisfied by the semigroup $\{S_\mu(t)\}$ since, as is shown in the proof of Lemma 8.4.4, $S_\mu f \le \widetilde{T}(t)f$ for any $t \ge 0$ and any nonnegative function f. We can thus apply Proposition 8.4.8 and conclude that $\|f\|^2_{L^{2N/(N-2)}(\mathbb{R}^N,\mu)} \le b_0\mathfrak{a}_\mu(f,f)$ for any $f \in D(\mathfrak{a}_\mu)$. The Hölder inequality then reveals that

$$\int_{\mathbb{R}^N} h|f|^2 d\mu \le b_0\|g\|_{L^{N/2}(\mathbb{R}^N,\mu)}\mathfrak{a}_\mu(f,f), \qquad f \in D(\mathfrak{a}_\mu),\ h \in L^{N/2}(\mathbb{R}^N,\mu). \tag{8.4.10}$$

Since $\|cf\|_{L^2(\mathbb{R}^N,\mu)} \le \mathfrak{a}_\mu(f,f)$ for any $f \in D(\mathfrak{a}_\mu)$ and $W(x) := |x|^\gamma \le \varepsilon(|c(x)| + \lambda_0) + C_{\beta,\gamma}\varepsilon^{-\frac{\gamma}{\beta-\gamma}} - \varepsilon\lambda_0$ for any $x \in \mathbb{R}^N$, $\varepsilon > 0$ and $\gamma \in ((\beta+2)/2, \beta)$, it follows straightforwardly that

$$\int_{\mathbb{R}^N} |x|^\gamma |f(x)|^2 \mu(dx) \le \varepsilon\mathfrak{a}_\mu(f,f) + C\varepsilon^{-\frac{\gamma}{\beta-\gamma}}\|f\|^2_{L^2(\mathbb{R}^N,\mu)}. \tag{8.4.11}$$

From this estimate we can infer that

$$\int_{\mathbb{R}^N} \log(\widetilde{\psi}_0)|f|^2\mu(dx) \ge -\varepsilon\mathfrak{a}_\mu(f,f) - C\Big(1 + \varepsilon^{-\frac{\gamma}{\beta-\gamma}}\Big)\|f\|^2_{L^2(\mathbb{R}^N,\mu)} \tag{8.4.12}$$

for any $f \in D(\mathfrak{a}_\mu)$. Indeed, there exists $R \ge |\lambda_0|^{\frac{1}{\beta}}$ such that $|\nabla W|^2 - \Delta W \ge |c| + \lambda_0$ in $\mathbb{R}^N \setminus B_R$ and, by Proposition 7.2.11, the function $\widetilde{\psi}_0$ is everywhere positive in $\mathbb{R}^N$. Therefore, we can determine a positive constant $\overline{C} = \overline{C}(R)$ such that $e^{-\overline{C}-W} \le \widetilde{\psi}_0$ in B_R. The function $\psi = e^{-\overline{C}-W}$ belongs to $C^2(\mathbb{R}^N)$ and solves the equation $q\Delta\psi + \widetilde{c}\psi = 0$ in $\mathbb{R}^N$, where $\widetilde{c} = q(|\nabla W|^2 - \Delta W)$. In particular, $q\Delta\psi \ge (|c| + \lambda_0)\psi$ in $\mathbb{R}^N \setminus B_R$. On the other hand, $q\Delta(\overline{C}\widetilde{\psi}_0) = (|c| - \lambda_0)\overline{C}\widetilde{\psi}_0$ in $\mathbb{R}^N$ Since $\psi \le \widetilde{\psi}_0$ on ∂B_R and both ψ and $\widetilde{\psi}_0$ vanish at ∞, the classical maximum principle shows that $\psi \le \widetilde{\psi}_0$ in $\mathbb{R}^N \setminus B_R$. This implies that $-\log(\widetilde{\psi}_0) \le -\log(\psi) = \overline{C} + W$ and estimate (8.4.11) yields (8.4.12).

Using (8.4.10) and (8.4.12) we can prove that

$$\int_{\mathbb{R}^N} (f^2\log(f))\widetilde{\psi}_0^2 d\mu \le \varepsilon\mathfrak{a}_\mu(\widetilde{\psi}_0 f, \widetilde{\psi}_0 f) + \beta(\varepsilon)\|f\widetilde{\psi}_0\|^2_{L^2(\mathbb{R}^N,\mu)} + \|f\widetilde{\psi}_0\|^2_{L^2(\mathbb{R}^N,\mu)}\log(\|\widetilde{\psi}_0 f\|^2_{L^2(\mathbb{R}^N,\mu)}) \tag{8.4.13}$$

for any nonnegative function $f \in L^1(\mathbb{R}^N,\mu) \cap L^\infty(\mathbb{R}^N)$ such that $\widetilde{\psi}_0 f \in D(\mathfrak{a}_\mu)$ and any $\varepsilon > 0$, where $\beta(\varepsilon) = C(1 + \varepsilon^{-\gamma/(\beta-\gamma)})$. (This estimate is known as Rosen's lemma.)

By a homogeneity argument, we can assume that $\|f\widetilde{\psi}_0\|_{L^2(\mathbb{R}^N,\mu)} = 1$. Let ς be a positive constant such that $\log_+(t) := (\log(t))^+ \le \varsigma t^{4/N}$ for any $t > 0$. Using this estimate we deduce that $\log_+(\beta f\widetilde{\psi}_0) \in L^{N/2}(\mathbb{R}^N,\mu)$ for any $\beta > 0$. Therefore, we can apply the weighted Nash inequality (8.4.10) and, since the function $f\widetilde{\psi}_0$ is in $L^2(\mathbb{R}^N,\mu)$, we conclude that

$$\int_{\mathbb{R}^N} \log_+(\beta f\widetilde{\psi}_0)f^2\widetilde{\psi}_0^2 d\mu \le \varsigma b_0\beta^{\frac{4}{N}}\mathfrak{a}_\mu(\widetilde{\psi}_0 f, \widetilde{\psi}_0 f).$$

Now, we choose $\beta > 0$ such that $\varsigma b_0 \beta^{\frac{4}{N}} = \varepsilon/2$ and use (8.4.12) to get

$$
\begin{aligned}
\int_{\mathbb{R}^N} \log(f) f^2 \widetilde{\psi}_0^2 d\mu \leq & \int_{\mathbb{R}^N} \log_+(f) f^2 \widetilde{\psi}_0^2 d\mu \\
\leq & \varsigma b_0 \beta^{\frac{4}{N}} \mathfrak{a}_\mu(\widetilde{\psi}_0 f, \widetilde{\psi}_0 f) - \int_{\mathbb{R}^N} \log(\widetilde{\psi}_0) \widetilde{\psi}_0^2 f^2 d\mu - \log(\beta) \\
\leq & \varepsilon \mathfrak{a}_\mu(\widetilde{\psi}_0 f, \widetilde{\psi}_0 f) + C\left(1 + \varepsilon^{-\frac{\gamma}{\beta-\gamma}}\right) - \log(\beta).
\end{aligned}
$$

Replacing β with its expression in terms of ε, estimate (8.4.13) follows at once.

From (8.4.13) we can infer that the form $f \mapsto \mathfrak{a}_\mu(\widetilde{\psi}_0 f, \widetilde{\psi}_0 f)$ satisfies the hypotheses of Proposition 8.4.8, with μ being replaced by the measure $\mu_{\widetilde{\psi}_0} = \widetilde{\psi}_0 \mu$. Its associated semigroup $\{\widetilde{S}(t)\}$ is defined by $\widetilde{S}(t) = \mathscr{I}_{\widetilde{\psi}_0}^{-1} S_\mu(t) \mathscr{I}_{\widetilde{\psi}_0}$ for any $t > 0$, where $\mathscr{I}_{\widetilde{\psi}_0}$ is the isometry between $L^2(\mathbb{R}^N, \mu_{\widetilde{\psi}_0})$ and $L^2(\mathbb{R}^N, \mu)$ defined by $\mathscr{I}_{\widetilde{\psi}_0} f = f\widetilde{\psi}_0$ for any $f \in L^2(\mathbb{R}^N, \mu_{\widetilde{\psi}_0})$. Therefore, from Proposition 8.4.8(ii) we deduce that

$$
\|\widetilde{S}(t)\|_{L(L^1(\mathbb{R}^N, \mu_{\widetilde{\psi}_0}), L^\infty(\mathbb{R}^N))} \leq K_1 e^{-t^{-K_2 \frac{\gamma}{\beta-\gamma}}}, \qquad t \in (0, 1],
$$

where $K_1 = K_1(\gamma)$ and $K_2 = K_2(\gamma)$. Since

$$
(\widetilde{S}(t)f)(x) = \int_{\mathbb{R}^N} (\widetilde{\psi}_0(x))^{-1} g_\mu(t, x, y) \widetilde{\psi}_0(y) f(y) dy, \qquad f \in L^2(\mathbb{R}^N, \mu_{\widetilde{\psi}_0}), \ x \in \mathbb{R}^N,
$$

estimate (8.4.8) follows observing that $\inf\left\{\frac{\gamma}{\beta-\gamma} : \gamma \in ((\beta+2)/2, \beta)\right\} = \frac{\beta+2}{\beta-2}$ and using the Dunford-Pettis theorem.

Step 2. To estimate the function $g_\mu(t, \cdot, \cdot)$ for $t > 1$, we use the formula (8.4.4) to infer that

$$
g_\mu(t, x, y) = \int_{\mathbb{R}^N} g_\mu(t - 1/2, x, z) g_\mu(1/2, y, z) \mu(dz), \qquad t > 1/2, \ x, y \in \mathbb{R}^N.
$$

By Step 1, the function $g_\mu(1/2, y, \cdot)$ belongs to $L^2(\mathbb{R}^N, \mu)$. Hence,

$$
g_\mu(t, x, y) = (S_\mu(t - 1/2) g_\mu(1/2, y, \cdot))(x), \qquad t > 1/2, \ x, y \in \mathbb{R}^N.
$$

From the formula (8.4.5) and estimate (7.2.36) we conclude that

$$
\begin{aligned}
g_\mu(t, x, x) = & \int_{\mathbb{R}^N} |g_\mu(t/2, x, y)|^2 \mu(dy) \\
\leq & M_2 e^{\lambda_0(t-1)} \|g_\mu(1/2, x, \cdot)\|_{L^2(\mathbb{R}^N, \mu)}^2 \\
= & M_2 e^{\lambda_0(t-1)} g_\mu(1, x, x) \\
\leq & K_3 e^{\lambda_0 t} (\widetilde{\psi}_0(x))^2
\end{aligned}
$$

for any $t > 1$, $x \in \mathbb{R}^N$ and some positive constant K_3, where in the last inequality we have taken advantage of (8.4.8) with $t = 1$ and $x = y$. Finally, using the inequality $g_\mu(t, x, y) \leq (g_\mu(t, x, x))^{1/2} (g_\mu(t, y, y))^{1/2}$, which holds true for any $t > 0$ and any $x, y \in \mathbb{R}^N$ (and follows from (8.4.4) and the Hölder inequality), we get (8.4.9). □

As a consequence of Theorem 8.4.6 we can estimate the behaviour as $|x|$ tends to $+\infty$ of the eigenfunction of the operator A_2.

Corollary 8.4.9 *If the assumptions of Theorem* 8.4.6 *hold, then the normalized eigenfunctions ψ_j of A_2 satisfy*

$$|\psi_j(x)| \le C_j |x|^{\frac{\alpha-\beta}{4}-\frac{N-1}{2}} \exp\left(-\int_1^{|x|} \frac{s^{\beta/2}}{(1+s^\alpha)^{1/2}}ds\right)$$

for any $x \in \mathbb{R}^N \setminus B_1$, $j \in \mathbb{N}$ and some positive constants C_j.

Proof Let λ_j be an eigenvalue of the realization A_2 in $L^2(\mathbb{R}^N)$ of the operator $\mathcal{A}$, with domain given by (7.0.2), and denote by ψ_j any normalized (i.e. $\|\psi_j\|_{L^2(\mathbb{R}^N)} = 1$) eigenfunction associated with λ_j. Then, by the Cauchy-Schwarz inequality and (8.4.5), with $2t$ instead of t, we get

$$e^{\lambda_j t}|\psi_j(x)| = \left|\int_{\mathbb{R}^N} g_\mu(t,x,y)\psi_j(y)\mu(dy)\right| \le (g_\mu(2t,x,x))^{\frac{1}{2}}, \qquad t>0,\ x\in\mathbb{R}^N.$$

Using estimate (8.4.6) the assertion follows at once. □

8.4.1 A more general class of elliptic operators

In [176] also the more general class of elliptic operators $\mathcal{A} = q\mathrm{div}(Q\nabla) + c$ has been considered under the following assumptions:

Hypotheses 8.4.10 (i) *$q(x) = 1 + |x|^\alpha$ for any $x \in \mathbb{R}^N$ and some $\alpha \in [0,2)$;*

(ii) *$q_{ij} = q_{ji}$ belongs to $C_b(\mathbb{R}^N) \cap W^{1,\infty}_{\rm loc}(\mathbb{R}^N)$ for any $i,j = 1,\ldots,N$ and there exists a positive constant κ_0 such that $\langle Q(x)\xi,\xi\rangle \ge \kappa_0|\xi|^2$ for any $x,\xi \in \mathbb{R}^N$;*

(iii) *$c \in L^1_{\rm loc}(\mathbb{R}^N)$ satisfies $c(x) \le -|x|^\beta$ for any $x \in \mathbb{R}^N$ and some $\beta > 2$ and $D_j q_{ij}(x) = \mathrm{o}(|x|^{\frac{\beta-\alpha}{2}})$ as $|x|$ tends to $+\infty$ for any $i,j = 1,\ldots,N$.*

For this class of operators the following result holds true.

Theorem 8.4.11 *Let Λ denote the sup-norm over $\mathbb{R}^N$ of the maximum eigenvalue of the matrix $Q(x)$. Then, for any $\theta \in (0,\Lambda^{-1/2})$ and $b > (\beta-2)^{-1}(\beta+2)$, there exist positive constants C and M such that*

$$\begin{aligned} g_\mu(t,x,y) \le & M e^{\lambda_{0,\theta}t} e^{Ct^{-b}} |xy|^{\frac{\alpha-\beta}{4}-\frac{N-1}{2}} \\ & \times \exp\left(-\theta\int_1^{|x|} \frac{s^{\beta/2}}{(1+s^\alpha)^{1/2}}ds - \theta\int_1^{|y|} \frac{s^{\beta/2}}{(1+s^\alpha)^{1/2}}ds\right) \end{aligned}$$

for any $t>0$ and $x,y \in \mathbb{R}^N \setminus B_1$. Here, $\lambda_{0,\theta}$ denotes the largest (negative) eigenvalue of the realization in $L^2(\mathbb{R}^N)$ of the operator $\Delta + \theta^2 c$, with domain being given by (7.0.2).

Chapter 9

The invariant measure μ and the semigroup in $L^p(\mathbb{R}^N, \mu)$

9.0 Introduction

In this chapter we deal with the invariant measure of the semigroup $\{T(t)\}$ associated with the operator $\mathcal{A}$ defined on smooth functions $u : \mathbb{R}^N \to \mathbb{R}$ by

$$\mathcal{A}u(x) = \sum_{i,j=1}^{N} q_{ij}(x)D_{ij}u(x) + \sum_{j=1}^{N} b_j(x)D_ju(x), \qquad x \in \mathbb{R}^N.$$

We assume that the operator $\mathcal{A}$ satisfies Hypotheses 1.0.1. According to Remark 1.2.11, $\{T(t)\}$ is a semigroup of contractions in $B_b(\mathbb{R}^N)$.

Throughout the chapter, for any measure μ we write L^p_μ for $L^p(\mathbb{R}^N, \mu)$ and we denote by $\|\cdot\|_p$ the norm of L^p_μ. Moreover, we write $W^{k,p}_\mu$ for $W^{k,p}(\mathbb{R}^N, \mu)$ for any $k \in \mathbb{N}$ and $p \in [1, +\infty]$.

By definition, a probability measure μ is an invariant measure for $\{T(t)\}$ if

$$\int_{\mathbb{R}^N} T(t)f d\mu = \int_{\mathbb{R}^N} f d\mu \tag{9.0.1}$$

for any $f \in B_b(\mathbb{R}^N)$ or, equivalently, for any $f \in C^\infty_c(\mathbb{R}^N)$ (see Lemma 9.1.2).

In Section 9.1 we show that, if there exists an invariant measure of $\{T(t)\}$, then the semigroup can be extended to a semigroup of bounded operators in the space L^p_μ for any $p \in [1, +\infty)$. Such an extension, which we still denote by $\{T(t)\}$, enjoys many interesting properties. First of all, it defines a strongly continuous semigroup for any $p \in [1, +\infty)$. Moreover, it is possible to describe the behaviour of the function $T(t)f$ in L^p_μ when t approaches $+\infty$. More precisely, for any $f \in L^p_\mu$, $T(t)f$ converges to the average of f with respect to the measure μ (which we denote by $\overline{f}$) in L^p_μ as t tends to $+\infty$. Such a result implies a Liouville type theorem, i.e., it implies that if $u \in D(L_p)$ satisfies $L_pu = 0$ for some $p \in (1, +\infty)$, then u is constant. Here, and in all the chapter, by L_p we denote the infinitesimal generator of the semigroup in L^p_μ.

Under Hypotheses 1.0.1, the invariant measure μ of $\{T(t)\}$, whenever existing, is unique and it is equivalent to the Lebesgue measure on the σ-algebra of the Borel sets of $\mathbb{R}^N$, in the sense that a Borel set is negligible with respect to μ if and only if it is negligible with respect to the Lebesgue measure. Moreover, the density $\rho = d\mu/dx$ satisfies $\operatorname{ess\,inf}_{x\in B_r} \rho(x) > 0$ for any $r > 0$.

We prove the existence of an invariant measure of $\{T(t)\}$ in three different ways and situations. First, we show a classical result by Khas'minskii, which guarantees the existence of the invariant measure in terms of a Lyapunov function related to the operator $\mathcal{A}$. Next we show that an invariant measure exists whenever $T(t)$ is compact in $C_b(\mathbb{R}^N)$ for any

$t>0$. Finally, we consider the case when the operator $\mathcal{A}$ is defined on smooth functions $u:\mathbb{R}^N\to\mathbb{R}$ by

$$\mathcal{A}u(x)=\Delta u(x)-\langle\nabla U(x)+G(x),\nabla u(x)\rangle,\qquad x\in\mathbb{R}^N, \tag{9.0.2}$$

and the functions U and G belong, respectively, to $C^{1+\alpha}_{\rm loc}(\mathbb{R}^N)$ and $C^1(\mathbb{R}^N,\mathbb{R}^N)$ for some $\alpha\in(0,1)$. We prove that, if $e^{-U}\in L^1(\mathbb{R}^N)$ and $\operatorname{div}G=\langle G,\nabla U\rangle$, then the measure $\mu(dx)=K^{-1}e^{-U(x)}dx$, where $K=\|e^{-U}\|_{L^1(\mathbb{R}^N)}$, is the invariant measure of $\{T(t)\}$. Moreover, the semigroup $\{T(t)\}$ and its infinitesimal generator L_2 are symmetric in L^2_μ, and the following formula holds:

$$\int_{\mathbb{R}^N}\langle\nabla f,\nabla g\rangle d\mu=-\int_{\mathbb{R}^N}gL_2f\,d\mu,\qquad f\in D(L_2),\;\; g\in W^{1,2}_\mu.$$

Notice that for functions $f\in C^\infty_c(\mathbb{R}^N)$ this formula follows immediately from an integration by parts. The proof that μ is an invariant measure of $\{T(t)\}$ is obtained using variational methods in L^p_μ.

In Section 9.2 we show some regularity properties of the invariant measure μ. We first show that, if the diffusion coefficients q_{ij} $(i,j=1,\dots,N)$ are continuously differentiable in $\mathbb{R}^N$, then μ has a density ρ which belongs to $W^{1,p}_{\rm loc}(\mathbb{R}^N)$. In particular, according to the Sobolev embedding theorems, ρ is a continuous function in $\mathbb{R}^N$. Next we specialize to the case when $\mathcal{A}u=\operatorname{div}(Q\nabla u)+\langle b,\nabla u\rangle$ on smooth functions. Allowing both the diffusion and the drift coefficients to be unbounded in $\mathbb{R}^N$, we show some global L^p- and $W^{1,p}$-regularity properties of ρ. In particular, we show that, if $b_i\in L^p_\mu$ for any $i=1,\dots,N$, then $\rho\in C_b(\mathbb{R}^N)$. Further, in the particular case when the diffusion coefficients belong to $C^1_b(\mathbb{R}^N)$, we prove some global $W^{1,q}$- and $W^{2,q}$-regularity properties of ρ. Finally, under some additional assumptions on the coefficients of $\mathcal{A}$, we show some pointwise bounds for the function ρ.

In Section 9.3 we prove some estimates for the derivatives of $T(t)f$ (up to the third-order) and for the gradient of the resolvent operator in the L^p_μ-norm. They follow integrating the pointwise estimates in Chapter 6.

In Section 9.4 we state the logarithmic Sobolev inequality

$$\int_{\mathbb{R}^N}|f|^p\log(|f|)d\mu\le\|f\|^p_{L^p_\mu}\log(\|f\|_{L^p_\mu})+\frac{p\Lambda}{2|r_0|}\int_{\mathbb{R}^N}|f|^{p-2}|\nabla f|^2\chi_{\{f\neq0\}}d\mu$$

for any $p\in(1,+\infty)$ and any function $f\in C^1_b(\mathbb{R}^N)$ (which can be extended to any $f\in W^{1,p}_\mu$ if $p\ge2$) and some of its remarkable consequences, such as the Poincaré inequality and hypercontractivity of the semigroup $\{T(t)\}$. Then, we consider other summability improving properties of the semigroup $\{T(t)\}$ such as its supercontractivity, ultraboundedness and ultracontractivity. Most of the results can be obtained simplifying the arguments of the non-autonomous case, addressed in Chapter 16. For this reason, here we provide the proofs only of the results which are not presented in Chapter 16. We stress that, in general, the Sobolev embedding theorems do not hold when the Lebesgue measure is replaced by an invariant measure. Hence, it is not immediate to conclude that $T(t)$ maps a function $f\in L^p_\mu$ (for some p) into a function belonging to L^q_μ for some $q>p$. Actually, without any further assumption on the coefficients of the operator $\mathcal{A}$, this property is false in general (see Remark 9.4.7). Roughly speaking the logarithmic Sobolev inequality is, for invariant measures, the counterpart of the Sobolev embedding theorems. Differently from the Sobolev embedding theorems, it is independent of the dimension of the Euclidean space $\mathbb{R}^N$.

In Section 9.5, under suitable conditions on the coefficients of the operator $\mathcal{A}$, which allow us to write explicitly the invariant measure μ, we prove that the semigroup $\{T(t)\}$ is also analytic in L^2_μ and characterize the domain of its infinitesimal generator as the set of all $u \in L^2_\mu$ such that the functions $|\sqrt{Q}\nabla u|$ and $|\sqrt{Q}D^2u\sqrt{Q}|$ are in L^2_μ. We also show that $D(L_2)$ coincides with the domain of the maximal realization of the operator $\mathcal{A}$ in L^2_μ and obtain further estimates for the derivatives of the resolvent $R(\lambda, L_2)f$. These results cover e.g. the case when $\mathcal{A} = \Delta - \langle \nabla U, \nabla\rangle$ and U is a convex function.

Then, we provide a sufficient condition for the embedding $W^{1,p}_\mu \subset L^p_\mu$ to be compact for any $p \in (1 + \infty)$. In the case when $\mathcal{A} = \Delta - \langle \nabla U, \nabla\rangle$, U being a convex function, such a condition requires that $|\nabla U(x)|$ diverges to $+\infty$ as $|x|$ tends to $+\infty$ and $|\Delta U| \leq \delta_1|\nabla U|^2 + \delta_2$ in $\mathbb{R}^N$, plus the additional condition $\langle D^2U\nabla U, \nabla U\rangle \geq (\delta_3|\nabla U|^2 + \delta_4)|\nabla U|^2$ in $\mathbb{R}^N$ if $p \in (1, 2)$. Here, $\delta_1, \ldots, \delta_4$ are suitable constants. The results of this section are taken from [173, 190].

Finally, in Section 9.6 we provide sufficient conditions on the coefficients of the operator $\mathcal{A}$ in (9.0.2) such that the associated semigroup is not analytic in L^p_μ.

9.1 Existence, uniqueness and general properties

In this section we deal with the problem of the existence and uniqueness of the invariant measure of $\{T(t)\}$ and prove some general properties of invariant measures. If not otherwise specified, we assume that the coefficients of the operator $\mathcal{A}$ satisfy Hypotheses 1.0.1(i), (ii).

9.1.1 General properties and uniqueness

To begin with we characterize the invariant measures of $\{T(t)\}$ in terms of the weak generator $\widehat{A}$ of the semigroup (see Section 1.3).

Proposition 9.1.1 *A Borel probability measure μ is an invariant measure of $\{T(t)\}$ if and only if*

$$\int_{\mathbb{R}^N} \widehat{A}f\,d\mu = 0, \qquad f \in D(\widehat{A}). \tag{9.1.1}$$

To prove the proposition, we need the following lemma.

Lemma 9.1.2 *A Borel probability measure μ such that*

$$\int_{\mathbb{R}^N} T(t)f\,d\mu = \int_{\mathbb{R}^N} f\,d\mu, \qquad t > 0, \tag{9.1.2}$$

for any $f \in C^\infty_c(\mathbb{R}^N)$ is an invariant measure of $\{T(t)\}$.

Proof Fix $f \in B_b(\mathbb{R}^N)$. We claim that we can approximate f with a sequence $\{f_n\} \subset C^\infty_c(\mathbb{R}^N)$ which is bounded with respect to the sup-norm and such that f_n tends to f almost everywhere in $\mathbb{R}^N$ as n tends to $+\infty$. Observe that a convolution argument with standard mollifiers proves the claim when $f \in C_b(\mathbb{R}^N)$. For a general $f \in B_b(\mathbb{R}^N)$ it suffices to observe that, according to [97, Proposition 3.4.2]), we can determine a sequence $\{f_n\} \subset C_b(\mathbb{R}^N)$ such that $\|f_n\|_\infty \leq C$ for any $n \in \mathbb{N}$ and $\lim_{n\to+\infty} f_n(x) = f(x)$ for almost every $x \in \mathbb{R}^N$.

Now, the assertion easily follows from the dominated convergence theorem, writing the formula (9.1.2) with f_n instead of f and letting n tend to $+\infty$. Indeed, by Remark 1.2.11, the sequence $\{T(\cdot)f_n\}$ converges to $T(\cdot)f$ pointwise in $(0, +\infty) \times \mathbb{R}^N$, as n tends to $+\infty$, and $\sup_{n\in\mathbb{N}} \|T(t)f_n\|_\infty < +\infty$. □

Proof of Proposition 9.1.1 According to (1.3.12), for any $f \in D(\widehat{A})$, $t > 0$ and $x \in \mathbb{R}^N$, we can write

$$(T(t)f)(x) - f(x) = \int_0^t (T(s)\widehat{A}f)(x)ds. \tag{9.1.3}$$

Let μ be an invariant measure of $\{T(t)\}$ and fix $f \in D(\widehat{A})$. Then

$$\int_{\mathbb{R}^N} \frac{T(t)f - f}{t} d\mu = 0, \qquad t > 0. \tag{9.1.4}$$

By (9.1.3), recalling that $\{T(t)\}$ is a semigroup of contractions since $c \equiv 0$ (see Theorem 1.2.5), it follows that $t^{-1}|(T(t)f)(x) - f(x)| \le \|\widehat{A}f\|_\infty$ for any $t > 0$ and $x \in \mathbb{R}^N$. Thus, (9.1.1) follows letting t tend to 0^+ in (9.1.4), using the dominated convergence theorem.

Conversely, assume that μ is a Borel probability measure satisfying (9.1.1) and fix $f \in C_c^\infty(\mathbb{R}^N)$. By Lemma 1.3.2 and Proposition 1.3.5, f and $T(s)f$ belong to $D(\widehat{A})$ and $T(s)\widehat{A}f = \widehat{A}T(s)f$ for any $s > 0$. Integrating (9.1.3) in $\mathbb{R}^N$ and using the Fubini theorem, we get

$$\int_{\mathbb{R}^N} (T(t)f - f)d\mu = \int_0^t ds \int_{\mathbb{R}^N} \widehat{A}T(s)f d\mu = 0, \qquad t > 0,$$

and, then, the conclusion follows from Lemma 9.1.2. □

Remark 9.1.3 A probability measure μ, which solves equation (9.1.1) for any $f \in C_c^\infty(\mathbb{R}^N)$, is usually called *infinitesimally invariant*. In general an infinitesimally invariant measure μ is not an invariant measure of the semigroup $\{T(t)\}$. Indeed, in [241, Example 1.12], W. Stannat shows that the measure $d\mu = e^{-x^2}dx$ is infinitesimally invariant for the one-dimensional elliptic operator defined by $\mathcal{A}u(x) = u''(x) - (2x + 6e^{x^2})u'(x)$ for any $x \in \mathbb{R}$, on smooth functions $u : \mathbb{R} \to \mathbb{R}$, but it is not invariant for the semigroup associated with the operator $\mathcal{A}$.

Checking that the measure μ is infinitesimally invariant for the operator $\mathcal{A}$ is an easy task. According to the forthcoming Proposition 9.1.9, to prove that it is not invariant for the associated semigroup $\{T(t)\}$ it suffices to show that $\{T(t)\}$ is not conservative. By Theorem 2.2.2, this is the case if $+\infty$ or $-\infty$ are accessible. In fact, we are going to prove that $+\infty$ is an entrance and $-\infty$ is an exit point. For this purpose, we show that the functions $\mathscr{P}, \mathscr{R} : \mathbb{R} \to \mathbb{R}$, defined by

$$\mathscr{P}(x) = \frac{1}{\mathscr{W}(x)} \int_0^x \mathscr{W}(t)dt, \qquad \mathscr{R}(x) = \mathscr{W}(x) \int_0^x \frac{1}{\mathscr{W}(t)}dt, \qquad x \in \mathbb{R},$$

where

$$\mathscr{W}(x) = \exp\left(x^2 + 6\int_0^x e^{t^2}dt\right), \qquad x \in \mathbb{R},$$

(see (2.1.3)-(2.1.5)) are such that $\mathscr{P} \in L^1((0, +\infty))$, $\mathscr{P} \notin L^1((-\infty, 0))$, $\mathscr{R} \in L^1((-\infty, 0))$, $\mathscr{R} \notin L^1((0, +\infty))$. This easily follows observing that

$$\lim_{x \to +\infty} x^4 \mathscr{P}(x) = \lim_{x \to -\infty} x^4 \mathscr{R}(x) = 0, \qquad \lim_{x \to -\infty} \mathscr{P}(x) = \lim_{x \to +\infty} \mathscr{R}(x) = +\infty.$$

Let us now exploit some properties of the invariant measures.

Proposition 9.1.4 *Let μ be an invariant measure of $\{T(t)\}$. Then, μ is equivalent*[1] *to the Lebesgue measure m on the σ-algebra of the Borel sets of $\mathbb{R}^N$. Besides, for any fixed $r > 0$, the density ρ of μ with respect to the Lebesgue measure satisfies $\operatorname{ess\,inf}_{x\in B_r} \rho(x) > 0$.*

Proof For any Borel set $B \subset \mathbb{R}^N$, any $t > 0$ and $x \in \mathbb{R}^N$ we can write

$$(T(t)\chi_B)(x) = \int_B g(t,x,y)dy,$$

see Theorem 1.2.5, and, by (9.0.1),

$$\mu(B) = \int_{\mathbb{R}^N} \chi_B d\mu = \int_{\mathbb{R}^N} T(t)\chi_B d\mu.$$

Therefore, if $m(B) = 0$, then $(T(t)\chi_B)(x) = 0$ for any $t > 0$ and $x \in \mathbb{R}^N$, since g is strictly positive in $(0,+\infty) \times \mathbb{R}^N \times \mathbb{R}^N$. Thus, $\mu(B) = 0$. Conversely, if $m(B) > 0$, then $T(\cdot)\chi_B$ is positive in $(0,+\infty) \times \mathbb{R}^N$ and, therefore, $\mu(B) > 0$. Thus, μ and m are equivalent.

Now, fix $R > r > 0$ and let g_R be the Green function of the semigroup associated with the realization of the operator $\mathcal{A}$ in $C(\overline{B}_R)$ with homogeneous Dirichlet boundary conditions on ∂B_R (see Proposition D.3.1). Since $g_R(t,x,y) \le g(t,x,y)$ for any $R > 0$, $t > 0$ and $x, y \in B_R$ (see the proof of Theorem 1.2.5), for any positive function $f \in B_b(\mathbb{R}^N)$, it holds that

$$\begin{aligned}\int_{\mathbb{R}^N} f(x)\rho(x)dx &= \int_{\mathbb{R}^N} (T(1)f)(x)\rho(x)dx = \int_{\mathbb{R}^N}\int_{\mathbb{R}^N} g(1,x,y)f(y)\rho(x)dy\,dx \\ &\ge \int_{\mathbb{R}^N} f(y)\int_{B_R} g_R(1,x,y)\rho(x)dx\,dy.\end{aligned}$$

The arbitrariness of f implies that $\rho(y) \ge F(y) = \|g_R(1,\cdot,y)\rho\|_{L^1(\mathbb{R}^N)}$ for almost every $y \in B_R$. Since $g_R(1,\cdot,\cdot)$ is positive, bounded and continuous in $B_R \times B_R$, the function F is positive, bounded and continuous in B_R. Therefore, $\rho(y) \ge \inf_{B_r} F > 0$ for almost every $y \in B_r$. □

As an immediate consequence of the results in Proposition 9.1.4, we get the following corollary.

Corollary 9.1.5 *Any invariant measure μ of $\{T(t)\}$ can be extended to a complete probability measure defined on the σ-algebra of all the Lebesgue measurable sets.*

Moreover, since the Lebesgue measure is regular, Proposition 9.1.4 immediately implies that μ is regular as well. Hence, applying [231, Theorem 3.14], we get the following useful result.

Corollary 9.1.6 *If μ is an invariant measure of $\{T(t)\}$, then $C_c^\infty(\mathbb{R}^N)$ is dense in L^p_μ for any $p \in [1,+\infty)$.*

We can now prove the following result.

Proposition 9.1.7 *Let μ be an invariant measure of $\{T(t)\}$. For any $p \in [1,+\infty)$, $\{T(t)\}$ extends to a strongly continuous semigroup of contractions in L^p_μ.*

---

[1]in the sense that μ and the Lebesgue measure have the same sets of zero measure.

Proof We begin the proof recalling that

$$(T(t)f)(x) = \int_{\mathbb{R}^N} f(y)p(t,x;dy), \qquad t>0,\ x\in\mathbb{R}^N,\ f\in C_b(\mathbb{R}^N),$$

where $p(t,x;dy)$ are positive measures satisfying $p(t,x;\mathbb{R}^N)\le 1$ for any $t>0$ and $x\in\mathbb{R}^N$.
Using the Hölder inequality we deduce that

$$|(T(t)f)(x)|^p = \left|\int_{\mathbb{R}^N} f(y)p(t,x;dy)\right|^p \le \int_{\mathbb{R}^N} |f(y)|^p p(t,x;dy) = (T(t)|f|^p)(x)$$

for any $f\in C_b(\mathbb{R}^N)$, $t>0$ and $x\in\mathbb{R}^N$. Therefore,

$$\int_{\mathbb{R}^N} |T(t)f|^p d\mu \le \int_{\mathbb{R}^N} T(t)(|f|^p)d\mu = \int_{\mathbb{R}^N} |f|^p d\mu, \qquad t>0.$$

Thus, since $C_b(\mathbb{R}^N)$ is dense in L^p_μ (due to Corollary 9.1.6), for any $t>0$ the operator $T(t)$ can be uniquely extended to a contraction in L^p_μ. The strong continuity of $\{T(t)\}$ in L^p_μ now is an immediate consequence of Propositions 1.2.8 and B.1.2. □

For any $p\in[1,+\infty)$, we denote by L_p the generator of $\{T(t)\}$ in L^p_μ. From the formula (9.0.1) it readily follows that

$$\int_{\mathbb{R}^N} L_p f d\mu = 0, \qquad f\in D(L_p),\ p\in[1,+\infty). \tag{9.1.5}$$

Besides, for any $p\in[1,+\infty)$ the resolvent $R(\lambda,L_p)$ is an extension of $R(\lambda)$ (see Theorem 1.1.3) to L^p_μ and $\|R(\lambda,L_p)f\|_p\le\lambda^{-1}\|f\|_p$ for any $f\in D(L_p)$, $\lambda>0$ and $p\in[1,+\infty)$. As the following propositions show, $D(\widehat{A})$ is a core of L_p and $\{T(t)\}$ is conservative.

Proposition 9.1.8 *Let $\{T(t)\}$ admit an invariant measure μ. Then, $D(\widehat{A})$ is a core of L_p for any $p\in[1,+\infty)$.*

Proof Let us begin by observing that $D(\widehat{A})\subset D(L_p)$. This follows easily recalling that $R(\lambda,L_p)_{|C_b(\mathbb{R}^N)} = R(\lambda)$ and $R(\lambda)(C_b(\mathbb{R}^N)) = D(\widehat{A})$.
Now, since $C_c^\infty(\mathbb{R}^N)\subset D(\widehat{A})$ and $C_c^\infty(\mathbb{R}^N)$ is dense in L^p_μ (see Proposition 1.3.5 and Corollary 9.1.6), $D(\widehat{A})$ is dense in L^p_μ as well. Besides, $T(t)(D(\widehat{A}))\subset D(\widehat{A})$ for any $t>0$ (see Lemma 1.3.2). Thus, the conclusion follows from Proposition C.1.14. □

Proposition 9.1.9 *Assume Hypotheses* 1.0.1(*i*), (*ii*). *If there exists an invariant measure μ of $\{T(t)\}$, then $\{T(t)\}$ is conservative.*

Proof Fix $t>0$. According to Theorem 1.2.5, $0<T(t)\mathbb{1}\le\mathbb{1}$ for any $t>0$. Besides, the formula (9.0.1), with $f\equiv\mathbb{1}$, yields that

$$\int_{\mathbb{R}^N} T(t)\mathbb{1}d\mu = \int_{\mathbb{R}^N}\mathbb{1}d\mu = 1, \qquad t>0,$$

and, therefore, $T(t)\mathbb{1}\equiv\mathbb{1}$ μ-almost everywhere in $\mathbb{R}^N$. Since $T(t)\mathbb{1}$ is a continuous function and μ is equivalent to the Lebesgue measure, then $T(t)\mathbb{1}\equiv\mathbb{1}$ everywhere in $\mathbb{R}^N$, for any $t>0$. □

Remark 9.1.10 For sufficient conditions for $C_c^\infty(\mathbb{R}^N)$ to be a core of L_p, we refer the reader to [9, 10, 11] and also to Subsection 17.1.1 which deals with the non-autonomous case.

Now, we show that an invariant measure of $\{T(t)\}$ is always ergodic, i.e., that

$$\lim_{t\to+\infty} t^{-1}\int_0^t T(s)fds = \int_{\mathbb{R}^N} fd\mu, \qquad f\in L^2_\mu,$$

where the limit is meant in L^2_μ.

Proposition 9.1.11 *Let μ be an invariant measure of $\{T(t)\}$. Then, μ is ergodic.*

To prove the proposition, we need some preliminary results.

Lemma 9.1.12 *Let H be a Hilbert space and let $T\in L(H)$ be a bounded operator such that $\|T^k\|_{L(H)}\le M$ for any $k\in\mathbb{N}$, some constant $M>0$ and any $k\in\mathbb{N}$. Then, the sequence of bounded operators $\{P_n\}$, defined by $P_nx = n^{-1}\sum_{k=0}^{n-1}T^kx$ for any $x\in H$ and $n\in\mathbb{N}$, converges pointwise to a bounded operator P_*, which is a projection on the kernel of the operator $I-T$.*

Proof A straightforward computation shows that P_nx converges to x in H, as n tends to $+\infty$, for any $x\in \mathrm{Ker}(I-T)$. Similarly, P_nx converges in H for any $x\in(I-T)(H)$. Since the norms of the operators T_n are equibounded by M, by Proposition B.1.2 P_nx converges in H for any $x\in\overline{(I-T)(H)}$.

Next, we prove that the sequence $\{P_nx\}$ converges for any $x\in H$. Since the previous sequence is bounded, up to a subsequence we can assume that P_nx converges weakly to some $y\in H$ as n tends to $+\infty$. As a consequence, $T(P_nx)$ converges weakly to Ty in H. Since

$$T(P_nx) = P_nx + \frac{1}{n}(T^nx - x), \qquad n\in\mathbb{N}, \tag{9.1.6}$$

we immediately deduce that $y\in \mathrm{Ker}\,(I-T)$. We now split $x=y+(x-y)$ and claim that $x-y\in\overline{(I-T)(X)}$. Once the claim is proved, we will conclude that P_nx converges in the strong topology of H. To prove the claim, we observe that $x-y$ is the weak limit of the sequence $x-P_nx$ and $x-P_nx\in(I-T)(X)$ for any $n\in\mathbb{N}$. Indeed,

$$x-P_nx = \frac{1}{n}\sum_{k=0}^{n-1}(I-T^k)(x) = (I-T)\left(\frac{1}{n}\sum_{k=0}^{n-1}\sum_{j=0}^{k-1}T^j\right)(x).$$

Since $(I-T)(H)$ is convex, its weak and strong closures coincide. The claim follows.

To conclude the proof, we set $P_*x=\lim_{n\to+\infty}P_nx$ for any $x\in H$ and show that P_* is a projection on $\mathrm{Ker}\,(I-T)$. Letting n tend to $+\infty$ in (9.1.6) we get $TP_*x=P_*x$. This implies that $T^kP_*x=P_*x$ for any $x\in H$, $k\in\mathbb{N}$ and, consequently, $P_*^2=P_*$, so that P_* is a projection. Moreover, the formula $TP_*=P_*$ shows that $P_*(H)\subset\mathrm{Ker}\,(I-T)$. Since we already know that $\mathrm{Ker}\,(I-T)\subset P_*(H)$, we conclude that $P_*(H)=\mathrm{Ker}\,(I-T)$ and we are done. □

Proposition 9.1.13 (Von Neumann) *The limit $\lim_{t\to+\infty}t^{-1}\int_0^tT(s)fds$ is well defined for any $f\in L^2_\mu$. Here, the integral is meant in the L^2_μ-norm. Moreover, if we denote by $P_\infty f$ the previous limit, then the operator P_∞ is a projection on the subspace $C:=\{f\in L^2_\mu : T(t)f=f\ \mu\text{-a.e. for any } t>0\}$.*

The set C enjoys the following properties:

(i) *if $f\in C$, then $|f|\in C$;*

(ii) *if $f,g\in C$, then $f\wedge g$ and $f\vee g$ belong to C. In particular, f^+ and f^- belong to C;*

(iii) *if $f \in C$, then, for any $\lambda \in \mathbb{R}$, the function $\chi_{\{x:\ f(x)>\lambda\}}$ belongs to C.*

Finally,

$$\int_{\mathbb{R}^N} P_\infty f d\mu = \int_{\mathbb{R}^N} f d\mu, \qquad f \in L^2_\mu. \tag{9.1.7}$$

Proof Let $P_t \in L(L^2_\mu)$ be the operator defined by $P_t f = t^{-1}\int_0^t T(s) f ds$ for any $f \in L^2_\mu$ and $t > 0$. Denoting, respectively, by $[t]$ and $\{t\}$ the integer and the fractional part of $t > 0$, we can write

$$\begin{aligned} P_t f =& \frac{1}{t}\sum_{k=0}^{[t]-1}\int_0^1 T(s+k) f ds + \frac{1}{t}\int_0^{\{t\}} T(s+[t]) f ds \\ =& \frac{[t]}{t}\frac{1}{[t]}\sum_{k=0}^{[t]-1} T^k(1)(P_1 f) + \frac{\{t\}}{t} T^{[t]}(1)(P_{\{t\}} f). \end{aligned} \tag{9.1.8}$$

Since the L^2_μ-norm of $T^k(1)$ is bounded by 1 for any $k \in \mathbb{N}$, Lemma 9.1.12 implies that there exists a projection P_* such that $\frac{1}{n}\sum_{k=0}^{n-1} T^k(1) f$ converges to $P_* f$, as n tends to $+\infty$, for any fixed $f \in L^2_\mu$. Therefore, letting t tend to $+\infty$ in (9.1.8), we deduce that P_t converges pointwise to $P_\infty = P_* \circ P_1$ as t tends to $+\infty$.

To prove that P_∞ is a projection it suffices to show that

$$T(r) \circ P_\infty = P_\infty, \qquad r > 0. \tag{9.1.9}$$

Indeed, this will imply that $P_1 \circ P_\infty = P_\infty$ and $P_* \circ P_\infty = P_\infty$, which, of course, gives $P_\infty^2 = P_\infty$. To prove (9.1.9) we observe that

$$\begin{aligned} T(r)P_t f =& \frac{1}{t}T(r)\int_0^t T(s) f ds = \frac{1}{t}\int_0^t T(r+s) f ds = \frac{1}{t}\int_r^{r+t} T(s) f ds \\ =& P_t f + \frac{1}{t}\int_0^r (T(t)-1)T(s) f ds. \end{aligned} \tag{9.1.10}$$

Hence, letting t tend to $+\infty$ in (9.1.10) we get (9.1.9).

Now, showing that P_∞ is a projection on C is immediate. The inclusion $C \subset P_\infty(L^2_\mu)$ is straightforward, whereas the other inclusion follows from (9.1.9). Indeed, since $P_\infty f = f$ for any $f \in C$, C is contained in $P_\infty(L^2_\mu)$. Hence, the first part of the assertion follows.

To prove the second part of the proposition, we begin by observing that, since $|T(t)g| \le T(t)|g|$ for any $t > 0$ and any function $g \in C_b(\mathbb{R}^N)$, by density, such an inequality can be extended to any $g \in L^2_\mu$. Fix now $f \in C$. Then, $|f| = |T(t)f| \le T(t)|f|$ for any $t > 0$. The invariance property of μ implies that $\|T(t)|f| - |f|\|_1 = 0$, so that $T(t)|f| = |f|$ μ-almost everywhere for any $t > 0$. Therefore, $|f|$ belongs to C. Now, since $f^+ = (f + |f|)/2$ and $f^- = (f - |f|)/2$, f^+ and f^- belong to C if f does. Finally, since $f \vee g = g + (f-g)^+$ and $f \wedge g = g + (f-g)^-$, $f \vee g$ and $f \wedge g$ belong to C if f and g do.

To prove property (iii), we fix $\lambda \in \mathbb{R}$, $f \in C$ and set $f_n = (n(f-\lambda)^+) \wedge 1$ for any $n \in \mathbb{N}$. As is immediately seen, f_n converges to pointwise to $\chi_{\{x:\ f(x)>\lambda\}}$ as n tends to $+\infty$. By Remark 1.2.11 it follows that $T(\cdot)f_n$ tends to $T(\cdot)\chi_{\{x:\ f(x)>\lambda\}}$ pointwise in $(0,+\infty)\times\mathbb{R}^N$. Since each function f_n belongs to C, the function $\chi_{\{x:\ f(x)>\lambda\}}$ is in C as well.

Finally, since

$$\int_{\mathbb{R}^N} P_t f d\mu = \frac{1}{t}\int_{\mathbb{R}^N} d\mu \int_0^t T(s) f ds = \frac{1}{t}\int_0^t ds \int_{\mathbb{R}^N} T(s) f d\mu = \int_{\mathbb{R}^N} f d\mu$$

for any $t > 0$, letting t tend to $+\infty$, the dominated convergence theorem yields (9.1.7). □

Proof of Proposition 9.1.11 We begin the proof observing that, if A is a Borel set, then χ_A belongs to the set $C = \{f \in L^2_\mu : T(t)f = f \ \mu\text{-}a.e. \ for \ any \ t > 0\}$ if and only if $\mu(A) = 0$ or $\mu(A) = 1$. Clearly, if $\mu(A) = 0$ or $\mu(A) = 1$, then $\chi_A \in C$. On the other hand, if $\mu(A) \in (0,1)$, then χ_A belongs to C. Indeed, due to Remark 1.2.14, the function $T(1)\chi_A$ is strictly positive in $\mathbb{R}^N$. Hence, $T(1)\chi_A \neq \chi_A$ on $\mathbb{R}^N \setminus A$ which is not μ-negligible.

We can now show that C contains only the constant functions. For this purpose, let $f \in C$ be a non constant function. Then, by Proposition 9.1.13(iii) the characteristic function of the set $A_\lambda = \{x \in \mathbb{R}^N : f(x) > \lambda\}$ is in C for any $\lambda \in \mathbb{R}$. Therefore, $\mu(A_\lambda) = 0$ or $\mu(A_\lambda) = 1$. Observe that there should exist $\lambda \in \mathbb{R}$ such that $\mu(A_\lambda) = 1$ since $\mathbb{R}^N = \{x \in \mathbb{R}^N : f(x) = \pm\infty\} \cup \bigcup_{n \in \mathbb{N}\cup\{+\infty\}} A_{-n}$. Similarly, there exists $\lambda \in \mathbb{R}$ such that $\mu(A_\lambda) = 0$, otherwise we would have $f(x) = +\infty$ for μ-almost every $x \in \mathbb{R}^N$, which can not be the case. Set $\lambda_0 = \sup\{\lambda \in \mathbb{R} : \mu(A_\lambda) = 1\}$. We infer that $f = \lambda_0$ μ-almost everywhere in $\mathbb{R}^N$. Indeed, by definition of λ_0, f is not less than λ_0 μ-almost everywhere in $\mathbb{R}^N$. Moreover, since $\mu(A_\lambda) = 0$ for any $\lambda > \lambda_0$, it follows that $\mu(A_{\lambda_0}) = 0$, so that f is μ-almost everywhere constant in $\mathbb{R}^N$.

Since C contains only the constant functions, its dimension is 1. Therefore, we can find a linear operator $S \in (L^2_\mu)'$ such that $P_\infty f = S(f)\mathbb{1}$ for any $f \in L^2_\mu$. According to the Riesz-Fisher representation theorem, there exists $g \in L^2_\mu$ such that $Sf = \int_{\mathbb{R}^N} fg d\mu$ for any $f \in L^2_\mu$. To complete the proof we just need to show that $g = \mathbb{1}$ μ-almost everywhere. For this purpose we observe that from (9.1.7) we get

$$\int_{\mathbb{R}^N} f d\mu = \int_{\mathbb{R}^N} P_\infty f d\mu = \int_{\mathbb{R}^N} fg d\mu, \qquad f \in L^2_\mu.$$

By the arbitrariness of f, it follows that $g = \mathbb{1}$ μ-almost everywhere and we are done. □

Remark 9.1.14 The results of Proposition 9.1.13 imply, in particular, that there exists a subsequence $\{t_n\}$ diverging to $+\infty$ such that $P_{t_n} f$ converges pointwise, as n tends to $+\infty$, to $\overline{f}$. As the Chacon-Ornstein theorem shows (see [214, Chapter 3, Section 8]), the pointwise limit is $\overline{f}$, for any $f \in L^1_\mu$, and we can take $t_n = n$ for any $n \in \mathbb{N}$.

Now we can prove that $\{T(t)\}$ admits at most one invariant measure. The uniqueness of the invariant measure has been proved in a more general context by J.L. Doob. We refer the reader to [80] for further details.

Theorem 9.1.15 *There exists at most one invariant measure of* $\{T(t)\}$.

Proof Let μ_1 and μ_2 be two invariant measures of $\{T(t)\}$. Let us prove that $\mu_1 = \mu_2$. According to Proposition 9.1.11, μ_1 and μ_2 are ergodic measures. Therefore, for any Borel set A, we can find two Borel sets M_1 and M_2 such that $\mu_j(M_j) = 1$ for $j = 1, 2$ and

$$\lim_{n\to+\infty} \frac{1}{n} \int_0^n (T(s)\chi_A)(x) ds = \int_{\mathbb{R}^N} \chi_A d\mu_j = \mu_j(A), \qquad x \in M_j, \ \ j = 1, 2,$$

(see Remark 9.1.14). But the measures μ_1 and μ_2 are equivalent (see Proposition 9.1.4). Hence, $M_1 \cap M_2 \neq \varnothing$, so that $\mu_1(A) = \mu_2(A)$. □

To conclude this subsection we deal with the behaviour of $T(t)$ as t tends to $+\infty$ and we prove a Liouville type theorem. The following result is due to Doob. We refer the reader to Chapter 18, where the non-autonomous case is considered. The results presented in that chapter hold also in the autonomous case, where the evolution systems of measures are replaced by the invariant measure.

Theorem 9.1.16 *For any $f \in L^p_\mu$ and $p \in [1,+\infty)$, the function $T(t)f$ converges to $\overline{f} = \int_{\mathbb{R}^N} f d\mu$ in L^p_μ as t tends to $+\infty$.*

As a straightforward consequence of Theorem 9.1.16 we get the following Liouville type theorem.

Theorem 9.1.17 *Suppose that the semigroup $\{T(t)\}$ admits an invariant measure μ. Moreover, let $p \in [1,+\infty)$ and $u \in D(L_p)$ be such that $L_p u = 0$. Then, u is constant.*

Proof Fix $u \in \mathrm{Ker}(L_p)$. Since $\{T(t)\}$ is a strongly continuous semigroup, the function $t \mapsto T(t)u$ is continuously Fréchet differentiable in $[0,+\infty)$ with values in L^p_μ and $D_t T(t)u = T(t)L_p u = 0$ for any $t > 0$ and $p \in [1,+\infty)$ (see Section C.1). This implies that the function $t \mapsto T(t)u$ is constant and, consequently, $T(t)u = u$ for any $t > 0$. Letting t tend to $+\infty$, from Theorem 9.1.16 we deduce that $u = \int_{\mathbb{R}^N} u d\mu$, so that u is constant. □

9.1.2 Existence by Khas'minskii theorem

The main result concerning the existence of an invariant measure of a Markov semigroup is the Khas'minskii theorem. To prove it, we need some preliminaries.

Theorem 9.1.18 (Prokhorov) *A family $\mathscr{F}$ of Borel probability measures on $\mathbb{R}^N$ is tight (see Definition 4.1.1) if and only if, for any sequence $\{\mu_n\} \subset \mathscr{F}$, there exists a subsequence weakly* convergent in the dual space $C_b(\mathbb{R}^N)'$ to a probability measure.*

Proof To begin the proof, we assume that the family $\mathscr{F}$ is tight, and we fix a sequence $\{\mu_n\} \subset \mathscr{F}$. For any $n \in \mathbb{N}$, let us consider the restriction $\mu_{n,1}$ of the measure μ_k to $\overline{B}_1$. Since $C(\overline{B}_1)$ is separable, the weak* topology of $(C(\overline{B}_1))'$ is metrizable. Therefore, up to a subsequence, $\mu_{n,1}$ weakly* converges to some measure μ_1. Applying the same argument to the restrictions of the sequence $\{\mu_n\}$ to the ball B_k ($k \in \mathbb{N}$) and using a diagonal procedure, we can determine a subsequence $\{\mu_{k_n}\} \subset \{\mu_n\}$ such that for any $m \in \mathbb{N}$ μ_{k_n} converges weakly* to a Borel measure $\mu^{(m)}$ on $C(\overline{B}_m)$. Then, for any positive function $f \in C_b(\mathbb{R}^N)$ and $m \in \mathbb{N}$, we have

$$\int_{\overline{B}_m} f d\mu^{(m)} = \lim_{n\to+\infty} \int_{\overline{B}_m} f d\mu_{k_n} \le \lim_{n\to+\infty} \int_{\overline{B}_{m+1}} f d\mu_{k_n}. \tag{9.1.11}$$

Now, fix a Borel set B and a bounded sequence $\{f_j\} \subset C_b(\mathbb{R}^N)$ of positive functions converging pointwise to χ_B as j tends to $+\infty$. Taking (9.1.11) into account, we can write

$$\begin{aligned}\mu^{(m)}(B\cap\overline{B}_m) &= \int_{\overline{B}_m} \chi_B d\mu^{(m)} = \lim_{j\to+\infty} \int_{\overline{B}_m} f_j d\mu^{(m)} \le \lim_{j\to+\infty} \int_{\overline{B}_{m+1}} f_j d\mu^{(m+1)} \\ &= \int_{\overline{B}_{m+1}} \chi_B d\mu^{(m+1)} = \mu^{(m+1)}(B\cap\overline{B}_m).\end{aligned}$$

Since the sequence $\{\mu^{(m)}(B\cap\overline{B}_m)\}$ is positive, increasing, and bounded from above by 1, we can define a function μ on the σ-algebra of the Borel sets of $\mathbb{R}^N$ by setting $\mu(B) := \lim_{m\to+\infty} \mu^{(m)}(B\cap\overline{B}_m)$ for any Borel set B. As is immediately seen, μ is a Borel measure.

Let us now prove that μ is a probability measure and μ_{k_n} converges weakly* to μ. For this purpose, we fix $\varepsilon > 0$ and $h \in \mathbb{N}$ such that $\mu_{k_n}(\mathbb{R}^N\setminus B_h) \le \varepsilon$ for any $n \in \mathbb{N}$. Moreover, for any $m > h$, we denote by g_m any continuous function such that $\chi_{\overline{B}_m\setminus B_{h+1}} \le g_m \le \chi_{\overline{B}_{m+1}\setminus B_h}$. Then,

$$\mu(\overline{B}_m\setminus B_{h+1}) \le \int_{\overline{B}_m} g_m d\mu = \lim_{n\to+\infty} \int_{\overline{B}_m} g_m d\mu_{k_n} \le \limsup_{n\to+\infty} \mu_{k_n}(\overline{\mathbb{R}}^N\setminus B_h) \le \varepsilon.$$

Letting m tend to $+\infty$ gives $\mu(\mathbb{R}^N \setminus B_{h+1}) \le \varepsilon$ and we are almost done. Indeed, for any $f \in C_b(\mathbb{R}^N)$ we can estimate

$$\begin{aligned}\left|\int_{\mathbb{R}^N} fd\mu - \int_{\mathbb{R}^N} fd\mu_{k_n}\right| &\le \left|\int_{\overline{B}_{h+1}} fd\mu - \int_{\overline{B}_{h+1}} fd\mu_{k_n}\right| + \left|\int_{\mathbb{R}^N\setminus\overline{B}_{h+1}} fd\mu - \int_{\mathbb{R}^N\setminus\overline{B}_{h+1}} fd\mu_{k_n}\right| \\ &\le \left|\int_{\overline{B}_{h+1}} fd\mu - \int_{\overline{B}_{h+1}} fd\mu_{k_n}\right| + 2\varepsilon\|f\|_\infty. \end{aligned} \tag{9.1.12}$$

Observing that $\mu(\overline{B}_{h+1}) = \mu^{(m)}(B_{h+1})$ for any $m > h+1$ and letting n tend to $+\infty$, (9.1.11) and (9.1.12) gives

$$\limsup_{n\to+\infty}\left|\int_{\mathbb{R}^N} fd\mu - \int_{\mathbb{R}^N} fd\mu_{k_n}\right| \le 2\varepsilon\|f\|_\infty. \tag{9.1.13}$$

The arbitrariness of $\varepsilon > 0$ and $f \in C_b(\mathbb{R}^N)$ allows us to conclude that μ_{k_n} weakly* converges to μ. Moreover, taking $f \equiv \mathbb{1}$ in (9.1.13), we obtain that μ is a probability measure.

Conversely, let us assume that $\mathscr{F}$ is relatively weak* compact and let us prove that it is also tight. By contradiction, suppose that there exist $\varepsilon > 0$ and a sequence $\{\mu_n\} \subset \mathscr{F}$ such that $\mu_n(B_n) < 1 - \varepsilon$ for any $n \in \mathbb{N}$. Since $\mathscr{F}$ is a relatively weak* compact set we can assume, up to a subsequence, that μ_n converges to some probability measure μ. Let us prove that

$$\mu(B_k) \le \liminf_{n\to+\infty} \mu_n(B_k) \le \liminf_{n\to+\infty} \mu_n(B_n) \le 1 - \varepsilon \tag{9.1.14}$$

for any k sufficiently large. Of course, (9.1.14) will lead us to a contradiction, since $\mu(B_k)$ converges to $\mu(\mathbb{R}^N) = 1$ as k tends to $+\infty$.

As is easily seen, it suffices to show that

$$\limsup_{n\to+\infty} \mu_n(\mathbb{R}^N \setminus B_k) \le \mu(\mathbb{R}^N \setminus B_k). \tag{9.1.15}$$

Fix $\varepsilon > 0$ and $m \in \mathbb{N}$ sufficiently large that $\mu(\mathbb{R}^N \setminus B_{k-1/m}) \le \mu(\mathbb{R}^N \setminus B_k) + \varepsilon$. Further, let $f : \mathbb{R}^N \to \mathbb{R}$ be the function defined by $f(x) = (m|x|+1-mk)\chi_{\{k-\frac{1}{m}<|x|<k\}}(x) + \chi_{\{|x|\ge k\}}(x)$ for any $x \in \mathbb{R}^N$ and observe that

$$\mu_n(\mathbb{R}^N \setminus B_k) = \int_{\mathbb{R}^N\setminus B_k} fd\mu_n \le \int_{\mathbb{R}^N\setminus B_{k-1/m}} fd\mu_n.$$

Letting n tend to $+\infty$ we deduce that

$$\limsup_{n\to+\infty} \mu_n(\mathbb{R}^N\setminus B_k) \le \int_{\mathbb{R}^N\setminus B_{k-1/m}} fd\mu \le \mu(\mathbb{R}^N\setminus B_{k-1/m}) \le \mu(\mathbb{R}^N \setminus B_k) + \varepsilon.$$

The arbitrariness of $\varepsilon > 0$ yields inequality (9.1.15), and we are done. □

Let $\{p(t, x; dy) : t > 0,\ x \in \mathbb{R}^N\}$ be the same transition family as in Theorem 1.2.5, and define the family of probability measures $\{r(t, x; dy) : t > 0,\ x \in \mathbb{R}^N\}$ by setting

$$r(t, x; B) = \frac{1}{t}\int_0^t p(s, x; B)ds, \qquad B \in \mathcal{B}(\mathbb{R}^N).$$

We recall that $p(s, x; B) = (T(s)\chi_B)(x)$ for any $s > 0$ and $x \in \mathbb{R}^N$.

Let us prove the following preliminary theorem ([138, Theorem 3.2.1]). In the sequel we will use only the first part of it.

Theorem 9.1.19 (Krylov-Bogoliubov) *If for some $t_0>0$ and some $x_0\in\mathbb{R}^N$ the family of measures $\{r(t,x_0;dy) : t>t_0\}$ is tight, then the semigroup $\{T(t)\}$ admits an invariant measure μ.*

Conversely, if there exists an invariant measure μ of the semigroup $\{T(t)\}$, then $\liminf_{t\to+\infty} r(t,\cdot;\mathbb{R}^N\setminus B_R) = 0$ converges to 0, as R tends to $+\infty$, almost everywhere in $\mathbb{R}^N$, with respect to the Lebesgue measure.

Proof Suppose that the family of measures $\{r(t,x_0;dy) : t>t_0\}$ is tight. By the Prokhorov theorem, there exists a sequence $\{t_n\}$, diverging to $+\infty$ as n tends to $+\infty$, such that $r(t_n,x_0;dy)$ converges weakly* to a probability measure μ. We claim that μ is an invariant measure of $\{T(t)\}$. Indeed, for any $f\in B_b(\mathbb{R}^N)$ and $t>0$, we have

$$\begin{aligned}\int_{\mathbb{R}^N}(T(t)f)(y)r(t_n,x_0;dy) =&\frac{1}{t_n}\int_0^{t_n}(T(t+s)f)(x_0)ds = \frac{1}{t_n}\int_t^{t+t_n}(T(s)f)(x_0)ds\\ =&\int_{\mathbb{R}^N} f(y)r(t_n,x_0;dy)+\frac{1}{t_n}\int_{t_n}^{t_n+t}(T(s)f)(x_0)ds\\ &-\frac{1}{t_n}\int_0^t (T(s)f)(x_0)ds.\end{aligned}$$

Since $\|T(t)\|_{L(B_b(\mathbb{R}^N))}\le 1$ for any $t\ge 0$, letting n tend to $+\infty$ we get

$$\int_{\mathbb{R}^N} T(t)f\,d\mu = \int_{\mathbb{R}^N} f\,d\mu,$$

i.e., μ is an invariant measure of $\{T(t)\}$.

Conversely, assume that μ is an invariant measure of $\{T(t)\}$. For any fixed $x\in\mathbb{R}^N$ the function $R\mapsto \liminf_{t\to+\infty} r(t,x;\mathbb{R}^N\setminus B_R)$ is decreasing. Therefore, we can define $r(x)=\lim_{R\to+\infty}\liminf_{t\to+\infty} r(t,x;\mathbb{R}^N\setminus B_R)$ for any $x\in\mathbb{R}^N$. Now, we observe that

$$\begin{aligned}\mu(\mathbb{R}^N\setminus B_R) =&\frac{1}{t}\int_0^t \mu(\mathbb{R}^N\setminus B_R)ds = \frac{1}{t}\int_0^t ds\int_{\mathbb{R}^N} T(s)\chi_{\mathbb{R}^N\setminus B_R}d\mu\\ =&\frac{1}{t}\int_0^t ds\int_{\mathbb{R}^N} p(s,x;\mathbb{R}^N\setminus B_R)d\mu = \int_{\mathbb{R}^N} r(t,x;\mathbb{R}^N\setminus B_R)d\mu\end{aligned}$$

for any $t>0$ and then, by Fatou's lemma, we get

$$\mu(\mathbb{R}^N\setminus B_R) = \liminf_{t\to+\infty}\int_{\mathbb{R}^N} r(t,x;\mathbb{R}^N\setminus B_R)d\mu \ge \int_{\mathbb{R}^N}\liminf_{t\to+\infty} r(t,x;\mathbb{R}^N\setminus B_R)d\mu.$$

Letting R tend to $+\infty$, by the monotone convergence theorem we deduce that

$$0\ge \lim_{R\to+\infty}\int_{\mathbb{R}^N}\liminf_{t\to+\infty} r(t,x;\mathbb{R}^N\setminus B_R)d\mu = \int_{\mathbb{R}^N} r\,d\mu,$$

which implies that $r=0$, μ-almost everywhere. Since μ is equivalent to the Lebesgue measure, the conclusion follows. □

We can now prove the main result of this section, that is, the Khas'minskii theorem, which gives a sufficient condition for the tightness of the measures $\{r(t,x_0;dy) : t>1\}$ and, hence, for the existence of an invariant measure for the semigroup $\{T(t)\}$.

Theorem 9.1.20 (Khas'minskii) *Suppose that there exists a nonnegative function $\varphi\in C^2(\mathbb{R}^N)$ such that $\mathcal{A}\varphi(x)$ diverges to $-\infty$ as $|x|$ tends to $+\infty$. Then, for any fixed $x_0\in\mathbb{R}^N$, the family of measures $\{r(t,x_0;dy) : t>1\}$ is tight. As a consequence, the semigroup $\{T(t)\}$ admits an invariant measure.*

Proof Let X be the Markov process associated with $\{T(t)\}$ (see Section 1.5), and let $\tau_n = \tau_{B_n}$ be the first exit time of X from the ball B_n, as defined in (1.5.5). Fix $t > 0$ and $x \in \mathbb{R}^N$; the formula (1.5.6) with $\tau' = t \wedge \tau_n$ shows that

$$\mathbb{E}_x \varphi(X_{t\wedge\tau_n}) = \varphi(x) + \mathbb{E}_x \int_0^{t\wedge\tau_n} \mathcal{A}\varphi(X_s)ds \tag{9.1.16}$$

for any $n \in \mathbb{N}$. Now, for any $R > 0$, let $-M_R = \sup_{|x|>R} \mathcal{A}\varphi(x)$ and let $K \geq 0$ be such that $\mathcal{A}\varphi(x) \leq K$ for any $x \in \mathbb{R}^N$. Then, clearly $\mathcal{A}\varphi \leq K - M_R \chi_{\mathbb{R}^N \setminus \overline{B}_R}$ in $\mathbb{R}^N$ and, by (9.1.16), we get

$$\mathbb{E}_x \varphi(X_{t\wedge\tau_n}) - \varphi(x) \leq Kt - M_R \mathbb{E}_x \int_0^{t\wedge\tau_n} \chi_{\{|X_s|>R\}} ds, \qquad t > 0, \ x \in \mathbb{R}^N.$$

Hence,

$$M_R \mathbb{E}_x \int_0^{t\wedge\tau_n} \chi_{\{|X_s|>R\}} ds \leq Kt + \varphi(x) - \mathbb{E}_x \varphi(X_{t\wedge\tau_n}) \leq Kt + \varphi(x) \tag{9.1.17}$$

for any $t > 0$ and $x \in \mathbb{R}^N$. Now, since $\{T(t)\}$ is conservative (see Remark 3.0.2 and Theorem 3.1.3), τ_n tends to $+\infty$ almost surely, and then

$$\begin{aligned}\lim_{n\to+\infty} \mathbb{E}_x \int_0^{t\wedge\tau_n} \chi_{\{|X_s|>R\}} ds =& \mathbb{E}_x \int_0^t \chi_{\{|X_s|>R\}} ds \\ =& \int_0^t \mathbb{E}_x \chi_{\{|X_s|>R\}} ds = t\, r(t, x; \mathbb{R}^N \setminus B_R).\end{aligned}$$

Thus, letting n tend to $+\infty$ in (9.1.17), we obtain $r(t, x; \mathbb{R}^N \setminus B_R) \leq (K + \varphi(x))M_R^{-1}$ for any $t > 1$ and $x \in \mathbb{R}^N$. Since $\lim_{R\to+\infty} M_R = +\infty$, the measures $\{r(t, x; dy) : t > 1\}$ are tight for any $x \in \mathbb{R}^N$. □

In [197] Khas'minskii theorem is proved in a similar way, taking advantage of the Krylov-Bogoliubov theorem, but without using the formula (1.5.6). For the seek of completeness we provide this proof. In Chapter 13 we will adapt such a proof, to prove the existence of an invariant measure for the semigroups therein considered.

A second proof of Theorem 9.1.20 We are going to prove that for any $t_0 > 0$ and $x_0 \in \mathbb{R}^N$ the family of measures $\{r(t, x_0; dy) : t > t_0\}$ is tight.

Let $\{\psi_n\} \in C^\infty([0, +\infty))$ be a sequence of smooth functions with the following properties: (i) $\psi_n(t) = t$ for any $t \in [0, n]$; (ii) $\psi_n(t) = n + \frac{1}{2}$ for any $t \geq n + 1$; (iii) $\psi_n'(t) \in [0, 1]$ and $\psi_n''(t) \leq 0$ for any $t \geq 0$. As is immediately seen, the function $\varphi_n = \psi_n \circ \varphi$ belongs to $D_{\max}(\mathcal{A})$ for any $n \in \mathbb{N}$ (see (1.0.1)). Since the semigroup $\{T(t)\}$ is conservative (see Remark 3.0.2 and Theorem 3.1.3), Propositions 1.3.4 and 1.3.5 imply that the function $u_n := T(\cdot)\varphi_n$ is differentiable with respect to t in $[0, +\infty) \times \mathbb{R}^N$ and

$$\begin{aligned}D_t u_n(t, x) =& (T(t)\mathcal{A}\varphi_n)(x) = \int_{\mathbb{R}^N} (\mathcal{A}\varphi_n) p(t, x; dy) \\ =& \int_{\mathbb{R}^N} ((\psi_n' \circ \varphi)\mathcal{A}\varphi + (\psi_n'' \circ \varphi)|\sqrt{Q}\nabla\varphi|^2) p(t, x; dy)\end{aligned} \tag{9.1.18}$$

for any $t \in [0, +\infty)$, $x \in \mathbb{R}^N$ and $n \in \mathbb{N}$. We now fix $t > 0$, $x \in \mathbb{R}^N$ and integrate (9.1.18) with respect to $s \in [0, t]$. Recalling that $u_n \geq 0$ in $[0, +\infty) \times \mathbb{R}^N$ and $\psi_n'' \leq 0$ in $[0, +\infty)$, we get

$$-\varphi_n(x) \leq \int_0^t ds \int_{\mathbb{R}^N} (\psi_n' \circ \varphi)\mathcal{A}\varphi\, p(s, x; dy)$$

$$= \int_0^t ds \int_E (\psi_n' \circ \varphi)\mathcal{A}\varphi\, p(s,x;dy) + \int_0^t ds \int_{\mathbb{R}^N\setminus E} (\psi_n' \circ \varphi)\mathcal{A}\varphi\, p(s,x;dy), \quad (9.1.19)$$

where $E = \{y \in \mathbb{R}^N : \mathcal{A}\varphi(y) \geq 0\}$. Since E is a bounded set, the dominated convergence theorem implies that the first integral term in the last side of (9.1.19) converges to $\int_0^t ds \int_E (\mathcal{A}\varphi)p(s,x;dy)$, as n tends to $+\infty$. Further, since the sequence $\{\psi_n'(s)\}$ is increasing to 1 for any $s \in [0,+\infty)$ and $\mathcal{A}\varphi \leq 0$ in $\mathbb{R}^N \setminus E$, using the monotone convergence theorem it is immediate to check that the second integral term converges to $\int_0^t ds \int_{\mathbb{R}^N\setminus E} \mathcal{A}\varphi\, p(s,x;dy)$, which is finite due to (9.1.19). Therefore, again from (9.1.19), we get

$$-\varphi(x) \leq \int_0^t ds \int_{\mathbb{R}^N} \mathcal{A}\varphi\, p(s,x;dy). \quad (9.1.20)$$

Now, fix $\varepsilon > 0$ and let $\rho > 0$ be such that $\mathcal{A}\varphi < -\varepsilon^{-1}$ outside the ball B_ρ. From (9.1.20) we obtain that

$$\begin{aligned} -\varphi(x) \leq & \int_0^t ds \int_{B_\rho} \mathcal{A}\varphi\, p(s,x;dy) + \int_0^t ds \int_{\mathbb{R}^N\setminus B_\rho} \mathcal{A}\varphi\, p(s,x;dy) \\ \leq & Mt - \frac{1}{\varepsilon}\int_0^t p(s,x;\mathbb{R}^N\setminus B_\rho)ds = Mt - \frac{t}{\varepsilon} r(t,x;\mathbb{R}^N\setminus B_\rho) \end{aligned}$$

for any $x \in \mathbb{R}^N$, where $M = \sup_{\mathbb{R}^N} \mathcal{A}\varphi$. Consequently, $r(t,x;\mathbb{R}^N\setminus B_\rho) \leq \varepsilon(M+\varphi(x)t^{-1})$, which, of course, implies that for any $t_0 > 0$ and $x \in \mathbb{R}^N$ the family of measures $\{r(t,x;dy) : t > t_0\}$ is tight. □

9.1.3 Existence by compactness in $C_b(\mathbb{R}^N)$

The compactness of $\{T(t)\}$ in $C_b(\mathbb{R}^N)$, which we studied in Section 4.1, implies the existence of an invariant measure as well as other remarkable properties. To begin with, we consider the following lemma.

Lemma 9.1.21 *If for some $t_0 > 0$ and $x_0 \in \mathbb{R}^N$ the family of measures $\{p(t,x_0;dy) : t > t_0\}$ is tight, then the family $\{r(t,x_0;dy) : t > t_0\}$ is tight, as well.*

Proof For any $t > t_0$ it holds that

$$\begin{aligned} r(t,x_0;B) = & \frac{1}{t}\int_0^{t_0} p(s,x_0;B)ds + \frac{1}{t}\int_{t_0}^t p(s,x_0;B)ds \\ \leq & \sup_{0\leq s\leq t_0} p(s,x_0;B) + \sup_{s>t_0} p(s,x_0;B). \end{aligned}$$

Therefore, it is sufficient to prove that the family of measures $\{p(t,x_0;dy) : t \in [0,t_0]\}$ is tight. This follows from Theorem 9.1.18. Indeed, consider a sequence $\{p(t_n,x_0;dy)\}$ where $t_n \in [0,t_0]$ for any $n \in \mathbb{N}$. Up to a subsequence, $\{t_n\}$ converges to some $t^* \in [0,t_0]$. Taking Theorem 1.2.5 into account, we get

$$\lim_{n\to+\infty} \int_{\mathbb{R}^N} f(y)p(t_n,x_0;dy) = \lim_{n\to+\infty} (T(t_n)f)(x_0) = (T(t^*)f)(x_0) = \int_{\mathbb{R}^N} f(y)p(t^*,x_0;dy)$$

for any $f \in C_b(\mathbb{R}^N)$. Therefore, the sequence $\{p(t_n,x_0;dy)\}$ weakly* converges to the measure $p(t^*,x_0;dy)$ in the dual space $C_b(\mathbb{R}^N)'$. □

Theorem 9.1.22 *If $T(t)$ is compact in $C_b(\mathbb{R}^N)$ for any $t > 0$, then there exists an invariant measure of $\{T(t)\}$. Moreover, for any $p \in (1, +\infty)$, $T(t)$ and $R(\lambda, L_p)$ are compact operators in L^p_μ for any $t > 0$ and $\lambda > 0$, and the spectrum of L_p consists of isolated eigenvalues.*

Proof Fix $t_0 > 0$, $x_0 \in \mathbb{R}^N$ and $\varepsilon > 0$. Since $\{T(t)\}$ is conservative, by Proposition 4.1.2 there exists $R > 0$ such that $p(t_0, y; \mathbb{R}^N \setminus B_R) \le \varepsilon$ for any $y \in \mathbb{R}^N$. Recalling that $p(t, x; dy)$ is a probability measure for any $t > 0$ and $x \in \mathbb{R}^N$ (see Theorem 1.2.5 and Proposition 9.1.9), it follows that

$$p(t, x_0; \mathbb{R}^N \setminus B_R) = \int_{\mathbb{R}^N} p(t_0, y; \mathbb{R}^N \setminus B_R) p(t - t_0, x_0; dy) \le \varepsilon$$

for any $t > t_0$, namely, the family of measures $\{p(t, x_0; dy) : t > t_0\}$ is tight. Then, the existence of an invariant measure of $\{T(t)\}$ follows from Theorem 9.1.19 and Lemma 9.1.21.

We now fix an arbitrary $t > 0$ and prove that $T(t)$ is compact in L^p_μ. For this purpose, we observe that, since μ is equivalent to the Lebesgue measure on the σ-algebra of the Borel sets of $\mathbb{R}^N$ (see Proposition 9.1.4), L^∞_μ is contained in $L^\infty(\mathbb{R}^N)$. Moreover, since $\{T(t)\}$ is strong Feller (see Proposition 1.2.13), it maps L^∞_μ into $C_b(\mathbb{R}^N)$. Therefore, writing $T(t) = T(t/2) \circ T(t/2)$ and taking Proposition B.1.1(ii) into account, it is immediate to check that $T(t)$ is compact from L^∞_μ into $C_b(\mathbb{R}^N)$ and, since $C_b(\mathbb{R}^N) \hookrightarrow L^\infty_\mu$, $T(t)$ is compact from L^∞_μ into itself. Applying Theorem B.3.11 we conclude that $T(t)$ is compact in L^p_μ for any $p \in (1, +\infty)$.

Now, we show that $R(\lambda, L_p)$ is compact for any $\lambda \in \rho(L_p)$. For this purpose, we prove that the semigroup $\{T(t)\}$ is norm-continuous in $(0, +\infty)$. Once this property is checked, taking (C.1.2) into account, we will get $R(\lambda, L_p) = \int_0^{+\infty} e^{-\lambda t} T(t) dt$ for any λ with positive real part, where the integral is meant in L^p_μ. Proposition B.1.1(i) will allow us to conclude that $R(\lambda, L_p)$ is compact for any λ as above. Then, the resolvent identity (see the formula (B.1.1)), together with Proposition B.1.1(ii) will yield the compactness of $R(\lambda, L_p)$ for any $\lambda \in \rho(L_p)$.

So, let us prove that $\{T(t)\}$ is norm-continuous in $(0, +\infty)$. Let us fix $t_0 > 0$ and set $K = \overline{T(t_0)(\mathcal{B}_1)}$, where now $\mathcal{B}_1$ denotes the open unit ball in L^p_μ with centre at 0. Since $T(t_0)$ is a compact operator, K is compact in L^p_μ. According to Remark C.1.3, $T(t)f$ tends to f as t tends to 0^+, uniformly with respect to f on compact subsets of L^p_μ. Therefore, $T(\cdot + t_0)f$ converges to $T(t_0)f$ as t tends to 0^+, uniformly with respect to $f \in \mathcal{B}_1$, and this implies that $T(t)$ tends to $T(t_0)$ in $L(L^p_\mu)$ from the right. Similarly, if $t < 0$, then we can fix $\delta \in (0, t_0)$ and write

$$\begin{aligned} \|T(t_0 + t)f - T(t_0)f\|_p \le& \|T(t_0 + t - \delta)(T(\delta)f - T(\delta - t)f)\|_p \\ \le& \|T(\delta)f - T(\delta - t)f\|_p. \end{aligned} \tag{9.1.21}$$

Letting t tend to 0^-, the last side of (9.1.21) vanishes. Hence, $T(\cdot)f$ is norm-continuous in $(0, +\infty)$.

Finally, since $R(1, L_p)$ is compact (and invertible), its spectrum consists of isolated eigenvalues. Therefore, the spectrum of L_p consists of isolated eigenvalues as well. □

Remark 9.1.23 Applying [17, Proposition 2.6], one can also show that, under the assumptions of Theorem 9.1.22, the spectrum of the operator L_p, is independent of p and it coincides with the spectrum of $(A, D_{\max}(\mathcal{A}))$ in $C_b(\mathbb{R}^N)$ (see (1.0.1)).

To conclude this subsection, we show that, under the assumptions of Theorem 9.1.22, the spectral gap for the operator L_p holds for any $p \in (1, +\infty)$, i.e., there exists $\delta > 0$ such that $\sigma(L_p) \setminus \{0\} \subset \{\lambda \in \mathbb{C} : \mathrm{Re}\,\lambda < -\delta\}$.

Theorem 9.1.24 *Under the assumptions of Theorem 9.1.22, the spectral gap holds for the operator L_p, for any $p \in (1, +\infty)$. Moreover, $T(t)f$ tends exponentially to $\overline{f} = \int_{\mathbb{R}^N} f d\mu$ in L^2_μ, as t tends to $+\infty$.*

Proof Fix $p \in (1, +\infty)$ and denote, as in Subsection 9.1.1, by P_∞ the operator defined in L^p_μ by $P_\infty f = \int_{\mathbb{R}^N} f d\mu$ for any $f \in L^p_\mu$. Since P_∞ is a projection, $L^p_\mu = H \oplus K$, where $H = \mathrm{Ker}(P_\infty)$ and $K = \mathrm{Ker}(I - P_\infty)$. As is immediately seen, $T(t)$ commutes with P_∞ for any $t > 0$. Therefore, H and K are both invariant under the action of the semigroup and the restrictions of the semigroup to H and K give rise to two strongly continuous semigroups, whose infinitesimal generators are the restrictions of L_p to H and K, respectively. In particular, $T(t)$ coincides with the identity operator on K for any $t > 0$. As a consequence, for any $f \in L^p_\mu$ such that $f = f_1 + f_2$ with $f_1 \in H$ and $f_2 \in K$, we can split $T(t)f = T(t)_{|H} f_1 + f_2$.

Now, we are going to prove that $T(t)_{|H}$ decreases exponentially as t tends to $+\infty$. As a first step, we claim that $\sigma(T(1)_{|H}) \subset \mathcal{B}_1$. Since $\{T(t)\}$ is a semigroup of contractions, $\sigma(T(1)) \subset \overline{\mathcal{B}}_1$. Let us show that, if $\lambda \in \partial\mathcal{B}_1$, then $\lambda \in \rho(T(1)_{|H})$. On the contrary, we assume that there exists $\lambda \in \sigma(T(1)_{|H}) \cap \partial\mathcal{B}_1$. Since $T(1)$ is a compact operator, λ belongs to the point spectrum of $T(1)$. Therefore, there exists $f \in H$, with $f \neq 0$, such that $T(1)f = \lambda f$. Using the semigroup rule, we obtain that $T(n)f = \lambda^n f$ for any $n \in \mathbb{N}$. From Theorem 9.1.16 it follows that $T(n)f$ converges to $P_\infty f = 0$ in L^p_μ, as n tends to $+\infty$. Therefore, $\lambda^n f$ should vanish as n tends to $+\infty$, but, of course, this can not be the case. The claim follows and, in particular, we deduce that there exists $\delta < 1$ such that $\lim_{n\to+\infty} \|T(1)^n\|^{1/n}_{L(H)} = \lim_{n\to+\infty} \|T(n)\|^{1/n}_{L(H)} = \delta$. Using the semigroup law it is now immediate to check that there exists $\omega < 0$ such that $\|T(t)\|_{L(H)} \le Ce^{\omega t}$ for any $t > 0$. Since $\{T(t)_{|H}\}$ is a strongly continuous semigroup in H, $\inf\{\mathrm{Re}\lambda : \lambda \in \sigma((L_p)_{|H})\} \le \omega$ (see Section C.1).

To conclude the proof of the first part of the theorem, it suffices to show that

$$\sigma(L_p) = \sigma((L_p)_{|H}) \cup \sigma((L_p)_{|K}) = \sigma((L_p)_{|H}) \cup \{0\}. \tag{9.1.22}$$

Note that the second equality in (9.1.22) is immediately checked since $T(t)_{|K} = I$ for any $t > 0$.

As a first step to prove (9.1.22), we observe that $0 \in \sigma(L_p)$ and $0 \notin \sigma((L_p)_{|H})$. Indeed, Theorem 9.1.17 implies that $L_p u = 0$ if and only if u is constant. Therefore, 0 belongs to $\sigma(L_p)$. But, since 0 is the only constant in H, 0 does not belong to $\sigma((L_p)_{|H})$.

Now, fix $0 \neq \lambda \in \rho(L)$ and $f \in H$. Set $u = R(\lambda, L_p)f$, so that $\lambda u - L_p u = f$. Since $\overline{f} = 0$ and $\overline{L_p u} = 0$, by (9.1.5), $\overline{u} = 0$ as well, so that $u \in H$. This implies that $\lambda \in \rho((L_p)_{|H})$ and $R(\lambda, (L_p)_{|H}) = R(\lambda, L_p)_{|H}$. Therefore, $\sigma((L_p)_{|H}) \cup \{0\}$ is contained in $\sigma(L_p)$.

To prove the other inclusion in (9.1.22), we fix $0 \neq \lambda \in \rho((L_p)_{|H})$ and show that $\lambda \in \rho(L_p)$. For this purpose, we observe that, for any $f \in L^p_\mu$, the function $u = R(\lambda, (L_p)_{|H})(f - \overline{f}) + \lambda^{-1}\overline{f}$ is a solution to the equation $\lambda u - L_p u = f$. Note that $\overline{f} \in D(L_p)$, since the constants belong to $D_{\max}(\mathcal{A})$, which is a core of L_p (see Propositions 1.3.5, 3.1.10 and 9.1.8). Hence, the operator $\lambda I - L_p$ is surjective in L^p_μ.

To prove that $\lambda I - L_p$ is also one to one, we observe that, since P_∞ commutes with $T(t)$ for any $t > 0$, it commutes with L_p on $D(L_p)$. Therefore, if $\lambda u = L_p u$ for some $u \in D(L_p)$, then both $u_1 = P_\infty u$ and $u_2 = (I - P_\infty)u$ solve the equation $\lambda v = L_p v$. Since $\lambda \notin \sigma((L_p)_{|H}) \cup \sigma((L_p)_{|K})$, it follows that $u_1 \equiv u_2 \equiv 0$ and, consequently, $u \equiv 0$.

To conclude the proof, we observe that the function $f - \overline{f}$ belongs to H for any $f \in L^2_\mu$. Therefore, $\|T(t)f - \overline{f}\|_2 = \|T(t)f - T(t)\overline{f}\|_2 \le 2Ce^{\omega t}\|f\|_2$ for any $t > 0$, and we are done. □

9.1.4 Existence by symmetry

In this subsection we study the case when $\mathcal{A} = \Delta - \langle \nabla U + G, \nabla \rangle$.

Hypotheses 9.1.25 (i) $U \in C^{1+\alpha}_{\rm loc}(\mathbb{R}^N)$, for some $\alpha \in (0,1)$, and $e^{-U} \in L^1(\mathbb{R}^N)$;

(ii) $G \in C^1(\mathbb{R}^N, \mathbb{R}^N)$, $\operatorname{div} G = \langle G, \nabla U\rangle$ and $\displaystyle\int_{\mathbb{R}^N} |G(x)|e^{-U(x)}dx < +\infty$.

Under Hypotheses 9.1.25 we can define the probability measure $\mu(dx) = K^{-1}e^{-U(x)}dx$, where $K = \|e^{-U}\|_{L^1(\mathbb{R}^N)}$. By means of an integration by parts we get

$$\int_{\mathbb{R}^N} \mathcal{A}f\, g\, d\mu = -\int_{\mathbb{R}^N} \langle \nabla f, \nabla g\rangle d\mu, \qquad f \in C^\infty_c(\mathbb{R}^N),\ g \in W^{1,p}_{\rm loc}(\mathbb{R}^N) \tag{9.1.23}$$

for any $p \in [1, +\infty)$. In particular,

$$\int_{\mathbb{R}^N} \mathcal{A}f d\mu = 0, \qquad f \in C^\infty_c(\mathbb{R}^N). \tag{9.1.24}$$

The main result of the section is the following theorem.

Theorem 9.1.26 *Assume Hypotheses* 9.1.25. *Then, the semigroup $\{T(t)\}$ associated with the operator $\mathcal{A}$ in $C_b(\mathbb{R}^N)$ is conservative, μ is the invariant measure of $\{T(t)\}$ and $C^\infty_c(\mathbb{R}^N)$ is a core of the infinitesimal generator L_p in L^p_μ for any $p \in [1, +\infty)$.*

Proof For any $p \in [1, +\infty)$ consider the operator $A_p : C^\infty_c(\mathbb{R}^N) \to L^p_\mu$ defined by $A_p u = \mathcal{A}u$ for any $u \in C^\infty_c(\mathbb{R}^N)$. As a first step, we prove that A_p is dissipative, that is, $\lambda\|u\|_p \le \|\lambda u - A_p u\|_p$ for any $\lambda > 0$ and $u \in D(A_p)$. For this purpose, let $u \in D(A_p)$ and set $f = \lambda u - A_p u$. Multiplying both sides of the previous equality by $\operatorname{sign}(u)|u|^{p-1}$ and integrating in $\mathbb{R}^N$, we get

$$\lambda\|u\|_p^p - \int_{\mathbb{R}^N} A_p u \operatorname{sign}(u)|u|^{p-1}d\mu = \int_{\mathbb{R}^N} f \operatorname{sign}(u)|u|^{p-1}d\mu \le \|f|u|^{p-1}\|_1. \tag{9.1.25}$$

Now, if $p \ge 2$, then integrating by parts we get

$$\begin{aligned}\int_{\mathbb{R}^N} A_p u \operatorname{sign}(u)|u|^{p-1}d\mu &= -\int_{\mathbb{R}^N} \langle \nabla u, \nabla(\operatorname{sign}(u)|u|^{p-1})\rangle d\mu \\ &= -(p-1)\int_{\mathbb{R}^N} |\nabla u|^2|u|^{p-2}d\mu \le 0.\end{aligned} \tag{9.1.26}$$

Hence, from (9.1.25) and (9.1.26) we deduce that A_p is dissipative.

In the case when $p \in [1,2)$ we consider the sequence $\{\psi_n\} \subset C^\infty(\mathbb{R})$ defined by $\psi_n(x) = x(x^2+n^{-1})^{(p-2)/2}$ for any $x \in \mathbb{R}^N$. Since any ψ_n is an increasing function in $\mathbb{R}^N$, converging pointwise to $\operatorname{sign}(x)|x|^{p-1}$ as n tends to $+\infty$, by dominated convergence and by (9.1.23) we obtain

$$\begin{aligned}\int_{\mathbb{R}^N} A_p u \operatorname{sign}(u)|u|^{p-1}d\mu &= \lim_{n\to+\infty}\int_{\mathbb{R}^N} A_p u(\psi_n \circ u)d\mu \\ &= -\lim_{n\to+\infty}\int_{\mathbb{R}^N} (\psi_n' \circ u)|\nabla u|^2 d\mu \le 0\end{aligned}$$

and, then, the dissipativity of A_p follows from (9.1.25).

As a second step we show that the range of $\lambda I - A_p$ is dense in L^p_μ for some $\lambda > 0$. Let $g \in L^q_\mu(\mathbb{R}^N)$ (where $1/p+1/q=1$, or $q=+\infty$ if $p=1$) be such that $\int_{\mathbb{R}^N}(\lambda u - A_p u)g d\mu = 0$ for any $u \in D(A_p)$. First we assume that $p \in (1,+\infty)$. By classical regularity results for weak solutions of elliptic equations (see Theorem D.1.4(ii)), g belongs to $W^{1,q}_{\rm loc}(\mathbb{R}^N)$. If $p=1$, then $g \in L^2_{\rm loc}(\mathbb{R}^N)$. Therefore, still from Theorem D.1.4(ii), we deduce that $g \in W^{1,2}_{\rm loc}(\mathbb{R}^N)$. Now, from the formula (9.1.23) we get

$$\lambda \int_{\mathbb{R}^N} ug\, d\mu + \int_{\mathbb{R}^N} \langle \nabla u, \nabla g\rangle d\mu = 0, \qquad u \in C^\infty_c(\mathbb{R}^N). \tag{9.1.27}$$

By density, (9.1.27) can be extended to any compactly supported function $u \in W^{1,p}(\mathbb{R}^N)$, if $p \in (1,+\infty)$, and to any compactly supported function $u \in W^{1,2}(\mathbb{R}^N)$, if $p=1$. Now, for any $n \in \mathbb{N}$, let $\eta_n \in C^\infty_c(\mathbb{R}^N)$ be a smooth function such that $\eta_n \equiv 1$ in B_n, $0 \le \eta_n \le \eta_{n+1}$, and $|\nabla \eta_n| \le 1/n$. If $1 < p \le 2$, then, taking $u = \eta_n^2 {\rm sign}(g)|g|^{q-1}$ in (9.1.27), we get

$$\begin{aligned}
&\lambda \int_{\mathbb{R}^N} \eta_n^2 |g|^q d\mu + (q-1)\int_{\mathbb{R}^N} \eta_n^2 |g|^{q-2}|\nabla g|^2 d\mu \\
=& -2\int_{\mathbb{R}^N} \eta_n {\rm sign}(g)|g|^{q-1}\langle \nabla \eta_n, \nabla g\rangle d\mu \le \frac{1}{n}\int_{\mathbb{R}^N} \eta_n^2 |\nabla g|^2 |g|^{q-2} d\mu + \frac{1}{n}\|g\|_q^q.
\end{aligned}$$

Hence, $\lambda\|\eta_n^2|g|^q\|_1 \le n^{-1}\|g\|_q^q$ for any $n \ge 2$. Letting n tend to $+\infty$, by monotone convergence we deduce that $g \equiv 0$.

The case when $p=1$ can be treated as the case when $p=2$. Suppose now that $p>2$ and, for any $n,m \in \mathbb{N}$, let $u_{n,m} = \eta_n^2(\psi_m \circ g)$, where the sequence $\{\psi_n\}$ is as above with p being replaced by q. Then, $u_{n,m} \in W^{1,p}(\mathbb{R}^N)$ and has compact support for any $n,m \in \mathbb{N}$. Writing (9.1.27) with $u_{n,m}$ instead of u, yields

$$\begin{aligned}
&\lambda \int_{\mathbb{R}^N} \eta_n^2 g^2 \left(g^2 + \frac{1}{m}\right)^{\frac{q-2}{2}} d\mu + (q-1)\int_{\mathbb{R}^N} \eta_n^2 g^2 \left(g^2 + \frac{1}{m}\right)^{\frac{q-4}{2}} |\nabla g|^2 d\mu \\
\le& \frac{2}{n}\int_{\mathbb{R}^N} \eta_n |g| \left(g^2 + \frac{1}{m}\right)^{\frac{q-2}{2}} |\nabla g| d\mu.
\end{aligned} \tag{9.1.28}$$

Letting m tend to $+\infty$ in (9.1.28) we get

$$\lambda \int_{\mathbb{R}^N} \eta_n^2 |g|^q d\mu + (q-1)\int_{\mathbb{R}^N} \eta_n^2 |g|^{q-2}|\nabla g|^2 d\mu \le \frac{2}{n}\int_{\mathbb{R}^N} \eta_n |g|^{q-1}|\nabla g| d\mu. \tag{9.1.29}$$

Since the right-hand side of (9.1.29) is finite, the second integral in the left-hand side is finite as well. The Cauchy-Schwarz inequality gives

$$\lambda\|\eta_n^2|g|^q\|_1 + (q-1)\|\eta_n^2|g|^{q-2}|\nabla g|^2\|_1 \le \frac{1}{n}\|g\|_q^q + \frac{1}{n}\|\eta_n^2|g|^{q-2}|\nabla g|^2\|_1.$$

Hence, for $n(q-1) \ge 1$, we get $\lambda \int_{\mathbb{R}^N} \eta_n^2 |g|^q d\mu \le n^{-1}\|g\|_q^q$ and, letting n tend to $+\infty$, we obtain $g \equiv 0$ also in this case.

Now, according to Proposition C.1.10, the closure $\overline{A_p}$ of the operator A_p in L^p_μ is the generator of a strongly continuous semigroup $\{S(t)\}$. By density, the formula (9.1.24) implies that $\int_{\mathbb{R}^N} \overline{A_p} f d\mu = 0$ for any $f \in D(\overline{A_p})$. Thus, for any $f \in D(\overline{A_p})$ it holds that

$$\frac{d}{dt}\int_{\mathbb{R}^N} S(t) f d\mu = \int_{\mathbb{R}^N} \overline{A_p} S(t) f d\mu = 0, \qquad t > 0,$$

so that

$$\int_{\mathbb{R}^N} S(t) f d\mu = \int_{\mathbb{R}^N} f d\mu, \qquad t > 0. \tag{9.1.30}$$

Since $D(\overline{A_p})$ is dense in L^p_μ, the equality (9.1.30) holds for any $f \in L^p_\mu$.

Next, we prove that the semigroup $\{T(t)\}$ can be extended to L^p_μ with a strongly continuous semigroup of contractions and, then, that the so extended semigroup coincides with $\{S(t)\}$. Note that $\{T(t)\}$ is well defined since the coefficients of the operator $\mathcal{A}$ satisfy Hypotheses 1.0.1(i), (ii). For this purpose, for any $R > 0$ we denote by $\{T_R(t)\}$ the semigroup associated with the realization of the operator $\mathcal{A}$ with homogeneous Dirichlet boundary conditions in $C(\overline{B}_R)$. Let us observe that μ is a subinvariant measure of $\{T_R(t)\}$ for any $R > 0$, that is

$$\int_{B_R} T_R(t) f d\mu \le \int_{B_R} f d\mu, \qquad t > 0, \tag{9.1.31}$$

for any positive function $f \in C_0(B_R)$. Indeed, differentiating the left-hand side of (9.1.31) with respect to t and integrating by parts we get

$$\int_{B_R} \frac{d}{dt} T_R(t) f d\mu = \int_{B_R} \mathcal{A} T_R(t) f d\mu = K^{-1} \int_{\partial B_R} \left(\frac{\partial}{\partial \nu} T_R(t) f \right) e^{-U} d\sigma,$$

where ν is the unit outward normal vector to ∂B_R, and $d\sigma$ is the surface measure on ∂B_R. Since $T_R(t) f$ is nonnegative in B_R and it vanishes on ∂B_R (see Proposition D.3.1), its normal derivative on ∂B_R is non positive and (9.1.31) follows.

Recalling that $T_R(t) f$ converges to $T(t) f$ in a dominated way, letting R tend to $+\infty$, inequality (9.1.31) holds also when $T_R(t) f$ is replaced with $T(t) f$, for any nonnegative function $f \in C_c(\mathbb{R}^N)$, and B_R is replaced by $\mathbb{R}^N$. Using an approximation argument we can show that (9.1.31) holds for any nonnegative function $f \in C_b(\mathbb{R}^N)$. Arguing as in the proof of Proposition 9.1.7, we can extend $\{T(t)\}$ to a strongly continuous semigroup in L^p_μ for any $p \in [1, +\infty)$.

Now, let L_p be the generator of $\{T(t)\}$ in L^p_μ. Since $\overline{A_p}$ and L_p coincide with $\mathcal{A}$ on $C^\infty_c(\mathbb{R}^N)$, which is a core of $\overline{A_p}$, L_p is an extension of $\overline{A_p}$. But $\overline{A_p}$ and L_p are both generators of strongly continuous semigroups of contractions; hence

$$L^p_\mu = (\lambda - \overline{A_p})(D(\overline{A_p})) = (\lambda - L_p)(D(\overline{A_p})) \subset (\lambda - L_p)(D(L_p)) = L^p_\mu$$

for any $\lambda > 0$, which implies that $(\lambda - L_p)(D(\overline{A_p})) = (\lambda - L_p)(D(L_p))$ or, equivalently, that $D(\overline{A_p}) = D(L_p)$, so that $\overline{A_p}$ and L_p coincide, and $S(t) \equiv T(t)$ in L^p_μ for any $t > 0$ (see Proposition C.1.5). Thus, by (9.1.30), μ is the invariant measure of $\{T(t)\}$. By Proposition 9.1.9, it follows that $\{T(t)\}$ is conservative. □

Proposition 9.1.27 *$D(L_2)$ is continuously and densely embedded in $W^{1,2}_\mu$. Further,*

$$\int_{\mathbb{R}^N} \langle \nabla g, \nabla h \rangle d\mu = - \int_{\mathbb{R}^N} h L_2 g d\mu, \qquad g \in D(L_2), \;\; h \in W^{1,2}_\mu. \tag{9.1.32}$$

Finally, $\|\nabla R(\lambda, L_2) f\|_2 \le \sqrt{2\lambda^{-1}} \|f\|_2$ for any $f \in L^2_\mu$ and $\lambda > 0$.

Proof The proof of the first part of the assertion is a straightforward consequence of the formula (9.1.23) and Theorem 9.1.26. Indeed, fix $\psi \in D(L_2)$. Since $C^\infty_c(\mathbb{R}^N)$ is a core of L_2, there exists a sequence $\{\psi_n\} \subset C^\infty_c(\mathbb{R}^N)$ such that ψ_n and $L_2 \psi_n$ tend, respectively, to ψ and $L_2 \psi$ as n tends to $+\infty$. Writing (9.1.23) with $f = g = \psi_n - \psi_m$ $(n, m \in \mathbb{N})$ we get $\|\nabla \psi_n - \nabla \psi_m\|_2^2 \le \|\psi_n - \psi_m\|_2 \|L_2 \psi_n - L_2 \psi_m\|_2$. Therefore, $\{\psi_n\}$ is a Cauchy sequence in

$W^{1,2}_\mu$. Since it converges to ψ in L^2_μ, we deduce that $\psi \in W^{1,2}_\mu$ and, then, writing (9.1.23) with $f = g = \psi$, we obtain that $D(L_2) \hookrightarrow W^{1,2}_\mu$.

To conclude the proof of the first part of the proposition, we have to show that $D(L_2)$ is dense in $W^{1,2}_\mu$. But this follows immediately from the forthcoming Lemma 16.3.2.

Now, the formula (9.1.32) can be easily proved by a density argument, recalling that $L_2\varphi = \mathcal{A}\varphi$ for any $\varphi \in C^\infty_c(\mathbb{R}^N)$.

Finally, to prove the estimate in the statement it suffices to apply the formula (9.1.32) with $g = h = R(\lambda, L_2)f$, recalling that $\|R(\lambda, L_2)f\|_2 \le |\lambda|^{-1}\|f\|_2$ and $\|L_2R(\lambda, L_2)f\|_2 \le 2\|f\|_2$. □

Remark 9.1.28 In Section 9.5 we will improve the results in Proposition 9.1.27, explicitly characterizing the domain of the operator L_2.

To conclude this section, let us recall the following result proved in [110, Lemma 1.3.3].

Proposition 9.1.29 *Under Hypotheses 9.1.25, $T(t)f$ belongs to $W^{1,2}_\mu$ for any $f \in L^2_\mu$ and $t > 0$. Moreover, $\|\nabla T(t)f\|_2^2 \le (2t)^{-1}(\|f\|_2^2 - \|T(t)f\|_2^2)$ for any $t > 0$.*

9.2 Regularity properties of invariant measures

In this section we provide some conditions on the coefficients of the operator $\mathcal{A}$ implying global regularity properties of the invariant measure μ.

In Proposition 9.1.4 we have seen that, if the semigroup $\{T(t)\}$ admits an invariant measure μ, then μ is absolutely continuous with respect to the Lebesgue measure and its density is positive. Now, we specialize to the case when $\mathcal{A}$ is given on smooth functions u by

$$\mathcal{A}u = \sum_{i,j=1}^N D_i(q_{ij}D_j)u + \sum_{i=1}^N b_iD_iu := \mathcal{A}_0u + \sum_{i=1}^N b_iD_iu. \tag{9.2.1}$$

Let us state the main hypotheses on the coefficients q_{ij} and b_i $(i, j = 1, \ldots, N)$ which will be assumed throughout this section.

Hypotheses 9.2.1 (i) $q_{ij} = q_{ji} \in C^{1+\alpha}_{\rm loc}(\mathbb{R}^N)$ for any $i, j = 1, \ldots, N$ $(N \ge 2)$ and some $\alpha \in (0, 1)$ and there exists a positive constant κ_0 such that $\langle Q(x)\xi, \xi\rangle \ge \kappa_0|\xi|^2$ for any $x, \xi \in \mathbb{R}^N$;

(ii) the functions $x \mapsto (1+|x|^2)^{-1}q_{ij}(x)$ and $x \mapsto (1+|x|)^{-1}D_iq_{ij}(x)$ belong to L^1_μ for any $i, j = 1, \ldots, N$;

(iii) $b_i \in C^\alpha_{\rm loc}(\mathbb{R}^N)$ for any $i = 1, \ldots, N$.

Remark 9.2.2 (i) Hypothesis 9.2.1(ii) is satisfied for instance in the case when the functions $x \mapsto (1 + |x|^2)^{-1}q_{ij}$, $x \mapsto (1 + |x|)^{-1}D_iq_{ij}(x)$ $(i, j = 1, \ldots, N)$ are bounded in $\mathbb{R}^N$. Another sufficient condition for the integrability of the previous functions with respect to the measure μ can be easily given under the assumptions of the forthcoming Proposition 9.2.14, which yield the integrability of the function $x \mapsto \exp(\delta|x|^\beta)$ for some constants $\beta, \delta > 0$.

(ii) The reason that, throughout this section, we decide to deal only with the case when $N \geq 2$, is based on the fact that, when $N = 1$, an explicit representation formula for the density of the infinitesimally invariant measure μ is available. Indeed, in such a case, a straightforward integration by parts shows that $\int_{\mathbb{R}}[q(x)\rho'(x) - b(x)\rho(x)]\psi'(x)dx = 0$ for any $\psi \in C_c^\infty(\mathbb{R})$. Hence, $q\rho' - b\rho$ is constant in $\mathbb{R}$.

In order to present the main results of this section (proved by G. Metafune, D. Pallara, and A. Rhandi in [193]) we recall the following theorem, which has been proved by V.I. Bogachev, N.V. Krylov, and M. Röckner in [39, 40] and gives a local regularity property of the function ρ.

Theorem 9.2.3 *Under Hypotheses* 9.2.1*(i) and (iii), the function ρ belongs to $W^{1,p}_{\rm loc}(\mathbb{R}^N)$ for any $p \in [1, +\infty)$. As a consequence, ρ is continuous in $\mathbb{R}^N$.*

Proof We split the proof into three steps.

Step 1. Here, we show that $\rho \in L^p_{\rm loc}(\mathbb{R}^N)$ for any $p < N/(N-1)$. For this purpose, we begin by observing that, according to Proposition 9.1.1,

$$\int_{\mathbb{R}^N} {\rm Tr}(QD^2\zeta)d\mu = -\int_{\mathbb{R}^N} \langle \overline{b}, \nabla\zeta\rangle d\mu, \qquad \zeta \in C_c^\infty(\mathbb{R}^N), \tag{9.2.2}$$

where $\overline{b} = (\overline{b}_1, \ldots, \overline{b}_N)$ and $\overline{b}_i := b_i + \sum_{j=1}^N D_i q_{ij}$ for any $i = 1, \ldots, N$. By density, it is immediate to extend (9.2.2) to any function $\zeta \in C_c^2(\mathbb{R}^N)$.

Now, we fix $R > 0$ and a function $\vartheta \in C_c^\infty(B_R)$ such that $\chi_{B_{R/2}} \leq \vartheta \leq \chi_{B_R}$. For any $\psi \in C^2(\overline{B}_R)$ we write (9.2.2) with $\zeta = \psi\vartheta$ and get

$$\begin{aligned}\left|\int_{B_R} {\rm Tr}(QD^2\psi)\vartheta d\mu\right| \leq & \bigg|2\int_{B_R}\langle Q\nabla\vartheta, \nabla\psi\rangle d\mu + \int_{B_R}\psi{\rm Tr}(QD^2\vartheta)d\mu \\ & + \int_{\mathbb{R}^N}\vartheta\langle\overline{b}, \nabla\psi\rangle d\mu + \int_{\mathbb{R}^N}\psi\langle\overline{b}, \nabla\vartheta\rangle d\mu\bigg| \leq C_1\|\psi\|_{C^1(B_R)},\end{aligned} \tag{9.2.3}$$

where C_1 is a positive constant, depending on R, but being independent of ψ.

Let $f \in C_c^\infty(B_R)$ be a smooth function. By [116, Theorem 6.14 and Lemma 9.17], the problem

$$\begin{cases} {\rm Tr}(Q(x)D^2u(x)) = f(x), & x \in B_R, \\ u(x) = 0, & x \in \partial B_R, \end{cases} \tag{9.2.4}$$

admits a unique solution $u \in C^2(\overline{B}_R)$ satisfying $\|u\|_{W^{2,q}(B_R)} \leq C_2\|f\|_{L^q(B_R)}$ for any $q > N$ and some positive constant $C_2 = C_2(q, R)$, independent of f. Using the Sobolev embedding theorems (see [3, Theorem 5.4]) we deduce that $\|u\|_{C^1(\overline{B}_R)} \leq C_3\|f\|_{L^q(B_R)}$, with C_3 being independent of f. Therefore, plugging $\psi = u$ into (9.2.3) yields

$$\left|\int_{B_R} f\,\vartheta\rho dx\right| \leq C_1C_3\|f\|_{L^q(B_R)}.$$

By the arbitrariness of $f \in C_c^\infty(B_R)$ and $q > N$, the function $\rho\vartheta$ belongs $L^p(B_R)$ for any $p \in [1, N/(N-1))$. Since $\rho\vartheta \equiv \rho$ in $B_{R/2}$ and R is arbitrary, ρ belongs to $L^p_{\rm loc}(\mathbb{R}^N)$ for any $p \in [1, N/(N-1))$.

Step 2. Now we show that ρ belongs to $W^{1,p}_{\rm loc}(\mathbb{R}^N)$ for any $p \in [1, N/(N-1))$. So, let us fix $p \in (1, N/(N-1))$ and $M \in \mathbb{N}$. For any $x_0 \in \overline{B}_M$ and $R > 0$ we fix two functions $\eta, \psi \in C^2(\overline{B}_R(x_0))$ such that $\chi_{B_{R/2}(x_0)} \leq \eta \leq \chi_{B_R(x_0)}$ and ψ vanishes on $\partial B_R(x_0)$. We are going to prove that we can fix R sufficiently small (depending only on M) such that

$\rho \in W^{1,p}(B_{R/2}(x_0))$ for any $x_0 \in B_M$. The arbitrariness of $x_0 \in B_M$ and $M > 0$ then will imply that $\rho \in W^{1,p}_{\rm loc}(\mathbb{R}^N)$. Without loss of generality, we can assume that $R < 1$.

Writing (9.2.2) with $\zeta = \psi\eta$ and using the Poincaré inequality, we get

$$\begin{aligned}\left|\int_{B_R(x_0)}\eta{\rm Tr}(QD^2\psi)d\mu\right| \le & 2\left|\int_{B_R(x_0)}\langle Q\nabla\eta,\nabla\psi\rangle d\mu\right| + \left|\int_{B_R(x_0)}\psi{\rm Tr}(QD^2\eta)d\mu\right| \\ & + \left|\int_{B_R(x_0)}\eta\langle\overline{b},\nabla\psi\rangle d\mu\right| + \left|\int_{B_R(x_0)}\psi\langle\overline{b},\nabla\eta\rangle d\mu\right| \\ \le & C_4\int_{B_R(x_0)}(|\psi|+|\nabla\psi|)d\mu \le C_5\|\nabla\psi\|_{L^{p/(p-1)}(B_R(x_0))} \end{aligned} \tag{9.2.5}$$

for some positive constants C_4 and C_5, depending on R, M, and the sup-norm of the coefficients of $\mathcal{A}$ in B_{M+1}, but independent of ψ and x_0.

For any set of functions $\{f_1,\ldots,f_n\} \subset C^\infty_c(B_R(x_0))$, let us denote by $u \in C^2(\overline{B}_R(x_0))$ the solution to the elliptic boundary problem (9.2.4) with $\sum_{i=1}^N D_if_i$ instead of f. Plugging $\psi = u$ into (9.2.5) we get

$$\left|\int_{B_R(x_0)}\eta\sum_{i=1}^N D_if_id\mu\right| \le C_5\|\nabla u\|_{L^{p/(p-1)}(B_R(x_0))}. \tag{9.2.6}$$

Let us show that

$$\|\nabla u\|_{L^{p/(p-1)}(B_R(x_0))} \le C_6\sum_{i=1}^N\|D_if_i\|_{W^{-1,p/(p-1)}(B_R(x_0))} \tag{9.2.7}$$

for some positive constant C_6, independent of R, x_0 and f, provided that R is sufficiently small. Here, $W^{-1,p/(p-1)}(B_R(x_0))$ denotes the dual of the space $W^{1,p}_0(B_R(x_0))$. From estimates (9.2.6) and (9.2.7) we will then deduce that $\eta\rho \in W^{1,p}_0(B_R(x_0))$ and, since $\eta = 1$ in $B_{R/2}(x_0)$, that $\rho \in W^{1,p}(B_{R/2}(x_0))$.

From now on, we denote by C_j positive constants, depending on M, but independent of x_0, R and f.

To prove (9.2.7) we begin by observing that

$$\|\nabla u\|_{L^{p/(p-1)}(B_R(x_0))} \le C_7(\|g_1\|_{W^{-1,p/(p-1)}(B_R(x_0))} + \|g_2\|_{W^{-1,p/(p-1)}(B_R(x_0))}), \tag{9.2.8}$$

where $\sum_{i,j=1}^N D_i(q_{ij}D_ju) = \sum_{i=1}^N D_if_i + \sum_{i,j=1}^N D_iq_{ij}\,D_ju := g_1 + g_2$, see e.g. [114, Section 4.3] or [208, Section 5.5].

To estimate the norm of g_2 we observe that, since $p < N$, $L^{pN/(N(p-1)+p)}(B_R(x_0)) \hookrightarrow W^{-1,p/(p-1)}(B_R(x_0))$. Therefore,

$$\begin{aligned}\|g_2\|_{W^{-1,p/(p-1)}(B_R(x_0))} \le & C_8\|g_2\|_{L^{pN/(N(p-1)+p)}(B_R(x_0))} \\ \le & C_9\|h\|_{L^N(B_R(x_0))}\|\nabla u\|_{L^{p/(p-1)}(B_R(x_0))} \\ \le & C_{10}R\|h\|_{L^\infty(B_{M+1})}\|\nabla u\|_{L^{p/(p-1)}(B_R(x_0))}. \end{aligned} \tag{9.2.9}$$

Here, $h^2 = \sum_{i,j=1}^N|D_iq_{ij}|^2$. From (9.2.8) and (9.2.9) we deduce that

$$\|\nabla u\|_{L^{p/(p-1)}(B_R(x_0))} \le C_{11}(R\|\nabla u\|_{L^{p/(p-1)}(B_R(x_0))} + \|g_1\|_{W^{-1,p/(p-1)}(B_R(x_0))}).$$

Taking R sufficiently small, we get estimate (9.2.7).

Step 3. In this step we conclude the proof, using a bootstrap argument. Since $\rho \in$

$W^{1,p}_{\rm loc}(\mathbb{R}^N)$ for any $p \in [1, N/(N-1))$, the Sobolev embedding theorems imply that $\rho \in L^p_{\rm loc}(\mathbb{R}^N)$ for any $p \in [1, N/(N-2))$. Repeating the argument in Step 2, we deduce that $\rho \in W^{1,p}_{\rm loc}(\mathbb{R}^N)$ for any $p \in [1, N/(N-2))$. Iterating this argument, we can show that $\rho \in L^p_{\rm loc}(\mathbb{R}^N)$ for any $p \in [1, +\infty)$.

We can adapt the arguments in Step 2 to prove that ρ belongs to $W^{1,p}_{\rm loc}(\mathbb{R}^N)$ also in the case when $p > N$: we just need to modify estimate (9.2.9). For this purpose, we observe that, if $p > N$, then $L^1(B_R(x_0)) \subset W^{-1,p/(p-1)}(B_R(x_0))$ and $\|k\|_{W^{-1,p/(p-1)}(B_R(x_0))} \le C_{12}R^{1-N/p}\|k\|_{L^1(B_R(x_0))}$ for any $k \in W^{-1,p/(p-1)}(B_R(x_0))$. Hence, in this case we can estimate

$$\begin{aligned}
\|g_2\|_{W^{-1,p/(p-1)}(B_R(x_0))} \le & C_{13}R^{\frac{p-N}{p}}\|g_2\|_{L^1(B_R(x_0))}\\
\le & C_{14}R^{\frac{p-N}{p}}\|h\|_{L^p(B_R(x_0))}\|\nabla u\|_{L^{p/(p-1)}(B_R(x_0))}\\
\le & C_{15}R\|h\|_{L^\infty(B_{M+1})}\|\nabla u\|_{L^{p/(p-1)}(B_R(x_0))}.
\end{aligned}$$

Taking R sufficiently small, we get (9.2.7) and, consequently, we obtain that $u \in W^{1,p}_{\rm loc}(\mathbb{R}^N)$ also in this case. □

Remark 9.2.4 (i) The previous theorem also applies to the case when μ is just an infinitesimally invariant measure. See Remark 9.1.3 for the definition of infinitesimally invariant measures.

(ii) The result in Theorem 9.2.3 holds also in the case when $N = 1$. To see it, we denote by x the variables in $\mathbb{R}$ and we introduce the operator $\mathcal{B}$ defined on smooth functions of two variables (x, y) by $\mathcal{B}u(x,y) = \mathcal{A}u(x,y) + D_{yy}u(x,y) - yD_yu(x,y)$ for any $(x,y) \in \mathbb{R}^2$. Moreover, we denote by $\widetilde{\mu}$ the measure in $\mathbb{R}^2$ whose density with respect to the Lebesgue measure is given by $\widetilde{\rho}(x,y) = (2\pi)^{-1/2}\rho(x)e^{-\frac{1}{2}y^2}$ for any $(x,y) \in \mathbb{R}^2$. Of course, the coefficients of the operator $\mathcal{B}$ satisfy the assumptions of Theorem 9.2.3. Moreover, for any $\psi \in C^\infty_c(\mathbb{R}^2)$, we have

$$\begin{aligned}
\int_{\mathbb{R}^2}\mathcal{B}\psi d\widetilde{\mu} =& \frac{1}{\sqrt{2\pi}}\int_{\mathbb{R}} e^{-\frac{1}{2}y^2}\bigg(\int_{\mathbb{R}}(\psi_{yy}(x,y) - y\psi_y(x,y))\rho(x)dx\bigg)dy\\
=& \frac{1}{\sqrt{2\pi}}\int_{\mathbb{R}}(D_{yy}e^{-\frac{1}{2}y^2} + D_y(ye^{-\frac{1}{2}y^2}))\bigg(\int_{\mathbb{R}^2}\psi(x,y)\rho(x)dx\bigg)dy = 0.
\end{aligned}$$

Therefore, from the results in Theorem 9.2.3, the function $\widetilde{\rho}$ belongs to $W^{1,p}_{\rm loc}(\mathbb{R}^2)$ and, hence, $\rho \in W^{1,p}_{\rm loc}(\mathbb{R})$ for any $p \in [1, +\infty)$.

9.2.1 Global L^q-regularity of the density ρ

In this subsection we are devoted to prove some L^q-regularity results for the function ρ.

Theorem 9.2.5 *Assume Hypotheses* 9.2.1. *Further, suppose that $b_i \in L^p_\mu$ for any $i = 1, \ldots, N$ and some $p \in [2, +\infty)$. Then, the following properties are satisfied:*

(i) if $N \ge 3$ and $p \in [2, N)$, then ρ belongs to $L^q(\mathbb{R}^N)$ for any $q \le \frac{N}{N-p}$;

(ii) if $p = N$, then ρ belongs to $L^q(\mathbb{R}^N)$ for any $q \in [1, +\infty)$;

(iii) if $p > N$, then ρ belongs to $C_b(\mathbb{R}^N)$.

Remark 9.2.6 As has been noticed in Remark 9.2.2(i), one can take advantage of the forthcoming Proposition 9.2.14 for a sufficient condition ensuring that the integrability assumptions on the coefficients b_i ($i = 1, \ldots, N$) are satisfied.

Proof of Theorem 9.2.5 To prove the assertion we use a bootstrap argument. Since the proof is rather long, we split it into three steps. First, in Step 1 we assume that $p = 2$ and prove that ρ belongs to $L^{N/(N-2)}(\mathbb{R}^N)$, if $N > 2$, and to $L^q(\mathbb{R}^N)$ for any $q \in [1, +\infty)$, if $N = 2$. For this purpose, according to the Sobolev embedding theorems (see [3, Theorem 5.4]), it suffices to show that $\sqrt{\rho} \in W^{1,2}(\mathbb{R}^N)$. Next, in Step 2, where we assume that $N > 2$, we prove that, if $\rho \in L^r(\mathbb{R}^N)$ for some $r \in (1, +\infty)$, then actually it belongs to $L^q(\mathbb{R}^N)$, where $q = N(r(p-2)+2)/(p(N-2))$. Finally, in Step 3 we conclude the proof. Throughout the proof, for notational convenience, we set

$$\mathfrak{q}(v, w) = \sum_{i,j=1}^{N} q_{ij} D_i v D_j w, \qquad \mathfrak{q}_0(v) = \mathfrak{q}(v, v),$$

for any smooth functions $v, w : \mathbb{R}^N \to \mathbb{R}$. We also stress that, throughout the proof, actually we do not require μ to be the invariant measure of $\{T(t)\}$. We just need that μ is infinitesimally invariant, i.e.,

$$\int_{\mathbb{R}^N} \mathcal{A}\psi d\mu = 0, \qquad \psi \in C_c^\infty(\mathbb{R}^N). \tag{9.2.10}$$

Of course, according to Proposition 9.1.1, if μ is the invariant measure of $\{T(t)\}$, then the previous condition is always satisfied.

Step 1. Integrating by parts (taking Theorem 9.2.3 into account), from the formula (9.2.10) we easily see that

$$\int_{\mathbb{R}^N} \mathfrak{q}(\rho, \varphi) dx = \int_{\mathbb{R}^N} \langle b, \nabla\varphi \rangle d\mu \tag{9.2.11}$$

for any $\varphi \in C_c^\infty(\mathbb{R}^N)$. Since ρ is continuous (see again Theorem 9.2.3), $b\rho$ belongs to $L^2_{\rm loc}(\mathbb{R}^N)$. Therefore, by density, we can extend (9.2.11) to any $\varphi \in W^{1,2}(\mathbb{R}^N)$ with compact support.

Now, for any ε, k such that $0 < \varepsilon < 1 < k$, we introduce the function $\rho_{\varepsilon,k} : \mathbb{R}^N \to \mathbb{R}$ defined by

$$\rho_{\varepsilon,k}(x) = (\rho(x) \vee \varepsilon) \wedge k, \qquad x \in \mathbb{R}^N. \tag{9.2.12}$$

Moreover, let the sequence of smooth functions $\{\vartheta_n\}$ be defined by $\vartheta_n(x) = \vartheta(x/n)$ for any $x \in \mathbb{R}^N$ and $n \in \mathbb{N}$, where $\vartheta \in C_c^\infty(\mathbb{R}^N)$ is such that $\chi_{B_{1/2}} \leq \vartheta \leq \chi_{B_1}$. A straightforward computation shows that there exists a positive constant C such that

$$|\nabla\vartheta_n(x)| \leq \frac{C}{1+|x|}, \qquad |D^2\vartheta_n(x)| \leq \frac{C}{1+|x|^2}, \qquad x \in \mathbb{R}^N. \tag{9.2.13}$$

As is easily seen, the function $\log(\rho_{\varepsilon,k})$ belongs to $W^{1,p}_{\rm loc}(\mathbb{R}^N) \cap L^\infty(\mathbb{R}^N)$ for any $n \in \mathbb{N}$. Plugging $\varphi = \vartheta_n^2 \log(\rho_{\varepsilon,k})$ in (9.2.11), we get

$$\begin{aligned}\int_{\mathbb{R}^N} \vartheta_n^2 \chi_{\{\varepsilon<\rho<k\}} \frac{\mathfrak{q}_0(\rho)}{\rho} dx = & -2\int_{\mathbb{R}^N} \vartheta_n \log(\rho_{\varepsilon,k}) \mathfrak{q}(\rho, \vartheta_n) dx + \int_{\mathbb{R}^N} \vartheta_n^2 \langle b, \nabla\rho \rangle \chi_{\{\varepsilon<\rho<k\}} dx \\ & + 2\int_{\mathbb{R}^N} \vartheta_n \rho \log(\rho_{\varepsilon,k}) \langle b, \nabla\vartheta_n \rangle dx := I_n + J_n + K_n.\end{aligned}$$

Let us estimate the term I_n. An integration by parts shows that

$$\begin{aligned}I_n =& 2\int_{\mathbb{R}^N}\log(\rho_{\varepsilon,k})\mathfrak{q}_0(\vartheta_n)d\mu+2\int_{\mathbb{R}^N}\vartheta_n\log(\rho_{\varepsilon,k})\mathrm{Tr}(QD^2\vartheta_n)d\mu\\&+2\int_{\mathbb{R}^N}\vartheta_n\log(\rho_{\varepsilon,k})\sum_{i,j=1}^N D_iq_{ij}D_j\vartheta_n d\mu+2\int_{\mathbb{R}^N}\vartheta_n\chi_{\{\varepsilon<\rho<k\}}\frac{\mathfrak{q}(\rho,\vartheta_n)}{\rho}d\mu.\end{aligned}$$

Using the Young inequality, we get

$$\begin{aligned}|I_n|\le& C_{\varepsilon,k}\int_{\mathbb{R}^N}\bigg(\mathfrak{q}_0(\vartheta_n)+\mathrm{Tr}(QD^2\vartheta_n)+\sum_{i,j=1}^N|D_iq_{ij}D_j\vartheta_n|\bigg)d\mu\\&+2\bigg(\int_{\mathbb{R}^N}\vartheta_n^2\chi_{\{\varepsilon<\rho<k\}}\frac{\mathfrak{q}_0(\rho)}{\rho}dx\bigg)^{\frac12}\bigg(\int_{\mathbb{R}^N}\mathfrak{q}_0(\vartheta_n)\rho dx\bigg)^{\frac12}\\ \le& C_{\varepsilon,k}\int_{\mathbb{R}^N}\bigg(\mathfrak{q}_0(\vartheta_n)+\mathrm{Tr}(QD^2\vartheta_n)+\sum_{i,j=1}^N|D_iq_{ij}D_j\vartheta_n|\bigg)d\mu\\&+\frac{1}{\delta^2}\int_{\mathbb{R}^N}\mathfrak{q}_0(\vartheta_n)d\mu+\delta\int_{\mathbb{R}^N}\vartheta_n^2\chi_{\{\varepsilon<\rho<k\}}\frac{\mathfrak{q}_0(\rho)}{\rho}dx,\end{aligned}\tag{9.2.14}$$

where $C_{\varepsilon,k}=|\log(\varepsilon)|\vee\log(k)$. Similarly, we can estimate

$$J_n+K_n\le\delta\int_{\mathbb{R}^N}\vartheta_n^2\chi_{\{\varepsilon<\rho<k\}}\frac{|\nabla\rho|^2}{\rho}dx+\frac{1}{4\delta^2}\int_{\mathbb{R}^N}|b|^2\rho dx+\frac{CC_{\varepsilon,k}}{n}\int_{\mathbb{R}^N}|b|d\mu\tag{9.2.15}$$

for any $\delta>0$. From (9.2.14) and (9.2.15) and Hypothesis 9.2.1(i) it follows that

$$\begin{aligned}&\bigg(1-\delta-\frac{\delta}{\kappa_0}\bigg)\int_{\mathbb{R}^N}\vartheta_n^2\chi_{\{\varepsilon<\rho<k\}}\frac{\mathfrak{q}_0(\rho)}{\rho}dx\\ \le& C_{\varepsilon,k}\int_{\mathbb{R}^N}\bigg(\mathfrak{q}_0(\vartheta_n)+\mathrm{Tr}(QD^2\vartheta_n)+\sum_{i,j=1}^N|D_iq_{ij}D_j\vartheta_n|\bigg)d\mu\\&+\frac{1}{\delta^2}\int_{\mathbb{R}^N}\mathfrak{q}_0(\vartheta_n)d\mu+\frac{CC_{\varepsilon,k}}{n}\int_{\mathbb{R}^N}|b|d\mu+\frac{1}{4\delta^2}\int_{\mathbb{R}^N}|b|^2d\mu.\end{aligned}$$

Letting n tend to $+\infty$ and taking Hypothesis 9.2.1(ii) and the condition (9.2.13) into account, from the dominated convergence theorem we deduce that

$$\bigg(1-\delta-\frac{\delta}{\kappa_0}\bigg)\int_{\mathbb{R}^N}\chi_{\{\varepsilon<\rho<k\}}\frac{\mathfrak{q}_0(\rho)}{\rho}dx\le\frac{1}{4\delta^2}\int_{\mathbb{R}^N}|b|^2\rho dx$$

for any $\delta>0$ and $0<\varepsilon<1<k$. Now, taking δ small enough and letting ε tend to 0^+ and k tend to $+\infty$, we deduce that $\rho^{-1}\mathfrak{q}_0(\rho)$ (and, hence, $\rho^{-1}|\nabla\rho|^2$) belongs to $L^1(\mathbb{R}^N)$ and

$$\kappa_0\int_{\mathbb{R}^N}\frac{|\nabla\rho|^2}{\rho}dx\le\int_{\mathbb{R}^N}\frac{\mathfrak{q}_0(\rho)}{\rho}dx\le\widetilde{C}\int_{\mathbb{R}^N}|b|^2d\mu\tag{9.2.16}$$

for a suitable positive constant $\widetilde{C}$. Hence, $\sqrt{\rho}$ belongs to $W^{1,2}(\mathbb{R}^N)$. Note that ρ belongs to $L^1(\mathbb{R}^N)$ according to Proposition 9.1.4.

Step 2. We assume that $N>2$, that $b_i\in L^p_\mu$ for any $i=1,\dots,N$, and that $\rho\in L^r(\mathbb{R}^N)$ for some $r\in(1,+\infty)$. We prove that $\rho^{(\beta-1)/2}|\nabla\rho|\in L^2(\mathbb{R}^N)$, where $\beta=(r-1)(1-2/p)$ and

$$\kappa_0^2\int_{\mathbb{R}^N}\rho^{\beta-1}|\nabla\rho|^2dx\le\int_{\mathbb{R}^N}|b|^2\rho^{\beta+1}dx.\tag{9.2.17}$$

Estimate (9.2.17) will imply that the function $\rho^{(\beta+1)/2}$ belongs to $W^{1,2}(\mathbb{R}^N)$ (note that, since $\beta+1<2r$ and $\rho\in L^1(\mathbb{R})$, ρ belongs to $L^s(\mathbb{R}^N)$ for any $s\in[1,r]$). Again, the Sobolev embedding theorems will yield $\rho\in L^q(\mathbb{R}^N)$ where $q=N(r(p-2)+2)/(p(N-2))$.

So, let us prove (9.2.17). The integral term in the right-hand side of (9.2.17) converges since, by the Hölder inequality, we can estimate

$$\int_{\mathbb{R}^N}|b|^2\rho^{\beta+1}dx\le\left(\int_{\mathbb{R}^N}|b|^pd\mu\right)^{\frac{2}{p}}\left(\int_{\mathbb{R}^N}\rho^r dx\right)^{1-\frac{2}{p}} \tag{9.2.18}$$

and the right-hand side of (9.2.18) is finite. Estimate (9.2.17) can be proved by applying the same arguments as in Step 1. Indeed, plugging $\varphi_n=\vartheta_n^2\rho_{\varepsilon,k}^{\beta}$ into (9.2.11) we get

$$\begin{aligned}\beta&\int_{\mathbb{R}^N}\vartheta_n^2\rho^{\beta-1}\chi_{\{\varepsilon<\rho<k\}}\mathfrak{q}_0(\rho)dx\\&=-2\int_{\mathbb{R}^N}\vartheta_n\rho_{\varepsilon,k}^{\beta}\mathfrak{q}(\rho,\vartheta_n)dx+\beta\int_{\mathbb{R}^N}\vartheta_n^2\rho^{\beta}\chi_{\{\varepsilon<\rho<k\}}\langle b,\nabla\rho\rangle dx\\&\quad+2\int_{\mathbb{R}^N}\vartheta_n\rho\rho_{\varepsilon,k}^{\beta}\langle b,\nabla\theta_n\rangle dx=:I_n+J_n+K_n.\end{aligned} \tag{9.2.19}$$

With the same computations as in Step 1 we can easily show that

$$\left[\beta\left(1-\frac{\delta}{\kappa_0}\right)-\delta\right]\int_{\mathbb{R}^N}\eta_n^2\rho^{\beta-1}\chi_{\{\varepsilon<\rho<k\}}\mathfrak{q}_0(\nabla\rho)dx\le\frac{1}{4\delta^2}\int_{\mathbb{R}^N}|b|^2\rho^{\beta+1}dx+C(\varepsilon,k,\delta,n), \tag{9.2.20}$$

where $\{C(\varepsilon,k,\delta,n)\}$ is a positive sequence which vanishes as n tends to $+\infty$, for any fixed ε,k,δ. Therefore, letting first n tend to $+\infty$ and, then, ε and k tend, respectively, to 0 and $+\infty$, we deduce that the function $\rho^{\beta-1}\mathfrak{q}_0(\nabla\rho)$ (and, consequently, the function $\rho^{\beta-1}|\nabla\rho|^2$) belongs to $L^1(\mathbb{R}^N)$. Now, from (9.2.19) the formula (9.2.17) easily follows.

Step 3. Here, we make the bootstrap argument work to prove the assertions (i)-(iii). We first assume that $N>2$. In such a situation the results of Steps 1 and 2 allow us to show that $\rho\in L^{q_n}(\mathbb{R}^N)$ for any $n\in\mathbb{N}$, where q_n is defined recursively by the following relation:

$$q_1=\frac{N}{N-2},\qquad q_{n+1}=\frac{N(p-2)}{p(N-2)}q_n+\frac{2N}{p(N-2)},\qquad n\ge1.$$

Therefore, $q_n=N^n(p-2)^np^{1-n}(N-2)^{-n}(p-N)^{-1}+N(N-p)^{-1}$ for any $n\ge2$, if $p\ne N$, whereas $q_n=1+2n(N-2)^{-1}$ for any $n\ge2$, if $p=N$. In the case when $p\in(2,N)$, q_n converges to $N/(N-p)$ as n tends to $+\infty$. Hence, to complete the proof of the assertion (i) it suffices to show that the sequence $\{\|\rho\|_{L^{q_n}(\mathbb{R}^N)}\}$ is bounded. Indeed, once this is proved, we will easily deduce that

$$\|\rho\|_{L^{N/(N-p)}(B_k)}=\lim_{n\to+\infty}\|\rho\|_{L^{q_n}(B_k)}\le\limsup_{n\to+\infty}\|\rho\|_{L^{q_n}(\mathbb{R}^N)}\le\sup_{n\in\mathbb{N}}\|\rho\|_{L^{q_n}(\mathbb{R}^N)}$$

for any $k\in\mathbb{N}$ and Fatou's lemma will imply that $\|\rho\|_{L^{N/(N-p)}(\mathbb{R}^N)}$ is finite.

Let us observe that according to the formulae (9.2.17) and (9.2.18) and the Sobolev embedding theorems we can estimate

$$\|\rho\|_{L^{q_{n+1}}(\mathbb{R}^N)}\le C_1^{\frac{2N}{q_{n+1}(N-2)}}\|\rho\|_{L^{q_n}(\mathbb{R}^N)}^{\frac{N}{N-2}\frac{p-2}{p}\frac{q_n}{q_{n+1}}},$$

where $C_1=C\|b\|_p$, C being the constant appearing in the Sobolev embedding theorems. Equivalently, setting $\alpha_n=\log(\|\rho\|_{L^{q_n}(\mathbb{R}^N)})$, we can rewrite the previous estimate as follows:

$$\alpha_{n+1}\le\frac{2N}{q_{n+1}(N-2)}|\log(C_1)|+\frac{N}{N-2}\frac{p-2}{p}\frac{q_n}{q_{n+1}}\alpha_n. \tag{9.2.21}$$

Since $N(p-2)/(p(N-2)) < 1$ and $\{q_n\}$ is increasing, from (9.2.21) it follows that the sequence $\{\alpha_n\}$ satisfies $\alpha_{n+1} \le \beta + \gamma\alpha_n$ for any $n \in \mathbb{N}$, some $\beta > 0$, $\gamma \in (0,1)$, and this implies that $\{\alpha_n\}$ and, consequently, $\{\|\rho\|_{L^{q_n}(\mathbb{R}^N)}\}$ are bounded. On the other hand, in the case when $p \ge N$ the sequence $\{q_n\}$ is increasing and it diverges to $+\infty$. Hence, the assertion (ii) easily follows. To prove that ρ belongs to $L^\infty(\mathbb{R}^N)$ when $p > N$, we will show that the L^{q_n}-norms of ρ are bounded by a constant C_2, independent of n. This will be enough for our purposes. Indeed, since ρ is continuous, for any $k \in \mathbb{N}$ we can estimate $\|\rho\|_{C(B_k)} = \lim_{n\to+\infty} \|\rho\|_{L^{q_n}(B_k)} \le \limsup_{n\to+\infty} \|\rho\|_{L^{q_n}(\mathbb{R}^N)} \le C_2$. The arbitrariness of k then will imply that $\rho \in L^\infty(\mathbb{R}^N)$.

To prove the boundedness of $\|\rho\|_{L^{q_n}(\mathbb{R}^N)}$ we argue as in the case when $p < N$. From (9.2.21) we easily deduce that

$$\alpha_n \le \beta_{n-1} + \sum_{k=1}^{n-2} \beta_{n-1-k} \prod_{l=n-k}^{n-1} \gamma_l + \alpha_1 \prod_{k=1}^{n-1} \gamma_k, \qquad n \ge 2,$$

where

$$\beta_j = \frac{2N}{q_{j+1}(N-2)} |\log(C_1)|, \qquad \gamma_j = \frac{N}{N-2}\frac{p-2}{p}\frac{q_j}{q_{j+1}} = 1 - \frac{2N}{p(N-2)}\frac{1}{q_{j+1}}$$

for any j. Observing that $\gamma_j \le 1$ for any j, we get $\alpha_n \le \beta_{n-1} + \sum_{k=1}^{n-2} \beta_k + \alpha_1$. Since $\sum_{k=1}^{+\infty} \beta_k < +\infty$, it follows that $\alpha_n \le \alpha_1 + \sum_{k=1}^{+\infty} \beta_k$ for any $n \in \mathbb{N}$ and we are done.

To conclude the proof, let us consider the case when $N = 2$. Property (ii) follows from the Sobolev embedding theorems since, as we have shown in Step 1, the function $\sqrt{\rho}$ belongs to $W^{1,2}(\mathbb{R}^N)$.

To prove property (iii) in the case when $N = 2$ we observe that, since $L^p_\mu(\mathbb{R}^N) \hookrightarrow L^2_\mu(\mathbb{R}^N)$, ρ belongs to $L^q(\mathbb{R}^N)$ for any $q \in [1,+\infty)$. Therefore, it belongs to $L^{r_0/(r_0-2)}(\mathbb{R}^N)$, where r_0 is such that $2 < r_0 < p$. Adapting the proof of Step 2 to this situation, we can show that, if $\rho \in L^r(\mathbb{R}^N)$ for some $r > 1$, then, actually, it belongs to $L^q(\mathbb{R}^N)$, where $q = r_0(r(p-2)+2)/(p(r_0-2))$. Applying the same bootstrap argument used in the case when $N \ge 3$, we can show that $\rho \in L^{q_n}(\mathbb{R}^N)$ for any $n \in \mathbb{N}$, where now the sequence $\{q_n\}$ is defined recursively by

$$q_1 = \frac{r_0}{r_0-2}, \qquad q_{n+1} = \frac{r_0(p-2)}{p(r_0-2)} q_n + \frac{2r_0}{p(r_0-2)}, \qquad n \ge 2.$$

Since $r_0(p-2)/(p(r_0-2)) > 1$, we conclude that $\rho \in C_b(\mathbb{R}^N)$ repeating step by step the same arguments used in the case when $N > 2$. □

Remark 9.2.7 In fact, the proof of Theorem 9.2.5 shows some differentiability properties of the density ρ. In particular, it shows that, if $b_i \in L^p_\mu$ for any $i = 1, \ldots, N$ and some $p \ge (N+2)/2$, $p > 2$, then $\rho \in W^{1,2}(\mathbb{R}^N)$. To check it, first suppose that $N \ge 3$. By Theorem 9.2.5(i), $\rho \in L^q(\mathbb{R}^N)$ for any $q \in [1, N/(N-p)]$; in particular, it belongs to $L^{2N/(N-2)}(\mathbb{R}^N)$. Therefore, we can apply Step 2 in the proof of Theorem 9.2.5, taking $\beta = 1$. This shows that $|\nabla\rho| \in L^2(\mathbb{R}^N)$, and we are done.

In the case when $N = 2$, Theorem 9.2.5 implies that $\rho \in L^q(\mathbb{R}^N)$ for any $q \in [1, +\infty]$. In particular, it belongs to $L^{2(p-1)/(p-2)}(\mathbb{R}^N)$ and, again, we can apply Step 2 in the proof of Theorem 9.2.5 to get the assertion.

A result similar to whose in Theorem 9.2.5(iii) can be proved, without assuming $p > N$, in the case when the function b is more regular. We refer the reader to the forthcoming Proposition 9.2.14 for a sufficient condition ensuring that the integrability assumptions on the coefficients b_i $(i = 1, \ldots, N)$ in the following theorem are satisfied.

Theorem 9.2.8 *Suppose that* $b_i \in W^{1,\infty}_{\rm loc}(\mathbb{R}^N)$ *and* $b_i, {\rm div}\, b \in L^p_\mu$ *for some* $p > N/2$, $p \ge 2$ *and any* $i = 1, \ldots, N$. *Then,* $\rho \in C_b(\mathbb{R}^N)$.

Proof The proof is similar to that of Theorem 9.2.5. To make the same arguments used therein work, we need to show that, if $\rho \in L^r(\mathbb{R}^N)$ for some $r > 1$, then the function $\rho^{(\beta+1)/2}$ belongs to $W^{1,2}(\mathbb{R}^N)$, where $\beta = (r-1)(1-1/p)$. For this purpose it suffices to show that

$$\kappa_0(\beta+1)\int_{\mathbb{R}^N}\rho^{\beta-1}|\nabla\rho|^2 dx \le -\int_{\mathbb{R}^N}\rho^{\beta+1}{\rm div} b\, dx, \tag{9.2.22}$$

since the integral in the right-hand side of (9.2.22) converges (compare with (9.2.18)). Multiplying the (distributional) identity $\mathcal{A}_0\rho = {\rm div}\,(\rho b)$ by $\vartheta_n^2\rho_{\varepsilon,k}^\beta$ and integrating by parts in $\mathbb{R}^N$, we obtain again (9.2.19).[2] The terms I_n and K_n can be estimated as in Step 1 in the proof of Theorem 9.2.5. As far as the term J_n is concerned, we observe that $\rho^\beta\chi_{\{\varepsilon<\rho<k\}}\nabla\rho = (\beta+1)^{-1}\nabla\rho_{\varepsilon,k}^{\beta+1}$. Therefore, since $\beta+1<r$, both the functions $\rho_{\varepsilon,k}^{\beta+1}$ and $\rho^{\beta+1}$ belong to $L^1(\mathbb{R}^N)$. Hence, integrating by parts, we can write

$$J_n = -\frac{\beta}{\beta+1}\int_{\mathbb{R}^N}\rho_{\varepsilon,k}^{\beta+1}\langle b, \nabla(\vartheta_n^2)\rangle dx - \frac{\beta}{\beta+1}\int_{\mathbb{R}^N}\rho_{\varepsilon,k}^{\beta+1}\vartheta_n^2{\rm div} b\, dx.$$

Letting n tend to $+\infty$, yields $\lim_{n\to+\infty} J_n = -\beta(\beta+1)^{-1}\int_{\mathbb{R}^N}\rho_{\varepsilon,k}^{\beta+1}{\rm div} b\, dx$. Therefore, arguing as in the proof of (9.2.20), we get

$$\beta(1-\delta)\int_{\mathbb{R}^N}\rho^{\beta-1}\chi_{\{\varepsilon<\rho<k\}}\langle Q\nabla\rho,\nabla\rho\rangle dx \le -\frac{\beta}{\beta+1}\int_{\mathbb{R}^N}\rho_{\varepsilon,k}^{\beta+1}{\rm div} b\, dx.$$

Finally, letting ε and δ tend to 0^+ and k tend to $+\infty$, estimate (9.2.22) follows. □

9.2.2 Global Sobolev regularity

In this subsection we assume the following more restrictive assumptions on the diffusion coefficients q_{ij} $(i,j = 1,\ldots,N)$.

Hypotheses 9.2.9 (i) Hypotheses 9.2.1(i) and 9.2.1(iii) are satisfied;

(ii) the functions q_{ij} belong to $C^1_b(\mathbb{R}^N)$ for any $i,j = 1,\ldots,N$.

Theorem 9.2.10 *Under Hypotheses* 9.2.9 *the following properties are satisfied:*

(i) *if* $b_i \in L^1_\mu$ $(i = 1,\ldots,N)$, *then* ρ *belongs to* $L^q(\mathbb{R}^N)$ *for any* $q \in \left[1, \frac{N}{N-1}\right)$;

(ii) *if* $b_i \in L^p_\mu$ $(i = 1,\ldots,N)$ *for some* $p \in (1,2)$, *then* ρ *belongs to* $W^{1,q}(\mathbb{R}^N)$ *for any* $q \in \left(1, \frac{N}{N-p+1}\right)$;

(iii) *if* $b_i \in L^p_\mu$ $(i = 1,\ldots,N)$ *for some* $p \in [2,N)$, *then* ρ *belongs to* $W^{1,q}(\mathbb{R}^N)$ *for any* $q \in \left[1, \frac{N}{N-p+1}\right]$;

(iv) *if* $b_i \in L^N_\mu$ $(i = 1,\ldots,N)$, *then* ρ *belongs to* $W^{1,q}(\mathbb{R}^N)$ *for any* $q \in [1,N)$;

[2] Here, $\mathcal{A}_0$ is given by (9.2.1), $\rho_{\varepsilon,k}$ is given by (9.2.12) and $\{\vartheta_n\}$ is the sequence of smooth functions defined by $\vartheta_n(x) = \vartheta(x/n)$ for any $x \in \mathbb{R}^N$ and $n \in \mathbb{N}$, $\vartheta \in C^\infty_c(\mathbb{R}^N)$ being any smooth function such that $\chi_{B_{1/2}} \le \vartheta \le \chi_{B_1}$.

(v) *if* $b_i \in L^p_\mu$ $(i = 1, \ldots, N)$ *for some* $p > N$, *then* ρ *belongs to* $W^{1,q}(\mathbb{R}^N)$ *for any* $q \in [1, p]$.

Proof We begin the proof checking property (i). For this purpose, we observe that, according to (9.2.10), we can write

$$\int_{\mathbb{R}^N} (\varphi - \mathcal{A}_0\varphi)d\mu = \int_{\mathbb{R}^N} (\varphi + \langle b, \nabla\varphi\rangle)d\mu$$

for any $\varphi \in C^\infty_c(\mathbb{R}^N)$. Here $\mathcal{A}_0$ is defined by (9.2.1). Then, using the Hölder inequality we get

$$\left|\int_{\mathbb{R}^N} (\varphi - \mathcal{A}_0\varphi)d\mu\right| \le C\|\varphi\|_{1,\infty} \tag{9.2.23}$$

for some positive constant C, independent of φ. Let us now show that (9.2.23) implies that

$$\left|\int_{\mathbb{R}^N} f d\mu\right| \le C\|f\|_{L^q(\mathbb{R}^N)} \tag{9.2.24}$$

for any $f \in L^q(\mathbb{R}^N)$ and $q > N$. This, of course, will lead us to the assertion thanks to the Riesz-Fisher representation theorem. Without loss of generality we assume that $f \in C^\infty_c(\mathbb{R}^N)$. According to [116, Theorems 9.15], the equation $w - \mathcal{A}_0 w = f$ admits a unique solution $w \in W^{2,q}(\mathbb{R}^N)$. Moreover, there exists a positive constant C, independent of f, such that $\|w\|_{W^{2,q}(\mathbb{R}^N)} \le C\|f\|_{L^q(\mathbb{R}^N)}$ (see [116, Lemma 9.17]). Since $q > N$, the Sobolev embedding theorems imply that $w \in C^1_b(\mathbb{R}^N)$ and $\|w\|_{C^1_b(\mathbb{R}^N)} \le \widetilde{C}\|f\|_{L^q(\mathbb{R}^N)}$ for some constant $\widetilde{C}$, independent of w. To get (9.2.24) we just need to show that we can plug w into (9.2.23). For this purpose, we use an approximation argument. Let $\{\vartheta_n\}$ be the sequence of smooth functions defined by $\vartheta_n(x) = \vartheta(x/n)$ for any $x \in \mathbb{R}^N$ and $n \in \mathbb{N}$, where $\vartheta \in C^\infty_c(\mathbb{R}^N)$ is such that $\chi_{B_{1/2}} \le \vartheta \le \chi_{B_1}$. The function $\vartheta_n w$ converges to w in $W^{2,q}(\mathbb{R}^N)$ as n tends to $+\infty$ and, consequently, $\mathcal{A}_0(\vartheta_n w)$ converges to $\mathcal{A}_0 w$ in $C_b(\mathbb{R}^N)$. We now regularize the sequence $\{\vartheta_n w\}$ by convolution in a standard way, introducing the sequence $\{w_{\varepsilon,n}\}$ defined by

$$w_{\varepsilon,n} = \int_{\mathbb{R}^N} \vartheta_n(y)w(y)\zeta_\varepsilon(x-y)dy, \qquad x \in \mathbb{R}^N,\ \varepsilon > 0,\ n \in \mathbb{N},$$

where $\zeta_\varepsilon(x) = \varepsilon^{-N}\zeta(\varepsilon^{-1}x)$ and $\zeta \in C^\infty_c(\mathbb{R}^N)$ is a smooth function with $\|\zeta\|_{L^1(\mathbb{R}^N)} = 1$. Of course, $w_{\varepsilon,n} \in C^\infty_c(\mathbb{R}^N)$ and it converges to $\vartheta_n w$ in $W^{2,q}(\mathbb{R}^N)$ as ε tends to 0^+. Moreover,

$$\begin{aligned}(\mathcal{A}_0 w_{\varepsilon,n})(x) = &\int_{\mathbb{R}^N} (\mathcal{A}_0(\vartheta_n w))(y)\zeta_\varepsilon(x-y)dy \\ &+ \sum_{i,j=1}^N \int_{\mathbb{R}^N} (q_{ij}(x) - q_{ij}(x-y))(D_{ij}(\vartheta_n w))(x-y)\zeta_\varepsilon(y)dy \\ &+ \sum_{i=1}^N \int_{\mathbb{R}^N} (c_i(x) - c_i(x-y))(D_i(\vartheta_n w))(x-y)\zeta_\varepsilon(y)dy\end{aligned} \tag{9.2.25}$$

for any $x \in \mathbb{R}^N$, where $c_i = \sum_{j=1}^N D_j q_{ij}$ for any $i = 1, \ldots, N$. Since $\mathcal{A}_0 w \in BUC(\mathbb{R}^N)$, $\mathcal{A}_0(\vartheta_n w)$ belongs to $BUC(\mathbb{R}^N)$ as well. Therefore, the first term in the right-hand side of (9.2.25) converges to $\mathcal{A}(\vartheta_n w)$ uniformly in $\mathbb{R}^N$, as ε tends to 0^+. Since $q_{ij} \in \mathrm{Lip}(\mathbb{R}^N)$ for any $i, j = 1, \ldots, N$, a similar argument shows that the second term in the right-hand side of (9.2.25) converges to zero uniformly in $\mathbb{R}^N$ as ε tends to 0^+. Finally, the third term converges to 0 locally uniformly in $\mathbb{R}^N$. It follows that $\mathcal{A}_0 w_{\varepsilon,n}$ converges to $\mathcal{A}_0(\vartheta_n w)$ locally uniformly in $\mathbb{R}^N$, as ε tends to 0^+. Moreover, there exists a positive constant $\overline{C}$, independent of ε and n, such that $\|\mathcal{A}_0 w_{\varepsilon,n}\|_\infty \le \overline{C}$ for any $\varepsilon > 0$ and $n \in \mathbb{N}$. We have so proved that there exists a sequence $\{w_n\} \subset C_c^\infty(\mathbb{R}^N)$ converging to w in $W^{2,q}(\mathbb{R}^N)$ such that $\mathcal{A}_0 w_n$ converges to $\mathcal{A}_0 w$ locally uniformly in $\mathbb{R}^N$, as n tends to $+\infty$, and the sequence $\{\mathcal{A}_0 w_n\}$ is bounded in $C_b(\mathbb{R}^N)$. As a consequence, writing (9.2.23) with w_n instead of φ and letting n tend to $+\infty$, easily yields (9.2.24).

Now we can prove property (ii). We argue as in Step 2 of the proof of Theorem 9.2.5. Suppose that $\rho \in L^q(\mathbb{R}^N)$ for some $q > 1$. Using (9.2.10) and the Hölder inequality, we can show that $|\int_{\mathbb{R}^N} \mathcal{A}_0\varphi d\mu| \le C\|\varphi\|_{1,r'}$, where $1/r + 1/r' = 1$ and $1/r = 1/p + (1-1/p)/q$. According to Theorem D.1.4(i), ρ belongs to $W^{1,r}(\mathbb{R}^N)$ and since $r < N$ (recall that $N \ge 2$) the Sobolev embedding theorems imply that $\rho \in L^{\widetilde{q}}(\mathbb{R}^N)$ where $\widetilde{q}^{-1} = p^{-1} - N^{-1} + (1 - p^{-1})q^{-1}$. Therefore, a bootstrap argument reveals that $\rho \in L^{q_n}(\mathbb{R}^N)$ for any $n \in \mathbb{N}$, where $s_n := 1/q_n$ $(n \in \mathbb{N})$ is defined recursively by $s_1 = \alpha$ and $s_{n+1} = (1/p - 1/N) + (1 - 1/p)s_n$ for any $n \in \mathbb{N}$, α being any real number belonging to the interval $((N-1)/N, 1)$. Since the sequence $\{s_n\}$ converges to $(N-p)/N$, q_n converges to $N/(N-p)$ as n tends to $+\infty$ Using the definition of r, the assertion follows.

Let us now prove property (iii). The arguments that we are going to use can also be adapted to prove properties (iv) and (v). According to Theorem 9.2.5(i) we know that $\rho \in L^{N/(N-p)}(\mathbb{R}^N)$. Therefore, the Hölder inequality and our assumptions show that the functions $b_i\rho$ belong to $L^{(N/(N-p+1)}(\mathbb{R}^N)$ for any $i = 1, \ldots, N$. Therefore, from (9.2.10) we deduce that $\left|\int_{\mathbb{R}^N} \mathcal{A}_0\varphi d\mu\right| \le C\|\varphi\|_{W^{1,N/(p-1)}(\mathbb{R}^N)}$ for any $\varphi \in C_c^\infty(\mathbb{R}^N)$. Theorem D.1.4(i) implies that the function ρ belongs to $W^{1,N/(N-p+1)}(\mathbb{R}^N)$. Since $\rho^{-1}|\nabla\rho|^2$ belongs to $L^1(\mathbb{R}^N)$ (see Step 1 in the proof of Theorem 9.2.5), using the Hölder inequality we conclude that $|\nabla\rho|$ belongs to $L^1(\mathbb{R}^N)$. Therefore, $\rho \in W^{1,1}(\mathbb{R}^N)$ and, hence, it belongs to $W^{1,q}(\mathbb{R}^N)$ for any $q \in [1, N/(N-p+1)]$. □

Remark 9.2.11 For a sufficient condition ensuring that the integrability assumptions on the coefficients b_i $(i = 1, \ldots, N)$ are satisfied, we refer the reader to the forthcoming Proposition 9.2.14.

We now deal with the regularity of the second-order derivatives of ρ.

Theorem 9.2.12 *In addition to Hypotheses 9.2.9, assume that b_i belongs to $W^{1,\infty}_{\rm loc}(\mathbb{R}^N)\cap L^p_\mu$ for any $i = 1, \ldots, N$ and $\operatorname{div} b \in L^p_\mu$ for some $p \ge 2$. Then, the following properties are satisfied:*

(i) *if $p \in (N/2, N)$, then ρ belongs to $W^{2,q}(\mathbb{R}^N)$ for any $q \in \left(1, \frac{pN}{pN-p^2+3p-2}\right]$;*

(ii) *if $p = N$, then ρ belongs to $W^{2,q}(\mathbb{R}^N)$ for any $q \in \left(1, \frac{N^2}{3N-2}\right)$;*

(iii) *if $p \in (N, 2N)$, then ρ belongs to $W^{2,q}(\mathbb{R}^N)$ for any $q \in \left(1, \frac{p^2}{3p-2}\right]$;*

(iv) *if $p = 2N$, then ρ belongs to $W^{2,q}(\mathbb{R}^N)$ for any $q \in (1, \frac{p}{2})$;*

(v) *if $p > 2N$, then ρ belongs to $W^{2,q}(\mathbb{R}^N)$ for any $q \in (1, \frac{p}{2}]$.*

Proof To prove the first three properties it suffices to show that, if $p > N/2$, $p \ge 2$, and $|\nabla\rho| \in L^r(\mathbb{R}^N)$ for some $r \in (1, +\infty)$, then ρ belongs to $W^{2,q}(\mathbb{R}^N)$ for any $q \in (1, s]$, where $1/s = (1 - 2/p)/r + 2/p$. Indeed, once this property is checked, conditions (i)-(iii) will immediately follow from Theorem 9.2.10.

Using (9.2.10) and integrating by parts, we can write

$$\int_{\mathbb{R}^N} (\mathcal{A}_0\psi)\rho dx = \int_{\mathbb{R}^N} (\rho \operatorname{div} b + \langle b, \nabla\rho\rangle)\psi dx, \qquad \psi \in C_c^\infty(\mathbb{R}^N),$$

where $\mathcal{A}_0$ is defined by (9.2.1). Since the function $f := \rho\operatorname{div} b + \langle b, \nabla\rho\rangle$ belongs to $L^m_{\rm loc}(\mathbb{R}^N)$ for any $m \in [1,+\infty)$ (see Theorem 9.2.3), according to Theorem D.1.4(iv) ρ belongs to $W^{2,m}_{\rm loc}(\mathbb{R}^N)$ for any $m \in [1,+\infty)$. Moreover, $\rho \operatorname{div} b$ belongs to $L^m(\mathbb{R}^N)$ for any $m \in [1,p]$, since $\rho \in C_b(\mathbb{R}^N)$ (see Theorem 9.2.8). Now, using the Hölder inequality we can write

$$\begin{aligned}\int_{\mathbb{R}^N} |b|^\alpha |\nabla\rho|^\alpha dx &= \int_{\mathbb{R}^N} |b|^\alpha |\nabla\rho|^{\alpha-\frac{2}{\beta}} |\nabla\rho|^{\frac{2}{\beta}} \rho^{-\frac{1}{\beta}} \rho^{\frac{1}{\beta}} dx\\ &\le \left(\int_{\mathbb{R}^N} \frac{|\nabla\rho|^2}{\rho} dx\right)^{\frac{1}{\beta}} \left(\int_{\mathbb{R}^N} |\nabla\rho|^{\frac{\alpha\beta-2}{\beta-2}} dx\right)^{\frac{\beta-2}{\beta}} \||b|^\alpha\|_{L^\beta_\mu} \end{aligned} \tag{9.2.26}$$

for any $\alpha \ge 1$ and $\beta \ge 2$ and, taking (9.2.16) into account, we conclude that the right-hand side of (9.2.26) is finite provided that $\frac{\alpha\beta-2}{\beta-2} = r$ and $\alpha\beta = r$. Hence, the best choice of α is s. Summing up, we have proved that $g := \langle b, \nabla\rho\rangle + \rho\operatorname{div} b \in L^s(\mathbb{R}^N)$. Since $\mathcal{A}_0\rho = g$, the function ρ belongs to $W^{2,s}(\mathbb{R}^N)$. Hence, the statement follows with $q = s$.

To prove that $\rho \in W^{2,q}(\mathbb{R}^N)$ for any $q \in (1,s)$, we observe that $|\nabla\rho| \in L^s(\mathbb{R}^N)$ since $\rho \in W^{2,s}(\mathbb{R}^N)$. Moreover, it belongs to $L^1(\mathbb{R}^N)$ by virtue of Theorem 9.2.10(iii) (recall that $p \ge 2$). Therefore, ρ belongs to $L^{r_*}(\mathbb{R}^N)$, where $r_* \in (1,s)$ is chosen so that $q^{-1} = (1-2/p)r_*^{-1} + 2/p$. Now, from the first part of the proof, we get $\rho \in W^{2,q}(\mathbb{R}^N)$.

To conclude the proof, let us check properties (iv) and (v). To begin with, we show that $|\nabla\rho| \in L^q(\mathbb{R}^N)$ for any $q < +\infty$, if $p = 2N$, and that $|\nabla\rho| \in L^\infty(\mathbb{R}^N)$, if $p > 2N$. Iterating the results in the first part of the proof, it is easy to check that $\rho \in W^{2,q_{n+1}}(\mathbb{R}^N)$ for any $n \in \mathbb{N}$ such that $q_n < N$. Here, $\tau_n := 1/q_n$ is defined recursively by $\tau_{n+1} = (1-2/p)\tau_n + (2N-p+2)(pN)^{-1}$ for any $n \ge 2$.

We claim that, if $p > 2N$, then there exists $n_0 \in \mathbb{N}$ such that $q_{n_0} > N$. The Sobolev embedding theorems then will imply that $|\nabla\rho| \in C_b(\mathbb{R}^N)$. To prove the claim, we observe that, if $q_n < N$ for any $n \le m$ and some $m \in \mathbb{N}$, then

$$\tau_n = \left(1-\frac{2}{p}\right)^n \left(\tau_1 - 1 + \frac{p-2}{2N}\right) + 1 + \frac{2-p}{2N}$$

for any $n \le m+1$. Since the sequence $\{\tau_n\}$ converges to $1 + (2-p)(2N)^{-1} < N^{-1}$ as n tends to $+\infty$, there should exist $n \in \mathbb{N}$ such that $q_n > N$. In the case when $p = 2N$, the sequence τ_n converges to N^{-1}. Therefore, $\rho \in W^{2,\alpha}(\mathbb{R}^N)$ for any $\alpha < N$ and, consequently, $|\nabla\rho| \in L^r(\mathbb{R}^N)$ for any $r \in (1,+\infty)$.

Now we are almost done. Indeed, to prove properties (iv) and (v) it suffices to show that the function $g = \sum_{i=1}^N b_i D_i\rho + \rho\operatorname{div} b$ belongs to $L^q(\mathbb{R}^N)$ for any $q \in (1,p/2)$, if $p = 2N$, and it belongs also to $L^{p/2}(\mathbb{R}^N)$, if $p > 2N$. Of course, without loss of generality we can limit ourselves to dealing with the function $|b\nabla\rho|$. Indeed, since $\rho \in C_b(\mathbb{R}^N)$, all the other terms in the definition of g belong to $L^q(\mathbb{R}^N)$ for any $q \in (1,p/2)$, if $p = 2N$, and for any $q \in (1,p/2]$, if $p > 2N$. From the Hölder inequality we get

$$\begin{aligned}\int_{\mathbb{R}^N} |b|^q |\nabla\rho|^q dx &= \int_{\mathbb{R}^N} |b|^q |\nabla\rho|^{q-\frac{2q}{p}} |\nabla\rho|^{\frac{2q}{p}} \rho^{-\frac{q}{p}} \rho^{\frac{q}{p}} dx\\ &\le \left(\int_{\mathbb{R}^N} |\nabla\rho|^{\frac{q(p-2)}{p-2q}} dx\right)^{\frac{p-2q}{p}} \left(\int_{\mathbb{R}^N} \frac{|\nabla\rho|^2}{\rho} dx\right)^{\frac{q}{p}} \|b\|^q_{L^p_\mu} \end{aligned} \tag{9.2.27}$$

for any $q \in (1,p/2)$, if $p = N/2$, and

$$\int_{\mathbb{R}^N} |b|^{\frac{p}{2}}|\nabla\rho|^{\frac{p}{2}}dx = \int_{\mathbb{R}^N} |b|^{\frac{p}{2}}|\nabla\rho|^{\frac{p}{2}-1}|\nabla\rho|\rho^{-\frac{1}{2}}\rho^{\frac{1}{2}}dx$$
$$\leq \|\nabla\rho\|_\infty^{\frac{p}{2}-1}\left(\int_{\mathbb{R}^N}\frac{|\nabla\rho|^2}{\rho}dx\right)^{\frac{1}{2}}\left(\int_{\mathbb{R}^N}|b|^p d\mu\right)^{\frac{1}{2}}, \tag{9.2.28}$$

if $p > 2N$. Since the last sides of (9.2.27) and (9.2.28) are finite, the assertion follows. □

To conclude this subsection we prove some pointwise bounds for the function ρ. For this purpose, we need to assume some additional assumptions on the measure μ and on the drift coefficients b_i $(i = 1,\ldots,N)$. For notational convenience, for any $\delta,\beta > 0$ we denote by $V_{\delta,\beta}$ the function defined by $V_{\delta,\beta}(x) = \exp(\delta|x|^\beta)$ for any $x \in \mathbb{R}^N$.

Theorem 9.2.13 *In addition to Hypotheses 9.2.9 assume that $V_{\delta,\beta} \in L^1_\mu$ for some $\beta,\delta > 0$. Then, the following properties are satisfied:*

(i) *if $|b(x)| \leq C\exp(|x|^\gamma)$ for some $C > 0$, $\gamma < \beta$ and any $x \in \mathbb{R}^N$, then there exist positive constants c_1 and c_2 such that $\rho(x) \leq c_1\exp(-c_2|x|^\beta)$ for any $x \in \mathbb{R}^N$;*

(ii) *if in addition to the previous set of hypotheses, $b_i \in W^{1,\infty}_{\rm loc}(\mathbb{R}^N)$ $(i = 1,\ldots,N)$ and $|b(x)| + |{\rm div}\, b(x)| \leq C\exp(|x|^\gamma)$ for some $C > 0$, some $\gamma < \beta$ and any $x \in \mathbb{R}^N$, then there exist positive constants c_1 and c_2 such that $|\nabla\rho(x)| \leq c_1\exp(-c_2|x|^\beta)$ for any $x \in \mathbb{R}^N$.*

Proof (i) Let Set $\zeta = V_{c_2,\beta}\psi$, where ψ is any function belonging to $C^\infty_c(\mathbb{R}^N)$ and c_2 is a positive constant to be fixed later on. The invariance property satisfied by the measure μ implies that

$$\int_{\mathbb{R}^N} \mathcal{A}_0\zeta d\mu = -\int_{\mathbb{R}^N}\langle b,\nabla\zeta\rangle d\mu \tag{9.2.29}$$

or, equivalently by differentiation,

$$\int_{\mathbb{R}^N} V_{c_2,\beta}\rho\mathcal{A}_0\psi dx = -\int_{\mathbb{R}^N}(\psi\mathcal{A}_0V_{c_2,\beta} + 2\langle Q\nabla\psi,\nabla V_{c_2,\beta}\rangle + V_{c_2,\beta}\langle b,\nabla\psi\rangle + \psi\langle b,\nabla V_{c_2,\beta}\rangle)d\mu. \tag{9.2.30}$$

Here, $\mathcal{A}_0$ is defined by (9.2.1). Now, we fix $p > N$ and prove that the right-hand side of (9.2.30) can be estimated by $C\|\psi\|_{W^{1,p'}(\mathbb{R}^N)}$ for a suitable positive constant C, independent of ψ. Here, $1/p + 1/p' = 1$. Theorem D.1.4(i) then will imply that $\rho V_{c_2,\beta} \in W^{1,p}(\mathbb{R}^N)$ and the Sobolev embedding theorems will yield $\rho V_{c_2,\beta} \in C_b(\mathbb{R}^N)$, which is the statement of property (i). So, let us fix $q > p$ and choose $c_2 < \delta/q$. According to our assumptions, the functions $V_{c_2,\beta}$, $\nabla_x V_{c_2,\beta}$ and $\mathcal{A}_0 V_{c_2,\beta}$ belong to L^q_μ and $\rho \in C_b(\mathbb{R}^N)$ (see Theorem 9.2.5(iii)). Therefore, the first two terms in the right-hand side of (9.2.30) can be estimated as wished. Indeed, let us consider, for instance, the first one. We can estimate

$$\left|\int_{\mathbb{R}^N}\psi\mathcal{A}_0V_{c_2,\beta}d\mu\right| = \left|\int_{\mathbb{R}^N}\psi(\mathcal{A}_0V_{c_2,\beta})\rho dx\right| \leq \|\rho\|_\infty^{1-\frac{1}{p}}\left|\int_{\mathbb{R}^N}\psi(\mathcal{A}_0V_{c_2,\beta})\rho^{\frac{1}{p}}dx\right|$$
$$\leq \|\rho\|_\infty^{1-\frac{1}{p}}\|\mathcal{A}_0V_{c_2,\beta}\|_p\|\psi\|_{L^{p'}(\mathbb{R}^N)}. \tag{9.2.31}$$

As far as the last two terms in the right-hand side of (9.2.30) are concerned, we observe that the functions $V_{c_2,\beta}b_i$ and $b_iD_iV_{c_2,\beta}$ belong to $L^p_\mu(\mathbb{R}^N)$ for any $i = 1,\ldots,N$. Indeed, according to our assumptions, the functions b_i $(i = 1,\ldots,N)$ belong to $L^r_\mu(\mathbb{R}^N)$ for any

$r \in [1, +\infty)$. Hence, in particular, they belong to $L^{pq/(q-p)}_\mu(\mathbb{R}^N)$. The Hölder inequality now implies that $V_{c_2,\beta}b_i$ and $b_iD_iV_{c_2,\beta}$ belong to $L^p_\mu(\mathbb{R}^N)$ for any $i = 1, \ldots, N$, and, repeating the same arguments as in the proof of (9.2.31), we are done.

(ii) According to Theorem 9.2.12(v), ρ belongs to $W^{2,p}(\mathbb{R}^N)$ for any $p \in (1, +\infty)$. Therefore, integrating by parts the identity (9.2.29), we easily see that

$$\int_{\mathbb{R}^N} (\mathcal{A}_0\rho)\zeta dx = \int_{\mathbb{R}^N} [\rho(\operatorname{div} b) + \langle b, \nabla\rho\rangle]\zeta dx, \qquad \zeta \in C^\infty_c(\mathbb{R}^N)$$

or, equivalently, $\mathcal{A}_0\rho = \rho(\operatorname{div} b) + \langle b, \nabla\rho\rangle$. It follows that

$$\mathcal{A}_0(\rho V_{c_2,\beta}) = V_{c_2,\beta}\rho \operatorname{div} b + V_{c_2,\beta}\langle b, \nabla\rho\rangle + \rho(\mathcal{A}_0 V_{c_2,\beta}) + 2\langle Q\nabla V_{c_2,\beta}, \nabla\rho\rangle. \qquad (9.2.32)$$

Now, we replace the constant c_2 with a smaller constant to make $\rho\nabla V_{c_2,\beta}$ bounded in $\mathbb{R}^N$. Our purpose consists in showing that the right-hand side of (9.2.32) belongs to $L^p(\mathbb{R}^N)$ for some $p > N$. This will imply that $\rho V_{c_2,\beta}$ belongs to $W^{2,p}(\mathbb{R}^N)$ and the Sobolev embedding theorems then will yield $\nabla(\rho V_{c_2,\beta}) \in C_b(\mathbb{R}^N)$ and, consequently, $V_{c_2,\beta}|\nabla\rho| \in C_b(\mathbb{R}^N)$. As in the proof of property (i), one can check that $V_{c_2,\beta}\rho\,(\operatorname{div} b), \rho(\mathcal{A}_0V_{c_2,\beta}) \in L^p(\mathbb{R}^N)$. As far as the terms containing $\nabla\rho$ are concerned, we observe that

$$\begin{aligned}\int_{\mathbb{R}^N} |V_{c_2,\beta}\langle b, \nabla\rho\rangle|^p dx \leq& \int_{\mathbb{R}^N} V^p_{c_2,\beta}|b|^p|\nabla\rho|^{p-1}|\nabla\rho|\rho^{-\frac{1}{2}}\rho^{\frac{1}{2}} dx\\ \leq&\|\nabla\rho\|^{p-1}_\infty\left(\int_{\mathbb{R}^N}\frac{|\nabla\rho|^2}{\rho}dx\right)^{\frac{1}{2}}\left(\int_{\mathbb{R}^N} V^{2p}_{c_2,\beta}|b|^{2p}d\mu\right)^{\frac{1}{2}}\end{aligned}$$

and the last side of the previous chain of inequalities is finite due to our assumptions and to (9.2.16). In a similar way one can estimate the term $|\nabla V_{c_2,\beta}||\nabla\rho|$. □

A situation in which $V_{\delta,\beta} \in L^1_\mu$, for suitable values of β and δ, is provided by the next proposition.

Proposition 9.2.14 *Assume that*

$$\limsup_{|x|\to+\infty}\left\{c\lambda_{\max}(x) + |x|^{-\beta}\left(\sum_{i=1}^N b_i(x)x_i + \sum_{i,j=1}^N D_jq_{ij}(x)x_i\right)\right\} < 0 \qquad (9.2.33)$$

for some positive constants c and β, where, for any $x \in \mathbb{R}^N$, $\lambda_{\max}(x)$ denotes the maximum eigenvalue of the matrix $Q(x)$. Then, the operator $\mathcal{A}$ admits a (unique) invariant measure μ. Moreover, the function $V_{\delta,\beta}$ is integrable with respect to μ for any $\delta < \beta^{-1}c$.

Proof As is easily seen,

$$\mathcal{A}V_{\delta,\beta}(x) \leq \delta\beta|x|^{\beta-2}e^{\delta|x|^\beta}\left[(\sigma + \delta\beta|x|^\beta)\lambda_{\max}(x) + \langle b(x), x\rangle + \sum_{i,j=1}^N D_jq_{ij}x_i\right]$$

for any $x \in \mathbb{R}^N$, where $\sigma = N + |\beta - 2|$ and, due to (9.2.33) and the choice of δ and β, the right-hand side of the previous estimate is negative for large $|x|$. This implies that $\mathcal{A}V_{\delta,\beta}(x)$ diverges to $-\infty$ as $|x|$ tends to $+\infty$. Therefore, according to Theorem 9.1.20, the semigroup associated with the operator $\mathcal{A}$ admits a (unique) invariant measure μ.

Let us now prove that the function $V_{\delta,\beta}$ belongs to L^1_μ. From the above pointwise estimate on $\mathcal{A}V_{\delta,\beta}$ we deduce that, if $\beta > 1$, then $V_{\delta,\beta}(x) \leq |\mathcal{A}V_{\delta,\beta}(x)|$ for $|x|$ sufficiently large, whereas, if $\beta \in (0, 1)$, then, for any $\varepsilon > 0$, $|V_{(1-\varepsilon)\delta,\beta}(x)| \leq |\mathcal{A}V_{\delta,\beta}(x)|$ still for sufficiently large $|x|$. Therefore, in both cases it suffices to show that $\mathcal{A}V_{\delta,\beta} \in L^1_\mu$ for any $\delta < \beta^{-1}c$ to deduce that $V_{\delta,\beta} \in L^1_\mu$.

To prove that $\mathcal{A}V_{\delta,\beta} \in L^1_\mu$ we observe that, by a simple approximation argument, we can show that $\int_{\mathbb{R}^N} \mathcal{A}\zeta d\mu = 0$ for any $\zeta \in C^2_b(\mathbb{R}^N)$ with compact support. Since $\mathcal{A}\mathbb{1} = 0$, we can extend the previous formula to any $\zeta \in C^2_b(\mathbb{R}^N)$, which is constant outside of a ball. Now, let $\{\psi_n\} \in C^2_b([0, +\infty))$ be a sequence of increasing functions such that $\psi_n(t) = t$ for any $t \in [0, n]$, $\psi(t) = n + 1/2$ for any $t \geq n + 1$, and $\psi_n''(t) \leq 0$ for any $t \geq 0$. Then, the function $V_n = \psi_n \circ V_{\delta,\beta}$ belongs to $C^2_b(\mathbb{R}^N)$ and is constant outside of a ball. Therefore,

$$\int_{\mathbb{R}^N} \mathcal{A}V_n d\mu = 0, \qquad n \in \mathbb{N}. \tag{9.2.34}$$

Fix now M sufficiently large that $\mathcal{A}V_{\delta,\beta}(x) < 0$ for any $x \notin B_M$ and, then, fix a large enough n so that $V_n = V_{\delta,\beta}$ in B_M. Taking (9.2.34) into account, we can write

$$\int_{\mathbb{R}^N \setminus B_M} |\mathcal{A}V_n| d\mu = -\int_{\mathbb{R}^N \setminus B_M} |\mathcal{A}V_n| d\mu = \int_{B_M} \mathcal{A}V_n d\mu = \int_{B_M} \mathcal{A}V_{\delta,\beta} d\mu.$$

According to Fatou's lemma, $\mathcal{A}V_{\delta,\beta}$ is integrable in $\mathbb{R}^N \setminus B_M$ (and, hence, in $\mathbb{R}^N$) with respect to the measure μ. □

Assuming more regularity on the diffusion coefficients and some bounds on the growth at infinity of the drift term, one can prove also lower bounds for ρ.

Theorem 9.2.15 ([193], Theorem 6.3) *Assume that $q_{ij} \in C^3_b(\mathbb{R}^N)$ and $b_i \in C^2(\mathbb{R}^N)$ $(i, j = 1, \ldots, N)$ satisfy $|b_i(x)| + |\nabla b_i(x)| + |D^2 b_i(x)| \leq C(1 + |x|^{\beta-1})$ for any $x \in \mathbb{R}^N$, any $i = 1, \ldots, N$, and some $\beta, C > 0$. Then, $\rho(x) \geq \exp(-M(1 + |x|^\beta))$ for any $x \in \mathbb{R}^N$ and some positive constant M, depending only on C, κ_0 and $\|q_{ij}\|_{C^3_b(\mathbb{R}^N)}$ $(i, j = 1, \ldots, N)$.*

9.3 Some consequences of the estimates in Chapter 6

In this section we show some interesting consequences of the estimates proved in Chapter 6, when the semigroup $\{T(t)\}$ admits an invariant measure.

Remark 9.3.1 Note that, if there exists a function φ, blowing up as $|x|$ tends to $+\infty$, such that $\mathcal{A}\varphi$ diverges to $-\infty$ as $|x|$ tends to $+\infty$, then, both the assumptions of Khas'minskii theorem and Hypothesis 5.1.1(ii) are satisfied.

The following proposition is a consequence of the results in Theorem 6.2.2 and Corollary 6.2.3.

Proposition 9.3.2 *Suppose that Hypotheses 5.1.1(i)-(iii), and Hypothesis 5.1.1(iv-k) or 6.1.3(ii-k) are satisfied for some $k \in \{1, 2, 3\}$, and let $p \in (1, +\infty)$. Then, for any $t > 0$, $T(t)$ maps L^p_μ into $W^{k,p}_\mu$ and*

$$\|D^k_x T(t) f\|_p \leq \left(\sum_{j=0}^{k-1} k^j \prod_{h=0}^{j} \frac{\widetilde{\omega}_{k-h,p\wedge 2} \psi_{k-h,p\wedge 2}(t/k)}{1 - e^{-\widetilde{\omega}_{k-h,p\wedge 2} t}} \right)^{\frac{1}{p} \vee \frac{1}{2}} \|f\|_p \tag{9.3.1}$$

for any $t > 0$ and $f \in L^p_\mu$. Here, for $r = 1, 2, 3$, $\widetilde{\omega}_{r,p} = 0$, under Hypothesis 5.1.1(iv-k), and

$\widetilde{\omega}_{r,p} = \omega_{k,p} < 0$ *is given by Theorem* 6.1.5, *if Hypothesis* 6.1.3(*ii-k*) *is satisfied. Moreover,* $\psi_{r,p}$ *is given by* (6.2.5). *In particular, the* L^p_μ*-norm of* $D^k_x T(t)f$ *behaves as* $t^{-1/2}$ *as* t *tends to* 0, *whereas, at infinity, it stays bounded, if Hypothesis* 5.1.1(*iv-k*) *is satisfied, and it decreases exponentially if Hypothesis* 6.1.3(*ii-k*) *is satisfied.*

Proof The proof is a straightforward consequence of Theorem 6.2.2 and Corollary 6.2.3. Indeed, let us first assume that $f \in C_b(\mathbb{R}^N)$; integrating (6.2.4) (if $k = 1$) or (6.2.20) (if $k = 2, 3$) in $\mathbb{R}^N$, we deduce that $T(t)f$ belongs to $W^{k,p}_\mu$ and it satisfies (9.3.1) for any $t > 0$. In the general case when $f \in L^p_\mu$, consider a sequence $\{f_n\} \subset C_b(\mathbb{R}^N)$ converging to f in L^p_μ. Writing (9.3.1) for $f_n - f_m$ it follows that $\{T(t)f_n\}$ is a Cauchy sequence in $W^{k,p}_\mu$. Therefore, $T(t)f$ belongs to $W^{k,p}_\mu$ and satisfies (9.3.1). □

Proposition 9.3.3 *Let Hypotheses* 5.1.1(*i*)-(*iii*) *and* 5.1.1(*iv*-1) *be satisfied and let* $p > 1$. *Then,* $D(L_p) \subset W^{1,p}_\mu$ *and, for any* $\omega > 0$, *there exists a positive constant* M_ω *such that*

$$\|\nabla f\|_p \le M_{\omega,p} \|f\|_p^{\frac{1}{2}} \|(L_p - \omega)f\|_p^{\frac{1}{2}}, \qquad f \in D(L_p). \tag{9.3.2}$$

Under Hypotheses 5.1.1(*i*)-(*iii*) *and* 6.1.3(*i*), *estimate* (9.3.2) *holds true with* $\omega = 0$.

Proof The proof is similar to that of Theorem 5.2.2. Indeed, fix $f \in D(L_p)$, $\lambda > 0$ and let ω be as in the statement of the proposition. Moreover, let us set $u = (\lambda + \omega)f - L_p f$. Then, we have

$$f = \int_0^{+\infty} e^{-(\lambda+\omega)t} T(t)u\,dt. \tag{9.3.3}$$

Now, from Proposition 9.3.2 we deduce that in correspondence of ω we can find a positive constant $C = C_\omega$ such that $\|\nabla_x T(t)f\|_p \le Ct^{-\frac{1}{2}} e^{\omega t}\|f\|_p$ for any $t > 0$. Using this estimate and (9.3.3) we conclude that $|\nabla f| \in L^p_\mu$ and

$$\begin{aligned}\|\nabla f\|_p \le& C_{\omega,p}\left(\int_0^{+\infty} t^{-\frac{1}{2}} e^{-\lambda t}dt\right)\|\lambda f - (L_p - \omega)f\|_p \\ \le& \sqrt{\pi} C_{\omega,p}\left(\sqrt{\lambda}\|f\|_p + \frac{1}{\sqrt{\lambda}}\|(L_p - \omega)f\|_p\right).\end{aligned} \tag{9.3.4}$$

Minimizing with respect to $\lambda \in (0, +\infty)$ the last side of (9.3.4), we get the assertion with $M_{\omega,p} = 2\sqrt{2}C_{\omega,p}$. □

In the remainder of this section, we consider the particular case when $\mathcal{A} = \Delta + \langle b, \nabla\rangle$ under the following assumptions on the function b.

Hypotheses 9.3.4 (i) the drift coefficient b belongs to $C^{1+\delta}_{\rm loc}(\mathbb{R}^N, \mathbb{R}^N)$ for some $\delta > 0$. Moreover, there exists $d_0 \in \mathbb{R}$ such that $\langle b(x) - b(y), x - y\rangle \le d_0|x - y|^2$ for any $x, y \in \mathbb{R}^N$;

(ii) Hypothesis 5.1.1(ii) is satisfied and the semigroup associated with the operator $\mathcal{A}$ admits an invariant measure.

Remark 9.3.5 (i) If the drift term b satisfies Hypotheses 9.3.4, then it satisfies condition (6.1.8). To check this latter condition, it suffices to write the inequality for b in Hypotheses 9.3.4 with $y = x + t\xi$, dividing, then, both the members by t^2 and letting t tend to 0. Therefore, the assumptions of Theorem 6.3.1 are satisfied and estimates

(6.3.3) and (6.3.4) apply. In particular, these latter two estimates imply that, for any $f\in L^p_\mu(\mathbb{R}^N)$ and $t>0$, it holds that

$$\|\nabla_x T(t)f\|_p\le \frac{2p^{1/p-1/2}d_0^{1/p}}{(p-1)^{1/2}(1-e^{-pd_0t})^{1/p}}t^{\frac{1}{p}-\frac{1}{2}}\|f\|_p$$

for any $p\in(1,2]$ and $\|\nabla_x T(t)f\|_p\le\sqrt{d_0}(1-e^{-2d_0t})^{-1/2}\|f\|_p$ for any $p\in(2,+\infty)$. Finally, Proposition 9.3.3 holds true.

(ii) Hypothesis 9.3.4(ii) is satisfied, for instance, under the assumptions of Remark 9.3.1. In particular, this is the case when $d_0<0$. Indeed, by the previous point, it suffices to take $\varphi(x)=|x|^2$ for any $x\in\mathbb{R}^N$.

Under Hypotheses 9.3.4 we can prove the following formulae.

Proposition 9.3.6 *Assume that Hypotheses 9.3.4 hold and let $p,q>1$ be such that $p^{-1}+q^{-1}=1$. For any $f\in D(L_p)$ and $g\in D(L_q)$, the function fg belongs to $D(L_1)$ and*

$$L_1(fg)=fL_qg+gL_pf+2\langle\nabla f,\nabla g\rangle. \tag{9.3.5}$$

In particular,

$$\int_{\mathbb{R}^N} gL_pf\,d\mu=-\int_{\mathbb{R}^N} fL_qg\,d\mu-2\int_{\mathbb{R}^N}\langle\nabla f,\nabla g\rangle d\mu \tag{9.3.6}$$

and, for any $f\in D(L_2)$, we have

$$\int_{\mathbb{R}^N} fL_2f\,d\mu=-\int_{\mathbb{R}^N}|\nabla f|^2d\mu. \tag{9.3.7}$$

Proof Assume first $f,g\in D(\widehat{A})$. Since $D(\widehat{A})=D_{\max}(\mathcal{A})$ (see Propositions 1.3.5 and 3.1.10), fg belongs to $D(\widehat{A})$ and

$$A(fg)=f\widehat{A}g+g\widehat{A}f+2\langle\nabla f,\nabla g\rangle. \tag{9.3.8}$$

Now, fix $f\in D(L_p)$ and $g\in D(L_q)$. Since $D(\widehat{A})$ is a core of L_p and L_q (see Proposition 9.1.8), there exist two sequences $\{f_n\},\{g_n\}\subset D(\widehat{A})$ such that f_n, L_pf_n converge, respectively, to f and L_pf in L^p_μ, whereas g_n, L_qg_n converge, respectively, to g and L_qg in L^q_μ, as n tends to $+\infty$. Then, f_ng_n converges to fg in L^1_μ; besides by (9.3.2), for any $j=1,\dots,N$, D_jf_n converges to D_jf in L^p_μ and D_jg_n converges to D_jg in L^q_μ as n tends to $+\infty$, so that $\langle\nabla f_n,\nabla g_n\rangle$ converges to $\langle\nabla f,\nabla g\rangle$ in L^1_μ. By (9.3.8) we conclude that $\widehat{A}(f_ng_n)=L_1(f_ng_n)$ converges to the right-hand side of (9.3.5). The closedness of L_1 implies that $fg\in D(L_1)$ and that (9.3.5) holds.

Formula (9.3.6) now follows integrating (9.3.5) in $\mathbb{R}^N$ and taking (9.1.5) into account. Finally, the formula (9.3.7) follows taking $f=g$ in (9.3.6). □

In the symmetric case we can prove the following improvement of the formula (9.3.6), which generalizes the formula (9.1.32).

Proposition 9.3.7 *Assume that Hypotheses 9.1.25 hold with $U\in C^{2+\alpha}_{\rm loc}(\mathbb{R}^N)$ and $G\in C^{1+\alpha}_{\rm loc}(\mathbb{R}^N)$. Further, assume that $\langle\nabla U(x)+G(x)-\nabla U(y)-G(y),y-x\rangle\le d_0|x-y|^2$ for any $x,y\in\mathbb{R}^N$ and some constant $d_0\in\mathbb{R}$. Finally, let $p,q>1$ be such that $p^{-1}+q^{-1}=1$. Then, for any $f\in D(L_p)$ and $g\in W^{1,q}_\mu$ it holds that*

$$\int_{\mathbb{R}^N} gL_pf\,d\mu=-\int_{\mathbb{R}^N}\langle\nabla f,\nabla g\rangle d\mu. \tag{9.3.9}$$

Proof If $f, g \in D(\widehat{A})$, then Formula (9.3.9) follows from (9.1.32). Indeed, it suffices to observe that, since $D(\widehat{A}) \subset D(L_2)$, we can write

$$\int_{\mathbb{R}^N} g L_p f d\mu = \int_{\mathbb{R}^N} g L_2 f d\mu = -\int_{\mathbb{R}^N} \langle \nabla f, \nabla g \rangle d\mu.$$

Now, we consider the general case when $f \in D(L_p)$ and $g \in W^{1,q}_\mu$. Let $\{f_n\} \subset D(\widehat{A})$ be a sequence such that f_n and $\widehat{A} f_n$ converge in L^p_μ, respectively, to f and $L_p f$ as n tends to $+\infty$, and let $\{g_n\} \subset D(\widehat{A})$ converge to g in $W^{1,q}_\mu$. Then, by estimate (9.3.2), f_n converges to f in $W^{1,q}_\mu$. Thus, (9.3.9) follows letting n tend to $+\infty$. □

9.4 Logarithmic Sobolev inequality, Poincaré inequality, summability improving properties

In this section we introduce the logarithmic Sobolev inequality, the Poincaré inequality, and analyze the summability improving properties of the semigroup $\{T(t)\}$. As we have already remarked in the introduction, most of the results are byproducts of the results in Chapter 16, so we skip the proofs[3] of those results which can be deduced from the ones in Chapter 16.

If not otherwise specified, we assume the following set of assumptions.

Hypotheses 9.4.1 (i) *The diffusion coefficients q_{ij} are constant and the matrix Q is strictly positive definite;*

(ii) *the coefficients b_i $(i = 1, \dots, N)$ and their first-order spatial derivatives belong to $C^\alpha_{\rm loc}(\mathbb{R}^N)$ for some $\alpha \in (0,1)$;*

(iii) *there exists a negative constant r_0 such that $\langle {\rm Jac}_x b(x)\xi, \xi\rangle \le r_0|\xi|^2$ for any $x, \xi \in \mathbb{R}^N$.*

To begin with, we state the logarithmic Sobolev inequality and some of its remarkable consequences such as the Poincaré inequality and the hypercontractivity of the semigroup $\{T(t)\}$.

Theorem 9.4.2 *For any $f \in C^1_b(\mathbb{R}^N)$ and $p \in (1, +\infty)$ it holds*

$$\int_{\mathbb{R}^N} |f|^p \log(|f|) d\mu \le \|f\|_p^p \log(\|f\|_p) + \frac{p\Lambda}{2|r_0|} \int_{\mathbb{R}^N} |f|^{p-2} |\nabla f|^2 \chi_{\{f \neq 0\}} d\mu, \tag{9.4.1}$$

where Λ denotes the maximum eigenvalue of the matrix Q and $0 \log 0 = 0$. Formula (9.4.1) can be extended to any $f \in W^{1,p}_\mu$ and $p \ge 2$. Moreover,

$$\int_{\mathbb{R}^N} |f|^p \log(|f|) d\mu \le \|f\|_p^p \log(\|f\|_p) - \frac{p\Lambda}{2(p-1)|r_0|} \int_{\mathbb{R}^N} |f|^{p-1} L_p f d\mu \tag{9.4.2}$$

for any $f \in D(L_p)$ and $p \in (1, +\infty)$.

As a byproduct of estimate (9.4.1) the following properties hold true.

[3] We stress that in many situations the proofs provided in Sections 16.4 to 16.7 can be simplified since in the autonomous case we do not need to introduce the cut-off sequence $\{\vartheta_n\}$.

(i) *$W^{1,p}_\mu$ is compactly embedded in L^p_μ for any $p \in [2, +\infty)$;*

(ii) *the operator $T(t)$ is compact in L^p_μ for any $t > 0$ and $p \in (1, +\infty)$;*

(iii) *for any $p \geq 2$, there exists a positive constant C_p such that*
$$\|f - \overline{f}\|_p \leq C_p \|\nabla f\|_p, \qquad f \in W^{1,p}_\mu. \tag{9.4.3}$$
In particular, if $p = 2$, then we can take $C_2 = \Lambda^{1/2}|r_0|^{-1/2}$;

(iv) *the semigroup $\{T(t)\}$ is hypercontractive, i.e. for any $t > 0$ and $p \in (1, +\infty)$, the operator $T(t)$ is a contraction from L^p_μ to L^q_μ, where $q = e^{2\kappa_0|r_0|\Lambda^{-1}t}(p-1)+1$, κ_0 being the minimum eigenvalue of the matrix Q.*

Proof We just prove estimate (9.4.2), since the other results are straightforward consequences of the results in Section 16.4. To check this estimate it suffices to show that
$$\int_{\mathbb{R}^N} |f|^{p-2}|\nabla f|^2 \chi_{\{f\neq 0\}} d\mu \leq -\frac{1}{p-1}\int_{\mathbb{R}^N} f|f|^{p-2} L_p f d\mu \tag{9.4.4}$$
for any $f \in D(\widehat{A})$. Indeed, from this estimate and (9.4.1), Formula (9.4.2) follows for functions in $D(\widehat{A})$. Since this space is a core of L_p (see Proposition 9.1.8), the formula (9.4.2) can be extended at once to any $f \in D(L_p)$ and any $p \in (1, +\infty)$.

First, we suppose that $p \geq 2$ and fix $f \in D(\widehat{A})$. Then, Proposition 1.3.5, Theorems 3.1.5, 5.2.2 and Remark 5.1.8 show that $D(\widehat{A}) \hookrightarrow C^1_b(\mathbb{R}^N)$. Since $\mathcal{A}(|f|^p) = p|f|^{p-2} f \mathcal{A} f + p(p-1)|f|^{p-2}|\sqrt{Q}\nabla f|^2$, the function $|f|^p$ belongs to $D(\widehat{A})$. Therefore, by Proposition 9.1.1 it follows that $\int_{\mathbb{R}^N} \mathcal{A}(|f|^p) d\mu = 0$, which yields (9.4.4) for functions in $D(\widehat{A})$, which is, in fact, an equality.

Now let $p \in (1, 2)$ and fix $f \in D(\widehat{A})$. For any $\varepsilon > 0$ we consider the function $g_\varepsilon = (f^2 + \varepsilon)^{p/2}$, which belongs to $D(\widehat{A})$. A straightforward computation reveals that
$$\begin{aligned}
&\int_{\mathbb{R}^N} |\sqrt{Q}\nabla f|^2 (f^2+\varepsilon)^{\frac{p-2}{2}} \chi_{\{f\neq 0\}} d\mu \\
=& -\frac{1}{p-1}\int_{\mathbb{R}^N} f(f^2+\varepsilon)^{\frac{p-2}{2}} L_p f d\mu + \frac{p-2}{p-1}\varepsilon \int_{\mathbb{R}^N} (f^2+\varepsilon)^{\frac{p-4}{2}} |\sqrt{Q}\nabla f|^2 \chi_{\{f\neq 0\}} d\mu \\
\leq& -\frac{1}{p-1}\int_{\mathbb{R}^N} f|f|^{p-2} L_p f d\mu
\end{aligned}$$
and this inequality yields the formula (9.4.4) letting ε tend[4] to 0^+. □

The following two examples show that the logarithmic Sobolev inequality (9.4.2) is precise.

Example 9.4.3 Let μ be the Gaussian measure $\mu(dx) = \pi^{-N/2}e^{-|x|^2}dx$. Moreover, for any $p > 1$ and $\varepsilon > 0$, let the function $u : \mathbb{R}^N \to \mathbb{R}$ be defined by $u(x) = \exp(2(2p+\varepsilon)^{-1}|x|^2)$ for any $x \in \mathbb{R}^N$. Then, $u \in L^p_\mu$, is continuously differentiable in $\mathbb{R}^N$ and $D^\alpha u = P_{|\alpha|}u$ in $\mathbb{R}^N$, where $P_{|\alpha|}$ is a real polynomial with degree $|\alpha|$. Since
$$\int_{\mathbb{R}^N} |D^\alpha u(x)|^p d\mu = \frac{1}{\pi^{N/2}}\int_{\mathbb{R}^N} |P_{|\alpha|}(x)|^p \exp\left(-\frac{\varepsilon}{2p+\varepsilon}|x|^2\right) dx < +\infty,$$
u belongs to $W^{k,p}(\mathbb{R}^N, \mu)$ for any $k \in \mathbb{N}$. However, u does not belong to $L^{p+\varepsilon}_\mu$. Indeed,
$$\int_{\mathbb{R}^N} |u(x)|^{p+\varepsilon} d\mu = \frac{1}{\pi^{N/2}}\int_{\mathbb{R}^N} \exp\left(\frac{\varepsilon}{2p+\varepsilon}|x|^2\right) dx = +\infty.$$

[4]Actually, also in this case, (9.4.4) is an equality. See the proof of Theorem 17.1.6.

This second example, due to L. Gross (see [119, p. 1074]), is still sharper than the previous one.

Example 9.4.4 Let $\mu(dx) = (2\pi)^{-1/2}\exp(-x^2/2)dx$ be the one-dimensional Gaussian measure. Further, let $f : \mathbb{R} \to \mathbb{R}$ be any smooth function which agrees with the function

$$x \mapsto g(x) = e^{x^2/4}[(x^2+2)\log(x^2+2)]^{-3/4} =: e^{x^2/4}h(x)$$

for any $|x| \geq M$, where M is such that $g(x) \geq 1$ for any $x \in (-\infty, M) \cup (M, +\infty)$. As is immediately seen, f belongs to L^2_μ, since at $\pm\infty$ h^2 is asymptotic to the function $x \mapsto |x|^{-3/2}(\log(|x|)^{-3/4}$, which is integrable in a neighborhood of $\pm\infty$. Moreover, f is continuously differentiable in $\mathbb{R}$ and

$$f'(x) = \frac{x}{2}\frac{(x^2-1)\log(x^2+2)-3}{((x^2+2)\log(x^2+2))^{\frac{7}{4}}}e^{\frac{x^2}{4}}, \qquad |x| > M.$$

Thus, $x \mapsto |f'(x)|^2 e^{-x^2/2}$ is asymptotic to the function $x \mapsto |x|^{-1}(\log(|x|)^{-3/2}$ at $\pm\infty$, which is integrable both in a neighborhood of $+\infty$ and in a neighborhood of $-\infty$. On the other hand, $x \mapsto |f(x)|^2 \log(f(x)) \log(\log(f(x)))e^{-x^2/2}$ is asymptotic to the function $x \mapsto (2|x|^2\log(|x|))^{-1/2}$ at $\pm\infty$, and this latter function is integrable neither in a neighborhood of $+\infty$ nor in a neighborhood of $-\infty$.

A refinement of the logarithmic Sobolev inequality in (9.4.1) leads to an improvement of the integrability properties of each operator $T(t)$.

Theorem 9.4.5 (Supercontractivity) *The following properties are equivalent.*

(i) *The semigroup $\{T(t)\}$ is supercontractive, i.e., each operator $T(t)$ maps $L^p(\mathbb{R}^N, \mu)$ into L^q_μ, for any $1 < p < q < +\infty$, and there exists a positive decreasing function $C_{p,q} : \mathbb{R}_+ \to \mathbb{R}_+$ such that $\lim_{r\to 0^+} C_{p,q}(r) = +\infty$ and $\|T(t)\|_{L(L^p_\mu, L^q_\mu)} \leq C_{p,q}(t)$ for any $t > 0$;*

(ii) *the family of logarithmic Sobolev inequalities*

$$\int_{\mathbb{R}^N} f^2 \log\left(\frac{|f|}{\|f\|_2}\right)d\mu \leq \varepsilon\|\nabla f\|_2^2 + \beta(\varepsilon)\|f\|_2^2$$

holds for any $f \in W^{1,2}_\mu$, $\varepsilon > 0$ and some decreasing function $\beta : (0, +\infty) \to \mathbb{R}$, blowing up as ε tends to 0^+;

(iii) *for any $\lambda > 0$, the function φ_λ belongs to L^1_μ.*

A sufficient condition for the supercontractivity of the semigroup $\{T(t)\}$ is provided in the following proposition.

Proposition 9.4.6 *Assume that there exist $K, \beta > 0$ and $R > 1$ such that $\langle b(x), x\rangle \leq -K|x|^2\log|x|$ for any $x \in \mathbb{R}^N \setminus B_R$. Then, the evolution operator $\{G(t,s)\}$ is supercontractive.*

Remark 9.4.7 The condition on the inner product $\langle b(t,x), x\rangle$ in Proposition 9.4.6 is quite optimal. Indeed, the Ornstein-Uhlenbeck operator $\mathcal{A}\zeta(x) = \Delta\zeta(x) - \langle x, \nabla\zeta(x)\rangle$ does not satisfy it and the associated semigroup is not supercontractive with respect to the Gaussian invariant measure $\mu(dx) = (2\pi)^{-N/2}e^{-|x|^2/2}dx$, as proved by E. Nelson in [210].

As the supercontractivity of the semigroup $\{T(t)\}$ its ultraboundedness can also be characterized in terms of the functions φ_λ.

Theorem 9.4.8 (Ultraboundedness) *The following properties are equivalent.*

(i) *The semigroup $\{T(t)\}$ is ultrabounded, i.e., each operator $T(t)$ maps L^p_μ into $L^\infty(\mathbb{R}^N)$ for any $p>1$ and there exists a decreasing function $C_{p,\infty}:\mathbb{R}_+\to\mathbb{R}_+$, diverging to $+\infty$ as r tends to 0^+, such that $\|T(t)\|_{L(L^p_\mu,L^\infty(\mathbb{R}^N))}\le C_{p,\infty}(t-s)$ for any $I\ni s<t$;*

(ii) *for any $\lambda>0$ and $t>s$, the function $T(t)\varphi_\lambda$ belongs to $L^\infty(\mathbb{R}^N)$ and $\sup\{\|T(t)\varphi_\lambda\|_\infty : t\ge\delta\}<+\infty$ for any $\delta,\lambda>0$.*

The following theorem provides a sufficient condition for the semigroup $\{T(t)\}$ to be ultrabounded.

Theorem 9.4.9 *Suppose that there exist constants $K>0$, $R>0$ and $\alpha>1$ such that $\langle b(x),x\rangle\le -K|x|^2|\log|x||^\alpha$ for any $x\in\mathbb{R}^N\setminus B_R$. Then, $\{T(t)\}$ is ultrabounded.*

Remark 9.4.10 The condition on the inner product $\langle b(x),x\rangle$ in Theorem 9.4.9 is rather sharp. Indeed the results proved by O. Kavian, G. Kerkyacharian and B. Roynette in [137] show that, if $\Phi:\mathbb{R}^N\to\mathbb{R}$ is a sufficiently smooth function satisfying $\Phi(x)\sim|x|^2\log|x|$ as $|x|$ tends to $+\infty$, then the semigroup $\{T(t)\}$ associated with the elliptic operator $\mathcal{A}=\Delta-\langle\nabla\Phi,\nabla\rangle$ is not ultrabounded in the L^p-spaces with respect to its invariant measure.

To conclude this section, we provide a sufficient condition for the semigroup $\{T(t)\}$ to be ultracontractive, where ultracontractive means that each operator $T(t)$ maps L^1_μ into $L^\infty(\mathbb{R}^N)$ and there exists a decreasing function $C_{1,\infty}:(0,+\infty)\to(0,+\infty)$ such that $\lim_{r\to 0^+}C_{1,\infty}(r)=+\infty$ and $\|T(t)\|_{L(L^1_\mu,L^\infty(\mathbb{R}^N))}\le C_{1,\infty}(t)$ for any $t>0$.

Theorem 9.4.11 *Besides the previous hypotheses, assume that there exist three constants $K>0$, $R>0$ and $\gamma>2$ such that $\langle b(t,x),x\rangle\le -K|x|^\gamma$ for any $x\in\mathbb{R}^N\setminus B_R$. Then, the evolution operator $\{T(t)\}$ is ultracontractive and*

$$C_{1,\infty}(r)=\exp[C(r^{-1}\vee r^{-\frac{\gamma}{\gamma-2}})],\qquad r>0,$$

where the constant C depends on γ, K, R, Λ (see (16.4.2)), N, κ_0.

9.4.1 Concluding remarks

Summability improving properties for Markov semigroup have been widely studied in the literature, starting from the pioneering papers by E. Nelson (see [210]) and L.Gross (see [119, 120]). E. Nelson was the first one who proved the hypercontractivity for the Ornstein Uhlenbeck semigroup, whereas L. Gross proved the equivalence between the logarithmic Sobolev inequality and hypercontractivity (see also the forthcoming Theorem 9.4.12). On the other hand, D. Bakry and M. Émery (see [28]) showed the hypercontractivity of the semigroup associated with the operator $\mathcal{A}=\Delta-\langle\nabla U,\nabla\rangle$ with U satisfying the estimate $\langle D^2U(x)\xi,\xi\rangle\ge\alpha|\xi|^2$ for any $x,\xi\in\mathbb{R}^N$ and some $\alpha>0$. Ultraboundedness and ultracontractivity have been studied mainly in the symmetric case (where they are equivalent). The first result in this direction is due to E.B. Davies and B. Simon [84, 85] that, following the idea of L. Gross and requiring some additional integrability conditions, connect ultracontractivity with a family of logarithmic Sobolev inequalities. Other different approaches to study ultracontractivity have been also suggested by E.A. Carlen, S. Kusuoka, and D.W.

Stroock in [52] and by F.-Y. Wang in [255]. For higher order logarithmic Sobolev inequalities we refer the reader to [100, 101], where G.F. Feissner proves that, if μ is a Gaussian measure and $f \in W^{2,2}(\mathbb{R}^N,\mu)$, then $f^2(\log(f))^2$ is in $L^1(\mathbb{R}^N,\mu)$.

For a very rich overview on Poincaré, logarithmic Sobolev inequalities and their deep consequences, we refer the reader to [13]. See also [121].

To conclude this subsection, we briefly comment on the logarithmic Sobolev inequality and the hypercontractivity.

Theorem 9.4.2 shows a link between the logarithmic Sobolev inequality and the hypercontractivity of the semigroup $\{T(t)\}$, under Hypotheses 16.2.1. Actually, the link between the logarithmic Sobolev inequality and the hypercontractivity is deeper. In the following theorem we just assume that minimal assumptions, i.e., Hypotheses 1.0.1 and the existence of an invariant measure.

Theorem 9.4.12 ([120], Thms. 3.7 and 3.12) *The following properties are equivalent.*

(i) *The logarithmic Sobolev inequality*

$$\int_{\mathbb{R}^N}|f|^p\log(|f|)d\mu \le \|f\|_p^p\log(\|f\|_p) - C_p\int_{\mathbb{R}^N}|f|^{p-2}fL_pf\,d\mu \tag{9.4.5}$$

holds true for any $f \in D(L_p)$, any $p \in (1,+\infty)$ and some constant $C_p > 0$;

(ii) *for any $p \in (1,+\infty)$ there exists a continuous increasing function $t \mapsto q(t,p)$, which is differentiable at $t=0$ with positive derivative, such that*

$$\|T(t)f\|_{q(t,p)} \le \|f\|_p, \qquad f \in L^p_\mu,\ t>0.$$

Proof Fix $p \in (1,+\infty)$, $\psi \in D(\widehat{A}) \subset D(L_p)$ (see Proposition 9.1.8) and denote by $\beta : [0,+\infty) \to [0,+\infty)$ the function defined by $\beta(t) = \|T(t)\psi\|_{q(t,p)}$ for any $t \ge 0$, where $q(\cdot,p)$ is a function with values in $[p,+\infty)$.

If $q(\cdot,p)$ is differentiable at some point $t \ge 0$, then the function β is differentiable as well at t and

$$\begin{aligned}\beta'(t) = \frac{D_tq(t,p)}{q(t,p)}\beta(t)^{1-q(t,p)}\bigg\{ &-\|T(t)\psi\|_{q(t,p)}^{q(t,p)}\log(\|T(t)\psi\|_{q(t,p)})\\ &+\int_{\mathbb{R}^N}|T(t)\psi|^{q(t,p)}\log(|T(t)\psi|)d\mu\\ &+\frac{q(t,p)}{D_tq(t,p)}\int_{\mathbb{R}^N}|T(t)\psi|^{q(t,p)-2}T(t)\psi\mathcal{A}T(t)\psi d\mu\bigg\}.\end{aligned} \tag{9.4.6}$$

(i) $\Rightarrow$ *(ii)* Take $q(t,p) = e^{tC_p^{-1}}$ for any $t \ge 0$. From the previous formula and estimate (9.4.5) we deduce that the function β is decreasing. Hence, in particular, $\beta(t) \le \beta(0)$ for any $t > 0$, i.e., $\|T(t)\psi\|_{q(t,p)} \le \|\psi\|_p$ for any $t > 0$. By density the previous estimate can be extended to any function in L^p_μ.

(ii) $\Rightarrow$ *(i)* By assumptions, $\beta(t) \le \beta(0)$ for any $t \ge 0$. Since $q(\cdot,p)$ is differentiable at zero and not less than p, β is differentiable at $t=0$ as well and its derivative is nonpositive. From (9.4.6) it thus follows that

$$\int_{\mathbb{R}^N}|\psi|^p\log(|\psi|)d\mu \le \|\psi\|_p^p\log(\|\psi\|_p) - \frac{p}{D_tq(0,p)}\int_{\mathbb{R}^N}|\psi|^{p-2}\psi\mathcal{A}\psi d\mu. \tag{9.4.7}$$

Since $D(\widehat{A})$ is a core for $D(L_p)$ (see again Proposition 9.1.8), inequality (9.4.7) can be extended straightforwardly to any function in $D(L_p)$. □

As the following proposition shows, the Poincaré inequality can be deduced, with a different argument, from the compactness of the semigroup $\{T(t)\}$ without using the Logarithmic-Sobolev inequality. As a matter of fact, in this case, we miss any information on the constant C_p. We prove it in the following theorem, where we do not assume Hypotheses 9.4.1.

Theorem 9.4.13 *Let the semigroup $\{T(t)\}$ admit an invariant measure μ and $W^{1,p}_\mu$ is compactly embedded in L^p_μ for some $p \in (1,+\infty)$. Then, the Poincaré inequality (9.4.3) holds true.*

Proof Fix $p \in [1,+\infty)$ and suppose, by contradiction, that (16.4.10) does not hold. Then, there exists a sequence $\{f_n\} \subset W^{1,p}_\mu$ such that

$$\overline{f_n} = 0, \qquad \|f_n\|_p = 1, \qquad \|\nabla f_n\|_p \le n^{-1} \tag{9.4.8}$$

for any $n \in \mathbb{N}$. Since $W^{1,p}_\mu$ is compactly embedded in L^p_μ, there exists a subsequence $\{f_{n_k}\}$ converging in L^p_μ to a function f^*, which satisfies

$$\overline{f^*} = 0, \qquad \|f^*\|_p = 1. \tag{9.4.9}$$

Now, fix $r > 0$; according to Proposition 9.1.4, $L^p(B_r,\mu) \subset L^p(B_r) := L^p(B_r,dx)$. Therefore, $\|f_{n_k} - f^*\|_{L^1(B_r)}$ vanishes as k tends to $+\infty$ and, moreover, from (9.4.8) it follows that $\|\nabla f_{n_k}\|_{L^1(B_r)} \le C n_k^{-1}$ for any $k \in \mathbb{N}$ and some positive constant C. The Poincaré inequality in $W^{1,1}(B_r)$ (see e.g. [98, Theorem 4.5.2]) shows that $\|f_{n_k} - (f_{n_k})_r\|_{L^1(B_r)} \le C'\|\nabla f_{n_k}\|_{L^1(B_r)}$ for any $k \in \mathbb{N}$ and some positive constant C', where $(f_{n_k})_r$ denotes the average, with respect to the Lebesgue measure, of f_{n_k} over the ball B_r. Letting k tend to $+\infty$ in this last estimate, it follows that $\|f^* - (f^*)_r\|_{L^1(B_r)} = 0$. Therefore, f^* is almost everywhere constant in B_r. Since $r > 0$ is arbitrary, f^* is almost everywhere constant in $\mathbb{R}^N$. But this contradicts (9.4.9). □

9.5 A class of analytic semigroups in $L^p(\mathbb{R}^N,\mu)$

In this section we consider the operator $\mathcal{A} = \sum_{i,j=1}^N q_{ij}D_{ij} + \sum_{j=1}^N b_jD_j$ under the following assumptions on its coefficients.

Hypotheses 9.5.1 (i) *the functions q_{ij} and b_i belong to $C^{1+\alpha}_{\rm loc}(\mathbb{R}^N)$ for some $\alpha \in (0,1)$ and there exists a function $\kappa : \mathbb{R}^N \to (0,+\infty)$, bounded from below by a positive constant κ_0, such that $\langle Q(x)\xi,\xi\rangle \ge \kappa(x)|\xi|^2$ for any $x,\xi \in \mathbb{R}^N$;*

(ii) *the diffusion coefficients grow at most quadratically at infinity;*

(iii) *there exists a function $\Phi : \mathbb{R}^N \to \mathbb{R}$ such that $Q^{-1}(\operatorname{div} Q - b) = \nabla\Phi$, where $(\operatorname{div} Q)_j := \sum_{i=1}^N D_i q_{ij}$ for any $j = 1,\dots,N$, and $e^{-\Phi} \in L^1(\mathbb{R}^N)$;*

(iv) *there exist two positive constants $k_1 > 0$ and $k_2 \in (0,1)$ such that*

$$\langle Q(\operatorname{Jac} b)^*\xi,\xi\rangle + \langle Q\xi,\nabla(\operatorname{Tr}(QS))\rangle - \operatorname{Tr}(((\operatorname{Jac} Q)^*\xi)(QS)^*) \le k_1|\sqrt{Q}\xi|^2 + k_2|\sqrt{Q}S\sqrt{Q}|^2 \tag{9.5.1}$$

for any symmetric matrix S and $\xi \in \mathbb{R}^N$.

Let μ be the probability measure whose density with respect to the Lebesgue measure is the function $K^{-1}e^{-\Phi}$, where $K = \|e^{-\Phi}\|_{L^1(\mathbb{R}^N)}$. Throughout this section we will use the spaces $W^{1,p}_{\mu,Q}$ and $W^{2,p}_{\mu,Q}$, defined as follows:

- $W^{1,p}_{\mu,Q}$ is the space of all functions $u \in L^p_\mu$ such that $|\sqrt{Q}\nabla u| \in L^p_\mu$;
- $W^{2,2}_{\mu,Q}$ is its subspace of all the functions in $W^{1,p}_{\mu,Q}$ such that $|\sqrt{Q}D^2u\sqrt{Q}| \in L^p_\mu$.

They are endowed with the norms $\|\cdot\|_{1,p,Q}$ and $\|\cdot\|_{2,p,Q}$, defined by $\|u\|^p_{1,p,Q} = \|u\|^p_p + \|\sqrt{Q}\nabla u\|^p_p$ for any $u \in W^{1,p}_{\mu,Q}$ and $\|v\|^p_{2,p,Q} = \|v\|^p_{1,p,Q} + \|\sqrt{Q}D^2v\sqrt{Q}\|^p_p$ for any $v \in W^{2,p}_{\mu,Q}$.

Remark 9.5.2 In the particular case when $Q = I$, Hypotheses 9.5.1 are satisfied if $b = -\nabla\Phi$ for some convex function $\Phi \in C^2(\mathbb{R}^N)$.

More generally, if $Q = \kappa I$ for some function κ, with positive infimum κ_0 and which grows at most quadratically at infinity, then Hypothesis 9.5.1(iii) is satisfied provided that $b = \kappa\nabla F$ for some function $F : \mathbb{R}^N \to \mathbb{R}$ such that e^F/κ is in $L^1(\mathbb{R}^N)$. Finally, since (9.5.1) reads as follows:

$$\langle \mathrm{Jac}\, b(x)\,\xi,\xi\rangle + \langle \nabla\kappa(x), (\mathrm{Tr}(S)I - S)\xi\rangle \le k_1|\xi|^2 + k_2\kappa(x)|S|^2, \qquad x,\xi \in \mathbb{R}^N,$$

Hypothesis 9.5.1(iv) is satisfied if $|\nabla\kappa| \le C\sqrt{\kappa}$ and $\langle \mathrm{Jac}\, b\,\xi,\xi\rangle \le C|\xi|^2$ in $\mathbb{R}^N$ for any $\xi \in \mathbb{R}^N$ and some positive constant C.

The following two lemmata will play a fundamental role in what follows.

Lemma 9.5.3 *Suppose that Hypotheses* 9.5.1*(i)-(ii) are satisfied. Then* $C^\infty_c(\mathbb{R}^N)$ *is dense in* $W^{1,p}_{\mu,Q}$ *and in* $W^{2,p}_{\mu,Q}$ *for any* $p \in [1,+\infty)$.

Proof Let us prove that $C^\infty_c(\mathbb{R}^N)$ is dense in $W^{2,p}_{\mu,Q}$. The same arguments show that $C^\infty_c(\mathbb{R}^N)$ is dense also in $W^{1,p}_{\mu,Q}$.

If $u \in W^{2,p}_{\mu,Q}$ has compact support in some ball B_R, then the assertion is straightforward since there exists a sequence $\{u_n\} \subset C^\infty_c(\mathbb{R}^N)$ compactly supported in B_{R+1}, which converges to u in $W^{2,p}(\mathbb{R}^N)$ as n tends to $+\infty$. Clearly, this sequence converges to u also in $W^{2,p}_{\mu,Q}$.

Now we show that any function $u \in W^{2,p}_{\mu,Q}$ can be approximated in the $W^{2,p}_{\mu,Q}$-norm by a sequence of compactly supported functions $u_n \in W^{2,p}_{\mu,Q}$. For any $n \in \mathbb{N}$, let $\vartheta_n \in C^\infty_c(\mathbb{R}^N)$ be defined by $\vartheta_n(x) = \vartheta(x/n)$ for any $x \in \mathbb{N}$, where $\chi_{B_{1/2}} \le \vartheta \le \chi_{B_1}$, and set $u_n(x) = u(x)\vartheta(x/n)$ for any $x \in \mathbb{R}^N$ and $n \in \mathbb{N}$. Each function u_n belongs to $W^{2,p}_{\mu,Q}$ has support contained in B_n and $u_n \equiv 1$ in $B_{n/2}$. Moreover, a straightforward computation reveals that

$$\begin{aligned}
\|u_n - u\|_{2,p,Q} \le & \left(\int_{\mathbb{R}^N} |u|^p|1-\vartheta_n|^p d\mu\right)^{\frac{1}{p}} + \left(\int_{\mathbb{R}^N} |\sqrt{Q}(x)\nabla u|^p|1-\vartheta_n|^p d\mu\right)^{\frac{1}{p}}\\
& + \left(\int_{\mathbb{R}^N} |\sqrt{Q}D^2u\sqrt{Q}|^p|1-\vartheta_n|^p d\mu\right)^{\frac{1}{p}}\\
& + \frac{1}{n}\left(\int_{\mathbb{R}^N} |\sqrt{Q}(x)\nabla\vartheta(x/n)|^p|u(x)|^p\mu(dx)\right)^{\frac{1}{p}}\\
& + \frac{2}{n}\left(\int_{\mathbb{R}^N} |\sqrt{Q}(x)\nabla u(x)|^p|\sqrt{Q}(x)\nabla\vartheta(x/n)|^p\mu(dx)\right)^{\frac{1}{p}}
\end{aligned}$$

$$+\frac{1}{n^2}\left(\int_{\mathbb{R}^N}|\sqrt{Q}(x)D^2\vartheta(x/n)\sqrt{Q}(x)|^p|u(x)|^p\mu(dx)\right)^{\frac{1}{p}}=:\sum_{j=1}^{6}I_{j,n}.$$

By dominated convergence, $I_{1,n}$, $I_{2,n}$ and $I_{3,n}$ converge to 0 as n tends to $+\infty$. On the other hand, taking into account that the diffusion coefficient may grow at most quadratically at infinity, we can estimate

$$\begin{aligned}I_{4,n}\leq&\frac{C_1}{n}\|\nabla\vartheta\|_\infty\left(\int_{B_n\setminus B_{n/2}}(1+|x|^2)^{\frac{p}{2}}|u(x)|^p\mu(dx)\right)^{\frac{1}{p}}\\ \leq&C_1\frac{\sqrt{1+n^2}}{n}\|\nabla\vartheta\|_\infty\left(\int_{\mathbb{R}^N\setminus B_{n/2}}|u(x)|^p\mu(dx)\right)^{\frac{1}{p}}\end{aligned}$$

for some positive constant C_1 and the last side of the previous inequality vanishes as n tends to $+\infty$. Similarly,

$$\begin{aligned}I_{5,n}+I_{6,n}\leq&C_2\frac{\sqrt{1+n^2}}{n}\|\nabla\vartheta\|_\infty\left(\int_{\mathbb{R}^N\setminus B_{n/2}}|\sqrt{Q}\nabla u|^pd\mu\right)\\ &+C_3\frac{1+n^2}{n^2}\|D^2\vartheta\|_\infty\left(\int_{\mathbb{R}^N\setminus B_{n/2}}|u|^pd\mu\right)^{\frac{1}{p}}\end{aligned}$$

for some positive constants C_2 and C_3, and also in this case the right-hand side vanishes as n tends to $+\infty$. □

Lemma 9.5.4 *For any $u\in W^{2,2}_{\rm loc}(\mathbb{R}^N)$ and $v\in W^{1,2}_{\rm loc}(\mathbb{R}^N)$, such that u or v has compact support, it holds that*

$$\int_{\mathbb{R}^N}v\mathcal{A}ud\mu=-\int_{\mathbb{R}^N}\langle Q\nabla u,\nabla v\rangle d\mu. \tag{9.5.2}$$

Proof An integration by parts reveals that

$$\begin{aligned}\int_{\mathbb{R}^N}\sum_{i,j=1}^N q_{ij}D_{ij}u\,vd\mu=&-\int_{\mathbb{R}^N}\sum_{i,j=1}^N D_i(q_{ij}ve^{-\Phi})D_judx\\ =&\int_{\mathbb{R}^N}\langle Q\nabla\Phi,\nabla u\rangle vd\mu-\int_{\mathbb{R}^N}\langle Q\nabla u,\nabla v\rangle d\mu-\int_{\mathbb{R}^N}\sum_{i,j=1}^N D_iq_{ij}vD_ju\,d\mu.\end{aligned}$$

This formula and Hypothesis 9.5.1(iii) yield the assertion. □

Next theorem is the main result of the section.

Theorem 9.5.5 *The measure μ is invariant for the semigroup $\{T(t)\}$. Moreover, the infinitesimal generator L_2 of the extension of $\{T(t)\}$ to L^2_μ has domain $D(L_2)=\{u\in W^{2,2}_{\mu,Q}:\langle b,\nabla u\rangle\in L^2_\mu\}$, is dissipative and self-adjoint in L^2_μ. Hence, $\{T(t)\}$ is a strongly continuous analytic semigroup of contractions in L^2_μ. Moreover, for any $\lambda>0$ and $u\in D(L_2)$ it holds that*

$$\|u\|_2\leq\frac{1}{\lambda}\|\lambda u-L_2u\|_2, \tag{9.5.3}$$

$$\|\sqrt{Q}\nabla u\|_2\leq\frac{1}{\sqrt{\lambda}}\|\lambda u-L_2u\|_2, \tag{9.5.4}$$

$$\|\sqrt{Q}D^2u\sqrt{Q}\|_2 \le C(\lambda)\|\lambda u - L_2 u\|_2, \tag{9.5.5}$$

where $C(\lambda)$ is a positive constant, independent of u. Finally, $D(L_2)$ coincides with the domain of the maximal realization of L_2 in L^2_μ, i.e., $D(\widehat{L}_2)=\{u\in L^2_\mu\cap W^{2,2}_{\rm loc}(\mathbb{R}^N): \mathcal{A}u\in L^2_\mu\}$.

Proof We split the proof into three steps. In the first step, we prove that the realization $\widehat{L}_2$ of the operator $\mathcal{A}$ in L^2_μ, with domain $\{u\in W^{2,2}_{\mu,Q}: \langle b,\nabla u\rangle\in L^2_\mu\}$, is dissipative and self-adjoint in L^2_μ. Then, in Step 2 we prove that $D(\widehat{L}_2)$ coincides with the domain of the maximal realization of $\mathcal{A}$ in L^2_μ. Finally, in Step 3, we prove that, actually, $\widehat{L}_2=L_2$ and conclude the proof.

Step 1. Using Lemma 9.5.3 we can straightforwardly extend the formula (9.5.2) to any $u,v\in D(\widehat{L}_2)$ and deduce that the operator $\widehat{L}_2$ is symmetric. It is also dissipative: indeed, for any $u\in D(\widehat{L}_2)$ and $\lambda>0$ we can write

$$\int_{\mathbb{R}^N}(\lambda u-\widehat{L}_2u)u\,d\mu=\lambda\int_{\mathbb{R}^N}u^2d\mu+\int_{\mathbb{R}^N}|\sqrt{Q}\nabla u|^2d\mu$$

and conclude that $\lambda\|u\|_2^2\le\|\lambda u-\widehat{L}_2u\|_2\|u\|_2$, from which the dissipativity of $\widehat{L}_2$ follows.

Let us now prove that the resolvent of the operator $\widehat{L}_2$ contains the halfline $(0,+\infty)$. In particular, this will imply that the resolvent of L_2 is not empty, so that L_2 is self-adjoint.

The Lax-Milgram lemma shows that, for any $f\in C^\infty_c(\mathbb{R}^N)$ and $\lambda>0$, the equation

$$\lambda\int_{\mathbb{R}^N}uv\,d\mu+\int_{\mathbb{R}^N}\langle Q\nabla u,\nabla v\rangle d\mu=\int_{\mathbb{R}^N}fv\,d\mu,\qquad v\in W^{1,2}_{\mu,Q}, \tag{9.5.6}$$

admits a unique solution $u\in W^{1,2}_{\mu,Q}$, which actually belongs to $C^3(\mathbb{R}^N)$, by local elliptic regularity (see Theorem D.1.4), and satisfies the equation $\lambda u-\mathcal{A}u=f$ pointwise. Taking $v=u$ in (9.5.6) yields

$$\lambda\|u\|_2\le\|f\|_2,\qquad \sqrt{\lambda}\|\sqrt{Q}\nabla u\|_2\le\|f\|_2. \tag{9.5.7}$$

To prove that u belongs to $W^{2,2}_{\mu,Q}$, we differentiate the equation $\lambda u-\mathcal{A}u=f$ with respect to any variable x_h, and obtain

$$\lambda D_hu-\mathcal{A}D_hu-\sum_{i,j=1}^N D_hq_{ij}D_{ij}u-\sum_{j=1}^N D_hb_jD_ju=D_hf. \tag{9.5.8}$$

Then, we fix $n_0\in\mathbb{N}$ such that $\mathrm{supp}(f)\subset B_{n_0}$ and, for any $n\ge 2n_0$ and $x\in\mathbb{R}^N$, we set $\vartheta_n(x)=\vartheta(x/n)$, where ϑ is any smooth function such that $\chi_{B_{1/2}}\le\vartheta\le\chi_{B_1}$. Multiplying both sides of (9.5.8) by $\vartheta_n^2\sum_{k=1}^N q_{hk}D_ku$, summing with respect to h and integrating in $\mathbb{R}^N$, gives

$$\begin{aligned}&\lambda\int_{\mathbb{R}^N}\vartheta_n^2|\sqrt{Q}\nabla u|^2d\mu-\int_{\mathbb{R}^N}\vartheta_n^2\sum_{h,k=1}^N q_{hk}D_ku\mathcal{A}(D_hu)d\mu\\&-\int_{\mathbb{R}^N}\vartheta_n^2\sum_{i,j,h,k=1}^N q_{hk}D_hq_{ij}D_kuD_{ij}u\,d\mu-\int_{\mathbb{R}^N}\vartheta_n^2\sum_{j,h,k=1}^N q_{hk}D_kuD_hb_jD_ju\,d\mu\\&=\int_{\mathbb{R}^N}\vartheta_n^2\sum_{h,k=1}^N q_{hk}D_hfD_ku\,d\mu=\int_{\mathbb{R}^N}\langle Q\nabla u,\nabla f\rangle d\mu=-\int_{\mathbb{R}^N}\mathcal{A}uf\,d\mu,\end{aligned} \tag{9.5.9}$$

where we have used the formula (9.5.2) in the last equality. To get rid of the third-order derivatives of u in the left-hand side of (9.5.9), we use again the formula (9.5.2) and the equality $\mathrm{Tr}(QD^2uQD^2u)=\mathrm{Tr}(\sqrt{Q}D^2u\sqrt{Q}\sqrt{Q}D^2u\sqrt{Q})=|\sqrt{Q}D^2u\sqrt{Q}|^2$ to write

$$\begin{aligned}
&\int_{\mathbb{R}^N}\vartheta_n^2\sum_{h,k=1}^N q_{hk}D_ku\mathcal{A}(D_hu)d\mu\\
&=-\sum_{h,k=1}^N\int_{\mathbb{R}^N}\langle Q\nabla(\vartheta_n^2 q_{hk}D_ku),\nabla(D_hu)\rangle d\mu\\
&=-\int_{\mathbb{R}^N}\vartheta_n^2\sum_{i,j,h,k=1}^N q_{ij}D_iq_{hk}D_kuD_{jh}u\,d\mu-\int_{\mathbb{R}^N}\vartheta_n^2|\sqrt{Q}D^2u\sqrt{Q}|^2d\mu\\
&\quad-2\int_{\mathbb{R}^N}\vartheta_n\sum_{i,j,h,k=1}^N q_{ij}D_i\vartheta_n\,q_{hk}D_kuD_{jh}u\,d\mu.
\end{aligned}\tag{9.5.10}$$

Combining (9.5.9) and (9.5.10) we get

$$\begin{aligned}
&\lambda\int_{\mathbb{R}^N}\vartheta_n^2|\sqrt{Q}\nabla u|^2d\mu+\int_{\mathbb{R}^N}\vartheta_n^2|\sqrt{Q}D^2u\sqrt{Q}|^2d\mu\\
=&-\int_{\mathbb{R}^N}f\mathcal{A}ud\mu+\int_{\mathbb{R}^N}\vartheta_n^2\sum_{i,j,h=1}^N(Q\nabla u)_hD_hq_{ij}D_{ij}u\,d\mu\\
&-\int_{\mathbb{R}^N}\vartheta_n^2\sum_{i,h,k=1}^N D_iq_{hk}D_ku(QD^2u)_{ih}d\mu+\int_{\mathbb{R}^N}\vartheta_n^2\sum_{j,h=1}^N(Q\nabla u)_hD_hb_jD_ju\,d\mu\\
&-2\int_{\mathbb{R}^N}\vartheta_n\langle D^2uQ\nabla\vartheta_n,Q\nabla u\rangle d\mu=:\sum_{k=1}^5 J_{k,n}.
\end{aligned}\tag{9.5.11}$$

Using the first formula in (9.5.7), the term $J_{1,n}$ can be estimated from above by $2\|f\|^2_{L^2_\mu}$. On the other hand, Hypothesis 9.5.1(iv), with $\xi=\nabla u$ and $S=D^2u$, yields

$$\begin{aligned}
&\sum_{i,j,h=1}^N(Q\nabla u)_hD_hq_{ij}D_{ij}u-\sum_{i,h,k=1}^N D_iq_{hk}D_ku(QD^2u)_{ih}+\sum_{j,h=1}^N(Q\nabla u)_hD_hb_jD_ju\\
&\le k_1|\sqrt{Q}\nabla u|^2+k_2|\sqrt{Q}D^2u\sqrt{Q}|^2.
\end{aligned}$$

Hence,

$$J_{2,n}+J_{3,n}+J_{4,n}\le k_1\int_{\mathbb{R}^N}\vartheta_n^2|\sqrt{Q}\nabla u|^2d\mu+k_2\int_{\mathbb{R}^N}\vartheta_n^2|\sqrt{Q}D^2u\sqrt{Q}|^2d\mu.$$

Finally, since $|\langle D^2uQ\nabla\vartheta_n,Q\nabla u\rangle|\le|\sqrt{Q}\nabla\vartheta_n||\sqrt{Q}D^2u\sqrt{Q}||\sqrt{Q}\nabla u|$ and $\sup_{\mathbb{R}^N}|\sqrt{Q}\nabla\vartheta_n|\le C_1$ for some positive constant C_1, using the Young inequality we can estimate

$$J_{6,n}\le\varepsilon\int_{\mathbb{R}^N}\vartheta_n^2|\sqrt{Q}D^2u\sqrt{Q}|^2d\mu+\frac{C_1^2}{4\varepsilon}\int_{\mathbb{R}^N}|\sqrt{Q}\nabla u|^2d\mu\tag{9.5.12}$$

for any $\varepsilon>0$. Hence, from (9.5.7), (9.5.11)-(9.5.12) and choosing $\varepsilon=(1-k_2)/4$ we get

$$\frac{1-k_2}{2}\int_{\mathbb{R}^N}\vartheta_n^2|\sqrt{Q}D^2u\sqrt{Q}|^2d\mu\le\left(2+\frac{2C_1^2}{(1-k_2)\lambda}+\frac{(k_1-\lambda)^+}{\lambda}\right)\|f\|_2^2,$$

so that, letting n tend to $+\infty$, we conclude that $u \in W^{2,2}_{\mu,Q}$ and $\|\sqrt{Q}D^2u\sqrt{Q}\|_2 \le C(\lambda)\|f\|_2$, where $C(\lambda) = \left(\frac{4}{1-k_2} + \frac{4C_1^2}{(1-k_2)^2\lambda} + \frac{2(k_1-\lambda)^+}{\lambda(1-k_2)}\right)^{1/2}$. We have so proved that $u \in D(\widehat{L}_2)$ and the formulae (9.5.3)-(9.5.5) hold true.

To prove that the equation $\lambda u - \mathcal{A}u = f$ is solvable in $D(\widehat{L}_2)$ for any $f \in L^2_\mu$ and estimates (9.5.3)-(9.5.5) are satisfied by any $u \in D(\widehat{L}_2)$, it suffices to approximate f by a sequence $\{f_n\} \subset C^\infty_c(\mathbb{R}^N)$, denote by u_n the solution of the equation $\lambda u_n - \mathcal{A}u_n = f_n$ in $D(L_2)$ and use the formulae (9.5.3)-(9.5.5) to infer that $\{u_n\}$ is a Cauchy sequence in $W^{2,2}_{\mu,Q}$ (and, hence, in $D(\widehat{L}_2)$ endowed with the graph norm).

Step 2. Here, we prove that $D(\widehat{L}_2) = \{u \in L^2_\mu \cap W^{2,2}_{\rm loc}(\mathbb{R}^N) : \mathcal{A}u \in L^2_\mu\}$. Clearly, we just need to prove the inclusion "$\supset$". For this purpose, we fix $\lambda > 0$ and $u \in L^2_\mu \cap W^{2,2}_{\rm loc}(\mathbb{R}^N)$ such that $\mathcal{A}u \in L^2_\mu$. Moreover, we set $f := \lambda u - \mathcal{A}u$. Then, the function $v := u - R(\lambda, L)f$ solves the equation $\lambda v - \mathcal{A}v = 0$. We are going to prove that, if λ is sufficiently large, then $v \equiv 0$. Let ϑ_n be as in Step 2. Multiplying the equation $\lambda v - \mathcal{A}v = f$ by $v\vartheta_n$ and using the formula (9.5.2), we get

$$0 = \lambda \int_{\mathbb{R}^N} v^2\vartheta_n^2 d\mu + \int_{\mathbb{R}^N} |\sqrt{Q}\nabla v|^2\vartheta_n^2 d\mu + 2\int_{\mathbb{R}^N} v\langle\sqrt{Q}\nabla v, \sqrt{Q}\nabla\vartheta_n\rangle\vartheta_n d\mu. \tag{9.5.13}$$

Using the Young inequality we can estimate

$$\begin{aligned}\left|\int_{\mathbb{R}^N} v\langle\sqrt{Q}\nabla v, \sqrt{Q}\nabla\vartheta_n\rangle\vartheta_n d\mu\right| \le & C_1\int_{\mathbb{R}^N} |v||\sqrt{Q}\nabla v|\vartheta_n d\mu\\ \le & \frac{C_1^2}{2}\int_{\mathbb{R}^N} v^2 d\mu + \frac{1}{2}\int_{\mathbb{R}^N} |\sqrt{Q}\nabla v|^2\vartheta_n^2 d\mu,\end{aligned}$$

which, replaced in (9.5.13), yields

$$0 \ge \lambda\int_{\mathbb{R}^N} v^2\vartheta_n^2 d\mu + \frac{1}{2}\int_{\mathbb{R}^N} |\sqrt{Q}\nabla v|^2\vartheta_n^2 d\mu - \frac{C_1^2}{2}\int_{\mathbb{R}^N} v^2 d\mu.$$

Letting n tend to $+\infty$ and taking $\lambda = C_1^2$, we conclude that $v \equiv 0$.

Step 3. Taking $v \equiv 1$ (which belongs to $D(\widehat{L}_2)$) in (9.5.2) it follows that

$$\int_{\mathbb{R}^N} \widehat{L}_2 f d\mu = 0, \qquad f \in D(\widehat{L}_2).$$

In particular, the previous formula is satisfied by any function $f \in D(\widehat{A})$ in view of Step 2. Proposition 9.1.1 allows us to conclude that μ is invariant for the semigroup $\{T(t)\}$ associated with the operator $\mathcal{A}$ in $C_b(\mathbb{R}^N)$. Moreover, by Propositions 9.1.7 and 9.1.8 this semigroup extends to a strongly continuous semigroup of contractions in L^2_μ and $D(\widehat{A})$ is a core of its infinitesimal generator L_2. If we denote by $\{\widehat{T}(t)\}$ the semigroup generated in L^2_μ by the operator $\widehat{L}_2$, then for any $f \in D(\widehat{A})$ and $t > 0$ the function $s \mapsto g(s) = \widehat{T}(t-s)T(s)f$ is differentiable in $[0,t]$ and $g'(s) = -\widehat{T}(t-s)\widehat{L}_2T(s)f + \widehat{T}(t-s)\widehat{A}T(s)f = -\widehat{T}(t-s)\widehat{L}_2T(s)f + \widehat{T}(t-s)\widehat{L}_2T(s)f = 0$. Hence, $g(t) = g(0)$ or, equivalently, $\widehat{T}(t)f = T(t)f$. The arbitrariness of $t > 0$ and the density of $D(\widehat{A})$ in L^2_μ show that the semigroups $\{\widehat{T}(t)\}$ and $\{T(t)\}$ coincide. □

We now prove further properties of the semigroup $\{T(t)\}$ and of L_2.

Proposition 9.5.6 *The domain of $\sqrt{-L_2}$ is $W^{1,2}_{\mu,Q}$. Therefore, the restriction of $\{T(t)\}$ to $W^{1,2}_{\mu,Q}$ is an analytic semigroup.*

Proof We recall that $D(\sqrt{-L_2})$ is the closure of $D(L_2)$ with respect to the norm induced by the inner product

$$\langle u, v\rangle := \int_{\mathbb{R}^N} uv\,d\mu - \int_{\mathbb{R}^N} L_2 uv d\mu.$$

Hence, according to the formula (9.5.2), it coincides with the inner product of $W^{1,2}_{\mu,Q}$. Since $D(L_2)$ contains $C^\infty_c(\mathbb{R}^N)$, which is dense in $W^{1,2}_{\mu,Q}$, $D(L_2)$ is dense in $W^{1,2}_{\mu,Q}$ and we thus conclude that $W^{1,2}_{\mu,Q} = D(\sqrt{-L})$. □

Remark 9.5.7 By Proposition 9.1.7, $\{T(t)\}$ is strongly continuous in L^p_μ for any $p \in [1,+\infty)$. Actually, if $p \in (1,+\infty)$, then $\{T(t)\}$ is also analytic in L^p_μ as it follows by a classical result. Indeed, this semigroup preserves positivity, its generator L_2 is self-adjoint and the quadratic form associated with L_2 is nonpositive. We refer the reader to [84, Theorem 1.4.2] for further details.

The following proposition gives a sufficient condition for the embedding of $W^{1,p}_{\mu,Q}$ in L^p_μ to be compact. As a byproduct, it shows that $D(L_2)$ is compactly embedded in L^2_μ.

Proposition 9.5.8 *Under the Hypotheses* 9.5.1, *assume that the diffusion coefficients* q_{ij} $(i,j = 1,\dots,N)$ *belong to* $C^2(\mathbb{R}^N)$ *and*

$$\sum_{i,j=1}^N D_{ij}q_{ij} - \mathrm{div} b \le \delta_1|Q^{-1/2}(\mathrm{div}Q - b)|^2 + \delta_2 \tag{9.5.14}$$

for some constants $\delta_1 \in (0,1)$ *and* $\delta_2 > 0$. *Further, suppose that* $|\sqrt{Q}\nabla\Phi|$ *blows up as* $|x|$ *tends to* $+\infty$. *Then,* $W^{2,p}_{\mu,Q}$ *is compactly embedded in* L^p_μ *for any* $p \in [2,+\infty)$ *and, hence,* $D(L_2)$ *is compactly embedded in* L^2_μ. *If further*

$$\sum_{j=1}^N \langle D_jQ\nabla\Phi + 2Q\nabla D_j\Phi, \nabla\Phi\rangle(Q\nabla\Phi)_j \ge (\delta_3|\sqrt{Q}\nabla\Phi|^2 + \delta_4)|\sqrt{Q}\nabla\Phi|^2 \tag{9.5.15}$$

in $\mathbb{R}^N$, *for some real constants* δ_3, *such that* $2\delta_1 - \delta_3^- < 2$, *and* δ_4, *then* $W^{1,p}_{\mu,Q}$ *is compactly embedded in* L^p_μ *also for any* $p \in (1,2)$.

Proof We split the proof into three steps. In the first two steps we prove that there exists a positive constant C_p such that

$$\int_{\mathbb{R}^N} |f|^p|\sqrt{Q}\nabla\Phi|^{p\wedge 2}d\mu \le C_p\left(\int_{\mathbb{R}^N}|f|^p d\mu + \int_{\mathbb{R}^N}|\sqrt{Q}\nabla f|^p d\mu\right), \qquad f \in W^{1,p}_{\mu,Q}. \tag{9.5.16}$$

Step 1. An integration by parts shows that

$$\begin{aligned}\int_{\mathbb{R}^N} u^2|\sqrt{Q}\nabla\Phi|^2 d\mu =& -\int_{\mathbb{R}^N} u^2\langle Q\nabla\Phi, \nabla e^{-\Phi}\rangle dx\\ =&2\int_{\mathbb{R}^N} u\langle Q\nabla\Phi, \nabla u\rangle d\mu + \int_{\mathbb{R}^N} u^2 \mathrm{div}(Q\nabla\Phi)d\mu\end{aligned} \tag{9.5.17}$$

for any $u \in C^\infty_c(\mathbb{R}^N)$. Since condition (9.5.14) is equivalent to $\mathrm{div}(Q\nabla\Phi) \le \alpha|\sqrt{Q}\nabla\Phi|^2 + \beta$, we can estimate

$$\int_{\mathbb{R}^N} u^2|\sqrt{Q}\nabla\Phi|^2 d\mu \le\delta_1\int_{\mathbb{R}^N} u^2|\sqrt{Q}\nabla\Phi|^2 d\mu + \delta_2\int_{\mathbb{R}^N} u^2 d\mu$$

$$
\begin{aligned}
&+2\left(\int_{\mathbb{R}^N}|\sqrt{Q}\nabla u|^2d\mu\right)^{\frac{1}{2}}\left(\int_{\mathbb{R}^N}u^2|\sqrt{Q}\nabla\Phi|^2d\mu\right)^{\frac{1}{2}}\\
\le&(\delta_1+\varepsilon)\int_{\mathbb{R}^N}u^2|\sqrt{Q}\nabla\Phi|^2d\mu+(\delta_2\vee\varepsilon^{-1})\|u\|_{1,2,Q}^2
\end{aligned}
\tag{9.5.18}
$$

for any $\varepsilon>0$. Replacing (9.5.18) in (9.5.17) and choosing ε such that $1-\delta_1-\varepsilon>0$, we deduce $\|u\sqrt{Q}\nabla\Phi\|_2\le C\|u\|_{1,2,Q}$ for any $u\in C_c^\infty(\mathbb{R}^N)$. By the density of $C_c^\infty(\mathbb{R}^N)$ in $W^{1,2}_{\mu,Q}$ (see Lemma 9.5.3), we can extend the previous estimate to any $u\in W^{1,2}_{\mu,Q}$.

Let us suppose that $p>2$ and fix $u\in C_c^\infty(\mathbb{R}^N)$. Then, the function $|u|^{p/2}$ belongs to $W^{1,2}_{\mu,Q}$. Hence, from the previous estimate and the Hölder inequality we can infer that

$$
\begin{aligned}
\int_{\mathbb{R}^N}|u|^p|\sqrt{Q}\nabla\Phi|^2d\mu\le&C^2\left(\int_{\mathbb{R}^N}|u|^pd\mu+\frac{p^2}{4}\int_{\mathbb{R}^N}|u|^{p-2}|\sqrt{Q}\nabla u|^2d\mu\right)\\
\le&C_p\left(\int_{\mathbb{R}^N}|u|^pd\mu+\int_{\mathbb{R}^N}|\sqrt{Q}\nabla u|^pd\mu\right).
\end{aligned}
$$

Formula (9.5.16) follows from the density of $C_c^\infty(\mathbb{R}^N)$ into $W^{1,p}_{\mu,Q}$.

Step 2. Let us now suppose that $p\in(1,2)$. As in the previous step, it suffices to prove estimate (9.5.16) for functions $u\in C_c^\infty(\mathbb{R}^N)$. Fix $g\in W^{1,p}_{\mu,Q}$ with compact support and $\beta\in(1,p]$. Since $L^q_\mu\hookrightarrow L^r_\mu$ for any $q>r$ and g is compactly supported, the function $|g|^\beta(1+|\sqrt{Q}\nabla\Phi|^2)$ belongs to L^1_μ. Integrating by parts we get

$$
\begin{aligned}
\int_{\mathbb{R}^N}|g|^\beta(1+|\sqrt{Q}\nabla\Phi|^2)d\mu=&\int_{\mathbb{R}^N}|g|^\beta d\mu-K^{-1}\int_{\mathbb{R}^N}|g|^\beta\langle Q\nabla\Phi,\nabla e^{-\Phi}\rangle dx\\
=&\int_{\mathbb{R}^N}|g|^\beta(1+\operatorname{div}(Q\nabla\Phi))d\mu+\beta\int_{\mathbb{R}^N}|g|^{\beta-2}g\langle\nabla g,Q\nabla\Phi\rangle d\mu,
\end{aligned}
\tag{9.5.19}
$$

where $K=\|e^{-\Phi}\|_{L^1(\mathbb{R}^N)}$. Letting β tend to 1^+ in (9.5.19), we get

$$
\int_{\mathbb{R}^N}|g|(1+|\sqrt{Q}\nabla\Phi|^2)d\mu=\int_{\mathbb{R}^N}[|g|(1+\operatorname{div}(Q\nabla\Phi))+\operatorname{sign}(g)\langle\nabla g,Q\nabla\Phi\rangle]d\mu,
\tag{9.5.20}
$$

where we set $\operatorname{sign}(0)=0$. Choosing $g=|u|^p(1+|\sqrt{Q}\nabla\Phi|^2)^{\frac{p-2}{2}}$ and observing that

$$
\begin{aligned}
D_jg=&p|u|^{p-2}uD_ju(1+|\sqrt{Q}\nabla\Phi|^2)^{\frac{p-2}{2}}\\
&+2^{-1}(p-2)|u|^p(1+|\sqrt{Q}\nabla\Phi|^2)^{\frac{p-4}{2}}(\langle D_jQ\nabla\Phi,\nabla\Phi\rangle+2\langle Q\nabla D_j\Phi,\nabla\Phi\rangle)
\end{aligned}
$$

for any $j=1,\dots,N$, from (9.5.20) we obtain that

$$
\begin{aligned}
&\int_{\mathbb{R}^N}|u|^p(1+|\sqrt{Q}\nabla\Phi|^2)^{\frac{p}{2}}d\mu\\
=&\int_{\mathbb{R}^N}|u|^p(1+|\sqrt{Q}\nabla\Phi|^2)^{\frac{p-2}{2}}(1+\operatorname{div}(Q\nabla\Phi))d\mu\\
&+p\int_{\mathbb{R}^N}|u|^{p-2}u(1+|\sqrt{Q}\nabla\Phi|^2)^{\frac{p-2}{2}}\langle\nabla u,Q\nabla\Phi\rangle d\mu\\
&+\frac{p-2}{2}\int_{\mathbb{R}^N}|u|^p(1+|\sqrt{Q}\nabla\Phi|^2)^{\frac{p-4}{2}}\sum_{j=1}^N(\langle D_jQ\nabla\Phi,\nabla\Phi\rangle+2\langle Q\nabla D_j\Phi,\nabla\Phi\rangle)(Q\nabla\Phi)_jd\mu.
\end{aligned}
$$

Denote by I_1, I_2 and I_3 the terms in the right-hand side of the previous inequality. Since $\mathrm{div}Q - b = Q\nabla\Phi$, condition (9.5.14) can be rewritten in the form

$$\mathrm{div}(Q\nabla\Phi) \le \delta_1|\sqrt{Q}\nabla\Phi|^2 + \delta_2. \tag{9.5.21}$$

The Hölder inequality and (9.5.21) yield

$$\begin{aligned}I_1 &\le \delta_1\int_{\mathbb{R}^N}|u|^p(1+|\sqrt{Q}\nabla\Phi|^2)^{\frac{p}{2}}d\mu + (1+\delta_2)\int_{\mathbb{R}^N}|u|^p(1+|\sqrt{Q}\nabla\Phi|^2)^{\frac{p-2}{2}}d\mu\\ &\le \delta_1\int_{\mathbb{R}^N}|u|^p(1+|\sqrt{Q}\nabla\Phi|^2)^{\frac{p}{2}}d\mu + (1+\delta_2)\int_{\mathbb{R}^N}|u|^p d\mu,\end{aligned}$$

$$\begin{aligned}I_2 &\le p\int_{\mathbb{R}^N}|u|^{p-1}(1+|\sqrt{Q}\nabla\Phi|^2)^{\frac{p-1}{2}}|\sqrt{Q}\nabla u|d\mu\\ &\le p\left(\int_{\mathbb{R}^N}|u|^p(1+|\sqrt{Q}\nabla\Phi|^2)^{\frac{p}{2}}d\mu\right)^{\frac{p-1}{p}}\left(\int_{\mathbb{R}^N}|\nabla u|^p d\mu\right)^{\frac{1}{p}}\\ &\le \varepsilon\int_{\mathbb{R}^N}|u|^p(1+|\sqrt{Q}\nabla\Phi|^2)^{\frac{p}{2}}d\mu + C_{\varepsilon,p}\int_{\mathbb{R}^N}|\sqrt{Q}\nabla u|^p d\mu\end{aligned}$$

for any $\varepsilon > 0$ and some positive constant $C_{\varepsilon,p}$. Moreover, using (9.5.15) we can estimate

$$\begin{aligned}I_3 &\le \frac{p-2}{2}\int_{\mathbb{R}^N}|u|^p(1+|\sqrt{Q}\nabla\Phi|^2)^{\frac{p-4}{2}}(\delta_3^-|\sqrt{Q}\nabla\Phi|^2+\delta_4^-)|\sqrt{Q}\nabla\Phi|^2 d\mu\\ &\le \frac{p-2}{2}\int_{\mathbb{R}^N}|u|^p(1+|\sqrt{Q}\nabla\Phi|^2)^{\frac{p-2}{2}}(\delta_3^-|\sqrt{Q}\nabla\Phi|^2+\delta_4^-)d\mu\\ &\le \frac{p-2}{2}\delta_3^-\int_{\mathbb{R}^N}|u|^p(1+|\sqrt{Q}\nabla\Phi|^2)^{\frac{p}{2}}d\mu + \frac{1}{2}|(p-2)(\delta_4^- - \delta_3^-)|\int_{\mathbb{R}^N}|u|^p d\mu.\end{aligned}$$

Since $2\delta_1 - \delta_3^- < 2$, choosing $\varepsilon > 0$ small enough, estimate (9.5.16) follows.

Step 3. From estimate (9.5.16) we can easily infer the compactness of the embedding $W^{1,p}_{\mu,Q} \subset L^p_\mu$. It suffices to prove that the set $E = \{u \in W^{1,p}_{\mu,Q} : \|u\|_{1,p,Q} \le 1\}$ is totally bounded in L^2_μ. We apply the arguments in [184, Proposition 3.4]. For this purpose, fix $r > 0$ and let $E_r = \{u_{|B_r} : u \in E\}$. Since Φ and Q are continuous, the norms of $L^p(B_r)$ and $L^p(B_r,\mu)$ are equivalent on the σ-algebra of the Borel subsets of B_r (in the sense that there exists a positive constant C such that $C^{-1}\mu(A) \le m(A) \le C\mu(A)$ for any Borel set A, where by m we denote the Lebesgue measure). Therefore, E_r is bounded in $W^{1,2}(B_r)$ and then, by Rellich-Kondrachov theorem (see [3, Theorem 6.2]), it is totally bounded in $L^p(B_r)$ and in $L^p(B_r,\mu)$ as well. Thus, for any $\varepsilon > 0$ there exists a finite number of functions $g_1,\ldots,g_n$ in $L^p(B_r,\mu)$ such that $E_r \subset \bigcup_{i=1}^n\{u : \|u-g_i\|_{L^p(B_r,\mu)} < \varepsilon\}$. Moreover, from the estimate $\|u|\sqrt{Q}\nabla\Phi|^{1\wedge(2/p)}\|_2 \le C_p$ it follows that

$$\int_{\mathbb{R}^N\setminus B_r}|u|^p d\mu \le \frac{1}{\inf\limits_{\mathbb{R}^N\setminus B_r}|\sqrt{Q}\nabla\Phi|^{p\wedge 2}}\int_{\mathbb{R}^N}|\sqrt{Q}\nabla\Phi|^{p\wedge 2}|u|^p d\mu \le \frac{C_p}{\inf\limits_{\mathbb{R}^N\setminus B_r}|\sqrt{Q}\nabla\Phi|^{p\wedge 2}}$$

for any $u \in E$. Choosing r large enough and recalling that $|\sqrt{Q}\nabla\Phi(x)|$ blows up as $|x|$ tends to $+\infty$, we deduce that $\|u\|_{L^p(\mathbb{R}^N\setminus B_r)} \le \varepsilon$ for any $f \in E$. Therefore, extending the functions g_j $(j = 1,\ldots,n)$ by zero to the whole $\mathbb{R}^N$ we obtain $E \subset \bigcup_{i=1}^n\{u : \|u-g_i\|_2 < 2\varepsilon\}$, that is, E is totally bounded in L^p_μ. □

Corollary 9.5.9 *Under the assumptions of Proposition* 9.5.8, *for any* $p \in (1,+\infty)$ *and* $t > 0$, *the operator* $T(t)$ *is compact in* L^p_μ.

Proof By Theorem 9.5.5 (see the formula (9.5.4)) the resolvent operator $R(\lambda, L_2)$ is bounded from L^2_μ into $W^{1,2}_{\mu,Q}$ for any $\lambda \in \rho(L_2)$. Hence, it is compact from L^2_μ into itself for any $\lambda \in \rho(L_2)$. Since $\{T(t)\}$ is analytic in L^2_μ, it can be expressed in terms of the resolvent operator $R(\cdot, L_2)$ through the formula (C.2.2). Being the limit of compact operators, each operator $T(t)$ is compact in L^2_μ.

To complete the proof it suffices to apply Theorem B.3.11. □

As the following example shows, in general the embedding of $W^{1,p}_\mu$ into L^p_μ is not compact.

Example 9.5.10 Let $N = 1$ and $U \in C^2(\mathbb{R})$ be any function such that $U(x) = x$ for any $x > 0$ and $U(x) = 0$ for any $x \le -1$. Moreover, let $\mu(dx) = e^{-U(x)}dx$. Consider the sequence of functions $\{f_n\}$, defined by $f_n(x) = \chi_{(0,+\infty)}x^n$ for any $x \in \mathbb{R}$ and $n \in \mathbb{N}$. Then, we have

$$\|f_n\|_1 = \int_0^{+\infty} x^n e^{-x}dx = n!, \qquad \|\nabla f_n\|_1 = n\int_0^{+\infty} x^{n-1}e^{-x}dx = n!.$$

Thus, the sequence of functions $g_n = (n!)^{-1}f_n$ $(n \in \mathbb{N})$ is bounded in $W^{1,1}_\mu$, but it is not relatively compact in L^1_μ because g_n vanishes pointwise in $\mathbb{R}^N$ as n tends to $+\infty$, whereas $\|g_n\|_{L^1_\mu} = 1$ for any $n \in \mathbb{N}$.

Remark 9.5.11 In [74] G. Da Prato and A. Lunardi prove some more refined results in the particular case when $\mathcal{A} = \Delta - \langle \nabla U, \nabla\rangle$ assuming that U is a convex function which diverges to $+\infty$ as $|x|$ tends to $+\infty$. No regularity conditions on U are assumed. When U is not differentiable at $x \in \mathbb{R}^N$, $\nabla U(x)$ is meant as the element with minimal norm in the subdifferential $\partial U(x)$ of U at x, where $\partial U(x) = \{y \in \mathbb{R}^N : U(\xi) \ge U(x) + \langle y, \xi - x\rangle,\ \forall \xi \in \mathbb{R}^N\}$. Under these rather weak assumptions on U, they prove that the realization L_2 in L^2_μ of the operator $\mathcal{A}$ with domain $D(L_2) = \{u \in W^{2,2}_\mu : \langle \nabla U, \nabla u\rangle \in L^2_\mu\}$ is a dissipative self-adjoint operator. Therefore, it generates a strongly continuous analytic semigroup in L^2_μ. Moreover, they show that $\|\nabla R(\lambda, L_2)f\|_{L^2_\mu} \le 2\lambda^{-1/2}\|f\|_{L^2_\mu}$ and $\|D^2R(\lambda, L_2)f\|_{L^2_\mu} \le 4\|f\|_{L^2_\mu}$ for any $\lambda > 0$ and $f \in L^2_\mu$.

To prove their results, the authors use an approximation argument. In the case when U is not smooth, they approximate U with a sequence of smooth functions by means of the Moreau-Yosida approximations U_α of U, which are defined by $U_\alpha(x) = \inf_{y\in\mathbb{R}^N}(U(y) + (2\alpha)^{-1}|x-y|^2)$ for any $x \in \mathbb{R}^N$ and $\alpha > 0$. Each function U_α is convex and differentiable. Moreover, $U_\alpha \le U$, $|\nabla U_\alpha| \le |\nabla U|$ in $\mathbb{R}^N$ and U_α, ∇U_α converge pointwise in $\mathbb{R}^N$ to U and ∇U, respectively, as α tends to 0^+. Finally, each function U_α is Lipschitz continuous in $\mathbb{R}^N$ and its Lipschitz constant is $1/\alpha$.

To the approximated operators $\mathcal{A}_\alpha := \Delta u - \langle \nabla U_\alpha, \nabla u\rangle$ $(\alpha > 0)$, the results in Theorem 9.5.5 apply. In particular, if A_α denotes the realization in $L^2_{\mu_\alpha}$ of the operator $\mathcal{A}_\alpha$ with domain $D(A_\alpha) = \{u \in W^{2,2}_{\mu_\alpha} : \langle \nabla U, \nabla u\rangle \in L^2_{\mu_\alpha}\}$, where $\mu_\alpha = \|e^{-U_\alpha}\|_{L^1(\mathbb{R}^N)}e^{-U_\alpha(x)}dx$, then the estimates $\|D^2u\|_{L^2_{\mu_\alpha}} \le \|\mathcal{A}_\alpha u\|_{L^2_{\mu_\alpha}}$ and $\|u\|_{W^{2,2}_{\mu_\alpha}} \le \frac{3}{2}(\|u\|_{L^2_{\mu_\alpha}} + \|\mathcal{A}_\alpha u\|_{L^2_{\mu_\alpha}})$ hold true for any $u \in D(A_\alpha)$. Taking advantage of this fact, the authors show that, if $f \in C^\infty_c(\mathbb{R}^N)$, then, as α tends to 0^+, $R(\lambda, A_\alpha)f$ converges weakly in $W^{2,2}_\mu(\mathbb{R}^N)$ to a function $u \in D(L_2)$, which turns out to be the unique solution of the equation $\lambda u - \mathcal{A}u = f$ in such a space and satisfies the estimates $\|D^2u\|_{L^2_\mu} \le \|L_2u\|_{L^2_\mu}$ and $\|u\|_{W^{2,2}_\mu} \le \frac{3}{2}(\|u\|_{L^2_\mu} + \|L_2u\|_{L^2_\mu})$.

Finally, using a density argument, they conclude that the operator L_2 generates a strongly continuous analytic semigroup in L^2_μ and it satisfies the above estimates.

9.6 Non-analytic semigroups in $L^p(\mathbb{R}^N,\mu)$

In this section we introduce a class of Markov semigroups $\{T(t)\}$, which admit an invariant measure μ, but which are not analytic in $L^p_\mu := L^p(\mathbb{R}^N,\mu)$. The results that we present here are due to G. Metafune and E. Priola and are taken from [198].

Let $\mathcal{A}$ be the second-order elliptic operator, defined by

$$\mathcal{A}\zeta = \Delta\zeta - \langle \nabla U + G, \nabla\zeta\rangle \tag{9.6.1}$$

on smooth functions $\zeta : \mathbb{R}^N \to \mathbb{R}$, where U and G satisfy the following hypotheses.

Hypotheses 9.6.1 (i) *$U \in C^2(\mathbb{R}^N)$ and $e^{-U} \in L^1(\mathbb{R}^N)$;*

(ii) *$G \in C^1(\mathbb{R}^N,\mathbb{R}^N)$ satisfies $|G|e^{-U} \in L^1(\mathbb{R}^N)$ and $\operatorname{div} G - \langle G, \nabla U\rangle = 0$.*

These conditions will be kept throughout this section and ensure that the semigroup $\{T(t)\}$ associated with $\mathcal{A}$ admits the measure $d\mu = Me^{-U(x)}dx$, where $M = \|e^{-U}\|^{-1}_{L^1(\mathbb{R}^N)}$, as its (unique) invariant measure (see Section 9.1.4). As has been shown in Proposition 9.1.7, the semigroup $\{T(t)\}$ extends to a strongly continuous semigroup of contractions in L^p_μ for any $p \in [1,+\infty)$.

Let us state the main result of this section.

Theorem 9.6.2 *Suppose that there exist some constants $k > 0$, $0 < \beta < \alpha$ and a sequence $\{\sigma_n\} \subset \mathbb{R}^N$, diverging to infinity, such that*

$$\begin{aligned} &(i) \quad |G(\sigma_n)| \geq k|\sigma_n|^\alpha, \qquad n \geq 1;\\ &(ii) \quad |\Delta U(x)| + |\nabla U(x)| + |\nabla G(x)| \leq k(1+|x|^\beta), \qquad x \in \mathbb{R}^N. \end{aligned} \tag{9.6.2}$$

Then, the semigroup $\{T(t)\}$ is not analytic in L^p_μ for any $p \in [1,+\infty)$.

Proof We first consider the case $p = 2$. Let us consider the isometry $R : L^2(\mathbb{R}^N) \to L^2_\mu$ defined by $Rf = f\psi$ for any $f \in L^2(\mathbb{R}^N)$, where $\psi(x) = M^{-\frac{1}{2}}e^{\frac{1}{2}U(x)}$ for any $x \in \mathbb{R}^N$. Further, let us define the operator $\widetilde{L}_2 : D(\widetilde{L}_2) \subset L^2(\mathbb{R}^N) \to L^2(\mathbb{R}^N)$, with domain $D(\widetilde{L}_2) = \{g \in L^2(\mathbb{R}^N) : \psi g \in D(L_2)\}$, by setting $\widetilde{L}_2 u = R^{-1}L_2Ru$ for any $u \in D(\widetilde{L}_2)$. Since, as has already been recalled, the semigroup $\{T(t)\}$ is strongly continuous in L^2_μ, $\widetilde{L}_2$ is the generator of the strongly continuous semigroup of contractions $\{\widetilde{T}(t)\}$ in $L^2(\mathbb{R}^n)$, defined by $\widetilde{T}(t)f = R^{-1}T(t)(Rf)$ for any $t \geq 0$ and $f \in L^2(\mathbb{R}^N)$. According to Propositions 1.3.5 and 9.1.8, it holds that $C^\infty_c(\mathbb{R}^N) \subset D(L_2)$. Therefore, $C^\infty_c(\mathbb{R}^N)$ is contained in $D(\widetilde{L}_2)$ and, since $L_2u = \mathcal{A}u$ for any $u \in C^\infty_c(\mathbb{R}^N)$,

$$(\widetilde{L}_2u)(x) = \Delta u(x) - \langle G(x), \nabla u(x)\rangle - c(x)u(x), \qquad x \in \mathbb{R}^N, \tag{9.6.3}$$

for any $u \in C^\infty_c(\mathbb{R}^N)$, where $c = \frac{1}{2}\left(\frac{1}{2}|\nabla U|^2 - \Delta U + \langle G, \nabla U\rangle\right)$. We claim that the operator $\widetilde{L}_2$ is not sectorial in $L^2(\mathbb{R}^N)$. Of course, this will imply that L_2 is not sectorial in L^2_μ as well.

To prove the claim, we adapt the technique in the proof of Theorem 1.4.1 to this situation. For this purpose, let us set $\gamma = \frac{1}{2}(\alpha+\beta)$ and introduce the sequences $\{\lambda_n\}$ and $\{r_n\}$ defined by $\lambda_n = (|\sigma_n|^\gamma + 1)^{-1}$ and $r_n = (|G(\sigma_n)|(|\sigma_n|^\gamma + 1))^{-1}$ for any $n \geq 1$.[5] Moreover,

[5] Up to replacing $\{\sigma_n\}$ with a suitable subsequence, we can assume that $G(\sigma_n) \neq 0$ for any $n \in \mathbb{N}$, so that the sequence $\{r_n\}$ is well defined.

for any $n\in\mathbb{N}$ we introduce the operators I_n, defined by $(I_nu)(x)=u(\lambda_n^{-1}(x-\sigma_n))$ for any $x\in\mathbb{R}^N$, $u\in L^2(\mathbb{R}^N)$, $n\geq 1$, and $\widetilde{L}_{2,n}=r_nI_n^{-1}\widetilde{L}_2I_n$ with domain $D(\widetilde{L}_{2,n})=I_n^{-1}(D(\widetilde{L}_2))$. The operator $\widetilde{L}_{2,n}$ is the infinitesimal generator of the strongly continuous semigroup of contractions $\{I_n^{-1}\widetilde{T}(r_nt)I_n\}$ in $L^2(\mathbb{R}^N)$.

Let us show that there exists $h\in\mathbb{R}^N$ such that, for any $u\in C_c^\infty(\mathbb{R}^N)$, the function $\widetilde{L}_{2,n}u$ converges to $\mathcal{B}_{-h}u:=\langle\nabla u,h\rangle$ as n tends to $+\infty$, locally uniformly in $\mathbb{R}^N$. For this purpose, we observe that, by (9.6.3), we can write

$$\widetilde{L}_{2,n}u=r_n\lambda_n^{-2}\Delta u-r_n\lambda_n^{-1}\langle G(\lambda_n\cdot+\sigma_n),\nabla u\rangle-r_nc(\lambda_n\cdot+\sigma_n)u$$

for any $n\in\mathbb{N}$ and $u\in C_c^\infty(\mathbb{R}^N)$.

Due to the choice of the sequences $\{\sigma_n\}$ and $\{r_n\}$, r_n/λ_n^2 vanishes as n tends to $+\infty$. Moreover, by the mean value theorem and (9.6.2)(ii) we can infer that

$$|G(\lambda_nx+\sigma_n)|\leq|G(\sigma_n)|+kR(1+(R+|\sigma_n|)^\beta) \tag{9.6.4}$$

for any $x\in\overline{B}_R$ and $R>0$. Hence, $|G(\sigma_n+\lambda_nx)-G(\sigma_n)||G(\sigma_n)|^{-1}$ vanishes as n tends to $+\infty$, locally uniformly in $\mathbb{R}^N$. Up to replacing $\{\sigma_n\}$ with a suitable subsequence, we can assume that $G(\sigma_n)/|G(\sigma_n)|$ converges to some vector $h\in\mathbb{R}^N$. Therefore, $\lambda_n^{-2}r_n\Delta u-\lambda_n^{-1}r_n\langle G(\lambda_n\cdot+\sigma_n),\nabla u\rangle$ converges to $\mathcal{B}_hu$ uniformly in $\mathbb{R}^N$ as n tends to $+\infty$ and, hence, in $L^2(\mathbb{R}^N)$, since the support of u is compact.

Let us now show that $c(\lambda_n\cdot+\sigma_n)u$ converges to 0 in $L^2(\mathbb{R}^N)$ as n tends to $+\infty$. For this purpose, we note that, thanks to (9.6.2)(ii), $|c(x)|\leq C((1+|x|^\beta)^2+(1+|x|^\beta)|G(x)|)$ for any $x\in\mathbb{R}^N$ and some positive constant C. This estimate and (9.6.4) imply that the sequence $\{r_n|c(\lambda_n\cdot+\sigma_n)|\}$ converges to 0 as n tends to $+\infty$, locally uniformly in $\mathbb{R}^N$, and we are done.

Now the proof follows immediately, arguing by contradiction. Indeed, if $\widetilde{L}_2$ were sectorial in $L^2(\mathbb{R}^N)$, then the operators $\widetilde{L}_{2,n}$ should be sectorial as well, and there should exist $K,C>0$, independent of n, such that[6] $\|R(\lambda,\widetilde{L}_{2,n})u\|_{L^2(\mathbb{R}^N)}\leq C|\lambda|^{-1}\|u\|_{L^2(\mathbb{R}^N)}$ for any $\lambda\in\mathbb{C}$, with $\operatorname{Re}\lambda>K$, and $u\in C_c^\infty(\mathbb{R}^N)$. Since the realization B_h of the operator $\mathcal{B}_h$ in $L^2(\mathbb{R}^N)$ generates a strongly continuous semigroup and $C_c^\infty(\mathbb{R}^N)$ is a core of B_h, using Trotter-Kato theorem (see Theorem C.1.15), it should follow that $\|R(\lambda,B_h)u\|_{L^2(\mathbb{R}^N)}\leq C|\lambda|^{-1}\|u\|_{L^2(\mathbb{R}^N)}$ for any $u\in L^2(\mathbb{R}^N)$, which, of course, is a contradiction.

To prove the assertion with a general $p\in[1,+\infty)$ it now suffices to observe that, if $\{T(t)\}$ were analytic in L^p_μ for some $p\neq 2$, then, by Stein interpolation theorem, it should be analytic in L^2_μ as well. For more details, we refer the reader to [84, Theorem 1.4.2]. □

Example 9.6.3 We now provide two concrete examples of differential operators in $\mathbb{R}^2$, such that the associated semigroups $\{T(t)\}$ are not analytic in L^2_μ. For this purpose, let $\mathcal{A}_1$, $\mathcal{A}_2$ be the operators defined on smooth functions $u:\mathbb{R}^2\to\mathbb{R}$ by

$$\begin{aligned}\mathcal{A}_1u(x,y)=&\Delta u(x,y)-\left(x+y(x^2+y^2)^{\frac{r}{2}}\right)D_xu(x,y)\\&-\left(y-x(x^2+y^2)^{\frac{r}{2}}\right)D_yu(x,y),\end{aligned}$$

$$\begin{aligned}\mathcal{A}_2u(x,y)=&\Delta u(x,y)-(x^2+y^2)\left(x-y(x^2+y^2)^{\frac{r}{2}}\right)D_xu(x,y)\\&-(x^2+y^2)\left(y-x(x^2+y^2)^{\frac{r}{2}}\right)D_yu(x,y)\end{aligned}$$

for some $r>1$ and any $(x,y)\in\mathbb{R}^2$. We can rewrite the operators $\mathcal{A}_1$ and $\mathcal{A}_2$ in the compact

[6] Observe that $R(\lambda,\widetilde{L}_{2,n})=r_n^{-1}I_n^{-1}R(r_n^{-1}\lambda,\widetilde{L}_2)I_n$.

form (9.6.1) taking $U_j(x,y) = (2j)^{-1}(x^2+y^2)^j$, $G_j(x,y) = (x^2+y^2)^{r/2+(j-1)}(y,-x)$ for any $(x,y) \in \mathbb{R}^2$ and $j = 1,2$. It is immediate to check that the condition $\operatorname{div} G - \langle G, \nabla U\rangle = 0$ is satisfied. Moreover, if we take $\sigma_n = (n,0)$ for any $n \in \mathbb{N}$, then also the assumptions of Theorem 9.6.2 are satisfied with $\alpha = r - 1 + 2j$ and $\beta = r - 2 + 2j$, $j = 1,2$. Therefore, the semigroup $\{T_j(t)\}$ associated with the operator $\mathcal{A}_j$ is not analytic in $L^p_{\mu_j}$ ($p \in [1,+\infty)$) either for $j = 1$ or for $j = 2$, where

$$d\mu_1 = (2\pi)^{-1}e^{-\frac{1}{2}(x^2+y^2)}dxdy, \qquad d\mu_2 = \left(\int_{\mathbb{R}^2} e^{-\frac{1}{4}(x^2+y^2)^2}dxdy\right)^{-1} e^{-\frac{1}{4}(x^2+y^2)^2}dxdy.$$

Chapter 10

The Ornstein-Uhlenbeck operator

10.0 Introduction

In this chapter we analyze the Ornstein-Uhlenbeck operator, which is the prototype of an elliptic operator with unbounded coefficients. It is defined on smooth functions $\zeta : \mathbb{R}^N \to \mathbb{R}$ by

$$\mathcal{A}\zeta(x) = \frac{1}{2}\mathrm{Tr}\,(QD^2\zeta)(x) + \langle Bx, \nabla\zeta(x)\rangle, \qquad x \in \mathbb{R}^N, \tag{10.0.1}$$

where Q and B are $N \times N$ constant matrices, with Q positive definite and $B \neq 0$. Here, if not otherwise specified, we consider only the case when Q is strictly positive. In any case, some of the results that we present hold also when Q is a degenerate positive definite matrix and the operator $\mathcal{A}$ in (10.0.1) is hypoelliptic, i.e., when the matrix $\int_0^t e^{sB}Qe^{sB^*}ds$ is strictly positive definite for any $t > 0$. Such a condition can also be expressed by saying that the kernel of Q does not contain any invariant subspace of B^*.

First, in Section 10.1, we show that an explicit formula for the Ornstein-Uhlenbeck semigroup is available both in the nondegenerate and in the degenerate case. Having such a formula simplifies the study of the main properties of the semigroup. For instance, one can prove uniform estimates for the spatial derivatives of any order of the function $T(t)f$ when $f \in C_b(\mathbb{R}^N)$ just differentiating under the integral sign. We do this in Section 10.2 in the nondegenerate case. The case of the degenerate Ornstein-Uhlenbeck operator is much more involved. It has been studied by A. Lunardi in [181]. Here, we limit ourselves to state the main results of [181]. We also refer the reader to Chapter 11, where a class of degenerate elliptic operators, which includes the degenerate Ornstein-Uhlenbeck operator, is considered.

The Ornstein-Uhlenbeck semigroup is neither analytic nor strongly continuous in $C_b(\mathbb{R}^N)$. In particular, $T(t)f$ tends to f in $C_b(\mathbb{R}^N)$ as t tends to 0^+, if and only if $f \in BUC(\mathbb{R}^N)$ and $f(e^{tB}\cdot)$ tends to f uniformly in $\mathbb{R}^N$.

In Section 10.3 we deal with the invariant measure of $\{T(t)\}$. We show that, when the spectrum of the matrix B is contained in the left halfplane, the Ornstein-Uhlenbeck semigroup admits the Gaussian measure $\mu(dx) = ((2\pi)^N \det Q_\infty)^{-1/2} e^{-\frac{1}{2}\langle Q_\infty^{-1}x, x\rangle}dx$ as the (unique) invariant measure, both in the nondegenerate and in the degenerate case. Here,

$$Q_\infty = \int_0^{+\infty} e^{sB}Qe^{sB^*}ds. \tag{10.0.2}$$

The assumptions on the location of the spectrum of the matrix B also turns out to be necessary to guarantee the existence of the invariant measure of $\{T(t)\}$.

From the results in Chapter 9 we know that the extension of the Ornstein-Uhlenbeck semigroup to the L^p-spaces associated with the invariant measure μ (in short L^p_μ) gives rise to a strongly continuous semigroup for any $p \in [1, +\infty)$. Moreover, for any $f \in L^p_\mu$, $T(t)f$ is still given by the same formula as in the case when $f \in C_b(\mathbb{R}^N)$.

In the nondegenerate case, we show that $\{T(t)\}$ is also analytic for any $p \in (1,+\infty)$. Also in this situation, having an explicit representation formula for $T(t)f$ is of much help. Indeed, we can quite easily show that $T(t)$ maps L^p_μ into $W^{k,p}_\mu := W^{k,p}(\mathbb{R}^N,\mu)$ (for any $k \in \mathbb{N}$) and we can also give precise estimates on the behaviour of the spatial derivatives of $T(t)f$ in L^p_μ when t approaches 0.

An important feature of the Ornstein-Uhlenbeck semigroup in L^p_μ is that a complete characterization of its infinitesimal generator L_p is available. More precisely, $D(L_p) = W^{2,p}_\mu$ for any $p \in (1,+\infty)$ and the graph norm is equivalent to the Euclidean norm of $W^{2,p}_\mu$. Such a result was firstly proved by A. Lunardi and G. Da Prato in the Hilbert case and, then, by G. Metafune, J. Prüss, A. Rhandi, and R. Schnaubelt for a general p.

Since the Ornstein-Uhlenbeck semigroup in L^p_μ is compact for any $p \in (1,+\infty)$, the spectrum of L_p is a discrete set. It has been completely characterized by G. Metafune, D. Pallara, and E. Priola in terms of the eigenvalues $\lambda_1,\ldots,\lambda_N$ of the matrix B. More precisely, they showed that

$$\sigma(L_p) = \left\{\lambda = \sum_{i=1}^{r} n_i\lambda_i : n_i \in \mathbb{N}\cup\{0\},\ i = 1,\ldots,r\right\}.$$

In particular, $\sigma(L_p)$ is independent of p and the eigenfunctions of L_p are polynomials with degree at most $(\mathrm{Re}\lambda)/s(B)$. Moreover, the eigenvalues of L_p are all semisimple if and only if B is diagonalizable in $\mathbb{C}$. In the particular case, when $Q = -B = I$, the Hermite polynomials are an orthonormal basis of $L^2(\mathbb{R}^N,\mu)$ consisting of eigenfunctions of the operator L_2.

The picture changes drastically in the case when $p = 1$. Indeed, in such a situation $\sigma(L_1)$ is the set of all the $\lambda \in \mathbb{C}$ with nonpositive real part and any λ with negative real part is an eigenvalue of L_1.

Since the Ornstein-Uhlenbeck semigroup is analytic in L^p_μ, the function $z \mapsto T(z)$ is analytic in a sector centered at the origin. A result by R. Chill, E. Fašangová, G. Metafune, and D. Pallara explicitly characterizes the amplitude of such a sector (see Theorem 10.3.26).

Finally, in Section 10.4, we consider the Ornstein-Uhlenbeck semigroup in the L^p-spaces associated with the Lebesgue measure, for any $p \in [1,+\infty)$. In such a setting $\{T(t)\}$ is strongly continuous but it is not analytic. The domain of its infinitesimal generator A_p is the set of all the functions $u \in W^{2,p}(\mathbb{R}^N)$ such that $\mathcal{A}u \in L^p(\mathbb{R}^N)$. Moreover, the spectrum of A_p consists of all the complex numbers with real parts not greater than $-\mathrm{Tr}(B)/p$. In particular, differently from what happens when the underline measure is the invariant measure, the spectrum depends explicitly on p.

We conclude this introduction by observing that Ornstein-Uhlenbeck operators have also been widely studied in the infinite dimensional setting. This analysis is beyond the scope of this book and we refer the interested reader e.g. to [24, 59, 60, 61, 67, 68, 71, 75, 109, 185, 186, 252].

10.1 The formula for $T(t)f$

One of the best features of the Ornstein-Uhlenbeck operator is that an explicit formula for the associated semigroup is available both in the nondegenerate and in the degenerate case. Such a formula is due to Kolmogorov (see [139]).

Let us introduce some notation. For any $t > 0$, we denote by $Q_t \in L(\mathbb{R}^N)$ the matrix defined by

$$Q_t = \int_0^t e^{sB} Q e^{sB^*} ds. \tag{10.1.1}$$

It is readily seen that Q_t ($t > 0$) is a positive definite matrix, and it is strictly positive definite whenever Q is. Moreover, for any strictly positive definite matrix K and any $a \in \mathbb{R}^N$, we denote by $\mathcal{N}(a, K)(dx)$ the Gaussian measure, defined by

$$\mathcal{N}(a, K)(dx) = \frac{1}{\sqrt{(2\pi)^N \det K}} e^{-\frac{1}{2}\langle K^{-1}(x-a), x-a\rangle} dx. \tag{10.1.2}$$

We can now prove the following theorem.

Theorem 10.1.1 *For any $f \in C_b(\mathbb{R}^N)$ the Cauchy problem*

$$\begin{cases} D_t u(t,x) = \mathcal{A}u(t,x), & t > 0, \quad x \in \mathbb{R}^N, \\ u(0,x) = f(x), & \qquad\quad x \in \mathbb{R}^N, \end{cases} \tag{10.1.3}$$

admits a unique classical solution[1] *u which is bounded in $[0,T] \times \mathbb{R}^N$ for any $T > 0$. In fact, such a solution is bounded in $[0,+\infty) \times \mathbb{R}^N$ and*

$$\sup_{t>0} \|u(t,\cdot)\|_{C_b(\mathbb{R}^N)} \le \|f\|_\infty. \tag{10.1.4}$$

Moreover, u is given by

$$u(t,x) = (T(t)f)(x) := \frac{1}{\sqrt{(2\pi)^N \det Q_t}} \int_{\mathbb{R}^N} e^{-\frac{1}{2}\langle Q_t^{-1}(e^{tB}x - y), e^{tB}x - y\rangle} f(y) dy \tag{10.1.5}$$

for any $t > 0$ and $x \in \mathbb{R}^N$. The semigroup $\{T(t)\}$ is called the Ornstein-Uhlenbeck semigroup associated with the operator $\mathcal{A}$ in $C_b(\mathbb{R}^N)$.

Proof First of all we observe that the existence of the classical solution to problem (10.1.3) as well as estimate (10.1.4) are guaranteed by Theorem 1.2.1, whereas its uniqueness is an immediate consequence of Theorem 3.1.3, where we can choose as a Lyapunov function, the function $\varphi : \mathbb{R}^N \to \mathbb{R}$, defined by $\varphi(x) = |x|^2$ for any $x \in \mathbb{R}^N$. Indeed, an explicit computation shows that $\mathcal{A}\varphi(x) = \operatorname{Tr} Q + 2\langle Bx, x\rangle \le \operatorname{Tr} Q + 2\|B\|_\infty |x|^2$ for any $x \in \mathbb{R}^N$. Hence, it suffices to take $\lambda = 2\|B\|_\infty$ to get $\sup_{\mathbb{R}^N}(\mathcal{A}\varphi - \lambda\varphi) < +\infty$ and, by Remark 3.0.2, this is enough to apply Theorem 3.1.3.

So, to complete the proof we just need to check the formula (10.1.5). For this purpose, we use stochastic calculus. Indeed, the stochastic differential equation associated with the operator $\mathcal{A}$ is $d\xi_t^x = B\xi_t^x dt + \sigma dW_t$ with the initial condition $\xi_0^x \equiv x$, where $\sigma = \sqrt{Q}$, W_t is a standard N-dimensional Brownian motion and $x \in \mathbb{R}^N$. The solution of this problem is the Gaussian process

$$\xi_t^x = e^{tB}x + \int_0^t \sigma e^{(t-s)B} dW_s, \qquad t > 0.$$

For any fixed $t > 0$ and $x \in \mathbb{R}^N$ the mean value of ξ_t^x is $e^{tB}x$ and the covariance matrix is the matrix Q_t defined in (10.1.1). Therefore, the distribution of ξ_t^x is the normal distribution $\mathcal{N}(e^{tB}x, Q_t)(dx)$. This implies that the semigroup $\{T(t)\}$ is given by the formula

$$(T(t)f)(x) = \mathbb{E} f(\xi_t^x) = \int_{\mathbb{R}^N} f(y) \mathcal{N}(e^{tB}x, Q_t)(dy), \qquad x \in \mathbb{R}^N,$$

[1] i.e., a function $u \in C([0,+\infty) \times \mathbb{R}^N) \cap C^{1,2}((0,+\infty) \times \mathbb{R}^N)$ which solves the Cauchy problem (10.1.3).

for any $f \in B_b(\mathbb{R}^N)$ and (10.1.5) follows.

Let us also give an analytical proof of the formula (10.1.5). For this purpose, we look for a solution u of (10.1.3) in the form $u(t,x) = v(t, M(t)x)$ for any $t > 0$ and $x \in \mathbb{R}^N$, where the matrix-valued function $M : (0, +\infty) \to L(\mathbb{R}^N)$ is to be properly chosen.

The function v solves the Cauchy problem

$$\begin{cases} D_t v(t,x) = \mathcal{A}(t)v(t,x), & t > 0, \quad x \in \mathbb{R}^N, \\ v(0,x) = f(M(0)x), & x \in \mathbb{R}^N, \end{cases} \tag{10.1.6}$$

where $\mathcal{A}(t)$ is given by

$$\mathcal{A}(t) = \frac{1}{2}\mathrm{Tr}(\Lambda(t)D_x^2 v(t,x)) + \langle (M(t)B - M'(t))(M(t))^{-1}x, \nabla_x v(t,x)\rangle, \tag{10.1.7}$$

with $\Lambda(t) = M(t)Q(M(t))^*$ for any $t > 0$.

Now, we choose $M(t)$ in order to eliminate the drift term in (10.1.7). This is possible if we choose M satisfying the differential equation $M' = MB$. For instance, we can take $M(t) = e^{tB}$ for any $t > 0$. With this choice of M the elliptic equation in (10.1.6) reduces to

$$\begin{cases} D_t v(t,x) = \dfrac{1}{2}\mathrm{Tr}(e^{tB}Qe^{tB^*}D_x^2 v(t,x)), & t > 0, \quad x \in \mathbb{R}^N, \\ u(0,x) = f(x), & x \in \mathbb{R}^N. \end{cases} \tag{10.1.8}$$

Taking the Fourier transform of both sides of the differential equation in (10.1.8) gives the ordinary differential equation $D_t \widehat{v}(t,\xi) = -\frac{1}{2}\langle \Lambda(t)\xi, \xi\rangle\, \widehat{v}(t,\xi)$ for any $t > 0$ and $\xi \in \mathbb{R}^N$. Hence, $\widehat{v}(t,\xi) = e^{-\frac{1}{2}\langle Q_t\xi,\xi\rangle}\widehat{f}(\xi)$ for any $t > 0$ and $\xi \in \mathbb{R}^N$, where Q_t is given by (10.1.1). Applying the inverse Fourier transform, we obtain that

$$v(t,x) = \frac{1}{\sqrt{(2\pi)^N \det Q_t}} \int_{\mathbb{R}^N} e^{-\frac{1}{2}\langle Q_t^{-1}(x-y), x-y\rangle} f(y)dy, \qquad t > 0, \; x \in \mathbb{R}^N.$$

Now, since $u(t,x) = v(t, M(t)x)$ for any $t > 0$ and $x \in \mathbb{R}^N$, the formula (10.1.5) follows at once. □

Remark 10.1.2 The same results as in Theorem 10.1.1 also hold true in the case when Q is singular (but still positive definite) provided that the matrix Q_t in (10.1.1) is nonsingular for any $t > 0$. The semigroup defined by the formula (10.1.5) is still called the (degenerate) Ornstein-Uhlenbeck semigroup associated with the operator $\mathcal{A}$ in (10.0.1). As has been stressed in the introduction to this chapter, the condition $\det Q_t > 0$, for any $t > 0$, is equivalent to the hypoellipticity of the operator $\mathcal{A}$ in the sense of Hörmander (see [125]). We refer the reader to [160] and Remark 11.1.1 for further details.

10.2 Properties of $\{T(t)\}$ in $C_b(\mathbb{R}^N)$

In this section we study some properties of $\{T(t)\}$ in the space $C_b(\mathbb{R}^N)$. To begin with, let us consider the following proposition.

Proposition 10.2.1 *The following properties are satisfied:*

(i) *$\{T(t)\}$ is conservative in $C_b(\mathbb{R}^N)$;*

(ii) *$C_0(\mathbb{R}^N)$ is invariant for $\{T(t)\}$;*

(iii) *for any $t > 0$, $T(t)$ is not compact in $C_b(\mathbb{R}^N)$.*

Proof Properties (i) and (ii) follow from Proposition 3.1.10 (since $T(t)\mathbb{1} = \mathbb{1}$ for any $t > 0$) and from Example 4.3.3.

Finally, property (iii) is a straightforward consequence of properties (i) and (ii) and Proposition 4.3.4. □

The estimates for the derivatives of $\{T(t)f\}$ can be easily obtained as a consequence of the results in Chapters 5 and 6. They can also be deduced differentiating under the integral sign the representation formula (10.1.5). This is the approach used in [73] that allows the authors to prove uniform (and pointwise) estimates for the spatial derivatives of $T(t)f$ of any order.

Proposition 10.2.2 *For $t > 0$ and $f \in C_b(\mathbb{R}^N)$, the function $T(t)f$ belongs to $C_b^\infty(\mathbb{R}^N)$. In particular, if $f \in C_b^1(\mathbb{R}^N)$, then*

$$(D_iT(t)f)(x) = \sum_{j=1}^N (e^{tB^*})_{ij}(T(t)D_jf)(x), \qquad t > 0,\ x \in \mathbb{R}^N,\ i,j = 1,\dots,N. \quad (10.2.1)$$

Moreover, for any $\varepsilon > 0$ and $h \in \mathbb{N}$, with $h \le k$, there exists a positive constant $C = C(\varepsilon, h, k)$ such that

$$\|T(t)f\|_{C_b^k(\mathbb{R}^N)} \le C(1 \vee t^{\frac{h-k}{2}})e^{k(s(B)+\varepsilon)t}\|f\|_{C_b^h(\mathbb{R}^N)}, \qquad t > 0, \quad (10.2.2)$$

for any $f \in C_b^h(\mathbb{R}^N)$, where $s(B)$ is the spectral bound of the matrix B.

Proof Estimate (10.2.2) can be proved by induction on k. Let us begin by proving it in the case when $h = 0$. For $k = 1$ estimate (10.2.2), as well as the formula (10.2.1), follow easily differentiating the formula (10.1.5) under the integral sign and observing that, for any $\delta > 0$, there exists a constant $\widetilde{C}_1 = \widetilde{C}_1(\delta)$ such that

$$\|e^{tB}\|_\infty \le \widetilde{C}_1 e^{(s(B)+\delta)t}, \qquad t > 0. \quad (10.2.3)$$

Let us suppose that $T(t)f$ belongs to $C_b^{m-1}(\mathbb{R}^N)$ for any $t > 0$ and (10.2.2) holds with $k = m-1$. We are going to prove that the m-th order derivative $D_{i_1,\dots,i_m}T(t)f$ exists in $C_b(\mathbb{R}^N)$ and (10.2.2) holds true with $k = m$. For this purpose using (10.2.1) we get

$$\begin{aligned}
D_{i_2,\dots,i_m}T(t)f =& D_{i_2,\dots,i_m}T(t/n)\circ\dots\circ T(t/n)f\\
=& D_{i_2,\dots,i_{m-1}}\left\{T((1-1/n)t)\sum_{j_m=1}^N (e^{\frac{n-1}{n}B^*t})_{i_mj_m}D_{j_m}T(t/n)f\right\}\\
=& T((1-(m-1)/n)t)\sum_{j_2,\dots,j_m=1}^N (e^{\frac{n-m+1}{n}tB^*})_{i_2j_2}\cdot\ldots\cdot(e^{\frac{n-1}{n}tB^*})_{i_mj_m}\\
&\times D_{j_2}T(t/n)\circ\dots\circ D_{j_m}T(t/n)f
\end{aligned} \quad (10.2.4)$$

in $\mathbb{R}^N$, for any $n > m$ and $t > 0$. Similarly,

$$D_{j_2}T(t/n)\circ\dots\circ D_{j_m}T(t/n)f$$

$$=T((m/n-2/n)t)\sum_{l_2,\ldots,l_{m-1}=1}^{N}(e^{\frac{1}{n}tB^*})_{j_{m-1}l_{m-1}}\cdot\ldots\cdot(e^{\frac{m-2}{n}tB^*})_{j_2l_2}D_{l_2,\ldots,l_{m-1},j_m}T(t/n)f \tag{10.2.5}$$

in $\mathbb{R}^N$, for any $t>0$. From (10.2.4), (10.2.5) and (10.2.2) (with $k=1$) we conclude that $D_{i_2,\ldots,i_m}T(t)\in C_b^1(\mathbb{R}^N)$. Moreover,

$$\begin{aligned}D_{i_1,\ldots,i_m}T(t)f=&\sum_{j_2,\ldots,j_m=1}^{N}(e^{\frac{n-m+1}{n}tB^*})_{i_2j_2}\cdot\ldots\cdot(e^{\frac{n-1}{n}tB^*})_{i_mj_m}\\&\times\sum_{l_1,\ldots,l_{m-1}=1}^{N}(e^{\frac{1}{n}tB^*})_{j_{m-1}l_{m-1}}\cdot\ldots\cdot(e^{\frac{m-2}{n}tB^*})_{j_2l_2}(e^{\frac{n-2}{n}tB^*})_{i_1l_1}\\&\times T((1-2/n)t)D_{l_1}T(t/n)D_{l_2,\ldots,l_{m-1},j_m}T(t/n)f\end{aligned} \tag{10.2.6}$$

in $\mathbb{R}^N$. Therefore, taking $\delta=1/n$ in (10.2.3) and using (10.2.2) with $k=1$, $k=m-1$ and $\varepsilon=1/n$, we get

$$\|D_{i_1,\ldots,i_m}T(t)f\|_\infty\le C_n\exp\left\{\left(m-\frac{1}{n}\right)\left(s(B)+\frac{1}{n}\right)\right\}\frac{1}{1\wedge t^{m/2}}\|f\|_\infty$$

in $\mathbb{R}^N$, for any $t>0$ and some positive constant C_n. Estimate (10.2.2) with $k=m$ follows, taking n sufficiently large.

Estimate (10.2.2) with $h=k$ follows immediately from (10.2.3). Indeed, a straightforward computation shows that

$$D_{i_1,\ldots,i_k}T(t)f=\sum_{j_1,\ldots,j_k=1}^{N}(e^{tB^*})_{i_1j_1}\cdot\ldots\cdot(e^{tB^*})_{i_kj_k}T(t)D_{j_1,\ldots,j_k}f \tag{10.2.7}$$

in $\mathbb{R}^N$, for any $i_1,\ldots,i_k\in\{1,\ldots,N\}$, $k\in\mathbb{N}$ and $t>0$.

Finally, in the case when $0<h<k$, estimate (10.2.2) follows from (10.2.6) (with $m=k$) observing that, thanks to (10.2.7), we can write

$$\begin{aligned}&D_{l_1}T(t/n)D_{l_2,\cdots,l_{k-1},j_k}T(t/n)f\\&=\sum_{r_1,\ldots,r_{h+1}=1}^{N}(e^{\frac{1}{n}tB^*})_{l_1r_1}(e^{\frac{1}{n}tB^*})_{l_{k-h+1}r_2}\cdot\ldots\cdot(e^{\frac{1}{n}tB^*})_{l_{k-1}r_h}(e^{\frac{1}{n}tB^*})_{j_kr_{h+1}}\\&\qquad\times T(t/n)(D_{r_1,l_2,\ldots,l_{k-h}}T(t/n)D_{r_2,\ldots,r_{h+1}}f)\end{aligned}$$

in $\mathbb{R}^N$, for any $t>0$. □

Remark 10.2.3 (i) From the proof of Proposition 10.2.2 it is clear that, if (10.2.3) holds with $\delta=0$, then estimate (10.2.2) can be written also with $\varepsilon=0$.

(ii) In view of Remark 6.2.4, when $k=1,2,3$, the uniform estimate (10.2.2) is useful only in the particular case when $s(B)<0$.

Remark 10.2.4 In the case when the matrix Q is singular and the operator $\mathcal{A}$ is hypoelliptic, uniform estimates of the Ornstein-Uhlenbeck semigroup are still available, but they differ from those in Proposition 10.2.2 since, as we will see in Chapter 11, the behaviour of the spatial derivatives of $T(t)f$ near $t = 0$ is anisotropic. Such estimates, which have been proved by A. Lunardi, can be written in an elegant way if the Ornstein-Uhlenbeck is expressed not in the usual Euclidean basis of $\mathbb{R}^N$ but in a suitable orthonormal basis associated with the matrix Q. In such a basis the estimates read as the ones in Chapter 11. We refer the reader to [181, Proposition 3.2] for further details.

As we have already stressed, both the Ornstein-Uhlenbeck semigroup and the degenerate Ornstein-Uhlenbeck semigroup are neither analytic nor strongly continuous in $C_b(\mathbb{R}^N)$ and in $BUC(\mathbb{R}^N)$. Now we provide a proof of these claims. To begin with we show that $\{T(t)\}$ is not analytic, adapting the proof of [73, Lemma 3.3]. This result holds also in the case when Q is singular.

Proposition 10.2.5 *The Ornstein-Uhlenbeck semigroup is analytic neither in $C_b(\mathbb{R}^N)$ nor in $BUC(\mathbb{R}^N)$.*

Proof If $Q = I$, the assertion follows from Remark 1.4.3.

Let us consider the general case. Clearly, we can limit ourselves to proving that $\{T(t)\}$ is not analytic in $BUC(\mathbb{R}^N)$. For this purpose, we prove that there exists $h \in \mathbb{R}^N$ such that

$$\sup_{h\in\mathbb{N}} \|\mathcal{A}T(1)f_h\|_\infty = +\infty, \tag{10.2.8}$$

where $f_h = \sin(\langle h, \cdot\rangle)$. Indeed, if $\{T(t)\}$ were analytic in $BUC(\mathbb{R}^N)$ then, by Theorem C.2.2, the above supremum would be finite. Since $(T(1)f_h)(x) = e^{-\frac{1}{2}\langle Q_1 h,h\rangle}\sin(\langle e^B x, h\rangle)$ for any $x, h \in \mathbb{R}^N$ (see also (10.3.6)), it follows that the function $\mathrm{Tr}(QD^2T(1)f_h)$ is bounded in $\mathbb{R}^N$ for any $h \in \mathbb{R}^N$. Thus, to prove (10.2.8) we need to show that the function $x \mapsto \langle Bx, (\nabla_x T(1)f_h)(x)\rangle$ is unbounded in $\mathbb{R}^N$ for some $h \in \mathbb{R}^N$. As is easily seen $(B^*(\nabla_x T(1)f_h)(x))_i = (B^*e^{B^*}h)_i e^{-\frac{1}{2}\langle Q_1 h,h\rangle}\cos(\langle e^B x, h\rangle)$ for any $x \in \mathbb{R}^N$ and $i = 1, \ldots, N$. Since $B \neq 0$, there exist $h_0 \in \mathbb{R}^N$ and $j \in \{1, \ldots, N\}$ such that $(B^*e^{B^*}h_0)_j \neq 0$. Thus, the function $r \mapsto \langle rBe_j, (\nabla_x T(1)f_{h_0})(re_j)\rangle = r(B^*e^{B^*}h_0)_j e^{-\frac{1}{2}\langle Q_1 h_0,h_0\rangle}\cos(r(e^{B^*}h_0)_j)$ is unbounded in $\mathbb{R}$ and, therefore, $\langle B\cdot, \nabla_x T(1)f_{h_0}\rangle$ is unbounded in $\mathbb{R}^N$. This completes the proof. □

The next proposition describes the set of strong continuity of $\{T(t)\}$. It holds for both nondegenerate and degenerate Ornstein-Uhlenbeck semigroups.

Proposition 10.2.6 *Let $f \in C_b(\mathbb{R}^N)$. Then, $\|T(t)f - f\|_\infty$ vanishes as t tends to 0 if and only if $f \in BUC(\mathbb{R}^N)$ and $\lim_{t\to 0^+}(f(e^{tB}x) - f(x)) = 0$, uniformly with respect to $x \in \mathbb{R}^N$.*

Proof First of all let us notice that, if $T(t)f$ tends to f uniformly in $\mathbb{R}^N$, then $f \in BUC(\mathbb{R}^N)$. Indeed, Proposition 10.2.2 ensures that, for any $f \in C_b(\mathbb{R}^N)$, the function $T(t)f$ belongs to $BUC(\mathbb{R}^N)$. Therefore, in the rest of the proof we assume that $f \in BUC(\mathbb{R}^N)$.

For any $t > 0$, let G_t be the linear operator in $BUC(\mathbb{R}^N)$ defined by

$$(G_t f)(x) = \frac{1}{\sqrt{(2\pi)^N \det Q_t}} \int_{\mathbb{R}^N} f(x-y)e^{-\frac{1}{2}\langle Q_t^{-1}y,y\rangle}dy, \qquad x \in \mathbb{R}^N.$$

Then, we can write $(T(t)f)(x) = (G_t f)(e^{tB}x)$ for any $t > 0$ and $x \in \mathbb{R}^N$. Let us observe that $G_t f$ tends to f uniformly in $\mathbb{R}^N$ as n tends to $+\infty$. Indeed, for any $n \in \mathbb{N}$, we can estimate

$$|(G_t f)(x) - f(x)| \leq \frac{1}{\sqrt{(2\pi)^N}} \int_{B_n} e^{-\frac{1}{2}|y|^2}|f(x+\sqrt{Q_t}y) - f(x)|dy$$

$$+\frac{2}{\sqrt{(2\pi)^N}}\|f\|_\infty\int_{\mathbb{R}^N\setminus B_n} e^{-\frac{1}{2}|y|^2}dy \tag{10.2.9}$$

for any $t>0$ and $x\in\mathbb{R}^N$. From estimate (10.2.9) it is clear that for any $\varepsilon>0$ we can find, first, n sufficiently large and, then, t sufficiently close to 0 such that both of the integral terms in the right-hand side of (10.2.9) are less than $\varepsilon/2$, uniformly with respect to $x\in\mathbb{R}^N$. Now, the conclusion follows writing $(T(t)f)(x)-f(x)=(G_tf)(e^{tB}x)-f(e^{tB}x)+f(e^{tB}x)-f(x)$ for any $t>0$ and $x\in\mathbb{R}^N$. □

To conclude this section we characterize the spectrum of the weak generator $\widehat{A}$ of the Ornstein-Uhlenbeck operator in $C_b(\mathbb{R}^N;\mathbb{C})$ when Q is not singular. For the proof we refer the reader to [190, Corollary 6.3], where the characterization of the spectrum of the restriction of $\widehat{A}$ to $BUC(\mathbb{R}^N;\mathbb{C})$ is given. But the spectra of $\widehat{A}$ and its restriction to $BUC(\mathbb{R}^N;\mathbb{C})$ actually coincide. Indeed, of course, $\sigma(\widehat{A}_{|BUC(\mathbb{R}^N;\mathbb{C})})\subset\sigma(\widehat{A})$. Moreover, since $D(\widehat{A})\subset BUC(\mathbb{R}^N;\mathbb{C})$, the point spectra of $\widehat{A}$ and $\widehat{A}_{|BUC(\mathbb{R}^N;\mathbb{C})}$ coincide. Further, if $\mathrm{Re}\,\lambda>0$, then $\lambda\in\rho(\widehat{A})$. Indeed, fix $f\in C_b(\mathbb{R}^N;\mathbb{C})$ and consider a bounded sequence $\{f_n\}\subset C_c^\infty(\mathbb{R}^N;\mathbb{C})$ converging locally uniformly in $\mathbb{R}^N$ to f. Taking the estimates in Proposition 10.2.2 into account it is easy to check that

$$(R(\lambda,\widehat{A}_{|BUC(\mathbb{R}^N;\mathbb{C})})f_n)(x)=\int_0^{+\infty}e^{-\lambda t}(T(t)f_n)(x)dt \tag{10.2.10}$$

for any $x\in\mathbb{R}^N$ and $n\in\mathbb{N}$, where we still denote by $\{T(t)\}$ the extension of the Ornstein-Uhlenbeck semigroup to complex-valued functions. From the formula (10.2.10) and Proposition 1.2.10, it follows that $u_n=R(\lambda,\widehat{A}_{|BUC(\mathbb{R}^N;\mathbb{C})})f_n$ converges locally uniformly to a function $u\in C_b(\mathbb{R}^N;\mathbb{C})$. Since $\widehat{A}u_n=\lambda u_n-f_n$ and $\|R(\lambda,\widehat{A}_{|BUC(\mathbb{R}^N;\mathbb{C})})f_n\|_\infty\le K\|f_n\|_\infty$ for any $n\in\mathbb{N}$ and some positive constant K, independent of n, $\{\widehat{A}u_n\}$ is bounded in $C_b(\mathbb{R}^N;\mathbb{C})$ and it converges locally uniformly in $\mathbb{R}^N$. Therefore, according to Proposition 1.3.4, $u\in D(\widehat{A})$ and $\lambda u-\widehat{A}u=f$. Moreover, by the previous estimate we get $\|u\|_\infty\le K\|f\|_\infty$ since we can assume that $\|f_n\|_\infty\le\|f\|_\infty$ for any $n\in\mathbb{N}$.

Theorem 10.2.7 *If $\sigma(B)\cap i\mathbb{R}=\varnothing$ then $\sigma(\widehat{A})=\{\lambda\in\mathbb{C}:\mathrm{Re}\,\lambda\le 0\}$.*

10.3 The invariant measure μ and the semigroup in $L^p(\mathbb{R}^N,\mu)$

In this section we assume that

$$\sigma(B)\subset\{\lambda\in\mathbb{C}:\mathrm{Re}\,\lambda<0\}. \tag{10.3.1}$$

As a straightforward consequence, the matrix Q_t (see (10.1.1)) can also be defined for $t=+\infty$. Q_∞ is a positive definite matrix and it is strictly positive definite if Q_t is strictly positive definite for some $t>0$.

We set

$$\mu(dx):=\mathcal{N}(0,Q_\infty)(dx)=\frac{1}{\sqrt{(2\pi)^N\det Q_\infty}}e^{-\frac{1}{2}\langle Q_\infty^{-1}x,x\rangle}dx=:\rho(x)dx. \tag{10.3.2}$$

As the following theorem shows, μ is the invariant measure of both the nondegenerate and the degenerate Ornstein-Uhlenbeck semigroups.

Proposition 10.3.1 *If condition* (10.3.1) *holds, then the Gaussian measure μ is the unique invariant measure of the semigroup $\{T(t)\}$.*

Proof In the case of the nondegenerate Ornstein-Uhlenbeck semigroup, the uniqueness of the invariant measure of $\{T(t)\}$ follows immediately from Theorem 9.1.15.

Actually, Theorem 9.1.15 also holds in the case when the diffusion matrix Q is singular. Indeed, to make the proof of that theorem work, one needs to be able to extend the semigroup with continuity to a strong Feller and irreducible semigroup in $B_b(\mathbb{R}^N)$ and to prove the convergence of the sequence $(T(t)f_n)(x)$ to $(T(t)f)(x)$, as n tends to $+\infty$, for any $t>0$ and $x\in\mathbb{R}^N$, whenever the sequence $\{f_n\}\subset B_b(\mathbb{R}^N)$ converges in a dominated way to f. Of course, these two properties are also satisfied by the degenerate Ornstein-Uhlenbeck semigroup, as can be checked using the representation formula (10.1.5).

So, let us prove that, in both the nondegenerate and the degenerate case, the Gaussian measure μ is the invariant measure of $\{T(t)\}$. Throughout the proof, we extend $\{T(t)\}$ to the space of bounded and continuous complex-valued functions in the natural way, and we still denote by $\{T(t)\}$ the so extended semigroup.

Denote by $\mathcal{E}$ the linear span of the set $\{f_h : h\in\mathbb{R}^N\}$, where $f_h:\mathbb{R}^N\to\mathbb{C}$ is defined by $f_h(x)=e^{i\langle x,h\rangle}$ for any $x,h\in\mathbb{R}^N$. We observe that $\mathcal{E}$ is dense in L^1_μ. To check this property, it suffices to show that, if $T\in(L^1_\mu)'$ vanishes on $\mathcal{E}$, then $T=0$. For this purpose, we recall that $(L^1_\mu)'=L^\infty_\mu\subset L^\infty(\mathbb{R}^N)$ (see [231, Theorem 6.16]). Fix $T\in(L^1_\mu)'$ vanishing on $\mathcal{E}$ and let $g\in L^\infty(\mathbb{R}^N)$ be such that $Tf=\int_{\mathbb{R}^N} fg\,d\mu$ for any $f\in L^1_\mu$. Then,

$$\int_{\mathbb{R}^N} g(x)e^{i\langle x,h\rangle}e^{-\frac{1}{2}\langle Q_\infty x,x\rangle}dx=0,\qquad h\in\mathbb{R}^N.$$

It follows that the Fourier transform of the function $x\mapsto g(x)e^{-\langle Q_\infty x,x\rangle/2}$ identically vanishes in $\mathbb{R}^N$. By the uniqueness of the Fourier transform, we deduce that $g\equiv 0$. Hence, to prove that μ is the invariant measure of the Ornstein-Uhlenbeck semigroup, it suffices to show that

$$\int_{\mathbb{R}^N} T(t)f_h\,d\mu=\int_{\mathbb{R}^N} f_h\,d\mu,\qquad t>0,\ \ h\in\mathbb{R}^N. \tag{10.3.3}$$

For this purpose, we fix $a\in\mathbb{R}^N$, a strictly positive definite matrix K and compute the Fourier transform of the probability measure $\mathcal{N}(a,K)(dx)$ (see (10.1.2)). Let M be an orthogonal matrix such that $MKM^*=\operatorname{diag}(\lambda_1,\ldots,\lambda_N)$. Then

$$\begin{aligned}\int_{\mathbb{R}^N} f_h\mathcal{N}(a,K)(dy) =&\frac{e^{i\langle a,h\rangle}}{\sqrt{(2\pi)^N\det K}}\int_{\mathbb{R}^N} e^{i\langle M^*y,h\rangle}\exp\left(-\sum_{i=1}^N\frac{y_i^2}{2\lambda_i}\right)dy\\ =&\frac{e^{i\langle a,h\rangle}}{\sqrt{(2\pi)^N\lambda_1\cdot\ldots\cdot\lambda_N}}\prod_{k=1}^N\int_{\mathbb{R}} e^{iy_k(Mh)_k}e^{-\frac{y_k^2}{2\lambda_k}}dy_k\\ =&e^{i\langle a,h\rangle}\prod_{k=1}^N e^{-\frac{1}{2}\lambda_k(Mh)_k^2}=e^{i\langle a,h\rangle-\frac{1}{2}\langle Kh,h\rangle}\end{aligned} \tag{10.3.4}$$

for any $h\in\mathbb{R}^N$. Therefore,

$$\int_{\mathbb{R}^N} f_h\,d\mu=\int_{\mathbb{R}^N} e^{i\langle h,y\rangle}\mathcal{N}(0,Q_\infty)(dy)=e^{-\frac{1}{2}\langle Q_\infty h,h\rangle},\qquad h\in\mathbb{R}^N. \tag{10.3.5}$$

On the other hand, it holds that

$$(T(t)f_h)(x)=\int_{\mathbb{R}^N} e^{i\langle h,y\rangle}\mathcal{N}(e^{tB}x,Q_t)(dy)=e^{i\langle e^{tB}x,h\rangle-\frac{1}{2}\langle Q_t h,h\rangle} \tag{10.3.6}$$

for any $t > 0$, any $x, h \in \mathbb{R}^N$ and then

$$\int_{\mathbb{R}^N} T(t) f_h d\mu = e^{-\frac{1}{2}\langle Q_t h, h\rangle} \int_{\mathbb{R}^N} e^{i\langle e^{tB^*}h, x\rangle} \mathcal{N}(0, Q_\infty)(dx) = e^{-\frac{1}{2}\langle (Q_t + e^{tB} Q_\infty e^{tB^*})h, h\rangle}. \tag{10.3.7}$$

Now, observing that

$$\begin{aligned} Q_t + e^{tB} Q_\infty e^{tB^*} &= \int_0^t e^{sB} Q e^{sB^*} ds + \int_0^{+\infty} e^{(t+s)B} Q e^{(t+s)B^*} ds \\ &= \int_0^t e^{sB} Q e^{sB^*} ds + \int_t^{+\infty} e^{sB} Q e^{sB^*} ds = Q_\infty \end{aligned} \tag{10.3.8}$$

for any $t > 0$, from (10.3.5) and (10.3.7) we get (10.3.3). □

Remark 10.3.2 As is shown in [79, Section 11.2.3], (10.3.1) is also a necessary condition to guarantee the existence of an invariant measure of $\{T(t)\}$.

In view of Remark 10.3.2, throughout the rest of this section we assume that condition (10.3.1) is satisfied.

According to Proposition 9.1.7, the semigroup $\{T(t)\}$ can be extended to a strongly continuous semigroup in $L^p_\mu(\mathbb{R}^N)$. As in Chapter 9, we denote by $\|\cdot\|_p$ the norm of L^p_μ. Moreover, we denote by $W^{k,p}_\mu$ the Sobolev space of the functions in L^p_μ with weak derivatives up to k-th-order in L^p_μ. Finally, we denote by $L_p : D(L_p) \subset L^p_\mu \to L^p_\mu$ ($p \in [1, +\infty)$) the infinitesimal generator of $\{T(t)\}$ in L^p_μ.

To prove the main result of this first part of the section (i.e., the analyticity of $\{T(t)\}$ in L^p_μ for any $p \in (1, +\infty)$), we need some preliminary results.

Lemma 10.3.3 *For any $p \in [1, +\infty)$ and $k \in \mathbb{N}$, $C^\infty_c(\mathbb{R}^N)$ is dense in L^p_μ and in $W^{k,p}_\mu$.*

Proof The proof can be obtained using the same truncation argument as in the proof of Lemma 9.5.3. □

Lemma 10.3.4 *Let $p \in [1, +\infty)$. Then, for any $u \in W^{1,p}_\mu$ and $j = 1, \ldots, N$, the function $x \mapsto x_j u(x)$ belongs to L^p_μ. Moreover, there exists a positive constant C_j such that*

$$\left(\int_{\mathbb{R}^N} |x_j|^p |u(x)|^p d\mu\right)^{\frac{1}{p}} \le C_j \left(\int_{\mathbb{R}^N} (|u|^p + |\nabla u|^p) d\mu\right)^{\frac{1}{p}}.$$

Proof Without loss of generality we can limit ourselves to proving the assertion in the case when $j = 1$. Moreover, we can assume that $Q_\infty = \mathrm{diag}(\lambda_1, \ldots, \lambda_n)$. Indeed, we can always reduce to this situation by means of a suitable linear change of variables as in the proof of Proposition 10.3.1. From the formula (10.1.5) we deduce that

$$\rho(x) = \frac{1}{\sqrt{(2\pi)^N \lambda_1 \cdot \ldots \cdot \lambda_n}} \exp\left(-\sum_{i=1}^N \frac{x_j^2}{2\lambda_j}\right), \qquad x \in \mathbb{R}^N. \tag{10.3.9}$$

Let us now fix $u \in C^\infty_c(\mathbb{R}^N)$ and assume that $p > 1$ (the case $p = 1$ is similar and much easier). Taking (10.3.9) into account and integrating by parts, we deduce the following chain of inequalities

$$\int_{\mathbb{R}^N} |x_1 u(x)|^p d\mu \le -\lambda_1 \int_{\mathbb{R}^N} x_1 |x_1|^{p-2} |u(x)|^p D_1 \rho(x) dx$$

$$
\begin{aligned}
&=\lambda_1 p\int_{\mathbb{R}^N} x_1u(x)|x_1u(x)|^{p-2}D_1u(x)d\mu\\
&\quad+\lambda_1(p-1)\int_{\mathbb{R}^N}|x_1|^{p-2}|u(x)|^pd\mu =: I_1+I_2. \qquad (10.3.10)
\end{aligned}
$$

Using the Hölder and Young inequalities we can estimate

$$
\begin{aligned}
I_1 \leq&\lambda_1 p\bigg(\int_{\mathbb{R}^N}|x_1u(x)|^pd\mu\bigg)^{\frac{p-1}{p}}\bigg(\int_{\mathbb{R}^N}|D_1u|^pd\mu\bigg)^{\frac{1}{p}}\\
\leq&\frac{1}{4}\int_{\mathbb{R}^N}|x_1u(x)|^pd\mu+K_1\int_{\mathbb{R}^N}|D_1u|^pd\mu \qquad (10.3.11)
\end{aligned}
$$

for any $\varepsilon>0$ and some positive constant $K_1=K_1(\varepsilon)$.

Then, we observe that, if $p\geq 2$, then by the Young inequality we can determine a positive constant K_2 such that $|x_1|^{p-2}\leq(4\lambda_1(p-1))^{-1}|x_1|^p+K_2$ for any $x_1\in\mathbb{R}$. Therefore,

$$
I_2\leq\frac{1}{4}\int_{\mathbb{R}^N}|x_1|^p|u(x)|^pd\mu+K_3\int_{\mathbb{R}^N}|u(x)|^pd\mu \qquad (10.3.12)
$$

for some positive constant K_3.

If $p<2$ we have to apply a different argument to estimate the term I_2. For notational convenience we split $x\in\mathbb{R}^N$ as $x=(x_1,z)$. Moreover, we denote by $d\mu_1$ and $d\mu_2$ the measures in $\mathbb{R}$ and $\mathbb{R}^{N-1}$ whose densities with respect to the Lebesgue measure are $\rho_1(x_1)=(2\pi\lambda_1)^{-1/2}e^{-x_1^2/(2\lambda_1)}$, for any $x_1\in\mathbb{R}$, and

$$
\rho_2(z)=\frac{1}{\sqrt{(2\pi)^{N-1}\lambda_2\cdot\ldots\cdot\lambda_N}}\exp\bigg(-\sum_{j=1}^{N-1}\frac{z_j^2}{2\lambda_{j+1}}\bigg),\qquad z\in\mathbb{R}^{N-1},
$$

respectively. Applying the Fubini-Tonelli theorem we deduce that

$$
\begin{aligned}
I_2\leq&2\lambda_1\int_{\mathbb{R}^{N-1}}d\mu_2\int_{|x_1|\geq\sqrt{8\lambda_1}}|x_1|^{p-2}|u(x_1,z)|^pd\mu_1\\
&+2\lambda_1\int_{\mathbb{R}^{N-1}}d\mu_2\int_{-\sqrt{8\lambda_1}}^{\sqrt{8\lambda_1}}|x_1|^{p-2}|u(x_1,z)|^pd\mu_1\\
\leq&\frac{1}{4}\int_{\mathbb{R}^{N-1}}d\mu_2\int_{\mathbb{R}}|x_1|^p|u(x_1,z)|^pd\mu_1+\int_{\mathbb{R}^{N-1}}d\mu_2\int_{-\sqrt{8\lambda_1}}^{\sqrt{8\lambda_1}}|x_1|^{p-2}|u(x_1,z)|^pd\mu_1.
\end{aligned}
\qquad (10.3.13)
$$

Let us estimate the second integral in the last side of (10.3.13). For this purpose we observe that, since $W^{1,p}((-\sqrt{8\lambda_1},\sqrt{8\lambda_1}))$ is continuously embedded in $L^\infty((-\sqrt{8\lambda_1},\sqrt{8\lambda_1}))$ (see [3, Theorem 5.4]), we can write

$$
\begin{aligned}
\int_{-\sqrt{8\lambda_1}}^{\sqrt{8\lambda_1}}|x_1|^{p-2}|u(x_1,z)|^pd\mu_1\leq&\Big(\sup_{|x_1|\leq\sqrt{8\lambda_1}}|u(x_1,z)|\Big)^p\int_{-\sqrt{8\lambda_1}}^{\sqrt{8\lambda_1}}|x_1|^{p-2}dx\\
\leq&K_4\int_{-\sqrt{8\lambda_1}}^{\sqrt{8\lambda_1}}(|u(x_1,z)|^p+|D_1u(x_1,z)|^p)dx_1\\
\leq&K_4\sup_{|x_1|\leq\sqrt{8\lambda_1}}|\rho_1(x_1)|^{-1}\int_{-\sqrt{8\lambda_1}}^{\sqrt{8\lambda_1}}(|u(x_1,z)|^p+|D_1u(x_1,z)|^p)d\mu_1
\end{aligned}
$$

for some positive constant K_4. Integrating in $\mathbb{R}^{N-1}$ with respect to the measure μ_2 gives

$$\int_{\mathbb{R}^{N-1}} d\mu_2 \int_{-\sqrt{8\lambda_1}}^{\sqrt{8\lambda_1}} |x_1|^{p-2}|u(x_1,z)|^p d\mu_1 \le K_5\|u\|_{1,p}^p$$

for some constant $K_5 > 0$. Therefore,

$$I_2 \le \frac{1}{4}\int_{\mathbb{R}^{N-1}} d\mu_2 \int_{\mathbb{R}} |x_1|^p|u(x_1,z)|^p d\mu_1 + K_5\|u\|_{1,p}^p. \tag{10.3.14}$$

Summing up, from (10.3.10)-(10.3.12) and (10.3.14) we deduce that there exists a positive constant K_6, independent of u, such that

$$\int_{\mathbb{R}^N} |x_1 u(x)|^p d\mu \le K_6\|u\|_{1,p}^p. \tag{10.3.15}$$

Since, by Lemma 10.3.3, $C_c^\infty(\mathbb{R}^N)$ is dense in $W_\mu^{1,p}$, estimate (10.3.15) yields the assertion. □

The following result is now a straightforward consequence of Lemma 10.3.4.

Corollary 10.3.5 *The realization of the Ornstein-Uhlenbeck operator $\mathcal{A}$ in L_μ^p with domain $W_\mu^{2,p}$ is a bounded linear operator.*

Lemma 10.3.6 *For any $f \in L_\mu^p$ ($p \in [1,+\infty)$) and $t > 0$, $T(t)f$ is given by the formula* (10.1.5).

Proof In order to avoid any misunderstanding, throughout the proof we denote by $\{T(t)\}$ the Ornstein-Uhlenbeck semigroup in $C_b(\mathbb{R}^N)$ and by $\{T_p(t)\}$ its extension to L_μ^p.
To begin with, let us observe that since, for any $t > 0$, Q_t and Q_∞ differ in a (strictly) positive definite matrix, $Q_t < Q_\infty$ in the sense of the positive matrices[2]. Hence, $Q_tQ_\infty^{-1} < Q_\infty Q_\infty^{-1} = I$ or, equivalently, $Q_\infty^{-1} = Q_t^{-1}Q_tQ_\infty^{-1} < Q_t^{-1}$. As a consequence,

$$\rho_t(x) := \frac{1}{\sqrt{(2\pi)^N \det Q_t}} e^{-\frac{1}{2}\langle Q_t^{-1}x,x\rangle} \le C_t\rho(x), \qquad x \in \mathbb{R}^N,$$

for some positive constant C_t, and this implies that $L_\mu^1 \subset L_{\mu_t}^1$ where $\mu_t(dx) = \rho_t(x)dx$. Therefore, if $\{f_n\}$ is a sequence of smooth functions converging to f in L_μ^p as n tends to $+\infty$, then, for any $t > 0$ and $x \in \mathbb{R}^N$, $f_n(e^{tB}x + \cdot)$ converges to $f(e^{tB}x + \cdot)$ in L_μ^1 and, hence,

$$\lim_{n\to+\infty}(T(t)f_n)(x) = \lim_{n\to+\infty}\int_{\mathbb{R}^N} f_n(e^{tB}x+y)\mu_t(dy) = \int_{\mathbb{R}^N} f(e^{tB}x+y)\mu_t(dy).$$

Now, the assertion follows if we recall that, by definition, $T(t)f_n$ converges to $T_p(t)f$ in L_μ^p (see Proposition 9.1.7). □

Remark 10.3.7 Note that the results in Corollary 10.3.5 and Lemma 10.3.6 hold true also in the case when the matrix Q is singular.

Arguing as in the proof of Proposition 10.2.2, we can prove the following estimates for the derivatives of the function $T(t)f$ in the L_μ^p-norm.

[2] i.e., $\langle Q_th,h\rangle < \langle Q_\infty h,h\rangle$ for any $h \in \mathbb{R}^N$.

Proposition 10.3.8 *For any $p\in(1,+\infty)$, $k\in\mathbb{N}$, $t>0$ and $f\in L^p_\mu$ the function $T(t)f$ belongs to $W^{k,p}_\mu$. Moreover, for any multi-index α and $\varepsilon>0$ there exists a positive constant $C=C(\varepsilon,|\alpha|,p)$ such that*

$$\|D^\alpha_x T(t)f\|_p\le Ce^{|\alpha|(s(B)+\varepsilon)t}t^{-\frac{|\alpha|}{2}}\|f\|_p,\qquad t>0,\tag{10.3.16}$$

for any $p\in(1,+\infty)$ and

$$\|D^\alpha_x T(t)f\|_p\le Ce^{|\alpha|(s(B)+\varepsilon)t}\sum_{|\beta|=|\alpha|}\|D^\beta f\|_p,\qquad t>0,\tag{10.3.17}$$

for any $p\in[1,+\infty)$. Further, the function $t\mapsto D^\alpha_x T(t)f$ is continuous in $(0,+\infty)$ with values in L^p_μ for any $f\in L^p_\mu$ ($p\in(1,+\infty)$).

Proof Since $C^\infty_c(\mathbb{R}^N)$ is dense in $W^{k,p}_\mu$ (see Lemma 10.3.3), it suffices to prove the assertions in the case when $f\in C^\infty_c(\mathbb{R}^N)$.

To show estimate (10.3.16) we can argue by induction on the length of α as in the proof of Proposition 10.2.2. We limit ourselves to checking (10.3.16) in the case when $|\alpha|=1$. For this purpose, we fix $f\in C^\infty_c(\mathbb{R}^N)$, $i\in\{1,\ldots,N\}$, $p\in(1,+\infty)$ and observe that

$$(D_iT(t)f)(x)=-\frac{1}{\sqrt{(2\pi)^N\det Q_t}}\int_{\mathbb{R}^N}(e^{tB^*}Q_t^{-1}y)_ie^{-\frac{1}{2}\langle Q_t^{-1}y,y\rangle}f(e^{tB}x-y)dy$$

for any $t>0$ and $x\in\mathbb{R}^N$. Since

$$|(e^{tB^*}Q_t^{-1}y)_i|=|\langle Q_t^{-1/2}y,Q_t^{-1/2}e^{tB}e_i\rangle|\le|Q_t^{-1/2}y||(Q_t^{-1/2}e^{tB})_i|$$

for any $t>0$, $y\in\mathbb{R}^N$ and $i=1,\ldots,N$, using the Hölder inequality and taking (10.2.3) into account, we conclude that

$$\begin{aligned}|D_iT(t)f|^p\le&\left(\int_{\mathbb{R}^N}|(e^{tB^*}Q_t^{-1}y)_i|^{\frac{p}{p-1}}e^{-\frac{1}{2}\langle Q_t^{-1}y,y\rangle}dy\right)^{p-1}T(t)|f|^p\\ \le&\|(Q_t^{-1/2}e^{tB})_i\|^{\frac{p}{p-1}}T(t)|f|^p\left(\frac{1}{\sqrt{(2\pi)^N\det Q_t}}\int_{\mathbb{R}^N}|Q_t^{-1/2}y|^{\frac{p}{p-1}}e^{-\frac{1}{2}\langle Q_t^{-1}y,y\rangle}dy\right)^{p-1}\\ \le&C_pt^{-\frac{1}{2}}e^{(s(B)+\varepsilon)t}T(t)|f|^p\end{aligned}\tag{10.3.18}$$

in $\mathbb{R}^N$, for any $t>0$ and some positive constant C_p. Integrating the first and the last term in (10.3.18) in $\mathbb{R}^N$, with respect to the measure μ, we get (10.3.16).

Estimate (10.3.17) is easier to prove: indeed, it suffices to iterate the formula (10.2.1).

Finally, we observe that for any multi-index α the function $D^\alpha_xT(\cdot)f$ is continuous in $(0,+\infty)$ with values in L^p_μ. Indeed, estimate (10.3.16) implies that $\|D^\alpha_xT(t)f-D^\alpha_xT(s)f\|_p\le Ce^{|\alpha|(s(B)+\varepsilon)t}s^{-\frac{|\alpha|}{2}}\|T(t-s)f-f\|_p$ for any $0<s<t$ and $p\in(1,+\infty)$. The strong continuity of $\{T(t)\}$ in L^p_μ yields the assertion. □

Now, we can show that the Ornstein-Uhlenbeck semigroup is analytic in L^p_μ for any $p\in(1,+\infty)$. Such a result has been proved in [73] and in [108, 117], also in infinite dimensional settings. Here, we present a simplified proof taken from [180].

Theorem 10.3.9 *For any $p\in(1,+\infty)$, the Ornstein-Uhlenbeck semigroup is analytic in L^p_μ.*

Proof Fix $p \in (1,+\infty)$. According to Theorem C.2.8 to prove the assertion it suffices to show that the function $T(\cdot)f$ is differentiable in $(0,+\infty)$, for any $f \in L^p_\mu$, and there exists a positive constant C, independent of f, such that

$$t\|D_tT(t)f\|_p \le C\|f\|_p, \qquad t \in (0,1). \tag{10.3.19}$$

Suppose first that $f \in C^\infty_c(\mathbb{R}^N)$. In such a case, taking Propositions 1.3.4, 1.3.5 and 3.1.10 into account, it is easy to check that the function $T(\cdot)f$ is differentiable in $[0,+\infty)$ with values in L^p_μ and $D_tT(t)f = \mathcal{A}T(t)f$ for any $t \ge 0$. Moreover, taking Corollary 10.3.5 and estimate (10.3.16) into account, it follows that

$$t\|\mathcal{A}T(t)f\|_p \le C_1\|f\|_p, \qquad t > 0, \tag{10.3.20}$$

for some positive constant C_1, independent of f, and (10.3.19) follows.

Now, suppose that $f \in L^p_\mu$ and let $\{f_n\} \in C^\infty_c(\mathbb{R}^N)$ be a sequence of smooth functions converging to f in L^p_μ (see Lemma 10.3.3). Since

$$\frac{T(t)f_n - T(t_0)f_n}{t-t_0} - \mathcal{A}T(t_0)f_n = \frac{1}{t-t_0}\int_{t_0}^t (\mathcal{A}T(s)f_n - \mathcal{A}T(t_0)f_n)ds \tag{10.3.21}$$

for any $t_0, t \in (0,+\infty)$ and $n \in \mathbb{N}$ (where the integral is meant in L^p_μ), letting n tend to $+\infty$, and observing that (10.3.20) is satisfied by f_n $(n \in \mathbb{N})$ and f, with the same constant C_1, we conclude that (10.3.21) holds true also with f instead of f_n. Now using the continuity of the function $t \mapsto \mathcal{A}T(t)f$ in $(0,+\infty)$ with values in L^p_μ (see Corollary 10.3.5 and Proposition 10.3.8), we can conclude that $T(\cdot)f$ is differentiable in $(0,+\infty)$ with values in L^p_μ and $D_tT(\cdot)f = \mathcal{A}T(\cdot)f$. Hence, estimate (10.3.19) follows and we are done. □

The next proposition characterizes the symmetric Ornstein-Uhlenbeck semigroups, i.e., the case when $T(t)$ coincides with its adjoint operator for any $t > 0$. The characterization of the symmetric Ornstein-Uhlenbeck semigroups has been studied in [61], also in the infinite dimensional setting.

Proposition 10.3.10 *The Ornstein-Uhlenbeck semigroup is symmetric in L^2_μ if and only if $BQ = QB^*$. In such a case its infinitesimal generator L_2 is self-adjoint. Further, if $B = -Q_\infty^{-1}$, then*

$$\int_{\mathbb{R}^N} gL_2f\,d\mu = -\int_{\mathbb{R}^N} \langle Q\nabla f, \nabla g\rangle d\mu, \qquad f,g \in D(L_2). \tag{10.3.22}$$

Proof As a first step we show that, for any $t > 0$, the adjoint operator of $T(t)$ is given by

$$(T(t)^*f)(x) = \int_{\mathbb{R}^N} f(\sqrt{Q}_\infty e^{tB^*_\infty}Q_\infty^{-1/2}x + Q_\infty^{1/2}(I - e^{tB^*_\infty}e^{tB_\infty})^{1/2}Q_\infty^{-1/2}y)d\mu \tag{10.3.23}$$

for any $f \in \mathcal{D} := \{g_h^{(j)} : h \in \mathbb{R}^N,\ j = 1,2\}$, where $g_h^{(1)}(x) = \cos(\langle h,x\rangle)$ and $g_h^{(2)}(x) = \sin(\langle h,x\rangle)$ for any x and h in $\mathbb{R}^N$. Here, $B_\infty = Q_\infty^{-1/2}BQ_\infty^{1/2}$. Note that the matrix $(I - e^{tB^*_\infty}e^{tB_\infty})^{1/2}$ is well defined. Indeed, using the formula (10.3.8), it is easy to check that

$$Q_t = Q_\infty^{1/2}(I - e^{tB_\infty}e^{tB^*_\infty})Q_\infty^{1/2}, \qquad t > 0. \tag{10.3.24}$$

As a consequence, the matrix $I - e^{tB_\infty}e^{tB^*_\infty}$ is strictly definite positive and this implies that $\|e^{tB_\infty}\|_{L(L^2_\mu)} = \|e^{tB^*_\infty}\|_{L(L^2_\mu)} < 1$ for any $t > 0$. Therefore, $I - e^{tB^*_\infty}e^{tB_\infty}$ is a strictly positive definite matrix as well.

Let us fix $t > 0$, $g \in \mathcal{D}$ and denote by $R(t)$ the operator defined by the right-hand side of (10.3.23). Since $\mathcal{D}$ is dense in L^2_μ (see the proof of Proposition 10.3.1), to prove that $T(t)^*g = R(t)g$ it suffices to show that

$$\int_{\mathbb{R}^N} gT(t)f d\mu = \int_{\mathbb{R}^N} fR(t)g d\mu \tag{10.3.25}$$

for any $f \in \mathcal{D}$. For this purpose, we begin by observing that, from the formula (10.3.6), we get

$$T(t)g_h^{(j)} = e^{-\frac{1}{2}\langle Q_t h,h\rangle} g^{(j)}_{e^{tB^*}h}, \qquad j = 1, 2. \tag{10.3.26}$$

Similarly, setting $P_{1,t} = Q_\infty^{1/2}(I - e^{tB_\infty^*}e^{tB_\infty})^{1/2}Q_\infty^{-1/2}$, $P_{2,t} = Q_\infty^{1/2}e^{tB_\infty^*}Q_\infty^{-1/2}$, we obtain that

$$\begin{aligned}(R(t)g_h^{(j)})(x) &= \int_{\mathbb{R}^N} g_h^{(j)}(P_{1,t}y)\mathcal{N}(P_{1,t}^{-1}P_{2,t}x, Q_\infty)(dy)\\ &= \int_{\mathbb{R}^N} g_h^{(j)}(y)\mathcal{N}(P_{2,t}x, P_{1,t}Q_\infty P_{1,t}^*)(dy)\\ &= e^{-\frac{1}{2}\langle Q_\infty^{1/2}(I-e^{tB_\infty^*}e^{tB_\infty})Q_\infty^{1/2}h,h\rangle} g^{(j)}_{Q_\infty^{-1/2}e^{tB_\infty}Q_\infty^{1/2}h}(x)\end{aligned} \tag{10.3.27}$$

for any $x \in \mathbb{R}^N$ and $j = 1, 2$, where we took advantage of (10.3.4) and $\mathcal{N}(a, K)$ is defined in (10.1.2) for any $a \in \mathbb{R}^N$ and any positive definite matrix K.

Since the density of μ (with respect to the Lebesgue measure) is even with respect to each variable, by (10.3.26) and (10.3.27) it suffices to check (10.3.25) in the case when $(f, g) = (g_h^{(j)}, g_k^{(j)})$ for $j = 1, 2$. This can easily be done, observing that $2g_h^{(j)}g_k^{(j)} = g_{h-k}^{(j)} + (-1)^{j-1}g_{h+k}^{(j)}$ for any $h, k \in \mathbb{R}^N$ and $j = 1, 2$. Indeed, we get

$$\begin{aligned}&\int_{\mathbb{R}^N} g_k^{(j)}T(t)g_h^{(j)}d\mu\\ &= \frac{1}{2}e^{-\frac{1}{2}\langle Q_t h,h\rangle}\left(e^{-\frac{1}{2}\langle Q_\infty(e^{tB^*}h-k), e^{tB^*}h-k\rangle} - (-1)^j e^{-\frac{1}{2}\langle Q_\infty(e^{tB^*}h+k), e^{tB^*}h+k\rangle}\right),\end{aligned}$$

whereas

$$\begin{aligned}\int_{\mathbb{R}^N} g_h^{(j)}R(t)g_k^{(j)}d\mu =& \frac{1}{2}e^{-\frac{1}{2}\langle Q_\infty^{1/2}(I-e^{tB_\infty^*}e^{tB_\infty})Q_\infty^{1/2}k,k\rangle}\\ &\times\left(e^{\frac{1}{2}|Q_\infty^{1/2}(h-Q_\infty^{-1/2}e^{tB_\infty}Q_\infty^{1/2}k)|^2} - (-1)^j e^{\frac{1}{2}|Q_\infty^{1/2}(h+Q_\infty^{-1/2}e^{tB_\infty}Q_\infty^{1/2}k)|^2}\right),\end{aligned}$$

from which the equality (10.3.26) follows observing that $e^{tB}Q_\infty e^{tB^*} + Q_t = Q_\infty$ for any $t > 0$. Now, the formulae (10.3.26) and (10.3.27) imply that $T(t)$ is symmetric for any $t > 0$ if and only

$$(i)\ Q_\infty^{-1/2}e^{tB_\infty}Q_\infty^{1/2} = e^{tB^*}, \qquad (ii)\ Q_\infty^{1/2}(I - e^{tB_\infty^*}e^{tB_\infty})Q_\infty^{1/2} = Q_t. \tag{10.3.28}$$

Differentiating both sides of (10.3.28)(i) at $t = 0$, we conclude that such a condition holds if and only if $B_\infty Q_\infty^{1/2} = Q_\infty^{1/2}B^*$ or, equivalently, if and only if B_∞ is symmetric. Taking (10.3.24) into account, condition (10.3.28)(ii) is now an immediate consequence of (10.3.28)(i). Summing up, we have shown that $T(t)$ is symmetric in L^2_μ for any $t > 0$ if and only if the matrix B_∞ is symmetric or, equivalently, if and only if $BQ_\infty = Q_\infty B^*$. In such a case, according to [213, Chapter 1, Corollary 10.6], L_2 is self-adjoint in L^2_μ.

To conclude the first part of the proof, let us show that $BQ_\infty = Q_\infty B^*$ if and only if

$BQ = QB^*$. Showing that $BQ = QB^*$ implies that $BQ_\infty = Q_\infty B^*$ is an easy task in view of (10.0.2). Conversely, let us assume that $BQ_\infty = Q_\infty B^*$. Using the formula (10.3.8) it is immediate to check that $Q_t B^* = BQ_t$ for any $t > 0$. Differentiating this formula at $t = 0$, gives $BQ = QB^*$.

We now assume that $B = -Q_\infty^{-1}$ and prove the formula (10.3.22). For this purpose, since $\nabla\rho(x) = -\rho(x)Q_\infty^{-1}x$, an integration by parts gives

$$\begin{aligned}\int_{\mathbb{R}^N}\langle Q\nabla f, \nabla g\rangle d\mu &= -\int_{\mathbb{R}^N} g(\mathrm{Tr}(QD^2 f)\rho + \langle Q\nabla f, \nabla\rho\rangle)dx\\ &= -\int_{\mathbb{R}^N} g(x)(\mathrm{Tr}\,(QD^2 f(x)) - \langle Q\nabla f(x), Q_\infty^{-1}x\rangle)\mu(dx)\end{aligned}$$

for any $f, g \in C_c^\infty(\mathbb{R}^N)$, that is (10.3.22). Since $C_c^\infty(\mathbb{R}^N)$ is a core of L_2 (see the forthcoming Lemma 10.3.11), the formula (10.3.22) follows for any $f, g \in D(L_2)$. □

10.3.1 The domain of the realization of $\{T(t)\}$ in $L^p(\mathbb{R}^N, \mu)$

In this subsection we give a complete characterization of the domain of the infinitesimal generator L_p of $\{T(t)\}$ in L^p_μ for any $p \in [1, +\infty)$. The results that we present are taken from [199]. We notice that in the case when $p = 2$ the characterization of the domain of L_2 has been proved in [180] and in [65] (also in the infinite dimensional case) and, then, it has been generalized to a more general context first in [78] and, then, in [74]. Here, we consider the case when Q is not singular.

The following lemma will be very useful to characterize the domain of L_p.

Lemma 10.3.11 *For any $p \in [1, +\infty)$, $C_c^\infty(\mathbb{R}^N)$ is a core of L_p.*

Proof We first prove that the Schwarz space $\mathcal{S}$, of all the smooth functions $f : \mathbb{R}^N \to \mathbb{R}$ such that $\lim_{|x|\to+\infty} |x|^k D^\alpha f(x) = 0$ for any $k \in \mathbb{N}$ and $\alpha \in \mathbb{N}^N$, is a core of L_p. For this purpose, let us notice that the same arguments as in the first part of the proof of Theorem 10.3.9 show that $\mathcal{S}$ is contained in $D(L_p)$ for any $p \in [1, +\infty)$. Since $T(t)$ maps $\mathcal{S}$ into itself for any $t > 0$, by Lemma 10.3.3 and Proposition C.1.14, $\mathcal{S}$ is a core of L_p.

Now, to prove that $C_c^\infty(\mathbb{R}^N)$ is a core of L_p, it suffices to show that, for any $f \in \mathcal{S}$, there exists a sequence $\{f_n\} \subset C_c^\infty(\mathbb{R}^N)$ such that f_n and $\mathcal{A}f_n$ converge, respectively, to f and $\mathcal{A}f$ in L^p_μ, as n tends to $+\infty$. So, fix $f \in \mathcal{S}$ and, for any $x \in \mathbb{R}^N$ and $n \in \mathbb{N}$, let $\vartheta_n(x) = \vartheta(x/n)$ where $\vartheta \in C_c^\infty(\mathbb{R}^N)$ satisfies the condition $\chi_{B_{1/2}} \le \vartheta \le \chi_{B_1}$. Then, the function $f_n = f\vartheta_n$ belongs to $C_c^\infty(\mathbb{R}^N)$ for any $n \in \mathbb{N}$ and it converges to f in $W^{2,p}_\mu$ as n tends to $+\infty$. Moreover, since $\langle x, \nabla f_n(x)\rangle = \vartheta_n(x)\langle x, \nabla f(x)\rangle + n^{-1}f(x)\langle x, \nabla\vartheta(n^{-1}x)\rangle$, for any $x \in \mathbb{R}^N$, $\mathcal{A}f_n$ converges to $\mathcal{A}f$ in L^p_μ. □

Now, we characterize the domain of L_p for any $p \in (1, +\infty)$, when the Ornstein-Uhlenbeck operator $\mathcal{A}$ is given by

$$(\mathcal{A}u)(x) = \frac{1}{2}\Delta u(x) - \sum_{i=1}^N \frac{x_i}{2\lambda_i} D_i u(x), \qquad x \in \mathbb{R}^N, \tag{10.3.29}$$

and λ_i $(i = 1, \ldots, N)$ are suitable positive constants. This is a crucial step in order to characterize the domain of the more general (nondegenerate) Ornstein-Uhlenbeck operator. In the case when $\mathcal{A}$ is given by (10.3.29), $\{T(t)\}$ is symmetric (see Proposition 10.3.10) and it is given by

$$(T(t)f)(x)=\frac{1}{\sqrt{(2\pi)^N\prod_{i=1}^N\lambda_i(1-e^{-\frac{t}{\lambda_i}})}}\int_{\mathbb{R}^N}\exp\bigg(-\sum_{i=1}^N\frac{(e^{-\frac{t}{2\lambda_i}}x_i-y_i)^2}{2\lambda_i(1-e^{-\frac{t}{\lambda_i}})}\bigg)f(y)dy$$

for any $t>0$, $x\in\mathbb{R}^N$ and $f\in L^p_\mu$. Moreover, the density ρ of the invariant measure of the semigroup $\{T(t)\}$, with respect to the Lebesgue measure, is given by

$$\rho(x)=\frac{1}{\sqrt{(2\pi)^N\lambda_1\cdot\ldots\cdot\lambda_N}}\exp\bigg(-\sum_{i=1}^N\frac{x_i^2}{2\lambda_i}\bigg),\qquad x\in\mathbb{R}^N.$$

Proposition 10.3.12 *Suppose that the operator $\mathcal{A}$ is given by* (10.3.29). *Then, for any $p\in(1,+\infty)$, it holds that $D(L_p)=W^{2,p}_\mu$.*

Proof Since the proof is rather long, we divide it into three steps.

Step 1. Let us introduce the semigroup $\{T_i(t)\}$ $(i=1,\ldots,N)$ defined in $L^p_\mu(\mathbb{R}^N)$ $(p\in(1,+\infty))$ by

$$(T_i(t)f)(x)=\frac{1}{\sqrt{2\pi\lambda_i(1-e^{-\frac{t}{\lambda_i}})}}\int_{\mathbb{R}}\exp\bigg(-\frac{y^2}{2\lambda_i(1-e^{-\frac{t}{\lambda_i}})}\bigg)f(E_i(t,x,y))dy$$

for any $t>0$, $x\in\mathbb{R}^N$, where $E_i(t,x,y)\in\mathbb{R}^N$ is defined by $\langle E_i(t,x,y),e_j\rangle=x_j$, if $i\neq j$ and $\langle E_i(t,x,y),e_j\rangle=e^{-t/(2\lambda_i)}x_i-y$, if $i=j$. Arguing as in the proof of Proposition 9.1.7 and Theorem 10.3.9, it can be shown that $\{T_i(t)\}$ is a strongly continuous analytic semigroup of positive contractions in L^p_μ. We denote by $L_{p,i}$ its infinitesimal generator and prove that it coincides with the operator $A_{p,i}$ defined by $A_{p,i}u(x)=\frac{1}{2}D_{ii}u(x)-(2\lambda_i)^{-1}x_iD_iu(x)$ for any $u\in D(A_{p,i})=\{u\in L^p_\mu : D_iu,\ D_{ii}u\in L^p_\mu\}$. For this purpose, we observe that, arguing as in the proof of Lemma 10.3.11, we can easily show that $C^\infty_c(\mathbb{R}^N)$ is a core of $L_{p,i}$ and that $L_{p,i}=A_{p,i}$ on $C^\infty_c(\mathbb{R}^N)$. Since, as the proof of Lemma 10.3.4 shows, $A_{p,i}$ is a bounded operator in $D(A_{p,i})$ endowed with the norm $\|u\|_{i,p}=\|u\|_p+\|D_iu\|_p+\|D_{ii}u\|_p$, $L_{p,i}$ is the closure of the operator $A_{p,i}$. Therefore, to prove the assertion it suffices to show that $A_{p,i}$ is closed in L^p_μ. This will follow immediately if we prove that there exists a positive constant C_1 such that $\|D_iu\|_p+\|D_{ii}u\|_p\le C_1(\|u\|_p+\|A_{i,p}u\|_p)$ for any $u\in C^\infty_c(\mathbb{R}^N)$ and $i=1,\ldots,N$. Equivalently, we can prove that

$$\int_{\mathbb{R}}|u'|^pd\mu_i+\int_{\mathbb{R}}|u''|^pd\mu_i\le C_2\int_{\mathbb{R}}(|u|^p+|\widetilde{L}_iu|^p)d\mu_i \tag{10.3.30}$$

for any $u\in C^\infty_c(\mathbb{R})$, $i=1,\ldots,N$, and some positive constant C_2, where $(\widetilde{L}_iu)(x)=\frac{1}{2}u''(x)-\frac{1}{2\lambda_i}xu'(x)$, for any $x\in\mathbb{R}$, and $\mu_i(dx)=(2\pi\lambda_i)^{-1/2}\exp(-x^2/(2\lambda_i))dx$.

Actually, we can limit ourselves to showing that there exists $C_3>0$ such that

$$\int_{\mathbb{R}}|xu'(x)|^pd\mu_i\le C_3\int_{\mathbb{R}}(|u|^p+|\widetilde{L}_iu|^p)d\mu_i,\qquad u\in C^\infty_c(\mathbb{R}). \tag{10.3.31}$$

Indeed, once (10.3.31) is checked, it will follow easily that the $L^p_{\mu_i}$-norm of u'' satisfies (10.3.30). Concerning the $L^p_{\mu_i}$-norm of u', we observe that

$$\frac{1}{\sqrt{2\pi\lambda_i}}|u'(x)|^pe^{-\frac{x^2}{2\lambda_i}}=\frac{1}{\sqrt{2\pi\lambda_i}}\int_{-\infty}^x\frac{d}{dt}\Big(|u'(t)|^pe^{-\frac{t^2}{2\lambda_i}}\Big)dt$$

$$
\begin{aligned}
&\leq p\int_{\mathbb{R}}|u'(x)|^{p-1}|u''(x)|d\mu_i+\frac{1}{\lambda_i}\int_{\mathbb{R}}|x||u'(x)|^p d\mu_i\\
&\leq p\left(\int_{\mathbb{R}}|u'(x)|^p d\mu_i\right)^{\frac{p-1}{p}}\left(\int_{\mathbb{R}}|u''(x)|^p d\mu_i\right)^{\frac{1}{p}}\\
&\quad+\frac{1}{4}\int_{\mathbb{R}}|u'(x)|^p d\mu_i+C_4\int_{\mathbb{R}}|x|^p|u'(x)|^p d\mu_i\\
&\leq\frac{1}{2}\int_{\mathbb{R}}|u'(x)|^p d\mu_i+C_4\int_{\mathbb{R}}|x|^p|u'(x)|^p d\mu_i+C_5\int_{\mathbb{R}}|u''(x)|^p d\mu_i
\end{aligned}
\tag{10.3.32}
$$

for any $x\in\mathbb{R}$ and some positive constants C_4 and C_5. Here, in the last two inequalities we have taken advantage of the Young inequality. Integrating the first and last sides of (10.3.32) in $\mathbb{R}$ with respect to the Lebesgue measure gives

$$
\int_{\mathbb{R}}|u'(x)|^p d\mu_i\leq 2C_4\int_{\mathbb{R}}|x|^p|u'(x)|^p d\mu_i+2C_5\int_{\mathbb{R}}|u''(x)|^p d\mu_i
$$

and the estimate for u' in (10.3.30) follows.

So, let us prove (10.3.31). For this purpose, we fix $u\in C_c^\infty(\mathbb{R})$, we multiply both sides of the equation $u-\widetilde{L}_1u=:f$ by $e^{-x^2/2\lambda_i}$ and, then, integrate in $(-\infty,x)$ and in $(x,+\infty)$, with respect to the measure $d\mu_i$. We get

$$
u'(x)=2\int_x^{\pm\infty}(f(y)-u(y))e^{\frac{x^2-y^2}{2\lambda_i}}dy,\qquad x\in\mathbb{R}^N.
$$

Therefore,

$$
u'(x)e^{-\frac{x^2}{2p\lambda_i}}=2\int_x^{\pm\infty}h(y)e^{\frac{x^2-y^2}{2p'\lambda_i}}dy,\qquad x\in\mathbb{R}^N,
$$

where $1/p+1/p'=1$ and we have set $h(y)=e^{-y^2/(2p\lambda_i)}(f(y)-u(y))$ for any $y\in\mathbb{R}$. Consequently,

$$
\begin{aligned}
\int_{\mathbb{R}}|xu'(x)|^p d\mu_i=&2\int_0^{+\infty}\left|x\int_x^{+\infty}h(y)e^{\frac{x^2-y^2}{2p'\lambda_i}}dy\right|^p d\mu_i+2\int_{-\infty}^0\left|x\int_{-\infty}^x h(y)e^{\frac{x^2-y^2}{2p'\lambda_i}}dy\right|^p d\mu_i\\
\leq&4\int_0^{+\infty}\left|x\int_x^{+\infty}(|h(y)|+|h(-y)|)e^{\frac{x^2-y^2}{2p'\lambda_i}}dy\right|^p d\mu_i.
\end{aligned}
$$

Now, observing that $e^{(x^2-y^2)/(2p'\lambda_i)}\leq e^{x(x-y)/(p'\lambda_i)}$ for any $0\leq x\leq y$ and using the Hölder inequality and then the Fubini-Tonelli theorem, we deduce that

$$
\begin{aligned}
\int_{\mathbb{R}}|xu'(x)|^p d\mu_i\leq&4\int_0^{+\infty}dx\left(x\int_x^{+\infty}(|h(y)|+|h(-y)|)^p e^{\frac{x(x-y)}{p'\lambda_i}}dy\right)\left(\int_x^{+\infty}xe^{\frac{x(x-y)}{p'\lambda_i}}dy\right)^{p-1}\\
=&4(p'\lambda_i)^{p-1}\int_0^{+\infty}(|h(y)|+|h(-y)|)^p dy\int_0^y xe^{\frac{x(x-y)}{p'\lambda_i}}dx.
\end{aligned}
\tag{10.3.33}
$$

Since

$$
\begin{aligned}
\int_0^y xe^{\frac{x(x-y)}{p'\lambda_i}}dx\leq&y^2\int_0^1 e^{-\frac{y^2t(1-t)}{p'\lambda_i}}dt=y^2\int_0^{\frac{1}{2}}e^{-\frac{y^2t(1-t)}{p'\lambda_i}}dt+y^2\int_{\frac{1}{2}}^1 e^{-\frac{y^2t(1-t)}{p'\lambda_i}}dt\\
=&2y^2\int_0^{\frac{1}{2}}e^{-\frac{y^2t(1-t)}{p'\lambda_i}}dt\leq 2y^2\int_0^{\frac{1}{2}}e^{-\frac{y^2t}{2p'\lambda_i}}dt\leq 4p'\lambda_i,
\end{aligned}
$$

from (10.3.33) we deduce that

$$\int_{\mathbb{R}} |xu'(x)|^p d\mu_i \le 2^{p+3}(p'\lambda_i)^p \|h\|_{L^p(\mathbb{R})} \le 2^{p+3}(p'\lambda_i)^p(\|u\|_p + \|\widetilde{L}_i u\|_p)$$

and (10.3.31) follows.

Step 2. Now, we prove that

$$D(L_p) = \{u \in L^p_\mu : D_i u,\ D_{ii}u \in L^p_\mu,\ i = 1,\dots,N\}. \tag{10.3.34}$$

By virtue of Step 1, we can assume that $N > 1$. A straightforward computation shows that $(T(t)f = (T_1(t) \circ \ldots \circ T_N(t))f$ for any $t > 0$ and $f \in L^p_\mu$. Since for any $i, j \in \{1,\dots,N\}$ the operators $L_{p,i}$ and $L_{p,j}$ commute in the resolvent sense, a general theorem for commuting analytic semigroups implies that L_p is the closure of the operator $\sum_{i=1}^N L_{p,i}$ with domain given by the right-hand side of (10.3.34) (see [72, Theorem 3.3]). Hence, to prove (10.3.34) it suffices to show that the operator $\sum_{i=1}^N L_{p,i}$ is closed. For this purpose, we use Theorem C.1.20. According to Proposition C.1.18(iii), the operators $I - L_{i,p}$ $(i = 1,\dots,N)$ admit bounded imaginary powers in L^p_μ with power angles $\theta_{i,p} := \theta_{I-L_{i,p}}$ which do not exceed $\pi/2$. Now, since $I - L_{i,2}$ is self-adjoint in L^2_μ (see Proposition 10.3.10), from Proposition C.1.18(v) we deduce that $\theta_{i,2} = 0$, and the Riesz-Thorin interpolation theorem (see Theorem B.3.10) implies that $\theta_{i,p} < \pi/2$ for any $p \in (1,+\infty)$. Hence, Theorem C.1.20 applies and yields the closedness of the sum $\sum_{i=1}^N (I - L_{i,p})$ (equivalently, the closedness of the operator $\sum_{i=1}^N L_{i,p}$) in $\{u \in L^p_\mu : D_i u,\ D_{ii}u \in L^p_\mu,\ i = 1,\dots,N\}$.

Step 3. To conclude the proof we need to show that $W^{2,p}_\mu = \{u \in L^p_\mu : D_i u,\ D_{ii}u \in L^p_\mu,\ i = 1,\dots,N\}$. For this purpose, we fix $u \in L^p_\mu$ such that $D_i u, D_{ii}u \in L^p_\mu$, for any $i = 1,\dots,N$, introduce the function $v \in L^p(\mathbb{R}^N)$, defined by

$$v(x) = u(x)\exp\left(-\sum_{i=1}^N \frac{x_i^2}{2p\lambda_i}\right), \qquad x \in \mathbb{R}^N,$$

and prove that $\Delta v \in L^p(\mathbb{R}^N)$. Since

$$D_{ii}v(x) = \left[D_{ii}u(x) - \frac{x_i}{p\lambda_i}D_i u(x) + \left(\frac{x_i^2}{(p\lambda_i)^2} - \frac{1}{p\lambda_i}\right)u(x)\right]\exp\left(-\sum_{j=1}^N \frac{x_j^2}{2p\lambda_j}\right)$$

for any $x \in \mathbb{R}^N$ and $i = 1,\dots,N$, according to the proof of Lemma 10.3.4 the function $x \mapsto x_i D_i u(x)$ belongs to $L^p_\mu(\mathbb{R}^N)$. Repeating the same arguments as in the proof of the quoted lemma, we can show that the function $x \mapsto x_i^2 u(x)$ belongs to L^p_μ as well. Therefore, $D_{ii}v \in L^p(\mathbb{R}^N)$ and, consequently, $\Delta v \in L^p(\mathbb{R}^N)$. By classical results for elliptic equations (see Theorem D.1.4(iii)), v belongs to $W^{2,p}(\mathbb{R}^N)$.

As a consequence of the previous results, we also deduce that the functions $x \mapsto x_i x_j v(x)$ and $x \mapsto x_i D_i v(x)$ belong to $L^p(\mathbb{R}^N)$ for any $i, j \in \{1,\dots,N\}$. Since

$$D_{ij}u(x) = \left[D_{ij}v(x) + \frac{x_i}{p\lambda_i}D_j v(x) + \frac{x_j}{p\lambda_j}D_i v(x) + \left(\frac{x_i x_j}{p^2\lambda_i\lambda_j} + \frac{\delta_{ij}}{(p\lambda_i)^2}\right)v(x)\right]\exp\left(\sum_{h=1}^N \frac{x_h^2}{2p\lambda_h}\right)$$

for any $x \in \mathbb{R}^N$ and $i, j \in \{1,\dots,N\}$, it follows that $D_{ij}u \in L^p_\mu$. □

In order to characterize the domain of the more general Ornstein-Uhlenbeck operator $\mathcal{A}$, let us prove the following proposition which allows us to transform the operator $\mathcal{A}$ into a perturbation of the operator defined in (10.3.29). With any non singular matrix M

we associate the operator Φ_M defined by $(\Phi_M u)(x) = u(Mx)$ for any $x \in \mathbb{R}^N$ and any $u : \mathbb{R}^N \to \mathbb{R}$. It is clearly seen that Φ_M is an isometry between $L^p_{\widetilde{\mu}}$ and L^p_μ ($p \in [1,+\infty)$), where

$$\widetilde{\mu}(dx) = \frac{1}{|\det M|\sqrt{(2\pi)^N \det Q_\infty}} e^{-\frac{1}{2}\langle (M^*)^{-1}Q_\infty^{-1}M^{-1}x,x\rangle} dx =: \widetilde{\rho}(x)dx. \tag{10.3.35}$$

Lemma 10.3.13 *There exists a non singular matrix $M \in L(\mathbb{R}^N)$ such that the operator $\widetilde{\mathcal{A}} = \Phi_{M^{-1}}\mathcal{A}\Phi_M$ is the Ornstein-Uhlenbeck operator defined on smooth functions u by $(\widetilde{\mathcal{A}}u)(x) = \mathcal{A}^0 u(x) + \langle B^0 x, \nabla u(x)\rangle$ for any $x \in \mathbb{R}^N$, where $(\mathcal{A}^0 u)(x) = \frac{1}{2}\Delta u(x) - \sum_{i=1}^N \frac{x_i}{2\lambda_i} D_i u(x)$ for any $x \in \mathbb{R}^N$ and $B^0 = MBM^{-1} + \frac{1}{2}\mathrm{diag}\,(\lambda_1^{-1},\ldots,\lambda_N^{-1})$. The measure*

$$\widetilde{\mu}(dx) = \frac{1}{\sqrt{(2\pi)^N \lambda_1 \cdot \ldots \cdot \lambda_N}} \exp\left(-\sum_{i=1}^N \frac{x_i^2}{2\lambda_i}\right) dx \tag{10.3.36}$$

is the invariant measure of both the semigroup $\{\widetilde{T}(t)\}$ associated with the operator $\widetilde{\mathcal{A}}$ and the semigroup $\{T^0(t)\}$ associated with the operator $\mathcal{A}^0$. Finally, denoting by $\widetilde{L}_p$ the generator of $\{\widetilde{T}(t)\}$ in $L^p_{\widetilde{\mu}}$ ($p \in (1,+\infty)$), it holds that $D(L_p) = \Phi_M(D(\widetilde{L}_p))$ and $\widetilde{L}_p := \Phi_{M^{-1}} L_p \Phi_M$.

Proof Since Q is strictly positive definite, the matrix $Q^{-1/2}$ is well defined. Now, let M_1 be an orthogonal matrix such that $M_1 Q^{-1/2} Q_\infty (Q^{-1/2}) M_1^* = \mathrm{diag}\,(\lambda_1,\ldots,\lambda_N)$. A straightforward computation shows that

$$(\widetilde{\mathcal{A}}u)(x) = \frac{1}{2}\Delta u(x) + \langle MBM^{-1}x, \nabla u(x)\rangle, \qquad x \in \mathbb{R}^N, \tag{10.3.37}$$

where $M = M_1 Q^{-1/2}$. Therefore, $\widetilde{\mathcal{A}}$ is still an Ornstein-Uhlenbeck operator. The decomposition of the operator $\widetilde{\mathcal{A}}$ now easily follows.

The associated Ornstein-Uhlenbeck semigroup is given by (10.1.5), where now the matrix Q_t is replaced with the matrix

$$\begin{aligned}\widetilde{Q}_t &= \int_0^t e^{sMBM^{-1}} e^{s(M^{-1})^* B^* M^*} ds \\ &= \int_0^t M e^{sB} M^{-1}(M^{-1})^* e^{sB^*} M^* ds = \int_0^t M e^{sB} Q e^{sB^*} M^* ds = MQ_t M^*\end{aligned} \tag{10.3.38}$$

for any $t > 0$. In particular, $\widetilde{Q}_\infty = MQ_\infty M^* = \mathrm{diag}(\lambda_1,\ldots,\lambda_N)$. Therefore, according to Proposition 10.3.1, the measure $\widetilde{\mu}$ in (10.3.36) is the invariant measure of both the semigroups $\{\widetilde{T}(t)\}$ and $\{T^0(t)\}$.

Now, let us denote by $\widetilde{L}_p$ the infinitesimal generator of $\{\widetilde{T}(t)\}$ in $L^p_{\widetilde{\mu}}$ ($p \in (1,+\infty)$). Since, by Lemma 10.3.11, $C_c^\infty(\mathbb{R}^N)$ is a core of both L_p and $\widetilde{L}_p$ and

$$\widetilde{L}_p = \widetilde{\mathcal{A}} = \Phi_{M^{-1}}\mathcal{A}\Phi_M = \Phi_{M^{-1}} L_p \Phi_M$$

in $C_c^\infty(\mathbb{R}^N)$, we easily deduce that $D(\widetilde{L}_p) = \Phi_{M^{-1}}(D(L_p))$ and $\widetilde{L}_p = \Phi_{M^{-1}} L_p \Phi_M$. This completes the proof. □

Lemma 10.3.14 *Let $\widetilde{\mu}$ and B^0 be the measure and the matrix defined, respectively, in (10.3.35) and in Lemma 10.3.13. Then, for any $p \in [1,+\infty)$, the family of bounded operators $\{S(t) : t \geq 0\}$ in $L^p_{\widetilde{\mu}}$, defined by $(S(t)f)(x) = f(e^{tB^0}x)$ for any $t \in [0,+\infty)$, $x \in \mathbb{R}^N$ and $f \in L^p_\mu$, is a semigroup of isometries. Moreover, $C_c^\infty(\mathbb{R}^N)$ is a core of the infinitesimal generator B_p^0 and $(B_p^0 u)(x) = \langle B^0 x, \nabla u(x)\rangle$ for any $u \in C_c^\infty(\mathbb{R}^N)$ and $x \in \mathbb{R}^N$.*

Proof A straightforward computation shows that

$$\int_{\mathbb{R}^N} |S(t)f|^p d\widetilde{\mu} = \frac{e^{-t\mathrm{Tr}\,B^0}}{\sqrt{(2\pi)^N\lambda_1\cdot\ldots\cdot\lambda_N}}\int_{\mathbb{R}^N} |f(y)|^p e^{-\frac{1}{2}\langle \widetilde{Q}_\infty^{-1}e^{-tB^0}y, e^{-tB^0}y\rangle}dy$$

for any $t>0$, where $\widetilde{Q}_\infty = \mathrm{diag}\,(\lambda_1,\ldots,\lambda_N)$. To prove that $\|S(t)f\|_p = \|f\|_p$, it suffices to show that $\mathrm{Tr}\,B^0 = 0$ and

$$\langle \widetilde{Q}_\infty^{-1}e^{-tB^0}h, e^{-tB^0}h\rangle = \langle \widetilde{Q}_\infty^{-1}h, h\rangle, \qquad t>0,\ h\in\mathbb{R}^N. \tag{10.3.39}$$

Applying (10.3.8) to the Ornstein-Uhlenbeck operator $\widetilde{\mathcal{A}}$ (see (10.3.37)), we get

$$\widetilde{Q}_t e^{-t(MBM^{-1})^*} + e^{tMBM^{-1}}\widetilde{Q}_\infty = \widetilde{Q}_\infty e^{-t(MBM^{-1})^*} \tag{10.3.40}$$

for any $t>0$, where $\widetilde{Q}_t$ is given by (10.3.38). Differentiating both sides of (10.3.40) at $t=0$ gives $I + MBM^{-1}\widetilde{Q}_\infty = -\widetilde{Q}_\infty(MBM^{-1})^*$ or, equivalently,

$$\widetilde{Q}_\infty^{-1}B^0\widetilde{Q}_\infty = -(B^0)^*. \tag{10.3.41}$$

From (10.3.41) we get $\mathrm{Tr}\,B^0 = 0$ and $\widetilde{Q}_\infty^{-1}e^{-tB^0} = e^{t(B^0)^*}\widetilde{Q}_\infty^{-1}$ for any $t>0$, which implies (10.3.39).

Let us now observe that $\{S(t)\}$ is a strongly continuous group. Indeed, for any $f\in C_c^\infty(\mathbb{R}^N)$, $f(e^{tB^0}\cdot)$ converges uniformly to f as t tends to 0. Therefore, $S(t)f$ converges to f as t tends to 0 in $L^p_{\widetilde{\mu}}$. Since $C_c^\infty(\mathbb{R}^N)$ is dense in $L^p_{\widetilde{\mu}}$ (see Lemma 10.3.3) and $\{S(t)\}$ is a semigroup of isometries in $L^p_{\widetilde{\mu}}$, using Proposition B.1.2 it is immediate to check that $S(t)f$ converges to f as t tends to 0, for any $f\in L^p_{\widetilde{\mu}}$.

Now, denote by B_p^0 the infinitesimal generator of the semigroup $\{S(t)\}$ in L^p_μ. To prove that $C_c^\infty(\mathbb{R}^N)$ is a core of B_p^0, one can argue as in the proof of Lemma 10.3.11. Indeed, the Schwarz space $\mathcal{S}$ is contained in $D(B_p^0)$ and $(B_p^0 f)(x) = \langle B^0x, \nabla f(x)\rangle$ for any $x\in\mathbb{R}^N$ and $f\in\mathcal{S}$. Moreover, $S(t)$ maps $\mathcal{S}$ into itself for any $t>0$. Therefore, by Proposition C.1.14, $\mathcal{S}$ is a core of $\mathcal{B}_p^0$. Then, the same approximation argument as in the proof of Lemma 10.3.11 shows that $C_c^\infty(\mathbb{R}^N)$ is a core of B_p^0 and $(B_p^0u)(x) = \langle Bx, \nabla u(x)\rangle$ for any $x\in\mathbb{R}^N$ and any $u\in W^{2,p}_{\widetilde{\mu}}$. □

Theorem 10.3.15 *For any $p\in(1,+\infty)$, $D(L_p) = W^{2,p}_\mu$ and $L_pu = \mathcal{A}u$ for any $u\in W^{2,p}_\mu$.*

Proof By virtue of Lemma 10.3.13, to characterize $D(L_p)$ it suffices to characterize $D(\widetilde{L}_p)$. In particular, since Φ_M is an isometry between $W^{2,p}_{\widetilde{\mu}}$ and $W^{2,p}_\mu$, if we show that $D(\widetilde{L}_p) = W^{2,p}_{\widetilde{\mu}}$ ($p\in(1,+\infty)$), then we will immediately deduce that $D(L_p) = W^{2,p}_\mu$.

We fix now, and for the rest of the proof, an arbitrary $p\in(1,+\infty)$. Moreover, we denote by L_0^p the infinitesimal generator of the Ornstein-Uhlenbeck semigroup $\{T^0(t)\}$ in L^p_μ, associated with the operator $\mathcal{A}^0$ in the statement of Lemma 10.3.13, and, by B_p^0, the operator in Lemma 10.3.14. By Proposition 10.3.12, $D(L_p^0) = W^{2,p}_{\widetilde{\mu}}$ and $L_p^0u = \mathcal{A}^0u$ for any $u\in W^{2,p}_{\widetilde{\mu}}$. Since $C_c^\infty(\mathbb{R}^N)$ is a core of $\widetilde{L}_p$ (see Lemma 10.3.11) and the operator $L_p^0 + B_p^0$, with domain $W^{2,p}_{\widetilde{\mu}}$, coincides with the bounded operator $\widetilde{\mathcal{A}}$ (see Corollary 10.3.5), $\widetilde{L}_p$ is the closure of the sum of the operators L_p^0 and B_p^0.

To complete the proof it suffices to show that the operator $L_p^0+B_p^0$, with domain $W^{2,p}_\mu$, is closed. For this purpose, we prove that all the assumptions of Theorem C.1.21 are satisfied.

As a first step we observe that, repeating the same arguments as in the proof of Proposition 10.3.12 and observing that the operator L_2^0 is self-adjoint in $L^2_{\tilde{\mu}}$ (see Proposition 10.3.10), we conclude that both the operators $I - L_p^0$ and $\lambda I - B_p^0$ admit bounded imaginary powers. Moreover, by Proposition C.1.18(iii), the power angle $\theta_{\lambda I - B_p^0} \le \pi/2$ of $\lambda I - B_p^0$ is not greater than $\pi/2$. On the other hand, since L_2^0 is self-adjoint, Proposition C.1.18(v) shows that $\theta_{I-L_2^0} = 0$. Therefore, $\theta_{I-L_2^0} < \pi/2$ for any $p \in (1, +\infty)$.

Now, we show that, if λ is sufficiently large, then the operators $I - L_p^0$ and $\lambda I - B_p^0$ satisfy the condition (C.1.3). For this purpose, for any $\lambda > 0$, $\mu \in \rho(B_p^0)$ and $\nu \in L_p^0$, we define the bounded operator $E(\lambda, \mu, \nu) : C_b^4(\mathbb{R}^N) \to L^p_{\tilde{\mu}}$ by setting

$$E(\lambda, \mu, \nu)u = (I - L_p^0)R(\nu + 1, L_p^0)[R(1, L_p^0), R(\lambda + \mu, B_p^0)]u, \qquad u \in W^{2,p}_{\tilde{\mu}}.$$

An explicit computation and Proposition 10.2.2 show that both $\{S(t)\}$ and $\{T^0(t)\}$ map $C_b^4(\mathbb{R}^N)$ into itself for any $t > 0$ and there exist two constants $M \ge 1$ and $\omega_0 \ge 0$ such that $e^{-\omega_0 t}\|S(t)\|_{L(C_b^4(\mathbb{R}^N))} + \|T^0(t)\|_{L(C_b^4(\mathbb{R}^N))} \le M$ for any $t > 0$. Thus, both the resolvent operators $R(\xi + \omega_0, B_p^0)$ and $R(\xi, L_p^0)$ map $C_b^4(\mathbb{R}^N)$ into itself for any $\xi \in \mathbb{C}$ with $\mathrm{Re}\,\xi > 0$. Therefore, if $\lambda > \omega_0$, then it holds that

$$[R(1, L_p^0), R(\lambda + \mu, B_p^0)] = R(\lambda + \mu, B_p^0)R(1, L_p^0)[L_p^0, B_p^0]R(1, L_p^0)R(\lambda + \mu, B_p^0).$$

Observing that

$$\begin{aligned}
&(I - L_p^0)R(\nu + 1, L_p^0)R(\lambda + \mu, B_p^0)R(1, L_p^0)u \\
=&- (I - L_p^0)R(\nu + 1, L_p^0)[R(1, L_p^0), R(\lambda + \mu, B_p^0)]u \\
&+ (I - L_p^0)R(\nu + 1, L_p^0)R(1, L_p^0)R(\lambda + \mu, B_p^0)u \\
=&- E(\lambda, \mu, \nu)u + (I - L_p^0)R(1, L_p^0)R(\nu + 1, L_p^0)R(\lambda + \mu, B_p^0)u \\
=&- E(\lambda, \mu, \nu)u + R(\nu + 1, L_p^0)R(\lambda + \mu, B_p^0)u
\end{aligned}$$

for any $u \in C_b^4(\mathbb{R}^N)$, we finally get

$$E(\lambda, \mu, \nu)u = \{E(\lambda, \mu, \nu) - R(\nu + 1, L_p^0)R(\lambda + \mu, B_p^0)\}[B_p, L_p^0]R(1, L_p^0)R(\lambda + \mu, B_p^0)u \tag{10.3.42}$$

for any $u \in C_b^4(\mathbb{R}^N)$ and λ sufficiently large.

Observe that the equality in (10.3.42) can be extended to all the functions $u \in L^p_\mu$. To see it, it suffices to prove that the operator $[B_p^0, L_p^0]R(1, L_p) : C_b^4(\mathbb{R}^N) \to L^p_\mu$ extends to a bounded operator in L^p_μ, but this is immediate, since an explicit computation and (10.3.41) show that

$$([B_p, L_p^0]u)(x) = -\mathrm{Tr}(B^0 D^2 u(x)) + \frac{1}{2}\langle (B^0 + (B^0)^*)\widetilde{Q}_\infty^{-1}x, \nabla u(x)\rangle$$

for any $x \in \mathbb{R}^N$ and $u \in C_b^4(\mathbb{R}^N)$. Lemma 10.3.4 and Proposition 10.3.12 then allow us to conclude.

Now, we observe that, according to Proposition C.1.18(ii), we can determine two angles φ_1 and φ_2, with $\varphi_1 > \max\{\theta_{B_p^0}, \pi/2\}$, $\varphi_2 \in (\theta_{L_p^0}, \pi/2)$ and $\varphi_1 + \varphi_2 < \pi$, such that

$$\|R(\mu + \lambda, B_p^0)\|_{L(L^p_{\tilde{\mu}})} \le \frac{C}{|\mu + \lambda|}, \qquad \|R(\nu + 1, L_p^0)\|_{L(L^p_{\tilde{\mu}})} \le \frac{C}{1 + |\nu|} \tag{10.3.43}$$

for any μ, ν, with $|\arg \mu| < \pi - \varphi_1$, $|\arg \nu| < \pi - \varphi_2$, and some positive constant C, independent of μ,ν. Note that we can take $|\nu|+1$ instead of $|\nu|$ in the second estimate in (10.3.43) since the operator $\nu \mapsto R(\nu+1,L_p^0)$ is bounded in a neighborhood of 0.

From (10.3.42) and (10.3.43) we get

$$\|E(\lambda,\mu,\nu)\|_{L(L^p_{\tilde{\mu}})} \le \frac{C_1}{|\mu+\lambda|}\|E(\lambda,\mu,\nu)\|_{L(L^p_{\tilde{\mu}})} + \frac{C_2}{(1+|\nu|)|\mu+\lambda|^2}$$

for some positive constants C_1 and C_2, independent of λ,μ,ν. Taking $\lambda > \max\{\omega_0, 2C_1\}$, we deduce that

$$\|E(\lambda,\mu,\nu)\|_{L(L^p_{\tilde{\mu}})} \le \frac{2C_2}{(1+|\nu|)|\mu+\lambda|^2} \le \frac{2C_2}{(1+|\nu|)|\mu|^2}$$

for any $\mu,\nu \in \mathbb{C}\setminus\{0\}$ such that $|\arg\mu| < \pi-\varphi_1$ and $|\arg\nu| < \pi-\varphi_2$. Theorem C.1.21 allows us to conclude that the operator $(\lambda I - B_p^0)+(I-L_p^0)$ (or, equivalently, the operator $L_p^0+B_p^0$) is closed in $W^{2,p}_{\tilde{\mu}}$. This concludes the proof. □

Remark 10.3.16 In the case when $p=1$, from Lemma 10.3.11 it follows easily that $D(L_1)$ is the closure of $C_c^\infty(\mathbb{R}^N)$ with respect to the graph norm of the operator L_1. On the other hand, the set equality $D(L_1)=W^{2,1}_\mu$ can not hold, since, in particular, $D(L_1)$ is not continuously embedded in $W^{1,1}_\mu$. Indeed, suppose on the contrary that $D(L_1)\hookrightarrow W^{1,1}_\mu$. Then, the operator $R(1,L_1): W^{1,1}_\mu \to W^{1,1}_\mu$ should be compact. Therefore the spectrum of the part of L_1 in $W^{1,1}_\mu$ (i.e., the restriction of the operator L_1 to the set $\{u\in D(L_1): L_1u \in W^{1,1}_\mu\}$) should be a discrete set. But this can not be the case, since, as is easily seen, it coincides with $\sigma(L_1)$, which, according to the forthcoming Theorem 10.3.22, is a halfplane.

10.3.2 The spectrum of L_p

In this subsection we explicitly characterize the spectrum of the operator L_p for any $p\in[1,+\infty)$. The results that we present here are taken from [192] and they also hold for degenerate Ornstein-Uhlenbeck operators.

First of all, we show that the operator $R(\lambda,L_p)$ is compact in L^p_μ for any $p\in(1,+\infty)$.

Theorem 10.3.17 *For any $p\in(1,+\infty)$, $W^{1,p}_\mu$ is compactly embedded in L^p_μ. As a consequence, for any p as above, $T(t)$ and $R(\lambda,L_p)$ are compact operators in L^p_μ for any $t>0$ and $\lambda\in\rho(L_p)$, respectively.*

Proof Of course, it suffices to show that $W^{1,p}_\mu$ is compactly embedded in L^p_μ for any $p\in(1,+\infty)$. Indeed, since $T(t)$ maps L^p_μ into $W^{1,p}_\mu$ for any $t>0$ (see Proposition 10.3.8), the compactness of $T(t)$ then will easily follow. Finally, since $R(\lambda,L_p)$ is the Laplace transform of the semigroup for any λ with positive real part (see the formula (C.1.2)), taking Proposition B.1.1 and the resolvent formula (B.1.1) into account, we will deduce that $R(\lambda,L_p)$ is compact as well.

To prove that $W^{1,p}_\mu$ is compactly embedded in L^p_μ, we adapt the technique in the proof of Proposition 9.5.8. For this purpose, it suffices to show that

$$\lim_{r\to+\infty}\int_{\mathbb{R}^N\setminus B_r}|f|^p d\mu = 0, \tag{10.3.44}$$

uniformly with respect to $f\in W^{1,p}_\mu$ with $\|f\|_{W^{1,p}_\mu}\le 1$. Taking Lemma 10.3.4 into account, we can write

$$\int_{\mathbb{R}^N\setminus B_r}|f|^p d\mu \le \frac{1}{r^p}\int_{\mathbb{R}^N}|xf(x)|^p d\mu \le \frac{C}{r^p}\|f\|_{W^{1,p}_\mu}$$

for any $f \in W^{1,p}_\mu$ and some positive constant C, independent of f, and (10.3.44) follows. □

As a consequence of Theorem 10.3.17, the spectrum of L_p consists of at most a sequence of eigenvalues for any $p \in (1,+\infty)$. In the remainder of this subsection, we characterize explicitly $\sigma(L_p)$. To begin with, we observe that 0 is an eigenvalue of the operator L_p for any $p \in (1,+\infty)$ and, according to the Liouville type Theorem 9.1.17 and Proposition 10.3.8, the corresponding eigenspace is one-dimensional and consists of all the constant functions. As the following proposition shows, if u is a generalized eigenfunction corresponding to the eigenvalue $\lambda \in \mathbb{C}$, then u is a polynomial. Of course, if u is a polynomial, then it belongs to $W^{2,p}_\mu = D(L_p)$ for any $p \in (1,+\infty)$ (see Theorem 10.3.15).

Proposition 10.3.18 *Fix $p \in (1,+\infty)$ and let $u \in D(L_p^r)$ satisfy $(\lambda - L_p)^r u = 0$ for some $\lambda \in \mathbb{C}$ and some $r \in \mathbb{N}$. Then, $\mathrm{Re}\,\lambda \le 0$ and u is a polynomial with degree at most equal to $\mathrm{Re}\,\lambda/s(B)$.*

Proof We prove the assertion by induction on r. Suppose that $r = 1$ and let $\lambda \in \mathbb{C}$ and $u \in D(L_p)$ be such that $\lambda u - L_p u = 0$. Then, $T(t)u = e^{\lambda t}u$ for any $t \ge 0$ and, from Proposition 10.3.8, we get $u \in W^{k,p}_\mu$ for any $k \in \mathbb{N}$. From the formula (10.3.17) we deduce that, for any $\varepsilon > 0$ and any multi-index α, there exists a positive constant $C = C(\varepsilon, |\alpha|)$ such that $e^{\mathrm{Re}\lambda\, t}\|D^\alpha u\|_p \le Ce^{|\alpha|(s(B)+\varepsilon)t}\sum_{|\beta|=|\alpha|}\|D^\beta u\|_p$ for any $t > 0$. From such an inequality we obtain that, if $\mathrm{Re}\lambda > |\alpha| s(B)$, then C should be zero. Therefore, the derivatives of u of order $k > \mathrm{Re}\lambda/s(B)$ should vanish and the assertion follows in this particular case.

Now, suppose that the assertion is true with $r = n > 1$ and let us prove it with $r = n+1$. For this purpose we observe that, if $u \in D(L_p^{n+1})$ and $(\lambda u - L_p u)^{n+1} = 0$, then the function $v = \lambda u - L_p u$ belongs to $D(L_p^n)$ and satisfies the equation $(\lambda I - L_p)^r v = 0$. Therefore, v is a polynomial of degree at most $\mathrm{Re}\,\lambda/s(B)$. Moreover, since $(\lambda I - L_p)^s v = 0$ for any $s \ge n+1$, it follows that

$$T(t)u = e^{\lambda t}\sum_{i=0}^{n}\frac{t^i}{i!}(\lambda I - L_p)^i u = e^{\lambda t}u + e^{\lambda t}\sum_{i=0}^{n-1}\frac{t^{i+1}}{(i+1)!}(\lambda I - L_p)^i v \qquad (10.3.45)$$

for any $t > 0$. From the formula (10.3.45), we obtain that

$$D^\alpha_x T(t)u = e^{\lambda t}D^\alpha u + e^{\alpha t}\sum_{i=0}^{n-1}\frac{t^{i+1}}{(i+1)!}D^\alpha(\lambda v - L_p v)^i, \qquad t > 0,$$

for any multi-index α. Since v is a polynomial with degree at most $\mathrm{Re}\lambda/s(B)$, for any $t > 0$ and any multi-index α with $|\alpha| > \mathrm{Re}\,\lambda/s(B)$ it holds that $D^\alpha_x T(t)u = e^{\lambda t}D^\alpha u$. Now the proof goes on as in the case when $r = 1$. □

We now show that the problem of determining the eigenvalues of the operator L_p can be stated in an equivalent (and simpler, as far as computations are concerned) way. For this purpose, we introduce the operator $\mathcal{B}_p : W^{2,p}_\mu \to L^p_\mu$ defined by $\mathcal{B}_p u(x) = \langle Bx, \nabla u(x)\rangle$ for almost any $x \in \mathbb{R}^N$ and any $u \in W^{2,p}_\mu$. According to Lemma 10.3.4, $\mathcal{B}_p$ is well defined and bounded in $W^{2,p}_\mu$.

Proposition 10.3.19 *Let $p \in (1,+\infty)$. The following statements are equivalent:*

(i) *$\lambda \in \mathbb{C}$ is an eigenvalue of the operator L_p;*

(ii) *λ is an eigenvalue of the operator $\mathcal{B}_p$ and the corresponding eigenspace contains a nonzero homogeneous polynomial.*

Proof Let $\lambda \in \mathbb{C}$ be an eigenvalue of the operator L_p. Further, let u be a corresponding eigenfunction so that $\lambda u - \mathcal{B}_p u = \mathrm{Tr}(QD^2u)$. By Proposition 10.3.18, u is a polynomial. We denote by m its degree. If $m = 0, 1$ then, clearly, λ is an eigenvalue of $\mathcal{B}_p$. So, let us assume that $m \geq 2$. Note that for any $n \in \mathbb{N}$ both $\mathcal{B}_p$ and L_p map the set $\mathcal{P}_n$, of all the polynomials with degree at most n, into itself. Suppose that $\lambda I - \mathcal{B}_p$ is invertible in $\mathcal{P}_{m-2}$. Then, there exists a polynomial v with degree $m-2$ such that $\lambda v - \mathcal{B}_p v = \mathrm{Tr}(QD^2u)$. The function $w = u - v$, which is a nontrivial polynomial with degree m, turns out to be an eigenfunction of the operator L_p corresponding to the eigenvalue λ. If the operator $\lambda I - \mathcal{B}_p$ is not invertible in $\mathcal{P}_{m-2}$, then, of course, λ is an eigenvalue of $\mathcal{B}_p$ and we can take as a corresponding eigenfunction a polynomial w with degree $m-2$.

To complete the proof of property (ii) we have to check that there exists a homogeneous polynomial among the eigenfunctions of the operator $\mathcal{B}_p$ corresponding to the eigenvalue λ. For this purpose, it suffices to notice that, for any $k \in \mathbb{N}$, $\mathcal{B}_p$ maps the space of homogeneous polynomials with degree k into itself. Therefore, the homogeneous part with maximum degree of the eigenfunction w is itself an eigenfunction.

Let us now assume that property (ii) holds and let us prove that property (i) holds as well. For this purpose, we assume that $\lambda \in \mathbb{C}$ is an eigenvalue of the operator $\mathcal{B}_p$ and that the homogeneous polynomial u is a corresponding eigenfunction. Of course, $\lambda u - L_p u = -\mathrm{Tr}(QD^2u)$. Repeating the same arguments used above, now applied to the operator L_p, we can easily show that λ is an eigenvalue of L_p. This concludes the proof. □

By virtue of Proposition 10.3.19, to characterize the spectrum of the operator A_p we can limit ourselves to characterize the spectrum of the operator $\mathcal{B}_p$, which is easier to handle.

Theorem 10.3.20 *Denote by $\lambda_1, \ldots, \lambda_r$ the r distinct eigenvalues of the matrix B. Then, for any $p \in (1, +\infty)$,*

$$\sigma(L_p) = \left\{\lambda = \sum_{i=1}^{r} n_i \lambda_i : n_i \in \mathbb{N} \cup \{0\},\ i = 1, \ldots, r\right\}. \tag{10.3.46}$$

In particular, $\sigma(L_p)$ is independent of p.

Proof A straightforward computation shows that $v : \mathbb{R}^N \to \mathbb{C}$ is an eigenfunction of the operator $\mathcal{B}_p$, corresponding to the eigenvalue $\lambda \in \mathbb{C}$, if and only if

$$v(e^{tB}x) = e^{\lambda t} v(x), \qquad t > 0,\ x \in \mathbb{R}^N. \tag{10.3.47}$$

According to Proposition 10.3.19, we can assume that v is a homogeneous polynomial. Therefore, the equality (10.3.47) can be extended to all $z \in \mathbb{C}$.

Denote by $\lambda_1, \ldots, \lambda_r$ the distinct eigenvalues of the matrix B. By a general result of linear algebra (see [162, Chapter 11, Section 6]), there exists a nonsingular $N \times N$ complex-valued matrix M such that $B = M^{-1}CM$, with $C = \mathrm{diag}(C_1, \ldots, C_r)$, where each block C_i is a Jordan block, which can be split into the sum $C_i = \lambda_i I_{n_i} + N_i$, n_i being the algebraic multiplicity of the eigenvalue λ_i and N_{λ_i} being a nihilpotent matrix. It follows that $e^{tB} = M^{-1}e^{tC}M$ and, consequently, setting $w(z) = v(M^{-1}z)$ for any $z \in \mathbb{C}^N$, we conclude that v satisfies (10.3.47) if and only if $w(e^{tC}z) = e^{\lambda t}w(z)$ for any $t > 0$ and $z \in \mathbb{C}^N$.

Now, we can prove the inclusion "$\subset$" in the formula (10.3.46). For this purpose, fix $\lambda \in \sigma(L_p)$ and let v be a corresponding eigenfunction. Since we have assumed that v is a homogeneous polynomial, w is a homogeneous polynomial as well, and we can write

$$w(z) = \sum_{|\alpha_1|+\ldots+|\alpha_n|=n} a_{\alpha_1,\ldots,\alpha_r} \prod_{j=1}^{r} z_j^{\alpha_j}, \qquad z \in \mathbb{C}^N,$$

for some $n \in \mathbb{N}$ and some $a_{\alpha_1,\ldots,\alpha_r} \in \mathbb{C}$ ($|\alpha_1| + \ldots + |\alpha_n| = n$), where $z = (z_1, \ldots, z_r)$ and $z_i \in \mathbb{C}^{n_i}$. Since $e^{tC}z = (e^{tC_1}z_1, \ldots, e^{tC_r}z_r)$ and N_i is nihilpotent, it holds that $e^{tC_i}y = e^{\lambda t}(P_{1,i}(t,y), \ldots, P_{r_i,i}(t,y))$ for any $t > 0$, $y \in \mathbb{C}^{n_i}$ and $i = 1, \ldots, r$, where $P_{k,i}$ ($i = 1, \ldots, r$) is a polynomial in both its entries. It follows that

$$e^{\lambda t}w(z) = w(e^{tC}z) = \sum_{|\alpha_1|+\ldots+|\alpha_n|=n} a_{\alpha_1,\ldots,\alpha_r} e^{(\lambda_1|\alpha_1|+\ldots+\lambda_n|\alpha_n|)t} P_{\alpha_1,\ldots,\alpha_r}(t,z)$$

for any $t > 0$ and $z \in \mathbb{C}^N$. Here, $P_{\alpha_1,\ldots,\alpha_r}$ is a polynomial. Fixing $z^* \in \mathbb{C}^N$ such that $w(z^*) \neq 0$ and letting t tend first to $+\infty$ and then to $-\infty$ in the previous formula, we easily see that $\lambda = \lambda_1|\alpha_1| + \ldots + \lambda_n|\alpha_n|$ for some $\alpha_1, \ldots, \alpha_n$, and the inclusion "$\subset$" in (10.3.46) follows.

We now show the other inclusion in (10.3.46). For this purpose, let $\lambda = \sum_{i=1}^r n_i\lambda_i$ for some $n_i \in \mathbb{N}$ ($i = 1, \ldots, r$) and let w be the polynomial, defined by $w(z) = \widehat{z}_1^{n_1} \cdot \widehat{z}_2^{n_2} \cdot \ldots \cdot \widehat{z}_r^{n_r}$ for any $z \in \mathbb{C}^N$, where $\widehat{z}_i$ is the last component of the vector z_i ($i = 1, \ldots, r$). Since the last row of the matrix N_i is 0 for any $i = 1, \ldots, r$, the last entry of the vector $N_i^k h$ is 0 for any $h \in \mathbb{C}^{n_i}$, $k \in \mathbb{N}$ and $i = 1, \ldots, r$. It follows that the last entry of the vector $e^{tC_i}z_i$ is $e^{\lambda t}\widehat{z}_i$ for any $i = 1, \ldots, r$ and, consequently, $w(e^{tC}z) = e^{\lambda t}w(z)$ for any $t > 0$ and $z \in \mathbb{C}^N$. Hence, λ is an eigenvalue of $\mathcal{B}_p$ and the (homogeneous) polynomial v, defined by $v(x) = w(Mx)$ for any $x \in \mathbb{R}^N$, is a corresponding eigenfunction. Proposition 10.3.19 allows us to conclude that $\lambda \in \sigma(L_p)$. □

Some more information is available on the eigenspaces corresponding to the eigenvalues of the operator L_p. In the following theorem we give a necessary and sufficient condition which guarantees that the eigenvalues of L_p are semisimple. For notational convenience, for any linear operator S with compact resolvent operator, we denote by $i_\lambda(S)$ the index of the eigenvalue λ, (i.e., the smallest integer k such that $\mathrm{Ker}\,(\lambda I - S)^k = \mathrm{Ker}\,(\lambda I - S)^{k+1}$). According to Proposition B.1.8, $i_\lambda(S)$ is well defined for any $\lambda \in \sigma(S)$ and $R(\cdot, S)$ has a pole of order $i_\lambda(S)$ at λ.

Theorem 10.3.21 *The index $i_\lambda(L_p)$ is one for any eigenvalue λ of L_p (equivalently, all the eigenvalues of L_p are semisimple) if and only if the matrix B is diagonalizable in $\mathbb{C}$.*

Proof To show the assertion, we prove that for any eigenvalue λ of the operator A_p, it holds that

$$1 + \max\left\{\sum_{i=1}^r n_i(i_{\lambda_i}(B) - 1) : \sum_{i=1}^r n_i\lambda_i = \lambda\right\} \le i_\lambda(L_p) \le 1 + \frac{\mathrm{Re}\lambda}{s(B)}(i(B) - 1), \qquad (10.3.48)$$

where $i(B)$ denotes the maximum of the indices $i_{\lambda_i}(B)$ ($i = 1, \ldots, r$).

To prove (10.3.48) we proceed in steps. As in the proof of Theorem 10.3.20, we split any $z \in \mathbb{C}^N$ into blocks as $z = (z_1, \ldots, z_r)$, where $z_i \in \mathbb{C}^{n_i}$ and n_i is the algebraic multiplicity of the eigenvalue λ_i of the matrix B. Moreover, we recall (see again the proof of Theorem 10.3.20) that we can determine a nonsingular matrix M such that $B = M^{-1}CM$, where $C = \mathrm{diag}\,(C_1, \ldots, C_r)$ and each block C_i is a Jordan block, which can be split into the sum $C_i = \lambda_i I_{n_i} + N_i$ for some nihilpotent matrix N_i. By well known results of linear algebra, $i_{\lambda_i}(B)$ coincides with the algebraic multiplicity of the eigenvalue λ, i.e., it coincides with n_i. As a consequence

$$e^{tC_i} = e^{t\lambda_i}p_i(t), \qquad t \in \mathbb{R}, \quad i = 1, \ldots, N, \qquad (10.3.49)$$

where p_i are suitable matrices whose entries are polynomials with degree at most $i_{\lambda_i}(B) - 1$.

Moreover, there exists $h \in \mathbb{R}^{n_i}$ such that at least one component of $p_i(t)x$ is a polynomial with degree $i_{\lambda_i}(B)-1$.

Step 1. Here, we prove that $i_\lambda(\mathcal{B}_p) \le i_\lambda(L_p)$. For this purpose, we show that, for any $\mu \in \rho(L_p)$, it holds that

$$R(\mu,\mathcal{B}_p)u = \sum_{i=1}^{+\infty} P_i R(\mu, L_p) P_i u \tag{10.3.50}$$

for any polynomial u. Here, P_i ($i \in \mathbb{N}\cup\{0\}$) is the canonical projection of the space of polynomials into the space $\mathcal{H}_i$ of the homogeneous polynomials with degree i. Once the formula (10.3.50) is checked, we can quite easily show that $R(\cdot,L_p)$ has a pole at λ, whose order is not less than the order of the pole of $R(\cdot,\mathcal{B}_p)$ at λ. Indeed, using such a formula we deduce that, for any polynomial u, the function $\mu \mapsto (\mu-\lambda)^{i_\lambda(L_p)}R(\mu,\mathcal{B}_p)u$ can be analytically extended to $B_r(\lambda)\subset\mathbb{C}$ for some $r>0$, independent of u. By density, the function $\mu\mapsto(\mu-\lambda)^{i_\lambda(L_p)}R(\mu,\mathcal{B}_p)u$ is analytic in $B_r(\lambda)$ for any $u\in L^p_\mu$, and the Banach-Steinhaus theorem (see [260, Chapter 2, Section 4]) implies that $\|(\mu-\lambda)^{i_\lambda(L_p)}R(\mu,\mathcal{B}_p)\|_{L(L^p_\mu)} \le C$ for any $\mu\in B_r(\lambda)$ and some positive constant C, independent of λ. Of course, this implies that $R(\mu,\mathcal{B}_p)u$ has a pole of order $m\le i_\lambda(L_p)$ at λ. Therefore, λ belongs to $\sigma(\mathcal{B}_p)$. According to the formula (B.1.2), $(\lambda I-\mathcal{B}_p)^m P=0$ and $(\lambda I-\mathcal{B}_p)^{m-1}P\neq 0$, where P is the spectral projection associated with λ, i.e., $P=\frac{1}{2\pi i}\int_\gamma R(\xi,\mathcal{B}_p)d\xi$, γ being the boundary of a suitable ball centered at λ and oriented counterclockwise (see Proposition B.1.5). Therefore, $P(L^p_\mu)$ is contained in $\mathcal{E}_\lambda$, where by $\mathcal{E}_\lambda$ we have denoted the space of generalized eigenfunctions of $\mathcal{B}_p$ corresponding to λ. Since the restriction of $\lambda I-\mathcal{B}_p$ to $(I-P)(L^p_\mu)$ is an invertible operator (see again Proposition B.1.5), it follows that $P(L^p_\mu)=\mathcal{E}_\lambda$. Now, it is easy to check that $i_\lambda(\mathcal{B}_p)$ coincides with m.

By linearity it suffices to check (10.3.50) in the case when $u\in\mathcal{H}_i$ ($i\in\mathbb{N}$). Set $v=R(\mu,L_p)u$. Since $\mathcal{B}_p$ maps $\mathcal{H}_n$ into itself, we easily get $u=P_iu=P_i(\mu v-L_pv)=P_i(\mu v-\mathcal{B}_pv)=(\mu I-\mathcal{B}_p)P_iv$. Formula (10.3.50) now follows.

Step 2. Now, we prove that

$$i_\lambda(\mathcal{B}_p)=1+\max\left\{\sum_{i=1}^r n_i(i_{\lambda_i}(B)-1):\sum_{i=1}^r n_i\lambda_i=\lambda\right\}. \tag{10.3.51}$$

Let v be a generalized eigenfunction of $\mathcal{B}_p$ corresponding to the eigenvalue $\lambda=m_1\lambda_1+\ldots+m_r\lambda_r$ such that $(\lambda I-\mathcal{B}_p)^{i_\lambda(\mathcal{B}_p)-1}v\neq 0$ and $(\lambda I-B_p)^{i_\lambda(\mathcal{B}_p)}v=0$. Let us show that v is a linear combination of polynomials $p_i\in\mathcal{H}_{m_i}$, $i=1,\ldots,r$, such that each p_i depends only on the block of variables z_i. For this purpose, we observe that $v(e^{tB}x)=(S(t)v)(x)$ for any $t\ge 0$ and $x\in\mathbb{R}^N$, where $\{S(t)\}$ is the strongly continuous group generated by the closure of the operator $\mathcal{B}_p$ in L^p_μ.[3] Since v is a generalized eigenfunction of $\mathcal{B}_p$, v is a polynomial. This can easily be checked by arguing as in the proof of Proposition 10.3.18, observing that $\|D^\alpha_x S(t)f\|_p\le Ce^{|\alpha|(s(B)+\varepsilon)t}\sum_{|\beta|=|\alpha|}\|D^\alpha f\|_p$ for any $t>0$, any multi-index α, any $\varepsilon>0$, any $f\in W^{k,p}_\mu$ and some positive constant $C=C(|\alpha|,\varepsilon)$. Moreover, by arguing as in the proof of (10.3.45), we can show that

$$(S(t)v)(x)=v(e^{tB}x)=e^{t\lambda}p(t,x),\qquad t>0,\ x\in\mathbb{R}^N, \tag{10.3.52}$$

where p is a polynomial in (t,x) whose degree in t does not exceed $i_B(\lambda)-1$. Since v is a polynomial, we can extend (10.3.52) to any $z\in\mathbb{C}^N$. We can also determine some constants $c_{\alpha_1,\ldots,\alpha_r}$ such that

$$v(z)=\sum_{|\alpha_1|+\ldots+|\alpha_r|\le n} c_{\alpha_1,\ldots,\alpha_r}\prod_{i=1}^r z_i^{\alpha_i},\qquad z\in\mathbb{C}^N,$$

[3] The strong continuity of $\{S(t)\}$ can be proved by arguing as in the proof of Lemma 10.3.14.

and, hence,

$$v(e^{tB}z) = \sum_{|\alpha_1|+\ldots+|\alpha_r|\le n} c_{\alpha_1,\ldots\alpha_r} e^{t(\lambda_1|\alpha_1|+\ldots+\lambda_r|\alpha_r|)} p_{\alpha_1,\ldots,\alpha_r}(t,z) \tag{10.3.53}$$

for any $t>0$ and $z\in\mathbb{C}^N$, where $p_{\alpha_1,\ldots,\alpha_r}$ are polynomials in (t,x) with degree at most $\sum_{i=1}^r |\alpha_i|(i_{\lambda_i}(B)-1)$ in t. Comparing (10.3.52) and (10.3.53) for large t, we easily deduce that $c_{\alpha_1,\ldots,\alpha_r}=0$, if $|\alpha_1|\lambda_1+\ldots+|\alpha_r|\lambda_r\neq\lambda$. It also follows that each polynomial $p_{\alpha_1,\ldots,\alpha_r}$ depends only on the block of variables x_i $(i=1,\ldots,r)$ for which $\alpha_i\neq 0$.

From (10.3.52) and (10.3.53) we now get that

$$p(t,z) = q(t,z) := \sum_{|\alpha_1|\lambda_1+\ldots+|\alpha_r|\lambda_r=\lambda} c_{\alpha_1,\ldots,\alpha_r} p_{\alpha_1,\ldots,\alpha_r}(t,z)$$

for any $t>0$ and $z\in\mathbb{C}^N$. Moreover, from the arguments before Step 1, we deduce that there exists $z\in\mathbb{C}^N$ such that the degree of $q(\cdot,z)$ is $\max\{\sum_{i=1}^r n_i(i_{\lambda_i}(B)-1):\sum_{i=1}^r n_i\lambda_i=\lambda\}$ and the degree of $p(\cdot,z)$ is $i_\lambda(\mathcal{B}_p)-1$. Formula (10.3.51) easily follows.

Step 3. Now we show that

$$i_\lambda(L_p) \le 1 + \frac{\mathrm{Re}\lambda}{s(B)}(i(B)-1). \tag{10.3.54}$$

For this purpose, let $u=\sum_{|\alpha|\le m} c_\alpha x^\alpha$ be a polynomial (with degree $m\le \mathrm{Re}\,\lambda/s(B)$, see Proposition 10.3.18) such that $(\lambda I - L_p)^{i_\lambda(L_p)-1}u\neq 0$ and $(\lambda I-L_p)^{i_\lambda(L_p)}u=0$. From the formula (10.3.45) with $n=i_\lambda(L_p)-1$, we deduce that $(T(t)u)(x)=e^{\lambda t}p(t,x)$ for any $t>0$ and $x\in\mathbb{R}^N$, and there exists $x_0\in\mathbb{R}^N$ such that the degree of $p(\cdot,x_0)$ is $i_\lambda(L_p)-1$.

A long but straightforward computation shows that

$$\begin{aligned}(T(t)u)(x) &= \sum_{|\alpha|\le m} c_\alpha \sum_{h\le\alpha}\binom{\alpha}{h}(e^{tB}x)^h \frac{1}{\sqrt{(2\pi)^N\det Q_t}}\int_{\mathbb{R}^N} y^{\alpha-h}e^{-\frac{1}{2}\langle Q_t^{-1}y,y\rangle}dy\\ &= \sum_{|\alpha|\le m} c_\alpha \sum_{\substack{h\le\alpha\\ |\alpha-h|\ \mathrm{even}}} d_{\alpha,h}\binom{\alpha}{h}(e^{tB}x)^h D_w^{\alpha-h}((Q_tw,w)^{\frac{|\alpha-h|}{2}})_{|w=0}\end{aligned}$$

for any $t>0$ and $x\in\mathbb{R}^N$, where $d_{\alpha,h}^{-1}=(2|\alpha-h|^{-1})!$. Taking (10.3.49) into account, it follows that $(Q_t)_{ij}=\sum_{h=1}^r e^{\lambda_h t}p_{ij}^{(h)}(t)+c_{ij}^{(h)}$ for any $t>0$ and $i,j=1,\ldots,N$, where $p_{ij}^{(h)}$ are suitable polynomials with degree at most $2i(B)-2$, and $c_{ij}^{(h)}$ are complex constants. Thus, $(D^{\alpha-h}((Q_tw,w)^{|\alpha-h|/2}))_{|w=0}=\sum_{i=1}^{l_1}e^{\tau_i t}q_i(t)$ for any $t>0$ and some $l_1\ge m$, where τ_i are integer multiples of $\lambda_1,\ldots,\lambda_r$ and q_i are polynomials with degree at most $(i(B)-1)(|\alpha|-|h|)$. Summing up, $(T(t)u)(x)=\sum_{i=1}^{l_2}e^{\xi_i t}q_i(t,x)$ for any $t>0$, $x\in\mathbb{R}^N$ and some $l_2\in\mathbb{N}$, where ξ_i $(i=1,\ldots,r)$ are again integer multiples of $\lambda_1,\ldots,\lambda_r$, and q_i, as functions of t, are polynomials with degree at most $m(i(B)-1)$. Now, estimate (10.3.54) easily follows. This completes the proof. □

Now, we consider the case when $p=1$. In such a situation the picture drastically changes. Indeed, the spectrum of L_1 is no more a discrete set but it is a halfplane. The proof of this property is based on the results of the forthcoming Section 10.4, where the spectrum of the Ornstein-Uhlenbeck operator in L^p-spaces associated with the Lebesgue measure is characterized.

Theorem 10.3.22 *The spectrum of the Ornstein-Uhlenbeck operator in L^1_μ is the halfplane $\{\lambda\in\mathbb{C}:\mathrm{Re}\,\lambda\le 0\}$. Moreover, any $\lambda\in\mathbb{C}$ with negative real part is an eigenvalue of L_1.*

Proof Set $\psi = 1/\rho$, where ρ is the density of the invariant measure μ with respect to the Lebesgue measure (see (10.3.2)). Denote by $I : L^1(\mathbb{R}^N) \to L^1_\mu$ the isometry defined by $If = f\psi$ for any $f \in L^1(\mathbb{R}^N)$, and by $\widetilde{L}_1$ the operator defined on $I^{-1}(D(L_1))$ by $\widetilde{L}_1 = I^{-1}L_1I$. As is immediately seen, $\widetilde{L}_1$ is the generator of a strongly continuous semigroup of contractions in $L^1(\mathbb{R}^N)$ and $\sigma(L_1) = \sigma(\widetilde{L}_1)$. Therefore, it suffices to characterize the spectrum of the operator $\widetilde{L}_1$ in $L^1(\mathbb{R}^N)$. For this purpose, we observe that, since $C^\infty_c(\mathbb{R}^N) \subset D(L_1)$ and I maps $C^\infty_c(\mathbb{R}^N)$ into itself, $C^\infty_c(\mathbb{R}^N)$ is contained in $D(\widetilde{L}_1)$. A straightforward computation shows that $\widetilde{L}_1 = \widetilde{\mathcal{A}}$ in $C^\infty_c(\mathbb{R}^N)$, where

$$\begin{aligned}\widetilde{\mathcal{A}}u(x) =&\mathcal{A}u(x) + \rho(x)\langle Q\nabla\psi(x), \nabla u(x)\rangle + \rho(x)u(x)\mathcal{A}\psi(x)\\ =&\frac{1}{2}\mathrm{Tr}\,(QD^2u(x)) + \langle (B + QQ_\infty^{-1})x, \nabla u(x)\rangle + (\rho\mathcal{A}\psi)(x)u(x)\end{aligned}$$

for any $x \in \mathbb{R}^N$ and $u \in C^\infty_c(\mathbb{R}^N)$. Now, we observe that differentiating the formula (10.3.8) at $t = 0$ gives

$$Q + BQ_\infty + Q_\infty B^* = 0. \tag{10.3.55}$$

Using (10.3.55) we get $2\langle Q_\infty^{-1}Bx, x\rangle = 2\langle BQ_\infty Q_\infty^{-1}x, Q_\infty^{-1}x\rangle = -\langle QQ_\infty^{-1}x, Q_\infty^{-1}x\rangle$ for any $x \in \mathbb{R}^N$ and, therefore, $\rho\mathcal{A}\psi = \frac{1}{2}\mathrm{Tr}\,(QQ_\infty^{-1}) =: k$. Using again the formula (10.3.55) to obtain

$$B + QQ_\infty^{-1} = -Q_\infty B^* Q_\infty^{-1}, \tag{10.3.56}$$

we finally conclude that

$$\widetilde{\mathcal{A}}u(x) = \frac{1}{2}\mathrm{Tr}\,(QD^2u(x)) - \langle Q_\infty B^* Q_\infty^{-1}x, \nabla u(x)\rangle + ku(x), \qquad x \in \mathbb{R}^N.$$

Since $\widehat{\mathcal{A}} := \widetilde{\mathcal{A}} - kI$ is an Ornstein-Uhlenbeck operator, according to the forthcoming Propositions 10.4.1 and 10.4.2, the realization $\widehat{L}_1$ of $\widehat{\mathcal{A}}$ in $L^1(\mathbb{R}^N)$ is the generator of a strongly continuous semigroup of contractions and $C^\infty_c(\mathbb{R}^N)$ is a core of $\widehat{L}_1$. We claim that $D(\widetilde{L}_1) = D(\widehat{L}_1)$. Indeed, since $(\widetilde{L}_1, D(\widetilde{L}_1))$ is a closed operator and $C^\infty_c(\mathbb{R}^N)$ is a core of $\widehat{L}_1$, $D(\widehat{L}_1)$ is contained in $D(\widetilde{L}_1)$ and $\widehat{L} = \widetilde{L} - k$ in $D(\widehat{L}_1)$. To prove the other inclusion, we observe that $\lambda I - \widetilde{L}_1$ and $\lambda I - \widehat{L}_1$ are both invertible if λ is sufficiently large. Therefore,

$$L^1(\mathbb{R}^N) = (\lambda - \widetilde{L}_1)(D(\widetilde{L}_1)) \supset (\lambda - \widetilde{L}_1)(D(\widehat{L}_1)) = (\lambda I - \widehat{L}_1)(D(\widehat{L}_1)) = L^1(\mathbb{R}^N),$$

which implies that $D(\widetilde{L}_1) = D(\widehat{L}_1)$, so that $(\widetilde{L}_1, D(\widetilde{L}_1))$ and $(\widehat{L}_1 - kI, D(\widehat{L}_1))$ do coincide. It follows that $\sigma(\widetilde{L}_1) = \{\lambda \in \mathbb{C} : \lambda + k \in \sigma(\widehat{L}_1)\}$.

Now, since the spectrum of the matrix $-Q_\infty B^* Q_\infty^{-1}$ is contained in the halfplane $\{\lambda \in \mathbb{C} : \mathrm{Re}\lambda < 0\}$, according to the forthcoming Theorem 10.4.3 the spectrum of $\widehat{L}_1$ is the halfplane $\{\lambda \in \mathbb{C} : \mathrm{Re}\lambda \le \mathrm{Tr}\,(Q_\infty B^* Q_\infty^{-1})\}$. Moreover, any λ with $\mathrm{Re}\lambda < \mathrm{Tr}\,(Q_\infty B^* Q_\infty^{-1})$ is an eigenvalue of $\widehat{L}_1$.

To complete the proof it suffices to show that $k = -\mathrm{Tr}\,(Q_\infty B^* Q_\infty^{-1})$ and this can be easily done by taking the traces of both sides of (10.3.56). □

10.3.3 Hermite polynomials

In this subsection we show some relations connecting the Hermite polynomials and the Ornstein-Uhlenbeck operators. We partially follow the approach in [188].

For any multi-index α, the N-dimensional Hermite polynomial $H^{(N)}_\alpha$ is defined by $H^{(N)}_\alpha(x) = (-1)^{|\alpha|}e^{\frac{1}{2}|x|^2}D^\alpha e^{-\frac{1}{2}|x|^2}$ for any $x \in \mathbb{R}^N$. Equivalently,

$$H^{(N)}_\alpha(x) = \prod_{k=1}^N H^{(1)}_{\alpha_k}(x_k), \qquad x \in \mathbb{R}^N, \tag{10.3.57}$$

where $\alpha = (\alpha_1, \ldots, \alpha_N)$.

Throughout the first part of this subsection we denote by $\mathcal{A}$ the Ornstein-Uhlenbeck operator $\mathcal{A}$ defined in $W^{2,2}_\mu$ by

$$\mathcal{A}u(x) = \Delta u(x) - \langle x, \nabla u(x)\rangle, \qquad x \in \mathbb{R}^N, \ u \in W^{2,2}_\mu. \tag{10.3.58}$$

Here, $\mu(dx) = \mathcal{N}(0, I)(dx) = (2\pi)^{-N/2}e^{-\frac{1}{2}|x|^2}dx$.

Let us first consider the one-dimensional case and introduce the operators in $W^{1,2}_\mu$ defined by $\partial f(x) = f'(x)$ and $\partial^* f(x) = -f'(x) + xf(x)$ for any $x \in \mathbb{R}$ and $f \in W^{1,2}_\mu$. The notation is justified by the identity

$$\int_{\mathbb{R}} \partial f\, g d\mu = \int_{\mathbb{R}} f \partial^* g d\mu, \qquad f, g \in W^{1,2}_\mu,$$

which follows integrating by parts, if $f, g \in C^\infty_c(\mathbb{R})$, and by approximation with compactly supported functions for a general pair of functions $f, g \in W^{1,2}_\mu$ (see Lemmata 10.3.3 and 10.3.4).

With the notation now introduced, we can rewrite the Ornstein-Uhlenbeck operator $\mathcal{A}$ in the more compact form $\mathcal{A}u = -\partial^*\partial u$ for any $u \in D(L_2)$.

We summarize some basic properties of the Hermite polynomials. In particular, we show that they are an orthonormal basis of L^2_μ consisting of eigenfunctions of L_2.

Proposition 10.3.23 *For any $n \in \mathbb{N} \cup \{0\}$ the following properties are satisfied:*

(i) *the Hermite polynomials $\{H^{(1)}_n\}$ can be equivalently defined by recurrence by $H^{(1)}_0 = 1$, $H^{(1)}_n = \partial^* H^{(1)}_{n-1}$ for any $n \geq 1$;*

(ii) *$H^{(1)}_n$ is a polynomial with degree n with leading term x^n;*

(iii) *$\partial H^{(1)}_{n+1} = (n + 1)H^{(1)}_n$;*

(iv) *$\mathcal{A}H^{(1)}_n = -nH^{(1)}_n$;*

(v) *$\{H^{(1)}_n/\sqrt{n!}\}$ is an orthonormal basis of L^2_μ. Here, we set $0! = 1$.*

Proof Properties (i), (ii) and (iii) can be proved by induction. In particular, to prove property (iii) one can take advantage of the fact that $[\partial, \partial^*] = I$ in $W^{2,2}_\mu$.

Property (iv) is now a straightforward consequence of (i) and (iii). Indeed, it holds that $\mathcal{A}H^{(1)}_n = -\partial^*\partial H^{(1)}_n = -\partial^*(nH^{(1)}_{n-1}) = -nH^{(1)}_n$.

Let us now prove property (v). For this purpose we observe that each polynomial of order $m \in \mathbb{N}$ belongs to span $(H^{(1)}_0, \ldots, H^{(1)}_m)$. Therefore, to prove that $\{H^{(1)}_n/\sqrt{n!}\}$ is a basis of L^2_μ, it suffices to show that the set of all the polynomials are a dense subspace of L^2_μ. So, let us fix a function $f \in L^2_\mu$ such that

$$0 = \int_{\mathbb{R}} f(x)x^n d\mu = \frac{1}{\sqrt{2\pi}} \int_{\mathbb{R}} f(x)x^n e^{-\frac{1}{2}x^2} dx, \qquad n \in \mathbb{N} \cup \{0\}.$$

Then, the Fourier transform of the function $x \mapsto f(x)e^{-x^2/2}$ is an entire function, whose derivatives at 0 vanishes. Therefore, $f(x)e^{-x^2/2} = 0$ for any $x \in \mathbb{R}$ and, consequently, $f \equiv 0$.

To conclude the proof, let us prove that $\{H^{(1)}_n/\sqrt{n!}\}_{n\in\mathbb{N}\cup\{0\}}$ is an orthonormal system in L^2_μ. Fix $m, n \in \mathbb{N} \cup \{0\}$ with $m \leq n$ and denote by $(\cdot, \cdot)_{2,\mu}$ the Euclidean inner product in L^2_μ. Then, it holds that

$$(H^{(1)}_m, H^{(1)}_n)_{2,\mu} = (H^{(1)}_m, \partial^* H^{(1)}_{n-1})_{2,\mu} = (\partial H^{(1)}_m, H^{(1)}_{n-1})_{2,\mu} = -n(H^{(1)}_{m-1}, H^{(1)}_{n-1})_{2,\mu}$$

and, iterating the argument, we finally get

$$(H_m^{(1)},H_n^{(1)})_{2,\mu}=\frac{n!}{(n-m)!}(H_0^{(1)},H_{n-m}^{(1)})_{2,\mu}.$$

Thus, if $n=m$, then we get $\|H_n^{(1)}\|_2=\sqrt{n!}$, while, if $m<n$, then we get $(H_0^{(1)},H_{n-m}^{(1)})_{2,\mu}=(\partial H_0^{(1)},H_{n-m-1}^{(1)})_{2,\mu}=0$, and, consequently, $(H_m^{(1)},H_n^{(1)})_{2,\mu}=0$. Property (v) now follows. □

We now consider the multidimensional case $N>1$. According to the one-dimensional case, for any $k=1,\dots,N$ we define the operators ∂_k and ∂_k^* in $W_\mu^{1,2}$ by setting $\partial_k f(x)=D_k f(x)$ and $\partial_k^* f(x)=-D_k f(x)+x_k f(x)$ for any $x\in\mathbb{R}^N$. With this notation the Ornstein-Uhlenbeck operator in (10.3.58) can be rewritten in the compact form $\mathcal{A}=-\sum_{k=1}^N\partial_k^*\partial_k$.

We can now prove the N-dimensional version of Proposition 10.3.23.

Proposition 10.3.24 *The following properties are satisfied:*

(i) $H_{\alpha+e_k}^{(N)}=\partial_k^* H_\alpha^{(N)}$ *for any multi-index* α *and* $k\in\{1,\dots,N\}$;

(ii) *for any multi-index* α, $H_\alpha^{(N)}$ *is a polynomial with degree* $|\alpha|$, *whose leading term is given by* $x^\alpha=\prod_{i=1}^N x^{\alpha_i}$, *and it depends only on the variables* x_j *such that* $\alpha_j\neq 0$;

(iii) $\partial_k H_\alpha^{(N)}=\alpha_k H_{\alpha-e_k}^{(N)}$;

(iv) $\mathcal{A}H_\alpha^{(N)}=-|\alpha|H_\alpha^{(N)}$;

(v) $\{H_\alpha^{(N)}/\sqrt{\alpha!}\}_{\alpha\in(\mathbb{N}\cup\{0\})^N}$ *is an orthonormal basis of* L_μ^2.

Proof Properties (i)-(iii) are immediate consequences of the results in Proposition 10.3.23 and the formula (10.3.57), whereas property (iv) can be checked by adapting the proof of property (v) in the quoted proposition. □

Propositions 10.3.23 and 10.3.24 can be easily extended to the more general Ornstein-Uhlenbeck operator

$$\mathcal{A}u(x)=\frac{1}{2}\mathrm{Tr}\,(QD^2u(x))+\langle Bx,\nabla u(x)\rangle,\qquad x\in\mathbb{R}^N, \tag{10.3.59}$$

with $\sigma(B)\subset\{\lambda\in\mathbb{C}:\mathrm{Re}\,\lambda<0\}$.

Corollary 10.3.25 *Let* μ *be the invariant measure of the Ornstein-Uhlenbeck semigroup associated with the operator* $\mathcal{A}$ *in* (10.3.59) *and let* R *be a nonsingular matrix such that* $RQ_\infty R^*=\mathrm{diag}\,(\lambda_1,\dots,\lambda_N)$. *Then, the family of polynomials* $\{K_\alpha^{(N)}\}$, *defined by* $K_\alpha^N(x)=\widetilde{H}_\alpha^{(N)}(Rx)$, *where* $\widetilde{H}_\alpha^{(N)}(x)=\prod_{k=1}^N H_{\alpha_k}^{(1)}\left(\frac{x_k}{\sqrt{\lambda_k}}\right)$ *for any* $x\in\mathbb{R}^N$, *is a complete orthonormal set in* L_μ^2.

Proof Set $D=\mathrm{diag}\,(\sqrt{\lambda_1},\dots,\sqrt{\lambda_N})$, $M=D^{-1}R$. From the remarks before Lemma 10.3.13, it follows that the operator $\Phi_{M^{-1}}$, defined by $(\Phi_{M^{-1}}f)(x)=f(M^{-1}x)$ for any $x\in\mathbb{R}^N$ and any function $f:\mathbb{R}^N\to\mathbb{R}$, is a linear isometry between L_μ^p and $L_{\widetilde{\mu}}^p$, where $\widetilde{\mu}(dx)=(2\pi)^{-N/2}e^{-|x|^2/2}dx$. The assertion now follows from property (iv) in Proposition 10.3.24. □

10.3.4 The sector of analyticity of L_p

In this subsection we present a result due to R. Chill, E. Fašangová, G. Metafune, and D. Pallara ([58]) which gives a precise characterization of the sector of analyticity of the Ornstein-Uhlenbeck operator in L^p_μ ($p \in (1,+\infty)$), that is of the largest sector $\Sigma_\theta = \{\lambda \in \mathbb{C} : |\arg\lambda| < \theta\}$ where the function $z \mapsto T(z)$ is analytic with values in $L(L^p_\mu)$. Such a result has been extended to more general elliptic operators by the same authors in [57]. We refer the reader to [167, 168] for related results.

Theorem 10.3.26 ([58], Theorems 1, 2 and Remark 6) *For any $p \in (1,+\infty)$ define $\theta_p \in (0,\pi/2]$ by*

$$\cot\theta_p = \frac{\sqrt{(p-2)^2 + p^2\|I + 2Q^{-\frac{1}{2}}Q_\infty B^* Q^{-\frac{1}{2}}\|_\infty^2}}{2\sqrt{p-1}}. \tag{10.3.60}$$

Then, Σ_{θ_p} is the biggest sector, with its centre at the origin, where the Ornstein-Uhlenbeck semigroup can be analytically extended. Moreover, the so extended semigroup is contractive in L^p_μ. Finally, for $p = 2$,

$$\{z \in \mathbb{C} : \|Q_\infty^{\frac{1}{2}} e^{zB} Q_\infty^{-\frac{1}{2}}\|_\infty \le 1\} \tag{10.3.61}$$

is the set of analyticity of the function $z \mapsto T(z)$ and $\{T(z)\}$ is contractive in it.

Remark 10.3.27 (i) Formula (10.3.60) provides us with a relation between the sector of analyticity of the semigroup $\{T(t)\}$ and the sector (centered at 0) Σ_{θ_0} of analyticity of the resolvent operator of the matrix $R := Q^{-1/2}Q_\infty B^* Q^{-1/2}$. Indeed, $\pi - \theta_0 = \arctan(\|I + 2Q^{-\frac{1}{2}}Q_\infty B^* Q^{-\frac{1}{2}}\|_\infty)$. To check this formula we begin by observing that, due to the formula (10.3.55), $\frac{1}{2}I + Q^{-1/2}Q_\infty B^* Q^{-1/2}$ is a skew-symmetric matrix and, consequently, R is normal. Therefore, $\|I + 2R\|_\infty$ coincides with the spectral radius[4] $r(I + 2R)$ of $I + 2R$. Still using (10.3.55) it is easy to check that all the eigenvalues of the matrix R lie on the straight line $\{\lambda \in \mathbb{C} : \operatorname{Re}\lambda = -1/2\}$ and, therefore, $\arctan(r(I + 2R)) = \pi - \theta_0$.

(ii) Formula (10.3.60) also shows that, when the Ornstein-Uhlenbeck semigroup is symmetric[5], the sector of analyticity of $\{T(t)\}$ depends only on p and it equals $\pi/2$ when $p = 2$. Indeed, in such a case, the formula (10.3.55) gives $Q^{-1/2}Q_\infty B^* Q^{-1/2} = -\frac{1}{2}I$.

Note that the symmetry of the Ornstein-Uhlenbeck semigroup in L^2_μ is also a necessary condition for its sector of analyticity to be $\Sigma_{\pi/2}$. Indeed, if $\{T(t)\}$ is analytic in $\Sigma_{\pi/2}$, then $\cot\theta_p = 0$. But this implies that $I + 2Q^{-1/2}Q_\infty B^* Q^{-1/2} = 0$ and, by virtue of (10.3.55), it follows that $Q_\infty B^* = B^* Q_\infty$, so that $\{T(t)\}$ is symmetric in L^2_μ, by Proposition 10.3.10.

(iii) In general, the domain of analyticity of the functions $z \mapsto T(z)$ is larger than the sector Σ_{θ_p}. For instance, this is the case of the one-dimensional Ornstein-Uhlenbeck semigroup and $p \ne 2$ (see [96]). But also in the multidimensional case, the domain of analyticity of $\{T(z)\}$ is, in general, larger than Σ_{θ_p}. We can easily see it when $p = 2$. Indeed, consider the two dimensional Ornstein-Uhlenbeck operator $\mathcal{A}$, defined by

$$\mathcal{A}u(x,y) = \frac{1}{2}\Delta u(x,y) - (ax - by)D_x u(x,y) - ayD_y u(x,y)$$

[4] i.e., the radius of the smallest ball centered at the origin which contains all the eigenvalues of the matrix $I + 2R$.

[5] i.e., when $BQ = QB^*$ or, equivalently, when $Q_\infty B^* = BQ_\infty$, see the proof of Proposition 10.3.10.

for any $(x,y) \in \mathbb{R}^2$ and any smooth function $u : \mathbb{R}^N \to \mathbb{R}$, where a and b are real constants, with $a > 0$ and $b \neq 0$. A straightforward computation shows that the associated Ornstein-Uhlenbeck semigroup is not symmetric in L^2_μ. Therefore, θ_2 is strictly less than $\pi/2$. Moreover, since

$$e^{zB} = \begin{pmatrix} e^{-az} & bze^{-az} \\ 0 & e^{-az} \end{pmatrix}, \qquad z \in \mathbb{C},$$

then $\lim_{\substack{|z|\to+\infty \\ z\in\Sigma_\theta}} e^{zB^*} = 0$ for any $\theta < \pi/2$, so that there exist points $z \in \mathbb{C}$ which satisfy (10.3.61) and do not belong to Σ_{θ_2}.

10.4 The Ornstein-Uhlenbeck operator in $L^p(\mathbb{R}^N)$

To conclude the analysis of the Ornstein-Uhlenbeck operator, in this section we present some results concerning the realization of the Ornstein-Uhlenbeck operator $\mathcal{A}$ in $L^p(\mathbb{R}^N)$. Here, B is any nonzero matrix and Q is strictly positive definite. The results that we illustrate here are taken from [190, 199].

Proposition 10.4.1 *For any $p \in [1,+\infty)$ the restriction of the Ornstein-Uhlenbeck semigroup $\{T(t)\}$ to $C_c(\mathbb{R}^N)$ can be extended to a strongly continuous semigroup (which we still denote by $\{T(t)\}$) to $L^p(\mathbb{R}^N)$. For any $f \in L^p(\mathbb{R}^N)$ and any $t > 0$, $T(t)f$ is given by the right-hand side of* (10.1.5). *Finally, $\|T(t)\|_{L(L^p(\mathbb{R}^N))} \le e^{-\frac{t}{p}\mathrm{Tr}\,B}$ for any $t > 0$.*

Proof Since $C_c^\infty(\mathbb{R}^N)$ is dense in $L^p(\mathbb{R}^N)$ for any $p \in [1,+\infty)$, it suffices to prove that for any $f \in C_c^\infty(\mathbb{R}^N)$ $T(t)f$ tends to f in $L^p(\mathbb{R}^N)$, as t tends to 0^+, and

$$\|T(t)f\|_{L^p(\mathbb{R}^N)} \le e^{-\frac{t}{p}\mathrm{Tr}\,B}\|f\|_{L^p(\mathbb{R}^N)}, \qquad t > 0. \tag{10.4.1}$$

Denote by $\{S(t)\}$ the semigroup in $L^p(\mathbb{R}^N)$ defined by $(S(t)f)(x) = f(e^{tB}x)$ for any $t > 0$, $x \in \mathbb{R}^N$ and $f \in L^p(\mathbb{R}^N)$. For any $f \in C_c^\infty(\mathbb{R}^N)$ and $t > 0$, we can write $T(t)f = S(t) \circ G_t$, where

$$(G_tf)(x) = \frac{1}{\sqrt{(2\pi)^N \det Q_t}}\int_{\mathbb{R}^N} e^{-\frac{1}{2}\langle Q_t^{-1}y,y\rangle} f(x-y)dy, \qquad t > 0,\ x \in \mathbb{R}^N.$$

Let us show that $\{S(t)\}$ is a strongly continuous (semi)group in $L^p(\mathbb{R}^N)$ and that

$$\|S(t)\|_{L(L^p(\mathbb{R}^N))} \le e^{-\frac{t}{p}\mathrm{Tr}\,B}, \qquad t > 0. \tag{10.4.2}$$

Estimate (10.4.2) easily follows by a change of variable in the integral defining $\|S(t)f\|_p$, recalling that $\det(e^{-tB}) = e^{-t\mathrm{Tr}\,B}$ for any $t \in \mathbb{R}$.

To show that $\{S(t)\}$ is a strongly continuous semigroup, we observe that, if $f \in C_c^\infty(\mathbb{R}^N)$, then $S(t)f$ converges to f uniformly in $\mathbb{R}^N$ as t tends to 0^+. Since there exists $R > 0$ such that $\mathrm{supp}(S(t)f) \subset B_R$ for any $t \in [0,1]$, $S(t)f$ converges to f also in $L^p(\mathbb{R}^N)$ as t tends to 0^+. Now, using the density of $C_c^\infty(\mathbb{R}^N)$ in $L^p(\mathbb{R}^N)$ and taking (10.4.2) into account, from Proposition B.1.2 we deduce that $S(t)f$ converges to f in $L^p(\mathbb{R}^N)$, as t tends to 0^+, for any $f \in L^p(\mathbb{R}^N)$.

As far as the sequence $\{G_t\}$ is concerned, the Young inequality implies that each operator

G_t is a contraction in $L^p(\mathbb{R}^N)$. Moreover, standard arguments show that $G_t f$ converges to f in $L^p(\mathbb{R}^N)$, as t tends to 0^+, for any $f \in L^p(\mathbb{R}^N)$. Hence, the semigroup $\{T(t)\}$ is strongly continuous in $L^p(\mathbb{R}^N)$. Finally, (10.4.1) follows from (10.4.2) and the contractiveness of $\{G_t\}$. □

In the rest of the section we denote by $A_p : D(A_p) \subset L^p(\mathbb{R}^N) \to L^p(\mathbb{R}^N)$ the infinitesimal generator of $\{T(t)\}$ in $L^p(\mathbb{R}^N)$. The following proposition gives a complete characterization of the operator A_p and of its domain.

Proposition 10.4.2 *The domain of the operator A_p is the set $D(A_p) = \{u \in W^{2,p}(\mathbb{R}^N) : \mathcal{A}u \in L^p(\mathbb{R}^N)\}$ and $A_p u = \mathcal{A}u$ for any $u \in D(A_p)$ and $p \in (1,+\infty)$. Moreover, there exist two positive constants C_1 and C_2 such that*

$$C_1(\|u\|_p + \|\mathcal{A}u\|_p) \le \|u\|_{W^{2,p}(\mathbb{R}^N)} + \|\langle B\cdot, \nabla u\rangle\|_p \le C_2(\|u\|_p + \|\mathcal{A}u\|_p) \tag{10.4.3}$$

for any $u \in D(A_p)$ and p as above. Finally, for any $p \in [1,+\infty)$, $C_c^\infty(\mathbb{R}^N)$ is a core of A_p.

Proof Using the same arguments as in Lemma 10.3.11, it can be easily checked that $C_c^\infty(\mathbb{R}^N)$ is a core of A_p for any $p \in [1,+\infty)$. In particular, this implies that A_p is the closure of $\mathcal{A} : C_c^\infty(\mathbb{R}^N) \to L^p(\mathbb{R}^N)$ for any $p \in [1,+\infty)$.

Now, let us prove the characterization of $D(A_p)$ and the formula (10.4.3). For this purpose, we adapt to the present situation the proof of Theorem 10.3.15. We introduce the operators $A_p^0 : W^{2,p}(\mathbb{R}^N) \to L^p(\mathbb{R}^N)$, defined by $A_p^0 u = \frac{1}{2}\mathrm{Tr}(QD^2u)$ for any $u \in W^{2,p}(\mathbb{R}^N)$, and $B_p^0 : \{u \in L^p(\mathbb{R}^N) : x \mapsto \langle Bx, \nabla u(x)\rangle \in L^p(\mathbb{R}^N)\} \to L^p(\mathbb{R}^N)$, defined by $B_p^0 u = \langle B\cdot, \nabla u\rangle$ for any $u \in D(B_p^0)$, where ∇u is meant in the sense of distributions. Note that B_p^0 is the infinitesimal generator of the strongly continuous group $\{S(t)\}$ defined in the proof of Proposition 10.4.1. Since $C_c^\infty(\mathbb{R}^N)$ is a core of A_p and $A_p = A_p^0 + B_p^0$ in $C_c^\infty(\mathbb{R}^N)$, A_p is the closure of the operator $A_p^0 + B_p^0 : D(A_p^0) \cap D(B_p^0) = \{u \in W^{2,p}(\mathbb{R}^N) : \mathcal{A}u \in L^p(\mathbb{R}^N)\} \to L^p(\mathbb{R}^N)$. Therefore, to complete the proof it suffices to show that the operator $A_p^0+B_p^0 : D(A_p^0)\cap D(B_p^0) \to L^p(\mathbb{R}^N)$ is closed. According to Proposition C.1.18(iv), the operator $\omega I - A_p^0$ admits bounded imaginary powers with power angle $\theta_{\omega I - A_p^0} = 0$ for any $\omega > 0$. Similarly, taking estimate (10.4.2) into account and applying Proposition C.1.18(iii), we deduce that the operator $\omega I - B_p^0$ admits bounded imaginary powers for any $\omega > -(\mathrm{Tr}\, B)/p$, with power angle $\theta_{\omega I - B_p^0} \le \pi/2$.

To apply Theorem C.1.21 we need to show that condition (C.1.3) is satisfied. Following the arguments in the proof of Theorem 10.3.15, we replace $C_b^4(\mathbb{R}^N)$ with the space $\mathcal{G}_k = \{u \in W^{k,2}(\mathbb{R}^N) : x \mapsto (1+|x|^2)^{k/2} D^\alpha u \in L^2(\mathbb{R}^N)$ for any $|\alpha| \le k\}$, where $k = k(p)$ is to be properly chosen in order that $\mathcal{G}_k$ is contained in the domain of the commutator $[A_p^0, B_p^0]$. It is then easy to check that $[A_p^0, B_p^0]u = \mathrm{Tr}\,(QB^* D^2 u)$ for any $u \in \mathcal{G}_k$. Therefore, the operator $[A_p^0, B_p^0]R(1, A_p^0)$ extends to a bounded linear operator in $L^p(\mathbb{R}^N)$.

To make the arguments in the proof of Theorem 10.3.15 work, it suffices to show that, for any $k \in \mathbb{N}$, $R(\lambda, A_p^0)$ and $R(\lambda, B_p^0)$ leave $\mathcal{G}_k$ invariant, if $\mathrm{Re}\lambda$ is sufficiently large. Showing that $\mathcal{G}_k$ is invariant for $R(\lambda, A_p^0)$ is an easy task, taking advantage of the Fourier transform. As far as $R(\lambda, B_p^0)$ is concerned, for any multi-index α we introduce the operator $\mathcal{E}_\alpha$, formally defined by $(\mathcal{E}_\alpha w)(x) = (1+|x|^2)^{k/2}(D_x^\alpha w)(x)$, and begin by showing that $\mathcal{E}_\alpha S(t)u$ belongs to $L^2(\mathbb{R}^N)$ for any multi-index with length at most k and any $u \in \mathcal{G}_k$. Since,

$$\nabla_x S(t) f = e^{tB^*} S(t) \nabla f, \qquad t > 0,\ f \in W^{1,p}(\mathbb{R}^N), \tag{10.4.4}$$

it suffices to prove that $\mathcal{E}_0 S(t)v \in L^2(\mathbb{R}^N)$ for any v such that $\mathcal{E}_0 v \in L^2(\mathbb{R}^N)$. For this purpose, we observe that

$$|x|^k |(S(t)v)(x)| \le \|e^{-tB}\|_\infty^k (S(t)\mathcal{E}_0 v)(x), \qquad t > 0,\ x \in \mathbb{R}^N. \tag{10.4.5}$$

From (10.4.5) we deduce that $\mathcal{E}_0 S(t)v \in L^2(\mathbb{R}^N)$. Using (10.4.4) in an iterative way and taking (10.4.5) into account, we can show (up to replacing C_k and ω_k with larger constants) that $\|\mathcal{E}_\alpha S(t)u\|_{L^2(\mathbb{R}^N)} \le C_k e^{\omega_k t}\|\mathcal{E}_0|D^{|\alpha|}u|\|_{L^2(\mathbb{R}^N)}$ for any $t > 0$ and any multi-index α with length k. Since $R(\lambda, B_p^0)$ is the Laplace transform of the semigroup $\{S(t)\}$ (see the formula (C.1.2)), from the previous estimate we deduce that, if $\mathrm{Re}\,\lambda$ is sufficiently large, then $R(\lambda, B_p^0)$ maps $\mathcal{G}_k$ into itself. This completes the proof. □

To conclude this section we characterize the spectrum of $(A_p, D(A_p))$. Since the proof is rather technical we prefer not to enter into detail, referring the reader to the paper [190].

Theorem 10.4.3 *Suppose that either the eigenvalues of the matrix B or the eigenvalues of the matrix $-B$ have all negative real parts. Then, for any $p \in [1, +\infty)$, it holds that $\sigma(A_p) = \{\lambda \in \mathbb{C} : \mathrm{Re}\lambda \le -(\mathrm{Tr}\,B)/p\}$.*

Proof See [190, Theorems 4.4, 4.7, 4.11 and 4.12]. □

The following result is now a straightforward consequence of Theorem 10.4.3.

Corollary 10.4.4 *For any $p \in [1, +\infty)$ the Ornstein-Uhlenbeck semigroup is not analytic in $L^p(\mathbb{R}^N)$.*

Remark 10.4.5 The results in Theorem 10.4.3 should be compared with the corresponding results in Theorem 10.3.20. In particular, it should be noticed that the spectrum of the Ornstein-Uhlenbeck operator in $L^p(\mathbb{R}^N)$ strictly depends on p (unless in the case when $\mathrm{Tr}\,B = 0$), whereas the spectrum in L^p_μ ($p \in (1, +\infty)$) is p-invariant.

Chapter 11

Degenerate Markov semigroups in $\mathbb{R}^N$

11.0 Introduction

The prototype of degenerate elliptic operators with unbounded coefficients is the degenerate Ornstein-Uhlenbeck, already considered in Chapter 10. As in the nondegenerate case, one of the main peculiarities of the associated Ornstein-Uhlenbeck semigroup is that an explicit representation formula is available (see (10.1.5)). As shown by A. Lunardi in [181] the behaviour of the spatial derivatives of $T(t)f$ depends on the variable along which one differentiates: a typical feature of hypoelliptic operators.

The results in [181] have been extended, both with analytic and probabilistic methods, in [99, 158, 170, 171, 218, 219, 232] to non-trivial perturbations of the degenerate Ornstein-Uhlenbeck operator to which an additional unbounded drift term is added. Degenerate operators modeled on the Ornstein-Uhlenbeck operator have been considered also in the L^p-setting in [44, 45].

In this chapter, we present the results in [99, 170, 171]. More precisely, we consider a class of elliptic operators that, up to a change of the coordinates, may be written on smooth functions $\zeta : \mathbb{R}^N \to \mathbb{R}$ as follows:

$$\mathcal{A}\zeta(x) = \sum_{i,j=1}^{p_0} q_{ij}(x)D_{ij}\zeta(x) + \sum_{i,j=1}^{N} b_{ij}x_jD_i\zeta(x) + \sum_{j=1}^{p_0} F_j(x)D_j\zeta(x) \quad x \in \mathbb{R}^N, \tag{11.0.1}$$

for some $p_0 < N$, under the following set of assumptions.

Hypotheses 11.0.1 (i) there exist $r \in \mathbb{N}$ and integers $p_0 \geq p_1 > \ldots \geq p_r$ such that the matrix B can be split into blocks as follows:[1]

$$B = \begin{pmatrix} \star & \star & \cdots & \cdots & \star \\ B_1 & \star & \cdots & \cdots & \star \\ 0 & B_2 & \star & \cdots & \star \\ \vdots & \ddots & \ddots & \ddots & \vdots \\ 0 & 0 & 0 & B_r & \star \end{pmatrix}, \tag{11.0.2}$$

where B_h is a $p_h \times p_{h-1}$ matrix with full rank p_h ($h = 1, \ldots, r$);

(ii) $Q(x)$ is a $p_0 \times p_0$ symmetric matrix, with entries that belong to $C^{\varsigma+\gamma}_{\rm loc}(\mathbb{R}^N)$ for some $3 \leq \varsigma \in \mathbb{N}$ and $\gamma \in (0,1)$, such that $\langle Q(x)\xi, \xi\rangle \geq \kappa(x)|\xi|^2$ for any $x \in \mathbb{R}^N$, $\xi \in \mathbb{R}^{p_0}$ and some function κ with positive infimum κ_0. Further, $|D^\alpha q_{ij}(x)| \leq C|x|^{(1-|\alpha|)^+}\sqrt{\kappa(x)}$ for any $x \in \mathbb{R}^N$, $i,j = 1, \ldots, p_0$, $|\alpha| \leq \varsigma_0$ and some positive constant C;

[1] Here, $\star$ denotes matrices that we are not interested in writing explicitly, and all the matrices appearing in a column below a matrix B_j are null matrices.

(iii) F belongs to $C^{\varsigma+\gamma}_{\rm loc}(\mathbb{R}^N,\mathbb{R}^{p_0})$, has at most linear growth and $|D^\alpha F(x)|\le C\sqrt{\kappa(x)}$ for any $x\in\mathbb{R}^N$ and $1\le|\alpha|\le\varsigma$.

Under the previous set of assumptions, it is possible to associate a semigroup $\{T(t)\}$ with the operator $\mathcal{A}$ in $C_b(\mathbb{R}^N)$ as in the nondegenerate case, and the following uniform estimates hold true:

$$\|D^\alpha_x T(t)f\|_\infty\le Ct^{-q_h(\widehat{\alpha})}\|f\|_{C^h_b(\mathbb{R}^N)},\qquad t\in(0,T_0],\ f\in C^h_b(\mathbb{R}^N),\tag{11.0.3}$$

for any $T_0>0$, any multi-index α with length at most ς, any $h\le|\alpha|$ and some positive constant $C=C(T_0)$. Here $\widehat{\alpha}\in(\mathbb{N}\cup\{0\})^{r+1}$ is obtained, writing $\alpha=(\gamma_0,\ldots,\gamma_r)$ with $\gamma_j\in\mathbb{N}_0^{p_j}$ for any $j=0,\ldots,r$, and setting $\widehat{\alpha}=(|\gamma_0|,\ldots,|\gamma_r|)$, where $|\gamma_j|$ denotes the length of the index γ_j. Finally,

$$q_h(\beta)=\begin{cases}\displaystyle\frac{1}{2}|\beta|-\frac{1}{2}h+\sum_{k=0}^{j(\beta)-1}k\beta_k+(j(\beta)-1)\Big(\sum_{k=j(\beta)}^{r}\beta_k-h\Big), & |\beta|>h,\\[2mm] 0, & \text{otherwise.}\end{cases}\tag{11.0.4}$$

Here, $j(\beta)\ge0$ is the smallest integer such that $\sum_{j=j(\beta)}^r\beta_j\le h$ and, if $\beta_r>h$, then $j(\beta)=r+1$. Moreover, we agree that $\sum_{k=0}^{-1}k\beta_k:=0$ and $\sum_{k=r+1}^r\beta_k:=0$.[2]
Suppose that $h<|\alpha|$. Theorem 5.1.7 shows that, in the nondegenerate case, estimate (11.0.3) holds true with $q_h(\alpha)$ being replaced by $(|\alpha|-h)/2$. Under Hypotheses (11.0.1), the additional term $\sum_{k=0}^{j(\widehat{\alpha})-1}k\gamma_k+(j(\widehat{\alpha})-1)(\sum_{k=j(\widehat{\alpha})}^r\gamma_k-h)$ appears. Suppose that $h=0$. Then, $q_h(\widehat{\alpha})=\frac{1}{2}\gamma_0+\frac{3}{2}\gamma_1+\ldots+\frac{2r+1}{2}\gamma_r$ Hence, if D^α_x contains only derivatives with respect to the first p_0 variables $x_1,\ldots,x_{p_0}$, then the behaviour of $\|D^\alpha_xT(t)f\|_\infty$ as t tends to 0^+ is the same as in the nondegenerate case. On the contrary, each derivative with respect to a variable x_j with $j\in\{p_{k-1}+1,\ldots,p_k\}$ ($k>0$), contributes to the behaviour of $\|D^\alpha_xT(t)f\|_\infty$ around $t=0$ as $t^{-1/2-k}$. If $h>0$, then $q_h(\widehat{\alpha})=q_0(\beta)$, where β is obtained from $\widehat{\alpha}$ by canceling, starting from the right, the entries γ_k for any $k=j(\widehat{\alpha}),\ldots,r$ and replacing $\gamma_{j(\widehat{\alpha})-1}$ with $\beta_{j(\widehat{\alpha})-1}=\gamma_{j(\widehat{\alpha})-1}-h+\sum_{k=j(\alpha)}^r\gamma_k$. In other words, $\|D^\alpha_xT(t)f\|_\infty$ behaves near $t=0$ as $\|D^{\alpha'}_xT(t)f\|_\infty$, where the index α', with length $|\alpha'|=|\alpha|-h$, is obtained removing h derivatives from $D^\alpha_x=D^{\alpha_1}_x\cdots D^{\alpha_N}_x$, starting from the right.

To associate a semigroup with the operator $\mathcal{A}$ in (11.0.1), for any $\varepsilon\in(0,1]$ we introduce the semigroup $\{T_\varepsilon(t)\}$ in $C_b(\mathbb{R}^N)$, associated with the non-degenerate elliptic operator $\mathcal{A}_\varepsilon:=\mathcal{A}+\varepsilon\Delta_\star$, where $\Delta_\star$ denotes the Laplacian with respect to the missing second-order derivatives, i.e., $\Delta_\star=D_{(p_0+1)(p_0+1)}+\cdots+D_{NN}$. In Section 11.2 we prove that estimates (11.0.3) are satisfied with $T(t)$ being replaced by $T_\varepsilon(t)$, with a constant independent of ε. For this purpose, as in Chapter 5, we adapt Bernstein method to our situation. Note that the degeneracy of the operator $\mathcal{A}$ gives rise to additional technical difficulties. The semigroup $\{T(t)\}$ is constructed in Section 11.3, where the uniform estimates (11.0.3) and some relevant continuity properties of the semigroup are also proved. As in the nondegenerate case (see Section 1.3) we show that a weak generator can be associated with the semigroup $\{T(t)\}$ and we characterize its domain. We use the uniform estimates (11.0.3), as we did

[2] If $r=2$, then we have for instance

$$q_0((2,1,1))=2\tfrac{1}{2}+\tfrac{3}{2}+\tfrac{5}{2}=5,\qquad q_1((2,1,1))=q_0((2,1,0))=2\tfrac{1}{2}+\tfrac{3}{2}=\tfrac{5}{2},$$
$$q_2((2,1,1))=q_0((2,0,0))=2\tfrac{1}{2}=1,\qquad q_3((2,1,1))=q_0((1,0,0))=\tfrac{1}{2},$$
$$q_4((2,1,1))=q_0((0,0,0))=0.$$

in Chapter 5, to prove anisotropic Hölder estimates for the distributional solutions to the elliptic equation

$$\lambda u(x) - \mathcal{A}u(x) = h(x), \qquad x \in \mathbb{R}^N,\ \lambda > 0, \tag{11.0.5}$$

and to the nonhomogeneous Cauchy problem

$$\begin{cases} D_t u(t,x) = \mathcal{A}u(t,x) + g(t,x), & t > 0, \quad x \in \mathbb{R}^N, \\ u(0,x) = f(x), & x \in \mathbb{R}^N, \end{cases} \tag{11.0.6}$$

when f, g and h are suitable continuous functions. In the case when $p_0 > N/2$ and B has only two blocks on the main diagonal, we prove more refined Hölder estimates for the solutions to the previous problems, which turn out to be optimal, as the degenerate Ornstein-Uhlenbeck operator reveals. This is the content of Section 11.4.

11.1 Remarks on the assumptions and technical results

We begin this section with some comments on Hypotheses 11.0.1.

Remark 11.1.1 (i) Hypotheses 11.0.1 imply that, for any $x \in \mathbb{R}^N$, the matrix $\mathcal{F}^{(r)}(x) = [Q(x), BQ(x), B^2Q(x), \ldots, B^rQ(x)]$ has full rank, i.e., the Kalman rank condition is satisfied at any $x \in \mathbb{R}^N$. Indeed, denote by $\mathcal{F}_j^{(r)}(x)$ $(j = 1, \ldots, N)$ the rows of $\mathcal{F}^{(r)}(x)$ and observe that $\sum_{j=1}^N \xi_j \mathcal{F}_j^{(r)}(x) = 0$ if and only if the vector $\xi = (\xi_1, \ldots, \xi_N)$ belongs to $W_r := \{\xi \in \mathbb{R}^N : Q(x)(B^*)^k\xi = 0 \text{ for any } k = 0, \ldots, r\}$. Hence, it suffices to prove that $W_r = \{0\}$. But this follows splitting the matrix $(B^*)^k$ into blocks according to the splitting in (11.0.2) and observing that the first block-row of $(B^*)^k$ has the following structure: $(\star_1, \ldots, \star_k, B_1B_2\cdots B_k, 0, \ldots, 0)$ for suitable matrices $\star_j$ $(j = 1, \ldots, k)$.
The previous property shows that the matrix

$$Q_t(x) = \int_0^t e^{sB}Q(x)e^{sB^*}\,ds$$

is positive definite for any $t > 0$ and $x \in \mathbb{R}^N$. Indeed, fix $t > 0$, $x \in \mathbb{R}^N$ and take $\xi \in \mathrm{Ker}(Q_t(x))$. Since $\langle Q_t(x)\xi, \xi\rangle = 0$, $\langle e^{sB}Q(x)e^{sB^*}\xi, \xi\rangle$ vanishes for any $s \in [0,t]$. Equivalently, $\sqrt{Q}(x)e^{sB^*} = 0$ for any $s \in (0,t)$ and, consequently, the function $s \mapsto Q(x)e^{sB^*}\xi$ identically vanishes in $[0,t]$. In particular, all the derivatives at $s = 0$ of this function vanish, so that $\xi \in W = \{\xi \in \mathbb{R}^N : Q(x)(B^*)^k\xi = 0 \text{ for any } k \in \mathbb{N}\cup\{0\}\}$. Since $W \subset W_r$, W reduces to the trivial vector, i.e., $\xi = 0$.

(ii) Actually, more general degenerate operators of the form $\mathcal{A} = \mathrm{Tr}(QD^2) + \langle Bx + F, \nabla\rangle$ can be reduced, by a change of variables, to operators which satisfy Hypotheses 11.0.1. This is the case when (i) the entries q_{ij} of the matrix Q belong to $C^{\varsigma+\gamma}_{\rm loc}(\mathbb{R}^N)$ $(i,j = 1, \ldots, N)$ for some $\varsigma \in \mathbb{N}$, $\varsigma \geq 3$ and some $\gamma \in (0,1)$, and there exists a positive constant C such that $|D^\alpha q_{ij}(x)| \leq C|x|^{(1-|\alpha|)^+}\sqrt{\kappa(x)}$ for any $x \in \mathbb{R}^N$, $i,j = 1, \ldots, N$ and any multi-index α with length at most ς; (ii) there exists a function $\kappa : \mathbb{R}^N \to (0,+\infty)$ such that $\kappa_0 := \inf_{x\in\mathbb{R}^N}\kappa(x) > 0$ and $\langle Q(x)\xi, \xi\rangle \geq \kappa(x)|\xi|^2$ for any $\xi \in (\ker(Q(0)))^\perp$ and $x \in \mathbb{R}^N$; (iii) the kernel of the matrix $Q(x)$ is independent of $x \in \mathbb{R}^N$, is a proper subspace of $\mathbb{R}^N$ and it does not contain non-trivial proper subspaces which are invariant for B^*; (iv) $F \in C^{\varsigma+\gamma}_{\rm loc}(\mathbb{R}^N, \mathbb{R}^{p_0})$ has at most linear growth and $|D^\alpha F(x)| \leq$

$C\sqrt{\kappa(x)}$ for any $x \in \mathbb{R}^N$ and $1 \le |\alpha| \le \varsigma$. Indeed, let $\{V_k : k \in \mathbb{N}\}$ be the sequence of nested vector spaces, defined by $V_k = W_k^{\perp}$ for any $k \in \mathbb{N}$ (where $W_k = \{\xi \in \mathbb{R}^N : Q(x)(B^*)^k\xi = 0$ for any $k = 0,\ldots,k\}$). Using property (iii), it can be proved that there exists a positive integer $p_0 < N$ such that $V_{p_0} = \mathbb{R}^N$ and V_k is properly contained in V_{k+1} if $k < p_0$. Let now $W_0 = V_0$ and W^k be the orthogonal complement of V_{k-1} in V_k, for any $k = 1,\ldots,p_0$. Let $p_k = \dim(W_k)$ for any $k \le p_0$. Of course, $\mathbb{R}^N = \bigoplus_{k=0}^r W_k$. In the orthonormal basis $\{e_1',\ldots,e_N'\}$ of $\mathbb{R}^N$ consisting of vectors of the spaces W_k $(k = 0,\ldots,r)$, the coefficients of the operator $\mathcal{A}$ satisfies Hypotheses 11.0.1. We refer the reader to [160, Proposition 2.1] for further details.

(iii) Hypotheses 11.0.1(ii) imply that the diffusion coefficients may grow at most quadratically at infinity.

11.1.1 Ordering the derivatives of smooth functions

Here, given a function $u : \mathbb{R}^N \to \mathbb{R}$, we introduce a splitting of the vector $D^k u$ of all its derivatives of order k ($k \in \mathbb{N}$) into sub-blocks. This splitting will be extensively used in Section 11.2.

Let $\{p_0,\ldots,p_r\}$ be as in Hypothesis 11.0.1(i). We split any multi-index $\alpha \in (\mathbb{N}\cup\{0\})^N$ with length k as $\alpha = (\alpha_0,\ldots,\alpha_r)$ with $\alpha_j \in \mathbb{N}_0^{p_j}$ $(j = 0,\ldots,r)$ and write $\widehat{\alpha} := (|\alpha_0|,\ldots,|\alpha_r|)$. Clearly, $\widehat{\alpha}$ has length k. Denote by $\mathcal{I}_k$ the set of all the elements of $(\mathbb{N}\cup\{0\})^{r+1}$ with length k. It consists of $c_k := \binom{k+r}{r}$ elements $i_1^{(k)} \preceq \cdots \preceq i_{c_k}^{(k)}$, ordered according to the following rule: $(m_0,\ldots,m_r) \preceq (m_0',\ldots,m_r')$ if there exists $h = 0,\ldots,r$ such that $m_j = m_j'$ for any $j = 0,\ldots,h-1$ and $m_h > m_h'$.

We can now order the elements of $D^k u$ as follows:

(i) we split $D^k u$ into blocks according to the rule: $D^k u = (\partial_1^k u,\ldots,\partial_{c_k}^k u)$, where $\partial_j^k u$ $(j = 1,\ldots,c_k)$ consists of all the derivatives $D^\alpha u$ of order k such that $\widehat{\alpha} = i_j^{(k)}$;

(ii) we order the entries of the vectors $\partial_j^k u$ $(j = 1,\ldots,c_{k,r})$ according to the following rule: if $D^\alpha u, D^\beta u \in \partial_j^k u$, then $D^\alpha u$ precedes $D^\beta u$ if there exists $h \le N$ such that $\alpha_j = \beta_j'$ for any $j = 0,\ldots,N-1$ and $\alpha_h > \beta_h$.

Example 11.1.2 Let $N = 4$, $k = 3$, $p_0 = 2$, $p_1 = p_2 = 1$ (i.e., $r = 2$) and denote by (x,y,z,w) the elements of $\mathbb{R}^4$. Then,

$$\mathcal{I}_3 = \{(3,0,0),(2,1,0),(2,0,1),(1,2,0),(1,1,1),(1,0,2),(0,3,0),(0,2,1),(0,1,2),(0,0,3)\},$$

where the elements are ordered according to the ordering $\preceq$. Hence, the vector D^3u is split into ten blocks as $D^3u = (\partial_1^3,\ldots,\partial_{10}^3 u)$ where the blocks ∂_j^3 $(j = 1,\ldots,10)$ with ordered entries are defined as follows:

$$
\begin{aligned}
&\partial_1^3 u = (D^3_{xxx}u, D^3_{xxy}u, D^3_{xyy}u, D^3_{yyy}u), && \partial_2^3 u = (D^3_{xxz}u, D^3_{xyz}u, D^3_{yyz}u),\\
&\partial_3^3 u = (D^3_{xxw}u, D^3_{xyw}u, D^3_{yyw}u), && \partial_4^3 u = (D^3_{xzz}u, D^3_{yzz}u),\\
&\partial_5^3 u = (D^3_{xzw}u, D^3_{yzw}u), && \partial_6^3 u = (D^3_{xww}u, D^3_{yww}u),\\
&\partial_7^3 u = (D^3_{zzz}u), && \partial_8^3 u = (D^3_{zzw}u),\\
&\partial_9^3 u = (D^3_{zww}u), && \partial_{10}^3 = (D^3_{www}u).
\end{aligned}
$$

11.1.2 Technical results

In this section we collect some preliminary results which play a crucial role in the proof of Theorem 11.2.2. To begin with, we introduce the function $\ell : \mathbb{N}\setminus\{1\} \to \mathbb{N}$. It is defined as follows: let $i_m^{(k)} = (0, \alpha_1, \ldots, \alpha_r)$, where $k = |\alpha|$, and let j be the smallest integer such that $\alpha_j > 0$. Then, $\ell(m)$ is the index such that $i_{\ell(m)}^{(k)} = (0, \ldots, 0, 1, \alpha_j - 1, \alpha_{j+1}, \ldots, \alpha_r)$. As is immediately seen, $\ell(m)$ is less than m.

Lemma 11.1.3 *Fix $k \geq 1$ and $m > c_{k-1}$. Then, for any function $w \in C_b^{k+1}(\mathbb{R}^N)$ it holds that*[3] *$[\partial_{\ell(m)}^k, \langle B\cdot, D\rangle]w = \sum_{n\in A_m^{(k)}} \mathcal{J}_n^{(k)} \partial_n^k w$, where the set $A_m^{(k)}$ is defined as follows: if $d_{j_1}, d_{j_2}, \ldots, d_{j_k}$ $(1 \leq j_1 < \ldots < j_k \leq r)$ are all the non-zero entries of the vector $i_m^{(k)} = (0, d_1, \ldots, d_r)$, then*

$$
\begin{aligned}
A_m^{(k)} = \Big\{ s : i_s^{(k)} = i_m^{(k)} - e_{j_1}^{(r+1)} + e_{j_1-1}^{(r+1)} - e_{j_i}^{(r+1)} + e_h^{(r+1)} \\
\text{for some } i = 2, \ldots, n \text{ and } h \leq j_i + 1 \Big\} \\
\cup \Big\{ s : i_s^{(k)} = i_m^{(k)} - e_{j_1}^{(r+1)} + e_h^{(r+1)} \text{ for some } h \leq j_1 \Big\} \\
\cup \Big\{ s : i_s^{(k)} = i_m^{(k)} - 2e_{j_1}^{(r+1)} + e_{j_1-1}^{(r+1)} + e_h^{(r+1)} \\
\text{for some } h \leq \min\{j_1 + 1, r\}, \text{ if } \alpha_{j_1} > 1 \Big\},
\end{aligned}
\tag{11.1.1}
$$

where $e_h^{(r+1)}$ denotes the hth-vector of the standard Euclidean basis of $\mathbb{R}^{r+1}$. The entries of the matrices $\mathcal{J}_n^{(k)}$ $(n \in A_m^{(k)})$ linearly depend only on the entries of the matrix B. Moreover, the matrix $\mathcal{J}_m^{(k)}$ has full rank.

Proof A long but straightforward computation and the structure of the matrix B in (11.0.2) show that

$$
\begin{aligned}
([D^\alpha, \langle B\cdot, \nabla\rangle]w)(x) &= \sum_{i,j=1}^{N} \sum_{s=0}^{r} \sum_{\tau\in\mathcal{I}_s} \langle \alpha, e_\tau^{(N)}\rangle b_{ij} D_\tau x_j D^{\alpha - e_\tau^{(N)} + e_i^{(N)}} w(x) \\
&= \sum_{s=0}^{r} \sum_{\tau\in\mathcal{I}_s} \sum_{h=0}^{(s+1)\wedge r} \sum_{i\in\mathcal{I}_h} \langle \alpha, e_\tau^{(N)}\rangle b_{i\tau} D^{\alpha - e_\tau^{(N)} + e_i^{(N)}} w(x)
\end{aligned}
$$

for any $\alpha \in (\mathbb{N}\cup\{0\})^N$ and $x \in \mathbb{R}^N$. Considering all the possible multi-indices $\alpha \in (\mathbb{N}\cup\{0\})^N$ with $\widehat{\alpha} = i_{\ell(m)}^{(k)} = (0, \ldots, 0, 1, d_{j_1}-1, \ldots, d_{j_2}, \ldots, d_{j_n}, 0, \ldots, 0)$, from this formula we conclude that $[\partial_{\ell(m)}^k, \langle B\cdot, D\rangle]w = \sum_{n\in A_m^{(k)}} \mathcal{J}_n^{(k)} \partial_n^k w$ for some matrices $\mathcal{J}_n^{(k)}$ $(n \in A_m)$ and $\mathcal{J}_m^{(n)}$. It remains to show that the matrix $\mathcal{J}_m^{(k)}$ has full rank which equals the number of its columns. We split the rest of the proof into two steps.

Step 1. Here, we perform some reductions to simplify the proof. Since the entries of the matrix $\mathcal{J}_m^{(k)}$ depend only on the matrices $B_1, \ldots, B_r$, we can assume that the only nontrivial blocks of the matrix B in (11.0.2) are $B_1, \ldots, B_r$. Hence,

$$
[D^\alpha, \langle B\cdot, \nabla\rangle]w = \sum_{s=0}^{r-1} \sum_{\tau\in\mathcal{I}_s} \sum_{i\in\mathcal{I}_{s+1}} \langle \alpha, e_\tau^{(N)}\rangle b_{i\tau} D^{\alpha - e_\tau^{(N)} + e_i^{(N)}} w. \tag{11.1.2}
$$

In particular, if $\widehat{\alpha} = i_{\ell(m)}^{(k)}$, then the only possibilities to obtain a multi-index $\alpha - e_\tau^{(N)} + e_i^{(N)}$ having the block form $i_m^{(k)}$, are the choices $\tau \in \mathcal{I}_{j_1-1}$, $i \in \mathcal{I}_{j_1}$. So, if we split $i_{\ell(m)}^{(k)} =$

[3] Here, the commutator is meant componentwise.

$(0,\dots,0,1,d_{j_1}-1,0,\dots,0)+(0,\dots,0,0,0,d_{j_1+1},\dots)$ and, accordingly, $\alpha=\beta+\gamma$, with $|\beta|=(0,\dots,0,1,d_{j_1}-1,0,\dots,0)$ and $|\gamma|=(0,\dots,0,0,0,d_{j_1+1},\dots)$, we get

$$[D^\alpha,\langle B\cdot,\nabla\rangle]w=\sum_{\tau\in\mathcal{I}_{j_1-1}}\sum_{i\in\mathcal{I}_{j_1}}\langle\alpha,e_\tau^{(N)}\rangle b_{i\tau}D^{\beta-e_\tau^{(N)}+e_i^{(N)}}D^\gamma w+\dots,$$

where we have not explicitly written the terms, which do not contribute to $\mathcal{J}_m^{(k)}$. It turns out that it suffices to argue on the function $D^\gamma w$ or, equivalently, we can assume that $i_{\ell(m)}^{(k)}=(0,\dots,0,1,d_{j_1}-1,0,\dots,0)$.

Step 2. Let us take a derivative D^α in the block $\partial^k_{\ell(m)}$. Then, there exists an index $\tau\in\mathcal{I}_{j_1-1}$ such that $\langle\alpha,e_\tau^{(N)}\rangle=1$. From the formula (11.1.2) we get

$$\begin{aligned}[D^\alpha,\langle B\cdot,\nabla\rangle]w=&\sum_{i\in\mathcal{I}_{j_1}}b_{i\tau}D^{\alpha-e_\tau^{(N)}+e_i^{(N)}}w+\dots\\ =&(B_{j_1}^*D_{j_1}^1D^{\alpha-e_\tau^{(N)}}w)_\tau+\dots=\mathcal{K}_\alpha(w)+\dots,\end{aligned}\tag{11.1.3}$$

where $(\,\cdot\,)_\tau$ and "..." denote, respectively, the τ-th component of the vector in brackets and terms which depend on the k-th derivatives of w and belong to a block different from ∂^k_m. Finally, we recall that $\partial^1_{j_1}z$ denotes the vector of the first-order derivatives D_hz of the function z, with $h\in\mathcal{I}_{j_1}$. We are interested exclusively in $\mathcal{K}_\alpha(w)$, because only this term contributes to $\mathcal{J}_m^{(k)}$.

Let us reorder the vectors ∂^k_m and $\partial^k_{\ell(m)}$ in such a way that the assertion about the rank of $\mathcal{J}_m^{(k)}$ will be clear. We order the set $\Gamma\subset(\mathbb{N}\cup\{0\})^{p_{j_1}}$ of multi-indices of length $d_{j_1}-1$ by $\prec$ anti-lexicographically.[4] Next, we introduce the set

$$\Lambda=\Big\{\lambda_{i,\gamma}=(0,\dots,0,e_i^{(p_{j_1-1})},\underbrace{\gamma}_{j_1-\text{th block}},0,\dots,0):\ i=1,\dots,p_{j_1-1},\ \gamma\in\Gamma\Big\},$$

which we order again anti-lexicographically, still denoting the ordering by $\prec$. The set Λ describes all the possible multi-indices having blocks from $i^{(k)}_{\ell(m)}$. We reorder the vector $\partial^k_{\ell(m)}w$ according to this ordering. Fix $\gamma\in\Gamma$. By considering multi-indices $\gamma^{+i}:=(0,\dots,0,\gamma+e_i^{(p_{j_1})},0,\dots,0)$ $(i=1,\dots,p_{j_1})$, where $\gamma+e_i^{(p_{j_1})}$ occupies the j_1-th block, we recover all the multi-indices of block form $i_m^{(k)}$, but most of them several times. For a multi-index $\gamma\in(\mathbb{N}\cup\{0\})^{p_{j_1}}$ let $n(\gamma)$ denote the smallest nonnegative integer n such that $\gamma_h=0$ for any $n+1<h\le p_{j_1}$.[5] Consider now a multi-index $\gamma\in\Gamma$ and all the multi-indices γ^{+i}, $i=1,\dots,p_{j_1}$. For $i=1,\dots,n(\gamma)$ we obtain multi-indices β which can be written both as γ^{+i} and $\gamma'^{+i'}$ for some $\gamma'\prec\gamma$ and for some $1\le i'\le p_{j_1}$. We set $\partial_m^\gamma w:=(D^{\gamma^{+(n(\gamma)+1)}}w,D^{\gamma^{+(n(\gamma)+2)}}w,\dots,D^{\gamma^{+p_{j_1}}}w)^T$. If $\gamma_1\prec\gamma_2\prec\dots$ is an enumeration of Γ, $(D_m^{\gamma_1}w,D_m^{\gamma_2}w,\dots)^T$ is a reordering of ∂^k_mw. Further, we set $\mathcal{H}_\gamma(w):=$

---

[4] i.e., $(d_{j_1}-1,0,\dots,0)\prec(d_{j_1}-2,1,\dots,0)\prec\dots\prec(0,\dots,0,d_{j_1}-1)$.

[5] For instance, $n((d_{j_1}-1,0,\dots,0))=0$ (only in this case is $n(\gamma)=0$), $n((d_{j_1}-1,1,0,\dots,0))=1$ and $n((0,\dots,0,d_{j_1}-1))=p_{j_1}-1$.

$(\mathcal{K}_{\lambda_{1,\gamma}}(w), \mathcal{K}_{\lambda_{2,\gamma}}(w), \dots, \mathcal{K}_{\lambda_{p_{j1}-1,\gamma}}(w))^T$. From (11.1.3) we conclude that

$$\begin{pmatrix} \mathcal{H}_{\gamma_1}(w) \\ \mathcal{H}_{\gamma_2}(w) \\ \vdots \\ \mathcal{H}_{\gamma_k}(w) \\ \vdots \end{pmatrix} = \begin{pmatrix} [B_{j1}^*]_{-n(\gamma_1)} & \dots & \dots & 0 & \dots & 0 \\ \star & [B_{j1}^*]_{-n(\gamma_2)} & 0 & \dots & \dots & 0 \\ \vdots & \ddots & \ddots & \ddots & & \\ \star & \dots & \star & [B_{j1}^*]_{-n(\gamma_k)} & 0 & \\ \vdots & \dots & \dots & \ddots & \ddots & \ddots \end{pmatrix} \begin{pmatrix} D_m^{\gamma_1} w \\ D_m^{\gamma_2} w \\ \vdots \\ D_m^{\gamma_k} w \\ \vdots \end{pmatrix},$$

where $[B_{j1}^*]_{-s}$ denotes the matrix obtained from B_{j1}^* by dropping out the first s columns. The block matrix above is block-lower triangular and has full rank, since all its blocks on the diagonal have full rank, and its rank equals the number of its columns. Thus, $\mathcal{J}_m^{(k)}$, which is similar to the block matrix above, has the asserted properties. □

The following lemma collects the main properties of the functions q_h ($h \in \mathbb{N} \cup \{0\}$) (see (11.0.4)).

Lemma 11.1.4 *The following properties hold.*

(i) $q_h(\alpha) \ge (|\alpha| - h)^+/2$ *and* $q_h(\alpha) = 0$ *if and only if* $|\alpha| \le h$.

(ii) *If* $|\alpha| \ge h$ *and* $\beta = (\alpha_0 + 1, \alpha_1, \dots, \alpha_r)$, *then* $q_h(\beta) = q_h(\alpha) + 1/2$.

(iii) *If* $\beta = \alpha - e_j^{(r+1)} + e_k^{(r+1)}$ *for some* $0 \le j, k \le r$ *such that* $\alpha_j > 0$ *and* $k \le j+1$, *then* $q_h(\alpha) \ge q_h(\beta) - 1$.

(iv) *Suppose that* $|\alpha| > h$, $\alpha_0, \dots, \alpha_{j_0-1} = 0$ *and* $\alpha_{j_0} > 0$. *Set* $\beta = \alpha - e_{j_0}^{(r+1)} + e_{j_0-1}^{(r+1)}$. *Then,* $q_h(\beta) = q_h(\alpha) - 1 > 0$.

(v) *Let* α *and* β *be two multi-indices such that* $\alpha_j \le \beta_j$ *for all* $j = 0, \dots, r$ *and* $\alpha_{j_0} < \beta_{j_0}$ *for some* j_0. *Further, let* $\gamma = 2e_1^{(r+1)} + \alpha$. *Then, it holds that* $q_h(\beta) \ge q_h(\gamma) - 1$.

(vi) *Fix* $k \le \varsigma$ *and* $m > c_{k-1}$. *Then, for any* $s \in A_m^{(k)}$ *(see (11.1.1)), it holds that* $q_h(i_s^{(k)}) \le q_h(i_{\ell(m)}^{(k)})$.

Proof (i) By definition, $q_h(\alpha) = 0$ if $|\alpha| \le h$. On the other hand, if $|\alpha| > h$ then we can estimate

$$\begin{aligned} q_h(\alpha) \ge& \frac{1}{2}|\alpha| - \frac{1}{2}h + (j(\alpha)-1)\beta_{j(\alpha)-1} + (j(\alpha)-1)\Big(\sum_{k=j(\alpha)}^{r} \beta_k - h\Big) \\ =& \frac{1}{2}|\alpha| - \frac{1}{2}h + (j(\alpha)-1)\Big(\sum_{k=j(\alpha)-1}^{r} \alpha_k - h\Big) \ge \frac{1}{2}(|\alpha| - h), \end{aligned}$$

since $\sum_{k=j(\alpha)-1}^{r} \alpha_k > h$.

(ii) If $|\alpha| = h$ then $j(\beta) = 1$. Therefore, $q_h(|\beta|) = (|\beta| - h)/2 = 1/2$. On the other hand, if $|\alpha| > h$ then, clearly, $1 \le j(\alpha) = j(\beta)$. Since $\alpha_k = \beta_k$ for any $k \ge 1$, it follows that $q_h(\beta) = q_h(\alpha) + 1/2$.

(iii) Of course, we can assume that $|\alpha| > h$, otherwise the assertion is trivial. First, suppose that $k < j$. Then, $\sum_{m=i}^{r} \beta_m = \sum_{m=i}^{r} \alpha_m$, if $i \le k$ or $i > j$, and $\sum_{m=i}^{r} \beta_m = \sum_{m=i}^{r} \alpha_m - 1$, otherwise. Therefore, if $j(\alpha) \le k$, then $j(\beta) = j(\alpha)$ and $q_h(\alpha) = q_h(\beta)$. On the other hand, if $j(\alpha) \in \{k+1, \dots, j\}$ (supposing that this set of indices is not empty), then there are two possibilities: either $\sum_{m=j(\alpha)-1}^{r}\alpha_m > h+1$ or $\sum_{m=j(\alpha)-1}^{r}\alpha_m = h+1$. In the first case, the above formulae show that $j(\alpha)=j(\beta)$, and $q_h(\beta)=q_h(\alpha)+k+1-j(\alpha)\le q_h(\alpha)$; in the second case, $\sum_{s=j(\alpha)-1}^{r}\beta_s = h$. Hence, $j(\beta)<j(\alpha)$. More precisely, $j(\beta)=j(\alpha)-p$, where $p\in\{1,\ldots,j(\alpha)-k-1\}$ is the first index such that $\alpha_{j(\alpha)-p-1}\neq 0$. Then, we can write

$$
\begin{aligned}
q_h(\beta) =&\frac{1}{2}|\alpha| - \frac{1}{2}h + \sum_{s=0}^{j(\alpha)-p-1} s\beta_s + (j(\alpha)-p-1)\Big(\sum_{s=j(\alpha)-p}^{r}\beta_s - h\Big)\\
=&\frac{1}{2}|\alpha| - \frac{1}{2}h + \sum_{s=0}^{j(\alpha)-1} s\beta_s - (j(\alpha)-1)\beta_{j(\alpha)-1}\\
=&\frac{1}{2}|\alpha| - \frac{1}{2}h + \sum_{s=0}^{j(\alpha)-1} s\beta_s = q_h(\alpha)+k-j(\alpha)+1 \le q_h(\alpha).
\end{aligned}
$$

The case when $j(\alpha)=j+1$ is similar to the previous one. It holds that $q_h(\beta)=q_h(\alpha)+k-j$, if $\sum_{s=j}^{r}\alpha_s > h+1$. If $\sum_{s=j}^{r}\alpha_s = h+1$, then $\sum_{s=j}^{r}\beta_s = h$. Thus, $j(\beta)=j+1-p\ge k$, where p is the smallest integer such that $\alpha_{j-p}\neq 0$. Consequently,

$$
\begin{aligned}
q_h(\beta) =&\frac{1}{2}|\alpha| - \frac{1}{2}h + \sum_{s=0}^{j-p} s\beta_s + (j-p)\Big(\sum_{s=j-p+1}^{r}\beta_s - h\Big)\\
=&\frac{1}{2}|\alpha| - \frac{1}{2}h + \sum_{s=0}^{j} s\beta_s + j\Big(\sum_{s=j+1}^{r}\beta_s - h\Big) = q_h(\alpha)+k-j.
\end{aligned}
$$

Finally, if $j(\alpha)>j+1$, then $j(\beta)=j(\alpha)$ and $q_h(\beta)=q_h(\alpha)+k-j<q_h(\alpha)$.

Now, we consider the case when $k=j+1$. Then, $\sum_{s=i}^{r}\beta_s=\sum_{s=i}^{r}\alpha_s$ if $i\le j$ or $i\ge j+2$. Therefore, if $j(\alpha)\le j$, then $j(\beta)=j(\alpha)$ and $q_h(\alpha)=q_h(\beta)$. Similarly, if $j(\alpha)\ge j+2$, then again $j(\beta)=j(\alpha)$. But in this situation $\sum_{s=0}^{j(\alpha)-1}s\beta_s = \sum_{s=0}^{j(\alpha)-1}s\alpha_s+1$. Hence, $q_h(\beta)=q_h(\alpha)+1$. Suppose now that $j(\alpha)=j+1$. If $\sum_{s=j+1}^{r}\alpha_s<h$, then $j(\beta)=j+1$ and, consequently, $q_h(\beta)=q_h(\alpha)$. On the other hand, if $\sum_{s=j+1}^{r}\alpha_s=h$, then $j(\beta)=j+p+2$, where $p\in\mathbb{N}$ is the smallest integer such that $\alpha_{j+p+1}>0$ (supposing that such an integer exists). In this case,

$$
\begin{aligned}
q_h(\beta) =&\frac{1}{2}|\beta| - \frac{1}{2}h + \sum_{s=0}^{j+p+1} s\beta_s + (j+p+1)\Big(\sum_{s=j+p+2}^{r}\beta_s - h\Big)\\
=&\frac{1}{2}|\alpha| - \frac{1}{2}h + \sum_{s=0}^{j+1} s\beta_s + (j+p+1)\Big(\sum_{s=j+p+1}^{r}\beta_s - h\Big)\\
=&\frac{1}{2}|\alpha| - \frac{1}{2}h + \sum_{s=0}^{j+1} s\alpha_s + 1 - (j+p+1)\alpha_{j+1}\\
=&q_h(\alpha)+1-p\alpha_{j+1}\le q_h(\alpha)+1,
\end{aligned}
$$

where we have used the condition $\sum_{s=j+1}^{r}\alpha_s=h$. We have so proved that $q_h(\beta)\le q_h(\alpha)$ unless in the case when $k=j+1$ and $j(\alpha)\ge j+2$ or $j(\alpha)=j+1$ and $\alpha_{j+1}\neq 0$. This completes the proof of the property.

(iv) As is immediately seen, $j(\alpha)\ge j_0+1$ and $j(\beta)=j(\alpha)$. Hence, arguing as in the proof of (iii) the assertion follows easily.

(v) If $|\alpha| \le h-2$, then $q_h(\alpha) = q_h(\gamma) = 0$ by definition. If $|\alpha| = h-1$, then $q_h(\alpha) = 0$, again by definition. Moreover, $j(\gamma) = 1$ and, hence, $q_h(\gamma) = 1/2$. Similarly, $j(\alpha) = j(\gamma)$ also if $|\alpha| \ge h$ and it holds that $q_h(\gamma) = q_h(\alpha) + 1$. So, in all the cases, we conclude that $q_h(\alpha) \ge q_h(\gamma) - 1$. The inequality $q_h(\alpha) \le q_h(\beta)$ is trivial, as can easily be seen by writing, for any $\widetilde{\alpha}$ with $|\widetilde{\alpha}| > h$,

$$q_h(\widetilde{\alpha}) = \frac{1}{2}|\widetilde{\alpha}| - \frac{1}{2}h + \sum_{k=0}^{j(\widetilde{\alpha})-2} k\widetilde{\alpha}_k + (j(\widetilde{\alpha})-1)\bigg(\sum_{k=j(\widetilde{\alpha})}^{r} \widetilde{\alpha}_k - h\bigg),$$

where now all the addends and factors are nonnegative, and observing that $j(\alpha) \le j(\beta)$ as well as $\alpha_k \le \beta_k$ for any $k = 0, \ldots, r$.

(vi) Since the proof is so similar to the previous ones, the details are omitted. □

Lemma 11.1.5 *For any $k, n \in \mathbb{N}$, with $n \le k$, there exist constants $\xi_m^{(n)}$, $\widetilde{\xi}_p^{(n)}$ ($l = 1, \ldots, k$, $m = 1, \ldots, c_n$, $p = c_{n-1}+1, \ldots, c_n$) such that $\xi_1^{(0)} = 1$ and*

$$\begin{cases} \text{(a)} \quad \xi_{\ell(m)}^{(n)} + \xi_m^{(n)} > 2\widetilde{\xi}_m^{(n)}, & m > c_{n-1} \\ \text{(b)} \quad 2\xi_m^{(n)} < \widetilde{\xi}_m^{(n)}, & c_{n-1} < m \le c_n, \\ \text{(c)} \quad \widetilde{\xi}_m^{(n)} < \widetilde{\xi}_p^{(n)}, & c_{n-1} < p < m \le c_n, \\ \text{(d)} \quad 2\max\limits_{m \le c_n} \xi_m^{(n)} < \min\limits_{m \le c_{n-1}} \xi_m^{(n-1)} =: \xi^{(n)}, & \\ \text{(e)} \quad \widetilde{\xi}_m^{(n)} < \xi_p^{(n)}, & p \le c_{n-1} < m, \\ \text{(f)} \quad 2\widetilde{\xi}_m^{(n)} < \widetilde{\xi}_{\ell(m)}^{(n)}, & m, \ell(m) > c_{n-1}. \end{cases} \tag{11.1.4}$$

Proof For any $n = 1, \ldots, k+1$ we consider a positive, strictly decreasing sequence $\{a_p^{(n)}\}$ with $a_p^{(n)} < 1$, and set $\xi_p^{(n)} := (a_p^{(n)})^2$, $\widetilde{\xi}_p^{(n)} := a_p^{(n)} a_{\ell(p)}^{(n)}$. By this restriction, condition (11.1.4)(a) is satisfied. Note that, if $c_{n-1} < m < p$, then $\ell(m) < \ell(p)$, so (11.1.4)(c) is automatically satisfied by monotonicity. Note also that $\ell(m) < m$ for such values of m. Hence, if we choose $a_{p+1}^{(n)} < \frac{1}{2}a_p^{(n)}$ for all $p \in \mathbb{N}$, then also conditions (11.1.4)(b) and (11.1.4)(f) are satisfied. Now, we turn to the actual construction keeping all the above requirements on $a_p^{(n)}$. First we choose $a_p^{(1)}$ for $p = 1, \ldots, c_1 = r+1$ according to the above. Then, for $n \ge 2$ we proceed by induction, first taking $a_1^{(n)} < \frac{1}{2}\sqrt{\xi^{(n)}}$ and, then, choosing $a_p^{(n)}$ ($p = 2, \ldots, c_n$) satisfying $a_p^{(n)} < \frac{1}{2}a_{p-1}^{(n)}$ for any $p \in \mathbb{N}$ and with $a_{c_{n-1}+1}^{(n)} < a_{c_{n-1}}^{(n)} a_{c_{n-1}}^{(n)}$. The first condition implies that (11.1.4)(d) is satisfied. On the other hand, since $a_p^{(n)} < 1$, the latter condition implies (11.1.4)(e). Indeed, for $m > c_{n-1}$ and $p \le c_{n-1}$ we can estimate $a_m a_{\ell(m)} \le a_{c_{n-1}+1} < a_{c_{n-1}} a_{c_{n-1}} < a_p^2$. □

11.2 Uniform estimates for $\mathcal{A}_\varepsilon$

To investigate the elliptic and parabolic problems associated with $\mathcal{A}$, we approximate this operator by the uniformly elliptic operators $\mathcal{A}_\varepsilon := \mathcal{A} + \varepsilon\sum_{i=p_0+1}^{N} D_{ii}$ ($\varepsilon > 0$). By the results in Chapter 1 we can associate a semigroup $\{T_\varepsilon(t)\}$ in $C_b(\mathbb{R}^N)$ with the operator $\mathcal{A}_\varepsilon$. This Markov semigroup is unique, as it follows from Theorem 3.1.3 taking $\varphi(x) = 1 + |x|^2$ for any $x \in \mathbb{R}^N$ in Hypothesis 3.0.1 (take Remark 11.1.1(iii) into account). By letting ε tend to 0^+ and applying a compactness argument, we will show the existence of a semigroup associated with the operator $\mathcal{A}$. For this purpose we need to prove estimates for the spatial derivatives of $\{T_\varepsilon(t)\}$ which do not depend on $\varepsilon \in (0,1]$. This section is devoted to the proof of such estimates.

The following theorem extends the results in Theorem 5.1.7 to the operators $\mathcal{A}_\varepsilon$ and is used in the proof of Theorem 11.2.2.

Theorem 11.2.1 *For any $\varepsilon > 0$, $h \in \mathbb{N}$, $t > 0$ and $f \in C_b^h(\mathbb{R}^N)$, the function $T_\varepsilon(t)f$ belongs to $C_b^\varsigma(\mathbb{R}^N)$ (where κ is as in Hypothesis 11.0.1(ii)). Moreover, for any $h \in \mathbb{N}\cup\{0\}$, $k \in \mathbb{N}$, with $h \le k \le \varsigma$, and $T_0 > 0$, the function $(t,x) \mapsto t^{(k-h)^+/2}(D^kT_\varepsilon(t)f)(x)$ is bounded and continuous in $[0,T_0]\times\mathbb{R}^N$ and, when $k > h$, it vanishes at $t = 0$.*

Proof We prove the assertion in the case when $h = 0$, the other cases being similar and even easier. Moreover, we split the proof into two steps. In the first one, we prove that, for any $T_0 > 0$, there exists a positive constant $C = C(T_0)$, independent of f, such that

$$\|D_x^kT_\varepsilon(t)f\|_\infty \le Ct^{-\frac{k}{2}}\|f\|_\infty, \qquad t \in (0,T_0],\ \ f \in C_b(\mathbb{R}^N). \tag{11.2.1}$$

Next, in Step 2, we prove that the function $(t,x) \mapsto t^{k/2}(D_x^kT_\varepsilon(t)f)(x)$ is continuous up to $t = 0$.

Step 1. We prove (11.2.1) for functions $f \in C_c^\infty(\mathbb{R}^N)$. To extend it to any $f \in C_b(\mathbb{R}^N)$, it suffices to approximate any such function by a sequence $\{f_n\} \subset C_c^\infty(\mathbb{R}^N)$, bounded with respect to the sup-norm and converging to f locally uniformly in $\mathbb{R}^N$, as n tends to $+\infty$. The interior Schauder estimates in Theorem D.1.5(ii) and Proposition 1.2.10 show that $D_x^kT_\varepsilon(\cdot)f_n$ converges to $D_x^kT_\varepsilon(\cdot)f$ locally uniformly in $(0,+\infty)\times\mathbb{R}^N$, as n tends to $+\infty$.

So, let us fix $f \in C_c^\infty(\mathbb{R}^N)$ and let $\vartheta \in C_c^\infty(\mathbb{R})$ be a non-increasing function which satisfies the condition $\chi_{(-1/2,1/2)} \le \vartheta \le \chi_{(-1,1)}$. For any $n \in \mathbb{N}$, we define the functions $\vartheta_n : \mathbb{R}^N \to \mathbb{R}$ and $v_n : [0,+\infty)\times\mathbb{R}^N \to \mathbb{R}$, respectively, by $\vartheta_n(x) := \vartheta(n^{-1}|x|)$ and $v_n(t,x) := \sum_{m=0}^k a^mt^m\vartheta_n^{2m}(x)|D^mu_n(t,x)|^2$ for any $t \in [0,+\infty)$ and $x \in B_n$, where u_n denotes the classical solution to the Cauchy-Dirichlet problem in the ball B_n with initial value f, and a is a positive parameter to be fixed later on. To simplify the notation, we drop out the index n, when there is no danger of confusion. Moreover, we arbitrarily fix $T_0 > 0$, $t \in (0,T_0]$ and denote by C_i positive constants, independent of ε, n and t.

Clearly, if $B_n \supset \text{supp}(f)$, then v is continuous in $[0,T_0]\times\overline{B}_R$. Moreover, v vanishes on $[0,T_0]\times\partial B_n$, $v(0,\cdot) = f^2$ in B_n and $D_tv \le \mathcal{A}v + g$ in $[0,T_0]\times B_n$, where $g = \sum_{j=1}^3 g_j$ with

$$g_1(t,\cdot) = -2\sum_{m=0}^k a^mt^m\vartheta^{2m}|\sqrt{Q}\nabla_xD_x^mu(t,\cdot)|^2 + \sum_{m=1}^k ma^mt^{m-1}\vartheta^{2m}|D_x^mu(t,\cdot)|^2$$

$$\begin{aligned} g_2(t,\cdot) = &-2\mathcal{A}\vartheta\sum_{m=1}^k ma^mt^m\vartheta^{2m-1}|D_x^mu(t,\cdot)|^2 \\ &-8\sum_{i=1}^N\sum_{m=1}^k ma^mt^m\vartheta^{2m-1}(Q\nabla\vartheta)_i\langle D_x^mu(t,\cdot),D_x^mD_iu(t,\cdot)\rangle, \end{aligned}$$

$$g_3(t,\cdot) = 2\sum_{m=1}^k a^mt^m\vartheta^{2m}\langle[D_x^m,\mathcal{A}]u(t,\cdot),D_x^mu(t,\cdot)\rangle.$$

Using the ellipticity assumption on q_{ij} we get

$$g_1(t,\cdot) \le -2\kappa\sum_{m=1}^{k+1}(at\vartheta^2)^{m-1}|D_{x,\star}^mu(t,\cdot)|^2 - 2\varepsilon\sum_{m=1}^{k+1}(at\vartheta^2)^{m-1}|D_{x,\star\star}^mu(t,\cdot)|^2$$

$$+k\sum_{m=1}^{k} a^m t^{m-1}\vartheta^{2m}|D_x^m u(t,\cdot)|^2,$$

where $D^m_{x,\star}u$ (respectively $D^m_{x,\star\star}u$) denotes the vector whose entries are the m-th order derivatives $D^m_{i_1,\dots,i_m}u$ with $i_j \le p_0$ for some $j=1,\dots,m$ (respectively $i_j > p_0$ for all $j=1,\dots,m$).

In view of Hypotheses 11.0.1, we can estimate $|\mathcal{A}\vartheta| \le C_1$, $|(Q\nabla\vartheta)_i| \le C_1\sqrt{\kappa}$, if $i \le p_0$, and $|(Q\nabla\vartheta)_i| \le C_1\varepsilon$, if $i > p_0$, which together with the Young inequality imply that $|a^m t^m\vartheta^{2m-1}(Q\nabla\vartheta)_i\langle D_x^m u(t,\cdot), D_x^m D_i u(t,\cdot)\rangle|$ can be bounded from above by $C_1 a^{m-\frac{1}{2}}t^{m-\frac{1}{2}}\vartheta^{2m-2}(|D_x^m u(t,\cdot)|^2 + \kappa a t\vartheta^2|D^{m+1}_{\star,x}u(t,\cdot)|^2)$, if $i \le p_0$, and by $C_1\varepsilon a^{m-\frac{1}{2}}t^{m-\frac{1}{2}}\vartheta^{2m-2}(|D_x^m u(t,\cdot)|^2 + at\vartheta^2|D_x^{m+1}u(t,\cdot)|^2)$, if $i > p_0$. Hence,

$$\begin{aligned}g_2(t,\cdot) \le& 2C_2\sum_{m=1}^{k}(at)^m\vartheta^{2m-1}|D_x^m u(t,\cdot)|^2 + C_2\sum_{m=1}^{k+1}(at)^{m-\frac{1}{2}}\vartheta^{2m-2}|D_x^m u(t,\cdot)|^2\\ &+C_2\sum_{m=1}^{k}(at)^{m+\frac{1}{2}}\vartheta^{2m}\kappa|D^{m+1}_{x,\star}u(t,\cdot)|^2.\end{aligned}$$

The function g_3 can be estimated similarly, taking Hypothesis 11.0.1(ii) into account, and we get

$$\begin{aligned}g_3(t,\cdot) \le& C_3\kappa\sum_{m=1}^{k+1} a^{m-\frac{1}{2}}t^{m-\frac{1}{2}}\vartheta^{2m}|D^m_{x,\star}u(t,\cdot)|^2 + C_3\sum_{m=1}^{k} a^m t^m\vartheta^{2m}|D_x^m u(t,\cdot)|^2\\ &+C_3\sum_{m=1}^{k} a^{m-\frac{1}{2}}t^{m-\frac{1}{2}}\vartheta^{2m}|D_x^m u(t,\cdot)|^2.\end{aligned}$$

From the previous estimate we can infer that

$$g(t,\cdot) \le \sum_{m=1}^{k+1} a^{m-1}t^{m-1}\vartheta^{2m-2}\big[M^\star(a,T_0)|D^m_{x,\star}u(t,\cdot)|^2 + M^{\star\star}|D^m_{x,\star\star}u(t,\cdot)|^2\big],$$

where $M^\star(a,T_0) := [-2 + C_4\sqrt{a}\sqrt{T_0}]\kappa + C_4\sqrt{a}\sqrt{T_0} + [k + (C_2 + C_4)T_0]a$, $M^{\star\star}(a,T_0) := -2\varepsilon + C_4\sqrt{a}\sqrt{T_0} + (k + 2C_1T_0 + C_3T_0)a$ and $C_4 = C_2 + C_3$. It follows that we can fix $a > 0$ small enough (and independent of n) such that $g \le 0$ in $[0,T_0]\times\overline{B}_n$. The classical maximum principle yields $|v_n(t,x)| \le \|f\|_\infty^2$ and, therefore, $t^m\kappa_n^{2m}(x)|D_x^m u_n(t,x)|^2 \le C_m\|f\|^2_\infty$ for any $(t,x)\in[0,T_0]\times\overline{B}_n$. Estimate (11.2.1) follows by letting n tend to $+\infty$. Indeed, as Theorem 1.2.1 shows, u_n tends to $T_\varepsilon(\cdot)f$ pointwise in $[0,+\infty)\times\mathbb{R}^N$, as n tends to $+\infty$.

Step 2. We now conclude the proof by showing that the function w_k, defined by $w_k(t,x) := t^{-k/2}D_x^k u(t,x)$ for any $(t,x)\in(0,+\infty)\times\mathbb{R}^N$ and $w_k(0,\cdot)\equiv 0$ in $\mathbb{R}^N$, is continuous in $[0,+\infty)\times\mathbb{R}^N$.

If $f\in C_c^\infty(\mathbb{R}^N)$, then this claim is easily checked. Indeed, for any $T_0 > 0$ and $M\in\mathbb{N}$, there exists a positive constant $C_k = C_k(M,T_0)$ such that $\|u_n\|_{C^{k+\theta/2,2k+\theta}((0,T_0)\times B_M)} \le C_k\|f\|_{C_c^{2k+\theta}(\mathbb{R}^N)}$ for any $n\in\mathbb{N}$ (see Theorem D.1.5(ii)). Hence, a compactness argument shows that u_n and its spatial derivatives, up to the order k, converge locally uniformly in $[0,+\infty)\times\mathbb{R}^N$, as n tends to $+\infty$, and the continuity of $D_x^k T_\varepsilon(\cdot)$ in $[0,+\infty)\times\mathbb{R}^N$ follows at once.

For a general function $f\in C_b(\mathbb{R}^N)$, we use the same approximating argument as in Step 1. By Proposition B.3.4 we can estimate $\|\psi\|_{C^k(\overline{B}_M)} \le C_k'\|\psi\|^{1/(k+1)}_{C(\overline{B}_M)}\|\psi\|^{k/(k+1)}_{C^{k+1}(\overline{B}_M)}$ for any $\psi \in C^{k+1}(\overline{B}_M)$ and some positive constant $C_k' = C_k'(M)$. Applying this inequality to the function $\psi = t^{k/2}T_\varepsilon(t)f_n - t^{k/2}T_\varepsilon(t)f$ and using inequality (11.2.1), we get $\|t^{k/2}D_x^kT_\varepsilon(t)f_n - t^{k/2}D_x^kT_\varepsilon(t)f\|_{C(\overline{B}_M)} \le C_k''\|T_\varepsilon(\cdot)f_n - T_\varepsilon(\cdot)f\|_{C([0,T_0]\times\overline{B}_M)}^{1/(k+1)}$ for some positive constant C_k'' and any $t \in (0, T_0]$. The right-hand side of this estimate vanishes as n tends to $+\infty$ thanks to Proposition 1.2.10. Since T_0 and M are arbitrarily fixed, the function $(t,x) \mapsto t^{k/2}(D_x^kT_\varepsilon(t)f)(x)$ is continuous in $[0,+\infty)\times\mathbb{R}^N$ and vanishes at $t=0$. □

Theorem 11.2.2 *For any $T_0 > 0$, any multi-index $\alpha \in (\mathbb{N}\cup\{0\})^N$ and $h \in \mathbb{N}\cup\{0\}$ such that $h \le |\alpha| \le \varsigma$, there exists a positive constant $\widetilde{C}$, depending on α, but being independent of $\varepsilon \in (0,1]$, such that, for any $f \in C_b(\mathbb{R}^N)$, it holds that*

$$\|D_x^\alpha T_\varepsilon(t)f\|_\infty \le \widetilde{C}t^{-q_h(\widehat{\alpha})}\|f\|_{C_b^h(\mathbb{R}^N)}, \qquad t \in (0,T_0],\ \varepsilon \in (0,1]. \tag{11.2.2}$$

Proof Throughout the proof, we denote by $s_p^{(n)}$ the cardinality of the set $\{\alpha \in (\mathbb{N}\cup\{0\})^N : |\alpha| = i_p^{(n)}\}$.

We fix $\varepsilon \in (0,1]$, $h,k \in \mathbb{N}$ with $h \le k \le \varsigma$, $f \in C_b^h(\mathbb{R}^N)$. Further, we consider the function $v_\varepsilon = u_\varepsilon^2 + \sum_{n=1}^k\langle \mathcal{H}^{(n)}D_x^nu_\varepsilon, D_x^nu_\varepsilon\rangle$, defined in $[0,+\infty)\times\mathbb{R}^N$, where $u_\varepsilon = T_\varepsilon(\cdot)f$ and the matrices $\mathcal{H}^{(n)}(t)$ $(n=1,\ldots,k)$ are split into c_n blocks $H_{m,p}^{(n)}(t)$ according to the splitting of the vector $D_x^nu_\varepsilon$ introduced in Subsection 11.1.1. More precisely,

$$H_{m,p}^{(n)}(t) = t^{q_h(i_m^{(n)})+q_h(i_p^{(n)})}\begin{cases} a^{\xi_p^{(n)}}I_{s_p^{(n)}}, & \text{if } m = p,\\ a^{\widetilde{\xi}_m^{(n)}}H_m^{(n)}, & \text{if } m > c_{n-1} \text{ and } p = \ell(m),\\ a^{\widetilde{\xi}_p^{(n)}}(H_p^{(n)})^*, & \text{if } p > c_{n-1} \text{ and } m = \ell(p),\\ 0, & \text{otherwise,}\end{cases}$$

for any $t > 0$ and some constant $s_m^{(n)} \times s_{\ell(m)}^{(n)}$-matrices $H_m^{(n)}$ with Euclidean norm equal to one and such that the matrix $-H_m^{(n)}\mathcal{J}_m^{(n)} - (H_m^{(n)}\mathcal{J}_m^{(n)})^*$ is positive definite for any $n = 1,\ldots,k$ and $m > c_{n-1}$. Here, the matrix $\mathcal{J}_m^{(n)}$ and the constants $\xi_m^{(n)}$ and $\widetilde{\xi}_m^{(n)}$ are given by Lemmata 11.1.3 and 11.1.5. For the moment, a is any positive parameter satisfying the following condition

$$a^{\xi_{\ell(m)}^{(n)}+\xi_m^{(n)}-2\widetilde{\xi}_m^{(n)}} > 2, \qquad n = 1,\ldots,k,\ m > c_{n-1}. \tag{11.2.3}$$

This condition and (11.1.4) guarantee that the matrices $\mathcal{H}^{(n)}$ are all positive definite.

Theorem 11.2.1 shows that the function v_ε is continuous in $[0,+\infty)\times\mathbb{R}^N$, $v_\varepsilon(0,\cdot) = \sum_{n=0}^k\langle\mathcal{H}^{(n)}(0)D^nf, D^nf\rangle$ and $D_tv_\varepsilon = \mathcal{A}_\varepsilon v_\varepsilon + g_\varepsilon$ in $(0,+\infty)\times\mathbb{R}^N$, where

$$\begin{aligned} g_\varepsilon = & -2\sum_{i,j=1}^N\sum_{n=0}^k q_{ij}^\varepsilon\langle\mathcal{H}^{(n)}D_x^nD_iu_\varepsilon, D_x^nD_ju_\varepsilon\rangle + 2\sum_{n=1}^k\langle\mathcal{H}^{(n)}[D_x^n,\langle B\cdot,\nabla_x\rangle]u_\varepsilon, D_x^nu_\varepsilon\rangle\\ & + \sum_{n=h+1}^k\langle\dot{\mathcal{H}}^{(n)}D_x^nu_\varepsilon, D_x^nu_\varepsilon\rangle + 2\sum_{n=1}^k\langle\mathcal{H}^{(n)}[D_x^n,\mathrm{Tr}(Q_\varepsilon D_x^2)]u_\varepsilon, D_x^nu_\varepsilon\rangle\\ & + 2\sum_{n=1}^k\langle\mathcal{H}^{(n)}[D_x^n,\langle F, D_1^1\rangle]u_\varepsilon, D_x^nu_\varepsilon\rangle := \sum_{j=1}^5 g_{j,\varepsilon}, \end{aligned} \tag{11.2.4}$$

the matrix $\dot{\mathcal{H}}^{(n)}$ is obtained by differentiating the matrix $\mathcal{H}^{(n)}$ componentwise with respect to time, and $D^0 D_i u_\varepsilon = D_i u_\varepsilon$. Here, the commutators are understood coordinatewise. When $h = k$ we agree that the first sum in the second line of (11.2.4) disappears.

We claim that, for $t \le 1$ and a satisfying (11.2.3), we can estimate the functions g_j ($j = 1, \dots, 5$) as follows:

$$g_{1,\varepsilon}(t,\cdot) \le -2\kappa|\partial_1^1 u_\varepsilon(t,\cdot)|^2 - 2\kappa \sum_{n=2}^{k+1}\sum_{m=1}^{c_{n-1}} (a^{\xi^{(n)}} + o(a^{\xi^{(n)}}))t^{(2q_h(i_m^{(n)})-1)^+}|\partial_m^n u_\varepsilon(t,\cdot)|^2, \tag{11.2.5}$$

$$\begin{aligned} g_{2,\varepsilon}(t,\cdot) \le & -\lambda^{(k)} \sum_{n=1}^{k}\sum_{m=c_{n-1}+1}^{c_n} (a^{\tilde{\xi}_m^{(n)}} + o(a^{\tilde{\xi}_m^{(n)}}))t^{(2q_h(i_m^{(n)})-1)^+}|\partial_m^n u_\varepsilon(t,\cdot)|^2 \\ & + \sum_{n=1}^{k}\sum_{m=1}^{c_{n-1}} o(a^{\xi^{(n)}})t^{(2q_h(i_m^{(n)})-1)^+}|\partial_m^n u_\varepsilon(t,\cdot)|^2, \end{aligned} \tag{11.2.6}$$

$$\begin{aligned} g_{3,\varepsilon}(t) \le & \sum_{n=h+1}^{k}\sum_{m=1}^{c_{n-1}} o(a^{\xi^{(n)}})t^{2q_h(i_m^{(n)})-1}|\partial_m^n u_\varepsilon(t,\cdot)|^2 \\ & + \sum_{n=h+1}^{k}\sum_{m=c_{n-1}+1}^{c_n} o(a^{\tilde{\xi}_m^{(n)}})t^{2q_h(i_m^{(n)})-1}|\partial_m^n u_\varepsilon(t,\cdot)|^2, \end{aligned} \tag{11.2.7}$$

$$\begin{aligned} g_{4,\varepsilon}(t,\cdot) \le & \kappa \sum_{n=1}^{k+1}\sum_{m=1}^{c_{n-1}} o(a^{\xi^{(n)}})t^{(2q(i_m^{(n)})-1)^+}|\partial_m^n u_\varepsilon(t,\cdot)|^2 \\ & + \sum_{n=1}^{k}\sum_{m=c_{n-1}+1}^{c_n} o(a^{\tilde{\xi}_m^{(n)}})t^{(2q(i_m^{(n)})-1)^+}|\partial_m^n u_\varepsilon(t,\cdot)|^2, \end{aligned} \tag{11.2.8}$$

$$\begin{aligned} g_{5,\varepsilon}(t,\cdot) \le & \kappa \sum_{n=1}^{k}\sum_{m=1}^{c_{n-1}} o(a^{\xi^{(n)}})t^{(2q_h(i_m^{(n)})-1)^+}|\partial_m^n u_\varepsilon(t,\cdot)|^2 \\ & + \sum_{n=1}^{k}\sum_{m=c_{n-1}+1}^{c_n} o(a^{\tilde{\xi}_m^{(n)}})t^{(2q_h(i_m^{(n)})-1)^+}|\partial_m^n u_\varepsilon(t,\cdot)|^2, \end{aligned} \tag{11.2.9}$$

where $\lambda^{(k)}$ is any positive constant less than the minimum eigenvalue of the matrix $-H_m^{(n)}\mathcal{J}_m^{(n)} - (H_m^{(n)}\mathcal{J}_m^{(n)})^*$ for $n \in \{1, \dots, k\}$ and $m \in \{c_{n-1}+1, \dots, c_n\}$. Finally, the Landau symbols are used as a tends to $+\infty$.

From (11.2.5)-(11.2.9) it follows that

$$\begin{aligned} g_\varepsilon(t,\cdot) \le & -2\kappa \sum_{n=1}^{k}\sum_{m=1}^{c_{n-1}} (a^{\xi^{(n)}} + o(a^{\xi^{(n)}}))t^{(2q_h(i_m^{(n)})-1)^+}|\partial_m^n u_\varepsilon(t,\cdot)|^2 \\ & - \lambda^{(k)} \sum_{n=1}^{k}\sum_{m=c_{n-1}+1}^{c_n} (a^{\tilde{\xi}_m^{(n)}} + o(a^{\tilde{\xi}_m^{(n)}}))t^{(2q_h(i_m^{(n)})-1)^+}|\partial_m^n u_\varepsilon(t,\cdot)|^2 \\ & - 2\kappa \sum_{m=1}^{c_k} (a^{\xi^{(k+1)}} + o(a^{\xi^{(k+1)}}))t^{(2q_h(i_m^{(k+1)})-1)^+}|\partial_m^{k+1} u_\varepsilon(t,\cdot)|^2 \end{aligned}$$

for $t \in (0,1]$. If we now fix the parameter a sufficiently large and t sufficiently small (both independent of ε), then condition (11.2.3) is satisfied and $g_\varepsilon \le 0$ in an interval

$(0,T_*]$. Theorem 3.1.3 shows that $v_\varepsilon \le \sum_{n=0}^h \langle \mathcal{H}^{(n)}(0)D^n f, D^n f\rangle$ in $(0,T_*]\times\mathbb{R}^N$; hence, $\langle \mathcal{H}^{(j)}D_x^j u_\varepsilon, D_x^j u_\varepsilon\rangle \le \widehat{C}\|f\|^2_{C_b^h(\mathbb{R}^N)}$ in $(0,T_*]\times\mathbb{R}^N$, for any $j=1,\ldots,k$ and some positive constant $\widehat{C}$. Since the matrices $\mathcal{H}^{(j)}(t)$ are positive definite for any j and t, estimate (11.2.2) follows in the time interval $(0,T_*]$. If $T_0 > T_*$, then the semigroup property allows us to extend this estimate to the interval $[0,T_0]$.

The rest of the proof is devoted to proving estimates (11.2.5)-(11.2.9). To simplify the notation, we do not stress the dependence on ε. Moreover, we assume that $t\le 1$ and a satisfies (11.2.3).

Proof of estimate (11.2.5). Using Hypothesis 11.0.1(ii) and the fact that the matrix whose entries are $\langle \mathcal{H}^{(n)}D_x^n D_i u_\varepsilon, D_x^n D_j u_\varepsilon\rangle$ in nonnegative definite, we get $g_1(t,\cdot)\le -2\kappa\sum_{j=1}^{p_0}\sum_{n=0}^k\langle \mathcal{H}^{(n)}(t)D_x^n D_j u(t,\cdot), D_x^n D_j u(t,\cdot)\rangle$. By the definition of the matrices $\mathcal{H}^{(n)}$ $(n=1,\ldots,k)$ we obtain

$$
\begin{aligned}
&\sum_{j=1}^{p_0}\sum_{n=1}^{k}\langle \mathcal{H}^{(n)}(t)D_x^n D_j u(t,\cdot), D_x^n D_j u(t,\cdot)\rangle\\
\ge&\sum_{j=1}^{p_0}\sum_{n=1}^{k}\sum_{m=1}^{c_n} a^{\xi_m^{(n)}} t^{2q_h(i_m^{(n)})}|\partial_m^n D_j u(t,\cdot)|^2\\
&-2\sum_{j=1}^{p_0}\sum_{n=1}^{k}\sum_{m=c_{n-1}+1}^{c_n} a^{\widetilde{\xi}_m^{(n)}} t^{q_h(i_m^{(n)})+q_h(i_{\ell(m)}^{(n)})}|\partial_{\ell(m)}^n D_j u(t,\cdot)||\partial_m^n D_j u(t,\cdot)|.
\end{aligned}
$$

Thanks to (11.1.4)(a), we can fix $\gamma_m^{(n)}$ and $\gamma_{\ell(m)}^{(n)}$ such that $\gamma_m^{(n)} < \xi_m^{(n)}$, $\gamma_{\ell(m)}^{(n)} < \xi_{\ell(m)}^{(n)}$ and $2\widetilde{\xi}_m^{(n)} = \gamma_{\ell(m)}^{(n)} + \gamma_m^{(n)}$. Using the Young inequality we get

$$
\begin{aligned}
&2a^{\widetilde{\xi}_m^{(n)}} t^{q_h(i_m^{(n)})+q_h(i_{\ell(m)}^{(n)})}|\partial_{\ell(m)}^n D_j u(t,\cdot)||\partial_m^n D_j u(t,\cdot)|\\
\le& a^{\gamma_{\ell(m)}^{(n)}} t^{2q_h(i_{\ell(m)}^{(n)})}|\partial_{\ell(m)}^n D_j u(t,\cdot)|^2 + a^{\gamma_m^{(n)}} t^{2q_h(i_m^{(n)})}|\partial_m^n D_j u(t,\cdot)|^2\\
=&o(a^{\xi_{\ell(m)}^{(n)}})t^{2q_h(i_{\ell(m)}^{(n)})}|\partial_{\ell(m)}^n D_j u(t,\cdot)|^2 + o(a^{\xi_m^{(n)}})t^{2q_h(i_m^{(n)})}|\partial_m^n D_j u(t,\cdot)|^2.
\end{aligned}
$$

Hence, we can infer that

$$
g_1(t,\cdot)\le -2\kappa|\partial_1^1 u(t,\cdot)|^2 - 2\kappa\sum_{j=1}^{p_0}\sum_{n=1}^{k}\sum_{m=1}^{c_n} a^{\xi^{(n)}}(1+o(1))t^{2q_h(i_m^{(n)})}|\partial_m^n D_j u(t,\cdot)|^2.
$$

Now, we observe that, by (11.1.4)(d), $a^{\xi_m^{(n)}}$ is greater than $a^{\xi^{(n)}}$. Fix $j\le p_0$, $n\le k$ and $m\le c_n$ and take any derivative D_x^α in the block ∂_m^n. Then, $D_x^\alpha D_j u$ belongs to some block $\partial_{m'}^{n+1}$ with $m'\le c_{n+1}-1$ and, by Lemma 11.1.4(ii), $q_h(i_{m'}^{(n+1)}) = (q_h(i_m^{(n)})+1/2)^+$. From these remarks, estimate (11.2.5) follows at once.

Proof of estimate (11.2.6). Lemma 11.1.3 and a straightforward computation show that

$$
\begin{aligned}
g_2(t,\cdot) =& 2\sum_{n=1}^{k}\sum_{m=c_{n-1}+1}^{c_n} a^{\widetilde{\xi}_m^{(n)}} t^{q_h(i_m^{(n)})+q_h(i_{\ell(m)}^{(n)})}\langle H_m^{(n)}\mathcal{J}_m^{(n)}\partial_m^n u(t,\cdot), \partial_m^n u(t,\cdot)\rangle\\
&+2\sum_{n=1}^{k}\sum_{m=c_{n-1}+1}^{c_n} a^{\widetilde{\xi}_m^{(n)}} t^{q_h(i_m^{(n)})+q_h(i_{\ell(m)}^{(n)})}\sum_{p\in A_m^{(n)}\setminus\{m\}}\langle \mathcal{M}_{m,p}^{(n)}\partial_p^n u(t,\cdot), \partial_m^n u(t,\cdot)\rangle\\
&+2\sum_{n=1}^{k}\sum_{m=1}^{c_n}\sum_{s\in B_m^{(n)}} a^{\xi_m^{(n)}} t^{2q_h(i_m^{(n)})}\langle \mathcal{N}_{m,s}^{(n)}\partial_s^n u(t,\cdot), \partial_m^n u(t,\cdot)\rangle
\end{aligned}
$$

$$+2\sum_{n=1}^{k}\sum_{m=c_{n-1}+1}^{c_n}\sum_{s\in B_m^{(n)}} a^{\widetilde{\xi}_m^{(n)}} t^{q_h(i_m^{(n)})+q_h(i_{\ell(m)}^{(n)})}\langle (H_m^{(n)})^*\mathcal{N}_{m,s}^{(n)}\partial_s^n u(t,\cdot),\partial_{\ell(m)}^n u(t,\cdot)\rangle,$$

(11.2.10)

where $A_m^{(n)}$ is given by (11.1.1) and, if $i_m^{(n)} = (\alpha_0,\ldots,\alpha_r)$, then $B_m^{(n)}$ is the set of all integers s such that $i_s^{(n)} = i_m^{(n)} - e_j^{(r+1)} + e_{j'}^{(r+1)}$ for some $j = 0,\ldots,r$, with $\alpha_j > 0$, and some $j' \le (j+1)\wedge r$. Moreover, the entries of the matrices $\mathcal{M}_{m,p}^{(n)}$ ($m = c_{n-1}+1,\ldots,c_n$, $p\in A_m^{(n)}$) and $\mathcal{N}_{m,s}^{(n)}$ ($m = 1,\ldots,c_n$, $s\in B_m^{(n)}$) depend only on the entries of B.

Denote by $g_{2,i}$ ($i = 1,\ldots,4$) the terms in the right-hand side of the formula (11.2.10). Since by properties (i), (iv) in Lemma 11.1.4 $q_h(i_{\ell(m)}^{(n)}) + q_h(i_m^{(n)}) = (2q_h(i_m^{(n)})-1)^+$ and $2\langle H_m^{(n)}\mathcal{J}_m^{(n)}\partial_m^n u(t,\cdot),\partial_m^n u(t,\cdot)\le -\lambda^{(k)}|\partial_m^n u(t,\cdot)|^2$, due to the choice of the matrices $H_m^{(n)}$, we can estimate

$$g_{2,1}(t,\cdot)\le -\lambda^{(k)}\sum_{n=1}^{k}\sum_{m=c_{n-1}+1}^{c_n} a^{\widetilde{\xi}_m^{(n)}} t^{(2q_h(i_m^{(n)})-1)^+}|\partial_m^n u(t,\cdot)|^2. \qquad (11.2.11)$$

Now, we consider the functions $g_{2,2}$ and $g_{2,3}$ and denote, respectively, by $\zeta_{m,p}^{(n)}$ and $\psi_{m,p}^{(n)}$ the terms that appear in the sums which define these two functions. We set $\beta_q^{(n)} = \xi^{(n)}$, if $q\le c_{n-1}$, and $\beta_q^{(n)} = \widetilde{\xi}_q^{(n)}$, otherwise, and claim that

$$\zeta_{m,p}^{(n)}(t,\cdot)\le o(a^{\beta_p^{(n)}})t^{(2q_h(i_p^{(n)})-1)^+}|\partial_p^n u(t,\cdot)|^2 + o(a^{\widetilde{\xi}_m^{(n)}})t^{(2q_h(i_m^{(n)})-1)^+}|\partial_m^n u(t,\cdot)|^2, \quad (11.2.12)$$

$$\psi_{m',s}^{(n)}(t,\cdot)\le o(a^{\beta_s^{(n)}})t^{(2q_h(i_s^{(n)})-1)^+}|\partial_s^n u(t,\cdot)|^2 + o(a^{\beta_{m'}^{(n)}})t^{(2q_h(i_{m'}^{(n)})-1)^+}|\partial_{m'}^n u(t,\cdot)|^2 \quad (11.2.13)$$

for $m\in\{c_{n-1}+1,\ldots,c_n\}$, $m'\le c_n$, $p\in A_m^{(n)}\setminus\{m\}$ and $s\in B_{m'}^{(n)}$. For this purpose, we arbitrarily fix (m,m',p,s) as above and observe that from Lemma 11.1.4(vi) we can infer that $q_h(i_p^{(n)})-1\le q_h(i_{\ell(m)}^{(n)})$ and $q_h(i_s^{(n)})-1\le q_h(i_{m'}^{(n)})$. Thus,

$$t^{q_h(i_m^{(n)})+q_h(i_{\ell(m)}^{(n)})}\le t^{\frac{(2q_h(i_m^{(n)})-1)^+}{2}+\frac{(2q_h(i_p^{(n)})-1)^+}{2}},$$

$$t^{2q_h(i_{m'}^{(n)})}\le t^{\frac{(2q_h(i_s^{(n)})-1)^+}{2}+\frac{(2q_h(i_{m'}^{(n)})-1)^+}{2}}.$$

We now distinguish some cases.

- $p,s\le c_{n-1}$. Using conditions (11.1.4)(d) and (11.1.4)(e), we obtain $2\widetilde{\xi}_m^{(n)} < \xi^{(n)} + \widetilde{\xi}_m^{(n)}$ and $2\xi_{m'}^{(n)} < \xi^{(n)} + \beta_{m'}^{(n)}$. Therefore, we conclude that

$$\zeta_{m,p}^{(n)}(t,\cdot)\le o(a^{\frac{1}{2}\xi^{(n)}})t^{\frac{(2q_h(i_p^{(n)})-1)^+}{2}}|\partial_p^n u(t,\cdot)|t^{\frac{(2q_h(i_m^{(n)})-1)^+}{2}}o(a^{\frac{1}{2}\widetilde{\xi}_m^{(n)}})|\partial_m^n u(t,\cdot)|$$

  and (11.2.12) follows by applying Cauchy-Schwarz inequality. Arguing in the same way, we get also (11.2.13).

- $m'\le c_{n-1}$ and $s > c_{n-1}$. We use (11.1.4)(d) to estimate $2\xi_m^{(n)} < \xi^{(n)} + \widetilde{\xi}_s^{(n)}$ and arguing as in the previous case, we still get (11.2.13).

- $m',p,s > c_{n-1}$. Observe that $p < m$ for $p\in A_m^{(n)}\setminus\{m\}$ and, hence, from condition (11.1.4)(c) we obtain $2\widetilde{\xi}_m^{(n)} < \widetilde{\xi}_m^{(n)} + \widetilde{\xi}_p^{(n)}$. It also holds that $2\xi_m^{(n)} < \widetilde{\xi}_m^{(n)} + \widetilde{\xi}_s^{(n)}$ by (11.1.4)(b). These inequalities and the same argument as above, yield to (11.2.12) and (11.2.13).

The function $g_{2,4}$ can be estimated similarly, using Lemma 11.1.4(iii) to infer that

$$t^{q_h(i_m^{(n)})+q_h(i_{\ell(m)}^{(n)})} \le t^{\frac{(2q_h(i_s^{(n)})-1)^+}{2}+\frac{(2q_h(i_{\ell(m)}^{(n)})-1)^+}{2}}, \qquad m > c_{n-1}\ s \in B_m^{(n)},$$

and observing that $2\widetilde{\xi}_m^{(n)} < \widetilde{\xi}_{\ell(m)}^{(n)}$, if $\ell(m) > c_{n-1}$ (see (11.1.4)(f)), and $2\widetilde{\xi}_m^{(n)} < \xi^{(n)}$, if $\ell(m) \le c_{n-1}$ (see (11.1.4)(d) and (11.1.4)(e)). We obtain

$$\begin{aligned}&|a^{\widetilde{\xi}_m^{(n)}} t^{q_h(i_m^{(n)})+q_h(i_{\ell(m)}^{(n)})}\langle (H_m^{(n)})^*\mathcal{N}_{m,s}^{(n)}\partial_s^n u(t,\cdot), \partial_{\ell(m)}^n u(t,\cdot)\rangle|\\ \le& o(a^{\beta_s^{(n)}})t^{(2q_h(i_s^{(n)})-1)^+}|\partial_s^n u(t,\cdot)|^2 + o(a^{\beta_{\ell(m)}^{(n)}})t^{(2q_h(i_{\ell(m)}^{(n)})-1)^+}|\partial_{\ell(m)}^n u(t,\cdot)|^2. \end{aligned} \tag{11.2.14}$$

Combining (11.2.10)-(11.2.14), estimate (11.2.6) follows.

Proof of estimate (11.2.7). This term occurs only if $h < k$. First of all we observe that

$$\begin{aligned} g_3(t,\cdot) =& 2\sum_{n=h+1}^{k}\sum_{m=1}^{c_n} q_h(i_m^{(n)})a^{\xi_m^{(n)}}t^{2q_h(i_m^{(n)})-1}|\partial_m^n u(t,\cdot)|^2\\ &+2\sum_{n=h+1}^{k}\sum_{m=c_{n-1}+1}^{c_n}\left(q_h(i_m^{(n)})+q_h(i_{\ell(m)}^{(n)})\right)a^{\widetilde{\xi}_m^{(n)}}t^{q_h(i_m^{(n)})+q_h(i_{\ell(m)}^{(n)})-1}\\ &\qquad\times\langle H_m D_{\ell(m)}^n u(t,\cdot), \partial_m^n u(t,\cdot)\rangle. \end{aligned} \tag{11.2.15}$$

Moreover, from Lemma 11.1.4(iv) it follows that $q_h(i_m^{(n)})+q_h(i_{\ell(m)}^{(n)})-1>0$ if $m > c_{n-1}$. By (11.1.4)(a), (d), (e), we have that $2\widetilde{\xi}_m^{(n)} < \xi^{(n)}+\widetilde{\xi}_m^{(n)}$, if $\ell(m)\le c_{n-1}$, and $2\widetilde{\xi}_m^{(n)} < \widetilde{\xi}_m^{(n)}+\widetilde{\xi}_{\ell(m)}^{(n)}$, if $\ell(m) > c_{n-1}$. Hencc, we can estimate

$$\begin{aligned}&a^{\widetilde{\xi}_m^{(n)}}t^{q_h(i_m^{(n)})+q_h(i_{\ell(m)}^{(n)})-1}|\partial_{\ell(m)}^n u(t,\cdot)||\partial_m^n u(t,\cdot)|\\ \le& o(a^{\beta_{\ell(m)}^{(n)}})t^{2q_h(i_{\ell(m)}^{(n)})-1}|\partial_{\ell(m)}^n u(t,\cdot)|^2 + o(a^{\widetilde{\xi}_m^{(n)}})t^{2q_h(i_m^{(n)})-1}|\partial_m^n u(t,\cdot)|^2,\end{aligned}$$

where $\beta_{\ell(m)}^{(n)}$ is as above. From this inequality and (11.2.15), estimate (11.2.7) follows at once.

Proof of estimates (11.2.8) *and* (11.2.9). We begin by proving (11.2.8) and observe that, for any $m = 1,\ldots,c_n$, it holds that $[\partial_m^n, \mathrm{Tr}(Q_\varepsilon D_x^2)]u = \sum_{i,j=1}^N[\partial_m^n, q_{ij}^\varepsilon D_{ij}]u = \sum_{q=2}^{n+1}\mathcal{P}_m^{(n,q)}D_x^q u$ for some matrices $\mathcal{P}_m^{(n,q)}$ whose entries linearly depend only on the derivatives (of order at most n) of the functions q_{ij} $(i,j = 1,\ldots,p_0)$. If we split the matrices $\mathcal{P}_m^{(n,q)}$ into sub-blocks $P_{m,s}^{(n,q)}$ $(m = 1,\ldots,c_n,\ s = 1,\ldots,c_q)$ according to the rule in Subsection 11.1.1, then it follows that $P_{m,s}^{(n,q)} = 0$ if $s > c_{q-1}$. Indeed, consider a derivative $D_x^\alpha \in \partial_m^n$. Then, $[D_x^\alpha, q_{ij}^\varepsilon D_{ij}]u$ contains only derivatives $D_x^\beta u$ of order at most $|\alpha|+1$. Moreover,[6] $\widehat{\beta} = (|\beta_0|,\ldots,|\beta_r|)$, with $1 < |\beta_0| \le |\alpha_0|+2$ and $|\beta_j| \le |\alpha_j|$ for any $j = 1,\ldots,r$. Since $|\beta_0| > 1$, it holds that $\widehat{\beta} = i_s^{(q)}$ with $s \le c_{q-1}$, where $q = |\beta|$. Denote by $C_m^{(n)}$ the indices $i_s^{(q)}$ obtained in this way from multi-indices α with $\widehat{\alpha} = i_m^{(n)}$. Lemma 11.1.4(v) shows that $q_h(i_m^{(n)}) \ge q_h(i_s^{(q)}) - 1$ for $i_s^{(q)} \in C_m^{(n)}$ and $q_h(i_{\ell(m)}^{(n)}) \ge q_h(i_s^{(q)}) - 1$ for $i_s^{(q)} \in C_{\ell(m)}^{(n)}$. These inequalities will allow us to split the powers of t by using the Young inequality. Hence, we can write

$$g_4(t,\cdot) = \sum_{n=1}^{k}\sum_{m=1}^{c_n}\sum_{q=2}^{n+1} a^{\xi_m^{(n)}}t^{2q_h(i_m^{(n)})}\langle \mathcal{P}_m^{(n,q)}D_x^q u(t,\cdot), \partial_m^n u(t,\cdot)\rangle$$

---

[6] See Subsection 11.1.1 for the definition of $\widehat{\beta}$.

$$
\begin{aligned}
&+ \sum_{(n,m,q)\in\mathcal{D}} a^{\widetilde{\xi}_m^{(n)}} t^{q_h(i_m^{(n)})+q_h(i_{\ell(m)}^{(n)})} \langle H_m^{(n)} \mathcal{P}_{\ell(m)}^{(n,q)} D_x^q u(t,\cdot), \partial_m^n u(t,\cdot)\rangle \\
&+ \sum_{(n,m,q)\in\mathcal{D}} a^{\widetilde{\xi}_m^{(n)}} t^{q_h(i_m^{(n)})+q_h(i_{\ell(m)}^{(n)})} \langle (H_m^{(n)})^* \mathcal{P}_m^{(n,q)} D_x^q u(t,\cdot), \partial_{\ell(m)}^n u(t,\cdot)\rangle, \qquad (11.2.16)
\end{aligned}
$$

where $\mathcal{D} = \{(n,m,q) \in \mathbb{N}^3 : n \le k,\ c_{n-1}+1 \le m \le c_n,\ 2 \le q \le n+1\}$. In the rest of the proof we denote by C positive constants, independent of κ, t and a, which may vary from line to line. Moreover, we denote by $g_{4,j}$ $(j = 1,\ldots,3)$ the terms which appear in the right-hand side of (11.2.16). Note that

$$
\begin{aligned}
g_{4,1}(t,\cdot) \le\, & C\sqrt{\kappa}\sum_{n=1}^{k}\sum_{m=1}^{c_n} a^{\xi_m^{(n)}} t^{2q_h(i_m^{(n)})} \sum_{q=2}^{n} \sum_{s:i_s^{(q)}\in C_m^{(n)}} |\partial_s^q u(t,\cdot)||\partial_m^n u(t,\cdot)| \\
&+ C\sqrt{\kappa}\sum_{n=1}^{k}\sum_{m=1}^{c_n} a^{\xi_m^{(n)}} t^{2q_h(i_m^{(n)})} \sum_{s:i_s^{(n+1)}\in C_m^{(n)}} |\partial_s^{n+1} u(t,\cdot)||\partial_m^n u(t,\cdot)| \\
=:\,& g_{4,1,1}(t,\cdot) + g_{4,1,2}(t,\cdot).
\end{aligned}
$$

Since $q \le n$, here we can use (11.1.4)(d) and estimate

$$
\begin{aligned}
g_{4,1,1}(t,\cdot) \le & \sum_{n=1}^{k}\sum_{m=1}^{c_n} o(a^{\beta_m^{(n)}}) t^{(2q_h(i_m^{(n)})-1)^+} |\partial_m^n u(t,\cdot)|^2 \\
&+ \kappa \sum_{n=1}^{k}\sum_{q=2}^{n}\sum_{s=1}^{c_{q-1}} o(a^{\xi^{(q)}}) t^{(2q_h(i_s^{(q)})-1)^+} |\partial_s^q u(t,\cdot)|^2.
\end{aligned}
$$

On the other hand, in the case when $q = n+1$ we use conditions (11.1.4)(b)-(d) to obtain $2\xi_m^{(n)} < \beta_m^{(n)} + \xi^{(l+1)}$. Thus, we can estimate

$$
\begin{aligned}
g_{4,1,2}(t,\cdot) \le & \sum_{n=1}^{k}\sum_{m=1}^{c_n} o(a^{\beta_m^{(n)}}) t^{(2q_h(i_m^{(n)})-1)^+} |\partial_m^n u(t,\cdot)|^2 \\
&+ \kappa \sum_{n=1}^{k}\sum_{s=1}^{c_n} o(a^{\xi^{(n+1)}}) t^{(2q_h(i_s^{(n+1)})-1)^+} |\partial_s^{n+1} u(t,\cdot)|^2.
\end{aligned}
$$

The functions $g_{4,2}$ and $g_{4,3}$ can be estimated similarly. Using (11.1.4)(d)-(f) we conclude that $\widetilde{\xi}_m^{(n)} < \xi^{(l+1)}$, $2\widetilde{\xi}_m^{(n)} < \xi^{(n)}$ and $2\widetilde{\xi}_m^{(n)} < 2\widetilde{\xi}_{\ell(m)}^{(n)}$ (this latter one for $\ell(m) > c_{n-1}$). Hence, by the Young inequality we can estimate

$$
\begin{aligned}
g_{4,2}(t,\cdot) \le & C\sqrt{\kappa} \sum_{(n,m,q)\in\mathcal{D}} \sum_{s:i_s^{(q)}\in C_{\ell(m)}^{(n)}} a^{\widetilde{\xi}_m^{(n)}} t^{q_h(i_m^{(n)})+q_h(i_{\ell(m)}^{(n)})} |\partial_s^q u(t,\cdot)||\partial_m^n u(t,\cdot)| \\
\le & \sum_{n=1}^{k}\sum_{m=c_{n-1}+1}^{c_n} o(a^{\widetilde{\xi}_m^{(n)}}) t^{(2q_h(i_m^{(n)})-1)^+} |\partial_m^n u(t,\cdot)|^2 \\
&+ \kappa \sum_{n=1}^{k}\sum_{q=2}^{l+1}\sum_{s=1}^{c_{q-1}} o(a^{\xi^{(q)}}) t^{(2q_h(i_s^{(q)})-1)^+} |\partial_s^q u(t,\cdot)|^2
\end{aligned}
$$

and

$$
g_{4,3}(t,\cdot) \le C\sqrt{\kappa} \sum_{(n,m,q)\in\mathcal{D}} \sum_{s:i_s^{(q)}\in C_m^{(n)}} a^{\widetilde{\xi}_m^{(n)}} t^{q_h(i_m^{(n)})+q_h(i_{\ell(m)}^{(n)})} |\partial_s^q u(t,\cdot)||\partial_{\ell(m)}^n u(t,\cdot)|
$$

$$
\begin{aligned}
&\le \sum_{n=1}^{k} \sum_{m=c_{n-1}+1}^{c_n} o(a^{\beta_{\ell(m)}^{(n)}}) t^{(2q_h(i_{\ell(m)}^{(n)})-1)^+} |\partial_{\ell(m)}^n u(t,\cdot)|^2 \\
&\quad + \kappa \sum_{n=1}^{k} \sum_{q=2}^{n+1} \sum_{s=1}^{c_{q-1}} o(a^{\xi^{(q)}}) t^{(2q_h(i_s^{(q)})-1)^+} |\partial_s^q u(t,\cdot)|^2.
\end{aligned}
$$

Putting everything together, we get (11.2.8). Arguing in the same way, we can prove (11.2.9). □

11.3 Construction of the semigroup

In this section we prove that, for any $f \in C_b(\mathbb{R}^N)$, the Cauchy problem

$$
\begin{cases}
D_t u(t,x) = \mathcal{A}u(t,x), & t > 0, \quad x \in \mathbb{R}^N, \\
u(0,x) = f(x), & \qquad\qquad x \in \mathbb{R}^N,
\end{cases}
\tag{11.3.1}
$$

admits a unique bounded classical solution u and, consequently, we can associate a semigroup of bounded operators in $C_b(\mathbb{R}^N)$ with the operator $\mathcal{A}$.

We start with a maximum principle for (degenerate) elliptic and parabolic equations, which leads to uniqueness of the distributional solutions to the elliptic equation $\lambda u - \mathcal{A}u = f \in C_b(\mathbb{R}^N)$ and the Cauchy problem (11.3.1).

Proposition 11.3.1 *Let $\mathcal{A}$ be as in* (11.0.1)*. Then, the following assertions hold true.*

(i) *Let $u \in C_b(\mathbb{R}^N)$ be a bounded distributional solution to the equation $\lambda u - \mathcal{A}u = f$, corresponding to some $f \in C_b(\mathbb{R}^N)$ and $\lambda > 0$, Further, suppose that $D_i u$ and $D_{ij}u$ exist in the classical sense for any $i,j = 1,\dots,p_0$. Then, $\lambda\|u\|_\infty \le \|f\|_\infty$.*

(ii) *Let $u : [0,T_0]\times\mathbb{R}^N \to \mathbb{R}$ $(T_0 > 0)$ be a bounded, continuous distributional solution of the Cauchy problem* (11.0.6) *corresponding to some $f \in C_b(\mathbb{R}^N)$ and $g \in C((0,T_0]\times\mathbb{R}^N)$. Further, assume that $D_i u$, $D_{ij}u$ $(i,j = 1,\dots,p_0)$ exist in the classical sense. If $g \le 0$ in $(0,T_0]\times\mathbb{R}^N$, then $\sup_{[0,T_0]\times\mathbb{R}^N} u \le \sup_{\mathbb{R}^N} f$. Similarly, if $g \ge 0$ in $(0,T_0]\times\mathbb{R}^N$, then $\inf_{[0,T_0]\times\mathbb{R}^N} u \ge \inf_{\mathbb{R}^N} f$. In particular, if $g \equiv 0$, then $\|u(t,\cdot)\|_\infty \le \|f\|_\infty$ for any $t \in [0,T_0]$.*

Proof (i) The proof is similar to that of Theorem 3.1.3. For any $n \in \mathbb{N}$, we set $u_n = u - n^{-1}\varphi$, where $\varphi(x) = 1 + |x|^2$ for any $x \in \mathbb{R}^N$. Note that, since the diffusion coefficients q_{ij} $(i,j = 1,\dots,p_0)$ may grow at most quadratically at infinity, there exists $\lambda_0 > 0$ such that $\mathcal{A}\varphi - \lambda_0\varphi < 0$ in $\mathbb{R}^N$.

Fix $\lambda > \lambda_0$, u and f as in the statement. Then, $\lambda u_n - \mathcal{A}u_n < f$ for any $n \in \mathbb{N}$. Let $x_n \in \mathbb{N}$ be such that $u_n(x_n) = \sup_{\mathbb{R}^N} u_n$. We claim that $\mathcal{A}u_n(x_n) \le 0$. From this inequality it follows that $\lambda u_n \le f$ for any $n \in \mathbb{N}$. Letting n tend to $+\infty$ we then conclude that $\lambda u \le f$ in $\mathbb{R}^N$. Replacing u with $-u$ we get the inequality $\lambda u \ge -f$ and the assertion follows for $\lambda > \lambda_0$. To prove it for $\lambda \in (0,\lambda_0)$, it suffices to rewrite the equation $\lambda u - \mathcal{A}u = f$ in the form $\lambda_0 u - \mathcal{A}u = (\lambda_0 - \lambda)u + f$ and apply the obtained result to infer that $\lambda\|u\|_\infty \le (\lambda_0 - \lambda)\|u\|_\infty + \|f\|_\infty$, and we are done.

So, let us prove the claim. Clearly, $\mathrm{Tr}(QD^2u_n(x_n)) \le 0$. To conclude that $\mathcal{A}u_n(x_n) \le 0$, we show that $\mathcal{B}u_n(x_n) = 0$, where $\mathcal{B}u_n(x) = \langle Bx, \nabla u(x)\rangle$ for any $x \in \mathbb{R}^N$. For this purpose, we adapt the argument in the second proof of Lemma 3.1.2. Without loss of generality we assume that $u_n(x_n)$ is positive.[7] Fix $R_n > 0$ such that $u_n \leq u_n(x_0)$ in $B_R(x_n)$ and let $\vartheta \in C_c^\infty(B_R(x_n))$ satisfy $\vartheta(x) < \vartheta(x_0)$ for any $x \in B_R(x_n) \setminus \{x_n\}$. Then, the function $v_n = u_n\vartheta$ is compactly supported in $B_R(x_n)$ and assumes its maximum value only at x_n. Moreover, $\mathcal{B}v_n \in BUC(\mathbb{R}^N)$. Let $v_{m,n} = v_n \star \varrho_m$, for any $m \in \mathbb{N}$, where $\{\varrho_m\}$ is a standard sequence of mollifiers supported in B_1 and $\star$ denotes convolution. Then, $v_{m,n}$ is smooth, compactly supported in $B_{R+1}(x_n)$ for any $m, n \in \mathbb{N}$ and it converges to v_n uniformly in $\mathbb{R}^N$ as m tends to $+\infty$. Moreover,

$$\mathcal{B}v_{m,n} = \mathcal{B}v_n \star \varrho_m + \mathrm{Tr}(B)v_n + v_n \star \mathcal{B}\varrho_m, \qquad m, n \in \mathbb{N}. \tag{11.3.2}$$

Indeed, an integration by parts shows that $\mathcal{B}v_{k,m,n} = \mathcal{B}v_{k,n} \star \varrho_m + \mathrm{Tr}(B)v_{m,n} + v_{k,n} \star \mathcal{B}\varrho_m$ for any $k \in \mathbb{N}$, where $v_{k,m,n} = v_{m,n} \star \varrho_k$. As k tends to $+\infty$, $v_{k,m,n}$, $\mathcal{B}v_{k,m,n}$ and $v_{k,n}$ converge, respectively, to $v_{m,n}$, $\mathcal{B}v_{m,n}$ and v_n, locally uniformly in $\mathbb{R}^N$. Moreover, since $\mathcal{B}v_{k,n}$ converges to $\mathcal{B}v_n$ in the sense of distributions, $\mathcal{B}v_{k,n} \star \varrho_m$ converges to $\mathcal{B}u_n \star \varrho_m$ pointwise in $\mathbb{R}^N$ as k tends to $+\infty$. Formula (11.3.2) follows.

Now, letting m tend to $+\infty$ in (11.3.2) it follows that $\mathcal{B}v_{m,n}$ converges to $\mathcal{B}v_n$, locally uniformly in $\mathbb{R}^N$, since $v_n \star \mathcal{B}\varrho_m$ and $\mathcal{B}v_n \star \varrho_m$ converge, respectively, to $-\mathrm{Tr}(B)v_n$ and $\mathcal{B}v_n$ uniformly in $\mathbb{R}^N$.

For any $m, n \in \mathbb{N}$, let $x_{m,n} \subset B_{R+1}(x_n)$ be any point where $v_{m,n}$ attains its maximum value. Up to a subsequence, $x_{m,n}$ converges, as m tends to $+\infty$, to a maximum point of v_n, which should be x_n. Clearly, $\mathcal{B}v_{m,n}(x_{m,n}) = 0$ for any $m \in \mathbb{N}$ and, thus, $\mathcal{B}v_n(x_n) = \mathcal{B}u_n(x_n) = 0$ as well.

(ii) The proof is similar to the previous one. Suppose that $g \leq 0$. The function $u_n : [0, T_0] \times \mathbb{R}^N \to \mathbb{R}$, defined by $u_n(t, x) = e^{-\lambda_0 t}(u(t, x) - \sup_{\mathbb{R}^N} f) - n^{-1}\varphi(x)$ for any $(t, x) \in [0, T_0] \times \mathbb{R}^N$, satisfies the inequality $D_t u_n - (\mathcal{A} - \lambda_0)u_n < 0$ in $(0, T_0) \times \mathbb{R}^N$ and $u_n(0, \cdot) \leq 0$. Moreover, it attains its maximum value at some point $(t_n, x_n) \in [0, T_0] \times \mathbb{R}^N$. We claim that $t_n = 0$. If this were not the case, we would have $(D_t u_n - \mathcal{A}u_n)(t_n, x_n) \geq 0$, as in the classical case, which is a contradiction. Indeed, clearly, $\mathrm{Tr}(QD_x^2 u_n(t_n, x_n)) \leq 0$. To prove that $(D_t u_n - \mathcal{B}u_n)(t_n, x_n) \geq 0$, it suffices to adapt the arguments in the proof of (i), replacing the functions ϑ and $u_{m,n}$ therein defined, respectively, by a cut-off function $\psi \in C_c^\infty((0, T_0) \times \mathbb{R}^N)$, with support in $[t_0 - R^{-1}, t_0 + R^{-1}] \times B_R(x_0)$, for some $R > 0$ sufficiently large, and such that $\psi(t, x) < \psi(t_0, x_0) = 1$ for any $(t_0, x_0) \neq (t, x) \in (0, +\infty) \times B_R(x_0)$, and by the function $v_{m,n} = \widetilde{v}_n \star \varrho_m$, where $\widetilde{v}_n$ is the even extension (with respect to t) of the function u_n to the whole of $\mathbb{R}^{N+1}$ and $\{\varrho_n\}$ is a standard sequence of mollifiers in $\mathbb{R}^{N+1}$.

To prove the assertion when $g \geq 0$, it suffices to apply the above argument to $-u$. Finally, the estimate $\|u(t, \cdot)\|_\infty \leq \|f\|_\infty$, for any $t \in [0, T_0]$, follows straightforwardly from these results. □

Theorem 11.3.2 *The following assertions hold true:*

(i) *for any $f \in C_b(\mathbb{R}^N)$ the Cauchy problem* (11.3.1) *admits a unique bounded classical solution;*[8]

(ii) *the family $\{T(t)\}$, defined by $T(\cdot)f := u$, where u is the bounded classical solution to problem* (11.3.1) *corresponding to the initial value f, is a positivity preserving semigroup of linear contractions in $C_b(\mathbb{R}^N)$;*

(iii) *if $f \in C_c(\mathbb{R}^N)$, then $T(t)f$ converges to f, as t tends to 0^+, uniformly in $\mathbb{R}^N$;*

[7] If $u_n(x_n) < 0$, then the function $v_n = u_n - 2u_n(x_n)$ has at x_n a positive maximum and $\mathcal{B}u_n = \mathcal{B}v_n$.
[8] i.e., a function $u \in C_b([0, +\infty) \times \mathbb{R}^N) \cap C^{1,2}((0, +\infty) \times \mathbb{R}^N)$ which solves problem (11.3.1).

(iv) *for any* $f \in C_b(\mathbb{R}^N)$ *and any multi-index* $\alpha \in (\mathbb{N}\cup\{0\})^N$, *with* $|\alpha| \le \varsigma-1$, *the derivative* $D_x^\alpha T(\cdot)f$ *exists in the classical sense in* $(0,+\infty)\times\mathbb{R}^N$ *and it is a continuous function. Moreover, for any* $\omega > 0$, *there exists a positive constant* C, *depending only on* ω, h *and* $|\alpha|$, *such that, for any* $f \in C_b^h(\mathbb{R}^N)$ *and any* α *as above, it holds that*

$$\|D_x^\alpha T(t)f\|_{C_b(\mathbb{R}^N)} \le Ct^{-q_h(\widehat{\alpha})}e^{\omega t}\|f\|_{C_b^h(\mathbb{R}^N)}, \qquad t > 0. \tag{11.3.3}$$

Proof (i)-(iii) The uniqueness of the bounded classical solution to problem (11.3.1) follows from Proposition 11.3.1, which also shows that the solutions to (11.3.1) corresponding to nonnegative initial data are nonnegative as well. To prove the existence part, we set $u_\varepsilon = T_\varepsilon(\cdot)f$ and observe that, since $D_t u_\varepsilon = \mathcal{A}_\varepsilon u_\varepsilon$ in $(0,+\infty)\times\mathbb{R}^N$, (11.2.2) and the smoothness of the coefficients of the operator $\mathcal{A}$ show that, for any $0 < T_1 < T_2$, $\sup_{\varepsilon\in(0,1)}\sup_{t\in[T_1,T_2]}\|u(t,\cdot)\|_{C_b^\varsigma(\mathbb{R}^N)} < +\infty$ and $u_\varepsilon \in \mathrm{Lip}([T_1,T_2];C(\overline{B}_R))$ with norm independent of $\varepsilon \in (0,1)$. Propositions B.3.4 and B.3.6(ii) imply that $u_\varepsilon \in C^{\theta/2,\varsigma-1+\theta}((T_1,T_2)\times B_R)$ for some $\theta \in (0,1)$, and $\sup_{\varepsilon\in(0,1)}\|u_\varepsilon\|_{C^{\theta/2,\varsigma-1+\theta}((T_1,T_2)\times B_R)} < +\infty$. As a byproduct, $\sup_{\varepsilon\in(0,1)}\|D_t u_\varepsilon\|_{C^{\theta/2,\varsigma-3+\theta}((T_1,T_2)\times B_R)} < +\infty$. Using these estimates and compactness and diagonal arguments, we can determine an infinitesimal sequence $\{\varepsilon_n\}$ such that $\{u_{\varepsilon_n}\}$ converges in $C^{1,\varsigma-1}(K)$, for any compact set $K \subset (0,+\infty)\times\mathbb{R}^N$, to a function u_f which belongs to $C^{1+\theta/2,\varsigma-1+\theta}_{\mathrm{loc}}((0,+\infty)\times\mathbb{R}^N)$ and solves the differential equation in (11.3.1) for $t > 0$. The continuity of u_f up to $t = 0$ and the condition $u_f(0,\cdot) = f$, are obtained in three steps. We observe that, as soon as we identify u_f with the unique bounded classical solution to (11.3.1), we can show that u_ε converges to u_f in $C^{1,\varsigma}_{\mathrm{loc}}([0,+\infty)\times\mathbb{R}^N)$ as ε tends to 0^+. Indeed, any sequence $\{u_{\varepsilon'_n}\}$ (with $\{\varepsilon'_n\}$ being positive and infinitesimal) which converges in $C^{1,\varsigma}_{\mathrm{loc}}([0,+\infty)\times\mathbb{R}^N)$, should converge to u_f.

Step 1. Suppose that f belongs to $C_c^2(\mathbb{R}^N)$. Using Proposition 1.3.4 we can estimate $\|u_{\varepsilon_n}(t,\cdot) - f\|_\infty \le t\sup_{s\in[0,+\infty)}\|T_{\varepsilon_n}(s)\mathcal{A}_{\varepsilon_n}f\|_\infty \le t\|\mathcal{A}_{\varepsilon_n}f\|_\infty \le Ct$ for any $t \ge 0$ and some positive constant C, independent of n. Letting first n tend to $+\infty$ and then t tend to 0^+, we conclude that u_f can be extended by continuity on $\{0\}\times\mathbb{R}^N$, by setting $u_f(0,\cdot) = f$. Hence, u_f is the unique classical solution to problem (11.3.1). Moreover, u_ε converges to u_f, as ε tends to 0^+, in $C^{1,\varsigma-1}([T_1,T_2]\times B_R)$ for any T_1, T_2, R as above.

Step 2. If $f \in C_0(\mathbb{R}^N)$, then the continuity of u_f at $t = 0$ follows from a density argument, approximating f by a sequence $\{f_n\} \subset C_c^2(\mathbb{R}^N)$, converging to f uniformly in $\mathbb{R}^N$, and observing that, by Proposition 11.3.1(ii), u_{f_n} converges to u_f uniformly in $(0,+\infty)\times\mathbb{R}^N$.

Step 3. Now we consider the general case when $f \in C_b(\mathbb{R})$. We fix $R > 0$ and a function $\psi \in C_c^\infty(\mathbb{R}^N)$ satisfying the condition $\chi_{B_R} \le \psi \le \chi_{B_{R+1}}$. Further, we split $f = \psi f + (1-\psi)f$ and write $T_{\varepsilon_n}(t)f = T_{\varepsilon_n}(t)(\psi f) + T_{\varepsilon_n}(t)((1-\psi)f)$. Since the semigroups $\{T_\varepsilon(t)\}$ preserve positivity (see Theorem 1.2.5), we can estimate $|T_{\varepsilon_n}(\cdot)((1-\psi)f)| \le \|f\|_\infty T_{\varepsilon_n}(\cdot)(\mathbf{1}-\psi) = \|f\|_\infty(\mathbf{1} - T_{\varepsilon_n}(\cdot)\psi)$ in $(0,+\infty)\times\mathbb{R}^N$. Letting n tend to $+\infty$ we conclude that $|u_f - u_{\psi f}| \le \|f\|_\infty(1-u_\psi)$ in $(0,+\infty)\times\mathbb{R}^N$. Since $\psi f \equiv f$ in B_R and $u_\psi(t,\cdot)$ converges to $\mathbf{1}$ as t tends to 0^+, locally uniformly in $\mathbb{R}^N$, u_f can be extended by continuity on $\{0\}\times B_R$ by setting $u_f(0,\cdot) = f$. The arbitrariness of $R > 0$ allows us to conclude that u_f can be extended to continuity to $\{0\}\times\mathbb{R}^N$.

(iv) The semigroup law allows us to extend (11.2.2) to all the positive times, showing that, for any $\omega > 0$, there exists a positive constant $\widehat{C} = \widehat{C}_\omega$, independent of ε, such that $\|D_x^\alpha T_\varepsilon(t)f\|_\infty \le \widehat{C}t^{-q_h(\widehat{\alpha})}e^{\omega t}\|f\|_{C_b^h(\mathbb{R}^N)}$ for any $t > 0$, $\varepsilon \in (0,1]$, $f \in C_b^h(\mathbb{R}^N)$ and $|\alpha| \le \varsigma$ (see the proof of Theorem 5.1.7 for further details). Letting ε tend to 0^+ we obtain (11.3.3) for any $|\alpha| \le \varsigma - 1$. □

Remark 11.3.3 We stress that estimate (11.3.3) agrees with the optimal estimates proved in [181] in the case of the degenerate Ornstein-Uhlenbeck operator, where an explicit formula for the associated semigroup is available (see (10.1.5)).

In the following theorem, we prove some additional smoothness of the function $T(t)f$.

Theorem 11.3.4 *For any $f \in C_b(\mathbb{R}^N)$ and any multi-index $\alpha \in (\mathbb{N}\cup\{0\})^N$, with $|\alpha| = \varsigma$ and $\alpha_j \neq 0$ for some $j \leq p_0 + p_1$, the derivative $D_x^\alpha T(\cdot)f$ exists in the classical sense in $(0,+\infty)\times\mathbb{R}^N$ and it is a continuous function. Moreover, for any $h \leq |\alpha|$ and $f \in C_b^h(\mathbb{R}^N)$, estimate (11.3.3) is satisfied.*

Proof Being rather long, we split the proof into several steps.

Step 1. We fix $R > 0$, $j \leq N$, $f \in C_b^{\varsigma-1}(\mathbb{R}^N)$ and set $u = T(\cdot)f$. Let us prove that for any $\zeta = \zeta_R \in C_c^\infty(\mathbb{R}^N)$, such that $\chi_{B_R} \leq \zeta_R \leq \chi_{B_{2R}}$, and any $\vartheta = \vartheta_R \in C_c^\infty((0,+\infty))$, such that $\chi_{(R^{-1},R)} \leq \vartheta_R \leq \chi_{((2R)^{-1},2R)}$, it holds that

$$D_j u(t,x) = \int_0^t (T(t-s)g_j(s,\cdot))(x)ds, \qquad t\in(R^{-1},R),\ x\in B_R, \tag{11.3.4}$$

where

$$\begin{aligned} g_j =& \vartheta\mathrm{Tr}((D_jQ_0)D_\star^2(\zeta u)) + \vartheta(B^*\partial_1^1(\zeta u))_j + \vartheta\langle D_jF, \partial_\star^1(\zeta u)\rangle - \vartheta D_j(u\mathcal{A}\zeta) \\ &- 2\vartheta\langle Q_0\partial_1^1 u, \partial_1^1 D_j\zeta\rangle - 2\vartheta\langle Q_0\partial_1^1 D_j u, \partial_1^1\zeta\rangle - 2\vartheta\langle (D_jQ_0)\partial_1^1 u, \partial_1^1\zeta\rangle + \vartheta' D_j(\zeta u). \end{aligned} \tag{11.3.5}$$

Here, Q_0 and $D_\star^2 u$ denote the quadratic submatrices obtained from Q and $D_x^2 u$, erasing the last $N-p_0$ rows and columns.

For any $\delta \in (-1,1)$, let τ_δ^j be the operator defined by $\tau_\delta^j\psi = \delta^{-1}(\psi(\cdot+\delta e_j)-\psi)$ for any $\psi \in C_b(\mathbb{R}^N)$ and set $w_{\varepsilon,\delta}^j = \vartheta\tau_\delta^j v_\varepsilon$, where $v_\varepsilon = \zeta u$. As is easily seen, $w_{\varepsilon,\delta}^j(0,\cdot)\equiv 0$ and $D_t w_{\varepsilon,\delta}^j = \mathcal{A}_\varepsilon w_{\varepsilon,\delta}^j + g_{j,\delta,\varepsilon}$ in $(0,+\infty)\times\mathbb{R}^N$, where

$$\begin{aligned} g_{j,\delta,\varepsilon} =& \vartheta\mathrm{Tr}(\tau_\delta^j Q_0 D_\star^2 v_\varepsilon) + \vartheta(B^*\nabla_x v_\varepsilon)_j + \vartheta\langle\tau_\delta^j F, \partial_1^1 v_\varepsilon\rangle - 2\vartheta\langle Q_0\partial_1^1 u_\varepsilon(\cdot,\cdot+\delta e_j), \partial_1^1\tau_\delta^j\zeta\rangle \\ &- \vartheta\tau_\delta^j(u_\varepsilon\mathcal{A}_\varepsilon\zeta) - 2\varepsilon\vartheta\langle\partial_\star^1\tau_\delta^j u_\varepsilon, \partial_\star^1\zeta\rangle - 2\vartheta\langle\tau_\delta^j Q_0\partial_1^1 u_\varepsilon(t,\cdot+\delta e_j), \partial_1^1\zeta(\cdot+\delta e_j)\rangle \\ &- 2\varepsilon\vartheta\langle\partial_\star^1 u_\varepsilon(\cdot,\cdot+\delta e_j), \partial_\star^1\tau_\delta^j\zeta\rangle - 2\vartheta\langle Q_0\partial_1^1\tau_\delta^j u_\varepsilon, \partial_1^1\zeta\rangle + \vartheta'\tau_\delta^j v_\varepsilon \end{aligned}$$

and $\partial_\star^1\psi$ denotes the vector of the first-order derivatives of the function $\psi:\mathbb{R}^N\to\mathbb{R}$ with respect to the last $N-p_0$ variables. Theorem 5.2.5 shows that we can write

$$w_{\varepsilon,\delta}^j(t,x) = \int_0^t (T_\varepsilon(t-s)g_{j,\delta,\varepsilon}(s,\cdot))(x)ds, \qquad t\geq 0,\ x\in\mathbb{R}^N. \tag{11.3.6}$$

Since $g_{j,\delta,\varepsilon}$ tends to the function

$$\begin{aligned} g_{j,\delta} =& \vartheta\mathrm{Tr}(\tau_\delta^j Q_0 D_\star^2(\zeta u)) + \vartheta(B^*\nabla(\zeta u))_j + \vartheta\langle\tau_\delta^j F, \partial_1^1(\zeta u)\rangle \\ &- \vartheta\tau_\delta^j(u\mathcal{A}\zeta) - 2\vartheta\langle Q_0\partial_1^1 u(\cdot,\cdot+\delta e_j), \partial_1^1\tau_\delta^j\zeta\rangle, \end{aligned}$$

uniformly in $\mathbb{R}^N$, as ε tends to 0^+, $\|T_\varepsilon(r)g_{j,\delta,\varepsilon}(s,\cdot)\|_\infty \leq \|g_{j,\delta,\varepsilon}\|_\infty$ and

$$\begin{aligned} &\|T_\varepsilon(r)g_{j,\delta,\varepsilon}(s,\cdot) - T(r)g_{j,\delta}(s,\cdot)\|_{C(K)} \\ \leq& \|g_{j,\delta,\varepsilon} - g_{j,\delta}\|_\infty + \|T_\varepsilon(r)g_{j,\delta}(s,\cdot) - T(r)g_{j,\delta}(s,\cdot)\|_{C(K)} \end{aligned}$$

for any $\varepsilon > 0$ and any compact set $K\subset\mathbb{R}^N$, we can let ε tend to 0^+ in (11.3.6) and, by dominated convergence, we get

$$v_\delta^j(t,x) := \zeta(x)(\tau_\delta^j u(t,\cdot))(x) = \int_0^t (T(t-s)g_{j,\delta}(s,\cdot))(x)ds, \qquad t>0,\ x\in\mathbb{R}^N. \tag{11.3.7}$$

By Theorem 11.3.2(iv) the function u is bounded in $[0, T_0]$ with values in $C_b^{\varsigma-1}(\mathbb{R}^N)$ for any $T_0 > 0$. Hence, $g_{j,\delta}(s,\cdot)$ converges to $g_j(s,\cdot)$ uniformly in $\mathbb{R}^N$ for any $s > 0$, as δ tends to 0^+. We can thus let δ tend to 0^+ in (11.3.7) and, still by dominated convergence, we get (11.3.4).

Step 2. Here and in Step 3 we assume that $f \in C_b^{\varsigma}(\mathbb{R}^N)$. Moreover, by C we denote a positive constant, independent of t, ε, which may vary from line to line. Finally, $\alpha = (\alpha_1, \ldots, \alpha_N) \in (\mathbb{N} \cup \{0\})^N$ is a multi-index with length ς, j is the largest integer such that $\alpha_j \neq 0$ and $\beta := \alpha - e_j^{(N)}$. We will prove that we can differentiate the function in (11.3.4) with respect to the multi-index β. Of course, this will show that the derivative $D_x^\alpha u$ exists in the classical sense.

In this step, we assume that $\alpha_k \neq 0$ for some $k \leq p_0$. Estimate (11.2.2) implies that the function $D_x^\gamma u_\varepsilon(t,\cdot)$ is Lipschitz continuous in $\mathbb{R}^N$ and its Lipschitz semi-norm can be bounded by $Ce^{\omega t}$ for any $\omega > 0$, any multi-index γ with length $\varsigma - 1$, any $t > 0$ and some positive constant $C = C(\omega)$. Since $u_\varepsilon(t,\cdot)$ converges to $u(t,\cdot)$ in $C_{\rm loc}^{\varsigma-1}(\mathbb{R}^N)$ (see the proof of Theorem 11.3.2), the function $D_x^\gamma u(t,\cdot)$ is Lipschitz continuous in $\mathbb{R}^N$ with Lipschitz-norm which can be bounded by C. This and Theorem 11.3.2 show that $\sup_{s\in(0,T_0)} \|g_j(s,\cdot)\|_{C_b^{\varsigma-3+\beta}(\mathbb{R}^N)} < +\infty$ for any $T_0 > 0$ and $\beta \in (0,1)$. Now, from (11.3.3) with $h = \varsigma - 3$ and $h = \varsigma - 2$, Proposition B.3.2 and Theorem B.3.8 it follows that $\|D_x^\beta T(t)\psi\|_\infty \leq Ct^{-\frac{1}{2}+(1-\theta)\frac{2i+1}{2}} e^{\omega t}\|\psi\|_{C_b^{\varsigma-3+\theta}(\mathbb{R}^N)}$ for any $t > 0$ and $\psi \in C_b^{\varsigma-3+\theta}(\mathbb{R}^N)$, i being the smallest integer such that $\widehat{\beta}'_i > 0$, where $\widehat{\beta}' = (\widehat{\beta}'_0, \widehat{\beta}'_1, \ldots, \widehat{\beta}'_r) := \widehat{\beta} - e_1^{(r+1)}$. Using this estimate with $\psi = g_j(s,\cdot)$ and $\theta > 2i/(2i+1)$, we deduce that we can differentiate under the integral sign in (11.3.4). This proves that $D_x^\alpha u$ exists in the classical sense and it is a continuous function of the variable x. Moreover, it satisfies estimate (11.3.3) since, as we have shown, the Lipschitz seminorm of $D_x^\beta u$ can be estimated from above by $Ce^{\omega t}$ for any $t > 0$ and $\omega > 0$.

Step 3. Here, we assume that $\alpha_j \neq 0$ for some $j \in \{p_0+1, \ldots, p_0+p_1\}$ and prove that g_j is bounded in $(0, T_0)$ with values in $C_b^{\varsigma-2+\theta}(\mathbb{R}^N)$, for some $\theta \in (\frac{2i-1}{2i+1}, 1)$. Indeed, once this property is proved, estimate (11.3.3) and the same interpolation argument used in Step 2 give

$$\|D_x^\beta T(t-s)g_j(s,\cdot)\|_\infty \leq C(t-s)^{-(1-\theta)\frac{2i+1}{2}} \sup_{s\in(0,T_0)} \|g_j(s,\cdot)\|_{C_b^{\varsigma-3+\theta}(\mathbb{R}^N)}$$

for any $0 < s < t \leq T_0$, any arbitrary $T_0 > 0$, and we can complete the proof differentiating under the integral sign in (11.3.4).

Since u is bounded in $[0, T]$ with values in $C_b^{\varsigma-2}(\mathbb{R}^N)$, an inspection of the formula (11.3.5) reveals that, to show that g_j is bounded in $(0, T_0)$ with values in $C_b^{\varsigma-2+\theta}(\mathbb{R}^N)$, it suffices to prove that, for any pair of indexes $h \leq p_0$ and $h' \leq p_0 + p_1$, with $h \leq h'$, the function $D_{hh'}u(t,\cdot)$ belongs to $C^{\varsigma-2+\theta}(\mathbb{R}^N)$ and $\sup_{t\in((2R)^{-1},R)} \|D_{hh'}u(t,\cdot)\|_{C^{\varsigma-2+\theta}(B_M)} < +\infty$.

Fix $T_0 > 0$, $\gamma \in (\mathbb{N} \cup \{0\})^N$ with $|\gamma| = \varsigma - 2$, denote by i_0 the largest integer such that $\gamma_{i_0} > 0$ and define $\gamma' = \gamma + e_h^{(N)} + e_{h'}^{(N)} - e_{i_0}^{(N)}$. From (11.3.3), the assertion already proved in Step 2, and an interpolation argument, we obtain that

$$\|D_x^{\gamma'} T(t)\psi\|_{C_b^\rho(\mathbb{R}^N)} \leq Ct^{-\frac{\rho\theta}{2}-(1-\theta)\frac{3\rho+1}{2}} e^{\omega t}\|\psi\|_{C_b^{\varsigma-2+\theta}(\mathbb{R}^N)}, \qquad t > 0, \tag{11.3.8}$$

for any $\theta, \rho \in [0,1]$. From (11.3.8) it now follows that

$$\|D_x^{\gamma'} T(t-s)g_{i_0}(s,\cdot)\|_{C_b^\rho(\mathbb{R}^N)} \leq C(t-s)^{-\frac{3\rho+1}{2}} \|g_{i_0}(s,\cdot)\|_{C_b^{\varsigma-2}(\mathbb{R}^N)} \leq C(t-s)^{-\frac{3\rho+1}{2}} \tag{11.3.9}$$

for any $0 < s < t \leq T_0$. Hence, if we fix $\delta \in (0,1)$ and take $\rho = \theta_1 = \delta/3$, then we deduce that the function in the last side of (11.3.9) is integrable in $(0,T_0)$. Thus, we can differentiate under the integral sign in (11.3.4) and conclude that the function $D_{hh'}u$ is bounded in $((2R)^{-1},2R)$ with values in $C^{\varsigma-2+\theta_1}(B_R)$. Due to the arbitrariness of R, $D_{hh'}u$ is bounded in J with values in $C^{\varsigma-2+\theta_1}(K)$ for any pair of compact sets $J\subset(0,+\infty)$ and $K\subset\mathbb{R}^N$.

Using (11.3.8) again, we deduce that $D_{ll'}u$ is bounded in J with values in $C^{\varsigma-2+\theta_2}(K)$ for any J and K as above, where $\theta_2=\delta\frac{1+\theta_1}{3-2\theta_1}$. Iterating this argument, we conclude that $D_{hh'}u$ is bounded in J with values in $C^{\varsigma-2+\theta_k}(K)$, where the sequence $\{\theta_k\}$ is defined by recurrence by setting $\theta_0=0$ and $\theta_{k+1}=\delta(1+\theta_k)(3-2\theta_k)^{-1}$ for any $k\le k_0$, where either $k_0=+\infty$ or k_0 is the largest integer such that $\theta_k<3/2$.

It easy to see that $\theta_k<\theta_{k+1}$ for any $k\le k_0$ and any choice $\delta\in(0,1)$. If $\delta=3/4$, then the equation $\ell=\delta\frac{1+\ell}{3-2\ell}$ has no real solutions. Hence, there exists k_1 such that $\theta_{k_1}>1$ and, consequently, $D_{hh'}u$ is bounded in J with values in $C^{\varsigma-2+\theta}(K)$ for any $\theta\in(0,1)$, any pair of compact sets $J\subset(0,+\infty)$ and $K\subset\mathbb{R}^N$. We have proved that g_i is locally bounded in $(0,+\infty)$ with values in $C_b^{\varsigma-2+\theta}(\mathbb{R}^N)$ for any $\theta\in(0,1)$. Finally, the same arguments as in Step 2 reveal that $D_x^\alpha T(t)f$ satisfies estimate (11.3.3).

Step 4. Here, we prove estimate (11.3.3) for a general $f\in C_b^h(\mathbb{R}^N)$ $(h<\varsigma)$ when α has length ς and $\alpha_j\neq 0$ for some $j\le p_0+p_1$. This follows immediately when $f\in BUC(\mathbb{R}^N)$ since we can determine a sequence $\{f_n\}\subset C_b^\varsigma(\mathbb{R}^N)$ converging to f uniformly in $\mathbb{R}^N$. We can use (11.3.3) to infer that $\{D_x^\alpha T(t)f_n\}$ is a Cauchy sequence in $C_b(\mathbb{R}^N)$ for any $t>0$ and, consequently, the derivative $D_x^\alpha T(t)f$ exists in the classical sense and it is bounded. In the general case when $f\in C_b(\mathbb{R}^N)$, we split $T(t)f=T(t/2)T(t/2)f$. Since $T(t/2)f\in C_b^1(\mathbb{R}^N)$, from the above arguments we conclude that $D_x^\alpha T(t)f=D_x^\alpha T(t/2)(T(t/2)f)$ exists in the classical sense for any $t>0$ and satisfies estimate (11.3.3).

Step 5. To complete the proof, we have to show that, for any multi-index α as in Step 4, the function $D_x^\alpha T(\cdot)f$ is continuous in $(0,+\infty)\times\mathbb{R}^N$. As usually, let j be the largest integer such that $\alpha_j>0$. Let us fix $y=(y_1,\ldots,y_{N-1})\in\mathbb{R}^{N-1}$ and introduce the function $\psi=D_x^\beta u(\cdot,y_1,\ldots,y_{j-1},\cdot,y_j,\ldots,y_{N-1})$ where, again, $\beta=\alpha-e_j^{(N)}$. From the results in Steps 2 to 4, ψ is bounded in (a,b) with values in $C^{1+\theta}(B_R)$ for some $\theta\in(0,1)$ and any $a,b,R>0$, with $a<b$. Applying Proposition B.3.4, with $(\theta,\alpha)=(1,1+\theta)$, to the function $\psi(t,\cdot)-\psi(s,\cdot)$ $(s,t,\in[a,b])$, yields that $\|\psi(t,\cdot)-\psi(s,\cdot)\|_{C^1(B_R)}\le C\|\psi(t,\cdot)-\psi(s,\cdot)\|_{C(B_R)}^{\theta/(1+\theta)}$ for some constant C, independent of y. Since $u\in C^{1,\varsigma-1}((0,+\infty)\times\mathbb{R}^N)$, the right-hand side of the previous estimate vanishes as $|t-s|$ tends to 0^+, implying that the function $D_x^\alpha u(\cdot,x)$ is continuous in $[a,b]$ uniformly with respect to $x\in\mathbb{R}^N$. This is enough to conclude that $D_x^\alpha u$ is continuous in $(0,+\infty)\times\mathbb{R}^N$. □

11.3.1 Properties of the semigroup

In this section we first state some continuity property of the semigroup $\{T(t)\}$ that will play a fundamental role in order to prove the anisotropic Hölder estimates of Section 11.4. Then, we characterize the domain of the weak generator of the semigroup.

Proposition 11.3.5 *The following assertions hold.*

(i) *If $\{f_n\}\subset C_b(\mathbb{R}^N)$ is a bounded sequence converging to $f\in C_b(\mathbb{R}^N)$, locally uniformly in $\mathbb{R}^N$, then $T(\cdot)f_n$ converges to $T(\cdot)f$ locally uniformly in $[0,+\infty)\times\mathbb{R}^N$ and in $C^{1,2}(K)$ for any compact set $K\subset(0,+\infty)\times\mathbb{R}^N$.*

(ii) *There exists a family of Borel probability measures $\{p(t,x;dy):t>0,x\in\mathbb{R}^N\}$ such* *that, for any $f \in C_b(\mathbb{R}^N)$,*

$$(T(t)f)(x) = \int_{\mathbb{R}^N} f(y)p(t,x;dy), \qquad t > 0, \ x \in \mathbb{R}^N. \tag{11.3.10}$$

Consequently, for any bounded sequence $\{f_n\} \subset C_b(\mathbb{R}^N)$ converging to $f \in C_b(\mathbb{R}^N)$, pointwise in $\mathbb{R}^N$, $T(\cdot)f_n$ converges to $T(\cdot)f$ locally uniformly in $(0,+\infty) \times \mathbb{R}^N$. Moreover, $\{T(t)\}$ can be extended to $B_b(\mathbb{R}^N)$ with a semigroup of positive contractions, which enjoys the strong Feller property.

Proof (i) Without loss of generality, we assume that $f \equiv 0$ and $\|f_n\|_\infty \le 1$ for any $n \in \mathbb{N}$. The arguments in the proof of Theorem 11.3.2 can be easily adapted to prove that there exists a subsequence $\{T(\cdot)f_{n_k}\}$ converging in $C^{1,2}(K)$, for any compact set $K \subset (0,+\infty) \times \mathbb{R}^N$, to some function $g \in C^{1,2}((0,+\infty) \times \mathbb{R}^N)$ which solves the differential equation $D_t g = \mathcal{A}g$. To prove that $g \equiv 0$, we show that it can be extended by continuity at $t = 0$ where it vanishes. Arguing as in Step 3 of the proof of Theorem 11.3.2 we can show that $|T(t)f_{n_k} - T(t)(\psi f_{n_k})| \le |\mathbf{1} - T(t)\psi|$ in $\mathbb{R}^N$ for any $t > 0$, any $R > 0$ and any smooth function ψ such that $\chi_{B_R} \le \psi \le \chi_{B_{R+1}}$. Since ψf_{n_k} tends to zero uniformly in $\mathbb{R}^N$ as k tends to $+\infty$, from the previous estimate we conclude that $|g| \le |\mathbf{1} - T(\cdot)\psi|$ in $(0,+\infty) \times \mathbb{R}^N$, so that g vanishes in B_R as t tends to 0^+. The arbitrariness of $R > 0$ implies that g can be extended by continuity at $t = 0$ and therein vanishes. We have so proved that $T(\cdot)f_{n_k}$ converges locally uniformly in $(0,+\infty) \times \mathbb{R}^N$ to the null function. Since we can repeat this argument starting from any arbitrarily fixed subsequence of $\{f_n\}$ and since the limit that we obtain is always the null function, the sequence $\{T(\cdot)f_n\}$ converges to zero locally uniformly in $(0,+\infty) \times \mathbb{R}^N$. To conclude that this sequence actually converges locally uniformly up to $t = 0$, we fix $R > 0$ and ψ as above, and estimate

$$|T(\cdot)f_n| \le |T(\cdot)f_n - T(\cdot)(\psi f_n)| + |T(\cdot)(\psi f_n)| \le |\mathbf{1} - T(\cdot)\psi| + \|\psi f_n\|_\infty$$

in $[0,+\infty) \times \mathbb{R}^N$. Therefore, for any $\varepsilon \in (0,T_0)$ and $n \in \mathbb{N}$, we can infer that

$$\begin{aligned}\|T(\cdot)f_n\|_{C([0,T]\times \overline{B}_R)} \le& \|T(\cdot)f_n\|_{C([0,\varepsilon]\times \overline{B}_R)} + \|T(\cdot)f_n\|_{C([\varepsilon,T]\times \overline{B}_R)}\\ \le& \|\mathbf{1} - T(\cdot)\psi\|_{C([0,\varepsilon]\times \overline{B}_R)} + \|\psi f_n\|_\infty + \|T(\cdot)f_n\|_{C([\varepsilon,T]\times \overline{B}_R)}.\end{aligned}$$

It follows that $\limsup_{n\to+\infty} \|T(\cdot)f_n\|_{C([0,T]\times \overline{B}_R)} \le \|\mathbf{1} - T(\cdot)\psi\|_{C([0,\varepsilon]\times \overline{B}_R)}$ and the right-hand side of the previous estimate vanishes as ε tends to 0^+. This completes the proof of (i).

(ii) By the Riesz representation theorem, for any $x \in \mathbb{R}^N$ there exists a positive finite Borel measure $p(t,x;dy)$ in $\mathbb{R}^N$ such that (11.3.10) holds for any $f \in C_0(\mathbb{R}^N)$. To extend it to any $f \in C_b(\mathbb{R}^N)$, we consider a bounded sequence $\{f_n\} \subset C_0(\Omega)$ which converges to f locally uniformly in $\mathbb{R}^N$. Writing (11.3.10) with f_n instead of f and letting n tend to $+\infty$, we obtain the desired formula, by dominated convergence and property (i).

From the formula (11.3.10) it is now clear that the semigroup $\{T(t)\}$ can be extended to $B_b(\mathbb{R}^N)$ with a semigroup of positive contractions. Moreover, if $\{f_n\} \subset C_b(\mathbb{R}^N)$ is a bounded sequence converging pointwise in $\mathbb{R}^N$ to a function $f \in C_b(\mathbb{R}^N)$, then again from (11.3.10) we deduce that $T(\cdot)f_n$ converges to $T(\cdot)f$ pointwise in $(0,+\infty) \times \mathbb{R}^N$. The arguments in the first part of the proof of (i) reveal that a subsequence $\{T(\cdot)f_{n_k}\}$ converges locally uniformly in $(0,+\infty) \times \mathbb{R}^N$ to a function, which, of course, is $T(\cdot)f$. Again, since the limit is independent of the subsequence that we consider, $T(\cdot)f_n$ converges to $T(\cdot)f$ locally uniformly in $(0,+\infty) \times \mathbb{R}^N$.

To prove that the semigroup $\{T(t)\}$ is strong Feller, we adapt the technique of [66, Theorem 3.2], which strongly relies on the gradient estimate (11.3.3). For this purpose, we fix $f \in B_b(\mathbb{R}^N)$ and introduce a bounded sequence $\{f_n\} \subset C_b(\mathbb{R}^N)$ converging pointwise to

f almost everywhere in $\mathbb{R}^N$. Estimate (11.3.3) shows that, for any $t > 0$, $T(t)f_n$ is a Lipschitz continuous function, uniformly with respect to n. Since $T(t)f_n$ pointwise converges to $T(t)f$, letting n tend to $+\infty$ we conclude that $T(t)f$ is a Lipschitz continuous function as well. Finally, since $p(t, x; \mathbb{R}^N) = 1$, $T(t)f$ is bounded. Hence, it belongs to $C_b(\mathbb{R}^N)$. □

As in Section 1.3, for any $\lambda > 0$ we can define the linear operator $R(\lambda) \in L(C_b(\mathbb{R}^N))$ by setting $(R(\lambda)f)(x) = \int_0^{+\infty} e^{-\lambda t}(T(t)f)(x)dt$ for any $x \in \mathbb{R}^N$ and $f \in C_b(\mathbb{R}^N)$. It is immediate to check that $\{R(\lambda) : \lambda > 0\}$ is the resolvent family associated with some closed operator $\widehat{A} : D(\widehat{A}) \subset C_b(\mathbb{R}^N) \to C_b(\mathbb{R}^N)$: the weak generator of the semigroup $\{T(t)\}$. From now on, we write $R(\lambda, \widehat{A})$ instead of $R(\lambda)$.

Proposition 11.3.6 *The following characterization holds true:*

$$D(\widehat{A}) = \Big\{ f \in C_b(\mathbb{R}^N) : \exists \{f_n\} \subset C_b^2(\mathbb{R}^N), \exists g \in C_b(\mathbb{R}^N) :$$
$$f_n \to f,\ \mathcal{A}f_n \to g \text{ loc. unif. in } \mathbb{R}^N,\ \sup_{n\in\mathbb{N}} (\|f_n\|_\infty + \|\mathcal{A}f_n\|_\infty) < +\infty \Big\}. \qquad (11.3.11)$$

Moreover, $g = \widehat{A}f = \mathcal{A}f$ for any $f \in D(A)$. Here and above, $\mathcal{A}f$ is meant in the sense of distributions.

Proof We adapt the technique of [181, Theorem 6.2]. For this purpose, let us denote by A_0 the realization of the operator $\mathcal{A}$ in $C_b(\mathbb{R}^N)$ with domain $D(A_0) = \{f \in C_b^2(\mathbb{R}^N) : \mathcal{A}f \in C_b(\mathbb{R}^N)\}$.

As a first step, we prove that

$$T(t)\mathcal{A}f = \mathcal{A}T(t)f, \qquad t > 0,\ f \in D(A_0). \qquad (11.3.12)$$

For this purpose, we observe that Lemma 1.3.2 and Propositions 1.3.5, 3.1.1(i) show that $T_\varepsilon(\cdot)\mathcal{A}_\varepsilon f = \mathcal{A}_\varepsilon T_\varepsilon(\cdot)f$ for any $\varepsilon > 0$ and $f \in C_b^2(\mathbb{R}^N)$ such that $\mathcal{A}f \in C_b(\mathbb{R}^N)$, where as usual $\{T_\varepsilon(t)\}$ is the semigroup associated in $C_b(\mathbb{R}^N)$ with the operator $\mathcal{A}_\varepsilon = \mathcal{A} + \varepsilon \sum_{j=p_0+1}^N D_{jj}$. To conclude the proof of (11.3.12), we observe that, by the proof of Theorem 11.3.2, $\mathcal{A}_\varepsilon T_\varepsilon(\cdot)f$ converges to $\mathcal{A}T(\cdot)f$ locally uniformly in $(0, +\infty) \times \mathbb{R}^N$. Moreover, $\|T_\varepsilon(t)\mathcal{A}_\varepsilon f - T(t)\mathcal{A}f\|_{C(K)} \le \|\mathcal{A}_\varepsilon f - \mathcal{A}f\|_\infty + \|(T_\varepsilon(t) - T(t))\mathcal{A}f\|_{C(K)}$ for any compact set $K \subset \mathbb{R}^N$ and any $t > 0$ and the right-hand side of the previous inequality vanishes as ε tends to 0^+, again by the proof of Theorem 11.3.2.

From (11.3.12) we deduce that

$$(R(\lambda, \widehat{A})\mathcal{A}f)(x) = \int_0^{+\infty} e^{-\lambda t}(T(t)\mathcal{A}f)(x)dt = \int_0^{+\infty} e^{-\lambda t}(\mathcal{A}T(t)f)(x)dt$$
$$= \int_0^{+\infty} e^{-\lambda t}\left(\frac{\partial}{\partial t}T(t)f\right)(x)dt = -f(x) + \lambda(R(\lambda, \widehat{A})f)(x)$$

for any $x \in \mathbb{R}^N$, $\lambda > 0$ and $f \in D(A_0)$, which shows that $f \in D(\widehat{A})$ and $\widehat{A}f = \mathcal{A}f$.

Now, denote by D_0 the function space defined by the right-hand side of (11.3.11) and fix $f \in D_0$. By the above results we know that $f_n = R(\lambda, \widehat{A})(\lambda f_n - \mathcal{A}f_n) =: R(\lambda, \widehat{A})h_n$ for any $n \in \mathbb{N}$ and $\lambda > 0$. Let us prove that $R(\lambda, \widehat{A})h_n$ converges to $R(\lambda, \widehat{A})(\lambda f - g)$ as n tends to $+\infty$, locally uniformly in $\mathbb{R}^N$. This is enough to infer that $f = R(\lambda, \widehat{A})(\lambda f - g)$, and the inclusion "$\supset$" in (11.3.11) follows with $g = \widehat{A}f$. Observe that

$$\|R(\lambda, \widehat{A})(h_n - \lambda f + g)\|_{C(K)} \le \int_0^{+\infty} e^{-\lambda t}\|T(t)(h_n - \lambda f + g)\|_{C(K)}dt$$

for any compact set $K \subset \mathbb{R}^N$. By Proposition 11.3.5(i), $\|T(t)(h_n - \lambda f + g)\|_{C(K)}$ vanishes as n tends to $+\infty$ for any $t \geq 0$. Moreover, $\|T(t)(h_n - \lambda f + g)\|_\infty \leq \sup_{n\in\mathbb{N}}(\lambda\|f_n\|_\infty + \|\mathcal{A}f_n\|_\infty)\|h_n\|_\infty + \lambda\|f\|_\infty + \|g\|_\infty < +\infty$ for any $n \in \mathbb{N}$. By the arbitrariness of K and the dominated convergence theorem we conclude that $R(\lambda, \widehat{A})h_n$ converges to $R(\lambda, \widehat{A})(\lambda f - g)$ locally uniformly in $\mathbb{R}^N$, as n tends to $+\infty$,

Next, we observe that for any $\varphi \in C_c^\infty(\mathbb{R}^N)$ and any $n \in \mathbb{N}$ it holds that $\int_{\mathbb{R}^N} \mathcal{A}f_n \varphi dx = \int_{\mathbb{R}^N} f_n \mathcal{A}^*\varphi dx$, where by $\mathcal{A}^*$ we denote the adjoint of the operator $\mathcal{A}$. Letting n tend to $+\infty$, we obtain that $g = \mathcal{A}f$ and, consequently, $\widehat{A}f = \mathcal{A}f$.

To complete the proof, let us show that $D(\widehat{A}) \subset D_0$. For this purpose, fix $u \in D(\widehat{A})$, $\lambda > 0$ and $f \in C_b(\mathbb{R}^N)$ such that $u = R(\lambda, \widehat{A})f$. By convolution, we can regularize f obtaining a sequence $\{f_n\} \subset C_b^2(\mathbb{R}^N)$, bounded in $C_b(\mathbb{R}^N)$ and converging locally uniformly to f as n tends to $+\infty$. Taking Theorem 11.3.4 into account, we deduce that $R(\lambda, \widehat{A})f_n$ belongs to $C_b^2(\mathbb{R}^N)$ for any $n \in \mathbb{N}$. Moreover,

$$(\mathcal{A}R(\lambda, \widehat{A})f_n)(x) = \int_0^{+\infty} e^{-\lambda t}(\mathcal{A}T(t)f_n)(x)dt = -f_n(x) + \lambda(R(\lambda, \widehat{A})f_n)(x)$$

for any $x \in \mathbb{R}^N$ and $n \in \mathbb{N}$. The above arguments show that $R(\lambda, \widehat{A})f_n$ converges to u, locally uniformly in $\mathbb{R}^N$. Therefore, letting n tend to $+\infty$ in the previous formula we deduce that $\mathcal{A}R(\lambda, \widehat{A})f_n$ converges to $\lambda u - f$, locally uniformly in $\mathbb{R}^N$. Since the sequences $\{R(\lambda, \widehat{A})f_n\}$ and $\{\mathcal{A}R(\lambda, \widehat{A})f_n\}$ are bounded in $C_b(\mathbb{R}^N)$, u belongs to D_0. □

11.4 Anisotropic Hölder estimates

In this section we prove Hölder estimates for the (distributional) solutions to the elliptic equation (11.0.5) and to the nonhomogeneous Cauchy problem (11.0.6). Throughout the section, we assume that Hypotheses 11.0.1 are satisfied with ς equal to the least common multiple of the odd numbers between 1 and $2r + 1$. Moreover, we use the spaces $\mathscr{C}_b^\theta(\mathbb{R}^N)$ and $\widetilde{\mathscr{C}}_b^\theta(\mathbb{R}^N)$ defined in Subsection A.1.2.

The main results of this section are collected in the following two theorems.

Theorem 11.4.1 *Let $\theta \in (0, 1)$ and $\lambda > 0$. Then, for any $h \in C_b^\theta(\mathbb{R}^N)$ there exists a function $u \in \mathscr{C}_b^{2+\theta}(\mathbb{R}^N)$ solving equation (11.0.5) in the sense of distributions. Moreover, there exists a positive constant C, independent of u and f, such that*

$$\|u\|_{\mathscr{C}_b^{2+\theta}(\mathbb{R}^N)} \leq C\|h\|_{C_b^\theta(\mathbb{R}^N)}. \tag{11.4.1}$$

Such a function u is the unique distributional solution to equation (11.0.5), which is bounded and continuous in $\mathbb{R}^N$, and it is twice continuously differentiable in $\mathbb{R}^N$ with respect to the first p_0 variables, with bounded derivatives.

Theorem 11.4.2 *Let $\theta \in (0, 1)$, $T_0 > 0$, $f \in C_b^{2+\theta}(\mathbb{R}^N)$ and $g \in C_b([0, T_0] \times \mathbb{R}^N)$ satisfy $\sup_{t\in[0,T_0]} \|g(t, \cdot)\|_{C_b^\theta(\mathbb{R}^N)} < +\infty$. Then, there exists a function $u \in C_b([0, T_0] \times \mathbb{R}^N)$, solution to problem (11.0.6) in the sense of distributions, such that*

$$\sup_{t\in[0,T_0]} \|u(t, \cdot)\|_{\mathscr{C}_b^{2+\theta}(\mathbb{R}^N)} \leq C\Big(\|f\|_{C_b^{2+\theta}(\mathbb{R}^N)} + \sup_{t\in[0,T_0]} \|g(t, \cdot)\|_{C_b^\theta(\mathbb{R}^N)}\Big) \tag{11.4.2}$$

for some positive constant C, independent of u, f, g. Moreover, u is the unique distributional solution to problem (11.0.6) which is bounded and continuous in $[0,T_0]\times\mathbb{R}^N$ and, therein, it is twice continuously differentiable with respect to the first p_0 spatial variables, with bounded derivatives.

To prove the previous theorems we need the following result, which is the counterpart of Theorem 5.1.9 of the nondegenerate setting.

Proposition 11.4.3 *For any $T_0>0$, $0<\beta\le\theta<3$ with $\beta,\theta\notin\mathbb{N}$, there exists a positive constant C_{T_0} such that, for any $f\in C_b^\beta(\mathbb{R}^N)$, the following inequality holds true:*

$$\|T(t)f\|_{\mathscr{C}_b^\theta(\mathbb{R}^N)}\le C_{T_0}t^{-\frac{\theta-\beta}{2}}\|f\|_{C_b^\beta(\mathbb{R}^N)},\qquad t\in(0,T_0]. \tag{11.4.3}$$

Proof The proof follows from an interpolation argument. To simplify the notation, in the sequel we fix $t\in(0,T_0]$ and denote by C a positive constant, possibly depending on T_0 but independent of t and f, which may vary from line to line. The results in Theorem 11.3.2 show that $\|T(t)f\|_{\mathscr{C}_b^m(\mathbb{R}^N)}\le Ct^{-\frac{m}{2}}\|f\|_{C_b(\mathbb{R}^N)}$. By applying Proposition B.3.2 with $X_1=X_2=Y_1=C_b(\mathbb{R}^N)$, $Y_2=\mathscr{C}_b^\varsigma(\mathbb{R}^N)$ and by taking this estimate, Theorem 11.3.2(ii) and Proposition B.3.9 into account, we obtain $\|T(t)\|_{L(C_b(\mathbb{R}^N),\mathscr{C}_b^{\theta_1}(\mathbb{R}^N))}\le Ct^{-\theta_1/2}$ for any $\theta_1\in(0,\varsigma)\setminus\mathbb{Q}$. Of course, this estimate holds also with $\theta_1=0$. Using again Proposition B.3.2, now with $X_1=C_b(\mathbb{R}^N)$, $X_2=C_b^\varsigma(\mathbb{R}^N)$, $Y_1=Y_2=\mathscr{C}_b^\varsigma(\mathbb{R}^N)$, we get $\|T(t)\|_{L(C_b^{\theta_2}(\mathbb{R}^N),\mathscr{C}_b^\varsigma(\mathbb{R}^N))}\le Ct^{-(\varsigma-\theta_2)/2}$ for any $\theta_2\in(0,\varsigma)\setminus\mathbb{N}$ and even for $\theta_2=0,\varsigma$. Using these last two estimates and Proposition B.3.2 we conclude that $\|T(t)\|_{\mathcal{L}(C_b^{\theta_3\theta_2}(\mathbb{R}^N),\mathscr{C}_b^{(1-\theta_3)\theta_1+\varsigma\theta_3}(\mathbb{R}^N))}\le Ct^{-[(1-\theta_3)\theta_1+(\varsigma-\theta_2)\theta_3]/2}$. Estimate (11.4.3) follows by taking $\theta_1,\theta_2,\theta_3$ such that $\theta_2\theta_3=\beta$ and $(1-\theta_3)\theta_1+\varsigma\theta_3=\theta$. □

Estimate (11.4.3) is the keystone in the proof of Theorems 11.4.1 and 11.4.2. The candidates to be the solutions to equation (11.0.5) and the nonhomogeneous Cauchy problem (11.0.6) are, respectively, the functions $R(\lambda,\widehat{A})f$ and u defined by

$$u(t,x)=(T(t)f)(x)+\int_0^t(T(t-s)g(s,\cdot))(x)ds,\qquad t\in[0,T_0],\ x\in\mathbb{R}^N. \tag{11.4.4}$$

The results in the following proposition are now a straightforward consequence of estimate (11.4.3) and the same arguments as in the proofs of Theorems 5.2.2 and 5.2.6. For this reason we skip the proof.

Proposition 11.4.4 *For any $\theta\in(0,1)$ and $T_0>0$ the following properties are satisfied.*

(i) *For any $h\in C_b^\theta(\mathbb{R}^N)$, the function $R(\lambda,\widehat{A})h$ belongs to $\mathscr{C}_b^{2+\theta}(\mathbb{R}^N)$ and estimate (11.4.1) is satisfied by some positive constant C, independent of f.*

(ii) *For any $f\in C_b^{2+\theta}(\mathbb{R}^N)$ and $g\in C([0,T_0]\times\mathbb{R}^N)$ such that $\sup_{t\in[0,T_0]}\|g(t,\cdot)\|_{C_b^\theta(\mathbb{R}^N)}<+\infty$, the function u in (11.4.4) is bounded and continuous in $[0,T_0]\times\mathbb{R}^N$. Moreover, $u(t,\cdot)\in\mathscr{C}_b^{2+\theta}(\mathbb{R}^N)$ for any $t\in[0,T_0]$ and estimate (11.4.2) is satisfied by some positive constant C, independent of f and g.*

Now, we can prove Theorems 11.4.1 and 11.4.2.

Proof of Theorem 11.4.1 By Proposition 11.3.6, $R(\lambda,\widehat{A})h$ is a distributional solution of equation (11.0.5). Moreover, by Proposition 11.4.4(i), $R(\lambda,\widehat{A})h\in\mathscr{C}_b^{2+\theta}(\mathbb{R}^N)$ and satisfies estimate (11.4.1). As a byproduct, Proposition 11.3.1(i) implies that $R(\lambda,\widehat{A})h$ is the unique distributional solution to equation (11.0.5) satisfying the properties of Theorem 11.4.1. □

Proof of Theorem 11.4.2 The uniqueness part of the statement follows immediately from the maximum principle in Proposition 11.3.1(ii). Moreover, by virtue of Proposition 11.4.4, it suffices to prove that the convolution term in (11.4.4), that we simply denote by v, is a distributional solution to the Cauchy problem (11.0.6), with $f \equiv 0$. Actually for smooth functions g with compact support, using estimate (11.3.3) with $|\alpha| = 2$ and $h = 2$, and arguing as in the proof of Theorems 5.2.5 and 5.2.6, we can show that v is a classical solution to problem (11.0.6), with $f \equiv 0$. In the general case we take a sequence $\{g_n\} \subset C_b^{1,2}([0,T_0] \times \mathbb{R}^N)$, bounded in the sup-norm and converging locally uniformly in $[0,T_0] \times \mathbb{R}^N$ to g, as n tends to $+\infty$. Moreover, for any $n \in \mathbb{N}$, we denote by v_n the function defined as v, with g being replaced by g_n. Each function v_n is smooth and satisfies the equation $D_t v_n = \mathcal{A} v_n + g_n$ in $[0,T] \times \mathbb{R}^N$. Moreover, its sup-norm can be bounded by a positive constant, independent of n and, by Proposition 11.3.5, v_n converges to v pointwise in $[0,T_0] \times \mathbb{R}^N$. Since

$$\int_{(0,T_0)\times\mathbb{R}^N} g_n \varphi dtdx = \int_{(0,T_0)\times\mathbb{R}^N} v_n(-D_t\varphi - \mathcal{A}^*\varphi)dtdx, \qquad n \in \mathbb{N},$$

where $\mathcal{A}^*$ is the formal adjoint of the operator $\mathcal{A}$, letting n tend to $+\infty$, we conclude that v is a distributional solution of the Cauchy problem (11.0.6) with $f \equiv 0$. □

11.4.1 The case $r = 1$

In the particular case when $r = 1$, the Schauder estimates in Theorems 11.4.1 and 11.4.2 can be improved. More precisely,

Theorem 11.4.5 (Optimal Schauder estimates) *Let $\theta \in (0,1)$. The following properties are satisfied.*

(i) *For any $\lambda > 0$ and $h \in \mathscr{C}_b^\theta(\mathbb{R}^N)$ there exists a function $u \in \mathscr{C}_b^{2+\theta}(\mathbb{R}^N)$ which solves equation (11.0.5) in the sense of distributions. Moreover, there exists a positive constant C, independent of u and f, such that*[9]

$$\|u\|_{\mathscr{C}_b^{2+\theta}(\mathbb{R}^N)} \leq C\|h\|_{\mathscr{C}_b^\theta(\mathbb{R}^N)}.$$

Such a function u is the unique distributional solution to equation (11.0.5), which is bounded and continuous in $\mathbb{R}^N$ and it is twice continuously differentiable in $\mathbb{R}^N$ with respect to the first p_0 variables, with bounded derivatives.

(ii) *For any $g \in C_b([0,T_0] \times \mathbb{R}^N)$ such that $\sup_{t\in[0,T_0]} \|g(t,\cdot)\|_{\mathscr{C}_b^\theta(\mathbb{R}^N)} < +\infty$ and any $f \in \mathscr{C}_b^{2+\theta}(\mathbb{R}^N)$, there exists a function $u \in C_b([0,T_0] \times \mathbb{R}^N)$, solution to the Cauchy problem (11.0.6) in the sense of distributions, such that*

$$\sup_{t\in[0,T_0]} \|u(t,\cdot)\|_{\mathscr{C}_b^{2+\theta}(\mathbb{R}^N)} \leq C\Big(\|f\|_{C_b^{2+\theta}(\mathbb{R}^N)} + \sup_{t\in[0,T_0]} \|g(t,\cdot)\|_{\mathscr{C}_b^\theta(\mathbb{R}^N)}\Big)$$

for some positive constant C, independent of u, f, g. Moreover, u is the unique distributional solution to problem (11.0.6) which is bounded and continuous in $[0,T_0] \times \mathbb{R}^N$ and therein twice continuously differentiable with respect to the first p_0 spatial variables, with bounded derivatives.

[9]For the definition of the spaces $\mathscr{C}_b^\alpha(\mathbb{R}^N)$ and $\widetilde{\mathscr{C}}_b^\alpha(\mathbb{R}^N)$, we refer the reader to Subsection A.1.2.

There are two main ingredients to prove Theorem 11.4.5: (i) the interpolation set equality

$$(C_b(\mathbb{R}^N), \mathscr{C}_b^3(\mathbb{R}^N))_{\theta,\infty} = \widetilde{\mathscr{C}}_b^{3\theta}(\mathbb{R}^N), \qquad \theta \in (0,1); \tag{11.4.5}$$

(ii) the estimate

$$\|T(t)\|_{L(\mathscr{C}_b^{3k}(\mathbb{R}^N), \mathscr{C}_b^3(\mathbb{R}^N))} \le Ce^{\omega t} t^{-\frac{3(1-k)}{2}}, \qquad t>0, \ k=0,1. \tag{11.4.6}$$

We begin by proving (i).

Proposition 11.4.6 *For any $\theta \in (0,1)$, the formula* (11.4.5) *holds true with equivalence of the corresponding norms.*

Proof Throughout the proof we identify $\mathbb{R}^N$ with $\mathbb{R}^{p_0} \times \mathbb{R}^{N-p_0}$. Moreover, we denote by C a positive constant, which is independent of s, t and $f \in C_b(\mathbb{R}^N)$, and may vary from line to line. Finally, since the proof is rather long, we split it into three steps.

Step 1. Let us denote by A_1 and A_2, respectively, the realizations in $C_b(\mathbb{R}^N)$ of the operators $\mathcal{A}_1 = \sum_{i=1}^{p_0} D_{ii}$ and $\mathcal{A}_2 = \sum_{i=p_0+1}^{N} D_{ii}$ with domains

$$D(A_1) = \left\{u : u, \mathcal{A}_1 u \in C_b(\mathbb{R}^N) : u(\cdot, y) \in \bigcap_{1\le p<+\infty} W^{2,p}_{\rm loc}(\mathbb{R}^{p_0}), \ \ \forall y \in \mathbb{R}^{N-p_0}\right\},$$

$$D(A_2) = \left\{u : u, \mathcal{A}_2 u \in C_b(\mathbb{R}^N) : u(x, \cdot) \in \bigcap_{1\le p<+\infty} W^{2,p}_{\rm loc}(\mathbb{R}^{N-p_0}), \ \ \forall x \in \mathbb{R}^{p_0}\right\}.$$

Further, let us set $A_3 = -A_1^2$. In this step, we show that

$$(C_b(\mathbb{R}^N), D(A_j))_{\beta,\infty} = \widetilde{\mathscr{C}}_b^{2(2-j)\beta, 2(j-1)\beta}(\mathbb{R}^N), \quad j=1,2, \tag{11.4.7}$$

$$(C_b(\mathbb{R}^N), D(A_3))_{\beta,\infty} = \widetilde{\mathscr{C}}_b^{4\beta,0}(\mathbb{R}^N) \tag{11.4.8}$$

for any $\beta \in (0,1)$, with equivalence of the corresponding norms. We will take advantage of the characterization of the interpolation spaces $(X, D(A))_{\beta,\infty}$ given in Proposition C.2.16, when A is the generator of an analytic semigroup in a Banach space X. Indeed, taking Theorem D.3.5(iv) into account, it is easy to check that A_1 and A_2 are sectorial operators. Moreover, since the realization of the Laplacian in $C_b(\mathbb{R}^N)$ satisfies (C.2.1) for any $\varphi \in (\pi/2, \pi)$, the operator A_1 satisfies (C.2.1) for any $\varphi \in (\pi/2,\pi)$ as well. Therefore, according to Theorem C.2.14, the operator A_3 is sectorial in $C_b(\mathbb{R}^N)$.

In the sequel, we simply write $D_{A_j}(\beta,\infty)$ and $D_{A_3}(\beta,\infty)$ to denote the interpolation spaces $(X, D(A_j))_{\beta,\infty}$ and $(X, D(A_3))_{\infty}$, $(j=1,2)$.

Let us check the formula (11.4.7) with $j=1$. For this purpose, we preliminarily observe that the analytic semigroup $\{T_1(t)\}$, associated with the operator A_1, is given by

$$(T_1(t)f)(x,y) = \frac{1}{(4\pi t)^{p_0/2}} \int_{\mathbb{R}^{p_0}} e^{-\frac{|x-z|^2}{4t}} f(z,y)dz, \qquad t>0, \ (x,y) \in \mathbb{R}^N,$$

for any $f \in C_b(\mathbb{R}^N)$. To prove the embedding "$\hookrightarrow$, we fix $f \in \widetilde{\mathscr{C}}_b^{2\beta,0}(\mathbb{R}^N)$ and observe that

$$\begin{aligned}&(T_1(t)f)(x,y) - f(x,y)\\ =&\frac{1}{2(4\pi)^{p_0/2}} \int_{\mathbb{R}^{p_0}} e^{-\frac{|z|^2}{4}} (f(x-\sqrt{t}z, y) - 2f(x,y) + f(x+\sqrt{t}z, y))dz\end{aligned}$$

for any $t>0$ and $(x,y)\in\mathbb{R}^N$. Therefore,

$$\|T_1(t)f-f\|_\infty\le\frac{t^\beta}{(4\pi)^{p_0/2}}[f]_{C_b^{2\beta,0}(\mathbb{R}^N)}\int_{\mathbb{R}^{p_0}}e^{-\frac{|z|^2}{4}}|z|^{2\beta}dz$$

and by Proposition C.2.16 we conclude that $\widetilde{\mathscr{C}}_b^{2\beta,0}(\mathbb{R}^N)\hookrightarrow D_{A_1}(\beta,\infty)$.

To prove the other inclusion in (11.4.7), we begin by observing that

$$\sqrt{t}\|\partial_1^1T_1(t)f\|_\infty+t\|A_1T_1(t)f\|_\infty\le C\|f\|_\infty,\qquad t>0\ f\in C_b(\mathbb{R}^N),\tag{11.4.9}$$

where, as usual $\partial_1^1T_1(t)f$ denotes the vector of all the first order derivatives $D_iT_1(t)f$, with $i\le p_0$. Fix $f\in D_{A_1}(\beta,\infty)$ and assume that $\beta\in(0,1/2)$. Then, for any $t>0$ and $x_1,x_2\in\mathbb{R}^{p_0}$, it holds that

$$\begin{aligned}|f(x_2,\cdot)-f(x_1,\cdot)|\le&|(T_1(t)f)(x_2,\cdot)-f(x_2,\cdot)|+|(T_1(t)f)(x_2,\cdot)-(T_1(t)f)(x_1,\cdot)|\\&+|(T_1(t)f)(x_1,y)-f(x_1,y)|\\\le&2[[f]]_{D_{A_1}(\beta,\infty)}t^\beta+\|\partial_1^1T_1(t)f\|_\infty|x_2-x_1|,\end{aligned}\tag{11.4.10}$$

where $[[f]]_{D_{A_1}(\beta,\infty)}$ is given by Proposition C.2.16. As is easily seen,

$$D_iT_1(n)f-D_iT_1(t)f=\int_t^n D_iA_1T_1(s)f\,ds,\qquad t\in(0,n),\ \ i=1,\dots,p_0.\tag{11.4.11}$$

From (11.4.9) and Proposition C.2.16 we deduce that $\|A_1T_1(t)f\|_\infty\le Ct^{\beta-1}\|f\|_{D_A(\beta,\infty)}$ for any $t>0$. From this estimate and (11.4.9) we get

$$\begin{aligned}\|D_iA_1T_1(s)f\|_\infty=&\|D_iT_1(s/2)A_1T_1(s/2)f\|_\infty\\\le&\|D_iT_1(s/2)\|_{L(C_b(\mathbb{R}^N))}\|A_1T_1(s/2)f\|_\infty\le Cs^{\beta-\frac{3}{2}}\|f\|_{D_{A_1}(\beta,\infty)}\end{aligned}\tag{11.4.12}$$

for any $s>0$ and $1\le i\le p_0$, so that we can let n tend to $+\infty$ in (11.4.11), concluding that $D_iT_1(t)f=-\int_t^{+\infty}D_iA_1T_1(s)fds$ for any $t>0$ and

$$\|\partial_1^1T_1(t)f\|_\infty\le Ct^{\beta-\frac{1}{2}}\|f\|_{D_{A_1}(\beta,\infty)},\qquad t>0.\tag{11.4.13}$$

Taking $t=|x_2-x_1|^2$ in estimates (11.4.10) and (11.4.13), we get $|f(x_2,y)-f(x_1,y)|\le C\|f\|_{D_{A_1}(\beta,\infty)}|x_2-x_1|^{2\beta}$ for any $|x_2-x_1|\le1$. Moreover, if $|x_2-x_1|>1$, then we can estimate $|f(x_2,y)-f(x_1,y)|\le2\|f\|_\infty\le2\|f\|_{D_{A_1}(\beta,\infty)}|x_2-x_1|^{2\beta}$, so that $D_{A_1}(\beta,\infty)\hookrightarrow\widetilde{\mathscr{C}}_b^{2\beta,0}(\mathbb{R}^N)$.

Now, suppose that $\beta>1/2$. In view of (11.4.11) and (11.4.12) we can infer that

$$\|D_iT_1(t_2)f-D_iT(t_1)f\|_\infty\le C|t_2-t_1|^{\beta-1/2}\|f\|_{D_{A_1}(\beta,\infty)}\tag{11.4.14}$$

for any $t_1,t_2>0$ and $i=1,\dots,p_0$. This implies that each derivative $D_iT_1(t)f$ converges uniformly in $\mathbb{R}^N$ as t tends to 0. Since $T_1(\cdot)f$ is continuous in $[0,+\infty)\times\mathbb{R}^N$ and $T_1(0)f=f$, we obtain that $f\in C_b^1(\mathbb{R}^N)$.

Now, taking estimates (11.4.9) and (11.4.13)[10] into account, we get

$$\begin{aligned}\|\partial_1^1D_iT_1(t)f\|_\infty=&\|\partial_1^1T(t/2)D_iT(t)f\|_\infty\le\|\partial_1^1T(t/2)\|_\infty\|D_iT(t)f\|_\infty\\\le&Ct^{\beta-1}\|f\|_{D_{A_1}(\beta,\infty)}.\end{aligned}\tag{11.4.15}$$

[10]which, of course, holds true also in the case when $\beta\in(1/2,1)$.

Hence, from (11.4.14) (where we take $t_1 = 0$), (11.4.15) and arguing as in the proof of (11.4.10), it follows that

$$\begin{aligned}|D_i f(x_2,y) - D_i f(x_1,y)| \le & 2\|D_i T_1(t)f - f\|_\infty + \|\partial_1^1 D_i T_1(t) f\|_\infty |x-y| \\ \le & C(t^{\beta-1/2}\|u\|_{D_{A_1}(\beta,\infty)} + t^{\beta-1}|x_2-x_1|\|f\|_{D_{A_1}(\beta,\infty)})\end{aligned}$$

for any $t > 0$, $x_1, x_2 \in \mathbb{R}^{p_0}$, $y \in \mathbb{R}^{N-p_0}$ and $i = 1, \ldots, p_0$. The same arguments as in the case when $\beta < 1/2$, allow us to conclude that $D_{A_1}(\beta,\infty) \hookrightarrow \widetilde{\mathscr{C}}_b^{2\beta,0}(\mathbb{R}^N)$.

Let us now consider the case when $\beta = 1/2$. For this purpose, we observe that the reiteration theorem shows that

$$D_{A_1}(1/2,\infty) = (D_{A_1}(1/4,\infty), D_{A_1}(3/4,\infty))_{1/2,\infty} = (\mathscr{C}_b^{1/2,0}(\mathbb{R}^N), \mathscr{C}_b^{3/2,0}(\mathbb{R}^N))_{1/2,\infty}.$$

Since $(C_b^{1/2}(\mathbb{R}^{p_0}), C_b^{3/2}(\mathbb{R}^{p_0}))_{1/2,\infty} = C_b^1(\mathbb{R}^{p_0})$ (see Theorem B.3.8) and the operator $T_y : C_b(\mathbb{R}^N) \to C_b(\mathbb{R}^{p_0})$, defined by $T_y f = f(\cdot, y)$ for any $f \in C_b(\mathbb{R}^N)$, is bounded from $\mathscr{C}_b^{(2k+1)/2,0}(\mathbb{R}^N)$ in $C_b^{(2k+1)/2}(\mathbb{R}^{p_0})$ $(k = 0, 1)$ with norms being independent of y, Proposition B.3.2 allows us to conclude that $D_{A_1}(1/2,\infty) \hookrightarrow \widetilde{\mathscr{C}}_b^{1,0}(\mathbb{R}^N)$. This completes the proof of (11.4.7), in the case when $j = 1$.

The proof of (11.4.7), with $j = 2$, is completely similar, so we skip it. Let us prove the formula (11.4.8). From the general interpolation theory, we know that, for any $\beta \in (0,1)$, $D_{A_3}(\beta/2,\infty) = (C_b(\mathbb{R}^N), D(A_1^2))_{\beta/2,\infty} = (D(A_1), D(A_1^2))_{\beta,\infty} = D_{A_1}(1+\beta,\infty)$, with equivalence of the corresponding norms. Hence, (11.4.8) follows immediately if we show that

$$D_{A_1}(1+\beta,\infty) = \widetilde{\mathscr{C}}_b^{2+2\beta,0}(\mathbb{R}^N), \tag{11.4.16}$$

with equivalence of the corresponding norms. The embedding $\widetilde{\mathscr{C}}_b^{2+2\beta,0}(\mathbb{R}^N) \hookrightarrow D_{A_1}(1+\beta,\infty)$ is an obvious consequence of (11.4.7) and the characterization of $D(A_1)$. To prove the other embedding, we fix $f \in D_{A_1}(1+\beta,\infty)$ and prove that f admits second-order derivatives with respect to the first r variables, which belong to $\widetilde{\mathscr{C}}_b^{2\beta,0}(\mathbb{R}^N)$. Observe that $f = \int_0^{+\infty} e^{-t}T_1(t)g\,dt$, where $g := f - \mathcal{A}_1 f$ belongs to $D_{A_1}(\beta,\infty) = \widetilde{\mathscr{C}}_b^{2\beta,0}(\mathbb{R}^N)$. Taking (11.4.13) into account, it is easy to check that f is once-continuously differentiable with respect to the first p_0 variables and $\|\partial_1^1 f\|_\infty \le C\|g\|_{D_{A_1}(\beta,\infty)}$. Splitting $D_{ij}T_1(t)g = D_j T_1(t/2) D_i T_1(t/2) g$ for any $t > 0$, $i, j = 1, \ldots, p_0$, and using (11.4.9), (11.4.13), we can show that

$$\|D_{ij}T_1(t)g\|_\infty \le Ct^{\beta-1}\|g\|_{D_{A_1}(\beta,\infty)}, \qquad i,j = 1,\ldots,p_0. \tag{11.4.17}$$

Hence, f is twice continuously differentiable in $\mathbb{R}^N$ with respect to the first r variables and $\|D_{ij}f\|_\infty \le C\|g\|_{D_{A_1}(\beta,\infty)} \le C\|f\|_{D_{A_1}(1+\beta,\infty)}$ for any $i,j = 1,\ldots,p_0$. This implies that f belongs to $C_b^{2,0}(\mathbb{R}^N)$ with norm bounded by $C\|f\|_{D_{A_1}(1+\beta,\infty)}$. Since we already know that $D_{A_1}(\beta,\infty) = \widetilde{\mathscr{C}}_b^{2\beta,0}(\mathbb{R}^N)$, it is sufficient to prove that $D_{ij}f$ belongs to $D_{A_1}(\beta,\infty)$. Using (11.4.9) and (11.4.17), we get

$$\begin{aligned}\|\xi^{1-\beta}A_1T_1(\xi)D_{ij}f\|_\infty \le & \left\|\int_0^{+\infty} \xi^{1-\beta}e^{-t}A_1T_1(\xi+t/2)D_{ij}T_1(t/2)g\,dt\right\|_\infty \\ \le & C\|g\|_{D_{A_1}(\beta,\infty)}\int_0^{+\infty}\frac{1}{(1+t)t^{1-\beta}}dt\end{aligned}$$

for any $\xi \in (0,1)$. Thus, $D_{ij}f$ belongs to $D_{A_1}(\beta,\infty) = \widetilde{\mathscr{C}}_b^{2\beta,0}(\mathbb{R}^N)$ for any $i,j \le p_0$ (see again Proposition C.2.16) and $\|D_{ij}f\|_{\widetilde{\mathscr{C}}_b^{2\beta,0}(\mathbb{R}^N)} \le C\|g\|_{D_{A_1}(\beta,\infty)} \le C\|f\|_{D_{A_1}(1+\beta,\infty)}$. Formula (11.4.16) now follows.

Step 2. Here, taking advantage of the results in Step 1, we show that

$$(C_b(\mathbb{R}^N), \widetilde{\mathscr{C}}_b^{3\alpha,\alpha}(\mathbb{R}^N))_{\gamma,\infty} = \widetilde{\mathscr{C}}_b^{3\alpha\gamma,\alpha\gamma}(\mathbb{R}^N) \tag{11.4.18}$$

for any $\alpha \in (0, 4/3)$ and $\gamma \in (0, 1)$. This can be done adapting the proof of [181, Theorem 2.2]. We begin by proving (11.4.18) in the case when $\alpha \in (0, 2/3)$. By (11.4.7) we can write $\widetilde{\mathscr{C}}_b^{3\alpha,\alpha}(\mathbb{R}^N) = D_{A_1}(3\alpha/2,\infty) \cap D_{A_2}(\alpha/2,\infty)$. The characterization (11.4.18) will follow immediately if we show that

$$(C_b(\mathbb{R}^N), D_{A_1}(3\alpha/2,\infty) \cap D_{A_2}(\alpha/2,\infty))_{\gamma,\infty} = D_{A_1}(3\gamma\alpha/2,\infty) \cap D_{A_2}(\gamma\alpha/2,\infty) \tag{11.4.19}$$

for any $\gamma \in (0, 1)$, with equivalence of the corresponding norms.

The embedding "$\hookrightarrow$" in (11.4.19) is immediate if we observe that, for any triplet of Banach spaces X, Y, Z such that $Z \subset Y \subset X$ with continuous embeddings, it holds that $(X, Z)_{\theta,\infty} \hookrightarrow (X, Y)_{\theta,\infty}$, for any $\theta \in (0, 1)$. To prove the other inclusion, we fix $f \in D_{A_1}(3\gamma\alpha/2,\infty) \cap D_{A_2}(\gamma\alpha/2,\infty)$ and set $u(t, \cdot) := T_1(t^{2/(3\alpha)})T_2(t^{2/\alpha})f$ for any $t \in (0, 1]$. Using Proposition C.2.16 yields

$$\begin{aligned}
\|u(t,\cdot)\|_{D_{A_1}(\frac{3\alpha}{2},\infty)} \leq& C\Big(\|T_1(t^{2/(3\alpha)})\|_{L(C_b(\mathbb{R}^N))}\|T_2(t^{2/\alpha})\|_{L(C_b(\mathbb{R}^N))}\|f\|_{C_b(\mathbb{R}^N)} \\
&+ \|T_2(t^{2/(3\alpha)})\|_{L(C_b(\mathbb{R}^N))} \sup_{0<\xi\leq 1} \|\xi^{1-\frac{2}{3\alpha}} A_1 T_1(\xi + t^{2/(3\alpha)})f\|_{C_b(\mathbb{R}^N)}\Big) \\
\leq& C\Big(\|f\|_{C_b(\mathbb{R}^N)} + \sup_{0<\xi\leq 1} \xi^{1-2/(3\alpha)}(\xi + t^{2/(3\alpha)})^{(3\alpha-2)/2}\|f\|_{D_{A_1}(\frac{3\alpha}{2},\infty)}\Big) \\
\leq& Ct^{(3\alpha-2)/2}\|f\|_{D_{A_1}(3\alpha/2,\infty)}
\end{aligned}$$

for any $t \in (0, 1]$. Similarly, $\|u(t, \cdot)\|_{D_{A_2}(\alpha/2,\infty)} \leq Ct^{(\alpha-2)/2}\|f\|_{D_{A_2}(\alpha/2,\infty)}$ and

$$\begin{aligned}
\|u(t,\cdot) - f\|_{C_b(\mathbb{R}^N)} \leq& \|T_1(t^{2/(3\alpha)})\|_{L(C_b(\mathbb{R}^N))}\|(T_2(t^{2/\alpha}) - 1)f\|_{C_b(\mathbb{R}^N)} \\
\leq& Ct^{\gamma}(\|f\|_{D_{A_1}(\frac{3/\alpha}{2},\infty)} + \|f\|_{D_{A_2}(\alpha/2,\infty)})
\end{aligned}$$

for any $t \in (0, 1]$. Splitting $f = u(t, \cdot) + (f - u(t, \cdot))$ and using these three last estimates, we get $t^{-\gamma}K(t, f) \leq C(\|f\|_{D_{A_1}(3\alpha/2,\infty)} + \|f\|_{D_{A_2}(\alpha/2,\infty)})$ for any $t \in (0, 1]$, which implies that $f \in (C_b(\mathbb{R}^N), D_{A_1}(3\alpha/2,\infty) \cap D_{A_2}(\alpha/2,\infty))_{\gamma,\infty}$. We have so proved both the set equality and the equivalence of the norms of the spaces in the left- and right-hand sides of (11.4.19).

Now, suppose that $\alpha \in [2/3, 4/3)$. According to (11.4.8), $\widetilde{\mathscr{C}}_b^{3\alpha,0}(\mathbb{R}^N) = D_{A_3}(3\alpha/4,\infty)$ with equivalence of the corresponding norms. Therefore, repeating the above arguments, with the operator A_1 being now replaced with A_3, we can prove (11.4.18).

Step 3. We can now conclude the proof checking the topological equality (11.4.5). We note that $\mathscr{C}_b^3(\mathbb{R}^N)$ belongs to the class $K_{1/\alpha}$ between $C_b(\mathbb{R}^N)$ and $\widetilde{\mathscr{C}}_b^{3\alpha,\alpha}(\mathbb{R}^N)$. Indeed, it is continuously embedded into $\widetilde{\mathscr{C}}_b^3(\mathbb{R}^N)$, which belongs to the class $K_{1/\alpha}(C_b(\mathbb{R}^N), \widetilde{\mathscr{C}}_b^{3\alpha,\alpha}(\mathbb{R}^N))$, as it follows from the formula (11.4.18), with $\alpha \in (1, 4/3)$ and $\gamma = 1/\alpha$. Since $C_b^3(\mathbb{R}^{p_0})$ and $C_b^1(\mathbb{R}^{N-p_0})$ belong, respectively, to the classes $J_{1/\alpha}(C_b(\mathbb{R}^{p_0}), C_b^{3\alpha}(\mathbb{R}^{p_0}))$ and $J_{1/\alpha}(C_b(\mathbb{R}^{N-p_0}), C_b^{\alpha}(\mathbb{R}^{N-p_0}))$ (see Proposition B.3.4), $\mathscr{C}_b^3(\mathbb{R}^N)$ belongs to the class $J_{1/\alpha}$ between $C_b(\mathbb{R}^N)$ and $\widetilde{\mathscr{C}}_b^{3\alpha,\alpha}(\mathbb{R}^N)$. The reiteration theorem (see Theorem B.3.7) and (11.4.18) imply that

$$(C_b(\mathbb{R}^N), \mathscr{C}_b^3(\mathbb{R}^N))_{\theta,\infty} = (C_b(\mathbb{R}^N), \widetilde{\mathscr{C}}_b^{3\alpha}(\mathbb{R}^N))_{\theta/\alpha,\infty} = \widetilde{\mathscr{C}}_b^{3\theta}(\mathbb{R}^N)$$

for any $\theta \in (0, 1)$, with equivalence of the corresponding norms. The set equality (11.4.5) now follows. □

Applying the reiteration Theorem B.3.7, we obtain the following corollary.

Corollary 11.4.7 *For $\alpha, \theta \in (0,1)$ and $\alpha \le \beta < 1$, it holds that $(\widetilde{\mathscr{C}}_b^{3\alpha}(\mathbb{R}^N), \mathscr{C}_b^3(\mathbb{R}^N))_{\theta,\infty} = \widetilde{\mathscr{C}}_b^{3(\alpha+(1-\alpha)\theta)}(\mathbb{R}^N)$ with equivalence of the corresponding norms.*

Remark 11.4.8 Adapting the techniques in the proof of Proposition 11.4.6 one can show that the formula (11.4.18) holds true for any $\alpha > 1$ and, hence by reiteration, that, for any $\alpha, \beta > 0$ and $\theta \in (0,1)$, $(\widetilde{\mathscr{C}}_b^{\alpha}(\mathbb{R}^N), \widetilde{\mathscr{C}}_b^{\beta}(\mathbb{R}^N))_{\theta,\infty} = \widetilde{\mathscr{C}}_b^{\alpha+\theta(\beta-\alpha)}(\mathbb{R}^N)$, with equivalence of the corresponding norms.

As far as (11.4.6) is concerned, we observe that, when $k = 0$, such an estimate has been already checked in the proof of Proposition 11.4.3. Hence, we just need to show it when $k = 1$. For this purpose, we adapt the Bernstein method to these anisotropic spaces.

Theorem 11.4.9 *Let $\varepsilon > 0$ and assume that Hypotheses 11.0.1 are satisfied. Then, there exist two positive constants C and ω such that estimate (11.4.6) holds true, with $k = 1$.*

Proof The proof is similar to that of Theorem 11.2.1. Hence, we just sketch the arguments that can be used to show the estimate $\|T_\varepsilon(t)\|_{L(\mathscr{C}_b^3(\mathbb{R}^N), \mathscr{C}_b^3(\mathbb{R}^N))} \le C$ for any $t \in (0, T_0]$ and some positive constant C, independent of ε, and $T_0 > 0$. We use the same notation as in the proof of the quoted theorem and we simply write u instead of $T_\varepsilon(\cdot)f$. An inspection of the proof of Lemma 11.1.3 reveals that

$$[\partial_k^2, \langle Bx, \nabla_x\rangle]u(t,x) = \sum_{j=0}^{3} L_{k,j}\partial_j^2 u(t,x), \qquad k = 0,1,2,3, \tag{11.4.20}$$

$$[\partial_k^3, \langle Bx, \nabla_x\rangle]u(t,x) = \sum_{j=0}^{4} M_{k,j}\partial_j^3 u(t,x), \qquad k = 0,1,2,4, \tag{11.4.21}$$

for any $(t,x) \in (0,+\infty) \times \mathbb{R}^N$, and some suitable matrices $L_{k,j}$ and $M_{k,j}$, whose entries linearly depend only on the entries of B. In particular, L_{13}, L_{31}, M_{13}, M_{14}, $M_{2,4}$, $M_{3,1}$, $M_{4,1}$ and $M_{4,2}$ are null matrices, whereas $L_{2,3} = \mathcal{I}_3^{(2)}$ and $M_{3,4} = \mathcal{I}_4^{(3)}$ (see Lemma 11.1.3)

Let us introduce the function $v_\varepsilon : [0,+\infty) \times \mathbb{R}^N \to \mathbb{R}$ defined by $v_\varepsilon(t,x) = \frac{1}{2}a^3(u(t,x))^2 + \sum_{n=1}^{3}\langle \mathcal{H}^{(1)} D_x^n u(t,x), D_x^n u(t,x)\rangle$ for any $t > 0$ and $x \in \mathbb{R}^N$, where the matrices $\mathcal{H}^{(j)}(t)$ ($j = 1,2$, $t \ge 0$) are the same as in the proof of Theorem 11.2.2 with the function q_h being replaced by the function $\widetilde{q}_h$ defined as follows: $\widetilde{q}_h(\alpha) = 0$ if $\alpha = (\alpha_1, 0, \ldots, 0)$ and $\widetilde{q}_h(\alpha) = q_h(\alpha)$, otherwise. Moreover, we do the following choices for the parameters $\xi_m^{(n)}$ and $\widetilde{\xi}_m^{(n)}$:

$$\begin{aligned}
&\xi_1^{(1)} = 1, \qquad \xi_2^{(1)} = 0, \qquad \widetilde{\xi}_2^{(1)} = 0,\\
&\xi_1^{(2)} = 0, \qquad \xi_2^{(2)} = -\tfrac{7}{16}, \qquad \xi_3^{(2)} = -\tfrac{7}{8}, \qquad \widetilde{\xi}_3^{(2)} = -\tfrac{4}{5},
\end{aligned}$$

and define the matrix $\mathcal{H}^{(3)}(t)$ as follows:

$$\mathcal{H}^{(3)}(t) = \begin{pmatrix} a^{-\frac{4}{3}} I_{s_1^{(3)}} & -a^{-\frac{4}{3}} t H_1^{(3)} & \frac{1}{4}t^2 a^{-\frac{4}{3}} H_2^{(3)} & -\frac{1}{12}a^{-\frac{4}{3}} t^3 H_3^{(3)} \\ -a^{-\frac{4}{3}} t (H_1^{(3)})^* & a^{-\frac{8}{9}} t^2 I_{s_2^{(3)}} & 0 & 0 \\ \frac{1}{4}t^2 \alpha^{-\frac{4}{3}} (H_2^{(3)})^* & 0 & a^{-1} t^4 I_{s_3^{(3)}} & a^{-\frac{13}{12}} t^5 (H_4^{(3)})^* \\ -\frac{1}{12} a^{-\frac{4}{3}} t^4 H_3^{(3)} & 0 & a^{-\frac{13}{12}} t^5 H_4^{(3)} & a^{-\frac{9}{8}} t^6 I_{s_4^{(3)}} \end{pmatrix}. \tag{11.4.22}$$

Here, $s_p^{(k)}$ denotes the cardinality of the set $\{\alpha \in (\mathbb{N} \cup \{0\})^N : |\alpha| = i_p^{(k)}\}$, $H_1^{(3)} = M_{1,2}$, $H_2^{(3)} = M_{1,2}M_{2,3}$, $H_3^{(3)} = M_{1,2}M_{2,3}M_{3,4}$ (see (11.4.20) and (11.4.21)).

As is easily seen, $v_\varepsilon(0,\cdot) = \frac{1}{2}a^3 f^2 + \langle \mathcal{H}_0^{(1)}\nabla f, \nabla f\rangle + |\partial_1^2 f|^2 + \alpha^{-\frac{4}{3}}|\partial_1^3 f|^2$ and $D_t v_\varepsilon = \mathcal{A}_\varepsilon v_\varepsilon + g_\varepsilon$ in $(0,+\infty)\times\mathbb{R}^N$, where g_ε is given by (11.2.4). To get the assertion, as in the proof of Theorem 11.2.2 it suffices to show that there exists $a>0$, independent of ε, such that $\mathcal{H}^{(k)}(t)$ $(k=1,2,3)$ are strictly positive definite matrices for any $t>0$ and $g_\varepsilon \le 0$ in $(0,T_0]\times\mathbb{R}^N$ for some T_0 independent of ε.

The first property is clearly satisfied if $a \ge a_0$ and a_0 is sufficiently large. Hence, it suffices to prove that $g_\varepsilon \le 0$ in $(0,T_0]\times\mathbb{R}^N$ for a suitable choice of T_0 and the parameter α. In the rest of the proof, we assume that t is arbitrarily fixed in $(0,1]$ and $a\ge a_0$. Arguing as in the proof of (11.2.5) we can prove that

$$\begin{aligned} g_1(t,\cdot) \le &- a^3\kappa|\partial_1^1 u(t,\cdot)|^2 - \kappa(2a - C\sqrt{a})|\partial_1^2 u(t,\cdot)|^2 - \kappa(2 - Ca^{-\frac{1}{2}})|\partial_2^2 u(t,\cdot)|^2\\ &- 2\kappa|\partial_1^3 u(t,\cdot)|^2 - \kappa t(2a^{-\frac{7}{16}} - Ca^{-\frac{3}{5}})|\partial_2^3 u(t,\cdot)|^2\\ &- \kappa t^3(2a^{-\frac{7}{8}} - Ca^{-1})|\partial_3^3 u(t,\cdot)|^2 - 2a^{-\frac{4}{3}}\kappa\partial_1^4 u(t,\cdot)|^2 - 2a^{-\frac{8}{9}}\kappa t^2|\partial_2^4 u(t,\cdot)|^2\\ &- \kappa t^4(2a^{-1} - Ca^{-\frac{49}{48}})|\partial_3^4 u(t,\cdot)|^2 - \kappa t^6(2a^{-\frac{9}{8}} - Ca^{-\frac{55}{48}})|\partial_4^4 u(t,\cdot)|^2, \end{aligned} \tag{11.4.23}$$

where, here and below, C is a constant independent of t, ε and a. The functions $g_{2,\varepsilon}$ and $g_{3,\varepsilon}$ can be estimated using the Young inequality properly and taking (11.4.20) and (11.4.21) into account.[11] We get

$$\begin{aligned} g_{2,\varepsilon}(t,\cdot)+g_{3,\varepsilon}(t,\cdot) \le &-(\lambda^{(1)} + o(1;+\infty))|\partial_2^1 u(t,\cdot)|^2\\ &- 2t^2[a^{-\frac{4}{5}}\lambda^{(2)} + o(a^{-\frac{4}{5}};+\infty) + o(1;0)]|\partial_3^2 u(t,\cdot)|^2\\ &- 2t^5[a^{-\frac{13}{12}}(\lambda^{(3)} + o(1;+\infty)) + o(1;0)]|\partial_4^3 u(t,\cdot)|^2\\ &+ C\Big\{a|\partial_1^1 u(t,\cdot)|^2 + (1+o(1;0))|\partial_1^2 u(t,\cdot)|^2 + [a^{-\frac{1}{16}} + o(1;0)]|\partial_2^2 u(t,\cdot)|^2\\ &\quad + [a^{-\frac{4}{3}} + o(1;0)]|\partial_1^3 u(t,\cdot)|^2 + t[a^{-\frac{8}{9}} + o(1;0)]|\partial_2^3 u(t,\cdot)|^2\\ &\quad + t^3[a^{-\frac{8}{9}} + o(1;0)]|\partial_3^3 u(t,\cdot)|^2\Big\} \end{aligned} \tag{11.4.24}$$

for any $t>0$, where $\lambda^{(k)}$ is any positive constant less than the minimum eigenvalue of the matrix $-H_m^{(n)}\mathcal{J}_m^{(n)} - (H_m^{(n)}\mathcal{J}_m^{(n)})^*$. Moreover, the Landau symbols have the following meaning: $o(a^\gamma;+\infty)$ $(\gamma\le 0)$ denotes a function h such that $a^{-\gamma}h(a)$ vanishes as a tends to $+\infty$, whereas $o(1;0)$ denotes a function of the variable t which vanishes as t tends to 0.

Let us now consider the function $g_{4,\varepsilon}$. Observe that

$$[\partial_h^2, q_{ij}D_{ij}]u = D_{ij}u\partial_h^2 q_{ij} + \sum_{j=0}^{4} N_{h,j}\partial_j^3 u,$$

$$[\partial_k^3, q_{ij}D_{ij}]u = D_{ij}u\partial_k^3 q_{ij} + \sum_{j=0}^{4} P_{k,j}\partial_j^3 u + \sum_{j=0}^{4} R_{k,j}\partial_j^4 u$$

for any $h=1,2,3$ and $k=1,2,3,4$, where the entries of the matrices $N_{k,j}(x)$ $P_{j,k}(x)$, $R_{j,k}(x)$ are linear combinations of the entries of the derivatives of the diffusion coefficients q_{ij} at x. In particular, $N_{1,j}$, $P_{1,j}$, $R_{1,j}$ $(j=2,3,4)$, $N_{2,j}$, $P_{2,j}$, $R_{2,j}$ $(j=3,4)$, $N_{3,j}$, $P_{3,j}$ $(j=1,3,4)$ $R_{3,j}$ $(j=1,4)$ and $R_{4,j}$ $(j=1,2,4)$ are null matrices. Moreover, there exists a positive constant C such that the norms of these matrices at any $x\in\mathbb{R}^N$ can be bounded from above by $C\sqrt{\kappa(x)}$. Taking these properties into account and applying Young inequality, it

[11] Here, the choice of the blocks on the first "line" of (11.4.22) is crucial to obtain terms which we can control with the negative terms in (11.4.23) and in (11.4.24).

is easy to check that all the terms in the definition of the function $g_{4,\varepsilon}$ are negligible (as t tends to 0^+) with respect to the terms in (11.4.23) and the leading terms in (11.4.24). Similarly, the terms in $g_{5,\varepsilon}$ are negligible with respect to the terms in (11.4.23) and the leading terms in (11.4.24). Summing up, we have proved that

$$\begin{aligned}
g_\varepsilon(t,\cdot) \leq & [-a^3 + o(a^3;+\infty)]\kappa|\partial_1^1 u(t,\cdot)|^2 + [-\lambda^{(1)} + o(1;+\infty)]|\partial_2^1 u(t,\cdot)|^2 \\
& + [-2a + o(a;+\infty) + o(1;0)]\kappa|\partial_1^2 u(t,\cdot)|^2 + [-2 + o(1;+\infty) + o(1;0)]\kappa|\partial_2^2 u(t,\cdot)|^2 \\
& + t^2[-2\lambda^{(2)} a^{-\frac{4}{5}} + o(a^{-\frac{4}{5}};+\infty) + o(1;0)]|\partial_3^2 u(t,\cdot)|^2 \\
& + [-2 + o(1;+\infty) + o(1;0)]\kappa|\partial_1^3 u(t,\cdot)|^2 \\
& + t[-2a^{-\frac{7}{16}} + (a^{-\frac{7}{16}};+\infty) + o(1;0)]\kappa|\partial_2^3 u(t,\cdot)|^2 \\
& + t^3[-2a^{-\frac{7}{8}} + o(a^{-\frac{7}{8}};+\infty) + o(1;0)]\kappa|\partial_3^3 u(t,\cdot)|^2 \\
& + t^5[-2a^{-\frac{13}{12}}\lambda^{(3)} + o(a^{-\frac{13}{12}};+\infty) + o(1;0)]|\partial_4^3 u(t,\cdot)|^2 \\
& - [2a^{-\frac{4}{3}} + o(a^{-\frac{4}{3}};+\infty) + o(1;0)]\kappa|\partial_1^4 u(t,\cdot)|^2 \\
& - t^2[2a^{-\frac{8}{9}} + o(a^{-\frac{8}{9}};+\infty) + o(1;0)]\kappa|\partial_2^4 u(t,\cdot)|^2 \\
& - t^4[2a^{-1} + o(a^{-1};+\infty) + o(1;0)]\kappa|\partial_3^4 u(t,\cdot)|^2 \\
& - t^6[2a^{-\frac{9}{8}} + o(a^{-\frac{9}{8}};+\infty)]\kappa|\partial_4^4 u(t,\cdot)|^2. \qquad (11.4.25)
\end{aligned}$$

It is now clear that we can fix T_0 sufficiently close to 0 and a large enough so that all the terms in the right-hand side of (11.4.25) are negative in $(0, T_0] \times \mathbb{R}^N$. □

Chapter 12

The Cauchy-Dirichlet problem

12.0 Introduction

In this chapter we consider the parabolic Cauchy-Dirichlet problem

$$\begin{cases} D_t u(t,x) = \mathcal{A}u(t,x), & t>0, \quad x\in\Omega, \\ u(t,x)=0, & t>0, \quad x\in\partial\Omega, \\ u(0,x)=f(x), & \quad\quad x\in\Omega, \end{cases} \tag{12.0.1}$$

where $\Omega \subset \mathbb{R}^N$ is an unbounded connected smooth open set with uniformly $C^{2+\alpha}$-boundary for some $\alpha \in (0,1)$, f is a bounded and continuous function on Ω, $\mathcal{A}$ is the second-order uniformly elliptic operator

$$\mathcal{A} = \sum_{i,j=1}^N q_{ij} D_{ij} + \sum_{i=1}^N b_i D_i + c, \tag{12.0.2}$$

with unbounded coefficients which satisfy the following conditions:

Hypotheses 12.0.1 (i) $q_{ij} \equiv q_{ji} \in C^1_b(\Omega)$ for any $i,j=1,\ldots,N$ and there exists $\kappa_0>0$ such that $\langle Q(x)\xi,\xi\rangle \geq \kappa_0|\xi|^2$ for any $x\in\Omega$ and $\xi\in\mathbb{R}^N$;

(ii) $c(x)\leq 0$ for any[1] $x\in\Omega$;

(iii) $q_{ij}, b_i, c \in C^{1+\alpha}(\Omega\cap B_R)$ for any $i,j=1,\ldots,N$ and $R>0$;

(iv) there exist a positive function $\varphi \in C^2(\overline{\Omega})$ and a positive constant λ_0 such that $\lim_{|x|\to+\infty}\varphi(x)=+\infty$ and $\sup_{x\in\Omega}(\mathcal{A}\varphi(x)-\lambda_0\varphi(x))<+\infty$.

In Section 12.1 we first show a maximum principle for the bounded classical solution[2] to problem (12.0.1), which is continuous in $[0,+\infty)\times\overline{\Omega}\setminus(\{0\}\times\partial\Omega)$. Then, by virtue of this maximum principle and some classical Schauder estimates, we prove the existence and uniqueness of the bounded classical solution u to the Cauchy-Dirichlet problem (12.0.1). As in Chapters 1 and 11, we use this result to define a semigroup $\{T(t)\}$ of bounded linear operators in $C_b(\Omega)$.

Next, under suitable additional assumptions on the coefficients b_j ($j=1,\ldots,N$) and c (see Hypothesis 12.2.1), we prove that the gradient estimate

$$\|\nabla_x T(t)f\|_\infty \leq Ct^{-\frac{1}{2}}e^{\omega t}\|f\|_\infty, \qquad t>0, \tag{12.0.3}$$

[1]This condition can be replaced by the condition $c(x)\leq c_0$ in $\mathbb{R}^N$ for some $c_0\in\mathbb{R}$, with slight changes in the proofs.

[2]i.e., a function $u\in C^{1,2}((0,+\infty)\times\Omega)$, bounded in $(0,+\infty)\times\Omega$ and continuous in $[0,+\infty)\times\overline{\Omega}\setminus(\{0\}\times\partial\Omega)$ which solves problem (12.0.1).

holds true for any $\omega > 0$, $f \in C_b(\Omega)$ and some positive constant $C = C(\omega)$. First, in Subsection 12.2.1 we prove (12.0.3) for classical solutions to problem (12.0.1) enjoying some more regularity properties. Unfortunately, it is not immediate to adapt the procedure in the proof of Theorem 5.1.7, due to the fact that no *a priori* information of the sign of $\nabla_x u$ on $(0, +\infty) \times \partial\Omega$ is available. To overcome such a difficulty, we first prove the gradient estimate (12.0.3) at the boundary of Ω by a comparison argument with a suitable one-dimensional parabolic problem. This result and the maximum principle in Proposition 12.1.1 then allow us to prove estimate (12.0.3).

Finally, in Subsections 12.2.2 and 12.2.3, adapting to our situation the approach introduced by P. Cannarsa and V. Vespri in [50, 51], we prove (12.0.3) for any classical solution to problem (12.0.1). For this purpose, the operator $\mathcal{A}$ is approximated by a family of uniformly elliptic operators $\mathcal{A}_\varepsilon$ such that, for any $p \geq 2$, their realizations in $L^p(\Omega)$ generate analytic semigroups $\{S_\varepsilon(t)\}$. Using the Sobolev embedding theorems, we show that, for any $f \in C_c^\infty(\Omega)$, the function $u_\varepsilon = S_\varepsilon(\cdot)f$ is the classical solution to problem (12.0.1), with $\mathcal{A}$ being replaced with the operator $\mathcal{A}_\varepsilon$. Then, letting ε tend to 0^+ and using a compactness argument based on classical Schauder estimates, we prove that u_ε converges to $T(t)f$, which satisfies all the additional assumptions needed, in Proposition 12.2.7, to prove the gradient estimate (12.0.3). Hence, (12.0.3) follows in this particular case. Finally, in the general case when $f \in C_b(\Omega)$, estimate (12.0.3) is obtained by approximating f with a suitable sequence of $C_c^\infty(\Omega)$-smooth functions converging, locally uniformly and in a dominated way, to f.

Unfortunately, to the best of our knowledge, estimates for the higher-order derivatives of the function $T(t)f$ when $f \in C_b(\Omega)$ seem to be available for an unbounded domain Ω only in some particular case (see Remark 12.2.16). As is readily seen, differently from what happens when $\Omega = \mathbb{R}^N$ (see Chapters 5 and 6), the Bernstein method can not be easily extended to estimate second- and third-order derivatives of the function $T(t)f$ and this is essentially due to the boundary conditions. This fact prevents us from making the machinery of Chapter 5 work to study the nonhomogeneous elliptic problem and the nonhomogeneous Cauchy-Dirichlet problem in Ω and to prove optimal Schauder estimates. To the best of our knowledge there are only a few results in this direction (see again Remark 12.2.16).

Differently from what happens in the whole of $\mathbb{R}^N$, there never exists an invariant measure μ associated with the semigroup $\{T(t)\}$ considered in this chapter. Indeed, if this were the case, then

$$\int_\Omega T(t)d\mu = \int_\Omega f d\mu \tag{12.0.4}$$

for any $f \in C_b(\Omega)$. But, according to Corollary 12.1.3, it holds that $0 \leq T(t)\mathbb{1} \leq \mathbb{1}$ for any $t > 0$. Hence, from (12.0.4), it should follow that $T(t)\mathbb{1} \equiv \mathbb{1}$ which is a contradiction.

The results of this chapter have been proved in [106].

12.1 Existence and uniqueness

Throughout this section we assume only Hypotheses 12.0.1.

To begin with, we introduce a few notation which will be used in the proof of the next proposition and, more generally, throughout this chapter. By $d : \overline{\Omega} \to \mathbb{R}$ and Ω_δ ($\delta > 0$) we denote, respectively, the distance function d, defined by $d(x) = \mathrm{dist}(x, \partial\Omega)$ for any $x \in \overline{\Omega}$, and the set $\Omega_\delta = \{x \in \overline{\Omega} : d(x) < \delta\}$. In view of the assumptions on Ω, the function d is C^2-smooth, with bounded second-order derivatives in Ω_δ, if $\delta > 0$ is sufficiently small. See [116, Lemma 14.16] and also Section A.4.

Next proposition provides us with a maximum principle for classical solutions to the Cauchy-Dirichlet problem (12.0.1).

Proposition 12.1.1 *Let $u \in C^{1,2}((0,T) \times \Omega) \cap C([0,T) \times \overline{\Omega} \setminus (\{0\} \times \partial\Omega))$ be bounded in $(0,T) \times \Omega$ and satisfy the inequality $D_t u - \mathcal{A}u \le 0$ in $(0,T) \times \Omega$. Then, $\sup_{(0,T)\times\Omega} u \le \|u(0,\cdot)\|_\infty + \sup_{(0,T)\times\partial\Omega} |u|$. In particular, if $u \le 0$ in $[0,T) \times \overline{\Omega} \setminus (\{0\} \times \partial\Omega)$, then $u \le 0$ in $(0,T) \times \Omega$. Finally, if u solves the Cauchy problem* (12.0.1) *for some $f \in C_b(\Omega)$, then $\|u\|_{C_b((0,T)\times\Omega)} \le \|f\|_\infty$.*

Proof We limit ourselves to proving that $u \le 0$ in $(0,T) \times \Omega$ if it is nonpositive in $[0,T) \times \overline{\Omega} \setminus (\{0\} \times \partial\Omega)$. The first part of the proposition then follows applying this result to the functions $u - K$, where $K = \sup_{(t,\xi)\in(0,T)\times\partial\Omega} |u(t,\xi)| + \|u(0,\cdot)\|_\infty$. The last statement follows applying the first part to u and $-u$.

To begin with, we observe that there exists a nonnegative function $g \in C^2(\overline{\Omega})$ such that $|\nabla g| = 1$ on $\partial\Omega = \{x \in \overline{\Omega} : g(x) = 0\}$. It suffices to take $g = \eta d + \mathbb{1} - \eta$, where η is a smooth function such that $0 \le \eta \le \mathbb{1}$ in Ω, $\eta \equiv \mathbb{1}$ in $\Omega_{\delta/2}$ and $\eta \equiv 0$ outside $\Omega_{3\delta/4}$.

Next, we introduce the function $v = e^{-\lambda_0 t}u$, where λ_0 is given by Hypothesis 12.0.1(iv), and prove that $v \le 0$ in $(0,T) \times \Omega$. For this purpose, we fix $R > 1$ and consider the set $\Omega_R = \Omega \cap B_R = \{g > 0\} \cap \{R^2 - |x|^2 > 0\}$. Moreover, we introduce the function $v_R = v - C_R^{-1}\|v\|_\infty \varphi$, where $C_R = \inf_{[0,T]\times(\partial B_R\cap\Omega)} \varphi$. Up to replacing λ_0 with a larger value, we can assume that $\mathcal{A}\varphi - \lambda_0\varphi \le 0$. By Hypothesis 12.0.1(iv), C_R tends to $+\infty$ as R tends to $+\infty$. Further, it is easy to see that $D_t v_R - \mathcal{A}v_R + \lambda_0 v_R \le 0$ in $(0,T) \times \Omega_R$ and $v_R \le 0$ in $(0,T) \times \partial\Omega_R$ and in $\{0\} \times \Omega_R$. We claim that $v_R \le 0$ in $(0,T) \times \Omega_R$. Once the claim is proved, letting R tend to $+\infty$, we will obtain that $v \le 0$ in $(0,T) \times \Omega$ and we will be done.

For any $\varepsilon, \lambda > 0$ let $\psi : (0,T) \times \Omega_R \to \mathbb{R}$ be the function defined by $\psi(t,x) = t^{-\varepsilon\kappa_0}[\exp(\lambda t - \varepsilon(g(x))^2 t^{-1}) + \exp(\lambda t - \varepsilon(h(x))^2 t^{-1})]$ for any $t > 0$ and $x \in \Omega_R$, where κ_0 is defined in Hypothesis 12.0.1(i) and $h(x) = R^2 - |x|^2$ for any $x \in \mathbb{R}^N$. A straightforward computation shows that

$$
\begin{aligned}
&D_t\psi(t,\cdot) - \mathcal{A}\psi(t,\cdot)\\
=&t^{-2}\{(\lambda - c)t^2 - \varepsilon[\kappa_0 - 2|\sqrt{Q}\nabla g|^2 - 2g(\mathcal{A} - c)g]t + \varepsilon g^2(1 - 4\varepsilon|\sqrt{Q}\nabla g|^2)\}\psi(t,\cdot)\\
&+t^{-2}\{(\lambda - c)t^2 - \varepsilon[\kappa_0 - 2|\sqrt{Q}\nabla h|^2 - 2h(\mathcal{A} - c)h]t + \varepsilon h^2(1 - 4\varepsilon|\sqrt{Q}\nabla h|^2)\}\psi(t,\cdot)\\
=&: t^{-2}[\Lambda_1(t,\cdot) + \Lambda_2(t,\cdot)]
\end{aligned}
$$

for any $t > 0$. Let us prove that we can fix $\varepsilon > 0$ sufficiently small and λ sufficiently large that $(D_t - \mathcal{A})\psi \ge 0$ in $(0,+\infty) \times \Omega_R$. Since $|\nabla g| \ge 1$ on $\partial\Omega$, we can fix $\delta > 0$ such that $|\nabla g(x)| \ge 3/4$ if $x \in \Omega_R$ and $g(x) \le \delta$. Then, we split $\Omega_R = \{x \in \Omega_R : g(x) > \delta\} \cup \{x \in \Omega_R : g(x) \le \delta\} =: \Omega_{R,1} \cup \Omega_{R,2}$. If $x \in \Omega_{R,1}$ then we can estimate

$$
\Lambda_1(t,x) \ge \lambda t^2 - \varepsilon\left(\kappa_0 + 2\sup_{\Omega_R}|g(\mathcal{A} - c)g|\right)t + \varepsilon\delta^2\left(1 - 4\varepsilon\|Q\|_\infty \sup_{\Omega_R}|\nabla g|^2\right). \tag{12.1.1}
$$

If we fix $\varepsilon = (8\|Q\|_\infty \sup_{\Omega_R}|\nabla g|^2)^{-1}$ and, then, $\lambda = \lambda(\varepsilon,\delta)$ large enough, then we can make the right-hand side of (12.1.1) nonnegative.

On the other hand,

$$
\begin{aligned}
\kappa_0 - 2|\sqrt{Q}\nabla g|^2 - 2g(x)(\mathcal{A} - c)g \le&\kappa_0 - 2\kappa_0|\nabla g|^2 + 2g\sup_{\Omega_R}|(\mathcal{A} - c)g|\\
\le&-\frac{1}{2}\kappa_0 + \frac{1}{2t}g^2 + \frac{t}{2}\sup_{\Omega_R}|(\mathcal{A} - c)g|^2
\end{aligned}
$$

in $\Omega_{R,2}$ Therefore,

$$\Lambda_1(t,x)\geq\left(\lambda-\frac{1}{2}\sup_{\Omega_R}|(\mathcal{A}-c)g|^2\right)t^2+\frac{1}{2}\varepsilon\kappa_0 t+\varepsilon g^2\left(\frac{1}{2}-4\varepsilon\|Q\|_\infty\sup_{\Omega_R}|\nabla g|^2\right).$$

Again, we can fix $\varepsilon=(16\|Q\|_\infty\sup_{\Omega_R}|\nabla g|^2)^{-1}$ and, then, $\lambda=\lambda(\varepsilon)$ large enough such that the right-hand side of the previous inequality is nonnegative. Similarly, we can argue with the function Λ_2. Indeed, the keystone to prove that Λ_1 is nonnegative in $(0,+\infty)\times\Omega_R$ was the fact that $|\nabla g|\geq 1$ when $g=0$. Now, it is easy to check that, since $R>1$, $|\nabla h|\geq 1$ on ∂B_R. Summing up, we have shown that we can fix the parameters λ and ε such that $D_t\psi-\mathcal{A}\psi\geq 0$ in $(0,+\infty)\times\Omega_R$.

Let us now suppose by contradiction that $0<M=\sup_{(0,T)\times\Omega_R}v_R$ for some $T>0$. For any $a>0$ we introduce the function $v_{a,R}$, defined by $v_{a,R}(t,x)=v_R(t,x)-Ma^{\varepsilon\kappa_0}\psi(t+a,x)$ for any $(t,x)\in[0,T]\times\overline{\Omega}_R\setminus(\{0\}\times\Omega\cap\partial B_R)$. Clearly, $(D_t-\mathcal{A}+\lambda_0)v_{a,R}$ is nonpositive in $(0,T)\times\Omega_R$. Take $\eta>0$ such that $\lambda a-\varepsilon\eta^2a^{-1}>0$ and introduce the set $I_{\eta,R}=\{x\in\overline{\Omega}_R: g(x)\wedge h(x)\leq\eta\}$. For any $x\in I_{\eta,R}$ it holds that $a^{\varepsilon\kappa_0}\psi(a,x)\geq 2\exp(\lambda a-a^{-1}\varepsilon\eta^2)>1$. By continuity, there exists $\delta>0$ such that $a^{\varepsilon\kappa_0}\psi(t+a,x)>1$ for any $t\in[0,\delta]$ and $x\in I_{\eta,R}$. It follows that $v_{a,R}<M-M=0$ in $[0,\delta]\times I_{\eta,R}$.

Since $v_R(0,\cdot)\leq 0$ in $\Omega_R\setminus I_{\eta,R}$, $v_{a,R}(0,\cdot)$ is negative in $\Omega_R\setminus I_{\eta,R}$. The boundedness of Ω_R and the continuity of u in $[0,T)\times\Omega_R\setminus I_{\eta,R}$ imply that $v_{a,R}\leq 0$ in $[0,\delta]\times(\Omega_R\setminus I_{\eta,R})$ for some $\delta>0$. Hence, $v_{a,R}$ is nonpositive in $[0,\delta]\times\overline{\Omega}_R\setminus(\{0\}\times\partial\Omega\cap B_R)$.

Applying the classical maximum principle in $[\delta,T)\times\overline{\Omega}_R$, we conclude that $v_{a,R}\leq 0$ in $[\delta,T)\times\overline{\Omega}_R$ and, consequently, $v_{a,R}$ is nonpositive in $[0,T)\times\overline{\Omega}_R\setminus(\{0\}\times\partial\Omega_R)$. Finally, letting a tend to 0^+, we deduce that $v_R\leq 0$ in $[0,T)\times\overline{\Omega}_R\setminus(\{0\}\times\partial\Omega_R)$, and, letting R tend to $+\infty$, we are done. □

Now we can prove that under Hypotheses 12.0.1, the Cauchy-Dirichlet problem (12.0.1) has a unique bounded classical solution u.

Theorem 12.1.2 *For any $f\in C_b(\Omega)$, the Cauchy-Dirichlet problem* (12.0.1) *admits a unique bounded classical solution u. Moreover, $\|u\|_\infty\leq\|f\|_\infty$ and $u\geq 0$ if $f\geq 0$. Further, if $f\in C_c^{2+\alpha}(\Omega)$, then u belongs to $C^{1+\alpha/2,2+\alpha}((0,T)\times(\Omega\cap B_R))$ for any $R,T>0$. Moreover, D_ju belongs to $C^{1+\alpha/2,2+\alpha}((\varepsilon,T)\times\Omega')$ for any $0<\varepsilon<T$, any bounded open set $\Omega'\Subset\Omega$ and $j=1,\ldots,N$. In particular, $D_ju\in C^{1,2}((0,+\infty)\times\Omega)$ for any $j=1,\ldots,N$.*

Proof Uniqueness follows from Proposition 12.1.1. So, let us prove that problem (12.0.1) actually admits a solution with the claimed regularity properties. We split the proof into two steps. First, in Step 1, we consider the case when $f\in C_c^{2+\alpha}(\Omega)$ and then, in Step 2, we deal with the general case. For notational convenience, in the rest of the proof we denote by u_g the classical solution to (12.0.1), which satisfies the condition $u(0,\cdot)=g$.

Step 1. Fix $f\in C_c^{2+\alpha}(\Omega)$. To prove that problem (12.0.1) admits a classical solution, we use an approximation argument different from the one used in the proof of Theorem 1.2.1. Instead of approximating the domain, we rather approximate the coefficients of the operator $\mathcal{A}$. More precisely, for any $n\in\mathbb{N}$ we introduce the uniformly elliptic operator $\mathcal{A}^{(n)}$ with bounded coefficients in $C^\alpha(\Omega)$, defined by $\mathcal{A}^{(n)}=\mathrm{Tr}(QD_x^2)+\langle b^{(n)},\nabla_x\rangle+c^{(n)}$, where the coefficients $b_i^{(n)}$ $(i=1,\ldots,N)$ and $c^{(n)}$ coincide, respectively, with b_i and c in $\Omega\cap B_n$, and $c^{(n)}\leq 0$ in Ω. Let $n_0\in\mathbb{N}$ be such that $\mathrm{supp}(f)\Subset\Omega\cap B_{n_0}$ and, for any $n\geq n_0$, let $u_n\in C^{1+\alpha/2,2+\alpha}((0,T)\times\Omega)\cap C([0,T]\times\overline{\Omega})$, for any $T>0$, be the classical solution of (12.0.1), with $\mathcal{A}^{(n)}$ instead of $\mathcal{A}$ (see Proposition D.3.1). Proposition 12.1.1 yields $\|u_n\|_\infty\leq\|f\|_\infty$ for any $n\geq n_0$. Let us fix $R>0$ and observe that, since Ω is unbounded and connected, $\mathrm{dist}(\Omega\setminus B_{R+1},\Omega\cap B_R)>0$. Noticing that $\mathcal{A}^{(n)}=\mathcal{A}^{(m)}=\mathcal{A}$ in $\Omega\cap B_{R+1}$ for $m,n\geq R+1$ and using the local Schauder estimates in Theorem D.1.6 (see (D.1.13)), we conclude that for any $T > 0$ there exists a positive constant C, independent of $n, m \geq n_0$, such that $\|u_n - u_m\|_{C^{1+\alpha/2,2+\alpha}((0,T)\times(\Omega\cap B_R))} \leq C\|u_n - u_m\|_{C((0,T)\times(\Omega\cap B_{R+1}))} \leq 2C\|f\|_\infty$. Now, using a compactness argument as in the proof of Theorem 1.2.1, we can determine a subsequence $\{u_{n_k}\}$ converging in $C^{1,2}(F)$, for any compact set $F \subset [0,T]\times\overline{\Omega}$, to a function $u \in C^{1+\alpha/2,2+\alpha}_{\rm loc}([0,T]\times\overline{\Omega})$, which solves the Cauchy-Dirichlet problem (12.0.1) and satisfies the estimate $\|u\|_\infty \leq \|f\|_\infty$. Moreover, by the maximum principle in Proposition 12.1.1, we deduce that u is nonnegative, whenever f is. Note that, applying the same arguments as above to any subsequence of $\{u_n\}$, we can actually show that the sequence $\{u_n\}$ itself converges to u in $C^{1,2}(F)$ for any compact set $F \subset (0,T)\times\overline{\Omega}$.

The further regularity properties of the function u follow from Theorem D.1.5(ii).

Step 2. Now, we consider the general case when $f \in C_b(\Omega)$. Let $\{f_n\} \subset C^{2+\alpha}_c(\Omega)$ be a sequence converging to f uniformly on compact subsets of Ω and such that $\sup_{n\in\mathbb{N}}\|f_n\|_\infty < +\infty$. According to Theorem D.1.6, for any $\varepsilon \in (0,T)$ we can determine a positive constant C, independent of n and m, such that $\|u_{f_n} - u_{f_m}\|_{C^{1+\alpha/2,2+\alpha}((\varepsilon,T)\times(\Omega\cap B_R))} \leq 2C\|f\|_\infty$ for any $m, n \in \mathbb{N}$. As in Step 1 there exists a subsequence $\{u_{f_{n_k}}\}$ converging to a function u which belongs to $C^{1+\alpha/2,2+\alpha}((\varepsilon,T)\times(\Omega\cap B_R))$ for any $R > 0$, $\varepsilon \in (0,T)$, solves the equation $D_t u = \mathcal{A}u$ in $(0,T)\times\Omega$ and vanishes in $(0,T)\times\partial\Omega$. Clearly, $u \geq 0$ if $f \geq 0$, since in this case we can assume that the functions f_n ($n \in \mathbb{N}$) are nonnegative in Ω and, by Step 1, $u_n \geq 0$ for any $n \in \mathbb{N}$.

To conclude the proof, it remains to show that $u(t,\cdot)$ converges to f as t tends to 0^+, uniformly on compact sets of Ω. If $f \in C_0(\Omega)$, then we can assume that the sequence $\{f_n\}$ converges to f uniformly in Ω. From Proposition 12.1.1 it follows that $\|u_{f_n} - u_{f_m}\|_{C_b([0,+\infty)\times\overline{\Omega})} \leq \|f_n - f_m\|_\infty$ for any $m, n \in \mathbb{N}$. Hence, $\{u_n\}$ converges uniformly in $[0,+\infty)\times\overline{\Omega}$ and, consequently, u can be extended by continuity to $\{0\}\times\overline{\Omega}$ by setting $u(0,\cdot) = f$.

Let us now suppose that $f \in C_b(\Omega)$. We fix a compact set $K \subset \Omega$ and a function $\vartheta \in C^{2+\alpha}_c(\Omega)$ such that $0 \leq \vartheta \leq \mathbb{1}$ in Ω and $\vartheta \equiv \mathbb{1}$ in K. Moreover, we split

$$u_{f_{n_k}} = u_{\vartheta f_{n_k}} + u_{(1-\vartheta)f_{n_k}}, \qquad k \in \mathbb{N}, \tag{12.1.2}$$

and we claim that there exists a positive constant C, independent of k, such that $|u_{(1-\vartheta)f_{n_k}}(t,\cdot)| \leq C(\mathbb{1} - u_\vartheta(t,\cdot))$ for any $t > 0$. To prove this inequality, it suffices to observe that, for any $m \in \mathbb{N}$ and $t > 0$, $|T_m(t)((\mathbb{1}-\vartheta)f_{n_k})| \leq \|f_{n_k}\|_\infty(\mathbb{1} - T_m(t)\vartheta) \leq C(\mathbb{1} - T_m(t)\vartheta)$, where $\{T_m(t)\}$ is the semigroup associated with the operator $\mathcal{A}^{(m)}$, and let m tend to $+\infty$. Now, subtracting ϑf from both sides of (12.1.2) and letting k tend to $+\infty$, give $|u(t,\cdot) - \vartheta f| \leq |u_{\vartheta f} - \vartheta f| + C(1 - u_\vartheta)$ for any $t > 0$. Since $\vartheta f, \vartheta \in C_c(\overline{\Omega})$, $u_{\vartheta f}$ and u_ϑ converge, respectively to ϑf and ϑ as t tends to 0, uniformly in $\overline{\Omega}$. Recalling that $\vartheta \equiv 1$ on K, we conclude that $u(t,\cdot)$ converges to f uniformly on K as t tends to 0^+. This shows that actually u is the classical solution to (12.0.1). □

Corollary 12.1.3 *The family $\{T(t)\}$, defined by $T(t)f = u(t,\cdot)$ for any $t \geq 0$ and any $f \in C_b(\Omega)$, where u is the classical solution to problem* (12.0.1), *is a semigroup of positive contractions in $C_b(\Omega)$.*

Proof It follows easily from Proposition 12.1.1 and Theorem 12.1.2. □

12.2 Gradient estimates

This section is devoted to proving that, under Hypotheses 12.0.1 and suitable additional growth assumptions on the coefficients of the operator $\mathcal{A}$, estimate (12.0.3) is satisfied by any $f \in C_b(\Omega)$. Of course, it suffices to prove (12.0.3) in some time-domain $(0,T)$ and then to use the semigroup rule to extend it to all the positive times (see the proof of Theorem 5.1.7). As mentioned at the very beginning of this chapter the arguments that we need to prove (12.0.3) differ from those used in Chapter 5 in the case when $\Omega = \mathbb{R}^N$.

Throughout this section, if not otherwise specified, we assume the following additional assumptions on the coefficients of the operator $\mathcal{A}$. We recall that $\Omega_\delta = \{x \in \overline{\Omega} : d(x) < \delta\}$.

Hypothesis 12.2.1 There exist some constants $k, M, \beta, d_1, d_2 \in \mathbb{R}$, $s < 1/2$ and $\delta > 0$ such that

$$\langle (\mathrm{Jac}\, b)\xi, \xi\rangle \le (-sc(x) + k)|\xi|^2, \qquad x \in \Omega,\ \xi \in \mathbb{R}^N, \tag{12.2.1}$$

$$\langle b(x), \nabla d(x)\rangle \le M, \qquad x \in \Omega_\delta, \tag{12.2.2}$$

$$|b(x)| \le d_1 e^{d_2|x|}, \qquad |\nabla c(x)| \le \beta(1 - c(x)), \qquad x \in \Omega. \tag{12.2.3}$$

Remark 12.2.2 Observe that, if $s = 0$ in (12.2.1) and Ω is star-shaped with respect to 0, then Hypothesis 12.0.1(iv) is a straightforward consequence of the positivity of $-c$ and the boundedness of q_{ij} $(i, j = 1, \ldots, N)$. Indeed, since

$$\langle b(x), x\rangle = \langle b(0), x\rangle + \int_0^1 \frac{d}{dt}\langle b(tx), x\rangle dt = \langle b(0), x\rangle + \int_0^1 \langle (\mathrm{Jac}\, b(tx))x, x\rangle dt$$

for any $x \in \Omega$, from (12.2.1) we get $\langle b(x), x\rangle \le \langle b(0), x\rangle + k|x|^2$ for any $x \in \Omega$. Hence, if we take $\varphi(x) = 1 + |x|^2$ for any $x \in \Omega$, then we get $\mathcal{A}\varphi(x) \le 2\mathrm{Tr}(Q(x)) + 2|b(0)||x| + k|x|^2 + c(x)(1 + |x|^2)$. Since $c \le 0$ and the q_{ij}'s are bounded for any $i, j = 1, \ldots, N$, we can determine λ_0 sufficiently large that $\mathcal{A}\varphi - \lambda_0\varphi \le 0$ and Hypothesis 12.0.1(iv) is satisfied taking φ as a Lyapunov function.

Remark 12.2.3 As we will see in Subsection 12.2.4, the gradient estimate (12.0.3) may fail if we do not assume Hypothesis 12.2.1.

Let us state the main result of this section. Its proof will be given in Subsection 12.2.3.

Theorem 12.2.4 *The bounded classical solution u to the Cauchy-Dirichlet problem* (12.0.1) *satisfies the gradient estimate* (12.0.3).

12.2.1 *A priori* gradient estimates

In this subsection we prove the gradient estimate (12.0.3) assuming some more regularity on the bounded classical solution u to problem (12.0.1). The following lemma will play a crucial role in the proof of the *a priori* gradient estimates.

Lemma 12.2.5 *Let a, b, δ be positive constants and let $g : [0, +\infty) \times [0, \delta] \to \mathbb{R}$ be the* *classical solution*[3] *to the Cauchy-Dirichlet problem*

$$\begin{cases} D_t g(t,r) = aD_{rr}g(t,r) + bD_r g(t,r), & t>0, \quad r\in(0,\delta), \\ g(t,0)=0, \;\; g(t,\delta)=1, & t>0, \\ g(0,r)=1, & \qquad\quad r\in(0,\delta). \end{cases} \tag{12.2.4}$$

Then, $D_r g \geq 0$*,* $D_{rr}g \leq 0$ *in* $(0,+\infty)\times(0,\delta)$ *and, for any* $T>0$*, there exists a positive constant* $C=C(T)$ *such that*

$$0 \leq g(t,r) \leq Ct^{-\frac{1}{2}}r, \qquad t\in(0,T),\ r\in(0,\delta). \tag{12.2.5}$$

Proof The uniqueness of the classical solution g to problem (12.2.4) follows from Proposition 12.1.1, where we replace Ω with $(0,\delta)$.

To prove the existence of a solution g and its properties, let us consider the differential operator $(B,D(B))$ defined by $Bu=au''+bu'$ for any $u\in D(B)=\{u\in C^2([0,\delta]) : u(0)=u(\delta)=0\}$. According to Theorem D.3.5(iv), $(B,D(B))$ generates an analytic semigroup $\{R(t)\}$ of positive contractions in $C([0,\delta])$. Let us introduce the function $\psi:[0,\delta]\to\mathbb{R}$, defined by $\psi(r)=q(1-e^{-br/a})$ for any $r\in[0,\delta]$, where $q=(1-e^{-b\delta/a})^{-1}$. As is immediately seen, $B\psi=0$, $\psi(0)=0$ and $\psi(\delta)=1$. A straightforward computation shows that $g(t,\cdot)=R(t)(\mathbb{1}-\psi)+\psi$ for any $t>0$. Since the function $t\mapsto R(t)(\mathbb{1}-\psi)$ is analytic in $(0,+\infty)$ with values in $D(B^n)$ for any $n\in\mathbb{N}$ (see Theorem C.2.2(iv)), g is a C^∞-smooth function in $(0,+\infty)\times[0,\delta]$. To show that $D_r g$ is nonpositive in $(0,+\infty)\times(0,\delta)$ we begin by observing that, according to the maximum principle, it holds that $0\leq g(t,\cdot)\leq\mathbb{1}$ for any $t>0$. Hence, $R(t)(\mathbb{1}-\psi)\leq\mathbb{1}-\psi$ for any $t>0$. We also deduce that $g(t+s,\cdot)=R(t)R(s)(\mathbb{1}-\psi)+\psi\leq R(t)(\mathbb{1}-\psi)+\psi=g(t,\cdot)$ for any $s,t>0$. Therefore, the function $g(\cdot,r)$ is decreasing for any $r\in(0,\delta)$. This implies that $D_t g(t,r)\leq 0$ for any $t\in(0,+\infty)$ and $r\in(0,\delta)$. From the differential equation solved by g, we now deduce that $aD_{rr}g(t,r)+bD_r g(t,r)\leq 0$ for any $t>0$ and $r\in(0,\delta)$, or, equivalently, that the function $r\mapsto e^{br/a}D_r g(t,r)$ is decreasing in $(0,\delta)$ for any $t>0$. Since $g(t,\delta)=1$ and $0\leq g\leq\mathbb{1}$, $D_r g(t,\delta)$ is nonnegative and, consequently, $D_r g(t,\cdot)\geq 0$ for any $t>0$.

Observing that $D_{rr}g=a^{-1}(D_t g-bD_r g)$, by difference, it follows that $D_{rr}g\leq 0$ in $(0,+\infty)\times(0,\delta)$.

To conclude the proof, let us check (12.2.5). Since $\{R(t)\}$ is an analytic semigroup, for any $T>0$ there exists a positive constant $\widetilde{C}=\widetilde{C}(T)$ such that

$$\|aD_{rr}g(t,\cdot)+bD_r g(t,\cdot)\|_\infty \leq \widetilde{C}t^{-1}, \qquad t\in(0,T). \tag{12.2.6}$$

Recalling that the set $C^1([0,\delta])$ belongs to the class $J_{1/2}$ between $C([0,\delta])$ and $C^2([0,\delta])$ (see Proposition B.3.4), we can estimate

$$\|v\|_{C^1([0,\delta])}\leq C\|v\|_\infty^{\frac{1}{2}}\left(\|v\|_{C^1([0,\delta])}^{\frac{1}{2}}+\|v''\|_\infty^{\frac{1}{2}}\right)\leq\frac{C}{2\varepsilon}\|v\|_\infty+C\varepsilon\|v\|_{C^1([0,\delta])}+C\varepsilon\|v''\|_\infty \tag{12.2.7}$$

for any $v\in C^2([0,\delta])$, $\varepsilon>0$ and some constant $C>0$. Choosing $C\varepsilon<1$ yields $\|v\|_{C^1([0,\delta])}\leq C(\varepsilon)\|v\|_\infty+C\varepsilon(1-C\varepsilon)^{-1}\|v''\|_\infty$ for some positive constant $C(\varepsilon)$, blowing up as ε tends to 0^+. Using this estimate, from (12.2.6) we deduce that

$$\|D_{rr}g(t,\cdot)\|_\infty\leq C_1\left(\frac{1}{t}+\|D_r g(t,\cdot)\|_\infty\right)\leq\frac{C_2}{t}+\frac{C\varepsilon}{1-C\varepsilon}\|D_{rr}g(t,\cdot)\|_\infty$$

for any $t\in(0,T)$, any $T>0$ and some positive constants $C_1=C_1(T)$ and $C_2=C_2(\varepsilon,T)$. Up to replacing ε with a smaller constant, we finally obtain that, for any $T>0$, there exists

[3] i.e., $g\in C([0,T]\times[0,\delta]\setminus(\{0\}\times\{0,\delta\}))\cap C^{1,2}((0,T)\times[0,\delta])$.

a positive constant $\overline{C} = \overline{C}(T)$ such that $\|D_{rr}g(t,\cdot)\|_\infty \le \overline{C}t^{-1}$ for any $t \in (0,T)$. Hence, using (12.2.7) with $C\varepsilon \le 1/2$ and $v = g(t,\cdot)$, estimate (12.2.5) follows. □

Taking advantage of Lemma 12.2.5, we can prove the following *a priori* gradient estimate on the boundary of Ω.

Proposition 12.2.6 *Assume that Hypotheses* 12.0.1 *and estimate* (12.2.2) *are satisfied. Then, there exists a positive constant C_0, depending on κ_0, M, δ, T and the sup-norm of the diffusion coefficients q_{ij} $(i,j = 1,\ldots,N)$, such that any bounded classical solution u of the Cauchy-Dirichlet problem* (12.0.1)*, which is differentiable with respect to the spatial variables in $(0,T)\times\overline{\Omega}$, satisfies the estimate*

$$|\nabla_x u(t,\xi)| \le C_0 t^{-\frac{1}{2}}\|f\|_\infty, \qquad t\in(0,T),\ \xi\in\partial\Omega. \tag{12.2.8}$$

Proof For any $x \in \overline{\Omega}$ such that $d(x) = d(x,\Omega) < \delta$, let $\xi(x)$ be the unique point on $\partial\Omega$ satisfying $|x-\xi(x)| = d(x)$. Then, we can write $x = \xi(x) - d(x)\nu(\xi(x))$, where $\nu(\xi)$ is the outer unit normal to $\partial\Omega$ at ξ. Moreover, $\nabla d(x) = -\nu(\xi(x))$ for any $x\in\Omega_\delta$. To proceed, we observe that, since $u=0$ on $\partial\Omega$, $|\nabla_x u(t,\xi)| = \left|\frac{\partial u}{\partial\nu}(t,\xi)\right|$ for any $t>0$ and $\xi\in\partial\Omega$. Hence, in order to prove the claim it is enough to show that

$$(T(t)\mathbb{1})(x) \le C_0 t^{-\frac{1}{2}} d(x), \qquad t\in(0,T),\ x\in\Omega_\delta, \tag{12.2.9}$$

for some constant C_0 as in the statement of the proposition, where $\{T(t)\}$ is the semigroup defined in Corollary 12.1.3. Indeed, once (12.2.9) is proved, using the fact that $\{T(t)\}$ is order preserving, for any $f\in C_b(\Omega)$, $\xi\in\partial\Omega$ and $x = \xi + d(x)\nu(\xi)$, we can write

$$|(T(t)f)(x) - (T(t)f)(\xi)| = |(T(t)f)(x)| \le \|f\|_\infty (T(t)\mathbb{1})(x) \le C_0 t^{-\frac{1}{2}} d(x)\|f\|_\infty$$

and (12.2.8) follows dividing both the first and the last side of the previous chain of inequalities by d and letting x tend to ξ.

To prove (12.2.9) we use a comparison argument. Let $z:(0,T)\times\Omega_\delta\to\mathbb{R}$ be the function defined by $z(t,x) = g(t,d(x))$ for any $t>0$ and $x\in\Omega_\delta$, $g:[0,+\infty)\times[0,\delta]\to\mathbb{R}$ being the solution to problem (12.2.4), where we take $a=\kappa_0$ and $b = M_0 := \sup_{x\in\Omega}(\mathcal{A}-c)d$ (which is finite, since the diffusion coefficients are bounded and the drift term b satisfies (12.2.2)). By (12.2.5) we know that $|z(t,x)| = g(t,d(x)) \le Ct^{-1/2}d(x)$ for any $T>0$, any $t\in(0,T)$, $x\in\Omega_\delta$ and some constant C depending on the same quantities as in the statement of the proposition. Thus, to prove (12.2.9) we have only to show that

$$(T(t)\mathbb{1})(x) \le z(t,x), \qquad t>0,\ x\in\Omega_\delta. \tag{12.2.10}$$

Let us consider the function $v = z - T(\cdot)\mathbb{1}$ in $(0,+\infty)\times\Omega_\delta$. In view of the assumptions on Ω, the function d is in $C^2(\Omega_\delta)$ and, consequently, $v\in C^{1,2}((0,+\infty)\times\Omega_\delta)$, is continuous in $[0,+\infty)\times\overline{\Omega}_\delta\setminus(\{0\}\times\partial\Omega_\delta)$, bounded in $(0,+\infty)\times\Omega_\delta$, and nonnegative in $(0,+\infty)\times\partial\Omega_\delta\cup(\{0\}\times\Omega_\delta)$. Moreover,

$$D_t v - \mathcal{A}v = (\kappa_0 - |\sqrt{Q}\nabla d|^2)D_{rr}g(\cdot,d(\cdot)) + (M_0 - \mathcal{A}d + c\,d)D_r g(\cdot,d(\cdot)) - cg(\cdot,d(\cdot)) \ge 0,$$

since g, $D_r g$, $-D_{rr}g$ are nonnegative. Proposition 12.1.1 yields (12.2.10). □

The following proposition provides us with an *a priori* estimate for $\nabla_x u$.

Proposition 12.2.7 *Let Hypotheses* 12.0.1 *and* 12.2.1*, but the first condition in* (12.2.3)*, be satisfied. Then, for any $T>0$, there exists a positive constant C, depending on κ_0, k, s, β, T and $\|\nabla Q\|_\infty$, such that*

$$\|\nabla_x u(t,\cdot)\|_\infty \le Ct^{-\frac{1}{2}}\|f\|_\infty, \qquad t\in(0,T), \tag{12.2.11}$$

for any bounded classical solution u of the Cauchy-Dirichlet problem (12.0.1) *such that* (*i*) D_ju *belongs to* $C^{1,2}((0,+\infty)\times\Omega)$ *for any* $j=1,\dots,N$;

(*ii*) $\sqrt{t}\,|\nabla_x u|$ *is continuous in* $[0,+\infty)\times\overline{\Omega}\setminus(\{0\}\times\partial\Omega)$, *bounded in* $(0,T)\times\Omega$ *for any* $T>0$, *it converges to* 0 *pointwise in* Ω, *as* t *tends to* 0^+, *and the gradient estimate* (12.2.8) *holds true.*

Proof Changing c to $c-1$ (hence, u to $e^{-t}u$) we may assume that $|\nabla c|\le-\beta c$ (see the second condition in (12.2.3)). To get (12.2.11) we apply the Bernstein method. For this purpose, we define the function $v:[0,+\infty)\times\overline{\Omega}\to\mathbb{R}$ by setting $v(t,x)=|u(t,x)|^2+a\,t\,|\nabla u(t,x)|^2$ for any $t>0$, $x\in\Omega$ and $v(0,\cdot)=f^2$, where the constant $a>0$ will be chosen later. Due to assumptions (i) and (ii), v belongs to $C^{1,2}((0,+\infty)\times\Omega)$, is continuous in $[0,+\infty)\times\overline{\Omega}\setminus(\{0\}\times\partial\Omega)$, bounded in $(0,+\infty)\times\Omega$ and $v(0,\cdot)=f^2$. We claim that we can fix $a>0$, depending only on κ_0,h,k,s,β,T and $\|\nabla Q\|_\infty$ (which is finite, since $q_{ij}\in C^1_b(\Omega)$ for any $i,j=1,\dots,N$), such that

$$D_tv(t,x)-\mathcal{A}v(t,x)\le 0,\qquad t\in(0,T),\ x\in\Omega. \tag{12.2.12}$$

Proposition 12.1.1 and estimate (12.2.8) will imply that

$$v(t,x)\le\|f\|_\infty+\sup_{(t,\xi)\in(0,T)\times\partial\Omega}at\,|\nabla_x u(t,\xi)|^2\le(1+aC_0^2)\|f\|_\infty^2$$

for any $t\in(0,T)$, $x\in\Omega$ and estimate (12.2.11) will follow with $C=(C_0^2+a^{-1})^{1/2}$.

To prove (12.2.12) we observe that

$$D_tv-\mathcal{A}v=a|\nabla_x u|^2-2|\sqrt{Q}\nabla u_x|^2+g_1+g_2,$$

where

$$g_1=at(2\langle(\mathrm{Jac}\,b)\nabla u_x,\nabla u_x\rangle+2u\langle\nabla c,\nabla u_x\rangle+c|\nabla_x u|^2)+cu^2$$

and

$$g_2=2at\sum_{i,j,h=1}^N D_hq_{ij}D_huD_{ij}u-2at\sum_{i=1}^N\langle Q\nabla_xD_iu,\nabla_xD_iu\rangle.$$

Using Hypothesis 12.2.1 and the inequality $2ab\le\varepsilon a^2+\varepsilon^{-1}b^2$, which holds for any $a,b,\varepsilon>0$, and recalling that $c\le0$, we can show that, for any $\varepsilon>0$,

$$\begin{aligned}D_tv-\mathcal{A}v\le&\{a-2\kappa_0+at[2k-(2s-1)c+\varepsilon^{-1}\|\nabla Q\|_\infty-\beta\varepsilon c]\}|\nabla_x u|^2\\&+c(1-\varepsilon^{-1}a\beta t)u^2+(\varepsilon\|\nabla Q\|_\infty-2\kappa_0)at\,|D_x^2u|^2.\end{aligned} \tag{12.2.13}$$

Since $2s<1$, choosing $\varepsilon<\min\{2\kappa_0\|\nabla Q\|_\infty,(1-2s)\beta^{-1}\}$, from (12.2.13) we get $D_tv-\mathcal{A}v\le c(1-\varepsilon^{-1}a\beta T)u^2+\{a-2\kappa_0+aT[2k+\varepsilon^{-1}\|\nabla Q\|_\infty]\}|\nabla_x u|^2$. Taking a small enough, we get immediately (12.2.12). □

12.2.2 An auxiliary problem

As a second step, in order to prove Theorem 12.2.4, in this subsection we introduce an auxiliary problem set in L^p-spaces. We still consider an elliptic operator $\mathcal{A}$ of the type (12.0.2), but we assume that its coefficients satisfy the following set of hypotheses.

Hypotheses 12.2.8 (i) the coefficients q_{ij}, b_j ($i,j=1,\dots,N$) and c satisfy Hypotheses 12.0.1(i) and 12.0.1(iii);

(ii) there exist $0 < \sigma < \sqrt{2\kappa_0}$ and $\beta > 0$ such that $c(x) \le -1$, $|\nabla c(x)| \le \beta |c(x)|$ *and* $|b(x)| \le \sigma\sqrt{|c(x)|}$ *for any* $x \in \Omega$.

We are going to show that, under the previous hypotheses, the operator $\mathcal{A}$ generates an analytic semigroup in $L^p(\Omega)$, for any $p \ge 2$. For this purpose we adapt the ideas of [50, 51, 200], where the case $\Omega = \mathbb{R}^N$ is considered.

We introduce the space $\mathscr{D}_p = \{u \in W^{2,p}(\Omega) \cap W_0^{1,p}(\Omega) : cu \in L^p(\Omega)\}$ and we endow it with the norm $\|u\|_{\mathscr{D}_p} = \|u\|_{W^{2,p}(\Omega)} + \|cu\|_{L^p(\Omega)}$ for any $u \in \mathscr{D}_p$. As is immediately seen, $\mathscr{D}_p$ is a Banach space when it is endowed with the previous norm. Moreover, we consider the dense subspace $\mathscr{D} = \{u \in C^\infty(\overline{\Omega}) : u_{|\partial\Omega} = 0, \ \mathrm{supp}\, u \Subset \overline{\Omega}\}$ of $\mathscr{D}_p$. Finally, we denote by $\mathcal{A}_0$ the operator defined on smooth functions ψ by $\mathcal{A}_0\psi = \mathrm{Tr}(QD^2\psi)$ and by A_p the realization of the operator $\mathcal{A}$ in $\mathscr{D}$, defined by $A_p u = \mathcal{A}u$ for any $u \in \mathscr{D}$.

In what follows, to shorten the notation we denote by $\|\cdot\|_p$ and $\|\cdot\|_{j,p}$ $(j = 1, 2)$ the usual norms of $L^p(\Omega)$ and $W^{j,p}(\Omega)$, respectively. We also set

$$\mathfrak{q}(\xi, \eta) = \langle Q\xi, \eta\rangle, \qquad \xi, \eta \in \mathbb{R}^N, \tag{12.2.14}$$

and introduce the function $b' : \overline{\Omega} \to \mathbb{R}^N$ with $b'_i = b_i - \sum_{j=1}^N D_j q_{ij}$ for any $i = 1, \ldots, N$. From Hypotheses 12.2.8 we obtain that

$$|b'(x)| \le \sigma\sqrt{|c(x)|} + K \tag{12.2.15}$$

for a positive constant K depending only on the sup-norm of the gradient of the diffusion coefficients.

Lemma 12.2.9 *For any $p \ge 2$ there exists a positive constant C, depending on N, p, β, R and the coefficients q_{ij} $(i, j = 1, \ldots, N)$, such that*

$$\|\sqrt{|c|}\nabla_x u\|_p \le \varepsilon \|\mathcal{A}_0 u\|_p + C\varepsilon^{-1}(\|u\|_p + \|cu\|_p) \tag{12.2.16}$$

for any $0 < \varepsilon < R$ and any $u \in \mathscr{D}$.

Proof Fix $u \in \mathscr{D}$. Integrating by parts and using the fact that u identically vanishes on $\partial\Omega$ and $p \ge 2$, we obtain

$$\begin{aligned}\int_\Omega |c|^{\frac{p}{2}} |\nabla u|^p dx =& \int_\Omega |c|^{\frac{p}{2}} |\nabla u|^{p-2} |\nabla u|^2 dx \\ =& \frac{p}{2}\int_\Omega |c|^{\frac{p}{2}-1} \langle \nabla c, \nabla u\rangle |\nabla u|^{p-2} u dx - \int_\Omega |c|^{\frac{p}{2}} u |\nabla u|^{p-2} \Delta u \, dx \\ & - (p-2)\int_\Omega |c|^{\frac{p}{2}} u |\nabla u|^{p-4} \langle D^2 u \nabla u, \nabla u\rangle dx.\end{aligned}$$

Using the Hölder inequality and observing that, by Hypothesis (12.2.8)(ii), $|\nabla c| \le \beta |c| \le \beta |c|^{\frac{3}{2}}$, we get

$$\begin{aligned}\|\sqrt{|c|}\nabla u\|_p^p \le& \frac{\beta p}{2}\int_\Omega |u||\nabla u|^{p-1} |c|^{\frac{p-1}{2}} |c| dx + (p-2+\sqrt{N})\int_\Omega |c|^{\frac{p}{2}-1} |\nabla u|^{p-2} |c||u||D^2 u| dx \\ \le& \frac{\beta p}{2}\|\sqrt{|c|}\nabla u\|_p^{p-1}\|cu\|_p + (p-2+\sqrt{N})\|\sqrt{|c|}\nabla u\|_p^{p-2}\|cu\|_p\|D^2 u\|_p\end{aligned}$$

or, equivalently,

$$\|\sqrt{|c|}\nabla u\|_p^2 \le \frac{\beta p}{2}\|cu\|_p\|\sqrt{|c|}\nabla u\|_p + (p-2+\sqrt{N})\|cu\|_p\|D^2 u\|_p.$$

It follows that

$$\|\sqrt{|c|}\nabla u\|_p \leq \frac{\beta p}{2}\|cu\|_p+\sqrt{p-2+\sqrt{N}}\,\|cu\|_p^{\frac{1}{2}}\|D^2u\|_p^{\frac{1}{2}} \leq C\varepsilon^{-1}\|cu\|_p+\varepsilon\|D^2u\|_p$$

for any $\varepsilon \in (0,1)$, with C depending on β, p, and the statement follows with $\|D^2u\|_p$ instead of $\|\mathcal{A}_0u\|_p$. To complete the proof it suffices to observe that

$$\|D^2u\|_p \leq C(\|u\|_p+\|\mathcal{A}_0u\|_p) \tag{12.2.17}$$

for some positive constant C, independent of u (see e.g. [116, Lemma 9.17]). □

Proposition 12.2.10 *For any $p \geq 2$ there exist two constants λ_0, λ_1 such that*

$$(\lambda-\lambda_0)\|u\|_p \leq \|\lambda u - A_pu\|_p, \qquad \lambda > \lambda_0, \tag{12.2.18}$$

$$C_1\|u\|_{\mathscr{D}_p} \leq \|\lambda u - A_pu\|_p \leq C_2\|u\|_{\mathscr{D}_p}, \qquad \lambda > \lambda_1, \tag{12.2.19}$$

for any $u \in \mathscr{D}$ and some positive constants C_1, C_2, depending only on λ, N, p, β, σ, κ_0 and the sup-norm of the coefficients q_{ij} $(i,j=1,\ldots,N)$.

Proof To shorten the notation, we denote by C any positive constant depending at most on $\lambda, N, p, \beta, \sigma$ and the coefficients q_{ij} $(i,j=1,\ldots,N)$, which may vary from line to line.

To prove (12.2.18) we multiply the identity $f=\lambda u - A_pu$ by $u|u|^{p-2}$ and integrate by parts over Ω. We get

$$\begin{aligned}&\int_\Omega(\lambda-c)|u|^pdx+(p-1)\int_\Omega|u|^{p-2}\mathfrak{q}(\nabla u,\nabla u)dx\\ &=\int_\Omega u|u|^{p-2}f dx+\int_\Omega u|u|^{p-2}\langle b',\nabla u\rangle dx.\end{aligned} \tag{12.2.20}$$

Using (12.2.15), the Cauchy-Schwarz and Young inequalities we deduce that

$$\begin{aligned}\int_\Omega u|u|^{p-2}\langle b',\nabla u\rangle dx \leq&\sigma\int_\Omega\sqrt{|c|}|u|^{p-1}|\nabla u|dx+K\int_\Omega|u|^{p-1}|\nabla u|dx\\ \leq&\sigma\bigg(\int_\Omega|c||u|^pdx\bigg)^{\frac{1}{2}}\bigg(\int_\Omega|u|^{p-2}|\nabla u|^2dx\bigg)^{\frac{1}{2}}\\ &+K\bigg(\int_\Omega|u|^{p-2}|\nabla u|^2dx\bigg)^{\frac{1}{2}}\bigg(\int_\Omega|u|^pdx\bigg)^{\frac{1}{2}}\\ \leq&\frac{1}{2}\int_\Omega|c||u|^pdx+\bigg(\frac{\sigma^2}{2}+\varepsilon\bigg)\int_\Omega|u|^{p-2}|\nabla u|^2dx+\frac{K^2}{4\varepsilon}\|u\|_p^p.\end{aligned} \tag{12.2.21}$$

Combining (12.2.20) and (12.2.21) gives

$$\begin{aligned}&\bigg(\lambda-\frac{K^2}{4\varepsilon}\bigg)\int_\Omega|u|^pdx+\frac{1}{2}\int_\Omega|c||u|^pdx\\ &+\bigg((p-1)\kappa_0-\frac{\sigma^2}{2}-\varepsilon\bigg)\int_\Omega|u|^{p-2}|\nabla u|^2dx \leq \|f\|_p\|u\|_p^{p-1}.\end{aligned} \tag{12.2.22}$$

Since $\sigma \in (0,\sqrt{2\kappa_0})$, $p \geq 2$ and $c \leq -1$, we can fix ε sufficiently small and, then, take $\lambda > \lambda_0 := K^2/(4\varepsilon)$ to make all the terms in the left-hand side of (12.2.22) nonnegative. Estimate (12.2.18) now easily follows.

Let us now prove the first inequality in (12.2.19). We begin by estimating cu. For this purpose, we multiply the identity $\lambda u - A_p u = f$ by $|c|^{p-1}u|u|^{p-2}$ and integrate by parts over Ω. Taking Hypothesis 12.0.1(i) into account, we obtain

$$
\begin{aligned}
&\int_\Omega (\lambda|c|^{p-1} + |c|^p)|u|^p dx + \kappa_0(p-1)\int_\Omega |c|^{p-1}|u|^{p-2}|\nabla u|^2 dx\\
\leq &\int_\Omega (\lambda|c|^{p-1} + |c|^p)|u|^p dx + (p-1)\int_\Omega |c|^{p-1}|u|^{p-2}\mathfrak{q}(\nabla u, \nabla u)dx\\
=&(p-1)\int_\Omega |c|^{p-2}u|u|^{p-2}\mathfrak{q}(\nabla u, \nabla c)dx + \int_\Omega |c|^{p-1}u|u|^{p-2}\langle b', \nabla u\rangle dx + \int_\Omega |c|^{p-1}u|u|^{p-2} f dx.
\end{aligned}
$$

Let us estimate the last side of this estimate. Arguing as in the proof of (12.2.21) we get

$$
\begin{aligned}
\left|\int_\Omega |c|^{p-1}u|u|^{p-2}\langle b', \nabla u\rangle dx\right| \leq &\left(\frac{\sigma^2}{2} + \varepsilon\right)\int_\Omega |c|^{p-1}|u|^{p-2}|\nabla u|^2 dx\\
&+ \frac{1}{2}\int_\Omega |c|^p|u|^p dx + \frac{K^2}{4\varepsilon}\int_\Omega |c|^{p-1}|u|^p dx
\end{aligned}
$$

for any $\varepsilon > 0$. Similarly, using Hypothesis 12.2.8(ii), we can estimate

$$
\begin{aligned}
\int_\Omega |c|^{p-2}|u|^{p-1}|q(\nabla u, \nabla c)|dx \leq &\|Q\|_\infty \int_\Omega |c|^{p-2}|\nabla c||u|^{p-1}|\nabla u| dx\\
\leq &\|Q\|_\infty \beta\left(\int_\Omega |c|^{p-1}|u|^{p-2}|\nabla u|^2 dx\right)^{\frac{1}{2}}\left(\int_\Omega |c|^{p-1}|u|^p dx\right)^{\frac{1}{2}}\\
\leq &\varepsilon\int_\Omega |c|^{p-1}|u|^{p-2}|\nabla u|^2 dx + \frac{\|Q\|_\infty^2\beta^2}{4\varepsilon}\int_\Omega |c|^{p-1}|u|^p dx.
\end{aligned}
$$

Summing up, from these last three estimates we conclude that

$$
\begin{aligned}
\|f\|_p\|cu\|_p^{p-1} \geq &\left(\lambda - \frac{K^2}{4\varepsilon} - (p-1)\frac{\|Q\|_\infty^2\beta^2}{4\varepsilon}\right)\int_\Omega |c|^{p-1}|u|^p dx + \frac{1}{2}\int_\Omega |c|^p|u|^p dx\\
&+ \left((p-1)(\kappa_0 - \varepsilon) - \frac{\sigma^2}{2} - \varepsilon\right)\int_\Omega |c|^{p-1}|u|^{p-2}|\nabla u|^2 dx.
\end{aligned}
$$

Choosing ε sufficiently small and, then, $\lambda > \lambda_1 := (4\varepsilon)^{-1}(K^2 + (p-1)\|Q\|_\infty^2\beta^2)$, we get

$$
\|cu\|_p \leq 2\|f\|_p. \tag{12.2.23}
$$

We now use Lemma 12.2.9 with $\varepsilon = (2\sigma)^{-1}$ to estimate the second-order derivatives of u. We obtain that

$$
\begin{aligned}
\|\langle b, \nabla u\rangle\|_p \leq &\sigma\|\sqrt{|c|}\nabla u\|_p \leq \frac{1}{2}\|\mathcal{A}_0 u\|_p + 2\sigma^2 C\|u\|_p + 2\sigma^2 C\|cu\|_p\\
\leq &\frac{1}{2}\|f\|_p + \frac{1}{2}\|\langle b, \nabla u\rangle\|_p + \left(\frac{1}{2} + 2\sigma^2 C\right)\|cu\|_p + \left(\frac{\lambda}{2} + 2\sigma^2 C\right)\|u\|_p.
\end{aligned}
$$

Hence, taking (12.2.18) and (12.2.23) into account, we get $\|\langle b, \nabla u\rangle\|_p \leq C\|f\|_p$ and, then, by difference, $\|\mathcal{A}_0 u\|_p \leq C\|f\|_p$. From (12.2.17) it follows that $\|D^2 u\|_p \leq C\|f\|_p$. Moreover, since $c \leq -1$, from (12.2.16) we deduce that $\|\nabla u\|_p \leq C\|f\|_p$. The first inequality in (12.2.19) follows.

To prove the other estimate in (12.2.19) it suffices to show that $\|\langle b, \nabla u\rangle\|_p \leq C\|u\|_{\mathscr{D}_p}$ for any $u \in \mathscr{D}$, and this can easily be done, taking (12.2.16), with $\varepsilon = 1$, and Hypothesis (12.2.8)(ii) into account. □

Remark 12.2.11 The estimate $\|\langle b, \nabla u\rangle\|_p \leq C\|u\|_{\mathscr{D}_p}$ and the density of $\mathscr{D}$ in $\mathscr{D}_p$ show that the operator A_p is well defined in $\mathscr{D}_p$. We still denote by A_p the operator so extended to $\mathscr{D}_p$. Of course, it satisfies all the results in Proposition 12.2.10. In particular, by (12.2.19), the graph norm of A_p is equivalent to the norm of $\mathscr{D}_p$.

Proposition 12.2.12 *For any $p \in [2, +\infty)$, $(A_p, \mathscr{D}_p)$ generates a strongly continuous analytic semigroup $\{S(t)\}$ in $L^p(\Omega)$.*

Proof Without loss of generality, we replace A_p with the operator $B_p = A_p - kI$, defined in $D(B_p) = \mathscr{D}_p$, where $k \geq 1$ is a suitable positive constant to be fixed.

We split the proof into two steps. First we prove that B_p generates a strongly continuous semigroup in $L^p(\Omega)$ and then, in Step 2, we show that such a semigroup is analytic.

Step 1. To begin with, we fix $\sigma' \in (\sigma, \sqrt{2\kappa_0})$ and $k > 0$ sufficiently large that

$$|b'(x)| \leq \sigma' \sqrt{|c'(x)|}, \qquad x \in \Omega, \tag{12.2.24}$$

where $c' = c - k$. Since now we can take $K = 0$ in (12.2.15), estimate (12.2.22) implies that B_p is dissipative in $L^p(\Omega)$.

Let us show that $\lambda - B_p$ is surjective for some $\lambda > 0$. The Lumer-Phillips theorem (see Theorem C.1.9) then will imply that B_p is the generator of a strongly continuous semigroup of contractions. For this purpose, for any $\varepsilon > 0$ we set $b_\varepsilon = b(1 - \varepsilon c)^{-1/2}$ and $c'_\varepsilon = c(1 - \varepsilon c)^{-1} - k$. The functions b_ε and c_ε satisfy Hypothesis 12.2.8, uniformly with respect to $\varepsilon > 0$, with the same constants σ and β, provided that $k > 1$. Since b_ε and c_ε are bounded, the realization $B_{\varepsilon,p}$ of the operator $\mathcal{B}_\varepsilon = \mathcal{A}_0 + \langle b_\varepsilon, \nabla\rangle + c'_\varepsilon$ in $L^p(\Omega)$, with domain $D(B_{\varepsilon,p}) = W^{2,p}(\Omega) \cap W^{1,p}_0(\Omega)$, generates an analytic semigroup (see Theorem D.3.5(ii)). Moreover, estimate (12.2.18) implies that there exists λ_0 such that $(\lambda - \lambda_0)\|u\|_p \leq \|\lambda u - B_{\varepsilon,p}u\|_p$ for any $u \in D(B_{\varepsilon,p})$, $\lambda > \lambda_0$ and $\varepsilon > 0$. Since $D(B_{\varepsilon,p})$ is dense in $L^p(\Omega)$, applying the Lumer-Phillips theorem to the operator $B_{\varepsilon,p} - \lambda_0 I$, we conclude that the resolvent of $B_{\varepsilon,p}$ contains the halfline $(\lambda_0, +\infty)$ for any $\varepsilon \in (0, 1)$.

Now, given $f \in L^p(\Omega)$, let $u_\varepsilon \in W^{2,p}(\Omega) \cap W^{1,p}_0(\Omega)$ satisfy the equation $(2\lambda_0 - B_{\varepsilon,p})u_\varepsilon = f$. According to Proposition 12.2.10 we know that $\|u_\varepsilon\|_{2,p} + \|c_\varepsilon u_\varepsilon\|_p \leq C\|f\|_p$ for some positive constant C, independent of ε. By a weak compactness argument we can determine an infinitesimal sequence $\{\varepsilon_n\}$ such that u_{ε_n} tends weakly to a function u in $W^{2,p}(\Omega) \cap W^{1,p}_0(\Omega)$. Moreover, we can assume that u_{ε_n} tends to u both strongly in $W^{1,p}_{\rm loc}(\Omega)$ and almost everywhere in Ω. Fatou's lemma yields $\|cu\|_p \leq C\|f\|_p$. Hence, u belongs to $D(B_p)$. Moreover, u solves the differential equation $\lambda u - B_p u = f$. Therefore, the operator $2\lambda_0 - B_p$ is surjective and we can apply the Lumer-Phillips theorem.

Step 2. Now, we prove that the semigroup $\{R(t)\}$ generated by B_p is analytic. For this purpose, we fix $u \in \mathscr{D}$ and set $u^* := \overline{u}|u|^{p-2}$. An integration by parts shows that

$$\begin{aligned}-\mathrm{Re}\int_\Omega u^*\, B_p u\, dx =& \mathrm{Re}\int_\Omega \left(\frac{p}{2}|u|^{p-2}\mathfrak{q}(\nabla u, \nabla\overline{u}) + \frac{p-2}{2}\overline{u}^2|u|^{p-4}\mathfrak{q}(\nabla u, \nabla u)\right) dx\\ &- \mathrm{Re}\int_\Omega \overline{u}|u|^{p-2}\,\langle b', \nabla u\rangle\, dx - \int_\Omega c'|u|^p dx,\end{aligned} \tag{12.2.25}$$

where $\mathfrak{q}$ is given by (12.2.14). Since

$$\begin{aligned}\mathrm{Re}\,(|u|^2\mathfrak{q}(\nabla u, \nabla\overline{u})) &= \mathfrak{q}(\mathrm{Re}(\overline{u}\nabla u), \mathrm{Re}(\overline{u}\nabla u)) + \mathfrak{q}(\mathrm{Im}(\overline{u}\nabla u), \mathrm{Im}(\overline{u}\nabla u)),\\ \mathrm{Re}\,(\overline{u}^2\mathfrak{q}(\nabla u, \nabla u)) &= \mathfrak{q}(\mathrm{Re}(\overline{u}\nabla u), \mathrm{Re}(\overline{u}\nabla u)) - \mathfrak{q}(\mathrm{Im}(\overline{u}\nabla u), \mathrm{Im}(\overline{u}\nabla u)),\end{aligned}$$

from (12.2.25) we deduce that

$$-\mathrm{Re}\int_\Omega u^* B_p u dx = (p-1)\int_\Omega |u|^{p-4}\mathfrak{q}(\mathrm{Re}(\overline{u}\nabla u), \mathrm{Re}(\overline{u}\nabla u))dx$$

$$+\int_\Omega |u|^{p-4}\mathfrak{q}(\mathrm{Im}(\overline{u}\nabla u),\mathrm{Im}(\overline{u}\nabla u))dx$$
$$-\int_\Omega |u|^{p-2}\langle b',\mathrm{Re}(\overline{u}\nabla u)\rangle dx-\int_\Omega c'|u|^p dx. \tag{12.2.26}$$

Condition (12.2.24) and the Hölder inequality imply that

$$\begin{aligned}\left|\int_\Omega |u|^{p-2}\langle b',\mathrm{Re}(\overline{u}\nabla u)\rangle dx\right| \leq &\sigma'\int_\Omega \sqrt{|c'|}|\mathrm{Re}(\overline{u}\nabla u)||u|^{p-2}dx\\ \leq &\sigma'\left(\int_\Omega |c'||u|^p dx\right)^{\frac{1}{2}}\left(\int_\Omega |u|^{p-4}|\mathrm{Re}(\overline{u}\nabla u)|^2dx\right)^{\frac{1}{2}}\\ \leq &\frac{\sigma'}{\sqrt{\kappa_0}}\left(\int_\Omega |c'||u|^p dx\right)^{\frac{1}{2}}\left(\int_\Omega |u|^{p-4}\mathfrak{q}(\mathrm{Re}(\overline{u}\nabla u),\mathrm{Re}(\overline{u}\nabla u))dx\right)^{\frac{1}{2}}.\end{aligned} \tag{12.2.27}$$

If we set

$$F^2:=\int_\Omega |u|^{p-4}\mathfrak{q}(\mathrm{Re}(\overline{u}\nabla u),\mathrm{Re}(\overline{u}\nabla u))dx,$$
$$G^2:=\int_\Omega |u|^{p-4}\mathfrak{q}(\mathrm{Im}(\overline{u}\nabla u),\mathrm{Im}(\overline{u}\nabla u))dx, \qquad H^2:=\int_\Omega |c'||u|^p dx,$$

then the formulae (12.2.26) and (12.2.27) yield

$$-\mathrm{Re}\int_\Omega u^*B_p u\,dx \geq \left(p-1-\frac{(\sigma')^2}{2\kappa_0}\right)F^2+G^2+\frac{1}{2}H^2. \tag{12.2.28}$$

Now, we consider the imaginary part of $\int_\Omega u^*B_p u dx$. Integrating by parts, we obtain

$$\begin{aligned}\mathrm{Im}\int_\Omega u^*B_p u\,dx =&-\mathrm{Im}\int_\Omega\left(\frac{p}{2}|u|^{p-2}\mathfrak{q}(\nabla u,\nabla\overline{u})+\frac{p-2}{2}\overline{u}^2|u|^{p-4}\mathfrak{q}(\nabla u,\nabla u)\right)dx\\ &+\mathrm{Im}\int_\Omega \overline{u}|u|^{p-2}\langle b',\nabla u\rangle\,dx.\end{aligned} \tag{12.2.29}$$

Since $\mathrm{Im}(\mathfrak{q}(\nabla u,\nabla\overline{u}))=0$ and $\mathrm{Im}(\mathfrak{q}(\overline{u}\nabla u,\overline{u}\nabla u))=2\mathfrak{q}(\mathrm{Re}(\overline{u}\nabla u),\mathrm{Im}(\overline{u}\nabla u))$, we get

$$\begin{aligned}\mathrm{Im}\int_\Omega u^*B_p u\,dx =&(p-2)\int_\Omega |u|^{p-4}\mathfrak{q}(\mathrm{Re}(\overline{u}\nabla u),\mathrm{Im}(\overline{u}\nabla u))dx\\ &+\mathrm{Im}\int_\Omega \overline{u}|u|^{p-2}\langle b',\nabla u\rangle\,dx.\end{aligned} \tag{12.2.30}$$

Using, first, the Cauchy-Schwarz inequality for the inner product induced by the matrix Q and, then, the Hölder inequality, we can show that

$$\left|\int_\Omega |u|^{p-4}\mathfrak{q}(\mathrm{Re}(\overline{u}\nabla u),\mathrm{Im}(\overline{u}\nabla u))dx\right|\leq FG,$$

whereas the other term in the right-hand side of (12.2.30) can be estimated as in (12.2.27), replacing everywhere the real part of $\overline{u}\nabla u$ with its imaginary part. Summing up, we get

$$\left|\mathrm{Im}\int_\Omega u^*B_p u dx\right|\leq (p-2)FG+\frac{\sigma'}{\sqrt{\kappa_0}}GH. \tag{12.2.31}$$

Combining (12.2.28) and (12.2.31) we can determine a positive constant C such that

$$\left|\mathrm{Im}\int_\Omega u^* B_p u dx\right| \le C\Big(-\mathrm{Re}\int_\Omega u^* B_p u dx\Big) \tag{12.2.32}$$

for any $u \in \mathscr{D}$. By density, (12.2.32) can be extended to any $u \in D(B_p)$.

Formula (12.2.32) implies that the numerical range $r(B_p)$ of B_p is contained in the sector $\{\lambda \in \mathbb{C} : |\mathrm{Im}\,\lambda| \le -C\,\mathrm{Re}\,\lambda\}$ so that, applying Theorem B.1.7, we can complete the proof. □

To conclude, we show some regularity properties of the function $S(\cdot)f$.

Proposition 12.2.13 *Let $p > N+1$. Then, for any $f \in C_c^\infty(\Omega)$ the function $u = S(\cdot)f$ is the bounded classical solution of the Cauchy-Dirichlet problem* (12.0.1). *Moreover, $u \in C^{1+\alpha/2,2+\alpha}((0,T)\times(\Omega\cap B_R))$ for any $R > 0$. Further, $D_j u$ belongs to $C_b([0,T]\times\overline{\Omega}) \cap C^{1+\alpha/2,2+\alpha}((\varepsilon,T)\times\Omega')$ for any $0<\varepsilon<T$, any $\Omega' \Subset \Omega$, any $R>0$ and $j=1,\dots,N$.*

Proof Fix $f \in C_c^\infty(\Omega) \subset D(A_p)$. Since $C_c^\infty(\Omega)$ is dense in $L^p(\Omega)$, according to Proposition 12.2.10 and Remark C.2.6 the function $t \mapsto S(t)f$ is continuous in $[0,+\infty)$ with values in $W^{2,p}(\Omega)$. Moreover, in view of the condition $p > N+1$, the Sobolev embedding theorems (see [3, Theorem 5.4]) and Remark C.2.6 imply that both u and $\nabla_x u$ are bounded and continuous in $[0,T]\times\overline{\Omega}$ for any $T>0$. To conclude that u is a classical solution to (12.0.1), we have to show that $u \in C^{1,2}((0,+\infty)\times\Omega)$. Since $\{S(t)\}$ is analytic, the function u is continuously differentiable in $[0,+\infty)$ with values in $W^{2,p}(\Omega)$ (see again Remark C.2.6). The Sobolev embedding theorems yield $D_t u \in C([0,+\infty);C_b(\overline{\Omega}))$.

Now, fix $\varepsilon \in (0,T)$ and set $\tau = \sup_{\varepsilon\le t\le T}(\|u(t,\cdot)\|_{W^{2,p}(\Omega)} + \|D_t u(t,\cdot)\|_{W^{2,p}(\Omega)})$. Since $u(t,\cdot) \in D(A_p^n)$ for any $t \in [\varepsilon,T]$ and $n \in \mathbb{N}$, the function $A_p u(t,\cdot) = D_t u(t,\cdot)$ belongs to $W^{2,p}(\Omega)$. Observing that $\mathrm{Tr}(QD_x^2 u(t,\cdot)) = D_t u(t,\cdot) - \langle b, \nabla_x u(t,\cdot)\rangle - cu(t,\cdot)$ for any $t \in (\varepsilon,T]$ and $u(t,\cdot) \in W^{2,p}(\Omega)$, by difference, we deduce that $\mathrm{Tr}(QD_x^2 u(t,\cdot)) \in W^{1,p}_{\mathrm{loc}}(\Omega)$. By local regularity results for elliptic equations in L^p-spaces, we deduce that $u(t,\cdot) \in W^{3,p}_{\mathrm{loc}}(\Omega)$ for any $t \in [\varepsilon,T]$ (see Theorem D.1.1). Moreover, since the coefficients of the operator $\mathcal{A}$ are locally bounded in Ω, from the definition of τ it follows that, for any pair of bounded open sets $\Omega_1 \Subset \Omega_2 \Subset \Omega$, it holds that

$$\|u(t,\cdot)\|_{W^{3,p}(\Omega_1)} \le C_1(\|A_p u(t,\cdot) - \langle b, \nabla u(t,\cdot)\rangle - cu(t,\cdot)\|_{W^{1,p}(\Omega_2)} + \|u\|_{L^p(\Omega_2)}) \le C_2\tau \tag{12.2.33}$$

for some positive constants $C_1 = C_1(\Omega_1,\Omega_2)$, $C_2 = C_2(\Omega_1,\Omega_2)$ and any $t \in [\varepsilon,T]$. We have thus proved that $D_x^2 u(t,\cdot) \in W^{1,p}(\Omega_1)$ for any $t \in [\varepsilon,T]$. Since $p > N+1$, the Sobolev embedding theorems imply that $D_x^2 u(t,\cdot) \in C^\theta(\Omega_1)$ for some $\theta \in (0,1)$ and, by virtue of (12.2.33), $\sup_{\varepsilon\le t\le T}\|D_x^2 u(t,\cdot)\|_{C^\theta(\Omega_1)} < +\infty$. Now, without loss of generality, we assume that Ω_1 has a smooth boundary. Therefore, applying Proposition B.3.6 with $X = C(\overline{\Omega}_1)$, $Y = C^{2+\theta}(\overline{\Omega}_1)$, $Z = C^2(\overline{\Omega}_1)$, $I = [\varepsilon,T]$ and $\theta = 0$ (and taking Proposition B.3.4 into account), we deduce that $D_{ij}u \in C([\varepsilon,T]\times\overline{\Omega}_1)$ for any $i,j = 1 = 1,\dots,N$ and the arbitrariness of ε, T and Ω_1 implies that the previous derivatives are continuous in $(0,+\infty)\times\Omega$. We have so proved that u is a classical solution to the Cauchy-Dirichlet problem (12.0.1).

The last part of the assertion follows from Proposition 12.1.1 and Theorem 12.1.2. □

12.2.3 Proof of Theorem 12.2.4

To prove that the bounded classical solution to the Cauchy-Dirichlet problem (12.0.1) satisfies the gradient estimate (12.0.3) we use an approximation argument. First, we prove the assertion in the case when $f \in C_c^\infty(\Omega)$ and, then, in the general case.

Step 1. Fix $f \in C_c^\infty(\Omega)$. For any $\varepsilon \in (0,1)$, let $c_\varepsilon : \mathbb{R}^N \to \mathbb{R}$ be defined by $c_\varepsilon(x) = -\varepsilon \exp\left(4d_2\sqrt{1+|x|^2}\right)$ for any $x \in \mathbb{R}^N$. Taking the first condition in (12.2.3) into account, it can be easily shown that for any $\sigma > 0$ there exists a positive constant $C_{\sigma,\varepsilon}$ such that

$$|b| \le \sigma\sqrt{|c_\varepsilon + c|} + C_{\sigma,\varepsilon}. \tag{12.2.34}$$

Moreover, using the second condition in (12.2.3), we get

$$|\nabla(c_\varepsilon + c)| \le \beta_0(1 - c_\varepsilon - c), \tag{12.2.35}$$

where $\beta_0 := \max(\beta, 4d_2)$, β and d_2 are given by (12.2.3).

Let $\mathcal{A}_\varepsilon = \mathcal{A}+c_\varepsilon-k$. Taking (12.2.34) and (12.2.35) into account, we can check that, if k is sufficiently large, then the operator $\mathcal{A}_\varepsilon$ satisfies Hypotheses 12.2.8 with β, σ being replaced by β_0 and some $\sigma' \in (\sigma_0, \sqrt{2\kappa_0})$. Hence, it generates an analytic semigroup $\{S_\varepsilon(t)\}$ in $L^p(\Omega)$ for any $p \ge 2$. By Proposition 12.2.13 we know that, if $p > N+1$, then the function $u_\varepsilon = S_\varepsilon(\cdot)f$ is the bounded classical solution to the Cauchy-Dirichlet problem (12.0.1), where we replace $\mathcal{A}$ with the operator $\mathcal{A}_\varepsilon$. Moreover, $D_j u_\varepsilon$ is bounded and continuous in $[0,T] \times \overline{\Omega}$, for any $T > 0$ and $j = 1, \ldots, N$, and it belongs to $C^{1,2}((0,+\infty) \times \Omega)$ Hence, u_ε satisfies the assumptions of Proposition 12.2.7. Since the coefficients of the operator $\mathcal{A}_\varepsilon$ satisfy Hypothesis 12.2.1 with constants being independent of $\varepsilon \in (0,1)$ (see (12.2.34) and (12.2.35)), for any $T > 0$ we can determine a positive constant $C = C(T)$, independent of ε, such that

$$\|\nabla_x u_\varepsilon(t,\cdot)\|_\infty \le Ct^{-\frac{1}{2}}\|f\|_\infty, \qquad t \in (0,T), \tag{12.2.36}$$

for any $T > 0$. For any $\alpha \in (0,1)$, the C^α-norms, in bounded subsets of Ω, of the coefficients of the operator $\mathcal{A}_\varepsilon$ are bounded uniformly with respect to ε. Therefore, arguing as in Step 1 of the proof of Theorem 12.1.2, we can show that, up to a subsequence, u_ε converges to $T(\cdot)f$ in $C^{1,2}([0,T] \times (\Omega \cap B_R))$, for any $R > 0$, as ε tends to 0^+. Thus, from (12.2.36) we get $\|\nabla_x T(t)f\|_\infty \le Ct^{-\frac{1}{2}}\|f\|_\infty$ for any $t \in (0,T)$ and any $T > 0$, and the assertion follows in this case.

Step 2. Given a function $f \in C_b(\Omega)$, we approximate it by a sequence $\{f_n\} \subset C_c^\infty(\Omega)$ of smooth functions converging to f uniformly on compact subsets of Ω and such that $\|f_n\|_\infty \le \|f\|_\infty$ for any $n \in \mathbb{N}$. Arguing as in Step 2 of the proof of Theorem 12.1.2, we conclude that, up to a subsequence, $T(\cdot)f_n$ converges to $T(\cdot)f$ in $C^{1,2}(F)$ for any compact set $F \subset (0,T)\times\Omega$ and any $T > 0$. By Step 1, we know that $\|\nabla_x T(t)f_n\|_\infty \le Ct^{-1/2}\|f_n\|_\infty \le Ct^{-1/2}\|f\|_\infty$ for any $t \in (0,T)$. Letting n tend to $+\infty$, we get the assertion. □

12.2.4 A counterexample to the gradient estimates

In this subsection we show that the gradient estimate (12.0.3) fails, in general, if Hypothesis 12.2.1 is not satisfied. The following result generalizes an example in [257].

Example 12.2.14 Let us consider the Cauchy-Dirichlet problem in $\mathbb{R}^2_+ = \{(x,y) \in \mathbb{R}^2, x > 0\}$:

$$\begin{cases} D_t u(t,x,y) = \Delta u(t,x,y) + \sqrt{1+y^2}D_x u(t,x,y), & t > 0, \quad (x,y) \in \mathbb{R}^2_+, \\ u(t,0,y) = 0, & t > 0, \quad y \in \mathbb{R}, \\ u(0,x,y) = 1, & \qquad\quad (x,y) \in \mathbb{R}^2_+. \end{cases} \tag{12.2.37}$$

Observe that (12.2.2) does not hold. Indeed, in this situation $d(x,y) = x$ and, consequently, the condition (12.2.2) reads as $\sqrt{1+y^2} \le M$ for any $(x,y) \in \Omega_\delta = \{(x,y) \in \mathbb{R}^2 :$

$x \in (0,\delta)\}$, which, of course, leads us to a contradiction, letting y tend to ∞. However, Theorem 12.1.2 guarantees that problem (12.2.37) admits a unique bounded classical solution u. To prove that the gradient estimate does not hold, it suffices to show that, for any $t > 0$, $u(t,\cdot)$ is not uniformly continuous in Ω. This can be done showing that, for any $t, x > 0$,

$$\sup_{y>0} u(t,x,y) = 1. \tag{12.2.38}$$

Indeed, if $u(t,\cdot)$ were uniformly continuous in $\mathbb{R}^2_+$, then, for any $t > 0$, the function $x \mapsto w_t(x) = \sup_{y>0} u(t,x,y)$ would be continuous in $[0,+\infty)$. Since $w_t(0) = 0$, we would get $w_t(x) < 1$ for any x in a sufficiently small neighborhood of $x = 0$, contradicting (12.2.38).

To prove (12.2.38), we fix $n > 0$ and take $r_n = \sqrt{n^2-1}$. Moreover, we denote by $v = v_n$ the classical solution to problem

$$\begin{cases} D_t v(t,x,y) = \Delta v(t,x,y) + nD_x v(t,x,y), & t>0, \quad (x,y) \in R_n,\\ v(t,x,y) = 0, & t>0, \quad (x,y)\in \partial R_n,\\ v(0,x,y) = 1, & \qquad\quad (x,y) \in R_n, \end{cases}$$

where $R_n = (0,+\infty) \times (r_n,+\infty)$. We are going to show that for any $(t,x) \in R_n$

$$(i)\ \lim_{n\to+\infty} \sup_{y>r_n} v_n(t,x,y) = 1, \qquad (ii)\ u(t,x,y) \ge v_n(t,x,y), \qquad n \in \mathbb{N}. \tag{12.2.39}$$

Clearly (i) and (ii) yield (12.2.38). Indeed, they yield $\sup_{y>0} u(t,x,y) \ge \mathbb{1}$ and the maximum principle in Proposition 12.1.1, implies that $u \le \mathbb{1}$.

To prove (12.2.39)(i), we look for the solution v_n in the form $v_n(t,x,y) = a_n(t,x)b_n(t,y)$, with $a = a_n$ and $b = b_n$ solving, respectively, the Cauchy problems

$$\begin{cases} D_t a = D_{xx}a + nD_x a, & \text{in } \mathbb{R}_+ \times \mathbb{R}_+,\\ a(\cdot,0) = 0, & \text{in } \mathbb{R}_+,\\ a(0,\cdot) = 1, & \text{in } \mathbb{R}_+, \end{cases} \qquad \begin{cases} D_t b = D_{yy} b, & \text{in } \mathbb{R}_+ \times (r_n,+\infty),\\ b(\cdot,r_n) = 0, & \text{in } \mathbb{R}_+,\\ b(0,\cdot) = 1, & \text{in } (r_n,+\infty). \end{cases} \tag{12.2.40}$$

To obtain an explicit formula for a_n, we first observe that $a_n(t,x) = a_1(n^2 t, nx)$. Next, we set $z(t,x) = e^{x/2}e^{t/4}a_1(t,x)$ and observe that z satisfies

$$\begin{cases} D_t z(t,x) = D_{xx} z(t,x), & t>0, \quad x>0,\\ z(t,0) = 0, & t > 0,\\ z(0,x) = e^{\frac{x}{2}}, & \qquad\quad x > 0. \end{cases} \tag{12.2.41}$$

Due to the well known representation formulae for the solutions to the Cauchy problem (12.2.41) and the second problem in (12.2.40), we can write

$$\begin{aligned} a_n(t,x) =& \frac{e^{-\frac{n^2t}{4}}}{n\sqrt{4\pi t}} \int_0^{+\infty} \left[\exp\left(-\frac{|nx-z|^2}{4n^2t}\right) - \exp\left(-\frac{|nx+z|^2}{4n^2t}\right)\right] \exp\left(\frac{z-nx}{2}\right) dz\\ =& \frac{1}{\sqrt{\pi}} \int_{-\frac{x+nt}{2\sqrt{t}}}^{+\infty} e^{-s^2}ds - \frac{e^{-nx}}{\sqrt{\pi}} \int_{\frac{x-nt}{2\sqrt{t}}}^{+\infty} e^{-s^2}ds, \end{aligned} \tag{12.2.42}$$

$$b_n(t,y) = \frac{1}{\sqrt{\pi t}} \int_0^{y-r_n} e^{-\frac{z^2}{4t}} dz$$

for any $t, x > 0$, $y > r_n$ and $n \in \mathbb{N}$. Hence, $\sup_{y>r_n} a_n(t,x,y) = a_n(t,x) \sup_{y>r_n} b_n(t,y) = a_n(t,x)$. Letting n tend to $+\infty$, from (12.2.42) we immediately get (12.2.39).

To prove (12.2.39)(ii) we use the maximum principle in Proposition 12.1.1. Set $w = u - v_n$. Clearly, w identically vanishes in $\{0\}\times R_n$. Moreover, w is nonnegative in $(0,+\infty)\times\partial R_n$. Indeed, the maximum principle implies that $u \ge 0$ in $(0,+\infty)\times\Omega$. To conclude that $u \ge v_n$ in $(0,+\infty)\times R_n$, it suffices to show that

$$D_t w(t,x,y) \ge \Delta w(t,x,y) + g(y) D_x w(t,x,y) \tag{12.2.43}$$

for any $t > 0$ and $(x,y) \in R_n$. For this purpose, we observe that $D_t w = \Delta w + g D_x w + (g-n) D_x a_n b_n$ in $(0,+\infty)\times R_n$, where we have set $g(y) = \sqrt{1+y^2}$ for any $y \in \mathbb{R}$. Formula (12.2.42) shows that $a_n(t,\cdot)$ is increasing in $(0,+\infty)$. Therefore, since $\sqrt{1+y^2} \ge n$ for any $y \in (r_n,+\infty)$, it follows that $(g-n)D_x a_n b_n \ge 0$ in $(0,+\infty)\times R_n$ and (12.2.43) follows.

Remark 12.2.15 Note that, if we replace $\sqrt{1+y^2}$ by $-\sqrt{1+y^2}$ in (12.2.37), then Hypotheses 12.0.1 and 12.2.1 are satisfied. In particular, we can take $\varphi(x,y) = \sqrt{1+x^2+y^2}$ for any $(x,y) \in \mathbb{R}^2_+$ in Hypothesis 12.0.1(iv). Therefore, the gradient estimate (12.0.3) holds. This shows that (12.2.1)-(12.2.3) are not merely growth conditions at infinity of the coefficients.

Remark 12.2.16 As it has been claimed in the introduction of this chapter, as far as we know, uniform estimates for higher order derivatives of the function $T(t)f$ seem be not available in a general domain Ω. This prevents us from performing the same techniques as in Chapter 5 to prove optimal Schauder estimates for both the nonhomogeneous elliptic problem with Dirichlet boundary conditions and the nonhomogeneous Cauchy-Dirichlet problem associated with the operator $\mathcal{A}$.

The situation is different when Ω is an exterior domain, where estimates for the higher-order derivatives of the function $T(t)f$ are available (see [123]).

In the case when Ω is the halfspace $\Omega = \{x \in \mathbb{R}^N : \langle x, v\rangle > 0\}$ (v being an unitary vector in $\mathbb{R}^N$) and $\mathcal{A}$ is the operator defined by $\mathcal{A} = \Delta u + \langle Bx, \nabla\rangle$ for any $x \in \Omega$, where B is a matrix such that v is an eigenvector of both B and B^*, optimal Schauder estimates for the elliptic problem

$$\begin{cases} \lambda u(x) - \mathcal{A}u(x) = f(x), & x \in \Omega, \\ u(x) = 0, & x \in \partial\Omega, \end{cases} \tag{12.2.44}$$

($\lambda \ge 0$), are available and they have been proved by E. Priola in [217]. To state them, we need to introduce some notation. We denote by Λ any orthogonal matrix such that $\Lambda(\Omega) := \mathbb{R}^N_+ = \{x = (x_1,\ldots,x_N) \in \mathbb{R}^N : x_1 > 0\}$. Moreover we set $\widetilde{B} = \Lambda B \Lambda^*$ and denote by $\widetilde{B}_1 \in L(\mathbb{R}^{N-1})$ the matrix obtained from $\widetilde{B}$ by erasing the first line and the first row.

Assuming that f belongs to $C^\alpha_b(\mathbb{R}^N_+)$ for some $\alpha \in (0,1)$, E. Priola shows that, for any $\lambda \ge 0$, problem (12.2.44) admits a unique solution $u \in C^{2+\alpha}_b(\mathbb{R}^N_+)$ if and only if $[[f]] := \sup\{t^{-\frac{\alpha}{2}}|f(\Lambda^{-1}(0,e^{t\widetilde{B}_1 y})) - f(\Lambda^{-1}(0,y))| : t > 0,\, y \in \mathbb{R}^{N-1}\} < +\infty$. In such a case, there exists a positive constant C, independent of f, such that $\|u\|_{C^{2+\alpha}_b(\Omega)} \le C([[f]] + \|f\|_{C^\alpha_b(\mathbb{R}^N)})$. This result shows that, differently from what happens when $\Omega = \mathbb{R}^N$, when Dirichlet boundary conditions are prescribed on $\partial\Omega$ (and Ω is not an exterior domain), we can not expect, in general, to prove optimal Schauder estimates for the solution to the elliptic equation in Ω, only assuming that the data are Hölder continuous in Ω: we need to assume some additional conditions on them.

Chapter 13

The Cauchy-Neumann problem

13.0 Introduction

Let $\Omega \subset \mathbb{R}^N$ be a uniformly unbounded domain uniformly of class $C^{2+\alpha}$ for some $\alpha \in (0,1)$. In this chapter we consider both the parabolic problem with homogeneous Neumann boundary conditions

$$\begin{cases} D_t u(t,x) = \mathcal{A}u(t,x), & t>0, \quad x \in \Omega, \\ \dfrac{\partial u}{\partial \nu}(t,x) = 0, & t>0, \quad x \in \partial\Omega, \\ u(0,x) = f(x), & \quad x \in \Omega, \end{cases} \tag{13.0.1}$$

and the elliptic problem

$$\begin{cases} \lambda u(x) - \mathcal{A}u(x) = f(x), & x \in \Omega, \\ \dfrac{\partial u}{\partial \nu}(x) = 0, & x \in \partial\Omega, \end{cases} \tag{13.0.2}$$

when $f \in C_b(\overline{\Omega})$. Here, $\mathcal{A}$ is given by

$$\mathcal{A} = \sum_{i,j=1}^{N} q_{ij}(x) D_{ij} + \sum_{j=1}^{N} b_j(x) D_j + c(x), \qquad x \in \Omega,$$

and $\nu(x)$ denotes the outer unit normal to $\partial\Omega$ at x.

Under suitable assumptions on the coefficients of the operator $\mathcal{A}$, which also contain a dissipativity assumption, we first show that any smooth enough solution to the Cauchy-Neumann problem (13.0.1) satisfies the gradient estimate

$$\|\nabla_x u(t,\cdot)\|_\infty \le C_T t^{-\frac{1}{2}} \|f\|_\infty, \qquad t \in (0,T), \tag{13.0.3}$$

for any $f \in C_b(\overline{\Omega})$, any $T>0$ and some positive constant C_T, independent of f. We also show that

$$\|\nabla_x u(t,\cdot)\|_\infty \le C\|f\|_{C^1_b(\overline{\Omega})}, \qquad t>0, \tag{13.0.4}$$

for any $f \in C^1_b(\overline{\Omega})$ with null normal derivative on $\partial\Omega$, where C is a positive constant independent of f. Such estimates are proved by using the Bernstein method as in Chapters 5, 11, and 12. Some additional technical difficulties arise in the nonconvex case since no information on the sign of the normal derivative on $\partial\Omega$ of $|\nabla_x u|^2$ is available. To overcome this difficulty, taking into account the smoothness of $\partial\Omega$, we introduce a suitable function m such that the normal derivative of $m|\nabla_x u|^2$ is nonpositive on $\partial\Omega$.

To prove that problem (13.0.1) actually admits a unique classical solution[1] u, which

[1] i.e., a function $u \in C([0,+\infty)\times\overline{\Omega}) \cap C^{0,1}((0,+\infty)\times\overline{\Omega}) \cap C^{1,2}((0,+\infty)\times\Omega)$ solving the Cauchy-Neumann problem (13.0.1).

is bounded in $(0,T)\times\Omega$ for any $T>0$ we first prove a suitable maximum principle and then we use different arguments depending on the geometry of Ω. For a general unbounded domain, we assume that the diffusion coefficients of the operator $\mathcal{A}$ belong to $C^1_b(\overline{\Omega})$ and we adapt to this situation the technique of Chapter 12. For this purpose, we consider auxiliary problems in $L^p(\Omega)$, which involve suitable new operators $\mathcal{A}_\varepsilon$ $(\varepsilon>0)$ and we show that, for p large enough and f smooth enough, there exists a unique classical solution u_ε to problem (13.0.1) with $\mathcal{A}$ being replaced with $\mathcal{A}_\varepsilon$, which is bounded in $(0,T)\times\Omega$ for any $T>0$. For f smooth enough $\{u_\varepsilon\}$ converges to a solution u of problem (13.0.1), which satisfies the gradient estimates (13.0.3) and (13.0.4). Finally, a standard approximation argument yields the existence of a classical solution to problem (13.0.1) and the gradient estimates in the general case when f is bounded and continuous. The boundedness of the diffusion coefficients and their gradients is essential to deal with the auxiliary problem in $L^p(\Omega)$.

When Ω is convex or an exterior domain, we can allow the diffusion coefficients to be unbounded. In the first case, as in Chapter 1, the solution to problem (13.0.1) is obtained by approximating this problem with a sequence of Cauchy-Neumann problems in (convex) bounded domains Ω_n $(n\in\mathbb{N})$. The Neumann boundary condition gives some problems. Indeed, differently to what happens in the case considered in Chapter 1, it is not immediate to show that the solutions u_n to the Cauchy-Neumann problems in the bounded sets Ω_n converge to a solution to problem (13.0.1). To overcome such a difficulty, we apply the gradient estimate (13.0.3) to the functions u_n in Ω_n, where the constant is independent of n. Finally, when Ω is an exterior domain, we approximate the operator $\mathcal{A}$ by a sequence of elliptic operators $\{\mathcal{A}_n\}$ with bounded diffusion coefficients to which we can apply the previous results, getting the existence of a unique classical solution u_n satisfying the gradient estimate (13.0.5) and being bounded in $(0,T)\times\Omega$ for any $T>0$. Then, letting n tend to $+\infty$, we see that u_n converges to the (unique) solution to problem (13.0.1), which, of course, satisfies (13.0.3).

The existence and uniqueness of a classical solution to problem (13.0.1), bounded in $(0,T)\times\Omega$ for any $T>0$, allow us to associate, in a natural way, a semigroup of bounded operators $\{T(t)\}$ in $C_b(\overline{\Omega})$ with such a problem. The semigroup rule can be used to extend estimate (13.0.3) to any $t>0$, showing that, for any $\omega>0$, there exists a positive constant $C=C(\omega)$ such that

$$\|\nabla_x T(t)f\|_\infty\le Ct^{-\frac{1}{2}}e^{\omega t}\|f\|_\infty,\qquad t>0. \tag{13.0.5}$$

In Section 13.3 we prove some remarkable properties of the semigroup such as the strong Feller property and its irreducibility. Then, in Section 13.4, we show that, for any $\lambda>c_0:=\sup_\Omega c(x)$ and $f\in C_b(\overline{\Omega})$, the elliptic problem (13.0.2) admits a unique solution $u\in D(\mathcal{A})$, where

$$D(\widehat{A})=\bigg\{f\in C_b(\overline{\Omega}):\ \sup_{t\in(0,1)}\frac{\|T(t)f-f\|_\infty}{t}<+\infty\ \text{ and }\ \exists g\in C_b(\overline{\Omega})\ \text{ s.t.}$$
$$\lim_{t\to 0^+}\frac{(T(t)f)(x)-f(x)}{t}=g(x)\ \text{ for any } x\in\overline{\Omega}\bigg\},$$

$$\widehat{A}f(x)=\lim_{t\to 0^+}\frac{(T(t)f)(x)-f(x)}{t},\qquad x\in\overline{\Omega},\qquad f\in D(\widehat{A}).$$

Moreover, using (13.0.3) and (13.0.4), we then show that $D(\mathcal{A})\hookrightarrow C^1_b(\mathbb{R}^N)$ and that $C^1_b(\overline{\Omega})$ is of class $J_{1/2}$ between $C_b(\overline{\Omega})$ and $D(\mathcal{A})$, that is $\|\nabla u\|_\infty\le M_\omega\|u\|_\infty^{1/2}\|(\mathcal{A}-\omega)u\|_\infty^{1/2}$ for any $f\in D(\mathcal{A})$ and $\omega>0$.

In Section 13.5 we come back to the parabolic Cauchy-Neumann problem and, under somewhat heavier assumptions on the growth of the coefficients at infinity, we prove the pointwise estimate $|\nabla_x T(t)f|^p \le C_p e^{M_p t} T(t)|\nabla f|^p$ in $\overline{\Omega}$ for any $f \in C^1_b(\overline{\Omega})$ with null normal derivative on $\partial\Omega$, and any $t > 0$, where $C_p > 0$ and $M_p \in \mathbb{R}$ are suitable constants, independent of f. But, here, we should have $T(t)\mathbb{1} \equiv \mathbb{1}$, otherwise this estimate fails for $f \equiv \mathbb{1}$. Thus, we suppose that $c \equiv 0$. Next, using the above pointwise estimate we prove that

$$|\nabla_x T(t)f|^p \le \widetilde{C}_p t^{-\frac{p}{2}} \left(\frac{M_2 t}{1 - e^{-M_2 t}}\right)^{\frac{p}{2}} T(t)|f|^p, \qquad p \in [2, +\infty), \tag{13.0.6}$$

$$|\nabla_x T(t)f|^p \le \widetilde{C}_p t^{-\frac{p}{2}} \frac{M_p t}{1 - e^{-M_p t}} T(t)|f|^p, \qquad p \in (1, 2), \tag{13.0.7}$$

in $\overline{\Omega}$, for any $t > 0$ and $f \in C_b(\overline{\Omega})$, where the ratio $M_p t/(1 - e^{-M_p t})$ is replaced with 1 when $M_p = 0$. Here, $\widetilde{C}_p$ is a positive constant independent of f. As in Chapter 6, such estimates lead to some interesting results. First, they allow to improve the uniform gradient estimate (13.0.3), since they show that the constant therein appearing can be taken independent of T and, when $M_p < 0$, they show that the sup-norm of $\nabla_x T(t)f$ decreases exponentially to 0 as t tends to $+\infty$. Next, when $M_p \le 0$, estimates (13.0.6) and (13.0.7) provide us with a Liouville type theorem. Namely, they imply that the constants are the only distributional solutions to $\mathcal{A}u = 0$ which belong to $D(\mathcal{A})$.

Assuming the same assumptions used to prove the uniform gradient estimates, we prove that $|\nabla_x T(t)f|^p \le C_p T(t)(f^2 + |\nabla f|^2)^{\frac{p}{2}}$ in $\overline{\Omega}$ for any $t > 0$ and any $f \in C^1_b(\overline{\Omega})$, with null normal derivative on $\partial\Omega$, any $p \in (1, +\infty)$ and some positive constant C_p.

Finally, in Section 13.6, we briefly generalize some results of Chapter 9 to the case of the invariant measure associated with the semigroup $\{T(t)\}$ considered in this chapter. As in the case when $\Omega = \mathbb{R}^N$, whenever an invariant measure exists, the semigroup can be extended to a strongly continuous semigroup defined in the L^p-space related to the measure μ, for any $p \in [1, +\infty)$. Using the above pointwise estimates, we prove gradient estimates for the semigroup in $L^p(\Omega, \mu)$. Such estimates allow us to give a partial characterization of the domain $D(L_p)$ of the infinitesimal generator of $\{T(t)\}$ in $L^p(\Omega, \mu)$.

In the last part of the section we deal with the case when Ω is convex and $\mathcal{A}\zeta = \frac{1}{2}\Delta\zeta - \langle\nabla U, \nabla\zeta\rangle$, on smooth functions ζ, where U is a convex function. In such a situation a complete description of $D(L_2)$ is available.

The results of this chapter have been proved in [31, 32, 33, 74].

13.1 A maximum principle and *a priori* estimates

To begin with, we observe that, since Ω is uniformly of class $C^{2+\alpha}$ for some $\alpha \in (0, 1)$, the distance function $x \mapsto d(x) = \mathrm{dist}(x, \partial\Omega)$, belongs to $C^2_b(\Gamma(\rho))$ for some $\rho > 0$, where

$$\Gamma(\rho) := \{x \in \overline{\Omega} : d(x) < \rho\}. \tag{13.1.1}$$

Moreover,

$$\nabla d(x) = -\nu(\xi(x)), \tag{13.1.2}$$

where $\xi(x)$ is the unique point on $\partial\Omega$ such that $d(x, \partial\Omega) = |x - \xi(x)|$ and $\nu(y)$ is the outer unit normal to $\partial\Omega$ at y. In particular, the formula (13.1.2) implies that $|\nabla d(x)| = 1$ for any $x \in \Gamma(\rho)$. We refer to Section A.4 for the proof of these properties concerning the distance function. Finally, according to Remark A.4.3,

$$+\infty > \omega_0 := \max\left\{0, -\inf_{x \in \partial\Omega}\left\{\frac{\partial\nu}{\partial\tau}(x) \cdot \tau,\ |\tau| = 1,\ \tau \cdot \nu(x) = 0\right\}\right\}. \tag{13.1.3}$$

Now, we can state the conditions on the coefficients of the operator $\mathcal{A}$ that we assume throughout this section and state a maximum principle, which can be proved adapting the arguments in the proof of Theorem 3.1.3.

Hypotheses 13.1.1 (i) the coefficients q_{ij} and b_j $(i,j=1,\ldots,N)$ and c belong to $C^{1+\alpha}_{\rm loc}(\overline{\Omega})$ and $\langle Q(x)\xi,\xi\rangle \geq \kappa(x)|\xi|^2$ for any $x\in\Omega$, $\xi\in\mathbb{R}^N$ and some function $\kappa:\overline{\Omega}\to\mathbb{R}$ with positive infimum κ_0;

(ii) there exist two constants c_0 and γ such that $c_0:=\sup_{x\in\overline{\Omega}}c(x)$ and $|\nabla c(x)|\leq\gamma(1+|c(x)|)$ for any $x\in\Omega$;

(iii) there exist $q_0>0$, $r_0\in\mathbb{R}$, $\tau\geq 0$ and β, with $4\beta+\tau^2N^2<2$, such that

$$|\nabla q_{ij}(x)|\leq q_0\kappa(x)+\tau|c(x)\kappa(x)|^{\frac{1}{2}}, \qquad x\in\Omega, \tag{13.1.4}$$

$$\langle \mathrm{Jac}\, b(x)\xi,\xi\rangle\leq(-\beta c(x)+r_0)|\xi|^2, \qquad x\in\Omega,\ \xi\in\mathbb{R}^N; \tag{13.1.5}$$

(iv) there exist a function $\varphi\in C^2(\overline{\Omega})$, blowing up as $|x|$ tends to $+\infty$, with nonnegative normal derivative on $\partial\Omega$, and a constant $\lambda_0>c_0$ such that $\sup_\Omega(\mathcal{A}\varphi-\lambda_0\varphi)<+\infty$;

(v) if $\omega_0>0$, then there exist constants $s_1\geq 0$, $r_1\in\mathbb{R}$ such that $4\beta+2\omega_0 s_1+\tau^2N^2<2$ and $\langle b(x),\nabla d(x)\rangle\geq s_1c(x)-r_1$ for any $x\in\Gamma(\varepsilon_0)$ for some $\varepsilon_0<\rho$. Moreover, the entries of the matrix Q are bounded in $\Gamma(\varepsilon_0)$.

Theorem 13.1.2 *If $z\in C_b([0,T]\times\overline{\Omega})\cap C^{0,1}((0,T]\times\overline{\Omega})\cap C^{1,2}((0,T]\times\Omega)$ satisfies*

$$\begin{cases} D_tz(t,x)-\mathcal{A}z(t,x)\leq 0, & t\in(0,T], \quad x\in\Omega,\\ \dfrac{\partial z}{\partial\nu}(t,x)\leq 0, & t\in(0,T], \quad x\in\partial\Omega,\\ z(0,x)\leq 0, & x\in\Omega, \end{cases}$$

then $z\leq 0$ in $[0,T]\times\overline{\Omega}$. In particular, the Cauchy-Neumann problem (13.0.1) *admits at most one classical solution, which is bounded in $[0,T]\times\overline{\Omega}$. Moreover, it satisfies $\|u(t,\cdot)\|_\infty\leq e^{c_0t}\|f\|_\infty$, where c_0 is given by Hypothesis* 13.1.1*(ii).*

We now prove the following *a priori* estimate

$$|\nabla_x u(t,x)|\leq C_Tt^{-\frac{1-k}{2}}\|f\|_{C^k_b(\overline{\Omega})}, \qquad t\in(0,T],\ x\in\overline{\Omega},\ k=0,1, \tag{13.1.6}$$

for any $T>0$ and any $f\in C^k_b(\overline{\Omega})$ (satisfying $\partial f/\partial\nu=0$ on $\partial\Omega$, if $k=1$), when u is a classical solution to problem (13.0.1), bounded in $(0,T)\times\Omega$ for any $T>0$, with the further property that $D_ju\in C_b([0,T]\times\overline{\Omega})\cap C^{0,1}((0,T]\times\overline{\Omega})\cap C^{1,2}((0,T]\times\Omega)$ for any $j=1,\ldots,N$. We will deal with the existence of such a solution in the next section.

In order to prove (13.1.6) we adapt to our situation the classical Bernstein method. For this purpose, we introduce the function $m:\overline{\Omega}\to\mathbb{R}$ defined by

$$m(x)=1+\omega_0\psi(d(x)), \qquad x\in\overline{\Omega}, \tag{13.1.7}$$

where $\psi\in C^2_b([0,+\infty))$ satisfies $\psi(0)=0\leq\psi\leq 2^{-1}\varepsilon_0\chi_{(\varepsilon_0,+\infty)}$, $0\leq\psi'\leq 1=\psi'(0)$, $-2\varepsilon_0^{-1}\leq\psi''\leq 0$ and ε_0 is defined in Hypothesis 13.1.1(v). As is readily seen, $1\leq m\leq 1+\omega_0\varepsilon_0/2$, $m(x)=1+\omega_0\varepsilon_0/2$ for any $x\in\Omega\setminus\Gamma(\varepsilon_0)$, and $m(x)=1$ for any $x\in\partial\Omega$.

Let us prove the following fundamental lemma.

Lemma 13.1.3 *Let $u \in C^2(\overline{\Omega})$ admits null normal derivative on $\partial\Omega$. Then, the function $v := m|\nabla u|^2$ has nonpositive normal derivative on $\partial\Omega$.*

Proof Since $\nabla u(x)\cdot\nu(x) = 0$ for any $x \in \partial\Omega$, differentiating along a unit tangent vector τ, we obtain

$$\frac{\partial}{\partial\tau}(\nabla u(x)\cdot\nu(x)) = D^2u(x)\tau\cdot\nu(x) + \frac{\partial\nu}{\partial\tau}(x)\cdot\nabla u(x) = 0, \qquad x \in \partial\Omega. \tag{13.1.8}$$

From (13.1.2), the definition of ω_0 and the properties of the function ψ, it follows that, if $x \in \partial\Omega$ and τ is a unit vector orthogonal to $\nu(x) = 0$, then

$$\frac{\partial m}{\partial\nu}(x) - 2\frac{\partial\nu}{\partial\tau}(x)\cdot\tau = -\omega_0 - 2\frac{\partial\nu}{\partial\tau}(x)\cdot\tau \le 0. \tag{13.1.9}$$

Therefore, if we choose, for any $x \in \partial\Omega$, $\tau = \tau(x) = |\nabla u(x)|^{-1}\nabla u(x)$, from (13.1.8) and (13.1.9) we get

$$\frac{\partial v}{\partial\nu}(x) = |\nabla u(x)|^2\left(\frac{\partial m}{\partial\nu}(x) - 2\frac{\partial\nu}{\partial\tau}(x)\cdot\tau\right) \le 0, \qquad x \in \partial\Omega,$$

and the assertion follows. □

We can now prove the *a priori* estimate (13.1.6) requiring, for the time being, a further regularity property to the function u, that will be removed later.

Proposition 13.1.4 *Let $u \in C_b([0,T]\times\overline{\Omega})\cap C^{0,1}((0,T]\times\overline{\Omega})\cap C^{1,2}((0,T]\times\Omega)$ be a solution of the Cauchy-Neumann problem (13.0.1). Further, assume that the functions D_ju ($j = 1,\ldots,N$) are bounded in $[0,T]\times\overline{\Omega}$ and belong to $C^{0,1}((0,T]\times\overline{\Omega})\cap C^{1,2}((0,T]\times\Omega)$, for any $T > 0$. Then, for any $T > 0$, there exists a constant $C_T > 0$ such that estimate (13.1.6) holds. Moreover, if $c \equiv 0$, then (13.1.6), with $k = 1$, holds with a constant being independent of T.*

Proof Without loss of generality, we can assume that $c \le 0$ in Ω. Indeed, if $c_0 = \sup_\Omega c > 0$, it suffices to consider the function $(t,x) \mapsto e^{-c_0t}u(t,x)$ which solves problem (13.0.1) with c being replaced by the nonpositive potential $c - c_0$. Hypotheses 13.1.1 are satisfied with the same constants β, τ and s_1 and possibly different constants γ, q_0, r_0 and r_1.

We begin by proving estimate (13.1.6) with $k = 0$. For this purpose, let us define the function $v : [0,+\infty)\times\overline{\Omega} \to \mathbb{R}$, by setting $v(t,x) = |u(t,x)|^2 + atm(x)|\nabla_xu(t,x)|^2$ for any $t \ge 0$ and $x \in \overline{\Omega}$, where $a > 0$ is a parameter that will be chosen later. By the assumptions, $v \in C_b([0,T]\times\overline{\Omega})\cap C^{0,1}((0,T]\times\overline{\Omega})\cap C^{1,2}((0,T]\times\Omega)$ and $v(0,\cdot) = f^2$.

We claim that there exists a suitable value of $a > 0$, independent of f, t, x, such that

$$\begin{cases} D_tv(t,x) - \mathcal{A}v(t,x) \le Mv, & t \in (0,T], \quad x \in \Omega,\\ \dfrac{\partial v}{\partial\nu}(t,x) \le 0, & t \in (0,T], \quad x \in \partial\Omega, \end{cases} \tag{13.1.10}$$

for some constant $M \ge 0$, independent of f, t, x as well. Then, by the maximum principle in Theorem 13.1.2, applied to the function $e^{-Mt}v$, we will get $v(t,x) \le e^{MT}\sup_{x\in\overline{\Omega}}v(0,x) = e^{MT}\|f\|_\infty^2$ for any $t \in [0,T]$ and $x \in \overline{\Omega}$, which yields estimate (13.1.6), with $k = 0$ and $C_T = e^{MT/2}a^{-1/2}$, since $m(x) \ge 1$ for any $x \in \overline{\Omega}$.

According to Lemma 13.1.3, the normal derivative of v on $\partial\Omega$ is nonpositive. Thus, we have only to prove that v satisfies the differential inequality in (13.1.10). A straightforward computation shows that v solves the equation $D_t v - \mathcal{A}v = am|\nabla_x u|^2 - 2|\sqrt{Q}\nabla_x u|^2 + g_1 + g_2 + g_3$ in $(0,T]\times\Omega$, where

$$\begin{aligned}g_1(t,\cdot) =&2atm\langle(\mathrm{Jac}\, b)\nabla_x u(t,\cdot), \nabla_x u(t,\cdot)\rangle + atcm|\nabla_x u(t,\cdot)|^2\\&+2atmu(t,\cdot)\langle\nabla c, \nabla_x u(t,\cdot)\rangle + c|u(t,\cdot)|^2,\end{aligned}$$

$$g_2(t,\cdot) =2atm\sum_{i,j,h=1}^N D_h q_{ij} D_h u(t,\cdot) D_{ij}u(t,\cdot) - 2atm\mathrm{Tr}(QD_x^2u(t,\cdot)D_x^2u(t,\cdot)),$$

$$\begin{aligned}g_3(t,\cdot) =&-\frac{\omega_0}{2}(\psi''\circ d)|\sqrt{Q}\nabla d|^2|\nabla_x u(t,\cdot)|^2 - \frac{\omega_0}{2}(\psi'\circ d)(\mathcal{A}d - cd)|\nabla_x u(t,\cdot)|^2\\&-2\omega_0(\psi'\circ d)\langle Q\nabla d, D_x^2 u(t,\cdot)\nabla_x u(t,\cdot)\rangle.\end{aligned}$$

Let us estimate each term in the definition of the functions g_1, g_2 and g_3. From (13.1.5) we get

$$m\langle(\mathrm{Jac}\, b)\nabla_x u, \nabla_x u\rangle \le r_0 m|\nabla_x u|^2 - \beta mc|\nabla_x u|^2 \tag{13.1.11}$$

and, from Hypothesis 13.1.1(ii), it follows that, for arbitrary $\delta > 0$,

$$mu\langle\nabla c, \nabla_x u\rangle \le \gamma m(1-c)|u||\nabla_x u| \le \gamma\delta m(1-c)|\nabla_x u|^2 + \frac{\gamma}{4\delta}m(1-c)u^2. \tag{13.1.12}$$

By (13.1.4) and the Hölder and Young inequalities, we can estimate

$$\begin{aligned}&m\sum_{i,j,h=1}^N D_h q_{ij} D_h u D_{ij} u\\\le& m\bigg(\sum_{i,j=1}^N |\langle\nabla q_{ij}, \nabla_x u\rangle|^2\bigg)^{\frac12}|D_x^2 u| \le mN(q_0\kappa + \tau|c|^{\frac12}\kappa^{\frac12})|\nabla_x u||D_x^2 u|\\\le& m\bigg(\frac{N^2}{4\delta}q_0\kappa - \varepsilon\tau^2N^2c\bigg)|\nabla_x u|^2 + m\kappa\bigg(\frac{1}{4\varepsilon} + \delta q_0\bigg)|D_x^2 u|^2\end{aligned} \tag{13.1.13}$$

for any $\delta, \varepsilon > 0$. Moreover, by the ellipticity condition in Hypothesis 13.1.1(i) it follows that

$$-m\mathrm{Tr}\,(QD_x^2uD_x^2u) \le -m\kappa|D_x^2 u|^2. \tag{13.1.14}$$

Using the properties of the function ψ'' and the Hölder inequality we can estimate

$$-\frac{\omega_0}{2}(\psi''\circ d)|\sqrt{Q}\nabla d|^2|\nabla_x u|^2 \le \frac{\omega_0}{\varepsilon_0}\|Q\|_{C_b(\Gamma(\varepsilon_0))}|\nabla_x u|^2. \tag{13.1.15}$$

Analogously, by Hypothesis 13.1.1(v) and the properties of the function ψ' we can infer that

$$\begin{aligned}-\frac{\omega_0}{2}(\psi'\circ d)(\mathcal{A}d - cd)|\nabla_x u|^2 =& -\frac{\omega_0}{2}(\psi'\circ d)\big(\mathrm{Tr}(QD^2d) + \langle b, \nabla d\rangle\big)|\nabla_x u|^2\\\le&\frac{\omega_0}{2}(\|Q\|_{C_b(\Gamma(\varepsilon_0))}\|D^2 d\|_\infty + r_1 - s_1 c)|\nabla_x u|^2.\end{aligned} \tag{13.1.16}$$

Finally,

$$\begin{aligned}-2\omega_0(\psi'\circ d)\langle Q\nabla d, D_x^2 u\nabla_x u\rangle \le& 2\omega_0\|Q\|_{C_b(\Gamma(\varepsilon_0))}|D_x^2 u||\nabla_x u|\\\le&\frac{\omega_0}{2\delta}\|Q\|_{C_b(\Gamma(\varepsilon_0))}|\nabla_x u|^2 + 2\delta\omega_0\|Q\|_{C_b(\Gamma(\varepsilon_0))}|D_x^2 u|^2.\end{aligned} \tag{13.1.17}$$

Now, combining estimates (13.1.11)-(13.1.17) and using the fact that $m \geq 1$, whenever it is necessary, we get

$$
\begin{aligned}
D_t v(t,\cdot) - \mathcal{A}v(t,\cdot) \leq & \left[am + \kappa\frac{N^2}{4\delta}aTq_0(2+\varepsilon_0\omega_0) - 2\kappa\right]|\nabla_x u(t,\cdot)|^2 + c(u(t,\cdot))^2 \\
& +2at\Bigg\{\left[\left(\delta q_0 + \frac{1}{4\varepsilon} - 1\right)\kappa + 2\delta\omega_0\|Q\|_{C_b(\Gamma(\varepsilon_0))}\right]m|D_x^2 u(t,\cdot)|^2 \\
& + Km|\nabla_x u(t,\cdot)|^2 + \frac{\gamma}{4\delta}m(1-c)(u(t,\cdot))^2 \\
& - \left(\beta + \gamma\delta + \frac{\omega_0}{2}s_1 - \frac{1}{2} + \varepsilon\tau^2 N^2\right)mc|\nabla_x u(t,\cdot)|^2\Bigg\}, \qquad (13.1.18)
\end{aligned}
$$

where we have set

$$
K := r_0 + \gamma\delta + \omega_0\|Q\|_{C_b(\Gamma(\varepsilon_0))}\left(\frac{1}{\varepsilon_0} + \frac{1}{2}\|D^2 d\|_\infty\right) + \frac{r_1\omega_0}{2} + \frac{\omega_0}{2\delta}\|Q\|_{C_b(\Gamma(\varepsilon_0))}.
$$

In view of the relation between β, ω_0 and s_1 in Hypothesis 13.1.1(v), we can choose δ small enough and $\varepsilon < 1/4$ to have $(\delta q_0 + (4\varepsilon)^{-1} - 1)\kappa + 2\delta\omega_0\|Q\|_{C_b(\Gamma(\varepsilon_0))} \leq 0$ and $\beta + \gamma\delta + 2^{-1}\omega_0 s_1 + 4^{-1}\tau^2 N^2 - 2^{-1} \leq 0$. Next, we choose a small enough such that the coefficient in the first square brackets in the right-hand side of (13.1.18) can be estimated from above by $am - \kappa_0$. Recalling that $m \leq 1 + \omega_0\varepsilon_0/2$, we thus obtain

$$
\begin{aligned}
D_t v - \mathcal{A}v \leq & \left[\left(1 + \frac{\omega_0\varepsilon_0}{2}\right)a(1+2TK) - \kappa_0\right]|\nabla_x u|^2 \\
& - \left[-1 + aT\left(1 + \frac{\omega_0\varepsilon_0}{2}\right)\frac{\gamma}{2\delta}\right]cu^2 + aT\left(1 + \frac{\omega_0\varepsilon_0}{2}\right)\frac{\gamma}{2\delta}u^2.
\end{aligned}
$$

Now, it is clear that, up to replacing a with a smaller value (which is independent of f, t, x), if needed, we conclude that v satisfies the differential inequality in (13.1.10) with $M = aT\gamma(2 + \omega_0\varepsilon_0)(4\delta)^{-1}$.

In order to prove (13.1.6) with $k = 1$, it suffices to replace v with the function $u^2 + am|\nabla_x u|^2$ and to argue as above. □

13.2 Existence and uniqueness of a classical solution to problem (13.0.1)

In this section we prove the existence and uniqueness of a classical solution u to the Cauchy problem (13.0.1), which is bounded $[0,T] \times \overline{\Omega}$ for any $T > 0$.

Throughout this section, we assume the following conditions, which depend on the geometry of Ω.

Hypotheses 13.2.1 (General case) (i) Hypotheses 13.1.1 are satisfied;

(ii) there exist two constants $d_1, d_2 \geq 0$ such that $|b(x)| \leq d_1 e^{d_2|x|}$ for any $x \in \Omega$;

(iii) the diffusion coefficients belong to $C_b^1(\Omega)$.

Hypothesis 13.2.2 (Convex case) *Hypotheses* 13.1.1*(i)-(iv) are satisfied.*

In the case of an exterior domain the boundedness of the diffusion coefficients in Hypotheses 13.2.1 can be relaxed. To state the assumption in this case, let $\Omega_\rho = \Omega \setminus \Gamma(\rho)$, where $\Gamma(\rho)$ has been defined in (13.1.1) and fix, now and for the rest of this section, a real number $\overline{R} > 0$ such that $\mathbb{R}^N \setminus \Omega_\rho \subseteq B_{\overline{R}}$. For any $n \in \mathbb{N}$ such that $n > \overline{R}$, we introduce the operator with bounded coefficients $\mathcal{A}^{(n)}$ defined by $\mathcal{A}^{(n)} = \mathrm{Tr}(Q^{(n)}D_x^2) + \langle b, \nabla_x\rangle + c^{(n)}$, where $q_{ij}^{(n)} = q_{ij} \circ \psi_n$ and $c^{(n)} = c \circ \psi_n$ for any $n \in \mathbb{N}$ and $\psi_n(x) = (\psi_{*,n}(x_1), \ldots, \psi_{*,n}(x_N))$ for any $x \in \mathbb{R}^N$. Here, the function $\psi_{*,n} \in C_b^2(\mathbb{R})$ is nondecreasing, odd and satisfies the conditions $y\chi_{[0,n]}(y) \le \psi_{*,n}(y) \le (n + 1/2)\chi_{[n+1,+\infty)}(y)$ for any $y \ge 0$ and $0 \le \psi'_{*,n} \le 1$ in $\mathbb{R}$.

Hypothesis 13.2.3 (Exterior domain) *Hypotheses* 13.2.1*(i)-(ii) are satisfied and there exist a positive function $\varphi_n \in C^2(\overline{\Omega})$, blowing up as $|x|$ tends to $+\infty$ and with nonnegative normal derivative on $\partial\Omega$, and a positive constant $\lambda_0^{(n)} > 0$ such that* $\sup_\Omega(\mathcal{A}^{(n)}\varphi_n - \lambda_0^{(n)}\varphi_n) < +\infty$.

Remark 13.2.4 A case in which Hypothesis 13.2.3 is satisfied is when there exist two constants $\lambda_0 > 0$ and $C \in \mathbb{R}$ such that $2\mathrm{Tr}(Q(x)) + 2\langle b(x), x\rangle + (c(x) - \lambda_0)(1 + |x|^2) \le C$ and $2\langle b(x), x\rangle - \lambda_0(1 + |x|^2) \le C$ for any x in a neighborhood of ∞. Indeed, in such a case the function $\varphi : \overline{\Omega} \to \mathbb{R}$, defined by $\varphi(x) = \psi(x) + (1 - \psi(x))(1 + |x|^2)$ for any $x \in \overline{\Omega}$, where $\psi \in C_c^2(B_{\overline{R}+1})$ is such that $\psi \equiv 1$ in $B_{\overline{R}}$, is a Lyapunov function for both the operators $\mathcal{A}^{(n)}$ $(n \in \mathbb{N})$ and $\mathcal{A}$.

The main result of this section is the following theorem.

Theorem 13.2.5 *Under either Hypotheses* 13.2.1 *or Hypothesis* 13.2.2 *or Hypothesis* 13.2.3*, for any $f \in C_b(\overline{\Omega})$ the Cauchy-Neumann problem* (13.0.1) *admits a unique classical solution u which is bounded in $(0, T) \times \Omega$ for any $t > 0$. The function u belongs to the space $C_{\rm loc}^{1+\alpha/2,2+\alpha}((0, +\infty) \times \overline{\Omega})$ and satisfies the gradient estimate* (13.1.6) *with $k = 0$. Moreover, $\|u(t, \cdot)\|_\infty \le e^{c_0 t}\|f\|_\infty$ for any $t > 0$. If, in addition, $f \in C_b^1(\overline{\Omega})$ and its normal derivative identically vanishes on $\partial\Omega$, then estimate* (13.1.6) *holds with $k = 1$.*

The uniqueness part of Theorem 13.2.5 and the estimate $\|u(t, \cdot)\|_\infty \le e^{c_0 t}\|f\|_\infty$, for any $t > 0$, follow from the maximum principle in Theorem 13.1.2. To prove the existence part of Theorem 13.2.5, we distinguish between three cases, depending on the geometry of Ω.

13.2.1 Proof of Theorem 13.2.5: the nonconvex case

To begin with, let us consider the following lemma.

Lemma 13.2.6 *Let f belong to $C_b(\overline{\Omega})$. Then, there exists a sequence of functions $\{f_n\} \subset C_c^{2+\alpha}(\overline{\Omega})$, bounded with respect to the sup-norm, with null normal derivative on $\partial\Omega$, and converging to f as n tends to $+\infty$, on the compact sets of $\overline{\Omega}$. In particular, if $f \in C_b^1(\overline{\Omega})$ has null normal derivative on $\partial\Omega$, then $\|f_n\|_{C_b^k(\overline{\Omega})} \le C\|f\|_{C_b^k(\overline{\Omega})}$ $(k = 0, 1)$ for any $n \in \mathbb{N}$ and some constant $C > 0$, independent of n, and ∇f_n converge to ∇f as n tends to $+\infty$, uniformly on compact sets of $\overline{\Omega}$.*

Proof Suppose that $f \in C_b(\overline{\Omega})$ and denote by $\{T(t)\}$ the semigroup associated with the Laplacian with homogeneous Neumann boundary conditions in Ω. Then, for any $n \in \mathbb{N}$, the function $T(1/n)f$ belongs to $C^2(\overline{\Omega})$, its normal derivative vanishes on $\partial\Omega$ and $\|T(1/n)f\|_\infty \le \|f\|_\infty$ for any $n \in \mathbb{N}$. Moreover, $T(1/n)f$ converges to f as n tends to $+\infty$, uniformly on compact sets of $\overline{\Omega}$. Now, to complete the proof of the first statement it suffices to prove that each function $T(1/n)f$ can be approximated with a sequence of functions as in the statement of the lemma, which is bounded with respect to the sup-norm. We are thus reduced to prove the second part of the statement.

Fix $f \in C_b^1(\overline{\Omega})$ with null normal derivative on $\partial\Omega$, let U_n and g_n be as in Definition A.4.1 and let $\{B_R(x_i)\}_{i\in\mathbb{N}}$ be a countable family of balls with the following properties: (i) the closure of $B_{2R}(x_i)$ is contained in Ω, for any $i \in \mathbb{N}$; (ii) $\bigcup_{i\in\mathbb{N}} B_R(x_i)$ covers $\Omega\setminus\Omega_\varepsilon$, where $\Omega_\varepsilon = \{x \in \mathbb{R}^N : d(x) < \varepsilon\}$ and $\varepsilon > 0$ is given by the condition (ii) of Definition A.4.1; (iii) there exists $s \in \mathbb{N}$ such that at most s among the balls $B_{2R}(x_i)$ overlap. We now introduce a covering $\{\Omega_k\}_{k\in\mathbb{N}}$ of $\overline{\Omega}$ defined as follows: $\Omega_{2k} = B_R(x_k)$, $\Omega_{2k+1} = V_k = g_k^{-1}(B_{1/2})$ and with this covering we associate a partition of the unity $\{\vartheta_k\}$ such that the C^2-norms of the functions ϑ_k are bounded by a constant independent of k and the normal derivative of the function ϑ_{2k+1} vanishes on $\Omega_{2k+1} \cap \partial\Omega$.

Let $f \in C_b^1(\overline{\Omega})$ admit null derivative on $\partial\Omega$. For any $k \in \mathbb{N}$, we set $F_k = \vartheta_k f$ and observe that $f = \sum_{k=1}^{+\infty} F_k$. At this point, the idea is to approximate each function F_k by regular functions. Fix $n \in \mathbb{N}$. If k is even, then F_k belongs to $C_c^1(B_R(x_{k/2}))$. By convolution, we can determine a function $f_{k,n} \in C_c^{2+\alpha}(B_{2R}(x_{k/2}))$ such that

$$\|F_k - f_{k,n}\|_{C_b^1(\mathbb{R}^N)} \leq \frac{1}{4^k n}. \tag{13.2.1}$$

If k is odd, then we consider the function $\widetilde{F}_k$ defined by $\widetilde{F}_k(y) = F_k(g_k^{-1}(y))$ for any $y \in B_{1/2}^+ = \{z \in B_{1/2} \mid z_N > 0\}$. Since the change of variables is of class $C^{2+\alpha}$, the function $\widetilde{F}_k$ belongs to $C^{2+\alpha}(B_{1/2}^+)$ and, if $y \in B_{1/2}$ with $y_N = 0$, then, taking Remark A.4.2 into account, we get

$$\begin{aligned}\frac{\partial \widetilde{F}_k}{\partial y_N}(y) =& \langle (\nabla F_k)(g_k^{-1}(y)), \mathrm{Jac}(g_k^{-1}(y))e_N\rangle \\ =& -\frac{1}{\alpha_n(g_k^{-1}(y))}\langle (\nabla F_k)(g_k^{-1}(y)), \nu(g_k^{-1}(y))\rangle = 0,\end{aligned}$$

since $\partial F_k/\partial\nu = 0$, thanks to the choice of ϑ_k. It follows that the extension of $\widetilde{F}_k$ with value zero in $\overline{\mathbb{R}^N_+}\setminus B_{1/2}^+$ belongs to $C_b^1(\overline{\mathbb{R}^N_+})$ and its normal derivative identically vanishes on $\{y \in \mathbb{R}^N : y_N = 0\}$. A standard technique based on a truncation argument and convolution with functions which are even with respect to the last variable, allows us to construct a function $\widetilde{f}_{k,n} \in C^{2+\alpha}(\overline{\mathbb{R}^N_+})$, with supp $\widetilde{f}_{k,n} \subseteq B_1$, $\partial\widetilde{f}_{k,n}/\partial y_N = 0$, if $y \in B_1$ and $y_N = 0$, such that $\|\widetilde{f}_{k,n} - \widetilde{F}_k\|_{C_b^1(B_1^+)} \leq 4^{-k}n^{-1}$. The function $f_{k,n} = \widetilde{f}_{k,n} \circ g_k$ belongs to $C^{2+\alpha}(U_n \cap \overline{\Omega})$, has support contained in U_n and its normal derivative vanishes on $U_n \cap \partial\Omega$. Since the $C^{2+\alpha}$-norm of g_k can be taken independent of n (by condition (iii) in Definition A.4.1), we can estimate

$$\|f_{k,n} - F_k\|_{C^1(U_n\cap\Omega)} < \frac{1}{4^k n}. \tag{13.2.2}$$

At this point, for any $n \in \mathbb{N}$, we set $f_n = \sum_{k=1}^n f_{k,n}$. Clearly, $f_n \in C_c^{2+\alpha}(\overline{\Omega})$ and $\partial f_n/\partial\nu = 0$ on $\partial\Omega$. Moreover, for any compact set $\Omega' \subset \overline{\Omega}$, using (13.2.1) and (13.2.2) we obtain $\|f - f_n\|_{C_b^1(\Omega')} \leq (3n)^{-1} + \|(1 - \sum_{k=1}^n \vartheta_k)u\|_{C_b^1(\Omega')}$. As n tends to $+\infty$, right-hand side of the previous estimate vanishes since $\sum_{k=1}^n \vartheta_k$ and $\sum_{k=1}^n \nabla\vartheta_k$ converge respectively to 1 and 0, uniformly in K. Finally, since the sequence $\{\vartheta_n\}$ is bounded in $C_b^1(\overline{\Omega})$, we can determine a constant C such that $\|f_n\|_{C_b^k(\overline{\Omega})} \leq C\|f\|_{C_b^k(\overline{\Omega})}$ for any $n \in \mathbb{N}$ and $k = 0, 1$. □

As the following proposition shows it suffices to prove Theorem 13.2.5 when f is smooth enough.

Proposition 13.2.7 *If Theorem* 13.2.5 *holds true for any $f \in C_c^{2+\alpha}(\overline{\Omega})$, with null derivative on $\partial\Omega$, then it holds for any $f \in C_b(\overline{\Omega})$. In particular, if $f \in C_b^1(\overline{\Omega})$ has a null derivative on $\partial\Omega$, then also estimate* (13.1.6) *with $k = 1$ is satisfied.*

Proof Let $\{f_n\}$ be a sequence of functions in $C_c^{2+\alpha}(\overline{\Omega})$ with null derivative on $\partial\Omega$, which converges to f uniformly on compact sets of $\overline{\Omega}$ and satisfies $\|f_n\|_\infty \leq C\|f\|_\infty$ with C being independent of n (see Lemma 13.2.6). Denote by u_n the solution to problem (13.0.1), with initial datum f_n. Arguing as in Step 2 of the proof of Theorem 12.2.4, we can show that there exists a subsequence $\{u_{n_k}\}$ which converges in $C^{1,2}(F)$, for any compact set $F \subset (0,+\infty) \times \overline{\Omega}$, to a function $u \in C^{1+\alpha/2,2+\alpha}_{\rm loc}((0,+\infty) \times \overline{\Omega})$ which solves the differential equation $D_t u = \mathcal{A}u$ and has null normal derivative on $\partial\Omega$.

To conclude that u is a classical solution to problem (13.0.1), which is bounded in each strip $(0,T) \times \Omega$, it remains to show that u is continuous on $\{0\} \times \overline{\Omega}$, where it equals the function f. This can be obtained by a localization argument. So, we fix $x_0 \in \overline{\Omega}$ and consider two bounded open neighborhoods $U_1 \subset U_0$ of x_0 such that $\Omega_0 := U_0 \cap \Omega$ is sufficiently smooth. Further, we introduce a smooth function $\vartheta \in C^\infty(\overline{\Omega}_0)$ such that $\vartheta \equiv 0$ in a neighborhood of $\Omega \cap \partial U_0$, $\vartheta \equiv 1$ in $U_1 \cap \Omega$ and $\partial\vartheta/\partial\nu = 0$ in $U_0 \cap \partial\Omega$. Finally, we set $v_{n_k} = \vartheta\, u_{n_k}$. Then, we can write

$$v_{n_k}(t,\cdot) = T_0(t)(\vartheta f_{n_k}) + \int_0^t T_0(t-s)\psi_{n_k}(s,\cdot)ds, \qquad t>0,$$

where $\{T_0(t)\}$ is the semigroup generated by the realization of $\mathcal{A}$ in $C(\overline{\Omega}_0)$ with homogeneous Neumann boundary conditions (see Theorem D.3.5(v)) and $\psi_{n_k} = -u_{n_k}(\mathcal{A}-c)\vartheta - 2\langle Q\nabla_x u_{n_k}, \nabla\vartheta\rangle$. Using the gradient estimate (13.1.6), satisfied by any function u_{n_k}, and the boundedness of the sequence $\{f_{n_k}\}$ in $C(\overline{\Omega}_0)$, it follows that $|v_{n_k}(t,\cdot)| \leq |T_0(t)(\vartheta f_{n_k})| + C\sqrt{t}$ in Ω_0 for any $t \in [0,T]$ and $k \in \mathbb{N}$, where $C > 0$ is a constant independent of $k \in \mathbb{N}$. Now, letting k tend to $+\infty$ we conclude that $|u(t,\cdot)| \leq |T_0(t)(\vartheta f)| + C\sqrt{t}$ in $\overline{\Omega} \cap U_1$ for any $t \in [0,T]$. From this formula it follows at once that $u(t,\cdot)$ converges to f uniformly in $U_1 \cap \Omega$, as t tends to 0^+. □

Remark 13.2.8 In the case when f belongs to $C_c^{2+\alpha}(\overline{\Omega})$, the Schauder estimates in Proposition 13.2.13, that we used to prove the convergence of the sequence u_n, show that $\nabla_x u_n$ is bounded in $[0,T] \times \overline{\Omega}$ for any $T > 0$ and it converges to $\nabla_x u$ in $[0,T] \times \overline{\Omega}'$ for any bounded open set $\Omega' \subset \Omega$. Therefore, the function $\nabla_x u$ is continuous in $[0,+\infty) \times \overline{\Omega}$ and is bounded in $[0,T] \times \overline{\Omega}$ for any $T > 0$. If $c \equiv 0$, then $\nabla_x u$ is bounded in $[0,+\infty) \times \overline{\Omega}$, since estimate (13.1.6) with $k = 1$ holds with $T = +\infty$ (see the end of the proof of Proposition 13.1.4).

Finally, by classical regularity results (see Theorem D.1.5), $D_j u$ belongs to the space $C^{1,2}((0,+\infty) \times \Omega)$ for any $j = 1, \ldots, N$.

To prove Theorem 13.2.5 for smooth functions $f \in C_c^{2+\alpha}(\overline{\Omega})$, we adapt the technique of Subsection 12.2.2 and approximate the operator $\mathcal{A}$ by new operators $\mathcal{A}_\varepsilon = \mathcal{A} + c_\varepsilon$, where the auxiliary potential c_ε is chosen in such a way that $\mathcal{A}_\varepsilon$ generates a strongly continuous analytic semigroup $\{T_\varepsilon(t)\}$ in $L^p(\Omega)$ for any $p \in [2,+\infty)$. We prove that, for any $\varepsilon > 0$, the function $u_\varepsilon = T_\varepsilon(\cdot)f$ solves problem (13.0.1), with $\mathcal{A}_\varepsilon$ replacing $\mathcal{A}$ and, for p large enough, it enjoys the regularity properties of Proposition 13.1.4. Hence, it verifies estimate (13.1.6). Taking the limit as ε tends to 0^+ allows us to prove the existence of a classical solution of the Cauchy-Neumann problem (13.0.1) as well as the validity of (13.1.6) for such a solution.

13.2.1.1 Generation of analytic semigroups in L^p-spaces

From the time being, we denote by $\|\cdot\|_p$ and $\|\cdot\|_{k,p}$ the usual norms of the spaces $L^p(\Omega)$ and $W^{k,p}(\Omega)$ ($k \in \mathbb{N}$), respectively. Besides Hypotheses 13.2.1 we assume the following more restrictive conditions on the coefficients of the operator $\mathcal{A}$.

Hypothesis 13.2.9 *$c \le -1$ and there exist $\sigma \in (0, \sqrt{2\kappa_0})$ and $\widetilde{\gamma} > 0$ such that $|\nabla c| \le \widetilde{\gamma}|c|$ and $|b| \le \sigma\sqrt{|c|}$ in Ω.*

For any $p \in [2, +\infty)$, we introduce the Banach space

$$\mathscr{D}_p = \left\{u \in W^{2,p}(\Omega) : \frac{\partial u}{\partial \nu} = 0 \text{ on } \partial\Omega, \quad cu \in L^p(\Omega)\right\}, \tag{13.2.3}$$

which is endowed with the norm $\|u\|_{\mathscr{D}_p} = \|u\|_{2,p} + \|cu\|_p$ for any $u \in \mathscr{D}_p$, and the subspace $\mathscr{D} = \left\{u \in C^2_c(\overline{\Omega}) : \frac{\partial u}{\partial \nu} = 0 \text{ on } \partial\Omega\right\}$.

Lemma 13.2.10 *For any $p \in [2, +\infty)$, $\mathscr{D}$ is dense in $\mathscr{D}_p$.*

Proof The proof is similar to that of Lemma 13.2.6. For this purpose, we limit ourselves to sketching it. Given $f \in \mathscr{D}_p$, we split it into the series $f = \sum_{k=1}^{+\infty} F_k$ and, using the same arguments as in the proof of the quoted lemma, we determine a function $f_{2k,n} \in C^2_c(B_{2R}(x_{k/2}))$ such that $\|F_{2k} - f_{2k,n}\|_{2,p} < (4^{2k}a_{2k}n)^{-1}$ for any $k, n \in \mathbb{N}$, where $a_{2k} = \|c\|_{L^\infty(B_{2R}(x_k))}$. As a byproduct, $\|cF_{2k} - cf_{2k,n}\|_p \le (4^{2k}n)^{-1}$. Similarly, for any $k \in \mathbb{N}$ we determine a function $f_{2k+1,n} \in C^2(U_k \cap \overline{\Omega})$ with null normal derivative on $U_n \cap \partial\Omega$, which satisfies the condition $\|F_{2k+1} - f_{2k+1,n}\|_{W^{2,p}(U_k\cap\Omega)} < (4^{2k+1}a_{2k+1}n)^{-1}$ and $\|cF_{2k+1} - cf_{2k+1,n}\|_{L^p(U_k\cap\Omega)} < (4^{2k+1}n)^{-1}$, for any $k, n \in \mathbb{N}$, where $a_{2k+1} = \|c\|_{L^\infty(V_k\cap\Omega)}$. The sequence $\{f_n\}$, defined by $f_n = \sum_{k=1}^{n} f_{k,n}$ for any $n \in \mathbb{N}$ is contained in $\mathscr{D}$ and converges to f in $\mathscr{D}_p$. □

Now, our purpose is to show that, under the additional Hypothesis 13.2.9, the realization A_p in $L^p(\Omega)$ of the operator $\mathcal{A}$ with domain $\mathscr{D}_p$ is well defined, for any $p \in [2, +\infty)$, and it generates an analytic semigroup $\{T_p(t)\}$ and that, for any $f \in C_b(\overline{\Omega})$ and any $p > N$, the function $T_p(t)f$ is a classical solution to the Cauchy-Neumann problem (13.0.1) and it satisfies the assumptions of Proposition 13.1.4.

For this purpose, we are going to adapt the techniques in Subsection 12.2.2, which are based on integration by parts. We note that, since the conormal vector associated with the operator Q needs not to coincide with the normal derivative on $\partial\Omega$, integrating by parts we are led to some surface integrals which we would avoid. The following lemma allows us to overcome this difficulty.

Lemma 13.2.11 *For any $x \in \overline{\Omega}$ there exists a matrix $Q'(x)$ such that*

(i) *$q'_{ij} \in C^1_b(\overline{\Omega})$ $(i, j = 1, \ldots, N)$ and $\langle Q'(x)\xi, \xi\rangle \ge \kappa_0|\xi|^2$ for any $x \in \Omega$ and $\xi \in \mathbb{R}^N$;*

(ii) *$\mathrm{Tr}(Q(x)A) = \mathrm{Tr}(Q'(x)A)$ for any symmetric matrix A and any $x \in \overline{\Omega}$;*

(iii) *$Q'(x)\nu(x) = \langle \nu(x), Q(x)\nu(x)\rangle\nu(x)$ for any $x \in \partial\Omega$.*

Moreover, for any $u \in \mathscr{D}_p$ and $v \in W^{1,q}(\Omega)$, with $1/p + 1/q = 1$, it holds

$$\int_\Omega (A_0 u)v dx = -\sum_{i,j=1}^{N} \int_\Omega v D_j q'_{ij} D_i u dx - \int_\Omega \langle Q'\nabla u, \nabla v\rangle dx. \tag{13.2.4}$$

Proof For any $x \in \Omega$, let $S(x)$ be the skew-symmetric matrix defined by

$$S(x)y = \vartheta(x)\{\langle y, Q(x)\nabla d(x)\rangle \nabla d(x) - \langle y, \nabla d(x)\rangle Q(x)\nabla d(x)\}, \qquad y \in \mathbb{R}^N,$$

where $\vartheta \in C_b^2(\mathbb{R}^N)$ satisfies the condition $\chi_{\Gamma(\rho/2)} \le \vartheta \le \chi_{\Gamma(\rho)}$ (see (13.1.1)). Then, we set $Q'(x) = Q(x) + S(x)$. Properties (i) and (ii) are immediate consequences of the skew-symmetry of the matrix $S(x)$. Similarly, the property (iii) follows easily taking (13.1.2) into account. In order to prove (13.2.4), it suffices to write $A_0 u = \mathrm{Tr}\,(Q'D^2u)$ and to integrate by parts using property (iii). □

For notational convenience, we now introduce the vector-valued function $b : \overline{\Omega} \to \mathbb{R}^N$ such that $b'_i = \sum_{j=1}^N D_j q'_{ij} - b_i$ for any $i = 1, \ldots, N$. Taking Hypothesis 13.2.9 into account, we immediately see that

$$|b'| \le |b| + \sqrt{N}\|\nabla Q'\|_\infty \le \sigma\sqrt{|c|} + \sqrt{N}\|\nabla Q'\|_\infty. \tag{13.2.5}$$

Proposition 13.2.12 *For any $p \ge 2$, $(A_p, \mathscr{D}_p)$ generates a strongly continuous analytic semigroup $\{T_p(t)\}$ in $L^p(\Omega)$. Moreover, there exists a positive constant λ_0, depending on κ_0, σ, p, N, $\|Q\|_\infty$, $\|\nabla Q\|_\infty$, $\tilde{\gamma}$, such that*

$$\|\lambda u - A_p\|_p \ge (\lambda - \lambda_0)\|u\|_p, \qquad \lambda > \lambda_0,\ u \in \mathscr{D}_p. \tag{13.2.6}$$

Finally, for any $\lambda > \lambda_0$, there exist two constants $C_1, C_2 > 0$ (depending also on λ), such that $C_1\|u\|_{\mathscr{D}_p} \le \|\lambda u - Au\|_p \le C_2\|u\|_{\mathscr{D}_p}$ for $u \in \mathscr{D}_p$.

Proof The estimates in the statement can be proved by arguing as in the proofs of Lemma 12.2.9 and Proposition 12.2.10, writing $\mathcal{A}u = \mathrm{Tr}(Q'D^2u) + \langle b, \nabla u\rangle + cu$. Indeed, according to Lemma 13.2.11, the conormal derivative relevant to the operator $\mathrm{Tr}(Q'D^2)$ coincides, up to a multiplication factor, with the normal derivative. Therefore, integrating by parts gives rise to the same integrals as in the proof of Proposition 12.2.10.

To show that $(A_p, D(A_p))$ generates a strongly continuous analytic semigroup for any $p \ge 2$, we adapt the proof of Proposition 12.2.12. For this purpose, we do not enter into too much detail. We fix $\sigma' \in (\sigma, \sqrt{2\kappa_0})$ and replace the operator A_p with the operator $B_p = A_p - k$ defined in $D(A_p)$, choosing k sufficiently large that $|b'(x)| \le \sigma'\sqrt{|c'(x)|}$ for any $x \in \overline{\Omega}$, where $c' = c - k$. This is possible since b' satisfies (13.2.5). Then, we approximate B_p by the operators $B_{p,\varepsilon} = \mathrm{Tr}(Q'D^2) + \langle b_\varepsilon, \nabla\rangle + c'_\varepsilon$ where $b_\varepsilon = b(1-\varepsilon c)^{-1/2}$ and $c'_\varepsilon = c(1-\varepsilon c)^{-1} + k$. Arguing as in the proof of Proposition 12.2.12, we can show that for any $\lambda > \lambda_0$ and $f \in L^p(\Omega)$ there exists a sequence of functions $u_{\varepsilon_n} \in W^{2,p}(\Omega)$, solutions to the equations $\lambda u_{\varepsilon_n} - B_{p,\varepsilon_n} u_{\varepsilon_n} = f$, such that $\partial u_{\varepsilon_n}/\partial\nu = 0$ on $\partial\Omega$ and $c_{\varepsilon_n} u_{\varepsilon_n} \in L^p(\Omega)$. Moreover, u_{ε_n} converges weakly in $W^{2,p}(\Omega)$ to a solution $u \in W^{2,p}(\Omega)$ of the equation $\lambda u - B_p u = f$ such that $cu \in L^p(\Omega)$. To conclude that u belongs to $\mathscr{D}_p$ it suffices to show that $\partial u/\partial\nu = 0$ on $\partial\Omega$. By the classical interior L^p-estimates, applied to the operator $\lambda - A_0$ (see Theorem D.1.2), for any pair of bounded and smooth open sets $\Omega' \subset \Omega'' \subset \Omega$ with $d(\Omega', \Omega \setminus \Omega'') > 0$, we can estimate $\|u_{\varepsilon_n} - u_{\varepsilon_m}\|_{W^{2,p}(\Omega')} \le C(\|u_{\varepsilon_n} - u_{\varepsilon_m}\|_{L^p(\Omega'')} + \|f_{\varepsilon_n} - f_{\varepsilon_m}\|_{L^p(\Omega'')})$, where $f_{\varepsilon_n} = f + \langle b_{\varepsilon_n}, \nabla u_{\varepsilon_n}\rangle + c'_{\varepsilon_n} u_{\varepsilon_n}$ and C is a suitable positive constant independent of n, m. Since u_{ε_n} and f_{ε_n} converge to u and $f + \langle b, \nabla u\rangle + cu$, respectively, in $L^p(\Omega \cap B_{R+1})$, u_{ε_n} converges to u in $W^{2,p}(\Omega')$. Therefore, by continuity, $\partial u_{\varepsilon_n}/\partial\nu$ converges to $\partial u/\partial\nu$ on $\Omega' \cap \partial\Omega$ and, consequently, $\partial u/\partial\nu = 0$ on $\Omega' \cap \partial\Omega'$ as well. From the arbitrariness of Ω', it follows that $\partial u/\partial\nu = 0$ on $\partial\Omega$. Hence, the operator $\lambda I - B_p$ is surjective for any $\lambda > \lambda_0$. Since, it is quasi-dissipative, due to (13.2.6), and it has dense domain in $L^p(\Omega)$, according to the Lumer-Phillips theorem (see Theorem C.1.9), the operator B_p generates a strongly continuous semigroup $\{R(t)\}$ in $L^p(\Omega)$.

To show that $\{R(t)\}$ is analytic, it suffices to prove that

$$\left|\mathrm{Im}\int_\Omega u^* B_p u dx\right| \le \varsigma\left(-\mathrm{Re}\int_\Omega u^* B_p u dx\right), \qquad u \in \mathscr{D}_p, \tag{13.2.7}$$

for some positive constant ς, where $u^* = |u|^{p-2}\overline{u}$. Indeed, estimate (13.2.7) implies that the numerical range of B_p is contained in the sector $\{\lambda \in \mathbb{C} : |\mathrm{Im}\,\lambda| \le -\varsigma\,\mathrm{Re}\,\lambda\}$ and Theorem B.1.7 yields the analyticity of $\{R(t)\}$.

An integration by parts shows that the real and imaginary parts of $\int_\Omega u^*\, B_p u\, dx$ are given, respectively, by (12.2.25) and (12.2.29), with $\mathfrak{q}'$ instead of $\mathfrak{q}$, where $\mathfrak{q}'$ is the bilinear form induced by the matrix Q'. Since Q and Q' differ only in the skew symmetric matrix S, it follows that $\mathrm{Re}(|u|^2\mathfrak{q}'(\nabla u, \nabla u)) = \mathrm{Re}(|u|^2\mathfrak{q}(\nabla u, \nabla u))$ and $\mathrm{Re}(\overline{u}^2\mathfrak{q}'(\nabla u, \nabla u)) = \mathrm{Re}(\overline{u}^2\mathfrak{q}(\nabla u, \nabla u))$. Hence, the real part of $\int_\Omega u^*\, B_p u\, dx$ can be estimated by

$$-\mathrm{Re}\int_\Omega u^* B_p u dx \ge \left(p - 1 - \frac{(\sigma')^2}{2\kappa_0}\right) F^2 + G^2 + \frac{1}{2}H^2, \tag{13.2.8}$$

where

$$F^2 = \int_\Omega |u|^{p-4}\mathfrak{q}(\mathrm{Re}(\overline{u}\nabla u), \mathrm{Re}(\overline{u}\nabla u))dx,$$

$$G^2 = \int_\Omega |u|^{p-4}\mathfrak{q}(\mathrm{Im}(\overline{u}\nabla u), \mathrm{Im}(\overline{u}\nabla u))dx, \qquad H^2 = \int_\Omega |c'||u|^p d\mu.$$

Observe that $\mathrm{Im}(|u|^2\mathfrak{q}'(\nabla u, \nabla \overline{u})) = -2\mathfrak{s}(\mathrm{Re}(\overline{u}\nabla u), \mathrm{Im}(\overline{u}\nabla u))$, where $\mathfrak{s}$ denotes the bilinear form induced by the matrix S, and $\mathrm{Im}(\mathfrak{q}'(\overline{u}\nabla u, \overline{u}\nabla u)) = 2\mathfrak{q}(\mathrm{Re}(\overline{u}\nabla u), \mathrm{Im}(\overline{u}\nabla u))$. Therefore,

$$\begin{aligned}\mathrm{Im}\int_\Omega u^* B_p u dx =& -p\int_\Omega |u|^{p-4}\mathfrak{s}(\mathrm{Re}(\overline{u}\nabla u), \mathrm{Im}(\overline{u}\nabla u))dx\\ &+ (p-2)\int_\Omega |u|^{p-4}\mathfrak{q}(\mathrm{Re}(\overline{u}\nabla u), \mathrm{Im}(\overline{u}\nabla u))dx - \mathrm{Im}\int_\Omega |u|^{p-2}\overline{u}\,\langle b', \nabla u\rangle\, dx,\end{aligned}$$

that differs from (12.2.30) only in the integral term containing the bilinear form $\mathfrak{s}$, that can be estimated by

$$\begin{aligned}&\left|\int_\Omega |u|^{p-4}\mathfrak{s}(\mathrm{Re}(\overline{u}\nabla_x u), \mathrm{Im}(\overline{u}\nabla u))dx\right|\\ \le&\|S\|_\infty\int_\Omega |u|^{p-4}|\mathrm{Re}(\overline{u}\nabla u)||\mathrm{Im}(\overline{u}\nabla u)|dx \le \frac{1}{\kappa_0}\|S\|_\infty FG.\end{aligned} \tag{13.2.9}$$

Using (12.2.31) we can write

$$\begin{aligned}&\left|(p-2)\int_\Omega |u|^{p-4}\mathfrak{q}(\mathrm{Re}(\overline{u}\nabla u), \mathrm{Im}(\overline{u}\nabla u))dx - \mathrm{Im}\int_\Omega |u|^{p-2}\overline{u}\,\langle b', \nabla u\rangle\, dx\right|\\ \le&(p-2)FG + \frac{\sigma'}{\sqrt{\kappa_0}}GH.\end{aligned} \tag{13.2.10}$$

Combining (13.2.9) and (13.2.10), we get

$$\left|\mathrm{Im}\int_\Omega u^* B_p u dx\right| \le \left(p - 2 + \frac{p}{\kappa_0}\|S\|_\infty\right) FG + \frac{\sigma'}{\sqrt{\kappa_0}}GH. \tag{13.2.11}$$

From (13.2.8) and (13.2.11) we obtain (13.2.7) provided we choose $\varsigma \ge 1$ large enough. The proof is now complete. □

13.2.1.2 Proof of Theorem 13.2.5 for smooth data

Proposition 13.2.13 *For any $p > N$ and $f \in C_c^{2+\alpha}(\overline{\Omega})$, with normal derivative which vanishes on $\partial\Omega$, the function $u = T_p(\cdot)f$ is the unique classical solution to problem* (13.0.1)*, which is bounded in $(0,T)\times\Omega$ for any $T>0$. Moreover, u belongs to $C_{\rm loc}^{1+\alpha/2,2+\alpha}([0,+\infty)\times\overline{\Omega})$ and D_ju belongs to $C_b([0,T]\times\overline{\Omega})\cap C^{1,2}((0,T]\times\Omega)$ for any $T>0$ and $j=1,\dots,N$.*

Proof Let f be as in the statement of the proposition. As is easily seen, f belongs to $\mathscr{D}_p$ (see (13.2.3)). Therefore, the density of $\mathscr{D}_p$ in $L^p(\Omega)$ implies that the function u is continuous from $[0,+\infty)$ in $\mathscr{D}_p$ (see Lemma 13.2.10, Theorem C.2.2 and Proposition C.2.5) and, hence, in $W^{2,p}(\Omega)$. Since $p>N$, according to the Sobolev embedding theorems (see [3, Theorem 5.4]), u belongs to $C([0,+\infty);C_b^1(\overline{\Omega}))$. In particular, u and $\nabla_x u$ are continuous functions in $[0,+\infty)\times\overline{\Omega}$ and they are bounded in $[0,T]\times\overline{\Omega}$ for any $T>0$.

Let us now prove that $u\in C^{1,2}((0,T]\times\Omega)$ for any $T>0$. For this purpose, we observe that, since $\{T_p(t)\}$ is an analytic semigroup, u belongs to $C^1([\varepsilon,T];\mathscr{D}_p)$ for any $0<\varepsilon<T$. Again using the inclusion $\mathscr{D}_p\subseteq W^{2,p}(\Omega)$, the Sobolev embedding theorems and the arbitrariness of ε, we conclude that $D_tu\in C((0,T]\times\overline{\Omega})$.

In order to show that the second-order spatial derivatives of u are continuous in $(0,T]\times\Omega$, we consider two bounded open sets Ω_1 and Ω_2, contained in Ω such that ${\rm dist}(\Omega_1,\Omega\backslash\Omega_2)>0$. For any fixed $t,t_0\in[\varepsilon,T]$, the function $v=u(t,\cdot)-u(t_0,\cdot)$ solves the equation ${\rm Tr}(QD^2v)=T_p(t)\mathcal{A}f-T_p(t_0)\mathcal{A}f-\langle b,\nabla_x v\rangle-cv:=g(t,t_0,\cdot)$, where the right-hand side belongs to $W^{1,p}(\Omega_2)$. By the classical L^p-regularity theory (see Theorem D.1.2), $v(t,\cdot)$ belongs to $W^{3,p}(\Omega_1)$ and

$$\begin{aligned}\|v(t,\cdot)\|_{W^{3,p}(\Omega_1)}\le& C_1\left(\|g(t,t_0,\cdot)\|_{W^{1,p}(\Omega_2)}+\|v(t,\cdot)\|_{W^{1,p}(\Omega_2)}\right)\\ \le& C_2\left(\|T(t)\mathcal{A}f-T(t_0)\mathcal{A}f\|_{W^{1,p}(\Omega)}+\|v(t,\cdot)\|_{W^{2,p}(\Omega)}\right)\end{aligned}$$

for some positive constants C_1 and C_2, independent of t,t_0. This estimate shows that u is continuous at t_0 with values in $W^{3,p}(\Omega_1)$. Indeed, the function $T(\cdot)\mathcal{A}f$ is continuous in $(0,+\infty)$ with values in $\mathscr{D}_p$ and, hence, with values in $W^{1,p}(\Omega)$. Moreover, we already know that u is continuous in $(0,+\infty)$ with values in $W^{2,p}(\Omega)$. By the arbitrariness of t_0, ε, T, we conclude that $u\in C((0,+\infty);W^{3,p}(\Omega_1))$ for any Ω_1 as above. Therefore, using once more the Sobolev embedding theorems, we deduce that $u\in C((0,+\infty);C^2(\overline{\Omega_1}))$, which implies that D_ju and $D_{ij}u$ belong to $C((0,+\infty)\times\overline{\Omega})$ for any $i,j=1,\dots,N$, by the arbitrariness of Ω_1. Thus, we have established that $u\in C^{1,2}((0,+\infty)\times\overline{\Omega})$, and, consequently, that u is a classical solution to the Cauchy-Neumann problem (13.0.1), which is bounded in $(0,T)\times\Omega$ for any $T>0$.

To show that u belongs to $C_{\rm loc}^{1+\alpha/2,2+\alpha}([0,+\infty)\times\overline{\Omega})$ and each first-order spatial derivative belongs to $C_{\rm loc}^{1+\alpha/2,2+\alpha}((0,+\infty)\times\overline{\Omega})$, we introduce a sequence $\{\Omega_n\}$ of smooth and bounded open sets such that (i) $\Omega_n\subset\Omega_{n+1}$ for any $n\in\mathbb{N}$, (ii) $\partial\Omega\subset\bigcup_{n\in\mathbb{N}}\partial\Omega_n$, (iii) $\Omega=\bigcup_{n\in\mathbb{N}}\Omega_n$. Moreover, we denote by $\{T_n(t)\}$ the analytic semigroup generated by the realization A_n of the operator $\mathcal{A}$ with homogeneous Neumann boundary conditions in $C(\overline{\Omega}_n)$ (see Theorem D.3.5(v)). If n is sufficiently large, then f belongs to $D_A(1+\alpha,\infty)=\{u\in C^{2+\alpha}(\overline{\Omega}_n):\partial u/\partial\nu=0 \text{ on } \partial\Omega_n\}$ (see Theorem D.3.5). This implies that for such values of n the function $u_n=T_n(\cdot)f$ belongs to $C_{\rm loc}^{1+\alpha/2,2+\alpha}([0,+\infty)\times\overline{\Omega}_n)$ and $D_ju_n\in C_{\rm loc}^{1+\alpha/2,2+\alpha}((0,+\infty)\times\Omega_n)$ for any $j=1,\dots,N$. Moreover, by the classical maximum principle, it follows that $\|u_n(t,\cdot)\|_\infty\le e^{c_0t}\|f\|_\infty$ for any $t\ge 0$ and $n\in\mathbb{N}$.

Now, we fix $k\in\mathbb{N}$ such that ${\rm supp}(f)\subset\overline{\Omega}_k$, and two bounded smooth open sets $\Omega'\subset\Omega''\Subset\Omega$, such that ${\rm dist}(\Omega',\Omega\setminus\Omega'')>0$. By the classical Schauder estimates in Theorems D.1.5, D.1.6 and by the local boundedness of the sequence $\{\|u_n(t,\cdot)\|_\infty\}$, we deduce that there exist two constants $C_1 = C_1(k)$ and $C_2 = C_2(\varepsilon, T, \Omega', \Omega'')$ such that

$$\|u_n - u_m\|_{C^{1+\alpha/2,2+\alpha}([0,T]\times\overline{\Omega}_k)} \le C_1\|u_n - u_m\|_{C([0,T]\times\overline{\Omega}_{k+1})} \le 2C_1\|f\|_\infty,$$
$$\|D_j u_n - D_j u_m\|_{C^{1+\alpha/2,2+\alpha}([\varepsilon,T]\times\overline{\Omega'})} \le C_2\|u_n - u_m\|_{C([0,T]\times\overline{\Omega''})} \le 2C_2\|f\|_\infty$$

for any $n, m > k$, $j = 1, \ldots, N$ and $T > 0$. By compactness and diagonal arguments, we can determine a subsequence $\{u_{n_k}\}$ converging to a function $v \in C^{1+\alpha/2,2+\alpha}_{\rm loc}([0,+\infty) \times \overline{\Omega})$ in $C^{1,2}(F)$, for any compact set $F \subset [0,+\infty) \times \overline{\Omega}$, such that $D_j u_{n_k}$ converges to $D_j v \in C^{1+\alpha/2,2+\alpha}_{\rm loc}((0,+\infty) \times \Omega)$ in $C^{1,2}([\varepsilon,T] \times \Omega')$ for any $\varepsilon \in (0,T)$, any $\Omega' \Subset \Omega$ and any $j = 1, \ldots, N$. As is immediately seen, v is a classical solution to the Cauchy-Neumann problem (13.0.1) and, consequently, it coincides with u, by Theorem 13.1.2. Thus, u admits continuous third-order spatial derivatives in $(0,+\infty) \times \Omega$. This completes the proof. □

To remove the additional conditions in Hypothesis 13.2.9, it suffices to argue as in the proof of Theorem 12.2.4. For this reason, we just sketch the arguments. For any $\varepsilon \in (0,1)$ we introduce the auxiliary potential c_ε defined by $c_\varepsilon(x) = -\varepsilon\exp(4d_2\sqrt{1+|x|^2})$ for any $x \in \Omega$, where d_2 is the constant in Hypothesis 13.2.1(i). As is immediately seen, there exists a positive constant K_ε such that the operator $\mathcal{A} + c_\varepsilon - K_\varepsilon$ satisfies the additional Hypothesis 13.2.9 with $\sigma = \sqrt{\kappa_0}$ and the constant γ being independent of ε. Proposition 13.2.13 shows that, for any $f \in C^{2+\alpha}_c(\overline{\Omega})$ with normal derivative which identically vanishes on $\partial\Omega$, there exists a unique bounded classical solution v_ε of the Cauchy-Neumann problem (13.0.1), with $\mathcal{A}$ being replaced with the operator $\mathcal{A} - c_\varepsilon - K_\varepsilon$. Further, $D_j v_\varepsilon$ belongs to $C_b([0,T]\times\overline{\Omega}) \cap C^{1,2}((0,T)\times\Omega)$ for any $T > 0$ and $j = 1, \ldots, N$. Applying Proposition 13.1.4 to the operator $\mathcal{A} + c_\varepsilon$ and the function u_ε, which is defined by $u_\varepsilon(t,x) = e^{-K_\varepsilon t}v_\varepsilon(t,x)$ for any $(t,x) \in [0,+\infty) \times \overline{\Omega}$ and has the same degree of smoothness as the function v_ε, we conclude that

$$\|\nabla_x u_\varepsilon(t,\cdot)\|_\infty \le C_T t^{-\frac{1-k}{2}}\|f\|_{C^k_b(\overline{\Omega})} \tag{13.2.12}$$

for any $t \in (0,T)$, $k = 0, 1$, $T > 0$ and some positive constant C_T. The classical Schauder estimates in Theorem D.1.6(ii) and the boundedness of u_ε in $(0,+\infty) \times \Omega$, which is uniform with respect to $\varepsilon \in (0,1)$, show that, for any $R > 0$, $\|u_\varepsilon\|_{C^{1+\alpha/2,2+\alpha}([0,T]\times(\overline{\Omega}\cap B_R))} \le C\|f\|_{C^{2+\alpha}_b(\overline{\Omega})}$ for some constant C, which depends on T and R, but is independent of ε. Hence, using the usual compactness and diagonal arguments, we can show that u_ε converges in $C^{1,2}([0,T] \times (\overline{\Omega} \cap B_R))$, for any $R > 0$, to the unique solution to problem (13.0.1). Further, estimate (13.1.6) follows letting ε tend to 0^+ in (13.2.12).

13.2.2 Proof of Theorem 13.2.5: the convex case

In the convex case, when the diffusion coefficients may be unbounded, we use a different technique to prove Theorem 13.2.5. More precisely, we approximate the domain Ω by a sequence $\{\Omega_n\}$ of bounded convex sets whose union is Ω and such that $B_n \cap \partial\Omega \subset \partial\Omega_n$ for any $n \in \mathbb{N}$. Moreover, we denote by $\{T_n(t)\}$ the strongly continuous analytic semigroup generated by the realization A_n of the operator $\mathcal{A}$ with homogeneous Neumann boundary conditions in Ω_n, i.e., by the operator $A_n : D(A_n) \subset C_b(\overline{\Omega}) \to C_b(\overline{\Omega})$ defined by $A_n u = \mathcal{A}u$ for any $u \in D(A_n)$, which is the set all the functions $u \in \bigcap_{1\le p<+\infty} W^{2,p}(\Omega_n)$ such that $\mathcal{A}u \in C(\overline{\Omega}_n)$ and the normal derivative of u vanishes on $\partial\Omega_n$ (see Theorem D.3.5(v)). According to Proposition D.3.2, for any $f \in C(\overline{\Omega}_n)$ the function $u_n = T_n(\cdot)f$ belongs to $C([0,+\infty) \times \overline{\Omega}_n) \cap C^{1+\alpha/2,2+\alpha}([\varepsilon,T] \times \overline{\Omega}_n)$ for any $0 < \varepsilon < T$ and it is the unique classical solution of the Cauchy-Neumann problem (13.0.1), with Ω being replaced by Ω_n. Moreover,

$$\|T_n(t)\|_{L(C(\overline{\Omega}_n))} \le e^{c_0 t}, \qquad t > 0, \tag{13.2.13}$$

and $T_n(t)f \geq 0$ for any $t \geq 0$ and any nonnegative function $f \in C(\overline{\Omega}_n)$.

Fix two open sets Ω' and Ω'', such that $\Omega' \subset \Omega''$ and $\text{dist}(\Omega', \Omega \setminus \Omega'') > 0$, and let $n_0 \in \mathbb{N}$ be such that $\Omega'' \subset \Omega_n$ for any $n \geq n_0$. According to Theorem D.1.6, there exists a positive constant C, independent of n, such that $\|u_n\|_{C^{1+\alpha/2,2+\alpha}([\varepsilon,T]\times\Omega')} \leq C\|f\|_\infty$ for any $n \geq n_0$. This estimate implies that there exist a subsequence $\{T_{n_k}(\cdot)f\}$ and a continuous function $u : (0,+\infty) \times \overline{\Omega} \to \mathbb{R}$ such that $T_{n_k}(\cdot)f$ converges to u pointwise in $(0,+\infty) \times \overline{\Omega}$, as k tends to $+\infty$. Moreover, $T_{n_k}(\cdot)f$ converges to u in $C^{1+\beta/2,2+\beta}([\varepsilon,T] \times \Omega')$ for any $\beta < \alpha$, any bounded set $\Omega' \subset \Omega$ and any $0 < \varepsilon < T$. As is immediately seen, u is a solution of the equation $D_t u = \mathcal{A}u$ in $(0,+\infty) \times \Omega$ and its normal derivative vanishes on $(0,+\infty) \times \partial\Omega$. Moreover, u belongs to $C^{1+\alpha/2,2+\alpha}_{\text{loc}}((0,+\infty) \times \overline{\Omega})$. By (13.2.13) we get $|u(t,x)| \leq e^{c_0 t}\|f\|_\infty$ for any $(t,x) \in (0,+\infty) \times \overline{\Omega}$. Hence, u is bounded in $(0,T] \times \overline{\Omega}$ for any $T > 0$.

The continuity of u on $\{0\} \times \overline{\Omega}$ can be obtained as in the proof of Proposition 13.2.7, replacing the function u_{n_k} therein defined with the function $T_{n_k}(\cdot)f$, observing that this function satisfies the assumptions of Proposition 13.1.4, with Ω being replaced by Ω_n. Hence, $\|\nabla_x T_{n_k}(t)f\|_\infty \leq C_T t^{-\frac{1}{2}}\|f\|_\infty$ for any $t \in (0,T]$, any $T > 0$ and some positive constant C_T, independent of k. Moreover, letting k tend to $+\infty$ in this estimate, we deduce that u satisfies (13.1.6) with $k = 0$.

Finally, let us prove that the sequence $T_n(\cdot)f$ itself converges to u in $C^{1,2}(F)$ for any compact set $F \subset (0,+\infty) \times \overline{\Omega}$. For this purpose, we observe that, applying all the above arguments to any subsequence $\{T_{n_h}(\cdot)f\}$, we can show that there exists a subsequence $\{T_{n_{h_k}}(\cdot)f\}$ converging to u in $C^{1,2}(F)$ for any compact set F as above. This, of course, implies that the whole sequence $\{T_n(\cdot)f\}$ converges to u in $C^{1,2}(F)$, and we are done.

13.2.3 Proof of Theorem 13.2.5: exterior domains

Fix $f \in C_b(\overline{\Omega})$. Since the operator $\mathcal{A}^{(n)}$ $(n > \overline{R})$[2] satisfies Hypotheses 13.1.1 with κ being replaced by $\kappa^{(n)}$ and with the same constants γ, q_0, r_0, β, s_1 and k_1, by Subsection 13.2.1 the parabolic Cauchy-Neumann problem (13.0.1), with $\mathcal{A}$ being replaced with $\mathcal{A}^{(n)}$, admits a unique classical solution u_n which belongs to $C^{1+\alpha/2,2+\alpha}_{\text{loc}}((0,+\infty) \times \overline{\Omega})$ and it is bounded in $(0,T) \times \Omega$ for any $T > 0$. Moreover, for any T as above there exists a positive constant C_T such that

$$|\nabla_x u_n(t,x)| \leq C_T t^{-\frac{1-k}{2}}\|f\|_{C_b(\overline{\Omega})}, \qquad t \in (0,T], \;\; x \in \overline{\Omega}. \tag{13.2.14}$$

In order to prove that there exists a subsequence of $\{u_n\}$, which converges to the classical solution to problem (13.0.1) (which is bounded in $(0,T)\times\Omega$ for any $T > 0$) in $C^{1,2}([\varepsilon,T]\times\Omega')$ for any $\varepsilon \in (0,T)$ and any $\Omega' \Subset \overline{\Omega}$, we argue as in the Subsection 13.2.2. Hence, in this context we may skip some details. First, we show that a subsequence $\{u_{n_k}\}$ converges to a solution u to the differential equation in (13.0.1), which satisfies the boundary condition $\partial u/\partial\nu = 0$ on $\partial\Omega$. For this purpose, we apply the Schauder estimates in Theorem D.1.6 to the function u_n, observing that the constants which appear can be taken independent of n, since for any bounded set $\widetilde{\Omega} \subset \Omega$, one has $\mathcal{A}^{(n)} = \mathcal{A}$ in $\widetilde{\Omega}$, if n is sufficiently large. Then, to prove the continuity of u at $t = 0$, we use a localization argument and the gradient estimate (13.2.14). For the same reason as above, also in this case all the constants involved can be taken independent of n. Finally, arguing as in the last part of Subsection 13.2.2, we conclude that u_n converges to u in $C^{1,2}(F)$ for any compact set $F \subset (0,+\infty) \times \overline{\Omega}$.

[2] We recall that $\overline{R}$ is defined before Hypothesis 13.2.3.

13.3 Some properties of the semigroup $\{T(t)\}$

For any $f \in C_b(\overline{\Omega})$, $t > 0$ and $x \in \overline{\Omega}$, we set $(T(t)f)(x) = u(t,x)$, where, according to Theorem 13.2.5, u is the unique classical solution to problem (13.0.1) which satisfies the estimate $\|u(t,\cdot)\|_\infty \le e^{c_0 t}\|f\|_\infty$ for any $t > 0$. Then, $\{T(t)\}$ is a semigroup of linear operators in $C_b(\overline{\Omega})$, such that $\|T(t)\|_{L(C_b(\overline{\Omega}))} \le e^{c_0 t}$ for any $t > 0$. Now we state some remarkable properties of this semigroup, which we need in Section 13.5.

Proposition 13.3.1 *If $\{f_n\} \subset C_b(\overline{\Omega})$ is a bounded sequence converging pointwise in Ω to a function $f \in C_b(\overline{\Omega})$, then $T(\cdot)f_n$ converges, as n tends to $+\infty$, to $T(\cdot)f$ in $C^{1,2}([\varepsilon,T]\times\overline{\Omega}')$ for any $0 < \varepsilon < T$ and any bounded set $\Omega' \subset \Omega$. Further, if $\{f_n\}$ converges to f uniformly on compact subsets of $\overline{\Omega}$, then $T(\cdot)f_n$ converges to $T(\cdot)f$ uniformly in $[0,T]\times\overline{\Omega}'$ for any $T > 0$ and any bounded set $\Omega' \subset \Omega$, as n tends to $+\infty$. Finally, $T(t)f$ can be represented in the form*

$$(T(t)f)(x) = \int_\Omega f(y)p(t,x;dy), \qquad t > 0,\ x \in \overline{\Omega}, \tag{13.3.1}$$

where $p(t,x;dy)$ is a positive finite Borel measure on Ω, which is a probability measure if $c \equiv 0$.

Proof To prove the first part of the proposition it suffices to adapt the arguments in the proof of Theorem 13.2.5. Without loss of generality, we can assume that $f \equiv 0$.

Let $\{f_n\}$ be a bounded sequence in $C_b(\overline{\Omega})$ converging pointwise to zero in Ω, and set $u_n = T(\cdot)f_n$. Applying the Schauder estimate (D.1.13) and the maximum principle to the sequence $\{u_n\}$, we can extract a subsequence $\{u_{n_k}\}$ converging in $C^{1,2}([\varepsilon,T]\times\overline{\Omega}')$ to a function $u \in C^{1,2}((0,+\infty)\times\overline{\Omega})$ for any $0 < \varepsilon < T$ and any bounded set $\Omega' \subset \Omega$. Moreover, the function u is a bounded solution of the equation $D_t u = \mathcal{A}u$ in $(0,+\infty)\times\Omega$, and its normal derivative vanishes on the boundary of Ω. To conclude that $u \equiv 0$, by Theorem 13.1.2 it suffices to show that u is continuous up to $t = 0$ and that $u(0,\cdot) \equiv 0$. This can be obtained by a localization argument. So, we fix $x_0 \in \overline{\Omega}$. Moreover, we consider two open neighborhoods $U_1 \subset U_0$ of x_0, such that $\Omega_0 := U_0 \cap \Omega$ is sufficiently smooth, and we introduce a smooth function $\vartheta \in C^\infty(\overline{\Omega}_0)$ such that $\vartheta \equiv 0$ in a neighborhood of $\Omega \cap \partial U_0$, $\vartheta \equiv 1$ in $U_1 \cap \Omega$ and $\partial\vartheta/\partial\nu = 0$ in $U_0 \cap \partial\Omega$. Finally, we set $v_{n_k} = \vartheta u_{n_k}$. Arguing as in the last part of the proof of Proposition 13.2.7, we can show that

$$|v_{n_k}(t,x)| \le |(T_0(t)\vartheta f_{n_k})(x)| + C\sqrt{t}, \qquad t \in [0,T],\ x \in \overline{\Omega}_0,\ k \in \mathbb{N}, \tag{13.3.2}$$

where $C > 0$ is a constant independent of $k \in \mathbb{N}$ and $\{T_0(t)\}$ is the semigroup generated by the realization of $\mathcal{A}$ in $C(\overline{\Omega}_0)$ with homogeneous Neumann boundary conditions (see Theorem D.3.5(v)). We now claim that $T_0(t)(\vartheta f_{n_k})$ vanishes as k tends to 0 for any $t \in [0,T]$. Once the claim is proved, letting k tend to $+\infty$ in (13.3.2), we will get $|u(t,x)| \le C\sqrt{t}$ and, consequently, u will turn out to be continuous at $(0,x_0)$, where it will vanish. To prove the claim, we take advantage of the L^p-theory. We observe that, for any $p \in (N,+\infty)$, the semigroup $\{T_0(t)\}$ extends to an analytic semigroup in $L^p(\Omega_0)$ (see Theorem D.3.5(iii)) and, by the Sobolev embedding theorem (see [3, Theorem 5.4]), the domain of its generator A_p is continuously embedded in $C(\overline{\Omega}_0)$. Since ϑf_{n_k} converges to zero in $L^p(\Omega_0)$ and $\|T_0(t)(\vartheta f_{n_k})\|_{D(A_p)} \le Ct^{-1}\|f_{n_k}\|_{L^p}$ for some positive constant C, independent of k, $T_0(t)(\vartheta f_{n_k})$ converges to 0 in $D(A_p)$ and, consequently, uniformly in $\overline{\Omega}$, as k tends to $+\infty$. So far, we have proved that the subsequence $\{u_{n_k}\}$ converges to zero in $C^{1,2}([\varepsilon,T]\times\overline{\Omega}')$ for any $0 < \varepsilon < T$ and any bounded set $\Omega' \subset \Omega$. As in Subsection 13.2.2 we can now prove that the sequence $\{u_n\}$ itself converges to zero in $C^{1,2}([\varepsilon,T]\times\overline{\Omega}')$ for any $0<\varepsilon<T$ and any bounded set $\Omega'\subset\Omega$, as stated.

Now suppose that $\{f_n\}$ converges to zero uniformly on compact subsets of $\overline{\Omega}$. By (13.3.2), with v_n instead of v_{n_k}, we get $|u_n(t,x)|\le\|T_0(t)(\vartheta f_n)\|_\infty+C\sqrt{t}\le e^{c_0T}\|\vartheta f_n\|_\infty+C\sqrt{t}$ for any $t\in[0,T]$, $x\in\overline{\Omega}_1$ and $n\in\mathbb{N}$, where $C>0$ is independent of n. Therefore, for any $\varepsilon>0$, we can estimate $\|u_n\|_{C([0,T]\times\overline{\Omega}_1)}\le e^{\varepsilon c_0}\|\vartheta f_n\|_\infty+C\sqrt{\varepsilon}+\|u_n\|_{C([\varepsilon,T]\times\overline{\Omega}_1)}$. Since u_n and ϑf_n vanish as n tends to $+\infty$, respectively in $[\varepsilon,T]\times\overline{\Omega}$ and in $\overline{\Omega}$, we get $\limsup_{n\to\infty}\|u_n\|_{C([0,T]\times\overline{\Omega}_1)}\le C\sqrt{\varepsilon}$, which implies that u_n converges to zero uniformly in $[0,T]\times\overline{\Omega}_1$. The arbitrariness of Ω_1 yields the conclusion.

Now we can prove (13.3.1). By the Riesz representation theorem, for any $x\in\overline{\Omega}$, there exists a positive finite Borel measure $p(t,x;dy)$ in Ω such that (13.3.1) is satisfied by any $f\in C_0(\Omega)$. If $f\in C_b(\overline{\Omega})$, then we consider a bounded sequence $\{f_n\}\subset C_0(\Omega)$ which converges to f uniformly on the compact sets of Ω. Writing (13.3.1) with f_n instead of f and letting n tend to $+\infty$, we obtain the desired formula, by dominated convergence.

To conclude the proof, we observe that, if $c\equiv 0$, then $T(t)\mathbb{1}\equiv\mathbb{1}$ and, consequently, $\int_\Omega p(t,x;dy)=1$ for any $t>0$ and $x\in\overline{\Omega}$. This implies that $p(t,x;dy)$ are all probability measures. □

Using the formula (13.3.1) we can extend the semigroup to the space $B_b(\Omega)$ of all the bounded Borel functions $f:\Omega\to\mathbb{R}$. We still denote by $\{T(t)\}$ such a semigroup. As the following proposition shows, the so obtained semigroup is irreducible[3] and strong Feller[4].

Proposition 13.3.2 *The semigroup $\{T(t)\}$ is irreducible and strong Feller.*

Proof To begin with, let us prove that $\{T(t)\}$ is irreducible. Fix an open set $U\subset\Omega$. We are going to prove that $(T(t)\chi_U)(x)>0$ for any $t>0$ and $x\in\overline{\Omega}$. For this purpose, we approximate χ_U by a sequence of nonnegative continuous functions f_n converging monotonically to χ_U and such that f_1 does not identically vanish in Ω. According to the representation formula (13.3.1), $(T(t)f_n)(x)$ converges monotonically to $(T(t)\chi_U)(x)$ as n tends to $+\infty$, for any $t>0$ and $x\in\overline{\Omega}$. To conclude that $T(\cdot)\chi_U>0$, it suffices to show that $T(t)f_1>0$ for any $t>0$. By contradiction, let us suppose that there exist $t_0>0$ and $x_0\in\Omega$ such that $(T(t_0)f_1)(x_0)=0$. The nonnegativity of the semigroup $\{T(t)\}$ immediately implies that (t_0,x_0) is a minimum point of the function $T(\cdot)f_1$. Therefore, according to Proposition D.2.2(ii), $T(\cdot)f_1$ vanishes on $(0,t_0]\times\Omega$. In particular, taking $t=0$ we get $f_1\equiv 0$ in Ω' and, hence, $f_1\equiv 0$ in Ω, which, of course can not be the case.

Similarly, if $(T(t_0)f_1)(x_0)=0$ at some point $(t_0,x_0)\in(0,+\infty)\times\partial\Omega$, then we can find a bounded open set Ω' of class $C^{2+\alpha}$ such that $x_0\in\partial\Omega$ and $\Omega'\cap(B_r(x_0))\subset\Omega\cap(B_r(x_0))$ for some $r>0$. Then, according to Proposition A.4.5, Ω' satisfies the interior sphere condition, so that $\partial T(t_0)f/\partial\nu>0$ at $x=x_0$ (see Proposition D.2.2(ii)), which, of course, can not be the case.

To prove that the semigroup $\{T(t)\}$ is strong Feller, we adapt the arguments in the proof of Proposition 11.3.5. We fix $f\in B_b(\Omega)$ and introduce a bounded sequence $\{f_n\}\in C_b(\overline{\Omega})$ converging pointwise to f, almost everywhere in Ω. We use estimate (13.1.6) with $k=0$ to conclude that, for any $t>0$, $T(t)f_n$ is a Lipschitz continuous function, uniformly with respect to n, on each smooth bounded subset of Ω. Letting n tend to $+\infty$, we deduce that the function $T(t)f$ is Lipschitz continuous as well in the same kind of subsets of $\overline{\Omega}$. Since we can cover $\overline{\Omega}$ with this kind of sets, $T(t)f$ is continuous in $\overline{\Omega}$. Finally, since $p(t,x;\Omega)\le e^{c_0t}$, we conclude that $T(t)f$ is bounded in $\overline{\Omega}$. Hence, it belongs to $C_b(\overline{\Omega})$. □

[3] i.e., $(T(t)\chi_U)(x)>0$ for any $t>0$, any $x\in\Omega$ and any open set $U\subset\Omega$.

[4] i.e., it maps $B_b(\Omega)$ into $C_b(\overline{\Omega})$.

Remark 13.3.3 In fact, in the proof of Proposition 13.3.2 we have shown that, if $f \in C_b(\overline{\Omega})$ is a nonnegative function which does not identically vanish in Ω, then $T(t)f > 0$ in $\overline{\Omega}$ for any $t > 0$.

Now we show that, if $f \in C_b^1(\overline{\Omega})$ has null derivative on $\partial\Omega$, then $\nabla_x T(\cdot)f$ is continuous at $t = 0$.

Proposition 13.3.4 *If $f \in C_b^1(\overline{\Omega})$ has null derivative on $\partial\Omega$, then*

$$\lim_{t \to 0^+} \|T(t)f - f\|_{C_b^1(\overline{\Omega'})} = 0 \tag{13.3.3}$$

for any bounded open set $\Omega' \subset \Omega$, i.e., $\nabla_x T(\cdot)f$ is continuous up to $\{0\} \times \overline{\Omega}$.

Proof We fix $x_0 \in \overline{\Omega}$ and introduce two open neighborhoods U_0, U_1 of x_0 with $U_1 \subset U_0$. Moreover, we set $\Omega_0 = U_0 \cap \Omega$, $\Omega_1 = U_1 \cap \Omega$ and assume that Ω_0 is sufficiently smooth. Finally, we denote by $\vartheta \in C^\infty(\overline{\Omega}_0)$ any function vanishing in a neighborhood of $\Omega \cap \partial U_0$ and such that $\vartheta \equiv 1$ in Ω_1, $\partial\vartheta/\partial\nu \equiv 0$ on $\partial\Omega \cap U_0$.

The same arguments as in the proof of Proposition 13.2.7 show that

$$T(t)f = T_0(t)(\vartheta f) + \int_0^t T_0(t-s)\psi(s,\cdot)ds$$

in $\Omega_1 \cap \Omega$, where $\{T_0(t)\}$ is the semigroup generated by the realization of the operator $\mathcal{A}$ in $C(\overline{\Omega}_0)$ with homogeneous Neumann boundary conditions, and $\psi = -(T(\cdot)f)(\mathcal{A} - c)\vartheta - 2\langle Q\nabla_x T(\cdot)f, \nabla\vartheta\rangle$. According to estimate (13.1.6), with $k = 1$, there exists a constant $C_1 > 0$ such that

$$\|\psi(t)\|_\infty \le C_1 \|f\|_{C_b^1(\overline{\Omega})}, \qquad t \in (0, 1]. \tag{13.3.4}$$

Moreover, since $\{u \in C^1(\overline{\Omega}_0) : \partial u/\partial\nu = 0 \text{ on } \partial\Omega_0\}$ belongs to the class $J_{1/2}$ between $C_b(\overline{\Omega}_0)$ and the domain of generator of $\{T_0(t)\}$ (see Theorem D.3.5(v)), we deduce that $\|\nabla_x T_0(t) g\|_\infty \le C_2 t^{-1/2} \|g\|_{C(\overline{\Omega}_0)} \le C_2 t^{-1/2}\|g\|_\infty$ for any $t \in (0, T]$, $g \in C_b(\overline{\Omega})$ and some positive constant C_2. Now, from this estimate and (13.3.4) we get

$$\begin{aligned}
&|\nabla_x T(t)f(x) - \nabla f(x_0)| \\
&= \left|(\nabla_x T_0(t)(\vartheta f))(x) + \int_0^t (\nabla_x T_0(t-s)\psi(s,\cdot))(x)ds - \nabla f(x_0)\right| \\
&\le |\nabla_x T_0(t)(\vartheta f)(x) - \nabla f(x_0)| + 2C_1C_2\|f\|_{C_b^1(\overline{\Omega})}\sqrt{t}
\end{aligned}$$

for any $t \in (0, T)$ and $x \in \Omega_1$. Since $\nabla_x T_0(\cdot)(\vartheta f)$ is continuous at $(0, x_0)$, letting (t, x) tend to $(0, x_0)$, the formula (13.3.3) follows. □

13.4 The weak generator of the semigroup and the elliptic equation (13.0.2)

Here, we use the gradient estimates to characterize the weak generator $\widehat{A}$ of the semigroup $\{T(t)\}$ and to solve the elliptic problem (13.0.2) when $f \in C_b(\overline{\Omega})$. As in the case of the problem in $\mathbb{R}^N$, the family of bounded operators $\{R(\lambda)\}$, defined by

$$(R(\lambda)f)(x) = \int_0^{+\infty} e^{-\lambda t}(T(t)f)(x)dt, \qquad x \in \overline{\Omega},$$

for any $f \in C_b(\overline{\Omega})$ and $\lambda > c_0$, is the resolvent family associated with some closed operator $\widehat{A} : D(\widehat{A}) \subset C_b(\overline{\Omega}) \to C_b(\overline{\Omega})$. So from now on, we write $R(\lambda, \widehat{A})$ instead of $R(\lambda)$. Adapting the proofs in Section 1.3, it can be shown that the weak generator $(\widehat{A}, D(\widehat{A}))$ is given by

$$D(\widehat{A}) = \Big\{ f \in C_b(\overline{\Omega}) : \sup_{t \in (0,1)} \frac{\|T(t)f - f\|_\infty}{t} < +\infty \text{ and } \exists g \in C_b(\overline{\Omega}) \text{ s.t.}$$
$$\lim_{t \to 0^+} \frac{(T(t)f)(x) - f(x)}{t} = g(x) \text{ for any } x \in \overline{\Omega} \Big\}, \tag{13.4.1}$$

$$\widehat{A}f(x) = \lim_{t \to 0^+} \frac{(T(t)f)(x) - f(x)}{t}, \qquad x \in \overline{\Omega}, \qquad f \in D(\widehat{A}).$$

Now our aim is to show that $D(\widehat{A}) = D(\mathcal{A})$ and $\widehat{A} = \mathcal{A}$, where

$$D(\mathcal{A}) = \Big\{ u \in C_b(\overline{\Omega}) \cap \bigcap_{1 \le p < +\infty} W^{2,p}(\Omega \cap B_R) \text{ for any } R > 0 :$$
$$\mathcal{A}u \in C_b(\overline{\Omega}), \ \frac{\partial u}{\partial \nu}(x) = 0 \text{ for any } x \in \partial\Omega \Big\}. \tag{13.4.2}$$

For this purpose, we first prove the following maximum principle for the elliptic problem (13.0.2).

Proposition 13.4.1 *Let $u \in C_b(\overline{\Omega}) \cap W^{2,p}(\Omega \cap B_R)$, for any $R > 0$ and $p \in [1, +\infty)$, be such that $\mathcal{A}u \in C_b(\overline{\Omega})$ and*

$$\begin{cases} \lambda u(x) - \mathcal{A}u(x) \le 0, & x \in \Omega, \\ \dfrac{\partial u}{\partial \nu}(x) \le 0, & x \in \partial\Omega, \end{cases}$$

for some $\lambda \ge \lambda_0$, where λ_0 is as in Hypothesis 13.1.1*(iv). Then, $u \le 0$. In particular, for any $\lambda \ge \lambda_0$ there exists at most one solution $u \in D(\mathcal{A})$ to problem* (13.0.2).

Proof Without loss of generality, we may assume that $\varphi \ge 0$ and $\mathcal{A}\varphi - \lambda_0 \varphi \le 0$. For any $n \in \mathbb{N}$, we introduce the function $u_n = u - n^{-1}\varphi$ and note that

$$\begin{cases} \lambda u_n(x) - \mathcal{A}u_n(x) \le 0, & x \in \Omega, \\ \dfrac{\partial u_n}{\partial \nu}(x) \le 0, & x \in \partial\Omega. \end{cases} \tag{13.4.3}$$

If we prove that $u_n \le 0$ for any $n \in \mathbb{N}$, then the conclusion will follow letting n tend to $+\infty$. By contradiction, suppose that u_n has a maximum point at $x_n \in \overline{\Omega}$. Adapting the second proof of Lemma 3.1.2 to this situation, we can show that, if $x_n \in \Omega$, then $\mathcal{A}u_n(x_n) - c(x_n)u(x_n) \le 0$. Thus, by (13.4.3) it follows that $(\lambda - c(x_n))u_n(x_n) \le 0$ and, hence, $u_n(x_n) \le 0$.

Suppose now that $x_n \in \partial\Omega$ is such that $u_n(x_n) > 0$ and $u_n(x) < u_n(x_n)$ for any $x \in \Omega$. These assumptions will lead us to a contradiction since they will imply that $(\partial u_n/\partial \nu)(x_n)$ is positive. Let $B_r(y) \subset \Omega$ be such that $\overline{B}_r(y) \cap \partial\Omega = \{x_n\}$ and assume that $u_n > 0$ in $B_r(y)$. Moreover, fix $\alpha > 0$ such that the function $z : \mathbb{R}^N \to \mathbb{R}$, defined by $z(x) = e^{-\alpha|x-y|^2} - e^{-\alpha r^2}$ for any $x \in \mathbb{R}^N$, satisfies $\mathcal{A}z > 0$ in $D = \{x \in \mathbb{R}^N : r/2 < |x - y| < r\}$. Then, set $w_n = u_n + \varepsilon z$, where $\varepsilon > 0$ is such that

$$w_n(x) < u_n(x_n), \qquad x \in \partial B_{r/2}(y). \tag{13.4.4}$$

Since $\mathcal{A}w_n = \mathcal{A}u_n + \varepsilon\mathcal{A}z > \lambda u_n > 0$ in D, the maximum of the function w_n in $\overline{D}$ is attained on ∂D and[5], due to (13.4.4), it is actually attained on $\partial B_r(y)$, where z vanishes. Hence, x_n is a maximum point of the function w_n in $\overline{D}$ as well. As a consequence, the normal derivative of w_n is nonpositive at x_n. Since the normal derivative of z is negative on $\partial\Omega$, it follows that the normal derivative of u_n is positive at x_n. □

Proposition 13.4.2 *For any $f \in C_b(\overline{\Omega})$ and $\lambda > c_0$ (see Hypothesis 13.1.1(iv)), the function $u = R(\lambda, \widehat{A})f$ belongs to $D(\mathcal{A})$ and solves problem (13.0.2). Moreover, $D(\widehat{A}) = D(\mathcal{A})$, $\widehat{A}v = \mathcal{A}v$, for any $v \in D(\mathcal{A})$, and $T(t)$ and $\mathcal{A}$ commute on $D(\mathcal{A})$. Finally, $D(\mathcal{A}) \hookrightarrow \{u \in C^1_b(\overline{\Omega}) : \partial u/\partial\nu \equiv 0$ on $\partial\Omega\}$ and, for any $\omega > c_0$, there exists a positive constant M_ω such that*

$$\|\nabla_x u\|_\infty \le M_\omega \|u\|_\infty^{\frac{1}{2}} \|(\mathcal{A}-\omega)\,u\|_\infty^{\frac{1}{2}}, \qquad u \in D(\mathcal{A}). \tag{13.4.5}$$

Proof Being rather long, we split the proof in some steps.

Step 1. Here, we prove the inclusion $D(\widehat{A}) \subset D(\mathcal{A})$ using an approximation argument, which depends on the geometry of Ω.

The convex case. We use the same notation from Subsection 13.2.2. Fix $f \in C_b(\overline{\Omega})$, $\lambda > c_0$ and set $u = R(\lambda, \widehat{A})f$ and $u_n = R(\lambda, A_n)f$ ($n \in \mathbb{N}$), where A_n is the generator of the semigroup $\{T_n(t)\}$ in $C_b(\overline{\Omega}_n)$. Of course, we can write

$$u_n(x) = \int_0^{+\infty} e^{-\lambda t}(T_n(t)f)(x)dt, \qquad x \in \overline{\Omega}_n. \tag{13.4.6}$$

Therefore, for any $\lambda > c_0$ and $n \in \mathbb{N}$, it holds that $\|u_n\|_\infty \le (\lambda - c_0)^{-1}\|f\|_\infty$ and $\|\mathcal{A}u_n\|_\infty \le (2\lambda - c_0)(\lambda - c_0)^{-1}\|f\|_\infty$. Since $T_n(\cdot)f$ tends to $T(\cdot)f$ locally uniformly in $(0,+\infty) \times \Omega$ (see Subsection 13.2.2), from (13.4.6) and taking (13.2.13) into account, we conclude that u_n converges to u pointwise in $\overline{\Omega}$ and in $L^p(\Omega_k)$, for any $k \in \mathbb{N}$, as n tends to $+\infty$. Furthermore, by the L^p-estimates in Theorem D.1.2, we can infer that $\|u_m - u_n\|_{W^{2,p}(\Omega_k)} \le c(p,k)\|u_m - u_n\|_{L^p(\Omega_{k+1})}$ for any $m, n > k$, $p \in (1,+\infty)$ and $k \in \mathbb{N}$, where $c(p,k) > 0$ is a constant independent of m, n. Therefore, u_n converges to u in $W^{2,p}(\Omega_k)$, for any $k \in \mathbb{N}$ and $p < +\infty$, as n tends to $+\infty$. Hence, u belongs to $W^{2,p}(\Omega \cap B_R)$, for any $R > 0$ and $p < +\infty$. By the Sobolev embedding theorems (see [3, Theorem 5.4]), u_n converges to u also in $C^1(\overline{\Omega}_k)$ for any $k \in \mathbb{N}$, and, therefore, $\partial u/\partial\nu = 0$ on $\partial\Omega$. Finally, since $\lambda u_n - \mathcal{A}u_n = f$ for any $n \in \mathbb{N}$, letting n tend to $+\infty$, it follows that $\lambda u - \mathcal{A}u = f$ in Ω. Therefore, u belongs to $D(\mathcal{A})$ and solves problem (13.0.2). Since $R(\lambda, \widehat{A})$ is surjective from $C_b(\overline{\Omega})$ onto $D(\widehat{A})$, we deduce that $D(\widehat{A}) \subset D(\mathcal{A})$ and $\widehat{A} = \mathcal{A}$ in $D(\widehat{A})$.

The general case. We first consider the case when $f \in C^{2+\alpha}_c(\overline{\Omega})$ is such that $\partial f/\partial\nu = 0$ on $\partial\Omega$. For any $n \in \mathbb{N}$, we introduce the operator $A^{(n)}$ with bounded coefficients belonging to $C^\alpha(\Omega \cap B_R)$ for any $R > 0$, defined by $\mathcal{A}^{(n)}u = \mathrm{Tr}(QD^2u) + \langle b^{(n)}, \nabla u\rangle + c^{(n)}u$ for any $u \in D(A^{(n)})$, where $D(A^{(n)})$ is defined as the right-hand side of (13.4.1), with the operator $\mathcal{A}$ being replaced by $\mathcal{A}^{(n)}$. We assume that $b_i^{(n)} \equiv b_i$, $c^{(n)} \equiv c$ in $\Omega \cap B_n$ and $c^{(n)} \le 0$ in $\overline{\Omega}$. According to Proposition D.3.2 and Theorem D.3.5(v), the operator $A^{(n)}$ generates an analytic semigroup of contractions $\{T_n(t)\}$ in $C_b(\overline{\Omega})$. Arguing as in Step 1 of the proof of Theorem 12.2.4, it turns out that $T_n(t)f$ converges to $T(t)f$ pointwise in $\overline{\Omega}$. Therefore, by dominated convergence, the sequence $R(\lambda, A^{(n)})f = \int_0^{+\infty} e^{-\lambda t}T_n(t)f dt$ converges to $u = R(\lambda, \widehat{A})f$ pointwise in $\overline{\Omega}$ and in $L^p(\Omega')$ for any bounded subset Ω' of Ω and any $p \in [1,+\infty)$. Now arguing as in the case when Ω is convex, we conclude that $u \in D(A)$.

In the general case when $f \in C_b(\overline{\Omega})$ we consider a sequence of functions $\{f_n\} \subset C^{2+\alpha}_c(\overline{\Omega})$,

[5]This is a classical result that can be obtained by arguing as in the proof of Lemma 3.1.2.

bounded in $C_b(\overline{\Omega})$, with null normal derivative on $\partial\Omega$, and converging to f pointwise in $\overline{\Omega}$ (see Lemma 13.2.6). By Proposition 13.3.1, $(T(t)f_n)(x)$ converges to $(T(t)f)(x)$, as n tends to $+\infty$, for any $t > 0$ and $x \in \overline{\Omega}$. Since $\|T(t)f_n\|_\infty \le \sup_{n\in\mathbb{N}} \|f_n\|_\infty < +\infty$, the dominated convergence theorem implies that $R(\lambda, \widehat{A})f_n$ tends to $R(\lambda, \widehat{A})f$ in $L^p(\Omega\cap B_R)$ for any $R > 0$ and $p \in [1, +\infty)$, as n tends to $+\infty$. Now, using the arguments of the convex case, with $u_n = T(t)f_n$, it can be shown that $R(\lambda, \widehat{A})f$ belongs to $D(\mathcal{A})$ and solves the differential equation $\lambda u - \mathcal{A}u = f$. Hence, $D(\widehat{A})$ is contained in $D(\mathcal{A})$ and $\widehat{A} = \mathcal{A}$ on $D(\widehat{A})$.

Exterior domains. In this case, we argue as in the general case, but since the diffusion coefficients need not to be bounded, we approximate them as we did for the drift and potential terms. Hence, we consider the operator $A^{(n)} = \mathrm{Tr}(Q^{(n)}D^2) + \langle b^{(n)}, \nabla\rangle + c^{(n)}$ where the coefficients $q_{ij}^{(n)}$ are bounded and smooth and $q_{ij}^{(n)} = q_{ij}$ in $\Omega\cap B_n$ for any $i, j = 1, \ldots, N$ and $n \in \mathbb{N}$.

Step 2. Here, we prove the inclusion $D(\mathcal{A}) \subset D(\widehat{A})$ and that the semigroup and the operator $\mathcal{A}$ commute on $D(\mathcal{A})$. We fix $u \in D(\mathcal{A})$ and set $f = \lambda u - \mathcal{A}u \in C_b(\overline{\Omega})$, where $\lambda = 2(\lambda_0 \vee c_0)$, and c_0, λ_0 are given, respectively, by Hypotheses 13.1.1(ii) and 13.1.1(iv). By the above results, the function $v = R(\lambda, \widehat{A})f$ is a bounded solution of problem (13.0.2) in $D(\mathcal{A})$, as well. Proposition 13.4.1 yields that $u \equiv v$, implying, in particular, that $u \in D(\widehat{A})$. Summing up, we conclude that $D(\mathcal{A}) = D(\widehat{A})$ and $\widehat{A} = \mathcal{A}$ on $D(\mathcal{A})$. Moreover, using Proposition 13.3.1 and (13.4.1), we deduce that $T(t)$ and $\mathcal{A}$ commute on $D(\mathcal{A})$, for any $t > 0$. We refer the reader to the proof of Lemma 1.3.2 for further details.

Step 3. Finally, we prove (13.4.5). We fix $u \in D(\mathcal{A})$, $\omega > c_0$, $\lambda > 0$ and set

$$u(x) = (R(\lambda+\omega, \widehat{A})f)(x) = \int_0^{+\infty} e^{-(\lambda+\omega)\,t}(T(t)f)(x)\,dt, \qquad x \in \overline{\Omega},$$

where $f = (\lambda+\omega)\,u - \mathcal{A}u$. By virtue of estimate (13.1.6), we may differentiate under the integral sign in the above formula and estimate

$$|\nabla u(x)| \le \frac{C_\omega}{\sqrt{\lambda}}\|f\|_\infty \le C_\omega\left(\sqrt{\lambda}\|u\|_\infty + \frac{1}{\sqrt{\lambda}}\|(\mathcal{A}-\omega)u\|_\infty\right), \qquad x \in \Omega,$$

for some positive constant C_ω. Minimizing over $\lambda > 0$, we get (13.4.5). □

13.5 Pointwise gradient estimates and their consequences

In this section we prove some pointwise gradient estimates for the function $T(t)f$. We start by proving the following estimate

$$|(\nabla_x T(t)f)(x)|^p \le C_p\, e^{M_p t}(T(t)|\nabla f|^p)(x), \qquad t > 0, \quad x \in \Omega, \tag{13.5.1}$$

in the case when the diffusion coefficients are bounded. As is immediately seen, $T(\cdot)\mathbf{1} \equiv \mathbf{1}$ is a necessary condition for (13.5.1) to hold; hence, we assume that $c \equiv 0$.

Theorem 13.5.1 *Let one among Hypotheses* 13.2.1, 13.2.2, *or* 13.2.3 *be satisfied, with* $c \equiv 0$. *Further, assume that*

$$\sum_{i,j=1}^N (\langle \nabla q_{ij}(x), \xi\rangle)^2 \le q_0\kappa(x)|\xi|^2, \qquad x \in \Omega, \ \xi \in \mathbb{R}^N, \tag{13.5.2}$$

for some positive constant[6] q_0. *Then, for any* $p \in (1, +\infty)$, *there exists a constant* $M_p \in \mathbb{R}$ *(see the proof) such that estimate* (13.5.1) *holds, with* $C_p = (1 + \omega_0\varepsilon_0/2)^{p/2}$, *for any*[7] $f \in C_b^1(\overline{\Omega})$ *with null normal derivative on* $\partial\Omega$. *In particular, if* $\omega_0 = 0$, *then we can take* $C_p = 1$.

Proof The proof is similar to and even simpler than those of Theorems 6.1.2 and 6.3.1. Hence, we just sketch it in the case when $p \in (1, 2]$. Moreover, we prove (13.5.1) for functions $f \in C_c^{2+\alpha}(\overline{\Omega})$ with null normal derivative on $\partial\Omega$, since the general case then will follow by density by virtue of Lemma 13.2.6 and Proposition 13.3.1.

Fix $\delta > 0$ and introduce the function $v = (m|\nabla_x u|^2 + \delta)^{p/2}$, where $u = T(\cdot)f$ and m is the function in (13.1.7). According to Remark 13.2.8, v is differentiable once with respect to the time variable and twice with respect to the spatial variables in $(0, +\infty) \times \Omega$ and it is bounded and continuous in $[0, +\infty) \times \overline{\Omega}$. Moreover, $D_t v - \mathcal{A}v = f_1 + f_2 + f_3$, where

$$f_1 = pv^{1-\frac{2}{p}} m\bigg(\sum_{i,j,h=1}^N D_h q_{ij} D_h u D_{ij} u + \langle (\mathrm{Jac}\, b)\nabla_x u, \nabla_x u\rangle - \mathrm{Tr}(QD_x^2 u D_x^2 u)\bigg),$$

$$f_2 = pv^{1-\frac{2}{p}}\bigg(-\frac{1}{2}|\nabla_x u|^2 \mathcal{A}m - 2\langle QD_x^2 u \nabla_x u, \nabla m\rangle\bigg),$$

$$f_3 = p(2-p)v^{1-\frac{4}{p}}\bigg|\sqrt{Q}\bigg(\frac{1}{2}|\nabla_x u|^2 \nabla m + m(D_x^2 u \nabla_x u)\bigg)\bigg|^2.$$

Using Hypothesis 13.1.1(i) and conditions (13.1.5), (13.5.2), we can estimate

$$f_1 \le pv^{1-\frac{2}{p}} m\bigg\{\bigg(\frac{q_0}{4\varepsilon} + r_0\bigg)|\nabla_x u|^2 + \varepsilon\kappa|D_x^2 u|^2 - \mathrm{Tr}(QD_x^2 u D_x^2 u)\bigg\}.$$

The function f_2 identically vanishes if $\omega_0 = 0$ (see (13.1.3)). On the other hand, if $\omega_0 > 0$, then, since $\mathcal{A}m = \omega_0(\psi' \circ d)\mathcal{A}d + \omega_0(\psi'' \circ d)|\sqrt{Q}\nabla d|^2$, from (13.1.15), (13.1.16) and (13.1.17), with ε instead of δ, we get

$$f_2 \le p\omega_0 v^{1-\frac{2}{p}}\bigg\{\bigg[q_1\bigg(\frac{1}{\varepsilon_0} + \frac{1}{2}\|D^2 d\|_\infty + \frac{1}{2\varepsilon}\bigg) + \frac{r_1}{2}\bigg]|\nabla_x u|^2 + 2\varepsilon q_1 |D_x^2 u|^2\bigg\} \tag{13.5.3}$$

for any $\varepsilon > 0$, where $q_1 = \|Q\|_{C_b(\Gamma(\varepsilon_0))}$. We now observe that

$$\begin{aligned}
|\sqrt{Q}D_x^2 u \nabla_x u|^2 &= \sum_{h,k=1}^N D_h u D_k u \langle Q\nabla_x D_h u, \nabla_x D_k u\rangle \\
&\le \bigg(\sum_{h=1}^N |D_h u||\sqrt{Q}\nabla_x D_h u|\bigg)^2 \\
&\le |\nabla_x u|^2 \sum_{h=1}^N |\sqrt{Q}\nabla_x D_h u|^2 = |\nabla_x u|^2 \mathrm{Tr}(QD_x^2 u D_x^2 u).
\end{aligned}$$

Recalling that $(\alpha + \beta)^2 \le (1 + \varepsilon_1)\alpha^2 + (1 + \varepsilon_1^{-1})\beta^2$ for any $\alpha, \beta, \varepsilon_1 > 0$ and using again the Cauchy-Schwarz inequality, we can estimate

$$\bigg|\sqrt{Q}\bigg(\frac{1}{2}|\nabla_x u|^2 \nabla m + m(D_x^2 u \nabla_x u)\bigg)\bigg|^2$$

[6] This latter condition replaces (13.1.4) in the convex case, and it is trivially satisfied in the general case.
[7] The constants ω_0 and ε_0 are defined in (13.1.3) and in Hypothesis 13.1.1(v).

$$\le\left[\frac{1}{2}|\sqrt{Q}\nabla m||\nabla_x u|^2+m\Big(\mathrm{Tr}(QD_x^2uD_x^2u)\Big)^{\frac{1}{2}}|\nabla_x u|\right]^2$$
$$\le m|\nabla_x u|^2\left[\frac{1}{2}|\sqrt{Q}\nabla m||\nabla_x u|+(m\mathrm{Tr}(QD_x^2uD_x^2u))^{\frac{1}{2}}\right]^2$$
$$\le v^{\frac{2}{p}}\left(m(1+\varepsilon_1)\mathrm{Tr}(QD_x^2uD_x^2u)+m\frac{1+\varepsilon_1}{4\varepsilon_1}|\nabla_x u|^2|\sqrt{Q}\nabla m|^2\right) \tag{13.5.4}$$

and conclude that

$$f_3\le p(2-p)v^{1-\frac{2}{p}}\left(m(1+\varepsilon_1)\mathrm{Tr}(QD_x^2uD_x^2u)+m\omega_0^2q_1\frac{1+\varepsilon_1}{4\varepsilon_1}|\nabla_x u|^2\right).$$

Note that in the case $\omega_0=0$, we can estimate $f_3\le p(2-p)v^{1-2/p}\mathrm{Tr}(QD_x^2uD_x^2u)$.

To fix the ideas, we complete the proof in the case $\omega_0\neq 0$. Choosing $\varepsilon_1>0$ small enough to have $q_2:=1-(1+\varepsilon_1)(2-p)>0$ and $\varepsilon<q_2$, from the previous estimate and (13.1.14) we get

$$D_tv-\mathcal{A}v\le pmv^{1-\frac{2}{p}}(M_p(\varepsilon)|\nabla_x u|^2+C(\varepsilon)m\kappa|D_x^2u|^2),$$

where $C(\varepsilon)=\varepsilon+2\varepsilon q_1\kappa_0^{-1}-q_2$ and

$$M_p(\varepsilon)=\left(1+\frac{1}{2}\omega_0\varepsilon_0\right)\left(\frac{q_0}{4\varepsilon}+r_0\right)+q_1\omega_0\left(\frac{1}{\varepsilon_0}+\frac{1}{2}\|D^2d\|_\infty+\frac{1}{2\varepsilon}\right)+\frac{r_1}{2}\omega_0+\frac{1+\varepsilon_1}{4\varepsilon_1}\omega_0^2q_1.$$

Let us now choose $\widetilde{\varepsilon}>0$ such that $C(\widetilde{\varepsilon})\le 0$ and set $M_p=M_p(\widetilde{\varepsilon})$ Then, we get $D_tv-\mathcal{A}v\le M_pv-(M_p\wedge 0)\delta^{\frac{p}{2}}$. Arguing as in the proof of Theorem 6.1.5, taking Theorem 13.1.2 into account, we deduce that $v(t,x)\le e^{M_pt}(T(t)(m|\nabla f|^2+\delta)^{p/2})(x)$ for any $t>0$ and $x\in\overline{\Omega}$, which leads to (13.5.1) letting δ tend to 0^+ and taking Proposition 13.3.1 into account. □

The assumption (13.5.2) is rather restrictive, since it does not allow us to consider diffusion coefficients of general polynomial type, when Ω is convex or an exterior domain. As the following proposition shows, this problem may be overcome assuming a dissipativity condition of negative type on the drift term, in order to balance the growth at infinity of the derivatives of the q_{ij}'s $(i,j=1,\dots,N)$.

Theorem 13.5.2 *Let either one between Hypotheses* 13.2.2 *and* 13.2.3 *be satisfied with* $c\equiv 0$ *and let condition* (13.1.5) *be satisfied with* r_0 *being replaced by a negative function* $r:\Omega\to\mathbb{R}$. *Moreover, assume that* $|\nabla q_{ij}|\le C\kappa^\alpha$ *in* Ω, *for any* $i,j=1,\dots,N$ *and some constants* $C>0$ *and* $\alpha\in(0,1]$. *Finally, fix* $p\in(1,2)$. *Then, the pointwise estimate* (13.5.1) *holds true if*

(i) $\widetilde{\sigma}_p:=\sup_\Omega\left(\frac{C^2N^2}{4(p-1)}\kappa_0^{\alpha-1}\kappa+r\right)<+\infty$ *in the convex case;*

(ii) $\overline{\sigma}_p:=\sup_\Omega\left(\frac{CN^2}{4p_0\kappa_0}(C\kappa_0^\alpha+2\omega_0q_1)\kappa^\alpha+r\right)<+\infty$ *for some* $0<p_0<p-1$, *where* $q_1=\|Q\|_{C_b(\Gamma(\varepsilon_0))}$ *(see Hypothesis* 13.1.1(*v*)*).*

Moreover, we can take $M_p=p\widetilde{\sigma}_p$ *if* Ω *is convex whereas* M_p *is given by* (13.5.8) *and* (13.5.9) *if* Ω *is an exterior domain.*

If Ω *is an exterior domain and* $\widehat{\sigma}=\sup_\Omega\left(\frac{CN^2}{4\kappa_0}(C\kappa_0^\alpha+2\omega_0q_1)\kappa^\alpha+r\right)<+\infty$, *then estimate* (13.5.1) *holds true for any* $p\ge 2$. *In this case,* M_p *is given by* (13.5.10) *and* (13.5.11). *Finally, if* Ω *is convex and* $\widetilde{\sigma}_2<+\infty$ *then estimate* (13.5.1), *with* $p\ge 2$, *holds true with* $M_p=p\widetilde{\sigma}_2$.

Proof It is similar to that of Theorem 13.5.1. The only difference is in the estimate of term $\sum_{i,j,h=1}^N D_h q_{ij} D_i u D_{jh} u$ which now should be estimated as follows:

$$\left|\sum_{i,j,h=1}^N D_h q_{ij} D_h u D_{ij} u\right| \le C\kappa^\alpha \sum_{i,j=1}^N |D_{ij} u| \le C\kappa^\alpha \left(\frac{N^2}{4\varepsilon}|\nabla_x u|^2 + \varepsilon|D_x^2 u|^2\right) \tag{13.5.5}$$

for any $\varepsilon > 0$. Suppose that $p < 2$. Then, for any $\varepsilon, \varepsilon_1 > 0$ we get

$$D_t v - \mathcal{A}v \le p v^{1-\frac{2}{p}} \left[\left(\frac{CN^2}{4\varepsilon}\kappa^\alpha + r\right)|\nabla_x u|^2 + (1-p)\mathrm{Tr}(QD_x^2 u D_x^2 u) + C\varepsilon\kappa^\alpha |D_x^2 u|^2\right], \tag{13.5.6}$$

when Ω is a convex set, and

$$\begin{aligned} D_t v - \mathcal{A}v \le & p v^{1-\frac{2}{p}} \left[m\left(\frac{CN^2}{4\varepsilon}\kappa^\alpha + r\right) + \omega_0 q_1\left(\frac{1}{\varepsilon_0} + \frac{1}{2}\|D^2 d\|_\infty + \frac{1}{2\varepsilon}\right) + \frac{\omega_0 r_1}{2}\right. \\ & \left. + (2-p)m\omega_0^2 q_1 \frac{1+\varepsilon_1}{4\varepsilon_1}\right]|\nabla_x u|^2 \\ & + p v^{1-\frac{2}{p}}\{m[1-p+(2-p)\varepsilon_1]\mathrm{Tr}(QD_x^2 u D_x^2 u) + \varepsilon(Cm\kappa^\alpha + 2\omega_0 q_1)|D_x^2 u|^2\}, \end{aligned} \tag{13.5.7}$$

when Ω is an exterior domain. In the first case, we take $\varepsilon = (p-1)\kappa_0^{1-\alpha}C^{-1}$ to make negative the sum of the two last terms in the right-hand side of (13.5.6), and (13.5.1) follows.

In the second case, we take $\varepsilon_1 = (p-1-p_0)(2-p)^{-1}$. This allows us to estimate the last line of (13.5.7) from above by $[(C\kappa_0^\alpha + 2\omega_0 q_1)\kappa_0^{-1}\varepsilon - p_0]m\kappa|D_x^2 u|^2$. Hence, the assertion follows choosing $\varepsilon > 0$ such that the coefficient in front of $|D_x^2 u|$ is negative. In particular, we can take

$$M_p = p\left[\overline{\sigma}_p + \omega_0 q_1\left(\frac{1}{\varepsilon_0} + \frac{1}{2}\|D^2 d\|_\infty + \frac{1}{2\varepsilon}\right) + \frac{\omega_0 r_1}{2} + (2-p)\left(1 + \frac{\omega_0\varepsilon_0}{2}\right)\omega_0^2 q_1 \frac{1+\varepsilon_1}{4\varepsilon_1}\right], \tag{13.5.8}$$

if $\overline{\sigma}_p \le 0$, and

$$M_p = p\left[\left(1 + \frac{\omega_0\varepsilon_0}{2}\right)\left(\overline{\sigma}_p + (2-p)\omega_0^2 q_1 \frac{1+\varepsilon_1}{4\varepsilon_1}\right) + \omega_0 q_1\left(\frac{1}{\varepsilon_0} + \frac{1}{2}\|D^2 d\|_\infty + \frac{1}{2\varepsilon}\right) + \frac{\omega_0 r_1}{2}\right], \tag{13.5.9}$$

otherwise

Finally, the case $p \ge 2$ is completely similar and even simpler since we can disregard the nonpositive function f_3. In particular, we can take

$$M_p = p\left[\widehat{\sigma} + \omega_0 q_1\left(\frac{1}{\varepsilon_0} + \frac{1}{2}\|D^2 d\|_\infty + \frac{1}{2\varepsilon}\right) + \frac{\omega_0 r_1}{2}\right], \tag{13.5.10}$$

if $\widehat{\sigma} \le 0$, and

$$M_p = p\left[\left(1 + \frac{\omega_0\varepsilon_0}{2}\right)\widehat{\sigma} + \omega_0 q_1\left(\frac{1}{\varepsilon_0} + \frac{1}{2}\|D^2 d\|_\infty + \frac{1}{2\varepsilon}\right) + \frac{\omega_0 r_1}{2}\right], \tag{13.5.11}$$

otherwise. □

Under the same assumptions used to prove the uniform gradient estimates, we can prove the following weaker pointwise gradient estimate.

Theorem 13.5.3 *Let either one between Hypotheses 13.2.2, 13.2.3 be satisfied with $c \equiv 0$. Then, for any $p \in (1,+\infty)$ there exists a constant $C_p > 0$ such that*

$$|(\nabla_x T(t)f)(x)|^p \le C_p\big(T(t)(f^2+|\nabla f|^2)^{\frac{p}{2}}\big)(x), \qquad t>0,\ x\in\overline{\Omega}, \tag{13.5.12}$$

for any $f \in C^1_b(\overline{\Omega})$ such that $\partial f/\partial\nu = 0$ on $\partial\Omega$.

Proof The proof is similar to that of Theorem 13.5.1. Hence, we limit ourselves to sketching it, pointing out the main differences. In the case when $p \in (1,2]$ and $f \in C^{2+\alpha}_c(\overline{\Omega})$, we introduce, for any $\delta > 0$, the function $v = (au^2 + m|\nabla_x u|^2 + \delta)^{p/2}$, where $a > 0$ is a real parameter to be fixed later on. By virtue of Remark 13.2.8, the function v is differentiable once with respect to the time variable and twice with respect to the spatial variables in $(0,+\infty)\times\Omega$, and it is continuous in $[0,+\infty)\times\overline{\Omega}$. Moreover, it satisfies the equation $D_t v - \mathcal{A}v = f_1 + f_2 + f_3$, where

$$\begin{aligned}
f_1 =& pv^{1-\frac{2}{p}}\bigg(m\sum_{i,j,h=1}^N D_h q_{ij} D_h u D_{ij}u + m\langle(\operatorname{Jac} b)\nabla_x u, \nabla_x u\rangle \\
&\qquad\qquad - a|\sqrt{Q}\nabla_x u|^2 - m\operatorname{Tr}(QD^2_x u D^2_x u)\bigg),\\
f_2 =& -p\,v^{1-\frac{2}{p}}\bigg(\frac{1}{2}|\nabla_x u|^2\mathcal{A}m + 2\langle Q\nabla m, D^2_x u\nabla_x u\rangle\bigg),\\
f_3 =& \frac{p}{2}\Big(1-\frac{p}{2}\Big)v^{1-\frac{4}{p}}|\sqrt{Q}\nabla_x(au^2+m|\nabla_x u|^2)|^2.
\end{aligned}$$

Taking (13.5.5) into account, we can estimate

$$f_1 \le pv^{1-\frac{2}{p}}\bigg[\frac{CN^2}{4\varepsilon}\kappa m|\nabla_x u|^2 - a|\sqrt{Q}\nabla_x u|^2 + C\varepsilon\kappa m|D^2_x u|^2 - m\operatorname{Tr}(QD^2_x u D^2_x u)\bigg].$$

On the other hand, the function f_2 can be estimated by (13.5.3). Finally, to estimate f_3 we observe that, arguing as in the proof of (13.5.4), we get

$$\begin{aligned}
&|\sqrt{Q}\nabla_x(au^2+m|\nabla_x u|^2)|^2\\
\le&\Big[|\sqrt{Q}\nabla m||\nabla_x u| + 2m(\operatorname{Tr}(QD^2_x u D^2_x u))^{\frac{1}{2}}|\nabla_x u| + 2a|u||\sqrt{Q}\nabla_x u|\Big]^2\\
\le& v^{\frac{2}{p}}\Big\{m\big[|\sqrt{Q}\nabla m||\nabla_x u| + 2(\operatorname{Tr}(QD^2_x u D^2_x u))^{\frac{1}{2}}\big]^2 + 4a|\sqrt{Q}\nabla_x u|^2\Big\}\\
\le& v^{\frac{2}{p}}\bigg(4m(1+\varepsilon_1)\operatorname{Tr}(QD^2_x uD^2_x u) + m\frac{1+\varepsilon_1}{\varepsilon_1}|\nabla_x u|^2|\sqrt{Q}\nabla m|^2 + 4a|\sqrt{Q}\nabla_x u|^2\bigg)
\end{aligned}$$

for any $\varepsilon_1 > 0$. If we take $\varepsilon_1 > 0$ such that $1-(1+\varepsilon_1)(2-p) > 0$, then we can estimate $D_t v - \mathcal{A}v \le pv^{1-\frac{2}{p}}[C_{1,p}(a,\varepsilon)|\nabla_x u|^2 + C_{2,p}(\varepsilon)m\kappa|D^2_x u|^2]$ for any $\varepsilon > 0$, where

$$\begin{aligned}
C_{1,p}(a,\varepsilon) =& \Big(1+\frac{\omega_0\varepsilon_0}{2}\Big)r_0 + \omega_0 q_1\Big(\frac{1}{\varepsilon_0} + \frac{1}{2}\|D^2 d\|_\infty + \frac{1}{2\varepsilon}\Big) + \frac{r_1\omega_0}{2}\\
&+ (2-p)\frac{1+\varepsilon_1}{4\varepsilon_1}\Big(1+\frac{\omega_0\varepsilon_0}{2}\Big) + \bigg[\Big(1+\frac{\omega_0\varepsilon_0}{2}\Big)\frac{CN^2}{4\varepsilon} + (1-p)a\bigg]\kappa,\\
C_{2,p}(\varepsilon) =& C\varepsilon + 2\varepsilon\omega_0 q_1\kappa_0^{-1} - 1 + (1+\varepsilon_1)(2-p)
\end{aligned}$$

and $q_1 = \|Q\|_{C_b(\Gamma(\varepsilon_0))}$. Now, we fix, first, $\varepsilon > 0$ such that $C_{2,p}(\varepsilon) = 0$ and, then, a sufficiently large that $C_{1,p}(a, \varepsilon) \leq 0$. We conclude that $D_t v - \mathcal{A}v \leq 0$. Therefore, using the same technique as in the proof of Theorem 13.5.1, we get (13.5.12) in the case when $p \in (1, 2]$. The other cases follow, just as in the proof of the quoted theorem. □

Now, we prove the second type of pointwise estimates.

Theorem 13.5.4 *Assume that the conditions of Theorem* 13.5.1 (*resp. Theorem* 13.5.2) *are satisfied. Then, for any $f \in C_b(\overline{\Omega})$ it holds that*

$$|(\nabla_x T(t)f)(x)|^p \leq K_{p,t}\, t^{-\frac{p}{2}} (T(t)|f|^p)(x), \qquad t > 0, \ x \in \overline{\Omega}, \tag{13.5.13}$$

where

$$K_{p,t} = \begin{cases} \dfrac{(2 + \omega_0\varepsilon_0)^{\frac{p}{2}}}{[2p(p-1)\kappa_0]^{\frac{p}{2}}} \dfrac{M_p t}{1 - e^{-M_p t}}, & \text{if } p \in (1, 2], \\ K_{2,t}^{\frac{p}{2}}, & \text{if } p > 2, \end{cases}$$

and M_p is defined in (*the proof of*) *Theorem* 13.5.1 (*resp. Theorem* 13.5.2). *In the case when $M_p = 0$, the term $M_p t(1 - e^{-M_p t})^{-1}$ should be replaced with* 1.

Proof To begin with, we observe that the case $p > 2$ follows from the case $p = 2$ using the Hölder inequality. Indeed, since $p(t, x; dy)$ are probability measures for any $t > 0$, $x \in \overline{\Omega}$, it follows that $(T(t)|f|^2)^{p/2} \leq T(t)(|f|^p)$. Writing $|\nabla_x T(t)f|^p = (|\nabla_x T(t)f|^2)^{p/2}$ and using estimate (13.5.13) with $p = 2$, we get it in the general case. Therefore, we confine ourselves to the case $p \in (1, 2]$ and, in view of Proposition 13.3.1, we limit ourselves to proving (13.5.13) for functions $f \in C_c^{2+\alpha}(\overline{\Omega})$ such that $\partial f/\partial\nu = 0$ on $\partial\Omega$.

The general case. This case is easier to deal with since the diffusion coefficients are bounded. For any $\delta > 0$, we introduce the function $\Phi : (0, t) \to C_b(\overline{\Omega})$, defined by $\Phi(s, \cdot) = T(s)(|T(t-s)f|^2 + \delta)^{p/2} =: T(s)g(s, \cdot)$ for any $s \in (0, \delta)$. We first prove that $g(s, \cdot)$ belongs to $D(\mathcal{A})$ for any $s \in (0, t)$, where $D(\mathcal{A})$ is defined in (13.4.2). This is not immediate, since the operator $\mathcal{A}$ has unbounded coefficients. By Proposition 13.3.4, for any $s \in (0, t)$ the function $g(s, \cdot)$ belongs to $C_b^1(\overline{\Omega}) \cap C^2(\Omega)$ and its normal derivative identically vanishes on $\partial\Omega$. To conclude that $g(s, \cdot) \in D(\mathcal{A})$, we just need to show that $\mathcal{A}g(s, \cdot) \in C_b(\overline{\Omega})$. For this purpose, we observe that

$$\begin{aligned} \mathcal{A}g(s, \cdot) =& p(g(s, \cdot))^{1-\frac{2}{p}} \left\{T(t-s)f\, \mathcal{A}T(t-s)f + |\sqrt{Q}\nabla_x T(t-s)f|^2\right\} \\ &+ p(p-2)(g(s, \cdot))^{1-\frac{4}{p}} |T(t-s)f|^2 |\sqrt{Q}\nabla_x T(t-s)f|^2. \end{aligned}$$

Since $f \in D(\widehat{A})$, by Proposition 13.4.2, $\mathcal{A}T(t-s)f = T(t-s)\mathcal{A}f$ and, therefore, the boundedness of the diffusion coefficients shows that $\mathcal{A}g(s, \cdot) \in C_b(\overline{\Omega})$. This implies that $\mathcal{A}\Phi(s, \cdot) = T(s)\mathcal{A}g(s, \cdot)$ for any $s \in (0, t)$.

Taking the positivity of the semigroup $\{T(t)\}$ into account, we now get

$$\begin{aligned} \Phi'(s, \cdot) =& T(s)[\mathcal{A}g(s, \cdot) - p(g(s, \cdot))^{1-\frac{2}{p}} T(t-s)f\, \mathcal{A}T(t-s)f] \\ =& T(s)[p(g(s, \cdot))^{1-\frac{4}{p}}((p-1)|T(t-s)f|^2 + \delta)|\sqrt{Q}\nabla_x T(t-s)f|^2] \\ \geq& p(p-1)\kappa_0 T(s)[(g(s, \cdot))^{1-\frac{2}{p}} |\nabla_x T(t-s)f|^2]. \end{aligned} \tag{13.5.14}$$

Fix $x \in \overline{\Omega}$ and $\varepsilon \in (0, t/2)$. Integrating the first and the last sides of (13.5.14) with respect to s from ε to $t - \varepsilon$ yields

$$\big(T(t-\varepsilon)(|T(\varepsilon)f|^2 + \delta)^{\frac{p}{2}}\big)(x)$$

$$\geq p(p-1)\kappa_0 \int_{\varepsilon}^{t-\varepsilon} \left(T(s)[(g(s,\cdot))^{1-\frac{2}{p}}|\nabla_x T(t-s)f|^2]\right)(x)ds.$$

Taking Proposition 13.3.1 into account, we can let ε tend to 0^+ and using the monotone convergence theorem we conclude that

$$(T(t)(|f|^2+\delta)^{\frac{p}{2}})(x) \geq p(p-1)\kappa_0 \int_0^t (T(s)((g(s,\cdot))^{1-\frac{2}{p}}|\nabla_x T(t-s)f|^2))(x)ds. \quad (13.5.15)$$

Now, using estimate (13.5.1) and the Hölder inequality, we can write

$$\begin{aligned}|\nabla_x T(t)f|^p \leq & C_p e^{M_p s} T_n(s)|\nabla_x T_n(t-s)f|^p \\ = & C_p e^{M_p s} T(s)[(g(s,\cdot))^{\frac{p}{2}-1}|\nabla_x T(t-s)f|^p\,(g(s,\cdot))^{1-\frac{p}{2}}] \\ \leq & C_p e^{M_p s}(T(s)[(g(s,\cdot))^{1-\frac{2}{p}}|\nabla_x T(t-s)f|^2])^{\frac{p}{2}}[T(s)g_n(s,\cdot)]^{1-\frac{p}{2}}.\end{aligned}$$

Then, using the Young and Jensen inequalities and recalling that $(a+b)^{p/2} \leq a^{p/2}+b^{p/2}$ for any $a, b > 0$, we get

$$\begin{aligned}|\nabla_x T(t)f|^p \leq C_p e^{M_p s}\bigg\{ & \frac{p}{2}\varepsilon^{\frac{2}{p}} T(s)\Big((g(s,\cdot))^{1-\frac{2}{p}}|\nabla_x T(t-s)f|^2\Big) \\ & + \frac{2-p}{2}\varepsilon^{\frac{2}{p-2}}\left(T(t)|f|^p+\delta^{\frac{p}{2}}\right)\bigg\}.\end{aligned}$$

Integrating from 0 to t and taking (13.5.15) into account, gives

$$\frac{1-e^{-M_p t}}{M_p}|\nabla_x T(t)f|^p \leq \frac{\varepsilon^{\frac{2}{p}}}{2(p-1)\kappa_0} C_p T(t)(|f|^2+\delta)^{\frac{p}{2}} + t\left(1-\frac{p}{2}\right)\varepsilon^{\frac{2}{p-2}} C_p\left(T(t)|f|^p+\delta^{\frac{p}{2}}\right), \quad (13.5.16)$$

where, if $M_p = 0$, then we replace the term $(1-e^{-M_p t})/M_p$ with t. Letting δ tend to 0 and, then, minimizing with respect to $\varepsilon \in (0, +\infty)$, we obtain (13.5.13).

Convex case. Since the diffusion coefficients could be unbounded, we can not guarantee that the function $g(s, \cdot)$ belongs to $D(\mathcal{A})$ for any $s \in (0, t)$. To overcome this difficulty, we replace the semigroup $\{T(t)\}$ with the family of semigroups $\{T_n(t)\}$ defined in Subsection 13.2.2 and, coherently, we replace the function g_δ with the function $g_{\delta,n} = (|T_n(t-\cdot)f|^2 + \delta)^{p/2}$. Since $g_{\delta,n}(s, \cdot)$ belongs $D(A_n)$ for[8] any $s \in (0, t)$, from the arguments of the general case we deduce estimate (13.5.16) with $\{T(t)\}$ being replaced by the semigroup $\{T_n(t)\}$. Note that this latter semigroup satisfies estimate (13.5.1) as it can be easily seen going through the proofs of Theorems 13.5.1 and 13.5.2. Finally, letting n tend to $+\infty$ we complete the proof.

Exterior domains. It suffices to use an approximation argument as in the case when Ω is convex, replacing the semigroup $\{T_n(t)\}$ therein introduced with the semigroup associated with the operator $\mathcal{A}^{(n)}$ defined before Hypothesis 13.2.3. This yields (13.5.13) with $T_n(t)$ instead of $T(t)$. Since $T_n(\cdot)f$ converges to $T(\cdot)f$ in $C^{1,2}(F)$ for any compact set $F \subset (0, +\infty) \times \overline{\Omega}$ (see Subsection 13.2.3), (13.5.13) follows. □

Remark 13.5.5 In the case when $p \in (1, 2)$ and $M_p > 0$ the function $t \mapsto K_{p,t}t^{-p/2}$ in

[8]We recall that $D(A_n)$ is the set all the functions $u \in \bigcap_{1\leq p<+\infty} W^{2,p}(\Omega_n)$ such that $\mathcal{A}u \in C(\overline{\Omega}_n)$ and the normal derivative of u vanishes on $\partial\Omega_n$.

(13.5.13) blows up as t tends to $+\infty$. This result can be refined when the potential c is nonpositive in Ω. Indeed, splitting $\nabla_x T(t)f = \nabla_x T(1)T(t-1)f$ for $t > 1$, we can estimate

$$|(\nabla_x T(t)f)(x)|^p = |(\nabla_x T(1)T(t-1)f)(x)|^p \le K_{p,1}(T(1)|T(t-1)f|^p)(x) \le (T(t)|f|^p)(x)$$

for any $x \in \overline{\Omega}$, where we have taken advantage of Theorem 13.2.5 and Proposition 13.3.1 to infer that $|T(t-1)f|^p \le T(t-1)|f|^p$, since $p(t-1,x;\mathbb{R}^N) \le e^{c_0(t-1)} \le 1$ for any $x \in \mathbb{R}^N$.

Corollary 13.5.6 *For any $f \in C_b(\overline{\Omega})$ and $t > 0$ it holds that*

$$\|\nabla_x T(t)f\|_\infty \le \|f\|_\infty \begin{cases} \dfrac{\sqrt{(2+\omega_0\varepsilon_0)M_2}}{2\sqrt{\kappa_0}} \dfrac{1}{\sqrt{1-e^{-M_2 t}}}, & M_2 \neq 0, \\ \dfrac{\sqrt{2+\omega_0\varepsilon_0}}{2\sqrt{\kappa_0 t}}, & M_2 = 0. \end{cases}$$

Corollary 13.5.6 gives uniform gradient estimates sharper than the ones in (13.0.5) since we can take $\omega = 0$. Moreover, under the assumptions of Theorem 13.5.4 it gives an explicit description of the constant C in terms of q_0 and k_0 (see (13.1.5) and (13.5.2)). Finally, in the case when $M_2 < 0$, the above estimate shows that the sup-norm of the gradient of $T(t)f$ vanishes exponentially as t tends to $+\infty$.

Another interesting consequence of the gradient estimate in Theorem 13.5.4 is shown in the following proposition which provides us a Liouville type theorem. We omit the proof since it can be obtained arguing as in the proof of Theorem 6.2.5.

Proposition 13.5.7 *Suppose that Theorem* 13.5.4 *holds with $M_2 \le 0$. If $f \in D(\mathcal{A})$ is such that $\mathcal{A}f \equiv 0$, then f is constant.*

13.6 The invariant measure of the semigroup

In this section we briefly generalize some results of Chapter 9. To simplify the notation, we simply write L^p_μ and $W^{k,p}_\mu$ ($k \in \mathbb{N}$, $p \in [1,+\infty)$) instead of $L^p(\Omega,\mu)$ and $W^{k,p}(\Omega,\mu)$. Moreover, we denote by $\|\cdot\|_p$ the usual norm of L^p_μ.

In view of the extension of the gradient estimates in Section 13.5 to this setting, we assume that c identically vanishes in $\overline{\Omega}$.

To begin with let us prove the following theorem.

Theorem 13.6.1 *Under the assumptions of Theorem* 13.2.5 *there exists at most one invariant measure of $\{T(t)\}$. Such an invariant measure is absolutely continuous with respect to the Lebesgue measure and its density ϱ is a positive function which belongs to $W^{1,p}_{\rm loc}(\Omega)$ for any $p \in (1,+\infty)$. In particular, ϱ is continuous in Ω (not necessarily bounded).*

Proof We split the proof in three steps.

Step 1. Here, we assume that μ is an invariant measure of $\{T(t)\}$. Repeating with slight changes the same arguments as in the proof of Theorem 9.2.3, we can show that μ admits a density ϱ with respect to the Lebesgue measure which belongs to $W^{1,p}_{\rm loc}(\Omega)$ for any $p \in (1,+\infty)$ (see also [40, Section 2] for more details). Moreover, the Harnack inequality (see [251, Corollary 5.3]), yields $\sup_{x\in K}\varrho(x) \le C\inf_{x\in K}\varrho(x)$ for any compact set K and some positive constant $C = C(K)$. Of course, this estimate and the continuity of ϱ in Ω imply that ϱ is everywhere positive.

To conclude this step of the proof, let us observe that $\varrho \in L^1(\mathbb{R}^N)$. Indeed, let $\{\Omega_n\}$ be a sequence of bounded sets such that $\Omega_n \subset \Omega_{n+1}$ for any $n \in \mathbb{N}$ and $\bigcup_{n\in\mathbb{N}} \Omega_n = \Omega$. By the above results we know that $\mu(\Omega_n) = \int_{\Omega_n} \varrho dx$ for any $n \in \mathbb{N}$. Fatou's lemma implies that $1 = \mu(\Omega) = \int_\Omega \varrho dx$. Therefore, ϱ belongs to $L^1(\Omega)$, i.e., μ is absolutely continuous with respect to the Lebesgue measure on the σ-algebra of the Borel sets of Ω.

Step 2. We now prove that there exists at most a unique invariant measure of $\{T(t)\}$. For this purpose, we begin by observing that the semigroup $\{T(t)\}$ is ergodic, i.e.,

$$\lim_{t\to+\infty} \frac{1}{t}\int_0^t T(s)f ds = \int_\Omega f d\mu, \qquad f \in L^2_\mu. \tag{13.6.1}$$

To prove the formula (13.6.1), one can argue as in the proof of Proposition 9.1.11 with minor changes. Indeed, Proposition 9.1.13 holds also in this situation as it can be easily seen repeating step by step the same proof (note in particular that, using the formula (13.3.1), one can show that, if $\{f_n\}$ is a bounded sequence of bounded and continuous functions converging to a function $f \in B_b(\mathbb{R}^N)$ pointwise almost everywhere in $\mathbb{R}^N$, then $T(\cdot)f_n$ converges to $T(\cdot)f$ pointwise in $(0, +\infty)\times\overline{\Omega}$ as n tends to $+\infty$). We need to modify the proof of Proposition 9.1.11 only to show that $\chi_A \in C = \{f \in L^2_\mu : T(t)f = f \ \mu\text{-a.e. for any } t > 0\}$ if and only if $\mu(A) = 0$ or $\mu(A) = 1$. For this purpose, suppose that $\chi_A \in C$ and $\mu(A) > 0$; then, $T(1)\chi_A = \chi_A$ μ-almost everywhere in Ω. Therefore, $T(1)\chi_A$ is positive μ-almost everywhere in A. Since the semigroup is strong Feller (see Proposition 13.3.2), $T(1)\chi_A$ belongs to $C_b(\overline{\Omega})$. Moreover, according to Remark 13.3.3, the function $T(2)\chi_A$ is everywhere positive in Ω. Therefore, the equality $T(2)\chi_A = \chi_A$ can hold μ-almost everywhere in Ω if and only if $\mu(\Omega \setminus A) = 0$, that is if and only if $\mu(A) = 1$.

Step 3. We can now conclude the proof. Suppose that μ_1 and μ_2 are two different invariant measures of $\{T(t)\}$. Then, according to the Chacon-Ornstein theorem (see [214, Chapter 3, Section 8]), for any Borel set A we can determine two sets M_1 and M_2, with $\mu_j(M_j) = 1$ $(j = 1, 2)$ such that

$$\lim_{n\to+\infty} \frac{1}{n}\int_0^n (T(s)\chi_A)(x)ds = \int_\Omega \chi_A d\mu_j = \mu_j(A) \tag{13.6.2}$$

for any $x \in M_j$ and $j = 1, 2$. Let us observe that $M_1 \cap M_2 \neq \varnothing$. Indeed, since the densities ϱ_1 and ϱ_2 of μ_1 and μ_2 (with respect to the Lebesgue measure) are positive in Ω, for any compact set $K \subset \Omega$ the restriction μ_j^K of the measure μ_j to K $(j = 1, 2)$ is equivalent to the Lebesgue measure (restricted to the Borel subsets of K). Fix a compact set K such that $\mu_j(M_j \cap K) > 0$ $(j = 1, 2)$ and suppose, by contradiction, that $M_1 \cap M_2 = \varnothing$. Since $\mu_1(\Omega \setminus M_1) = 0$, it follows that $\mu_1^K(M_1 \cap K) = \mu_1^K(K)$. Moreover,

$$\mu_1^K((M_1 \cup M_2) \cap K) = \mu_1^K(M_1 \cap K) + \mu_1^K(M_2 \cap K) \le \mu_1^K(K) = \mu_1^K(M_1 \cap K).$$

Hence, $\mu_1^K(M_2 \cap K) = 0$. Since, μ_1^K and μ_2^K are equivalent to the Lebesgue measure, $\mu_2^K(M_2 \cap K) = 0$, which, of course, can not be the case.

Now, taking $x \in M_1 \cap M_2$, from (13.6.2) we easily deduce that $\mu_1(A) = \mu_2(A)$ and the arbitrariness of A allows us to conclude that $\mu_1 = \mu_2$. □

As in the case when $\Omega = \mathbb{R}^N$ the main existence result of an invariant measure of $\{T(t)\}$ is the Khas'minskii theorem.

Theorem 13.6.2 (Khas'minskii) *Suppose that there exists a nonnegative function $\varphi \in C^2(\overline{\Omega})$, with null normal derivative on $\partial\Omega$, such that $\mathcal{A}\varphi(x)$ diverges to $-\infty$ as $|x|$ tends to $+\infty$. Then, the semigroup $\{T(t)\}$ admits an invariant measure μ.*

Proof It can be obtained by arguing as in the proof of Theorem 9.1.20, adapting to our situation the proofs of Theorems 9.1.18 and 9.1.19. □

From now on, we always assume that the semigroup $\{T(t)\}$ admits an invariant measure. In such a situation, as the following proposition shows, the semigroup can be extended to a strongly continuous semigroup of contractions defined in L^p_μ for any $p \in [1,+\infty)$. We will still denote by $\{T(t)\}$ such a semigroup.

Proposition 13.6.3 *Let μ be the invariant measure of $\{T(t)\}$. Then, the semigroup can be extended to a strongly continuous semigroup of contractions in L^p_μ for any $p \in [1,+\infty)$. Moreover, the set $D(\mathcal{A})$ (see (13.4.2)) is a core of the infinitesimal generator L_p of the semigroup in L^p_μ.*

Proof It can be obtained by arguing as in the proofs of Propositions 9.1.7 and 9.1.8. □

Now, we use the pointwise gradient estimates in Section 13.5 to derive some L^p_μ-gradient estimates and a partial characterization of the domain of the infinitesimal generator L_p of the semigroup $\{T(t)\}$ in L^p_μ.

Proposition 13.6.4 *Let the assumptions of Theorem 13.5.1 (resp. Theorem 13.5.2) be satisfied and assume that the semigroup $\{T(t)\}$ admits an invariant measure μ. Then, $T(t)$ maps L^p_μ into $W^{1,p}_\mu$, for any $t > 0$ and $p \in (1,+\infty)$ and*

$$\|\nabla_x T(t)f\|_p \le \begin{cases} \dfrac{(2+\omega_0\varepsilon_0)^{\frac{1}{2}}}{[2p(p-1)\kappa_0]^{\frac{1}{2}}}\left(\dfrac{M_p t}{1-e^{-M_p t}}\right)^{\frac{1}{p}} t^{-\frac{1}{2}}\|f\|_p, & p \in (1,2), \\[2ex] \left(\dfrac{2+\omega_0\varepsilon_0}{4\kappa_0}\dfrac{M_2 t}{1-e^{-M_2 t}}\right)^{\frac{1}{2}} t^{-\frac{1}{2}}\|f\|_p, & p \ge 2, \end{cases} \tag{13.6.3}$$

for $t > 0$ and $p \in (1,+\infty)$. In the case when $M_p = 0$, the term $M_p t(1-e^{-M_p t})^{-1}$ should be replaced with 1.

Moreover, $D(L_p) \hookrightarrow W^{1,p}_\mu$ and for any $\omega > 0$ there exists a positive constant M_ω such that $\|\nabla f\|_p \le M_{\omega,p}\|f\|_p^{1/2}\|(L_p-\omega)f\|_p^{1/2}$ for any $f \in D(L_p)$. In the case when the estimate (13.6.3) is satisfied with an exponential term of negative type, we can take $\omega = 0$ in the previous estimate.

Proof The gradient L^p-estimate follows straightforwardly from (13.5.13), On the other hand, the second part of the statement can be obtained by arguing as in the proof of Proposition 9.3.3. Hence, we skip the details. □

To conclude this section, we consider the case when Ω is convex and the operator $\mathcal{A}$ is given by

$$\mathcal{A}\psi = \Delta\psi - \langle \nabla U, \nabla\psi\rangle, \tag{13.6.4}$$

on smooth functions $\psi : \Omega \to \mathbb{R}$, under the following hypothesis on U:

Hypothesis 13.6.5 *The function U belongs to $C^2(\mathbb{R}^N)$, diverges to $+\infty$ as $|x|$ tends to $+\infty$ and $\langle D^2U(x)\xi,\xi\rangle \ge 0$ for any $x, \xi \in \mathbb{R}^N$.*

Denote by μ the measure defined by $\mu(dx) = K^{-1}e^{-U(x)}dx$, where $K = \|e^{-U}\|_{L^1(\Omega)}$.

Lemma 13.6.6 *μ is the invariant measure of $\{T(t)\}$.*

Proof Let us fix $f \in C_b(\overline{\Omega})$ and denote by $\{T_n(t)\}$ $(n \in \mathbb{N})$ the approximating semigroup defined in Subsection 13.2.2. We know that the function $T_n(t)f$ belongs to $C^2(\overline{\Omega}_n)$ for any $t>0$. Moreover, for any $t>0$, $\partial T_n(t)f/\partial\nu = 0$ on $\partial\Omega_n$ and $T_n(t)f$ converges pointwise to $T(t)f$ as n tends to $+\infty$. An integration by parts shows that $\int_{\Omega_n} \mathcal{A}T_n(t)f d\mu = 0$ for any $t>0$. Therefore,

$$\frac{d}{dt}\int_{\Omega_n} T_n(t)f d\mu = \int_{\Omega} \mathcal{A}T_n(t)f d\mu = 0, \qquad t>0,$$

which implies that $\int_{\Omega_n} T_n(t)f d\mu = \int_{\Omega_n} f d\mu$ for any $t>0$ and $n \in \mathbb{N}$. Letting n tend to $+\infty$, from the dominated convergence theorem we deduce that $\int_\Omega T(t)f d\mu = \int_\Omega f d\mu$ for any $t>0$, which, of course, implies that μ is the invariant measure of $\{T(t)\}$. □

As has been proved by G. Da Prato and A. Lunardi in [74], a precise characterization of the domain of the infinitesimal generator of $\{T(t)\}$ in L^2_μ is available, when $\mathcal{A}$ is given by (13.6.4).

Theorem 13.6.7 *Under Hypothesis* 13.6.5, *the resolvent set* $\rho(L_2)$ *contains the halfline* $(0,+\infty)$ *and*

$$D(L_2) = \left\{u \in W^{2,2}_\mu : \langle \nabla U, \nabla_x u\rangle \in L^2_\mu,\ \frac{\partial u}{\partial \nu} = 0 \text{ on } \partial\Omega\right\}. \tag{13.6.5}$$

Moreover,

$$\|D^k R(\lambda, L_2)\|_{L(L^2_\mu)} \le 2^k \lambda^{\frac{k}{2}-1}, \qquad k = 0,1,2. \tag{13.6.6}$$

Proof We split the proof into two steps.
Step 1. Here, we prove that, for any $f \in L^2_\mu$ and $\lambda>0$, the Neumann problem

$$\begin{cases} \lambda u(x) - \mathcal{A}u(x) = f(x), & x \in \Omega,\\ \dfrac{\partial u}{\partial \nu}(x) = 0, & x \in \partial\Omega, \end{cases} \tag{13.6.7}$$

admits a unique solution u belonging to the space defined by the right-hand side of (13.6.5) and $\|D^k u\|_2 \le 2^k \lambda^{k/2-1}\|f\|_2$.

Let us first prove the uniqueness part. For this purpose, we fix $\lambda>0$ and denote by u a solution to problem (13.6.7) (corresponding to $f \equiv 0$) with the claimed regularity properties. An integration by parts shows that

$$\lambda \int_\Omega u^2 d\mu = \lambda \int_\Omega u\mathcal{A}u d\mu = -\frac{1}{2}\int_\Omega |\nabla_x u|^2 d\mu$$

and, of course, this implies that $u=0$.

To prove the existence part, we fix $f \in L^2_\mu$ and approximate the operator $\mathcal{A}$ with the sequence of operators $\{\mathcal{A}^{(n)}\}$ defined as the operator $\mathcal{A}$, with U being replaced by the function U_n given by $U_n(x) = U(x) + n(\text{dist}(x,\Omega))^2/2$ for any $x \in \mathbb{R}^N$ and $n \in \mathbb{N}$. As is immediately seen, the function U_n diverges to $+\infty$ as $|x|$ tends to $+\infty$. Moreover, since Ω is convex, U_n is convex in $\mathbb{R}^N$. Therefore, according to Remark 9.5.11, the realization $L_2^{(n)}$ in $L^2_{\mu_n}$ of the operator $\mathcal{A}^{(n)}$ with domain $D(L_2^{(n)}) = \{u \in W^{2,2}(\mathbb{R}^N,\mu_n) : \langle \nabla U_n, \nabla u\rangle \in L^2(\mathbb{R}^N,\mu_n)\}$ generates a strongly continuous semigroup in $L^2(\mathbb{R}^N,\mu_n)$, where $\mu_n(dx) = K_n^{-1}e^{-U_n(x)}dx$, with $K_n = \|e^{-U_n}\|_{L^1(\mathbb{R}^N)}$. Let us extend the function f to the whole of $\mathbb{R}^N$

by zero and let us still denote by f such an extended function. The equation $\lambda u - \mathcal{A}^{(n)}u = f$ admits a unique solution $u \in D(L_2^{(n)})$ which satisfies

$$\|D^k u_n\|_{L^2(\mathbb{R}^N,\mu_n)} \le 2^k \lambda^{\frac{k}{2}-1}\|f\|_{L^2(\mathbb{R}^N,\mu_n)}, \qquad k=0,1,2, \tag{13.6.8}$$

see again Remark 9.5.11.

We now use a compactness argument to prove that, up to a subsequence, u_n converges to a solution of problem (13.6.7) which belongs to $W^{2,2}_\mu$ and satisfies the condition $\langle \nabla U, \nabla u\rangle \in L^2_\mu$. For this purpose we begin by observing that

$$\|f\|_{L^2(\mathbb{R}^N,\mu_n)} = \left(\frac{1}{K_n}\int_\Omega f^2 e^{-U}dx\right)^{\frac{1}{2}} = \left(\frac{\int_\Omega e^{-U}dx}{\int_{\mathbb{R}^N} e^{-U_n}dx}\right)^{\frac{1}{2}}\|f\|_2.$$

Therefore, the sequence $\{\|f\|_{L^2(\mathbb{R}^N,\mu_n)}\}$ is bounded. From (13.6.8) we now deduce that the restriction of the functions u_n to Ω gives rise to a bounded sequence in $W^{2,2}_\mu$. Indeed, $\|D^k u\|_{L^2(\Omega,\mu)} \le \|D^k u\|_{L^2(\mathbb{R}^N,\mu_n)}$. Hence, there exists a subsequence $\{u_{n_k}\} \subset \{u_n\}$ which converges weakly in $W^{2,2}_\mu$ and strongly in $W^{3/2,2}(\Omega\cap B_R)$, for any $R>0$, to a function $u \in W^{2,2}_\mu$. Moreover, we can assume that u_n converges to u pointwise almost everywhere in Ω. This implies that the function u solves the differential equation in (13.6.7). Since u and Δu belong to L^2_μ, the function $\langle \nabla U, \nabla_x u\rangle$ belongs to L^2_μ as well. Further, (13.6.6) follows by virtue of (13.6.8).

Finally, let us prove that the normal derivative of u identically vanishes on $\partial\Omega$. Using the formula (9.1.23) we can show that

$$\begin{aligned}
\int_{\mathbb{R}^N}\langle \nabla_x u_n, \nabla\psi\rangle d\mu &= -\int_{\mathbb{R}^N\setminus\Omega}(\mathcal{A}_n u_n)\psi d\mu_n - \int_\Omega (\mathcal{A}_n u_n)\psi d\mu_n\\
&= -\lambda\int_{\mathbb{R}^N\setminus\Omega} u_n\psi d\mu_n - \int_\Omega (\mathcal{A}_n u_n)\psi d\mu_n\\
&= -\lambda\int_{\mathbb{R}^N\setminus\Omega} u_n\psi d\mu_n + \int_\Omega \langle\nabla_x u_n, \nabla\psi\rangle d\mu_n\\
&\quad - \frac{1}{K_n}\int_{\partial\Omega}\frac{\partial u_n}{\partial\nu}\psi e^{-U_n}d\sigma
\end{aligned}$$

for any $\psi \in C^\infty_c(\mathbb{R}^N)$, where $d\sigma$ is the surface measure related to $\partial\Omega$. Thus,

$$\frac{1}{K_n}\int_{\partial\Omega}\frac{\partial u_n}{\partial\nu}\psi e^{-U_n}d\sigma = -\int_{\mathbb{R}^N\setminus\Omega}\langle\nabla_x u_n, \nabla\psi\rangle d\mu_n - \lambda\int_{\mathbb{R}^N\setminus\Omega} u_n\psi d\mu_n \tag{13.6.9}$$

for any $\psi \in C^\infty_c(\mathbb{R}^N)$. Let us observe that the right-hand side of (13.6.9) vanishes as n tends to $+\infty$. Indeed, the sequence $\{u_n\}$ is bounded in $W^{2,2}_\mu$ and, as is easily seen, the $W^{1,2}(\mathbb{R}^N\setminus\Omega,\mu_n)$-norm of ψ vanishes as n tends to $+\infty$. As far as the left-hand side of (13.6.9) is concerned, we observe that, since u_n converges to u in $W^{3/2,2}(\Omega\cap \operatorname{Supp}\psi)$, $\partial u_n/\partial\nu$ converges to $\partial u/\partial\nu$ in $L^2(\partial\Omega\cap\operatorname{Supp}\psi)$ as n tends to $+\infty$. As a consequence,

$$\int_{\partial\Omega}\frac{\partial u}{\partial\nu}\psi e^{-U}d\sigma = \lim_{n\to+\infty}\int_{\partial\Omega}\frac{\partial u_n}{\partial\nu}\psi e^{-U_n}d\sigma,$$

from which we deduce that $\partial u/\partial\nu = 0$ on $\partial\Omega$.

Finally, the formula (13.6.6) follows from (13.6.8), letting n tend to $+\infty$.

Step 2. Denote by $\widehat{L}_2$ the realization of the operator $\mathcal{A}$ in L^2_μ with domain given by the right-hand side of (13.6.5).

Since the operator $\widehat{L}_2$ is dissipative and self-adjoint in L^2_μ, according to Theorem C.2.10 it generates a strongly continuous analytic semigroup $\{S(t)\}$ in L^2_μ. To conclude the proof, we need to show that $T(t) \equiv S(t)$ for any $t > 0$. For this purpose, we begin by observing that $\widehat{L}_2$ and L_2 coincide in $C^\infty_c(\Omega)$. To check it, it suffices to observe that $C^\infty_c(\Omega)$ is properly contained in the space defined by the right-hand side of (13.6.5). Moreover, it is also contained in $D(L_2)$. Indeed, according to Proposition 13.6.3, the space defined by (13.4.2) is a core of L_2.

Now, fix $f \in C^\infty_c(\Omega)$ and $t > 0$. Moreover, let $\psi : [0, t] \to L^2_\mu$ be the function defined by $\psi(s) = S(t-s)T(s)f$ for any $s \in [0, t]$. Since $\{S(t)\}$ and $\{T(t)\}$ are both strongly continuous semigroups and $\{T(t)\}$ is also analytic, the function ψ is continuously differentiable in $[0, t)$ with values in L^2_μ and $\psi' = S(t - \cdot)T(\cdot)\widehat{L}_2 f - S(t - \cdot)T(\cdot)L_2 f = 0$. Therefore, $\psi(s) = \psi(0)$ for any $s \in (0, t)$. Since ψ is continuous up to $s = t$, we deduce that $\psi(t) = \psi(0)$ or, equivalently, that $T(t)f \equiv S(t)f$. This completes the proof. □

13.7 Final remarks

To conclude the chapter, we observe that under assumptions comparable with those in Hypothesis 5.1.1, uniform estimates up to the third-order have been proved in [123] for the semigroup associated with the operator $\mathcal{A} = \mathrm{Tr}(QD^2) + \langle b, \nabla\rangle$, with rather general homogeneous boundary conditions, including the Dirichlet and Neumann boundary conditions.

Such estimates have been the keystone to prove optimal Schauder estimates for both the elliptic equation

$$\begin{cases} \lambda u - \mathcal{A}u = f, & \text{in } \Omega, \\ \mathcal{B}u = 0, & \text{on } \partial\Omega, \end{cases}$$

(for λ positive) and the parabolic problem

$$\begin{cases} D_t u = \mathcal{A}u + f, & \text{in } (0,T) \times \Omega, \\ \mathcal{B}u = 0, & \text{on } (0,T) \times \partial\Omega, \\ u(0,\cdot) = u_0, & \text{in } \overline{\Omega}, \end{cases}$$

adapting the arguments used in Chapter 5.

Part II

Non-autonomous Kolmogorov equations

Chapter 14

The evolution operator and the evolution semigroup in the space of bounded and continuous functions

14.0 Introduction

In this chapter we set the basis of the analysis on non-autonomous Kolmogorov equations in $\mathbb{R}^N$ associated with the family of elliptic operators $\mathcal{A}(t)$, defined on smooth functions $\psi : \mathbb{R}^N \to \mathbb{R}$ by

$$\mathcal{A}(t)\psi(x) = \sum_{i,j=1}^{N} q_{ij}(t,x)D_{ij}\psi(x) + \sum_{i=1}^{N} b_i(t,x)D_i\psi(x) + c(t,x)\psi(x) \tag{14.0.1}$$

for any $t \in I$ and $x \in \mathbb{R}^N$. In view of the use of these results in the next chapters, we assume that I is either $\mathbb{R}$ or a right halfline, but most of the results that we present in this chapter also hold true in the case when I is a bounded interval.

As in the autonomous case such operators occur naturally in the study of stochastic differential equations. Indeed, consider the stochastic differential equation

$$\begin{cases} dX_t = \mu(t,X_t)dt + \sigma(t,X_t)dW_t, & t > s, \\ X_s = x, \end{cases} \tag{14.0.2}$$

where W_t is a standard N-dimensional Brownian motion and μ (resp. σ) are regular $\mathbb{R}^N$ (resp. $\mathbb{R}^{N\times N}$) valued coefficients. If problem (14.0.2) admits a solution[1] $X_t = X(t,s,x)$ for all $x \in \mathbb{R}^N$, Itô's formula reveals that, for $\zeta \in C_b^2(\mathbb{R}^N)$ and $t \in \mathbb{R}$, the function $u(s,x) := \mathbb{E}(\zeta(X(t,s,x)))$, where $\mathbb{E}$ denotes expectation, solves the Cauchy problem

$$\begin{cases} D_s u(s,x) = -\mathcal{A}(s)u(s,x), & (s,x) \in (-\infty,t) \times \mathbb{R}^N, \\ u(t,x) = \zeta, & x \in \mathbb{R}^N, \end{cases}$$

where $\mathcal{A}$ is given by (14.0.1) with $Q = \sigma\sigma^*/2$ and $b = \mu$.

To be consistent with the first part of the book, in this and in the forthcoming chapters we prefer to deal with forward Cauchy problems, but of course, this difference is immaterial since reversing time we can shift from forward to backward Cauchy problems and vice versa.

First, in Section 14.1 we prove the well posedness of the Cauchy problem

$$\begin{cases} D_t u(t,x) = \mathcal{A}(t)u(t,x), & (t,x) \in (s,+\infty) \times \mathbb{R}^N, \\ u(s,x) = f(x), & x \in \mathbb{R}^N, \end{cases} \tag{14.0.3}$$

for $f \in C_b(\mathbb{R}^N)$ and $s \in I$. The following are our standing assumptions:

---

[1] In the case of Lipschitz continuous coefficients there are many results providing conditions on μ and σ such that this problem is well posed. See e.g. [126, 134, 154].

Hypotheses 14.0.1 (i) q_{ij}, b_i $(i,j=1,\ldots,N)$ and c belong to $C^{\alpha/2,\alpha}_{\rm loc}(I\times\mathbb{R}^N)$ for some $\alpha\in(0,1)$;

(ii) the matrix $Q(t,x)=(q_{ij}(t,x))$ is symmetric for any $(t,x)\in I\times\mathbb{R}^N$ and there exists a function $\kappa: I\times\mathbb{R}^N\to\mathbb{R}$ such that $0<\kappa_0:=\inf_{I\times\mathbb{R}^N}\kappa$ and $\langle Q(t,x)\xi,\xi\rangle\ge\kappa(t,x)|\xi|^2$ for any $\xi\in\mathbb{R}^N$ and $(t,x)\in I\times\mathbb{R}^N$;

(iii) $c_0:=\sup_{I\times\mathbb{R}^N}c<+\infty$;

(iv) for any bounded interval $J\subset I$ there exist a positive function $\varphi=\varphi_J\in C^2(\mathbb{R}^N)$ and a real number $\lambda=\lambda_J$ such that $\varphi(x)$ diverges to $+\infty$ as $|x|$ tends to $+\infty$ and $\mathcal{A}\varphi-\lambda\varphi\le 0$ in $J\times\mathbb{R}^N$.

Note that Hypotheses 14.0.1 are the natural generalization of Hypotheses 1.0.1 to the non-autonomous setting. In particular, Hypotheses 14.0.1 allow to prove a maximum principle, which yields the uniqueness of the classical solution[2] to problem (14.0.3), which is bounded in $[s,T]\times\mathbb{R}^N$ for any $T>s$.

In the autonomous case, the solutions to (14.0.3) are governed by a semigroup $\{T(t)\}$. In the non autonomous setting, the semigroup is replaced by an evolution family $\{G(t,s)\}$, which admits a Green function, i.e., there exists a function g such that

$$(G(t,s)f)(x)=\int_{\mathbb{R}^N}g(t,s,x,y)f(y)dy,\qquad s,t\in I,\ s<t,\ x\in\mathbb{R}^N,$$

for any $f\in C_b(\mathbb{R}^N)$. For any fixed s and almost every $y\in\mathbb{R}^N$, $g(\cdot,s,\cdot,y)$ is smooth and solves the equation $D_tg=\mathcal{A}(t)g$. The previous formula allows us to extend each operator $G(t,s)$ to the space $B_b(\mathbb{R}^N)$ of all bounded and Borel measurable functions $f:\mathbb{R}^N\to\mathbb{R}$. The so extended operators turn out to be strong Feller and irreducible.

The regularity of the function $s\mapsto(G(t,s)f)(x)$ is an important issue, treated in the literature in the case of bounded coefficients because of its importance in several applications such as control theory. For functions $f\in C^\infty_c(\mathbb{R}^N)$ this property follows from the formula

$$(G(t,s_1)f)(x)-(G(t,s_0)f)(x)=-\int_{s_0}^{s_1}(G(t,\sigma)\mathcal{A}(\sigma)f)(x)d\sigma,$$

which holds true for any $I\ni s_0\le s_1\le t$, $x\in\mathbb{R}^N$, and for functions $f\in C_0(\mathbb{R}^N)$ it then follows from a density argument. For a general $f\in C_b(\mathbb{R}^N)$, the continuity of the function $s\mapsto(G(t,s)f)(x)$ (more precisely, of the function $(t,s,x)\mapsto(G(t,s)f)(x)$) is proved assuming that $\mathcal{A}\varphi$ is upperly bounded in $J\times\mathbb{R}^N$ for any bounded interval $J\subset I$.

Next, in Sections 14.2 and 14.3 we provide sufficient conditions for the compactness of the operator $G(t,s)$ in $C_b(\mathbb{R}^N)$ and for the invariance of $C_0(\mathbb{R}^N)$ under the action of the evolution operator. In particular, under the assumptions that guarantee that $C_0(\mathbb{R}^N)$ is invariant under the action of the evolution operator $\{G(t,s)\}$, we show that the restriction of $\{G(t,s)\}$ to $C_0(\mathbb{R}^N)$ gives rise to a strongly continuous evolution operator. All the results are in the spirit of Chapter 4. We stress that, already in the classical case when the coefficients of the operator $\mathcal{A}$ are bounded, the operator $G(t,s)$ in general is not compact. This justifies the need for additional assumptions on the coefficients of the operator $\mathcal{A}$.

Pointwise and uniform gradient estimates can be proved as in the autonomous case: the dependence of the coefficients on t does not cause any additional difficulty. Under the non-autonomous version of the assumptions of Chapters 5 and 6 we state such estimates, which will play a crucial role in the next chapters to study further properties of the evolution operator.

[2] i.e., a function $u\in C([s,+\infty)\times\mathbb{R}^N)\cap C^{1,2}((s,+\infty)\times\mathbb{R}^N)$ which solves the Cauchy problem (14.0.3).

Even if we do not provide the proofs, referring the reader to the original paper, we also state the results from [172] which provide us with optimal Schauder estimate for the solution to the nonhomogeneous Cauchy problem associated with the operator $\mathcal{A}$, when t runs in a bounded interval, also allowing some situations where the coefficients of the operator $\mathcal{A}$ are not smooth in the pair (t,x).

In the case when $I=\mathbb{R}$, one can introduce the so-called evolution semigroup $\{\mathscr{T}(t)\}$ associated with $\{G(t,s)\}$, which is defined in $C_b(\mathbb{R}^{N+1})$ by

$$(\mathscr{T}(t)f)(s,x)=(G(s,s-t)f(s-t,\cdot))(x),\qquad (s,x)\in\mathbb{R}^{N+1},\ t\geq 0,\ f\in C_b(\mathbb{R}^{N+1}).$$

In the classical case when the coefficients of the operator $\mathcal{A}$ are bounded, this semigroup is known to be a useful tool to determine several qualitative properties of the evolution family (see e.g. [56] and the references therein). Likewise, in our analysis it will play a crucial role since it will be used to study the asymptotic behaviour of the evolution operator $\{G(t,s)\}$ as t tends to $+\infty$ (see Chapter 18). For this purpose, in Section 14.5 we begin by studying some of its relevant properties in the general case and in the case when the coefficients of the operator $\mathcal{A}$ are T-periodic with respect to the variable t for some $T>0$. In this latter situation, $G(t+T,s+T)f=G(t,s)f$ for any $I\ni s<t$ and $f\in C_b(\mathbb{R}^N)$. Hence, the semigroup $\{\mathscr{T}(t)\}$ leaves invariant the set $C_\sharp(\mathbb{R}^{N+1})$ of all functions $f\in C_b(\mathbb{R}^{N+1})$ such that $f(\cdot+T,\cdot)=f$. We refer the reader to Chapter 17 for an analysis of the semigroup $\{\mathscr{T}(t)\}$ in L^p-spaces.

The results in this chapter are taken from [15, 155, 177, 183].

14.1 The evolution operator and its continuity properties

To begin with, we state a maximum principle. Its proof can be obtained by arguing as in the proof of Theorem 3.1.3. Hence, we skip the details.

Proposition 14.1.1 *Fix $T>s\in I$, $R>0$ and let $\Omega=\mathbb{R}^N$ or $\Omega=\mathbb{R}^N\setminus B_R$. If $u\in C_b([s,T]\times\overline{\Omega})\cap C^{1,2}((s,T]\times\Omega)$ is nonpositive in $\{s\}\times\overline{\Omega}$ (and in $[s,T]\times\partial\Omega$, if $\Omega\neq\mathbb{R}^N$) and $D_tu-\mathcal{A}u\leq 0$ in $(s,T]\times\Omega$, then $u\leq 0$ in $[s,T]\times\overline{\Omega}$. As a byproduct, if $u\in C_b([s,T]\times\mathbb{R}^N)\cap C^{1,2}((s,T]\times\mathbb{R}^N)$ solves the Cauchy problem (14.0.3) in the time domain $(s,T]$, then $\|u(t,\cdot)\|_\infty\leq e^{c_0(t-s)}\|f\|_\infty$ for any $t>s$.*

For any $f\in C_b(\mathbb{R}^N)$ and $n\in\mathbb{N}$ we introduce the Cauchy problems

$$\begin{cases} D_tu_n(t,x)=\mathcal{A}(t)u_n(t,x), & (t,x)\in(s,+\infty)\times B_n,\\ u_n(t,x)=0, & (t,x)\in(s,+\infty)\times\partial B_n,\\ u_n(s,x)=f(x), & x\in B_n,\end{cases}\tag{14.1.1}$$

and

$$\begin{cases} D_tu_n(t,x)=\mathcal{A}(t)u_n(t,x), & (t,x)\in(s,+\infty)\times B_n,\\ \dfrac{\partial u_n}{\partial\nu}(t,x)=0, & (t,x)\in(s,+\infty)\times\partial B_n,\\ u_n(s,x)=f(x), & x\in B_n,\end{cases}\tag{14.1.2}$$

where $\nu=\nu(x)$ denotes the exterior unit normal at $x\in\partial B_n$. We further denote by $G_n^{\mathcal{D}}(\cdot,s)$ (resp. $G_n^{\mathcal{N}}(\cdot,s)$) the bounded operator on $C_b(B_n)$ (resp. on $C_b(\overline{B}_n)$) which associates with any $f\in C_b(B_n)$ (resp. $f\in C_b(\overline{B}_n)$) the unique classical solution to problem (14.1.1) (resp. (14.1.2)).

Theorem 14.1.2 *For any $f \in C_b(\mathbb{R}^N)$ and $s \in I$ the Cauchy problem (14.0.3) admits a unique solution $u_f \in C([s,+\infty) \times \mathbb{R}^N) \cap C^{1+\alpha/2,2+\alpha}_{\rm loc}((s,+\infty) \times \mathbb{R}^N)$ (α being given by Hypothesis 14.0.1(i)), which is bounded in $[s,T] \times \mathbb{R}^N$ for any $T > s$. If for any $t > s$ and $f \in C_b(\mathbb{R}^N)$ we set $G(t,s)f := u_f(t,\cdot)$, then $G(t,s)$ is a bounded, order preserving linear operator in $C_b(\mathbb{R}^N)$ and*

$$\|G(t,s)\|_{L(C_b(\mathbb{R}^N))} \le e^{c_0(t-s)}, \qquad t \ge s. \tag{14.1.3}$$

Moreover, the following properties hold true:

(i) *$G_n^{\mathcal{N}}(\cdot,s)f$ converges to $G(\cdot,s)f$ in $C^{1,2}(D)$ for any compact set $D \subset (s,+\infty) \times \mathbb{R}^N$ and any $f \in C_b(\mathbb{R}^N)$;*

(ii) *$G_n^{\mathcal{D}}(\cdot,s)f$ converges to $G(\cdot,s)f$ in $C^{1,2}(D)$ for any compact set $D \subset (s,+\infty)\times\mathbb{R}^N$, any $s \in I$ and any $f \in C_b(\mathbb{R}^N)$. Moreover, if f is nonnegative, then $G_n^{\mathcal{D}}(t,s)f$ increases to $G(t,s)f$ for any $(t,s) \in \Delta_I := \{(t,s) \in I \times I : t \ge s\}$.*

Proof Fix $s \in I$ and $f \in C_b(\mathbb{R}^N)$. The uniqueness of the classical solution to problem (14.0.3) follows from Proposition 14.1.1. As far as the existence part is concerned, we observe that, replacing estimate (11.3.3) with the Schauder estimates in Theorem D.1.5(i) and arguing as in the proof of Theorem 11.3.2, it can be shown that both the sequences $\{G_n^{\mathcal{D}}(\cdot,s)\}$ and $\{G_n^{\mathcal{N}}(\cdot,s)\}$ converge in $C^{1,2}(D)$, for any compact set $D \subset (s,+\infty) \times \mathbb{R}^N$, to a classical solution to problem (14.0.3), which satisfies the estimate $\|u(t,\cdot)\|_\infty \le e^{c_0(t-s)}\|f\|_\infty$ and the regularity properties in the statement. Finally, if $f \ge 0$, then the classical maximum principle shows that the sequence $\{(G_n^{\mathcal{D}}(t,s)f)(x)\}$ is positive and increasing for any $t > s$, $x \in \mathbb{R}^N$, and this accomplishes the proof. □

From Theorem 14.1.2 the evolution law $G(t,s)G(s,r)= G(t,r)$ easily follows for any $r \le s \le t$. For this reason, the family $\{G(t,s) : (t,s) \in \Delta_I\}$ is called the evolution operator associated with problem (14.0.3). To simplify the notation, from now on we will write $\{G(t,s)\}$ to denote the evolution operator.

The autonomous case (see Theorem 1.2.5) suggests that any operator $G(t,s)$ could be associated with a Green function g, which gives rise to a transition family if $c \le 0$ in $\mathbb{R}^N$. The following proposition states that this is indeed the case; in fact, the Green function g forms the non autonomous equivalent of a stochastically continuous transition function.

Proposition 14.1.3 (Green function) *The following properties are satisfied.*

(i) *The evolution operator $\{G(t,s)\}$ can be represented in the form*

$$(G(t,s)f)(x) = \int_{\mathbb{R}^N} g(t,s,x,y)f(y)dy, \qquad s < t, \;\; x \in \mathbb{R}^N, \tag{14.1.4}$$

for any $f \in C_b(\mathbb{R}^N)$, where $g : \Delta_I \times \mathbb{R}^N \times \mathbb{R}^N \to \mathbb{R}$ is a positive function. For any $s \in I$ and almost every $y \in \mathbb{R}^N$, $g(\cdot,s,\cdot,y)$ belongs to $C^{1+\alpha/2,2+\alpha}_{\rm loc}((s,+\infty) \times \mathbb{R}^N)$ and solves the equation $D_t g = \mathcal{A}(t)g$ in $(s,+\infty) \times \mathbb{R}^N$. Moreover, $\|g(t,s,x,\cdot)\|_{L^1(\mathbb{R}^N)} = 1$ for any $t > s \in I$ and $x \in \mathbb{R}^N$, if $c \equiv 0$, and $\|g(t,s,x,\cdot)\|_{L^1(\mathbb{R}^N)} \le e^{c_0(t-s)}$ for any s, t and x as above, otherwise. The function g is called the Green function associated with the evolution operator $\{G(t,s)\}$.

(ii) *$G(t,s)$ can be extended to $L^\infty(\mathbb{R}^N)$ through the formula (14.1.4). Moreover, each operator $G(t,s)$ is strong Feller (i.e., $G(t,s)f \in C_b(\mathbb{R}^N)$ for any $f \in B_b(\mathbb{R}^N)$ and any $t > s$) and irreducible (i.e., $G(\cdot,\cdot)\chi_U > 0$ in $\Delta_I \times \mathbb{R}^N$ for any open set $U \subset \mathbb{R}^N$).*

(iii) *For any $t > s \in I$ and $x \in \mathbb{R}^N$, the measure $p(t,s,x;dy) = g(t,s,x,y)dy$ is equivalent to the Lebesgue measure, i.e., they have the same sets of zero measure. Moreover, if $c \equiv 0$, each measure $p(t,s,x;dy)$ is a probability measure.*

(iv) *For any fixed Borel set $A \subset \mathbb{R}^N$ and any $(t,s) \in \Delta_I$ the map $x \mapsto p(t,s,x;A)$ is Borel measurable. Moreover, for any open set $U \subset \mathbb{R}^N$ containing x, $p(t,s,x;U)$ converges to 1 as t tends to s^+.*

(v) *For any $t \geq s \geq r$, any $x \in \mathbb{R}^N$ and any Borel set $A \subset \mathbb{R}^N$ it holds that*

$$p(t,r,x;A) = \int_{\mathbb{R}^N} p(s,r,y;A)p(t,s,x;dy). \tag{14.1.5}$$

Proof It suffices to adapt the proofs of Theorem 1.2.5 and Proposition 1.2.13, taking [246, Theorem 1] into account. □

Now, taking advantage of the representation formula (14.1.4) and adapting the arguments in the proof of Proposition 11.3.5, we can prove the following proposition, which, as in the autonomous case, plays a crucial role in the analysis of the evolution operator $\{G(t,s)\}$.

Proposition 14.1.4 *Let $\{f_n\} \subset C_b(\mathbb{R}^N)$ be a bounded sequence. The following properties are satisfied.*

(i) *If f_n tends to f pointwise in $\mathbb{R}^N$, then $G(\cdot,s)f_n$ converges to $G(\cdot,s)f$ in $C^{1,2}(F)$ for any compact set $F \subset (s,+\infty) \times \mathbb{R}^N$.*

(ii) *If f_n converges uniformly to f in any compact set $K \subset \mathbb{R}^N$, then $G(\cdot,s)f_n$ converges to $G(\cdot,s)f$ uniformly in $[s,T] \times K$, for any $T > s$ and any compact set $K \subset \mathbb{R}^N$.*

Up to now we have just rephrased some results already established in the autonomous case. An issue, which of course does not arise in the autonomous case, is the smoothness of the function $G(\cdot,\cdot)f$ with respect to the variable s (and, hence, with respect to the triplet (t,s,x)). This is not a trivial task and we address it here. We begin by proving a very useful generalization of a well known formula that holds in the case of bounded coefficients. For this purpose, we introduce the following strengthening of Hypothesis 14.0.1(iv), in the case when the coefficient c does not identically vanish.

Hypotheses 14.1.5 *For any bounded interval $J \subset I$ there exist a positive function $\varphi = \varphi_J \in C^2(\mathbb{R}^N)$, blowing up as $|x|$ tends to $+\infty$, and a real number $\lambda = \lambda_J$ such that $\mathcal{A}(t)\varphi(t,x) - c(t,x)\varphi(x) - \lambda\varphi(x) \leq 0$ for any $(t,x) \in J \times \mathbb{R}^N$.*

Lemma 14.1.6 *The following properties are satisfied.*

(i) *Suppose that f belongs to $C^2_c(\mathbb{R}^N)$. Then,*

$$(G(t,s_1)f)(x) - (G(t,s_0)f)(x) = -\int_{s_0}^{s_1} (G(t,\sigma)\mathcal{A}(\sigma)f)(x)d\sigma \tag{14.1.6}$$

for any $s_0 \leq s_1 \leq t$ and $x \in \mathbb{R}^N$. In particular, the function $(G(t,\cdot)f)(x)$ is differentiable in $(-\infty,t] \times I$ for any $x \in \mathbb{R}^N$ and $(D_sG(t,s)f)(x) = -(G(t,s)\mathcal{A}(s)f)(x)$ for any $I \ni s \leq t$.

(ii) *Let $f \in C_b^2(\mathbb{R}^N)$ be constant and positive outside of a ball and assume Hypothesis 14.1.5. Then, for any $x \in \mathbb{R}^N$, the function $(G(t,\cdot)\mathcal{A}(\cdot)f)(x)$ is locally integrable in $(-\infty,t]\cap I$ and*

$$(G(t,s_1)f)(x) - (G(t,s_0)f)(x) \geq -\int_{s_0}^{s_1} (G(t,\sigma)\mathcal{A}(\sigma)f)(x)d\sigma \tag{14.1.7}$$

for any $s_0 \leq s_1 \leq t$. If c is bounded, then (14.1.7) is an equality.

Proof (i) Fix $f \in C_c^2(\mathbb{R}^N)$ and let n be sufficiently large that $\mathrm{supp}(f) \subset B_n$. It is well known that[3]

$$(G_n^{\mathcal{N}}(t,s_1)f)(x) - (G_n^{\mathcal{N}}(t,s_0)f)(x) = -\int_{s_0}^{s_1} (G_n^{\mathcal{N}}(t,r)\mathcal{A}(r)f)(x)dr \tag{14.1.8}$$

for any $s_0 \leq s_1 \leq t$ and $x \in \mathbb{R}^N$, where we recall that $\{G_n^{\mathcal{N}}(t,s)\}$ is the evolution operator associated with the Cauchy-Neumann problem (14.1.1). Since the function $\mathcal{A}f$ is bounded and continuous in $[s_0,s_1]\times\mathbb{R}^N$, taking Theorem 14.1.2(i) into account, we can let n tend to $+\infty$ in (14.1.8) and obtain (14.1.6).

(ii) Since any function, which is constant and positive in a neighborhood of ∞, can be split into the sum of a compactly supported function and a positive constant, due to the above result we just need to consider the case when $f = \mathbb{1}$. Being rather long, we split the proof into two steps.

Step 1. We first assume that the potential c is bounded in $J \times \mathbb{R}^N$ for any bounded set $J \subset \overline{I}$. Formula (14.1.8) actually holds true for any $f \in C_b^2(\mathbb{R}^N)$ with null normal derivative on ∂B_n. Therefore, we can write

$$(G_n^{\mathcal{N}}(t,s_1)\mathbb{1})(x) - (G_n^{\mathcal{N}}(t,s_0)\mathbb{1})(x) = -\int_{s_0}^{s_1} (G_n^{\mathcal{N}}(t,\tau)c(\tau,\cdot))(x)d\tau \tag{14.1.9}$$

for any s_0, s_1, t and x as above. Taking the limit as n tends to $+\infty$ in both sides of (14.1.9) yields, by dominated convergence,

$$(G(t,s_1)\mathbb{1})(x) - (G(t,s_0)\mathbb{1})(x) = -\int_{s_0}^{s_1} (G(t,\tau)c(\tau,\cdot))(x)d\tau \tag{14.1.10}$$

for any $s_0 \leq s_1 < t$, $x \in \mathbb{R}^N$ and the assertion follows.

Step 2. Let us now suppose that c is unbounded and set $c_n(s,x) = (c(s,x)-c_0)\vartheta_n(x)$ for any $(s,x) \in I \times \mathbb{R}^N$, where c_0 is the supremum of the function c and $\vartheta_n \in C_c(\mathbb{R}^N)$ satisfies the condition $\chi_{B_n} \leq \vartheta_n \leq \chi_{B_{n+1}}$ for any $n \in \mathbb{N}$. Clearly, each function c_n is nonnegative and belongs to $C(I; C_c(\mathbb{R}^N))$. Moreover, $c_n(s,x) \geq c(s,x) - c_0$ for any $(s,x) \in I \times \mathbb{R}^N$, $n \in \mathbb{N}$ and the sequence $\{c_n(s,x)\}$ is decreasing for any $(s,x) \in I \times \mathbb{R}^N$.

For any $n \in \mathbb{N}$, let $\mathcal{A}_n$ be the operator defined on smooth functions $\psi : \mathbb{R}^N \to \mathbb{R}$ by $\mathcal{A}_n(t)\psi(x) = \mathrm{Tr}(Q(t,x)D^2\psi(x)) + \langle b(t,x), \nabla\psi(x)\rangle + (c_n(t,x)+c_0)\psi(x)$ for any $(t,x) \in I \times \mathbb{R}^N$. Since each operator $\mathcal{A}_n$ satisfies Hypotheses 14.0.1, we can associate an evolution operator $\{G_n(t,s)\}$ with $\mathcal{A}_n$. Note that, for any nonnegative $f \in C_b(\mathbb{R}^N)$ and $m,n \in \mathbb{N}$ such that $n < m$, the function $v = G_m(\cdot,s)f - G_n(\cdot,s)f$ satisfies the differential inequality $D_t v - \mathcal{A}_m v \leq 0$ and vanishes at $t = s$. Proposition 14.1.1 then implies that $v \leq 0$ in $[s,+\infty)\times\mathbb{R}^N$, i.e., $G_m(\cdot,s)f \leq G_n(\cdot,s)f$. This shows that the sequence $\{(G_n(t,s)f)(x)\}$ is decreasing. Hence, $G_n(\cdot,s)f$ converges to some function u as n tends to $+\infty$, pointwise in $[s,+\infty)\times\mathbb{R}^N$. To show that $u = G(\cdot,s)f$ it suffices to use the same arguments as in the proof of Theorem 14.1.2.

---

[3] See [1, Theorem 2.3(ix)].

Now, we can complete the proof. Writing (14.1.10) with the evolution operator $\{G_n(t,s)\}$ replacing $\{G(t,s)\}$, we get

$$\begin{aligned}(G_n(t,s_1)\mathbb{1})(x) - (G_n(t,s_0)\mathbb{1})(x) &= -\int_{s_0}^{s_1} (G_n(t,\tau)(c_n(\tau,\cdot)+c_0))(x)d\tau \\ &\geq -\int_{s_0}^{s_1} (G(t,\tau)c_n(\tau,\cdot))(x)d\tau - c_0\int_{s_0}^{s_1} (G_n(t,\tau)\mathbb{1})(x)d\tau\end{aligned}$$

for any $s_0, s_1 \in I$, such that $s_0 \leq s_1 \leq t$, and any $x \in \mathbb{R}^N$. Since the sequence $\{c_n\}$ is decreasing and $G(t,\tau)$ is a positive operator, letting n tend to $+\infty$ we get $(G(t,s_1)\mathbb{1})(x) - (G(t,s_0)\mathbb{1})(x) \geq -\int_{s_0}^{s_1}(G(t,\tau)c(\tau,\cdot))(x)d\tau$ and (14.1.7) follows. □

Corollary 14.1.7 *For any $f \in C_0(\mathbb{R}^N)$ the following properties are satisfied:*

(i) *the function $G(t,\cdot)f$ belongs to $C((-\infty,t]\cap I; C_b(\mathbb{R}^N))$ for any $t \in I$;*

(ii) *the function $G(\cdot,\cdot)f$ is continuous in $\Delta_I \times \mathbb{R}^N$.*

Proof It suffices to prove the statements when $f \in C_c^\infty(\mathbb{R}^N)$. Indeed, the general case follows by density, approximating $f \in C_0(\mathbb{R}^N)$ uniformly in $\mathbb{R}^N$ by a sequence $\{f_n\} \subset C_c^\infty(\mathbb{R}^N)$ and taking into account that $G(\cdot,\cdot)f_n$ converges to $G(\cdot,\cdot)f$ uniformly in $K \times \mathbb{R}^N$ for any compact set $K \subset \Delta_I$.

(i) Formulae (14.1.3) and (14.1.6) show that

$$\|G(t,s_1)f - G(t,s_0)f\|_\infty \leq \sup_{r\in[s_0,s_1]} (e^{c_0(t-r)}\|\mathcal{A}(r)f\|_\infty)|s_1-s_0|$$

for any $t \leq s_0 < s_1$ and this implies that the function $G(t,\cdot)f$ is locally Lipschitz continuous in $(-\infty,t]$ with values in $C_b(\mathbb{R}^N)$.

(ii) Using the classical Schauder estimates in Theorem D.1.5(i), we can show that, for any pair of compact sets $[a,b] \subset I$, $K \subset \mathbb{R}^N$ and any $m \in \mathbb{N}$, $\|G(\cdot,s)f\|_{C^{1+\alpha/2,2+\alpha}([s,s+m]\times K)}$ is bounded from above by a constant C_1 independent of $s \in [a,b]$. In particular, it follows that

$$|(G(t_2,s)f)(x) - (G(t_1,s)f)(x_0)| \leq C_1(|t_2-t_1| + |x-x_0|) \tag{14.1.11}$$

for any $t_1, t_2 \in [s,s+m]$, $x, x_0 \in K$ and $s \in [a,b]$.

Fix (t,s,x), $(t_0,s_0,x_0) \in \Delta_I \times \mathbb{R}^N$ with $s, s_0 \in [a,b]$ for some $[a,b] \subset I$. If $s < s_0$, then by (14.1.6) and (14.1.11) we can estimate

$$\begin{aligned}|(G(t,s)f)(x) - (G(t_0,s_0)f)(x_0)| &\leq |(G(t,s)f)(x) - (G(t_0,s)f)(x_0)| \\ &\quad + |(G(t_0,s)f)(x_0) - (G(t_0,s_0)f)(x_0)| \\ &\leq C_1(|t-t_0| + |x-x_0|) + C_2|s-s_0|,\end{aligned}$$

where $C_2 = \sup_{r\in[a,b]}(e^{c_0(t_0-r)}\|\mathcal{A}(r)f\|_\infty)$. On the other hand, if $s \geq s_0$ and $|t-t_0| \leq 1$, then $(t,s_0) \in \Delta_I$ and

$$\begin{aligned}&|(G(t,s)f)(x) - (G(t_0,s_0)f)(x_0)| \\ \leq&|(G(t,s)f)(x) - (G(t,s_0)f)(x)| + |(G(t,s_0)f)(x) - (G(t_0,s_0)f)(x_0)| \\ \leq&C_3|s-s_0| + |(G(t,s_0)f)(x) - (G(t_0,s_0)f)(x_0)|,\end{aligned}$$

where $C_3 = e^{c_0(t_0+1-a)}\max_{r\in[a,b]}\|\mathcal{A}(r)f\|_\infty$. From these two estimates it follows that $(G(t,s)f)(x)$ converges to $(G(t_0,s_0)f)(x_0)$ as (t,s,x) tends to (t_0,s_0^+,x_0), and the proof is completed. □

To prove that the function $G(\cdot,\cdot)f$ is continuous in $\Delta_I \times \mathbb{R}^N$ for any $f \in C_b(\mathbb{R}^N)$, we need the following hypothesis which replaces Hypothesis 14.0.1(iv).

Hypothesis 14.1.8 *For any bounded interval $J \subset I$ there exist a function $\varphi = \varphi_J \in C^2(\mathbb{R}^N)$, diverging to $+\infty$ as $|x|$ tends to $+\infty$, and a positive constant M_J such that $\mathcal{A}(t)\varphi(x) \le M_J$ for any $t \in J$ and $x \in \mathbb{R}^N$.*

Lemma 14.1.9 *Assume that Hypotheses 14.0.1(i)-(iii) and 14.1.8 are satisfied. Fix a bounded interval $J \subset I$ and let $\varphi = \varphi_J$ be as in Hypothesis 14.1.8. Then, φ belongs to $L^1(\mathbb{R}^N, p(t,s,x;dy))$ for any[4] $(t,s,x) \in \Delta_J \times \mathbb{R}^N$. Moreover, the function $G(\cdot,\cdot)\varphi$, defined through the formula (14.1.4), is bounded in $\Delta_J \times B_\varrho$ for any $\varrho > 0$.*

Proof The proof is an adaption to the non-autonomous setting of the proof of Lemma 4.1.3 and it strongly relies on Lemma 14.1.6. For this purpose, we just sketch it.

Fix a bounded interval $J \subset I$. Possibly adding a constant, we can assume that $\varphi = \varphi_J$ is nonnegative. For any $n \in \mathbb{N}$ let $\psi_n \in C^\infty([0,+\infty))$ satisfy the conditions $\psi_n(t) = t$ for $t \in [0,n]$, $\psi_n(t) \equiv n + 1/2$ for $t \ge n+1$, $0 \le \psi_n' \le 1$ and $\psi_n'' \le 0$ in $[0,+\infty)$. Then, the function $\varphi_n := \psi_n \circ \varphi$ belongs to $C^2_b(\mathbb{R}^N)$ and is constant outside of a compact set. In view of Lemma 14.1.6 we can estimate

$$\varphi_n(x) \ge \varphi_n(x) - (G(t,s)\varphi_n)(x) \ge -\int_s^t dr \int_{\mathbb{R}^N} (\psi_n' \circ \varphi)\mathcal{A}(r)\varphi\, p(t,r,x;dy) \tag{14.1.12}$$

for any $s,t \in \Delta_I$ and $x \in \mathbb{R}^N$. To let n tend to $+\infty$ in (14.1.12), we split the integral term in the right-hand side of (14.1.12) into the sum

$$\begin{aligned}&\int_s^t dr \int_{\mathbb{R}^N} (\psi_n' \circ \varphi)\mathcal{A}(r)\varphi\, p(t,r,x;dy)\\ =&M_J \int_s^t dr \int_{\mathbb{R}^N} (\psi_n' \circ \varphi) p(t,r,x;dy) - \int_s^t dr \int_{\mathbb{R}^N} (\psi_n' \circ \varphi)(M_J - \mathcal{A}(r)\varphi) p(t,r,x;dy).\end{aligned} \tag{14.1.13}$$

Since the sequence $\{\psi_n' \circ \varphi)\}$ pointwise increases to $\mathbb{1}$ in $\mathbb{R}^N$, both the integrals in the right-hand side of (14.1.13) converge by the monotone convergence theorem. We can thus let n tend to $+\infty$ in (14.1.12) and conclude that

$$(G(t,s)\varphi)(x) \le \varphi(x) + \int_s^t (G(t,r)\mathcal{A}(r)\varphi)(x)dr \tag{14.1.14}$$

for any $(t,s,x) \in \Delta_J \times \mathbb{R}^N$. Moreover, from (14.1.13) we also deduce that

$$\int_s^t (G(t,r)\mathcal{A}(r)\varphi)(x)\,dr \le M_J(t-s), \qquad (t,s,x) \in \Delta_J \times \mathbb{R}^N. \tag{14.1.15}$$

Estimates (14.1.14) and (14.1.15) imply that $(G(t,s)\varphi)(x) \le \varphi(x) + M_J(t-s)$ for any $s,t \in J$, with $s \le t$ and any $x \in \mathbb{R}^N$. The assertion follows. □

Thanks to Lemma 14.1.9 and arguing as in the second proof of Theorem 9.1.20, we can prove the following result.

Lemma 14.1.10 *Let Hypotheses 14.0.1(i)-(iii) and 14.1.8 be satisfied. Then, for each bounded interval $J \subset I$ and for each $r > 0$ the family of measures $\{p(t,s,x;dy) : (t,s,x) \in \Delta_J \times \overline{B}_r\}$ is tight.*[5]

[4] The measure $p(t,s,x,dy)$ is defined in Proposition 14.1.3(iii).

[5] When $\{\mu_t : t \in F\}$ is a family of finite but not probability measures on $B(\mathbb{R}^N)$, by tight we mean that, for any $\varepsilon > 0$, there exists $R > 0$ such that $\mu_t(\mathbb{R}^N \setminus B_R) \le \varepsilon$ for any $t \in F$.

Proof Fix $\varepsilon > 0$ and consider the function $\varphi = \varphi_J$ in Hypothesis 14.1.8. As in the proof of Lemma 14.1.9, we assume that φ is nonnegative. Since $\varphi(x)$ blows up as $|x|$ tends to $+\infty$, there exists $\varrho > 0$ such that $\varphi \geq M_{J,r}\varepsilon^{-1}\chi_{\mathbb{R}^N\setminus B_\varrho}$ in $\mathbb{R}^N$, where $M_{J,r}$ denotes the supremum of the function $G(\cdot,\cdot)\varphi$ over $\Delta_J \times B_r$. Thus, for $(t,s) \in \Delta_J$ we can estimate

$$p(t,s,x;\mathbb{R}^N \setminus B_\varrho) \leq \frac{\varepsilon}{M_{J,r}}\int_{\mathbb{R}^N}\varphi(y)p(t,s,x;dy) = \frac{\varepsilon}{M_{J,r}}(G(t,s)\varphi)(x) \leq \varepsilon,$$

so that $p(t,s,x;\mathbb{R}^N \setminus B_\varrho) \leq \varepsilon$ for any $(t,s,x) \in \Delta_J \times \overline{B}_r$ and the statement follows. □

The tightness yields some convergence result.

Proposition 14.1.11 *Assume that Hypotheses* 14.0.1(i), (iii) *and* 14.1.8 *are satisfied. Further, let* $\{f_n\}$ *be a bounded sequence in* $C_b(\mathbb{R}^N)$ *converging to* $f \in C_b(\mathbb{R}^N)$ *locally uniformly in* $\mathbb{R}^N$*, as* n *tends to* $+\infty$*. Then, the function* $G(\cdot,\cdot)f_n$ *converges to* $G(\cdot,\cdot)f$ *locally uniformly in* $\Delta_I \times \mathbb{R}^N$*.*

Proof Fix any bounded interval $J \subset I$, $r,\rho > 0$. Using the representation formula (14.1.4), for $(t,s,x) \in \Delta_J \times \overline{B}_r$ we split $G(t,s)f_n - G(t,s)f$ in the sum of two integral terms, one over B_ϱ and the other one over $\mathbb{R}^N \setminus B_\rho$. We can thus estimate

$$\begin{aligned}|(G(t,s)f_n)(x) - (G(t,s)f)(x)| \leq& \sup_{B_\varrho}|f_n - f|\int_{\mathbb{R}^N}p(t,s,x;dy)\\ &+\left(\sup_{n\in\mathbb{N}}\|f_n\|_\infty + \|f\|_\infty\right)\int_{\mathbb{R}^N\setminus B_\varrho}p(t,s,x;dy)\\ \leq& e^{c_0(t-s)}\sup_{B_\varrho}|f_n - f| + 2Mp(t,s,x;\mathbb{R}^N\setminus B_\rho).\end{aligned}$$

Taking the supremum with respect to $(t,s,x) \in \Delta_J \times \overline{B}_r$ and letting first n and then ρ tend to $+\infty$, we conclude that $G(\cdot,\cdot)f_n$ converges to $G(\cdot,\cdot)f$ uniformly in $\Delta_J \times \overline{B}_r$ thanks to Lemma 14.1.9. □

Now, we have all the tools to prove that the function $G(\cdot,\cdot)f$ is continuous in $\Delta_I \times \mathbb{R}^N$ for each $f \in C_b(\mathbb{R}^N)$.

Theorem 14.1.12 *Under the assumptions of Proposition* 14.1.11*, the function* $G(\cdot,\cdot)f$ *is continuous in* $\Delta_I \times \mathbb{R}^N$ *for any* $f \in C_b(\mathbb{R}^N)$*.*

Proof Fix $f \in C_b(\mathbb{R}^N)$ and let $\{f_n\} \in C_c^\infty(\mathbb{R}^N)$ be a sequence of smooth functions, bounded with respect to the sup-norm and converging to f locally uniformly in $\mathbb{R}^N$. By Corollary 14.1.7, each function $G(\cdot,\cdot)f_n$ is continuous in $\Delta_I \times \mathbb{R}^N$. Moreover, $G(\cdot,\cdot)f_n$ converges to $G(\cdot,\cdot)f$ as n tends to $+\infty$, locally uniformly in $\Delta_I \times \mathbb{R}^N$, in view of Proposition 14.1.11. From these facts, the assertion follows at once. □

14.2 Compactness of the evolution operator in $C_b(\mathbb{R}^N)$

We now give sufficient conditions ensuring that the operator $G(t,s)$ is compact. For notational convenience, for any interval $J \subset I$, we set

$$\widetilde{\Delta}_J := \{(t,s) \in J \times J : t > s\}. \tag{14.2.1}$$

Using the same arguments as in the proof of Proposition 4.1.2, where now $S_nf = G(t,r)(\chi_{B_n}G(r,s)f)$ and r is any point arbitrarily fixed in (s,t), we can prove the following result.

Proposition 14.2.1 *Let $J \subset I$ be an interval. Then, $G(t,s)$ is compact in $C_b(\mathbb{R}^N)$ for any $(t,s) \in \widetilde{\Delta}_J$ if and only if the family of measures $\{p(t,s,x;dy) : x \in \mathbb{R}^N\}$, defined in Proposition 14.1.3(iii), is tight for any $(t,s) \in \widetilde{\Delta}_J$.*

Proposition 14.1.3(iii) shows that $p(t,s,x;\mathbb{R}^N) > 0$ for any $t > s \in I$ and $x \in \mathbb{R}^N$. If $c \not\equiv 0$, then it is not clear whether $\inf_{x\in\mathbb{R}^N} p(t,s,x;\mathbb{R}^N) > 0$ for any $t > s \in I$ or not. The positivity of the previous infimum, which is crucial for the proof of Theorem 14.2.3, can be guaranteed under an additional assumption.

Proposition 14.2.2 *Assume that Hypotheses 14.0.1, 14.1.5 hold and $c \not\equiv 0$. Let $J \subset I$ be an interval and suppose that there exist $\mu \in \mathbb{R}$, $R > 0$ and a positive and bounded function $W \in C^2(\mathbb{R}^N \setminus B_R)$ with positive infimum and such that*

$$\mathcal{A}W - \mu W \geq 0, \qquad \text{in } J \times (\mathbb{R}^N \setminus B_R). \tag{14.2.2}$$

Then, for any $s_0, T \in J$, with $T > s_0$, there exists a positive constant C_{T,s_0} such that $p(t,s,x;\mathbb{R}^N) \geq C_{T,s_0}$ for any $s,t \in \mathbb{R}$, with $s_0 \leq s \leq t \leq T$, and any $x \in \mathbb{R}^N$.

Proof We first assume that $c \leq 0$ and introduce the function v defined by $v(t,x) = e^{-\mu(t-s_0)}(G(t,s_0)\mathbb{1})(x)$ for any $t \geq s_0$ and $x \in \mathbb{R}^N$. Since the evolution operator $\{G(t,s)\}$ is strong Feller (see Proposition 14.1.3(ii)), $G(t,s_0)\mathbb{1}$ is everywhere positive in $\mathbb{R}^N$. Hence, the minimum of v over $[s_0,T] \times \overline{B}_R$ is a positive constant, which we denote by M.

The function $z = v - \gamma W$, where $\gamma = M/\sup_{\mathbb{R}^N\setminus B_R} W$, belongs to $C_b([s_0,T] \times \mathbb{R}^N) \cap C^{1,2}((s_0,T] \times \mathbb{R}^N)$ and solves the following problem:

$$\begin{cases} D_t z(t,x) \geq \mathcal{A}(t)z(t,x) - \mu z(t,x), & t \in (s_0,T], \quad x \in \mathbb{R}^N \setminus \overline{B}_R, \\ z(t,x) \geq 0, & t \in [s_0,T], \quad x \in \partial B_R, \\ z(s_0,x) \geq 0, & \qquad\qquad\quad x \in \mathbb{R}^N \setminus B_R. \end{cases}$$

Proposition 14.1.1 implies that $z \geq 0$ in $[s_0,T] \times \mathbb{R}^N \setminus B_R$, i.e.,

$$e^{-\mu(t-s_0)}(G(t,s_0)\mathbb{1})(x) \geq \gamma W(x) \geq \gamma \inf_{y\in\mathbb{R}^N\setminus B_R} W(y), \qquad t \in [s_0,T], \; x \in \mathbb{R}^N \setminus B_R.$$

It thus follows that $G(t,s_0)\mathbb{1}$ is not less than $C_{T,s_0} = e^{\mu^-(T-s_0)}(M \wedge \inf_{\mathbb{R}^N\setminus B_R}(\gamma W))$ in $\mathbb{R}^N$.

Let us now fix $s \in (s_0,t)$. From the formula (14.1.10) we infer that the function $(G(t,\cdot)\mathbb{1})(x)$ is increasing. Therefore, $(G(t,s)\mathbb{1})(x) \geq (G(t,s_0)\mathbb{1})(x) \geq C_{T,s_0}$ for any $x \in \mathbb{R}^N$, and this accomplishes the proof in the case when $c \geq 0$, since by the representation formula (14.1.4) and (14.1.5), $p(t,s,x;\mathbb{R}^N) = (G(t,s)\mathbb{1})(x)$ for any $s,t \in I$, with $s < t$, and $x \in \mathbb{R}^N$.

In the general case when $c_0 > 0$, let $\{\widetilde{G}(t,s)\} = \{e^{-c_0(t-s)}G(t,s)\}$ be the evolution operator associated with the second-order elliptic operator $\mathcal{A}_{-c_0} = \mathrm{Tr}(QD_x^2) + \langle b, \nabla_x\rangle + (c - c_0)$. Clearly, the operator $\mathcal{A}_{-c_0}$ satisfies Hypotheses 14.0.1(iv) and 14.1.5 with the same λ_J and φ_J. Moreover, it fulfills also assumption (14.2.2) with μ replaced with $\mu - c_0$. Hence, from the above arguments we conclude that for any s_0, T there exists a positive constant C'_{T,s_0} such that $(\widetilde{G}(t,s)\mathbb{1})(x) \geq C'_{T,s_0}$ for any $x \in \mathbb{R}^N$ and $T \geq t \geq s \geq s_0$, and the assertion follows with $C_{T,s_0} = C'_{T,s_0}$. □

Adapting to our situation the technique in Theorem 4.1.4, we provide a sufficient condition which ensures the compactness of the evolution operator.

Theorem 14.2.3 *Assume that there exist $R > 0$, $d_1, d_2 \in I$, with $d_1 < d_2$, a positive function $\zeta \in C^2(\mathbb{R}^N)$, blowing up as $|x|$ tends to $+\infty$, and a convex increasing function $h : [0, +\infty) \to \mathbb{R}$ such that $1/h \in L^1((a, +\infty))$ for large values of a and $\mathcal{A}(s)\zeta(x) \leq -h(\zeta(x))$ for any $s \in [d_1, d_2]$ and $|x| \geq R$. Finally, when $c \not\equiv 0$, let Hypothesis 14.1.5 and the assumptions of Proposition 14.2.2 hold true with $J = [d_1, d_2]$. Then, $G(t, s)$ is compact in $C_b(\mathbb{R}^N)$ for any $(t, s) \in \{(t, s) \in \Delta_I : s \leq d_2, t \geq d_1, t \neq s\}$.*

Proof Of course we can limit ourselves to proving the compactness of $G(t, s)$ for $(t, s) \in \widetilde{\Delta}_{[d_1,d_2]}$ (see (14.2.1)) since for the other values of (t, s) it suffices to use the evolution law. Moreover, without loss of generality, we assume that h is nonnegative, $\mathcal{A}\zeta \leq -h \circ \zeta$ in $[d_1, d_2] \times \mathbb{R}^N$ and $c \leq 0$. We reduce to this situation replacing $\{G(t, s)\}$ and h, respectively, by the evolution operator $\{e^{K(t-s)}G(t, s)\}$ and the function $h - 0 \wedge h(0)$, where $K = K_{d_1,d_2} = (0 \wedge h(0) - K_1)(\inf_{\mathbb{R}^N} \zeta)^{-1} - (c_0 + (0 \wedge h(0) - K_1)(\inf_{\mathbb{R}^N} \zeta)^{-1})^+$, $K_1 = K_{1,d_1,d_2}$ being any positive constant such that $\mathcal{A}\zeta + h \circ \zeta \leq K_1$ in $[d_1, d_2] \times B_R$ and c_0 is the supremum of c. Clearly, the operator $G(t, s)$ is compact in $C_b(\mathbb{R}^N)$ if and only if the operator $e^{K(t-s)}G(t, s)$ is compact in $C_b(\mathbb{R}^N)$. Moreover, the evolution operator $\{e^{K(t-s)}G(t, s)\}$ is associated with the operator $\mathcal{A} + K$ which satisfies the inequality $(\mathcal{A} + K)\zeta \leq h \circ \zeta - 0 \wedge h(0)$ and has a nonpositive potential.

In view of Proposition 14.2.1, we prove that $\{p(t, s, x; dy) : x \in \mathbb{R}^N\}$ is a tight family of measures for any $(t, s) \in \Delta_{[d_1,d_2]}$. Since h is increasing, $\mathcal{A}\zeta$ is bounded from above in $[d_1, d_2] \times \mathbb{R}^N$. Therefore, Lemma 14.1.9 shows that the function $G(\cdot, \cdot)\zeta$ is bounded in $\Delta_{[d_1,d_2]} \times B_r$ for any $r > 0$. Moreover, the same arguments in the proof of the quoted lemma allow us to prove that

$$(G(t, s)\zeta)(x) - (G(t, r)\zeta)(x) \geq -\int_r^s (G(t, \sigma)\mathcal{A}(\sigma)\zeta)(x)d\sigma \tag{14.2.3}$$

for any $r, s \in [d_1, d_2]$, such that $r < s < t$, and any $x \in \mathbb{R}^N$. Let us prove that the function $G(\cdot, \cdot)\zeta$ is actually bounded in $\Delta_{[d_1,d_2]} \times \mathbb{R}^N$. For this purpose, we begin by observing that $A := p(t, s, x; \mathbb{R}^N) \leq 1$ and $A^{-1}p(t, s, x; dy)$ is a probability measure. Hence, recalling that h is an increasing convex function and applying the Jensen inequality, we can estimate

$$\begin{aligned} h((G(t, s)\zeta)(x)) \leq & h\bigg(\int_{\mathbb{R}^N} \zeta A^{-1}p(t, s, x; dy)\bigg) \leq \int_{\mathbb{R}^N} (h \circ \zeta)\, A^{-1}p(t, s, x; dy) \\ = & \frac{(G(t, s)(h \circ \zeta))(x)}{p(t, s, x; \mathbb{R}^N)} \end{aligned}$$

for any $(t, s, x) \in \Delta_{[d_1,d_2]} \times \mathbb{R}^N$, where, with a slight abuse of notation, we have denoted by $(G(t, s)(h \circ \zeta))(x)$ the last integral on the first line of the previous chain of inequalities. This term is finite since it is bounded from above by $-(G(t, s)\mathcal{A}(s)\zeta)(x)$, which is finite for almost every $s \in [d_1, t]$: indeed, it belongs to $L^1((d_1, d_2))$, as the proof of Lemma 14.1.9 shows.

Fix $s_0 < T$. Then, by Proposition 14.2.2 it follows that $G(\cdot, \cdot)(h \circ \zeta) \geq C_{d_1,d_2}h(G(\cdot, \cdot)\zeta)$ in $\Delta_{[d_1,d_2]} \times \mathbb{R}^N$.

Now, let us fix $x \in \mathbb{R}^N$, $t \in [d_1, d_2]$ and define the function $\beta : [0, r_0) \to \mathbb{R}$ by setting $\beta(r) = (G(t, t - r)\zeta)(x)$ for any $r \in [0, r_0)$, where $r_0 = t - \inf I$. Function β is measurable since it is the limit of the sequence of the continuous functions $(G(t, t-\cdot)\zeta_n)(x)$ (see Corollary 14.1.7).

Fix $b = t - d_1$. From estimate (14.2.3), the condition $\mathcal{A}\zeta \leq -h \circ \zeta$ in $[d_1, d_2] \times \mathbb{R}^N$ and all the above remarks, we deduce that

$$\beta(b) - \beta(0) \leq -\int_{t-b}^t (G(t, \sigma)h(\zeta))(x)d\sigma \leq -C_{d_1,d_2}\int_0^b h(\beta(\sigma))d\sigma.$$

Let $y(\cdot;x)$ be the unique solution to the differential equation $y' = -C_{d_1,d_2} h \circ y$, which satisfies the condition $y(0) = \zeta(x)$. Since h is increasing, $y(\cdot;x)$ is defined for all the positive times. Moreover, recalling that the function $1/h$ is integrable in $[\zeta(x), +\infty)$, we can estimate

$$\int_{y(t;x)}^{+\infty} \frac{dz}{h(z)} = \int_{\zeta(x)}^{+\infty} \frac{dz}{h(z)} - \int_{\zeta(x)}^{z(t,x)} \frac{dz}{h(z)} \geq - \int_{\zeta(x)}^{y(t;x)} \frac{dz}{h(z)} = C_{d_1,d_2} t.$$

Fix $\delta > 0$ and let $\bar{y} = \bar{y}(\delta) > 0$ be the unique point such that $\int_{\bar{y}}^{+\infty} \frac{1}{h(s)} ds = C_{d_1,d_2}\delta$. Then, $y(r;x) \leq \bar{y}$ for any $r \geq \delta$. Let us now show that $\beta \leq y(\cdot;x)$. This will allow us to conclude that $\beta \leq \bar{y}$ in $[\delta, t-d_1]$ for any $\delta < t - d_1$ or, equivalently, that $\sup_{x\in\mathbb{R}^N} \sup_{s\in[d_1,t-\delta]} (G(t,s)\zeta)(x) \leq \bar{y}$. Arguing as in the proof of Lemma 14.1.10, we can then conclude the proof. By contradiction, we suppose that $\beta(s_0) > y(s_0;x)$ for some $s_0 \in (0,b)$. Then, there exists an interval J containing s_0 where $\beta > y$. It suffices to observe that the inequality

$$\beta(s_2) - \beta(s_1) \leq -C_{d_1,d_2} \int_{s_1}^{s_2} h(\beta(\sigma)) d\sigma, \qquad s_1, s_2 \in [0, r_0), \;\; s_1 \leq s_2,$$

implies that the function β is decreasing. Thus,

$$\lim_{s\to s_0^-} \beta(s) \geq \beta(s_0) > y(s_0;x) = \lim_{s\to s_0^-} y(s;x),$$

so that $\beta > y(\cdot;x)$ in a left neighborhood of s_0. If we set $a = \inf J$, then $\beta(a) \leq y(a;x)$. We get to a contradiction observing that

$$\beta(s) \leq \beta(a) - C_{d_1,d_2} \int_a^s (h\circ\beta) d\sigma, \qquad y(s;x) = y(a;x) - C_{d_1,d_2} \int_a^s (h\circ\beta) d\sigma$$

for any $s \in J$, which yield

$$\beta(s) - y(s;x) \leq C_{d_1,d_2} \int_a^s (h(y(\sigma;x)) - h(\beta(\sigma))) d\sigma, \qquad s \in J.$$

This is a contradiction since the left-hand side is positive while the right-hand side is negative. □

14.3 Invariance of $C_0(\mathbb{R}^N)$

As in the autonomous case, we can provide sufficient conditions for $C_0(\mathbb{R}^N)$ to be or not to be preserved by the action of the evolution operator $\{G(t,s)\}$.

Theorem 14.3.1 *Let the assumptions of Theorem* 14.2.3 *be satisfied. Further, assume that, for some $s, t \in J$ with $s < t$, the operator $G(t,s)$ is compact in $C_b(\mathbb{R}^N)$. Then, $G(t,s)$ does not preserve $C_0(\mathbb{R}^N)$.*

Proof Let $\{f_n\}$ be a sequence of smooth functions such that $\chi_{B_n} \leq f_n \leq \chi_{B_{2n}}$ for any $n \in \mathbb{N}$, and fix $s, t \in I$ with $s < t$. From the formula (14.1.4) and the dominated convergence theorem it follows immediately that $G(t,s)f_n$ converges pointwise in $\mathbb{R}^N$ to $G(t,s)\mathbb{1}$ as n tends to $+\infty$. Since $G(t,s)$ is a compact operator, $G(t,s)f_n$ actually converges uniformly in $\mathbb{R}^N$ to $G(t,s)\mathbb{1}$. If $G(t,s)$ preserved $C_0(\mathbb{R}^N)$, then the function $G(t,s)\mathbb{1}$ would vanish as $|x|$ tends to $+\infty$, but this is not the case. Indeed, Proposition 14.2.2 shows that $G(t,s)\mathbb{1}$ is bounded from below by a positive constant. □

Now we provide a sufficient condition for $C_0(\mathbb{R}^N)$ to be preserved. This is the counterpart of Theorem 4.3.2.

Theorem 14.3.2 *Fix $I \ni a < b$ and assume that there exist a strictly positive function $V \in C^2(\mathbb{R}^N)$ and $\lambda_0 > 0$ such that $\lim_{|x|\to+\infty} V(x) = 0$ and $\lambda_0 V - \mathcal{A}V \geq 0$ in $[a,b] \times \mathbb{R}^N$. Then, $G(t,s)$ preserves $C_0(\mathbb{R}^N)$ for any $(t,s) \in \Delta_{[a,b]}$.*

Proof Fix $s \in [a,b]$. It suffices to prove that $G(t,s)f \in C_0(\mathbb{R}^N)$ for any $(t,s) \in \Delta_{[a,b]}$ and $f \in C_c(\mathbb{R}^N)$; the general case then follows from a density argument, approximating an arbitrary $f \in C_0(\mathbb{R}^N)$ by a sequence $\{f_n\} \subset C_c(\mathbb{R}^N)$ with respect to the sup-norm in $\mathbb{R}^N$. It is not restrictive to suppose $f \geq 0$ otherwise we consider its positive and negative part. Fix $R > 0$ such that $\mathrm{supp}(f) \subset B_R$, let $\delta = \inf_{x\in B_R} V(x) > 0$ and $z(t,x) = e^{-\lambda_0(t-s)}(G(t,s)f)(x) - \delta^{-1}\|f\|_\infty V(x)$ for any $(t,x) \in [s,+\infty) \times \mathbb{R}^N$. The function z belongs to $C_b([s,b] \times \mathbb{R}^N) \cap C^{1,2}((s,b] \times \mathbb{R}^N)$ and

$$\begin{cases} D_t z(t,x) - \mathcal{A}(t)z(t,x) + \lambda_0 z(t,x) \leq 0, & (t,x) \in (s,b] \times \mathbb{R}^N, \\ z(s,x) \leq 0, & x \in \mathbb{R}^N. \end{cases}$$

Therefore, applying Proposition 14.1.1 (with $\mathcal{A}$ replaced with $\mathcal{A} - \lambda_0$) we get $z \leq 0$, i.e.,

$$0 \leq u(t,x) \leq e^{\lambda_0(t-s)}\delta^{-1}\|f\|_\infty V(x), \qquad s \leq t \leq b, \ x \in \mathbb{R}^N, \tag{14.3.1}$$

which implies that $u \in C_0(\mathbb{R}^N)$. □

In the autonomous case, when the coefficients of the operator $\mathcal{A}$ do not depend on t, it is known that, under Hypotheses 14.0.1, $T(t)f$ converges to f uniformly in $\mathbb{R}^N$ as t tends to 0 from the right, for any $f \in C_0(\mathbb{R}^N)$, where $\{T(t)\}$ is the semigroup in $C_b(\mathbb{R}^N)$ associated with the operator $\mathcal{A}$ (see Proposition 1.2.8). The proof of such a proposition is based on the formula

$$(T(t)f)(x) - f(x) = \int_0^t (T(s)\mathcal{A}f)(x)ds, \qquad t > 0, \ x \in \mathbb{R}^N, \ f \in C_c^2(\mathbb{R}^N),$$

which has not a natural counterpart in the non-autonomous case, since $\{G(t,s)\}$ does not commute with $\mathcal{A}$. This difficulty can be overcome by taking advantage of Theorem 14.3.2.

Corollary 14.3.3 *Let the assumptions of Theorem 14.3.2 be satisfied. Then, the function $G(\cdot,\cdot)f$ is continuous in $\Delta_{[a,b]}$ with values in $C_0(\mathbb{R}^N)$, for any $f \in C_0(\mathbb{R}^N)$. In particular, if for any $I \ni a < b$ there exist $\lambda = \lambda_{a,b}$ and $V = V_{a,b}$, with the properties in Theorem 14.3.2, then the restriction of $\{G(t,s)\}$ to $C_0(\mathbb{R}^N)$ gives rise to a strongly continuous evolution operator.*

Proof We split the proof into two steps.

Step 1. Here, we prove that the map $G(\cdot,s_0)f$ is continuous in $[s_0,b]$ for any $s_0 \in [a,b)$ and $f \in C_0(\mathbb{R}^N)$. We first assume that $f \in C_c(\mathbb{R}^N)$ and prove that the function $G(\cdot,s_0)f$ is continuous at t_0. It is not restrictive to assume that f is nonnegative. Indeed, if this is not the case, then it suffices to argue on the positive and negative parts of f.

Let R and δ be as in the proof of Theorem 14.3.2, fix $\varepsilon > 0$ and choose $\widetilde{R} > 0$ such that $V(x) \leq \varepsilon\delta e^{-\lambda_0(b-a)}\|f\|_\infty^{-1}$ for any $|x| \geq \widetilde{R}$. Then, estimate (14.3.1) implies that

$(G(t,s_0)f)(x) \le \varepsilon$ for any $|x| \ge \widetilde{R}$ and $t \in [s_0, b]$. Therefore, $\|G(t,s_0)f - G(t_0,s_0)f\|_\infty \le \|G(t,s_0)f - G(t_0,s_0)f_0\|_{C(\overline{B}_{\widetilde{R}})} + 2\varepsilon$. Since the function $G(\cdot, s_0)f$ is continuous in $[s_0,+\infty) \times \mathbb{R}^N$, from the previous estimate we obtain that $\limsup_{t\to t_0} \|G(t,s_0)f - G(t_0,s_0)f\|_\infty \le 2\varepsilon$. The arbitrariness of $\varepsilon > 0$ allows us to conclude that $G(t,s_0)f$ converges to $G(t_0,s_0)f$ as t tends to t_0, uniformly in $\mathbb{R}^N$.

The continuity of $G(\cdot,s_0)f$ when $f \in C_0(\mathbb{R}^N)$ follows approximating f by a sequence of compactly supported functions and taking (14.1.3) into account.

Step 2. Here, we complete the proof. We assume that $f \in C^2_c(\mathbb{R}^N)$. As in Step 1, the general case then will follow by a standard approximation argument and estimate (14.1.3).

So, let us fix $f \in C^2_c(\mathbb{R}^N)$, $(s_0,t_0), (s,t) \in \Delta_{[a,b]}$ and prove that $G(\cdot,\cdot)f$ is continuous at (t_0,s_0). If $t > t_0$, then we split

$$G(t,s)f - G(t_0,s_0)f = G(t,t_0)(G(t_0,s)f - G(t_0,s_0)f) + G(t,t_0)G(t_0,s_0)f - G(t_0,s_0)f.$$

Therefore, using (14.1.3) we can estimate

$$\begin{aligned}\|G(t,s)f - G(t_0,s_0)f\|_\infty \le & e^{c_0(t-t_0)}\|G(t_0,s)f - G(t_0,s_0)f\|_\infty \\ & + \|G(t,t_0)G(t_0,s_0)f - G(t_0,s_0)f\|_\infty.\end{aligned}$$

On the other hand if $t < t_0$, then we can estimate

$$\begin{aligned}\|G(t,s)f - G(t_0,s_0)f\| \le & \|G(t,s)f - G(t,s_0)f\|_\infty + \|G(t_0,t)G(t_0,s_0)f - G(t_0,s_0)f\|_\infty \\ & + e^{c_0(t-t_0)}\|G(t,s_0)f - G(t_0,s_0)f\|_\infty + \|G(t,s_0)f - G(t_0,s_0)f\|_\infty.\end{aligned}$$

Using Step 1 and Corollary 14.1.7(i), we conclude that the right-hand sides of the previous two inequalities vanish as s tends to s_0 and t tends to t_0, from the right and the left, respectively. This completes the proof. □

14.4 Gradient estimates

In this section we prove both uniform and pointwise gradient estimates for the function $G(t,s)f$ when f belongs to $C_b(\mathbb{R}^N)$ or to $C^1_b(\mathbb{R}^N)$. Besides being interesting in their own right, we will need them in the next chapters to prove some remarkable results. In view of the applications of these estimates, we assume that $c \equiv 0$.

Throughout the section we assume Hypotheses 14.0.1 and the following additional conditions on the coefficients of the operator $\mathcal{A}$.

Hypotheses 14.4.1 (i) *The coefficients q_{ij} and b_i $(i,j = 1,\dots,N)$ and their first-order spatial derivatives belong to $C^{\alpha/2,\alpha}_{\rm loc}(I \times \mathbb{R}^N)$ and $c \equiv 0$;*

(ii) *there exists a continuous function $k_0 : I \to \mathbb{R}$ such that $\langle {\rm Jac}_x b(t,x)\xi, \xi\rangle \le k_0(t)|\xi|^2$ for any $(t,x) \in I \times \mathbb{R}^N$ and $\xi \in \mathbb{R}^N$;*

(iii) *there exists a continuous function $q_0 : I \to \mathbb{R}_+$ such that, $|D_k q_{ij}(t,x)| \le q_0(t)\kappa(t,x)$ for any $(t,x) \in I \times \mathbb{R}^N$ and $i,j,k \in \{1,\dots,N\}$.*

First, we prove uniform gradient estimates.

Theorem 14.4.2 *Let $I \ni s < T$. Then, there exist positive constants C_1, C_2, depending on s and T, such that*

$$\|\nabla_x G(t,s)f\|_\infty \le \frac{C_1}{\sqrt{t-s}}\|f\|_\infty, \qquad s < t \le T, \; f \in C_b(\mathbb{R}^N), \tag{14.4.1}$$

$$\|\nabla_x G(t,s)f\|_\infty \le C_2\|f\|_{C_b^1(\mathbb{R}^N)}, \qquad s < t \le T, \; f \in C_b^1(\mathbb{R}^N). \tag{14.4.2}$$

Further, if the functions q_0 and k_0 in Hypothesis 14.4.1(ii), (iii) are bounded from above in I, then the constant C_1 is independent of s and T. Moreover, the pointwise gradient estimate

$$|\nabla_x G(t,s)f|^p \le C_1^p((t-s)^{-\frac{p}{2}} \vee 1)G(t,s)|f|^p, \qquad t > s \in I, \; f \in C_b(\mathbb{R}^N), \tag{14.4.3}$$

holds true for any $p \ge 2$ and some positive constant C_3, independent of s, t and f.

Proof First of all, we stress that it suffices to prove the statement when $f \in C_c^{2+\alpha}(\mathbb{R}^N)$. For a general $f \in C_b(\mathbb{R}^N)$ we consider a sequence $\{f_n\} \subset C_c^{2+\alpha}(\mathbb{R}^N)$, bounded with respect to the sup-norm and converging to f locally uniformly in $\mathbb{R}^N$. Proposition 14.1.4(i) shows that $\nabla_x G(\cdot,s)f_n$ converges to $\nabla_x G(\cdot,s)f$ pointwise in $(s,T] \times \mathbb{R}^N$. Hence, we can write (14.4.2) and (14.4.1), with f being replaced by f_n, and let n tend to $+\infty$ to get the assertions in the general case.

So, let us fix $f \in C_c^{2+\alpha}(\mathbb{R}^N)$ and $n_0 \in \mathbb{N}$ such that the support of f is contained in B_{n_0}. Then, for any $n \ge n_0$ the function $u_n = G_n^{\mathcal{N}}(\cdot,s)f$ has the regularity properties in the statement of Proposition 13.1.4. Moreover, even if the quoted proposition is concerned with the autonomous case, its proof applies also in our situation since Hypotheses 13.1.1 hold true in $\Omega = B_n$ with constants independent of $t \in [s,T]$ and $n \in \mathbb{N}$. We thus deduce that there exists a positive constant C_1, independent of n, t, s and f, such that

$$(G_n^{\mathcal{N}}(t,s)f)(x))^2 + C_1^{-2}(t-s)^{1-k}|(\nabla_x G_n^{\mathcal{N}}(t,s)f)(x)| \le G_n^{\mathcal{N}}(t,s)(f^2 + C_1^{-2}\delta_1(k)|\nabla f|^2) \tag{14.4.4}$$

for any $t \in (s,T)$, $x \in B_n$, $n \in \mathbb{N}$ and $k = 0,1$, where δ_1 denotes the Kronecker delta centered at 1. Hence, it follows that there exists a constant $C_2 > 0$ such that $|(\nabla_x G_n^{\mathcal{N}}(t,s)f)(x)| \le C_{k+1}(t-s)^{-(1-k)/2}\|f\|_{C_b^k(\mathbb{R}^N)}$ for t, x and k as above. By Theorem 14.1.2 we know that $G_n^{\mathcal{N}}(\cdot,s)f$ converges to $G(\cdot,s)f$ in $C^{1,2}(D)$ for any compact set $D \subset (s,+\infty) \times \mathbb{R}^N$, as n tend to $+\infty$, and this allows us to conclude the proof of estimates (14.4.1) and (14.4.2).

To complete the proof we observe that, if the functions q_0 and k_0 are bounded from above in I, then from (14.4.4) it follows that

$$|\nabla_x G_n^{\mathcal{N}}(t,s)f|^2 \le C_1^2(t-s)^{-1}G_n^{\mathcal{N}}(t,s)f^2, \qquad s \in I, \; t \in (s,s+1),$$

and letting n tend to $+\infty$ yields

$$|(\nabla_x G(t,s)f)(x)|^2 \le C_1^2(t-s)^{-1}G(t,s)f^2, \qquad s \in I, \; t \in (s,s+1). \tag{14.4.5}$$

Using the Hölder inequality and taking Proposition 14.1.3 into account, which shows that $\|g(t,s,x,\cdot)\|_{L^1(\mathbb{R}^N)} = 1$, from (14.4.5) we get

$$\begin{aligned}|\nabla_x G(t,s)f|^p &= (|\nabla_x G(t,s)f|^2)^{\frac{p}{2}} \\ &\le (C_1^2(t-s)^{-1}G(t,s)|f|^2)^{\frac{p}{2}} \\ &\le C_1^p(t-s)^{-\frac{p}{2}}G(s+t,s)|f|^p\end{aligned}$$

for s, t and f as above. To extend this estimate to $t > s+1$, it suffices to split $\nabla_x G(s+t,s)f = \nabla_x G(s+t,s+t-1)G(s+t-1,s)f$. The proof is now complete. □

Corollary 14.4.3 *For any $f \in C_b^1(\mathbb{R}^N)$ and $s \in I$, the function $\nabla_x G(\cdot, s)f$ is continuous in $[s, +\infty) \times \mathbb{R}^N$.*

Proof We have to show only the continuity on $\{s\} \times \mathbb{R}^N$. For any $n \in \mathbb{N}$, let $\zeta \in C_c^\infty(B_n)$ satisfy $\chi_{B_{n-1}} \le \zeta \le 1$. Set $u = G(\cdot, s)f$ and $v = \zeta u$. The function v satisfies the equation $D_t v = \mathcal{A}v + \psi$ in B_n, where $\psi = -u\mathcal{A}\zeta - 2\langle Q\nabla\zeta, \nabla_x u\rangle$. From Theorems 14.1.2 and 14.4.2, it follows that the functions u and $\nabla_x u$ are bounded and continuous in $(s, T] \times \mathbb{R}^N$ for any $T > s$. Since ζ is compactly supported in B_n, ψ belongs to $C((s, s+1], C_0(B_n))$. Moreover, Theorem 14.4.2 yields that $\|\psi\|_\infty \le C\|f\|_{C_b^1(\mathbb{R}^N)}$ for some $C > 0$.

Let $\{G_n^{\mathcal{D}}(t, s)\}$ be the evolution family associated with problem (14.1.1). By the variation-of-constants formula (e.g. [1, Proposition 3.2]) we can write

$$v(t, \cdot) = G_n^{\mathcal{D}}(t, s)(\zeta f) + \int_s^t G_n^{\mathcal{D}}(t, \sigma)\psi(\sigma, \cdot)d\sigma, \qquad s < t < s+1. \tag{14.4.6}$$

By classical gradient estimates,[6] we get

$$\|\nabla_x G_n^{\mathcal{D}}(t, \sigma)\psi(\sigma, \cdot)\|_\infty \le \frac{C_1}{\sqrt{t-\sigma}}\|\psi(\sigma, \cdot)\|_\infty \le \frac{C_2}{\sqrt{t-\sigma}}\|f\|_{C_b^1(\mathbb{R}^N)}$$

for any $s < \sigma < t \le s+1$ and some positive constants C_1 and C_2. Hence, we can differentiate (14.4.6) obtaining

$$\nabla_x v(t, \cdot) = \nabla_x G_n^{\mathcal{D}}(t, s)(\zeta f) + \int_s^t \nabla_x G_n^{\mathcal{D}}(t, \sigma)\psi(\sigma, \cdot)d\sigma, \qquad s < t < s+1.$$

The previous formula shows that

$$|\nabla_x u(t, \cdot) - \nabla f(x_0)| \le |\nabla_x G_n(t, s)(\zeta f) - \nabla f(x_0)| + 2C_2\|f\|_{C_b^1(\mathbb{R}^N)}\sqrt{t-s}$$

in B_{n-1}, for any $x_0 \in B_{n-1}$, and this implies that $\nabla_x G(\cdot, s)f$ is continuous at the point (s, x_0) since the function $\nabla_x G_n^{\mathcal{D}}(\cdot, s)(\zeta f)$ is continuous in $\{s\} \times \overline{B}_n$ by classical results. The statement follows from the arbitrariness of $n \in \mathbb{N}$. □

To extend the validity of estimate (14.4.3) and prove another kind of pointwise gradient estimate, we assume the following additional assumptions.

Hypotheses 14.4.4 (i) *There exist two functions $\widehat{q} : I \times \mathbb{R}^N \to \mathbb{R}$ and $r : I \times \mathbb{R}^N \to \mathbb{R}$, with finite supremum r_0, and $p \in (1, +\infty)$ such that $\langle(\mathrm{Jac}_x b)\xi, \xi\rangle \le r|\xi|^2$ in $I \times \mathbb{R}^N$ for any $\xi \in \mathbb{R}^N$ and*

$$\ell_p := \sup_{(t,x)\in I\times\mathbb{R}^N} \left(r(t, x) + \frac{N^2(\widehat{q}(t, x))^2}{4\min\{p-1, 1\}\kappa(t, x)}\right) < +\infty; \tag{14.4.7}$$

(ii) *Hypothesis* 14.4.1(*ii*) *holds true with the function k being replaced by a real constant k_0. Moreover, there exists a positive constant ζ_0 such that, for any $i, j, k = 1, \ldots, N$, we have $|D_k q_{ij}(t, x)| \le \zeta_0\sqrt{\kappa(t, x)}$ for any $(t, x) \in I \times \mathbb{R}^N$.*

Theorem 14.4.5 *The following properties are satisfied.*

(i) *Assume that Hypotheses* 14.0.1, 14.4.1(*i*), 14.4.1(*iii*) *and* 14.4.4(*i*) *are satisfied. Then, for any $f \in C_b^1(\mathbb{R}^N)$ it holds that*

$$|(\nabla_x G(t, s)f)(x)|^p \le e^{p\ell_p(t-s)}(G(t, s)|\nabla f|^p)(x), \qquad t \ge s,\ x \in \mathbb{R}^N. \tag{14.4.8}$$

[6] See e.g. [159, Chapter IV, Theorem 17].

(ii) *Under Hypotheses* 14.0.1, 14.4.1(*i*) *and* 14.4.4(*ii*), *estimate* (14.4.8) *holds true for any* $p \in (1, +\infty)$, *with*

$$\ell_p = r_0 + \frac{N^2 \zeta_0^2}{4\min\{p-1, 1\}}. \tag{14.4.9}$$

In particular, if the coefficients q_{ij} $(i, j = 1, \dots, N)$ *do not depend on* x *then estimate* (14.4.8) *holds true for* $p = 1$ *too, with* $\ell_p = r_0$ *for any* $p \geq 1$.

(iii) *Under the assumptions in* (*i*) *or* (*ii*) *there exists a positive constant* C_p *such that*

$$|(\nabla_x G(t,s)f)(x)|^p \leq C_p^p (1 \vee (t-s)^{-\frac{p}{2}}) e^{p\ell_p^-(t-s)} (G(t,s)|f|^p)(x) \tag{14.4.10}$$

for any $f \in C_b(\mathbb{R}^N)$, $t > s$ *and* $x \in \mathbb{R}^N$. *Here,* p *is as in* (14.4.7) *under the assumptions in* (*i*), *whereas it is arbitrary in* $(1, +\infty)$ *under the assumptions in* (*ii*).

Proof (i), (ii) By the proof of Theorem 13.5.1, which also works in the non-autonomous case with no particular differences,[7] it follows that the gradient estimate (14.4.8) holds true with $\{G(t,s)\}$ being replaced by $\{G_n^{\mathcal{N}}(t,s)\}$ for any $n \in \mathbb{N}$.

Letting n tend to $+\infty$ and taking Theorem 14.1.2(i) into account, we get (14.4.8). In the particular case when the diffusion coefficients are independent of x, we obtain that $\ell_p = \sup_{I \times \mathbb{R}^N} r = r_0$ for any $p \in (1, +\infty)$. Hence, letting p tend to 1 from the right we can extend such estimate to any $p \in [1, +\infty)$.

(iii) We adapt the proof of Theorem 13.5.4, pointing out the main differences in the case when $p \in (1, 2]$ and referring to the proof of the quoted theorem, to Remark 13.5.5 and the proof of Theorem 14.4.2 for further details. As in the proof of Theorem 14.4.2, it suffices to show (14.4.10) with the evolution operator $\{G(t,s)\}$ being replaced by $\{G_n^{\mathcal{N}}(t,s)\}$ for any $n \in \mathbb{N}$. For this purpose, we fix $f \in C_c^{2+\alpha}(\mathbb{R}^N)$, $n \in \mathbb{N}$ such that $\text{supp}(f) \subset B_n$, $\delta \in (0,1)$, $r, s \in \mathbb{R}$, with $r < s$. Next, we introduce the function $\Phi_{n,\delta} = G_n^{\mathcal{N}}(s, \cdot)(|G_n^{\mathcal{N}}(\cdot, r)f|^2 + \delta)^{\frac{p}{2}}$, defined in $[r, s]$. To simplify the notation, throughout the rest of the proof we set $\varphi_n^\tau := G_n^{\mathcal{N}}(\tau, r)f$ for any $\tau \in [r, s]$.

The function $\Phi_{n,\delta}$ is differentiable with respect to $\tau \in (r, s)$ for any fixed $x \in \mathbb{R}^N$ and[8]

$$\begin{aligned}\Phi'_{n,\delta}(\tau, \cdot) =& G_n^{\mathcal{N}}(s,\tau)\big[D_\tau(|\varphi_n^\tau|^2 + \delta)^{\frac{p}{2}} - \mathcal{A}(\tau)(|\varphi_n^\tau|^2 + \delta)^{\frac{p}{2}}\big] \\ \leq & -c_p G_n^{\mathcal{N}}(s,\tau)\big[\,(|\varphi_n^\tau|^2 + \delta)^{\frac{p}{2}-1} |\nabla_x \varphi_n^\tau|^2\big],\end{aligned} \tag{14.4.11}$$

where $c_p = p(p-1)\kappa_0$. By integrating (14.4.11) in $[r+\varepsilon, s-\varepsilon]$ $(0 < 2\varepsilon < s-r)$, we get

$$(G_n^{\mathcal{N}}(s, r+\varepsilon)|G_n^{\mathcal{N}}(r+\varepsilon, r)f|^p)(x) \geq c_p \int_{r+\varepsilon}^{s-\varepsilon} [G_n^{\mathcal{N}}(s,\tau)((|\varphi_n^\tau|^2 + \delta)^{\frac{p}{2}-1}|\nabla_x \varphi_n^\tau|^2)](x)d\tau$$

for any $x \in \mathbb{R}^N$. Letting ε tend to 0^+ in this inequality, yields

$$c_p \int_r^s (G_n^{\mathcal{N}}(s,\tau)[(|\varphi_n^\tau|^2 + \delta)^{\frac{p}{2}-1}|\nabla_x \varphi_n^\tau|^2])(x)d\tau \leq (G_n^{\mathcal{N}}(s,r)|f|^p)(x) \tag{14.4.12}$$

for any $x \in \mathbb{R}^N$. Using estimate (14.4.8), the Hölder and Young inequalities, from the previous estimate we deduce that

$$|\nabla_x G_n^{\mathcal{N}}(s,r)f|^p \leq e^{p\ell_p(s-\tau)} \Big\{\frac{p}{2}\varepsilon^{\frac{2}{p}} G_n^{\mathcal{N}}(s,\tau)\big[(|\varphi_n^\tau|^2 + \delta)^{\frac{p}{2}-1}|\nabla_x \varphi_n^\tau|^2\big]$$

---

[7]The only relevant difference is the estimate of the term $\left|\sum_{i,j,h=1}^N D_h q_{ij} D_h u D_{ij} u\right|$, which now can be bounded from above by $(1-p)\kappa|D_x^2 u|^2 + (4\min\{p-1,1\}\kappa)^{-1}\widehat{q}^2 N^2$.

[8]See also [1, Theorem 2.3(ix)].

$$+\left(1-\frac{p}{2}\right)\varepsilon^{\frac{2}{p-2}}(G_n^{\mathcal{N}}(s,r)|f|^p+\delta^{\frac{p}{2}})\Big\}\tag{14.4.13}$$

in $\mathbb{R}^N$, for any $\varepsilon>0$. To complete the proof we multiply both sides of (14.4.13) by $e^{-p\ell_p(s-\tau)}$ and integrate with respect to $\tau\in(r,s)$. Taking (14.4.12) into account and letting δ tend to 0^+, estimate (14.4.10) follows. □

The following result is a straightforward consequence of Theorem 14.4.5 and it can be obtained by arguing as in the last part of the proof of Theorem 14.4.2.

Corollary 14.4.6 *Under the hypotheses of Theorem* 14.4.5(*iii*) *suppose that* $\ell_p<0$ *for some* p. *Then, there exists a constant* $\widetilde{C}_p$ *such that*

$$\|\nabla_x G(t,s)f\|_\infty\le\widetilde{C}_p e^{\ell_p(t-s)}\|f\|_\infty,\qquad f\in C_b(\mathbb{R}^N),\ \ s\in I,\ \ t\ge s+1.$$

Proof Fix $t,s\in I$ with $t>s+1$. By Theorem 14.4.2, for any $f\in C_b(\mathbb{R}^N)$ the function $G(s+1,s)f$ belongs to $C_b^1(\mathbb{R}^N)$, and its C^1-norm does not exceed $C_1\|f\|_\infty$ for some $C_1>0$, independent of f. Therefore, if $\ell_p<0$, then by Theorem 14.4.5 we can estimate

$$\begin{aligned}|(\nabla_x G(t,s)f)(x)|^p=&|(\nabla_x G(t,s+1)G(s+1,s)f)(x)|\\ \le& e^{\ell_p(t-(s+1))}(G(t,s+1)|\nabla_x G(s+1,s)f|^p)(x)\\ \le& e^{\ell_p(t-(s+1))}\|\nabla_x G(s+1,s)f\|_\infty^p\end{aligned}$$

for any $x\in\mathbb{R}^N$. On the other hand, if $\ell_p\ge 0$ then we split $\nabla_x G(t,s)f=\nabla_x G(t,t-1)G(t-s,s)f$ and use estimate (14.4.10) and the contractivity of the evolution operator $\{G(t,s)\}$. Thus, the statement follows. □

Remark 14.4.7 If estimate (14.4.8) hold with $p=1$, then the diffusion coefficients should satisfy the following condition $D_iq_{jk}+D_jq_{ik}+D_kq_{ij}=0$ in $I\times\mathbb{R}^N$, for any $i,j,k=1,\dots,N$, (see [254, 14]). This is the reason why in the proof of (14.4.8) with $p=1$ we confine ourselves to the case when the diffusion coefficients are independent of x.

14.4.1 Schauder estimates for nonhomogeneous parabolic problems

The nonhomogeneous Cauchy problem

$$\begin{cases}D_tu(t,x)=\mathcal{A}u(t,x)+g(t,x), & t\in[0,T],\quad x\in\mathbb{R}^N,\\ u(0,x)=f(x), & \qquad\qquad\quad x\in\mathbb{R}^N,\end{cases}\tag{14.4.14}$$

has been studied in [172] under the following conditions on the coefficients of the operator $\mathcal{A}$.

Hypotheses 14.4.8 (i) *the coefficients* q_{ij},b_j $(i,j=1,\dots,N)$ *and* c *are thrice continuously differentiable with respect to the spatial variables in* $[0,T]\times\mathbb{R}^N$ *and they belong to* $C^{\alpha/2,\alpha}_{\rm loc}([0,T]\times\mathbb{R}^N)$ *for some* $\alpha\in(0,1)$, *together with their first-, second- and third-order spatial derivatives;*

(ii) $q_{ij}(t,x)=q_{ji}(t,x)$ *for any* $i,j=1,\dots,N$ *and* $(t,x)\in[0,T]\times\mathbb{R}^N$. *Moreover,* $\langle Q(t,x)\xi,\xi\rangle\ge\kappa(t,x)|\xi|^2$ *for any* $t\in[0,T]$, $\xi,x\in\mathbb{R}^N$ *and some function* $\kappa:(0,T)\times\mathbb{R}^N\to\mathbb{R}$ *with positive infimum* κ_0;

(iii) *there exists a positive constant* C_1 *such that*

$$|Q(t,x)x| + \mathrm{Tr}(Q(t,x)) \le C_1(1+|x|^2)\kappa(t,x), \qquad \langle b(t,x),x\rangle \le C_1(1+|x|^2)\kappa(t,x)$$

for any $t\in[0,T]$ *and any* $x\in\mathbb{R}^N$;

(iv) $c(t,x)\le c_0$ *for some real constant* c_0 *and any* $(t,x)\in[0,T]\times\mathbb{R}^N$;

(v) *there exists a positive constant* C_2 *such that* $|D_x^\beta q_{ij}|\le C_2\kappa$ *in* $[0,T]\times\mathbb{R}^N$, *for any* $i,j=1,\ldots,N$, *and*

$$\sum_{i,j,h,k=1}^N D_{hk}q_{ij}(t,x)m_{ij}m_{hk} \le C_2\kappa(t,x)\|M\|^2$$

for any $|\beta|=1,3$, *any* $(N\times N)$*-symmetric matrix* M *and any* $(t,x)\in[0,T]\times\mathbb{R}^N$;

(vi) *there exist three functions* $r_1,r_2,r_3:[0,T]\times\mathbb{R}^N\to\mathbb{R}$ *and three positive constants* L_1,L_2,L_3 *such that*

$$\langle(\mathrm{Jac}_x b)\xi,\xi\rangle\le r_1|\xi|^2, \qquad |D_x^\beta b_j|\le r_2, \qquad |D_x^\gamma c|\le r_3, \qquad r_1+L_1r_2+L_2r_3^2\le L_3\kappa$$

in $[0,T]\times\mathbb{R}^N$, *for any* $|\beta|=2,3$, *any* $|\gamma|=1,2,3$, *any* $j=1,\ldots,N$ *and any* $\xi\in\mathbb{R}^N$;

(vii) *there exist a positive function* $\varphi:\mathbb{R}^N\to\mathbb{R}$, *blowing up as* $|x|$ *tends to* $+\infty$, *and* $\lambda>0$ *such that* $\mathcal{A}\varphi-\lambda\varphi$ *is bounded from above in* $[0,T]\times\mathbb{R}^N$.

The following optimal Schauder estimates have been proved in [172, Theorem 2.7].

Theorem 14.4.9 *Fix* $\theta\in(0,1)$, $g\in C^{0,\theta}([0,T]\times\mathbb{R}^N)$ *and* $f\in C_b^{2+\theta}(\mathbb{R}^N)$. *Then, problem* (5.2.2) *admits a unique bounded classical solution. Moreover,* $u(t,\cdot)\in C_b^{2+\theta}(\mathbb{R}^N)$ *for any* $t\in[0,T]$ *and there exists a positive constant* C_0 *such that*

$$\sup_{t\in[0,T]}\|u\|_{C_b^{2+\theta}(\mathbb{R}^N)} \le C_0\Big(\|f\|_{C_b^{2+\theta}(\mathbb{R}^N)} + \sup_{t\in[0,T]}\|g\|_{C_b^\theta(\mathbb{R}^N)}\Big). \tag{14.4.15}$$

The proof of Theorem 14.4.9 is obtained by arguing as in Section 5.2 and it strongly relies on the uniform estimates

$$\|G(t,s)f\|_{C_b^\beta(\mathbb{R}^N)} \le C(t-s)^{-\frac{\beta-\alpha}{2}}\|f\|_{C_b^\alpha(\mathbb{R}^N)}, \qquad t\in(s,T),\ f\in C_b^\alpha(\mathbb{R}^N),$$

which hold true for any $\alpha,\beta\in[0,3]$, with $\alpha\le\beta$, and some positive constant $C=C(\alpha,\beta,T)$.
The results of the previous theorem have been also extended to some situations where the coefficients of the operator $\mathcal{A}$ are not smooth with respect to the variable t, which are made clear by the following set of assumptions.

Hypotheses 14.4.10 (i) *the coefficients* $q_{ij}=q_{ji}$, b_j $(i,j=1,\ldots,N)$ *and* c *are measurable functions in* $[0,T]\times\mathbb{R}^N$. *Moreover,*

$$\sup_{t\in[0,T]}\big(\|q_{ij}(t,\cdot)\|_{C^{3+\alpha}(B_R)} + \|b_j(t,\cdot)\|_{C^{3+\alpha}(B_R)} + \|c(t,\cdot)\|_{C^{3+\alpha}(B_R)}\big) < +\infty$$

for any $i,j=1,\ldots,N$, *any* $R>0$ *and some* $\alpha\in(0,1)$;

(ii) *Hypotheses* 14.4.8*(ii)-(vii) are satisfied with* $[0,T]$ *being replaced by* $\mathcal{D}$, *where* $[0,T]\setminus\mathcal{D}$ *is a negligible set. Moreover, for any* $x\in\mathbb{R}^N$, *the functions* $r_j(\cdot,x)$ $(j=1,2,3)$ *are bounded and measurable in* $(0,T)$.

The following theorem extends the results in [142, 143, 144].

Theorem 14.4.11 *Let Hypotheses* 14.4.10 *be satisfied. Fix* $\theta \in (0,1)$ *and suppose that* $f \in C_b^{2+\theta}(\mathbb{R}^N)$ *and* $g : [0,T] \times \mathbb{R}^N \to \mathbb{R}$ *is a bounded measurable function such that* $\sup_{t\in[0,T]} \|g(t,\cdot)\|_{C_b^\theta(\mathbb{R}^N)} \leq +\infty$. *Then, the Cauchy problem* (14.4.14) *admits a unique solution* u *with the following properties:*

(i) u *is Lipschitz continuous in* $(0,T) \times B_R$ *for any* $R > 0$, *its first- and second-order space derivatives are bounded and continuous functions in* $[0,T] \times \mathbb{R}^N$;

(ii) $u(0,x) = f(x)$ *for any* $x \in \mathbb{R}^N$;

(iii) *there exists a set* $F \subset [0,T] \times \mathbb{R}^N$, *with negligible complement, such that* $D_t u(t,x) = \mathcal{A}u(t,x) + g(t,x)$ *for any* $(t,x) \in F$. *Moreover, for any* $x \in \mathbb{R}^N$, *the set* $F_x = \{t \in [0,T] : (t,x) \in F\}$ *is measurable with measure* T.

Further, the function u *satisfies estimate* (14.4.15).

To conclude this subsection, we remark that, in the case when $[0,T]$ is replaced by a right-halfline I, global Schauder estimates for the solutions of the parabolic equation $D_t u = \mathcal{A}u + g$ in $I \times \mathbb{R}^N$ have been proved by N.V. Krylov and E. Priola in [149] assuming that the diffusion coefficients are bounded and the drift coefficients may grow at most linearly at infinity.

14.5 The evolution semigroup $\{\mathscr{T}(t)\}$

In this section we assume that $I = \mathbb{R}$ and, besides Hypotheses 14.0.1, we assume Hypothesis 14.1.8.

The evolution semigroup $\{\mathscr{T}(t)\}$ associated with the evolution family $\{G(t,s)\}$ in $C_b(\mathbb{R}^{N+1})$ is defined by $(\mathscr{T}(t)f)(s,x) = (G(s,s-t)f(s-t,\cdot))(x)$ for any $(s,x) \in \mathbb{R}^{N+1}$, $t \geq 0$ and $f \in C_b(\mathbb{R}^{N+1})$.

Theorem 14.5.1 $\{\mathscr{T}(t)\}$ *is a positive semigroup in* $C_b(\mathbb{R}^{N+1})$ *which satisfies the estimate* $\|\mathscr{T}(t)\|_{L(C_b(\mathbb{R}^{N+1}))} \leq e^{c_0 t}$ *for any* $t \geq 0$, *where* c_0 *is defined in Hypothesis* 14.0.1*(iii). In particular, if* $c \equiv 0$, *then each operator* $\mathscr{T}(t)$ *is a contraction. Finally,* $\mathscr{T}(t)f$ *converges to* f *locally uniformly in* $\mathbb{R}^{N+1}$ *as* t *tends to* 0^+, *for any* $f \in C_b(\mathbb{R}^{N+1})$.

Proof By Theorem 14.1.2 we know that $|(\mathscr{T}(t)f)(s,x)| \leq e^{c_0 t}\|f\|_{C_b(\mathbb{R}^{N+1})}$ for any $t \geq 0$ and $(s,x) \in \mathbb{R}^{N+1}$. Let us now prove that, for any $t > 0$, the operator $\mathscr{T}(t)$ maps $C_b(\mathbb{R}^{N+1})$ into itself. Fix (s_0,x_0), (s,x) in $\mathbb{R}^{N+1}$ and estimate

$$\begin{aligned}|(\mathscr{T}(t)f)(s,x) - (\mathscr{T}(t)f)(s_0,x_0)| \leq & |(G(s,s-t)f(s,\cdot))(x) - (G(s,s-t)f(s_0,\cdot))(x)| \\ & + |(G(s,s-t)f(s_0,\cdot))(x) - (G(s_0,s_0-t)f(s_0,\cdot))(x_0)|.\end{aligned}$$

Using Proposition 14.1.11 and Theorem 14.1.12, it follows that the right-hand side of the previous estimate vanishes as (s,x) tends to (s_0,x_0).

The semigroup property and the positivity of $\{\mathscr{T}(t)\}$ are immediate consequences of the fact that $\{G(t,s)\}$ is a positive evolution operator.

Finally, to prove that $\mathscr{T}(t)f$ converges to f locally uniformly in $\mathbb{R}^{N+1}$ as t tends to 0^+ for any $f \in C_b(\mathbb{R}^{N+1})$, it suffices to observe that the same arguments in first part of the proof show that the function $(t,s,x) \mapsto (\mathscr{T}(t)f)(s,x)$ is continuous in $[0,+\infty) \times \mathbb{R}^{N+1}$. $\square$

Remark 14.5.2 In general, $\{T(t)\}$ has no smoothing improving properties in the s variable. In particular, it is not strong Feller. This is straightforward to see when $c \equiv 0$ and $G(s, s-t)\mathbb{1} = \mathbb{1}$ for each $t > s$. Indeed, in this case, if f depends only on time then $(\mathscr{T}(t)f)(s,x) = f(s-t)$, i.e., $\{\mathscr{T}(t)\}$ acts as a semigroup of translations.

14.5.1 The weak generator G_∞ of $\{\mathscr{T}(t)\}$

Let us introduce the operator $\mathcal{G} = \mathcal{A} - D_s$ and the set

$$D_{\max}(\mathcal{G}) = \left\{\psi \in C_b(\mathbb{R}^{N+1}) \cap \bigcap_{p<+\infty} W^{1,2}_{p,\mathrm{loc}}(\mathbb{R}^{N+1}) : \mathcal{G}\psi \in C_b(\mathbb{R}^{N+1})\right\}.$$

To begin with, we consider the following generalization of the classical maximum principle.

Proposition 14.5.3 (Maximum principle) *The following properties are satisfied.*

(i) *Suppose that $u \in C_b(\mathbb{R}^{N+1}) \cap \bigcap_{p\in[1,+\infty)} W^{1,2}_{p,\mathrm{loc}}(\mathbb{R}^{N+1})$ and $\mathcal{G}u \in C(\mathbb{R}^{N+1})$. If u satisfies the inequality $\mathcal{G}u - \lambda u \geq 0$ in $\mathbb{R}^{N+1}$, for some $\lambda \geq \lambda_0 + 1$, then, $u \leq 0$.*

(ii) *Fix $f \in C_b(\mathbb{R}^{N+1})$ and let $u \in D_{\max}(\mathcal{G})$ solve the equation $\mathcal{G}u - \lambda u = f$ in $\mathbb{R}^{N+1}$ for some $\lambda > 0$. Then, $\|u\|_\infty \leq \lambda^{-1}\|f\|_\infty$.*

Proof To begin with, we observe that, if v is a function as in the statement of property (i), which has a maximum (resp. minimum point) at some point $(t_0, x_0) \in \mathbb{R}^{N+1}$, then $\mathcal{G}v(t_0,x_0) \leq 0$ (resp $\mathcal{G}v(t_0,x_0) \geq 0$). This property can be checked arguing e.g. as in the proof of Lemma 3.1.2. Now, the proof of (i) can be obtained using the arguments in the proof of Theorem 3.1.6, replacing the function φ, therein used, with the function $\overline{\varphi}$, defined by $\overline{\varphi}(s,x) = (1+s^2)\varphi(x)$ for any $(s,x) \in \mathbb{R}^{N+1}$, where φ is given by Hypothesis 14.1.8. This function satisfies $\mathcal{G}\overline{\varphi} \leq (\lambda_0+1)\overline{\varphi}$ in $\mathbb{R}^{N+1}$. The same arguments can be also adapted to prove (ii) for $\lambda \geq \lambda_0+1$. To get it for $\lambda < \lambda_0+1$, it suffices to split $\mathcal{G}u-(\lambda_0+2)u = (\lambda-\lambda_0-2)u+f$ and apply the above result to deduce that $(\lambda_0+2)\|u\|_\infty \leq \|(\lambda-\lambda_0-2)u+f\|_\infty \leq (\lambda_0 + 2-\lambda)\|u\|_\infty + \|f\|_\infty$. □

Now, thanks to Proposition 14.5.3 we can prove the following result.

Theorem 14.5.4 *For any $f \in C_b(\mathbb{R}^{N+1})$ and $\lambda > 0$, there exists a unique function $u \in D_{\max}(\mathcal{G})$ which solves the equation $\mathcal{G}u - \lambda u = f$. Moreover,*

$$\|u\|_\infty \leq \lambda^{-1}\|f\|_\infty. \tag{14.5.1}$$

Finally, if we denote by $R(\lambda)$ the operator that, with any $f \in C_b(\mathbb{R}^{N+1})$ associates the unique solution of the equation $\mathcal{G}u - \lambda u = f$ in $D_{\max}(\mathcal{G})$, then

$$R(\lambda)f - R(\mu)f = (\mu-\lambda)R(\lambda)R(\mu)f, \qquad \lambda,\mu > 0. \tag{14.5.2}$$

Proof The uniqueness part and estimate (14.5.1) follow from Proposition 14.5.3.
To prove the existence part, we fix $\lambda > 0$, $n \in \mathbb{N}$ and consider the problem

$$\begin{cases} \mathcal{G}u(s,x) - \lambda u(s,x) = f(s,x), & s \in (-n,n), \quad x \in B_n, \\ u(s,x) = 0, & s \in (-n,n), \quad x \in \partial B_n, \\ u(-n,x) = 0, & \qquad\qquad\quad x \in B_n. \end{cases}$$

By classical results (see e.g. [159, Chapter IV, Theorem 9.1]), this problem admits a unique solution u_n which belongs to $C((-n,n)\times B_n)\cap W^{1,2}_p((-n,n)\times B_n)$ for any $p\in(1,+\infty)$. Further, it satisfies the estimate $\|u_n\|_\infty\le\lambda^{-1}\|f\|_\infty$. Finally, for any $\sigma<\tau<n$ and $p\in(1,+\infty)$, it holds that

$$\|u_n\|_{W^{1,2}_p((-\sigma,\sigma)\times B_\sigma)}\le C_1\|u_n\|_{L^p((-\tau,\tau)\times B_\tau)}\le C_2\|f\|_\infty \tag{14.5.3}$$

for some positive constants C_1 and C_2, independent of n. Estimate (14.5.3) and the Sobolev embedding theorem show that the sequence $\{u_n\}$ is eventually bounded in $C^\gamma((-\sigma,\sigma)\times B_\sigma)$ for any $\gamma\in(0,1)$ and $\sigma>0$. Hence, by Arzelà-Ascoli theorem, up to a subsequence, u_n converges to some function $u\in C^\gamma_{\rm loc}(\mathbb{R}^{N+1})$, locally uniformly in $\mathbb{R}^{N+1}$.

Estimate (14.5.3) and compactness and diagonal arguments can be used to prove that (still up to a subsequence) the time derivative and the first- and second-order spatial derivatives of u_n weakly converge in $L^p((-\sigma,\sigma)\times B_\sigma)$ for any $\sigma>0$ and $p>1$. Therefore, u belongs to $W^{1,2}_{p,{\rm loc}}(\mathbb{R}^{N+1})$ for any $\sigma>0$ and satisfies the differential equation $\mathcal{G}u-\lambda u=f$. We have so proved that u belongs to $D_{\max}(\mathcal{G})$.

Finally, the formula (14.5.2) follows from Proposition 14.5.3. □

Remark 14.5.5 The identity (14.5.2) implies that $\{R(\lambda):\lambda>0\}$ is the resolvent family associated with some closed operator G_∞. Moreover, the uniqueness part of the proof of Theorem 14.5.4 implies that, actually, $D(G_\infty)=D_{\max}(\mathcal{G})$ and $G_\infty=\mathcal{G}$ on $D(G_\infty)$.

14.5.2 An equivalent characterization of $D(G_\infty)$

Lemma 14.5.6 *Let $\{f_n\}\in C_b(\mathbb{R}^{N+1})$ be a bounded sequence, converging locally uniformly in $\mathbb{R}^{N+1}$ to a bounded and continuous function f as n tends to $+\infty$. Then, $R(\lambda,G_\infty)f_n$ and $\mathscr{T}(t)f_n$ converge to $R(\lambda,G_\infty)f$ and $\mathscr{T}(t)f$, respectively, for any $\lambda>0$ and $t>0$, locally uniformly in $\mathbb{R}^{N+1}$.*

Proof Let us fix a bounded sequence $\{f_n\}$ as in the statement. By (14.5.1) the sequence $\{R(\lambda,G_\infty)f_n\}$ is bounded in $C_b(\mathbb{R}^{N+1})$. Hence, as in the proof of Theorem 14.5.4, interior L^p-estimates, show that this sequence is bounded in $W^{1,2}_p((-\sigma,\sigma)\times B_\sigma)$ for any $\sigma>0$. Now, diagonal and compactness arguments allow to conclude that $R(\lambda,G_\infty)f_n$ converges locally uniformly in $\mathbb{R}^{N+1}$) to a solution v of the equation $\mathcal{G}v-\lambda v=f$, which belongs to $D_{\max}(\mathcal{G})$. By Proposition 14.5.3, we deduce that $v=R(\lambda,G_\infty)f$.

To prove that the function $\mathscr{T}(\cdot)f_n$ converges to $\mathscr{T}(\cdot)f$ locally uniformly in $\mathbb{R}^{N+2}$, it suffices to argue as in the proof of Proposition 14.1.11. □

Proposition 14.5.7 *For any $\lambda>c_0$ and $f\in C_b(\mathbb{R}^{N+1})$ it holds that*

$$(R(\lambda,G_\infty)f)(s,x)=\int_0^{+\infty}e^{-\lambda t}(\mathscr{T}(t)f)(s,x)dt,\qquad (s,x)\in\mathbb{R}^{N+1}. \tag{14.5.4}$$

Proof By virtue of Lemma 14.5.6 it suffices to prove the formula (14.5.4) for functions $f\in C^\infty_c(\mathbb{R}^{N+1})$. Hence, throughout the proof, f is an arbitrarily fixed function in $C^\infty_c(\mathbb{R}^{N+1})$.

We split the proof into two steps. In the first one we prove that, for any $t>0$, $\mathscr{T}(t)f$ belongs to $D_{\max}(\mathcal{G})$, that the function $(\mathscr{T}(\cdot)f)(s,x)$ is differentiable in $[0,+\infty)$ for any $(s,x)\in\mathbb{R}^{N+1}$ and

$$D_t\mathscr{T}(t)f=\mathscr{T}(t)\mathcal{G}f=\mathcal{G}\mathscr{T}(t)f,\qquad t\ge 0. \tag{14.5.5}$$

Finally, we show $\mathscr{T}(\cdot)f\in C([0,+\infty);C_b(\mathbb{R}^{N+1}))$. Next, in Step 2, we conclude the proof.

Step 1. From Lemma 14.1.6(i) it follows that

$$\frac{(\mathscr{T}(t)f)(s,x)-f(s,x)}{t}=\frac{1}{t}\int_{s-t}^{s}[(G(s,r)\mathcal{A}(r)f(s-t,\cdot))(x)-D_tf(r,x)]dr. \qquad (14.5.6)$$

Note that the function $(s,r,t)\mapsto G(s,r)\mathcal{A}(r)f(t,\cdot)$ is continuous in $\Delta_{\mathbb{R}}\times\mathbb{R}$ with values in $C(\overline{B}_R)$ for any $R>0$. Indeed, for (r,s,t), (r_0,s_0,t_0) in $\Delta_{\mathbb{R}}\times\mathbb{R}$, we can infer that

$$\begin{aligned}|G(s,r)\mathcal{A}(r)f(t,\cdot)-G(s_0,r_0)\mathcal{A}(r_0)f(t_0,\cdot)|\le& e^{c_0(s-r)}\|\mathcal{A}(r)f(t,\cdot)-\mathcal{A}(r_0)f(t_0,\cdot)\|_\infty\\&+|G(s,r)\mathcal{A}(r_0)f(t_0,\cdot)-G(s_0,r_0)\mathcal{A}(r_0)f(t_0,\cdot)|\end{aligned}$$

in $\mathbb{R}^N$ and, using Theorem 14.1.12, we conclude that the right-hand side of the previous estimate vanishes as (r,s,t) tends to (r_0,s_0,t_0). Hence, we can let t tend to 0^+ in (14.5.6) and obtain that $t^{-1}(\mathscr{T}(t)f-f)$ converges to $\mathcal{G}f$ locally uniformly in $\mathbb{R}^{N+1}$. The semigroup property and Lemma 14.5.6 show that, for any $(s,x)\in\mathbb{R}^{N+1}$, the function $(\mathscr{T}(\cdot)f)(s,x)$ is differentiable from the right at any $t>0$, with the function $(\mathscr{T}(\cdot)\mathcal{G}f)(s,x)$ as right-derivative. Since this latter function is continuous in $(0,+\infty)$, the function $(\mathscr{T}(\cdot)f)(s,x)$ is differentiable at t and $D_t\mathscr{T}(t)f=\mathscr{T}(t)\mathcal{G}f$ for any $t>0$.[9] As a byproduct, the function $\mathscr{T}(\cdot)f$ turns out to be continuous in $[0,+\infty)$ with values in $C_b(\mathbb{R}^{N+1})$.

Let us now show that, for any $t>0$, $\mathscr{T}(t)f$ belongs to $D_{\max}(\mathcal{G})$ and (14.5.5) holds true. By Theorem 14.1.2, the derivatives $D_j\mathscr{T}(t)f$ and $D_{ij}\mathscr{T}(t)f$ exist in the classical sense in $\mathbb{R}^{N+1}$ for any $i,j=1,\ldots,N$. Moreover, they are continuous functions in $\mathbb{R}^{N+1}$. Indeed, the classical interior Schauder estimates imply that, for any $M>0$, $\varepsilon>0$ and $s\in[-M,M]$, there exists a constant $C=C(\varepsilon,M)>0$ such that $\|G(\cdot,s)\psi\|_{C^{1+\alpha/2,2+\alpha}([s+\varepsilon,M+2\varepsilon]\times B_M)}\le C\|\psi\|_\infty$ for any $s\in[-M,M-\varepsilon]$ and $\psi\in C_b(\mathbb{R}^N)$. This shows that $\|(\mathscr{T}(t)f)(s,\cdot)\|_{C^{2+\alpha}(B_M)}\le C\|f\|_\infty$ for any $s\in[-M,M]$ and some positive constant C, independent of s. We now recall that, for any $M\in\mathbb{N}$, there exists a positive constant C_M such that $\|\psi\|_{C^2(\overline{B_M})}\le C_M\|\psi\|_{C(\overline{B_M})}^{\alpha/(2+\alpha)}\|\psi\|_{C^{2+\alpha}(B_M)}^{2/(2+\alpha)}$ for any $\psi\in C^{2+\alpha}(B_M)$ (see e.g. [178, Proposition 1.1.3]). Applying this estimate with $\psi=(\mathscr{T}(t)f)(s,\cdot)-(\mathscr{T}(t)f)(s_0,\cdot)$, for arbitrarily fixed $s_0,s\in\mathbb{R}$, and letting s tend to s_0, the continuity in $\mathbb{R}^{N+1}$ of the spatial derivatives of the function $\mathscr{T}(t)f$ follows at once.

To prove that $\mathscr{T}(t)f$ is differentiable with respect to s in $\mathbb{R}^{N+1}$, it suffices to split

$$\begin{aligned}&\frac{(\mathscr{T}(t)f)(s+h,x)-(\mathscr{T}(t)f)(s,x)}{h}\\&=\left(G(s+h,s+h-t)\frac{f(s+h-t,\cdot)-f(s-t,\cdot)}{h}\right)(x)\\&\quad+\left(\frac{G(s+h,s+h-t)-G(s+h,s-t)}{h}f(s-t,\cdot)\right)(x)\\&\quad+\left(\frac{G(s+h,s-t)-G(s,s-t)}{h}f(s-t,\cdot)\right)(x)\end{aligned}$$

for any $(s,x)\in\mathbb{R}^{N+1}$ and $h\in(-t,t)\setminus\{0\}$, and argue as in the first part of the proof. We deduce that $D_s\mathscr{T}(t)f=-\mathscr{T}(t)\mathcal{G}f+\mathcal{A}(s)\mathscr{T}(t)f$ in $\mathbb{R}^{N+1}$. From this formula it follows that $D_s\mathscr{T}(t)f$ is continuous in $\mathbb{R}^{N+1}$ and that $\mathcal{G}$ and $\{\mathscr{T}(t)\}$ commute. In particular, $\mathscr{T}(t)f$ belongs to $D_{\max}(\mathcal{G})$. for any $t\ge 0$.

Step 2. Since the function $t\mapsto e^{-\lambda t}\mathscr{T}(t)f$ is continuous in $[0,+\infty)$ with values in $C_b(\mathbb{R}^{N+1})$ and $\|e^{-\lambda t}\mathscr{T}(t)f\|_\infty\le e^{-(\lambda-c_0)t}\|f\|_\infty$, for any $t>0$, the right-hand side of (14.5.4) defines a bounded operator in $C_b(\mathbb{R}^{N+1})$, which we denote by R_λ. Since also the function $t\mapsto$

[9] See e.g. [260, p. 239].

$e^{-\lambda t}G_\infty \mathscr{T}(t)f = e^{-\lambda t}\mathscr{T}(t)\mathcal{G}f$ is continuous[10] and summable in $[0,+\infty)$, by the closedness of the operator G_∞ we conclude that $G_\infty R_\lambda f = R_\lambda \mathcal{G}f$. Hence, from Step 1 we get

$$G_\infty R_\lambda = \int_0^{+\infty} e^{-\lambda t}(\mathcal{G}\mathscr{T}(t)f)(s,x)dt = \int_0^{+\infty} e^{-\lambda t}\frac{d}{dt}(\mathscr{T}(t)f(s,x))dt$$
$$= \lambda\int_0^{+\infty} e^{-\lambda t}(\mathscr{T}(t)f)(s,x)dt - f(s,x)$$

for any $(s,x)\in\mathbb{R}^{N+1}$. Hence, $(\lambda I - G_\infty)R_\lambda = I$, which yields the assertion. □

Remark 14.5.8 Actually, the proof of Proposition 14.5.7 shows that, for any $t>0$ and $f\in C_b(\mathbb{R}^{N+1})$, the function $\mathscr{T}(t)f$ admits first- and second-order spatial derivatives which are continuous functions in the whole of $\mathbb{R}^{N+1}$.

We can now prove the following characterization of $D(G_\infty)$.

Theorem 14.5.9 *A function f belongs to $D(G_\infty)$ if and only if the function $t\mapsto t^{-1}(\mathscr{T}(t)f-f)$ is bounded in $(0,1]$ with values in $C_b(\mathbb{R}^{N+1})$ and there exists $g\in C_b(\mathbb{R}^{N+1})$ such that $t^{-1}(\mathscr{T}(t)f-f)$ converges to g, locally uniformly in $\mathbb{R}^{N+1}$, as t tends to 0 from the right. In such a case $g=\mathcal{G}f$.*

Proof Denote by $\widehat{D}$ the space of all functions $f\in C_b(\mathbb{R}^{N+1})$ such that the function $t\mapsto t^{-1}(\mathscr{T}(t)f-f)$ is bounded in $(0,1]$ with values in $C_b(\mathbb{R}^{N+1})$ and it converges locally uniformly in $\mathbb{R}^{N+1}$, as t tends to 0^+, to some function, which we denote by $\widehat{G}f$. Arguing as in the proof of Proposition 1.3.1, we can show that $D(G_\infty)\subset\widehat{D}$ and $t^{-1}(\mathscr{T}(t)f-f)$ converges locally uniformly in $\mathbb{R}^{N+1}$ to $\mathcal{G}f$, as t tends to 0^+, for any $f\in D(G_\infty)$.

Now, fix $f\in\widehat{D}$. The proof of Lemma 1.3.2 and the above result show that $\widehat{G}R(\lambda,G_\infty)f = R(\lambda,G_\infty)\widehat{G}f = G_\infty R(\lambda,G_\infty)f$. Hence, we can write

$$f=(\lambda-G_\infty)R(\lambda,G_\infty f) = R(\lambda,G_\infty)(\lambda f-\widehat{G}f)$$

and conclude that $f\in D(G_\infty)$. □

Corollary 14.5.10 *For any $t>0$, $\mathscr{T}(t)$ maps $D(G_\infty)$ into itself.*

Proof The assertion is a straightforward consequence of the characterization of $D(G_\infty)$ in Theorem 14.5.9 and Lemma 14.5.6. □

We conclude this section with the following proposition which gives a partial characterization of $D(G_\infty)$.

Proposition 14.5.11 *Let the assumptions of Theorem 14.4.2 be satisfied, with the functions q_0 and r being bounded from above. Then, for any $t>0$ and $f\in C_b(\mathbb{R}^{N+1})$ it holds that*

$$\|\nabla_x\mathscr{T}(t)f\|_\infty \le C_1(t^{-\frac{1}{2}}\vee 1)\|f\|_\infty. \tag{14.5.7}$$

Moreover, $D(G_\infty)$ is continuously embedded into $C_b^{0,1}(\mathbb{R}^{N+1})$.

[10] The function $\mathcal{G}f$ belongs to $C_c(\mathbb{R}^{N+1})$. Hence, it suffices to approximate it uniformly in $\mathbb{R}^{N+1}$ by a sequence $\{f_n\}\subset C_c^\infty(\mathbb{R}^{N+1})$, converging uniformly in $\mathbb{R}^{N+1}$.

Proof Estimate (14.5.7) follows immediately from (14.4.3).

To prove that $D(G_\infty) \hookrightarrow C_b^{0,1}(\mathbb{R}^{N+1})$ we recall that $D(G_\infty) = R(\lambda, G_\infty)(C_b(\mathbb{R}^{N+1}))$ for any $\lambda > 0$, where $R(\lambda, G_\infty)f$ is given by (14.5.4) for any $f \in C_b(\mathbb{R}^{N+1})$. Estimate (14.5.7) implies that we can differentiate (14.5.4) under the integral sign and the moduli of the derivatives $D_i R(\lambda, G_\infty)f$ $(i = 1, \ldots, N)$ are bounded in $\mathbb{R}^{N+1}$ by $C\|f\|_\infty$ for some $C > 0$. Their continuity follows from the continuity of the spatial derivatives of $\mathscr{T}(t)f$ (see Remark 14.5.8) through the dominated convergence theorem. □

14.5.3 The periodic case

In this subsection we assume the following set of hypotheses.

Hypotheses 14.5.12 (i) *Hypotheses* 14.0.1(*i*), 14.0.1(*ii*) *and* 14.1.8 *are satisfied;*

(ii) *there exists $T > 0$ such that the coefficients q_{ij} and b_i are T-periodic with respect to the s variable for any $i, j = 1, \ldots, N$.*

We denote by $C_\sharp(\mathbb{R}^{N+1})$ and $D_\sharp(\mathcal{G})$ the subsets of $C_b(\mathbb{R}^{N+1})$ and $D_{\max}(\mathcal{G})$, respectively, consisting of T-periodic functions in the s-variable, i.e.,

$$D_\sharp(\mathcal{G}) = \left\{\psi \in C_\sharp(\mathbb{R}^{N+1}) \cap \bigcap_{p<+\infty} W^{1,2}_{p,\mathrm{loc}}(\mathbb{R}^{N+1}) : \mathcal{G}\psi \in C_\sharp(\mathbb{R}^{N+1})\right\}. \tag{14.5.8}$$

Proposition 14.5.13 *The following properties are satisfied.*

(i) *For any $t > 0$, $\mathscr{T}(t)$ leaves the spaces $C_\sharp(\mathbb{R}^{N+1})$ and $D_\sharp(\mathcal{G})$ invariant.*

(ii) *For any $f \in C_\sharp(\mathbb{R}^{N+1})$ and $\lambda > 0$, there exists a unique $u \in D_\sharp(\mathcal{G})$ which solves the equation $\lambda u - \mathcal{G}u = f$. The function u satisfies estimate* (14.5.1)*. Moreover, a function f belongs to $D_\sharp(\mathcal{G})$ if and only if $t^{-1}(\mathscr{T}(t)f - f)$ is bounded in* $(0, 1]$ *with values in $C_\sharp(\mathbb{R}^{N+1})$ and it converges to some function $g \in C_\sharp(\mathbb{R}^{N+1})$, locally uniformly in $\mathbb{R}^{N+1}$, as t tends to* 0 *from the right. In such a case, $g = \mathcal{G}f$.*

Proof (i) The T-periodicity of the coefficients of the operator $\mathcal{A}$ implies that, for any $s \in \mathbb{R}$ and $\psi \in C_b(\mathbb{R}^N)$, the functions $G(\cdot, r)\psi$ and $G(\cdot + T, r + T)\psi$ solve the same Cauchy problem. Hence, Proposition 14.1.1 yields $G(\cdot, r)\psi = G(\cdot{+}T, r{+}T)\psi$. This fact and Theorem 14.5.1 show that $\mathscr{T}(t)f \in C_\sharp(\mathbb{R}^{N+1})$ for any $f \in C_\sharp(\mathbb{R}^{N+1})$. Finally, using Corollary 14.5.10, we conclude that $\mathscr{T}(t)$ maps $D_\sharp(\mathcal{G})$ into itself for any $t > 0$.

(ii) To prove that the equation $\lambda u - \mathcal{G}u = f$ admits a unique solution in $D_\sharp(\mathcal{G})$ for any $f \in C_\sharp(\mathbb{R}^{N+1})$, it suffices to observe that it admits a unique solution u in $D_{\max}(\mathcal{G})$ given by the formula (14.5.4). Since $\mathscr{T}(t)f \in C_\sharp(\mathbb{R}^{N+1})$ for any $t > 0$, u belongs to $C_\sharp(\mathbb{R}^{N+1})$ as well. Thus, u belongs to $D_\sharp(\mathcal{G})$. The other assertions are now straightforward consequences of Theorem 14.5.9. □

Remark 14.5.14 By Proposition 14.5.13(i), $\{\mathscr{T}(t)\}$ gives rise to a semigroup of bounded operators in $C_\sharp(\mathbb{R}^{N+1})$, denoted by $\{\mathscr{T}^\sharp(t)\}$. Under the hypothesis of Theorem 14.4.2, $D_\sharp(\mathcal{G})$ is continuously embedded into $C_\sharp^{0,1}(\mathbb{R}^{N+1}) := C_\sharp(\mathbb{R}^{N+1}) \cap C_b^{0,1}(\mathbb{R}^{N+1})$. Moreover, estimate (14.5.7) holds with $\{\mathscr{T}(t)\}$ being replaced by the semigroup $\{\mathscr{T}^\sharp(t)\}$. These properties follow arguing as in the proof of Proposition 14.5.11.

Chapter 15

Estimates for Green functions

15.0 Introduction

In this chapter we prove some estimates for the Green function of the evolution operator $\{G(t,s)\}$, associated in $C_b(\mathbb{R}^N)$ ($N \geq 2$) with the non-autonomous elliptic operators $\mathcal{A}$ in (14.0.1). For notational convenience, throughout the chapter we denote by (s,y) the variable in $I \times \mathbb{R}^N$. We assume the following conditions on the coefficients of the operator $\mathcal{A}$.

Hypotheses 15.0.1 (i) *The coefficients of the operator $\mathcal{A}$ satisfy Hypotheses* 14.0.1(*i*), (*ii*). *Moreover $c \leq 0$ and $q_{ij} \in C^{0,1}((t_1,t_2) \times \mathbb{R}^N)$ for any $i,j = 1,\dots,N$ and some $I \ni t_1 < t_2$;*

(ii) *For any bounded interval $J \subset I$ there exists a positive function $\varphi \in C^2(\mathbb{R}^N)$, blowing up as $|x|$ tends to $+\infty$, such that $\mathcal{A}\varphi - c\varphi \leq \lambda\varphi$ in $J \times \mathbb{R}^N$ for some positive constant $\lambda = \lambda_J$;*

(iii) *there exist a nonnegative function $V \in C^2(\mathbb{R}^N)$, blowing up as $|x|$ tends to $+\infty$, and a constant $M \geq 0$ such that $\mathcal{A}(t)V(x) \leq M$ and $\kappa_0\Delta V(x) + \langle b(t,x), \nabla V(x)\rangle + c(t,x)V(x) \leq M$ for any $t \in [t_1,t_2]$ and $x \in \mathbb{R}^N$. Here, κ_0 is as in Hypothesis* 14.0.1(*ii*);

(iv) *for any $t \in (t_1,t_2]$ there exist two nonnegative functions $W_1, W_2 \in C([t_1,t] \times \mathbb{R}^N) \cap C^{1,2}((t_1,t) \times \mathbb{R}^N)$ and two nonnegative functions $h_1, h_2 \in L^1((t_1,t))$ such that*

(a) *$W_j(s,x) \leq \rho_0 V(x)^{1-\beta}$ for any $(s,x) \in [t_1,t] \times \mathbb{R}^N$, some constants $\rho_0 > 0$, $\beta \in (0,1)$ and for $j = 1,2$;*

(b) *$\lim_{|x|\to+\infty} W_j(s,x) = +\infty$ ($j = 1,2$), uniformly for s varying in compact subsets of $[t_1,t)$;*

(c) *$D_s W_j - \mathcal{A}W_j \geq -h_j W_j$ and $D_s W_j - \kappa_0\Delta W_j - \langle b, \nabla_y W_j\rangle - cW_j \geq -h_j W_j$ in $(t_1,t_2) \times \mathbb{R}^N$ for $j = 1,2$;*

(v) *for any $t \in (t_1,t_2]$ there exist a function $1 \leq w \in C^{1,2}((t_1,t) \times \mathbb{R}^N)$, such that the function $w^{-2}D_s w$ and $w^{-2}\nabla_y w$ are bounded in $[\sigma_0,\tau_0] \times \mathbb{R}^N$ for any interval $[\sigma_0,\tau_0] \subset (t_1,t)$, and constants $C_1,\dots,C_9$, possibly depending on the interval (σ_0,τ_0), such that*

(a) $w \leq C_1 W_1$, (b) $|Q\nabla_y w| \leq C_2 w^{\frac{q-1}{q}} W_1^{\frac{1}{q}}$,

(c) $|\mathrm{Tr}(QD_y^2 w)| \leq C_3 w^{\frac{q-2}{q}} W_1^{\frac{2}{q}}$, (d) $|D_s w| \leq C_4 w^{\frac{q-2}{q}} W_1^{\frac{2}{q}}$,

(e) $|\sum_{i=1}^N D_i q_{ij}| \leq C_5 w^{-\frac{1}{q}} W_2^{\frac{1}{q}}$, (f) $|b| \leq C_6 w^{-\frac{1}{q}} W_2^{\frac{1}{q}}$,

(g) $c \leq C_7 w^{-\frac{2}{q}} W_2^{\frac{2}{q}}$, (h) $|\Delta_y w| \leq C_8 w^{\frac{q-2}{q}} W_1^{\frac{2}{q}}$,

(k) $|Q\nabla_y W_1| \leq C_9 w^{-\frac{1}{q}} W_1 W_2^{\frac{1}{q}}$

in $[\sigma_0,\tau_0] \times \mathbb{R}^N$ for some $q > N+2$.

We note that Hypotheses 15.0.1(i), (ii) imply that it is possible to associate evolution operators $\{G(t,s)\}$ and $\{G_0(t,s)\}$ in $C_b(\mathbb{R}^N)$ with the operators $\mathcal{A}$ and $\mathcal{A}-c$, respectively. Moreover, there exist functions $g, g_0 : \{(t,s) \in I \times I : t > s\} \times \mathbb{R}^N \times \mathbb{R}^N \to \mathbb{R}$ such that

$$(G(t,s)f)(x) = \int_{\mathbb{R}^N} f(y)g(t,s,x,y)dy, \qquad (G_0(t,s)f)(x) = \int_{\mathbb{R}^N} f(y)g_0(t,s,x,y)dy \tag{15.0.1}$$

for any $f \in C_b(\mathbb{R}^N)$, $I \ni s < t$ and $x, y \in \mathbb{R}^N$. By the maximum principle we can infer that $g \le g_0$. We refer the reader to Chapter 14 for the proofs of the previous results. Sometimes, in the chapter we find it convenient to use the measure $p(t,s,x;dy) = g(t,s,x,y)dy$.

We remark that the existence of functions W_j as in Hypotheses 15.0.1 have been considered also in [8, 239] to prove estimates of Green functions.

The chapter is organized as follows. First, in Section 15.1, we establish Sobolev regularity of the Green function $g(t,\cdot,x,\cdot)$ in the case of bounded diffusion coefficients adapting the arguments in [103, 195]. Then, in Section 15.2 we prove the main result of this chapter (Theorem 15.2.2) and apply it to a class of non-autonomous nondegenerate elliptic operators. We first address the case when the diffusion coefficients are bounded in $[t_1,t_2] \times \mathbb{R}^N$ together with their spatial gradients and, then, by an approximation argument we remove this additional assumption.

The results of this chapter are taken from [156, 157].

15.1 Integrability properties of Green functions

To begin with, we consider the following crucial remark.

Remark 15.1.1 By Lemma 14.1.6(i), for any $t_1 \le \sigma < \tau \le t_2$ and $\psi \in C_c^{1,2}((\sigma,\tau) \times \mathbb{R}^N)$ it holds that $D_sG(t,\cdot)\psi = G(t,\cdot)(D_s\psi - \mathcal{A}\psi)$ or, equivalently,

$$\int_\sigma^\tau ds \int_{\mathbb{R}^N} (D_s\psi - \mathcal{A}\psi)g(t,\cdot,x,\cdot)dy = 0 \tag{15.1.1}$$

and the forthcoming Theorem 16.2.4 shows that the function $g(t,\cdot,x,\cdot)$ is continuous (actually, locally Hölder continuous) in $(t_1,t) \times \mathbb{R}^N$.

We now give the following definition.

Definition 15.1.2 (Time dependent Lyapunov function) *Let $t \in (t_1,t_2]$. A nonnegative function $W \in C([t_1,t] \times \mathbb{R}^N) \cap C^{1,2}((t_1,t) \times \mathbb{R}^N)$ is a time dependent Lyapunov function on $[t_1,t]$ (with respect to the function $h \in L^1((t_1,t))$) if*

(i) *$W \le V$ in $[t_1,t] \times \mathbb{R}^N$;*

(ii) *$\lim_{|y|\to+\infty} W(s,y) = +\infty$ uniformly with respect to s varying in $[t_1,\tau)$ for any $\tau \in (t_1,t_2)$;*

(iii) *$D_sW - \mathcal{A}W \ge -hW$ and $D_sW - \kappa_0\Delta W - \langle b, \nabla_y W\rangle - cW \ge -hW$ in $(t_1,t) \times \mathbb{R}^N$.*

The following lemma will be crucial in the proof of Theorem 15.2.2.

Lemma 15.1.3 *Fix $t \in (t_1, t_2]$ and let W be a time dependent Lyapunov function on $[t_1, t]$ (with respect to h). Then, the function $G(t,s)W(s,\cdot)$ is well defined in $\mathbb{R}^N$ and*

$$(G(t,s)W(s,\cdot))(x) \le \exp\left(\int_s^t h(\tau)d\tau\right)W(s,x), \qquad t_1 \le s < t,\ x \in \mathbb{R}^N. \tag{15.1.2}$$

Proof Arguing as in the proof of Lemma 14.1.9 and using Lemma 14.1.6(ii), we can show that $(G(t,s)V)(x) \le V(x) + M(t-s)$ for any $t_1 \le s < t \le t_2$ and $x \in \mathbb{R}^N$, where M is defined in Hypothesis 15.0.1(iii). Since $W(s,\cdot) \le V$ for any $s \in [t_1, t]$, $G(t,s)W(s,\cdot)$ is well defined for any $s \in [t_1, t]$ and $G(t,s)W(s,\cdot) \le V + M$ in $\mathbb{R}^N$.

Now, we fix $t_1 \le s < r < t$, $x \in \mathbb{R}^N$ and, for any $n \in \mathbb{N}$, a function $\psi_n \in C^\infty([0,+\infty))$ such that $\tau\chi_{[0,n]}(\tau) \le \psi_n(\tau) \le (n+1/2)\chi_{[0,n+1]}(\tau)$, $0 \le \psi_n'(\tau) \le 1$ and $\psi_n''(\tau) \le 0$ for any $\tau \in [0,+\infty)$. Since $W(\sigma, x)$ diverges to $+\infty$ as $|x|$ tends to $+\infty$, uniformly with respect to $\sigma \in [0,r]$, the restriction to $[0,r] \times \mathbb{R}^N$ of the function $W_n := \psi_n \circ W$ can be split into the sum of a function in $C^{1,2}_c((0,r) \times \mathbb{R}^N)$ and a positive constant. By Lemma 14.1.6(ii) it follows that

$$\begin{aligned}
&(G(t,t-1/k)W_n(t-1/k,\cdot))(x) - (G(t,s)W_n(s,\cdot))(x)\\
&= \int_s^{t-1/k} (G(t,r)[\psi_n'(W(r,\cdot))(D_rW(r,\cdot) - \mathcal{A}(r)W(r,\cdot))])(x)dr\\
&\quad - \int_s^{t-1/k} (G(t,r)[\psi_n''(W(r,\cdot))|\sqrt{Q}(r,\cdot)\nabla_y W(r,\cdot)|^2])(x)dr\\
&\quad + \int_s^{t-1/k} (G(t,r)[c(r,\cdot)W(r,\cdot)\psi_n'(W(r,\cdot)) - c(r,\cdot)\psi_n(W(r,\cdot))])(x)dr\\
&\ge -\int_s^{t-1/k} h(r)(G(t,r)[\psi_n'(W(r,\cdot))W(r,\cdot)])(x)dr
\end{aligned} \tag{15.1.3}$$

for any $k \in \mathbb{N}$ large and $n \in \mathbb{N}$, where we have used the condition $\tau\psi_n'(\tau) - \psi_n(\tau) \le 0$ (which follows since ψ_n'' is nonpositive) to infer that the integral containing the potential c is nonnegative. Since $(\psi_n' \circ W)hW$ is compactly supported in $[t_1, t] \times \mathbb{R}^N$, the function $r \mapsto h(r)G(t,r)[\psi_n'(W(r,\cdot))W(r,\cdot)]$ is continuous in $[t_1,t] \times \mathbb{R}^N$ for any $n \in \mathbb{N}$, by Theorem 14.1.12. Hence, as k tends to $+\infty$, the last side of (15.1.3) converges to $-h(r)\int_s^t (G(t,r)[\psi_n'(W(r,\cdot))W(r,\cdot)])(x)dr$. As far as the first side is concerned, we observe that

$$\begin{aligned}
(G(t,t-1/k)W_n(t-1/k,\cdot))(x) - W_n(x) =&(G(t,t-1/k)[W_n(t-1/k,\cdot) - W_n(t,\cdot)])(x)\\
&+ (G(t,t-1/k)W_n(t,\cdot))(x) - W_n(t,x)
\end{aligned} \tag{15.1.4}$$

for any $n \in \mathbb{N}$ and k large enough. Both the terms in the right-hand side of (15.1.4) vanish. Indeed, the condition $(G(t,s)V)(x) \le V(x) + M(t-s)$ for any $s \in [t_1, t]$ implies that

$$p(t,s,x;\mathbb{R}^N \setminus B_R) \le \frac{1}{\inf_{\mathbb{R}^N\setminus B_R} V}\int_{\mathbb{R}^N} V(y)p(t,s,x;dy) \le \frac{V(x)+M}{\inf_{\mathbb{R}^N\setminus B_R} V},$$

where $p(t,s,x;dy) = g(t,s,x,y)dy$, which allows us to estimate

$$\begin{aligned}
&|(G(t,t-1/k)[W_n(t-1/k,\cdot) - W_n(t,\cdot)])(x)|\\
&\le \int_{B_R} |W_n(t-1/k,y) - W_n(t,y)|p(t,s,x;dy)
\end{aligned}$$

$$
\begin{aligned}
&+\int_{\mathbb{R}^N\setminus B_R}|W_n(t-1/k,y)-W_n(t,y)|p(t,s,x;dy)\\
\le&\|W_n(t-1/k,\cdot)-W_n(t,\cdot)\|_{L^\infty(B_R)}+(2n+1)p(t,s,x;\mathbb{R}^N\setminus B_R)\\
\le&\|W_n(t-1/k,\cdot)-W_n(t,\cdot)\|_{L^\infty(B_R)}+(2n+1)\frac{V(x)+M}{\inf_{\mathbb{R}^N\setminus B_R}V}
\end{aligned}
$$

for any $R>0$. Letting k tend to $+\infty$ yields that

$$
\limsup_{k\to+\infty}|(G(t,t-1/k)[W_n(t-1/k,\cdot)-W_n(t,\cdot)])(x)|\le(2n+1)\frac{V(x)+M}{\inf_{\mathbb{R}^N\setminus B_R}V}.
$$

Then, letting R tend to $+\infty$, we conclude that $(G(t,t-1/k)[W_n(t-1/k,\cdot)-W_n(t,\cdot)])(x)$ vanishes as k tends to $+\infty$. On the other hand, Theorem 14.1.12 shows that $(G(t,t-1/k)W_n(t,\cdot))(x)-W_n(t,x)$ vanishes as k tends to $+\infty$. Hence, from (15.1.4) we obtain that

$$
W_n(t,x)-(G(t,s)W_n(s,\cdot))(x)\ge-\int_s^t h(r)(G(t,r)[\psi_n'(W(r,\cdot))W(r,\cdot)])(x)dr. \tag{15.1.5}
$$

Since $\left|(\psi_n'\circ W)hW\right|\le hV$ in $[s,t]\times\mathbb{R}^N$ and

$$
\int_s^t h(r)(G(t,r)V)(x)dr\le(V(x)+M)\int_s^t h(r)dr<+\infty,
$$

we can let n tend to $+\infty$ in (15.1.5) and by monotone and dominated convergence conclude that

$$
W(t,x)-(G(t,s)W(s,\cdot))(x)\ge-\int_s^t h(r)(G(t,r)W(r,\cdot))(x)dr. \tag{15.1.6}
$$

To complete the proof, we now introduce the continuous function $\Phi_x:[s,t]\to\mathbb{R}$, defined by

$$
\Phi_x(\tau):=\left(W(t,x)+\int_\tau^t h(\sigma)(G(t,\sigma)W(\sigma,\cdot))(x)d\sigma\right)\exp\left(\int_s^\tau h(\sigma)d\sigma\right),
$$

which is weakly differentiable and, due to (15.1.6), Φ_x' is nonnegative almost everywhere in $[s,t]$. Hence, Φ_x is increasing and, using (15.1.6), we deduce that

$$
(G(t,s)W(s,\cdot))(x)\le\Phi_x(s)\le\Phi_x(t)=\exp\left(\int_s^t h(\sigma)d\sigma\right)W(t,x),
$$

which gives (15.1.2). □

To study the integrability properties of the function $g(t,\cdot,x,\cdot)$, we adapt the procedure in [195]. For this purpose, for $0\le\sigma<\tau\le1$, $k\ge1$, $t\in(t_1,t_2]$ and $x\in\mathbb{R}^N$, we set

$$
\Gamma_1(k,\sigma,\tau,t,x):=\left(\int_\sigma^\tau ds\int_{\mathbb{R}^N}|b(s,y)|^k g(t,s,x,y)dy\right)^{\frac1k},
$$

$$
\Gamma_2(k,\sigma,\tau,t,x):=\left(\int_\sigma^\tau ds\int_{\mathbb{R}^N}|c(s,y)|^k g(t,s,x,y)dy\right)^{\frac1k}.
$$

Proposition 15.1.4 *Let $c\equiv0$ and let the diffusion coefficients q_{ij} $(i,j=1,\dots,N)$ belong to $C_b^{0,1}((t_1,t_2)\times\mathbb{R}^N)$. Further, suppose that $\Gamma_1(1,\sigma,\tau,t,x)<+\infty$ for some $t\in(t_1,t_2]$, $x\in\mathbb{R}^N$ and $t_1<\sigma<\tau<t$. Then, $g(t,\cdot,x,\cdot)$ belongs to $L^r((\sigma,\tau)\times\mathbb{R}^N)$ for any $r\in[1,(N+2)/(N+1))$ and $\|g(t,\cdot,x,\cdot)\|_{L^r((\sigma,\tau)\times\mathbb{R}^N)}\le C(1+\Gamma_1(1,\sigma,\tau,t,x))$ for some constant C, depending only on the coefficients q_{ij} $(i,j=1,\dots,N)$ and N.*

Proof Fix t and x as in the statement. By Remark 15.1.1 it follows that

$$\int_{(\sigma,\tau)\times\mathbb{R}^N}(D_s\psi - \mathrm{Tr}(QD_y^2\psi))g(t,\cdot,x,\cdot)dsdy$$
$$=(G(t,\tau)\psi(\tau,\cdot))(x) - (G(t,\sigma)\psi(\sigma,\cdot))(x) + \int_\sigma^\tau ds\int_{\mathbb{R}^N}(\langle b,\nabla_y\psi\rangle + c\psi)g(t,\cdot,x,\cdot)dy$$

for any $\psi\in C_c^{1,2}([t_1,t_2]\times\mathbb{R}^N)$. Since $\|g(t,s,x,\cdot)\|_{L^1(\mathbb{R}^N)}\le 1$ for any $s\in(0,t)$ (see Proposition 14.1.3(i)), we deduce that

$$\left|\int_{(\sigma,\tau)\times\mathbb{R}^N}(D_s\psi - \mathrm{Tr}(QD_y^2\psi))g(t,\cdot,x,\cdot)dsdy\right|$$
$$\le\Gamma_1(1,\sigma,\tau,t,x)\|\psi\|_{W^{0,1}_\infty((\sigma,\tau)\times\mathbb{R}^N)} + \Gamma_2(1,\sigma,\tau,t,x)\|\psi\|_\infty + 2\|\psi\|_\infty$$
$$\le(2+\Gamma_1(1,\sigma,\tau,t,x)+\Gamma_2(1,\sigma,\tau,t,x))\|\psi\|_{W^{0,1}_\infty((\sigma,\tau)\times\mathbb{R}^N)}. \tag{15.1.7}$$

Fix $\zeta\in C_c^\infty((\sigma,\tau)\times\mathbb{R}^N)$ and r as in the statement. By the Schauder theory, there exists a unique function $u\in C^{1+\alpha/2,2+\alpha}([t_1,t_2]\times\mathbb{R}^N)\cap W^{1,2}_{r'}((t_1,t_2)\times\mathbb{R}^N)$ such that $D_tu=\mathrm{Tr}(QD_y^2u)+\zeta$ in $(t_1,t_2)\times\mathbb{R}^N$, $u(0,\cdot)=0$ and $\|u\|_{W^{1,2}_{r'}((t_1,t_2)\times\mathbb{R}^N)}\le C\|g\|_{L^{r'}((\sigma,\tau)\times\mathbb{R}^N)}$, where $1/r+1/r'=1$. Since $r'>N+2$, from the Sobolev embedding theorem we deduce that[1]

$$\|u\|_{W^{0,1}_\infty((t_1,t_2)\times\mathbb{R}^N)}\le C\|\zeta\|_{L^{r'}((\sigma,\tau)\times\mathbb{R}^N)}. \tag{15.1.8}$$

Now, we apply (15.1.7) with ψ being replaced by the function ψ_n ($n\in\mathbb{N}$), defined by $\psi_n(s,x)=\vartheta(x/n)u(s,x)$ for any $(s,x)\in[t_1,t_2]\times\mathbb{R}^N$, where $\vartheta\in C_c^\infty(\mathbb{R}^N)$ satisfy $\chi_{B_1}\le\vartheta\le\chi_{B_2}$. Taking (15.1.8) into account, we obtain

$$\left|\int_{(\sigma,\tau)\times\mathbb{R}^N}(D_s\psi_n - \mathrm{Tr}(QD_y^2\psi_n))(t,\cdot,x,\cdot)dsdy\right|$$
$$\le C[1+\Gamma_1(1,\sigma,\tau,t,x)+\Gamma_2(1,\sigma,\tau,t,x)]\|\zeta\|_{L^{r'}((\sigma,\tau)\times\mathbb{R}^N)}$$

for any $n\in\mathbb{N}$ and some positive constant C, independent of n. Letting n tend to $+\infty$, we conclude that

$$\left|\int_{(\sigma,\tau)\times\mathbb{R}^N}\zeta g(t,\cdot,x,\cdot)dsdy\right|\le C[1+\Gamma_1(1,\sigma,\tau,t,x)+\Gamma_2(1,\sigma,\tau,t,x)]\|\zeta\|_{L^{r'}((\sigma,\tau)\times\mathbb{R}^N)}.$$

The arbitrariness of ζ yields the assertion. □

Lemma 15.1.5 *Let $t_1<\sigma_0<\sigma<\tau<\tau_0<t_2$ and suppose that both $\Gamma_1(q,\sigma_0,\tau_0,t,x)$ and $\Gamma_2(q,\sigma_0,\tau_0,t,x)$ are finite for some $q>1$, $t\in(t_1,t_2]$ and $x\in\mathbb{R}^N$, and $g(t,\cdot,x,\cdot)$ belongs to $L^r((\sigma_0,\tau_0)\times\mathbb{R}^N)$ for some $r\in(1,+\infty)$. Then, $g(t,\cdot,x,\cdot)$ belongs to $\mathscr{H}^{p,1}((\sigma,\tau)\times\mathbb{R}^N)$ for $p=rq/(r+q-1)$.*

Proof Throughout the proof C denotes a constant depending on q, σ_0, σ, τ, τ_0, N and the coefficients q_{ij}, which may vary from line to line.

Let ϑ be a smooth function such that $\chi_{[\sigma,\tau]}\le\vartheta\le\chi_{[\sigma_0,\tau_0]}$ and fix $\psi\in C_c^{1,2}([t_1,t_2]\times\mathbb{R}^N)$. Since $D_sG(t,\cdot)(\vartheta\psi)=G(t,\cdot)(\vartheta'\psi+\vartheta D_s\psi-\vartheta\mathcal{A}\psi)$, using (15.1.1) we can write

$$\int_{\sigma_0}^{\tau_0}ds\int_{\mathbb{R}^N}\vartheta g(t,\cdot,x,\cdot)(D_s\psi-\mathrm{Tr}(QD_y^2\psi))dy$$

[1] For further details, see e.g. [145, Theorem 9.2.3], [159, Lemma II.3.3 and Theorem IV.9.1].

$$= \int_{\sigma_0}^{\tau_0} ds \int_{\mathbb{R}^N} g(t,\cdot,x,\cdot)(\langle b, \nabla_y\psi\rangle\vartheta + c\psi\vartheta - \psi\vartheta')dy. \tag{15.1.9}$$

Applying the Hölder inequality we get

$$\|bg(t,\cdot,x,\cdot)\|_{L^p((\sigma_0,\tau_0)\times\mathbb{R}^N)} \le \Gamma_1(q,\sigma_0,\tau_0,t,x)\|g(t,\cdot,x,\cdot)\|_{L^r((\sigma_0,\tau_0)\times\mathbb{R}^N)}^{\frac{q-1}{q}} \tag{15.1.10}$$

and

$$\|cg(t,\cdot,x,\cdot)\|_{L^p((\sigma_0,\tau_0)\times\mathbb{R}^N)} \le \Gamma_2(q,\sigma_0,\tau_0,t,x)\|g(t,\cdot,x,\cdot)\|_{L^r((\sigma_0,\tau_0)\times\mathbb{R}^N)}^{\frac{q-1}{q}}. \tag{15.1.11}$$

Arguing similarly, we can show that $g(t,\cdot,x,\cdot) \in L^p((\sigma_0,\tau_0)\times\mathbb{R}^N)$ and

$$\|g(t,\cdot,x,\cdot)\|_{L^p((\sigma_0,\tau_0)\times\mathbb{R}^N)} \le \|g(t,\cdot,x,\cdot)\|_{L^r((\sigma_0,\tau_0)\times\mathbb{R}^N)}^{\frac{q-1}{q}}. \tag{15.1.12}$$

Hence, from (15.1.9) we deduce that

$$\begin{aligned}
&\left|\int_{\sigma_0}^{\tau_0} ds \int_{\mathbb{R}^N} \vartheta g(t,\cdot,x,\cdot)(D_s\psi - \mathrm{Tr}(QD_y^2\psi))dy\right| \\
\le & C\|g(t,\cdot,x,\cdot)\|_{L^r((\sigma_0,\tau_0)\times\mathbb{R}^N)}^{\frac{q-1}{q}}\|\psi\|_{W^{0,1}_{p'}((0,1)\times\mathbb{R}^N)},
\end{aligned} \tag{15.1.13}$$

where $1/p+1/p'=1$. To simplify the notation, in what follows we set $\widetilde{g}(t,\cdot,x,\cdot) = \vartheta g(t,\cdot,x,\cdot)$ and extend this function in the trivial way to $(t_1,t_2)\times\mathbb{R}^N$.

Now fix $j \in \{1,\ldots,N\}$ and set $\Delta_h\psi(t,x) := h^{-1}\big[\psi(t,x+he_j) - \psi(t,x)\big]$ for any $(t,x) \in [t_1,t_2]\times\mathbb{R}^N$. An easy computation shows that

$$\begin{aligned}
&\int_{t_1}^{t_2} ds \int_{\mathbb{R}^N} \widetilde{g}(t,\cdot,x,\cdot)[D_s(\Delta_{-h}\psi) - \mathrm{Tr}(QD_y^2\Delta_{-h}\psi)]dy \\
=&\int_{t_1}^{t_2} ds \int_{\mathbb{R}^N} \widetilde{g}(t,s,x,y+he_j)\sum_{i,j=1}^N (\Delta_h q_{ij})(s,y)D_{ij}\psi(s,y)dy \\
&-\int_{t_1}^{t_2} ds \int_{\mathbb{R}^N} (\Delta_h\widetilde{g}(t,\cdot,x,\cdot))(D_s\psi - \mathrm{Tr}(QD_y^2\psi))dy.
\end{aligned}$$

Since the first-order derivatives of the diffusion coefficients are bounded functions on $(t_1,t_2)\times\mathbb{R}^N$, it follows that

$$\begin{aligned}
&\left|\int_{t_1}^{t_2} ds \int_{\mathbb{R}^N} \widetilde{g}(t,s,x,y+he_j)\sum_{i,j=1}^N (\Delta_h q_{ij})(s,y)D_{ij}\psi(s,y)dy\right| \\
\le & C\|g(t,\cdot,x,\cdot)\|_{L^r((\sigma_0,\tau_0)\times\mathbb{R}^N)}^{\frac{q-1}{q}}\|\psi\|_{W^{1,2}_{p'}((t_1,t_2)\times\mathbb{R}^N)}.
\end{aligned} \tag{15.1.14}$$

On the other hand, $\|\Delta_h\psi\|_{W^{0,1}_{p'}((t_1,t_2)\times\mathbb{R}^N)} \le \|\psi\|_{W^{1,2}_{p'}((t_1,t_2)\times\mathbb{R}^N)}$. Hence, from (15.1.13) and (15.1.14) we obtain that

$$\begin{aligned}
&\left|\int_{t_1}^{t_2} ds \int_{\mathbb{R}^N} (\Delta_h\widetilde{g}(t,\cdot,x,\cdot))(D_s\psi - \mathrm{Tr}(QD_y^2\psi))dy\right| \\
\le&\left|\int_{t_1}^{t_2} ds \int_{\mathbb{R}^N} \widetilde{g}(t,s,x,y+he_j)\sum_{i,j=1}^N (\Delta_h q_{ij})(s,y)D_{ij}\psi(s,y)dy\right|
\end{aligned}$$

$$
\begin{aligned}
&+\left|\int_{t_1}^{t_2} ds\int_{\mathbb{R}^N}\widetilde{g}(t,\cdot,x,\cdot)[D_s(\Delta_{-h}\psi)-\mathrm{Tr}(QD^2_y\Delta_{-h}\psi)]dy\right|\\
&\le C\|g(t,\cdot,x,\cdot)\|^{\frac{q-1}{q}}_{L^r((\sigma_0,\tau_0)\times\mathbb{R}^N)}\|\psi\|_{W^{1,2}_{p'}((t_1,t_2)\times\mathbb{R}^N)}.
\end{aligned}\tag{15.1.15}
$$

Observing that $\widetilde{g}(t,\cdot,x,\cdot)$ (and, hence, $\Delta_h\widetilde{g}(t,\cdot,x,\cdot)$) belongs to $L^p((t_1,t_2)\times\mathbb{R}^N)$, by density we can extend (15.1.15) to any $\psi\in W^{1,2}_{p'}((t_1,t_2)\times\mathbb{R}^N)$.

Since the function $|\Delta_h\widetilde{g}(t,\cdot,x,\cdot)|^{p-2}\Delta_h\widetilde{g}(t,\cdot,x,\cdot)$ belongs to $L^{p'}((t_1,t_2)\times\mathbb{R}^N)$, there exists a unique function[2] $u\in W^{1,2}_{p'}((t_1,t_2)\times\mathbb{R}^N)$ such that

$$
D_s u=\mathrm{Tr}(QD^2_y u)+|\Delta_h\widetilde{g}(t,\cdot,x,\cdot)|^{p-2}\Delta_h\widetilde{g}(t,\cdot,x,\cdot)
$$

almost everywhere in $(0,1)\times\mathbb{R}^N$, $\|u\|_{W^{1,2}_{p'}((t_1,t_2)\times\mathbb{R}^N)}\le C\||\Delta_h\widetilde{g}(t,\cdot,x,\cdot)|^{p-1}\|_{L^{p'}((t_1,t_2)\times\mathbb{R}^N)}$ and $u(t_1,\cdot)=0$. Replacing $u=\psi$ in (15.1.15) yields

$$
\|\Delta_h\widetilde{g}(t,\cdot,x,\cdot)\|_{L^p((t_1,t_2)\times\mathbb{R}^N)}\le C\|g(t,\cdot,x,\cdot)\|^{\frac{q-1}{q}}_{L^r((\sigma_0,\tau_0)\times\mathbb{R}^N)}.
$$

It thus follows that $D_j\widetilde{g}(t,\cdot,x,\cdot)$ exists in $L^p((t_1,t_2)\times\mathbb{R}^N)$ and, consequently, $\widetilde{g}(t,\cdot,x,\cdot)\in W^{0,1}_p((t_1,t_2)\times\mathbb{R}^N)$ and

$$
\|\nabla_y\widetilde{g}(t,\cdot,x,\cdot)\|_{L^p((t_1,t_2)\times\mathbb{R}^N)}\le C\|g(t,\cdot,x,\cdot)\|^{\frac{q-1}{q}}_{L^r((\sigma_0,\tau_0)\times\mathbb{R}^N)}.\tag{15.1.16}
$$

Now, let us consider the distributional time derivative of $g(t,\cdot,x,\cdot)$. From (15.1.9) we deduce that

$$
\begin{aligned}
&\left|\int_{(t_1,t_2)\times\mathbb{R}^N}\widetilde{g}(t,\cdot,x,\cdot)D_s\psi dsdy\right|\\
=&\left|\int_{(t_1,t_2)\times\mathbb{R}^N}[\widetilde{g}(t,\cdot,x,\cdot)\mathrm{Tr}(QD^2_y\psi)+\langle b,\nabla_y\psi\rangle\widetilde{g}(t,\cdot,x,\cdot)+c\psi\widetilde{g}(t,\cdot,x,\cdot)-g(t,\cdot,x,\cdot)\psi\vartheta']dy\right|
\end{aligned}
$$

for any $\psi\in C^{1,2}_c([t_1,t_2]\times\mathbb{R}^N)$. Integrating by parts we find

$$
\begin{aligned}
\int_{(t_1,t_2)\times\mathbb{R}^N}\widetilde{g}(t,\cdot,x,\cdot)\mathrm{Tr}(QD^2_y\psi)dy=&-\int_{(t_1,t_2)\times\mathbb{R}^N}\langle Q\nabla_y\widetilde{g}(t,\cdot,x,\cdot),\nabla_y\psi\rangle dy\\
&-\int_{(t_1,t_2)\times\mathbb{R}^N}\widetilde{g}(t,\cdot,x,\cdot)\sum_{i,j=1}^N D_iq_{ij}D_j\psi dy.
\end{aligned}
$$

Using estimates (15.1.12), (15.1.16) and recalling that the gradient of the diffusion coefficients is bounded, we deduce that

$$
\left|\int_{t_1}^{t_2}ds\int_{\mathbb{R}^N}\widetilde{g}(t,\cdot,x,\cdot)\mathrm{Tr}(QD^2_y\psi)dy\right|\le C\|g(t,\cdot,x,\cdot)\|^{\frac{q-1}{q}}_{L^r((\sigma_0,\tau_0)\times\mathbb{R}^N)}\|\nabla_y\psi\|_{L^{p'}((t_1,t_2)\times\mathbb{R}^N)}.
$$

From this estimate and (15.1.10)-(15.1.12) we conclude that

$$
\left|\int_{t_1}^{t_2}ds\int_{\mathbb{R}^N}\widetilde{g}(t,\cdot,x,\cdot)D_s\psi dy\right|\le C\|g(t,\cdot,x,\cdot)\|^{\frac{q-1}{q}}_{L^r((\sigma_0,\tau_0)\times\mathbb{R}^N)}\|\psi\|_{W^{0,1}_{p'}((t_1,t_2)\times\mathbb{R}^N)},
$$

which shows that $D_s\widetilde{g}$ belongs to $(W^{0,1}_{p'}((t_1,t_2)\times\mathbb{R}^N))'$. We have so proved that $\widetilde{g}(t,\cdot,x,\cdot)$ belongs to $\mathscr{H}^{p,1}((t_1,t_2)\times\mathbb{R}^N)$. Since $\widetilde{g}(t,\cdot,x,\cdot)\equiv g(t,\cdot,x,\cdot)$ in $(\sigma,\tau)\times\mathbb{R}^N$, the assertion follows. □

We can now use a bootstrap argument to improve the summability properties of g.

[2]See e.g. [159, Theorem IV.9.1].

Proposition 15.1.6 *Suppose that $c \equiv 0$ and $q_{ij} \in C_b^{0,1}((t_1,t_2)\times\mathbb{R}^N)$ $(i,j=1,\dots,N)$. Fix $(t,x)\in(t_1,t_2]\times\mathbb{R}^N$ and $t_1<\sigma_0<\sigma<\tau<\tau_0<t_2$. Then, the following properties are satisfied.*

(i) *If $\Gamma_1(q,\sigma_0,\tau_0,t,x)<+\infty$ for some $1<q\le N+2$, then $g(t,\cdot,x,\cdot)$ belongs to $L^r((\sigma,\tau)\times\mathbb{R}^N)\cap\mathscr{H}^{p,1}((\sigma,\tau)\times\mathbb{R}^N)$ for any $p\in(1,(N+2)/(N+3-q))$ and $r\in[1,(N+2)/(N+2-q))$.*

(ii) *If $\Gamma_1(q,\sigma_0,\tau_0,t,x)<+\infty$ for some $q>N+2$, then $g(t,\cdot,x,\cdot)$ belongs to $\mathscr{H}^{p,1}((\sigma,\tau)\times\mathbb{R}^N)$ for some $p>N+2$, and also to $L^\infty((\sigma,\tau)\times\mathbb{R}^N)$.*

Proof (i) As a first step we observe that, for any $d_*>(N+1)/(N+2)$, the sequence

$$\begin{cases} d_{n+1}=d_n\left(1-\dfrac{1}{q}\right)+\dfrac{1}{q}-\dfrac{1}{N+2}, & n\ge 1,\\ \sigma_1=d_* \end{cases}$$

is decreasing and it converges to $(N+2-q)/(N+2)$. As a consequence, the sequence $\{e_n\}$, defined by $e_n=d_n+(N+2)^{-1}$ for any $n\in\mathbb{N}$, is decreasing and converges to $(N+2)/(N+3-q)$ as n tends to $+\infty$.

We now fix $r\in(1,(N+2)/(N+2-q))$ and $p\in(1,(N+2)/(N+3-q))$, set $d_*=1/r$ and, for any $n\le m$, $\sigma_n=\sigma_0+n(\sigma-\sigma_0)/m$ and $\tau_n=\tau_0-n(\tau_0-\tau)/m$, where m is the smallest integer such that $\sigma_m>r$ and $\tau_m>p$. Since $\Gamma_1(1,\sigma_0,\tau_0,t,x)\le\Gamma_1(q,\sigma_0,\tau_0,t,x)$, Proposition 15.1.4 shows that $g(t,\cdot,x,\cdot)\in L^r((\sigma_1,\tau_1)\times\mathbb{R}^N)$. Now, we can invoke Lemma 15.1.5, with $(\sigma_0,\sigma,\tau,\tau_0)=(\sigma_1,\sigma_2,\tau_2,\tau_1)$, to infer that $g(t,\cdot,x,\cdot)\in\mathscr{H}^{p_2,1}((\sigma_2,\tau_2)\times\mathbb{R}^N)$, where $p_2=rq/(r+q-1)$. Since this space is continuously embedded in $L^{1/d_2}((\sigma_2,\tau_2)\times\mathbb{R}^N)$ (see Section A.3), we conclude that $g(t,\cdot,x,\cdot)\in L^{1/d_2}((\sigma_2,\tau_2)\times\mathbb{R}^N)$. Iterating this argument shows that $g(t,\cdot,x,\cdot)\in L^{1/d_n}((\sigma_n,\tau_n)\times\mathbb{R}^N)\cap\mathscr{H}^{1/e_n,1}((\sigma_n,\tau_n)\times\mathbb{R}^N)$ for any $n\le m$. Taking $n=m$, we complete the proof.

(ii) Since $\Gamma_1(N+2,\sigma_0,\tau_0)\le\Gamma_1(q,\sigma_0,\tau_0,t,x)<+\infty$, from property (i) it follows that $g(t,\cdot,x,\cdot)\in\mathscr{H}^{p,1}((\sigma,\tau)\times\mathbb{R}^N)$ for any $p\in(1,k)$. Recalling that $\mathscr{H}^{p,1}((\sigma,\tau)\times\mathbb{R}^N)\hookrightarrow L^\infty((\sigma,\tau)\times\mathbb{R}^N)$ for any $p>N+2$ (see Section A.3), the assertion follows at once. □

We now provide an estimate of the sup-norm of weak solutions to the nonhomogeneous parabolic equation $D_s u=\mathcal{A}u+f$, with a constant that does not depend on the sup-norm of the diffusion coefficients q_{ij} of the operator $\mathcal{A}$. The proof is the parabolic version of [103, Theorem A.1] and is based on De Giorgi's technique.

To begin with, we recall the following embedding result and, then, we prove an integration by parts formula.

Lemma 15.1.7 ([159], Chapter 2, Section 3) *Suppose that $p\in(2,+\infty]$, $q\in[2,+\infty)$, if $N=2$, and $p\in[2,+\infty]$, $q\in[2,2N/(N-2)]$, if $N\ge 3$. Further, suppose that $\frac{1}{p}+\frac{N}{2q}=\frac{N}{4}$. Then, there exists a positive constant K, which is independent of σ,τ varying in bounded subsets of $\mathbb{R}$, such that $\|u\|_{L^p((\sigma,\tau);L^q(\mathbb{R}^N))}\le K(\|u\|_{L^\infty((\sigma,\tau);L^2(\mathbb{R}^N))}+\|\nabla_y u\|_{L^2((\sigma,\tau)\times\mathbb{R}^N)})$ for any $u\in W_2^{0,1}((\sigma,\tau)\times\mathbb{R}^N)\cap L^\infty((\sigma,\tau);L^2(\mathbb{R}^N))$.*

Lemma 15.1.8 *Let u belong to $\mathscr{H}^{p,1}((\sigma,\tau)\times\mathbb{R}^N)\cap C_b([\sigma,\tau]\times\mathbb{R}^N)$ for some $p\in[2,+\infty)$ and $\sigma,\tau\in\mathbb{R}$ with $\sigma<\tau$. Further, let $\ell>0$ and $\vartheta:\mathbb{R}^N\to\mathbb{R}$ be a nonnegative, smooth and compactly supported function. Then, $\vartheta(u-\ell)_+$ belongs to $W_{p'}^{0,1}((\sigma,\tau)\times\mathbb{R}^N)$ and*

$$(D_t u,\vartheta(u-\ell)_+)_p=\frac{1}{2}\left(\int_{\mathbb{R}^N}\vartheta((u(\tau,\cdot)-\ell)_+)^2dy-\int_{\mathbb{R}^N}\vartheta((u(\sigma,\cdot)-\ell)_+)^2dy\right),\tag{15.1.17}$$

where $(\cdot,\cdot)_p$ denotes the pairing between $(W_{p'}^{0,1}((\sigma,\tau)\times\mathbb{R}^N))'$ and $W_{p'}^{0,1}((\sigma,\tau)\times\mathbb{R}^N)$, and $1/p+1/p'=1$.

Proof As a first step, we observe that, since $p' \le p$ and ϑ is compactly supported in $\mathbb{R}^N$, the function $\vartheta(u-\ell)$ belongs to $W^{0,1}_{p'}((\sigma,\tau)\times\mathbb{R}^N)$. As a straightforward consequence, $\vartheta(u-\ell)_+$ belongs to $W^{0,1}_{p'}((\sigma,\tau)\times\mathbb{R}^N)$, as well.

To prove the formula (15.1.17) we approximate u by a sequence $\{u_n\}\in C^\infty_c(\mathbb{R}^{N+1})$ converging to u in the $\mathscr{H}^{p,1}$-norm, see Section A.3. As is easily seen,

$$\begin{aligned}&(D_t u, \vartheta(u_n-\ell)_+)_p\\ =&\frac{1}{2}(D_t((u_n-\ell)_+^2,\vartheta)_p + (D_t u - D_t u_n, \vartheta(u_n-\ell)_+)_p\\ =&\frac{1}{2}\int_{\mathbb{R}^N}\vartheta[(u_n(\sigma,\cdot)-\ell)_+^2-(u_n(\tau,\cdot)-\ell)_+^2]dx + (D_t u - D_t u_n, \vartheta(u_n-\ell)_+)_p. \end{aligned}\tag{15.1.18}$$

To let n tend to $+\infty$ in the first and last side of the previous formula, we observe that Lemma A.3.1 shows that u_n converges to u uniformly in $(\sigma,\tau)\times\mathbb{R}^N$. Moreover, the sequence $\{\vartheta(u_n-\ell)_+\}$ converges to $\vartheta(u-\ell)_+$ in $W^{0,1}_{p'}((\sigma,\tau)\times\mathbb{R}^N)$ as n tends to $+\infty$. Hence, all the terms in the first and last side of (15.1.18) converge to the corresponding terms with u_n being replaced by u and (15.1.17) follows. □

Now, we can prove the following crucial result.

Theorem 15.1.9 *Let $t_1\le\sigma<\tau\le t_2$, $r>N+2$ and let the functions f and h belong to $L^{r/2}((\sigma,\tau)\times\mathbb{R}^N)$ and $L^r((\sigma,\tau)\times\mathbb{R}^N;\mathbb{R}^N)$, respectively. Further, assumc that u belongs to $\mathscr{H}^{p,1}((\sigma,\tau)\times\mathbb{R}^N)\cap C_b([\sigma,\tau]\times\mathbb{R}^N)$, for some $p>2$, satisfies $u(\tau,\cdot)\equiv 0$ and*

$$\begin{aligned}&\int_{(\sigma,\tau)\times\mathbb{R}^N}\langle Q\nabla_y u,\nabla_y\psi\rangle dsdy-(D_s u,\psi)_p\\ =&\int_{(\sigma,\tau)\times\mathbb{R}^N} f\psi dsdy+\int_{(\sigma,\tau)\times\mathbb{R}^N}\langle h,\nabla_y\psi\rangle dsdy\end{aligned}\tag{15.1.19}$$

for any $\psi\in C^\infty_c([\sigma,\tau]\times\mathbb{R}^N)$, where $(\cdot,\cdot)_p$ denotes the pairing between $(W^{0,1}_{p'}((\sigma,\tau)\times\mathbb{R}^N))'$ and $W^{0,1}_{p'}((\sigma,\tau)\times\mathbb{R}^N)$. Then, u is bounded and there exists a positive constant C, depending only on κ_0, N, r, t_1 and t_2, such that

$$\|u\|_\infty\le C\big(\|u\|_{L^\infty((\sigma,\tau);L^2(\mathbb{R}^N))}+\|f\|_{L^{r/2}((\sigma,\tau)\times\mathbb{R}^N)}+\|h\|_{L^r((\sigma,\tau)\times\mathbb{R}^N)}\big).$$

Proof A density argument shows that (15.1.19) is satisfied by any function $\psi\in W^{0,1}_2((\sigma,\tau)\times\mathbb{R}^N)$ vanishing in $(\sigma,\tau)\times(\mathbb{R}^N\setminus B_R)$ for some $R>0$. By replacing u by u/K_0, where $K_0=\|u\|_{L^\infty((\sigma,\tau);L^2(\mathbb{R}^N))}+\|f\|_{L^{r/2}((\sigma,\tau)\times\mathbb{R}^N)}+\|h\|_{L^r((\sigma,\tau)\times\mathbb{R}^N)}$, we can assume that $\|u\|_{L^\infty((\sigma,\tau);L^2(\mathbb{R}^N))}$, $\|f\|_{L^{r/2}((\sigma,\tau)\times\mathbb{R}^N)}$, $\|h\|_{L^r((\sigma,\tau)\times\mathbb{R}^N)}$ are all not greater than 1.

With this reduction in mind, we fix $\ell>1$ and a standard sequence $\{\vartheta_n\}\in C^\infty_c(\mathbb{R}^N)$ of cut-off functions. Replacing $\psi:=\vartheta_n^2(u-\ell)_+$ in (15.1.19) and taking Lemma 15.1.8 into account we get

$$\begin{aligned}&\frac{1}{2}\int_{\mathbb{R}^N}\vartheta_n^2(u(\sigma,\cdot)-\ell)_+^2dx+\int_{(\sigma,\tau)\times\mathbb{R}^N}\vartheta_n^2|\sqrt{Q}\nabla_y(u-\ell)_+|^2dsdy\\ =&-2\int_{(\sigma,\tau)\times\mathbb{R}^N}\vartheta_n\langle Q\nabla_y(u-\ell)_+,\nabla\vartheta_n\rangle(u-\ell)_+dsdy\\ &+\int_{(\sigma,\tau)\times\mathbb{R}^N}f\vartheta_n^2(u-\ell)_+dsdy+\int_{(\sigma,\tau)\times\mathbb{R}^N}\vartheta_n^2\langle h,\nabla_y(u-\ell)_+\rangle dsdy\\ &+2\int_{(\sigma,\tau)\times\mathbb{R}^N}\vartheta_n(u-\ell)_+\langle h,\nabla\vartheta_n\rangle dsdy=:I_{1,n}+I_{2,n}+I_{3,n}+I_{4,n}\end{aligned}$$

for any $n \in \mathbb{N}$. The Cauchy-Schwarz inequality shows that

$$|I_{1,n}| \leq \frac{1}{2}\int_{(\sigma,\tau)\times\mathbb{R}^N} \vartheta_n^2|\sqrt{Q}\nabla_y(u-\ell)_+|^2 dsdy + \frac{C\|Q\|_\infty}{n^2}\int_{(\sigma,\tau)\times\mathbb{R}^N}(u-\ell)_+^2 dsdy$$

for some positive constant C, independent of n.

To estimate $I_{2,n}$ we use the Hölder inequality, Lemma 15.1.7 and obtain that

$$\begin{aligned}|I_{2,n}| \leq&\|\vartheta_n f\|_{L^{\frac{r}{2}}((\sigma,\tau)\times\mathbb{R}^N)}\|(u-\ell)_+\|_{L^{2+\frac{4}{N}}((\sigma,\tau)\times\mathbb{R}^N)}|A_\ell|^{\frac{1}{2}-\frac{2}{r}+\frac{1}{N+2}}\\ \leq&K(\|(u-\ell)_+\|_{L^\infty((\sigma,\tau);L^2(\mathbb{R}^N))} + \|\nabla_y(u-\ell)_+\|_{L^2((\sigma,\tau)\times\mathbb{R}^N)})|A_\ell|^{\frac{1}{2}-\frac{2}{r}+\frac{1}{N+2}}\end{aligned}$$

for any $n \in \mathbb{N}$, where $|A_\ell|$ denotes the $(N+1)$-dimensional Lebesgue measure of the set $A_\ell = \{(s,y) \in (\sigma,\tau)\times\mathbb{R}^N : u(s,y) \geq \ell\}$. Using the Hölder inequality again, we get

$$|I_{3,n}| \leq \|h\|_{L^r((\sigma,\tau)\times\mathbb{R}^N)}\|\nabla_y(u-\ell)_+\|_{L^2((\sigma,\tau)\times\mathbb{R}^N)}|A_\ell|^{\frac{1}{2}-\frac{1}{r}}, \qquad n \in \mathbb{N}.$$

The previous two estimates show that the functions $f(u-\ell)_+$ and $\langle h, \nabla(u-\ell)_+\rangle$ belong to $L^1((\sigma,\tau)\times\mathbb{R}^N)$. Finally, we estimate

$$|I_{4,n}| \leq \frac{C}{n}\|h\|_{L^r((\sigma,\tau)\times\mathbb{R}^N)}\|(u-\ell)_+\|_{L^2((\sigma,\tau)\times\mathbb{R}^N)}|A_\ell|^{\frac{1}{2}-\frac{1}{r}}, \qquad n \in \mathbb{N}.$$

Collecting everything together and letting n tend to $+\infty$, we obtain

$$\begin{aligned}&\frac{1}{2}\int_{\mathbb{R}^N}(u(\sigma,\cdot)-\ell)_+^2 dy + \frac{1}{2}\int_{(\sigma,\tau)\times\mathbb{R}^N}|\sqrt{Q}\nabla_y(u-\ell)_+|^2 dsdy\\ &\leq \int_{(\sigma,\tau)\times\mathbb{R}^N}|f|(u-\ell)_+ dsdy + \int_{(\sigma,\tau)\times\mathbb{R}^N}|h||\nabla_y(u-\ell)_+|dsdy\\ &\leq K\big(\|(u-\ell)_+\|_{L^\infty((\sigma,\tau);L^2(\mathbb{R}^N))} + \|\nabla_y(u-\ell)_+\|_{L^2((\sigma,\tau)\times\mathbb{R}^N)}\big)|A_\ell|^{\frac{1}{2}-\frac{2}{r}+\frac{1}{N+2}}\\ &\quad + \|\nabla_y(u-\ell)_+\|_{L^2((\sigma,\tau)\times\mathbb{R}^N)}|A_\ell|^{\frac{1}{2}-\frac{1}{r}}.\end{aligned}$$

Clearly, the previous estimate also holds true with σ being replaced by an arbitrary $\sigma' \in (\sigma,\tau)$. We now write this estimate with σ' instead of σ and take the supremum with respect to $\sigma' \in (\sigma,\tau)$. Using the ellipticity assumption on Q and recalling that $\|h\|_{L^r((\sigma,\tau)\times\mathbb{R}^N)} \leq 1$, we obtain

$$\begin{aligned}&(1\wedge\kappa_0)\big(\|(u-\ell)_+\|_{L^\infty((\sigma,\tau);L^2(\mathbb{R}^N))}^2 + \|\nabla_y(u-\ell)_+\|_{L^2((\sigma,\tau)\times\mathbb{R}^N)}^2\big)\\ \leq&2K\big(\|(u-\ell)_+\|_{L^\infty((\sigma,\tau);L^2(\mathbb{R}^N))}+\|\nabla_y(u-\ell)_+\|_{L^2((\sigma,\tau)\times\mathbb{R}^N)}\big)|A_\ell|^{\frac{1}{2}-\frac{2}{r}+\frac{1}{N+2}}\\ &+ 2\|\nabla_y(u-\ell)_+\|_{L^2((\sigma,\tau)\times\mathbb{R}^N)}|A_\ell|^{\frac{1}{2}-\frac{1}{r}}.\end{aligned}$$

Since $\frac{1}{2}-\frac{1}{r} < \frac{1}{2}-\frac{2}{r}+\frac{1}{N+2}$ and $|A_\ell| \leq 1$, from the assumptions on r and ℓ it follows that

$$\|(u-\ell)_+\|_{L^\infty((\sigma,\tau);L^2(\mathbb{R}^N))} + \|\nabla_y(u-\ell)_+\|_{L^2((\sigma,\tau)\times\mathbb{R}^N)} \leq L|A_\ell|^{\frac{1}{2}-\frac{1}{r}} \tag{15.1.20}$$

for some positive constant L. For $m > \ell$ we find

$$\begin{aligned}(m-\ell)^2|A_m| \leq& \int_{A_m}(u-\ell)_+^2 dsdy \leq \int_{A_\ell}(u-\ell)_+^2 dsdy\\ \leq& \|(u-\ell)_+\|_{L^{2+\frac{4}{d}}((\sigma,\tau)\times\mathbb{R}^N)}^2\|\chi_{A_\ell}\|_{L^{\frac{N+2}{2}}((\sigma,\tau)\times\mathbb{R}^N)}\\ \leq& K\big(\|(u-\ell)_+\|_{L^\infty((\sigma,\tau);L^2(\mathbb{R}^N))} + \|\nabla_y(u-\ell)_+\|_{L^2((\sigma,\tau)\times\mathbb{R}^N)}\big)^2|A_\ell|^{\frac{2}{N+2}}\end{aligned}$$

$$\leq LK|A_\ell|^{1-\frac{2}{r}+\frac{2}{N+2}}. \tag{15.1.21}$$

Here, we have used the Hölder inequality with exponents $1+2/d$ and $(N+2)/2$, Lemma 15.1.7 and estimate (15.1.20).

Now, we can complete the proof. We fix $\overline{\ell} \geq 1$ and define $\ell_n = 2\overline{\ell} - 2^{-n}\overline{\ell}$ and $y_n = |A_{\ell_n}|$ for any $n \in \mathbb{N}$. From (15.1.21), with $m = \ell_{n+1}$ and $\ell = \ell_n$, it follows that $y_{n+1} \leq 4LK\overline{\ell}^{-2}2^{2n}y_n^{1+\alpha}$, where $\alpha = \frac{2}{N+2} - \frac{2}{r} > 0$. If $y_0 \leq (4LK\overline{\ell}^{-2})^{-1/\alpha}4^{-\alpha^{-2}}$ then y_n vanishes as n tends to $+\infty$ and this implies that $|A_l| = 0$ for $l \geq 2\overline{\ell}$, i.e., $u \leq 2\overline{\ell}$. As $y_0 = |A_{\overline{\ell}}| \leq 1$, this estimate holds if we take $\overline{\ell} = \max\{1, 2^{1+\frac{1}{\alpha}}\sqrt{LK}\} =: C/2$. Thus, $u \leq C$. Replacing u with $-u$, we obtain that $-u \leq C$ and, consequently, $\|u\|_\infty \leq C$. □

15.2 Kernel estimates

To state the main result of this chapter, we need to introduce some approximate operators $\mathcal{A}^{(n)} = \mathrm{Tr}(Q^{(n)}D^2) + \langle b, \nabla\rangle$ $(n \in \mathbb{N})$. For this purpose, we denote by ϑ any smooth function such that $\chi_{(-1,1)} \leq \vartheta \leq \chi_{(-2,2)}$ and $|r\vartheta'(r)| \leq 2$ for any $r \in \mathbb{R}$. Then, we fix $t_* \in [t_1, t_2)$ and set $q_{ij}^{(n)} = \vartheta_n q_{ij} + (1-\vartheta_n)\kappa_0\delta_{ij}$ $(i,j = 1,\ldots,N)$, where $\vartheta_n(s,y) = \vartheta(n^{-1}W_1(s,y))$ for any $(s,y) \in [t_1,t_*] \times \mathbb{R}^N$, $n \in \mathbb{N}$ and W_1 is as in Hypothesis 15.0.1(iv), with t being replaced by t_*. Clearly, each function $q_{ij}^{(n)}$ belongs to $C_b^{0,1}((t_1,t) \times \mathbb{R}^N)$.

The following lemma collects some properties of the operator $\mathcal{A}^{(n)}$ which we need in the proof of Theorem 15.2.2.

Lemma 15.2.1 *For any $n \in \mathbb{N}$ and $t_* \in (t_1, t_2]$, the operator $\mathcal{A}^{(n)}$ satisfies Hypotheses 15.0.1(i) and (iv) (with the same functions W_k and h_k $(k = 1,2)$) in $[t_1,t_*]$ and Hypotheses 15.0.1(v)-(b), (c), (e) and (k) in any interval $[\sigma_0,\tau_0] \subset (t_1,t_*)$. In particular, we can associate an evolution operator $\{G_n(t,s) : t_1 < s < t < t_*\}$ in $C_b(\mathbb{R}^N)$ with the operator $\mathcal{A}^{(n)}$ for any $n \in \mathbb{N}$. Such an evolution operator admits a Green function g_n and, for any $x \in \mathbb{R}^N$, $g_n(t_*,\cdot,x,\cdot)$ converges to $g(t_*,\cdot,x,\cdot)$ locally uniformly in $(t_1,t_*) \times \mathbb{R}^N$, as n tends to $+\infty$.*

Proof As is immediately seen, the diffusion coefficients of each operator $\mathcal{A}^{(n)}$ belong to $C_b^{0,1}((t_1,t_*) \times \mathbb{R}^N)$ and $\langle Q^{(n)}(t,x)\xi,\xi\rangle \geq \kappa_0|\xi|^2$ for any $(t,x) \in (t_1,t_*) \times \mathbb{R}^N$ and $\xi \in \mathbb{R}^N$. Hence, Hypothesis 15.0.1(i) is trivially satisfied. On the other hand, $\mathcal{A}^{(n)}V = \vartheta_n\mathcal{A}V + (1-\vartheta_n)(\kappa_0\Delta V + \langle b, \nabla_x V\rangle + cV)$ in $[t_1,t_*] \times \mathbb{R}^N$. Hence, by Hypothesis 15.0.1(iii) it follows that $\mathcal{A}^{(n)}V \leq M$ in $(t_1,t_*) \times \mathbb{R}^N$. In view of Theorem 14.1.2, with $I = (t_1,t_*)$, we can associate an evolution operator $\{G_n(t,s) : t_1 < s < t < t_*\}$ with the operator $\mathcal{A}^{(n)}$ in $C_b(\mathbb{R}^N)$, for any $n \in \mathbb{N}$. By Proposition 14.1.3 there exists a Green function g_n such that

$$(G_n(t,s)f)(x) = \int_{\mathbb{R}^N} g_n(t,s,x,y)f(y)dy, \qquad t_1 < s < t < t_*, \;\; f \in C_b(\mathbb{R}^N), \;\; n \in \mathbb{N}.$$

A straightforward computation reveals that the functions W_1 and W_2 in Hypothesis 15.0.1(iv) are time dependent Lyapunov functions for the each operator $\mathcal{A}^{(n)}$ with the same L^1-functions h_1 and h_2.

Next, we show that the operator $\mathcal{A}^{(n)}$ $(n \in \mathbb{N})$ satisfies also Hypotheses 15.0.1(v)-(b), (c), (e) and (k). For this purpose, we note that $|\nabla w| = |Q^{-1}Q\nabla w| \leq \kappa_0^{-1}C_2 w^{\frac{q-1}{q}}W_1^{\frac{1}{q}}$. Similarly, using Hypothesis 15.0.1(v)-(k), we get $|\nabla W_1| \leq \kappa_0^{-1}C_9 w^{\frac{q-1}{q}}W_2^{\frac{1}{q}}$. From these estimates it follows at once that $|Q^{(n)}\nabla w| \le 2C_2 w^{\frac{q-1}{q}} W_1^{\frac{1}{q}}$ and $|Q^{(n)}\nabla W_1| \le 2C_9 w^{-\frac{1}{q}} W_1 W_2^{\frac{1}{q}}$. Moreover, we can estimate $|\mathrm{Tr}(Q^{(n)}D_y^2 w)| \le |\mathrm{Tr}(QD_y^2 w)| + \kappa_0|\Delta w| \le (C_3+\kappa_0 C_8)w^{1-2/q}W_1^{2/q}$. Finally, it holds that

$$\begin{aligned}\left|\sum_{i=1}^N D_i q_{ij}^{(n)}\right| &\le \left|\sum_{i=1}^N D_i q_{ij}\right| + \frac{1}{W_1}|n^{-1}W_1\varphi'(n^{-1}W_1)|(|Q\nabla_y W_1| + \kappa_0|\nabla_y W_1|)\\ &\le \left|\sum_{i=1}^N D_i q_{ij}\right| + \frac{2}{W_1}(|Q\nabla_y W_1| + \kappa_0|\nabla_y W_1|) \le (C_5+4C_9)w^{-\frac{1}{q}}W_2^{\frac{1}{q}}.\end{aligned}$$

Hence, each operator $\mathcal{A}^{(n)}$ satisfies Hypotheses 15.0.1(v)-(b), (c), (e) and (k) in any closed interval contained in (t_1, t_*).

To complete the proof, let us prove that $g_n(t_*, \cdot, x, \cdot)$ converges to $g(t_*, \cdot, x, \cdot)$ locally uniformly in $(t_1, t_*) \times \mathbb{R}^N$, as n tends to $+\infty$. As a first step, adapting the arguments in the proof of Theorem 11.3.2, we prove that, for any $f \in C_b(\mathbb{R}^N)$ and $s \in [t_1, t_*)$, the function $G_n(\cdot, s)f$ converges to $G(\cdot, s)f$ locally uniformly in $\mathbb{R}^N$. This result is obtained in some steps.

First, when $f \in C_c^{2+\alpha}(\mathbb{R}^N)$ the classical Schauder estimates and the contractivity of each operator $G_n(t, s)$ in $C_b(\mathbb{R}^N)$ show that the sequence $\{G_n(\cdot, s)\}$ is bounded in $C^{1+\alpha/2,2+\alpha}([s, r] \times B_R)$ for any $r \in (s, t_*)$ and $R > 0$. Compactness and diagonal arguments allow us to determine a subsequence of $\{G_n(\cdot, s)\}$ converging locally uniformly in $[s, t_*) \times \mathbb{R}^N$ to a function u which solves the equation $D_t u = \mathcal{A}u$ in $(s, t_*) \times \mathbb{R}^N$ and satisfies the condition $u(s, \cdot) = f$. The maximum principle in Proposition 14.1.1 reveals that $u = G(\cdot, s)f$. Since the limit is independent of the subsequence $\{G_n(\cdot, s)f\}$, which converges locally uniformly in $[s, t_*) \times \mathbb{R}^N$, the sequence $\{G_n(\cdot, s)f\}$ itself converges to $G(\cdot, s)f$ locally uniformly in $[s, t_*) \times \mathbb{R}^N$, as n tends to $+\infty$. Next, the contractivity of $G_n(t, s)$ in $C_b(\mathbb{R}^N)$ (see (14.1.3)) and the density of $C_c^{2+\alpha}(\mathbb{R}^N)$ in $C_0(\mathbb{R}^N)$ can be used to prove that $G_n(\cdot, s)f$ converges to $G(\cdot, s)f$ locally uniformly in $[s, t_*) \times \mathbb{R}^N$, as n tends to $+\infty$, for any $f \in C_0(\mathbb{R}^N)$. Finally, in the general case when $f \in C_b(\mathbb{R}^N)$, the interior Schauder estimates and a diagonal argument show that a subsequence $\{G_{n_k}(\cdot, s)f\}$ converges locally uniformly in $(s, t_*) \times \mathbb{R}^N$ to a solution u of the equation $D_t u = \mathcal{A}u$ which is bounded in $(s, t_*) \times \mathbb{R}^N$. To identify u with $G(\cdot, s)f$ we use a localization argument. For any $R > 0$, we fix a smooth function ϑ such that $\chi_{B_R} \le \vartheta \le \chi_{B_{2R}}$. Then, we split $G_{n_k}(\cdot, s)f = G_{n_k}(\cdot, s)(f\vartheta) + G_{n_k}(\cdot, s)(f(1-\vartheta))$ and observe that the maximum principle shows that $\|G_{n_k}(t, s)(f(1-\vartheta))\|_\infty \le \|f\|_\infty(1 - G_{n_k}(t, s)\vartheta)$ for any $s < t < t_*$ and $k \in \mathbb{N}$. Since ϑ and $f\vartheta$ belong to $C_0(\mathbb{R}^N)$, $G_{n_k}(\cdot, s)\vartheta$ and $G_{n_k}(\cdot, s)(f\vartheta)$ converge, respectively, to $G(\cdot, s)\vartheta$ and $G(\cdot, s)(f\vartheta)$ locally uniformly in $\mathbb{R}^N$, as k tends to $+\infty$. Hence, $|u(t, \cdot) - G(t, s)(f\vartheta)| \le \|f\|_\infty(1 - G(t, s)\vartheta)$ for any $s < t < t_*$. Letting t tend to s^+, it thus follows that $u(t, \cdot)$ converges to f in B_R and, by the arbitrariness of R, we conclude that u can be extended by continuity to $\{s\} \times \mathbb{R}^N$ by setting $u(s, \cdot) = f$. Hence, $u = G(\cdot, s)f$. As in the case when $f \in C_c^{2+\alpha}(\mathbb{R}^N)$, we can now infer that all the sequence $\{G_n(\cdot, s)f\}$ converges to $G(\cdot, s)f$ locally uniformly in $(s, t_*) \times \mathbb{R}^N$, as n tends to $+\infty$.

The convergence of $(G_n(t, s)f)(x)$ to $(G(t, s)f)(x)$ can be rephrased by saying that the measures $p_n(t, s, x, dy) = g_n(t, s, x, y)dy$ converges weakly* to the measure $p(t, s, x; dy) = g(t, s, x; y)dy$ as n tends to $+\infty$. Since Remark 15.1.1 (see also the forthcoming Theorem 16.2.4) can be applied to the measure $g_n(t, s, x, y)dsdy$, for any compact set $K \subset \mathbb{R}^N$ and any compact interval $J \subset (t_1, t_*)$ there exist a positive constant $C > 0$ and $\gamma \in (0, 1)$ such that $\|g_n(t, \cdot, x, \cdot)\|_{C^\gamma(J\times K)} \le C$. In particular, the constant C can be taken independent of n since the coefficients $q_{ij}^{(n)}$ coincide with the coefficients q_{ij} $(i, j = 1, \dots, N)$ in the compact set $J \times K$ if n is sufficiently large. Therefore, using a diagonal argument, we can determine a subsequence of $\{g_n(t, \cdot, x, \cdot)\}$ which converges locally uniformly in $(t_1, t_*) \times \mathbb{R}^N$ to a continuous function $\psi_{t,x}$. Since $p_n(t, s, x, dy)$ converges weakly* to $p(t, s, x; dy)$, we easily infer that $\psi_{t,x} = g(t,\cdot,x,\cdot)$. Finally, observing that the limit is independent of the subsequence of $\{g_n(t,\cdot,x,\cdot)\}$ which converges, we conclude that the sequence $\{g_n(t,\cdot,x,\cdot)\}$ converges locally uniformly in $(t_1,t_*) \times \mathbb{R}^N$ to $g(t,\cdot,x,\cdot)$, as n tends to $+\infty$. $\square$

In view of Lemma 15.1.3 and Hypothesis 15.0.1(iv), for any $s \in [t_1,t]$ and $i = 1,2$ the function $G(t,s)W_i(s,\cdot)$ is well defined. With this remark in mind we can state and prove the main result of this chapter.

Theorem 15.2.2 *Fix $t_1 < \sigma_0 < \sigma < \tau < \tau_0 < t \le 1$. The following properties are satisfied.*

(i) *If $c \equiv 0$, then there exists a positive constant M_0, depending only on κ_0, t_1, t_2, q and N, such that*

$$\begin{aligned}
&w(s,y)g(t,s,x,y)\\
\le& M_0\Bigg[C_1\sup_{s\in(\sigma_0,\tau_0)}(G(t,s)W_1(s,\cdot))(x)\\
&+\left(\frac{C_1}{(\tau_0-\tau)^{q/2}}+C_2^q+C_3^{\frac{q}{2}}+C_4^{\frac{q}{2}}+C_8^{\frac{q}{2}}\right)\int_{\sigma_0}^{\tau_0}(G(t,s)W_1(s,\cdot))(x)ds\\
&+(C_2^{\frac{q}{2}}C_6^{\frac{q}{2}}+C_5^q+C_6^q+C_9^q)\int_{\sigma_0}^{\tau_0}(G(t,s)W_2(s,\cdot))(x)ds\Bigg]
\end{aligned}\tag{15.2.1}$$

for any $s \in (\sigma,\tau)$ and $x,y \in \mathbb{R}^N$, where $C_j = C_j(\sigma_0,\tau_0)$ are the constants in Hypothesis 15.0.1(v). In the particular case when the diffusion coefficients are bounded in $[t_1,t_2]$ together with the first-order spatial derivatives, Hypotheses 15.0.1(v)-(h), (k) are not required. Hence, in (15.2.1) we can take $C_8 = C_9 = 0$.

(ii) *Assume that $c \not\equiv 0$ and fix $x \in \mathbb{R}^N$. If the diffusion coefficients of the operator $\mathcal{A}$ are bounded together with their spatial gradients and $g_0(t,\cdot,x,\cdot)$ is bounded in $[\sigma_0,\tau_0]\times\mathbb{R}^N$ for any $[\sigma_0,\tau_0] \subset (t_1,t)$ (where g_0 is the Green function associated with the evolution operator $\{G_0(t,s)\}$, see (15.0.1)), then there exists a positive constant M_1, depending only on κ_0, q, t_1, t_2 and N, such that*

$$\begin{aligned}
&w(s,y)g(t,s,x,y)\\
\le& M_1\Bigg[C_1\sup_{s\in(\sigma_0,\tau_0)}(G(t,s)W_1(s,\cdot))(x)\\
&+\left(\frac{C_1}{(\tau_0-\tau)^{q/2}}+C_2^q+C_3^{\frac{q}{2}}+C_4^{\frac{q}{2}}+C_8^{\frac{q}{2}}\right)\int_{\sigma_0}^{\tau_0}(G(t,s)W_1(s,\cdot))(x)ds\\
&+(C_2^{\frac{q}{2}}C_6^{\frac{q}{2}}+C_5^q+C_6^q+C_7^{\frac{q}{2}}+C_9^q)\int_{\sigma_0}^{\tau_0}(G(t,s)W_2(s,\cdot))(x)ds\Bigg]
\end{aligned}\tag{15.2.2}$$

for any $s \in (\sigma,\tau)$ and $y \in \mathbb{R}^N$. The same estimate holds true when the diffusion coefficients or their spatial gradients are unbounded in $[t_1,t_2]\times\mathbb{R}^N$ if we assume in addition that the Green function of the operator $\mathcal{A}^{(n)}$ is bounded in $[\sigma_0,\tau_0]\times\mathbb{R}^N$ for any $n \in \mathbb{N}$ and $[\sigma_0,\tau_0]$ as above.

Proof We split the proof into four steps. The first three steps deal with the case when the diffusion coefficients are bounded in $[t_1,t_2]$ together with their first-order spatial derivatives and we do not assume Hypotheses 15.0.1(v)-(h), (k). In the first two steps, in addition we assume that w is bounded together with its first-order partial derivatives and consider separately the cases when $c \equiv 0$ and $c \not\equiv 0$. The assumption on the boundedness of w and its derivatives is removed in Step 3, using an approximation argument. The general case when the diffusion coefficients do not belong to $C_b^{0,1}((t_1,t_2)\times\mathbb{R}^N)$ is addressed in Step 4.

Throughout the proof, σ_0, σ_1, τ_0 and τ_1 are arbitrarily fixed as in the statement.

Step 1. Assume that $c\equiv 0$ and w is bounded together with its first-order partial derivatives. Then, thanks to conditions (f) and (g) in Hypothesis 15.0.1(v) and Lemma 15.1.3, we can estimate

$$
\begin{aligned}
(\Gamma_1(q,\sigma_0,\tau_0,t,x))^q &\le \int_{(\sigma_0,\tau_0)\times\mathbb{R}^N} w(s,y)|b(s,y)|^q g(t,s,x,y)dsdy\\
&\le C_6\int_{\sigma_0}^{\tau_0}(G(t,s)W_2(s,\cdot))(x)ds<+\infty
\end{aligned}
$$

for any $t\in(t_1,t_2]$ and $x\in\mathbb{R}^N$. Hence, by Proposition 15.1.6 we conclude that $g(t,\cdot,x,\cdot)\in\mathscr{H}^{p,1}((\sigma_1,\tau_1)\times\mathbb{R}^N)$ for some $p>N+2$, any $(\sigma_1,\tau_1)\subset(\sigma_0,\tau_0)$ and any $(t,x)\in(t_1,t_2]\times\mathbb{R}^N$. Further, $g(t,\cdot,x,\cdot)$ belongs to $C_b([\sigma_1,\tau_1]\times\mathbb{R}^N)$ for the same values of t and x.

Fix $\sigma_0<\sigma_1<\sigma<\tau<\tau_1<\tau_0$, $\psi\in C_c^{1,2}((\sigma_1,\tau_1)\times\mathbb{R}^N)$ and $\vartheta\in C_b^\infty(\mathbb{R})$ such that $\chi_{(\sigma,\tau)}\le\vartheta\le\chi_{[\tau_1,+\infty)}$ and $|\vartheta'|\le 2(\tau_1-\tau)^{-1}$ in $\mathbb{R}$. From Remark 15.1.1, with ψ being replaced by $h=\vartheta^{\frac{q}{2}}w\psi$, it follows that

$$
\int_{(\sigma_1,\tau_1)\times\mathbb{R}^N}(D_sh-\mathcal{A}h)g(t,\cdot,x,\cdot)dsdy=0. \tag{15.2.3}
$$

We write $\widetilde{g}(t,\cdot,x,\cdot):=\vartheta^{\frac{q}{2}}g(t,\cdot,x,\cdot)$ and observe that the function $w\widetilde{g}(t,\cdot,x,\cdot)$ belongs to $\mathscr{H}^{p,1}((\sigma_1,\tau_1)\times\mathbb{R}^N)$ since w and its first-order derivatives are bounded. Thus,

$$
\begin{aligned}
\int_{(\sigma_1,\tau_1)\times\mathbb{R}^N}D_shg(t,\cdot,x,\cdot)dsdy =&\frac{q}{2}\int_{(\sigma_1,\tau_1)\times\mathbb{R}^N}\vartheta^{\frac{k-2}{2}}\vartheta' w\psi g(t,\cdot,x,\cdot)dsdy\\
&+\int_{(\sigma_1,\tau_1)\times\mathbb{R}^N}(D_sw)\psi\widetilde{g}(t,\cdot,x,\cdot)dsdy-(D_s(w\widetilde{g}(t,\cdot,x,\cdot)),\psi)_p,
\end{aligned}
$$

where $(\cdot,\cdot)_p$ denotes the pairing between $(W_{p'}^{0,1}((\sigma_1,\tau_1)\times\mathbb{R}^N))'$ and $W_{p'}^{0,1}((\sigma_1,\tau_1)\times\mathbb{R}^N)$. Moreover, since $\mathcal{A}(\psi w)=\mathcal{A}(w)\psi+w\langle b,\nabla_y\psi\rangle+\mathrm{Tr}(QD_y^2\psi)w+2\langle Q\nabla_y\psi,\nabla_yw\rangle$, integrating by parts the term $\mathrm{Tr}(QD^2_y\psi)w$ gives

$$
\begin{aligned}
&\int_{(\sigma_1,\tau_1)\times\mathbb{R}^N}\mathcal{A}(\vartheta^{\frac{q}{2}}\psi w)g(t,\cdot,x,\cdot)dsdy=\int_{(\sigma_1,\tau_1)\times\mathbb{R}^N}\mathcal{A}(\psi w)\widetilde{g}(t,\cdot,x,\cdot)dsdy\\
=&\int_{(\sigma_1,\tau_1)\times\mathbb{R}^N}\mathcal{A}(w)\psi\widetilde{g}(t,\cdot,x,\cdot)dsdy+\int_{(\sigma_1,\tau_1)\times\mathbb{R}^N}\langle b,\nabla_y\psi\rangle\, w\widetilde{g}(t,\cdot,x,\cdot)dsdy\\
&+2\int_{(\sigma_1,\tau_1)\times\mathbb{R}^N}\langle Q\nabla_yw,\nabla_y\psi\rangle\widetilde{g}(t,\cdot,x,\cdot)dsdy-\int_{(\sigma_1,\tau_1)\times\mathbb{R}^N}\mathrm{div}(Q\nabla_y\psi)w\widetilde{g}(t,\cdot,x,\cdot)dsdy\\
&-\int_{(\sigma_1,\tau_1)\times\mathbb{R}^N}\langle Q\nabla_y(w\widetilde{g}(t,\cdot,x,\cdot)),\nabla_y\psi\rangle dsdy.
\end{aligned}
$$

Replacing these last two formulae in (15.2.3), we deduce that

$$
\begin{aligned}
&\int_{(\sigma_1,\tau_1)\times\mathbb{R}^N}\langle Q\nabla_y(w\widetilde{g}(t,\cdot,x,\cdot)),\nabla_y\psi\rangle dsdy-(D_s(w\widetilde{g}(t,\cdot,x,\cdot)),\psi)_p\\
=&\int_{(\sigma_1,\tau_1)\times\mathbb{R}^N}f\psi dsdy+\int_{(\sigma_1,\tau_1)\times\mathbb{R}^N}\widetilde{g}(t,\cdot,x,\cdot)\langle\zeta,\nabla\psi\rangle dsdy,
\end{aligned}
$$

where $f = \widetilde{g}(t,\cdot,x,\cdot)(\mathrm{Tr}(QD_y^2w) + \langle b, \nabla_y w\rangle - D_s w) - 2^{-1}qw\vartheta^{\frac{q-2}{q}}\vartheta' g(t,\cdot,x,\cdot)$ and $\zeta_j = 2(Q\nabla_y w)_j - w\sum_{i=1}^N D_i q_{ij} + wb_j$ for any $j = 1,\dots,N$. We may thus invoke Theorem 15.1.9 and obtain that

$$\begin{aligned}\|w\widetilde{g}(t,\cdot,x,\cdot)\|_\infty \le C(&\|w\widetilde{g}(t,\cdot,x,\cdot)\|_{L^\infty((\sigma_1,\tau_1);L^2(\mathbb{R}^N))}\\ &+\|\zeta\|_{L^q((\sigma_1,\tau_1)\times\mathbb{R}^N)} + \|f\|_{L^{\frac{q}{2}}((\sigma_1,\tau_1)\times\mathbb{R}^N)}),\end{aligned} \tag{15.2.4}$$

where C is a constant depending only on q, N, t_1, t_2 and κ_0. Let us estimate the right-hand side of (15.2.4). From now on we use the constants in Hypothesis 15.0.1(v) related to the interval (σ_0,τ_0). Using condition (a) in such an hypothesis, we obtain that

$$\begin{aligned}\|w\widetilde{g}(t,\cdot,x,\cdot)\|_{L^\infty((\sigma_1,\tau_1);L^2(\mathbb{R}^N))} &\le \|w\widetilde{g}(t,\cdot,x,\cdot)\|_\infty^{\frac{1}{2}} \sup_{s\in(\sigma_1,\tau_1)}\left(\int_{\mathbb{R}^N}\widetilde{g}(t,s,x,\cdot)w(s,\cdot)dy\right)^{\frac{1}{2}}\\ &\le \sqrt{C_1}\mathscr{Y}^q\left(\sup_{s\in(\sigma_1,\tau_1)}(G(t,s)W_1(s,\cdot))(x)\right)^{\frac{1}{2}},\end{aligned}$$

where $\mathscr{Y} = \|w\widetilde{g}(t,\cdot,x,\cdot)\|_\infty^{1/q}$. Moreover, using condition (b) in Hypothesis 15.0.1(v), we can estimate

$$\begin{aligned}\|\widetilde{g}(t,\cdot,x,\cdot)Q\nabla_y w\|_{L^q((\sigma_1,\tau_1)\times\mathbb{R}^N)} &\le C_2\left(\int_{(\sigma_1,\tau_1)\times\mathbb{R}^N}(\widetilde{g}(t,\cdot,x,\cdot))^q w^{q-1}W_1 ds\,dy\right)^{\frac{1}{q}}\\ &\le C_2\mathscr{Y}^{2q-2}\left(\int_{a_1}^{b_1}(G(t,s)W_1(s,\cdot))(x)ds\right)^{\frac{1}{q}}.\end{aligned}$$

The other terms can be estimated likewise, and we get

$$\begin{aligned}\|w\widetilde{g}(t,\cdot,x,\cdot)b\|_{L^q((\sigma_1,\tau_1)\times\mathbb{R}^N)} &\le C_6\mathscr{Y}^{q-1}\left(\int_{\sigma_1}^{\tau_1}(G(t,s)W_2(s,\cdot))(x)ds\right)^{\frac{1}{q}},\\ \left\|w\widetilde{g}(t,\cdot,x,\cdot)\sum_{i=1}^N D_i q_{ij}\right\|_{L^q((\sigma_1,\tau_1)\times\mathbb{R}^N)} &\le C_5\mathscr{Y}^{q-1}\left(\int_{\sigma_1}^{\tau_1}(G(t,s)W_2(s,\cdot))(x)ds\right)^{\frac{1}{q}},\\ \|w\widetilde{g}(t,\cdot,x,\cdot)\vartheta^{\frac{k-2}{k}}\|_{L^{\frac{q}{2}}((\sigma_1,\tau_1)\times\mathbb{R}^N)} &\le C_1^{\frac{2}{q}}\mathscr{Y}^{q-2}\left(\int_{\sigma_1}^{\tau_1}(G(t,s)W_1(s,\cdot))(x)ds\right)^{\frac{2}{q}},\\ \|\widetilde{g}(t,\cdot,x,\cdot)\mathrm{Tr}(QD_y^2\psi)\|_{L^{\frac{q}{2}}((\sigma_1,\tau_1)\times\mathbb{R}^N)} &\le C_3\mathscr{Y}^{q-2}\left(\int_{\sigma_1}^{\tau_1}(G(t,s)W_1(s,\cdot))(x)ds\right)^{\frac{2}{q}},\\ \|\widetilde{g}(t,\cdot,x,\cdot)D_s w\|_{L^{\frac{q}{2}}((\sigma_1,\tau_1)\times\mathbb{R}^N)} &\le C_4\mathscr{Y}^{q-2}\left(\int_{\sigma_1}^{\tau_1}(G(t,s)W_1(s,\cdot))(x)ds\right)^{\frac{2}{q}},\\ \|\widetilde{g}(t,\cdot,x,\cdot)\langle b,\nabla_y w\rangle\|_{L^{\frac{q}{2}}((\sigma_1,\tau_1)\times\mathbb{R}^N)} &\le C_2C_6\kappa_0^{-1}\mathscr{Y}^{q-2}\left(\int_{\sigma_1}^{\tau_1}(G(t,s)W_2(s,\cdot))(x)ds\right)^{\frac{2}{q}}.\end{aligned}$$

Replacing the previous estimates in (15.2.4), we obtain that

$$\mathscr{Y}^q \le \gamma_1\mathscr{Y}^{q/2} + \gamma_2\mathscr{Y}^{q-1} + \gamma_3\mathscr{Y}^{q-2},$$

where

$$\gamma_1 = CC_1^{\frac{1}{2}}\left(\sup_{s\in(\sigma_1,\tau_1)}(G(t,s)W_1(s,\cdot))(x)\right)^{\frac{1}{2}},$$

$$\gamma_2 = CC_2\bigg(\int_{\sigma_1}^{\tau_1}(G(t,s)W_1(s,\cdot))(x)ds\bigg)^{\frac{1}{q}} + C(C_5N+C_6)\bigg(\int_{\sigma_1}^{\tau_1}(G(t,s)W_2(s,\cdot))(x)ds\bigg)^{\frac{1}{q}},$$

$$\gamma_3 = C\bigg(\frac{C_1^{\frac{2}{q}}}{\tau_1-\tau}+C_3+C_4\bigg)\bigg(\int_{\sigma_1}^{\tau_1}(G(t,s)W_1(s,\cdot))(x)ds\bigg)^{\frac{2}{q}}$$
$$+ C(\kappa_0^{-1}C_2C_6)\bigg(\int_{\sigma_1}^{\tau_1}(G(t,s)W_2(s,\cdot))(x)ds\bigg)^{\frac{2}{q}}.$$

Since $\gamma_1\mathscr{Y}^{q/2}\le\frac{1}{4}\mathscr{Y}^q+\gamma_1^2$, we can estimate

$$\mathscr{Y}^q \le \frac{4}{3}\gamma_1^2+\frac{4}{3}\gamma_2\mathscr{Y}^{q-1}+\frac{4}{3}\gamma_3\mathscr{Y}^{q-2}. \tag{15.2.5}$$

Denote by χ the function defined by

$$\chi(r) = r^q-\frac{4}{3}\gamma_2r^{q-1}-\frac{4}{3}\gamma_3r^{q-2}-\frac{4}{3}\gamma_1^2 = r^{q-2}\bigg(r^2-\frac{4}{3}\gamma_2r-\frac{4}{3}\gamma_3\bigg)-\frac{4}{3}\gamma_1^2$$

and observe that condition (15.2.5) implies that $\chi(\mathscr{Y})\le 0$. A straightforward computation reveals that χ is increasing in the interval $\Big(\frac{4}{3}\gamma_2+\sqrt{\frac{4}{3}\gamma_3}+(\frac{4}{3}\gamma_1^2)^{\frac{1}{q}},+\infty\Big)$. Moreover, it vanishes at the left point of the previous interval. We thus conclude that $\chi(r)>0$ for any $r>\frac{4}{3}\gamma_2+\sqrt{\frac{4}{3}\gamma_3}+(\frac{4}{3}\gamma_1^2)^{\frac{1}{q}}$ and, consequently, $\mathscr{Y}\le\frac{4}{3}\gamma_2+\sqrt{\frac{4}{3}\gamma_3}+(\frac{4}{3}\gamma_1^2)^{\frac{1}{q}}$. Recalling the definition of $\mathscr{Y}$ we can infer that $\|w\widetilde{g}(t,\cdot,x,\cdot)\|_\infty\le K_1(\gamma_1^2+\gamma_2^q+\gamma_3^{q/2})$ for some positive constant K_1, and estimate (15.2.1) follows letting τ_1 tend to τ_0, with some constant M_0 as in the statement and with $C_7=C_8=C_9=0$.

Step 2. Here we remove the assumption that c identically vanishes in $\mathbb{R}^N$ still keeping the condition on the boundedness of w and its derivatives. As is immediately seen,

$$(\Gamma_1(q/2,\sigma_0,\tau_0,t,x))^{\frac{q}{2}} \le C_1^{\frac{1}{2}}C_6^{\frac{q}{2}}\int_{\sigma_0}^{\tau_0}(G(t,s)W_2(s,\cdot))(x)ds$$

$$(\Gamma_2(q/2,\sigma_0,\tau_0,t,x))^{\frac{q}{2}} \le C_7^{\frac{1}{2}}\int_{\sigma_0}^{\tau_0}(G(t,s)W_2(s,\cdot))(x)ds<+\infty.$$

Hence, both $\Gamma_1(q/2,\sigma_0,\tau_0,t,x)$ and $\Gamma_2(q/2,\sigma_0,\tau_0,t,x)$ are finite. Lemma 15.1.5 and the boundedness of $g(t,\cdot,x,\cdot)$ (which follows from the boundedness of the Green function g_0 since $g\le g_0$) show that $g(t,\cdot,x,\cdot)\in L^r((\sigma_1,\tau_1)\times\mathbb{R}^N)$ for any $r\in[1,+\infty]$, where σ_1 and τ_1 are in Step 1. Hence, from Lemma 15.1.5 we deduce that $g(t,\cdot,x,\cdot)\in\mathscr{H}^{p,1}((\sigma_1,\tau_1)\times\mathbb{R}^N)$ for any $p\in(1,q/2)$. Since $q>N+2$, we can apply Theorem 15.1.9 and conclude that estimate (15.2.4) still holds, with f being replaced by $f+\widetilde{g}(t,\cdot,x,\cdot)cw$. Thus, we have to estimate the $L^{q/2}((\sigma_1,\tau_1)\times\mathbb{R}^N)$-norm of $\widetilde{g}(t,\cdot,x,\cdot)cw$. This can easily be done using the condition (g) in Hypothesis 15.0.1(v) which shows that

$$\|\widetilde{g}(t,\cdot,x,\cdot)cw\|_{L^{\frac{q}{2}}((\sigma_1,\tau_1)\times\mathbb{R}^N)} \le C_7\mathscr{Y}^{2q-4}\bigg(\int_{\sigma_1}^{\tau_1}(G(t,s)W_2(s,\cdot))(x)ds\bigg)^{\frac{2}{q}}.$$

Now, proceeding as in Step 1, we can prove estimate (15.2.2) with $C_8=C_9=0$.

Step 3. Now, we remove the additional assumption that w is bounded. For any $\varepsilon>0$ we introduce the function $w_\varepsilon:=(1+\varepsilon w)^{-1}w$, which is bounded together with its first-order partial derivatives. Moreover, since $w^{-2}D_sw$ and $w^{-2}\nabla_yw$ are bounded in $[\sigma_0,\tau_0]\times\mathbb{R}^N$ for any interval $[\sigma_0,\tau_0]\subset(t_1,t)$, the functions $w_\varepsilon^{-2}D_sw_\varepsilon$ and $w_\varepsilon^{-2}\nabla_yw_\varepsilon$ are bounded in the same set as well. This shows that ω_ε satisfies Hypothesis 15.0.1(iv). We claim that it also satisfies Hypothesis 15.0.1(v) with the same constants C_j, $j\neq 3$, and with C_3 being replaced by $\widetilde{C}_3 := C_3 + 2\kappa_0^{-1}C_2^2$. We limit ourselves to checking condition (c) in Hypothesis 15.0.1(v) since the other conditions are immediate to prove. For this purpose we observe that

$$\mathrm{Tr}(QD_y^2 w_\varepsilon) = (1+\varepsilon w)^{-2}\mathrm{Tr}(QD_y^2 w) - 2\varepsilon(1+\varepsilon w)^{-3}|\sqrt{Q}\nabla_y w|^2,$$

so that $|\mathrm{Tr}(QD_y^2 w_\varepsilon)| \le (1+\varepsilon w)^{-2}|\mathrm{Tr}(QD_y^2 w)| + 2\kappa_0^{-1}\varepsilon(1+\varepsilon w)^{-3}|Q\nabla_y w|^2$. Using conditions (b) and (c) in Hypothesis 15.0.1(v), we get

$$|\mathrm{Tr}(QD_y^2 w_\varepsilon)| \le \frac{w^{\frac{q-2}{q}}}{(1+\varepsilon w)^3} W_1^{\frac{2}{q}}(C_3(1+\varepsilon w) + 2\varepsilon\kappa_0^{-1}C_2^2 w) \le (C_3 + 2\kappa_0^{-1}C_2^2) w_\varepsilon^{\frac{q-2}{q}} W_1^{\frac{2}{q}},$$

which proves that w_ε satisfies condition (c) in Hypothesis 15.0.1(v). By Steps 1 and 2 we conclude that (15.2.1) (with $C_7 = C_8 = C_9 = 0$) and (15.2.2) (with $C_8 = C_9 = 0$) hold if we replace w with w_ε and C_3 with $\widetilde{C}_3$. Letting ε tend to 0^+, we get the previous estimate with w instead of w_ε.

Step 4. Since $g_n(t,\cdot,x,y)$ is bounded in $[\sigma_*,\tau_*]\times\mathbb{R}^N$ for any $[\sigma_*,\tau_*]\subset(t_1,t)$, in view of (the proof of) Lemma 15.2.1 and the previous steps we conclude that estimates (15.2.1) and (15.2.2) are satisfied by $g_n(t,\cdot,x,\cdot)$ with $G(t,s)W_i$ $(i=1,2)$ being replaced by $G_n(t,s)W_i$. To conclude the proof, we need to let n tend to $+\infty$ in such estimates. The quoted lemma shows that $g_n(t,\cdot,x,\cdot)$ converges to $g(t,\cdot,x,\cdot)$ locally uniformly in $(t_1,t)\times\mathbb{R}^N$. Hence, we just need to address the convergence of the sequence $\{(G_n(t,\cdot)W_i(\cdot,\cdot))(x)\}$. For this purpose, we observe that

$$\begin{aligned}
&|(G_n(t,s)W_i(s,\cdot))(x) - (G(t,s)W_i(s,\cdot))(x)|\\
\le&\int_{B_R} W_i(s,\cdot)|g_n(t,s,x,\cdot) - g(t,s,x,\cdot)|dy\\
&+\int_{\mathbb{R}^N\setminus B_R} W_i(s)p_n(t,s,x,dy) + \int_{\mathbb{R}^N\setminus B_R} W_i(s,\cdot)p(t,s,x;dy)\\
\le&\|W_i\|_{L^\infty((\sigma_0,\tau_0)\times B_R)}\|g_n(t,\cdot,x,\cdot) - g(t,\cdot,x,\cdot)\|_{L^\infty((\sigma_0,\tau_0)\times B_R)}\omega_N R^N\\
&+\rho_0\bigg(\int_{\mathbb{R}^N\setminus B_R} Vp_n(t,s,x,dy)\bigg)^{1-\beta}(p_n(t,s,x,\mathbb{R}^N\setminus B_R))^\beta\\
&+\rho_0\bigg(\int_{\mathbb{R}^N\setminus B_R} Vp(t,s,x;dy)\bigg)^{1-\beta}(p(t,s,x;\mathbb{R}^N\setminus B_R))^\beta\\
\le&\|W_i\|_{L^\infty((\sigma_0,\tau_0)\times B_R)}\|g_n(t,\cdot,x,\cdot) - g(t,\cdot,x,\cdot)\|_{L^\infty((\sigma_0,\tau_0)\times B_R)}\omega_N R^N\\
&+\rho_0((G_n(t,s)V)(x))^{1-\beta}(p_n(t,s,x,\mathbb{R}^N\setminus B_R))^\beta\\
&+\rho_0((G(t,s)V)(x))^{1-\beta}(p(t,s,x;\mathbb{R}^N\setminus B_R))^\beta
\end{aligned}\tag{15.2.6}$$

for any $s\in(\sigma_0,\tau_0)$, $n\in\mathbb{N}$, $i=1,2$ and $R>0$, where ω_N denotes the measure of the ball B_1 and we have used the condition $W_j \le \rho_0 W^{1-\beta}$ in Hypothesis 15.0.1(iv)-(a). Since $\mathcal{A}^{(n)}V \le M$ in $[0,1]\times\mathbb{R}^N$, where M is the constant in Hypothesis 15.0.1(iii), by the proof of Lemma 14.1.9 it follows that $(G_n(t,s)V)(x) \le V(x) + M$ for any $t_1\le s\le t\le t_2$, $x\in\mathbb{R}^N$. Moreover, Lemma 14.1.10 shows that the family of measures $\{p_n(t,s,x,dy) : n\in\mathbb{N}, s\in(\sigma_0,\tau_0)\}$ is tight for any $x\in\mathbb{R}^N$. Hence, for any $\varepsilon>0$ and $x\in\mathbb{R}^N$ there exists $R>0$ such that $(p_n(t,s,x,\mathbb{R}^N\setminus B_R))^\sigma + (p(t,s,x;\mathbb{R}^N\setminus B_R))^\sigma \le \varepsilon$ for any $n\in\mathbb{N}$ and $s\in(\sigma_0,\tau_0)$. Letting n tend to $+\infty$ in (15.2.6) yields

$$\limsup_{n\to+\infty}\sup_{s\in(\sigma_0,\tau_0)} |(G_n(t,s)W_i(s,\cdot))(x) - (G(t,s)W_i(s,\cdot))(x)| \le \rho_0(V(x)+M)^{1-\sigma}\varepsilon.$$

The arbitrariness of $\varepsilon > 0$ allows us to conclude that the sequence $\{G_n(t,\cdot)W_i(\cdot,\cdot))(x)\}$ converges to $(G(t,\cdot)W_i(\cdot,\cdot))(x)$, uniformly in (σ_0,τ_0), as n tends to $+\infty$. In particular, this shows that

$$\lim_{n\to+\infty}\int_{\sigma_0}^{\tau_0}(G_n(t,s)W_i(s,\cdot))(x)ds=\int_{\sigma_0}^{\tau_0}(G(t,s)W_i(s,\cdot))(x)ds.$$

Thus, we can let n tend to $+\infty$ in the estimates satisfied by $wg_n(t,\cdot,x,\cdot)$ and complete the proof. □

15.2.1 A concrete application of Theorem 15.2.2

We now make the results of Theorem 15.2.2 more concrete by assuming the following conditions on the operator $\mathcal{A}$.

Hypotheses 15.2.3 (i) *The coefficients q_{ij}, b_j $(i,j=1,\ldots,d)$ and c satisfy Hypothesis* 15.0.1*(i);*

(ii) *there exist $m\geq 0$ and, for any bounded interval $J\subset I$, a positive constant K_J such that $|Q(t,x)|\leq K_J(1+|x|^m)$ and $|D_iq_{ij}(t,x)|\leq K_J(1+|x|^m)$ for any $(t,x)\in J\times\mathbb{R}^N$ and $i,j=1,\ldots,N$;*

(iii) *there exist $p>1\vee(m-1)$ and, for any J as above, a positive constant $b_{0,J}$ such that $|b(t,x)|\leq b_{0,J}|x|^p$ and $\langle b(t,x),x\rangle\leq -b_{0,J}|x|^{p+1}$ for any $(t,x)\in J\times(\mathbb{R}^N\setminus B_1)$;*

(iv) *for any J as above, there exists a positive constant $c_{0,J}$ such that $|c(t,x)|\leq c_{0,J}|x|^r$ for any $(t,x)\in J\times(\mathbb{R}^N\setminus B_1)$. Here, r is either zero or is positive and larger than $r>m-2$.*

Theorem 15.2.4 *Let Hypotheses* 15.2.3 *be satisfied, fix $q>N+2$ and $I\ni t_1<t_2$. Then, there exists a positive constant C, depending on κ_0, q, m, p, r, $K_{[t_1,t_2]}$, $b_{0,[t_1,t_2]}$, $c_{0,[t_1,t_2]}$, γ, ε, t_1, t_2, such that following properties are satisfied.*

(i) *if $p\geq\frac{1}{2}(m+r)$, then for any $\gamma>\frac{p+1-m}{p-1}$ and $\varepsilon<\frac{1}{(p+1-m)}\left(1\wedge\frac{b_{0,[t_1,t_2]}}{K_{[t_1,t_2]}}\right)(t_2-t_1)^{-\gamma}$ it holds that*

$$g(t,s,x,y)\leq C(t-s)^{1-\frac{\gamma(m\vee p)q}{p+1-m}}e^{-\varepsilon(t-s)^\gamma|y|^{p+1-m}},\qquad t_1<s<t<t_2,\ x,y\in\mathbb{R}^N;\tag{15.2.7}$$

(ii) *if $p<\frac{1}{2}(m+r)$, then, for any $\varepsilon<\frac{2}{(r+2-m)}\left(1\wedge\frac{c_{0,[t_1,t_2]}^{1/2}}{K_{[t_1,t_2]}^{1/2}}\right)(t_2-t_1)^{-\gamma}$ and any $\gamma>\frac{r-m+2}{r+m-2}$, if $r+m>2$, and any $\gamma>\frac{r+2-m}{2(p-1)}$, if $r+m\leq 2$, it holds that*

$$g(t,s,x,y)\leq C(t-s)^{1-\frac{\gamma(2m\vee 2p\vee r)q}{r+2-m}}e^{-\varepsilon(t-s)^\gamma|y|^{\frac{1}{2}(r+2-m)}},\qquad t_1<s<t<t_2,\ x,y\in\mathbb{R}^N.\tag{15.2.8}$$

Proof We fix $I\ni t_1<t_2$ and set $J=[t_1,t_2]$. Moreover, throughout the proof, we fix $(\beta,\delta_{0,J})=(p+1-m,b_{0,J}(K_J\beta)^{-1})$ or $(\beta,\delta_{0,J})=((r+2-m)/2,|c_{0,J}|^{1/2}K_J^{-1/2}\beta^{-1})\}$ (this latter case requires that c is unbounded), and denote by $v_\beta\in C^2(\mathbb{R}^N)$ any function which satisfies $v_\beta(x)=|x|^\beta$ for $x\in\mathbb{R}^N\setminus B_1$.

Being rather long, we split the proof into three steps.

Step 1. To begin with, we prove that for any $\delta < \delta_{0,J}$ the function $V = e^{\delta v_\beta}$ satisfies Hypothesis 15.0.1(iii). A straightforward computation reveals that

$$\begin{aligned}(V(x))^{-1}\mathcal{A}(t)V(x) =&\beta\delta|x|^{\beta-2}\mathrm{Tr}(Q(t,x)) + \beta(\beta-2)|x|^{\beta-4}|\sqrt{Q}(t,x)x|^2\\ &+\beta^2\delta^2|x|^{2\beta-4}|\sqrt{Q}(t,x)x|^2 + \beta\delta|x|^{\beta-2}\langle b(t,x),x\rangle + c(t,x)\\ \le&\beta[\delta N + (\beta-2)^+K_J]|x|^{\beta-2}(1+|x|^m)\\ &+\beta^2\delta^2K_J|x|^{2\beta-2}(1+|x|^m) - \beta\delta b_{0,J}|x|^{\beta+p-1} - c_{0,J}|x|^r \end{aligned}\tag{15.2.9}$$

for any $t \in [t_1,t_2]$ and $x \in \mathbb{R}^N$. Since the function in the last side of (15.2.9) diverges to $-\infty$ as $|x|$ tends to $+\infty$, with both the choices of β, and the coefficients of the operator $\mathcal{A}$ are continuous functions in $I \times \mathbb{R}^N$, we can infer that there exists a positive constant M_J such that $\mathcal{A}(t)V \le M_J$ in $\mathbb{R}^N$ for any $t \in J$. Arguing similarly we can show that the function $\kappa_0\Delta V + \langle b, \nabla V\rangle + cV$ is bounded from above in $[t_1,t_2]\times\mathbb{R}^N$ and V satisfies Hypothesis 15.0.1(iii). Finally, taking $\beta = p+1-m$, from the previous computations it follows also that the function $\varphi = V$ satisfies Hypothesis 15.0.1(ii).

Next, we set $\gamma_0 = \beta/(\beta+m-2)$ if $\beta+m-2>0$ and $\gamma_0 = \beta/(p-1)$ otherwise, and show that, for any $t \in (t_1,t_2]$, $\varepsilon \in (0,(\beta^{-1}\wedge\delta)(t_2-t_1)^{-\gamma})$ and $\gamma > \gamma_0$, the function $Z_\varepsilon : [t_1,t]\times\mathbb{R}^N \to \mathbb{R}$, defined by $Z_\varepsilon(s,x) := \exp(\varepsilon(t-s)^\gamma v_\beta(x))$ for any $(t,x) \in [t_1,t]\times\mathbb{R}^N$, is a time dependent Lyapunov function for the operator $\mathcal{A}$. For this purpose, for any $x \in \mathbb{R}^N$ we set

$$\zeta_a(x) - (1+|x|^m)|x|^{\beta-2}[N + (\beta-2)^+K_J + \beta\delta(t-s)^\gamma K_J|x|^\beta] - |x|^{\beta+p-1}b_{0,J} + c_{0,J}|x|^r.$$

The same computations as above show that

$$\begin{aligned}&(Z_\varepsilon(s,x))^{-1}(D_sZ_\varepsilon(s,x) - \mathcal{A}(s)Z_\varepsilon(s,x))\\ \ge& -\gamma\varepsilon(t-s)^{\gamma-1}|x|^\beta - c_{0,J}|x|^r\\ &-\varepsilon\beta(t-s)^\gamma\Big[(1+|x|^m)|x|^{\beta-2}[N+(\beta-2)^+K_J + \beta\varepsilon(t-s)^\gamma K_J|x|^\beta] - |x|^{\beta+p-1}b_{0,J}\Big].\end{aligned}$$

Since $\varepsilon \in (0,\beta^{-1}(t_2-t_1)^{-\gamma})$, it follows that $-c_{0,J}|x|^r \ge -\varepsilon\beta(t-s)^\gamma c_{0,J}|x|^r$. Hence, from the previous estimate we get

$$(Z_\varepsilon(s,x))^{-1}(D_sZ_\varepsilon(s,x) - \mathcal{A}(s)Z_\varepsilon(s,x)) \ge -\gamma\varepsilon(t-s)^{\gamma-1}|x|^\beta - \varepsilon\beta(t-s)^\gamma\zeta_\varepsilon(x) \tag{15.2.10}$$

or even

$$\begin{aligned}&(Z_\varepsilon(s,x))^{-1}(D_sZ_\varepsilon(s,x) - \mathcal{A}(s)Z_\varepsilon(s,x))\\ \ge&\varepsilon(t-s)^{\gamma-1}|x|^\beta[\beta^2(\delta-\varepsilon)(t-s)K_J|x|^{\beta+m-2} - \gamma] - \varepsilon\beta(t-s)^\gamma\zeta_\delta(x)\end{aligned}\tag{15.2.11}$$

for any $x \in \mathbb{R}^N$. We now distinguish between the cases $\beta+m-2>0$ and $\beta+m-2\le 0$. In the first case, we set $C := K_1 \vee (\beta^2(\delta-\varepsilon)K_J/\gamma)^{-1/(\beta+m-2)}$, where K_1 is such that $\zeta_\delta \le 0$ in $\mathbb{R}^N \setminus B_{K_1}$. From (15.2.11) it follows immediately that $(D_sZ_\varepsilon - \mathcal{A}Z_\varepsilon)/Z_\varepsilon$ is nonnegative in $[t_1,t)\times(\mathbb{R}^N\setminus B_{C_t})$, where $C_t = C(t_2-t_1)^{1/(\beta+m-2)}(t-s)^{-1/(\beta+m-2)}$. On the other hand, if $x \in B_{C_t}\setminus B_{K_1}$, then the function ζ_ε is nonpositive in $\mathbb{R}^N\setminus B_{K_1}$, since $\varepsilon<\delta$. Hence, using (15.2.10) we can estimate

$$D_sZ_\varepsilon(s,x) - \mathcal{A}(s)Z_\varepsilon(s,x) \ge -M_1(t-s)^{\gamma-\frac{2\beta+m-2}{\beta+m-2}}Z_\varepsilon(s,x), \qquad (s,x)\in[t_1,t)\times\mathbb{R}^N,$$

for some positive constant M_1. Clearly, this estimate can be extended also to any $x \in B_{K_1}$. The condition on γ show that the function $s \mapsto (t-s)^{\gamma-\frac{2\beta+m-2}{\beta+m-2}}$ belongs to $L^1((t_1,t))$.

Let us now suppose that $\beta+m-2\le 0$. Then, $\beta=(r+2-m)/2$ and $m+r-2\le 0$. From (15.2.10), disregarding the term $c_{0,J}|x|^r$, we get

$$
\begin{aligned}
&(Z_\varepsilon(s,x))^{-1}(D_sZ_\varepsilon(s,x)-\mathcal{A}(s)Z_\varepsilon(s,x))\\
&\ge -\varepsilon(t-s)^{\gamma-1}|x|^\beta(\gamma+\beta^2\varepsilon(t_2-t_1)^{\gamma-1}K_J-\beta(t-s)|x|^{p-1}b_{0,J})-L_{1,\varepsilon},
\end{aligned}
$$

where $L_{1,\varepsilon}=\varepsilon\beta(t_2-t_1)^\gamma(N+(\beta-2)K_J)$. Arguing as above, we distinguish between the cases $|x|\ge((\gamma+\beta^2\varepsilon(t_2-t_1)^{\gamma-1}K_J)/(\beta b_{0,J}(t-s)))^{1/(p-1)}$ and $1<|x|<((\gamma+\beta^2\varepsilon(t_2-t_1)^{\gamma-1}K_J)/(\beta b_{0,J}(t-s)))^{1/(p-1)}$. In the first case, $(Z_\varepsilon(s,x))^{-1}(D_sZ_\varepsilon(s,x)-\mathcal{A}(s)Z_\varepsilon(s,x))\ge -L_{1,\varepsilon}$, whereas in the second case

$$
(Z_\varepsilon(s,x))^{-1}(D_sZ_\varepsilon(s,x)-\mathcal{A}(s)Z_\varepsilon(s,x))\ge -L_{2,\varepsilon}(t-s)^{\gamma-1-\frac{\beta}{p-1}}-L_{1,\varepsilon}
$$

for some positive constant $L_{2,\varepsilon}$ and the function $s\mapsto(t-s)^{\gamma-1-\frac{\beta}{p-1}}$ belongs to $L^1((t_1,t))$. Summing up, we have proved that Z_ε is a time dependent Lyapunov function for $\mathcal{A}$.

The analogous estimate for $\kappa_0\Delta+\langle b,\nabla\rangle+c$ follows observing that $\kappa_0\le K_J$ and this operator satisfies the inequalities in (15.2.10) and (15.2.11), with K_J being replaced by κ_0. Finally, each operator $\mathcal{A}^{(n)}$ satisfies the same estimate as $\mathcal{A}$, with possibly different constants.

Step 2. We fix $0<\varepsilon_0<\varepsilon_1<\varepsilon_2<(t_2-t_1)^{-\gamma}(\beta^{-1}\wedge\delta)$ and define the functions $w=Z_{\varepsilon_0}$, $W_1=Z_{\varepsilon_1}$, $W_2=Z_{\varepsilon_2}$. Clearly, $w\le W_1\le W_2$ and, by Step 1, W_1 and W_2 are time dependent Lyapunov functions such that $W_j\le\rho_0W_j^{1-\beta}$ for $j=1,2$ and suitable constants ρ_0 and $\beta\in(0,1)$. Hence, Hypothesis 15.0.1(iv) is satisfied. We further fix $t_1<\sigma_0<\sigma<\tau<\tau_0<t$ and claim that the functions w, W_1 and W_2 satisfy Hypothesis 15.0.1(v). At the end of the proof, we will make a suitable choice of the constants σ_0, σ, τ_0 and τ.

As is easily seen, the functions $w^{-2}D_sw$ and $w^{-2}\nabla_yw$ are bounded in $[\sigma_0,\tau_0]\times\mathbb{R}^N$. Hence, Hypothesis 15.0.1(iv) is satisfied.

Let us now prove that also Hypothesis 15.0.1(v) is satisfied. Since $\varepsilon_0<\varepsilon_1$, we can clearly take $C_1=1$ in condition (a) of Hypothesis 15.0.1(v). As far as the condition (b) is concerned, we observe that

$$
\begin{aligned}
\frac{|Q(s,y)\nabla_yw(s,y)|}{(w(s,y))^{1-1/q}(W_1(s,y))^{1/q}}\le&\,\varepsilon_0\beta K_J(t-s)^\gamma|y|^{\beta-1}(1+|y|^m)e^{-\frac{1}{q}(\varepsilon_1-\varepsilon_0)(t-s)^\gamma|y|^\beta}\\
\le&\,2\varepsilon_0\beta K_J(t-s)^\gamma|y|^{\beta+m-1}e^{-\frac{1}{q}(\varepsilon_1-\varepsilon_0)(t-s)^\gamma|y|^\beta}
\end{aligned}
$$

for any $y\in\mathbb{R}^N\setminus B_1$. To go further in the estimate, we note that

$$
|x|^\omega e^{-r|x|^\lambda}\le r^{-\frac{\omega}{\lambda}}C(\omega,\lambda),\qquad x\in\mathbb{R}^N,\ r,\omega,\lambda>0,\tag{15.2.12}
$$

where $C(\omega,\lambda)=[\omega(\lambda e)^{-1}]^{\omega/\lambda}$. Applying this estimate with $\lambda=\beta$, $r=q^{-1}(\varepsilon_1-\varepsilon_0)(t-s)^\gamma$ and $\omega=\beta-1+m$, we get

$$
\begin{aligned}
\frac{|Q(s,y)\nabla_yw(s,y)|}{(w(s,y))^{1-1/q}(W_1(s,y))^{1/q}}&\le 2\varepsilon_0\beta K_J(t-s)^{-\frac{\gamma(m-1)}{\beta}}\left(\frac{\varepsilon_1-\varepsilon_0}{q}\right)^{-\frac{\beta-1+m}{\beta}}C(\beta-1+m,\beta)\\
&=:\overline{C}_2(t-s)^{-\frac{\gamma(m-1)}{\beta}}\le\overline{C}_2(t-\tau_0)^{\frac{-\gamma(m-1)_+}{\beta}}
\end{aligned}
$$

for any $y\notin B_1$. Since $|Q\nabla_yw|w^{1/q-1}W_1^{-1/q}$ is bounded in $[t_1,t]\times B_1$, the previous estimate holds true in $[\sigma_0,\tau_0]\times\mathbb{R}^N$ at the cost of a possibly larger constant $\overline{C}_2$.

The other constants $C_3,\dots,C_9$ are obtained in a similar way. It turns out that they are always of the form $C_j=\overline{C}_j(t-\tau_0)^{-r_j}$ for a certain constant $\overline{C}_j$ (which may depend on N,

p, m, q, ε_0, ε_1, ε_2, K_J and the behaviour of the function v_β on B_1) and a certain exponent r_j. We are interested only in the exponent r_j. For this purpose, in the following estimates we do not write explicitly the constants $\overline{C}_j$. We simply denote them by C, where C may vary from line to line.

Concerning r_3, using (15.2.12) we get

$$\begin{aligned}\frac{|\mathrm{Tr}(Q(s,y)D_y^2 w(s,y))|}{(w(s,y))^{1-2/k}(W_1(s,y))^{2/k}} \leq & C\big[(t-s)^\gamma |y|^{\beta+m-2} + (t-s)^{2\gamma}|y|^{2\beta+m-2}\big]e^{-\frac{2}{q}(\varepsilon_1-\varepsilon_0)(t-s)^\gamma|y|^\beta} \\ \leq & C(t-s)^{2\gamma}(t-s)^{-\gamma\frac{2\beta+m-2}{\beta}} \leq C(t-\sigma_0)^{-\frac{\gamma(m-2)_+}{\beta}},\end{aligned}$$

so that $r_3 = \gamma(m-2)_+/\beta$. This estimate with $m = 0$ shows that $r_7 = 0$. Moreover, the estimates

$$\frac{|D_s w(s,y)|}{(w(s,y))^{1-2/q}(W_1(s,y))^{2/q}} \leq C(t-s)^{\gamma-1}|y|^\beta e^{-\frac{2}{q}(\varepsilon_1-\varepsilon_0)(t-s)^\gamma|y|^\beta} \leq C(t-\sigma_0)^{-1},$$

$$\frac{|\sum_{j=1}^N D_j q_{ij}(s,y)|}{(w(s,y))^{-1/q}(W_2(s,y))^{1/q}} \leq C|y|^m e^{-\frac{1}{q}(\varepsilon_2-\varepsilon_0)(t-s)^\gamma|y|^\beta} \leq C(t-s)^{-\frac{\gamma m}{\beta}} \leq (t-\sigma_0)^{-\frac{\gamma m}{\beta}},$$

$$\frac{(w(s,y))^{1/q}|b(s,y)|}{(W_2(s,y))^{1/q}} \leq C|y|^p e^{-\frac{1}{q}(\varepsilon_2-\varepsilon_0)(t-s)^\gamma|y|^\beta} \leq C(t-s)^{-\frac{\gamma p}{\beta}} \leq C(t-\sigma_0)^{-\frac{\gamma p}{\beta}},$$

$$\frac{(w(s,y))^{2/q}}{(W(s,y))^{2/q}}|c(s,y)| \leq C|y|^r e^{-\frac{2}{q}(\varepsilon_2-\varepsilon_0)(t-s)^\gamma|y|^\beta} \leq C(t-\sigma_0)^{-\frac{\gamma r}{\beta}},$$

yield $r_4 = 1$, $r_5 = \gamma m/\beta$, $r_6 = \gamma p/\beta$ and $r_7 = \gamma r/\beta$. Similarly, arguing as in the estimate of c_2 and c_3, we can show that $r_8 = 0$ and $r_9 = \gamma(m-1)_+/\beta$.

Step 3. Now, we can complete the proof. From Step 1, we know that $\mathcal{A}W_j \geq -hW_j$ for $j = 1, 2$, where, for any $s \in (t_1, t)$, $h(s) = L_{2,\varepsilon}(t-s)^{\gamma-\frac{2\beta+m-2}{\beta+m-2}} + L_{1,\varepsilon}$, if $\beta+m-2 > 0$, and $h(s) = L_{2,\varepsilon}(t-s)^{\gamma-1-\frac{\beta}{p-1}} + L_{1,\varepsilon}$ otherwise.

We first assume that $c \equiv 0$. In this case, we take $r = 0$ and $\beta = p+1-m$ (Note that $\beta+m-2 > 0$.) We fix $t \in (t_1, t_2]$, $s \in (t_1, t)$ and choose $\sigma_0 = (s/2) \vee (s-(t-s)/2)$ and $\tau_0 = s + (t-s)/2$ so that $t - \tau_0 = (t-s)/2$ and $\tau_0 - \sigma_0 \leq t - s$. From Lemma 15.1.3 we can infer that $(G(t,r)W_i(r,\cdot))(x) \leq \exp(\|h\|_{L^1((t_1,t))})$ for any $t_1 < \xi < t$, $x \in \mathbb{R}^N$ and $i = 1, 2$. Hence,

$$\int_{\sigma_0}^{\tau_0}(G(t,s)W_i(\xi,\cdot))(x)d\xi \leq K_1(\tau_0-\sigma_0) = K_2(t-s),$$

where the constants K_1 and K_2 are independent of σ_0, τ_0, $t \in (t_1, t_2]$ and $x \in \mathbb{R}^N$.

Thus, we can apply Theorem 15.2.2 and, taking into account that $t - s \leq t_2 - t_1$, $p > m - 1$ as well as the choice of γ, we get

$$w(s,y)g(t,s,x,y) \leq K_3(t-s)^{1-\frac{\gamma(m\vee p)q}{p+1-m}}, \qquad s \in (t_1,t),\ x,y \in \mathbb{R}^N,$$

for some positive constant K_3 depending on t_1 and t_2 but being independent of s, t and x. Recalling the definitions of w, estimates (15.2.7) and (15.2.8) follow at once in the case when $c \equiv 0$.

We now address the general case. For this purpose, we observe that, taking the previous steps and Lemma 15.2.1 into account, we deduce that each operator $\mathcal{A}^{(n)} - c$ satisfies Hypotheses 15.0.1(i)-(v) with the same functions V, w, W_1 and W_2. Hence, we can associate an evolution operator $\{G_{n,0}(t,s)\}$ with the operator $\mathcal{A}^{(n)} - c$, and its Green function $g_{n,0}$

is such that $g_{n,0}(t,\cdot,x,\cdot)$ is bounded in $[\sigma_0,\tau_0]\times\mathbb{R}^N$. By the maximum principle, $g_n\le g_{n,0}$ and $g_n(t,\cdot,x,\cdot)$ is bounded in $[\sigma_0,\tau_0]\times\mathbb{R}^N$ as well. The assumptions of Theorem 15.2.2(ii) are thus satisfied and, as a consequence, we deduce that

$$w(s,y)g(t,s,x,y)\le K_4(t-s)^{1-\frac{\gamma(m\vee p\vee\frac{r}{2})q}{\beta}},\qquad s\in(t_1,t),\ \ x,y\in\mathbb{R}^N,$$

for some positive constant K_4 depending on t_1 and t_2 but being independent of s, t and x. To conclude the proof, we take

$$\beta=p+1-m,\qquad \gamma>\frac{p-1+m}{p-1},\qquad \varepsilon<\frac{1}{(p+1-m)}\left(1\wedge\frac{b_{0,[t_1,t_2]}}{K_{[t_1,t_2]}}\right)(t_2-t_1)^{-\gamma},$$

if $p\ge(m+r)/2$. On the other hand, if $p<(m+r)/2$, then we take

$$\beta=\frac{r-m+2}{2},\qquad \varepsilon<\frac{2}{(r+2-m)}\left(1\wedge\frac{c_{0,[t_1,t_2]}^{1/2}}{K_{[t_1,t_2]}^{1/2}}\right)(t_2-t_1)^{-\gamma},$$

$$\gamma>\begin{cases}\dfrac{r-m+2}{r+m-2}, & \text{if } r+m>2,\\[2mm] \dfrac{r+2-m}{2(p-1)}, & \text{otherwise.}\end{cases}$$

The statement follows. □

15.3 Concluding remarks

Kernel estimates in the non-autonomous case have also been proven by S.V. Shaposhnikov in [235]. In that paper, no differentiability conditions are assumed on the diffusion coefficients of the operator $\mathcal{A}$. Even if in [235] Lyapunov functions are also used, the approach is different since it relies on local L^p-estimates and it is not easy to compare his results to the ones presented here. For the class of operators considered in Theorem 15.2.4 the decay of the Green function with respect to the spatial variable in such a theorem agrees with the one obtained in [235]. On the other hand, the behaviour with respect to t is different. We refer the interested reader to [235, Example 3.10].

Equation (15.1.1) can be rephrased as an equation for the measure $g(t,s,x,y)dsdy$. Equations for measures of this kind have been widely studied in the literature, assuming a final condition which the solution should satisfy. We refer the interested reader e.g. to the papers [36, 37, 38, 41, 233, 234] and the literature therein quoted.

Chapter 16

The evolution operator in L^p-spaces

16.0 Introduction

In this chapter we study the evolution operator $\{G(t,s)\}$, associated with the non-autonomous elliptic operator $\mathcal{A}$ defined in (14.0.1), in L^p-spaces. Throughout the chapter, I is a right-halfline (possibly $I = \mathbb{R}$).

We first consider the case when the underlying measure is the Lebesgue measure and we briefly extend the results in Section 7.1, providing sufficient conditions for $L^p(\mathbb{R}^N)$ ($p \in [1, +\infty)$) to be preserved by the action of the evolution operator $\{G(t,s)\}$. As the autonomous case shows (see e.g. Example 7.1.1) the Lebesgue measure in not the measure which fits best the properties of elliptic operators with unbounded coefficients.

Next, in Section 16.2 we introduce the natural counterpart of the invariant measure of the autonomous case, which is a one-parameter family of probability measures $\{\mu_t : t \in I\}$ such that

$$\int_{\mathbb{R}^N} G(t,s)f\,d\mu_t = \int_{\mathbb{R}^N} f\,d\mu_s, \qquad I \ni s < t, \;\; f \in B_b(\mathbb{R}^N). \tag{16.0.1}$$

As in the autonomous case, we assume that the potential c of the operator $\mathcal{A}$ identically vanishes in $\mathbb{R}^N$.

Whenever it exists, the family $\{\mu_t : t \in I\}$, which is called *evolution system of measures* in [77] and *entrance laws at* $-\infty$ in [93], consists of measures which are equivalent to the Lebesgue measure. As a matter of fact, in general the measures μ_t are not known explicitly and the L^p-spaces related to μ_t and μ_s differ for $t \neq s$. Moreover, when they exist, evolution systems of measures are infinitely many in general, see Proposition 16.2.8 which deals with the case when $\mathcal{A}$ is the non-autonomous Ornstein-Uhlenbeck operator.

Hypotheses 16.4.3 guarantee the existence of an eventually tight evolution system of measures $\{\mu_t : t \in I\}$, i.e., an evolution system of measures with the following property: for any $\varepsilon > 0$ there exists $R > 0$ such that $\mu_t(B_R) \geq 1 - \varepsilon$ for any $t \geq t_0$ and some $t_0 \in I$ (see Theorem 16.2.7). In Section 16.3, using the formula (16.0.1) we show that each operator $G(t,s)$ can be extended to a contraction mapping $L^p(\mathbb{R}^N, \mu_s)$ into $L^p(\mathbb{R}^N, \mu_t)$ for any $p \in [1, +\infty)$. When the pointwise gradient estimates in Theorem 14.4.5 hold true, the operator $G(t,s)$ maps $L^p(\mathbb{R}^N, \mu_s)$ into the Sobolev space $W^{1,p}(\mathbb{R}^N, \mu_t)$ and estimates on the behaviour of the norm $\|\nabla_x G(t,s)\|_{L^p(\mathbb{R}^N, \mu_t)}$ as t tends to s^+ can be obtained.

Under suitable assumptions (see e.g. Theorem 16.2.11), the eventually tight evolution system of measures is unique. Moreover it has a distinguished role among all the evolution systems of measures, as we will see in Chapter 18. For this reason, in Sections 16.4 to 16.7, where we study the summability improving properties of the evolution operator $\{G(t,s)\}$, we assume that $\{\mu_t : t \in I\}$ is the unique eventually tight evolution system of measures for $\{G(t,s)\}$. More precisely, in those sections we study the hypercontractivity, supercontractivity, ultraboundedness and ultracontractivity of the operators $G(t,s)$ (see Definitions 16.4.1, 16.5.1, 16.6.1 and 16.7.1). As is well known, already in the autonomous case the semigroups associated with elliptic operators with unbounded coefficients do not improve, in general, the summability properties, since the Sobolev embedding theorems fail to hold, when the Lebesgue measure is replaced by a general (probability) measure (see Example 9.4.3). Hence, additional assumptions are needed to guarantee that the operator $G(t,s)$ maps a function $f \in L^p(\mathbb{R}^N,\mu_s)$ into a function belonging to $L^q(\mathbb{R}^N,\mu_t)$ for some $q>p$. The main condition assumed in Hypotheses 16.4.3 is a dissipativity condition on the drift term b, which implies the validity of the pointwise gradient estimate

$$|\nabla_x G(t,s)f| \le e^{r_0(t-s)}G(t,s)|\nabla f|^p, \qquad t\ge s\in I,\ f\in C^1_b(\mathbb{R}^N), \tag{16.0.2}$$

for some $r_0<0$. This estimate, which has been proved in Theorem 14.4.5, is crucial and extensively used throughout this chapter. Due to Remark 14.4.7, this estimate to hold requires that the diffusion coefficients satisfy the condition $D_iq_{jk}+D_jq_{ik}+D_kq_{ij}=0$ in $I\times\mathbb{R}^N$ for any $i,j,k=1,\dots,N$, (see [14]). For this reason, we confine ourself to the case of diffusion coefficients depending only on t.

The starting point of our analysis is the following estimate

$$G(t,s)(|f|^p\log|f|^p) \le C_p(1-e^{2r_0(t-s)})G(t,s)(|f|^{p-2}|\nabla f|^2)+(G(t,s)|f|^p)\log(G(t,s)|f|^p),$$

which holds true in $\mathbb{R}^N$ for any $I\ni s<t$, $p\in[1,+\infty)$, $f\in C_b(\mathbb{R}^N)$ (with positive infimum, if $p<2$) and some positive constant C_p. Integrating both sides of the previous estimate with respect to the measure μ_t and using the invariance property of the family $\{\mu_t : t\in I\}$, leads to the logarithmic Sobolev inequality (LSI in short)

$$\int_{\mathbb{R}^N}|f|^p\log(|f|)d\mu_t \le \|f\|^p_{L^p(\mathbb{R}^N,\mu_t)}\log(\|f\|_{L^p(\mathbb{R}^N,\mu_t)})+C'_p\int_{\mathbb{R}^N}|f|^{p-2}|\nabla f|^2\chi_{\{f\neq 0\}}d\mu_t,$$

satisfied by any $f\in C^1_b(\mathbb{R}^N)$, $p\in(1,+\infty)$, $t\in I$ and some positive constant C'_p. Roughly speaking the LSI is, for evolution systems of measures, the counterpart of the Sobolev embedding theorems. As Examples 9.4.3 and 9.4.4 show, this estimate is rather precise: in general, the function $f^2\log(|f|)\log(\log(|f|))$ is not in $L^1(\mathbb{R}^N,\mu_t)$.

The hypercontractivity of $G(t,s)$ is then obtained as a consequence of the LSI, which is also used to prove both a Poincaré inequality satisfied by the functions in $W^{1,p}(\mathbb{R}^N,\mu_t)$ and the compactness of the embedding of $W^{1,p}(\mathbb{R}^N,\mu_t)$ in $L^p(\mathbb{R}^N,\mu_t)$ for any $t\in I$ and $p\ge 2$.

Next, in Section 16.5, we first prove that the supercontractivity property of the evolution operator $\{G(t,s)\}$ is equivalent to the validity of the following family of logarithmic Sobolev inequalities (in short LSI$_\varepsilon$)

$$\int_{\mathbb{R}^N} f^2\log\left(\frac{|f|}{\|f\|_{L^2(\mathbb{R}^N,\mu_t)}}\right)d\mu_t \le \varepsilon\|\nabla f\|^2_{L^2(\mathbb{R}^N,\mu_t)}+\beta(\varepsilon)\|f\|^2_{L^2(\mathbb{R}^N,\mu_t)} \tag{16.0.3}$$

for any $f\in W^{1,2}(\mathbb{R}^N,\mu_s)$, $t\in I$, $\varepsilon>0$ and some decreasing function $\beta:(0,+\infty)\to\mathbb{R}$, blowing up as ε tends to 0^+. The proof of estimate (16.0.3) is obtained by adapting the arguments in [229].

Another important consequence of (16.0.2) is a Harnack type estimate, satisfied by the function $G(\cdot,s)f$ for any $f\in C_b(\mathbb{R}^N)$, which, together with LSI$_\varepsilon$, allows to prove (see Theorem 16.5.4) that the supercontractivity of $\{G(t,s)\}$ is equivalent to the condition $\sup_{t\in I}\|\varphi_\lambda\|_{L^1(\mathbb{R}^N,\mu_t)}<+\infty$ for any $\lambda>0$, where $\varphi_\lambda(x):=e^{\lambda|x|^2}$ for any $x\in\mathbb{R}^N$ and $\lambda>0$. The Harnack type estimate is also the key tool to provide a necessary and sufficient condition for the ultraboundedness to hold (see Theorem 16.6.2). Sufficient conditions on the drift term b are provided, which guarantee the supercontractivity and the ultraboundedness of $\{G(t,s)\}$. The comments in Section 9.4 show that these conditions are rather precise.

Since the ultracontractivity implies the ultraboundedness of $\{G(t,s)\}$, the main step to prove that the evolution operator $\{G(t,s)\}$ is ultracontractive consists of showing that each operator $G(t,s)$ maps $L^1(\mathbb{R}^N,\mu_s)$ into $L^2(\mathbb{R}^N,\mu_t)$. We provide a sufficient condition (see Hypothesis 16.7.2) for the previous property to hold.

16.1 The evolution operator in $L^p(\mathbb{R}^N)$

As the autonomous case shows (see Example 7.1.1), the usual L^p-spaces are not the L^p-spaces which fit best the properties of semigroups/evolution operators associated with elliptic operators with unbounded coefficients.

Adapting the arguments in the proof of Theorem 7.1.2, we can provide some sufficient conditions which guarantee that the restriction to $C_c^\infty(\mathbb{R}^N)$ of the evolution operator $\{G(t,s)\}$ can be extended to $L^p(\mathbb{R}^N)$. For this purpose, we set $\beta_i = b_i - \sum_{j=1}^N D_j q_{ij}$ for any $i = 1, \ldots, N$.

Theorem 16.1.1 *Fix $a, b \in I$, with $a < b$. Suppose that Hypotheses 14.0.1 hold true and the diffusion coefficients q_{ij} $(i,j = 1,\ldots,N)$ are continuously differentiable in $[a,b] \times \mathbb{R}^N$ with respect to the spatial variables. Moreover, assume that one of the following additional conditions is satisfied:*

(i) *the drift coefficients b_j are continuously differentiable in $[a,b] \times \mathbb{R}^N$ with respect to the spatial variables, the derivatives $D_{ij}q_{ij}$ $(i,j = 1,\ldots,N)$ exist in $[a,b] \times \mathbb{R}^N$ and there exists a positive constant K such that $c(t,x) + \mathrm{div}_x\beta(t,x) \geq -K$ for any $(t,x) \in [a,b] \times \mathbb{R}^N$;*

(ii) *there exists a positive constant K_p' such that*

$$\frac{|\beta(t,x)|^2}{4(p-1)\kappa(t,x)} - c(t,x) \leq K_p', \qquad (t,x) \in [a,b] \times \mathbb{R}^N.$$

Then, $L^p(\mathbb{R}^N)$ is invariant under $G(t,s)$ for any $(t,s) \in \Delta_{[a,b]}$ and $p \in (1,+\infty)$ (and even for $p = 1$ under condition (i)), and, for these values of p, $\|G(t,s)f\|_{L^p(\mathbb{R}^N)} \leq e^{K_p(t-s)}\|f\|_{L^p(\mathbb{R}^N)}$ for any $a \leq s \leq t \leq b$, where $K_p = p^{-1}(K - (p-1)c_0)$, under condition (i), and $K_p = K_p'$ under condition (ii). Finally, $G(t,\cdot)f$ belongs to $C([a,t]; L^p(\mathbb{R}^N))$ for any $t \in (a,b]$.

16.2 Evolution systems of measures

Throughout this section we assume the following set of assumptions on the coefficients of the operator $\mathcal{A}$ in (14.0.1)

Hypotheses 16.2.1 (i) *Hypotheses 14.0.1(i), (ii) and (iv) are satisfied and $c \equiv 0$;*

(ii) *there exist a function $\varphi \in C^2(\mathbb{R}^N)$ with positive values, diverging to $+\infty$ as $|x|$ tends to $+\infty$, and constants $a, d > 0$ and $t_0 \in I$ such that $\mathcal{A}(t)\varphi(x) \leq a - d\varphi(x)$ for any $t \geq t_0$ and $x \in \mathbb{R}^N$.*

Definition 16.2.2 *Let $\{G(t,s)\}$ be the evolution operator associated with the operator $\mathcal{A}$ in $B_b(\mathbb{R}^N)$. A family $\{\mu_t : t \in I\}$ of probability measures on $\mathbb{R}^N$ is an evolution system of measures for $\{G(t,s)\}$ if*

$$\int_{\mathbb{R}^N} G(t,s)f \, d\mu_t = \int_{\mathbb{R}^N} f \, d\mu_s \qquad (16.2.1)$$

for any $f \in B_b(\mathbb{R}^N)$ and $I \ni s < t$.

Formula (16.2.1) may be rewritten as $G^*(t,s)\mu_t = \mu_s$, where $G^*(t,s)$ denotes the adjoint of the operator $G(t,s)$. The following lemma, which will be crucial in the proof of Theorem 16.2.7, shows that each evolution system of measures is uniquely determined by any subfamily $\{\mu_{t_n} : n \in \mathbb{N}\}$ corresponding to a sequence $\{t_n\} \subset I$ which diverges to $+\infty$.

Lemma 16.2.3 *Suppose that $\{t_n\} \subset I$ is an increasing sequence diverging to $+\infty$. Further, let $\mathscr{F} = \{\mu_{t_n} : n \in \mathbb{N}\}$ be a family of probability measures such that $G^*(t_n,t_m)\mu_{t_n} = \mu_{t_m}$ for any $n,m \in \mathbb{N}$ with $n > m$. Then, there exists, and it is unique, an evolution system of measures $\{\mu_t : t \in I\}$ for the evolution operator $\{G(t,s)\}$ such that $\{\mu_{t_n} : n \in \mathbb{N}\} \subset \{\mu_t : t \in I\}$.*

Proof As a first step, we show that, for any probability measure μ, defined on the σ-algebra of all the Borel subsets of $\mathbb{R}^N$, and $t > s \in I$, $G^*(t,s)\mu$ is a probability measure. For this purpose, we introduce the map $\nu_{t,s}$ defined by

$$\nu_{t,s}(A) = \int_{\mathbb{R}^N} G(t,s)\chi_A d\mu$$

for any Borel set $A \subset \mathbb{R}^N$. Clearly, $\nu_{t,s}(\mathbb{R}^N) = 1$ since $G(t,s)\mathbb{1} \equiv \mathbb{1}$. Moreover, $\nu_{t,s}(A) \geq 0$ for any Borel set A due to Proposition 14.1.3, which shows that the function $G(t,s)\chi_A$ is nonnegative in $\mathbb{R}^N$ for any A as above. To conclude that $\nu_{t,s}$ is a measure, we fix a sequence $\{A_n\}$ of pairwise disjoint Borel sets. Then, $\chi_{\bigcup_{j\in\mathbb{N}} A_j} = \sum_{j=1}^{+\infty} \chi_{A_j}$ and using both the representation formula (14.1.4) and Beppo-Levi theorem yields that $G(t,s)\chi_{\bigcup_{j\in\mathbb{N}} A_j} = \sum_{j=1}^{+\infty} G(t,s)\chi_{A_j}$. Thus,

$$\nu_{t,s}\bigg(\bigcup_{j\in\mathbb{N}} A_j\bigg) = \int_{\mathbb{R}^N} G(t,s)\chi_{\bigcup_{j\in\mathbb{N}} A_j} d\mu = \sum_{j=1}^{+\infty} G(t,s)\chi_{A_j} d\mu = \sum_{j=1}^{+\infty} \nu_{t,s}(A_j).$$

Next, we observe that, by definition,

$$\int_{\mathbb{R}^N} f d\nu_{t,s} = \int_{\mathbb{R}^N} G(t,s) f d\mu \tag{16.2.2}$$

for any f which is the characteristic function of a Borel set A. By linearity, we extend the previous equality to any simple Borel measurable function f. Since any bounded and continuous function is the uniform limit of a sequence of linear combinations of characteristic functions, we can extend (16.2.2) to any $f \in B_b(\mathbb{R}^N)$. Formula (16.2.2) reveals that $\nu_{t,s} = G(t,s)^*\mu$ is a positive measure.

Now, we can complete the proof defining the family $\{\mu_t : t \in I\}$ as follows: for any $t \in I$, which is not an element of the sequence $\{t_n\}$, we set $\mu_t = G^*(t,t_k)\mu_{t_k}$ where k is the smallest integer such that $t_k > t$. To check the invariance property, we fix $t > s \in I$, $f \in B_b(\mathbb{R}^N)$ and let k_0 and k_1 be the smallest integers such that $t_{k_0} \geq s$ and $t_{k_1} \geq t$. Since $G^*(t_{k_1},t_{k_0})\mu_{t_{k_1}} = \mu_{k_0}$ we can write

$$\begin{aligned}\int_{\mathbb{R}^N} f d\mu_s &= \int_{\mathbb{R}^N} G(t_{k_0},s) f d\mu_{t_{k_0}} = \int_{\mathbb{R}^N} G(t_{k_1},t_{k_0})G(t_{k_0},s) f d\mu_{t_{k_1}} \\ &= \int_{\mathbb{R}^N} G(t_{k_1},t)G(t,s) f d\mu_{t_{k_1}} = \int_{\mathbb{R}^N} G(t,s) f d\mu_t.\end{aligned}$$

Clearly, the family $\{\mu_t : t \in I\}$ is the unique evolution system of measures for $\{G(t,s)\}$ which contains the family $\mathscr{F}$. □

Before providing a sufficient condition for the existence of an evolution system of measures, which generalizes the condition in Theorem 9.1.20, we show that whenever an evolution system of measures exists, each measure μ_t is equivalent to the Lebesgue measure. For this purpose, we need the following result, proved in [40] by V.I. Bogachev, N.V. Krylov, and M. Röckner under weaker assumptions than Hypotheses 16.2.1 .

Theorem 16.2.4 ([40], Thm. 3.8 and Cor. 3.1, 3.9) *Fix $I \ni a < b$, $R > 0$ and let $\widehat{\mu}$ be a Borel signed measure on $(a,b) \times B_R$ such that*

$$\int_{(a,b)\times B_R} \mathcal{G}\psi d\widehat{\mu} = 0, \qquad \psi \in C_c^\infty((a,b)\times B_R), \tag{16.2.3}$$

where $\mathcal{G} = \mathcal{A} - D_t$. Then, μ admits a density with respect to the Lebesgue measure which belongs to $L^r_{\rm loc}((a,b)\times B_R))$ for any $r < (N+2)'$. Further, if the diffusion coefficients q_{ij} $(i,j = 1,\ldots,N)$ are differentiable with respect to the spatial variables in $(a,b)\times B_R$ and the partial derivatives belong to $C_b((a,b); L^p(B_R))$ for some $p > N+2$, then the density of $\widehat{\mu}$ belongs to $C^\beta_{\rm loc}((a,b)\times B_R)\cap W^{0,1}_{p,{\rm loc}}(\Omega)$ for some $\beta \in (0,1)$ and any open set $\Omega \Subset (a,b)\times B_R$. In particular, if $\widehat{\mu}$ is a positive measure, then the density of $\widehat{\mu}$ is positive in $(a,b)\times B_R$. Finally, if the coefficients of the operator $\mathcal{G}$ belong to $C_b^{\alpha/2,\alpha}((a,+\infty)\times B_R)$ and the first-order spatial derivatives of the diffusion coefficients exist and belong to $C_b((a,+\infty); L^p(B_R))$ for any $R > 0$ and some $p > N+2$, then the density of $\widehat{\mu}$ belongs to $C^\beta((a,+\infty)\times B_R)$ for any[1] $R > 0$.

Proposition 16.2.5 *If $\{\mu_t : t \in I\}$ is an evolution system of measures for $\{G(t,s)\}$ then each measure μ_t is equivalent to the Lebesgue measure. More precisely, $\mu_s = \rho(s,\cdot)dx$ for some strictly positive and locally Hölder continuous function $\rho : I\times\mathbb{R}^N \to \mathbb{R}$.*

Proof Let μ be the unique measure, defined on the σ-algebra of all the Borel subsets of $I\times\mathbb{R}^N$ such that

$$\mu(A\times B) = \int_A \mu_s(B)ds$$

for any pair of Borel sets $A \subset I$ and $B \subset \mathbb{R}^N$. This measure is not finite since $\mu(I\times\mathbb{R}^N)$ equals the Lebesgue measure of I, which is infinite. Anyway, to some extent it acts as an invariant measure. In particular, it satisfies (16.2.3). Indeed, fix $(a,b) \Subset I$, $\psi \in C_c^\infty((a,b)\times\mathbb{R}^N)$, and consider the function $\mathscr{T}(t)\psi$ introduced in Section 14.5, which is defined in $\mathbb{R}^{N+1}$ provided we set $\mathscr{T}(t)\psi \equiv 0$ outside $(a+t,b+t)\times\mathbb{R}^N$. The invariance property of the family $\{\mu_t : t\in I\}$ implies that

$$\begin{aligned}\int_{I\times\mathbb{R}^N} \mathscr{T}(t)\psi d\nu &= \int_I ds\int_{\mathbb{R}^N} G(s,s-t)\psi(s-t,\cdot)d\mu_s\\ &= \int_I ds\int_{\mathbb{R}^N}\psi(s-t,\cdot)d\mu_{s-t} = \int_{I\times\mathbb{R}^N}\psi d\mu.\end{aligned}$$

Hence,

$$\int_{I\times\mathbb{R}^N}\frac{\mathscr{T}(t)\psi-\psi}{t}d\mu = \int_{(a,b+1)\times\mathbb{R}^N}\frac{\mathscr{T}(t)\psi-\psi}{t}d\mu = 0, \qquad t\in(0,1].$$

Arguing as in Step 1 in the proof of Proposition 14.5.7, with minor changes, we can show

[1] This last statement follows from a deep inspection of the proofs of the Theorem 3.8 and Corollary 3.9 in [40].

that the ratio $t^{-1}(\mathscr{T}(t)\psi - \psi)$ converges to $\mathcal{A}\psi - D_s\psi$ locally uniformly in $[a,b]\times\mathbb{R}^N$. Moreover, its modulus can be bounded from above in $(a,b+1)\times\mathbb{R}^N$ by a positive constant times the $C^{1,2}$-norm of ψ. Letting t tend to 0^+, the formula (16.2.3) follows by dominated convergence. Hence, by Theorem 16.2.4 and the arbitrariness of (a,b), the measure μ admits a density ρ with respect to the Lebesgue measure, which is positive in $\mathrm{Int}(I)\times\mathbb{R}^N$, where $\mathrm{Int}(I)$ denotes the set of all the interior points of I.

To complete the proof, we observe that, for any Borel set $A\subset I$ and any $f\in C_c^\infty(\mathbb{R}^N)$, it holds that

$$\int_A ds\int_{\mathbb{R}^N} f d\mu_s = \int_{I\times\mathbb{R}^N}\chi_A f d\mu = \int_A ds\int_{\mathbb{R}^N} f\rho(s,\cdot)dx.$$

The arbitrariness of A yields

$$\int_{\mathbb{R}^N} f d\mu_s = \int_{\mathbb{R}^N} f\rho(s,\cdot)dx \tag{16.2.4}$$

for almost every $s\in I$. To extend (16.2.4) to any $s\in I$, we observe that both its sides are continuous with respect to s. The right-hand side is continuous since ρ is. On the other hand, by (16.2.1) the left-hand side is equal to $\int_{\mathbb{R}^N} G(r,s)f d\mu_r$ for any $r>s$, and the function $G(r,\cdot)f$ is continuous in $I\cap(-\infty,r]$, by Lemma 14.1.6(i). This shows that the function $s\mapsto\int_{\mathbb{R}^N} f d\mu_s$ is continuous in $I\cap(-\infty,r]$ and, hence, in I due to the arbitrariness of r.

We are almost done. Indeed, let $B\subset\mathbb{R}^N$ be any Borel set, consider a sequence $\{f_n\}\subset C_c^\infty(\mathbb{R}^N)$ converging to χ_B locally uniformly in $\mathbb{R}^N$ and write (16.2.4) with f being replaced by f_n. Letting n tend to $+\infty$ the right-hand side tends to $\int_B\rho(s,\cdot)dx$. On the other hand, $\int_{\mathbb{R}^N} f_n d\mu_s = \int_{\mathbb{R}^N} G(s+1,s)f_n d\mu_{s+1}$ and, by Proposition 14.1.3, the right-hand side of this formula converges to $\int_{\mathbb{R}^N} G(s+1,s)\chi_B d\mu_{s+1} = \mu_s(B)$ as n tends to $+\infty$. We thus conclude that $\mu_s=\rho(s,\cdot)dx$ for any $s\in I$. This shows that μ_s is equivalent to the Lebesgue measure for any $s\in\mathrm{Int}(I)$. Finally, suppose that I is a closed interval and s_0 denotes its left endpoint. Then, $\mu_{s_0}(B)=\int_{\mathbb{R}^N} G(s_0+1,s_0)\chi_B d\mu_{s_0+1}$ for any Borel set $B\subset\mathbb{R}^N$ and, by Proposition 14.1.3, $G(s_0+1,s_0)\chi_B$ is continuous in $\mathbb{R}^N$. Moreover, it is everywhere positive in $\mathbb{R}^N$ if and only if B has positive Lebesgue measure. This shows that $\mu_{s_0}(B)=0$ if and only if B is negligible with respect to the Lebesgue measure. As a byproduct $\rho(s_0,\cdot)>0$ in $\mathbb{R}^N$ and the proof is complete. □

To prove the existence of evolution systems of measures we use a procedure similar to the Krylov-Bogoliubov theorem which, in the autonomous case, states that the existence of an invariant measure is equivalent to the tightness of a certain set of probability measures. The following theorem improves the result in Lemma 14.1.9.

Lemma 16.2.6 *Assume that Hypotheses* 16.2.1 *are satisfied. Then, $G(t,s)\varphi$ is well defined for any $t_0\le s\le t\in I$. Moreover, for any fixed $x\in\mathbb{R}^N$, it holds that*

$$(G(t,s)\varphi)(x)\le\varphi(x)+d^{-1}a,\qquad t_0\le s\le t. \tag{16.2.5}$$

Proof Lemma 14.1.9 implies that the function $G(\cdot,\cdot)\varphi$ is well defined and locally bounded in $\{(t,s,x)\in I\times I\times\mathbb{R}^N : t_0\le s<t\}$. To complete the proof, we fix $t>t_0$, $x\in\mathbb{R}^N$ and observe that the function $(G(t,\cdot)\varphi)(x)$ is measurable being the pointwise limit of the functions $(G(t,\cdot)\varphi_n)(x)$ in the proof of Lemma 14.1.9, that are continuous with respect to s by Theorem 14.1.12. The same arguments used in the proof of Lemma 14.1.9 reveal that

$$(G(t,r)\varphi)(x)-(G(t,s)\varphi)(x)\ge\int_s^r (d(G(t,\sigma)\varphi)(x)-a)\,d\sigma \tag{16.2.6}$$

for any $t_0 \le s \le r \le t$.

Let us now introduce the function Φ, defined by

$$\Phi(r) := \left((G(t,s)\varphi)(x) - \frac{a}{d} + \int_s^r (d(G(t,\sigma)\varphi)(x) - a)\, d\sigma \right) e^{-dr}$$

for any $r \in [s,t]$, where $s \ge t_0$ has been arbitrarily fixed. This function is continuous in $[s,t]$ and therein weakly differentiable with $\Phi'(r) \ge 0$. In particular, $\Phi(s) \le \Phi(t)$, which, combined with (16.2.6), yields to (16.2.5). □

Theorem 16.2.7 *Under Hypotheses* 16.2.1 *there exist an evolution system of measures* $\{\mu_t : t \in I\}$ *for* $\{G(t,s)\}$ *and a constant* $M > 0$ *such that*

$$\int_{\mathbb{R}^N} \varphi \, d\mu_t \le M, \qquad t \ge t_0. \tag{16.2.7}$$

Moreover, the family $\{\mu_t : t \ge t_0\}$ *is tight.*

Proof Fix $x_0 \in \mathbb{R}^N$ and, for any $t > s \in I$, let $\mu_{t,s}$ be the measure defined by

$$\mu_{t,s}(A) = \frac{1}{t-s}\int_s^t g_{\tau,s,x_0}(A)\, d\tau = \frac{1}{t-s}\int_s^t (G(\tau,s)\chi_A)(x_0)\, d\tau.$$

Arguing as in the proof of Lemma 14.1.10 and taking Lemma 16.2.6 into account, we can easily show that the family $\{\mu_{t,s} : t > s \ge t_0\}$ is tight. The Prokhorov theorem (see Theorem 9.1.18) and a diagonal argument show that there exist a sequence $\{t_k\}$ diverging to $+\infty$ and a family $\{\mu_n : t_0 < n \in \mathbb{N}\}$ such that $\mu_{t_k,n}$ weakly* converges to μ_n. Now, we want to use Lemma 16.2.3. For this purpose we need to show that $G^*(n,m)\mu_n = \mu_m$ for $t_0 < m < n$. Fix an arbitrary Borel set $A \subset \mathbb{R}^N$ and observe that

$$\begin{aligned}
G^*(n,m)\mu_n(A) &= \int_{\mathbb{R}^N} (G(n,m)\chi_A)(y)\, \mu_n(dy)\\
&= \lim_{k\to+\infty} \frac{1}{t_k - n}\int_n^{t_k} (G(\tau,n)G(n,m)\chi_A)(x_0)\, d\tau\\
&= \lim_{k\to+\infty} \frac{1}{t_k - n}\int_n^{t_k} (G(\tau,m)\chi_A)(x_0)\, d\tau\\
&= \lim_{k\to+\infty} \frac{1}{t_k - m}\int_m^{t_k} (G(\tau,m)\chi_A)(x_0)\, d\tau = \mu_m(A)
\end{aligned}$$

and the equality $G^*(n,m)\mu_n = \mu_m$ follows.

Let us now prove (16.2.7). Fix $t \ge t_0$ and, for any $n \in \mathbb{N}$, let $\varphi_n : \mathbb{R}^N \to \mathbb{R}$ be defined by $\varphi_n(x) = n \wedge \varphi(x)$ for any $x \in \mathbb{R}^N$. By Lemma 16.2.6 we can estimate $0 < (G(t+1,t)\varphi_n)(x_0) \le \varphi(x_0) + a/d$ for any $n \in \mathbb{N}$. Integrating this inequality with respect to the measure μ_{t+1} we obtain

$$\int_{\mathbb{R}^N} \varphi_n d\mu_t = \int_{\mathbb{R}^N} G(t+1,t)\varphi_n d\mu_{t+1} \le \varphi(x_0) + \frac{a}{d}.$$

Letting n tend to $+\infty$, estimate (16.2.7) follows with $M = \varphi(x_0) + a/d$.

The tightness of the family $\{\mu_t : t \ge t_0\}$ follows from observing that, for any $\varepsilon > 0$, there exists $R > 0$ such that $\varphi \ge M\varepsilon^{-1}\chi_{\mathbb{R}^N\setminus B_R}$. Hence,

$$\mu_t(\mathbb{R}^N \setminus B_R) = \int_{\mathbb{R}^N\setminus B_R} d\mu_t \le \frac{\varepsilon}{M}\int_{\mathbb{R}^N} \varphi\mu_t \le \varepsilon, \qquad t \ge t_0.$$

This completes the proof. □

It should be noted that the evolution system of measures constructed in Theorem 16.2.7 could depend on x_0. Indeed, in general, evolution systems of measures are not unique as the following result from [112] shows in the case of the non-autonomous Ornstein-Uhlenbeck operator $\mathcal{A}$ defined on smooth functions $u : \mathbb{R}^N \to \mathbb{R}$ by

$$(\mathcal{A}u)(x) = \mathrm{Tr}(Q(t)D^2u(x)) + \langle A(t)x, \nabla u(x)\rangle, \qquad t \in \mathbb{R}, \ x \in \mathbb{R}^N. \tag{16.2.8}$$

In the case when the functions $t \mapsto A(t)$ and $t \mapsto Q(t)$ are continuous and there exists a positive constant κ_0 such that $\langle Q(t)\xi, \xi\rangle \geq \kappa_0|\xi|^2$ for any $t \in \mathbb{R}$ and $\xi \in \mathbb{R}^N$, an explicit formula for the associated evolution operator, the non-autonomous Ornstein-Uhlenbeck evolution operator, is available. More precisely,

$$(G(t,s)f)(x) = \int_{\mathbb{R}^N} f(y)\mathcal{N}_{U(s,t)x,Q(t,s)}(dy), \qquad t > s \in \mathbb{R}, \ x \in \mathbb{R}^N,$$

for any $f \in C_b(\mathbb{R}^N)$, where $\mathcal{N}_{U(s,t)x,Q(t,s)}(dy)$ is the Gaussian measure in $\mathbb{R}^N$ with mean $U(s,t)x$ and covariance matrix

$$Q(t,s) = \int_s^t U(s,r)Q(r)U^*(s,r)dr.$$

Here, $\{U(t,s)\}$ is the evolution operator associated with the matrix-valued function $-A(\cdot)$ and $\{U^*(t,s)\}$ is its adjoint operator. More precisely, for any $r \in \mathbb{R}$, $U(\cdot, r)$ solves the Cauchy problem

$$\begin{cases} D_tU(t,r) = -A(t)U(t,r), & t \in \mathbb{R}, \\ U(t,t) = I. \end{cases}$$

If $\|U(t,s)\| \leq Me^{\omega(t-s)}$ for some positive constants ω, M and any $t \geq s \in \mathbb{R}$, then the matrix

$$Q_t = \int_t^{+\infty} U(t,r)Q(r)U^*(t,r)dr \tag{16.2.9}$$

is well defined (and positive definite) for any $t \in \mathbb{R}$.

Proposition 16.2.8 *Let all the above conditions be satisfied. For any $t_0 \in \mathbb{R}$ and any Borel probability measure ν in $\mathbb{R}^N$, let ν_t ($t \in \mathbb{R}$) be the probability measure, whose Fourier transform is the function $\widehat{\nu}_t$, defined by $\widehat{\nu}_t(h) = \widehat{\nu}(U^*(t,t_0)h)$. Then, the family $\{\mu_t : t \in \mathbb{R}\}$, where*

$$\mu_t = \mathcal{N}_{0,Q_t} \star \nu_t, \qquad t \in \mathbb{R}, \tag{16.2.10}$$

and $\star$ denotes convolution, is an evolution system of measures for $\{G(t,s)\}$. Moreover, all the evolution systems of measures for $\{G(t,s)\}$ are of this type.

Proof As is well known, each probability measure μ is uniquely determined by its Fourier transform $\widehat{\mu}$ defined by

$$\widehat{\mu}(h) = \int_{\mathbb{R}^N} e^{i\langle h,\cdot\rangle}d\mu, \qquad h \in \mathbb{R}^N.$$

Hence, a family $\{\mu_t : t \in \mathbb{R}^N\}$ is an evolution system for $\{G(t,s)\}$ if and only if

$$\int_{\mathbb{R}^N} G(t,s)f_h d\mu_t = \int_{\mathbb{R}^N} f_h d\mu_s, \qquad t > s \in \mathbb{R}, \ h \in \mathbb{R}^N, \tag{16.2.11}$$

where $f_h(x) = e^{i\langle h,x\rangle}$ for any $x, h \in \mathbb{R}^N$. For any $x \in \mathbb{R}^N$, $(G(t,s)f_h)(x)$ is easily computed, observing that it is the Fourier transform of the Gaussian measure $\mathcal{N}_{U(s,t)x,Q(t,s)}(dy)$, and we get $G(t,s)f_h = e^{-\langle Q(t,s)h,h\rangle/2}e^{i\langle\cdot,U^*(s,t)h\rangle}$ for any $t \geq s$. Hence, (16.2.11) holds if and only if

$$e^{-\frac{1}{2}\langle Q(t,s)h,h\rangle}\widehat{\mu}_t(U^*(s,t)h) = \widehat{\mu}_s(h), \qquad h \in \mathbb{R}^N \; s < t. \tag{16.2.12}$$

Suppose that the family $\{\mu_t : t \in \mathbb{R}\}$ is defined by (16.2.10). Then, for each $h \in \mathbb{R}^N$ and $t > s \in \mathbb{R}$, it holds that

$$\begin{aligned}
e^{-\frac{1}{2}\langle Q(t,s)h,h\rangle}\widehat{\mu}_t(U^*(s,t)h) =& e^{-\frac{1}{2}\langle Q(t,s)h,h\rangle}\widehat{\mathcal{N}}_{0,Q_t}(U^*(s,t)h)\widehat{\nu}_t(U^*(s,t)h)\\
=& e^{-\frac{1}{2}\langle Q(t,s)h,h\rangle}e^{-\frac{1}{2}\langle Q_tU^*(s,t)h,U^*(s,t)h\rangle}\widehat{\nu}(U^*(t,t_0)U^*(s,t)h)\\
=& e^{-\frac{1}{2}\langle Q_sh,h\rangle}\widehat{\nu}(U^*(s,t_0)h) = \widehat{\mu}_s(h)
\end{aligned}$$

and (16.2.12) follows.

On the other hand, suppose that $\{\mu_t : t \in \mathbb{R}\}$ is an evolution system of measures for $\{G(t,s)\}$. Then, it satisfies (16.2.12) and, letting t tend to $+\infty$, we deduce that, $\widehat{\mu}_t(U^*(s,t)h)$ converges to $e^{\langle Q_sh,h\rangle/2}\widehat{\mu}_s(h)$, for any $h \in \mathbb{R}^N$ and $s \in \mathbb{R}$. Hence, by the Bochner theorem, for any $s \in \mathbb{R}$ there exists a probability measure η_s such that $\widehat{\eta}_s(h) = e^{\langle Q_sh,h\rangle/2}\widehat{\mu}_s(h)$. To conclude the proof, we observe that a straightforward computation reveals that $\widehat{\eta}_s(h) = \widehat{\eta}_{t_0}(U^*(s,t_0)h)$ for any $s, t_0 \in \mathbb{R}$. Fix such a t_0 and set $\nu = \eta_{t_0}$. Then, from the previous formulae, it follows that $\widehat{\mu}_s(h) = e^{\langle Q_sh,h\rangle/2}\widehat{\nu}_s(h)$ for any $s \in \mathbb{R}$. □

Remark 16.2.9 Proposition 16.2.8 shows that the spaces $L^p(\mathbb{R}^N, \mu_t)$ and $L^p(\mathbb{R}^N, \mu_s)$ differ, in general, for different values of t and s and any $p \in [1, +\infty)$.

Even if the evolution systems of measures for $\{G(t,s)\}$ are, in general, infinitely many, however, uniqueness may be achieved among all systems of measures which satisfy an additional condition.

Definition 16.2.10 *An evolution system of measures $\{\mu_t : t \in I\}$ for $\{G(t,s)\}$ is called eventually tight if there exists $t_1 \in I$ such that the family $\{\mu_t : t \geq t_1\}$ is tight.*

Let $\{\mu_t : t \in I\}$ be an evolution system of measures for $\{G(t,s)\}$. For any $t \in I$ and $f \in L^1(\mathbb{R}^N, \mu_t)$ we denote by $m_t(f)$ the average of f with respect to μ_t, i.e.,

$$m_t(f) = \int_{\mathbb{R}^N} f\, d\mu_t.$$

We can now provide a sufficient condition for the uniqueness of an eventually tight evolution system of measures for $\{G(t,s)\}$.

Theorem 16.2.11 *Assume that there exists $\omega < 0$ such that*

$$\|\nabla_x G(t,s)f\|_\infty \leq Ce^{\omega(t-s)}\|f\|_\infty, \qquad t-1 \geq s \in I, \tag{16.2.13}$$

for any $f \in C_b(\mathbb{R}^N)$ and some positive constant C and let $\{\mu_t : t \in I\}$ be an eventually tight evolution system of measures for $\{G(t,s)\}$. Then, for any $s \in I$ and $f \in C_b(\mathbb{R}^N)$, $G(t,s)f$ converges to $m_s(f)$ locally uniformly in $\mathbb{R}^N$. In particular, the previous is the unique eventually tight evolution system of measures for $\{G(t,s)\}$.

Proof Fix $f \in C_b(\mathbb{R}^N)$, $t - 1 > s \in I$ and $x \in \mathbb{R}^N$. The invariance property of $\{\mu_t : t \in I\}$ shows that $m_s(f) = m_t(G(t,s)f)$. Hence, using the gradient estimate (16.2.13) we get

$$\begin{aligned}|(G(t,s)f)(x) - m_s(f)| \leq &\int_{\mathbb{R}^N \setminus B_{e^{-\omega t/2}}} |(G(t,s)f)(x) - (G(t,s)f)(y)|\mu_t(dy)\\ &+ \int_{B_{e^{-\omega t/2}}} |(G(t,s)f)(x) - (G(t,s)f)(y)|\mu_t(dy)\\ \leq& 2\|f\|_\infty \mu_t(\mathbb{R}^N \setminus B_{e^{-\omega t/2}}) + \|\nabla_x G(t,s)f\|_\infty \int_{B_{e^{-\omega t/2}}} |x-y|\mu_t(dy)\\ \leq& 2\|f\|_\infty \mu_t(\mathbb{R}^N \setminus B_{e^{-\omega t/2}}) + \|\nabla f\|_\infty e^{-\omega s}(e^{\omega t}|x| + e^{\frac{1}{2}\omega t}), \quad (16.2.14)\end{aligned}$$

where we have used estimate (16.2.13). Since the family $\{\mu_t : t \in I\}$ is eventually tight, $\mu_t(\mathbb{R}^N \setminus B_{e^{-\omega t/2}})$ vanishes as t tends to $+\infty$. Hence, (16.2.14) shows that $G(t,s)f$ converges to $m_s(f)$ as t tends to $+\infty$, locally uniformly in $\mathbb{R}^N$.

Now, suppose that $\{\widetilde{\mu}_t : t \in I\}$ is another eventually tight evolution system of measures for $\{G(t,s)\}$ and denote by $\widetilde{m}_t(f)$ the average of $f \in C_b(\mathbb{R}^N)$ with respect to the measure $\widetilde{\mu}_t$ for any $t \in I$. By the first part of the proof, for any $s \in I$ and $f \in C_b(\mathbb{R}^N)$, $G(t,s)f$ converges both to $m_s(f)$ and to $\widetilde{m}_s(f)$ as t tends to $+\infty$. Hence, $m_s(f) = \widetilde{m}_s(f)$ for any $s \in I$ and $f \in C_b(\mathbb{R}^N)$. Since the characteristic functions of Borel sets in $\mathbb{R}^N$ are the almost everywhere limit of bounded sequences of functions in $C_b(\mathbb{R}^N)$, we conclude that $\mu_s = \widetilde{\mu}_s$ for any $s \in I$. □

Remark 16.2.12 (i) Condition (16.2.13) is satisfied, for instance, under Hypotheses 14.4.4 provided that the supremum in (14.4.7) is negative (see Theorem 14.4.5).

(ii) The convergence in Theorem 16.2.11 is not uniform in $\mathbb{R}^N$ in general, even in the autonomous case. For instance, let $\mathcal{A}$ be any Ornstein-Uhlenbeck operator $\mathcal{A}$, defined by $\mathcal{A}\psi = \frac{1}{2}\mathrm{Tr}(QD^2\psi) + \langle Bx, \nabla\psi\rangle$ for any smooth enough function $\psi : \mathbb{R}^N \to \mathbb{R}$, where Q is a symmetric and positive definite matrix and all the eigenvalues of B have negative real part. Then, by Proposition 10.3.1, the associated Ornstein-Uhlenbeck semigroup $\{T(t)\}$ has a unique invariant measure μ, which is the Gaussian measure with zero mean and covariance operator $Q_\infty = \int_0^{+\infty} e^{sB}Qe^{sB^*}ds$. Clearly, $G(t,s) = T(t-s)$ and $\mu_t = \mu$ for any $\mathbb{R} \ni s < t$. Let $f = e^{i\langle\cdot,h\rangle}$ for some $h \in \mathbb{R}$. Then, $T(t)f = \exp(-\frac{1}{2}\langle Q_t h, h\rangle + i\langle\cdot, e^{tB^*}h\rangle)$ for any $t \geq 0$, where $Q_t := \int_0^t e^{sB}Qe^{sB^*}ds$, and $\int_{\mathbb{R}^N} f d\mu = e^{-\langle Q_\infty h,h\rangle/2}$. Therefore,

$$\begin{aligned}T(t)f - \int_{\mathbb{R}^N} f d\mu =& \{\exp(-\tfrac{1}{2}\langle Q_t h, h\rangle) - \exp(-\tfrac{1}{2}\langle Q_\infty h, h\rangle)\}e^{i\langle\cdot, e^{tB^*}h\rangle}\\ &+ \exp(-\tfrac{1}{2}\langle Q_\infty h, h\rangle)(\exp(i\langle\cdot, e^{tB^*}h\rangle) - 1)\end{aligned}$$

for any $t > 0$. The sup norm of the first addendum in the right-hand side of the previous formula vanishes as t tends to $+\infty$ but the second one does not, since $\sup_{x\in\mathbb{R}^N} |\exp(i\langle x, e^{tB^*}h\rangle) - 1| = \sup_{r\in\mathbb{R}} |\exp(ir) - 1| = 2$ for any $t > 0$.

(iii) In Chapter 18 (see Theorem 18.1.6) we will prove the local uniform convergence of $G(t,s)f$ to $m_s(f)$ under weaker assumptions.

To conclude this section, we prove the following lemma which will play a crucial role in the next section and in the next chapters.

Lemma 16.2.13 *Let Hypotheses* 16.2.1 *hold and let* $\{\mu_t : t \in I\}$ *be an evolution system of measures for* $\{G(t,s)\}$. *Then, the following properties are satisfied.*

(i) *For any* $f \in C_b(\mathbb{R}^{N+1})$ *the function* $s \mapsto m_s(f)$ *is continuous in* I.

(ii) *If* $\{\mu_t : t \in I\}$ *is eventually tight, then, for any* $f \in C_b(\mathbb{R}^{N+1})$, *the function* $t \mapsto m_t(f(t,\cdot))$ *is continuous in* $[t_1, +\infty)$, *where* t_1 *is as in Definition* 16.2.10.

(iii) *If* $\{\mu_t : t \in I\}$ *is eventually tight, the diffusion coefficients belong to* $C^{0,1}([a,b] \times \mathbb{R}^N)$ *for some* $[a,b] \subset [t_1, +\infty)$, *where* t_1 *is as in Definition* 16.2.10, $f \in C_b^{1,2}([a,b] \times \mathbb{R}^N)$ *and* $f(t,\cdot)$ *is constant outside of a compact set* $K \subset \mathbb{R}^N$ *for any* $t \in [a,b]$, *then the function* $t \mapsto m_t(f(t,\cdot))$ *is continuously differentiable in* $[a,b]$ *and*

$$\frac{d}{dt} m_t(f(t,\cdot)) = \int_{\mathbb{R}^N} D_t f(t,\cdot) d\mu_t - \int_{\mathbb{R}^N} \mathcal{A}(t) f(t,\cdot) d\mu_t$$

for any $t \in [a,b]$.

Proof (i) Fix $f \in C_b(\mathbb{R}^N)$, $s_0 \in I$ and observe that the invariance property of $\{\mu_t : t \in I\}$ yields

$$m_s(f) - m_{s_0}(f) = \int_{\mathbb{R}^N} (G(s_0+1,s)f - G(s_0+1,s_0)f)\, d\mu_{s_0+1} \tag{16.2.15}$$

for any $s \in (s_0-1, s_0+1) \cap I$. By Theorem 14.1.12, $G(t,s)f$ converges to $G(t,s_0)f$ pointwise in $\mathbb{R}^N$ as s tends to s_0. Moreover, thanks to Proposition 14.1.3 it follows that $\|G(t,s)f - G(t,s_0)f\|_\infty \le 2\|f\|_\infty$ in $\mathbb{R}^N$. Hence, we can let s tend to s_0 in (16.2.15) and conclude that $m_s(f)$ converges to $m_{s_0}(f)$ by dominated convergence.

(ii) Fix $f \in C_b(\mathbb{R}^{N+1})$, $s_0 \ge t_1$ and observe that

$$|m_s(f(s,\cdot)) - m_{s_0}(f(s_0,\cdot))| \le |m_s(f(s,\cdot)) - m_s(f(s_0,\cdot))| + |m_s(f(s_0,\cdot)) - m_{s_0}(f(s_0,\cdot))|$$

for any $s \in I$. The second term in the right-hand side of the previous inequality vanishes as s tends to s_0 by the first part of the proof. As far as the other term is concerned, we estimate

$$\begin{aligned}|m_s(f(s,\cdot)) - m_s(f(s_0,\cdot))| \le& \int_{B_R} |f(s,\cdot) - f(s_0,\cdot)| d\mu_s + \int_{\mathbb{R}^N \setminus B_R} |f(s,\cdot) - u(f_0,\cdot)| d\mu_s \\ \le& \|f(s,\cdot) - f(s_0,\cdot)\|_{L^\infty(B_R)} + 2\mu_s(\mathbb{R}^N \setminus B_R)\|f\|_\infty\end{aligned}$$

for any $R > 0$. Hence,

$$\limsup_{s \to s_0} |m_s(f(s,\cdot)) - m_s(f(s_0,\cdot))| \le 2 \sup_{s \ge t_1} \mu_s(\mathbb{R}^N \setminus B_R)\|f\|_\infty$$

for any $R > 0$. Finally, letting R tend to $+\infty$, from the tightness of the family $\{\mu_t : t \ge t_1\}$ we conclude that $\limsup_{s \to s_0} |m_s(f(s,\cdot)) - m_s(f(s_0,\cdot))| = 0$ and the assertion follows.

(iii) We first assume that f is independent of t. Fix $t \ge t_1$ and $|h| \le 1$ such that $t+h \ge t_1$. By (14.1.7) (which is an equality since $c \equiv 0$) we can write

$$(G(t+1,t+h)f)(x) - (G(t+1,t)f)(x) = -\int_t^{t+h} (G(t+1,r)\mathcal{A}(r)f)(x) dr.$$

Integrating over $\mathbb{R}^N$ with respect to μ_{t+1} and using the invariance property of the family $\{\mu_t : t \in I\}$, we get

$$m_{t+h}(f) - m_t(f) = -\int_t^{t+h} m_r(\mathcal{A}(r)f) dr.$$

Since $\mathcal{A}f \in C_b(I \times \mathbb{R}^N)$, by property (ii) the function $r \mapsto m_r(\mathcal{A}(r)f)$ is continuous in I. Hence, the claim follows dividing both sides by h and letting h tend to 0.

Now, we consider the general case. We fix $t \in [a,b]$, $|h| \le 1$ and split

$$\frac{m_{t+h}(f(t+h,\cdot)) - m_t(f(t,\cdot))}{h} = \int_{\mathbb{R}^N} \frac{f(t+h,\cdot) - f(t,\cdot)}{h} d\mu_{t+h} + \frac{m_{t+h}(f(t,\cdot)) - m_t(f(t,\cdot))}{h}.$$

As h tends to 0^+, the second term of the previous formula converges to $-m_t(\mathcal{A}(t)f)$, whereas the first one converges to $m_t(D_t f)$ by dominated convergence, taking into account Proposition 16.2.5. □

16.3 The evolution operator in L^p-spaces related to evolution systems of measures: basic properties

Throughout this section $\{\mu_t : t \in I\}$ is an arbitrarily fixed evolution system of measures for $\{G(t,s)\}$.

Proposition 16.3.1 *For any $p \in [1,+\infty)$ and $t > s \in I$, the operator $G(t,s)$ can be extended with a bounded operator, which we still denote by $G(t,s)$, mapping $L^p(\mathbb{R}^N, \mu_s)$ into $L^p(\mathbb{R}^N, \mu_t)$. Moreover,*

$$\|G(t,s)f\|_{L^p(\mathbb{R}^N,\mu_t)} \le \|f\|_{L^p(\mathbb{R}^N,\mu_s)}, \qquad f \in L^p(\mathbb{R}^N, \mu_s),\ t > s \in I. \tag{16.3.1}$$

Proof From Proposition 14.1.3 and the Hölder inequality it follows easily that $|G(t,s)f|^p \le G(t,s)|f|^p$ for any $f \in C_b(\mathbb{R}^N)$ and $t > s \in I$. Integrating this inequality with respect to the measure μ_t and using the invariance property of the family $\{\mu_t : t \in I\}$, estimate (16.3.1) follows at once. The density of $C_b(\mathbb{R}^N)$ in $L^p(\mathbb{R}^N, \mu_s)$ allows us to extend each operator $G(t,s)$ to a bounded operator from $L^p(\mathbb{R}^N, \mu_s)$ to $L^p(\mathbb{R}^N, \mu_t)$ which satisfies estimate (16.3.1). □

The pointwise gradient estimates in Theorem 14.4.5 can be extended to the L^p-setting and show some smoothing effect of $\{G(t,s)\}$ in the L^p-spaces related to evolution systems of measures.

The following lemma will be crucial in the proof of Proposition 16.3.3 and in other parts of the chapter. It can be checked by arguing as in the proof of Lemma 9.5.3. For this purpose, we omit the details.

Lemma 16.3.2 *For any $s \in I$, $C_c^\infty(\mathbb{R}^N)$ is dense in $W^{1,p}(\mathbb{R}^N, \mu_s)$.*

Proposition 16.3.3 *Let Hypotheses 14.4.4(i) and 16.2.1 be satisfied. Then, the estimates*

$$\|\nabla_x G(t,s)f\|_{L^p(\mathbb{R}^N,\mu_t)} \le K_p\Big(1 \vee (t-s)^{-\frac{1}{2}}\Big) e^{\ell_p^-(t-s)} \|f\|_{L^p(\mathbb{R}^N,\mu_s)} \tag{16.3.2}$$

for any $f \in L^p(\mathbb{R}^N, \mu_s)$ and

$$\|\nabla_x G(t,s)f\|_{L^p(\mathbb{R}^N,\mu_t)} \le e^{\ell_p(t-s)} \|\nabla f\|_{L^p(\mathbb{R}^N,\mu_s)}, \qquad f \in W^{1,p}(\mathbb{R}^N, \mu_s), \tag{16.3.3}$$

hold true, for any $t > s$, where p is defined in Hypothesis 14.4.4(*i*), *ℓ_p is given by* (14.4.7) *and K_p is a positive constant.*

If Hypothesis 14.4.4(*i*) *is replaced by Hypothesis* 14.4.4(*ii*), *then the previous two estimates hold true for any $p \in (1, +\infty)$, with ℓ_p being given by* (14.4.9). *Finally, if the diffusion coefficients are independent of x and there exists a constant $r_0 \in \mathbb{R}$ such that $\langle \mathrm{Jac}_x b(t,x)\xi, \xi\rangle \le r_0|\xi|^2$ for any $(t,x) \in I \times \mathbb{R}^N$ and $\xi \in \mathbb{R}^N$, then the estimate*

$$\|\nabla_x G(t,s)f\|_{L^1(\mathbb{R}^N,\mu_t)} \le e^{r_0(t-s)}\|\nabla f\|_{L^1(\mathbb{R}^N,\mu_s)}, \qquad f \in W^{1,p}(\mathbb{R}^N,\mu_s), \tag{16.3.4}$$

is satisfied by any $t > s \in I$.

Proof Fix $t > s \in I$. Integrating (14.4.8) with respect to μ_t and using the invariance property of the family $\{\mu_t : t \in \mathbb{R}\}$, estimates (16.3.3) and (16.3.4) follow immediately for functions $f \in C^1_b(\mathbb{R}^N)$. To extend them to any $f \in W^{1,p}(\mathbb{R}^N,\mu_s)$ it suffices to use Lemma 16.3.2.

Concerning estimate (16.3.2), we integrate (14.4.10) and get

$$\|\nabla_x G(t,s)f\|_{L^p(\mathbb{R}^N,\mu_t)} \le C_{3,p}\left(1 \vee (t-s)^{-\frac{1}{2}}\right)e^{\ell_p(t-s)}\|f\|_{L^p(\mathbb{R}^N,\mu_s)} \tag{16.3.5}$$

for any $f \in C_b(\mathbb{R}^N)$. By density, we extend this estimate to any $f \in L^p(\mathbb{R}^N,\mu_s)$ and we get (16.3.2) if $\ell_p \le 0$. On the other hand, if $\ell_p > 0$ and $t > s+1$, then we split $G(t,s)f = G(t,t-1)G(t-1,s)f$ and use (16.3.1), (16.3.5) to estimate

$$\|\nabla_x G(t,x)f\|_{L^p(\mathbb{R}^N,\mu_t)} \le C_{3,p}e^{\ell_p}\|G(t-1,s)f\|_{L^p(\mathbb{R}^N,\mu_{t-1})} \le C_{3,p}e^{\ell_p}\|f\|_{L^p(\mathbb{R}^N,\mu_s)}. \tag{16.3.6}$$

From (16.3.6) and (16.3.5), we get (16.3.2) also in this case. □

16.4 Logarithmic Sobolev inequality, Poincaré inequality and hypercontractivity

In this section we prove the logarithmic Sobolev inequality and the Poincaré inequality, and we study the hypercontractivity of the evolution operator $\{G(t,s)\}$.

Definition 16.4.1 *The evolution operator $\{G(t,s)\}$ is called hypercontractive if, for any $p > 1$ and $t > s$, there exists $q(t,s) > p$ such that $G(t,s)$ maps $L^p(\mathbb{R}^N,\mu_s)$ into $L^{q(t)}(\mathbb{R}^N,\mu_t)$ and $\|G(t,s)\|_{L(L^p(\mathbb{R}^N,\mu_s),L^{q(t,s,p)}(\mathbb{R}^N,\mu_t))} \le 1$ for any $I \ni s < t$.*

Remark 16.4.2 Since each μ_t ($t \in I$) is a probability measure, if $\{G(t,s)\}$ is hypercontractive, then, for any $I \ni s < t$, $G(t,s)$ is a contraction from $L^p(\mathbb{R}^N,\mu_s)$ to $L^r(\mathbb{R}^N,\mu_t)$ for any $r \in (p, q(t,s,p))$.

Throughout this and the next sections, if not otherwise specified, we assume the following set of assumptions.

Hypotheses 16.4.3 (i) *Hypotheses* 16.2.1 *are satisfied with the diffusion coefficients q_{ij} ($i,j = 1,\dots,N$) being bounded in I and independent of x, and the condition $\mathcal{A}(t)\varphi \le a - d\varphi$ holds true in $\mathbb{R}^N$ for any $t \in I$;*

(ii) *the first-order spatial derivatives of* b_i *exist, belong to* $C^{\alpha/2,\alpha}_{\rm loc}(I\times\mathbb{R}^N)$ *for any* $i=1,\ldots,N$, *and* $\langle {\rm Jac}_x b(t,x)\xi,\xi\rangle\le r_0|\xi|^2$ *for any* $(t,x)\in I\times\mathbb{R}^N$, $\xi\in\mathbb{R}^N$ *and some negative constant* r_0.

Remark 16.4.4 The condition on the Jacobian matrix of b in Hypothesis 16.4.3(ii) is equivalent to the condition $\langle b(t,x)-b(t,y),x-y\rangle\le r_0|x-y|^2$ for any $t\in I$ and $x,y\in\mathbb{R}^N$. This implies that for any $[a,b]\subset I$ there exists a positive constant $C_{a,b}$ such that

$$\langle b(t,x),x\rangle\le C_{a,b}|x|,\qquad t\in[a,b],\;\; x\in\mathbb{R}^N. \tag{16.4.1}$$

It suffices to take $y=0$ and set $C_{a,b}=\|b(\cdot,0)\|_{L^\infty((a,b))}$.

Remark 16.4.5 Due to Hypotheses 16.4.3 and Theorems 16.2.7 and 16.2.11, there exists a unique tight evolution system of measures for $\{G(t,s)\}$.

In view of this remark, here and in the next sections, $\{\mu_t:t\in I\}$ will denote the unique tight evolution system of measure for $\{G(t,s)\}$.

As in the forthcoming sections, we set $0\log 0:=0$ and

$$\Lambda=\sup_{t\in I}\sup_{\xi\in\partial B_1}|\sqrt{Q}(t)\xi|^2. \tag{16.4.2}$$

16.4.1 Logarithmic Sobolev inequality and consequences

To begin with, we prove a logarithmic Sobolev inequality (LSI for short) for the measures $p(t,s,x;dy)$ defined in Proposition 14.1.3. The ideas of the proof come from [86, Theorem 6.2.42], which deals with the autonomous case.

Proposition 16.4.6 *For any* $p\in[1,+\infty)$, $t\ge s\in I$ *and* $f\in C^1_b(\mathbb{R}^N)$ *(such that* $|f|$ *has positive infimum, if* $p\in[1,2)$*), it holds that*

$$\begin{aligned}G(t,s)(|f|^p\log|f|^p)\le&\frac{p^2\Lambda}{|r_0|}(1-e^{2r_0(t-s)})G(t,s)(|f|^{p-2}|\nabla f|^2)\\&+(G(t,s)|f|^p)\log(G(t,s)|f|^p),\end{aligned} \tag{16.4.3}$$

where r_0 *and* Λ *are defined by Hypothesis* 16.4.3*(ii) and* (16.4.2).

Proof It suffices to prove (16.4.3) for $p=2$. Indeed, for any $p\in[1,+\infty)$ and $f\in C^1_b(\mathbb{R}^N)$ (such that $|f|$ has positive infimum, if $p<2$), the claim follows from applying estimate (16.4.3) with $p=2$ to the function $|f|^{p/2}$. By homogeneity we can also assume that $\|f\|_\infty\le 1$.

Fix $I\ni s<t$, $x\in\mathbb{R}^N$, a function $0\le f\in C^1_b(\mathbb{R}^N)$ such that $0<\|f\|_\infty\le 1$, and, for $n\in\mathbb{N}$, consider the function $F_n=\{G(t,\cdot)[\vartheta_n(G(\cdot,s)f)^2\log(G(\cdot,s)f)^2]\}(x)$ with $[s,t]$ as a domain, where $\vartheta_n(x)=\vartheta(|x|/n)$ for any $x\in\mathbb{R}^N$ and $\vartheta\in C^\infty(\mathbb{R})$ satisfies the condition $\chi_{(-\infty,1]}\le\vartheta\le\chi_{(-\infty,2]}$. Each function F_n is well defined. Indeed, the formula (14.1.4) shows that the evolution operator transforms nonnegative and not identically vanishing functions into positive functions. In particular, $(G(r,s)f)^2\log(G(r,s)f)^2$ belongs to $C_b(\mathbb{R}^N)$ for any $r\ge s$. From Proposition 14.1.4, we thus infer that F_n converges to $F=\{G(t,\cdot)[(G(\cdot,s)f)^2\log(G(\cdot,s)f)^2]\}(x)$ pointwise in $[s,t]$, as n tends to $+\infty$. Moreover, since $\vartheta_n(G(r,s)f)^2\log(G(r,s)f)^2\in C^2_c(B_{2n})$, by Lemma 14.1.6(i) each function F_n is differentiable in (s,t), with

$$F_n'=-2(G(t,\cdot)[\vartheta_n(3+\log(G(\cdot,s)f)^2)|\sqrt{Q}\nabla_x G(\cdot,s)f|^2])(x)$$

$$
\begin{aligned}
&- 4(G(t,\cdot)[(G(\cdot,s)f)(\log(G(\cdot,s)f)^2+1)\langle Q\nabla\vartheta_n, \nabla_x G(\cdot,s)f\rangle])(x)\\
&- (G(t,\cdot)[(G(\cdot,s)f)^2\log(G(\cdot,s)f)^2\mathcal{A}\vartheta_n])(x) =: I_n + J_n + H_n.
\end{aligned}
$$

The terms I_n, K_n tend, respectively, to $-2\{G(t,\cdot)[(3+\log(G(\cdot,s)f)^2)|\sqrt{Q}\nabla_x G(\cdot,s)f|^2]\}(x)$ and 0, as n tends to $+\infty$, pointwise in $[s,t]$. On the other hand, by (16.4.1), we can estimate $\mathcal{A}\vartheta_n \geq n^{-1}(C_{s,t}\|\vartheta'\|_\infty - \Lambda N\|\vartheta\|_{C^2_b(\mathbb{R})})$ in $[s,T]\times\mathbb{R}^N$ for any $n\in\mathbb{N}$. Thus, since $(G(\cdot,s)f)^2\log(G(\cdot,s)f)^2 \leq 0$ (recall that $\|f\|_\infty \leq 1$), it follows that $\liminf_{n\to+\infty} J_n \geq 0$ in $[s,t]$. We conclude that

$$
\begin{aligned}
\liminf_{n\to+\infty} F_n' \geq& -2\{G(t,\cdot)[(3+\log(G(\cdot,s)f)^2)|\sqrt{Q}\nabla_x G(\cdot,s)f|^2]\}(x)\\
\geq& -6\Lambda(G(t,\cdot)|\nabla_x G(\cdot,s)f|^2)(x),
\end{aligned}
$$

pointwise in $[s,t]$. Since the sequence $\{F_n'\}$ is bounded from below thanks to estimate (14.4.8), using first Fatou's lemma, and then again (14.4.8) with $p=1$, we get

$$
F(t)-F(s) \geq \int_s^t \liminf_{n\to+\infty} F_n'(\sigma)\,d\sigma \geq 3|r_0|^{-1}(e^{2r_0(t-s)}-1)(G(t,s)|\nabla f|^2)(x)
$$

and (16.4.3) follows. □

As a byproduct of Proposition 16.4.6, we can now prove a logarithmic Sobolev inequality, in short LSI.

Theorem 16.4.7 *For any $f\in C^1_b(\mathbb{R}^N)$, $p\in(1,+\infty)$ and $s\in I$, it holds that*

$$
\int_{\mathbb{R}^N}|f|^p\log(|f|)d\mu_s \leq \|f\|^p_{L^p(\mathbb{R}^N,\mu_s)}\log(\|f\|_{L^p(\mathbb{R}^N,\mu_s)}) + \frac{p\Lambda}{2|r_0|}\int_{\mathbb{R}^N}|f|^{p-2}|\nabla f|^2\chi_{\{f\neq 0\}}d\mu_s. \tag{16.4.4}
$$

Formula (16.4.4) *can be extended to any $f\in W^{1,p}(\mathbb{R}^N,\mu_s)$ and $p\geq 2$.*

Proof We split the proof into two steps.

Step 1. Fix $t>s\in I$ and $f\in C^1_b(\mathbb{R}^N)$ with positive infimum. Integrating estimate (16.4.3) with respect to μ_t and taking the invariance property of $\{\mu_t : t\in I\}$ into account, we obtain

$$
\begin{aligned}
\int_{\mathbb{R}^N}|f|^p\log(|f|^p)d\mu_s \leq& \frac{p^2\Lambda}{|r_0|}(1-e^{2r_0(t-s)})\int_{\mathbb{R}^N}|f|^{p-2}|\nabla f|^2 d\mu_s\\
&+\int_{\mathbb{R}^N}(G(t,s)|f|^p)\log(G(t,s)|f|^p)d\mu_s.
\end{aligned} \tag{16.4.5}
$$

Now, we observe that the second integral term in the right-hand side of (16.4.5) converges to $m_s(f^p)\log(m_s(f^p))$ as t tends to $+\infty$. Indeed, since the function $y\mapsto y\log y$ is 1/2-Hölder continuous in bounded sets of $(0,+\infty)$, we can estimate

$$
\begin{aligned}
&\left|\int_{\mathbb{R}^N}(G(t,s)f^p)\log(G(t,s)f^p)d\mu_t - m_s(f^p)\log(m_s(f^p))\right|\\
=&\left|\int_{\mathbb{R}^N}[(G(t,s)f^p)\log(G(t,s)f^p) - m_s(f^p)\log(m_s(f^p))]d\mu_t\right|\\
\leq& C\int_{\mathbb{R}^N}|G(t,s)f^p - m_s(f^p)|^{\frac12}d\mu_t \leq C\|G(t,s)f^p - m_s(f^p)\|^{\frac12}_{L^1(\mathbb{R}^N,\mu_t)}
\end{aligned}
$$

for some positive constant C, and the last side of the previous chain of inequalities vanishes as t tends to $+\infty$, by dominated convergence taking Theorem 16.2.11 into account. Now, since $r_0 < 0$, letting t tend to $+\infty$ in (16.4.5), estimate (16.4.4) follows at once.

To get (16.4.4) in the general case, it suffices to write it with f being replaced by the function $f_n = (f^2 + n^{-1})^{1/2}$ and let n tend to $+\infty$. Indeed, since $0 < f_n^p \le \|f^2 + 1\|_\infty^{p/2}$, $\||f_n|^p \log(|f_n|)\|_{L^1(\mathbb{R}^N,\mu_s)}$ and $m_s(f_n^p)\log(m_s(f_n^p))$ converge to $\||f|^p \log(|f|)\|_{L^1(\mathbb{R}^N,\mu_s)}$ and $m_s(|f|^p)\log(m_s(|f|^p))$, respectively, as n tends to $+\infty$. Moreover, observing that $|\nabla f_n|^2 \le |\nabla f|^2$ for any $n \in \mathbb{N}$, we can infer that

$$\lim_{n\to+\infty} \int_{\mathbb{R}^N} f_n^{p-2}|\nabla f_n|^2 d\mu_s = \int_{\mathbb{R}^N} |f|^{p-2}|\nabla f|^2 \chi_{\{f\neq 0\}} d\mu_s,$$

by monotone convergence, if $p < 2$, and by dominated convergence otherwise.

Step 2. Fix a function $f \in W^{1,p}(\mathbb{R}^N, \mu_s)$, $p \ge 2$ and let $\{f_n\} \in C_c^\infty(\mathbb{R}^N)$ be a sequence of smooth functions converging to f in $W^{1,p}(\mathbb{R}^N, \mu_s)$ as n tends to $+\infty$ (see Lemma 16.3.2). Writing (16.4.4) with f being replaced by f_n yields

$$\begin{aligned}\int_{\mathbb{R}^N} |f_n|^p \log_+(|f_n|) d\mu_s \le& \|f_n\|_{L^p(\mathbb{R}^N,\mu_s)}^p \log(\|f_n\|_{L^p(\mathbb{R}^N,\mu_s)})\\ &+ \frac{p\Lambda}{2|r_0|}\int_{\mathbb{R}^N} |f_n|^{p-2}|\nabla f_n|^2 d\mu_s + \int_{\mathbb{R}^N} |f_n|^p \log_-(|f_n|) d\mu_s,\end{aligned}$$

where $\log_+ t = 0 \vee \log(t)$, $\log_-(t) = 0 \vee (-\log(t))$ for any $t > 0$. Without loss of generality, we can also assume that f_n converges to f pointwise μ_s-almost everywhere in $\mathbb{R}^N$. Therefore, Fatou's lemma implies that

$$\begin{aligned}\int_{\mathbb{R}^N} |f|^p \log_+(|f|) d\mu_s \le& \|f\|_{L^p(\mathbb{R}^N,\mu_s)}^p \log(\|f\|_{L^p(\mathbb{R}^N,\mu_s)}) + \frac{p\Lambda}{2|r_0|}\liminf_{n\to+\infty}\int_{\mathbb{R}^N} |f_n|^{p-2}|\nabla f_n|^2 d\mu_s\\ &+ \liminf_{n\to+\infty}\int_{\mathbb{R}^N} |f_n|^p \log_-(|f_n|) d\mu_s.\end{aligned}$$

So, to conclude the proof, we just need to show that

$$\liminf_{n\to+\infty}\int_{\mathbb{R}^N} |f_n|^{p-2}|\nabla f_n|^2 d\mu_s = \int_{\mathbb{R}^N} |f|^{p-2}|\nabla f|^2 d\mu_s \tag{16.4.6}$$

and

$$\liminf_{n\to+\infty}\int_{\mathbb{R}^N} |f_n|^p \log_-(|f_n|) d\mu_s = \int_{\mathbb{R}^N} |f|^p \log_-(|f|) d\mu_s. \tag{16.4.7}$$

For this purpose, we observe that

$$\begin{aligned}&\left|\int_{\mathbb{R}^N} |f_n|^{p-2}|\nabla f_n|^2 d\mu_s - \int_{\mathbb{R}^N} |f|^{p-2}|\nabla f|^2 d\mu_s\right|\\ &\le \int_{\mathbb{R}^N} |f_n|^{p-2}||\nabla f_n|^2 - |\nabla f|^2| d\mu_s + \int_{\mathbb{R}^N} ||f_n|^{p-2} - |f|^{p-2}||\nabla f|^2 d\mu_s =: I_n + J_n.\end{aligned}$$

The Hölder inequality implies that

$$\begin{aligned}I_n \le& \|f_n\|_{L^p(\mathbb{R}^N,\mu_s)}^{p-2} \||\nabla f_n|^2 - |\nabla f|^2\|_{L^{p/2}(\mathbb{R}^N,\mu_s)}\\ \le& \|f_n\|_{L^p(\mathbb{R}^N,\mu_s)}^{p-2}(\|\nabla f_n\|_{L^p(\mathbb{R}^N,\mu_s)} + \|\nabla f\|_{L^p(\mathbb{R}^N,\mu_s)})\|\nabla f_n - \nabla f\|_{L^p(\mathbb{R}^N,\mu_s)},\end{aligned}$$

so that it vanishes as n tends to $+\infty$. As far as I_2 is concerned, we observe that, if $p \in [2,3]$, then $||f_n|^{p-2} - |f|^{p-2}| \le |f_n - f|^{p-2}$ for any $n \in \mathbb{N}$. Therefore, the Hölder inequality yields

$$J_n \le \int_{\mathbb{R}^N} |f_n - f|^{p-2}|\nabla f|^2 d\mu_s \le \|f_n - f\|_{L^p(\mathbb{R}^N,\mu_s)}^{p-2}\|\nabla f\|_{L^p(\mathbb{R}^N,\mu_s)}^2. \tag{16.4.8}$$

On the other hand, if $p>3$, then we can estimate

$$\begin{aligned}||f_n|^{p-2}-|f|^{p-2}| \le&(p-2)||f_n|^{p-3}+|f|^{p-3}||f_n-f|\\ \le&(p-2)(2^{4-p}\vee 1)(|f_n|+|f|)^{p-3}|f_n-f|\end{aligned}$$

for any $n\in\mathbb{N}$. Therefore,

$$\begin{aligned}J_n \le& C_p\int_{\mathbb{R}^N}(|f_n|+|f|)^{p-3}|f_n-f||\nabla f|^2 d\mu_s\\ \le& C_p\bigg(\int_{\mathbb{R}^N}|f_n-f|^{\frac{p}{p-2}}(|f_n|+|f|)^{\frac{p(p-3)}{p-2}}d\mu_s\bigg)^{\frac{p-2}{p}}\|\nabla f\|^2_{L^p(\mathbb{R}^N,\mu_s)}\\ \le& C_p\|f_n-f\|_{L^p(\mathbb{R}^N,\mu_s)}(\|f_n\|_{L^p(\mathbb{R}^N,\mu_s)}+\|f\|_{L^p(\mathbb{R}^N,\mu_s)})^{\frac{p-3}{p}}\|\nabla f\|^2_{L^p(\mathbb{R}^N,\mu_s)} \qquad (16.4.9)\end{aligned}$$

for any $n\in\mathbb{N}$, where $C_p=(2^{4-p}\vee 1)(p-2)$. From (16.4.8) and (16.4.9) we deduce that, in both cases $p\in[2,3]$ and $p>3$, J_n vanishes as n tends to $+\infty$. Summing up, (16.4.6) follows.

Finally, to prove (16.4.7) it suffices to observe that the function $t\mapsto t^p\log_-(t)$ is Lipschitz continuous in $(0,+\infty)$. Hence,

$$\begin{aligned}\bigg|\int_{\mathbb{R}^N}|f_n|^p\log_-(|f_n|)d\mu_s-\int_{\mathbb{R}^N}|f|^p\log_-(|f|)d\mu_s\bigg| \le& K_p\int_{\mathbb{R}^N}|f_n-f|d\mu_s\\ \le& K_p\int_{\mathbb{R}^N}|f_n-f|^p d\mu_s\end{aligned}$$

for some positive constant K_p and any $t>0$, and (16.4.7) follows. □

The logarithmic Sobolev inequality (16.4.4) yields some compactness results.

Theorem 16.4.8 *Fix $s\in I$. Then:*

(i) *$W^{1,p}(\mathbb{R}^N,\mu_s)$ is compactly embedded in $L^p(\mathbb{R}^N,\mu_s)$ for any $p\in[2,+\infty)$;*

(ii) *for any $t>s$ and $p\in(1,+\infty)$, the operator $G(t,s):L^p(\mathbb{R}^N,\mu_s)\to L^p(\mathbb{R}^N,\mu_t)$ is compact.*

Proof (i) We prove that the unit ball $\mathcal{B}_1\subset W^{1,p}(\mathbb{R}^N,\mu_s)$ centered at zero, is totally bounded in $L^p(\mathbb{R}^N,\mu_s)$. The main step of the proof consists of proving that, for any $\varepsilon>0$, there exists $R>0$ such that $\|f\|_{L^p(\mathbb{R}^N\setminus B_R,\mu_s)}\le\varepsilon$ for any $f\in\mathcal{B}_1$. Once this property is checked, we can apply the same arguments as in the proof of Proposition 9.5.8, recalling that, by Proposition 16.2.5, $L^p(B_R,\mu_s)=L^p(B_R)$, with equivalence of the corresponding norms, for any $R>0$.

For any $f\in\mathcal{B}_1$ and $k\in\mathbb{N}$ we set $F_k=\{x\in\mathbb{R}^N:|f(x)|\le k\}$ and observe that

$$\begin{aligned}\int_{\mathbb{R}^N\setminus B_R}|f|^p d\mu_s=&\int_{F_k\cap(\mathbb{R}^N\setminus B_R)}|f|^p d\mu_s+\int_{\mathbb{R}^N\setminus(B_R\cup F_k)}|f|^p d\mu_s\\ \le& k^p\mu_s(\mathbb{R}^N\setminus B_R)+\frac{1}{\log(k)}\int_{\mathbb{R}^N}|f|^p\log(|f|)d\mu_s.\end{aligned}$$

By the logarithmic Sobolev inequality (16.4.4) (which can be extended to any $g\in W^{1,p}(\mathbb{R}^N,\mu_s)$ by Lemma 16.3.2 since $p\ge 2$) and the Hölder inequality we obtain

$$\int_{\mathbb{R}^N}|f|^p\log(|f|)d\mu_s\le\frac{p\Lambda}{2|r_0|}\|\nabla f\|^2_{L^p(\mathbb{R}^N,\mu_s)}\|f\|^{p-2}_{L^p(\mathbb{R}^N,\mu_s)}\le\frac{p\Lambda}{2|r_0|},$$

since $\|f\|_{W^{1,p}(\mathbb{R}^N,\mu_s)}\le 1$. Therefore,

$$\int_{\mathbb{R}^N\setminus B_R}|f|^p d\mu_s\le k^p\mu_s(\mathbb{R}^N\setminus B_R)+\frac{p\Lambda}{2|r_0|\log(k)},\qquad f\in\mathcal{B}_1.$$

The claim follows choosing properly k and R.

(ii) Since each operator $G(t,s)$ is bounded from $L^p(\mathbb{R}^N,\mu_s)$ to $L^p(\mathbb{R}^N,\mu_t)$ by Proposition 16.3.1, the assertion follows immediately from (i) in the case $p\ge 2$. The proof in the case when $p\in(1,2)$ follows by interpolation since $G(t,s)$ is bounded from $L^1(\mathbb{R}^N,\mu_s)$ to $L^1(\mathbb{R}^N,\mu_t)$ and compact from $L^2(\mathbb{R}^N,\mu_s)$ to $L^2(\mathbb{R}^N,\mu_t)$. □

The logarithmic Sobolev inequality also yields the Poincaré inequality.

Theorem 16.4.9 *For any $p\ge 2$ there exists a positive constant C_p such that*

$$\|f-m_s(f)\|_{L^p(\mathbb{R}^N,\mu_s)}\le C_p\|\nabla f\|_{L^p(\mathbb{R}^N,\mu_s)},\qquad f\in W^{1,p}(\mathbb{R}^N,\mu_s),\ \ s\in I.\tag{16.4.10}$$

In particular, if $p=2$, then we can take $C_2=\Lambda^{1/2}|r_0|^{-1/2}$, where Λ and r_0 are given by (16.4.2) *and Hypothesis* 16.4.3(*ii*).

Proof We split the proof into two steps. First, we prove (16.4.10) for $p=2$ using the arguments in [230]. Then, we use a bootstrap argument to extend it to any $p\ge 2$. Throughout the proof, $s\in I$ is arbitrarily fixed.

Step 1. Fix $f\in C^1_b(\mathbb{R}^N)$ and for any $\varepsilon>0$ set $f_\varepsilon:=1+\varepsilon(f-m_s(f))$. The logarithmic Sobolev inequality (16.4.4) yields

$$\int_{\mathbb{R}^N}f_\varepsilon^2\log(f_\varepsilon)d\mu_s-\|f_\varepsilon\|^2_{L^2(\mathbb{R}^N,\mu_s)}\log(\|f_\varepsilon\|_{L^2(\mathbb{R}^N,\mu_s)})\le\varepsilon^2\frac{\Lambda}{|r_0|}\int_{\mathbb{R}^N}|\nabla f|^2 d\mu_s.\tag{16.4.11}$$

As ε tends to 0^+ the left-hand side of (16.4.11) can be written as $2\varepsilon^2\|f-m_s(f)\|^2_{L^2(\mathbb{R}^N,\mu_s)}+o(\varepsilon^2)$, since $m_s(f)=0$. Hence, dividing both sides of (16.4.11) by ε^2 and letting ε tend to 0^+, estimate (16.4.10) follows for any $f\in C^1_b(\mathbb{R}^N)$. Since $C^1_b(\mathbb{R}^N)$ is dense in $W^{1,2}(\mathbb{R}^N,\mu_s)$ (see Proposition 16.2.5 and Lemma 16.3.2), the proof of (16.4.10), with $p=2$, is complete.

Step 2. Now, we extend the validity of (16.4.10) to any $p\in(2,+\infty)$. To begin with, we observe that if $g\in W^{1,p}(\mathbb{R}^N,\mu_s)$, then the function $|g|^{p/2}$ belongs to $W^{1,2}(\mathbb{R}^N,\mu_s)$. This property can be proved by approximating g by a sequence $\{g_n\}\subset C^\infty_c(\mathbb{R}^N)$, which converges to g in $W^{1,p}(\mathbb{R}^N,\mu_s)$. Hence, applying (16.4.10), with $p=2$, to the function $|g|^{p/2}$ and using the Hölder inequality we obtain

$$\|g\|^p_{L^p(\mathbb{R}^N,\mu_s)}\le\frac{p^2\Lambda}{4|r_0|}\|g\|^{p-2}_{L^p(\mathbb{R}^N,\mu_s)}\|\nabla g\|^2_{L^p(\mathbb{R}^N,\mu_s)}+\|g\|^p_{L^{p/2}(\mathbb{R}^N,\mu_s)}.\tag{16.4.12}$$

The Young inequality $a^{1-2/p}b^{2/p}\le\varepsilon a+C_{\varepsilon,p}b$ with $\varepsilon=2|r_0|p^{-2}\Lambda^{-1}$, $a=\|g\|^p_{L^p(\mathbb{R}^N,\mu)}$ and $b=\|\nabla g\|^p_{L^p(\mathbb{R}^N,\mu)}$ shows that

$$\|g\|^{p-2}_{L^p(\mathbb{R}^N,\mu_s)}\|\nabla g\|^2_{L^p(\mathbb{R}^N,\mu_s)}\le 2|r_0|p^{-2}\Lambda^{-1}\|g\|^p_{L^p(\mathbb{R}^N,\mu_s)}+K_p\|\nabla g\|^p_{L^p(\mathbb{R}^N,\mu_s)}$$

for some positive constant K_p, which replaced in (16.4.12) leads to the crucial estimate

$$\|g\|^p_{L^p(\mathbb{R}^N,\mu_s)}\le\widetilde{K}_p\|\nabla g\|^p_{L^p(\mathbb{R}^N,\mu_s)}+2\|g\|^p_{L^{p/2}(\mathbb{R}^N,\mu_s)},\qquad g\in W^{1,p}(\mathbb{R}^N,\mu_s),\tag{16.4.13}$$

for some positive constant $\widetilde{K}_p$, independent of g. Using (16.4.13) we immediately extend the validity of (16.4.10) to any $f\in W^{1,p}(\mathbb{R}^N,\mu_s)$ when $p\in(2,4]$. It suffices to take into account that $p/2 \le 2$ and $\|h\|_{L^q(\mathbb{R}^N,\mu_s)} \le \|h\|_{L^r(\mathbb{R}^N,\mu_s)}$ if $h \in L^r(\mathbb{R}^N,\mu_s)$ and $q < r$, and apply (16.4.13) with $g = f - m_s(f)$ to deduce that

$$\begin{aligned}\|f - m_s(f)\|^p_{L^p(\mathbb{R}^N,\mu_s)} \le& \widetilde{K}_p\|\nabla f\|^p_{L^p(\mathbb{R}^N,\mu_s)} + 2\|f - m_s(f)\|^p_{L^2(\mathbb{R}^N,\mu_s)}\\ \le& \widetilde{K}_p\|\nabla f\|^p_{L^p(\mathbb{R}^N,\mu_s)} + 2\Lambda^{p/2}|r_0|^{-p/2}\|\nabla f\|_{L^2(\mathbb{R}^N,\mu_s)}\\ \le&(\widetilde{K}_p + 2\Lambda^{p/2}|r_0|^{-p/2})\|\nabla f\|_{L^p(\mathbb{R}^N,\mu_s)}.\end{aligned}$$

To get estimate (16.4.10) for $p \in (4, 8]$, we argue similarly, starting from (16.4.13) and using (16.4.10) with $p/2$, which belongs to $(2, 4]$. Iterating this procedure, we get (16.4.10) for any $p > 2$. □

Remark 16.4.10 In Chapter 18 we will prove the Poincaré inequality in some cases when the diffusion coefficients also depend on the spatial variables. See Remark 18.1.15.

16.4.2 Hypercontractivity

Now, using the LSI, we can prove that the evolution operator $\{G(t, s)\}$ is hypercontractive.

Theorem 16.4.11 *The evolution operator $\{G(t, s)\}$ is hypercontractive, with $q(t, s, p) = e^{2\kappa_0|r_0|\Lambda^{-1}(t-s)}(p-1)+1$, where κ_0 is the infimum of the function κ in Hypothesis* 14.0.1*(ii) and r_0 is defined in Hypothesis* 16.4.3*(ii).*

Proof It suffices to prove that

$$\|G(t, s)f\|_{L^{q(t,s,p)}(\mathbb{R}^N,\mu_t)} \le \|f\|_{L^p(\mathbb{R}^N,\mu_s)}, \qquad t > s \in I, \tag{16.4.14}$$

for positive functions $f \in C^\infty_b(\mathbb{R}^N)$, which are constant outside of a compact set. Indeed, for a general function $f \in C^\infty_c(\mathbb{R}^N)$ we write this estimate, with f being replaced by $f_n = \sqrt{f^2 + n^{-1}}$, and use Proposition 14.1.4 to let n tend to $+\infty$. This gives (16.4.14) with $|f|$ instead of f. Since $|G(t, s)f| \le G(t, s)|f|$ (see Proposition 14.1.3), we conclude that f itself satisfies estimate (16.4.14). Finally, we can conclude the proof using the density of $C^\infty_c(\mathbb{R}^N)$ in $L^p(\mathbb{R}^N, \mu_s)$ for any $s \in I$.

So, let us fix $p > 1$, $s \in I$ and a function $0 < f \in C^\infty_b(\mathbb{R}^N)$, constant outside of a compact subset of $\mathbb{R}^N$. Then, $G(t, s)f \ge \inf_{\mathbb{R}^N} f > 0$ in $\mathbb{R}^N$ for any $t > s$. Estimate (16.4.14) follows if we prove that the function $t \mapsto \beta(t) = \|G(t, s)f\|_{L^{q(t,s,p)}(\mathbb{R}^N,\mu_t)}$ is decreasing in $[s, +\infty)$. Since this function might not be differentiable, we use an approximation argument. To simplify the notation, in the rest of the proof we simply write $q(t)$ instead of $q(t, s, p)$.

For any $n \in \mathbb{N}$ and $t \ge s$, let $\beta_n(t) = \|\vartheta_n^{1/q(t)}G(t, s)f\|_{L^{q(t)}(\mathbb{R}^N,\mu_t)}$, where $\vartheta_n(x) = \vartheta(|x|/n)$ for any $x \in \mathbb{R}^N$ and $\vartheta \in C^\infty(\mathbb{R})$ satisfies the condition $\chi_{(-\infty,1]} \le \vartheta \le \chi_{(-\infty,2]}$. By dominated convergence, β_n converges pointwise in $[s, +\infty)$ to β as n tends to $+\infty$. Moreover, by Lemma 16.2.13(iii) each function β_n is differentiable in $[s, +\infty)$.

Fix $T > s$. Clearly, $\|\nabla_x\vartheta_n\|_\infty \le C_1n^{-1}$ and, by Remark 16.4.4, it follows that $\mathcal{A}\vartheta_n \ge -C_2n^{-1}$ in $[s, T] \times \mathbb{R}^N$, where C_1 and C_2 are positive constants independent of n. Hence, a straightforward computation shows that $\beta_n'(t) \le \gamma_n(t)$ for any $t \in [s, T]$, where

$$\begin{aligned}\gamma_n(t) = \beta_n(t)\bigg\{& -\frac{q'(t)}{(q(t))^2}\log\left(\int_{\mathbb{R}^N}\vartheta_n(G(t, s)f)^{q(t)}d\mu_t\right)\\ &+\frac{q'(t)}{q(t)(\beta_n(t))^{q(t)}}\int_{\mathbb{R}^N}\vartheta_n(G(t, s)f)^{q(t)}\log(G(t, s)f)d\mu_t\end{aligned}$$

$$
\begin{aligned}
&-\frac{q(t)-1}{(\beta_n(t))^{q(t)}}\int_{\mathbb{R}^N}\vartheta_n(G(t,s)f)^{q(t)-2}|\sqrt{Q(t)}\nabla_x G(t,s)f|^2 d\mu_t\\
&+\frac{C_1}{n(\beta_n(t))^{q(t)}}\int_{\mathbb{R}^N}(G(t,s)f)^{q(t)-1}|Q(t)\nabla_x G(t,s)f|d\mu_t\\
&+\frac{C_2}{n^2 q(t)(\beta_n(t))^{q(t)}}\int_{\mathbb{R}^N}(G(t,s)f(x))^{q(t)}d\mu_t\bigg\}. \qquad (16.4.15)
\end{aligned}
$$

It follows that

$$\beta_n(t_2)-\beta_n(t_1)\le\int_{t_1}^{t_2}\gamma_n(s)ds, \qquad s\le t_1<t_2\le T. \qquad (16.4.16)$$

Since $q(t)\in[1,e^{2\kappa_0|r_0|\Lambda^{-1}(T-s)}(p-1)+1]$ for any $t\in[s,T]$ and $G(t,s)f$ has positive infimum in $\mathbb{R}^N$ for any $t>s$, taking Theorem 14.4.2 into account, which shows that the function $\nabla_x G(\cdot,s)f$ is bounded in $[s,T]\times\mathbb{R}^N$, it is easy to check that the sup norm of γ_n in $[s,t]$ can be bounded from above by a constant independent of n and that γ_n converges pointwise in $[s,T]$ to the function γ defined by the first three lines of (16.4.15), with ϑ_n and β_n being replaced, respectively, by 1 and β. This function is nonpositive in $(s,+\infty)$ as can be checked by applying the logarithmic Sobolev inequality (16.4.4) to $G(t,s)f$ and, then, estimating $|\sqrt{Q}\nabla_x G(\cdot,s)f|^2\ge\kappa_0|\nabla_x G(\cdot,s)f|^2$ in $[s,+\infty)\times\mathbb{R}^N$. Hence, we can let n tend to $+\infty$ in (16.4.16), by the dominated convergence theorem, and conclude that $\beta(t_2)-\beta(t_1)\le 0$. □

16.5 Supercontractivity and LSI$_\varepsilon$

In this section we study the supercontractivity of the evolution operator $\{G(t,s)\}$.

Definition 16.5.1 *The evolution operator $\{G(t,s)\}$ is called supercontractive if each operator $G(t,s)$ maps $L^p(\mathbb{R}^N,\mu_s)$ into $L^q(\mathbb{R}^N,\mu_t)$, for any $1<p<q<+\infty$, and there exists a positive decreasing function $C_{p,q}:(0,+\infty)\to(0,+\infty)$ such that $\lim_{r\to 0^+}C_{p,q}(r)=+\infty$ and $\|G(t,s)\|_{L(L^p(\mathbb{R}^N,\mu_s),L^q(\mathbb{R}^N,\mu_t))}\le C_{p,q}(t-s)$ for any $I\ni s<t$.*

As the following theorem shows, the supercontractivity of the evolution operator $\{G(t,s)\}$ can be characterized in terms of the validity of a family of logarithmic Sobolev inequalities (which we will denote by LSI$_\varepsilon$).

Theorem 16.5.2 *The evolution operator $\{G(t,s)\}$ is supercontractive if and only if the family of logarithmic Sobolev inequalities*

$$\int_{\mathbb{R}^N}f^2\log\left(\frac{|f|}{\|f\|_{L^2(\mathbb{R}^N,\mu_s)}}\right)d\mu_s\le\varepsilon\|\nabla f\|^2_{L^2(\mathbb{R}^N,\mu_s)}+\beta(\varepsilon)\|f\|^2_{L^2(\mathbb{R}^N,\mu_s)} \qquad (16.5.1)$$

holds for any $f\in W^{1,2}(\mathbb{R}^N,\mu_s)$, $s\in I$, $\varepsilon>0$ and some decreasing function $\beta:(0,+\infty)\to\mathbb{R}$, blowing up as ε tends to 0^+.

The proof of Theorem 16.5.2 is based on two propositions. The first one shows that the boundedness of $G(t,s)$ from $L^p(\mathbb{R}^N,\mu_s)$ into $L^q(\mathbb{R}^N,\mu_t)$, for any $t>s$, yields a family of logarithmic Sobolev inequalities satisfied by the evolution system of measures $\{\mu_t:t\in I\}$. The key tools used in the proof are estimate (16.4.3) and the Riesz-Thorin interpolation theorem.

Proposition 16.5.3 *Assume that the evolution operator $\{G(t,s)\}$ is supercontractive. Then,*

$$\int_{\mathbb{R}^N} f^2 \log\left(\frac{|f|}{\|f\|_{L^2(\mathbb{R}^N,\mu_s)}}\right) d\mu_s \leq \frac{2\Lambda p(q-1)}{|r_0|(q-p)}(1-e^{2r_0(t-s)})\|\nabla f\|^2_{L^2(\mathbb{R}^N,\mu_s)} + \frac{pq}{2(q-p)}\log(C_{p,q}(t-s))\|f\|^2_{L^2(\mathbb{R}^N,\mu_s)} \qquad (16.5.2)$$

for any $t > s \in I$, $1 < p < q < +\infty$ and $f \in W^{1,2}(\mathbb{R}^N,\mu_s)$, where r_0 is the constant in Hypothesis 16.4.3*(ii)*.

Proof We split the proof into two steps. In the first one we show that it suffices to prove (16.5.2) for functions $f \in C^1_c(\mathbb{R}^N)$. In the second step, we prove estimate (16.5.2) for such functions.

Step 1. For notational convenience we denote by $M_{p,q}(t-s)$ and $N_{p,q}(t-s)$ the constants appearing in the right-hand side of (16.5.2) in front of $\|\nabla f\|^2_{L^2(\mathbb{R}^N,\mu_s)}$ and $\|f\|^2_{L^2(\mathbb{R}^N,\mu_s)}$, respectively.

Suppose that inequality (16.5.2) is satisfied by any function $f \in C^1_c(\mathbb{R}^N)$ and fix $f \in W^{1,2}(\mathbb{R}^N,\mu_s)$. Let $\{f_n\} \subset C^1_c(\mathbb{R}^N)$ be a sequence converging to f in $W^{1,2}(\mathbb{R}^N,\mu_s)$ and almost everywhere. Then,

$$\int_{\mathbb{R}^N} f_n^2 \log_+(|f_n|)d\mu_s \leq M_{p,q}(t-s)\int_{\mathbb{R}^N} |\nabla f_n|^2 d\mu_s + N_{p,q}(t,s)\|f_n\|^2_{L^2(\mathbb{R}^N,\mu_s)} + \int_{\mathbb{R}^N} f_n^2|\log(|f_n|)|\chi_{\{|f_n|\leq 1\}}d\mu_s + \|f_n\|^2_{L^2(\mathbb{R}^N,\mu_s)}\log(\|f_n\|_{L^2(\mathbb{R}^N,\mu_s)}) \qquad (16.5.3)$$

for any $n \in \mathbb{N}$, where $\log_+(r) = 0 \vee \log(r)$ for any $r > 0$. As n tends to $+\infty$, the terms in the right-hand side of (16.5.3) converge to the corresponding integral terms, with f_n being replaced by f, and they are both finite. Applying Fatou's lemma, we conclude that the function $f^2 \log_+ |f|$ also belongs to $L^1(\mathbb{R}^N,\mu_s)$ and

$$\int_{\mathbb{R}^N} f^2 \log_+ |f|\, d\mu_s \leq M_{p,q}(t-s)\int_{\mathbb{R}^N} |\nabla f|^2\, d\mu_s + N_{p,q}(t-s)\|f\|^2_{L^2(\mathbb{R}^N,\mu_s)} + \int_{\mathbb{R}^N} f^2|\log|f|\,|\chi_{\{|f|\leq 1\}}\, d\mu_s,$$

which leads to (16.5.2).

Step 2. Fix $1 < p < q < +\infty$ and $f \in C^1_c(\mathbb{R}^N)$. By an homogeneity argument we can assume that $\|f\|_{L^2(\mathbb{R}^N,\mu_s)} = 1$. Integrating the formula (16.4.3), with $p = 2$, in $\mathbb{R}^N$ with respect to the measure μ_t and using the invariance property of the family $\{\mu_t : t \in I\}$, we get

$$\int_{\mathbb{R}^N} f^2 \log f^2 d\mu_s \leq \frac{4\Lambda}{|r_0|}(1-e^{2r_0(t-s)})\int_{\mathbb{R}^N} |\nabla f|^2 d\mu_s + \int_{\mathbb{R}^N} (G(t,s)f^2)\log(G(t,s)f^2)d\mu_t. \qquad (16.5.4)$$

Let us prove that

$$\int_{\mathbb{R}^N} (G(t,s)f^2)\log(G(t,s)f^2)\, d\mu_t \leq \frac{q(p-1)}{p(q-1)}\int_{\mathbb{R}^N} f^2 \log f^2\, d\mu_s + \frac{q}{q-1}\log(C_{p,q}(t-s)). \qquad (16.5.5)$$

This estimate replaced in (16.5.4) will yield (16.5.2). Since $\{G(t,s)\}$ is supercontractive and $\|G(t,s)\|_{L(L^1(\mathbb{R}^N,\mu_s))} \le 1$ for any $I \ni s < t$, from the Riesz-Thorin interpolation theorem we get

$$\|G(t,s)g\|_{L^{q_h}(\mathbb{R}^N,\mu_t)} \le (C_{p,q}(t-s))^{r_h}\|g\|_{L^{p_h}(\mathbb{R}^N,\mu_s)} \tag{16.5.6}$$

for any $g \in L^{p_h}(\mathbb{R}^N,\mu_s)$ and $h \in (0,1-1/p)$, where $r_h = ph(p-1)^{-1} \in (0,1)$, $p_h = (1-h)^{-1}$ and $q_h^{-1} = 1 - r_h + q^{-1}r_h$. From (16.5.6) it follows that

$$\int_{\mathbb{R}^N} (G(t,s)|f|^{2(1-h)})^{q_h} d\mu_t \le (C_{p,q}(t-s))^{r_h q_h}, \qquad t > s,$$

which holds also for $h = 0$. Consequently,

$$\frac{1}{h}\bigg(\int_{\mathbb{R}^N} (G(t,s)|f|^{2(1-h)})^{q_h} d\mu_t - \int_{\mathbb{R}^N} G(t,s)|f|^2 d\mu_t\bigg) \le \frac{1}{h}(C_{p,q}(t-s))^{r_h q_h} - 1). \tag{16.5.7}$$

Since the functions $h \mapsto \|G(t,s)|f|^{2(1-h)}\|^{q_h}_{L^{q_h}(\mathbb{R}^N,\mu_s)}$ and $h \mapsto (C_{p,q}(t,s))^{r_h q_h}$ are differentiable at $h = 0$, letting h tend to 0^+ in (16.5.7) and again using the invariance property of the family $\{\mu_t : t \in I\}$ estimate (16.5.5) follows. □

Proof of Theorem 16.5.2 Suppose that $\{G(t,s)\}$ is supercontractive and fix $s \in I$. By density, it suffices to prove (16.5.1) for functions $f \in C_b^1(\mathbb{R}^N)$. For such functions and $\varepsilon \in (0, 4\Lambda|r_0|^{-1}]$, estimate (16.5.1) follows from (16.5.2), with $p = 2$, $q = 3$ and $t > s$ such that $8\Lambda|r_0|^{-1}(1 - e^{2r_0(t-s)}) = \varepsilon$. In this case we obtain that

$$\beta(\varepsilon) = 3\log\left[C_{2,3}\left(\frac{1}{2r_0}\log\left(1 + \frac{r_0}{8\Lambda}\varepsilon\right)\right)\right]. \tag{16.5.8}$$

Finally, for $\varepsilon > 4\Lambda|r_0|^{-1}$ estimate (16.5.1) follows taking $\beta(\varepsilon) = \beta(4\Lambda|r_0|^{-1})$.

Vice versa, let us assume that estimate (16.5.1) holds. Fix $f \in C_b^1(\mathbb{R}^N)$ with positive infimum, $q > p > 1$ and $t > s \in I$. Further, set $\varepsilon = 2\kappa_0(t-s)(\log((q-1)/(p-1)))^{-1}$ and denote by $H : [s,+\infty) \to \mathbb{R}$ the function defined by $H(r) = e^{-m(r)}\|G(r,s)f\|_{L^{q(r)}(\mathbb{R}^N,\mu_r)}$ for any $r \ge s$, where $q(r) = e^{2\kappa_0\varepsilon^{-1}(r-s)}(p-1) + 1$ and $m(r) = 2\beta(\varepsilon)(p^{-1} - (q(r))^{-1})$. The main step of the proof consists of showing that $H(t) \le H(s)$. Indeed, this estimate yields $\|G(t,s)f\|_{L^q(\mathbb{R}^N,\mu_t)} \le C_{p,q}(t-s)\|f\|_{L^p(\mathbb{R}^N,\mu_s)}$, with

$$C_{p,q}(r) = \exp\left[\frac{2(q-p)}{pq}\beta\left(2\kappa_0 r\left(\log\frac{q-1}{p-1}\right)^{-1}\right)\right], \qquad r > 0,$$

which is a decreasing function since β is decreasing as well. Then, using a density argument we can extend the previous estimate to any $f \in L^p(\mathbb{R}^N,\mu_s)$ which is almost everywhere nonnegative in $\mathbb{R}^N$ (with respect to the measure μ_s and, hence, with respect to the Lebesgue measure, due to Proposition 16.2.5). Finally, the estimate $|G(t,s)f| \le G(t,s)|f|$, which holds true for any $f \in L^p(\mathbb{R}^N,\mu_s)$, allows us to complete the proof.

To check the inequality $H(s) \le H(t)$, for any $n \in \mathbb{N}$ we introduce the function H_n, defined by $H_n(t) = e^{-m(t)}\|\vartheta_n^{1/q(t)}G(t,s)f\|_{L^{q(t)}(\mathbb{R}^N,\mu_t)}$ for any $t \ge s$, where $\vartheta_n(x) = \vartheta(|x|/n)$, for any $x \in \mathbb{R}^N$, and $\vartheta \in C^\infty(\mathbb{R})$ satisfies the condition $\chi_{(-\infty,1]} \le \vartheta \le \chi_{(-\infty,2]}$. The function H_n is differentiable in (s,t) and

$$H_n(t) - H_n(s) = \int_s^t H_n(\sigma)(\psi_n(\sigma) - m'(\sigma))d\sigma, \qquad n \in \mathbb{N}, \tag{16.5.9}$$

for some function ψ_n, which satisfies the condition $\limsup_{n\to+\infty}\psi_n(\sigma) \le \psi(\sigma)$ for any $\sigma > s$, where

$$\begin{aligned}\psi(\sigma) \le &\left(\int_{\mathbb{R}^N}(G(\sigma,s)f)^{q(\sigma)}d\mu_\sigma\right)^{-1}\\ &\times\left\{\frac{q'(\sigma)}{q(\sigma)}\int_{\mathbb{R}^N}(G(\sigma,s)f)^{q(\sigma)}\log\left(\frac{G(\sigma,s)f}{\|G(\sigma,s)\|_{L^{q(\sigma)}(\mathbb{R}^N,\mu_\sigma)}}\right)d\mu_\sigma\right.\\ &\left.\quad -(q(\sigma)-1)\int_{\mathbb{R}^N}(G(\sigma,s)f)^{q(\sigma)-2}|\sqrt{Q(\sigma)}\nabla_x G(\sigma,s)f|^2 d\mu_\sigma\right\}\end{aligned}$$

for any $\sigma > s$. Letting n tend to $+\infty$ in (16.5.9) yields $H(t) - H(s) \le \int_s^t(\psi - m')H d\sigma$. Now, we show that $\psi - m'$ is nonpositive in (s,t). For this purpose, we write (16.5.1) with $|G(\sigma,s)f|^{p/2}$, $q(\sigma)$ and μ_σ in place of f, p and μ_s, respectively, and use Hypotheses 16.4.3 to estimate $|\sqrt{Q}\nabla_x G(\cdot,s)f|^2 \ge \kappa_0|\nabla_x G(\cdot,s)f|^2$. We thus get

$$\begin{aligned}&\psi(\sigma)-m'(\sigma)\\ \le&[e^{m(\sigma)}H(\sigma)]^{-q(\sigma)}\left(\varepsilon\frac{q'(\sigma)}{2}-\kappa_0(q(\sigma)-1)\right)\int_{\mathbb{R}^N}(G(\sigma,s)f)^{q(\sigma)-2}|\nabla_x G(\sigma,s)f|^2 d\mu_\sigma\\ &+[e^{m(\sigma)}H(\sigma)]^{-q(\sigma)}\frac{q'(\sigma)}{q(\sigma)}\left(2\frac{\beta(\varepsilon)}{q(\sigma)}-m'(\sigma)\frac{q(\sigma)}{q'(\sigma)}\right)\int_{\mathbb{R}^N}(G(\sigma,s)f)^{q(\sigma)}d\mu_\sigma\end{aligned}$$

for any $\sigma \in (s,t)$ and the right-hand side of the previous inequality identically vanishes in (s,t) due to the definitions of q and m. □

For notational convenience, throughout the rest of this section, by φ_λ ($\lambda > 0$) we denote the function defined by

$$\varphi_\lambda(x) = e^{\lambda|x|^2}, \qquad x \in \mathbb{R}^N.$$

The second characterization of the supercontractivity is contained in the following theorem.

Theorem 16.5.4 *The evolution operator $\{G(t,s)\}$ is supercontractive if and only if, for any $\lambda > 0$, the function φ_λ belongs to $L^1(\mathbb{R}^N,\mu_s)$ for any $s \in I$ and $\sup_{s\in I}\|\varphi_\lambda\|_{L^1(\mathbb{R}^N,\mu_s)} < +\infty$.*

For its proof we need some preliminary results.

Proposition 16.5.5 *The function $x \mapsto e^{\lambda|x|}$ belongs to $L^1(\mathbb{R}^N,\mu_s)$ for any $\lambda > 0$. More precisely,*

$$\sup_{s\in I}\int_{\mathbb{R}^N} e^{\lambda|x|}\mu_s(dx) < +\infty, \qquad \lambda > 0.$$

Moreover, if inequality (16.5.1) *holds true, then $\varphi_\lambda \in L^1(\mathbb{R}^N,\mu_s)$ for any $\lambda > 0$ and $\sup_{s\in I}\|\varphi_\lambda\|_{L^1(\mathbb{R}^N,\mu_s)} < +\infty$.*

Proof We adapt to our setting the arguments in [164]. Fix $\lambda > 0$ and, for any $n \in \mathbb{N}$, a smooth increasing function $\psi_n : [0,+\infty) \to \mathbb{R}$ such that $\psi_n(t) = t$ for any $t \in [0,n]$, $\psi_n(t) = n+1$ for any $t \ge n+2$ and $0 \le \psi_n'(t) \le 1$ for any $t \ge 0$. The functions f_n ($n \in \mathbb{N}$), defined by $f_n(x) := \psi_n(|x|)$ for any $x \in \mathbb{R}^N$, are bounded and $\|\nabla f_n\|_\infty \le 1$ for any $n \in \mathbb{N}$. Moreover, $f_n(x)$ increases to $f(x) := |x|$ for any $x \in \mathbb{R}^N$, as n tends to $+\infty$. For any $s \in I$, $n \in \mathbb{N}$ and $r > 0$, let $H_{n,\lambda,s}(r) = \|e^{\lambda r f_n}\|_{L^1(\mathbb{R}^N,\mu_s)}$. Applying the logarithmic

Sobolev inequality (16.4.4) to the function $e^{\lambda r f_n/2}$ and then dividing by $r^2 H_{n,\lambda,s}(r)$, we conclude that the derivative of the function $r \mapsto r^{-1}\log H_{n,\lambda,s}(r)$ can be bounded from above by $C\lambda^2/2$, where C is the constant in (16.4.4). From this result, the crucial estimate $H_{n,\lambda,s}(2) \le e^{C\lambda^2}(H_{n,\lambda,s}(1))^2$ follows.

Since the family $\{\mu_t : t \in I\}$ is tight, we can fix $M = M(\lambda) > 0$ such that $\mu_s(\mathbb{R}^N \setminus B_{M\lambda^{-1}}) \le (4e^{C\lambda^2})^{-1}$ for any $s \in I$. This estimate implies that $\mu_s(\{\lambda f_n \ge M\}) \le \mu_s(\{\lambda f \ge M\}) = \mu_s(\mathbb{R}^N \setminus B_{M\lambda^{-1}}) \le (4e^{C\lambda^2})^{-1}$ for any $s \in I$. Hence, we can estimate

$$\begin{aligned} H_{n,\lambda,s}(1) &= \int_{\{\lambda f_n \ge M\}} e^{\lambda f_n} d\mu_s + \int_{\{\lambda f_n < M\}} e^{\lambda f_n} d\mu_s \\ &\le (\mu_s(\{\lambda f_n \ge M\}))^{\frac{1}{2}} \|e^{\lambda f_n}\|_{L^2(\mathbb{R}^N,\mu_s)} + e^M \\ &\le (4e^{C\lambda^2})^{-\frac{1}{2}} \left(H_{n,\lambda}(2)\right)^{\frac{1}{2}} + e^M \le 2^{-1} H_{n,\lambda,s}(1) + e^M \end{aligned}$$

and conclude that $\|e^{\lambda f_n}\|_{L^1(\mathbb{R}^N,\mu_s)} \le 2e^M$ for any $s \in I$. Letting n tend to $+\infty$, we get the first part of the claim.

In order to prove the second part of the claim, we begin by observing that

$$\int_{\mathbb{R}^N} e^{\lambda f_n^2} d\mu_s = \frac{1}{2\sqrt{\lambda\pi}} \int_{\mathbb{R}^N} d\mu_s \int_{\mathbb{R}} e^{\sigma f_n - \frac{\sigma^2}{4\lambda}} d\sigma = \frac{1}{2\sqrt{\lambda\pi}} \int_{\mathbb{R}} H_{n,1,s}(\sigma) e^{-\frac{\sigma^2}{4\lambda}} d\sigma. \tag{16.5.10}$$

From the results of the first part of the proof we infer that $H_{n,1,s}(\sigma) \le 2e^{M(1)}$ for any $\sigma \in (-\infty, 1]$, $n \in \mathbb{N}$ and $s \in I$. To estimate the function $H_{n,1,s}$ in $(1, +\infty)$, we use (16.5.1), with $\varepsilon = (4\lambda)^{-1}$, instead of the logarithmic Sobolev inequality and by arguing as above we can show that

$$\frac{d}{dr}\left(\frac{1}{r}\log H_{n,1,s}(r)\right) \le \frac{1}{8\lambda} + \frac{c_\lambda}{r^2}, \qquad r > 0, \tag{16.5.11}$$

for any $n \in \mathbb{N}$ and some positive constant c_λ. Integrating (16.5.11) from 1 to $\sigma > 1$, we deduce that

$$H_{n,1,s}(\sigma) \le \exp\left(\frac{\sigma^2}{8\lambda} + \sigma\left(M(1) + \log(2) - \frac{1}{8\lambda} + c_\lambda\right) - c_\lambda\right).$$

We have so proved that the functions $\sigma \mapsto H_{n,1,s}(\sigma)e^{-\frac{\sigma^2}{4\lambda}}$ are in $L^1((1, +\infty))$ and their L^1-norms can be bounded from above by a positive constant, independent of n and $s \in I$. Letting n tend to $+\infty$ in (16.5.10), we complete the proof. □

Remark 16.5.6 Actually, just assuming the validity of estimate (16.4.4) and repeating the arguments in the second part of the proof of Proposition 16.5.5, one can show that $\sup_{s\in I} \|\varphi_\lambda\|_{L^1(\mathbb{R}^N,\mu_s)} < +\infty$ for any $\lambda < (2C)^{-1}$ where C is the constant in (16.4.4).

Besides its own interest, the following Harnack-type estimate will play a crucial role in the proof of Theorem 16.5.4. It is obtained by extending the method used by F.-Y Wang in [253] to the non-autonomous case.

Proposition 16.5.7 *For any $f \in C_b(\mathbb{R}^N)$, $p > 1$, $t > s$ and $x, y \in \mathbb{R}^N$ it holds that*

$$|(G(t,s)f)(x)|^p \le (G(t,s)|f|^p)(y) \exp\left(\frac{p|x-y|^2}{4(p-1)\kappa_0(t-s)}\right). \tag{16.5.12}$$

Proof In view of Proposition 14.1.4 it suffices to prove (16.5.12) for functions $f \in C_b^1(\mathbb{R}^N)$. Since, $|G(t,s)f| \le G(t,s)|f|$ in $\mathbb{R}^N$, for any $I \ni s < t$ and $f \in C_b(\mathbb{R}^N)$, we can assume that f is nonnegative. So, we fix a nonnegative function $f \in C_b^1(\mathbb{R}^N)$, $t > s \in I$, $x, y \in \mathbb{R}^N$ and, for any $n \in \mathbb{N}$, we consider the function $\Phi_n = \{G(t,\cdot)[\vartheta_n(G(\cdot,s)f)^p]\} \circ \psi$, where $\psi(r) = (t-s)^{-1}(t-r)y + (t-s)^{-1}(r-s)x$, for any $r \in [s,t]$, and $\vartheta_n(z) = \vartheta(|z|/n)$ for any $z \in \mathbb{R}^N$, $n \in \mathbb{N}$ and some function $\vartheta \in C^\infty(\mathbb{R})$ satisfying the condition $\chi_{(-\infty,1]} \leq \vartheta \leq \chi_{(-\infty,2]}$. By Proposition 14.1.4 and Lemma 14.1.6(i), the function $G(t,\cdot)[\vartheta_n(G(\cdot,s)f)^p]$ is differentiable in $[s,t]$. Moreover, if $n \geq n_0$, where $n_0 \geq [\,|x|\,] + 1$ is the smallest integer such that f does not identically vanish in B_{n_0}, then Φ_n has positive infimum. Indeed, fix $n \geq n_0$. Proposition 14.1.3 shows that the function $(G(\sigma,s)f)^p$ is strictly positive in $\mathbb{R}^N$ for any $\sigma \in (s,t]$. Applying Proposition 14.1.3 once more , we conclude that $\Phi_n > 0$ in (s,t). Moreover, $\Phi_n(t) = \vartheta_n(x)((G(t,s)f)(x))^p = ((G(t,s)f)(x))^p > 0$ and, since $G(t,s)(\vartheta_n f^p)$ is strictly positive in $\mathbb{R}^N$, $\Phi_n(s)$ is positive as well. We thus conclude that Φ_n has a positive infimum. Such infimum is not less than the infimum C_Φ of the function Φ_{n_0} due to the fact that $\Phi_n \geq \Phi_{n_0}$.

Differentiating the function Φ_n we obtain

$$\begin{aligned}\Phi_n(\log\Phi_n)' = &- p(p-1)\{G(t,\cdot)[(G(\cdot,s)f)^{p-2}|\sqrt{Q}\nabla_x G(\cdot,s)f|^2]\} \circ \psi \\ &- \{G(t,\cdot)[(\mathcal{A}\vartheta_n)(G(\cdot,s)f)^p]\} \circ \psi \\ &- 2p\{G(t,\cdot)[(G(\cdot,s)f)^{p-1}\langle Q\nabla\vartheta_n, \nabla_x G(\cdot,s)f\rangle]\} \circ \psi \\ &+ \frac{1}{t-s}\langle[\nabla_x G(t,\cdot)(\vartheta_n(G(\cdot,s)f)^p)] \circ \psi, x-y\rangle \end{aligned} \tag{16.5.13}$$

in (s,t). Using estimate (14.4.8), with $p=1$ and $\ell_p = r_0$, and Hypothesis 16.4.3(i), we can estimate

$$\begin{aligned}&|\nabla_x G(t,r)(\vartheta_n(G(r,s)f)^p)| \\ \leq& G(t,r)[|\nabla\vartheta_n|(G(r,s)f)^p + p\kappa_0^{-1/2}\vartheta_n(G(r,s)f)^{p-1}|\sqrt{Q}(r)\nabla_x G(r,s)f|]\end{aligned}$$

for any $r \in (s,t)$. Hence, from (16.5.13) and Remark 16.4.4, which shows that $\langle b, \nabla\vartheta_n\rangle \geq -C_{s,t}n^{-1}\|\vartheta'\|_\infty$ in $[s,t]$, we get

$$\Phi_n(\log\Phi_n)' \leq \left\{G(t,\cdot)\left[g_n - p\vartheta_n(G(\cdot,s)f)^p\left((p-1)\zeta^2 - \frac{|x-y|}{\sqrt{\kappa_0}(t-s)}\zeta\right)\right]\right\} \circ \psi,$$

where

$$\begin{aligned}g_n =&(G(\cdot,s)f)^p(|\mathrm{Tr}(QD^2\vartheta_n)| - \langle b, \nabla\vartheta_n\rangle) \\ &+ 2p(G(\cdot,s)f)^{p-1}|\langle Q\nabla\vartheta_n, \nabla_x G(\cdot,s)f\rangle| + \frac{|x-y|}{t-s}|\nabla\vartheta_n|(G(\cdot,s)f)^p \leq \frac{C_1}{n}\end{aligned}$$

in $[s,t]$ and $\zeta = (G(\cdot,s)f)^{-1}|\sqrt{Q}\nabla_x G(\cdot,s)f|$. Here, C_1 is a positive constant depending on $\|f\|_{C_b^1(\mathbb{R}^N)}$, p, ϑ, Λ, $|x-y|$ and $t-s$. Recalling that $\delta^2 - \beta\delta \geq -\beta^2/4$ for any $\beta, \delta \in \mathbb{R}$ and that $\{G(t,s)\}$ is order preserving (see Proposition 14.1.3(i)), from the estimate for g_n proved above we deduce that

$$(\log\Phi_n)' \leq \frac{C_1}{C_\Phi n} + \frac{p|x-y|^2}{4(p-1)\kappa_0(t-s)^2}$$

in $[s,t]$, for any $n > n_0$. Integrating both sides of the previous inequality with respect to r in $[s,t]$ and letting n tend to $+\infty$, estimate (16.5.12) follows. □

Now, we can prove Theorem 16.5.4.

Proof of Theorem 16.5.4 If $\{G(t,s)\}$ is supercontractive, then, Theorem 16.5.2 and Proposition 16.5.5 show that $\sup_{s\in I}\|\varphi_\lambda\|_{L^1(\mathbb{R}^N,\mu_s)} < +\infty$ for any $\lambda > 0$.

Vice versa, let us assume that $M_\lambda := \sup_{s\in I} \|\varphi_\lambda\|_{L^1(\mathbb{R}^N,\mu_s)} < +\infty$ for any $\lambda > 0$. Integrating the Harnack inequality (16.5.12) with respect to the measure μ_t yields

$$\int_{\mathbb{R}^N} |f|^p\,d\mu_s = \int_{\mathbb{R}^N} G(t,s)|f|^p\,d\mu_t \geq |(G(t,s)f)(x)|^p \int_{\mathbb{R}^N} \exp\left(-\frac{p\,|x-y|^2}{4\kappa_0(p-1)(t-s)}\right)\mu_t(dy)$$
$$\geq |(G(t,s)f)(x)|^p \mu_t(B_r)\exp\left(-\frac{p(r^2+|x|^2)}{2\kappa_0(p-1)(t-s)}\right)$$

for any $t > s$, $r > 0$, $x \in \mathbb{R}^N$ and $f \in C_b(\mathbb{R}^N)$. Hence,

$$|(G(t,s)f)(x)| \leq 2\exp\left(\frac{R^2+|x|^2}{2\kappa_0(p-1)(t-s)}\right)\|f\|_{L^p(\mathbb{R}^N,\mu_s)}, \qquad t > s,\ x \in \mathbb{R}^N, \quad (16.5.14)$$

where R is such that $\mu_t(B_R) > 2^{-p}$ for any $t \in I$. Let us now fix $q > p$ and set $\lambda_0 = (2\kappa_0(p-1)(t-s))^{-1}q$. By (16.5.14) we can estimate

$$\int_{\mathbb{R}^N} |G(t,s)f|^q d\mu_t \leq 2^q \exp\left(\frac{qR^2}{2\kappa_0(p-1)(t-s)}\right) M_{\lambda_0}\|f\|^q_{L^p(\mathbb{R}^N,\mu_s)}$$
$$=: (C_{p,q}(t-s))^q \|f\|^q_{L^p(\mathbb{R}^N,\mu_s)} \quad (16.5.15)$$

for any $I \ni s < t$. By density, we can extend the previous inequality to any $f \in L^p(\mathbb{R}^N,\mu_s)$. This completes the proof. □

Now, we provide a concrete condition on the coefficients of the operator $\mathcal{A}$ which yields the supercontractivity of the evolution operator $\{G(t,s)\}$.

Proposition 16.5.8 *Let Hypotheses* 16.2.1(*i*) *and* 16.4.3(*ii*) *be satisfied. Further, assume that there exist $K, \beta > 0$ and $R > 1$ such that $\langle b(t,x), x\rangle \leq -K|x|^2\log(|x|)$ for any $t \in I$ and $x \in \mathbb{R}^N \setminus B_R$. Then, the evolution operator $\{G(t,s)\}$ is supercontractive.*

Proof Since $\mathcal{A}(t)\varphi_\lambda(x) \leq 2\lambda[2\lambda\Lambda|x|^2 + \Lambda N - 2K|x|^2\log(|x|)]\varphi_\lambda(x)$ for any $t \in I$, $x \in \mathbb{R}^N \setminus B_R$ and $\lambda > 0$, taking Remark 16.4.4 into account we deduce that there exist two positive constants $a = a(\lambda,\delta)$ and $d = d(\lambda,\delta)$ such that $\mathcal{A}\varphi_\lambda \leq a - d\varphi_\lambda$ in $I \times \mathbb{R}^N$. The proof of Theorem 16.2.7 shows that $\sup_{s\in I} \varphi_\lambda d\mu_s < +\infty$, for any $\lambda > 0$. Hence, by Theorem 16.5.4, the evolution operator $\{G(t,s)\}$ is supercontractive. □

16.6 Ultraboundedness

Definition 16.6.1 *The evolution operator $\{G(t,s)\}$ is called ultrabounded if $G(t,s) \in L(L^p(\mathbb{R}^N,\mu_s), L^\infty(\mathbb{R}^N))$ for any $t > s \in I$, $p > 1$ and $\|G(t,s)\|_{L(L^p(\mathbb{R}^N,\mu_s),L^\infty(\mathbb{R}^N))} \leq C_{p,\infty}(t-s)$ for some decreasing function $C_{p,\infty} : \mathbb{R}_+ \to \mathbb{R}_+$, such that $\lim_{r\to 0^+} C_{p,\infty}(r) = +\infty$, and any $I \ni s < t$.*

To begin with, we prove the following equivalent characterization of ultraboundedness. As in [76, 229, 253, 256], which deal with the autonomous case, a key tool to prove such a result is the Harnack type estimate (16.5.12).

Theorem 16.6.2 *The evolution operator $\{G(t,s)\}$ is ultrabounded if and only if, for any $\lambda > 0$ and $t > s$, the function $G(t,s)\varphi_\lambda$ belongs to $L^\infty(\mathbb{R}^N)$ and $\sup\{\|G(t,s)\varphi_\lambda\|_\infty : t,s \in I,\ t \geq s+\delta\} < +\infty$ for any $\delta, \lambda > 0$.*

Proof Suppose that $M_{\delta,\lambda} := \sup\{\|G(t,s)\varphi_\lambda\|_\infty : t,s \in I,\ t \geq s+\delta\} < +\infty$ for any $\delta, \lambda > 0$. Then, $\sup_{s\in I} \|\varphi_\lambda\|_{L^1(\mathbb{R}^N,\mu_s)} < +\infty$ for any $\lambda > 0$ and, by Theorem 16.5.4, $\{G(t,s)\}$ is supercontractive. Indeed, for any $n \in \mathbb{N}$, let us set $\varphi_{\lambda,n} = \varphi_\lambda \wedge n$. Then, for any $s \in I$,

$$\int_{\mathbb{R}^N} \varphi_{\lambda,n} d\mu_s = \int_{\mathbb{R}^N} G(s+1,s)\varphi_{\lambda,n} d\mu_{s+\delta} \leq \|G(s+\delta,s)\varphi_{\lambda,n}\|_\infty \leq M_{1,\lambda}.$$

Letting n tend to $+\infty$, we conclude that $\|\varphi_\lambda\|_{L^1(\mathbb{R}^N,\mu_s)} \leq M_{1,\lambda}$. In view of this remark, to prove that $\{G(t,s)\}$ is ultrabounded it suffices to show that there exists a non-increasing function $C_{2,\infty}$ such that $\|G(t,s)\|_{L(L^2(\mathbb{R}^N,\mu_s),L^\infty(\mathbb{R}^N))} \leq C_{2,\infty}(t-s)$ for any $t > s \in I$. Indeed, since $\|f\|_{L^2(\mathbb{R}^N,\mu_s)} \leq \|f\|_{L^p(\mathbb{R}^N,\mu_s)}$ for $s \in I$, $f \in L^p(\mathbb{R}^N,\mu_s)$ and $p > 2$, we conclude that $\|G(t,s)\|_{L(L^p(\mathbb{R}^N,\mu_s),L^\infty(\mathbb{R}^N))} \leq C_{2,\infty}(t-s)$ for any $t > s \in I$. On the other hand, if $p \in (1,2)$ then, using the supercontractivity of the evolution operator and splitting $G(t,s)f = G(t,(t+s)/2)G((t+s)/2,s)f$, we can estimate

$$\begin{aligned}\|G(t,s)f\|_\infty \leq & \|G(t,(t+s)/2)\|_{L(L^2(\mathbb{R}^N,\mu_{(t+s)/2}),L^\infty(\mathbb{R}^N))}\|G((t+s)/2,s)f\|_{L^2(\mathbb{R}^N,\mu_{(t+s)/2})}\\ \leq & C_{2,\infty}((t-s)/2)C_{p,2}((t-s)/2)\|f\|_{L^p(\mathbb{R}^N,\mu_s)}\end{aligned}$$

for any $f \in L^p(\mathbb{R}^N,\mu_s)$ and we are done.

So, let us consider the case $p = 2$. Using (16.5.14), with $p = 2$ and the same splitting of $G(t,s)f$ as above, we get

$$\begin{aligned}\|G(t,s)f\|_\infty =& \|G(t,(t+s)/2)G((t+s)/2,s)f\|_\infty\\ \leq & 2e^{\frac{R}{2\kappa_0(t-s)}}\|f\|_{L^2(\mathbb{R}^N,\mu_s)}\|G(t,(t+s)/2)\varphi_{\frac{1}{2\kappa_0(t-s)}}\|_\infty\\ \leq & 2e^{\frac{R}{2\kappa_0(t-s)}}M_{\frac{t-s}{2},\frac{1}{2\kappa_0(t-s)}}\|f\|_{L^2(\mathbb{R}^N,\mu_s)} =: C_{2,\infty}(t-s)\|f\|_{L^2(\mathbb{R}^N,\mu_s)}\end{aligned}$$

for any $f \in C_b(\mathbb{R}^N)$. By density we can extend the previous estimate to any $f \in L^2(\mathbb{R}^N,\mu_s)$. Note that the function $r \mapsto C_{2,\infty}(r)$ is decreasing since $M_{\delta_2,\lambda_1} \leq M_{\delta_1,\lambda_2}$ for any $0 < \delta_1 \leq \delta_2$ and $0 < \lambda_1 \leq \lambda_2$.

Conversely, suppose that $\{G(t,s)\}$ is ultrabounded. Then, it is supercontractive and, by Propositions 16.5.3 and 16.5.5, $\sup_{s\in I}\|\varphi_\lambda\|_{L^1(\mathbb{R}^N,\mu_s)} < +\infty$ for any $\lambda > 0$ or, equivalently, $\sup_{s\in I}\|\varphi_\lambda\|_{L^2(\mathbb{R}^N,\mu_s)} < +\infty$. Therefore, $\|G(t,s)\varphi_\lambda\|_\infty \leq C_{2,\infty}(t-s)\sup_{r\in I}\|\varphi_\lambda\|_{L^2(\mathbb{R}^N,\mu_r)}$ for any $t > s$, $\lambda > 0$. Since the function $r \mapsto C_{2,\infty}(r)$ is decreasing, it follows that $M_{\lambda,\delta} \leq C_{2,\infty}(\delta)\sup_{r\in I}\|\varphi_\lambda\|_{L^2(\mathbb{R}^N,\mu_r)}$ and the proof is complete. □

The following theorem provides a sufficient condition for the evolution operator $\{G(t,s)\}$ to be ultrabounded.

Theorem 16.6.3 *Let Hypotheses* 16.2.1(*i*) *and* 16.4.3(*ii*) *be satisfied. Further, suppose that there exist constants $K > 0$, $R > 0$ $\alpha > 1$ such that $\langle b(t,x),x\rangle \leq -K|x|^2|\log|x||^\alpha$ for any $t \in I$ and $x \in \mathbb{R}^N$. Then, $\{G(t,s)\}$ is ultrabounded.*

Proof As a first step we prove that for any $\lambda > 0$ there exists a convex increasing function $h_\lambda : [0,+\infty) \to \mathbb{R}$ such that $1/h_\lambda$ belongs to $L^1((c,+\infty))$ for large $c = c(\lambda)$ and $\mathcal{A}(t)\varphi_\lambda(x) \leq -h_\lambda(\varphi_\lambda(x))$ for any $t \in I$ and $x \in \mathbb{R}^N$. Taking Remark 16.4.4 into account, we can estimate $\langle b(t,x),x\rangle \leq K_1 - K|x|^2|\log|x||^\alpha$ for any $(t,x) \in I \times \mathbb{R}^N$ and some positive constant $K_1 = K_1(R,\alpha)$. It follows that

$$\mathcal{A}(t)\varphi_\lambda(x) \leq -2\lambda\varphi_\lambda(x)\big(K|x|^2|\log(|x|)|^\alpha - 2\lambda\Lambda|x|^2 - \Lambda N\big), \qquad t \in I,\ x \in \mathbb{R}^N.$$

Hence, if we fix $C_1 = C_1(\alpha,\lambda) > 0$ such that $4\lambda\Lambda y^2 \leq Ky^2(\log y)^\alpha + C_1$ for any $y \geq 0$, then $\mathcal{A}\varphi_\lambda \le -\zeta_\lambda \circ \varphi_\lambda$ in $I \times \mathbb{R}^N$, where $\zeta_\lambda(r) = r(2^{-\alpha}K \log r|\log(\lambda^{-1}\log r)|^\alpha - C_2)$ for any $r \ge 1$ and $C_2 = C_2(\alpha, \lambda, N)$ is a positive constant. This function is convex in $[e^\lambda, +\infty)$ and $1/\zeta_\lambda$ is integrable in a neighborhood of $+\infty$. Moreover, ζ_λ attains its minimum value at some point $y_{0,\lambda} > e^\lambda$ since $\zeta_\lambda'(e^\lambda) = -C_2 < 0$. The function h_λ that we are looking for is defined by $h_\lambda := \zeta_\lambda(y_{0,\lambda})\chi_{[0,y_{0,\lambda}]} + \zeta_\lambda\chi_{(y_{0,\lambda},+\infty)} - (4\lambda^2\Lambda y_{0,\lambda}^2 + 2\Lambda N + |\zeta_\lambda(y_{0,\lambda})|)$. This function is convex and increasing in $[0, +\infty)$. It is clearly integrable in a neighborhood of $+\infty$ and $\mathcal{A}\varphi_\lambda \le -h_\lambda \circ \varphi_\lambda$ for any $t \in I$ and $x \in \mathbb{R}^N$.

Fix $t \in I$, $0 \le s_1 < s_2$ such that $t - s_2 \in I$ and $x \in \mathbb{R}^N$. Using this inequality we can show that $(G(t, t-r)\varphi_\lambda)(x)$ is finite for any $r \in [s_1, s_2]$ and

$$(G(t, t-s_2)\varphi_\lambda)(x) - (G(t, t-s_1)\varphi_\lambda)(x) \le -\int_{s_1}^{s_2} h_\lambda((G(t, t-r)\varphi_\lambda)(x))dr. \tag{16.6.1}$$

For this purpose, for any $n \in \mathbb{N}$, let $\psi_n \in C^\infty([0, +\infty))$ satisfy the conditions $\psi_n(\tau) = \tau$ for $\tau \in [0, n]$, $\psi_n(\tau) \equiv n + 1/2$ for $\tau \ge n+1$, $0 \le \psi_n' \le 1$ and $\psi_n'' \le 0$ in $[0, +\infty)$. Then, the function $\varphi_{n,\lambda} := \psi_n \circ \varphi_\lambda$ belongs to $C_b^2(\mathbb{R}^N)$ and is constant outside of a compact set. By Lemma 14.1.6, we have

$$\begin{aligned}
&(G(t, t-s_2)\varphi_{n,\lambda})(x) - (G(t, t-s_1)\varphi_{n,\lambda})(x)\\
\le& \int_{s_1}^{s_2} dr \int_{\mathbb{R}^N} (\psi_n'(\varphi_\lambda(y))\mathcal{A}(t-r)\varphi_\lambda(y)g(t, r, x, y)dy\\
\le& -\int_{s_1}^{s_2} dr \int_{\mathbb{R}^N} (\psi_n'(\varphi_\lambda(y))h_\lambda(\varphi_\lambda(y))g(t, t-r, x, y)dy.
\end{aligned} \tag{16.6.2}$$

Since h_λ is bounded from above and ψ_n' pointwise increases to 1 as n tends to $+\infty$, the last side of (16.6.2) converges to the corresponding integral with $\psi_n' \circ \varphi_\lambda$ being replaced by 1. Writing (16.6.2) with $s_1 = 0$ and $s_2 = \sigma > 0$, we get

$$(G(t, t-\sigma)\varphi_{n,\lambda})(x) \le \varphi_\lambda(x) - \int_0^\sigma dr \int_{\mathbb{R}^N} (\psi_n'(\varphi_\lambda(y))h_\lambda(\varphi_\lambda(y))g(t, t-r, x, y)dy.$$

Again by monotone convergence we conclude that

$$(G(t, \sigma)\varphi_\lambda)(x) \le \varphi_\lambda(x) - \int_\sigma^t dr \int_{\mathbb{R}^N} h_\lambda(\varphi_\lambda(y))g(t, r, x, y)dy$$

and this shows that $(G(t, \sigma)\varphi_\lambda)(x)$ and the integral term are both finite. Hence, we can now let n tend to $+\infty$ in the first and last side of (16.6.2) and conclude that

$$(G(t, t-s_2)\varphi_\lambda)(x) - (G(t, t-s_1)\varphi_\lambda)(x) \le -\int_{s_1}^{s_2} dr \int_{\mathbb{R}^N} h_\lambda(\varphi_\lambda(y))g(t, t-r, x, y)dy. \tag{16.6.3}$$

Since $p(t, r, x; dy) = g(t, r, x, y)dy$ is a probability measure (see Proposition 14.1.3), we can apply Jensen inequality and obtain (16.6.1).

Next, we observe that the function $(G(t, t-\cdot)\varphi_\lambda)(x)$ is bounded from above by the solution $z(\cdot; x)$ of the Cauchy problem

$$\begin{cases} z'(s) = -h_\lambda(z(s)),\\ z(0) = \varphi_\lambda(x), \end{cases}$$

which is implicitly defined by the equation

$$\int_{\varphi_\lambda(x)}^{z(s;x)} \frac{1}{h_\lambda(\sigma)} d\sigma = -s, \qquad s > 0.$$

Suppose by contradiction, that there exists a point $s_0 > 0$ such that $(G(t,t-s_0)\varphi_\lambda)(x) > z(s_0;x)$. Then, we can determine an interval J containing s_0 where $(G(t,t-\cdot)\varphi_\lambda)(x) > z(\cdot;x)$. Indeed, by inequality (16.6.3), the function $s \mapsto (G(t,t-s)\varphi_\lambda)(x) + \zeta_\lambda(y_{0,\lambda})s$ is decreasing. Thus, we can estimate

$$\begin{aligned}\lim_{s\to s_0^-}[(G(t,t-s)\varphi_\lambda)(x)+\zeta_\lambda(y_{0,\lambda})s] \geq &(G(t,t-s_0)\varphi_\lambda)(x)+\zeta_\lambda(y_{0,\lambda})s\\ >&z(s_0;x)+\zeta_\lambda(y_{0,\lambda})s = \lim_{s\to s_0^-}[z(s;x)+\zeta_\lambda(y_{0,\lambda})s],\end{aligned}$$

so that $(G(t,t-\cdot)\varphi_\lambda)(x) > z(\cdot;x)$ in a left neighborhood of s_0. Denote by a the infimum of J. Then, $(G(t,t-a)\varphi_\lambda)(x) \leq z(a;x)$. But using (16.6.3) with $s_2 = s$ and $s_1 = a$, and observing that

$$z(s;x) = z(a;x) - \int_a^s h_\lambda((G(t,t-\sigma)\varphi_\lambda)(x))d\sigma, \qquad s \in J,$$

we conclude that

$$(G(t,t-s)\varphi_\lambda)(x) - z(s;x) \leq \int_a^s (h_\lambda(z(\sigma;x)) - h_\lambda((G(t,t-\sigma)\varphi_\lambda)(x))d\sigma < 0, \qquad s \in J.$$

This is a contradiction, since the left-hand side is positive.

In view of the above remarks, it suffices to bound from above the function $z(\cdot;x)$. For this purpose, we denote by $\widehat{y}_\lambda$ the positive zero of the function h_λ. Clearly, if $\varphi_\lambda(x) \leq \widehat{y}_\lambda$ then $z(s;x) \leq \widehat{y}_\lambda$ for any $s > 0$. On the other hand, if $\varphi_\lambda(x) > \widehat{y}_\lambda$, then $z(\cdot;x)$ is decreasing in $(0,+\infty)$ and

$$s = -\int_{\varphi_\lambda(x)}^{z(s;x)} \frac{1}{h_\lambda(\sigma)}d\sigma = -\int_{\varphi_\lambda(x)}^{+\infty} \frac{1}{h_\lambda(\sigma)}d\sigma + \int_{z(s;x)}^{+\infty} \frac{1}{h_\lambda(\sigma)}d\sigma \leq \int_{z(s;x)}^{+\infty} \frac{1}{h_\lambda(\sigma)}d\sigma$$

for any $s > 0$. In particular, if $s \geq \delta$, then from the previous chain of inequalities we get

$$\int_{z(s;x)}^{+\infty} \frac{1}{h_\lambda(\sigma)}d\sigma \geq \delta$$

and, from the integrability of the function $1/h_\lambda$ in a neighborhood of $+\infty$, it follows that $z(\cdot,x)$ is bounded in $[\delta,+\infty)$, uniformly with respect to $x \in \mathbb{R}^N$. Hence, $\sup\{\|G(t,s)\varphi_\lambda\|_\infty : t,s \in I,\ t \geq s+\delta\} < +\infty$ for any $\delta, \lambda > 0$ and the assertion follows from (16.6.2). □

16.7 Ultracontractivity

In this section we provide sufficient conditions for the evolution operator $\{G(t,s)\}$ to be ultracontractive.

Definition 16.7.1 *The evolution operator $\{G(t,s)\}$ is called ultracontractive if each operator $G(t,s)$ maps $L^1(\mathbb{R}^N,\mu_s)$ into $L^\infty(\mathbb{R}^N)$ and there exists a decreasing function $C_{1,\infty} : (0,+\infty) \to (0,+\infty)$ such that $\lim_{r\to 0^+} C_{1,\infty}(r) = +\infty$ and $\|G(t,s)\|_{L(L^1(\mathbb{R}^N,\mu_s),L^\infty(\mathbb{R}^N))} \leq C_{1,\infty}(t-s)$ for any $I \ni s < t$.*

Throughout this section, we assume the following additional assumption on the drift term of the operator $\mathcal{A}$.

Hypothesis 16.7.2 *There exist constants $K, R > 0$ and $\gamma > 2$ such that $\langle b(t,x), x\rangle \le -K|x|^\gamma$ for any $t \in I$ and $x \in \mathbb{R}^N \setminus B_R$.*

Theorem 16.7.3 *The evolution operator $\{G(t,s)\}$ is ultracontractive. Moreover,*

$$C_{1,\infty}(r) = \exp\left[C(r^{-1} \vee r^{-\frac{2\gamma}{\gamma-2}})\right], \qquad r > 0, \tag{16.7.1}$$

where the constant C depends on γ, K, R, Λ (see (16.4.2)), N, κ_0.

Proof The proof consists of two main ingredients: the boundedness of each operator $G(t,s)$ from $L^1(\mathbb{R}^N,\mu_s)$ into $L^2(\mathbb{R}^N,\mu_t)$, which is proved by adapting the basic ideas in [187, Theorem 3.4], and the estimate on the constant $M_{\delta,\lambda} = \sup\{\|G(t,s)\varphi_\lambda\|_\infty : t \ge s+\delta,\ s \in I\}$. Being rather long, we split the proof into three steps.

Step 1. Here we prove that the function $\varepsilon \mapsto \varepsilon^{\frac{\gamma}{\gamma-2}}\beta(\varepsilon)$ is bounded as ε tends to 0^+, where β is the function in (16.5.8). To begin with, we claim that the function $x \mapsto \varphi_{\delta,\gamma}(x) = e^{\delta|x|^\gamma}$ belongs to $L^1(\mathbb{R}^N,\mu_s)$ for any $s \in I$ and $\delta < K/(\gamma\Lambda)$, where Λ is the constant defined in (16.4.2), and $\sup_{s\in I}\|\varphi_{\delta,\gamma}\|_{L^1(\mathbb{R}^N,\mu_s)} < +\infty$. To prove the claim we observe that Hypothesis 16.7.2 and Remark 16.4.4 show that there exists a positive constant $K_1 = K_1(R,\gamma)$ such that

$$\langle b(t,x), x\rangle \le K_1 - K|x|^\gamma, \qquad t \in I,\ x \in \mathbb{R}^N. \tag{16.7.2}$$

Using this estimate it is easy to show that

$$\mathcal{A}(t)\varphi_{\delta,\gamma}(x) \le \delta\gamma\varphi_{\delta,\gamma}(x)[(\delta\gamma\Lambda - K)|x|^{2\gamma-2} + (\Lambda N + \Lambda(\gamma-2) + K_1)|x|^{\gamma-2}] = g_1(x)\varphi_{\delta,\gamma}(x)$$

for any $(t,x) \in I \times \mathbb{R}^N$. Since $g_1(x)$ diverges to $-\infty$ as $|x|$ tends to $+\infty$, there exist two positive constants a_1 and c_1 such that $\mathcal{A}(t)\varphi_{\delta,\gamma}(x) \le a_1 - c_1\varphi_{\delta,\gamma}(x)$ for any $(t,x) \in I \times \mathbb{R}^N$ and the claim follows from Theorem 16.2.7.

Now, let us denote by $c_1 = c_1(\gamma)$ a positive constant such that $\delta t^\gamma - \lambda t^2 \ge -c_1\lambda^{\frac{\gamma}{\gamma-2}}$ for any $\lambda > 0$ and $t \ge 0$. Using this inequality we get

$$\|\varphi_\lambda\|_{L^1(\mathbb{R}^N,\mu_s)} \le e^{c_1\lambda^{\frac{\gamma}{\gamma-2}}}\sup_{r\in I}\|\varphi_{\delta,\gamma}\|_{L^1(\mathbb{R}^N,\mu_r)} =: c_2 e^{c_1\lambda^{\frac{\gamma}{\gamma-2}}},$$

which, together with (16.5.15), allows us to infer that

$$C_{2,3}(\tau) = c_3\exp\left(c_4\tau^{-1} + c_5\tau^{-\frac{\gamma}{\gamma-2}}\right), \qquad \tau > 0,$$

for some positive constants c_3, c_4 and c_5. Replacing this formula in (16.5.8) we can estimate

$$\beta(\varepsilon) \le c_6 + c_7\left[\frac{1}{2r_0}\log\left(1 + \frac{r_0}{8\Lambda}\varepsilon\right)\right]^{-\frac{\gamma}{\gamma-2}} + c_8\left[\frac{1}{2r_0}\log\left(1 + \frac{r_0}{8\Lambda}\varepsilon\right)\right]^{-1}, \qquad \varepsilon \le \frac{4\Lambda}{|r_0|},$$

for some positive constants c_6, c_7, c_8. The assertion follows at once.

Step 2. Here, taking Step 1 into account, we prove that, for any $t > s \in I$, the operator $G(t,s)$ is bounded from $L^1(\mathbb{R}^N,\mu_s)$ into $L^2(\mathbb{R}^N,\mu_t)$. To begin with, we fix $t > s \in I$ and a nontrivial nonnegative function $\zeta \in C_b(\mathbb{R}^N)$. Applying Jensen inequality to the function $z = \|\zeta\|^{-1}_{L^1(\mathbb{R}^N,\mu_s)}\zeta$, the probability measure $d\nu_s = z\,d\mu_s$ and the convex function $x \mapsto x\log x$, we can infer that

$$2\|\zeta\|^2_{L^2(\mathbb{R}^N,\mu_s)}\log\|\zeta\|_{L^2(\mathbb{R}^N,\mu_s)} - \|\zeta\|^2_{L^2(\mathbb{R}^N,\mu_s)}\log\|\zeta\|_{L^1(\mathbb{R}^N,\mu_s)} \le \int_{\mathbb{R}^N}\zeta^2\log\zeta\,d\mu_s. \tag{16.7.3}$$

We now fix a positive function $f \in C_c^\infty(\mathbb{R}^N)$, with $\|f\|_{L^1(\mathbb{R}^N,\mu_s)} = 1$, and apply the logarithmic Sobolev inequality (16.5.1), with f and μ_s replaced respectively by $\vartheta_n G(r,s)f$ and μ_r. Here, $\vartheta_n(x) = \vartheta(|x|/n)$ for any $x \in \mathbb{R}^N$ and $n \in \mathbb{N}$, where $\vartheta \in C^\infty(\mathbb{R})$ satisfies the condition $\chi_{(-\infty,1]} \le \vartheta \le \chi_{(-\infty,2]}$. Using estimate (16.7.3) with $\zeta = \vartheta_n G(r,s)f$ we obtain

$$\begin{aligned}
&\|\vartheta_n G(r,s)f\|^2_{L^2(\mathbb{R}^N,\mu_r)} \log\left(\frac{\|\vartheta_n G(r,s)f\|_{L^2(\mathbb{R}^N,\mu_r)}}{\|\vartheta_n G(r,s)f\|_{L^1(\mathbb{R}^N,\mu_r)}}\right)\\
\le&\varepsilon\|\nabla_x(\vartheta_n G(r,s)f)\|^2_{L^2(\mathbb{R}^N,\mu_r)} + \beta(\varepsilon)\|\vartheta_n G(r,s)f\|^2_{L^2(\mathbb{R}^N,\mu_r)}
\end{aligned} \tag{16.7.4}$$

for $I \ni s \le r$. Further, we fix $t > s$ and set $\zeta_n(r) := \log \|\vartheta_n G(r,s)f\|^2_{L^2(\mathbb{R}^N,\mu_r)}$ for any $r \in [s,t]$ and $n \ge n_0$, where n_0 is chosen such that $\inf_{r\in I}\mu_r(B_{n_0}) \ge 1/2$. Formula (14.1.4) shows that the continuous function $\vartheta_n G(\cdot,\cdot)f$ is positive in $\Delta_I \times B_{2n}$, where $\Delta_I = \{(t,s) \in I\times I : t \ge s\}$. If δ denotes its positive infimum, then $\|\vartheta_n G(r,s)f\|_{L^2(\mathbb{R}^N,\mu_r)} \ge \delta/\sqrt{2}$ for any $r \in [s,t]$ and $n \ge n_0$. By Lemma 14.1.6(i) the function ζ_n is differentiable in $[s,t]$ and

$$\begin{aligned}
\|\vartheta_n G(r,s)f\|^2_{L^2(\mathbb{R}^N,\mu_r)}\zeta_n'(r) =&-2\int_{\mathbb{R}^N}|\sqrt{Q}(r)\nabla_x(\vartheta_n G(r,s)f)|^2 d\mu_r\\
&-4\int_{\mathbb{R}^N}\vartheta_n(G(r,s)f)\langle Q(r)\nabla\vartheta_n, \nabla_x G(r,s)f\rangle d\mu_r\\
&-2\int_{\mathbb{R}^N}\vartheta_n(G(r,s)f)^2\mathcal{A}(r)\vartheta_n d\mu_r.
\end{aligned}$$

Using Hypothesis 16.7.2 we get $-\mathcal{A}\vartheta_n \le 2\Lambda N n^{-2}\|\vartheta\|_{C^2_b(\mathbb{R})}$ in $I\times\mathbb{R}^N$ (see (16.4.2)), which replaced in the above formula gives

$$\begin{aligned}
\|\vartheta_n G(r,s)f\|^2_{L^2(\mathbb{R}^N,\mu_r)}\zeta_n'(r) \le&-2\kappa_0\|\nabla_x(\vartheta_n G(r,s)f)\|^2_{L^2(\mathbb{R}^N,\mu_r)}\\
&+\frac{4\Lambda}{n^2}\|\vartheta\|_{C^2_b(\mathbb{R})}\|f\|_\infty(n\|\nabla f\|_\infty + N\|f\|^2_\infty)
\end{aligned}$$

for any $r \ge s$ and $n \in \mathbb{N}$. Since $\|\vartheta_n G(r,s)f\|_{L^1(\mathbb{R}^N,\mu_r)} \le \|f\|_{L^1(\mathbb{R}^N,\mu_s)} = 1$, from (16.7.4) we conclude that

$$\zeta_n'(r) \le -\kappa_0\varepsilon^{-1}\zeta_n(r) + 2\kappa_0\varepsilon^{-1}\beta(\varepsilon) + 2Cn^{-1}\delta^{-2}, \qquad r \in [s,t], \tag{16.7.5}$$

where C is a positive constant depending on Λ, $\|\vartheta\|_{C^2_b(\mathbb{R})}$, $\|f\|_{C^1_b(\mathbb{R}^N)}$ and N. Fix $m > 2(\gamma-2)^{-1}$ and take $\varepsilon = \kappa_0(r-s)(m+1)^{-1}$ in the previous inequality. Multiplying both sides of (16.7.5) by $(r-s)^{m+1}$ and integrating between s and t we get

$$\begin{aligned}
\int_s^t (r-s)^m\zeta_n(r)dr \le&-\frac{1}{m+1}\int_s^t (r-s)^{m+1}\zeta_n'(r)dr\\
&+2\int_s^t (r-s)^m\beta\left(\frac{\kappa_0(r-s)}{m+1}\right)dr + \frac{2C(t-s)^{m+2}}{n\delta^2(m+1)(m+2)}.
\end{aligned} \tag{16.7.6}$$

Note that the second integral term in the right-hand side of (16.7.6) is finite, thanks to Step 1. Integrating by parts in the first integral term in the right-hand side of (16.7.6) and changing variable in the second one, we conclude that

$$(t-s)^{m+1}\zeta_n(t) \le \frac{2(m+1)^{m+2}}{\kappa_0^{m+1}}\int_0^{\frac{\kappa_0(t-s)}{m+1}}\sigma^m\beta(\sigma)d\sigma + \frac{2C}{n\delta^2}(t-s)^{m+2}.$$

Then, letting n tend to $+\infty$ it follows that

$$(t-s)^{m+1}\log(\|G(t,s)f\|^2_{L^2(\mathbb{R}^N,\mu_t)}) \le \frac{2(m+1)^{m+2}}{\kappa_0^{m+1}}\int_0^{\frac{\kappa_0(t-s)}{m+1}}\sigma^m\beta(\sigma)d\sigma$$

$$\leq C(\gamma, \kappa_0)(t-s)^{m+1-\frac{\gamma}{\gamma-2}}$$

for some positive constant $C(\gamma, \kappa_0)$, i.e.,

$$\|G(t,s)f\|_{L^2(\mathbb{R}^N,\mu_t)} \leq \exp\left(\frac{C(\gamma,\kappa_0)}{2(t-s)^{\gamma/(\gamma-2)}}\right) = \exp\left(\frac{C(\gamma,\kappa_0)}{2(t-s)^{\gamma/(\gamma-2)}}\right)\|f\|_{L^1(\mathbb{R}^N,\mu_s)}. \tag{16.7.7}$$

By homogeneity, we can extend estimate (16.7.7) to any $0 < f \in C_c^\infty(\mathbb{R}^N)$. For a general $f \in C_c^\infty(\mathbb{R}^N)$, we write (16.7.7) with $f_n = (f^2+n^{-1})^{1/2}$ and let n tend to $+\infty$ to get (16.7.7) with $|f|$ instead of f. To replace $|f|$ by f, it suffices to recall that $|G(t,s)f| \leq G(t,s)|f|$. Finally, we can extend (16.7.7) to any $f \in L^1(\mathbb{R}^N, \mu_s)$ by density.

Step 3. By Hypothesis 16.7.2 and Proposition 16.5.8, $\{G(t,s)\}$ is ultrabounded. Hence, splitting $G(t,s) = G(t,(t+s)/2)G((t+s)/2,s)$, using Step 1 and the definition of the function $C_{2,\infty}$ in the proof of Theorem 16.6.2, we can estimate

$$\begin{aligned}
\|G(t,s)\|_{L(L^1(\mathbb{R}^N,\mu_s),L^\infty(\mathbb{R}^N))} \leq &\|G(t,(t+s)/2)\|_{L(L^1(\mathbb{R}^N,\mu_{(t+s)/2}),L^\infty(\mathbb{R}^N))}\\
&\times \|G((t+s)/2,s)\|_{L(L^1(\mathbb{R}^N,\mu_s),L^2(\mathbb{R}^N,\mu_{(t+s)/2}))}\\
\leq &2\exp\left(\frac{R}{\kappa_0(t-s)} + \frac{C(\gamma,\kappa_0)}{(t-s)^{\frac{\gamma}{\gamma-2}}}\right)M_{\frac{t-s}{4},\frac{1}{\kappa_0(t-s)}}
\end{aligned}$$

for $t \in (s,+\infty)$, where c_1 is a positive constant depending on γ and κ_0, and $M_{\delta,\lambda} = \sup\{\|G(t,s)\varphi_\lambda\|_\infty : t \geq s+\delta,\ s \in I\}$.

Let us estimate $M_{(t-s)/4,(\kappa_0(t-s))^{-1}}$. Using (16.7.2) we can show that

$$(\mathcal{A}(t)\varphi_\lambda)(x) \leq 2\lambda(-K|x|^\gamma + 2\lambda\Lambda|x|^2 + K_1 + \Lambda N)\varphi_\lambda(x), \qquad t \in I,\ x \in \mathbb{R}^N.$$

Writing $2\lambda\Lambda|x|^2 = (\varepsilon|x|^2)(2\varepsilon^{-1}\lambda\Lambda)$, using the well known inequality $ab \leq a^p/p+b^q/q$ (which holds true for any $a,b \geq 0$ and $p,q \in (1,+\infty)$ such that $1/p+1/q = 1$) with $p = \gamma/2$ and choosing properly $\varepsilon > 0$, we can estimate $2\lambda\Lambda|x|^2 \leq K|x|^\gamma/2 + C_1\lambda^{\gamma/(\gamma-2)}$ for any $x \in \mathbb{R}^N$ and some positive constant $C_1 = C_1(\gamma,\Lambda)$. This estimate allows us to conclude that

$$(\mathcal{A}(t)\varphi_\lambda)(x) \leq \lambda(-K|x|^\gamma + 2C_1\lambda^{\frac{\gamma}{\gamma-2}} + 2\Lambda N + 2K_1)\varphi_\lambda(x), \qquad t \in I,\ x \in \mathbb{R}^N.$$

Hence, the condition $\mathcal{A}\varphi_\lambda \leq -h_\lambda \circ \varphi_\lambda$ is satisfied with $h_\lambda = \zeta_\lambda(y_{0,\lambda})\chi_{[0,y_{0,\lambda}]} + \zeta_\lambda\chi_{(y_{0,\lambda},+\infty)}$, where $y_{0,\lambda} > 1$ denotes the minimum of the function ζ_λ, defined by

$$\zeta_\lambda(y) = \lambda^{1-\frac{\gamma}{2}}y\Big(K(\log y)^{\frac{\gamma}{2}} - \lambda^{\frac{\gamma^2}{2(\gamma-2)}}C_1 - \lambda^{\frac{\gamma}{2}}C_2\Big), \qquad y \geq 1,$$

where $C_2 = 2(\Lambda N + K_1)$. Clearly, h_λ is a nondecreasing convex function in $[0,+\infty)$. Moreover, it is integrable in a neighborhood of $+\infty$.

Repeating the same arguments as in the proof of Theorem 16.6.3, we can easily show that the function $(G(t,t-\cdot)\varphi_\lambda)(x)$ is bounded from above by the solution $z(\cdot;x)$ of the Cauchy problem

$$\begin{cases} z'(s) = -h_\lambda(z(s)),\\ z(0) = \varphi_\lambda(x),\end{cases}$$

which is implicitly defined by the equation

$$\int_{\varphi_\lambda(x)}^{z(s;x)} \frac{1}{h_\lambda(\sigma)}d\sigma = -s, \qquad s > 0.$$

Hence, it suffices to bound from above the function $z(\cdot;x)$. For this purpose, we denote by $\widehat{y}_\lambda$ the positive zero of h_λ. Clearly, if $\varphi_\lambda(x) \le \widehat{y}_\lambda$ then $z(s;x) \le \widehat{y}_\lambda$ for any $s > 0$. On the other hand, if $\varphi_\lambda(x) > \widehat{y}_\lambda$ and $s \ge \delta$, then we can estimate

$$\int_{z(s;x)}^{+\infty} \frac{1}{h_\lambda(\sigma)} d\sigma \ge \delta, \tag{16.7.8}$$

(see again the proof of Theorem 16.6.3) for further details. We now observe that

$$K(\log y)^{\frac{\gamma}{2}} - \lambda^{\frac{\gamma^2}{2(\gamma-2)}} C_1 - \lambda^{\frac{\gamma}{2}} C_2 \ge \frac{1}{2} K(\log y)^{\frac{\gamma}{2}}$$

for any $y \ge y_\lambda := \exp[(\widetilde{C}_1\lambda^{\gamma^2/(2\gamma-4)} + \widetilde{C}_2\lambda^{\gamma/2})^{2/\gamma}]$, where $\widetilde{C}_1 = 2C_1/K$, $\widetilde{C}_2 = 2C_2/K$, and ζ_λ is positive in $[y_{0,\lambda}, +\infty)$. Hence, $y_\lambda \ge \widehat{y}_\lambda$ and

$$h_\lambda(y) = \zeta_\lambda(y) \ge \frac{1}{2} K \lambda^{1-\frac{\gamma}{2}} y(\log y)^{\frac{\gamma}{2}}, \qquad y \ge y_\lambda.$$

Fix $\delta > 0$ and set $\tau_\delta = \exp[K_2\delta^{2/(2-\gamma)}\lambda - (\widetilde{C}_1\lambda^{\gamma^2/(2\gamma-4)} + \widetilde{C}_2\lambda^{\gamma/2})^{2/\gamma}] \vee 1$, where $K_2 = [(\gamma-2)K/4]^{2/(2-\gamma)}$. Then, we can estimate

$$\int_{\tau_\delta y_\lambda}^{+\infty} \frac{1}{h_\lambda(s)} ds \le \frac{2}{K} \lambda^{\frac{\gamma-2}{2}} \int_{r y_\lambda}^{+\infty} \frac{1}{y(\log y)^{\frac{\gamma}{2}}} dy \le \delta$$

and, taking (16.7.8) into account, it thus follows that

$$\int_{z(s;x)}^{+\infty} \frac{1}{h_\lambda(\sigma)} d\sigma \ge \int_{\tau_\delta y_\lambda}^{+\infty} \frac{1}{h_\lambda(s)} ds.$$

Therefore, $z(\cdot;x)$ is bounded from above by $\tau_\delta y_\lambda$ in $[\delta, +\infty)$. Summing up, we have proved that $M_{\delta,\lambda} \le \tau_\delta y_\lambda$. Taking $\delta = (t-s)/4$ and $\lambda = (\kappa_0(t-s))^{-1}$, estimate (16.7.1) follows. □

Chapter 17

The evolution semigroups $\{\mathscr{T}(t)\}$ and $\{\mathscr{T}^\sharp(t)\}$ in L^p-spaces

17.0 Introduction

In Chapter 14 the evolution semigroup $\{\mathscr{T}^\sharp(t)\}$ associated in $C_b(\mathbb{R}^{N+1})$ with the operator $\mathcal{A}$, defined on smooth functions ψ by

$$\mathcal{A}(t)\psi(t,x) = \sum_{i,j=1}^{N} q_{ij}(t,x)D_{ij}\psi(t,x) + \sum_{j=1}^{N} b_j(t,x)D_j\psi(t,x), \qquad (t,x) \in \mathbb{R}^{N+1},$$

has been introduced and its basic properties have been established. Then, in Chapter 16 (see the proof of Proposition 16.2.5), assuming the existence of an evolution system of measures $\{\mu_t : t \in \mathbb{R}\}$ for the evolution operator $\{G(t,s)\}$ it has been proved that the measure μ, which is the unique positive Borel measure on the σ-algebra of all the Borel subsets of $\mathbb{R}^{N+1}$ which satisfies the property

$$\mu(A \times B) = \int_A \mu_s(B)ds, \qquad A \subset \mathbb{R},\ B \subset \mathbb{R}^N, \tag{17.0.1}$$

is infinitesimally invariant for the semigroup $\{\mathscr{T}^\sharp(t)\}$ i.e.,

$$\int_{\mathbb{R}^{N+1}} \mathcal{G}\psi d\mu = 0, \qquad \psi \in C_c^\infty(\mathbb{R}^{N+1}), \tag{17.0.2}$$

where $\mathcal{G} = \mathcal{A} - D_s$, and

$$\int_{\mathbb{R}^{N+1}} \mathscr{T}(t) f d\mu = \int_{\mathbb{R}^{N+1}} f d\mu$$

for any bounded and continuous function f whose support is contained in $[-R,R] \times \mathbb{R}^N$ for some $R > 0$. This formula is the keystone used to define the strongly continuous semigroup $\{\mathscr{T}_p(t)\}$ in the space $L^p(\mathbb{R}^{N+1},\mu)$ for any $p \in [1,+\infty)$. The characterization of the domain of its infinitesimal generator G_p is known when $\mathcal{A}$ is a non-autonomous Ornstein-Uhlenbeck operator (see Theorem 17.1.3). For a more general non-autonomous operator $\mathcal{A}$ the characterization of $D(G_p)$ is not obvious (in Theorem 17.1.3 we partially characterize it taking advantage of the pointwise gradient estimates in Theorem 14.4.5) and even determining whether a given smooth function belongs to $D(G_p)$ is not obvious. For this reason, it is important to determine suitable sets which are cores of G_p.

The problem of the existence/uniqueness of nonnegative measures satisfying (17.0.2), and related issues have been extensively studied by several authors (see e.g. [37, 38, 41, 42]). In all these papers the authors are mainly concerned with the case when the $\mathbb{R}^{N+1}$ is replaced by $(0,1)\times\mathbb{R}^N$ or, more generally, by[1] $(a,b)\times\mathbb{R}^N$ for some $a,b\in\mathbb{R}$ such that $a<b$. They consider measures μ defined by (17.0.1), with the family $\{\mu_s : s\in(a,b)\}$ consisting of probability measures satisfying the equation $\mathcal{G}^*\mu=0$ as well as the initial condition

$$\lim_{t\to a}\int_{\mathbb{R}^N}\zeta d\mu_t=\int_{\mathbb{R}^N}\zeta d\mu_a$$

for any $\zeta\in C_c^\infty(\mathbb{R}^N)$ and some probability measure μ_a.

In Section 17.1, under suitable assumptions on the coefficients of the operator $\mathcal{A}$, we provide two different cores of the operator G_p ($p\in[1,+\infty)$). The first core is the set

$$D_{\rm comp}(\mathcal{G})=\bigg\{\psi\in C_b(\mathbb{R}^{N+1})\cap\bigcap_{p<+\infty}W^{1,2}_{p,{\rm loc}}(\mathbb{R}^{N+1}):$$
$$\mathcal{G}\psi\in C_b(\mathbb{R}^{N+1}),\ {\rm supp}(\psi)\subset[-M,M]\times\mathbb{R}^N \text{ for some } M>0\bigg\}. \quad (17.0.3)$$

This result generalizes a well known property holding in the autonomous case (see e.g. Proposition 9.1.8).

As a byproduct of this result, we prove the inequality

$$\int_{\mathbb{R}^{N+1}}|u|^{p-2}|\sqrt{Q}\nabla_x u|^2\chi_{\{u\neq 0\}}d\mu\le\frac{1}{1-p}\int_{\mathbb{R}^{N+1}}u|u|^{p-2}G_p u d\mu, \quad (17.0.4)$$

satisfied by any function $u\in D(G_p)$ and $p\in(1,+\infty)$. Under stronger assumptions on the inner product $\langle b(t,x),x\rangle$ or assuming that the diffusion coefficients are bounded in $\mathbb{R}^{N+1}$, the previous inequality is, in fact an equality (the so called time dependent *identité de carré du champs*) satisfied by any $u\in D(G_p)$ for any $p\ge 2$. Formula (17.0.4) will play a crucial role in the study of the asymptotic behaviour of the evolution operator $\{G(t,s)\}$ addressed in Chapter 18.

Next, we provide a sufficient condition for $C_c^\infty(\mathbb{R}^{N+1})$ to be a core of G_p (see Theorem 17.1.8). For this purpose, we need to refine the smoothness assumptions on the coefficients of the operator $\mathcal{A}$. Under such additional assumptions we improve the regularity result in Theorem 16.2.4, proving that the density ρ of μ with respect to the Lebesgue measure belongs to $C^{1+\alpha/2,2+\alpha}_{\rm loc}(\mathbb{R}^{N+1})$. The criterion stated in Theorem 17.1.8 requires that suitable functions are integrable in $\mathbb{R}^{N+1}$ with respect to the measure μ. Sufficient conditions for the previous condition to hold are provided in Theorem 17.1.13. They also cover some situations in which the drift coefficients of the operator $\mathcal{A}$ may grow with an exponential rate at infinity.

In Section 17.2 we consider the case when the coefficients of the operator $\mathcal{A}$ are T-periodic with respect to the variable s, for some $T>0$. In this situation, under Hypotheses 17.2.1 (which are the natural counterpart of the assumptions of the first section) there exists a unique probability measure $\mu^\sharp$ which satisfies the condition

$$\int_{\mathbb{R}^{N+1}}\mathscr{T}^\sharp(t)f d\mu^\sharp=\int_{\mathbb{R}^{N+1}}f d\mu^\sharp$$

for any continuous function $f:\mathbb{R}^{N+1}\to\mathbb{R}$, such that $f(\cdot+T,\cdot)=f$, and this measure is induced by the unique T-periodic evolution system of measures for $\{G(t,s)\}$. Also in this situation, we provide a partial characterization of the domain of the infinitesimal generator $G_p^\sharp$ ($p\in(1,+\infty)$) of the strongly continuous semigroup $\{\mathscr{T}_p^\sharp(t)\}$, which is the extension of

[1]but some of the obtained results apply also to the case of the whole of $\mathbb{R}^{N+1}$.

$\{\mathscr{T}^\sharp(t)\}$ to the space $L^p_\sharp(\mathbb{R}^{N+1},\nu^\sharp)$ of all functions $f:\mathbb{R}^{N+1}\to\mathbb{R}$, which are T-periodic with respect to the time variable and satisfy the condition $\int_{(0,T)\times\mathbb{R}^N}|f|^p d\mu^\sharp<+\infty$. Moreover, we prove that the semigroups $\{\mathscr{T}_p^\sharp(t)\}$, $\{\mathscr{T}_p(t)\}$ satisfies the spectral mapping theorem (see Theorem 17.2.6 and Remark 17.2.7). Next, we prove that the set

$$D_\sharp(\mathcal{G})=\left\{\psi\in C_\sharp(\mathbb{R}^{N+1})\cap\bigcap_{p<+\infty}W^{1,2}_{p,\mathrm{loc}}(\mathbb{R}^{N+1}):\mathcal{G}\psi\in C_\sharp(\mathbb{R}^{N+1})\right\},$$

which is the domain of the weak generator of the semigroup $\{\mathscr{T}^\sharp(t)\}$ in $C_\sharp(\mathbb{R}^{N+1})$ (see Subsection 14.5.3), is a core of $G_p^\sharp$ for any $p\in(1,+\infty)$. This result can be obtained by arguing as in the proof of Theorem 17.1.5. Here, we rather prefer to obtain it as a consequence of the fact that another set of smooth functions is a core of $G_p^\sharp$ (see Theorem 17.2.8). Using this result we show the counterpart of inequality (17.0.4) in the periodic setting, i.e., we prove that

$$\int_{(0,T)\times\mathbb{R}^N}|u|^{p-2}|\sqrt{Q}\nabla_x u|^2\chi_{\{u\neq 0\}}d\mu^\sharp\le\frac{1}{1-p}\int_{(0,T)\times\mathbb{R}^N}u|u|^{p-2}G_p^\sharp u\,d\mu^\sharp$$

for any $u\in D(G_p^\sharp)$. We also prove the *identité de carré du champs* in this periodic setting.

A sufficient condition for the space $C^\infty_{\sharp,c}(\mathbb{R}^{N+1}):=\{f\in C^\infty(\mathbb{R}^{N+1})\cap C_\sharp(\mathbb{R}^{N+1}):\mathrm{supp}(f)\subset\mathbb{R}\times B_R$ for some $R>0\}$ to be a core is also provided.

Finally, in Subsection 17.2.2 we prove a logarithmic Sobolev type inequality satisfied by all the functions $f\in D_\sharp(\mathcal{G})$ and we use it to show that the domains $D(G_p^\sharp)$ are compactly embedded in $L^p_\sharp(\mathbb{R}^{N+1},\mu^\sharp)$ for $p\in(1,+\infty)$. In particular, this result implies that the spectrum of each operator $G_p^\sharp$ consists of eigenvalues and it is independent of p.

The results of this chapter are taken from [175, 177].

17.1 The general case

Throughout this section we assume the following conditions on the coefficients of the operator $\mathcal{A}$.

Hypotheses 17.1.1 (i) *q_{ij} and b_i $(i,j=1,\ldots,N)$ belong to $C^{\alpha/2,\alpha}_{\mathrm{loc}}(I\times\mathbb{R}^N)$ for some $\alpha\in(0,1)$;*

(ii) *the matrix $Q(t,x)=(q_{ij}(t,x))$ is symmetric for any $(t,x)\in I\times\mathbb{R}^N$ and there exists a function $\kappa:I\times\mathbb{R}^N\to\mathbb{R}$ such that $0<\kappa_0:=\inf_{I\times\mathbb{R}^N}\kappa$ and $\langle Q(t,x)\xi,\xi\rangle\ge\kappa(t,x)|\xi|^2$ for any $\xi\in\mathbb{R}^N$ and $(t,x)\in I\times\mathbb{R}^N$;*

(iii) *there exist a positive function $\varphi\in C^2(\mathbb{R}^N)$, diverging to $+\infty$ as $|x|$ tends to $+\infty$, and a positive number λ_0 such that $\mathcal{A}\varphi-\lambda_0\varphi\le 0$ in $\mathbb{R}^{N+1}$. Moreover, for any $R>0$ there exist two positive constants $a=a_R$ and $d=d_R$ such that $\mathcal{A}\varphi\le a-d\hat{\varphi}$ in $(-R,+\infty)\times\mathbb{R}^N$.*

In view of Theorem 16.2.7, the previous hypotheses guarantee that there exists at least one evolution system of measures $\{\mu_t:t\in\mathbb{R}\}$ associated with the evolution operator $\{G(t,s)\}$. Throughout this section, we consider any of such systems.

In Theorem 14.5.1 we have shown that $\{\mathscr{T}(t)\}$ is a semigroup of positive contractions in $C_b(\mathbb{R}^{N+1})$, which, in general, is not strongly continuous. Each operator $\mathscr{T}(t)$ preserves the space $C_{\rm comp}(\mathbb{R}^{N+1})$ of all the functions $u \in C_b(\mathbb{R}^{N+1})$ whose support is contained $[-R, R] \times \mathbb{R}^N$ for some $R > 0$. Indeed, the support of the function $u(\cdot - t, \cdot)$ is contained in $[-R+t, R+t] \times \mathbb{R}^N$. Hence, the support of the function $\mathscr{T}(t)u$ is contained in $[-R+t, R+t] \times \mathbb{R}^N$. Moreover, if μ is the measure defined by (17.0.1), then the proof of Proposition 16.2.5 shows that

$$\int_{\mathbb{R}^{N+1}} \mathscr{T}(t)u d\mu = \int_{\mathbb{R}^{N+1}} u d\mu \tag{17.1.1}$$

for any function u as above.[2]

By Proposition 14.1.3 and the Jensen inequality, it follows that $|G(t,s)\psi|^p \le G(t,s)|\psi|^p$ for any $t > s$ and $\psi \in C_b(\mathbb{R}^N)$, so that $|\mathscr{T}(t)u|^p \le \mathscr{T}(t)|u|^p$ for any $u \in C_b(\mathbb{R}^{N+1})$ and $p \in [1, +\infty)$. Using (17.1.1) we conclude that

$$\int_{\mathbb{R}^{N+1}} |\mathscr{T}(t)u|^p d\mu \le \int_{\mathbb{R}^{N+1}} |u|^p d\mu, \qquad u \in C_{\rm comp}(\mathbb{R}^{N+1}) \tag{17.1.2}$$

and this formula allows us to extend each operator $\mathscr{T}(t)$ with a contraction $\{\mathscr{T}_p(t)\}$ in $L^p(\mathbb{R}^{N+1}, \mu)$ for any $p \in [1, +\infty)$. As the proof of Proposition 14.5.7 shows, $\mathscr{T}(\cdot)u$ is continuous in $[0, +\infty)$ with values in $C_b(\mathbb{R}^N)$ and $D_t\mathscr{T}(\cdot)u = \mathscr{T}(\cdot)\mathcal{G}u$ for any $u \in C_c^\infty(\mathbb{R}^{N+1})$, where $\mathcal{G} = \mathcal{A} - D_s$. From the formula (17.1.2), the density of $C_c^\infty(\mathbb{R}^{N+1})$ in $L^p(\mathbb{R}^{N+1}, \mu)$ and all the above remarks, we conclude that $\{\mathscr{T}_p(t)\}$ is a strongly continuous semigroup. Finally, by density, we can extend (17.1.1) to any $u \in D(G_1)$ (the domain of the infinitesimal generator of the semigroup $\{\mathscr{T}_1(t)\}$) and, as a consequence, we deduce that, for any $u \in D(G_1)$, the function $G_1 u$ has zero average with respect to the measure μ. We summarize all these results in the following proposition.

Proposition 17.1.2 *For any $p \in [1, +\infty)$ the semigroup $\{\mathscr{T}_p(t)\}$ is strongly continuous in $L^p(\mathbb{R}^{N+1}, \mu)$. The domain of its infinitesimal generator G_p contains $C_c^\infty(\mathbb{R}^{N+1})$ and $G_p u = \mathcal{G}u$ for any $u \in C_c^\infty(\mathbb{R}^{N+1})$. Finally,*

$$\int_{\mathbb{R}^{N+1}} \mathscr{T}_1(t)u d\mu = \int_{\mathbb{R}^{N+1}} u d\mu, \qquad u \in L^1(\mathbb{R}^{N+1}, \mu),$$

and

$$\int_{\mathbb{R}^{N+1}} G_1 u d\mu = 0, \qquad u \in D(G_1).$$

In the case of the Ornstein-Uhlenbeck operator $\mathcal{A}$ defined in (16.2.8) and under the same assumptions as in Proposition 16.2.8 the domain of G_p has been characterized in [111, 112], when the measure μ is obtained, as explained in the introduction, starting from the Gaussian measures $\mu_t = \mathcal{N}(0, Q_t)$ $(t \in \mathbb{R})$, where Q_t is the covariance matrix defined in (16.2.9).

Theorem 17.1.3 *Under the same assumptions as in Proposition* 16.2.8, *for any $p \in (1, +\infty)$ let G_p be the infinitesimal generator of the evolution operator $\{\mathscr{T}_p(t)\}$, associated in $L^p(\mathbb{R}^{N+1}, \mu)$ with the Ornstein-Uhlenbeck operator in* (16.2.8)*. Then, $D(G_p) = W_p^{1,2}(\mathbb{R}^{N+1}, \mu)$.*

[2]Note that in the proof of Proposition 16.2.5 the assumption on the differentiability of the diffusion coefficients is not needed to prove (17.1.1).

As has already been stressed, for a more general non-autonomous elliptic operator $\mathcal{A}$ the characterization of the domain $D(G_p)$ of G_p is not obvious, and even determining whether a given smooth function f belongs to $D(G_p)$ is not obvious and we do not expect the same result in the general case.

Fortunately, the explicit knowledge of $D(G_p)$ is not necessary in several circumstances, provided we have a good core of G_p at our disposal. In Subsection 17.1.1 we provide some cores of $D(G_p)$. Here, using the estimates on the spatial derivatives of $G(t,s)f$ in Proposition 16.3.3, we provide a useful embedding result for $D(G_p)$.

Proposition 17.1.4 *The following properties are satisfied.*

(i) *Let Hypotheses* 14.4.1 *be satisfied, with the functions k_0 and q_0 be bounded from above in $\mathbb{R}$, and fix $p \in [2,+\infty)$. Then, $\mathscr{T}_p(t)$ maps $L^p(\mathbb{R}^{N+1},\mu)$ into $W^{0,1}_p(\mathbb{R}^{N+1},\mu)$ for any $t>0$ and*

$$\|\nabla_x \mathscr{T}_p(t)u\|_{L^p(\mathbb{R}^{N+1},\mu)} \le C_1(t^{-\frac{1}{2}} \vee 1)\|u\|_{L^p(\mathbb{R}^{N+1},\mu)}, \qquad u \in L^p(\mathbb{R}^{N+1},\mu), \quad (17.1.3)$$

where C_1 is the constant in (14.4.3). *Moreover, $D(G_p)$ is continuously embedded into $W^{0,1}_p(\mathbb{R}^{N+1},\mu)$ and*

$$\|\nabla_x u\|_{L^p(\mathbb{R}^{N+1},\mu)} \le C\|u\|^{\frac{1}{2}}_{L^p(\mathbb{R}^{N+1},\mu)}\|G_p u - \omega u\|^{\frac{1}{2}}_{L^p(\mathbb{R}^{N+1},\mu)} \quad (17.1.4)$$

for any $u \in D(G_p)$, $\omega > 0$ and some positive constant C, depending on ω but being independent of u.

(ii) *Let Hypotheses* 14.4.1*(i) and either of Hypotheses* 14.4.1*(iii) and* 14.4.4*(i) or* 14.4.4*(ii) be satisfied. Then, the properties in (i) hold true for any $p \in (1,+\infty)$. In particular,*

$$\|\nabla_x \mathscr{T}_p(t)u\|_{L^p(\mathbb{R}^{N+1},\mu)} \le C_p e^{\ell_p^- t}(t^{-\frac{1}{2}} \vee 1)\|u\|_{L^p(\mathbb{R}^{N+1},\mu)}, \qquad t>0, \ u \in L^p(\mathbb{R}^{N+1},\mu), \quad (17.1.5)$$

and

$$\|\nabla_x \mathscr{T}_p(t)u\|_{L^p(\mathbb{R}^{N+1},\mu)} \le e^{\ell_p t}\|\nabla_x u\|_{L^p(\mathbb{R}^{N+1},\mu)}, \qquad t>0, \ u \in W^{0,1}_p(\mathbb{R}^{N+1},\mu), \quad (17.1.6)$$

where C_p is the constant in (14.4.10) *and ℓ_p is given by* (14.4.7) *and* (14.4.9). *In particular, if $\ell_p < 0$, then we can take $\omega = 0$ in* (17.1.4).

Proof Estimate (17.1.6) follows immediately integrating both sides of estimate (14.4.8) and using the density of $C^\infty_c(\mathbb{R}^N)$ in $W^{0,1}_p(\mathbb{R}^{N+1},\mu)$. Similarly, estimate (17.1.3) (resp. (17.1.5)) follows integrating (14.4.3)) (resp. (14.4.10)) and using the density[3] of $C^\infty_c(\mathbb{R}^{N+1})$ in $L^p(\mathbb{R}^{N+1},\mu)$. Finally, estimate (17.1.4) follows by arguing as in the proof of Proposition 9.3.3. □

17.1.1 Cores of the operator G_p

In this subsection we provide two different cores for the operator G_p. The first core is contained in the next theorem.

Theorem 17.1.5 *Let Hypotheses* 17.1.1 *be satisfied. Then, $D_{\rm comp}(\mathcal{G})$ (see* (17.0.3)*) is a core of the operator G_p for any $p \in [1,+\infty)$.*

[3]See the proof of Lemma 16.3.2 and take the proof of Proposition 16.2.5 into account.

Proof Fix $p \in [1,+\infty)$. In view of Proposition C.1.14 it suffices to prove that $D_{\rm comp}(\mathcal{G})$ is dense in $L^p(\mathbb{R}^{N+1})$, is contained in $D(G_p)$ and the semigroup $\{\mathscr{T}_p(t)\}$ leaves it invariant. Since it contains $C_c^\infty(\mathbb{R}^{N+1})$, $D_{\rm comp}(\mathcal{G})$ is dense in $L^p(\mathbb{R}^{N+1},\mu)$. Moreover, $D_{\rm comp}(\mathcal{G}) \subset D(G_p)$. Indeed, if $u \in D_{\rm comp}(\mathcal{G})$, then its support is contained in some strip $(a,b)\times\mathbb{R}^N$. Therefore, for any $t \in (0,1)$, the support of the function $t^{-1}(\mathscr{T}(t)f-f)$ is contained in $(a,b+1)\times\mathbb{R}^N$. Since $D_{\rm comp}(\mathcal{G}) \subset D(G_\infty)$, the characterization of this latter space, given in Theorem 14.5.9 shows that $t^{-1}(\mathscr{T}(t)f-f)$ pointwise converges to $\mathcal{G}u$ in $\mathbb{R}^{N+1}$ as t tends to 0^+ and $\mathcal{G}u$ is supported in $(a,b+1)\times\mathbb{R}^N$. Hence, by dominated convergence we conclude that $u \in D(G_p)$ and $G_pu = \mathcal{G}u$. Finally, since $\mathscr{T}(t) = \mathscr{T}_p(t)$ in $C_{\rm comp}(\mathbb{R}^{N+1})$ for any $t>0$, by Corollary 14.5.10 it follows that $\{\mathscr{T}_p(t)\}$ leaves $D_{\max}(\mathcal{G})$ invariant. □

As a consequence of this theorem, we can prove the following results.

Theorem 17.1.6 *Let Hypotheses* 14.4.1 *and* 17.1.1 *be satisfied, with the functions k_0 and q_0 being bounded. Further assume that $\|Q(s,x)\|_{L(\mathbb{R}^N)} \le C(|x|+1)\varphi(x)$ for any $(s,x)\in\mathbb{R}^{N+1}$ and some positive constant C. Then, the following properties are satisfied.*

(i) *If $\langle b(s,x),x\rangle \le C(|x|^2+1)\varphi(x)$ for any $(s,x)\in\mathbb{R}^{N+1}$, then $|u|^{p-2}|\sqrt{Q}\nabla_x u|^2\chi_{\{u\neq 0\}}$ belongs to $L^1(\mathbb{R}^{N+1},\mu)$ for any $p\in(1,+\infty)$, $u\in D(G_p)$ and*

$$\int_{\mathbb{R}^{N+1}}|u|^{p-2}|\sqrt{Q}\nabla_x u|^2\chi_{\{u\neq 0\}}d\mu \le \frac{1}{1-p}\int_{\mathbb{R}^{N+1}}u|u|^{p-2}G_pu\,d\mu. \tag{17.1.7}$$

(ii) *If $|\langle b(s,x),x\rangle| \le C(|x|^2+1)\varphi(x)$ for any $(s,x)\in\mathbb{R}^{N+1}$, then* (17.1.7) *is an equality for any $p\ge 2$ and $u\in D(G_p)$.*

(iii) *If the diffusion coefficients q_{ij} are bounded, then the equality in* (17.1.7) *holds true for any $p\in(1,+\infty)$ and $u\in D_{\rm comp}(\mathcal{G})$ and even for any $u\in D(G_p)$ if $p\ge 2$.*

(iv) *Under either the condition in (i) or in (iii), $D(G_p)$ is contained in $W_p^{0,1}(\mathbb{R}^{N+1},\mu)$ for any $p\in(1,+\infty)$ and the map $u\mapsto|\sqrt{Q}\nabla_x u|$ is continuous from $D(G_p)$ into $L^p(\mathbb{R}^{N+1},\mu)$ for any $1<p\le 2$.*

(v) *If $u,v\in D(G_2)$, then $uv\in D(G_1)$, $G_1(uv) = vG_2u+uG_2v+2\langle Q\nabla_x u,\nabla_x v\rangle$ and*

$$\int_{\mathbb{R}^{N+1}}(vG_2u+uG_2v)d\mu = -2\int_{\mathbb{R}^{N+1}}\langle Q\nabla_x u,\nabla_x v\rangle d\mu. \tag{17.1.8}$$

Proof Being rather long, we split the proof into four steps. In the first two steps we prove properties (i) to (iii) for functions in $D_{\rm comp}(\mathcal{G})$. More precisely, in Step 1 we deal with the case when $p\ge 2$ and, in Step 2, with the case when $p\in(1,2)$. In Step 3, we extend the validity of properties (i) and (ii) to any function of $D(G_p)$ and prove property (iv). Finally, in Step 4 we prove property (v).

Step 1. Fix $u\in D_{\rm comp}(\mathcal{G})$, $p\ge 2$. Then, $\mathcal{G}(|u|^p) = p(p-1)|u|^{p-2}|\sqrt{Q}\nabla_x u|^2+pu|u|^{p-2}\mathcal{G}u$. Since u, $\nabla_x u$ and $\mathcal{G}u$ belong to $C_{\rm comp}(\mathbb{R}^{N+1})$ (see Proposition 14.5.11), if the diffusion coefficients are bounded then $\mathcal{G}(|u|^p)\in C_{\rm comp}(\mathbb{R}^{N+1})$. Therefore, $|u|^p\in D_{\rm comp}(\mathcal{G})\subset D(G_1)$, so that $\int_{\mathbb{R}^{N+1}}\mathcal{G}(|u|^p)d\mu = 0$, by Proposition 17.1.2, and the equality in (17.1.7) holds.

Let us now suppose that the diffusion coefficients are unbounded. Since $\mathcal{G}(|u|^p)$ splits into the sum of a bounded function and a positive function, $\int_{\mathbb{R}^{N+1}}\mathcal{G}(|u|^p)d\mu$ is either finite or equal to $+\infty$. We will prove that $\mathcal{G}(|u|^p)$ belongs to $L^1(\mathbb{R}^{N+1},\mu)$ and the previous integral is nonpositive (resp. vanishes) if the assumption in property (i) (resp. (ii)) holds.

Let $\vartheta \in C^\infty(\mathbb{R})$ be a nonincreasing function such that $\chi_{[0,1]} \le \vartheta \le \chi_{[0,2]}$. For $n \in \mathbb{N}$ and $x \in \mathbb{R}^N$, let $\vartheta_n(x) = \vartheta(|x|/n)$ and observe that

$$\int_{\mathbb{R}^{N+1}} \mathcal{G}(|u|^p)d\mu = \lim_{n\to+\infty}\int_{\mathbb{R}^{N+1}} \mathcal{G}(|u|^p)\vartheta_n d\mu. \tag{17.1.9}$$

This formula follows by applying the monotone convergence theorem to the positive and increasing sequence $\{\vartheta_n|u|^{p-2}|\sqrt{Q}\nabla_x u|^2\}$ and the dominated convergence theorem to the sequence $\{p\vartheta_n u|u|^{p-2}\mathcal{G}u\}$.

Let us estimate the integrals in the right-hand side of (17.1.9). Since the function $|u|^p\vartheta_n$ belongs to $D(G_\infty)$ for any $n \in \mathbb{N}$, the mean value of $\mathcal{G}(|u|^p\vartheta_n)$ vanishes, i.e.,

$$\int_{\mathbb{R}^{N+1}} \mathcal{G}(|u|^p)\vartheta_n\, d\mu = -\int_{\mathbb{R}^{N+1}} (|u|^p\mathcal{G}(\vartheta_n) + 2pu|u|^{p-2}\langle Q\nabla_x u, \nabla\vartheta_n\rangle)d\mu. \tag{17.1.10}$$

A straightforward computation shows that

$$\begin{aligned}\mathcal{G}\vartheta_n(s,x) = {}& \vartheta''\left(\frac{|x|}{n}\right)\frac{|\sqrt{Q}(s,x)x|^2}{|x|^2n^2} + \vartheta'\left(\frac{|x|}{n}\right)\frac{\mathrm{Tr}(Q(s,x))}{|x|n}\\ & - \vartheta'\left(\frac{|x|}{n}\right)\frac{|\sqrt{Q}(s,x)x|^2}{|x|^3n} + \vartheta'\left(\frac{|x|}{n}\right)\frac{\langle b(s,x),x\rangle}{|x|n}.\end{aligned} \tag{17.1.11}$$

Since ϑ', ϑ'' vanish outside the interval $(1,2)$, the modulus of the sum of the first three terms in the right-hand side of (17.1.11) can be bounded from above by $C_1n^{-1}\varphi(x)$ for any $(s,x) \in \mathbb{R}^{N+1}$ and some positive constant C_1. Moreover, $\langle Q\nabla_x u, \nabla\vartheta_n\rangle$ vanishes pointwise in $\mathbb{R}^{N+1}$, as n tends to $+\infty$, and $|\langle Q\nabla_x u, \nabla\vartheta_n\rangle| \le 3C\|\vartheta'\|_\infty\|\nabla_x u\|_\infty\varphi$ in $\mathbb{R}^{N+1}$. Thus, for $(s,x) \in \mathbb{R}\times(\mathbb{R}^N\setminus\{0\})$ we can write

$$-(|u|^p\mathcal{G}(\vartheta_n) + 2pu|u|^{p-2}\langle Q\nabla_x u, \nabla\vartheta_n\rangle)(s,x) = f_n(s,x) - \vartheta'\left(\frac{|x|}{n}\right)\frac{\langle b(s,x),x\rangle}{|x|n} \tag{17.1.12}$$

for some infinitesimal sequence $\{f_n\} \subset L^1(\mathbb{R}^{N+1},\mu)$. Let us split $\langle b(s,x),x\rangle = \langle b(s,x),x\rangle^+ - \langle b(s,x),x\rangle^-$. Then, $\vartheta'(|x|/n)(n|x|)^{-1}\langle b(s,x),x\rangle^- \le 0$, while $\vartheta'(|x|/n)(n|x|)^{-1}\langle b(s,x),x\rangle^+$ vanishes pointwise in $\mathbb{R}^{N+1}$ as n tends to $+\infty$, and by the assumption in property (i)

$$\left|\vartheta'\left(\frac{|x|}{n}\right)\frac{\langle b(s,x),x\rangle^+}{|x|n}\right| \le 5C\|\vartheta'\|_\infty\varphi(x), \qquad (s,x) \in \mathbb{R}^{N+1}, \tag{17.1.13}$$

so that, by dominated convergence,

$$\lim_{n\to+\infty}\int_{\mathbb{R}^{N+1}}\left|\vartheta'\left(\frac{|x|}{n}\right)\right|\frac{\langle b(s,x),x\rangle^+}{|x|n}d\mu = 0. \tag{17.1.14}$$

Formulae (17.1.10), (17.1.12) and (17.1.14) show that $\int_{\mathbb{R}^{N+1}}\mathcal{G}(|u|^p)\vartheta_n d\mu$ converges to a nonpositive limit as n tends to $+\infty$. Taking (17.1.9) into account, we thus conclude that $\int_{\mathbb{R}^{N+1}}\mathcal{G}(|u|^p)d\mu \le 0$, so that $\mathcal{G}(|u|^p) \in L^1(\mathbb{R}^{N+1},\mu)$ and (17.1.7) follows. If, in addition, $|\langle b(s,x),x\rangle| \le C(|x|^2+1)\varphi(x)$ for any $(s,x) \in \mathbb{R}^{N+1}$, then estimates (17.1.13) and (17.1.14) hold with $\langle b(s,x),x\rangle^+$ replaced by $\langle b(s,x),x\rangle$, so that $\int_{\mathbb{R}^{N+1}}\mathcal{G}(|u|^p)\vartheta_n d\mu$ vanishes as n tends to $+\infty$ and the equality in (17.1.7) follows.

Step 2. Fix $p \in (1,2)$, $u \in D_{\rm comp}(\mathcal{G})$ and $\delta > 0$. Then, the function $u_\delta := (u^2+\delta)^{\frac{p}{2}} - \delta^{\frac{p}{2}}$ belongs to $C_{\rm comp}(\mathbb{R}^{N+1})$. A straightforward computation shows that $\mathcal{G}u_\delta = pu(u^2+\delta)^{\frac{p}{2}-1}\mathcal{G}u + p(u^2+\delta)^{\frac{p}{2}-2}[(p-1)u^2+\delta]|\sqrt{Q}\nabla_x u|^2$. If the assumption in the statement of property (i) is satisfied, then (17.1.9) holds true with $|u|^p$ being replaced by u_δ. Now, we can repeat the same arguments as in Step 1 and conclude that $\mathcal{G}u_\delta \in L^1(\mathbb{R}^{N+1},\mu)$ and it has nonpositive mean value, i.e.,

$$\int_{\mathbb{R}^{N+1}} (u^2+\delta)^{\frac{p}{2}-2}[(p-1)u^2+\delta]|\sqrt{Q}\nabla_x u|^2 d\mu \le -\int_{\mathbb{R}^{N+1}} u(u^2+\delta)^{\frac{p}{2}-1}\mathcal{G}u\, d\mu. \qquad (17.1.15)$$

If the diffusion coefficients are bounded, then $\mathcal{G}u_\delta \in C_{\rm comp}(\mathbb{R}^{N+1})$. Hence, $u_\delta \in D_{\rm comp}(\mathcal{G}) \subset D(G_1)$ and the inequality in (17.1.15) can be replaced by an equality.

By dominated convergence, the right-hand side of (17.1.15) converges, as δ tends to 0^+, to the corresponding integral with $\delta = 0$. Indeed, for any $\delta > 0$ the function $|u|(u^2+\delta)^{\frac{p}{2}-1}|G_p u|$ is bounded from above by $|u|^{p-1}|G_p u|$ since $p < 2$, and the μ-almost everywhere pointwise convergence is obvious. On the other hand, the functions $v_\delta = (u^2+\delta)^{\frac{p}{2}-2}[(p-1)u^2+\delta]|\sqrt{Q}\nabla_x u|^2$ converge pointwise almost everywhere (with respect to the Lebesgue measure), in the set where u does not vanish, to the function $(p-1)|u|^{p-2}|\sqrt{Q}\nabla_x u|^2$, and $\nabla_x u = 0$ almost everywhere (with respect to the Lebesgue measure) in the set where u vanishes. Taking the proof of Proposition 16.2.5 into account, we conclude that the function v_δ converges to 0 μ-almost everywhere in the set where u vanishes. This shows that v_δ converges pointwise to $(p-1)|u|^{p-2}|\sqrt{Q}\nabla_x u|^2\chi_{\{u\neq 0\}}$ μ-almost everywhere in $\mathbb{R}^{N+1}$. Fatou's lemma shows that $|u|^{p-2}\langle Q\nabla_x u, \nabla_x u\rangle\chi_{\{u\neq 0\}}$ belongs to $L^1(\mathbb{R}^{N+1},\mu)$ and (17.1.7) follows.

To complete the proof of property (iii) we note that $v_\delta \le (p-1)|u|^{p-2}|\sqrt{Q}\nabla_x u|^2\chi_{\{u\neq 0\}}$, since $p < 2$. Hence, we can apply the dominated convergence theorem to both sides of (17.1.15) (which is an equality if the diffusion coefficients are bounded) and the equality in (17.1.7) follows.

Step 3. Let $p \in (1,2]$ and $u \in D_{\rm comp}(\mathcal{G})$. Using the Hölder inequality, (17.1.7) and then the Hölder inequality again, we get

$$\begin{aligned}\|\sqrt{Q}\nabla_x u\|^2_{L^p(\mathbb{R}^{N+1},\mu)} \le &\|u\|^{2-p}_{L^p(\mathbb{R}^{N+1},\mu)}\||u|^{p-2}|\sqrt{Q}\nabla_x f|^2\chi_{\{u\neq 0\}}\|_{L^1(\mathbb{R}^{N+1},\mu)}\\ \le &\frac{1}{p-1}\|u\|_{L^p(\mathbb{R}^{N+1},\mu)}\|G_p u\|_{L^p(\mathbb{R}^{N+1},\mu)},\end{aligned}$$

i.e., $\|\sqrt{Q}\nabla_x u\|_{L^p(\mathbb{R}^{N+1},\mu)} \le (2\sqrt{p-1})^{-1}\|u\|_{D(G_p)}$. Since $D_{\rm comp}(\mathcal{G})$ is dense in $D(G_p)$ by Theorem 17.1.5, the map $u \mapsto |\sqrt{Q}\nabla_x u|$ is bounded from $D(G_p)$ into $L^p(\mathbb{R}^{N+1},\mu)$ and, since $Q(s,x) \ge \kappa_0 I$ for any $(s,x) \in \mathbb{R}^{N+1}$, also the map $u \mapsto |\nabla_x u|$ is bounded from $D(G_p)$ into $L^p(\mathbb{R}^{N+1},\mu)$, i.e., $D(G_p) \hookrightarrow W^{0,1}_p(\mathbb{R}^{N+1},\mu)$.

If $p > 2$ then the embedding $D(G_p) \hookrightarrow W^{0,1}_p(\mathbb{R}^{N+1},\mu)$ follows by interpolation. More precisely, for any $i = 1,\dots,N$ and $\lambda > 0$ the mapping $u \mapsto \int_0^{+\infty} e^{-\lambda t}D_i\mathscr{T}_2(t)u dt$ is bounded in $L^2(\mathbb{R}^{N+1},\mu)$ by the first part of this section. Similarly, the mapping $u \mapsto \int_0^{+\infty} e^{-\lambda t}D_i\mathscr{T}(t)u dt$ is bounded in $L^\infty(\mathbb{R}^{N+1})$, by Proposition 14.5.11, observing that, by Proposition 14.1.3, each operator $G(t,s)$ can be extended to $L^\infty(\mathbb{R}^N)$. Hence, by the Riesz-Thorin theorem it follows that the operator $u \mapsto D_i R(\lambda, G_p)u = \int_0^{+\infty} e^{-\lambda t}D_i\mathscr{T}_p(t)u dt$ is bounded in $L^p(\mathbb{R}^{N+1},\mu)$ for any $i = 1,\dots,N$. This shows that the range of $R(\lambda, G_p)$, which is the domain of G_p, is continuously embedded in $W^{0,1}_p(\mathbb{R}^{N+1},\mu)$.

Now, we can extend the formula (17.1.7) to any function in $D(G_p)$. Fix $u \in D(G_p)$ and let $\{u_n\} \subset D_{\rm comp}(\mathcal{G})$ be such that u_n and $G_p u_n$ converge, respectively, to u and $G_p u$ as n tends to $+\infty$. Clearly, $\int_{\mathbb{R}^{N+1}} u_n|u_n|^{p-2}G_p u_n d\mu$ converges to $\int_{\mathbb{R}^{N+1}} u|u|^{p-2}G_p u d\mu$, as n tends to $+\infty$, for any $p \in (1,+\infty)$. On the other hand, for $p \in (1,2)$, $|\nabla_x u_n|$ converges to $|\nabla_x u|$ in $L^p(\mathbb{R}^{N+1},\mu)$. Hence, up to a subsequence, we can assume that $|u_n|^{p-2}|\sqrt{Q}\nabla_x u_n|^2$ converges to $|u|^{p-2}|\sqrt{Q}\nabla_x u|^2$ μ-almost everywhere in $\mathbb{R}^{N+1}$, as n tends to $+\infty$. Fatou's lemma shows that

$$\int_{\mathbb{R}^{N+1}} |u|^{p-2}|\sqrt{Q}\nabla_x u_n|^2\chi_{\{u\neq 0\}}d\mu \le \liminf_{n\to+\infty}\int_{\mathbb{R}^{N+1}} |u_n|^{p-2}|\sqrt{Q}\nabla_x u_n|^2\chi_{\{u_n\neq 0\}}d\mu.$$

Hence, we can write (17.1.7) with u being replaced by u_n and let n tend to $+\infty$, to get it with u. In particular, this is an equality under the assumption in (ii).

Step 4. Let us fix $u, v \in D(G_2)$. Then, there exist two sequences $\{u_n\}$ and $\{v_n\}$ in $D_{\rm comp}(\mathcal{G})$ such that $(u_n, \mathcal{G}u_n)$ and $(v_n, \mathcal{G}v_n)$ converge, respectively, to (u, G_2u) and (v, G_2v) in $(L^2(\mathbb{R}^{N+1}, \mu))^2$, as n tends to $+\infty$. Clearly, $\mathcal{G}(u_nv_n) = v_n\mathcal{G}u_n + u_n\mathcal{G}v_n + 2\langle Q\nabla_x u_n, \nabla_x v_n\rangle$ for any $n \in \mathbb{N}$, and, by property (iv), the right-hand side of this formula converges to $v\mathcal{G}u + u\mathcal{G}v + 2\langle Q\nabla_x u, \nabla_x v\rangle$ in $L^1(\mathbb{R}^{N+1}, \mu)$. Similarly, u_nv_n tends to uv in $L^1(\mathbb{R}^{N+1}, \mu)$. Since the operator G_1 is closed, this means that $uv \in D(G_1)$ and $G_1(uv) = vG_2u + uG_2v + 2\langle Q\nabla_x u, \nabla_x v\rangle$. Integrating this formula in $\mathbb{R}^{N+1}$ and letting n tend to $+\infty$, we get (17.1.8). □

We now provide sufficient conditions for $C_c^\infty(\mathbb{R}^{N+1})$ to be a core of the operator G_p for any $p \in [1, +\infty)$. For this purpose, we refine the smoothness assumptions on the coefficients of the operator $\mathcal{A}$.

Hypothesis 17.1.7 *The coefficients q_{ij} and b_i $(i, j = 1, \dots, N)$ belong to $C^{\alpha/2,2+\alpha}_{\rm loc}(\mathbb{R}^{N+1})$ and to $C^{\alpha/2,1+\alpha}_{\rm loc}(\mathbb{R}^{N+1})$, respectively, for some $\alpha \in (0, 1)$.*

Theorem 17.1.8 *Let $p \in [1, +\infty)$ and assume that Hypotheses 17.1.1(ii), (iii) and Hypothesis 17.1.7 are satisfied. Further, assume that there exist a function $V \in C^2(\mathbb{R}^N)$, with infimum greater than 1 and blowing up as $|x|$ tends to $+\infty$, and a constant $c > 0$ such that the functions*

$$\mathscr{E}_c(s, x) := e^{-c|s|}\frac{\mathcal{A}(s)V(x)}{V(x)\log V(x)}, \qquad \mathscr{F}_c(s, x) := e^{-c|s|}\frac{|\sqrt{Q}(s, x)\nabla V(x)|^2}{(V(x))^2\log V(x)} \tag{17.1.16}$$

belong to $L^p(\mathbb{R}^{N+1}, \mu)$. Then, $C_c^\infty(\mathbb{R}^{N+1})$ is a core of the operator G_q for any $1 \le q \le p$.

To prove Theorem 17.1.8, we adapt the techniques in [9]. To this aim, we need the following preliminary result.

Theorem 17.1.9 *Suppose that Hypotheses 17.1.1(ii), (iii) and Hypothesis 17.1.7 are satisfied and let $\widehat{\mu}$ be a signed measure on $(a, b) \times B_R$ (for some $a, b \in \overline{\mathbb{R}}$, such that $a < b$, and $R > 0$) satisfying the condition*

$$\int_{(a,b)\times\mathbb{R}^N} \mathcal{G}\psi d\widehat{\mu} = 0, \qquad \psi \in C_c^\infty((a, b) \times B_R). \tag{17.1.17}$$

Then, $\widehat{\mu}$ is absolutely continuous with respect to the Lebesgue measure and its density $\widehat{\rho}$ belongs to $C^{1+\alpha/2,2+\alpha}_{\rm loc}((a, b) \times B_R)$, where α is the same as in Hypothesis 17.1.7. In particular, if Hypothesis 17.1.1(ii) is satisfied, then each invariant measure μ of $\{\mathscr{T}(t)\}$ admits a density $\rho \in C^{1+\alpha/2,2+\alpha}_{\rm loc}(\mathbb{R}^{N+1})$ with respect to the Lebesgue measure.

Proof [4] By Proposition 17.1.2, the measure μ satisfies the formula (17.1.17) with $a = -\infty$ and $b = +\infty$. Hence, we just need to prove the first part of the statement. By Proposition 16.2.5, $\widehat{\mu}$ admits a density $\widehat{\rho}$ with respect to the Lebesgue measure, which belongs to $C^\beta_{\rm loc}((a, b) \times \mathbb{R}^N) \cap W^{0,1}_{p,{\rm loc}}((a, b) \times \mathbb{R}^N)$ for some $\beta \in (0, 1)$ and any $p \in [1, +\infty)$. This function is a weak solution of the equation $D_s\widehat{\rho} + \mathcal{A}^*\widehat{\rho} = 0$, where $\mathcal{A}^*$ denotes the adjoint of the operator $\mathcal{A}$, i.e.,

$$\int_{(a,b)\times B_R} (\widehat{\rho}{\rm div}_x(Q\nabla_x\psi) + \langle Q\nabla_x\widehat{\rho}, \nabla_x\psi\rangle + ({\rm div}_x\, b)\widehat{\rho}\psi + \psi\langle b, \nabla_x\widehat{\rho}\rangle + \widehat{\rho}D_s\psi)dsdx = 0$$

[4] For the classical results used in the proof, we refer the reader, e.g., to [159, Chapters 3, 4], [87, Appendix], [178, Chapter 1].

for any $\psi \in C_c^\infty((a,b)\times B_R)$ or, equivalently, for any $\psi \in C_c^{1,2}((a,b)\times B_R)$.

For any $n\in\mathbb{N}$ greater than $2(b-a)^{-1}$, we set $a_n = a+1/n$, $b_n = b-1/n$, $R_n = R-1/n$ and $\Omega_n = (-b_n,-a_n)\times B_{R_n}$. Further, we fix a cut-off function ϑ_n such that $\chi_{\Omega_n}\le\vartheta_n\le\chi_{\Omega_{n+1}}$. The function $v := \rho^-\vartheta_{n+1}$, where $\rho^-(s,x) = \widehat{\rho}(-s,x)$ for any $s\in(-b,-a)$, has the same degree of smoothness as ρ^- and is a weak solution of the Cauchy problem

$$\begin{cases} D_s v(s,x) = (\mathcal{A}^-)^* v(s,x) + f(s,x), & (s,x)\in\Omega_{n+2},\\ v(s,x) = 0, & (s,x)\in(-b_{n+2},-a_{n+2})\times\partial B_{R_{n+2}},\\ v(-b_{n+2},x) = 0, & x\in B_{R_{n+2}},\end{cases}\tag{17.1.18}$$

where the operator $\mathcal{A}^-$ is defined as the operator $\mathcal{A}$ with the coefficients q_{ij} and b_j being replaced by q_{ij}^- and b_j^- $(i,j=1,\ldots,N)$, which are obtained from the q_{ij}'s and b_j's as ρ^- is obtained from ρ, and

$$\begin{aligned} f =& -2\rho^-\mathrm{div}_x(Q_-\nabla\vartheta_{n+1}) - 2\langle Q^-\nabla_x\rho^-, \nabla_x\vartheta_{n+1}\rangle\\ &-\rho^-\mathrm{Tr}(Q^-D_x^2\vartheta_{n+1}) + \rho^-\langle b^-, \nabla_x\vartheta_{n+1}\rangle + \rho^- D_s\vartheta_{n+1}.\end{aligned}$$

Since $f\in L^p(\Omega_{n+2})$ for any $p\in[1,+\infty)$, by classical results there exists a unique function w which belongs to $W_p^{1,2}(\Omega_{n+2})$ for any $p\ge1$ and solves the Cauchy problem (17.1.18). We now observe that, in fact, $w=v$. Indeed, there exists a unique function $z\in V_2(\Omega_{n+2})$ such that[5]

$$\begin{aligned}\int_{\Omega_{n+2}} f\psi dsdx =& \int_{\Omega_{n+2}}\big[-zD_s\psi + z\mathrm{div}_x(Q^-\nabla_x\psi) + \langle Q^-\nabla_x z, \nabla_x\psi\rangle\big]dsdx\\ &+\int_{\Omega_{n+2}}\big[(\mathrm{div}_x b^-)z\psi + \psi\langle b^-, \nabla_x z\rangle\big]dsdx\end{aligned}\tag{17.1.19}$$

for any $\psi\in W_2^{1,1}(\Omega_{n+2})$ which vanishes on $(-b_{n+2},-a_{n+2})\times\partial B_{R_{n+2}}$ and on $\{-a_{n+2}\}\times B_{R_{n+2}}$. The Sobolev embedding theorem implies that both the functions v and w belong to $V_2(\Omega_{n+2})$ and solve the variational equation (17.1.19). Hence, they do coincide.

Now, we can check that v belongs to $C^{\sigma,1+\sigma}(\Omega_{n+2})$ for any $\sigma\in(0,1)$. Indeed, v belongs to $L^p((-b_{n+2},-a_{n+2});W^{2\gamma,p}(B_{R_{n+2}}))\cap W_0^{1,p}((-b_{n+2},-a_{n+2});L^p(B_{R_{n+2}}))$ for any $\gamma\in(0,1)$. Since

$$\{g\in W^{2\delta,p}(B_{R_{n+2}}) : g_{|\partial B_{R_{n+2}}} = 0\} = (L^p(B_{R_{n+2}}), W^{2,p}(B_{R_{n+2}})\cap W_0^{1,p}(B_{R_{n+2}}))_{\delta,p}$$

for any $\delta\in(0,1)\setminus\{1/2\}$ and $p>(2\delta)^{-1}$, v is in $W^{\varepsilon,p}((-b_{n+2},-a_{n+2});W^{2\vartheta-2\varepsilon,p}(B_{R_{n+2}}))$ for any $\varepsilon>0$ such that $2(\vartheta-\varepsilon)>1$. Let us fix $\gamma,\vartheta\in(0,1)$ and $\varepsilon\in(0,\vartheta)$ such that $2\vartheta-2\varepsilon>1+\gamma$. Taking $p>1/\varepsilon$ it follows that $v\in C([-b_{n+2},-a_{n+2}];W^{2\vartheta-2\varepsilon,p}(B_{R_{n+2}}))$. Up to replacing p with a larger index, we can assume that $2\vartheta-2\varepsilon-N/p>1+\gamma$. Hence, by the Sobolev embedding theorem, v belongs to $C([-b_{n+2},-a_{n+2}];C^{1+\gamma}(\overline{B}_{R_{n+2}}))$. Next, we observe that, since $v\in W^{1,p}(\Omega_{n+2})$ for any $p>N+1$, v belongs to $C^\beta(\Omega_{n+2})$ for any $\beta\in(0,1)$. Now, we are almost done. Indeed, there exists a positive constant K_γ such that $\|z\|_{C^1(\overline{B}_{R_{n+2}})}\le K_\gamma\|z\|_\infty^{\frac{\gamma}{1+\gamma}}\|z\|_{C^{1+\gamma}(B_{R_{n+2}})}^{\frac{1}{1+\gamma}}$ for any $z\in C^{1+\gamma}(B_{R_{n+2}})$. Plugging the function $z=v(s,\cdot)-v(r,\cdot)$ into this estimate (where r,s are arbitrarily fixed in $(-b_{n+2},-a_{n+2})$), gives

$$\|v(s,\cdot)-v(r,\cdot)\|_{C^1(\overline{B}_{R_{n+2}})}\le K[v]_{C^\gamma(B_{n+2})}^{\frac{\gamma}{1+\gamma}}|t-s|^{\frac{\beta\gamma}{1+\gamma}}(2\|u\|_{C([-b_{n+2},-a_{n+2}];C^{1+\gamma}(B_{R_{n+2}}))})^{\frac{1}{1+\gamma}}.$$

[5] $V_2(\Omega_{n+2})$ is the set of all functions $z\in W_2^{0,1}(\Omega_{n+2})$ vanishing on $(-b_{n+2},-a_{n+2})\times\partial B_{n+2}$ and such that $\sup_{t\in(-b_{n+2},-a_{n+2})}\|z(t,\cdot)\|_{L^2(B_{n+2})}<+\infty$.

A suitable choice of the exponents β and γ shows that $v \in C^{\sigma,1+\sigma}(\Omega_{n+2})$.

We are almost done. Indeed, since $v = \rho^-$ in Ω_{n+1}, $\widehat{\rho}^-$ belongs to $C^{\alpha,1+\alpha}(\Omega_{n+1})$. Hence, f is in $C^{\alpha/2,\alpha}(\Omega_{n+1})$. Consider problem (17.1.18) with ϑ_{n+1} being replaced by ϑ_n. By classical Hölder regularity results for parabolic equations we conclude that $v \in C^{1+\alpha/2,2+\alpha}(\Omega_{n+1})$. Hence, $\widehat{\rho} \in C^{1+\alpha/2,2+\alpha}((a_n,b_n)\times B_{R_n})$ and the proof is complete. □

Proof of Theorem 17.1.8 Fix $p \in [1,+\infty)$. By Theorem 17.1.5, the operator $\lambda I - G_p$ is bijective from $D(G_p)$ onto $L^p(\mathbb{R}^{N+1},\mu)$ for any $\lambda > 0$. Hence, to prove the assertion it suffices to show that $(\lambda - \mathcal{G})(C_c^\infty(\mathbb{R}^{N+1}))$ is dense in $L^p(\mathbb{R}^{N+1},\mu)$ for some $\lambda > 0$, i.e., we need to show that, if $\psi \in L^{p'}(\mathbb{R}^{N+1},\mu)$, where p' is the conjugate index of p, is such that

$$\int_{\mathbb{R}^{N+1}} (\lambda\zeta - \mathcal{G}\zeta)\psi d\mu = 0 \tag{17.1.20}$$

for some $\lambda > 0$ and any $\zeta \in C_c^\infty(\mathbb{R}^{N+1})$, then $\psi \equiv 0$ μ-almost everywhere. From now on we assume that (17.1.20) holds with $\lambda > 0$ being arbitrarily fixed. At the end of the proof, we will choose λ.

By Theorem 17.1.9, μ has a density $\rho \in C^1(\mathbb{R}^{N+1})$. For our purposes we need to show that $\rho\psi$ also belongs to $C^1(\mathbb{R}^{N+1})$. Observe that, writing (17.1.20) with ζ being replaced by the function $(s,x) \mapsto e^{-\lambda s}\zeta(x)$, it follows that $\int_{\mathbb{R}^{N+1}} (\mathcal{G}\zeta)(s,x)\, e^{-\lambda s} d\widehat{\mu} = 0$ for any $\zeta \in C_c^\infty(\mathbb{R}^{N+1})$, where $d\widehat{\mu} = e^{-\lambda s}\psi(s,x)dsdx$ is a signed Borel measure in $(-R,R)\times B_R$. Theorem 17.1.9 implies that the function $(s,x) \mapsto e^{-\lambda s}\rho(s,x)\psi(s,x)$ belongs to $C^1(\mathbb{R}^{N+1})$ and, consequently, also the function $\rho\psi$ does. Since $\rho > 0$ in $\mathbb{R}^{N+1}$, by Theorem 17.1.9 the function ψ belongs to $C^1(\mathbb{R}^{N+1})$ as well.

Integrating by parts the identity $\lambda \int_{\mathbb{R}^{N+1}} \zeta\psi d\mu = \int_{\mathbb{R}^{N+1}} \mathcal{G}\zeta\psi d\mu$ yields

$$\lambda \int_{\mathbb{R}^{N+1}} \zeta\psi d\mu = \int_{\mathbb{R}^{N+1}} (-\langle Q\nabla_x(\psi\rho), \nabla_x\zeta\rangle + \psi\rho\langle \widehat{b}, \nabla_x\zeta\rangle - \psi\rho D_s\zeta) dsdx \tag{17.1.21}$$

for any $\zeta \in C_c^\infty(\mathbb{R}^{N+1})$, where $\widehat{b}_i = b_i - \sum_{j=1}^N D_j q_{ij}$ for any $i = 1,\dots,N$. By density, the formula (17.1.21) holds for any $\zeta \in C_c^1(\mathbb{R}^{N+1})$.

Let $G : \mathbb{R} \to [-1,1]$ be an increasing C^∞-function such that $G(t) = 0$ if $|t| \le 1$, $G(t) = -1$, if $t \le -2$, $G(t) = 1$, if $t \ge 2$, and $\|G'\|_\infty \le 2$. For any $n \in \mathbb{N}$ and $(s,x) \in \mathbb{R}^{N+1}$, set $\psi_n(s,x) := G(n\psi(s,x)\rho(s,x))$. The function ψ_n belongs to $C^1(\mathbb{R}^{N+1})$, $|\psi_n| \le 1$ and ψ_n converges to $\operatorname{sign}(\psi)$ as n tends to $+\infty$, pointwise in $\mathbb{R}^{N+1}$. Fix a positive function $\vartheta \in C_c^\infty(\mathbb{R}^{N+1})$. Taking $\zeta = \psi_n\vartheta$ in (17.1.21), integrating by parts the term $\int_{\mathbb{R}^{N+1}} \psi_n\langle Q\nabla_x(\psi\rho), \nabla_x\vartheta\rangle dsdx$ and observing that $n\vartheta G'(n\psi\rho)|\sqrt{Q}\nabla_x(\psi\rho)|^2 \ge 0$ in $\mathbb{R}^{N+1}$, we get

$$\begin{aligned}\lambda \int_{\mathbb{R}^{N+1}} \vartheta\psi_n\psi d\mu \le & \int_{\mathbb{R}^{N+1}} [\psi_n\psi\rho \operatorname{div}_x(Q\nabla_x\vartheta) + n\psi\rho G'(n\psi\rho)\langle Q\nabla_x(\psi\rho), \nabla_x\vartheta\rangle] dsdx \\ & + \int_{\mathbb{R}^{N+1}} [\psi_n\psi\rho\langle \widehat{b}, \nabla_x\vartheta\rangle + n\vartheta\psi\rho G'(n\psi\rho)\langle \widehat{b}, \nabla_x(\psi\rho)\rangle] dsdx \\ & - \int_{\mathbb{R}^{N+1}} [n\vartheta\psi\rho G'(n\psi\rho) D_s(\psi\rho) + \psi_n\psi\rho D_s\vartheta] dsdx.\end{aligned} \tag{17.1.22}$$

Recalling that $G'(ny) \ne 0$ if and only if $1 \le |ny| \le 2$, we conclude that $|nyG'(ny)| \le 4$ for any $n \in \mathbb{N}$ and $y \in \mathbb{R}$. Moreover, $nyG'(ny)$ converges to 0 as n tends to $+\infty$ for any $y \in \mathbb{R}$. By passing to the limit and taking into account that the supports of all the involved functions are contained in the support of ϑ, from (17.1.22) we get

$$\lambda \int_{\mathbb{R}^{N+1}} \vartheta|\psi| d\mu \le \int_{\mathbb{R}^{N+1}} |\psi|\mathcal{G}\vartheta d\mu. \tag{17.1.23}$$

Let $H:\mathbb{R}\to[0,1]$ be a decreasing C^∞-function such that $H(t)=1$, if $t\le 1$, and $H(t)=0$, if $t\ge 2$, and let $g:\mathbb{R}\to\mathbb{R}$ be a positive C^∞-function such that $g(t)=1$, if $|t|\le 1$, $g(t)=e^{-2c|t|/p}$, if $|t|\ge 2$, where $c>0$ is as in the statement of the theorem, and $\|g\|_\infty\le 1$. Define the function $\vartheta_n:\mathbb{R}^{N+1}\to\mathbb{R}$ by setting $\vartheta_n(s,x)=g(s)H(n^{-1}\Psi(s,x))$ for any $(s,x)\in\mathbb{R}^{N+1}$, where $\Psi(s,x)=(1+s^2)\log V(x)$. Since $V(x)$ blows up as $|x|$ tends to $+\infty$ and $\inf_{x\in\mathbb{R}^N}V(x)>1$, Ψ is a positive function diverging to $+\infty$ as $|(s,x)|$ tends to $+\infty$. Therefore, $\vartheta_n\in C^\infty_c(\mathbb{R}^{N+1})$ and it converges to $g(s)$ pointwise in $\mathbb{R}^{N+1}$ as n tends to $+\infty$. Moreover, since

$$
\begin{aligned}
\mathcal{G}\vartheta_n(s,\cdot)=&g(s)H'\left(\frac{\Psi(s,\cdot)}{n}\right)\frac{\Psi(s,\cdot)}{n}\left\{\frac{\mathcal{A}V(s,\cdot)}{V\log V}-\frac{|\sqrt{Q}(s,\cdot)\nabla V|^2}{V^2\log V}-\frac{2s}{1+s^2}\right\}\\
&+g(s)H''\left(\frac{\Psi(s,\cdot)}{n}\right)\frac{(\Psi(s,\cdot))^2}{n^2}\frac{|\sqrt{Q}(s,\cdot)\nabla V|^2}{(V\log V)^2}-g'(s)H\left(\frac{\Psi(s,\cdot)}{n}\right)
\end{aligned}
$$

for any $s\in\mathbb{R}$, $\mathcal{G}\vartheta_n$ converges to g' pointwise in $\mathbb{R}^{N+1}$ as n tends to $+\infty$. Finally, for any $n\in\mathbb{N}$ we can estimate

$$
|\mathcal{G}\vartheta_n|\le C_1 g\frac{|\mathcal{A}V|}{V\log V}+C_2 g\frac{|\sqrt{Q}(s,\cdot)\nabla V|^2}{V^2\log V}+C_3 g+|g'|
$$

for some positive constants C_j $(j=1,2,3)$ depending on the infimum of V and $\|H\|_{C^2(\mathbb{R})}$. Since the right-hand side of the previous inequality belongs to $L^p(\mathbb{R}^{N+1},\mu)$, we can write (17.1.23), with ϑ_n replacing ϑ, and take the limit as n tends to $+\infty$ by the dominated convergence theorem. This gives $\int_{\mathbb{R}^{N+1}}(\lambda g+g')|\psi|d\mu\le 0$. Since the function g belongs to $C^\infty(\mathbb{R})$ and $|g'(s)|^{-1}g(s)=2p^{-1}c$ if $|s|\ge 2$, $g_0:=\sup_{s\in\mathbb{R}}(g'(s))^{-1}g(s)$ is finite and taking $\lambda>g_0$, the previous estimate implies that $\psi\equiv 0$ μ-almost everywhere in $\mathbb{R}^{N+1}$. □

Remark 17.1.10 In the case when $\mathbb{R}^{N+1}$ is replaced by $(0,T)\times\mathbb{R}^N$, W. Stannat in [242] proved that $C^\infty_c((0,T)\times\mathbb{R}^N)$ is a core of the realization of $\mathcal{G}$ in $L^1((0,T)\times\mathbb{R}^N)$, under somewhat weaker regularity assumptions on the coefficients of the operator $\mathcal{A}$ in divergence form, but assuming that the diffusion coefficients are bounded in $[a,b]\times\mathbb{R}^N$ for any $[a,b]\subset\mathbb{R}^N$.

We now provide concrete conditions on the operator $\mathcal{A}$, which imply the validity of the assumptions of Theorem 17.1.8. For this purpose, for any $\beta,\delta>0$ we introduce the function $V_{\beta,\delta}:\mathbb{R}^N\to\mathbb{R}$ defined by

$$
V_{\beta,\delta}(x)=\begin{cases}2\exp(\delta|x|^\beta), & \text{if }\beta\ge 2,\\ 2\vartheta(|x|)+2(1-\vartheta(|x|))\exp(\delta|x|^\beta), & \text{if }\beta\in(0,2),\end{cases}
$$

for any $x\in\mathbb{R}^N$, where $\vartheta\in C^2([0,+\infty))$ is a decreasing function such that $\chi_{[0,1]}\le\vartheta\le\chi_{[0,2]}$.

Lemma 17.1.11 *Besides Hypotheses 17.1.1(i), (ii) and Hypothesis 17.1.7 assume that there exist positive constants k, β and γ such that*

$$
b_0:=\sup_{(s,x)\in\mathbb{R}^{N+1}}\langle b(s,x),x\rangle<+\infty, \tag{17.1.24}
$$

$$
\lim_{|x|\to+\infty}\sup_{s\in\mathbb{R}}\left(\gamma\Lambda(s,x)|x|^\beta+\langle b(s,x),x\rangle\right)=-\infty, \tag{17.1.25}
$$

$$
\sup_{(s,x)\in\mathbb{R}\times B_M}\left(|q_{ij}(s,x)|+e^{-k|s|}|b_j(s,x)|\right)<+\infty \tag{17.1.26}
$$

for any $M>0$ and $i,j=1,\dots,N$, where $\Lambda(s,x)$ denotes the maximum eigenvalue of the *matrix $Q(s,x)$. Then, for any $\delta>0$ such that $\beta\delta<\gamma$, the function $V_{\beta,\delta}$ satisfies Hypothesis 17.1.1(iii). Hence, there exists a tight evolution system of measures $\{\mu_s : s\in\mathbb{R}\}$ associated with the operator $\{G(t,s)\}$.*

Proof As a straightforward computation shows

$$\mathcal{A}(s)V_{\beta,\delta}(x)\leq \beta\delta|x|^{\beta-2}V_{\beta,\delta}(x)[(\beta\delta|x|^{\beta}+K_{\beta,\delta})\Lambda(s,x)+\langle b(s,x),x\rangle] \tag{17.1.27}$$

for any $(s,x)\in\mathbb{R}\times(\mathbb{R}^N\setminus B_2)$, where $K_{\beta,\delta}:=N+\beta\delta(\beta-2)^+$.

Now, let us fix $R>2$ such that $\beta\delta|x|^{\beta}+K_{\beta,\delta}\leq\gamma|x|^{\beta}$ in $\mathbb{R}^N\setminus B_R$. Up to replacing R with a larger value if necessary, we can assume that $\sup_{s\in\mathbb{R}}\{\Lambda(s,x)(\beta\delta|x|^{\beta}+K_{\beta,\delta})+\langle b(s,x),x\rangle\}\leq -1$ for any $x\in\mathbb{R}^N\setminus B_R$ and, consequently, from (17.1.27) it follows that $\mathcal{A}V_{\beta,\delta}\leq-\beta\delta V_{\beta,\delta}$ in $\mathbb{R}\times(\mathbb{R}^N\setminus B_R)$. On the other hand, if $\beta\geq 2$ then (17.1.24) and (17.1.26) imply that $\mathcal{A}V_{\beta,\delta}\leq\beta\delta a$ in $\mathbb{R}\times B_R$, where $a=R^{\beta-2}e^{\delta R^{\beta}}[(\beta\delta R^{\delta}+K_{\beta,\delta})\|\Lambda\|_{C_b(\mathbb{R}\times B_R)}+b_0^+]$. Combining these two estimates, we deduce that $\mathcal{A}V_{\beta,\delta}\leq\beta\delta a'-\beta\delta V_{\beta,\delta}$ in $\mathbb{R}^{N+1}$, where $a'=a+2e^{\delta R^{\beta}}$. It is now clear that the function $V_{\beta,\delta}$ satisfies Hypothesis 17.1.1(iii).

If $\beta\in(0,2)$ then the terms in $\mathcal{A}V_{\beta,\delta}$ which contain the diffusion coefficients are clearly bounded in $\mathbb{R}\times B_2$. since the diffusion coefficients are bounded in $\mathbb{R}\times B_2$ by condition (17.1.26). On the other hand, the terms which contain the drift coefficients b_j are $2\beta\delta|x|^{\beta-2}(1-\vartheta(|x|))e^{\delta|x|^{\beta}}\langle b(s,x),x\rangle$ and $2|x|^{-1}\vartheta'(|x|)(1-e^{\delta|x|^{\beta}})\langle b(s,x),x\rangle$. In both the terms the coefficient in front of $\langle b(s,x),x\rangle$ is nonnegative. Hence, thanks to (17.1.24), they can be bounded from above by a positive constant, independent of $(s,x)\in\mathbb{R}\times B_2$. Hence, also in this case, $\mathcal{A}V_{\beta,\delta}\leq\overline{a}-\beta\delta V_{\beta,\delta}$ in $\mathbb{R}^{N+1}$ for some constant $\overline{a}>0$.

The last assertion follows from (the proof of) Theorem 16.2.7. □

In order to apply Theorem 17.1.8, we have to show that both the functions $\mathscr{E}_c$ and $\mathscr{F}_c$ are in $L^p(\mathbb{R}^{N+1},\mu)$ for some $c>0$. The main obstruction in this direction is the fact that in most cases the measure μ is not explicitly known. To overcome this difficulty, we will provide some functions which are integrable with respect to the measure μ and we use them to apply a comparison argument to show the integrability of $\mathscr{E}_c$ and $\mathscr{F}_c$. For this purpose, we need the following result.

Lemma 17.1.12 *Let the assumptions of Lemma* 17.1.11 *be satisfied. Then, the following properties hold true.*

(i) *The function $fV_{\beta,\delta}$ belongs to $L^1(\mathbb{R}^N,\mu)$ for any $f\in L^1(\mathbb{R})$.*

(ii) *The function $f\mathcal{A}V_{\beta,\delta}$ belongs to $L^1(\mathbb{R}^N,\mu)$ for any $f:\mathbb{R}\to\mathbb{R}$ such that the function $s\mapsto e^{\lambda|s|}|f(s)|$ is bounded and measurable for some $\lambda>k$, where k is as in* (17.1.26).

Proof (i) The proofs of Theorem 16.2.7 and Lemma 17.1.11 show that

$$\sup_{s\in\mathbb{R}}\|V_{\beta,\delta}\|_{L^1(\mathbb{R}^N,\mu_s)}<+\infty.$$

The integrability of the function $fV_{\beta,\delta}$ thus follows from the Fubini-Tonelli theorem.

(ii) We split the proof into two steps. To ease the notation, throughout the proof, we simply write V instead of $V_{\beta,\delta}$.

Step 1. Here, we prove that $\int_{\mathbb{R}^{N+1}}\mathcal{G}(\psi\zeta)\,d\mu=0$ for any $\psi\in C^1(\mathbb{R})$, such that $\psi(s)=e^{-\lambda|s|}$ for any $|s|\geq 1$ and some $\lambda>k$, and any $\zeta\in C^2(\mathbb{R}^N)$ which is constant outside of a compact set.

Proposition 17.1.2 shows that $\int_{\mathbb{R}^{N+1}}\mathcal{G}ud\mu=0$ for any $u\in C_c^{\infty}(\mathbb{R}^{N+1})$ and this estimate can be extended straightforwardly to any $u\in C_c^{1,2}(\mathbb{R}^{N+1})$.

Let $\zeta \in C^2(\mathbb{R}^N)$ be constant outside of a compact set and, for any $n \in \mathbb{N}$, set $\psi_n = \psi\vartheta_n$, where $\vartheta_n \in C_c^\infty(\mathbb{R})$ satisfies the conditions $\chi_{(-n,n)} \le \vartheta_n \le \chi_{(-2n,2n)}$ and $\|\vartheta_n'\|_\infty \le 2n^{-1}$. Clearly, ψ_n converges to ψ locally uniformly in $\mathbb{R}$ and $|\psi_n(s,x)| + |\psi_n'(s,x)| \le C_1 e^{-\lambda|s|}$ for any $(s,x) \in \mathbb{R}^{N+1}$, $n \in \mathbb{N}$ and some positive constant C_1. Moreover, $\int_{\mathbb{R}^{N+1}} \mathcal{G}(\psi_n\zeta)d\mu = 0$ for any $n \in \mathbb{N}$. Indeed, $\mathcal{G}(\psi_n\zeta) = \mathcal{G}(\psi_n(\zeta - k))$ for any $k \in \mathbb{R}$ and we can choose k such that the function $\psi_n(\zeta - k)$ is compactly supported in $\mathbb{R}^{N+1}$. Since, by (17.1.26), $|\mathcal{A}\zeta(s,x)| \le C_2 e^{k|s|}$ for any $(s,x) \in \mathbb{R}^{N+1}$ and some positive constant C_2, by dominated convergence we get $\int_{\mathbb{R}^{N+1}} \mathcal{G}(\psi\zeta)d\mu = 0$.

Step 2. Now, we can prove that the function $\psi\mathcal{A}\zeta$ belongs to $L^1(\mathbb{R}^{N+1},\mu)$. For this purpose, for any $n \in \mathbb{N}$ let $V_n = \zeta_n \circ V$, where $\zeta_n \in C^2([0,+\infty))$ is any function such that $\zeta_n(\sigma) = \sigma$ for any $\sigma \in [0,n]$, $\zeta_n = n+1$ in $[n+2,+\infty)$, $-4 \le \zeta_n'' \le 0 \le \zeta_n' \le 1$. Each function V_n belongs to $C^2(\mathbb{R}^N)$ and is constant outside of a compact set. Moreover, $V_n \le V$ for any $n \in \mathbb{N}$ and V_n converges to V pointwise in $\mathbb{R}^N$ as n tends to $+\infty$. Since $\mathcal{A}V_n = \zeta_n'(V)\mathcal{A}V + \zeta_n''(V)|\sqrt{Q}\nabla V|^2$, $\mathcal{A}V_n$ converges to $\mathcal{A}V$ pointwise in $\mathbb{R}^{N+1}$ as n tends to $+\infty$. Moreover, the proof of Lemma 17.1.11 shows that $\mathcal{A}V \le a - dV$ in $\mathbb{R}^{N+1}$ for some positive constants a, d. Hence, $\mathcal{A}V_n \le \zeta_n'(V)(a - dV)$ for any $n \in \mathbb{N}$ and this implies that $\mathcal{A}V_n \le 0$ in $\mathbb{R} \times (\mathbb{R}^N \setminus B_R)$ for some $R > 0$, independent of n. The identity $\int_{\mathbb{R}^{N+1}} \mathcal{G}(\psi V_n)d\mu = 0$ yields $\int_{\mathbb{R}\times(\mathbb{R}^N\setminus B_R)} \mathcal{G}(\psi V_n)d\mu = -\int_{\mathbb{R}\times B_R} \mathcal{G}(\psi V_n)d\mu$, from which we deduce, first, that

$$\begin{aligned}\int_{\mathbb{R}\times(\mathbb{R}^N\setminus B_R)} |\psi\mathcal{A}V_n|d\mu &= -\int_{\mathbb{R}\times(\mathbb{R}^N\setminus B_R)} \mathcal{G}(\psi V_n)d\mu - \int_{\mathbb{R}\times(\mathbb{R}^N\setminus B_R)} \psi' V_n d\mu\\ &= \int_{\mathbb{R}\times B_R} \psi\mathcal{A}V_n d\mu - \int_{\mathbb{R}^{N+1}} \psi' V_n d\mu\end{aligned}$$

and, then, that

$$\int_{\mathbb{R}^{N+1}} |\psi\mathcal{A}V_n|d\mu \le 2\int_{\mathbb{R}\times B_R} |\psi\mathcal{A}V_n|d\mu - \int_{\mathbb{R}^{N+1}} \psi' V_n d\mu. \tag{17.1.28}$$

Since $|\mathcal{A}V_n| \le |\mathcal{A}V| + 4|\sqrt{Q}\nabla V|^2$ for any $n \in \mathbb{N}$, from (17.1.26) and a straightforward computation it follows that, if $\beta \ge 2$, then

$$\begin{aligned}|\mathcal{A}V_n(s,x)| \le &\beta\delta|x|^{\beta-2}V(x)[\Lambda(s,x)(\beta\delta|x|^\beta + K_{\beta,\delta}) + |\langle b(s,x),x\rangle|]\\ &+ (\beta\delta)^2\Lambda(s,x)|x|^{2\beta-2}|V(x)|^2\\ \le &2\beta\delta R^{\beta-2}e^{k|s|}\bigg\{\|\Lambda\|_{C_b(\mathbb{R}\times B_R)}(2\beta\delta|R|^\beta e^{\delta R^\beta} + \beta\delta|R|^\beta + K_{\beta,\delta})\\ &+ \sup_{(s,x)\in\mathbb{R}\times B_R}(e^{-k|s|}|\langle b(s,x),x\rangle|)\bigg\}e^{\delta R^\beta} =: C_3 e^{k|s|}\end{aligned}$$

for any $(s,x) \in \mathbb{R}\times B_R$, $n \in \mathbb{N}$ and some positive constant C_3, where $K_{\beta,\delta} = N + \beta\delta(\beta-2)^+$ and $\Lambda(s,x)$ denotes the maximum eigenvalue of the matrix $Q(s,x)$. If $R > 2$ then this estimate holds true in $\mathbb{R} \times (B_R \setminus B_2)$ also in the case $\beta < 2$. On the other hand, if $|x| \le 2$ then, by (17.1.26) we can show that $|\mathcal{A}(s)V_n(x)| \le C_4 e^{k|s|}$ for any $(s,x) \in \mathbb{R} \times B_2$ and, consequently, $|\mathcal{A}(s)V_n(x)| \le C_5 e^{k|s|}$ for any $(s,x) \in \mathbb{R} \times B_R$, C_4 and C_5 being positive constants. Recalling that $V_n \le V$ and taking property (i) into account, from estimate (17.1.28) we deduce that $\sup_{n\in\mathbb{N}} \|\psi\mathcal{A}V_n\|_{L^1(\mathbb{R}^{N+1},\mu)} < +\infty$. Applying Fatou's lemma, we conclude that the function $\psi\mathcal{A}V$ belongs to $L^1(\mathbb{R}^{N+1},\mu)$. □

Theorem 17.1.13 *Let the assumptions of Lemma* 17.1.11 *be satisfied and fix $\delta > 0$ such that $\beta\delta < \gamma$. Then, the following properties hold true.*

(i) *If* $|q_{ij}(s,x)| < M_1|x|^{2-\beta}e^{\delta|x|^\beta}$ *for any* $s \in \mathbb{R}$, $x \in \mathbb{R}^N \setminus B_1$, $i,j = 1,\ldots,N$ *and some positive constant* M_1, *then* $C_c^\infty(\mathbb{R}^{N+1})$ *is a core of* G_1.

(ii) *Suppose that*

$$|q_{ij}(s,x)| \le M_2|x|^{2-\beta}e^{\frac{\delta}{p}|x|^\beta}e^{k|s|} \tag{17.1.29}$$

and

$$|\langle b(s,x),x\rangle| \le M_2|x|^{2+\frac{\beta}{p-1}}e^{\frac{\delta}{p-1}|x|^\beta}e^{k|s|} \tag{17.1.30}$$

for any $s \in \mathbb{R}$, $x \in \mathbb{R}^N \setminus B_1$, $i,j = 1,\ldots,N$ *and some positive constant* M_2, *where* p' *is the conjugate index of* p. *Then,* $C_c^\infty(\mathbb{R}^{N+1})$ *is a core of* G_q *for any* $1 \le q \le p$.

Proof In view of Theorem 17.1.8 it suffices to show that the functions $\mathscr{E}_{2k}$ and $\mathscr{F}_{2k}$ in (17.1.16) are in $L^p(\mathbb{R}^{N+1},\mu)$. Throughout the proof we denote by C a positive constant, independent of s and x, which may vary from line to line. Moreover, to ease the notation we simply write V instead of $V_{\beta,\delta}$.

(i) Since $V \ge 2$ in $\mathbb{R}^N$, we can estimate $|\mathscr{E}_{2k}(s,x)| \le e^{-2k|s|}|\mathcal{A}(s)V(x)|$ for any $(s,x) \in \mathbb{R}^{N+1}$. Hence, $\mathscr{E}_{2k}$ belongs to $L^1(\mathbb{R}^{N+1},\mu)$, due to Lemma 17.1.12(ii).

To prove that also $\mathscr{F}_{2k} \in L^1(\mathbb{R}^{N+1},\mu)$, we note that, if $\beta \ge 2$, then $|\sqrt{Q}(s,x)\nabla V(x)|^2 \le \beta^2\delta^2|x|^{2\beta-2}(V(x))^2\Lambda(s,x)$ for any $(s,x) \in \mathbb{R}^{N+1}$, where $\Lambda(s,x)$ denotes the maximum eigenvalue of the matrix $Q(s,x)$. This estimate holds true also in $\mathbb{R} \times (\mathbb{R}^N \setminus B_2)$, if $\beta \in (0,2)$. Moreover, for these values of β, the function $|\sqrt{Q}\nabla V|^2$ is bounded in $\mathbb{R} \times B_2$. It thus follows that for any $\beta > 0$ there exists a positive constant C, depending on β, such that

$$|\sqrt{Q}(s,x)\nabla V(x)|^2 \le \beta^2\delta^2|x|^{2\beta-2}\chi_{\mathbb{R}^N\setminus B_2}(x)(V(x))^2\Lambda(s,x) + C \tag{17.1.31}$$

for any $(s,x) \in \mathbb{R}^{N+1}$. As a byproduct,

$$\mathscr{F}_{2k}(s,x) \le \beta^2\delta e^{-2k|s|}|x|^{\beta-2}\chi_{\mathbb{R}^N\setminus B_2}(x)\Lambda(s,x) + C \le C(e^{-2k|s|}V(x) + 1)$$

for any $(s,x) \in \mathbb{R}^{N+1}$. Therefore, Lemma 17.1.12(i) implies that $\mathscr{F}_{2k} \in L^1(\mathbb{R}^{N+1},\mu)$.

(ii) We begin by showing that $\mathscr{E}_{k(p-1)/p}$ belongs to $L^p(\mathbb{R}^{N+1},\mu)$. Since the function $(s,x) \mapsto e^{-\sigma|s|}|\mathcal{A}(s)V(x)|^{1/p}$ belongs to $L^p(\mathbb{R}^N,\mu)$ for any $\sigma > k/p$, we can limit ourselves to proving that $\mathscr{F}_{k(p-1)/p} \le C|\mathcal{A}V|^{1/p}$ in $\mathbb{R}^{N+1}$, or, equivalently, that

$$|\mathcal{A}(s)V(x)| \le Ce^{k|s|}(V(x)\log V(x))^{p'}, \qquad (s,x) \in \mathbb{R}^{N+1}. \tag{17.1.32}$$

Taking estimate (17.1.26) into account and observing that the function $V\log V$ has positive infimum in $\mathbb{R}^N$, it suffices to prove (17.1.32) for any $s \in \mathbb{R}$ and x outside some ball B_R. Arguing as in the proof of (17.1.27), we get

$$\frac{|\mathcal{A}(s)V(x)|}{(V(x)\log V(x))^{p'}} \le \beta\delta^{1-p'}|x|^{\beta(1-p')-2}(V(x))^{1-p'}[\Lambda(s,x)(\beta\delta|x|^\beta + K_{\beta,\delta}) + |\langle b(s,x),x\rangle|]$$

for any $(s,x) \in \mathbb{R}^{N+1}$ such that $x \ne 0$, where $K_{\beta,\delta} = N + \beta\delta(\beta-2)^+$. Hence, taking R sufficiently large, recalling that $\beta\delta < \gamma$ and taking conditions (17.1.29) and (17.1.30) into account, we get

$$\frac{|\mathcal{A}(s)V(x)|}{(V(x)\log V(x))^{p'}} \le Ce^{k|s|}[(V(x))^{-\frac{1}{p(p-1)}} + 1] \le Ce^{k|s|}$$

for any $s \in \mathbb{R}$, and $x \in \mathbb{R}^N \setminus B_R$.

As far as the function $\mathscr{F}_{2k}$ is concerned, we observe that estimates (17.1.29) and (17.1.31) imply that $\mathscr{F}_{2k}(s,x) \le Ce^{-k|s|}(V(x))^{1/p}$ for any $(s,x) \in \mathbb{R}^{N+1}$. This shows that $\mathscr{F}_{2k} \in L^p(\mathbb{R}^{N+1},\mu)$ and completes the proof. □

17.2 The periodic case

Throughout this section we assume the following assumptions on the coefficients of the operator $\mathcal{A}$.

Hypotheses 17.2.1 (i) *The coefficients q_{ij} and b_j $(i,j=1,\dots,N)$ are T-periodic with respect to the variable t and satisfy Hypotheses* 17.1.1*(i) and (ii);*

(ii) *there exist a nonnegative function $\varphi\in C^2(\mathbb{R}^N)$, diverging to $+\infty$ as $|x|$ tends to $+\infty$, and two positive constants a and d such that $\mathcal{A}\varphi\le a-d\varphi$ in $(0,T)\times\mathbb{R}^N$.*

In view of the results in Subsection 14.5.3, the evolution operator $\{G(t,s)\}$ associated in $C_b(\mathbb{R}^N)$ with the operator $\mathcal{A}$ satisfies the condition $G(t+T,s+T)=G(t,s)$ for any $\mathbb{R}\ni s<t$. Moreover, the restriction of the evolution operator $\{\mathscr{T}(t)\}$ to the subspace $C_\sharp(\mathbb{R}^{N+1})$ of $C_b(\mathbb{R}^{N+1})$ gives rise to a semigroup, denoted by $\{\mathscr{T}^\sharp(t)\}$.

To extend the results of the general case to the evolution semigroup $\{\mathscr{T}^\sharp(t)\}$, the following proposition is crucial.

Proposition 17.2.2 *There exists a tight evolution system of measures $\{\mu_t^\sharp : t\in\mathbb{R}\}$ for $\{G(t,s)\}$, which is periodic of period T, i.e., $\mu_{t+T}^\sharp=\mu_t^\sharp$ for any $t\in\mathbb{R}$.*

Proof The proof of Theorem 16.2.7 shows that the evolution system of measures therein constructed satisfies the assertion of the proposition. Indeed, fix $x_0\in\mathbb{R}^N$ and $\{t_k\}$ as in the proof of the quoted theorem. Using the condition $G(t+T,s+T)=G(t,s)$ for $t\ge s$, we get

$$\begin{aligned}\mu_{(n+1)T}(A)&=\lim_{k\to+\infty}\frac{1}{t_k-(n+1)T}\int_{(n+1)T}^{t_k}(G(\tau,(n+1)T)\chi_A)(x_0)d\tau\\&=\lim_{k\to+\infty}\frac{1}{t_k-nT}\int_{(n+1)T}^{t_k+T}(G(\tau,(n+1)T)\chi_A)(x_0)d\tau\\&=\lim_{k\to+\infty}\frac{1}{t_k-nT}\int_{nT}^{t_k}(G(\tau,nT)\chi_A)(x_0)d\tau=\mu_{nT}(A)\end{aligned}$$

for any $n\in\mathbb{N}$ and any Borel set $A\subset\mathbb{R}^N$. Let now fix $t\in\mathbb{R}$ and let $n\in\mathbb{N}$ be such that $nT>s$. Then, $\mu_{t+T}(A)=(G^*(nT+T,t+T)\mu_{(n+1)T})(A)=(G^*(nT,t)\mu_{nT})(A)=\mu_t(A)$. So, it suffices to set $\mu_t^\sharp=\mu_t$ for any $t\in\mathbb{R}$ to complete the proof. □

As in the nonperiodic case, we use the evolution system of measures $\{\mu_t^\sharp : t\in\mathbb{R}\}$ to define a measure $\mu^\sharp$ on $\mathcal{B}((0,T)\times\mathbb{R}^N)$. More precisely, $\mu^\sharp$ is the unique measure such

$$\mu^\sharp(A\times B)=\frac{1}{T}\int_0^T\mu_s^\sharp(B)ds$$

for any pair of Borel sets $A\subset(0,T)$ and $B\subset\mathbb{R}^N$. Differently from the measure μ considered in the previous section, $\mu^\sharp$ is a probability measure. The arguments in the proof of Proposition 16.2.5 reveal[6] that $\mu^\sharp$ is an invariant measure for the semigroup $\{\mathscr{T}^\sharp(t)\}$, i.e.,

$$\int_{(0,T)\times\mathbb{R}^N}\mathscr{T}^\sharp(t)u\,d\mu^\sharp=\int_{(0,T)\times\mathbb{R}^N}u\,d\mu^\sharp.$$

[6] We stress that the smoothness assumption on the diffusion coefficients is not needed here

Starting from this formula and observing that $|\mathscr{T}^\sharp(t)u|^p \le \mathscr{T}^\sharp(t)|u|^p$ for any $t>0$, $p<+\infty$ and $u\in C_\sharp(\mathbb{R}^{N+1})$, we deduce that $\int_{(0,T)\times\mathbb{R}^N}|\mathscr{T}^\sharp(t)u|^p\,d\mu^\sharp \le \int_{(0,T)\times\mathbb{R}^N}|u|^p\,d\mu^\sharp$. This estimate allows us to extend the semigroup $\{\mathscr{T}^\sharp(t)\}$ to a strongly continuous semigroup of contractions in $L^p_\sharp(\mathbb{R}^{N+1}),\mu^\sharp)$ for[7] any $p\in[1,+\infty)$. We denote it by $\{\mathscr{T}^\sharp_p(t)\}$.

Actually, as the following proposition shows, $\{\mu^\sharp_t : t\in\mathbb{R}\}$ and $\mu^\sharp$ are, respectively, the unique T-periodic evolution system of measures for $\{G(t,s)\}$ and the unique invariant measure of $\{\mathscr{T}^\sharp(t)\}$. The proof is inspired by the proofs of [69, Proposition 3.3 and Theorem 4.3].

Proposition 17.2.3 *Under Hypotheses* 17.2.1 *and assuming that the diffusion coefficients are continuously differentiable in* $\mathbb{R}^{N+1}$ *with respect to the spatial variables, the evolution operator* $\{G(t,s)\}$ *has a unique* T*-periodic evolution system of measures, and* $\{\mathscr{T}^\sharp(t)\}$ *has a unique invariant measure.*

Proof We split the proof into three steps. In the first one we prove that each invariant measures of $\{\mathscr{T}^\sharp(t)\}$ is associated with a T-periodic evolution system of measures for $\{G(t,s)\}$. In Step 2, using this result, we prove that $\{\mathscr{T}(t)\}$ admits a unique invariant measure. Finally, in Step 3 we prove the uniqueness of the T-periodic evolution system of measures for $\{G(t,s)\}$.

Step 1. Let $\mu^\sharp$ be an invariant measure for $\{\mathscr{T}^\sharp(t)\}$. For any $u\in C^\infty_b((0,T)\times\mathbb{R}^N)$ with compact support, let $u^\sharp$ be its T-periodic extension to $\mathbb{R}^{N+1}$. Since $u^\sharp\in D_\sharp(\mathcal{G})$ (see (14.5.8)), Proposition 14.5.13 implies that $\int_{(0,T)\times\mathbb{R}^N}\mathcal{G}u d\mu^\sharp=0$. We can thus apply Theorem 16.2.4 which shows that $\mu^\sharp$ is absolutely continuous with respect to the Lebesgue measure and its density ρ is a locally Hölder continuous function and, since $\mu^\sharp$ is a probability measure, there exists a set $B\in\mathcal{B}((0,T))$, with negligible complement in $(0,T)$ such that $\rho(s,\cdot)\in L^1(\mathbb{R})$ for any $s\in B$. Using the function ρ we define the measures $\mu^\sharp_s$ $(s\in\mathbb{R})$. More precisely, we set $\mu^\sharp_s = T\rho(s,\cdot)dx$, for any $s\in(0,T)$, $\mu^\sharp_0=G^*(s_*,0)\mu^\sharp_{s_*}$, where s_* is arbitrarily fixed in B and, then, we extend the function $s\mapsto\mu^\sharp_s$ by periodicity to $\mathbb{R}$. If $s-kT\in B$ for some $k\in\mathbb{Z}$, then the measure $\mu^\sharp_s$ is finite. Actually, $\mu^\sharp_s$ is a probability measure. Indeed, denote by ν_* the marginal of $\mu^\sharp$, i.e., the measure defined by $\nu_*(A)=\mu^\sharp(A\times\mathbb{R}^N)$ for any $A\in\mathcal{B}((0,T))$. Further, let the family of probability measures $\{\nu_t : t\in(0,T)\}$ be a disintegration[8] of $\mu^\sharp$. Then,

$$\int_0^T u d\mu^\sharp = \int_0^T \nu_*(dt)\int_{\mathbb{R}^N}u(t,\cdot)d\nu_t,\qquad u\in C_\sharp(\mathbb{R}^{N+1}). \tag{17.2.1}$$

Take a function u independent of x and periodic with period T. Then, $\mathscr{T}^\sharp(t)u=u(\cdot-t)$. This last formula, (17.2.1) and the invariance of the measure $\mu^\sharp$, show that $\int_0^T u(\cdot-t)d\nu_* = \int_0^T u d\nu_*$ for any $t>0$. Hence, up to the multiplicative factor T^{-1}, ν_* coincides with the restriction of the Lebesgue measure to $\mathcal{B}((0,T))$. It follows that

$$\frac{1}{T}\int_{(0,T)\times\mathbb{R}^N}u d\mu^\sharp = \frac{1}{T}\int_0^T u(s)ds = \frac{1}{T}\int_0^T u(s)\mu^\sharp_s(\mathbb{R}^N)ds.$$

By the arbitrariness of u we conclude that $\mu^\sharp_s(\mathbb{R}^N)=1$ for almost every $s\in(0,T)$. Hence, without loss of generality, we can assume that $\mu^\sharp_s$ is a probability measure for any $s\in B$.

[7] $L^p_\sharp(\mathbb{R}^{N+1},\mu^\sharp)$ denotes the set of all the equivalence classes of functions u such that $u(s+T,x)=u(s,x)$ for almost every $(s,x)\in\mathbb{R}^{N+1}$ and it is normed by setting $\|u\|^p_{L^p_\sharp(\mathbb{R}^{N+1},\mu^\sharp)}=\int_{(0,T)\times\mathbb{R}^N}|u|^p d\mu^\sharp$ for any $u\in L^p_\sharp(\mathbb{R}^{N+1},\mu^\sharp)$.

[8] We refer the reader e.g. to [90] for more details on the disintegration of a measure.

If $s \in (0,T) \setminus B$, then $\mu_s^\sharp(\mathbb{R}^N) \le 1$ as can be easily checked approximating s by a sequence $\{s_n\} \in B$ and applying Fatou's lemma.

With all these remarks above in mind, we can prove that $\{\mu_t^\sharp : t \in \mathbb{R}\}$ is an evolution system of measures for $\{G(t,s)\}$. Fix $t \in (0,T)$, two functions $g \in C([0,T])$, with support contained in $(0,T)$, $f \in C_b(\mathbb{R}^N)$ and extend g to $\mathbb{R}$ with a T-periodic function $g^\sharp$. Finally, set $u = g^\sharp(\cdot - t)f$. The invariance of $\mu^\sharp$ and the periodicity of $g^\sharp$ and of the family of measures $\{\mu_s^\sharp : s \in \mathbb{R}\}$ yield

$$\int_0^T g(s)ds \int_{\mathbb{R}^N} G(s,s-t)f d\mu_s^\sharp = \int_0^T g^\sharp(s-t)ds \int_{\mathbb{R}^N} f d\mu_s^\sharp = \int_0^T g(s)ds \int_{\mathbb{R}^N} f d\mu_{s-t}^\sharp.$$

From the arbitrariness of g we conclude that

$$\int_{\mathbb{R}^N} f d\mu_{s-t}^\sharp = \int_{\mathbb{R}^N} G(s,s-t) f d\mu_s^\sharp, \qquad s \in C_{t,f}, \tag{17.2.2}$$

where $C_{t,f}$ is contained in B and has negligible complement in $(0,T)$. If we set $C_f = \bigcap_{t\in\mathbb{Q}\cap(0,T)} C_{t,f}$, then C_f has a negligible complement in $(0,T)$ and the formula (17.2.2) holds true, with $C_{t,f}$ being replaced by C_t, for any $t \in \mathbb{Q} \cap (0,T)$. We use this formula to prove that each $\mu_s^\sharp$ is a probability measure. We fix a sequence $\{f_m\} \subset C_c^\infty(\mathbb{R}^N)$, bounded with respect to the sup-norm and converging to $\mathbb{1}$ locally uniformly in $\mathbb{R}^N$, and we set $D = \bigcap_{m\in\mathbb{N}} C_{f_m}$. Let $r \in (0,T) \setminus C_f$ and fix $s \in D$ such that $s > r$. Further, fix $m \in \mathbb{N}$ and consider a sequence $\{t_n\} \subset \mathbb{Q}$ converging to $s - r$ as n tends to $+\infty$. Write (17.2.2) with t being replaced by t_n and f being replaced by f_m. Since the function ρ is continuous in $(0,T) \times \mathbb{R}^N$, $\int_{\mathbb{R}^N} f_m d\mu_{s-t_n}^\sharp$ converges to $\int_{\mathbb{R}^N} f d\mu_r^\sharp$ as n tends to $+\infty$. Similarly, the bounded sequence $\{G(s,s-t_n)f_m\}$ converges to $G(s,s-r)f_m$ locally uniformly in $\mathbb{R}^N$ as n tends to $+\infty$, by Theorem 14.1.12. Hence, the dominated convergence theorem reveals that $\int_{\mathbb{R}^N} G(s+t_n,s) f_m d\mu_s^\sharp$ converges to $\int_{\mathbb{R}^N} G(s+t,s) f_m d\mu_s^\sharp$ as n tends to $+\infty$ and we conclude that $\int_{\mathbb{R}^N} f_m d\mu_r^\sharp = \int_{\mathbb{R}^N} G(s,s-t) f_m d\mu_s^\sharp$. Since the measure $\mu_r^\sharp$ is finite, we can let m tend to $+\infty$ and, again by dominated convergence, we infer that $\mu_r^\sharp(\mathbb{R}^N) = \int_{\mathbb{R}^N} G(s,r)\mathbb{1} d\mu_s^\sharp = 1$. The arbitrariness of r shows that $\mu_s^\sharp$ is a probability measure for any $r \in \mathbb{R}$.

Finally, we can extend (17.2.2) to any $s, r \in (0,T)$, with $r < s$, and $f \in C_b(\mathbb{R}^N)$. The above arguments can be repeated verbatim to prove that, for any $f \in C_b(\mathbb{R}^N)$, $s \in C_f$ and $r \in (0,s)$, it holds that

$$\int_{\mathbb{R}^N} f d\mu_r^\sharp = \int_{\mathbb{R}^N} G(s,r) f d\mu_s^\sharp. \tag{17.2.3}$$

We now fix $s_1 \in (s,T)$ and write

$$\int_{\mathbb{R}^N} f d\mu_r^\sharp = \int_{\mathbb{R}^N} G(s_1,r) f d\mu_{s_1}^\sharp = \int_{\mathbb{R}^N} G(s_1,s)G(s,r) f d\mu_{s_1}^\sharp = \int_{\mathbb{R}^N} G(s,r) f d\mu_s^\sharp, \tag{17.2.4}$$

which shows that the formula (17.2.3) holds true for any $f \in C_b(\mathbb{R}^N)$ and any $0 < s < t < T$. It holds also with $s = 0$. Indeed, by definition $\int_{\mathbb{R}^N} f d\mu_0^\sharp = \int_{\mathbb{R}^N} G(s_*,0) f d\mu_{s*}^\sharp$ for any $f \in C_b(\mathbb{R}^N)$. If $t \in (s_*,T)$, then we write

$$\int_{\mathbb{R}^N} G(t,0) f d\mu_t^\sharp = \int_{\mathbb{R}^N} G(t,s_*)G(s_*,0) f d\mu_t^\sharp = \int_{\mathbb{R}^N} G(s_*,0) f d\mu_{s_*}^\sharp = \int_{\mathbb{R}^N} f d\mu_0^\sharp.$$

On the other hand, if $t \in (0,s_*)$, then we argue as in (17.2.4) and still conclude that $\int_{\mathbb{R}^N} f d\mu_0^\sharp = \int_{\mathbb{R}^N} G(t,s) d\mu_t^\sharp$. We have so proved that the formula (17.2.3) holds true for any $0 \le r < s < T$. Since $G(t+T,s+T) = G(t,s)$ and $\mu_{s+T}^\sharp = \mu_s^\sharp$ for any $s < T$, we conclude that $\{\mu_s^\sharp :\in \mathbb{R}\}$ is a T-periodic evolution system of measures for $\{G(t,s)\}$.

Step 2. The arguments in the proof of Theorem 9.1.15 can be applied to our situation provided that we show that, if $\mu^\sharp$ is an invariant measure for $\{\mathscr{T}^\sharp(t)\}$ and $A \in \mathcal{B}((0,T)\times\mathbb{R}^N)$ is such that $\mathscr{T}^\sharp(t)\chi_A^\sharp = \chi_A^\sharp$, $\mu^\sharp$-almost everywhere[9] in $(0,T)\times\mathbb{R}^N$ for any $t>0$, then $\mu^\sharp(A)\in\{0,1\}$. This is not immediate as in the proof of Theorem 9.1.15 since $\{\mathscr{T}(t)\}$ is not irreducible. To overcome this difficulty, we fix $t>0$ and let A be as above. For any set $C\in\mathcal{B}((0,T)\times\mathbb{R}^N)$ and $s\in(0,T)$, we set $C(s)=\{x\in\mathbb{R}^N:(s,x)\in C\}$. Moreover, we denote by $\{\mu_s^\sharp : s\in\mathbb{R}\}$ the periodic evolution system of measures for $\{G(t,s)\}$ associated with $\mu^\sharp$.

Let $C=C_t\subset(0,T)\times\mathbb{R}^N$ be the Borel set where $\mathscr{T}^\sharp(t)\chi_A^\sharp\neq\chi_A^\sharp$. Then, $C(s)$ is negligible[10] for almost every $s\in(0,T)$ since C is negligible in $(0,T)\times\mathbb{R}^N$. Denote by E the set of all $s\in(0,T)$ such that $\mu_s^\sharp(C(s))=0$. For any $s\in E$, $(\mathscr{T}^\sharp(t)\chi_A^\sharp)(s-t,\cdot)=\chi_A^\sharp(s,\cdot)$ almost everywhere, i.e., $G(s,s-t)\chi_{A(s-t)}=\chi_{A(s)}$ almost everywhere in $\mathbb{R}^N$. Suppose that $A(s)$ is not negligible for some $s\in E$. Then, the previous formula implies that $A(s-t)$ is not negligible as well, otherwise the equality $G(s,s-t)\chi_{A(s-t)}=\chi_{A(s)}$ could not hold almost everywhere in $\mathbb{R}^N$ (see Proposition 14.1.3). The same proposition shows that $G(s,s-t)\chi_{A(s-t)}$ is strictly positive in $\mathbb{R}^N$. Hence, $\chi_{A(s)}=1$ almost everywhere in $\mathbb{R}^N$, i.e. $\mu_s(A_s)=1$. We have so proved that, for any $s\in(0,T)$, either $\mu_s(A(s))=0$ or $\mu_s(A(s))=1$. Let D the set of all s such that $\mu_s(A(s))=1$. Then, $\mu^\sharp(A)=T^{-1}|D|$, where $|D|$ denotes the Lebesgue measure of D. Hence, if D is negligible, then A is negligible as well. Suppose that $|D|>0$ and consider the set $F=D\times\mathbb{R}^N$. It is easy to check that $\chi_F=\chi_A$ almost everywhere in $(0,T)\times\mathbb{R}^N$. Indeed, $F\setminus A=\{(s,x)\in(0,T)\times\mathbb{R}^N : s\in D,\ x\in\mathbb{R}^N\setminus A(s)\}$ and $A\setminus F=\{(s,x)\in A : s\notin D\}$. Hence,

$$\mu^\sharp(F\setminus A)=\int_D\mu_s(\mathbb{R}^N\setminus A(s))ds=\mu^\sharp(A\setminus F)=\int_{(0,T)\setminus D}\mu_s(A(s))ds=0,$$

so that $\mu^\sharp(A\cap F)=1$ and the functions χ_A and χ_F coincide almost everywhere in $(0,T)\times\mathbb{R}^N$. As a consequence, $\mathscr{T}(t)\chi_F^\sharp=\chi_F^\sharp$ almost everywhere in $\mathbb{R}^{N+1}$ for any $t>0$ or, equivalently, $\chi_D^\sharp=\chi_D^\sharp(\cdot-t)$ almost everywhere in $\mathbb{R}$. By the arbitrariness of $t>0$, the function $\chi_D^\sharp$ belongs to the kernel of the infinitesimal generator of the semigroup of translations in $L^1_\sharp(\mathbb{R})$. Since this operator is the realization in $W^{1,1}_\sharp(\mathbb{R})$ of the first-order derivative, $\chi_D^\sharp$ is constant almost everywhere in $\mathbb{R}$. The assumption $|D|>0$ then implies that D has negligible complement in $(0,T)$ so that $\mu^\sharp(A)=1$.

Step 3. To complete the proof, let us suppose that $\{\mu_t^\sharp : t\in\mathbb{R}\}$ and $\{\nu_t^\sharp : t\in\mathbb{R}\}$ are two T-periodic evolution systems of measures for $\{G(t,s)\}$. By Step 2, the measures $\mu^\sharp$ and $\nu^\sharp$, defined by $\mu^\sharp(A\times B)=\int_A\mu_s^\sharp(B)ds$ and $\nu^\sharp(A\times B)=\int_A\nu_s^\sharp(B)ds$ for any $A\in\mathcal{B}((0,T))$ and $B\in\mathcal{B}(\mathbb{R}^N)$, coincide. Therefore, by Step 1, $\mu_s^\sharp=\rho(s,\cdot)dx=\nu_s^\sharp$ for any $s\in(0,T)$, where ρ is the density of $\nu^\sharp$ with respect to the Lebesgue measure. This suffices to conclude that $\mu_s^\sharp=\nu_s^\sharp$ for any $s\in\mathbb{R}$. □

Remark 17.2.4 Let $\{\mu_t^\sharp : t\in\mathbb{R}\}$ be the unique (tight) evolution system of measures for $\{G(t,s)\}$ such that $\mu_{t+T}^\sharp=\mu_t^\sharp$ for any $t\in\mathbb{R}$. Then, the continuous function $\rho^\sharp:\mathbb{R}^{N+1}\to\mathbb{R}$ such that $\mu_t^\sharp(dx)=\rho^\sharp(t,\cdot)dx$ for any $t\in\mathbb{R}$ (see Proposition 16.2.5) is T-periodic. Moreover, the function $T^{-1}\rho^\sharp$ is the density of the measure $\mu^\sharp$ with respect to the Lebesgue measure in $(0,T)\times\mathbb{R}^N$. Finally, by the proof of Theorem 16.2.7 we conclude that $\sup_{s\in\mathbb{R}}\|\varphi\|_{L^1(\mathbb{R}^N,\mu_s^\sharp)}\leq\min_{\mathbb{R}^N}\varphi+ad^{-1}$.

[9] Here, $\chi_A^\sharp$ is the T-periodic extension of the function χ_A to $\mathbb{R}^{N+1}$.

[10] Since the measures $\mu_s^\sharp$ (resp. $\mu^\sharp$) are all equivalent to the Lebesgue measure in $\mathbb{R}^N$ (resp. in $(0,T)\times\mathbb{R}^N$) (see Proposition 16.2.5), when we say that a subset of $\mathbb{R}^N$ is negligible, we do not stress if the measure that we are considering is $\mu_s^\sharp$ (resp. $\mu^\sharp$) or the Lebesgue measure.

The estimates on the spatial derivatives of $G(t,s)f$ yield a useful embedding result for the domain of the infinitesimal generator $G_p^\sharp$ of $\{\mathscr{T}_p^\sharp(t)\}$.

Proposition 17.2.5 *The following properties are satisfied.*

(i) *Let Hypotheses* 14.4.1 *be satisfied and fix $p \in [2,+\infty)$. Then, $\mathscr{T}_p^\sharp(t)$ maps $L^p_\sharp(\mathbb{R}^{N+1},\mu^\sharp)$ into $W^{0,1}_{p,\sharp}(\mathbb{R}^{N+1},\mu^\sharp)$ for any $t>0$ and*

$$\|\nabla_x \mathscr{T}_p^\sharp(t)u\|_{L^p_\sharp(\mathbb{R}^{N+1},\mu^\sharp)} \le C_1\left(t^{-\frac{1}{2}}\vee 1\right)\|u\|_{L^p(\mathbb{R}^{N+1},\mu^\sharp)}, \qquad u\in L^p(\mathbb{R}^{N+1},\mu^\sharp),$$

where C_1 is the constant in (14.4.3). *Moreover, $D(G_p^\sharp)$ is continuously embedded into $W^{0,1}_{p,\sharp}(\mathbb{R}^{N+1},\mu^\sharp)$ and*

$$\|\nabla_x u\|_{L^p_\sharp(\mathbb{R}^{N+1},\mu^\sharp)} \le C\|u\|^{\frac{1}{2}}_{L^p_\sharp(\mathbb{R}^{N+1},\mu^\sharp)}\|G_p^\sharp u-\omega u\|^{\frac{1}{2}}_{L^p_\sharp(\mathbb{R}^{N+1},\mu^\sharp)}$$

for any $u\in D(G_p^\sharp)$, $\omega>0$ and some positive constant C, depending on ω but being independent of u.

(ii) *Let Hypotheses* 14.4.1*(i) and either of Hypotheses* 14.4.1*(iii) and* 14.4.4*(i) or* 14.4.4*(ii) be satisfied. Then, the properties in (i) hold true for any $p\in(1,+\infty)$. In particular,*

$$\|\nabla_x \mathscr{T}_p^\sharp(t)u\|_{L^p_\sharp(\mathbb{R}^{N+1},\mu^\sharp)} \le C_p e^{\ell_p^- t}\left(t^{-\frac{1}{2}}\vee 1\right)\|u\|_{L^p_\sharp(\mathbb{R}^{N+1},\mu)}, \qquad t>0,\ u\in L^p_\sharp(\mathbb{R}^{N+1},\mu^\sharp) \tag{17.2.5}$$

and

$$\|\nabla_x \mathscr{T}_p^\sharp(t)u\|_{L^p_\sharp(\mathbb{R}^{N+1},\mu^\sharp)} \le e^{\ell_p t}\|\nabla_x u\|_{L^p_\sharp(\mathbb{R}^{N+1},\mu^\sharp)}, \qquad t>0,\ u\in W^{0,1}_{p,\sharp}(\mathbb{R}^{N+1},\mu), \tag{17.2.6}$$

where C_p is the constant in (14.4.10) *and ℓ_p is given by* (14.4.7) *and* (14.4.9). *In particular, if $\ell_p<0$, then we can take $\omega=0$ in* (17.1.4).

We omit the proof, since it can be obtained by arguing as in the proof of Proposition 17.1.4.

To conclude this section, we prove the following spectral mapping theorem. The proof is obtained arguing as in the proof of [113, Theorem 2.1], which is a variant of [56, Theorem 3.13] and uses the arguments in the proof of [163, Theorem 2.5].

Theorem 17.2.6 *For any $p<+\infty$ and $t>0$ it holds that $\sigma(\mathscr{T}_p^\sharp(t))\setminus\{0\}=e^{t\sigma(G_p^\sharp)}$.*

Proof The spectral mapping theorem is a classical result in the theory of evolution semigroups associated with evolution operators in fixed Banach spaces, which is not our case (see Remark 16.2.9). To overcome this difficulty, we use a trick, adding one more variable, to reduce to the classical case of evolution semigroups in fixed Banach spaces. For this purpose, for any $p<+\infty$ we introduce two semigroups $\{\mathscr{S}_p^\sharp(t)\}$ and $\{\mathscr{U}_p^\sharp(t)\}$ in the space $L^p_\sharp(\mathbb{R};L^p_\sharp(\mathbb{R}^{N+1},\mu^\sharp))$ of all Bochner measurable functions $f:\mathbb{R}\to L^p_\sharp(\mathbb{R}^{N+1},\mu^\sharp)$ such that $f(\sigma+T)=f(\sigma)$ for almost every $\sigma\in\mathbb{R}$ and $\|f\|^p_{L^p_\sharp(\mathbb{R};L^p_\sharp(\mathbb{R}^{N+1},\mu^\sharp))}=\int_0^T\|f(\sigma)\|^p_{L^p_\sharp(\mathbb{R}^{N+1},\mu^\sharp)}d\sigma<+\infty$. They are defined as follows:

$$(\mathscr{S}_p^\sharp(t)f)(\theta)=\mathscr{T}_p^\sharp(t)f(\theta-t), \qquad (\mathscr{U}_p^\sharp(t)f)(\theta)=\mathscr{T}_p^\sharp(t)f(\theta)$$

for almost every $\theta\in\mathbb{R}$, any $t>0$ and $f\in L^p_\sharp(\mathbb{R};L^p_\sharp(\mathbb{R}^{N+1},\mu^\sharp))$. The first one is the evolution semigroup in $L^p_\sharp(\mathbb{R};L^p_\sharp(\mathbb{R}^{N+1},\mu^\sharp))$ associated with the semigroup $\{\mathscr{T}_\sharp(t)\}$. Hence, it satisfies the spectral mapping theorem and $\rho(\mathscr{S}_p^\sharp(t)) = e^{t\rho(G_p^\sharp)}$ for any $t > 0$, where $G_p^\sharp$ denotes the infinitesimal generator of the semigroup. To characterize the resolvent set of the operator $G_p^\sharp$, we identify $L_\sharp^p(\mathbb{R}; L_\sharp^p(\mathbb{R}^{N+1}, \mu^\sharp))$ with the space $L_\sharp^p(\mathbb{R}^{N+2}, \nu)$ of all the Lebesgue measurable functions f defined in $\mathbb{R}^{N+2}$ such that $f(\cdot + T, \cdot, \cdot) = f(\cdot, T + \cdot, \cdot) = f$ and $\|f\|^p_{L_\sharp^p(\mathbb{R}^{N+2},\nu)} = \int_{(0,T)^2\times\mathbb{R}^N} |f|^p d\nu < +\infty$, where $d\nu = T^{-1}\rho^\sharp(s,x)d\sigma ds dx$ and $\rho^\sharp$ is the density of $\mu^\sharp$ with respect to the Lebesgue measure. Then, we introduce the isometry $\mathscr{I}$ defined by $(\mathscr{I}f)(\sigma, t, x) = f(\sigma - t, t, x)$ for almost every $(\sigma, t, x) \in \mathbb{R}^{N+2}$ and any $f \in L_\sharp^p(\mathbb{R}^{N+2}, \nu)$. Since $\mathscr{S}_p^\sharp(t) = \mathscr{I}^{-1}\mathscr{U}_p^\sharp(t)\mathscr{I}$ for any $t > 0$, $G_p^\sharp$ and the infinitesimal generator $H_p^\sharp$ of $\{\mathscr{U}_p^\sharp(t)\}$ have the same resolvent set. As is easily seen, $D(H_p^\sharp) = \{f \in L_\sharp^p(\mathbb{R}; L_\sharp^p(\mathbb{R}^{N+1}, \mu^\sharp)) : f(\sigma) \in D(G_p^\sharp) \text{ for a.e. } \sigma \in \mathbb{R}\}$ and $H_p^\sharp f(\sigma) = G_p^\sharp f(\sigma)$ for almost every $\sigma \in \mathbb{R}$ and any $f \in D(H_p^\sharp)$. Moreover, $\rho(H_p^\sharp) = \rho(G_p^\sharp)$. Now, the inclusion $\sigma(\mathscr{T}_p^\sharp(t)) \setminus \{0\} \subset e^{t\sigma(G_p^\sharp)}$ follows. Since the other inclusion is obvious from the general theory of semigroups (see e.g. [95, Section 3.6]), the proof is complete. □

Remark 17.2.7 The spectral mapping theorem holds also in nonperiodic setting with obvious changes in the proof.

17.2.1 Cores

Arguing as in the proof of Theorem 17.1.5 and taking Proposition 14.5.13 into account, we can easily prove that $D_\sharp(\mathcal{G})$ is a core of the infinitesimal generator $G_p^\sharp$ of the semigroup $\{\mathscr{T}_p^\sharp(t)\}$ for any $p \in [1, +\infty)$. This result can be also obtained as a byproduct of the following theorem. To state it, let us introduce a few notation. For any $\tau \in \mathbb{R}$, $\chi \in C_c^\infty(\mathbb{R}^N)$ and $\alpha \in C_c^1(\mathbb{R})$ with support contained in $(\sigma, \sigma+T)$ for some $\sigma \geq \tau$, we define the function $u_{\tau,\chi,\alpha}$ as the T-periodic (with respect to s) extension of the function $(s, x) \mapsto \alpha(s)(G(s, \tau)\chi)(x)$ defined in $[\sigma, \sigma + T) \times \mathbb{R}^N$. Finally, by $\mathscr{C}$ we denote the span of the functions $u_{\tau,\chi,\alpha}$.

Theorem 17.2.8 *The space $\mathscr{C}$ is a core of $G_p^\sharp$, for any $p \in [1, +\infty)$.*

Proof First of all, we prove that $\mathscr{C}$ is contained in $D(G_p^\sharp)$ for any $p \in [1, +\infty)$. For this purpose, we fix $\tau \in \mathbb{R}$, $\chi \in C_c^\infty(\mathbb{R}^N)$ and $\alpha \in C_c^1(\mathbb{R})$ with support contained in $(\sigma, \sigma+T)$ for some $\sigma \geq \tau$. Clearly, the function $u_{\tau,\chi,\alpha}$ belongs to $C^{1,2}(\mathbb{R}^{N+1})$ by (the proof of) Theorem 14.1.2 and it belongs to $L_\sharp^p(\mathbb{R}^{N+1}, \mu^\sharp)$ for any $p \in [1, +\infty)$. Moreover, $\mathcal{G}u_{\tau,\chi,\alpha}$ is the periodic extension to $\mathbb{R}^{N+1}$ of the function $-\alpha' G(\cdot, \tau)\chi$, defined in $[\sigma, \sigma + T) \times \mathbb{R}^N$. Therefore, $\mathcal{G}u$ belongs to $C_\sharp(\mathbb{R}^{N+1})$ and $u_{\tau,\chi,\alpha} \in D_\sharp(\mathcal{G})$ (see Subsection 14.5.3). The characterization of this space given in Proposition 14.5.13 reveals that $u_{\tau,\chi,\alpha} \in D(G_p^\sharp)$ and the inclusion $\mathscr{C} \subset D(G_p^\sharp)$ follows.

Let us now prove that $\mathscr{C}$ is invariant under the action of the semigroup $\{\mathscr{T}_p^\sharp(t)\}$. For this purpose, fix $t > 0$ and, for any $s \in \mathbb{R}$, let $k \in \mathbb{Z}$ be such that $s - t \in [\sigma + kT, \sigma + (k + 1)T)$. Then, $u_{\tau,\chi,\alpha}(s - t, \cdot) = \alpha(s - t - kT)G(s - t - kT, \tau)\chi$ and we can write

$$\begin{aligned}
\mathscr{T}_p^\sharp(t)u_{\tau,\chi,\alpha}(s, \cdot) &= \alpha(s - t - kT)G(s, s - t)G(s - t - kT, \tau)\chi \\
&= \alpha(s - t - kT)G(s - kT, s - t - kT)G(s - t - kT, \tau)\chi \\
&= \alpha(s - t - kT)G(s - kT, \tau)\chi.
\end{aligned}$$

Hence, $\mathscr{T}_p^\sharp(t)u$ is the T-periodic extension of the function $\alpha(\cdot - t)G(\cdot, \tau)\chi$ defined in $[\sigma + t, \sigma + t + T) \times \mathbb{R}^N$, which belongs to $\mathscr{C}$. We have so proved that $\mathscr{T}_p^\sharp(t)$ leaves $\mathscr{C}$ invariant.

Let us finally prove that $\mathscr{C}$ is dense in the domain of $G_p^\sharp$. Since $\mathscr{T}_p(t)$ maps $\mathscr{C}$ into itself for any $t > 0$, it suffices to prove that $\mathscr{C}$ is dense in $L_\sharp^p(\mathbb{R}^{N+1}, \mu^\sharp)$. Note that the linear span of the functions $(s,x)\mapsto\beta(s)\chi(x)$, with $\beta\in C^1(\mathbb{R})$ being periodic of period T and $\chi\in C_c^\infty(\mathbb{R}^N)$, is dense in $L^p_\sharp(\mathbb{R}^{N+1},\mu^\sharp)$. Hence, we just prove that any function $\beta\chi$ can be approximated by elements of $\mathscr{C}$. Fix $\beta\in C^1_\sharp(\mathbb{R})$, $\chi\in C_c^\infty(\mathbb{R}^N)$, $\varepsilon>0$ and $\tau\in\mathbb{R}$. Corollary 14.1.7 shows that $G(s,\tau)\chi$ converges to χ, uniformly in $\mathbb{R}^N$, as s tends to τ^+. Therefore, there exists $\tau'\in(\tau,\tau+T)$ such that $\|G(s,\tau)\chi-\chi\|_\infty\le\varepsilon$ for any $s\in[\tau,\tau']$, which implies that $\|G(s,\tau)\chi-\chi\|_{L^p(\mathbb{R}^N,\mu^\sharp_s)}\le\varepsilon$ for any $\tau\le s\le\tau'$. Let us cover $(0,T)$ by a finite number of such intervals (τ_k,τ_k'), $k=1,\ldots,n$ and let $\{\alpha_k:k=1,\ldots,n\}$ be an associated partition of unity. Set $u_k=\beta\alpha_kG(\cdot,\tau_k)\chi$ in $[\tau_k,\tau_k+T)\times\mathbb{R}^N$ and still denote by u_k its T-periodic extension. Then, the function $u=\sum_{k=1}^n u_k$ belongs to $\mathscr{C}$ and

$$\|g(s,\cdot)-u(s,\cdot)\|_{L^p(\mathbb{R}^N,\mu^\sharp_s)}\le\|\beta\|_\infty\sum_{k=1}^n\alpha_k(s)\|G(s,\tau_k)\chi-\chi\|_{L^p(\mathbb{R}^N,\mu^\sharp_s)}\le\varepsilon\|\beta\|_\infty$$

for any $s\in[0,T]$. Integrating this estimate in $(0,T)$ we get $\|g-u\|_{L^p_\sharp(\mathbb{R}^{N+1},\mu^\sharp)}\le\varepsilon\|\beta\|_\infty$ and the statement follows. □

Corollary 17.2.9 *For $1<p<+\infty$, $D(G^\sharp_p)$ is contained in $W^{1,2}_{p,\mathrm{loc}}(\mathbb{R}^{N+1})$ and for any $r>0$ the restriction mapping $\mathcal{R}:D(G^\sharp_p)\to W^{1,2}_p((0,T)\times B_r)$, defined by $\mathcal{R}u=u_{|(0,T)\times B_r}$ for any $u\in D(G^\sharp_p)$, is continuous.*

Proof Each function $u\in\mathscr{C}$ belongs to $C^{1,2}_\sharp(\mathbb{R}^{N+1})$ and $G^\sharp_pu(s,x)=\mathcal{A}(s)u(s,x)-D_su(s,x)$, so that, by classical regularity results for parabolic equations in L^p-spaces with respect to the Lebesgue measure, there exists a positive constant $C_1=C_1(r)$ such that

$$\|u\|_{W^{1,2}_p((0,T)\times B_r)}\le C_1(\|u\|_{L^p((0,T)\times B_{2r})}+\|G^\sharp_pu\|_{L^p((0,T)\times B_{2r})}).$$

On the other hand, since $\mu^\sharp(ds,dx)=T\rho^\sharp(s,x)dsdx$ for a positive continuous function $\rho^\sharp$ (see Remark 17.2.4), there exists $C_2=C_2(r)$ such that $\|\cdot\|_{L^p((0,T)\times B_{2r})}\le C_2\|\cdot\|_{L^p_\sharp(\mathbb{R}^{N+1},\mu^\sharp)}$. Thus, $\mathcal{R}$ is continuous from $\mathscr{C}$ (endowed with the $D(G^\sharp_p)$-norm) to $W^{1,2}_p((0,T)\times B_r)$ and, since $\mathscr{C}$ is dense in $D(G^\sharp_p)$, the statement follows. □

Now, arguing as in the proof of Theorem 17.1.6 we can prove the following results.

Theorem 17.2.10 *Let Hypotheses 14.4.1 and 17.2.1 be satisfied. Further assume that $\|Q(s,x)\|_{L(\mathbb{R}^N)}\le C(|x|+1)V(x)$ for any $(s,x)\in\mathbb{R}^{N+1}$ for some positive constant C. Then, the following properties are satisfied.*

(i) *If $\langle b(s,x),x\rangle\le C(|x|^2+1)\varphi(x)$ for any $(s,x)\in\mathbb{R}^{N+1}$, then $|u|^{p-2}|\sqrt{Q}\nabla_xu|^2\chi_{\{u\neq0\}}$ belongs to $L^1_\sharp(\mathbb{R}^{N+1},\mu^\sharp)$ for any $p\in(1,+\infty)$, $u\in D(G^\sharp_p)$ and*

$$\int_{(0,T)\times\mathbb{R}^N}|u|^{p-2}|\sqrt{Q}\nabla_xu|^2\chi_{\{u\neq0\}}d\mu^\sharp\le\frac{1}{1-p}\int_{(0,T)\times\mathbb{R}^N}u|u|^{p-2}G^\sharp_pu\,d\mu^\sharp.\qquad(17.2.7)$$

(ii) *If $|\langle b(s,x),x\rangle|\le C(|x|^2+1)\varphi(x)$ for any $(s,x)\in(0,T)\times\mathbb{R}^N$, then (17.2.7) is an equality for any $p\ge2$ and $u\in D(G^\sharp_p)$.*

(iii) *If the diffusion coefficients q_{ij} $(i,j=1,\ldots,N)$ are bounded, then the equality in (17.1.7) holds true for any $p\in(1,+\infty)$ and $u\in D_\sharp(\mathcal{G})$ and even for any $f\in D(G^\sharp_p)$ if $p\ge2$.*

(iv) *Under either of the conditions in (i) or in (iii), $D(G^\sharp_p)\hookrightarrow W^{0,1}_{p,\sharp}(\mathbb{R}^{N+1},\mu^\sharp)$ for any $p\in(1,+\infty)$ and the map $f\mapsto|\sqrt{Q}\nabla_xf|$ is continuous from $D(G^\sharp_p)$ into $L^p_\sharp(\mathbb{R}^{N+1},\mu)$ for $1<p\le2$.*

The following theorem is the counterpart of Theorem 17.1.8 and provides sufficient conditions ensuring that the space $C^\infty_{\sharp,c}(\mathbb{R}^{N+1}) := \{f \in C^\infty(\mathbb{R}^{N+1}) \cap C_\sharp(\mathbb{R}^{N+1}) : \operatorname{supp}(f) \subset \mathbb{R} \times B_R$ for some $R > 0\}$ is a core of the infinitesimal generator of $\{\mathscr{T}_p(t)\}$ for any $p \in [1, +\infty)$.

Theorem 17.2.11 *Assume that Hypotheses* 17.1.7 *and* 17.2.1 *are satisfied. Further, assume that there exists a positive function $V \in C^2(\mathbb{R}^N)$, blowing up as $|x|$ tends to $+\infty$, such that the functions $(V \log V)^{-1}\mathcal{A}V$ and $(V^2 \log V)^{-1}|\sqrt{Q}\nabla V|^2$ belong to $L^p_\sharp(\mathbb{R}^{N+1}, \mu^\sharp)$ for any $p \in [1, +\infty)$. Then, $C^\infty_{\sharp,c}(\mathbb{R}^{N+1})$ is a core of the operator G_q for any $1 \le q \le p$.*

Proof The proof is very close to that of Theorem 17.1.8. Hence, we just sketch it, pointing out the main differences. For a fixed $p \in [1, +\infty)$ we need to prove that, if $\psi \in L^{p'}_\sharp(\mathbb{R}^{N+1}, \mu^\sharp)$ (where p' is the conjugate index of p) satisfies the condition

$$\int_{(0,T)\times\mathbb{R}^N} \psi(\lambda\zeta - \mathcal{G}\zeta)d\mu^\sharp = 0 \tag{17.2.8}$$

for some $\lambda > 0$ and any $\zeta \in C^\infty_{\sharp,c}(\mathbb{R}^{N+1})$, then $\psi = 0$ $\mu^\sharp$-almost everywhere. The main step to apply the same argument in the proof of Theorem 17.1.8 consists of showing that the function $\psi\rho^\sharp$ belongs to $C^1_c(\mathbb{R}^{N+1}) \cap C_\sharp(\mathbb{R}^{N+1})$, where $T\rho^\sharp$ is the density of $\mu^\sharp$ with respect to the Lebesgue measure (see Remark 17.2.4). To check this property we observe that, since ψ, $\rho^\sharp$ and ζ are T-periodic with respect to s, the formula (17.2.8) holds true also with $(0, T)$ being replaced by $(a, a + T)$, for any $a \in \mathbb{R}$. Arguing as in the very first part of the proof of Theorem 17.1.8, we deduce that $\psi\rho^\sharp \in C^1((a, a + T) \times \mathbb{R}^N)$ for any $a \in \mathbb{R}$. Hence, $\psi\rho^\sharp$ belongs to $C^1(\mathbb{R}^{N+1})$ and is T-periodic with respect to s, since both ψ and $\rho^\sharp$ are.

Now, we can repeat the same arguments as in the proof of Theorem 17.1.8. First, we infer that

$$\lambda \int_{(0,T)\times\mathbb{R}^N} \vartheta|\psi|d\mu^\sharp \le \int_{(0,T)\times\mathbb{R}^N} (\mathcal{G}\vartheta)|\psi|d\mu^\sharp$$

for any $\vartheta \in C^\infty_c(\mathbb{R}^N)$. Then, taking $\vartheta_n := H(n^{-1} \log V)$, where H is the same function as in the proof of the quoted theorem, we conclude that $\psi = 0$ $\mu^\sharp$-almost everywhere in $(0, T) \times \mathbb{R}^N$, just letting n tend to $+\infty$. □

17.2.2 A logarithmic Sobolev type inequality and compactness

Throughout this whole subsection, besides Hypotheses 17.2.1, we assume the following additional conditions on the coefficients of the operator $\mathcal{A}$.

Hypotheses 17.2.12 (i) *The diffusion coefficients q_{ij} $(i, j = 1, \dots, N)$ are independent of x;*

(ii) *the first-order spatial derivatives of the coefficients b_i $(i = 1, \dots, N)$ exist, belong to $C^{\alpha/2,\alpha}_{\rm loc}(\mathbb{R}^{N+1})$ and there exists a negative constant $r_0 < 0$ such that $\langle(\mathrm{Jac}_x b)\xi, \xi\rangle \le r_0|\xi|^2$ in $(0, T) \times \mathbb{R}^N$, for any $\xi \in \mathbb{R}^N$.*

As a straightforward consequence of Theorem 16.4.7, we can prove the following logarithmic Sobolev inequality for functions in $W^{0,1}_{p,\sharp}(\mathbb{R}^{N+1}, \mu^\sharp)$.

Proposition 17.2.13 *For any $p\in(1,+\infty)$ and $u\in D_\sharp(\mathcal{G})$ it holds that*

$$\int_{(0,T)\times\mathbb{R}^N}|u|^p\log(|u|^p)d\mu^\sharp \leq \frac{1}{T}\int_0^T(\Pi|u|^p)\log(\Pi|u|^p)ds + \frac{p^2\Lambda}{2|r_0|}\int_{(0,T)\times\mathbb{R}^N}|u|^{p-2}|\nabla_x u|^2 d\mu^\sharp, \tag{17.2.9}$$

where $\Pi(|u|^p)$ denotes the average of the function $|u|^p$ with respect to the measure $\mu^\sharp$. In addition, if $u\in D_\sharp(\mathcal{G})$ is such that $u^{-1}|\nabla_x u|^2\in L^1_\sharp(\mathbb{R}^{N+1},\mu^\sharp)$, then estimate (17.2.9) *holds also for $p=1$. Finally, if $p\geq 2$ then estimate* (17.2.9) *holds true in $D(G^\sharp_p)$.*

Proof Writing estimate (16.4.4) for a function in $D_\sharp(\mathcal{G})$, which belongs to $C^{0,1}_b(\mathbb{R}^{N+1})\cap C_\sharp(\mathbb{R}^{N+1})$ by Remark 14.5.14, and integrating with respect to $s\in(0,T)$, we get (17.2.9) for $p\in(1,+\infty)$ with $u\in D_\sharp(\mathcal{G})$. Note that the last integral in such an estimate is finite due to Theorem 17.2.10. If $u^{-1}|\nabla_x u|^2\in L^1_\sharp(\mathbb{R}^{N+1},\mu^\sharp)$, then we can let p tend to 1 from the right in (17.2.9) and conclude that this estimate holds true also for $p=1$.

To conclude the proof, we extend the validity of (17.2.9) to any $u\in D(G^\sharp_p)$ in the case when $p\geq 2$. For this purpose, we fix p and u as above and consider a sequence $\{u_n\}\subset D_\sharp(\mathcal{G})$ converging to u in the graph norm of $G^\sharp_p$. Possibly replacing $\{u_n\}$ with a subsequence, we may assume that u_n converges to u pointwise almost everywhere in $(0,T)\times\mathbb{R}^N$ (with respect to the Lebesgue measure, thanks to Remark 17.2.4) as n tends to $+\infty$. By the first part of the proof, it holds that

$$\int_{(0,T)\times\mathbb{R}^N}|u_n|^p\log_+(|u_n|^p)d\mu^\sharp \leq \frac{1}{T}\int_0^T(\Pi|u_n|^p)\log(\Pi|u_n|^p)ds + \frac{p^2\Lambda}{2|r_0|}\int_{(0,T)\times\mathbb{R}^N}|u_n|^{p-2}|\nabla_x u_n|^2 d\mu^\sharp + \int_{(0,T)\times\mathbb{R}^N}|u_n|^p\log_-(|u_n|^p)d\mu^\sharp \tag{17.2.10}$$

for any $n\in\mathbb{N}$, where $\log_+(y):=0\vee\log(y)$, $\log_-(y):=0\vee(-\log(y))$ for any $y>0$.

To let n tend to $+\infty$ in such inequality, as a first step we prove that $|u|^p$ belongs to $D(G^\sharp_1)$ and $|u_n|^p$ converges to $|u|^p$ in $D(L^1_\sharp)$, as n tends to $+\infty$. For this purpose, it suffices to observe that the same arguments as in the proof of Theorem 17.1.6 show that, for any $n\in\mathbb{N}$, $|u_n|^p$ belongs to $D_\sharp(\mathcal{G})$ and $\mathcal{G}|u_n|^p=pu_n|u_n|^{p-2}\mathcal{G}u_n+p(p-1)|u_n|^{p-2}|\sqrt{Q}\nabla_x u_n|^2$. By Proposition 17.2.5, $D(G^\sharp_p)$ is continuously embedded in $W^{0,1}_{p,\sharp}(\mathbb{R}^{N+1},\mu^\sharp)$. Hence, $|\sqrt{Q}\nabla_x u_n|^2$ converges to $|\sqrt{Q}\nabla_x u_n|^2$ in $L^{p/2}_\sharp(\mathbb{R}^{N+1},\mu^\sharp)$. Since the maps $v\mapsto v|v|^{p-2}$ and $v\mapsto|v|^{p-2}$ are continuous in $L^p_\sharp(\mathbb{R}^{N+1},\mu^\sharp)$ with values in $L^{p/(p-1)}_\sharp(\mathbb{R}^{N+1},\mu^\sharp)$ and $L^{p/(p-2)}_\sharp(\mathbb{R}^{N+1},\mu^\sharp)$, respectively, $\mathcal{G}|u_n|^p$ converges in $L^p_\sharp(\mathbb{R}^{N+1},\mu^\sharp)$, as n tends to $+\infty$. This is enough to conclude that $|u|^p\in D(G^\sharp_1)$ and $|u_n|^p$ converges to $|u|^p$ in $D(G^1_\sharp)$, as n tends to $+\infty$.

Now, it follows easily that $\Pi|u_n|^p$ converges to $\Pi|u|^p$ in $D(G^\sharp_1)$, as n tends to $+\infty$. Since the part of G_1 in $\Pi(L^1_\sharp(\mathbb{R}^{N+1},\mu^\sharp))$ is the time derivative $-D_s$ with domain isomorphic to[11] $W^{1,1}_\sharp(\mathbb{R},T^{-1}ds)$, $\Pi(|u_n|^p)$ converges to $\Pi(|u|^p)$ in $W^{1,1}_\sharp(\mathbb{R},T^{-1}ds)$, as n tends to $+\infty$, and hence in $C([0,T])$. Using the 1/2-Hölder continuity of the function $y\mapsto y\log y$ on bounded

[11] $W^{1,1}_\sharp(\mathbb{R},T^{-1}ds)$ is the set of all functions $f:\mathbb{R}\to\mathbb{R}$ which are T-periodic and such that their distributional derivative f' belongs to $L^1_{\rm loc}(\mathbb{R})$ and $\int_0^T(|f|+|f'|)ds<+\infty$.

sets of $[0,+\infty)$, we get

$$\frac{1}{T}\int_0^T |\Pi|u_n|^p\log(\Pi|u_n|^p)-\Pi|u|^p\log(\Pi|u|^p)|ds\le C_1\|\Pi|u_n|^p-\Pi|u|^p\|_\infty^{\frac{1}{2}}$$

for some positive constant C_1. Hence, the first term in the right-hand side of (17.2.10) converges to $T^{-1}\int_0^T(\Pi|u|^p)\log(\Pi|u|^p)ds$ as n tends to $+\infty$. The previous arguments show that $\int_{(0,T)\times\mathbb{R}^N}|u_n|^{p-2}|\nabla_x u_n|^2d\mu^\sharp$ converges to the corresponding integral with u instead of u_n. Next, since the function $y\mapsto y^p\log_-(y^p)$ is Lipschitz continuous, we can estimate

$$\||u|^p\log_-(|u|^p)-|u_n|^p\log_-(|u_n|^p)\|_{L^1_\sharp(\mathbb{R}^{N+1},\mu^\sharp)}\le C_2\||u|^p-|u_n|^p\|_{L^1_\sharp(\mathbb{R}^{N+1},\mu^\sharp)}$$

for some constant $C_2>0$, so that the last term in the right-hand side of (17.2.10) also converges to the corresponding term with u instead of u_n.

In view of the above results, to complete the proof it suffices to apply Fatou's lemma to the left-hand side of (17.2.10). □

To conclude this subsection, we show that the log-Sobolev inequality implies that $D(G_p^\sharp)$ is compactly embedded in $L^p_\sharp(\mathbb{R}^{N+1},\mu^\sharp)$, for any $p\in(1,+\infty)$.

Theorem 17.2.14 *$D(G_p^\sharp)$ is compactly embedded into $L^p_\sharp(\mathbb{R}^{N+1},\mu^\sharp)$ for any $p\in(1,+\infty)$. Moreover, the spectrum of $G_p^\sharp$ consists of isolated eigenvalues independent of $p\in(1,+\infty)$. The associated spectral projections are independent of p, too.*

Proof We first prove that $D(G_2^\sharp)$ is compactly embedded in $L^2_\sharp(\mathbb{R}^{N+1},\mu^\sharp)$. Arguing as in the proof of Theorem 16.4.8, we can prove that for any $\varepsilon>0$ there exists $R>0$ such that $\|u\|_{L^2((0,T)\times(\mathbb{R}^N\setminus B_R))}\le\varepsilon$ for any u in the unit ball $\mathcal{B}$ of $D(G_2^\sharp)$, centered at zero, and this is enough to apply a standard argument and conclude that $\mathcal{B}$ is totally bounded in $L^2_\sharp(\mathbb{R}^{N+1},\mu^\sharp)$. The main ingredients are the logarithmic Sobolev inequality (17.2.9), Proposition 17.2.5 to infer that $\int_{(0,T)\times\mathbb{R}^N}|\nabla_x u|^2d\mu^\sharp$ is bounded from above by a constant, independent of $u\in\mathcal{B}$, and the proof of Proposition 17.2.13, to show that the modulus of $\int_0^T(\Pi u^2)\log(\Pi u^2)ds$ can be bounded by a constant independent of $u\in\mathcal{B}$. Indeed, the proof of the quoted proposition shows that $u^2\in D(G_1^\sharp)$ for any $u\in\mathcal{B}$ and $\|u^2\|_{D(G_1^\sharp)}\le C$, with C being independent of u. Hence, Πu^2 belongs to the domain of the part of $G_1^\sharp$ in $\Pi(L^1_\sharp(\mathbb{R}^{N+1},\mu^\sharp))$, which is isomorphic to $W^{1,1}_\sharp(\mathbb{R},T^{-1}ds)$. By the classical Sobolev embedding theorems, $\|\Pi u^2\|_{L^\infty((0,T))}$ is bounded by a constant independent of u.

Since $D(G_2^\sharp)$ is compactly embedded into $L^2_\sharp(\mathbb{R}^{N+1},\mu^\sharp)$, the resolvent operator $u\mapsto R(\lambda,G_2^\sharp)u=\int_0^{+\infty}e^{-\lambda t}\mathscr{T}_2^\sharp(t)u\,dt$ is compact in $L^2_\sharp(\mathbb{R}^{N+1},\mu^\sharp)$ for any $\lambda>0$. Moreover, each operator $R(\lambda,G_p^\sharp)$ is bounded in $L^p_\sharp(\mathbb{R}^{N+1},\mu^\sharp)$ ($p\in[1,+\infty)\cup\{+\infty\}$) and $R(\lambda,G_p^\sharp)=R(\lambda,G_2^\sharp)$ on $L^2_\sharp(\mathbb{R}^{N+1},\mu^\sharp)\cap L^p_\sharp(\mathbb{R}^{N+1},\mu^\sharp)$. Hence, by Theorem B.3.11 the operator $R(\lambda,G_p^\sharp)$ is compact in $L^p_\sharp(\mathbb{R}^{N+1},\mu^\sharp)$ for any $p\in(1,+\infty)$.

To complete the proof, we observe that, by the general spectral theory, the spectrum of $G_p^\sharp$ consists of isolated eigenvalues. Moreover, arguing as in Step 2 of the proof of Theorem 7.2.10, we can show that the spectrum of $G_p^\sharp$ is independent of p. It also follows that the spectral projections are independent of p. □

Chapter 18

The asymptotic behaviour of the evolution operator and the evolution semigroup

18.0 Introduction

In this chapter we consider the family of second-order elliptic operators $\mathcal{A}$, defined on smooth functions ψ by

$$\mathcal{A}(t)\psi(x) = \sum_{i,j=1}^{N} q_{ij}(t,x)D_{ij}\psi(x) + \sum_{i=1}^{N} b_i(t,x)D_i\psi(x), \qquad t \in I, \ x \in \mathbb{R}^N,$$

where $I = [\sigma_*, +\infty)$, for some $\sigma_* \in \mathbb{R}$, in Section 18.1, and $I = \mathbb{R}$ in Section 18.2. Under suitable assumptions on the coefficients of the operator $\mathcal{A}$ (see Hypotheses 18.1.1), in Section 18.1 we first prove that, as t tends to $+\infty$, $\|G(t,s)f - m_s(f)\|_{L^p(\mathbb{R}^N,\mu_t)}$ converges to zero for any $f \in L^p(\mathbb{R}^N, \mu_s)$ and $G(t,s)f$ converges to $m_s(f)$ locally uniformly in $\mathbb{R}^N$ for any $f \in C_b(\mathbb{R}^N)$. As usual, $m_s(f)$ denotes the average of f with respect to the measure μ_s and $\{\mu_t : t \geq \sigma_*\}$ is the unique tight evolution system of measures for $\{G(t,s)\}$. To get such a result we extend the coefficients of the operator $\mathcal{A}$ to $\mathbb{R}^{N+1}$ in a suitable way and, coherently, we extend the evolution operator $\{G(t,s)\}$, with an evolution semigroup defined for any $\mathbb{R} \ni s \leq t$, and the measures μ_t to any $t \in \mathbb{R}$. Then, we introduce the evolution semigroup $\{\mathscr{T}_p(t)\}$ in $L^p(\mathbb{R}^{N+1}, \mu)$, where μ is the unique positive Borel measure such that

$$\mu(A \times B) = \int_A \mu_s(B)ds$$

for any pair of Borel sets $A \subset \mathbb{R}$ and $B \subset \mathbb{R}^N$, and, using semigroups arguments that seem not to have counterparts for evolution operators, we prove that

$$\lim_{t\to+\infty} \|\nabla_x \mathscr{T}_p(t)u\|_{L^p(\mathbb{R}^{N+1},\mu)} = 0, \qquad u \in L^p(\mathbb{R}^{N+1}, \mu), \ p \geq 2.$$

To prove this formula we use the estimate

$$\int_{\mathbb{R}^{N+1}} |\sqrt{Q}\nabla_x u|^2 d\mu \leq -\int_{\mathbb{R}^{N+1}} uG_2u d\mu, \qquad u \in D(G_2),$$

where G_2 is the infinitesimal generator of the strongly continuous semigroup $\{\mathscr{T}_2(t)\}$, which is obtained extending to $L^2(\mathbb{R}^{N+1}, \mu)$ the restriction of the semigroup $\{\mathscr{T}(t)\}$ to $C_c(\mathbb{R}^{N+1})$. This is the first crucial step in the proof of the asymptotic behaviour of $G(t,s)f$. The second crucial step consists of showing that the infimum of the density $\rho(t,\cdot)$ of the measure μ_t over any ball B_R is bounded from below by a positive constant as t tends to $+\infty$. This brief summary of the techniques used to study the convergence of $G(t,s)f$ to $m_s(f)$ sheds a light on the strongly relations between the evolution operator $\{G(t,s)\}$ and the evolution semigroup $\{\mathscr{T}(t)\}$. These relations will be made still much more clear in the periodic setting analyzed in the second part of this chapter.

Due to the nature of the convergence result of $G(t,s)f$ to $m_s(f)$ when $f \in L^p(\mathbb{R}^N, \mu_s)$, it is important to get information of the limit point (at infinity) of the family $\{\mu_t : t \geq \sigma_*\}$. If the coefficients q_{ij} and b_j of the operator $\mathcal{A}$ belong to $C_b^{\alpha/2,\alpha}(I \times B_R)$ for any $i,j \in \{1,\dots,N\}$ and $R > 0$ and they converge pointwise in $\mathbb{R}^N$ as t tends to $+\infty$ to functions q_{ij}^∞ and b_j^∞, then the measure μ_t weakly* converges to the invariant measure μ_∞ of the autonomous elliptic operator $\mathcal{A}_\infty = \mathrm{Tr}(Q^\infty D^2) + \langle b^\infty, \nabla\rangle$. More precisely, $\rho(t,\cdot)$ converges to the density ρ_∞ of μ_∞ in $L^1(\mathbb{R}^N)$ and locally uniformly in $\mathbb{R}^N$. These results imply the convergence of $G(t,s)f$ to $m_s(f)$ in a fixed L^p-space. More precisely, $G(t,s)$ converges to $m_s(f)$ in $L^p(\mathbb{R}^N, \mu_\infty)$ for any $p \in [1,+\infty)$ and $f \in C_b(\mathbb{R}^N)$.

Next, in Subsection 18.1.1, we study the exponential decay to zero of $G(t,s)f$ to $m_s(f)$. We first show that $\|\mathscr{T}_p(t)(I-\Pi)\|_{L(L^p(\mathbb{R}^{N+1},\mu))} \leq M_0 e^{\omega t}$ for any $t > 0$ and some constant $\omega \in \mathbb{R}$ and $M_0 > 0$ if and only if $\|G(t,s) - m_s\|_{L(L^p(\mathbb{R}^N,\mu_s),L^p(\mathbb{R}^N,\mu_t))} \leq M_1 e^{\omega(t-s)}$ for any $t > s$ and some positive constant M_1. Here, $(\Pi f)(s,x) = m_s(f)$ for almost every $(s,x) \in \mathbb{R}^{N+1}$.

Moreover, assuming that the diffusion coefficients are bounded and

$$|D_k q_{ij}(t,x)| \leq \widehat{q}(t,x), \qquad \langle \mathrm{Jac}_x b(t,x)\xi, \xi\rangle \leq r(t,x)|\xi|^2, \qquad t \in [\sigma_*,+\infty),\ x,\xi \in \mathbb{R}^N,$$

for some upperly bounded functions $r : [\sigma_*,+\infty) \times \mathbb{R}^N \to \mathbb{R}$ and $\widehat{q} : [\sigma_*,+\infty) \times [0,+\infty) \to \mathbb{R}$, such that

$$\ell_2 := \sup_{(t,x)\in[\sigma_*,+\infty)\times\mathbb{R}^N} \left(r(t,x) + \frac{N^2(\widehat{q}(t,x))^2}{4\kappa(t,x)} \right)$$

is negative, the decay rate to zero of the function $G(t,s)f - m_s(f)$ (as t tends to $+\infty$ and as s tends to $-\infty$) is of exponential type even if the diffusion coefficients depend on the spatial variables. More precisely, it holds that

$$\|G(t,s)f - m_s(f)\|_{L^p(\mathbb{R}^N,\mu_t)} \leq M_p e^{\theta_p(t-s)} \|f\|_{L^p(\mathbb{R}^N,\mu_s)}, \qquad t > s \geq \sigma_*, \tag{18.0.1}$$

for any $p \in (1,+\infty)$ where $\theta_p = 2\ell_2(1-(p\wedge 2)^{-1})$. A fundamental tool to prove this result is the Poincaré inequality

$$\|u - \Pi u\|_{L^p(\mathbb{R}^{N+1},\mu)} \leq C_p \|\nabla_x u\|_{L^p(\mathbb{R}^{N+1},\mu)}, \qquad u \in W_p^{0,1}(\mathbb{R}^{N+1},\mu).$$

When estimate (18.0.1) holds, each operator G_p has a spectral gap, i.e., the $\sup\{\mathrm{Re}\,\lambda : \lambda \in \sigma(G_p) \setminus i\mathbb{R}\} = \omega_p < 0$. In the particular case when the diffusion coefficients are independent of x and

$$\langle \mathrm{Jac}_x b(t,x)\xi, \xi\rangle \leq r_0|\xi|^2, \qquad t \in [\sigma_*,+\infty),\ x,\xi \in \mathbb{R}^N, \tag{18.0.2}$$

for some negative constant r_0, the Poincaré inequality, with $p = 2$, and the hypercontractivity of $G(t,s)$ allow us to prove that the sets

$$\begin{aligned}
\mathfrak{C}_p = \{&\omega \in \mathbb{R} : \|G(t,s)f - m_s(f)\|_{L^p(\mathbb{R}^N,\mu_t)} \leq M_{p,\omega} e^{\omega(t-s)} \|f\|_{L^p(\mathbb{R}^N,\mu_s)},\\
&\text{for any } \sigma_* \leq s < t, \text{ any } f \in L^p(\mathbb{R}^N,\mu_s) \text{ and some } M_{p,\omega} > 0\},\\
\mathfrak{D}_p = \{&\omega \in \mathbb{R} : \|\nabla_x G(t,s)f\|_{L^p(\mathbb{R}^N,\mu_t)} \leq N_{p,\omega} e^{\omega(t-s)} \|f\|_{L^p(\mathbb{R}^N,\mu_s)},\\
&\text{for any } s,t \geq \sigma_*,\ t-s \geq 1, \text{ any } f \in L^p(\mathbb{R}^N,\mu_s) \text{ and some } N_{p,\omega} > 0\}
\end{aligned}$$

coincide for any $p \in (1,+\infty)$ and are independent of p. Then, Theorem 14.4.5(ii) implies that $r_0 \in \mathfrak{D}_p$, and therefore $\|G(t,s) - m_s\|_{L(L^p(\mathbb{R}^N,\mu_s),L^p(\mathbb{R}^N,\mu_t))}$ decays exponentially to zero, as t tends to $+\infty$. This is a refinement of estimate (18.0.1).

In the second part of the chapter, we deal with the periodic case where the diffusion coefficients of the operator $\mathcal{A}$ are defined in $\mathbb{R}^{N+1}$ and are T-periodic with respect to the variable t for some $T > 0$. In this setting, Proposition 17.2.2 shows that the unique tight evolution system of measures of $\{G(t,s)\}$ is T-periodic, i.e., $\mu_{t+T} = \mu_t$ for any $t \in \mathbb{R}$.

As has already been noticed, in this periodic setting the interplay between the evolution operator $\{G(t,s)\}$ and the periodic evolution semigroup $\{\mathscr{T}_p^\sharp(t)\}$ is still much more clear, as Theorem 18.2.2 shows. In fact, under suitable assumptions, the following properties

$$\lim_{t\to+\infty} \|\mathscr{T}_p^\sharp(t)(u - \Pi u)\|_{L^p_\sharp(\mathbb{R}^{N+1},\mu^\sharp)} = 0, \qquad u \in L^p(\mathbb{R}^{N+1}, \mu^\sharp),$$

$$\lim_{t\to+\infty} \|G(t,s)f - m_s(f)\|_{L^p(\mathbb{R}^N,\mu_t)} = 0, \qquad f \in L^p(\mathbb{R}^N, \mu_s^\sharp),\ \ s \in \mathbb{R},$$

$$\lim_{s\to-\infty} \|G(t,s)f - m_s(f)\|_{L^p(\mathbb{R}^N,\mu_t)} = 0, \qquad f \in C_b(\mathbb{R}^N),\ \ t \in \mathbb{R},$$

$$\lim_{t\to+\infty} \|G(t,s)f - m_s(f)\|_{C_b(B_R)} = 0, \qquad f \in C_b(\mathbb{R}^N),\ \ s \in \mathbb{R},\ \ R > 0,$$

$$\lim_{s\to-\infty} \|G(t,s)f - m_s(f)\|_{C_b(B_R)} = 0, \qquad f \in C_b(\mathbb{R}^N),\ \ t \in \mathbb{R},\ \ R > 0,$$

are all equivalent to each other. Here, $\mu^\sharp$ is the probability measure on $\mathcal{B}((0,T) \times \mathbb{R}^N)$ given by $\mu^\sharp = T^{-1}\mu$, where μ is as above. Under suitable assumptions, the first property is satisfied, so that all the others are too. In the particular case when the diffusion coefficients are independent of x and condition (18.0.2) is satisfied by some negative constant $r_0 < 0$, $\sigma(G_p^\sharp) \cap i\mathbb{R} = \{2\pi k i/T : k \in \mathbb{Z}\}$ consists of simple eigenvalues and ω_p is independent of $p \in (1,+\infty)$. It turns out that the exponential decay rate of $\|G(t,s)f - m_s(f)\|_{L^p(\mathbb{R}^N,\mu_t)}$ as t tends to $+\infty$ is independent of $p \in (1,+\infty)$ and, in the case when $p \in (1,2)$, the decay rate in (18.0.1) can be improved.

The results of this chapter are inspired by [16, 174, 175].

18.1 The general case

In the first part of this section we assume the following standing assumptions about the coefficients of the operator $\mathcal{A}$.

Hypotheses 18.1.1 (i) *$q_{ij}, b_i \in C^{\alpha/2,\alpha}_{\rm loc}([\sigma_*,+\infty) \times \mathbb{R}^N)$ for some $\alpha \in (0,1)$ and any $i,j = 1,\ldots,N$. Moreover, $q_{ij} \in C_b([\sigma_*,+\infty) \times B_R)$ and $b_i \in C_b([\sigma_*,+\infty); L^p(B_R))$ for any $R > 0$, $i,j = 1,\ldots,N$ and some $p > N+2$;*

(ii) *the matrix $Q(t,x)$ is symmetric for any $(t,x) \in [\sigma_*,+\infty) \times \mathbb{R}^N$ and there exist a function $\kappa : [\sigma_*,+\infty) \times \mathbb{R}^N \to \mathbb{R}$, with positive infimum κ_0, such that $\langle Q(t,x)\xi,\xi\rangle \geq \kappa(t,x)|\xi|^2$ for any $t \geq \sigma_*$ and $x,\xi \in \mathbb{R}^N$;*

(iii) *there exist a function $\varphi \in C^2(\mathbb{R}^N)$, blowing up as $|x|$ tends to $+\infty$, and two positive constants a and d such that $\mathcal{A}\varphi \leq a - d\varphi$ in $[\sigma_*,+\infty) \times \mathbb{R}^N$;*

(iv) *the first-order spatial derivatives of the coefficients of the operator $\mathcal{A}$ exist, belong to* $C^{\alpha/2,\alpha}_{\rm loc}([\sigma_*,+\infty)\times\mathbb{R}^N)$ *and there exist two constants* $q_0>0$, $r_0\in\mathbb{R}$ *and two functions* $\widehat{q}_0:[\sigma_*,+\infty)\to[0,+\infty)$, *bounded from above by* q_0, *and* $r:[\sigma_*,+\infty)\times\mathbb{R}^N\to\mathbb{R}$, *bounded from above by* r_0, *such that* $|\nabla_x q_{ij}(t,x)|\le\widehat{q}_0(t)\kappa(t,x)$ *and* $\langle {\rm Jac}_x b(t,x)\xi,\xi\rangle\le r(t,x)|\xi|^2$ *for any* $t\ge\sigma_*$ *and* $x,\xi\in\mathbb{R}^N$. *Moreover,* D_kq_{ij} *belongs to* $C_b([\sigma_*,+\infty);L^p(B_R))$ *for any* $i,j,k\in\{1,\dots,N\}$, $R>0$ *and some* $p>N+2$;

(v) *there exists a constant* $C>0$ *such that one of the following conditions is satisfied:*

(a) $|Q(t,x)|\le C(1+|x|)\varphi(x)$, $\langle b(t,x),x\rangle\le C(1+|x|^2)\varphi(x)$ *for any* $(t,x)\in[\sigma_*,+\infty)\times\mathbb{R}^N$;

(b) *the diffusion coefficients are bounded in* $[\sigma_*,+\infty)\times\mathbb{R}^N$.

Remark 18.1.2 (i) Hypothesis 18.1.1(iii) and Theorem 16.2.7 show that there exists a tight evolution system of measures $\{\mu_t:t\ge\sigma_*\}$ associated with the evolution operator $\{G(t,s)\}$. If we extend the diffusion coefficients q_{ij} and b_i $(i,j=1,\dots,N)$ to $\mathbb{R}^{N+1}$ by setting $q_{ij}(t,\cdot)=q_{ij}(\sigma_*,\cdot)$ and $b_i(t,\cdot)=b_i(\sigma_*,\cdot)$ for any $t<\sigma_*$, then the operators $\mathcal{A}(t)$ can be defined for any $t\in\mathbb{R}$ and Hypotheses 18.1.1 hold with I replaced by $\mathbb{R}$. Moreover, if we set $\mu_s:=G^*(\sigma_*,s)\mu_0$ for $s<\sigma_*$, where $G^*(0,s)$ is the adjoint of the operator $G(0,s)$ in $C_b(\mathbb{R}^N)$, then the arguments in the proof of Lemma 16.2.3 show that μ_s is a probability measure for any $s<\sigma_*$ and the set $\{\mu_t:t\in\mathbb{R}\}$ is a tight evolution system of measures for $G(t,s)$ on $\mathbb{R}$ (see the proof of Theorem 16.2.7 for further details).

(ii) Hypotheses 18.1.1(iv), the above remark and Theorem 14.4.2 show that the estimate

$$\|\nabla_x G(t,s)f\|_\infty\le C_1\|f\|_\infty$$

is satisfied by any $f\in C_b(\mathbb{R}^N)$, $t,s\in\mathbb{R}$, with $t>s+1$, and some positive constant C_1.

(iii) The first part of Hypothesis 18.1.1(iv) is used to guarantee the gradient estimate $|\nabla_x G(t,s)f|^p\le K_p(t\vee 1)^{-p/2}G(t,s)|f|^p$ for any $t>s\in\mathbb{R}$, any $p\in[2,+\infty)$ and a suitable constant K_p. Actually, the condition on the growth at infinity of the gradient of the diffusion coefficients q_{ij} $(i,j=1,\dots,N)$ can be relaxed assuming more restrictive conditions on the function r (see Theorem 14.4.5).

To prove the first crucial result, which we need in the study of the asymptotic behaviour of the function $G(t,s)f$, we introduce the evolution semigroup $\{\mathscr{T}_p(t)\}$ associated in $L^p(\mathbb{R}^{N+1},\mu)$ $(p\ge2)$ with the evolution operator $\{G(t,s):t\ge s\in\mathbb{R}\}$ (obtained from the evolution operator $\{G(t,s):t\ge s\ge\sigma_*\}$ as explained in the previous remark). Here, μ is the measure on the σ-algebra of all the Borel subsets of $\mathbb{R}^{N+1}$ such that

$$\mu(A\times B)=\int_A\mu_s(B)ds$$

for any pair of Borel sets $A\subset\mathbb{R}$ and $B\subset\mathbb{R}^N$, obtained by starting from the extension of a tight evolution system of measures $\{\mu_t:t\ge\sigma_*\}$. *A priori* the measure μ might depend on the choice of the evolution system of measures $\{\mu_t:t\ge\sigma_*\}$. As the forthcoming Corollary 18.1.7 shows, this is not the case.

The proof of the following proposition is an extension to the evolution semigroup of a similar proof for Markov semigroups generated by elliptic operators (see e.g. [70]).

Proposition 18.1.3 *For any $p \in [2,+\infty)$ and $f \in L^p(\mathbb{R}^{N+1},\mu)$ it holds that*

$$\lim_{t\to+\infty} \|\nabla_x \mathscr{T}_p(t)f\|_{L^p(\mathbb{R}^{N+1},\mu)} = 0. \tag{18.1.1}$$

Proof We begin by observing that it suffices to prove the assertion in the case $p = 2$. Indeed, for $p > 2$, the Hölder inequality shows that

$$\begin{aligned}\|\nabla_x \mathscr{T}_p(t)f\|_{L^p(\mathbb{R}^{N+1},\mu)} \leq & \|\nabla_x \mathscr{T}(t)f\|_\infty^{1-\frac{2}{p}} \|\nabla_x \mathscr{T}_2(t)f\|_{L^2(\mathbb{R}^{N+1},\mu)}^{\frac{2}{p}} \\ \leq & C\|f\|_\infty^{1-\frac{2}{p}} \|\nabla_x \mathscr{T}_2(t)f\|_{L^2(\mathbb{R}^{N+1},\mu)}^{\frac{2}{p}}\end{aligned}$$

for any $f \in C_c(\mathbb{R}^{N+1})$ and some positive constant C. Hence, if the formula (18.1.1) holds with $p = 2$, then it holds for any $p > 2$ and $f \in C_c(\mathbb{R}^{N+1})$. To extend its validity to any $f \in L^p(\mathbb{R}^{N+1},\mu)$, it suffices to recall that $C_c(\mathbb{R}^{N+1})$ is dense in $L^p(\mathbb{R}^{N+1},\mu)$ and $\sup_{t\geq 1} \|\nabla_x \mathscr{T}_p(t)\|_{L(L^p(\mathbb{R}^{N+1},\mu),(L^p(\mathbb{R}^{N+1},\mu))^N)} < +\infty$ by Proposition 17.1.4.

So let us prove (18.1.1) with $p = 2$. As a first step, we prove that, for any $g \in D(G_2)$, the function $\chi_g : [0,+\infty) \to \mathbb{R}$, defined by $\chi_g(t) = \|\nabla_x \mathscr{T}_2(t)g\|^2_{L^2(\mathbb{R}^{N+1},\mu)}$ for any $t \geq 0$, belongs to $L^1((0,+\infty))$. For this purpose, we observe that, since the function $t \mapsto \|\mathscr{T}_2(t)g\|^2_{L^2(\mathbb{R}^{N+1},\mu)}$ is differentiable in $[0,+\infty)$, we can write

$$\begin{aligned}\|\mathscr{T}_2(t)g\|^2_{L^2(\mathbb{R}^{N+1},\mu)} - \|g\|^2_{L^2(\mathbb{R}^{N+1},\mu)} =& \int_0^t \frac{d}{ds}\|\mathscr{T}_2(s)g\|^2_{L^2(\mathbb{R}^{N+1},\mu)} ds \\ =& 2\int_0^t ds \int_{\mathbb{R}^{N+1}} \mathscr{T}_2(s)g\, G_2\mathscr{T}_2(s)g d\mu\end{aligned}$$

for any $t > 0$. Using (17.1.7) we can estimate

$$\|\mathscr{T}_2(t)g\|^2_{L^2(\mathbb{R}^{N+1},\mu)} + 2\int_0^t ds \int_{\mathbb{R}^{N+1}} |\sqrt{Q}\nabla_x \mathscr{T}_2(s)g|^2 d\mu \leq \|g\|^2_{L^2(\mathbb{R}^{N+1},\mu)}, \qquad t > 0.$$

Thus, the function χ_g belongs to $L^1((0,+\infty))$.

Now, we can prove (18.1.1) for functions $f \in D(G_2^2)$. For any such function f, χ_f is differentiable in $[0,+\infty)$ and

$$\chi_f'(t) = \int_{\mathbb{R}^{N+1}} 2\langle \nabla_x \mathscr{T}_2(t)f, \nabla_x \mathscr{T}_2(t)G_2 f\rangle d\mu, \qquad t \geq 0.$$

Thus, we can estimate $|\chi_f'| \leq \chi_f + \chi_{G_2 f}$ in $[0,+\infty)$, by the Hölder inequality, and conclude that χ_f' belongs to $L^1((0,+\infty))$. This implies that $\chi_f(t)$ vanishes as t tends to $+\infty$, and (18.1.1) holds.

To extend the validity of (18.1.1) to any $f \in L^2(\mathbb{R}^{N+1},\mu)$ it suffices to approximate any such f with a sequence of functions in $D(G_2^2)$ and use estimate (17.1.5), which implies that $\nabla_x \mathscr{T}_2(\cdot)$ is bounded in $(1,+\infty)$ with values in $L(L^2(\mathbb{R}^{N+1},\mu),(L^2(\mathbb{R}^{N+1},\mu))^N)$. □

By (the proof of) Proposition 16.2.5, each measure μ_t ($t \in \mathbb{R}$) is equivalent to the Lebesgue measure. Hence, it admits a density $\rho(t,\cdot)$ which is locally Hölder continuous in $\mathbb{R}^N$, and the function ρ is the density of the measure μ with respect to the $(N+1)$-th dimensional Lebesgue measure.

Using Hypothesis 18.1.1(i) we can prove a crucial lower bound on the densities $\rho(t,\cdot)$.

Lemma 18.1.4 *For any $k \in \mathbb{N}$ there exists a number $\delta_k > 0$ such that $\rho(\tau,x) \geq \delta_k$ for any $\tau \geq \sigma_*$ and $x \in \overline{B}_k$.*

Proof Let $\mathscr{D}_k=\{(t,s,x,y)\in[\sigma_*,+\infty)\times[\sigma_*,+\infty)\times\overline{B}_k\times\overline{B}_k: t>s\}$ for any $k\in\mathbb{N}$. By $g_k:\mathscr{D}_k\to[0,+\infty)$ we denote the Green function associated with the parabolic problem

$$\begin{cases} D_t u(t,x)=\mathcal{A}(t)u(t,x), & t>s, \quad x\in B_k,\\ u(t,x)=0, & t>s, \quad x\in\partial B_k,\\ u(s,x)=f(x), & \qquad\quad x\in B_k.\end{cases}$$

The proof of Theorem 14.1.2 shows that $g\geq g_k$ on $\mathscr{D}_k$ for any $k\in\mathbb{N}$, where g is the Green function of the evolution operator $\{G(t,s)\}$ (see Proposition 14.1.3(i)). By the tightness of the family $\{\mu_t: t\geq\sigma_*\}$, we can determine a radius $k_0\in\mathbb{N}$ such that $\mu_t(\overline{B}_{k_0})\geq 1/2$ for any $t\geq\sigma_*$. We fix now and for the rest of the proof an arbitrary integer $k\geq k_0$.

The core of the proof consists of showing that there exists a number $\delta_k>0$ such that

$$g_{k+2}(\tau+1,\tau,x,y)\geq 2\delta_k, \qquad \tau\geq\sigma_*,\ x,y\in\overline{B}_k. \tag{18.1.2}$$

Indeed, suppose that estimate (18.1.2) is satisfied, and take a Borel set $B\subset\overline{B}_k$ and $\tau\geq\sigma_*$. Then, using the invariance property of the family $\{\mu_t: t\in\mathbb{R}\}$ and Proposition 14.1.3, we can estimate

$$\begin{aligned}\int_B \rho(\tau,x)dx &= \int_{\mathbb{R}^N}\chi_B d\mu_\tau = \int_{\mathbb{R}^N} G(\tau+1,\tau)\chi_B d\mu_{\tau+1}\\ &= \int_{\mathbb{R}^N} dx\int_B g(\tau+1,\tau,x,y)\rho(\tau+1,x)dy\\ &\geq \int_{\overline{B}_k} dx\int_B g_{k+2}(\tau+1,\tau,x,y)\rho(\tau+1,x)dy\\ &\geq 2\delta_k\lambda(B)\int_{\overline{B}_k}\rho(\tau+1,x)dx = 2\delta_k\lambda(B)\mu_{\tau+1}(B_k)\geq\delta_k\lambda(B)\end{aligned}$$

and the assertion follows from the arbitrariness of B.

So, let us prove estimate (18.1.2). For this purpose, we take advantage of a classical estimate for the Green functions of parabolic problems associated with elliptic operators in divergence form. For any $\tau\geq\sigma_*$, $g_{k+2}(\cdot+\tau,\tau,\cdot,\cdot)$ is the Green function of the operator $D_t-\operatorname{div}(\widetilde{Q}(t+\tau,\cdot)\nabla_x)-\langle\widetilde{b}(t+\tau),\nabla_x\rangle$ in $(0,1]\times B_{k+2}$ (see e.g., [22, Theorem 6]), where $\widetilde{q}_{ij},\widetilde{b}_i\in C_b(\mathbb{R}_+\times\mathbb{R}^N)$ satisfy $\widetilde{q}_{ij}=q_{ij}$ $\widetilde{b}_i=b_i-\sum_{j=1}^N D_iq_{ij}$ in $\mathbb{R}_+\times B_{k+2}$ $(i,j\in\{1,\dots,N\})$ as well as the ellipticity condition $\langle\widetilde{Q}(t,x)\xi,\xi\rangle\geq\kappa_0/2|\xi|^2$ for any $t\in\mathbb{R}_+$, $x,\xi\in\mathbb{R}^N$ and $\widetilde{b}_i\equiv 0$ in $\mathbb{R}_+\times\mathbb{R}\setminus B_{k+3}$ for any $i\in\{1,\dots,N\}$. To this operator, we can apply [22, Theorem 9(iii)] with $\Omega'=B_{k+1}$, $\Omega=B_{k+2}$ and $T=8$ and conclude that

$$g_{k+2}(t+\tau,\tau,x,y)\geq C_1t^{-\frac{N}{2}}\exp(-C_2t^{-1}|x-y|^2) \tag{18.1.3}$$

for any $x,y\in B_{k+1}$, $t\in(0,\min\{8,(d(y,\partial B_{k+1}))^2\}]$ and $\tau\geq\sigma_*$. The constants C_1, C_2 depend on $\sup_{t\geq\sigma_*}\|q_{ij}(t,\cdot)\|_{L^\infty(B_{k+2})}$, $\sup_{t\geq\sigma_*}\|b_j(t,\cdot)\|_{L^p(B_{k+2})}$, $\sup_{t\geq\sigma_*}\|D_iq_{ij}(t,\cdot)\|_{L^p(B_{k+2})}$ $(i,j=1,\dots,N)$, where p is as in Hypothesis 18.1.1(i), and on κ_0. If we take $y\in B_k$, then $d(y,\partial B_{k+1})\geq 1$ and taking $t=1$ in (18.1.3), estimate (18.1.2) follows. □

Also the following lemma will be used in the proof of Theorem 18.1.6.

Lemma 18.1.5 *Let $\{\nu_n: n\in\mathbb{N}\}$ be a tight sequence of probability measures on $\mathcal{B}(\mathbb{R}^N)$ and let $\{f_n\}\subset C_b(\mathbb{R}^N)$ be a bounded sequence converging to zero locally uniformly in $\mathbb{R}^N$ as n tends to $+\infty$. Then,*

$$\lim_{n\to+\infty}\int_{\mathbb{R}^N} f_n d\nu_n = 0.$$

If ν_n weakly converges to a probability measure ν and the bounded sequence $\{f_n\} \subset C_b(\mathbb{R}^N)$ converges to some function $f \in C_b(\mathbb{R}^N)$ locally uniformly in $\mathbb{R}^N$ as n tends to $+\infty$, then it holds that*

$$\lim_{n\to+\infty}\int_{\mathbb{R}^N} f_n d\nu_n = \int_{\mathbb{R}^N} f d\nu.$$

Proof We first suppose that ν_n weakly* converges to the probability measure ν and f_n converges to some $f \in C_b(\mathbb{R}^N)$. We fix $\varepsilon > 0$ and let $R > 0$ be such that $\nu_n(\mathbb{R}^N \setminus B_R) \le \varepsilon$ and $\nu(\mathbb{R}^N \setminus B_R) \le \varepsilon$ for any $n \in \mathbb{N}$. Moreover, we set $M = \sup_{n\in\mathbb{N}} \|f_n\|_\infty$. Then, we can estimate

$$\begin{aligned}\left|\int_{\mathbb{R}^N} f_n d\nu_n - \int_{\mathbb{R}^N} f d\nu\right| \le& \int_{B_R} |f_n - f| d\nu_n + \int_{\mathbb{R}^N\setminus B_R} |f_n - f| d\nu_n + \left|\int_{\mathbb{R}^N} f_n d\nu - \int_{\mathbb{R}^N} f d\nu\right|\\ \le& \|f_n - f\|_{C_b(B_R)} + 2M\varepsilon + \left|\int_{\mathbb{R}^N} f_n d\nu - \int_{\mathbb{R}^N} f d\nu\right|.\end{aligned}$$

Since f_n converges to f locally uniformly in $\mathbb{R}^N$ and ν_n weakly* converges to ν as n tends to $+\infty$, from the previous formula it follows that

$$\limsup_{n\to+\infty}\left|\int_{\mathbb{R}^N} f_n d\nu_n - \int_{\mathbb{R}^N} f d\nu\right| \le 2M\varepsilon$$

and the arbitrariness of $\varepsilon > 0$ yields the assertion.

The other statement follows from the estimate

$$\left|\int_{\mathbb{R}^N} f_n d\nu_n\right| \le \int_{B_R} |f_n| d\nu_n + 2M\varepsilon.$$

The proof is complete. □

Now, we can prove the main result of this section.

Theorem 18.1.6 *Let $\{\mu_t : t \ge \sigma_*\}$ be a tight evolution system of measures for $\{G(t,s)\}$. The following assertions are true.*

(i) *$\|G(t,s)f - m_s(f)\|_{L^p(\mathbb{R}^N,\mu_t)}$ converges to 0 as t tends to $+\infty$ for any $f \in L^p(\mathbb{R}^N, \mu_s)$, $s \ge \sigma_*$ and $p \in [1,+\infty)$.*

(ii) *$G(t,s)f$ converges to $m_s(f)$ locally uniformly in $\mathbb{R}^N$ as t tends to $+\infty$ for any $s \ge \sigma_*$ and $f \in C_b(\mathbb{R}^N)$.*

Proof (i) First of all we note that it suffices to prove the assertion for $p > N$, since

$$\|G(t,s)f - m_s(f)\|_{L^p(\mathbb{R}^N,\mu_t)} \le \|G(t,s)f - m_s(f)\|_{L^q(\mathbb{R}^N,\mu_t)},$$

for any $t > s \ge \sigma_*$, $p \le N < q$ and $f \in L^q(\mathbb{R}^N, \mu_s)$ (which is dense in $L^p(\mathbb{R}^N, \mu_s)$ since it contains $C_c^\infty(\mathbb{R}^N)$). Being rather long, we split the rest of the proof into two steps.

Step 1. Here, we prove that, for any $f \in C_c^\infty(\mathbb{R}^N)$ there exists a null set $\mathcal{N}(f)$ such that $\|G(t,s)f - m_s(f)\|_{L^p(\mathbb{R}^N,\mu_t)}$ vanishes as t tends to $+\infty$ for any $s \in (\sigma_*, +\infty) \setminus \mathcal{N}$.

We fix $f \in C_c^\infty(\mathbb{R}^N)$, $p > 2N$, a positive sequence $\{t_n\}$ diverging to $+\infty$, and a sequence $\{\alpha_m\} \subset C_c^\infty(\mathbb{R})$ such that $\chi_{(-m,m)} \le \alpha_m \le 1$. Further, we extend the evolution operator $\{G(t,s)\}$ and the evolution system of measures $\{\mu_t : t \ge \sigma_*\}$ as explained in Remark 18.1.2(i). By Proposition 18.1.3

$$\int_{\mathbb{R}} |\alpha_m(s)|^p \|\rho(s+t_n,\cdot)|\nabla_x G(s+t_n,s)f|^p\|_{L^1(\mathbb{R}^N)} ds = \|\nabla_x \mathscr{T}_p(t_n)(\alpha_m f)\|^p_{L^p(\mathbb{R}^{N+1},\mu)}$$

vanishes as n tends to $+\infty$ for any $m \in \mathbb{N}$. Thus, there exist null sets $\mathcal{N}_m \subset [-m,m]$ and subsequences $\{t_n^{(m)}\}$ diverging to $+\infty$, with $t_k^{(m+1)} \in \{t_n^{(m)}\}$ for all $k,m \in \mathbb{N}$, such that

$$\lim_{n\to+\infty}\int_{\mathbb{R}^N}\rho(s+t_n^{(m)},\cdot)|\nabla_x G(s+t_n^{(m)},s)f|^p dx = 0, \qquad s\in(\sigma_*,+\infty)\setminus\mathcal{N}_m,\ \ m\in\mathbb{N}.$$

With a diagonal procedure we can construct a sequence $\{t_{n_j}\}$ such that

$$\lim_{j\to+\infty}\int_{\mathbb{R}^N}\rho(s+t_{n_j},\cdot)|\nabla_x G(s+t_{n_j},s)f|^p dx, \qquad s\in(\sigma_*,+\infty)\setminus\mathcal{N}, \tag{18.1.4}$$

where $\mathcal{N}=\bigcup_{m\in\mathbb{N}}\mathcal{N}_m$ is a null set.

Fix $s\in(\sigma_*,+\infty)\setminus\mathcal{N}$. Lemma 18.1.4, with $\tau = s+t_n$, shows that, for any $k\in\mathbb{N}$, there exists a positive constant δ_k such that $\rho(s+t_n,x)\geq\delta_k$ for any $n\in\mathbb{N}$ and $x\in\overline{B}_k$. This lower bound and (18.1.4) yield

$$\lim_{j\to+\infty}\|\nabla_x G(s+t_{n_j},s)f\|_{L^p(B_k)} = 0 \tag{18.1.5}$$

for any $k\in\mathbb{N}$. In particular, the sequence $|\nabla_x G(s+t_{n_j},s)f|$ is bounded in $L^p(B_k)$ for any $k\in\mathbb{N}$. Since $L^\infty(B_k)\hookrightarrow L^p(B_k)$ and each operator $G(s+t_{n_j},s)$ is a contraction in $C_b(\mathbb{R}^N)$, the sequence $\{G(s+t_{n_j},s)f\}$ is bounded in $W^{1,p}(B_k)$ for any $k\in\mathbb{N}$. Since $p>N$, $W^{1,p}(B_k)$ is compactly embedded in $C(\overline{B}_k)$. Again, a diagonal argument shows that, up to a subsequence, $G(s+t_{n_j},s)f$ converges to some function $g(s,\cdot)\in C_b(\mathbb{R}^N)$, locally uniformly in $\mathbb{R}^N$. Actually, the formula (18.1.5) shows that g depends only on s since its weak gradient vanishes. To identify $g(s)$ with $m_s(f)$, we use the invariance property of the evolution system of measures to write

$$m_s(f)-g(s)=\int_{\mathbb{R}^N}(f-g(s))d\mu_s=\int_{\mathbb{R}^N}G(s+t_{n_j},s)(f-g(s))\,d\mu_{s+t_{n_j}}$$

and, then, Lemma 18.1.5(i), with $\widetilde{\mu}_n=\mu_{s+t_{n_j}}$, to conclude that the last side of the previous formula vanishes as j tends to $+\infty$. Thus, we have proved that $G(s+t_{n_j},s)f$ converges to $m_s(f)$ locally uniformly in $\mathbb{R}^N$, as j tends to $+\infty$, for $s\in\mathbb{R}_+\setminus\mathcal{N}$. Since $G(s+t_{n_j},s)\mathbf{1}=\mathbf{1}$, again Lemma 18.1.5(i) shows that

$$\lim_{j\to+\infty}\|G(s+t_{n_j},s)(f-m_s(f))\|_{L^p(\mathbb{R}^N,\mu_{s+t_{n_j}})}=0.$$

To conclude that $\lim_{t\to+\infty}\|G(t,s)f-m_s(f)\|_{L^p(\mathbb{R}^N,\mu_t)}=0$ it suffices to observe that the function $t\mapsto h(t)=\|G(t,s)f-m_s(f)\|_{L^p(\mathbb{R}^N,\mu_t)}$ is decreasing in $[s,+\infty)$ since

$$\begin{aligned}h(t_2)&=\|G(t_2,s)(f-m_s(f))\|_{L^p(\mathbb{R}^N,\mu_{t_2})}=\|G(t_2,t_1)G(t_1,s)(f-m_s(f))\|_{L^p(\mathbb{R}^N,\mu_{t_2})}\\&\leq\|G(t_1,s)(f-m_s(f))\|_{L^p(\mathbb{R}^N,\mu_{t_1})}=h(t_1)\end{aligned}$$

for $s\leq t_1<t_2$.

Step 2. In our construction, the set $\mathcal{N}(f)$ depends on f. To remove the dependence of the previous set on f, here we prove that there exists a countable set $\mathcal{I}\subset C_c^\infty(\mathbb{R}^N)$, which is dense in $L^p(\mathbb{R}^N,\mu_s)$ for any $s\geq\sigma_*$. Note that, we can limit ourselves to proving that there exists a set $\mathcal{I}$ as above such that for any $s\geq\sigma_*$ and any function $f\in L^p(\mathbb{R}^N,\mu_s)$, which is nonnegative and bounded, there exists a sequence in $\mathcal{I}$ which converges to f in $L^p(\mathbb{R}^N,\mu_s)$. Indeed, a general $f\in L^p(\mathbb{R}^N,\mu_s)$ can be approximated in the $L^p(\mathbb{R}^N,\mu_s)$-norm by the sequence $\{g_n\}$, where $g_n=f^+\wedge n-f^-\wedge n$ for any $n\in\mathbb{N}$, and all the functions $f^+\wedge n$ and $f^-\wedge n$ are nonnegative and bounded.

So, let us fix a nonnegative, bounded and compactly supported function $f \in L^p(\mathbb{R}^N, \mu_s)$ and denote by $\{f_n\}$ the sequence of functions defined by

$$f_n = \sum_{j=0}^{n2^n} \frac{j}{2^n} \chi_{f^{-1}([2^{-n}j, 2^{-n}(j+1)))\cap B_n}, \qquad n \in \mathbb{N}.$$

As n tends to $+\infty$ f_n pointwise increases to f. Hence, it converges to f in $L^p(\mathbb{R}^d, \mu_s)$ for $s \geq \sigma_*$. As is well known, there exists a sequence of simple functions, which are linear combinations (with rational coefficients) of characteristic functions of dyadic cubes with vertices in $\mathbb{Q}^N$ and all have support contained in the same cube, which converges to $\chi_{f^{-1}([2^{-n}j, 2^{-n}(j+1)))\cap B_n}$ in $L^p(\mathbb{R}^N)$ (and, hence, in $L^p(\mathbb{R}^N, \mu_s)$ by Theorem 16.2.4).

Since this set of functions is countable, we can order its elements into a sequence $\{h_m\}$. Next, we consider a standard sequence $\{\varrho_k\}$ of mollifiers and regularize each function h_m by convolution with ϱ_k. Clearly, each function $\varrho_k \star h_m$ belongs to $C_c^\infty(\mathbb{R}^N)$. Moreover, for any $m \in \mathbb{N}$, $\varrho_k \star h_m$ converges to h_m in $L^p(\mathbb{R}^N)$ as k tends to $+\infty$ and has support into a ball which is independent of k. Hence, taking again Theorem 16.2.4 into account, we conclude that $\varrho_k \star h_m$ converges to h_m in $L^p(\mathbb{R}^N, \mu_s)$ for any $s \geq \sigma_*$. The set $\mathcal{I} = \{\varrho_k \star h_m : k, m \in \mathbb{N}\}$ is the set that we are looking for.

Step 3. We now complete the proof of property (i). For this purpose, we set

$$\mathcal{N} := \bigcup_{k,m\in\mathbb{N}} \mathcal{N}(\varrho_k \star h_m)$$

and begin by proving that, for any $s \in (\sigma_*, +\infty) \setminus \mathcal{N}$ and any $f \in L^p(\mathbb{R}^N, \mu_s)$, the function $t \mapsto \|G(t,s)f - m_s(f)\|_{L^p(\mathbb{R}^N,\mu_t)}$ vanishes as t tends to $+\infty$. For this purpose, we fix such a function f and consider a sequence $\{\psi_n\} \subset \mathcal{I}$, converging to f in $L^p(\mathbb{R}^N, \mu_s)$ as n tends to $+\infty$. Moreover, we split

$$\begin{aligned}\|G(t,s)f - m_s(f)\|_{L^p(\mathbb{R}^N,\mu_t)} \leq& \|G(t,s)(f-\psi_n)\|_{L^p(\mathbb{R}^N,\mu_t)} + \|G(t,s)\psi_n - m_s(\psi_n)\|_{L^p(\mathbb{R}^N,\mu_t)} \\ &+ |m_s(\psi_n) - m_s(f)| \\ \leq& 2\|f-\psi_n\|_{L^p(\mathbb{R}^N,\mu_s)} + \|G(t,s)\psi_n - m_s(\psi_n)\|_{L^p(\mathbb{R}^N,\mu_t)}\end{aligned}$$

for any $t > s$ and any $n \in \mathbb{N}$. Letting t tend to $+\infty$ in the first and last side of the previous chain inequalities we get

$$\limsup_{t\to+\infty} \|G(t,s)f - m_s(f)\|_{L^p(\mathbb{R}^N,\mu_t)} \leq 2\|f-\psi_n\|_{L^p(\mathbb{R}^N,\mu_s)}, \qquad n \in \mathbb{N}.$$

Finally, letting n tend to $+\infty$, we conclude that $\limsup_{t\to+\infty} \|G(t,s)f - m_s(f)\|_{L^p(\mathbb{R}^N,\mu_t)} = 0$ so that $\|G(t,s)f - m_s(f)\|_{L^p(\mathbb{R}^N,\mu_t)}$ converges to 0 as t tends to $+\infty$.

Now, we can extend the convergence of $\|G(t,s)f - m_s(f)\|_{L^p(\mathbb{R}^N,\mu_t)}$ also to the values of s in $\mathcal{N}$. For this purpose, we fix $s \in \mathcal{N}$, $f \in L^p(\mathbb{R}^N, \mu_s)$ and take $s < \overline{s} \notin \mathcal{N}$. Further, we split $G(t,s)f - m_s(f) = G(t,\overline{s})g - m_{\overline{s}}(g)$, with $g = G(\overline{s}, s)f$. Since g belongs to $L^p(\mathbb{R}^d, \mu_{\overline{s}})$, from the above results we conclude that $\|G(t,\overline{s})g - m_{\overline{s}}(g)\|_{L^p(\mathbb{R}^N,\mu_t)}$ (i.e., $\|G(t,s)f - m_s(f)\|_{L^p(\mathbb{R}^N,\mu_t)}$) vanishes as t tends to $+\infty$ and the proof of property (i) is complete.

(ii) Fix $f \in C_b(\mathbb{R}^N)$, $s \in [\sigma_*, +\infty)$, $R > 0$ and $p > N$. Since $C_b(\mathbb{R}^N) \hookrightarrow L^p(\mathbb{R}^N, \mu_s)$ and

$$\|G(t+s,s)f - m_s(f)\|_{L^p(B_R)} \leq \delta_R^{-1/p} \|G(t+s,s)f - m_s(f)\|_{L^p(\mathbb{R}^N,\mu_{t+s})}, \qquad t > 0,$$

for some positive constant δ_R, by Lemma 18.1.4, from (i) it follows that $\|G(t+s,s)f - m_s(f)\|_{L^p(B_R)}$ vanishes as t tends to $+\infty$. In particular, there exists a positive constant

$C_1 = C_1(R)$ such that $\|G(t+s,s)f - m_s(f)\|_{L^p(B_R)} \le C_1$ for any $t \ge 1$. Moreover, the gradient estimate (14.4.3) implies that $\|\nabla_x G(t+s,s)f\|_{L^p(B_R)} \le C_2\|f\|_\infty$ for any $t \ge 1$ and some positive constant $C_2 = C_2(R)$. These two last estimates show that the family of functions $\{G(t+s,s)f - m_s(f) : t \ge 1\}$ is bounded in $W^{1,p}(B_R)$ and, by the Sobolev embedding theorems, in $C^\beta(B_R)$ for some $\beta \in (0,1)$, since $p > N$. Applying the Arzelà-Ascoli theorem, any sequence $\{t_n\}$ diverging to $+\infty$ admits a subsequence $\{t_{n_k}\}$ such that $G(t_{n_k}+s,s)f - m_s(f)$ converges uniformly in B_R to zero, as k tends to $+\infty$. This is enough to conclude that $G(t+s,s)f - m_s(f)$ converges to 0, uniformly in B_R, as t tends to $+\infty$. □

Corollary 18.1.7 *The evolution operator $\{G(t,s)\}$ has exactly one tight evolution system of measures.*

We skip the proof, since it can be obtained by arguing as in the last part of the proof of Theorem 16.2.11.

To conclude this section, we refine the results in Theorem 18.1.6(i) under the following additional assumptions on the coefficients of the operator $\mathcal{A}$.

Hypothesis 18.1.8 *The coefficients q_{ij} and b_i belong to $C_b^{\alpha/2,\alpha}(I \times B_R)$ for any $i,j \in \{1,\ldots,N\}$, $R > 0$ and converge pointwise in $\mathbb{R}^N$, as t tends to $+\infty$.*

Let us consider the operator $\mathcal{A}_\infty$, defined on smooth functions $\psi : \mathbb{R}^N \to \mathbb{R}$ by

$$\mathcal{A}_\infty\psi(x) = \sum_{i,j=1}^N q_{ij}^\infty D_{ij}\psi(x) + \sum_{j=1}^N b_j^\infty(x) D_j\psi(x), \qquad x \in \mathbb{R}^N,$$

where, for any $i,j = 1,\ldots,N$, q_{ij}^∞ and b_j^∞ are the pointwise limit as t tends to $+\infty$ of the coefficients $q_{ij}(t,\cdot)$ and $b_j(t,\cdot)$. Hypotheses 18.1.1 and 18.1.8 imply that the coefficients of $\mathcal{A}_\infty$ belong to $C^\alpha_{\rm loc}(\mathbb{R}^N)$ and satisfy the t-independent analogues of Hypotheses 18.1.1(ii) and (iii). By the results in Chapter 1, Proposition 9.1.4 and Theorems 9.1.15, 9.1.20, we can associate a semigroup $\{T(t)\}$ in $C_b(\mathbb{R}^N)$ with the elliptic operator $\mathcal{A}_\infty$ and this semigroup admits a unique invariant measure μ_∞ having a positive density ρ_∞ with respect to the Lebesgue measure.

Under the additional Hypothesis 18.1.8 we show that the densities $\rho(t,\cdot)$ converge to ρ_∞ and we derive a variant of Theorem 18.1.6.

Theorem 18.1.9 *The densities $\rho(t,\cdot)$ converge to ρ_∞ locally uniformly in $\mathbb{R}^N$ and in $L^1(\mathbb{R}^N)$ as t tends to $+\infty$.*

Proof The main core of the proof consists of showing that μ_t weakly* converges to μ, as t tends to $+\infty$.

Since the evolution system of measures $\{\mu_t : t \ge \sigma_*\}$ is tight, by Theorem 9.1.18 it admits weak* limits as t tends to $+\infty$. We will prove that μ is its only weak* limit. For this purpose, let ν_0 be a weak* limit point of the sequence $\{\mu_t : t \ge \sigma_*\}$ and let $\{s_n\} \subset [\sigma_*, +\infty)$ be a sequence diverging to $+\infty$ such that

$$\lim_{n\to+\infty} \int_{\mathbb{R}^N} f\, d\mu_{s_n} = \int_{\mathbb{R}^N} f\, d\nu_0, \qquad f \in C_b(\mathbb{R}^N). \tag{18.1.6}$$

To prove that $\nu_0 = \mu$ we proceed in three steps.

Step 1. Here, we prove that the sequence $\{G(\cdot + s_n, s_n)f\}$ converges to $T(\cdot)f$ as n tends to $+\infty$, locally uniformly in $[0,+\infty) \times \mathbb{R}^N$, for any $f \in C_c^\infty(\mathbb{R}^N)$. For this purpose, we observe that, for each $n \in \mathbb{N}$, the function $v_n = G(\cdot + s_n, s_n)f \in C_b([0,+\infty) \times \mathbb{R}^N) \cap C^{1,2}([0,+\infty) \times \mathbb{R}^N)$ solves the Cauchy problem

$$\begin{cases} D_t v_n(t,x) = \mathcal{A}(t+s_n)v_n(t,x), & (t,x) \in [0,+\infty) \times \mathbb{R}^N, \\ v_n(0,x) = f(x), & x \in \mathbb{R}^N. \end{cases}$$

Due to Hypotheses 18.1.8 and to the classical Schauder estimates, for any $r, T > 0$ there exists a positive constant C (which is independent of n) such that $\|v_n\|_{C^{1+\alpha/2,2+\alpha}([0,T]\times B_r)} \le C\|f\|_{C_b^{2+\alpha}(\mathbb{R}^N)}$. The Arzelà-Ascoli theorem and a diagonal argument show that there exist a function $u \in C_b([0,+\infty)\times\mathbb{R}^N) \cap C^{1+\alpha/2,2+\alpha}_{\rm loc}([0,+\infty)\times\mathbb{R}^N)$ and a subsequence $\{s_{n_k}\} \subset \{s_n\}$ such that v_{n_k} converges to u in $C^{1,2}([0,T] \times B_r)$ for any $T, r > 0$, as k tends to $+\infty$, and $D_t u = \mathcal{A}u$ in $[0,+\infty) \times \mathbb{R}^N$. Since, clearly, $v_{n_k}(0,\cdot) = f$ in $\mathbb{R}^N$ for any $k \in \mathbb{N}$, it follows that $u(0,\cdot) = f$. Hence, $u = T(\cdot)f$. Actually, all the sequence $\{G(\cdot + s_n, s_n)f\}$ converges locally uniformly in $[0,+\infty) \times \mathbb{R}^N$ to $T(\cdot)f$, since all its converging subsequences have the same limit.

Step 2. Since each system of measures $\{\mu_{k+s_n} : n \in N\}$ is tight, for any $k \in \mathbb{N} \cup \{0\}$ there exist a subsequence $\{s_n^k\}$ of $\{s_n\}$ and a probability measure ν_k such that

$$\lim_{n\to+\infty} \int_{\mathbb{R}^N} f\, d\mu_{k+s_n^k} = \int_{\mathbb{R}^N} f\, d\nu_k, \qquad f \in C_b(\mathbb{R}^N),$$

for any $k \in \mathbb{N}$. Using again a diagonal argument we can extract a subsequence $\{t_n\}$ of $\{s_n\}$ such that the previous formula holds true for any $k \in \mathbb{N} \cup \{0\}$, with $\{s_n^k\}$ being replaced by $\{t_n\}$.

The family $\{\nu_k : k \in \mathbb{N} \cup \{0\}\}$ is tight. To check this property, we fix $\varepsilon > 0$ and let $R > 0$ be such that $\mu_t(B_R) \ge 1 - \varepsilon$ for any $t \ge \sigma_*$. Moreover, we fix a function $\psi \in C_b(\mathbb{R}^N)$ such that $\chi_{B_R} \le \psi \le \chi_{B_{R+1}}$. Then,

$$\inf_{k\in\mathbb{N}\cup\{0\}} \int_{\mathbb{R}^N} \psi d\mu_{k+s_n} \ge \inf_{t\ge\sigma_*} \int_{B_R} \mathbf{1} d\mu_t = \inf_{t\ge\sigma_*} \mu_t(B_R) \ge 1 - \varepsilon.$$

Letting n tend to $+\infty$ we conclude that $\inf_{k\in\mathbb{N}\cup\{0\}} \nu_k(B_{R+1}) \ge 1 - \varepsilon$ and the family $\{\nu_k : k \in \mathbb{N} \cup \{0\}\}$ is tight.

Next, we observe that

$$\int_{\mathbb{R}^N} T(k)f\, d\nu_k = \int_{\mathbb{R}^N} f\, d\nu_0, \qquad f \in C_c^\infty(\mathbb{R}^N),\ k \in \mathbb{N} \cup \{0\}. \tag{18.1.7}$$

Indeed, Lemma 18.1.5, the invariance property of the family $\{\mu_t : t \ge \sigma_*\}$ and (18.1.6) yield

$$\int_{\mathbb{R}^N} T(k)f d\nu_k = \lim_{n\to+\infty} \int_{\mathbb{R}^N} G(k+t_n, t_n)f\, d\mu_{k+t_n} = \lim_{n\to+\infty} \int_{\mathbb{R}^N} f\, d\mu_{t_n} = \int_{\mathbb{R}^N} f\, d\nu_0$$

for any $k \in \mathbb{N}$.

Step 3. Now, we complete the proof, letting k tend to $+\infty$ in (18.1.7). For this purpose, we observe that, by Theorem 9.1.16, $T(k)f$ converges to $\int_{\mathbb{R}^N} f\, d\mu$, locally uniformly in $\mathbb{R}^N$, for any $f \in C_c^\infty(\mathbb{R}^N)$. Hence, applying Lemma 18.1.5, we conclude that

$$\lim_{k\to+\infty} \int_{\mathbb{R}^N} T(k)f\, d\nu_k = \int_{\mathbb{R}^N} f\, d\mu. \tag{18.1.8}$$

Formulae (18.1.7) and (18.1.8) yield

$$\int_{\mathbb{R}^N} f\, d\mu = \int_{\mathbb{R}^N} f\, d\nu_0, \qquad f \in C_c^\infty(\mathbb{R}^N).$$

This shows that $\nu_0 = \mu$.

Now, we can prove that $\rho(t,\cdot)$ converges to ρ_∞ locally uniformly in $\mathbb{R}^N$. Clearly, it suffices to show that any sequence $\{s_n\}$ diverging to $+\infty$ admits a subsequence such that $\rho(s_{n_j},\cdot)$ converges to ρ_∞ locally uniformly in $\mathbb{R}^N$ as j tends to $+\infty$.

In view of Hypothesis 18.1.8(i), Theorem 16.2.4 shows that ρ belongs to $C_b^\beta([0,+\infty)\times B_R)$ for any $R>0$. By the Arzelà-Ascoli theorem there exists a sequence $\{t_n\}$ diverging to $+\infty$ such that $\rho(t_n,\cdot)$ converges to a function $g\in C(\mathbb{R}^N)$ locally uniformly in $\mathbb{R}^N$ as n tends to $+\infty$. The weak* convergence of μ_t to μ_∞ proved above yields

$$\int_{\mathbb{R}^N} f\rho_\infty\,dx = \int_{\mathbb{R}^N} f d\mu_\infty = \lim_{k\to+\infty}\int_{\mathbb{R}^N} f d\mu_{t_k} = \lim_{k\to+\infty}\int_{\mathbb{R}^N} f\rho(t_k,\cdot)dx = \int_{\mathbb{R}^N} fgdx$$

for any $f\in C_c^\infty(\mathbb{R}^N)$. Hence, $\rho_\infty = g$ and the local uniform convergence is shown.

To prove the L^1-convergence, we fix $\varepsilon>0$. By the tightness of the family $\{\mu_t : t\geq\sigma_*\}$, there exists a radius $R>0$ such that $\mu_t(\mathbb{R}^N\setminus B_R)$, $\mu_\infty(\mathbb{R}^N\setminus B_R)\leq\varepsilon$ for any $t\geq\sigma_*$. Thus we obtain that

$$\begin{aligned}\limsup_{t\to+\infty}\|\rho(t,\cdot)-\rho_\infty\|_{L^1(\mathbb{R}^N)} &= \limsup_{t\to+\infty}\left[\|\rho(t,\cdot)-\rho_\infty\|_{L^1(B_R)}+\|\rho(t,\cdot)-\rho_\infty\|_{L^1(\mathbb{R}^N\setminus B_R)}\right]\\ &\leq \limsup_{t\to+\infty}\mu_t(\mathbb{R}^N\setminus B_R)+\mu_\infty(\mathbb{R}^N\setminus B_R)\leq 2\varepsilon.\end{aligned}$$

The arbitrariness of $\varepsilon>0$ shows that $\rho(t,\cdot)$ converges to ρ_∞ in $L^1(\mathbb{R}^N)$ as t tends to $+\infty$. □

As a byproduct, we get the following convergence result.

Theorem 18.1.10 *Let $s\geq\sigma_*$, $p\in[1,+\infty)$ and $f\in C_b(\mathbb{R}^N)$. Then, $G(t,s)f$ converges to $m_s(f)$ in $L^p(\mathbb{R}^N,\mu_\infty)$ as t tends to $+\infty$.*

Proof For any $f\in C_b(\mathbb{R}^N)$ we can estimate.

$$\begin{aligned}&\|G(t,s)f-m_s(f)\|^p_{L^p(\mathbb{R}^N,\mu_\infty)}\\ \leq&\int_{\mathbb{R}^N}|\rho_\infty-\rho(t,\cdot)||G(t,s)(f-m_s(f))|^p\,dx+\|G(t,s)f-m_s(f)\|^p_{L^p(\mathbb{R}^N,\mu_t)}\\ \leq&2^p\|f\|^p_\infty\|\rho_\infty-\rho(t,\cdot)\|_{L^1(\mathbb{R}^N)}+\|G(t,s)f-m_s(f)\|^p_{L^p(\mathbb{R}^N,\mu_t)}.\end{aligned}$$

Hence, taking Theorems 18.1.6(i) and 18.1.9 into account and letting t tend to $+\infty$, the assertion follows. □

18.1.1 Exponential decay to zero

In this subsection we study the exponential decay to zero of the function $G(t,s)f-m_s(f)$ as t tends to $+\infty$, when $f\in L^p(\mathbb{R}^N,\mu_s)$. The standing assumptions that we assume here are the following.

Hypotheses 18.1.11 (i) *$q_{ij}, b_i\in C^{\alpha/2,\alpha}_{\rm loc}([\sigma_*,+\infty)\times\mathbb{R}^N)$ for some $\alpha\in(0,1)$ and any $i,j=1,\dots,N$.*

(ii) *the matrix $Q(t,x)$ is symmetric for any $(t,x)\in[\sigma_*,+\infty)\times\mathbb{R}^N$ and there exist a function $\kappa:[\sigma_*,+\infty)\times\mathbb{R}^N\to\mathbb{R}$, with positive infimum κ_0, such that $\langle Q(t,x)\xi,\xi\rangle\geq\kappa(t,x)|\xi|^2$ for any $t\geq\sigma_*$ and $x,\xi\in\mathbb{R}^N$;*

(iii) *there exist a function* $\varphi \in C^2(\mathbb{R}^N)$, *blowing up as* $|x|$ *tends to* $+\infty$, *and two positive constants* a *and* d *such that* $\mathcal{A}\varphi \le a - d\varphi$ *in* $[\sigma_*, +\infty) \times \mathbb{R}^N$.

Also in this subsection, we extend the coefficients q_{ij} and b_j $(i, j = 1, \ldots, N)$ to $\mathbb{R}^{N+1}$ as explained in Remark 18.1.2(i).

Theorem 18.1.12 *Let Hypotheses* 18.1.11 *hold. Fix* $1 \le p \le +\infty$, $\omega \in \mathbb{R}$. *Then, the following conditions are equivalent:*

(i) *there exists a positive constant* M_0 *such that* $\|\mathscr{T}_p(t)(I - \Pi)\|_{L(L^p(\mathbb{R}^{N+1},\mu))} \le M_0 e^{\omega t}$ *for any* $t > 0$;

(ii) *there exists a positive constant* M_1 *such that* $\|G(t, s) - m_s\|_{L(L^p(\mathbb{R}^N,\mu_s),L^p(\mathbb{R}^N,\mu_t))} \le M_1 e^{\omega(t-s)}$ *for any* $t > s$.

Similarly, the properties:

(iii) *there exists a positive constant* M_2 *such that* $\|\nabla_x \mathscr{T}_p(t)\|_{L(L^p(\mathbb{R}^{N+1},\mu),(L^p(\mathbb{R}^{N+1},\mu))^N)} \le M_2 e^{\omega t}$ *for any* $t \ge 1$;

(iv) *there exists a positive constant* M_3 *such that* $\|\nabla_x G(t, s)\|_{L(L^p(\mathbb{R}^N,\mu_s),(L^p(\mathbb{R}^N,\mu_t))^N)} \le M_3 e^{\omega(t-s)}$ *for any* $t \ge s + 1$

are equivalent.

Proof For $p = +\infty$ the equivalences $(i) \Leftrightarrow (ii)$ and $(iii) \Leftrightarrow (iv)$ are immediate. Similarly, for $p < +\infty$ the implications $(ii) \Rightarrow (i)$ and $(iv) \Rightarrow (iii)$ are straightforward.

Let us prove that $(i) \Rightarrow (ii)$ for $p < +\infty$. We fix $f \in C_c^\infty(\mathbb{R}^N)$, $\sigma_* \le s < t$, $\zeta \in C_c^\infty(\mathbb{R})$ such that $\zeta(s) = 1$ and consider the function $u \in C_\sharp(\mathbb{R}^{N+1})$, defined by $u(r, x) = \zeta(r) f(x)$ for any $(r, x) \in \mathbb{R}^{N+1}$. Then, $(\mathscr{T}_p(t)(I - \Pi)u)(s, x) = \zeta(s - t)[(G(s, s - t)f)(x) - m_{s-t}f]$ for any $(s, x) \in \mathbb{R}^{N+1}$ and $t > 0$. From this formula and Lemmata 14.1.6(ii), 16.2.13(ii) it follows easily that the function $s \mapsto \|(\mathscr{T}_p(t)(I - \Pi)u)(s, \cdot)\|^p_{L^p(\mathbb{R}^N,\mu_s)}$ is continuous in $\mathbb{R}$ for any $t \ge 0$.

Using property (i) we can estimate

$$
\begin{aligned}
\|G(t, s)f - m_s(f)\|^p_{L^p(\mathbb{R}^N,\mu_t)} =& \|(\mathscr{T}_p(t - s)(I - \Pi)u)(t, \cdot)\|^p_{L^p(\mathbb{R}^N,\mu_t)} \\
=& \lim_{\varepsilon \to 0^+} \frac{1}{\varepsilon} \int_{\mathbb{R}} \chi_{(t,t+\varepsilon)}(r) \|(\mathscr{T}_p(t - s)(I - \Pi)u)(r, \cdot)\|^p_{L^p(\mathbb{R}^N,\mu_r)} dr \\
=& \lim_{\varepsilon \to 0^+} \frac{1}{\varepsilon} \|\mathscr{T}_p(t - s)(\chi_{(s,s+\varepsilon)}(I - \Pi)u)\|^p_{L^p(\mathbb{R}^{N+1},\mu)} \\
\le& M_0^p e^{\omega p(t-s)} \lim_{\varepsilon \to 0^+} \frac{1}{\varepsilon} \|\chi_{(s,s+\varepsilon)}(I - \Pi)u\|^p_{L^p(\mathbb{R}^{N+1},\mu)} \\
=& M_0^p e^{\omega p(t-s)} \lim_{\varepsilon \to 0^+} \frac{1}{\varepsilon} |\zeta(r)|^p \int_s^{s+\varepsilon} \|f - m_r(f)\|^p_{L^p(\mathbb{R}^N,\mu_r)} dr \\
=& M_0^p e^{\omega p(t-s)} \|f - m_s(f)\|^p_{L^p(\mathbb{R}^N,\mu_s)}.
\end{aligned}
$$

Since $\|f - m_s(f)\|_{L^p(\mathbb{R}^N,\mu_s)} \le 2\|f\|_{L^p(\mathbb{R}^N,\mu_s)}$ the estimate

$$
\|G(t, s)f - m_s(f)\|_{L^p(\mathbb{R}^N,\mu_t)} \le 2M_0 e^{\omega(t-s)} \|f\|_{L^p(\mathbb{R}^N,\mu_s)}, \qquad t > s,
$$

follows for any $f \in C_c^\infty(\mathbb{R}^N)$. By density, we can extend it to any $f \in L^p(\mathbb{R}^N, \mu_s)$.

The proof of the property $(ii) \Rightarrow (i)$ is straightforward. Hence, we skip the details.

To prove that $(iii) \Rightarrow (iv)$ it suffices to argue as in the proof of $(i) \Rightarrow (ii)$. Indeed, if f, ζ are as above and $u = \zeta f$, then we can write

$$\begin{aligned}\|\nabla_x G(t,s)f\|^p_{L^p(\mathbb{R}^N,\mu_t)} &= \lim_{\varepsilon\to 0^+}\frac{1}{\varepsilon}\int_{\mathbb{R}}\chi_{(t,t+\varepsilon)}(r)\|(\nabla_x \mathscr{T}_p(t-s)u)(r,\cdot)\|^p_{L^p(\mathbb{R}^N,\mu_r)}dr\\ &= \lim_{\varepsilon\to 0^+}\frac{1}{\varepsilon}\|\nabla_x \mathscr{T}_p(t-s)(\chi_{(s,s+\varepsilon)}u)\|^p_{L^p(\mathbb{R}^{N+1},\mu)}\\ &\le M_2^p e^{\omega p(t-s)}\lim_{\varepsilon\to 0^+}\frac{1}{\varepsilon}\|\chi_{(s,s+\varepsilon)}u\|^p_{L^p(\mathbb{R}^{N+1},\mu)}\\ &= M_2^p e^{\omega p(t-s)}\|f\|^p_{L^p(\mathbb{R}^N,\mu_s)}\end{aligned}$$

for any $t \ge s+1$. By density, property (iv) follows. □

Next, we prove the Poincaré inequality. For this purpose, we need to strengthen the Hypotheses 18.1.11.

Hypotheses 18.1.13 (i) *Hypotheses* 18.1.11 *are satisfied;*

(ii) *the first-order spatial derivatives of the coefficients of the operator $\mathcal{A}$ exist, belong to $C^{\alpha/2,\alpha}_{\rm loc}([\sigma_*,+\infty)\times\mathbb{R}^N)$ and there exist two functions $\widehat{q}:[\sigma_*,+\infty)\times\mathbb{R}^N\to[0,+\infty)$ and $r:[\sigma_*,+\infty)\times\mathbb{R}^N\to\mathbb{R}$ such that $|\nabla_x q_{ij}(t,x)|\le\widehat{q}(t,x)$ and $\langle {\rm Jac}_x b(t,x)\xi,\xi\rangle\le r(t,x)|\xi|^2$ for any $t\ge\sigma_*$ and $x,\xi\in\mathbb{R}^N$. Moreover,*

$$\ell_2 := \sup_{(t,x)\in[\sigma_*,+\infty)\times\mathbb{R}^N}\left(r(t,x)+\frac{N^2(\widehat{q}(t,x))^2}{4\kappa(t,x)}\right)<0. \tag{18.1.9}$$

Proposition 18.1.14 *Let Hypotheses* 18.1.13 *be satisfied and assume that the diffusion coefficients are bounded. Then, for any $p\in[2,+\infty)$, there exists a positive constant C_p such that*

$$\|u-\Pi u\|_{L^p(\mathbb{R}^{N+1},\mu)}\le C_p\|\nabla_x u\|_{L^p(\mathbb{R}^{N+1},\mu)},\qquad u\in W^{0,1}_p(\mathbb{R}^{N+1},\mu). \tag{18.1.10}$$

In particular, we can take $C_2=\Lambda^{1/2}|\ell_2|^{-1/2}$, where Λ is the supremum over $(s,x)\in\mathbb{R}^{N+1}$ of the maximum eigenvalue of the matrix $Q(s,x)$.

Proof We split the proof into three steps.

Step 1. Here we prove that, for any $u\in L^2(\mathbb{R}^{N+1},\mu)$, $\|\mathscr{T}_2(t)(I-\Pi)u\|_{L^2(\mathbb{R}^{N+1},\mu)}$ vanishes as t tends to $+\infty$. We first assume that $u(s,x)=\alpha(s)f(x)$ for any $(s,x)\in\mathbb{R}^{N+1}$ and some functions $\alpha\in C^\infty_c(\mathbb{R})$ and $f\in C^\infty_c(\mathbb{R}^N)$. Then, we can write

$$\int_{\mathbb{R}^{N+1}}|\mathscr{T}_2(t)(I-\Pi)u|^2d\mu=\int_{\mathbb{R}}|\alpha(s-t)|^2\|G(s,s-t)f-m_{s-t}(f)\|^2_{L^2(\mathbb{R}^N,\mu_s)}ds,\qquad t>0.$$

Theorem 18.1.6(i) and the dominated convergence theorem show that the right-hand side of the previous formula vanishes as t tends to $+\infty$. The density of the functions u as above in $L^2(\mathbb{R}^{N+1},\mu)$ allows to conclude that $\|\mathscr{T}_2(t)(I-\Pi)u\|_{L^2(\mathbb{R}^{N+1},\mu)}$ vanishes as t tends to $+\infty$ for any $u\in L^2(\mathbb{R}^{N+1},\mu)$.

Step 2. Here we prove the Poincaré inequality with $p=2$. For this purpose, we fix $u\in C^\infty_c(\mathbb{R}^{N+1})\subset D(G_2)$. Then, the function $\|\mathscr{T}_2(\cdot)(I-\Pi)u\|^2_{L^2(\mathbb{R}^{N+1},\mu)}$ is differentiable in $[0,+\infty)$ and, using Theorem 17.1.6 and estimate (17.1.6), we obtain

$$\|\mathscr{T}_2(t)(I-\Pi)u\|^2_{L^2(\mathbb{R}^{N+1},\mu)}-\|u\|^2_{L^2(\mathbb{R}^{N+1},\mu)}=2\int_0^t ds\int_{\mathbb{R}^N}\mathscr{T}_2(s)u\,G_2\mathscr{T}_2(s)u d\mu$$

$$\begin{aligned}
&= -2\int_0^t \|\sqrt{Q}\mathscr{T}_2(s)u\|^2_{L^2(\mathbb{R}^{N+1},\mu)}ds\\
&\geq -2\Lambda\int_0^t \|\nabla_x\mathscr{T}_2(s)u\|^2_{L^2(\mathbb{R}^{N+1},\mu)}ds\\
&\geq -2\Lambda\int_0^t e^{2\ell_2 s}\|\nabla_x u\|^2_{L^2(\mathbb{R}^{N+1},\mu)}ds.
\end{aligned}$$

Letting t tend to $+\infty$ in the first- and last-side of the previous chain of inequalities, the assertion follows at once for functions in $C_c^\infty(\mathbb{R}^{N+1})$. Using the density of $C_c^\infty(\mathbb{R}^{N+1})$ in $W^{1,2}(\mathbb{R}^{N+1},\mu)$ the Poincaré inequality follows.

Step 3. To complete the proof we prove the Poincaré inequality for any $p\in(2,+\infty)$. We fix $r\in\mathbb{R}$, $\varepsilon>0$ and a function $f\in C_c^\infty(\mathbb{R}^N)$. Then, the function $u=\chi_{(r,r+\varepsilon)}f$ belongs to $W^{0,1}(\mathbb{R}^{N+1},\mu)$. Applying estimate (18.1.10), with $p=2$, gives

$$\int_r^{r+\varepsilon}ds\int_{\mathbb{R}^N}|f-m_s(f)|^2d\mu_s\leq\int_r^{r+\varepsilon}ds\int_{\mathbb{R}^N}|\nabla f|^2d\mu_s.$$

By Lemma 16.2.13(ii), the functions $s\mapsto\int_{\mathbb{R}^N}|f-m_s(f)|^2d\mu_s$ and $s\mapsto\int_{\mathbb{R}^N}|\nabla f|^2d\mu_s$ are continuous in $\mathbb{R}$. Hence, dividing by ε both sides of the previous formula and letting ε tend to 0 we conclude that

$$\int_{\mathbb{R}^N}|f-m_r(f)|^2d\mu_r=\int_{\mathbb{R}^N}|\nabla f|^2d\mu_r. \tag{18.1.11}$$

By density we can extend (18.1.11) to any $f\in L^2(\mathbb{R}^N,\mu_r)$.

Now, using the same arguments as in Step 2 of the proof of Theorem 16.4.9 we can prove that

$$\int_{\mathbb{R}^N}|f-m_s(f)|^pd\mu_s=\int_{\mathbb{R}^N}|\nabla f|^pd\mu_s,\qquad f\in L^p(\mathbb{R}^N,\mu_s),\ \ p\in[2,+\infty),$$

and deduce (18.1.10) for functions u which are the product of a function $\alpha\in C_c^\infty(\mathbb{R})$ and a function $f\in C_c^\infty(\mathbb{R}^N)$. Since these functions are dense in $W_p^{0,1}(\mathbb{R}^{N+1},\mu)$, the Poincaré inequality (18.1.10) follows in its generality. □

Remark 18.1.15 In fact, in the proof of the previous theorem we have proved that the Poincaré inequality

$$\|f-m_s(f)\|_{L^p(\mathbb{R}^N,\mu_s)}\leq C_p\|\nabla f\|_{L^p(\mathbb{R}^N,\mu_s)},\qquad f\in L^p(\mathbb{R}^N,\mu_s),\ \ s\geq\sigma_*.$$

holds true for some positive constant C_p and any $p\in[2,+\infty)$. The assumptions which yield to the previous estimate are weaker than the ones in Theorem 16.4.9.

For any $p\in[1,+\infty)$ let us define the right halflines

$$\mathfrak{A}_p:=\{\omega\in\mathbb{R}:\exists M_\omega>0\ \text{ s.t. }\|\mathscr{T}_p(t)(I-\Pi)\|_{L(L^p(\mathbb{R}^{N+1},\mu))}\leq M_\omega e^{\omega t}\ \ \forall t\geq 0\},$$

$$\mathfrak{B}_p:=\{\omega\in\mathbb{R}:\exists N_\omega>0\ \text{ s.t. }\|\nabla_x\mathscr{T}_p(t)\|_{L(L^p(\mathbb{R}^{N+1},\mu);(L^p(\mathbb{R}^{N+1},\mu))^N)}\leq N_\omega e^{\omega t}\ \ \forall t\geq 1\}$$

and their infima denoted, respectively, by ω_p and γ_p. Then, ω_p, the growth bound of the part of $\mathscr{T}(t)$ in $(I-\Pi)(L^p(\mathbb{R}^{N+1},\mu^\sharp))$, is nonpositive.

Theorem 18.1.16 *Let Hypotheses* 18.1.13 *hold. Then,* $\mathfrak{A}_p\subset\mathfrak{B}_p$ *for any* $p\geq 2$*. The previous inclusion holds also for*

(i) $p \in (1,2)$ *if*

$$\ell_p := \sup_{(t,x)\in[\sigma_*,+\infty)\times\mathbb{R}^N} \left(r(t,x) + \frac{N^2(\widehat{q}(t,x))^2\kappa(t,x)}{4(p-1)\kappa(t,x)}\right) < +\infty \tag{18.1.12}$$

(ii) *any* $p \in (1,2)$ *if* $|\nabla_x q_{ij}(t,x)| \le \zeta_0\sqrt{\kappa(t,x)}$ *for any* $(t,x) \in [\sigma_*,+\infty)\times\mathbb{R}^N$.

Finally, if the diffusion coefficients are bounded, then $\mathfrak{B}_p \subset \mathfrak{A}_p$ *for any* $p \ge 2$.

Proof The inclusion $\mathfrak{A}_p \subset \mathfrak{B}_p$ follows from (17.1.3) and (17.1.5) which show that

$$\begin{aligned}\|\nabla_x \mathscr{T}_p(t)u\|_{L^p(\mathbb{R}^{N+1},\mu)} =&\|\nabla_x \mathscr{T}_p(t)(u-\Pi u)\|_{L^p(\mathbb{R}^{N+1},\mu)}\\ =&\|\nabla_x \mathscr{T}_p(1)\mathscr{T}_p(t-1)(u-\Pi u)\|_{L^p(\mathbb{R}^{N+1},\mu)}\\ \le& C_p\|\mathscr{T}_p(t-1)(u-\Pi u)\|_{L^p(\mathbb{R}^{N+1},\mu)}\end{aligned}$$

for any $t \ge 1$ and $u \in L^p(\mathbb{R}^{N+1},\mu)$.

The other inclusion follows from the Poincaré inequality in Proposition 18.1.14, which, together with Proposition 17.1.4, shows that

$$\|\mathscr{T}_p(t)(u-\Pi u)\|_{L^p(\mathbb{R}^{N+1},\mu)} \le C_p\|\nabla_x \mathscr{T}_p(t)\|_{L^p(\mathbb{R}^{N+1},\mu)}$$

for any $u \in L^2(\mathbb{R}^{N+1},\mu)$ and $t > 0$. □

Theorem 18.2.6 has two important consequences. The first one concerns the spectral gap of G_p and the solvability of the equation $\lambda u - G_p u = f$; the second one is about the asymptotic behaviour of the evolution operator $\{G(t,s)\}$.

Corollary 18.1.17 *Let Hypotheses* 18.1.13 *be satisfied. Further, assume that the diffusion coefficients are bounded. Then, for any* $p \in (1,+\infty)$, $i\mathbb{R} \subset \sigma(G_p)$ *and* G_p *has a spectral gap. More precisely,* $\sup\{\mathrm{Re}\,\lambda : \lambda \in \sigma(G_p)\setminus i\mathbb{R}\} \le 2\ell_2(1-(p\wedge 2)^{-1})$ *for any* $p \in (1,+\infty)$.

Proof By Proposition 17.1.4 and Theorem 18.1.16 it follows that $\gamma_2 \le \ell_2$. Since

$$\|\mathscr{T}_1(t)(I-\Pi)f\|_{L^1(\mathbb{R}^{N+1},\mu)} \le 2\|f\|_{L^1(\mathbb{R}^{N+1},\mu)}, \qquad t>0,\ \ f \in L^1(\mathbb{R}^{N+1},\mu),$$

using (17.1.5), with $p = 2$, and interpolating between $L^1(\mathbb{R}^{N+1},\mu)$ and $L^2(\mathbb{R}^{N+1},\mu)$ we get

$$\|\mathscr{T}_p(t)(I-\Pi)f\|_{L^p(\mathbb{R}^{N+1},\mu)} \le M_p e^{2\ell_2(1-1/p)t}\|f\|_{L^p(\mathbb{R}^{N+1},\mu)}, \qquad f \in L^p(\mathbb{R}^{N+1},\mu).$$

Therefore, the spectrum of the part of G_p in $(I-\Pi)(L^p(\mathbb{R}^{N+1},\mu))$ is contained in the halfplane $\mathrm{Re}\,\lambda \le \ell_2$, if $p \ge 2$, and in the halfplane $\mathrm{Re}\,\lambda \le 2\ell_2(1-1/p)$, if $1 < p < 2$. For the other values of λ, it is convenient to write the equation $\lambda u - G_p u = f$ as the system

$$\begin{cases}\lambda\Pi u - G_p\Pi u = \Pi f,\\ \lambda(I-\Pi)u - G_p(I-\Pi)u = (I-\Pi)f,\end{cases}$$

where the second equation is uniquely solvable. As far as the first equation is concerned, we are led to solve in $W^{1,p}(\mathbb{R})$ the equation $\lambda v(s) + v'(s) = m_s(f(s,\cdot))$ for any $s \in \mathbb{R}$. This equation is uniquely solvable if and only if $\lambda \notin i\mathbb{R}$. □

Corollary 18.1.18 *Let Hypotheses* 18.1.13 *be satisfied. Further, assume that the diffusion coefficients are bounded. Then, for any* $p > 1$ *there exists a positive constant* M_p *such that, for any* $f \in L^p(\mathbb{R}^N,\mu_s)$, *it holds that*

$$\|G(t,s)f - m_s(f)\|_{L^p(\mathbb{R}^N,\mu_t)} \le M_p e^{\theta_p(t-s)}\|f\|_{L^p(\mathbb{R}^N,\mu_s)}, \quad t > s, \tag{18.1.13}$$

with $\theta_p = 2\ell_2(1-(p\wedge 2)^{-1})$.

Proof By estimate (17.1.5), ℓ_2 belongs to $\mathfrak{B}_p$ for $p \geq 2$, and (18.1.13) follows applying Theorems 18.1.12, 18.1.16. For $1 < p < 2$ and $s \geq \sigma_*$ the estimate

$$\|G(t,s)f - m_s(f)\|_{L^p(\mathbb{R}^N,\mu_t)} \leq M_p e^{2\ell_2(1-1/p)(t-s)}\|f\|_{L^p(\mathbb{R}^N,\mu_s)}, \qquad f \in L^p(\mathbb{R}^N,\mu_s),\ t > s,$$

follows interpolating between $L^1(\mathbb{R}^{N+1},\mu)$ and $L^2(\mathbb{R}^{N+1},\mu)$, recalling that

$$\|G(t,s)f - m_s(f)\|_{L^1(\mathbb{R}^N,\mu_t)} \leq 2\|f\|_{L^1(\mathbb{R}^N,\mu_s)}$$

for $t > s$ and $f \in L^1(\mathbb{R}^N,\mu_s)$. □

By Theorems 18.1.12 and 18.1.16, the sets

$$\begin{aligned}\mathfrak{C}_p =& \big\{\omega \in \mathbb{R} : \|G(t,s)f - m_s(f)\|_{L^p(\mathbb{R}^N,\mu_t)} \leq M_{p,\omega} e^{\omega(t-s)}\|f\|_{L^p(\mathbb{R}^N,\mu_s)} \\ & \text{for any } \sigma_* \leq s < t \text{ any } f \in L^p(\mathbb{R}^N,\mu_s) \text{ and some } M_{p,\omega} > 0\big\}, \\ \mathfrak{D}_p =& \big\{\omega \in \mathbb{R} : \|\nabla_x G(t,s)f\|_{L^p(\mathbb{R}^N,\mu_t)} \leq N_{p,\omega} e^{\omega(t-s)}\|f\|_{L^p(\mathbb{R}^N,\mu_s)} \\ & \text{for any } s,t \geq \sigma_*,\ t-s \geq 1, \text{ any } f \in L^p(\mathbb{R}^N,\mu_s) \text{ and some } N_{p,\omega} > 0\big\}\end{aligned}$$

coincide with the sets $\mathfrak{A}_p$ and $\mathfrak{B}_p$, respectively, for any $p \in (1,+\infty)$. Moreover, $\mathfrak{C}_p \subset \mathfrak{D}_p$ for any $p \in (1,+\infty)$ and, if the diffusion coefficients are bounded, then $\mathfrak{D}_p \subset \mathfrak{C}_p$ for any $p \geq 2$. Assuming the following conditions on the coefficients of the operator $\mathcal{A}$, such a result can be refined.

Hypothesis 18.1.19 *Hypotheses* 18.1.1 *are satisfied with the diffusion coefficients being independent of x and with the constant r_0 in Hypothesis* 18.1.1*(iv) being negative.*

Theorem 18.1.20 *The sets $\mathfrak{C}_p$ and $\mathfrak{D}_p$ are independent of $p \in (1,+\infty)$ and they coincide.*

Proof As has been stressed in the introduction, the main tool of the proof is the hypercontractivity of the evolution operator $\{G(t,s)\}$ guaranteed by Theorem 16.4.11. In view of the remark above the statement of the theorem, we just need to show that $\mathfrak{C}_p$ and $\mathfrak{D}_p$ are independent of $p \in (1,+\infty)$. For this purpose, we fix $p_1, p_2 \in (1,+\infty)$, with $p_1 < p_2$, $\omega \in \mathfrak{C}_{p_1}$, $s \geq \sigma_*$ and $f \in L^{p_2}(\mathbb{R}^N,\mu_s)$. Since $m_t(G(t,s)f) = m_s(f)$ for any $t > s$, we can split $G(t,s)f - m_s(f) = G(t,t-\tau)(G(t-\tau,s)f - m_s(f))$ and, using (16.4.14), we deduce that

$$\begin{aligned}\|G(t,s)f - m_s(f)\|_{L^{p_2}(\mathbb{R}^N,\mu_t)} \leq& \|G(t-\tau,s)f - m_s(f)\|_{L^{p_1}(\mathbb{R}^N,\mu_{t-\tau})} \\ \leq& M_{p_1,\omega} e^{-\omega\tau} e^{\omega(t-s)}\|f\|_{L^{p_1}(\mathbb{R}^N,\mu_s)} \\ \leq& M_{p_1,\omega} e^{-\omega\tau} e^{\omega(t-s)}\|f\|_{L^{p_2}(\mathbb{R}^N,\mu_s)}\end{aligned}$$

for any $t > s+\tau$, where $\tau > 0$ is such that $p_2 = e^{2\kappa_0|r_0|\Lambda^{-1}\tau}(p_1-1)+1$. This estimate can be extended to any $t \in (s,s+\tau]$, with a possibly larger constant $M_{p_1,\omega}$, by recalling that the operators $G(t,s)$ and m_s are contractions from $L^p(\mathbb{R}^N,\mu_s)$ to $L^p(\mathbb{R}^N,\mu_t)$ for any $t > s$ and $p \in [1,+\infty)$. Hence, $\omega \in \mathfrak{C}_{p_2}$ and the inclusion $\mathfrak{C}_{p_1} \subset \mathfrak{C}_{p_2}$ follows.

To prove the other inclusion, we fix $\omega \in \mathfrak{C}_{p_2}$ and $f \in L^{p_1}(\mathbb{R}^N,\mu_s)$. Using again estimate (16.4.14) and splitting $G(t,s)f - m_s(f) = G(t,s+\tau)G(s+\tau,s)f - m_{s+\tau}(G(s+\tau,s)f)$, we conclude that

$$\begin{aligned}\|G(t,s)f - m_s(f)\|_{L^{p_1}(\mathbb{R}^N,\mu_t)} \leq& \|G(t,s)f - m_s(f)\|_{L^{p_2}(\mathbb{R}^N,\mu_t)} \\ \leq& M_{p_2,\omega} e^{\omega(t-s-\tau)}\|G(s+\tau,s)f\|_{L^{p_2}(\mathbb{R}^N,\mu_{s+\tau})}\end{aligned}$$

$$\leq M_{p_2,\omega}e^{\omega(t-s-\tau)}\|f\|_{L^{p_1}(\mathbb{R}^N,\mu_s)}$$

for any $t > s+\tau$. As above, this is sufficient to infer that $\omega \in \mathfrak{C}_{p_1}$ so that $\mathfrak{C}_{p_2} \subset \mathfrak{C}_{p_1}$. The arbitrariness of p_1 and p_2 shows that $\mathfrak{C}_p$ is independent of $p \in (1,+\infty)$.

Let us prove that also $\mathfrak{D}_p$ is independent of $p \in (1,+\infty)$. We fix p_1, p_2, τ as above, $\omega \in \mathfrak{D}_{p_1}$ and $t \geq s+\tau+1$, Estimates (14.4.8), with $p=1$ and $\ell_p = r_0$, and (16.4.14) imply that

$$\begin{aligned}\|\nabla_x G(t,s)f\|_{L^{p_2}(\mathbb{R}^N,\mu_t)} =&\|\nabla_x G(t,t-\tau)G(t-\tau,s)f\|_{L^{p_2}(\mathbb{R}^N,\mu_t)}\\ \leq& e^{r_0\tau}\|G(t,t-\tau)|\nabla_x G(t-\tau,s)f|\|_{L^{p_2}(\mathbb{R}^N,\mu_t)}\\ \leq& e^{r_0\tau}\|\nabla_x G(t-\tau,s)f\|_{L^{p_1}(\mathbb{R}^N,\mu_{t-\tau})}\\ \leq& N_{p_1,\omega}e^{\tau(r_0-\omega)}e^{\omega(t-s)}\|f\|_{L^{p_2}(\mathbb{R}^N,\mu_s)}\end{aligned}$$

for any $f \in C^1_b(\mathbb{R}^N)$. The density of $C^1_b(\mathbb{R}^N)$ in $L^p(\mathbb{R}^N,\mu_s)$ for any $p \in (1,+\infty)$ and (16.3.2) allow us to extend the previous estimate to any $f \in L^{p_2}(\mathbb{R}^N,\mu_s)$ and $t \geq s+1$, with a possibly larger constant $N_{p_1,\omega}$. We thus conclude that $\omega \in \mathfrak{D}_{p_2}$ and the inclusion $\mathfrak{D}_{p_1} \subset \mathfrak{D}_{p_2}$ follows. On the other hand, if $\omega \in \mathfrak{D}_{p_2}$ and $t \geq s+\tau+1$, then we can estimate

$$\begin{aligned}\|\nabla_x G(t,s)f\|_{L^{p_1}(\mathbb{R}^N,\mu_t)} \leq&\|\nabla_x G(t,s)f\|_{L^{p_2}(\mathbb{R}^N,\mu_t)}\\ \leq& N_{p_2,\omega}e^{\omega(t-s-\tau)}\|G(s+\tau,s)f\|_{L^{p_2}(\mathbb{R}^N,\mu_{s+\tau})}\\ \leq& N_{p_2,\omega}e^{-\omega\tau}e^{\omega(t-s)}\|f\|_{L^{p_1}(\mathbb{R}^N,\mu_s)}\end{aligned}$$

for any $f \in L^{p_1}(\mathbb{R}^N,\mu_s)$ and, as above, this is enough to infer that ω belongs to $\mathfrak{D}_{p_1}$. Thus, $\mathfrak{D}_{p_1} = \mathfrak{D}_{p_2}$ and we conclude that $\mathfrak{D}_p$ is independent of $p \in (1,+\infty)$. □

Theorem 18.1.20 and Proposition 16.3.3 yield the following important corollary.

Corollary 18.1.21 *For any $p > 1$ there exists a positive constant C_p such that*

$$\|G(t,s)f - m_s(f)\|_{L^p(\mathbb{R}^N,\mu_t)} \leq C_p e^{r_0(t-s)}\|f\|_{L^p(\mathbb{R}^N,\mu_s)}, \qquad t > s \geq \sigma_*,\ f \in L^p(\mathbb{R}^N,\mu_s).$$

Remark 18.1.22 Since $r_0 = \ell_2$, the decay to zero estimate in Theorem 18.2.7(*ii*) is much more precise than the one in Corollary 18.1.18 for $1 < p < 2$.

Remark 18.1.23 (i) Using the same arguments as in the proof of Theorem 18.1.20 and taking Proposition 17.1.4 into account, one can show that $\mathfrak{C}_p = \mathfrak{E}_p$ for any $p \in (1,+\infty)$, where

$$\begin{aligned}\mathfrak{E}_p = \big\{\omega \in \mathbb{R} : \|\nabla_x G(t,s)f\|_{L^p(\mathbb{R}^N,\mu_t)} \leq K_{p,\omega}e^{\omega(t-s)}\|f\|_{W^{1,p}(\mathbb{R}^N,\mu_s)}\\ \text{for any } t > s \geq \sigma_*,\ f \in W^{1,p}(\mathbb{R}^N,\mu_s) \text{ and some } K_{p,\omega} > 0\big\}.\end{aligned}$$

(ii) For $p = 1$ the sets $\mathfrak{C}_1$ and $\mathfrak{D}_1$ may not coincide, even in the autonomous case. For instance, in the case of the Ornstein-Uhlenbeck operator $(\mathcal{A}\zeta)(x) := \zeta''(x) - x\zeta'(x)$, the tight evolution system of measures $\{\mu_t : t \in \mathbb{R}\}$ of the associated evolution operator is defined by $\mu_t(dx) = (2\pi)^{-1/2}e^{-x^2/2}dx$ for any $t \in \mathbb{R}$. Moreover, any $\lambda < 0$ is an eigenvalue of the realization of $\mathcal{A}$ in $L^1(\mathbb{R},\mu)$ as shown in Theorem 10.3.22. This implies that $\mathfrak{C}_1$ can not contain negative numbers, so that $\mathfrak{C}_1 = [0,+\infty)$. On the other hand, in this case $r_0 = -1$ belongs to $\mathfrak{D}_1$ by Proposition 16.3.3.

18.2 The periodic case

Throughout this section we assume the following conditions on the coefficients of the operator $\mathcal{A}$.

Hypotheses 18.2.1 (i) *The coefficients q_{ij} and b_i $(i,j = 1,\ldots,N)$ are T-time periodic and belong to $C^{\alpha/2,\alpha}_{\rm loc}(\mathbb{R}^{N+1})$ for some $\alpha \in (0,1)$;*

(ii) *the matrix $Q(s,x)$ is symmetric for any $(s,x) \in \mathbb{R}^{N+1}$ and there exists a T-time periodic function $\kappa : \mathbb{R}^{N+1} \to \mathbb{R}$, with positive infimum κ_0, such that $\langle Q(t,x)\xi,\xi\rangle \geq \kappa(t,x)|\xi|^2$ for any $\xi \in \mathbb{R}^N$ and $(t,x) \in (0,T)\times\mathbb{R}^N$;*

(iii) *there exist a positive function $\varphi \in C^2(\mathbb{R}^N)$, blowing up as $|x|$ tends to $+\infty$, and positive constants a, d such that $\mathcal{A}\varphi \leq a - d\varphi$ in $(0,T)\times\mathbb{R}^N$;*

(iv) *the first-order spatial derivatives of the coefficients q_{ij} and b_i $(i,j = 1,\ldots,N)$ exist and belong to $C^{\alpha/2,\alpha}_{\rm loc}(\mathbb{R}^{N+1})$;*

(v) *there exist two functions $r : \mathbb{R}^{N+1} \to \mathbb{R}$, $\widehat{q}_0 : \mathbb{R} \to \mathbb{R}$, bounded from above respectively by the constants r_0 and $\widehat{q}_0$, such that $\langle ({\rm Jac}_x b)\xi,\xi\rangle \leq r|\xi|^2$ and $|D_k q_{ij}| \leq \widehat{q}_0\kappa$ in $\mathbb{R}^{N+1}$ for any $\xi \in \mathbb{R}^N$ and $i,j,k = 1,\ldots,N$.*

Clearly, Hypotheses 18.2.1 are a refinement of Hypotheses 18.1.1. Hence, from Propositions 17.2.2, 17.2.3 and Corollary 18.1.7 it follows that the unique (eventually) tight evolution system of measures for $\{G(t,s)\}$ is T-periodic. Differently from Section 17.2 we denote simply by μ_t (instead that by $\mu_t^\sharp$) each measure of this system. On the other hand, we still denote by $\mu^\sharp$ the unique measure on $\mathcal{B}((0,T)\times\mathbb{R}^N)$ such that

$$\mu^\sharp(A\times B) = \frac{1}{T}\int_0^T \mu_t(B)dt, \qquad A \in \mathcal{B}((0,T)),\ \ B \in \mathcal{B}(\mathbb{R}^N).$$

We introduce the projection Π defined by $(\Pi f)(s,x) := m_s(f(s,\cdot))$ for any $(s,x) \in \mathbb{R}^{N+1}$ and $f \in C_\sharp(\mathbb{R}^{N+1})$. It is easy to see that $\|\Pi\|_{L(C_\sharp(\mathbb{R}^{N+1}))} = \|\Pi\|_{L(L^p_\sharp(\mathbb{R}^{N+1},\mu^\sharp))} = 1$. The ranges of $\Pi(C_\sharp(\mathbb{R}^{N+1}))$ and of $\Pi(L^p_\sharp(\mathbb{R}^{N+1},\mu^\sharp))$ may be identified with $C_\sharp(\mathbb{R})$ and $L^p_\sharp(\mathbb{R},T^{-1}ds)$, respectively. The semigroups $\{\mathscr{T}^\sharp(t)\}$ and $\{\mathscr{T}^\sharp_p(t)\}$ (see Section 17.2) leave, respectively, $C_\sharp(\mathbb{R})$ and $L^p_\sharp(\mathbb{R},T^{-1}ds)$ invariant, and their part in such spaces is just the translation semigroup $f \mapsto f(\cdot - t)$. Although $\{\mathscr{T}^\sharp(t)\}$ is not strongly continuous in $C_\sharp(\mathbb{R}^{N+1})$, the part of $\{\mathscr{T}^\sharp(t)\}$ in $C_\sharp(\mathbb{R})$ is strongly continuous. The infinitesimal generators of the parts of $\{\mathscr{T}^\sharp(t)\}$ and $\{\mathscr{T}^\sharp_p(t)\}$ in $C_\sharp(\mathbb{R})$ and in $L^p_\sharp(\mathbb{R},T^{-1}ds)$ have domains (isomorphic to) $C^1_\sharp(\mathbb{R})$ and $W^{1,p}_\sharp(\mathbb{R},T^{-1}ds)$, respectively, and coincide with $-D_s$.

In the next theorem we relate the asymptotic behaviour of $\{\mathscr{T}^\sharp(t)\}$ to the asymptotic behaviour of $G(t,s)$. To state it, we introduce some definitions. We say that a function f satisfies property

- $(\mathcal{P}_{-\infty,p,t})$ if $\|G(t,s)f - m_s(f)\|_{L^p(\mathbb{R}^N,\mu_t)}$ vanishes as s tends to $-\infty$;
- $(\mathcal{P}_{+\infty,p,s})$ if $\|G(t,s)f - m_s(f)\|_{L^p(\mathbb{R}^N,\mu_t)}$ vanishes as t tends to $+\infty$;
- $(\mathcal{P}_{-\infty,\infty,R,t})$ if $\|G(t,s)f - m_s(f)\|_{L^\infty(B_R)}$ vanishes as s tends to $-\infty$;
- $(\mathcal{P}_{+\infty,\infty,R,s})$ if $\|G(t,s)f - m_s(f)\|_{L^\infty(B_R)}$ vanishes as t tends to $+\infty$.

Theorem 18.2.2 *Let Hypotheses* 18.2.1(*i*)-(*iii*) *hold. For any* $p \in [1,+\infty)$ *consider the following statements:*

(i) *for any* $u \in L^p_\sharp(\mathbb{R}^{N+1},\mu^\sharp)$ *it holds that*

$$\lim_{t\to+\infty} \|\mathscr{T}^\sharp_p(t)(u-\Pi u)\|_{L^p_\sharp(\mathbb{R}^{N+1},\mu^\sharp)} = 0; \tag{18.2.1}$$

(ii) *for any* $f \in C_b(\mathbb{R}^N)$ *there exists* $t \in \mathbb{R}$ *such that property* $(\mathcal{P}_{-\infty,p,t})$ *is satisfied;*

(ii′) *property* $(\mathcal{P}_{-\infty,p,t})$ *is satisfied by any* $f \in C_b(\mathbb{R}^N)$ *and* $t \in \mathbb{R}$*;*

(iii) *there exists* $s \in \mathbb{R}$ *such that property* $(\mathcal{P}_{+\infty,p,s})$ *is satisfied by any* $f \in L^p(\mathbb{R}^N,\mu_s)$*;*

(iii′) *property* $(\mathcal{P}_{+\infty,p,s})$ *is satisfied by any* $f \in L^p(\mathbb{R}^N,\mu_s)$ *and* $s \in \mathbb{R}$*;*

(iv) *for any* $f \in C_b(\mathbb{R}^N)$ *and* $R > 0$ *there exists* $t \in \mathbb{R}$ *such that property* $(\mathcal{P}_{-\infty,\infty,R,t})$ *is satisfied;*

(iv′) *property* $(\mathcal{P}_{-\infty,\infty,R,t})$ *is satisfied by any* $f \in C_b(\mathbb{R}^N)$, $R > 0$ *and* $t \in \mathbb{R}$*;*

(v) *there exists* $s \in \mathbb{R}$ *such that, for any* $f \in C_b(\mathbb{R}^N)$ *and* $R > 0$, *property* $(\mathcal{P}_{+\infty,\infty,R,s})$ *is satisfied;*

(v′) *property* $(\mathcal{P}_{+\infty,\infty,R,s})$ *is satisfied by any* $f \in C_b(\mathbb{R}^N)$, $R > 0$ *and* $s \in \mathbb{R}$.

For any $p \in [1,+\infty)$, *statements* (*i*), (*ii*), (*ii′*), (*iii*) *and* (*iii′*) *are equivalent, and they are implied by statements* (*iv*), (*iv′*), (*v*) *and* (*v′*). *If in addition Hypotheses* 18.2.1(*iv*), (*v*) *hold, then for any* $p \in [1,+\infty)$ *statements* (*i*) *to* (*v′*) *are equivalent.*

Proof The proof is split into several steps.

Step 1. Here, we prove that (*ii*) ⇔ (*ii′*), (*iii*) ⇔ (*iii′*), (*iv*) ⇔ (*iv′*) and (*v*) ⇔ (*v′*). Clearly the implication "⇐" is obvious, hence we just need to prove the other implication.

To begin with, let us prove that (*ii*) ⇒ (*ii′*). Fix $f \in C_b(\mathbb{R}^N)$ and let $t_0 \in \mathbb{R}$ be such that $\|G(t_0,s)f - m_s(f)\|_{L^p(\mathbb{R}^N,\mu_{t_0})}$ vanishes as s tends to $-\infty$. For $t > t_0$, we split $G(t,s)f - m_s(f) = G(t,t_0)(G(t_0,s)f - m_s(f))$ and, recalling that $G(t,t_0)$ is a contraction from $L^p(\mathbb{R}^N,\mu_{t_0})$ to $L^p(\mathbb{R}^N,\mu_t)$, we conclude that property $(\mathcal{P}_{-\infty,p,t})$ is satisfied for any $t > t_0$. To prove that this property is satisfied also by any $t < t_0$ we use the T-time periodicity of the coefficients of the operator $\mathcal{A}$. We fix $k \in \mathbb{N}$ such that $t + kT \ge t_0$ and observe that $G(t,s)f = G(t+kT,s+kT)$ and $m_s(f) = m_{s+kT}(f)$ (see Proposition 17.2.2). Hence, $\|G(t,s)f - m_s(f)\|_{L^p(\mathbb{R}^N,\mu^\sharp_t)} = \|G(t+kT,s+kT)f - m_{s+kT}f\|_{L^p(\mathbb{R}^N,\mu^\sharp_{t+kT})}$ and letting s tend to $-\infty$ we conclude that f satisfies property $(\mathcal{P}_{-\infty,p,t})$.

The same arguments can be used to prove that (*iv*) ⇒ (*iv′*). Indeed, fix $f \in C_b(\mathbb{R}^N)$ and let $t_0 \in \mathbb{R}$ be such that $\|G(t_0,s)f - m_s(f)\|_{L^\infty(B_R)}$ vanishes as s tends to $-\infty$ for any $R > 0$. For $t > t_0$ we write $G(t,s)f - m_s(f) = G(t,t_0)f_s$, where $f_s := G(t_0,s)f - m_s(f)$, and estimate

$$\begin{aligned}|(G(t,t_0)f_s)(x)| \le& \int_{B_r} |f_s(y)|p(t,s,x;dy) + \int_{\mathbb{R}^N\setminus B_r} |f_s(y)|p(t,s,x;dy)\\ \le& \|f_s\|_{L^\infty(B_R)} + \sup_{x\in B_R} p(t,s,x;\mathbb{R}^N\setminus B_r)\sup_{s\in I}\|f_s\|_\infty\end{aligned}$$

for any $r, R > 0$. By assumptions $\|f_s\|_{L^\infty(B_R)}$ vanishes as s tends to $-\infty$. On the other hand, $\sup_{s\in I}\|f_s\|_\infty$ is finite and, by Lemma 14.1.10, $\sup_{x\in B_R} p(t,s,x;\mathbb{R}^N\setminus B_r)$ vanishes as r tends to $+\infty$, uniformly with respect to $s < t$. From these remarks and the previous chain of inequalities it follows that $G(t,t_0)f_s$ vanishes locally uniformly in $\mathbb{R}^N$, as s tends to $-\infty$. Hence, f satisfies property $(\mathcal{P}_{-\infty,\infty,R,t})$ for any $t>t_0$. If $t<t_0$ then it suffices to argue as in the last part of the proof of the implication $(ii)\Rightarrow(ii')$.

To prove that $(iii)\Rightarrow(iii')$ we fix $s_0\in\mathbb{R}$ such that property $(\mathcal{P}_{+\infty,p,s_0})$ is satisfied by any function in $L^p(\mathbb{R}^N,\mu_{s_0})$. Moreover, we consider $s\neq s_0$ and $f\in L^p(\mathbb{R}^N,\mu_s)$. If $s<s_0$ then if holds that $\|G(t,s)f-m_s(f)\|_{L^p(\mathbb{R}^N,\mu_t)}=\|G(t,s_0)\psi-m_{s_0}(\psi)\|_{L^p(\mathbb{R}^N,\mu_t)}$, where $\psi=G(s_0,s)f$. Since $\psi\in L^p(\mathbb{R}^N,\mu_{s_0})$ the right-hand side of the previous formula vanishes as t tends to $+\infty$. Hence, f satisfies property $(\mathcal{P}_{+\infty,p,s})$. On the other hand, if $s>s_0$, then for any $k\in\mathbb{N}$ such that $s-kT<s_0$ it holds that

$$\|G(t,s)f-m_s(f)\|_{L^p(\mathbb{R}^N,\mu_t)}=\|G(t-kT,s-kT)f-m_{s-kT}f\|_{L^p(\mathbb{R}^N,\mu_{t-kT})}.$$

This formula shows that $\|G(t,s)f-m_s(f)\|_{L^p(\mathbb{R}^N,\mu_t)}$ vanishes as t tends to $+\infty$.

Finally, the same arguments can be used to prove that $(v)\Rightarrow(v')$.

Step 2. Here we prove that $(i)\Rightarrow(ii)$. For this purpose, let us fix $f\in C_c^\infty(\mathbb{R}^N)$. By Step 1, it suffices to prove that $\|G(0,-s)f-m_{-s}(f)\|_{L^p(\mathbb{R}^N,\mu_0)}$ vanishes as s tends to $+\infty$. We will prove that any sequence $\{s_n\}$ diverging to $+\infty$ admits a subsequence $\{s_{n_k}\}$ such that $\|G(0,-s_{n_k})f-m_{-s_{n_k}}(f)\|_{L^p(\mathbb{R}^N,\mu_0)}$ vanishes as k tends to $+\infty$.

The function u, defined by $u(s,x)=f(x)$ for any $(s,x)\in\mathbb{R}^{N+1}$, belongs to $C_\sharp(\mathbb{R}^{N+1})$ and $(\mathscr{T}^\sharp(t)(I-\Pi)u)(s,x)=(G(s,s-t)f)(x)-m_{s-t}(f)$ for any $t>0$, $s\in\mathbb{R}$ and $x\in\mathbb{R}^N$. Since $G(0,s-t)f(x)-m_{s-t}(f)=G(0,s)[G(s,s-t)f-m_{s-t}(f)]$ for any $s\in[-T,0]$ and $t\geq 0$, and $G(0,s)$ is a contraction from $L^p(\mathbb{R}^N,\mu_s)$ to $L^p(\mathbb{R}^N,\mu_0)$, from the formula (18.2.1) it follows that

$$\lim_{t\to+\infty}\int_{-T}^0\|G(0,s-t)f-m_{s-t}(f)\|^p_{L^p(\mathbb{R}^N,\mu_0)}ds=0. \tag{18.2.2}$$

Thus, from any sequence $\{s_n\}$ diverging to $+\infty$, we can extract a subsequence $\{s_{n_k}\}$ and a set $\Gamma\subset[-T,0]$, with negligible complement, such that

$$\lim_{k\to+\infty}\|G(0,s-s_{n_k})f-m_{s-s_{n_k}}(f)\|_{L^p(\mathbb{R}^N,\mu_0)}=0,\qquad s\in\Gamma.$$

To show that $0\in\Gamma$, we fix a sequence $\{r_j\}\subset\Gamma$, converging to 0 as m tends to $+\infty$, and estimate

$$\begin{aligned}\|G(0,-s_{n_k})f-m_{-s_{n_k}}(f)\|_{L^p(\mathbb{R}^N,\mu_0)}\leq&\|G(0,-s_{n_k})f-G(0,r_j-s_{n_k})(f)\|_{L^p(\mathbb{R}^N,\mu_0)}\\&+\|G(0,r_j-s_{n_k})f-m_{r_j-s_{n_k}}(f)\|_{L^p(\mathbb{R}^N,\mu_0)}\\&+|m_{r_j-s_{n_k}}(f)-m_{-s_{n_k}}(f)|\end{aligned}\tag{18.2.3}$$

for any $j,k\in\mathbb{N}$. Since $C_b(\mathbb{R}^N)\hookrightarrow L^p(\mathbb{R}^N,\mu_0)$ and the function $G(0,\cdot)f$ is uniform continuous in $[-\infty,0]$ with values in $C_b(\mathbb{R}^N)$ (see Lemma 14.1.6(i)), the first term in the right-hand side of (18.2.3) vanishes as j tends to $+\infty$, uniformly with respect to $k\in\mathbb{N}$. Similarly, the third term in the right-hand side of (18.2.3) vanishes as j tends to $+\infty$, uniformly with respect to $k\in\mathbb{N}$, due to Lemma 16.2.13. Hence, taking the limsup as k tends to $+\infty$ in both sides of (18.2.3) and then letting j tend to $+\infty$, we conclude that $0\in\Gamma$ and property $(\mathcal{P}_{-\infty,p,0})$ is satisfied by any $f\in C_c^\infty(\mathbb{R}^N)$. To extend its validity to any $f\in C_b(\mathbb{R}^N)$ we observe that any such function is the local uniform limit in $\mathbb{R}^N$ of a bounded sequence $\{f_n\}\subset C_c^\infty(\mathbb{R}^N)$. From Lemma 18.1.5 it follows that $\sup_{s\in\mathbb{R}}\|f_n-f\|_{L^p(\mathbb{R}^N,\mu_s)}$ vanishes as n tends to $+\infty$. Hence, estimating

$$\begin{aligned}\|G(t,s)f-m_s(f)\|_{L^p(\mathbb{R}^N,\mu_t)}\leq&\|G(t,s)(f-f_n)\|_{L^p(\mathbb{R}^N,\mu_t)}+\|G(t,s)f_n-m_s(f_n)\|_{L^p(\mathbb{R}^N,\mu_t)}\\&+|m_s(f_n)-m_s(f)|\end{aligned}$$

$$\leq 2\sup_{s\in\mathbb{R}}\|f-f_n\|_{L^p(\mathbb{R}^N,\mu_s)}+\|G(t,s)f_n-m_s(f_n)\|_{L^p(\mathbb{R}^N,\mu_t)} \tag{18.2.4}$$

for any $n\in\mathbb{N}$ and letting n tend to $+\infty$, we deduce that f satisfies property $(\mathcal{P}_{-\infty,p,0})$.

Step 3. Here we prove that *(i)* $\Rightarrow$ *(iii)*. Clearly, it suffices to prove that property $(\mathcal{P}_{+\infty,p,s})$ is satisfied, for any $s\in\mathbb{R}$, by any $f\in C_b(\mathbb{R}^N)$ since, once it is proved for such functions, it can be extended to any $f\in L^p(\mathbb{R}^N,\mu_s)$, approximating any such function, in the $L^p(\mathbb{R}^N,\mu_s)$-norm, by a sequence of functions in $C_b(\mathbb{R}^N)$ and using the contractivity of the operators $G(t,s)$ and m_s from $L^p(\mathbb{R}^N,\mu_s)$ to $L^p(\mathbb{R}^N,\mu_t)$.

Fix $f\in C_b(\mathbb{R}^N)$. Changing variable in (18.2.2) we get

$$\lim_{t\to+\infty}\int_0^T ds\int_{\mathbb{R}^N}|G(s+t,s)f-m_s(f)|^p d\mu_{s+t}=0,$$

so that there exists a sequence $\{t_n\}$ diverging to $+\infty$ such that, for almost every $s\in(0,T)$ (and, by periodicity, for almost every $s\in\mathbb{R}$), it holds that

$$\lim_{n\to+\infty}\|G(s+t_n,s)f-m_s(f)\|_{L^p(\mathbb{R}^N,\mu_{s+t_n})}=0. \tag{18.2.5}$$

Let Γ' be the set of all $s\in\mathbb{R}$ such that (18.2.5) is satisfied. For $s\in\Gamma'$ and $t\in[t_n,t_{n+1})$ we split $G(s+t,s)f-m_s(f)=G(s+t,s+t_n)[G(s+t_n,s)f-m_s(f)]$ and estimate

$$\|G(s+t,s)f-m_s(f)\|_{L^p(\mathbb{R}^N,\mu_{s+t})}\leq\|G(s+t_n,s)f-m_s(f)\|_{L^p(\mathbb{R}^N,\mu_{s+t_n})}.$$

From this estimate we easily deduce that property $(\mathcal{P}_{+\infty,p,s})$ is satisfied by f for any $s\in\Gamma'$. To extend its validity to any $s\in\mathbb{R}$, it suffices to argue as at the end of Step 2, replacing f_n with $G(s,r_n)f$, where the sequence $\{r_n\}\subset\Gamma'$ increases to s as n tends to $+\infty$. Indeed, by Lemma 14.1.6(i), $G(s,r_n)f$ converges to f locally uniformly in $\mathbb{R}^N$ and $\|G(s,r_n)f\|_\infty\leq\|f\|_\infty$ for any $n\in\mathbb{N}$. Estimate (18.2.4) implies the statement.

Step 4. Here, we prove that, if Hypotheses 18.2.1(iv), (v) hold, then *(ii′)* $\Rightarrow$ *(iv)*. Similar arguments can be used to prove that *(iii′)* $\Rightarrow$ *(v)*.

Fix $f\in C_b(\mathbb{R}^N)$ and $t\in\mathbb{R}$. Then, the family of functions $\{G(t,s)f-m_s(f):s\leq t-1\}$ is bounded in $C_b(\mathbb{R}^N)$ and consists of equicontinuous functions due to the estimate in Remark 18.1.2(ii). We can thus invoke the Arzelà-Ascoli theorem, which shows that for any $R>0$ there exist a sequence $\{s_n\}$ diverging to $-\infty$ and a function $g\in C_b(B_R)$ such that $\|G(t,s_n)f-m^\sharp_{s_n}(f)-g\|_{L^\infty(B_R)}$ converges to 0 as n tends to $+\infty$.

Let $\rho^\sharp$ be the continuous positive version of the density of $\mu^\sharp$ with respect to the Lebesgue measure. Proposition 16.2.5 shows that $\mu_s=\rho^\sharp(s,x)dx$ for any $s\in\mathbb{R}$. Recalling that $L^p(\mathbb{R}^N,\mu_t)\hookrightarrow L^1(\mathbb{R}^N,\mu_t)$ and $\|\psi\|_{L^1(\mathbb{R}^N,\mu_t)}\leq\|\psi\|_{L^p(\mathbb{R}^N,\mu_t)}$ for any $\psi\in L^p(\mathbb{R}^N,\mu_t)$, we can estimate

$$\inf_{\mathbb{R}\times B_R}\rho^\sharp\int_{B_R}|(G(t,s_n)f)(x)-m_{s_n}(f)|dx\leq\|G(t,s_n)f-m_{s_n}(f)\|_{L^p(\mathbb{R}^N,\mu_t)}$$

for any $n\in\mathbb{N}$, and the right-hand side of the previous estimate vanishes as $n\to+\infty$ since property (ii) holds. On the other hand, the left-hand side converges to $\inf_{\mathbb{R}\times B_R}\rho^\sharp\|g\|_{L^1(B_R)}$, so that $g\equiv 0$, by the arbitrariness of $R>0$.

Step 5. Here, we prove that any of properties (ii), (iii), (iv), (v) implies property (i). We fix $\alpha\in C_\sharp(\mathbb{R})$, $f\in C_b(\mathbb{R}^N)$, consider the function $u\in C_\sharp(\mathbb{R}^{N+1})$, defined by $u(s,x)=\alpha(s)f(x)$ for any $(s,x)\in\mathbb{R}^{N+1}$, and observe that

$$\|\mathscr{T}^\sharp_p(t)(u-\Pi u)\|^p_{L^p_\sharp(\mathbb{R}^{N+1},\mu^\sharp)}=\frac{1}{T}\int_0^T|\alpha(s-t)|^p ds\int_{\mathbb{R}^N}|G(s,s-t)f-m^\sharp_{s-t}(f)|^p d\mu_s. \tag{18.2.6}$$

If property (ii) holds, then $\|G(s,s-t)f - m_{s-t}(f)\|_{L^p(\mathbb{R}^N,\mu_s)}$ converges to zero as t tends to $+\infty$ for each $s \in (0,T)$. Moreover, $\|G(s,s-t)f - m_{s-t}(f)\|_{L^p(\mathbb{R}^N,\mu_s)} \le 2\|f\|_\infty$ for any $s \in (0,T)$. Similarly, if property (iv) holds, then $\alpha(s-t)[(G(s,s-t)f)(x) - m_{s-t}(f)]$ converges to zero pointwise in $\mathbb{R}^{N+1}$, as t tends to $+\infty$, and it does not exceed $2\|\alpha\|_\infty\|f\|_\infty$ for any $(s,x) \in \mathbb{R}^{N+1}$ and $t>0$. In both cases we can apply the dominated convergence theorem to the right-hand side of (18.2.6) and conclude that $\mathscr{T}_p^\sharp(t)(u-\Pi u)$ vanishes in $L^p_\sharp(\mathbb{R}^{N+1},\mu^\sharp)$ as t tends to $+\infty$.

If (iii) or (v) holds, then, by a change of variables, we rewrite (18.2.6) as

$$\|\mathscr{T}_p^\sharp(t)(u-\Pi u)\|^p_{L^p_\sharp(\mathbb{R}^{N+1},\mu^\sharp)} = \frac{1}{T}\int_0^T |\alpha(s)|^p ds \int_{\mathbb{R}^N} |G(s+t,s)f - m_s(f)|^p d\mu^\sharp_{s+t}.$$

Using the dominated convergence theorem we conclude that $\mathscr{T}_p^\sharp(t)(u-\Pi u)$ vanishes in $L^p_\sharp(\mathbb{R}^{N+1},\mu^\sharp)$ as t tends to $+\infty$.

Since the linear span of the functions, which are the product of a function in $C_\sharp(\mathbb{R})$ and a function in $C_b(\mathbb{R}^N)$, is dense in $L^p_\sharp(\mathbb{R}^{N+1},\mu^\sharp)$, property (i) follows. □

Corollary 18.2.3 *Under Hypotheses* 18.2.1 *assume that there exists a constant $C>0$ such that one of the following conditions is satisfied:*

(i) $|Q(t,x)| \le C(1+|x|)\varphi(x)$ *and* $\langle b(t,x),x\rangle \le C(1+|x|^2)\varphi(x)$ *for any* $(t,x) \in (0,T)\times\mathbb{R}^N$*;*

(ii) *the diffusion coefficients are bounded in* $(0,T)\times\mathbb{R}^N$*.*

Then, all the properties in Theorem 18.2.2 *hold true.*

Proof By the assumptions, Hypotheses 18.1.1 are satisfied. Hence, property $(\mathcal{P}_{+\infty,p,s})$ is satisfied by any $f \in L^p(\mathbb{R}^N,\mu_s)$, $p \in (1,+\infty)$ and $s \in \mathbb{R}$, thanks to Theorem 18.1.6. This completes the proof. □

As in Proposition 18.1.14, the results in the previous theorem, the gradient estimate (17.2.6) and Theorem 17.2.10 allow to prove a Poincaré inequality.

Proposition 18.2.4 *Assume that the diffusion coefficients are bounded and the constant ℓ_2 in* (18.1.9) *is negative. Then, for any $p \in [2,+\infty)$, there exists a positive constant C_p such that*

$$\|u-\Pi u\|_{L^p_\sharp(\mathbb{R}^{N+1},\mu^\sharp)} \le C_p\|\nabla_x u\|_{L^p_\sharp(\mathbb{R}^{N+1},\mu^\sharp)}, \qquad u \in W^{0,1}_{p,\sharp}(\mathbb{R}^{N+1},\mu^\sharp).$$

In particular, we can take $C_2 = \Lambda^{1/2}|\ell_2|^{-1/2}$, where Λ is the supremum over $(s,x) \in (0,T)\times\mathbb{R}^N$ of the maximum eigenvalue of the matrix $Q(s,x)$.

We now deal with the asymptotic decay to zero of the function $G(t,s)f - m_s(f)$.

Theorem 18.2.5 *Let Hypotheses* 18.2.1*(i)-(iii) hold. Fix $1 \le p \le +\infty$, $\omega \in \mathbb{R}$. Then, the following conditions are equivalent:*

(i) *there exists a positive constant M_0 such that $\|\mathscr{T}_p^\sharp(t)(I-\Pi)\|_{L(L^p_\sharp(\mathbb{R}^{N+1},\mu^\sharp))} \le M_0e^{\omega t}$ for any $t>0$;*

(ii) *there exists a positive constant M_1 such that $\|G(t,s)-m_s\|_{L(L^p(\mathbb{R}^N,\mu_s),L^p(\mathbb{R}^N,\mu_t))} \le M_1e^{\omega(t-s)}$ for any $t>s$.*

Proof The proof is very close to the one of Theorem 18.1.12. Hence, we just sketch the proof of $(i) \Rightarrow (ii)$ for $p < +\infty$ pointing out the differences. We fix $f \in C_c^\infty(\mathbb{R}^N)$ and consider the function $u \in C_\sharp(\mathbb{R}^{N+1})$ defined by $u(s,x) = f(x)$ for any $(s,x) \in \mathbb{R}^{N+1}$. Next we fix $s, t \in \mathbb{R}$, with $s < t$, $k_0 \in \mathbb{Z}$ such that $t - k_0T \in [0,T)$, $0 < \varepsilon < (k_0+1)T - t$ and denote by $\chi^\sharp_{(t,t+\varepsilon)}$ the T-periodic extension of the function $\chi_{(t-k_0T, t-k_0T+\varepsilon)}$. From the T-time periodicity of the function $s \mapsto \|(\mathscr{T}_p^\sharp(t)(I-\Pi)u)(s,\cdot)\|_{L^p(\mathbb{R}^N,\mu_s)}$ and property (i), it follows that

$$
\begin{aligned}
\|G(t,s)f - m_s(f)\|^p_{L^p(\mathbb{R}^N,\mu_t)} &= \lim_{\varepsilon\to0^+}\frac{1}{\varepsilon}\int_{t-k_0T}^{t-k_0T+\varepsilon} \|(\mathscr{T}_p^\sharp(t-s)(I-\Pi)u)(r,\cdot)\|^p_{L^p(\mathbb{R}^N,\mu_r)}dr\\
&= \lim_{\varepsilon\to0^+}\frac{1}{\varepsilon}\int_0^T \chi^\sharp_{(t,t+\varepsilon)}(r)\|(\mathscr{T}_p^\sharp(t-s)(I-\Pi)u)(r,\cdot)\|^p_{L^p(\mathbb{R}^N,\mu_r)}dr\\
&= \lim_{\varepsilon\to0^+}\frac{T}{\varepsilon}\|\mathscr{T}_p^\sharp(t-s)(\chi^\sharp_{(s,s+\varepsilon)}(I-\Pi)u)\|^p_{L^p_\sharp(\mathbb{R}^{N+1},\mu^\sharp)}.
\end{aligned}
$$

Now, the proof can be concluded with the same arguments as in the proof of Theorem 18.1.12. □

For any $p \in [1,+\infty)$ let us define the right halflines

$$\mathfrak{A}_p^\sharp := \{\omega \in \mathbb{R} : \exists M_\omega > 0 \ \text{s.t.}\ \|\mathscr{T}_p^\sharp(t)(I-\Pi)\|_{L(L^p_\sharp(\mathbb{R}^{N+1},\mu^\sharp))} \le M_\omega e^{\omega t}\ \forall t \ge 0\},$$

$$\mathfrak{B}_p^\sharp := \{\omega \in \mathbb{R} : \exists N_\omega > 0 \ \text{s.t.}\ \|\nabla_x\mathscr{T}_p^\sharp(t)\|_{L(L^p_\sharp(\mathbb{R}^{N+1},\mu^\sharp);(L^p_\sharp(\mathbb{R}^{N+1},\mu^\sharp))^N)} \le N_\omega e^{\omega t}\ \forall t \ge 1\}$$

and their infima denoted, respectively, by ω_p and γ_p. Then, ω_p, the growth bound of the part of $\mathscr{T}_p^\sharp(t)$ in $(I-\Pi)(L^p_\sharp(\mathbb{R}^{N+1},\mu^\sharp))$, is nonpositive.

Theorem 18.2.6 *Let Hypotheses* 18.2.1 *hold. Then* $\mathfrak{A}_p^\sharp \subset \mathfrak{B}_p^\sharp$ *for any* $p \ge 2$. *The inclusion can also be extended to* $p \in (1,2)$ *if* $\ell_p < +\infty$ *(see* (18.1.12)*). If the diffusion coefficients are bounded and* $\ell_2 < 0$, *then* $\mathfrak{B}_p^\sharp \subset \mathfrak{A}_p^\sharp$ *for any* $p \ge 2$. *In this latter case the following properties are satisfied:*

(i) $\sigma(G_p^\sharp) \cap i\mathbb{R} = \{2\pi k i/T : k \in \mathbb{Z}\}$ *consists of simple isolated eigenvalues for any* $p \in (1,+\infty)$. *In particular, for any* $f \in L^p_\sharp(\mathbb{R}^{N+1},\mu^\sharp)$ *the parabolic problem* $G_p^\sharp u = f$ *is solvable if and only if* $\int_{(0,T)\times\mathbb{R}^N} f\,d\mu^\sharp = 0$. *In this case, it has infinite solutions and the difference of two solutions is constant in* $\mathbb{R}^{N+1}$.

(ii) *For any* $p \in (1,+\infty)$ $G_p^\sharp$ *has a spectral gap. More precisely,*

$$\sup\{\mathrm{Re}\,\lambda : \lambda \in \sigma(G_p^\sharp)\setminus i\mathbb{R}\} \le 2\ell_2(1-(p\wedge 2)^{-1}), \qquad p \in (1,+\infty).$$

Proof The first part of the theorem can be checked by arguing as in the proof of Theorem 18.1.16.

To complete the proof, let us prove properties (i) and (ii). Arguing as in the proof of Corollary 18.1.17 it can be shown that the spectrum of the part of $G_p^\sharp$ in $(I-\Pi)(L^p_\sharp(\mathbb{R}^{N+1},\mu^\sharp))$ is contained in the halfplane $\mathrm{Re}\,\lambda \le \ell_2$, if $p \ge 2$, and in the halfplane $\mathrm{Re}\,\lambda \le 2\ell_2(1-1/p)$, if $1 < p < 2$.

Splitting the equation $\lambda u - G_p^\sharp u = f$ along $\Pi(L^p_\sharp(\mathbb{R}^{N+1},\mu^\sharp))$ and $(I-\Pi)(L^p_\sharp(\mathbb{R}^{N+1},\mu^\sharp))$ we immediately deduce that, if λ does not belong to the spectrum of the part of $G_p^\sharp$ in $(I-\Pi)(L^p_\sharp(\mathbb{R}^{N+1},\mu^\sharp))$, then the previous equation is uniquely solvable in $D(G_p^\sharp)$ if and only if the equation $\lambda v + v' = \Pi f$ is uniquely solvable in $W^{1,p}_\sharp(\mathbb{R},T^{-1}ds)$. This equation is uniquely solvable if and only if $\lambda \neq 2\pi ik/T$ for any $k \in \mathbb{Z}$. Since the eigenvalues $2\pi ik/T$ of the realization of the first-order derivative in $L^p_\sharp(\mathbb{R}, T^{-1}ds)$ are simple, the eigenvalues $2\pi ik/T$ of $G^\sharp_p$ are simple too. In particular, for $\lambda = 0$ the above equation is solvable if and only if $\int_0^T m_s(f(s,\cdot))ds = 0$, which means that $\int_{(0,T)\times\mathbb{R}^N} f\,d\mu^\sharp = 0$ and, in this case, the solutions differ by constants. The statements follow. □

To conclude this section, we prove a refinement of Theorem 18.2.6 in the case of diffusion coefficients independent of the spatial variables.

Theorem 18.2.7 *Let Hypotheses* 18.2.1 *be satisfied with the diffusion coefficients being independent of x and the constant r_0 being negative. Then,* $\sup\{\mathrm{Re}\,\lambda : \lambda \in \sigma(G^\sharp_p) \setminus i\mathbb{R}\}$ *is independent of $p \in (1,+\infty)$. Denoting by $\omega_0 < 0$ this supremum, for any $\omega > \omega_0$ and $p \in (1,+\infty)$ there exists a positive constant M such that*

$$\|G(t,s)f - m_s(f)\|_{L^p(\mathbb{R}^N,\mu_t)} \le Me^{\omega(t-s)}\|f\|_{L^p(\mathbb{R}^N,\mu_s)}, \qquad t > s,\ \ f \in L^p(\mathbb{R}^N,\mu_s). \quad (18.2.7)$$

Proof To begin with, let us prove that ω_p is independent of $p \in (1,+\infty)$. For this purpose, we fix $p \in (1,+\infty)$ and denote by $G^\sharp_{p,\Pi}$ (resp. $\mathscr{T}^\sharp_{p,\Pi}(t)$) the part of $G^\sharp_p$ (resp. $\mathscr{T}^\sharp_p(t)$) in $\Pi(L^p_\sharp(\mathbb{R}^{N+1},\mu^\sharp))$, and by $G^\sharp_{p,I-\Pi}$ (resp. $\mathscr{T}^\sharp_{p,I-\Pi}(t)$) the part of $G^\sharp_p$ (resp. $\mathscr{T}^\sharp_p(t)$) in $(I-\Pi)(L^p_\sharp(\mathbb{R}^{N+1},\mu^\sharp))$. Since Π commutes with $\mathscr{T}^\sharp(t)$, it follows that $\sigma(G^\sharp_p) = \sigma(G^\sharp_{p,\Pi}) \cup \sigma(G^\sharp_{p,I-\Pi})$. The spectrum of $G^\sharp_{p,\Pi}$ is the set $\{2k\pi i/T : k \in \mathbb{Z}\}$ since $\Pi(L^p_\sharp(\mathbb{R}^{N+1},\mu^\sharp))$ is isometric to $L^p_\sharp(\mathbb{R}, T^{-1}ds)$ and $G^\sharp_{p,\Pi} = -D_s$ on $D(G^\sharp_{p,\Pi}) = \Pi(D(G^\sharp_p))$. Therefore,

$$\sup\{\mathrm{Re}\,\lambda : \lambda \in \sigma(G^\sharp_p),\ \mathrm{Re}\,\lambda < 0\} = \sup\{\mathrm{Re}\,\lambda : \lambda \in \sigma(G^\sharp_{p,I-\Pi})\}.$$

Let us prove that such suprema coincide with ω_p. This will imply that ω_p is independent of p, because the left-hand supremum is independent of p. For this purpose, we remark that, although the operators $\mathscr{T}^\sharp_p(t)$ are not compact, $\sigma(\mathscr{T}^\sharp_p(t))\setminus\{0\}$ consists of eigenvalues. Indeed, by the spectral mapping Theorem 17.2.6, $\sigma(\mathscr{T}^\sharp_p(t))\setminus\{0\} = e^{t\sigma(G^\sharp_p)}$, and by the general theory of strongly continuous semigroups (see Proposition C.1.12) $P\sigma(\mathscr{T}^\sharp_p(t)) \setminus \{0\} = e^{tP\sigma(G^\sharp_p)}$, where $P\sigma$ denotes the point spectrum. Since $\sigma(G^\sharp_p) = P\sigma(G^\sharp_p)$, the set equality $\sigma(\mathscr{T}^\sharp_p(t)) \setminus \{0\} = P\sigma(\mathscr{T}^\sharp_p(t)) \setminus \{0\}$ holds true. As a consequence, also $\sigma(\mathscr{T}^\sharp_{p,I-\Pi}(t)) \setminus \{0\}$ consists of eigenvalues, because the elements of $\sigma(\mathscr{T}_{p,I-\Pi}(t))$, which are not eigenvalues, are contained in $\sigma(\mathscr{T}^\sharp_p(t)) \setminus P\sigma(\mathscr{T}^\sharp_p(t))$, which does not contain nonzero elements. Again by the spectral mapping theorem for the point spectrum, it follows that $\sigma(\mathscr{T}^\sharp_{p,I-\Pi}(t)) \setminus \{0\} = e^{t\sigma(G^\sharp_{p,I-\Pi})}$ i.e., the semigroup $\{\mathscr{T}^\sharp_{p,I-\Pi}(t)\}$ satisfies the spectral mapping theorem. Taking Proposition C.1.2 into account (which shows that ω_p coincides with the logarithm of the spectral radius of $\mathscr{T}^\sharp_{p,I-\Pi}(1)$), this implies that $\omega_p = \sup\{\mathrm{Re}\,\lambda : \lambda \in \sigma(G^\sharp_{p,I-\Pi})\}$.

To conclude the proof, we apply Corolary 18.1.21 and Theorem 18.2.5 to infer that $\omega_0 \le r_0 < 0$. Estimate (18.2.7) now follows from Theorem 18.2.5. □

Part III

Appendices

Appendix A

Function spaces and smooth domains

In this appendix we collect the function spaces that we consider in this book and prove some properties of the distance function $x \mapsto d(x, \partial\Omega)$ when Ω is a sufficiently smooth (unbounded) domain.

A.1 Spaces of continuous or Hölder continuous functions

A.1.1 Isotropic spaces

Given an open set $\Omega \subset \mathbb{R}^N$ (even $\Omega = \mathbb{R}^N$), $C_b(\Omega)$ (resp. $C_b(\overline{\Omega})$, omitting the subscript "b" when Ω is bounded) denotes the space of all bounded and continuous functions $f : \Omega \to \mathbb{R}$ (resp. $f : \overline{\Omega} \to \mathbb{R}$) and it is endowed with the sup-norm, i.e., $\|f\|_\infty = \sup_{x\in\Omega} |f(x)|$ for any $f \in C_b(\Omega)$ (resp. $f \in C_b(\overline{\Omega})$).

If Ω is bounded then $C_0(\Omega)$ denotes the subspace of $C_b(\overline{\Omega})$ of all functions $f : \overline{\Omega} \to \mathbb{R}$ which vanish on $\partial\Omega$. If Ω is unbounded then $C_0(\Omega)$ denotes the subset of $C_b(\overline{\Omega})$ of all functions $f : \overline{\Omega} \to \mathbb{R}$ which vanish both on $\partial\Omega$ and as $|x|$ tends to $+\infty$. Next, for any $k > 0$, $C_b^k(\overline{\Omega})$ denotes the space of all functions $f : \overline{\Omega} \to \mathbb{R}$ which are continuously differentiable in Ω up to the $[k]$-th order with $D^\alpha f \in C_b(\overline{\Omega})$ for any $|\alpha| \le [k]$ ($[k]$ denoting the integer part of k) and with $D^\alpha f$ being (k-$[k]$)-Hölder continuous for any $|\alpha| = [k]$. This space is endowed with the norm

$$\|f\|_{C_b^k(\overline{\Omega})} := \sum_{|\alpha|\le[k]} \|D^\alpha f\|_\infty + \sum_{|\alpha|=[k]} [D^\alpha f]_{C^{k-[k]}(\overline{\Omega})}, \qquad f \in C_b^k(\overline{\Omega}), \tag{A.1.1}$$

where

$$[D^\alpha f]_{C^{k-[k]}(\overline{\Omega})} = \sup_{\substack{x,y\in\Omega \\ x\neq y}} \frac{|D^\alpha f(x) - D^\alpha f(y)|}{|x-y|^{k-[k]}}.$$

Similarly, $C^k_{\rm loc}(\Omega)$ ($k \in \mathbb{R}_+$) is the set of all functions $f : \Omega \to \mathbb{R}$ which are continuously differentiable up to the $[k]$-th order in Ω with $[k]$-th-order derivatives which are (k-$[k]$)-Hölder continuous in K for any compact set $\Omega \subset \mathbb{R}^N$.

A function f belongs to $BUC^k(\mathbb{R}^N)$ ($k \in \mathbb{N}$) if $f \in C_b^k(\mathbb{R}^N)$ and $D^\alpha f$ is uniformly continuous in $\mathbb{R}^N$ for any $|\alpha| = k$. We endow $BUC^k(\mathbb{R}^N)$ with the norm in (A.1.1).

Moreover, for any open set $\Omega \subset \mathbb{R}^N$ and any $k \ge 0$, $C_c^k(\Omega)$ (resp. $C_c^k(\overline{\Omega})$) denotes the subset of $C_b^k(\overline{\Omega})$ of all functions $f : \overline{\Omega} \to \mathbb{R}$ which are compactly supported in Ω (resp. in $\overline{\Omega}$). We endow both $C_c^k(\Omega)$ and $C_c^k(\overline{\Omega})$ with the norm of $C_b^k(\overline{\Omega})$. A function f belongs to $C_c^\infty(\Omega)$ if $f \in C_c^k(\Omega)$ for any $k \in \mathbb{N}$.

Sometimes we also consider spaces of vector-valued continuous functions. For any $k \in [0, +\infty)$, and $m \in \mathbb{N}$, we denote by $C^k(\Omega, \mathbb{R}^m)$ (resp. $C^k_{\rm loc}(\Omega, \mathbb{R}^m)$) the set of all functions $f : \Omega \to \mathbb{R}^m$ such that the components f_j belongs to $C^k(\Omega)$ (resp. to $C^k_{\rm loc}(\Omega)$) for any $j = 1, \ldots, m$.

A.1.2 Anisotropic Hölder spaces in $\mathbb{R}^N$

Throughout this subsection, we fix $p_0, \ldots, p_r \in \mathbb{N}$ satisfying $p_0 + \ldots + p_r = N$ and, to simplify the notation, we split any $x \in \mathbb{R}^N$ as $x = (x_0, \ldots, x_r)$ with $x_j \in \mathbb{R}^{p_j}$ for any $j = 0, \ldots, r$.

For any $\theta_0, \theta_1, \ldots, \theta_r > 0$ we define the space $\mathscr{C}_b^{\theta_0,\ldots,\theta_r}(\mathbb{R}^N)$ as the set of all the bounded functions $f : \mathbb{R}^N \to \mathbb{R}$ such that, for any $\widehat{x}_j := (x_0, \ldots, x_{j-1}, x_{j+1}, \ldots, x_r)$ in $\mathbb{R}^{N-p_j}$, the function $f(x_0, \ldots, x_{j-1}, \cdot, x_{j+1}, \ldots, x_r)$ belongs to the space $C_b^{\theta_j}(\mathbb{R}^{p_j})$ and

$$\|f\|_{\theta_j} := \sup_{\widehat{x}_j \in \mathbb{R}^{N-p_j}} \|f(x_1, \ldots, x_{j-1}, \cdot, x_{j+1}, \ldots, x_r)\|_{C_b^{\theta_j}(\mathbb{R}^{p_j})} < +\infty.$$

We norm it by $\|f\|_{\mathscr{C}_b^{\theta}(\mathbb{R}^N)} := \sum_{j=0}^{r} \|f\|_{\theta_j}$ for any $f \in \mathscr{C}_b^{\theta_0,\ldots,\theta_r}(\mathbb{R}^N)$. When $\theta_j \in \mathbb{N}$ for some $j = 0, \ldots, r$, we assume that all the existing derivatives of $f \in \mathscr{C}_b^{\theta_0,\ldots,\theta_r}(\mathbb{R}^N)$ are continuous in $\mathbb{R}^N$.

We also use the space $\widetilde{\mathscr{C}}_b^{\theta_0,\ldots,\theta_r}(\mathbb{R}^N)$, defined as $\mathscr{C}_b^{\theta_0,\ldots,\theta_r}(\mathbb{R}^N)$ with the space $C_b^{\theta_j}(\mathbb{R}^{p_j})$ being replaced by the Zygmund space $\mathcal{C}_b^{\theta_j}(\mathbb{R}^{p_j})$, where, for any $\gamma \in (0,1]$ and $m \in \mathbb{N}$, $\mathcal{C}_b^{\gamma}(\mathbb{R}^m)$ is the set of all functions $f \in C_b(\mathbb{R}^m)$ such that

$$[f]_{\mathcal{C}_b^{\gamma}(\mathbb{R}^m)} = \sup_{\substack{x,y \in \mathbb{R}^m \\ x \neq y}} \frac{|f(x) - 2f\left(\frac{x+y}{2}\right) + f(y)|}{|x-y|^{\gamma}} < +\infty.$$

$\mathcal{C}_b^{\gamma}(\mathbb{R}^m)$ is a Banach space when it is endowed with the norm $\|f\|_{\mathcal{C}_b^{\gamma}(\mathbb{R}^m)} := \|f\|_{\infty} + [f]_{\mathcal{C}_b^{\gamma}(\mathbb{R}^m)}$ for any $f \in \mathcal{C}_b^{\gamma}(\mathbb{R}^m)$. For $\gamma > 1$ such that $\gamma \notin \mathbb{N}$,

$$\mathcal{C}_b^{\gamma}(\mathbb{R}^m) = \{f \in C_b^{[\gamma]}(\mathbb{R}^m) : D^{\alpha} f \in \mathcal{C}_b^{\gamma-[\gamma]}(\mathbb{R}^m) \text{ for any } |\alpha| = [\gamma]\}$$

and we endow it with the norm $\|f\|_{\mathcal{C}_b^{\gamma}(\mathbb{R}^m)} := \|f\|_{C_b^{[\gamma]}(\mathbb{R}^m)} + \sum_{|\alpha|=[\gamma]} [D^{\alpha} f]_{\mathcal{C}_b^{\gamma-[\gamma]}(\mathbb{R}^m)}$ for any $f \in \mathcal{C}_b^{\gamma}(\mathbb{R}^m)$. Finally, if $\gamma \in \mathbb{N} \setminus \{1\}$ then

$$\mathcal{C}_b^{\gamma}(\mathbb{R}^m) = \{f \in C_b^{\gamma-1}(\mathbb{R}^m) : D^{\alpha} f \in \mathcal{C}_b^{1}(\mathbb{R}^m), \text{ for any } |\alpha| = \gamma - 1\}$$

and we endow it with the norm $\|f\|_{\mathcal{C}_b^{\gamma}(\mathbb{R}^m)} := \|f\|_{C_b^{\gamma-1}(\mathbb{R}^m)} + \sum_{|\alpha|=\gamma-1} [D^{\alpha} f]_{\mathcal{C}_b^{1}(\mathbb{R}^m)}$ for any $f \in \mathcal{C}_b^{\gamma}(\mathbb{R}^m)$.

Remark A.1.1 It is well known (see e.g. [178, Proposition 0.2.2]) that the spaces $C_b^{\gamma}(\mathbb{R}^m)$ and $\mathcal{C}_b^{\gamma}(\mathbb{R}^m)$ coincide when $\gamma \notin \mathbb{N}$ and the corresponding norms are equivalent, whereas, if $\gamma \in \mathbb{N}$, $C_b^{\gamma}(\mathbb{R}^m)$ is properly continuously embedded in $\mathcal{C}_b^{\gamma}(\mathbb{R}^m)$. It follows that, if $\theta_j \notin \mathbb{N}$ for any $j = 0, \ldots, r$, then $\mathscr{C}_b^{\theta_0,\ldots,\theta_r}(\mathbb{R}^N) = \widetilde{\mathscr{C}}_b^{\theta_0,\ldots,\theta_r}(\mathbb{R}^N)$ with equivalence of the corresponding norms, otherwise $\mathscr{C}_b^{\theta_0,\ldots,\theta_r}(\mathbb{R}^N)$ is properly and continuously embedded in $\widetilde{\mathscr{C}}_b^{\theta_0,\ldots,\theta_r}(\mathbb{R}^N)$.

When $\theta_j = \theta/(2j+1)$ for any $j = 0, \ldots, r$ and some $\theta > 0$, we simply write $\mathscr{C}_b^{\theta}(\mathbb{R}^N)$ and $\widetilde{\mathscr{C}}_b^{\theta}(\mathbb{R}^N)$ instead of $\mathscr{C}_b^{\theta_0,\ldots,\theta_r}(\mathbb{R}^N)$ and $\widetilde{\mathscr{C}}_b^{\theta_0,\ldots,\theta_r}(\mathbb{R}^N)$, respectively.

514

A.2 Parabolic Hölder spaces

Given any interval $I \subset (0, +\infty)$, any set $D \subset \mathbb{R}^N$, with nonempty interior part and two integers $h, k \in \mathbb{N} \cup \{0\}$, we denote by $C^{h,k}(I \times D)$, the set of all functions $f : I \times D \to \mathbb{R}$

which are continuous in $I \times D$, continuously differentiable in the interior of $I \times D$ up to the h-th order with respect to the time variable and up to the k-th order with respect to the spatial variables and with all the derivatives which can be extended by continuity to $I \times D$. When $I \times D$ is bounded, we norm $C^{h,k}(I \times D)$ by setting

$$\|f\|_{C^{h,k}(I\times D)} = \sum_{j=0}^{h} \|D_t^j f\|_\infty + \sum_{0<|\alpha|\le k} \|D^\alpha f\|_\infty. \tag{A.2.1}$$

Moreover, when $I \times D$ is unbounded we write $C_b^{h,k}(I \times D)$ to denote the set of all functions belonging to $C^{h,k}(I \times D)$ which are bounded together with their derivatives. We endow $C_b^{h,k}(I \times D)$ with the norm in (A.2.1).

We denote by $C_c^\infty(I \times D)$ the set of all functions $f : I \times D \to \mathbb{R}$ compactly supported in $I \times D$, which admit classical derivatives $D_t^j D_x^\alpha f$ of any order in $I \times D$.

Next, for any $\alpha \in (0,1)$ and $k \in \mathbb{N} \cup \{0\}$, we denote by $C^{1+\alpha/2,2+\alpha}(I \times D)$ the subspace of $C^{1,2}(I \times \Omega)$ of all functions $f : I \times D \to \mathbb{R}$ such that the functions $D_t f, D_x^\beta f$ ($|\beta| \le 2$) are bounded and satisfy the condition

$$\|f\|_{C^{1+\alpha/2,2+\alpha}(I\times D)} := \sup_{x\in D} \|f(\cdot,x)\|_{C_b^{1+\alpha/2}(I)} + \sup_{t\in I} \|f(t,\cdot)\|_{C_b^{2+\alpha}(D)} < +\infty. \tag{A.2.2}$$

It is a Banach space when endowed with the norm in (A.2.2).

Similarly, we say that $f \in C_{\rm loc}^{1+\alpha/2,2+\alpha}(I \times D)$ if $f \in C^{1+\alpha/2,2+\alpha}(F)$ for any open set $F := (a,b) \times \Omega'$ with compact closure in $I \times D$.

Let Y be a subspace of $C_b(\Omega)$ (resp. of $C_b(\overline{\Omega})$). By $B(I;Y)$ and $C_b^k(I;Y)$ ($k \ge 0$) we denote the set of all functions $f : I \to Y$ which are, respectively, bounded in I and differentiable in I (with values in Y) up to the $[k]$-th order, with bounded derivatives, $f^{([k])}$ being Hölder continuous in I with exponent $k - [k]$. We norm these spaces by setting $\|f\|_{B(I;Y)} := \sup_{t\in I} \|f(t,\cdot)\|_Y$ and

$$\|f\|_{C_b^k(I;Y)} := \sum_{j=0}^{[k]} \|f^{(j)}\|_{B(I;Y)} + \sup_{\substack{s,t\in I\\ s\neq t}} \frac{\|f(t) - f(s)\|_Y}{|t-s|^{k-[k]}}.$$

Sometimes, we identify a bounded and uniformly continuous function $f : I \times \Omega \to \mathbb{R}$ with an element of the space $C_b(I; C_b(\Omega))$ in the natural way. Modulo this identification, we can say that $f \in C^{1+\alpha/2,2+k+\alpha}(I \times \Omega)$ ($k \in \mathbb{N} \cup \{0\}$) if and only if $f \in C^{1+\alpha/2}(I; C_b(\Omega)) \cap C_b(I; C_b^{2+k+\alpha}(\Omega))$.

A.3 L^p- and Sobolev spaces

Given an open set $\Omega \subset \mathbb{R}^N$ (possibly $\Omega = \mathbb{R}^N$), we denote by $B_b(\Omega)$ the space of bounded Borel measurable functions defined in Ω.

Let μ be a positive measure defined on the σ-algebra of all the Borel sets of Ω. We denote by $L^p(\Omega,\mu)$ ($p \in [1,+\infty)$) the space of Borel measurable functions $f : \Omega \to \mathbb{R}$ such that

$$\|f\|_{L^p(\Omega,\mu)} := \left(\int_\Omega |f|^p d\mu\right)^{\frac{1}{p}} < +\infty, \qquad p < +\infty,$$

$$\|f\|_\infty = \min\{\Lambda > 0 \text{ s.t. } |f(x)| \le \Lambda,\ \mu\text{-}a.e. \text{ in } \Omega\}.$$

By $L^p(\Omega)$ ($p \in [1,+\infty]$), we denote the usual L^p-spaces related to the Lebesgue measure defined on the σ-algebra of the Lebesgue measurable sets.

By $L^p_{\rm loc}(\Omega)$ ($p \in [1,+\infty]$), we denote the space of all functions f which belong to $L^p(F)$ for any open set $F \Subset \mathbb{R}^N$.

The Sobolev space $W^{k,p}(\Omega,\mu)$ ($k \in \mathbb{N}$, $p \in [1,+\infty]$) is the subspace of $L^p(\Omega,\mu)$ of all functions f which admits distributional derivatives up to the k-th-order in $L^p(\Omega,\mu)$ It is a Banach space when endowed with the norm $\|f\|_{W^{k,p}(\Omega,\mu)} = \sum_{|\alpha|\le k}\|D^\alpha f\|_{L^p(\Omega,\mu)}$ for any $f \in W^{k,p}(\Omega,\mu)$. As above we simply write $W^{k,p}(\Omega)$ when the underlying measure is the Lebesgue measure.

We say that $u \in W^{k,p}_{\rm loc}(\Omega)$ ($k \in \mathbb{N}$, $p \in [1,+\infty]$) if $u \in W^{k,p}(F)$ for any open set $F \Subset \mathbb{R}^d$.

For any $p \in [1,+\infty]$, $h,k \in \mathbb{N}\cup\{0\}$ and $\Omega \subset \mathbb{R}^{N+1}$, the parabolic Sobolev space $W^{h,k}_p(\Omega,\mu)$ is defined as the set of the functions $u \in L^p(\Omega,\mu)$ having distributional derivatives $D^m_t u$ ($m \le h$) and $D^m_x u$ ($m \le k$) in $L^p(\Omega,\mu)$. It is endowed with the norm $\|u\|_{W^{h,k}_p(\Omega,\mu)} = \|u\|_{L^p(\Omega,\mu)} + \sum_{m=1}^h \|D^m_t u\|_{L^p(\Omega,\mu)} + \sum_{m=1}^k \|D^m_x u\|_{L^p(\Omega,\mu)}$ for any $u \in W^{h,k}_p(\Omega,\mu)$. We say that $u \in W^{h,k}_{p,{\rm loc}}(\Omega,\mu)$ if u belongs to $W^{h,k}_p(\Omega',\mu)$ for any $\Omega' \Subset \Omega$.

For more details on L^p and Sobolev spaces we refer the reader e.g. to [3, 148, 165, 189]

A.3.1 The spaces $\mathscr{H}^{p,1}((a,b)\times\mathbb{R}^N)$

By $\mathscr{H}^{p,1}((a,b)\times\mathbb{R}^N)$ ($a,b \in \mathbb{R}$, $a<b$, $p \in (1,+\infty)$) we denote the space of all functions $u \in W^{0,1}_p((a,b)\times\mathbb{R}^N)$ such that the distributional derivative $D_t u$ belongs to the space $(W^{0,1}_{p'}((a,b)\times\mathbb{R}^N))'$, where $1/p+1/p'=1$. It is endowed with the norm

$$\|f\|_{\mathscr{H}^{p,1}((a,b)\times\mathbb{R}^N)} = \|D_t f\|_{(W^{0,1}_{p'}((a,b)\times\mathbb{R}^N))'} + \|f\|_{W^{0,1}_p((a,b)\times\mathbb{R}^N)}.$$

As the following proposition shows, any function $u \in \mathscr{H}^{p,1}((a,b)\times\mathbb{R}^N)$ is the limit of smooth functions.

Lemma A.3.1 *Let $u \in \mathscr{H}^{p,1}((a,b)\times\mathbb{R}^N)$ for some $p \in (1,+\infty)$. Then, there exists a sequence $\{u_n\} \subset C^\infty_c(\mathbb{R}^{N+1})$ of smooth functions converging to u in $W^{0,1}_p(Q(a,b))$ as n tends to $+\infty$. Moreover, $D_t u_n$ converges to $D_t u$ weakly* in $(W^{0,1}_{p'}((a,b)\times\mathbb{R}^N))'$. Finally, if $u \in C_b([a,b]\times\mathbb{R}^N)$, then the sequence $\{u_n\}$ converges to u locally uniformly in $[a,b]\times\mathbb{R}^N$.*

Proof We split the proof into two steps: first, by a convolution argument, we prove the statement when $(a,b)\times\mathbb{R}^N$ is replaced by $\mathbb{R}^{N+1}$. Then, we use this result to complete the proof.

Step 1. Let $\vartheta \in C^\infty_c(\mathbb{R})$ be a smooth function such that $\chi_{(-1,1)} \le \vartheta \le \chi_{(-2,2)}$. For any $\sigma > 0$, $t \in \mathbb{R}$ and $x \in \mathbb{R}^N$, set $\vartheta_\sigma(t,x) = \vartheta(|t|/\sigma)\vartheta(|x|/\sigma)$ and use these functions to regularize u by setting

$$\begin{aligned} u_n(t,x) &= n^{N+1}\vartheta_n(t,x)\int_{\mathbb{R}^{N+1}} u(s,y)\vartheta_{1/n}(t-s,x-y)dsdy \\ &=: n^{N+1}\vartheta_n(t,x)(u\star\vartheta_{1/n})(t,x) \end{aligned}$$

for any $(t,x) \in \mathbb{R}^{N+1}$ and $n \in \mathbb{N}$. Each function u_n belongs to $C^\infty_c(\mathbb{R}^{N+1})$ and converges to u in $W^{0,1}_p(\mathbb{R}^{N+1})$. If, in addition, $u \in C_b([a,b]\times\mathbb{R}^N)$, then u_n converges to u locally uniformly in $[a,b]\times\mathbb{R}^N$.

Let us fix a function $\psi \in W^{0,1}_{p'}(\mathbb{R}^{N+1})$. Applying the Fubini-Tonelli theorem and taking into account that $\vartheta_{1/n}(r,z)=\vartheta_{1/n}(-r,-z)$ for any $(r,z)\in\mathbb{R}^{N+1}$, it can be shown that $(D_tu_n,\psi)_p=(D_tu,\psi_n)_p$ for any $n\in\mathbb{N}$, where $\psi_n=n^{N+1}\vartheta_{1/n}\star(\vartheta_n\psi)$ and $(\cdot,\cdot)_p$ denotes the duality pairing between $(W^{0,1}_{p'}(\mathbb{R}^{N+1}))'$ and $W^{0,1}_{p'}(\mathbb{R}^{N+1})$. Since ψ_n converges to ψ in $W^{0,1}_{p'}(\mathbb{R}^{N+1})$ as n tends to $+\infty$, it follows that $(D_tu,\psi_n)_p$ converges to $(D_tu,\psi)_p$ as n tends to $+\infty$. The arbitrariness of ψ shows that D_tu_n weakly* converges to D_tu in $(W^{0,1}_{p'}(\mathbb{R}^{N+1}))'$ as n tends to $+\infty$.

Step 2. Let us suppose that $a,b\in\mathbb{R}$. We extend u to $(3a-2b,2b-a)\times\mathbb{R}^N$, by symmetry, first with respect to $t=b$ and then with respect to $t=a$. The so obtained function v belongs to $W^{0,1}_p((3a-2b,2b-a)\times\mathbb{R}^N)\cap C_b((3a-2b,2b-a)\times\mathbb{R}^N)$. Moreover, the distributional derivative D_tv belongs to $(W^{0,1}_{p'}((3a-2b,2b-a)\times\mathbb{R}^N))'$, so that $u\in\mathscr{H}^{p,1}((3a-2b,2b-a)\times\mathbb{R}^N)$. Indeed, fix $\varphi\in C^\infty_c((3a-2b,2b-a)\times\mathbb{R}^N)$ and observe that

$$\int_{(3a-2b,2b-a)\times\mathbb{R}^N} vD_t\varphi dtdx=\int_{(a,b)\times\mathbb{R}^N} uD_t\Phi dtdx, \tag{A.3.1}$$

where $\Phi=\varphi-\varphi(2b-\cdot,\cdot)-\varphi(2a-\cdot,\cdot)+\varphi(2a-2b+\cdot,\cdot)$ belongs to $W^{0,1}_{p'}((a,b)\times\mathbb{R}^N)$. It follows immediately that $(D_tv,\varphi)_p=(D_tu,\Phi)_p$ and $|(D_tv,\varphi)_p|\le C_1\|\Phi\|_{W^{0,1}_{p'}((a,b)\times\mathbb{R}^N)}\le C_2\|\varphi\|_{W^{0,1}_{p'}((a,b)\times\mathbb{R}^N)}$ for some positive constants C_1 and C_2, independent of φ. The density of $C^\infty_c((a,b)\times\mathbb{R}^N)$ in $W^{0,1}_{p'}((3a-2b,2b-a)\times\mathbb{R}^N)$ implies that $D_tv\in(W^{0,1}_{p'}((3a-2b,2b-a)\times\mathbb{R}^N))'$.

Now we fix a function $\zeta\in C^\infty_c((3a-2b,2b-a))$ such that $\zeta\equiv1$ in $[a,b]$. Step 1, applied to the function ζv, which belongs to $\mathscr{H}^{p,1}(\mathbb{R}^{N+1})\cap C_b(\mathbb{R}^{N+1})$, yields the existence of a sequence $\{u_n\}\subset C^\infty_c(\mathbb{R}^{N+1})$, converging to ζv locally uniformly in $\mathbb{R}^{N+1}$ and in $W^{0,1}_p(\mathbb{R}^{N+1})$, and such that D_tu_n converges weakly* in $(W^{0,1}_{p'}(\mathbb{R}^{N+1}))'$ to $D_t(\zeta v)$. Clearly, u_n converges to u locally uniformly in $[a,b]\times\mathbb{R}^N$ and in $W^{0,1}_p((a,b)\times\mathbb{R}^N)$.

Fix $\varphi\in W^{0,1}_{p'}((a,b)\times\mathbb{R}^N)$ and denote by $\overline{\varphi}\in W^{0,1}_{p'}(\mathbb{R}^{N+1})$ the null extension of φ to the whole of $\mathbb{R}^{N+1}$. Since

$$\int_a^b\int_{\mathbb{R}^N} D_tu_n\varphi dtdx=\int_{\mathbb{R}^{N+1}} D_tu_n\overline{\varphi}dtdx,\qquad n\in\mathbb{N},$$

letting n tend to $+\infty$ and taking the formula (A.3.1) into account, we conclude that

$$\begin{aligned}\lim_{n\to+\infty}\int_{Q(a,b)} D_tu_n\varphi dtdx&=(D_t(\zeta v),\overline{\varphi})_p=\int_{Q(a,b)} v\zeta'\overline{\varphi}dtdx+(D_tv,\zeta\overline{\varphi})_p\\&=(D_tv,\overline{\varphi})_p=(D_tu,\varphi)_p.\end{aligned}$$

This completes the proof. □

Next, we recall some embedding theorems for the spaces $\mathscr{H}^{p,1}((a,b)\times\mathbb{R}^N)$. The original proofs are due to N.V. Krylov (see [147]) who proves the above embeddings in a more general situation. The proofs that we present here are taken from [195, Appendix A].

Theorem A.3.2 *The following properties are satisfied.*

(i) *If* $1<p<N+2$ *then* $\mathscr{H}^{p,1}((a,b)\times\mathbb{R}^N)$ *is continuously embedded in* $L^q((a,b)\times\mathbb{R}^N)$, *where* $q=p(N+2)/(N+2-p)$.

(ii) *If* $p=N+2$ *then* $\mathscr{H}^{p,1}((a,b)\times\mathbb{R}^N)$ *is continuously embedded in* $L^q((a,b)\times\mathbb{R}^N)$ *for* $N+2\le r<+\infty$.

(iii) *If $p > N+2$ then $\mathscr{H}^{p,1}((a,b)\times\mathbb{R}^N)$ is continuously embedded in $C_0((a,b)\times\mathbb{R}^N)$.*

Proof Fix $g \in C_c^\infty((a,b)\times\mathbb{R}^N)$. Then the function $v : \mathbb{R}^{N+1} \to \mathbb{R}$, defined by

$$v(t,x) = \int_{\mathbb{R}^{N+1}} G(t-\tau, x-y)g(\tau,y)d\tau dy =: (G\star g)(t,x), \qquad (t,x)\in\mathbb{R}^{N+1},$$

where

$$G(t,x) = \begin{cases} (4\pi t)^{-\frac{N}{2}} \exp\left(-\dfrac{1}{4t}|x|^2\right), & t>0,\\[2mm] 0, & t\le 0,\end{cases}$$

belongs to $C^2(\mathbb{R}^{N+1})$ and solves the equation $D_t v - \Delta v = g$ (see e.g. [147, Theorem 8.4.2]).

We consider the cases $p < N+2$, $p = N+2$ and $p > N+2$ separately. Throughout the rest of the proof, we denote by C a positive constant, independent of the functions that we consider, that may vary from line to line.

(i) Let $p < N+2$ and $q = p(N+2)/(N+2-p)$. Then, we can estimate

$$\|v\|_{W^{2,1}_{q'}((a,b)\times\mathbb{R}^N)} \le C\|g\|_{L^{q'}((a,b)\times\mathbb{R}^N)}.$$

Moreover, by the classical parabolic Sobolev embedding theorems, $W^{2,1}_{q'}((a,b)\times\mathbb{R}^N)$ is continuously embedded in $W^{0,1}_{p'}((a,b)\times\mathbb{R}^N)$ (we refer the reader e.g. to [159, Lemma II.3.3 and Theorem 9.2.3] for further details). Hence, for any $u\in C_c^\infty(\mathbb{R}^{N+1})$ we can estimate

$$\begin{aligned}\left|\int_{(a,b)\times\mathbb{R}^N} ugdtdx\right| &= \left|\int_{(a,b)\times\mathbb{R}^N} u(D_t v - \Delta v)dtdx\right|\\ &= \left|\int_{(a,b)\times\mathbb{R}^N} (uD_t v + \langle\nabla_x u, \nabla_x v\rangle)dtdx\right|\\ &\le C\|u\|_{\mathscr{H}^{p,1}((a,b)\times\mathbb{R}^N)}\|v\|_{W^{1,0}_{p'}((a,b)\times\mathbb{R}^N)}\\ &\le C\|u\|_{\mathscr{H}^{p,1}((a,b)\times\mathbb{R}^N)}\|g\|_{L^{q'}((a,b)\times\mathbb{R}^N)}.\end{aligned}$$

From this estimate we deduce that $\|u\|_{L^q((a,b)\times\mathbb{R}^N)} \le C\|u\|_{\mathscr{H}^{p,1}((a,b)\times\mathbb{R}^N)}$. Since $C_c^\infty(\mathbb{R}^{N+1})$ is dense in $\mathscr{H}^{p,1}((a,b)\times\mathbb{R}^N)$ by Lemma A.3.1, from the previous estimate we get (i) by a straightforward density argument.

(ii) Since the support of g is contained in $(a,b)\times\mathbb{R}^N$, it follows that $v = G_{a,b}\star g$ on $(a,b)\times\mathbb{R}^N$, where $G_{a,b} = G\chi_{[0,b-a]}$. This function belongs to $W^{0,1}_s(\mathbb{R}^{N+1})$ for any $s\in[1,(N+2)/(N+1))$. Hence, from the Young theorem we deduce that $v\in W^{1,0}_r((a,b)\times\mathbb{R}^N)$ and $\|v\|_{W^{1,0}_r((a,b)\times\mathbb{R}^N)} \le C\|g\|_{L^1((a,b)\times\mathbb{R}^N)}$ for any $r\in[1,(N+2)/(N+1))$.

In view of this remark, for any $q\in(N+2,+\infty)$ we fix $r\in(1,(N+2)/(N+1))$ such that $p'^{-1} = r^{-1} - q^{-1}$. The Young theorem yields $\|v\|_{W^{1,0}_{p'}((a,b)\times\mathbb{R}^N)} \le C\|g\|_{L^{q'}((a,b)\times\mathbb{R}^N)}$. Hence,

$$\begin{aligned}\left|\int_{(a,b)\times\mathbb{R}^N} ugdtdx\right| &\le C\|u\|_{\mathscr{H}^{p,1}((a,b)\times\mathbb{R}^N)}\|v\|_{W^{1,0}_{p'}((a,b)\times\mathbb{R}^N)}\\ &\le C\|u\|_{\mathscr{H}^{p,1}((a,b)\times\mathbb{R}^N)}\|g\|_{L^{r'}((a,b)\times\mathbb{R}^N)}\end{aligned}$$

for any $u\in C_c^\infty(\mathbb{R}^{N+1})$. Hence, $\|u\|_{L^q((a,b)\times\mathbb{R}^N)} \le C\|u\|_{\mathscr{H}^{p,1}((a,b)\times\mathbb{R}^N)}$ and, again by density, we get (ii).

(iii) Since $p > N+2$, $p' < (N+2)/(N+1)$ we can estimate

$$\left|\int_{(a,b)\times\mathbb{R}^N} ugdtdx\right| \le C\|u\|_{\mathscr{H}^{p,1}((a,b)\times\mathbb{R}^N)}\|v\|_{W^{1,0}_{p'}((a,b)\times\mathbb{R}^N)}$$

$$\le C\|u\|_{\mathscr{H}^{p,1}((a,b)\times\mathbb{R}^N)}\|g\|_{L^1((a,b)\times\mathbb{R}^N)}$$

and, arguing as above, we conclude that $\mathscr{H}^{p,1}((a,b)\times\mathbb{R}^N)\hookrightarrow C_b([a,b]\times\mathbb{R}^N)$. Finally, we observe that the estimate for $\partial_t u$ implies that u vanishes at $t=a$ and $t=b$. This completes the proof. □

For more details on the spaces $\mathscr{H}^{p,1}((a,b)\times\mathbb{R}^N)$ we refer the reader to [147].

A.4 Some properties of the distance function

In this appendix we collect some regularity results of the distance function $d(x) = \mathrm{dist}(x,\partial\Omega)$, when $\partial\Omega$ is the boundary of an unbounded smooth open subset Ω of $\mathbb{R}^N$.

Definition A.4.1 *We say that an open subset Ω of $\mathbb{R}^N$ is uniformly of class $C^{2+\alpha}$, for some $\alpha\in[0,1)$, if there exist a countable open covering $\{U_n\}$ of $\partial\Omega$ and a family of diffeomorphisms $g_n:\overline{U}_n\to\overline{B}_1$ of class $C^{2+\alpha}$ such that $g_n(U_n\cap\Omega)=\{x\in B_1 : x_N>0\}$, $g_n(U_n\cap\partial\Omega)=\{x\in B_1 : x_N=0\}$ (where x_N denotes the N-th component of the vector x) and*

(i) *there exists $k\in\mathbb{N}$ such that $\bigcap_{n\in J}U_n=\varnothing$ if $\mathrm{card}\,(J)>k$;*

(ii) *there exists $\varepsilon>0$ such that $\{x\in\Omega : d(x,\partial\Omega)<\varepsilon\}\subset\bigcup_{n\in\mathbb{N}}V_n$, where $V_n=g_n^{-1}(B_{1/2})$;*

(iii) *there exists $C>0$ such that $\|g_n\|_{C^{2+\alpha}}+\|g_n^{-1}\|_{C^{2+\alpha}}\le C$ for any $n\in\mathbb{N}$.*

Remark A.4.2 If Ω is uniformly of class $C^{2+\alpha}$ then the diffeomorphism g_n $(n\in\mathbb{N})$ can be chosen so that $(\mathrm{Jac}\,g_n)(x)\nu(x)=-\alpha_n(x)e_N$ for any $x\in U_n\cap\partial\Omega$, $n\in\mathbb{N}$ and some positive function α_n, where $\nu(x)$ denotes the unit outward normal vector to $\partial\Omega$ at x and $e_N=(0,\ldots,0,1)$. Roughly, speaking the diffeomorphisms g_n leave the normal direction unchanged. We refer the reader to [259, Chapter 1, Section 2.4] for the proof of this property.

Remark A.4.3 As a consequence of the condition (iii) in Definition A.4.1, the principal curvatures $\lambda_1,\ldots,\lambda_{N-1}$ of $\partial\Omega$ are bounded both from above and below with respect to x. Hence,

$$\inf_{x\in\partial\Omega}\left\{\frac{\partial\nu}{\partial\tau}(x)\cdot\tau,\ |\tau|=1,\ \tau\cdot\nu(x)=0\right\}>-\infty,$$

since, locally, $\partial\nu/\partial\tau\cdot\tau=\lambda_1\tau_1^2+\cdots+\lambda_{N-1}\tau_{N-1}^2$.

Definition A.4.4 *The open set Ω is said to satisfy the uniform interior sphere condition if, for any point $z_0\in\partial\Omega$, there exists a ball $B_{r_{z_0}}(z_0)\subset\Omega$ such that $\overline{B}_{r_{z_0}}(z_0)\cap\partial\Omega=\{z_0\}$ and $\inf_{z_0\in\Omega}r_{z_0}>0$.*

Proposition A.4.5 ([106], Prop. B.2) *If $\partial\Omega$ is uniformly of class C^2, then it satisfies a uniform interior sphere condition.*

Proof To begin with, we fix $n\in\mathbb{N}$ and observe that, if $y\in V_n$, then $B_{(2C)^{-1}}(y)\subset U_n$, where C is the constant in Definition A.4.1.

Next we fix $x_0 \in \partial\Omega$ and let $n \in \mathbb{N}$ be such that $y_0 \in V_n$. By the above remark, for any $t \in [0,(2C)^{-1})$ the point $x_0 - t\nu(x_0)$ belongs to U_n. Using the Taylor expansion of the N-th component $g_{n,N}$ of the function g_n around the point x_0, we get

$$g_{n,N}(x_0 - t\nu(x_0)) = -t\langle \nabla g_{n,N}(x_0), \nu(x_0)\rangle + r(t),$$

where $|r(t)| \le Ct^2/2$ for any $t \in (0,(2C)^{-1})$. By assumptions $g_{n,N}(x_0)$ vanishes on $U_n \cap \partial\Omega$. Hence, the tangential derivatives of $g_{n,N}$ at x_0 vanish and $\nabla g_{n,N}(x_0) = k\nu(x_0)$ for some constant $k \le -C^{-1}$ (still by Definition A.4.1). It thus follows that $g_{n,N}(x_0 - t\nu(x_0)) \ge tC^{-1} - Ct^{-2}/2$ for any $t \in (0, 2/C^3)$. Hence, if we set $\delta = 2/C^3$, then we conclude that the point $x_0 - t\nu(x_0)$ belongs to Ω for any $t \in (0,\delta)$.

Now, we are almost done. Indeed, if we set $x = x_0 - \nu(x_0)\delta/2$, then the ball $B_{\delta/2}(x)$ is contained in Ω and $x_0 \in \partial B_{\delta/2}(x)$. If x_0 is the unique point which belongs to $\partial\Omega \cap \partial B_{\delta/2}(x)$ we are done, otherwise we replace this ball with the ball centered at $x_0 - \nu(x_0)\delta/4$ with radius $\delta/4$. □

Theorem A.4.6 ([106], Proposition B.3) *Let $\partial\Omega$ be uniformly of class C^2, Further, fix $\delta > 0$ such that at each point $x_0 \in \partial\Omega$ there exists a ball, with radius not less than δ, which satisfies the interior sphere condition at x_0. Further, set $\Omega_\delta = \{y \in \overline{\Omega} \,|\, d(y) < \delta\}$. Then, $d \in C_b^2(\Omega_\delta)$ and $\nabla d(x) = -\nu(y(x))$ for any $x \in \Omega_\delta$, where $y(x) \in \partial\Omega$ is the only point which realizes the distance between x and $\partial\Omega$.*

Proof We limit ourselves to proving that for any $x \in \Omega_\delta$ there exists a unique point $y(x) \in \partial\Omega$ such that $|x - y| = d(x)$. Indeed, the proof of the other properties relies on this fact and the implicit function theorem, and it is completely similar to that given in [116, Section 14.6] in the case when Ω is bounded.

Of course, the existence of a point $y(x)$ which satisfies the equation $|x - y| = d(x)$ is obvious. To prove the uniqueness of the point $y(x)$ for any $x \in \Omega_\delta$, let $y \in \partial\Omega$ be such that $d(x) = |x - y|$. According to Proposition A.4.5, there exists a ball $B = B_\rho(\xi)$ such that $B \subset \Omega$ and $\overline{B} \cap \partial\Omega = \{y\}$. Moreover from the definition of δ it follows that $x \in B$. It is easy to check that x and ξ lie on the normal direction $\nu(y)$ and that the balls $B_{d(x)}(x)$, $B_\rho(\xi)$ are tangent at y. Then, $B_{d(x)}(x)$ still verifies the interior sphere condition at y. Hence, for any $z \in \partial\Omega \setminus \{y\}$, z does not belong to $B_{d(x)}(x)$, so that y is actually the unique point on $\partial\Omega$ such that $|x - y| = d(x)$. □

Appendix B

Basic notions of functional analysis in Banach spaces

In this appendix we collect a few basic results on linear operators, on elementary spectral theory, on Riemann integral for vector-valued functions and on interpolation theory, that we use in this book.

B.1 Linear operators, spectrum and resolvent

For more details and for the proofs of the results that we present here, we refer the reader e.g. to [19, 135, 260].

Proposition B.1.1 ([260], Chapter 10, Section 2) *Let X, Y, Z be three Banach spaces. Then, the following properties are satisfied:*

(*i*) *the set of all the compact linear operators from X to Y is a closed subspace of $L(X, Y)$;*

(*ii*) *for any compact linear operator $L : X \to Y$ and any bounded linear operator $M \in L(Y, Z)$ (resp. $M \in L(Z, X)$), the operator $M \circ L$ (resp. $L \circ M$) is a compact operator.*

A very useful criterion for the convergence of a given sequence of bounded linear operators is provided by the following proposition.

Proposition B.1.2 *Let $\{M_n\}$ be a sequence of bounded linear operators on a Banach space X such that $\sup_{n\in\mathbb{N}} \|M_n\|_{L(X)} < +\infty$. Further, suppose that $M_n x$ converges in X, as n tends to $+\infty$, for any x in a dense subspace of X. Then, M_n converges pointwise in X to a bounded linear operator M which satisfies the estimate $\|M\|_{L(X)} \leq C$.*

Now, we recall some simple properties of the resolvent $\rho(A)$ (which is an open subset of $\mathbb{C}$) and the spectrum $\sigma(A)$ of a linear operator A defined on (a subset of) a Banach space X. First of all, since the inverse of a bounded operator is a closed operator, it is clear that, if $\rho(A) \neq \varnothing$, then A is closed. For any $\lambda \in \rho(A)$ we set $R(\lambda, A) = (\lambda I - A)^{-1} \in L(X)$. The family of operators $\{R(\lambda, A) : \lambda \in \rho(A)\}$ satisfies the so-called *resolvent identity*

$$R(\lambda, A) - R(\mu, A) = (\mu - \lambda)R(\lambda, A)R(\mu, A), \qquad \lambda, \mu \in \rho(A). \tag{B.1.1}$$

Such a formula characterizes the resolvent operators, as specified in the following proposition.

Proposition B.1.3 ([260], Section 8.4) *Let $\Omega \subset \mathbb{C}$ be an open set and let $\{F(\lambda) : \lambda \in \Omega\} \subset L(X)$ be a family of bounded linear operators satisfying the resolvent identity $F(\lambda) - F(\mu) = (\mu - \lambda)F(\lambda)F(\mu)$ for any $\lambda, \mu \in \Omega$. If the operator $F(\lambda_0)$ is invertible for some*

$\lambda_0 \in \Omega$, then there exists a closed linear operator $A : D(A) \subset X \to X$ such that $\rho(A) \supset \Omega$ and $R(\lambda, A) = F(\lambda)$ for any $\lambda \in \Omega$.

Some properties of the resolvent operator are listed in the next proposition.

Proposition B.1.4 *The function $R(\cdot, A)$ is holomorphic in $\rho(A)$. Moreover, the domain of analyticity of the function $\lambda \mapsto R(\lambda, A)$ is $\rho(A)$ and*

$$\|R(\lambda, A)\|_{L(X)} \geq \frac{1}{\text{dist}(\lambda, \sigma(A))}, \qquad \lambda \in \rho(A).$$

As a consequence of Proposition B.1.4 we obtain the following Laurent expansion of $R(\cdot, A)$ in a neighborhood of an isolated point of $\sigma(A)$.

Proposition B.1.5 ([135], Chapter 3, Section 6.5) *Suppose that $\lambda \in \mathbb{C}$ is an isolated point in $\sigma(A)$. Then, there exists $r > 0$ such that*

$$R(\mu, A) = \sum_{n=0}^{+\infty} (\mu - \lambda)^n S^{n+1} + P + \sum_{n=1}^{+\infty} \frac{D^n}{(\mu - \lambda)^{n+1}} \tag{B.1.2}$$

for any $0 < |\mu - \lambda| < r$. Here, P is the projection defined by

$$P = \frac{1}{2\pi i} \int_\gamma R(\xi, A) d\xi, \tag{B.1.3}$$

where γ is the boundary of any ball $B_r(\lambda)$ (oriented counterclockwise), such that $\overline{B}_r(\lambda) \cap \sigma(A) = \{\lambda\}$. The operators D and S are defined, respectively, by $D = (\lambda I - A)P$ and $S = \lim_{\xi \to \lambda} R(\xi, A)(I - P)$. In particular, λ is the unique element in the spectrum of the restriction of the operator A to $P(X)$, whereas the spectrum of the restriction of A to $(I - P)(X)$ is $\sigma(A) \setminus \{\lambda\}$. Finally, $X = P(X) \oplus (I - P)(X)$.

With the help of the formula (B.1.3) the following result can be proved.

Proposition B.1.6 ([19], Prop. A-III.2.5(iii)) *Let $A : D(A) \subset X \to X$ be a densely defined linear operator with non-empty resolvent set. Then, λ_* is an isolated element in $\sigma(A)$ if and only if $(\lambda_0 - \lambda_*)^{-1}$ is an isolated element in the spectrum of $R(\lambda_0, A)$ for any $\lambda_0 \in \rho(A)$. In this case the residue (resp. the order of the pole) of $R(\cdot, A)$ at λ_* coincides with the residue (resp. the order of the pole) of $R(\cdot, R(\lambda_0))$ at $(\lambda_0 - \lambda_*)^{-1}$.*

The next theorem provides an estimate from above of the norm of the resolvent set of a densely defined linear operator $A : D(A) \subset X \to X$, in terms of its numerical range $r(A) = \{\langle x', Ax\rangle : x \in D(A),\ \|x\| = 1,\ x' \in X',\ \|x'\| = 1,\ \langle x', x\rangle = 1\}$.

Theorem B.1.7 ([213], Chapter 1, Theorem 3.9) *Let $A : D(A) \subset X \to X$ be a densely defined closed linear operator in X. If $\lambda \in \mathbb{C} \setminus \overline{r(A)}$, then the operator $\lambda I - A$ is one to one and has closed range. Moreover, if Σ is a component of $\mathbb{C} \setminus \overline{r(A)}$ which intersects $\rho(A)$, then the spectrum of A is contained in $\mathbb{C} \setminus \Sigma$ and $\|R(\lambda, A)\|_{L(X)} \leq (\text{dist}(\lambda, \overline{r(A)}))^{-1}$ for any $\lambda \in \mathbb{C} \setminus \Sigma$.*

In the case when $R(\lambda, A)$ is a compact operator for some (and, hence, any) $\lambda \in \rho(A)$, $\sigma(A)$ has a very easy structure, as the following proposition shows.

Proposition B.1.8 *Let $A : X \to X$ be a closed operator with $\sigma(A) \neq \varnothing$ and compact resolvent operator. Then, the following properties are satisfied:*

(i) *$\sigma(A)$ consists of an at most countable set of points. Any $\lambda \in \sigma(A)$ is an eigenvalue of A and the corresponding eigenspace is finite dimensional;*

(ii) *for any $\lambda \in \sigma(A)$ the vector space of the generalized eigenvectors*[1] *has finite dimension;*

(iii) *any $\lambda \in \sigma(A)$ is a pole of the function $R(\cdot, A)$ whose order coincides with the dimension of the vector space of all the generalized eigenvectors associated with the eigenvalue λ. Finally, the order of the pole coincides also with the index*[2] *$i_\lambda(A)$ of the eigenvalue λ.*

B.2 Vector-valued Riemann integral

In this section we define the Riemann integral for vector-valued functions. For more details and for the proof of the results that we present here, we refer the reader to [88, Chapter 2], [91, Chapter 3] and [124, Chapter 3]. Throughout the section, X denotes a Banach space.

Definition B.2.1 *A bounded function $f : [a, b] \to X$ $(-\infty < a < b < +\infty)$ is said to be integrable on $[a, b]$ if there exists $x \in X$ with the following property: for any $\varepsilon > 0$ there exists $\delta > 0$ such that, for any partition $\{a = t_0 < t_1 < \ldots < t_n = b\}$ of $[a, b]$, with $\max_{i=1,\ldots,n}(t_i - t_{i-1}) < \delta$, and any choice of the points $\xi_i \in [t_{i-1}, t_i]$, it holds that*

$$\left\| x - \sum_{i=1}^{n} f(\xi_i)(t_i - t_{i-1}) \right\| < \varepsilon.$$

In this case we define

$$\int_a^b f(t)dt = x.$$

Arguing as in the real-valued case, one can check that the set of functions which are integrable in $[a, b]$ is a vector space and that the integral over $[a, b]$ is a linear operator in this vector space. In particular, if $f : [a, b] \to X$ is continuous, then it is integrable. Moreover, if f is integrable in $[a, b]$, then the map $t \mapsto \|f(t)\|$ is integrable in $[a, b]$ as well. Finally, if f is integrable in $[a, b]$, then it is integrable in any $[c, d] \subset [a, b]$ and

$$\int_a^b f(t)dt = \int_a^c f(t)dt + \int_c^d f(t)dt,$$

for any $c \in (a, b)$.

As in the real-valued case, the definition of Riemann integral can be easily extended to unbounded intervals or unbounded functions.

Definition B.2.2 *Let $J \subset \mathbb{R}$ be an interval with endpoints a and b $(-\infty \le a < b \le +\infty)$ with a and b not necessarily in J. Moreover, let $f : J \to X$ be Riemann integrable in $[c, d]$ for any $a < c < d < b$. We say that the function f admits an improper integral in J if the limits*

$$\lim_{c \to a^+} \int_c^{t_0} f(t)dt, \qquad \lim_{d \to b^-} \int_{t_0}^d f(t)dt$$

[1] i.e., the vector space of all $x \in X$ such that $(\lambda I - A)^k x = 0$ for some $k \in \mathbb{N}$.

[2] i.e., with the smallest integer k such that $\mathrm{Ker}((\lambda I - A)^k) = \mathrm{Ker}((\lambda I - A)^{k+1})$.

exist in X for some (equivalently, for any) $t_0 \in J$. In this case we set

$$\int_J f(x)dx = \lim_{c\to a^+}\int_c^{t_0} f(t)dt + \lim_{d\to b^-}\int_{t_0}^{d} f(t)dt.$$

Note that the previous definition is independent of t_0.

We now recall the definition of the integral of vector-valued functions of a complex variable, along a smooth curve γ.

Definition B.2.3 *Let Ω be an open subset of $\mathbb{C}$, $f:\Omega\to X$ be a continuous function and $\gamma:[a,b]\to\Omega$ be a piecewise C^1-curve. The integral of f along γ is defined as follows:*

$$\int_\gamma f(z)dz = \int_a^b f(\gamma(t))\gamma'(t)dt.$$

As in the case of vector-valued functions defined in a real interval, we can introduce the *improper complex integral* in an obvious way.

Definition B.2.4 *Let $\Omega\subset\mathbb{C}$ be a (possibly) unbounded open set. Moreover, let $J=(a,b)$ be a (possibly unbounded) interval and $\gamma:J\to\mathbb{C}$ be a (piecewise) C^1 curve in Ω. We say that f admits an improper integral along γ if for any $t_0\in(a,b)$ the limits*

$$\lim_{s\to a^+}\int_s^{t_0} f(\gamma(\tau))\gamma'(\tau)d\tau \qquad \text{and} \qquad \lim_{s\to b^-}\int_{t_0}^{b} f(\gamma(\tau))\gamma'(\tau)d\tau$$

exist in X. In such a case, we set

$$\int_\gamma f(z)dz = \lim_{s\to a^+}\int_s^{t_0} f(\gamma(\tau))\gamma'(\tau)d\tau + \lim_{s\to b^-}\int_{t_0}^{s} f(\gamma(\tau))\gamma'(\tau)d\tau.$$

Note that the definition of the improper integral is independent of the choice of t_0. Moreover, if J is closed and bounded, and the integral of f along γ exists, then f admits an improper integral along γ and the two integrals coincide.

B.3 Some results from interpolation theory

In this section we recall some results from the interpolation theory. If not otherwise specified, for the proofs we refer the reader to [178, 250].

Definition B.3.1 *Let X, Y be two Banach spaces with $Y\hookrightarrow X$. Moreover, let $K:(0,+\infty)\times X\to\mathbb{R}$ be the function defined by $K(\xi,x)=\inf_{x=a+b}(\|a\|_X+\xi\|b\|_Y)$ for any $\xi>0$ and $x\in X$. For any $\theta\in[0,1]$ the interpolation space $(X,Y)_{\theta,\infty}$ is defined by*

$$(X,Y)_{\theta,\infty}=\left\{x\in X:\ \sup_{0<\xi<1}\xi^{-\theta}K(\xi,x)<+\infty\right\}$$

and it is normed by

$$\|x\|_{\theta,\infty}:=\sup_{0<\xi<1}\xi^{-\theta}K(\xi,x),\qquad x\in(X,Y)_{\theta,\infty}. \tag{B.3.1}$$

For any $\theta \in [0,1]$ $(X,Y)_{\theta,\infty}$ is a Banach space when it is endowed with the norm in (B.3.1) and

$$Y \hookrightarrow (X,Y)_{\theta,\infty} \hookrightarrow X. \tag{B.3.2}$$

The topological inclusions in (B.3.2) and the following proposition explain the reason why $(X,Y)_{\theta,\infty}$ are called interpolation spaces.

Proposition B.3.2 *Let X_1, X_2, Y_1, Y_2 be four Banach spaces with $Y_j \hookrightarrow X_j$ for $j = 1,2$. Moreover, let T belong to $L(X_1,X_2) \cap L(Y_1,Y_2)$. Then, for any $\theta \in (0,1)$, T belongs to $L((X_1,Y_1)_{\theta,\infty},(X_2,Y_2)_{\theta,\infty})$. Moreover,*

$$\|T\|_{L((X_1,Y_1)_{\theta,\infty},(X_2,Y_2)_{\theta,\infty})} \leq \|T\|_{L(X_1,X_2)}^{1-\theta}\|T\|_{L(Y_1,Y_2)}^{\theta}.$$

Now, we introduce the definition of spaces of class J_θ and K_θ ($\theta \in [0,1]$) which play a central role in the interpolation theory.

Definition B.3.3 *Let X,Y,Z be three Banach spaces such that $Y \hookrightarrow Z \hookrightarrow X$. Moreover, let $\theta \in [0,1]$. We say that Z belongs to the class J_θ between X and Y (in short $Z \in J_\theta(X,Y)$) if there exists a positive constant $C = C(\theta)$ such that $\|x\|_Z \leq C\|x\|_X^{1-\theta}\|x\|_Y^{\theta}$ for any $x \in Z$. Similarly, we say that Z belongs to the class K_θ between X and Y (in short $Z \in K_\theta(X,Y)$) if $Z \hookrightarrow (X,Y)_{\theta,\infty}$.*

Abstract and concrete examples of spaces of class J_θ are given in the following two propositions.

Proposition B.3.4 *Let $0 < \theta < \alpha$ and let $\Omega \subset \mathbb{R}^N$ ($N \geq 1$) be an open set with boundary uniformly of class C^α (possibly $\Omega = \mathbb{R}^N$). Then, $C_b^\theta(\overline{\Omega})$ belongs to the class $J_{\theta/\alpha}$ between $C_b(\overline{\Omega})$ and $C_b^\alpha(\overline{\Omega})$.*

Proposition B.3.5 *For any $\theta \in [0,1]$ and any pair of Banach spaces X and Y, with $Y \hookrightarrow X$, $(X,Y)_{\theta,\infty}$ belongs to both the classes $J_\theta(X,Y)$ and $K_\theta(X,Y)$.*

The next proposition shows interesting properties of the spaces of class J_α.

Proposition B.3.6 *Let $Y \hookrightarrow Z \hookrightarrow X$ be three Banach spaces such that $Z \in J_\alpha(X,Y)$ for some $\alpha \in (0,1)$. Further, let $I \subset \mathbb{R}$ be an interval. Then, the following properties are satisfied:*

(i) *if $u \in B(I;Y) \cap C_b^\theta(I;X)$ for some $\theta \in [0,1)$, then u belongs to $C_b^{\theta(1-\alpha)}(I;Z)$. Moreover, there exists a positive constant C such that $\|u\|_{C_b^{\theta(1-\alpha)}(I;Z)} \leq C\|u\|_{C_b^\theta(I;X)}^{1-\alpha}\|u\|_{B(I;Y)}^{\alpha}$;*

(ii) *if $u \in B(I;Y) \cap \mathrm{Lip}(I;X)$, then u belongs to $C_b^{1-\alpha}(I;Z)$. Moreover, there exists a positive constant C such that $\|u\|_{C_b^{1-\alpha}(I;Z)} \leq C\|u\|_{\mathrm{Lip}(I;X)}^{1-\alpha}\|u\|_{B(I;Y)}^{\alpha}$.*

We now state one of the most useful tools from interpolation theory, the so-called *reiteration theorem.*

Theorem B.3.7 *Let $0 \leq \theta_1 < \theta_2 \leq 1$. Suppose that Z_j belongs to $K_{\theta_j}(X,Y)$ ($j = 1,2$). Then, for any $\theta \in (0,1)$, it holds that $(Z_1,Z_2)_{\theta,\infty} \hookrightarrow (X,Y)_{(1-\theta)\theta_1+\theta\theta_2,\infty}$. Similarly, if $Z_j \in J_{\theta_j}(X,Y)$ ($j = 1,2$), then, for any $\theta \in (0,1)$, it holds that $(X,Y)_{(1-\theta)\theta_1+\theta\theta_2,\infty} \hookrightarrow (Z_1,Z_2)_{\theta,\infty}$. In particular, if $Z_j \in J_{\theta_j}(X,Y) \cap K_{\theta_j}(X,Y)$ ($j = 1,2$), then it holds that $(X,Y)_{(1-\theta)\theta_1+\theta\theta_2,\infty} = (Z_1,Z_2)_{\theta,\infty}$, with equivalence of the corresponding norms.*

Using the reiteration theorem some spaces of Hölder continuous functions can be characterized as interpolation spaces.

Theorem B.3.8 *For any $0 \le \alpha < \beta$ and $\theta \in (0,1)$ it holds that*

$$(C_b^{\alpha}(\mathbb{R}^N), C_b^{\beta}(\mathbb{R}^N))_{\theta,\infty} = C_b^{\alpha+\theta(\beta-\alpha)}(\mathbb{R}^N), \tag{B.3.3}$$

with equivalence of the corresponding norms. Moreover,

$$(BUC(\mathbb{R}^N), C_b^{\beta}(\mathbb{R}^N))_{\theta,\infty} = C_b^{\theta\beta}(\mathbb{R}^N),$$

with equivalence of the corresponding norms.

Proof For the proof of (B.3.3) see [178, Theorem 1.2.17 and Corollary 1.2.18]. The last part of the theorem follows from the previous one through the reiteration theorem, recalling that $BUC(\mathbb{R}^N)$ belongs to $J_0(C_b(\mathbb{R}^N), C_b^{\beta}(\mathbb{R}^N)) \cap K_0(C_b(\mathbb{R}^N), C_b^{\beta}(\mathbb{R}^N))$. □

Taking advantage of the previous results we can now prove the following characterization of the anisotropic spaces $\mathscr{C}_b^{\alpha}(\mathbb{R}^N)$ and $\widetilde{\mathscr{C}}_b^{\alpha}(\mathbb{R}^N)$ introduced in Subsection A.1.2.

Proposition B.3.9 *Fix $m \in \mathbb{N}$, $\beta \in [0,m)$, such that $\beta/(2j+1) \notin \mathbb{N}$ for any $j = 0, \dots, r$, and $\theta \in (0,1)$. Then, it holds that $(\mathscr{C}_b^{\beta}(\mathbb{R}^N), \mathscr{C}_b^{m}(\mathbb{R}^N))_{\theta,\infty} = \widetilde{\mathscr{C}}_b^{(1-\theta)\beta+m\theta}(\mathbb{R}^N)$ with equivalence of the corresponding norms. Here, $\mathscr{C}_b^{0}(\mathbb{R}^N) = C_b(\mathbb{R}^N)$.*

Proof We first prove the assertion in the case when $\beta = 0$. For this purpose, we recall that, by [181, Theorem 2.2], $(BUC(\mathbb{R}^N), \widetilde{\mathscr{C}}_b^{\gamma}(\mathbb{R}^N))_{\theta,\infty} = \widetilde{\mathscr{C}}_b^{\gamma\theta}(\mathbb{R}^N)$ for any $\gamma > 0$ and $\theta \in (0,1)$, with equivalence of the corresponding norms. Since $BUC(\mathbb{R}^N)$ belongs to both the classes $J_0(C_b(\mathbb{R}^N), \widetilde{\mathscr{C}}_b^{\gamma}(\mathbb{R}^N))$ and $K_0(C_b(\mathbb{R}^N), \widetilde{\mathscr{C}}_b^{\gamma}(\mathbb{R}^N))$, Theorem B.3.7 implies that

$$(C_b(\mathbb{R}^N), \widetilde{\mathscr{C}}_b^{\gamma}(\mathbb{R}^N))_{\theta,\infty} = \widetilde{\mathscr{C}}_b^{\gamma\theta}(\mathbb{R}^N), \tag{B.3.4}$$

with equivalence of the corresponding norms.

Fix $\gamma \in \mathbb{R} \setminus \mathbb{Q}$ such that $\gamma > m$. Then, Remark A.1.1 implies that $\mathscr{C}_b^{\gamma}(\mathbb{R}^N) = \widetilde{\mathscr{C}}_b^{\gamma}(\mathbb{R}^N)$. Therefore, the formula (B.3.4) with $\theta = m/\gamma$ yields the equality $(C_b(\mathbb{R}^N), \mathscr{C}_b^{\gamma}(\mathbb{R}^N))_{m/\gamma,\infty} = \widetilde{\mathscr{C}}_b^{m}(\mathbb{R}^N)$ with equivalence of the corresponding norms. Since $\mathscr{C}_b^{m}(\mathbb{R}^N) \hookrightarrow \widetilde{\mathscr{C}}_b^{m}(\mathbb{R}^N)$, $\mathscr{C}_b^{m}(\mathbb{R}^N)$ is continuously embedded in $(C_b(\mathbb{R}^N), \mathscr{C}_b^{\gamma}(\mathbb{R}^N))_{m/\gamma,\infty}$, or, equivalently, $\mathscr{C}_b^{m}(\mathbb{R}^N)$ belongs to the class $K_{m/\gamma}(C_b(\mathbb{R}^N), \mathscr{C}_b^{\gamma}(\mathbb{R}^N))$.

Let us prove that $\mathscr{C}_b^{m}(\mathbb{R}^N)$ belongs also to the class $J_{m/\gamma}(C_b(\mathbb{R}^N), \mathscr{C}_b^{\gamma}(\mathbb{R}^N))$. We fix $f \in \mathscr{C}_b^{\gamma}(\mathbb{R}^N)$ and apply estimate (B.3.3), with $\alpha = 0$, $\beta = \gamma/(2j+1)$ $(0 \le j \le r)$ and $\theta = m/\gamma$, to the function $\psi = f(x_0, \dots, x_{j-1}, \cdot, x_{j+1}, \dots, x_r)$ (where we have split $x \in \mathbb{R}^N$ as $x = (x_0, \dots, x_r)$, with $x_i \in \mathbb{R}^{p_i}$ $(i = 0, \dots, r)$). Taking the supremum when $(x_0, \dots, x_{j-1}, x_{j+1}, \dots, x_r)$ runs over $\mathbb{R}^{N-p_j}$ and summing with respect to j from 0 to r, we get $\|f\|_{\mathscr{C}_b^{m}(\mathbb{R}^N)} \le C\|f\|_{\infty}^{1-m/\gamma}\|f\|_{\mathscr{C}_b^{\gamma}(\mathbb{R}^N)}^{m/\gamma}$, i.e., $\mathscr{C}_b^{m}(\mathbb{R}^N) \in J_{m/\gamma}(C_b(\mathbb{R}^N), \mathscr{C}_b^{\gamma}(\mathbb{R}^N))$. Since $C_b(\mathbb{R}^N)$ belongs to both the classes $J_0(C_b(\mathbb{R}^N), \mathscr{C}_b^{\gamma}(\mathbb{R}^N))$ and $K_0(C_b(\mathbb{R}^N), \mathscr{C}_b^{\gamma}(\mathbb{R}^N))$, Theorem B.3.7 shows that $(C_b(\mathbb{R}^N), \mathscr{C}_b^{m}(\mathbb{R}^N))_{\theta,\infty} = (C_b(\mathbb{R}^N), \mathscr{C}_b^{\gamma}(\mathbb{R}^N))_{\theta m/\gamma,\infty}$ (with equivalence of the corresponding norms) that, combined with the formula (B.3.4), yields the assertion with $\beta = 0$.

The general case when $\beta \in (0,m)$ is such that $\beta/(2j+1) \notin \mathbb{N}$ for any $j = 0, \dots, r$ now follows from Theorem B.3.7. Indeed,

$$\begin{aligned}(\mathscr{C}_b^{\beta}(\mathbb{R}^N), \mathscr{C}_b^{m}(\mathbb{R}^N))_{\theta,\infty} &= ((C_b(\mathbb{R}^N), \mathscr{C}_b^{m}(\mathbb{R}^N))_{\beta/m,\infty}, \mathscr{C}_b^{m}(\mathbb{R}^N))_{\theta,\infty}\\ &= (C_b(\mathbb{R}^N), \mathscr{C}_b^{m}(\mathbb{R}^N))_{(1-\theta)\beta/m+\theta,\infty}.\end{aligned}$$

The proof is complete. □

To conclude this section we state the Riesz-Thorin interpolation theorem and a remarkable consequence. For the proof of the Riesz-Thorin interpolation theorem we refer the reader e.g. to [227, Theorem IX.17].

Theorem B.3.10 *Let $(\Omega_1, \mathcal{F}_1, \mu_1)$, $(\Omega_2, \mathcal{F}_2, \mu_2)$ be measure spaces, with μ_1 and μ_2 being σ-finite measures. Moreover, let $T \in L(L^{p_0}(\Omega_1, \mu_1), L^{q_0}(\Omega_2, \mu_2)) \cap L(L^{p_1}(\Omega_1, \mu_1), L^{q_1}(\Omega_2, \mu_2))$ for some $p_0, p_1, q_0, q_1 \in [1, +\infty]$. Then, T belongs to $L(L^{p_\theta}(\Omega_1, \mu_1), L^{q_\theta}(\Omega_2, \mu_2))$ for any $\theta \in (0, 1)$, where $p_\theta^{-1} = (1-\theta)p_0^{-1} + \theta p_1^{-1}$ and $q_\theta^{-1} = (1-\theta)q_0^{-1} + \theta q_1^{-1}$. Moreover,*

$$\|T\|_{L(L^{p_\theta}(\Omega_1,\mu_1),L^{q_\theta}(\Omega_2,\mu_2))} \le \|T\|^{1-\theta}_{L(L^{p_0}(\Omega_1,\mu_1),L^{q_0}(\Omega_2,\mu_2))} \|T\|^{\theta}_{L(L^{p_1}(\Omega_1,\mu_1),L^{q_1}(\Omega_2,\mu_2))}.$$

Theorem B.3.11 *Let $(\Omega_1, \mathcal{F}_1, \mu_1)$ and $(\Omega_2, \mathcal{F}_2, \mu_2)$ be two measure spaces, with the measures μ_1 and μ_2 being σ-finite. Further, suppose that $T \in L(L^{p_0}(\Omega, \mu), L^{q_0}(\Omega, \mu)) \cap L(L^{p_1}(\Omega, \mu), L^{q_1}(\Omega, \mu))$ for some $p_0, p_1, q_0, q_1 \in [1, +\infty]$. If T is compact from $L^{p_0}(\Omega, \mu)$ into $L^{q_0}(\Omega, \mu)$ (resp. from $L^{p_1}(\Omega, \mu)$ into $L^{q_1}(\Omega, \mu)$), then T is compact from $L^{p_\theta}(\Omega, \mu)$ into $L^{q_\theta}(\Omega, \mu)$ for any $\theta \in (0, 1)$.*

Proof To fix the ideas, let us suppose that the operator T is compact from $L^{p_0}(\Omega, \mu)$ into $L^{q_0}(\Omega, \mu)$ and denote by $\mathcal{B}$ the unit ball centered at 0 in $L^{p_0}(\Omega, \mu)$. Then, the set $T(\mathcal{B})$ is totally bounded in $L^{q_0}(\Omega, \mu)$. Since the simple functions are dense in $L^{q_0}(\Omega, \mu)$, for any $\varepsilon > 0$ we can determine $n \in \mathbb{N}$ and simple functions $f_1, \ldots, f_n$ such that

$$\overline{T(\mathcal{B})} \subset \bigcup_{i=1}^{n} \{g \in L^{q_0}(\Omega, \mu) : \|g - f_i\|_{L^{q_0}(\Omega,\mu)} < \varepsilon\}.$$

For any $i = 1, \ldots, n$, let us write $f_i = \sum_{j=1}^{M} c_{ij}\chi_{A_j}$, where A_j $(j = 1, \ldots, M)$ are pairwise disjoint measurable sets with positive measures. Let us set $E = \text{span}\{\chi_{A_j} : j = 1, \ldots, M\}$ and, for any $1 \le p \le +\infty$, let us consider the operator $P_\varepsilon : L^p(\Omega, \mu) \to E$, defined by

$$P_\varepsilon f = \sum_{j=1}^{M} \left(\frac{1}{\mu(A_j)} \int_{A_j} f d\mu\right) \chi_{A_j}, \qquad f \in L^p(\Omega, \mu).$$

Using the Jensen inequality, it follows easily that $P_\varepsilon \in L(L^p(\Omega, \mu))$ and its norm does not exceed one. Since P_ε has a finite dimensional rank, it is a compact operator which leaves E invariant.

Fix $f \in \mathcal{B}$ and let $i \in \{1, \ldots, n\}$ be such that $\|Tf - f_i\|_{L^{q_0}(\Omega,\mu)} < \varepsilon$. Then, we can estimate

$$\begin{aligned}\|Tf - P_\varepsilon Tf\|_{L^{q_0}(\Omega,\mu)} \le& \|Tf - f_i\|_{L^{q_0}(\Omega,\mu)} + \|f_i - P_\varepsilon Tf\|_{L^{q_0}(\Omega,\mu)} \\ =& \|Tf - f_i\|_{L^{q_0}(\Omega,\mu)} + \|P_\varepsilon f_i - P_\varepsilon Tf\|_{L^{q_0}(\Omega,\mu)} \\ \le& 2\|Tf - f_i\|_{L^{q_0}(\Omega,\mu)} < 2\varepsilon.\end{aligned}$$

Hence, $\|T - P_\varepsilon T\|_{L(L^{p_0}(\Omega,\mu),L^{q_0}(\Omega,\mu))} \le 2\varepsilon$. On the other hand $\|T - P_\varepsilon T\|_{L(L^{p_1}(\Omega,\mu),L^{q_1}(\Omega,\mu))} \le 2\|T\|_{L(L^{p_1}(\Omega,\mu),L^{q_1}(\Omega,\mu))}$. Applying the Riesz-Thorin interpolation theorem to the operator $T - P_\varepsilon T$, we conclude that

$$\|T - P_\varepsilon T\|_{L(L^{p_\theta}(\Omega,\mu),L^{q_\theta}(\Omega,\mu))} \le 2\varepsilon^{1-\theta} \|T\|^{\theta}_{L(L^{p_1}(\Omega,\mu),L^{q_1}(\Omega,\mu))}$$

for any $\theta \in (0, 1)$. The arbitrariness of $\varepsilon > 0$ shows that we can approximate $T : L^{p_\theta}(\Omega, \mu) \to L^{q_\theta}(\Omega, \mu)$ with operators with finite dimensional rank. Hence, $T : L^{p_\theta}(\Omega, \mu) \to L^{q_\theta}(\Omega, \mu)$ is compact for any $\theta \in (0, 1)$. □

Appendix C

An overview of strongly continuous and analytic semigroups

In this appendix we recall the definitions and the main properties of strongly continuous and analytic semigroups that we use in this book. If not otherwise specified, we refer the reader to [95, 124, 178, 213] for the proofs of the results here presented.

To begin with, we give the definition of semigroup of bounded linear operators. Throughout this appendix, if not otherwise specified, X denotes a Banach space.

Definition C.0.1 *A one-parameter family $\{T(t) : t \geq 0\}$ of linear bounded operators in X is said to be a semigroup of bounded linear operators if $T(t+s) = T(t) \circ T(s)$ for any $s, t \geq 0$.*[1]

We simply write $T(t)T(s)$ for $T(t) \circ T(s)$ when no confusion may arise. Moreover, we denote the semigroup simply by $\{T(t)\}$.

C.1 Strongly continuous semigroups

Definition C.1.1 *A semigroup $\{T(t)\}$ of bounded linear operators defined in X is strongly continuous if $T(t)x$ converges to x as t tends to 0 from the right, for any $x \in X$.*

Using the semigroup rule, one can see that, if $\{T(t)\}$ is a strongly continuous semigroup, then the function $t \mapsto T(t)x$ is continuous in $[0, +\infty)$ for any $x \in X$. Moreover, there exist two constants $M \geq 1$ and $\omega \in \mathbb{R}$ such that

$$\|T(t)\|_{L(X)} \leq Me^{\omega t}, \qquad t > 0. \tag{C.1.1}$$

When $\omega = 0$ and $M = 1$, the semigroup $\{T(t)\}$ is called a *strongly continuous semigroup of contractions in X*.

For any strongly continuous semigroup we define its *growth bound* as the infimum of the set of all $\omega \in \mathbb{R}$ such that estimate (C.1.1) holds true for some $M \geq 1$.

The growth bound can also be characterized in terms of the *spectral radius* of each operator $T(t)$, where the spectral radius of a bounded operator S is the radius of the smallest ball centered at the origin which contains the spectrum of S.

Proposition C.1.2 *Let $\{T(t)\}$ be a strongly continuous semigroup on X. Then, for any $t > 0$ the spectral radius of the operator $T(t)$ coincides with $e^{t\omega_0}$.*

Remark C.1.3 If $\{T(t)\}$ is a strongly continuous semigroup, then the map $(t, x) \mapsto T(t)x$

[1] This condition is usually referred to as the semigroup rule.

is locally uniformly continuous in $[0, +\infty) \times X$. See e.g. [95, Chapter 1, Lemma 5.2]. As a consequence, for any compact set $K \subset X$ and any $T > 0$, the map $t \mapsto T(t)x$ is continuous in $[0, T]$, uniformly with respect to $x \in K$.

Now we give the definition of the infinitesimal generator of a strongly continuous semigroup.

Definition C.1.4 *Let $\{T(t)\}$ be a strongly continuous semigroup in X. Its infinitesimal generator is the operator $A : D(A) \subset X \to X$ defined by*

$$D(A) = \left\{ x \in X : \exists \lim_{t\to 0^+} \frac{T(t)x - x}{t} \in X \right\}, \qquad Ax = \lim_{t\to 0^+} \frac{T(t)x - x}{t}.$$

It is easy to see that the function $t \mapsto T(t)x$ is continuously Fréchet differentiable in $[0, +\infty)$ with values in X if and only if $x \in D(A)$. In this case $D_t T(t)x = AT(t)x = T(t)Ax$ for any $t \geq 0$.

As the following proposition shows, the infinitesimal generator completely characterizes the strongly continuous semigroup.

Proposition C.1.5 *Let $\{S(t)\}$ and $\{T(t)\}$ be two strongly continuous semigroups in X and denote by A and B their infinitesimal generators. If $A = B$, then $S(t) \equiv T(t)$ for any $t > 0$.*

The Hille-Yosida theorem is a keystone in the theory of strongly continuous semigroups since it allows us to give a complete characterization of the operators which are the infinitesimal generators of strongly continuous semigroups.

Theorem C.1.6 (Hille-Yosida) *A linear operator $A : D(A) \subset X \to X$ is the infinitesimal generator of a strongly continuous semigroup $\{T(t)\}$, satisfying estimate* (C.1.1) *for some $M \geq 1$ and $\omega \in \mathbb{R}$, if and only if*

(i) *A is closed and its domain is dense in X;*

(ii) *the resolvent set $\rho(A)$ contains the halfline $(\omega, +\infty)$ and the estimate $\|R(\lambda, A)^n\|_{L(X)} \leq M(\lambda - \omega)^{-n}$ holds true for any $\lambda > \omega$ and $n \in \mathbb{N}$.*

The proof of the Hille-Yosida theorem shows a connection between the resolvent operator and the semigroup. In fact, the resolvent operator is the Laplace transform of the semigroup, i.e.,

$$R(\lambda, A)x = \int_0^{+\infty} e^{-\lambda t} T(t)x dt, \qquad \mathrm{Re}\,\lambda > \omega, \quad x \in X. \tag{C.1.2}$$

Hence, many properties of $R(\lambda, A)$ can be deduced from the corresponding properties of the semigroup. Moreover, as the following theorem shows, the semigroup $\{T(t)\}$ can be obtained through the resolvent operator, using the so-called exponential formula.

Theorem C.1.7 *Let $\{T(t)\}$ be a strongly continuous semigroup in X and let A be its infinitesimal generator. Then,*

$$T(t)x = \lim_{n\to+\infty} \left[\frac{n}{t} R\left(\frac{n}{t}, A\right)\right]^n x$$

for any $x \in X$ and the limit is uniform with respect to t varying in bounded intervals.

Another important result in the theory of strongly continuous semigroups is the Lumer-Phillips theorem. It provides an alternative characterization of the infinitesimal generators of strongly continuous semigroups of contractions.

Definition C.1.8 *A linear operator $A : D(A) \subset X \to X$ is called dissipative if the estimate $\|\lambda x - Ax\|_X \geq \lambda \|x\|_X$ is satisfied by any $\lambda > 0$ and $x \in D(A)$.*

Theorem C.1.9 (Lumer-Phillips) *Let $A : D(A) \subset X \to X$ be a linear operator with dense domain. Then, the following properties are satisfied:*

(i) *if A is dissipative and there exists $\lambda_0 > 0$ such that the range of the operator $\lambda_0 I - A$ is X, then A is the infinitesimal generator of a strongly continuous semigroup of contractions;*

(ii) *if A is the infinitesimal generator of a strongly continuous semigroup of contractions in X, then the range of the operator $\lambda I - A$ is X for any $\lambda > 0$. Moreover, A is dissipative.*

The following proposition is a straightforward consequence of the Lumer-Phillips theorem.

Proposition C.1.10 *Let $A : D(A) \subset X \to X$ be a dissipative operator. Then, the following properties are satisfied:*

(i) *if A is closable then the closure $\overline{A}$ is also dissipative;*

(ii) *if $D(A)$ is dense in X and $(\lambda I - A)(D(A))$ is dense in X for some (and hence all $\lambda > 0$), then the closure $\overline{A}$ is the infinitesimal generator of a strongly continuous semigroup of contractions.*

For strongly continuous semigroups the spectral mapping theorem fails to hold in general. Only the the inclusion $e^{t\sigma(A)} \subset \sigma(T(t))$ can be guaranteed for any $t \geq 0$. A spectral mapping theorem can be stated for the point spectrum as the following proposition shows.

Definition C.1.11 *Let $S : D(S) \subset X \to X$ be a closed operator. The point spectrum $P\sigma(S)$ of S is the set of all $\lambda \in \mathbb{C}$ such that the map $\lambda I - S$ is not injective.*

Proposition C.1.12 *Let $\{T(t)\}$ be a strongly continuous semigroup of X with infinitesimal generator A. Then,*

$$P\sigma(T(t)) \setminus \{0\} = e^{tP(\sigma(A))}.$$

Now, we recall the definition of a core of a closed operator A and we provide a useful sufficient condition guaranteeing that a set E is a core of A, when A is the infinitesimal generator of a strongly continuous semigroup.

Definition C.1.13 *Let $A : D(A) \subset X \to X$ be a closed linear operator. A subspace E of $D(A)$ is a core of A if E is dense in $D(A)$ in the graph norm.*

Proposition C.1.14 *Let $A : D(A) \subset X \to X$ be the generator of a strongly continuous semigroup of contractions $\{S(t)\}$ in X. Moreover, let E be a subspace of $D(A)$ dense in X such that $S(t)(E) \subset E$ for any $t > 0$. Then, E is a core of A.*

To conclude this section, we state the famous Trotter-Kato theorem and a remarkable result on the spectrum of the generator of an irreducible, positive and strongly continuous semigroup.

Theorem C.1.15 (Trotter-Kato) *Let $\{T(t)\}$, $\{T_n(t)\}$ ($n \in \mathbb{N}$) be strongly continuous semigroups of contractions in X, with infinitesimal generators $A : D(A) \subset X \to X$ and $A_n : D(A_n) \subset X \to X$, respectively. Further, let E be a core of the operator A such that $E \subset D(A_n)$ for any $n \in \mathbb{N}$. If $A_n x$ converges to Ax in X for any $x \in E$, as n tends to $+\infty$, then $R(\lambda, A_n)x$ tends to $R(\lambda, A)x$, as n tends to $+\infty$, for any $\lambda \in \mathbb{C}$ with positive real part.*

Theorem C.1.16 ([19], Prop. C-III.3.5) *Let A be the infinitesimal generator of an irreducible and positive semigroup on a Banach lattice X. If the spectral bound $s(A)$ of A is a pole of $R(\cdot, A)$, then its algebraic and geometric multiplicities are equal to one. Moreover, the kernel of $s(A)I - A$ is spanned by a positive eigenvector of A.*

C.1.1 On the closure of the sum of generators of semigroups

Here, we briefly recall some results on the sum of closed operators. Throughout the subsection we make the following hypotheses.

Hypotheses C.1.17 $A : D(A) \subset X \to X$ is a closed operator such that $-A$ is the infinitesimal generator of a strongly continuous semigroup of contractions in X and $0 \in \rho(A)$.

The analyticity of the function $\lambda R(\lambda, -A)$ in $\rho(A)$ implies that $\rho(A)$ contains a sector $\Sigma_\varphi := \{\lambda \in \mathbb{C} \setminus \{0\} : |\arg \lambda| < \varphi\}$ for some $\varphi \in [0, \pi)$ and that there exists a positive constant C_φ such that $\|\lambda R(\lambda, -A)\|_{L(X)} \le C_\varphi$ for any $\lambda \in \Sigma_\varphi$. Therefore, one can introduce the *spectral angle* φ_A of A, defined by $\varphi_A = \inf\{\varphi > 0 : \Sigma_{\pi-\varphi} \subset \rho(-A) \text{ and } C_\varphi < +\infty\}$. Moreover, for $s \in \mathbb{R} \setminus \{0\}$, one can define the operator A^{is} through the Komatsu formula

$$A^{is}x = -\frac{\sin(i\pi s)}{\pi}\left\{i\frac{x}{s} + \frac{1}{1+is}A^{-1}x + \int_0^1 \lambda^{is+1}R(-\lambda, A)A^{-1}x\,d\lambda \right.$$
$$\left. + \int_1^{+\infty} \lambda^{is-1}R(-\lambda, A)Ax\,d\lambda\right\}$$

for any $x \in D(A) \cap R(A)$ (see e.g. [122, 140, 141]). A^{is} is a closed and densely defined operator for any $s \neq 0$ but, in general, is not bounded. If the operators A^{is} are bounded for any $s \neq 0$ and the map $s \mapsto A^{is}$ ($A^0 := I$) is locally bounded in $\mathbb{R}$, then we say that A admits bounded imaginary powers or simply that $A \in \mathcal{BIP}(X)$.

If $A \in \mathcal{BIP}(X)$, then the limit $\theta_A := \lim_{|s|\to+\infty} |s|^{-1} \log(\|A^{is}\|_{L(X)})$ is well defined and it is called the *power angle* of A.

We summarize in the following proposition a few properties of the class $\mathcal{BIP}(X)$. For a proof, we refer the reader to e.g. [62, Theorem 5.8], [63], [224] and [225, Theorem 3].

Proposition C.1.18 *The following properties are satisfied:*

(i) *if $A \in \mathcal{BIP}(X)$, then $\omega I + A \in \mathcal{BIP}(X)$ for any $\omega > 0$ and $\theta_{\omega I + A} = \theta_A$;*

(ii) *if $A \in \mathcal{BIP}(X)$, then $\theta_A \ge \varphi_A$;*

(iii) *if $A : D(A) \subset L^p(\mathbb{R}^N, \mu) \to L^p(\mathbb{R}^N, \mu)$ (μ being a σ-finite measure, and $p \in (1, +\infty)$) is the generator of a strongly continuous semigroup of contractions $\{T(t)\}$ such that $T(t)f \ge 0$ for any $f \ge 0$, then, for any $\omega > 0$, $\omega I - A \in \mathcal{BIP}(L^p(\mathbb{R}^N, \mu))$ and $\theta_{\omega I - A} \le \pi/2$;*

(iv) *if A is the realization in $L^p(\mathbb{R}^N)$ ($p \in (1,+\infty)$) of the uniformly elliptic operator $\mathcal{A}u = \mathrm{Tr}(QD^2u)$ with constant coefficients, then $\theta_{\omega I - A} = 0$ for any $\omega > 0$;*

(v) *if X is a Hilbert space and $A : D(A) \subset X \to X$ is a self-adjoint operator such that $-A$ generates a strongly continuous semigroup of contractions in X, then $A \in \mathcal{BIP}(X)$ and $\theta_A = 0$.*

Property (iii) in Proposition C.1.18 is known as the *transference principle*. Property (iv) holds for a more general class of elliptic operators in divergence form than those that we consider here. We refer the reader to [225, Theorem C] for more details.

We now introduce a class of Banach spaces, the so-called ξ-convex spaces.

Definition C.1.19 *A Banach space X is said to be ξ-convex if there exists a function $\xi : X \times X \to \mathbb{R}$ such that: (i) ξ is convex with respect to both the variables, (ii) $\xi(0,0) > 0$, (iii) $\xi(x+y) \le |x+y|$ for any $x, y \in \partial B_1$.*

Examples of ξ-convex Banach spaces are the L^p-spaces ($p \in [1,+\infty)$) with respect to σ-finite measures in $\mathbb{R}^N$. In particular $L^p(\mathbb{R}^N)$ is a ξ-convex Banach space for any $p \in [1,+\infty)$. For further results on the operators which admit bounded imaginary powers and on ξ-convex Banach spaces, we refer the reader to [250, Chapter 1, Section 15], [89, Appendix], the survey paper [46] and the monograph [222].

To conclude this subsection we state two theorems which guarantee the closedness of the sum of sectorial operators which admit bounded imaginary powers. The first one deals with the sum of n operators A_i commuting in the resolvent sense[2]. It extends a well known result by G. Dore and A. Venni (see [89]).

Theorem C.1.20 ([225], Corollary 4) *Let X be a ξ-convex space and let the operators $A_i : D(A_i) \subset X \to X$ ($i = 1, \ldots, N$) belong to $\mathcal{BIP}(X)$ and commute in the resolvent sense. Further, suppose that the power angles θ_i satisfy $\theta_i + \theta_j < \pi$ for any $i, j = 1, \ldots, n$, $i \neq j$, and there exists only one index j such that $\theta_j = \max_{i=1,\ldots,n} \theta_i$. Then, the operator $A = \sum_{i=1}^n A_i$, with domain $D(A) = \bigcap_{i=1}^n D(A_i)$, is closed, belongs to $\mathcal{BIP}(X)$ and there exists a positive constant C such that $\sum_{i=1}^n \|A_i x\|_X \le C\|Ax\|_X$ for any $x \in D(A)$.*

Theorem C.1.21 ([207], Corollary 2) *Let X be a ξ-convex Banach space and let $A, B \in \mathcal{BIP}(X)$. Further, suppose that the power angles θ_A and θ_B satisfy $\theta_A + \theta_B < \pi$ and that there exist $C > 0$, α, β, satisfying $0 \le \alpha < \beta \le 1$, and φ_A, φ_B, with $\varphi_A > \theta_A$, $\varphi_B > \theta_B$ and $\varphi_A + \varphi_B < \pi$, such that*

$$\|AR(\lambda, -A)[A^{-1}, R(\mu, -B)]\|_{L(X)} \le \frac{C}{(1+|\lambda|^{1-\alpha})|\mu|^{1+\beta}} \tag{C.1.3}$$

for any $\lambda, \mu \in \mathbb{C} \setminus \{0\}$ such that $|\arg \lambda| < \pi - \varphi_A$ and $|\arg \mu| < \pi - \varphi_B$. Then, the operator $A + B$ with domain $D(A+B) = D(A) \cap D(B)$ is closed in X.

C.2 Analytic semigroups

In this section we introduce the analytic semigroups and describe their main properties.

[2] i.e., $R(\lambda, A_i)R(\mu, A_j) = R(\mu, A_j)R(\lambda, A_i)$ for any $\lambda \in \rho(A_i)$, $\mu \in \rho(A_j)$ and $i, j = 1, \ldots, N$.

Definition C.2.1 *A linear operator $A : D(A) \subset X \to X$ is said to be sectorial if there exist three constants $\omega \in \mathbb{R}$, $\theta \in (\pi/2, \pi)$, $M > 0$ such that*

$$\begin{cases} (i) & \rho(A) \supset S_{\theta,\omega} = \{\lambda \in \mathbb{C} : \lambda \neq \omega,\ |\arg(\lambda - \omega)| < \theta\}, \\ (ii) & \|R(\lambda, A)\|_{L(X)} \le M|\lambda - \omega|^{-1}, \qquad \lambda \in S_{\theta,\omega}. \end{cases} \tag{C.2.1}$$

For any $t > 0$, conditions (C.2.1) allow us to define a bounded linear operator $T(t)$ on X as follows. For any $r > 0$ and $\eta \in (\pi/2, \theta]$, let $\gamma_{r,\eta}$ be the curve $\{\lambda \in \mathbb{C} : |\arg\lambda| = \eta,\ |\lambda| \ge r\} \cup \{\lambda \in \mathbb{C} : |\arg\lambda| \le \eta,\ |\lambda| = r\}$, oriented counterclockwise. We set $T(0) = I$ and

$$T(t) = \frac{1}{2\pi i}\int_{\omega+\gamma_{r,\eta}} e^{t\lambda} R(\lambda, A)\, d\lambda, \qquad t > 0. \tag{C.2.2}$$

It is easy to check that the integral in (C.2.2) is well defined and is independent of $r > 0$ and $\eta \in (\pi/2, \theta)$.

In the following theorem we summarize the main properties of $\{T(t)\}$.

Theorem C.2.2 *Let A be a sectorial operator and let $T(t)$ be given by (C.2.2) for any $t > 0$. Then, the following properties are satisfied:*

(i) *the family of bounded operators $\{T(t)\}$ is a semigroup;*

(ii) *$T(t)x$ belongs to $D(A^k)$ for any $t > 0$, $x \in X$ and $k \in \mathbb{N}$. Moreover, if x belongs to $D(A^k)$ then $A^kT(t)x = T(t)A^kx$ for any $t \ge 0$;*

(iii) *there exists a constant $M > 0$ such that $\|T(t)\|_{L(X)} \le Me^{\omega t}$ for any $t > 0$, where ω is the number in (C.2.1). Moreover, for any $\varepsilon > 0$ and $k \in \mathbb{N}$, there exists a positive constant $C_{k,\varepsilon}$ such that $\|t^kA^kT(t)\|_{L(X)} \le C_{k,\varepsilon}e^{(\omega+\varepsilon)t}$ for any $t > 0$;*

(iv) *the function $t \mapsto T(t)$ is analytic in $(0+\infty)$ with values in $L(X)$ and $D_t^kT(t) = A^kT(t)$ for any $t > 0$ and $k \in \mathbb{N}$.*

We can now give the following definitions.

Definition C.2.3 *Let A be a sectorial operator. The semigroup $\{T(t)\}$ defined by (C.2.2) is called the analytic semigroup generated by A (in X).*

Definition C.2.4 *A semigroup of bounded linear operators $\{T(t)\}$ in X is said to be norm-continuous in the interval $I \subset [0, +\infty)$ if the map $t \mapsto T(t)$ is continuous in I with respect to the topology of $L(X)$.*

Since any analytic semigroup, is norm-continuous by Theorem C.2.2(iv), thanks to Theorem C.2.2(iii) we can write $R(\lambda, A) = \int_0^{+\infty} e^{-\lambda t}T(t)dt$ for any $\lambda \in \mathbb{C}$ with $\mathrm{Re}\lambda > \omega$, where the integral converges in the topology of $L(X)$.

In the following proposition we exploit the behaviour of the function $t \mapsto T(t)x$ at $t = 0$.

Proposition C.2.5 *The following properties are satisfied:*

(i) *the function $T(\cdot)x$ admits limit as t tends to 0 from the right if and only if $x \in \overline{D(A)}$. In such a case the limit is x;*

(ii) *the function $T(\cdot)x$ is differentiable at $t = 0$ if and only if $x \in D(A)$ and $Ax \in \overline{D(A)}$. In such a case, the derivative at zero of the function $T(\cdot)x$ is Ax.*

Remark C.2.6 Let $x \in X$ and set $u := T(\cdot)x$. Theorem C.2.2 and Proposition C.2.5 show that u belongs to $C([0,+\infty);X)$ if and only if $x \in \overline{D(A)}$. Similarly, u belongs to $C^1([0,+\infty);X)$ if and only if $x \in D(A)$ and $Ax \in \overline{D(A)}$. Since $u' = Au$ in $(0,+\infty)$, we conclude that $u \in C([0,+\infty);D(A))$ (i.e., both u and Au belong to $C([0,+\infty);X)$) if and only if $x \in D(A)$ and $Ax \in \overline{D(A)}$, whereas $u \in C^1([0,+\infty);D(A))$ if and only if $x \in D(A^2)$ and $A^2x \in \overline{D(A)}$.

As a straightforward consequence of Proposition C.2.5(i) we deduce that $\{T(t)\}$ fails to be strongly continuous in X unless $D(A)$ is dense in X.

Useful criteria to establish if an operator $A : D(A) \subset X \to X$ is sectorial are given by the following theorems, whose proofs can be found e.g. in [95, 178, 212, 247].

Theorem C.2.7 *Let $A : D(A) \subset X \to X$ be a linear operator such that $\rho(A) \supset \{\lambda \in \mathbb{C} : \mathrm{Re}\lambda \geq K\}$ for some $K \geq 0$, and $\|\lambda R(\lambda,A)\|_{L(X)} \leq M$ for any $\lambda \in \mathbb{C}$, with $\mathrm{Re}\lambda \geq K$, and some $M \geq 1$. Then, A is sectorial.*

Theorem C.2.8 *Let $\{T(t)\}$ be a strongly continuous semigroup in X such that the function $t \mapsto T(t)x$ is differentiable in $(0,+\infty)$ with values in X for any $x \in X$ and there exists a positive constant M such that $t\|T'(t)\|_{L(X)} \leq M$ for any $t \in (0,1)$. Then, $\{T(t)\}$ is analytic.*

Theorem C.2.9 *Let $A : D(A) \subset X \to X$ be a closed operator with $\overline{D(A)} = X$. If λ belongs to $\mathbb{C} \setminus \overline{r(A)}$ then the operator $\lambda - A$ is one-to-one and has closed range. Further, if Σ is a component of $\mathbb{C} \setminus \overline{r(A)}$ which intersects the resolvent set of A, then the spectrum of A is contained in $\mathbb{C} \setminus \Sigma$ and*

$$\|R(\lambda,A)\|_{L(X)} \leq \frac{1}{d(\lambda,\overline{r(A)})}, \qquad \lambda \in \mathbb{C} \setminus \Sigma.$$

Theorem C.2.10 *Let X be a Hilbert space and let $A : D(A) \subset X \to X$ be a self-adjoint dissipative operator. Then, A is sectorial in X and satisfies conditions (C.2.1) with $\omega = 0$ and arbitrary $\theta < \pi$.*

Theorem C.2.11 *Suppose that H and V are Hilbert spaces such that $V \hookrightarrow H$. Further, let $\mathfrak{a} : V \times V \to \mathbb{C}$ be a continuous and coercive form. Define the operator A by setting*

$$\begin{cases} D(A) = \{v \in V : \exists y \in H \text{ s.t. } \mathfrak{a}(v,w) = (y,w)_H \text{ for any } w \in V\}, \\ Av = f, \end{cases}$$

where $(\cdot,\cdot)_H$ denotes the inner product in H. Then, the operator $-A$ generates a strongly continuous analytic semigroup in H.

Now we observe that the estimate in Theorem C.2.2 can be sharpened. To state the next result, we need to introduce the definition of (semi)simple eigenvalue.

Definition C.2.12 *An isolated eigenvalue λ_0 of the sectorial operator $A : D(A) \subset X \to X$ is said to be a semisimple eigenvalue if it is a simple pole of the resolvent operator $\lambda \mapsto R(\lambda,A)$. A semisimple eigenvalue of A is called a simple eigenvalue if its algebraic multiplicity is one.*

Theorem C.2.13 *Let $A : D(A) \subset X \to X$ be a sectorial operator and let $\{T(t)\}$ be the associated analytic semigroup. Denote by ω_A the supremum of the real parts of all the elements of $\sigma(A)$ and suppose that $\sigma(A) \cap \{\lambda \in C : \mathrm{Re}\lambda = \omega_A\}$ is a spectral set consisting of a finite number of semisimple eigenvalues. Then, for any $n \in \mathbb{N} \cup \{0\}$ there exists a positive constant M_n such that $\|t^n A^n T(t)\|_{L(X)} \leq M_n e^{\omega_A t}$ for any $t > 0$.*

Next, we recall some classical perturbation results for analytic semigroups.

Theorem C.2.14 *Let $A : D(A) \subset X \to X$ and $B : D(B) \subset X \to X$ be, respectively, a sectorial operator and a linear operator such that (i) $D(A) \subset D(B)$, (ii) there exist $\theta \in (0,1)$ and $C > 0$ such that $\|Bx\| \le C\|x\|_{D(A)}^{\theta}\|x\|_X^{1-\theta}$ for any $x \in D(A)$. Then, $A+B : D(A+B) := D(A) \to X$ is sectorial. Moreover, if A satisfies* (C.2.1) *with $\theta > 3\pi/4$, then the operator $-A^2$ is sectorial as well.*

We now introduce the interpolation spaces $D_A(\alpha,\infty)$ ($\alpha \in (0,1)$), which play a crucial role in the theory of analytic semigroups.

Definition C.2.15 *For any $\alpha \in (0,1)$ we set $D_A(\alpha,\infty) = (X, D(A))_{\alpha,\infty}$ (see Section* B.3) *and we call it the interpolation space of order α.*

As the following proposition shows, the function $t \mapsto t^{-\alpha}(T(t)x - x)$ is bounded near $t = 0$ if and only if $x \in D_A(\alpha,\infty)$.

Proposition C.2.16 *The interpolation space $D_A(\alpha,\infty)$ is the set of all the $x \in X$ such that*

$$[[x]]_{D_A(\alpha,\infty)} := \sup_{0<t\le 1} t^{-\alpha}\|T(t)x - x\|_X < +\infty.$$

Moreover, the norm $x \mapsto \|x\|_X + [[x]]_{D_A(\alpha,\infty)}$ is equivalent to the norm $\|\cdot\|_{(X,D(A))_{\alpha,\infty}}$ defined in (B.3.1). *Equivalently, $D_A(\alpha,\infty)$ can be defined as the set of all the $x \in X$ such that*

$$[[[x]]]_{D_A(\alpha,\infty)} := \sup_{0<\xi\le 1} \|\xi^{1-\theta} AT(\xi)x\|_X < +\infty.$$

Moreover, the norm $x \mapsto \|x\|_X + [[[x]]]_{D_A(\alpha,\infty)}$ is equivalent to the norm $\|\cdot\|_{(X,D(A))_{\alpha,\infty}}$.

As a consequence of Theorem C.2.2 and Proposition C.2.16 it follows that the function $u = T(\cdot)x$ belongs to $C^\alpha([0,T];X)$ for some (and hence any) $T > 0$ if and only if $x \in D_A(\alpha,\infty)$. Similarly, $u \in C^\alpha([0,T];D(A))$ (or, equivalently, $u \in C^{1+\alpha}([0,T];X)$) if and only if $x \in D_A(1+\alpha,\infty)$, where $D_A(1+\alpha,\infty) := \{x \in D(A) : Ax \in D_A(\alpha,\infty)\}$.

To conclude this section, let us consider the abstract Cauchy problem

$$\begin{cases} u'(t) = Au(t) + f(t), & t \in (0,T], \\ u(0) = x, \end{cases} \tag{C.2.3}$$

where $f : [0,T] \to X$ is a given continuous function and $x \in X$.

Definition C.2.17 *A function $u : [0,T] \to X$ is a strict (resp. classical) solution to* (C.2.3) *if $u \in C^1([0,T];X) \cap C([0,T];D(A))$ (resp. $u \in C^1((0,T];X) \cap C((0,T];D(A)) \cap C([0,T];X)$) and satisfies the Cauchy problem* (C.2.3).

If f belongs to $C([0,T];X)$ then the Cauchy problem (C.2.3) admits at most a unique strict (resp. classical) solution, which is given by the variation-of-constants formula

$$u(t) = T(t)x + \int_0^t T(t-s)f(s)ds, \qquad t \in [0,T]. \tag{C.2.4}$$

Whenever the integral in (C.2.4) makes sense (which is the case if $f \in C([0,T];X)$), the function u defined by (C.2.4) is said to be the *mild solution* of the abstract Cauchy problem (C.2.3).

Appendix D

PDE's and analytic semigroups

In this appendix we recall some classical results on PDE's of elliptic and parabolic type that we use in this book. If not otherwise specified, by $\mathcal{A}$ we denote the non-autonomous elliptic operator defined on smooth functions $\psi : \Omega \to \mathbb{R}$ by

$$\mathcal{A}(t)\psi(x) = \sum_{i,j=1}^{N} q_{ij}(t,x)D_{ij}\psi(x) + \sum_{i=1}^{N} b_i(t,x)D_i\psi(x) + c(t,x)\psi(x),$$

with real coefficients q_{ij}, b_i $(i,j = 1,\ldots,N)$ and c defined in the closure of an open set $I \times \Omega$, where I is an interval. We do not assume any smoothness assumption on the domain $\Omega \subset \mathbb{R}^N$ unless it is explicitly mentioned.

Throughout this appendix, we assume that $q_{ij} \equiv q_{ji}$ for any $i,j = 1,\ldots,N$ and $\langle Q(t,x)\xi,\xi\rangle \geq \kappa_0|\xi|^2$ for any $(t,x) \in \overline{I \times \Omega}$, $\xi \in \mathbb{R}^N$ and some positive constant κ_0.

D.1 *A priori* estimates

Here, $\Omega \subset \mathbb{R}^N$ is a connected bounded open set or even $\mathbb{R}^N$. We begin with the elliptic case and, for this reason, we assume that the coefficients of the operator $\mathcal{A}$ are independent of t. We recall well-known interior L^p-estimates and L^p-estimates up to a part of the boundary of Ω, where homogeneous Dirichlet or Neumann boundary conditions are prescribed.

Theorem D.1.1 ([116], Theorems 9.11 and 9.19) *Fix $p \in (1,+\infty)$. Then, the following properties are satisfied:*

(i) *if the coefficients of the operator $\mathcal{A}$ are bounded and continuous in Ω, then for any open set $\Omega' \Subset \Omega$ there exists a positive constant C, depending on p, Ω, Ω', κ_0 and the moduli of continuity of the coefficients q_{ij} in Ω' $(i,j = 1,\ldots,N)$, such that $\|u\|_{W^{2,p}(\Omega')} \leq C(\|u\|_{L^p(\Omega)} + \|\mathcal{A}u\|_{L^p(\Omega)})$ for any $u \in L^p(\Omega) \cap W^{2,p}_{\rm loc}(\Omega)$ such that $\mathcal{A}u \in L^p(\Omega)$;*

(ii) *if the coefficients of the operator $\mathcal{A}$ are locally Lipschitz continuous in Ω, $u \in W^{2,p}_{\rm loc}(\Omega)$ and $\mathcal{A}u \in W^{1,p}_{\rm loc}(\Omega)$, then $u \in W^{3,p}_{\rm loc}(\Omega)$ and for any $\Omega' \Subset \Omega'' \Subset \Omega$ there exists a positive constant $\widetilde{C}$, depending on p, Ω', Ω'' and the coefficients of the operator $\mathcal{A}$, such that $\|u\|_{W^{3,p}(\Omega')} \leq \widetilde{C}(\|u\|_{L^p(\Omega'')} + \|\mathcal{A}u\|_{W^{1,p}(\Omega'')})$.*

Theorem D.1.2 *Let $p \in (1,+\infty)$ and let Ω be an open set of class C^2. Moreover, let Ω_0 and Ω_1 be two bounded subsets of Ω, with Ω_1 of class C^2, such that their boundaries intersect the boundary of Ω and $\text{dist}(\Omega_0, \Omega \setminus \Omega_1) > 0$. Further, assume that the coefficients of the operator $\mathcal{A}$ are continuous in $\overline{\Omega}$. Then, there exists a positive constant C, depending on $p, N, \mu, \Omega_0, \Omega_1$, the sup-norm of the coefficients of the operator $\mathcal{A}$ and the moduli of continuity of the diffusion coefficients in Ω_1, such that $\|u\|_{W^{2,p}(\Omega_0)} \leq C(\|\mathcal{A}u\|_{L^p(\Omega_1)} + \|u\|_{L^p(\Omega_1)})$ for any function $u \in W^{2,p}(\Omega)$ with normal derivative which vanishes on $\partial\Omega \cap \partial\Omega_1$.*

We also recall the well-known Calderón-Zygmund and Nash inequalities that we extensively use in Chapters 7 and 8. We refer the reader, e.g., to [116, Theorem 9.19], [209] and [228, Chapter 3, Section 3] for the proofs.

Theorem D.1.3 *The following properties hold true.*

(i) *There exists a positive constant $C = C(N,p)$ such that*

$$\|D^2u\|_{L^p(\mathbb{R}^N)} \le C(\|u\|_{L^p(\mathbb{R}^N)} + \|\Delta u\|_{L^p(\mathbb{R}^N)}), \qquad u \in W^{2,p}(\mathbb{R}^N). \tag{D.1.1}$$

(ii) *Let $\Omega \subset \mathbb{R}^N$ be an open set (possibly $\Omega = \mathbb{R}^N$). Then, there exists a positive constant C_N such that*

$$\|u\|_{L^2(\Omega)}^{1+2/N} \le C_N \|u\|_{L^1(\Omega)}^{2/N} \|\nabla u\|_{L^2(\Omega)}, \qquad u \in W_0^{1,2}(\Omega) \cap L^1(\Omega). \tag{D.1.2}$$

To conclude the analysis of the elliptic case, we present some results concerning the regularity of distributional solutions to autonomous elliptic equations. They are essentially due to S. Agmon (see [4] and [5], in the case when $p = 2$), who deals with the case when $\mathbb{R}^N$ is replaced with a bounded set Ω. Since, as far as we know they seem not to be written, at least in the most used textbooks, in the form that we need, for the reader's convenience we provide a proof of them, which is a slight variant of the one due to G. Metafune, D. Pallara, and A. Rhandi.

Theorem D.1.4 *Let $p \in (1,+\infty)$, denote by p' the conjugate exponent to p (i.e. $1/p' = 1 - 1/p$) and set $\mathcal{A} = \mathrm{Tr}(QD^2) + \langle b, \nabla\rangle$. Then, the following properties are satisfied:*

(i) *if $q_{ij} \in C_b^1(\mathbb{R}^N)$, $b_i \in C_b(\mathbb{R}^N)$ $(i,j = 1,\ldots,N)$ and $u \in L^p(\mathbb{R}^N)$ satisfies the estimate*

$$\left|\int_{\mathbb{R}^N} u\mathcal{A}\varphi dx\right| \le C\|\varphi\|_{W^{1,p'}(\mathbb{R}^N)}, \qquad \varphi \in C_c^\infty(\mathbb{R}^N), \tag{D.1.3}$$

then u belongs to $W^{1,p}(\mathbb{R}^N)$;

(ii) *if $q_{ij} \in C^1(\mathbb{R}^N)$, $b_i \in C(\mathbb{R}^N)$ $(i,j = 1,\ldots,N)$ and $u \in L^p_{\rm loc}(\mathbb{R}^N)$ satisfies* (D.1.3) *for any $\varphi \in C_c^\infty(\mathbb{R}^N)$, then $u \in W^{1,p}_{\rm loc}(\mathbb{R}^N)$;*

(iii) *if $q_{ij} \in C_b^2(\mathbb{R}^N)$, $b_i \in C_b^1(\mathbb{R}^N)$ $(i,j = 1,\ldots,N)$, $\mathrm{div}\, b \in L^p(\mathbb{R}^N)$ and $f, u \in L^p(\mathbb{R}^N)$ are such that*

$$\int_{\mathbb{R}^N} u\mathcal{A}\varphi dx = \int_{\mathbb{R}^N} f\varphi dx, \qquad \varphi \in C_c^\infty(\mathbb{R}^N), \tag{D.1.4}$$

then u belongs to $W^{2,p}(\mathbb{R}^N)$. If $\mathcal{A} = \mathrm{div}(Q\nabla) + \langle b, \nabla\rangle$ then the same result holds true under the weaker assumptions $q_{ij}, b_i \in C_b^1(\mathbb{R}^N)$ $(i,j = 1,\ldots,N)$;

(iv) *if $q_{ij} \in C^2(\mathbb{R}^N)$, $b_i \in C^1(\mathbb{R}^N)$ $(i,j = 1,\ldots,N)$ and $f, u \in L^p_{\rm loc}(\mathbb{R}^N)$ satisfy the formula* (D.1.4) *for any $\varphi \in C_c^\infty(\mathbb{R}^N)$, then u belongs to $W^{2,p}_{\rm loc}(\mathbb{R}^N)$. If $\mathcal{A} = \mathrm{div}(Q\nabla) + \langle b, \nabla\rangle$ then the same result holds true under the weaker assumptions $q_{ij}, b_i \in C^1(\mathbb{R}^N)$ $(i,j = 1,\ldots,N)$.*

Proof (i) Let $A_1 : W^{2,p'}(\mathbb{R}^N) \to L^{p'}(\mathbb{R}^N)$ be the operator defined by $A_1 v = \mathrm{Tr}(QD^2 v)$ for any $v \in W^{2,p'}(\mathbb{R}^N)$. From (D.1.3) we deduce that

$$\left|\int_{\mathbb{R}^N} uA_1\varphi dx\right| \le C\|\varphi\|_{W^{1,p'}(\mathbb{R}^N)}, \qquad \varphi \in C_c^\infty(\mathbb{R}^N). \tag{D.1.5}$$

For any $h \in \mathbb{R}^N$, set $\tau_h u = |h|^{-1}(u(\cdot + h) - u)$. An explicit computation shows that

$$\int_{\mathbb{R}^N} \tau_h u\, A_1\varphi\, dx = \int_{\mathbb{R}^N} uA_1(\tau_{-h}\varphi)\, dx + \int_{\mathbb{R}^N} u(\tau_{-h}q_{ij})D_{ij}\varphi(\cdot - h)dx$$

for any $\varphi \in C_c^\infty(\mathbb{R}^N)$. Using the boundedness of the diffusion coefficients q_{ij} in $\mathbb{R}^N$ $(i,j = 1,\ldots,N)$ and (D.1.5), we obtain

$$\left|\int_{\mathbb{R}^N} \tau_h u(\lambda\varphi - A_1\varphi)dx\right| \le C_1\|\varphi\|_{W^{2,p'}(\mathbb{R}^N)} \tag{D.1.6}$$

for any $\varphi \in C_c^\infty(\mathbb{R}^N)$, any $\lambda > 0$ and some positive constant $C_1 = C_1(\lambda)$, independent of φ and h. Since $u \in L^p(\mathbb{R}^N)$, by density (D.1.6) can be extended to any function in $W^{2,p'}(\mathbb{R}^N)$. We fix a large λ so that the operator $\lambda - A_1$ is invertible from $W^{2,p'}(\mathbb{R}^N)$ to $L^{p'}(\mathbb{R}^N)$ (see the forthcoming Theorem D.3.5(i)) and set $\varphi = R(\lambda, A_1)(\tau_h u|\tau_h u|^{p-2})$. The function φ satisfies the estimate

$$\|\varphi\|_{W^{2,p'}(\mathbb{R}^N)} \le C_2\|\tau_h u\|_{L^p(\mathbb{R}^N)}^{p-1} \tag{D.1.7}$$

for some positive constant C_2, independent of h. Plugging this particular φ into (D.1.6) and taking (D.1.7) into account, we obtain that $\|\tau_h u\|_{L^p(\mathbb{R}^N)} \le C_3$, with C_3 being independent of h. From this estimate it follows that $u \in W^{1,p}(\mathbb{R}^N)$.

(ii) Let us fix $r > 0$ and let $\varphi, \psi_r \in C_c^\infty(\mathbb{R}^N)$ be two smooth functions with $\chi_{B_r} \le \psi_r \le \chi_{B_{2r}}$. Since

$$\psi_r \mathcal{A}\varphi = \mathcal{A}(\psi_r\varphi) - \varphi\mathcal{A}\psi_r - 2\langle Q\nabla\varphi, \nabla\psi_r\rangle, \tag{D.1.8}$$

it follows that

$$\int_{\mathbb{R}^N} u\psi_r\mathcal{A}\varphi\, dx = \int_{\mathbb{R}^N} u\mathcal{A}(\varphi\psi_r)dx - \int_{\mathbb{R}^N} u\varphi\mathcal{A}\psi_r dx - 2\int_{\mathbb{R}^N} u\langle Q\nabla\varphi, \nabla\psi_r\rangle dx.$$

Hence, from (D.1.3) we infer that

$$\left|\int_{\mathbb{R}^N} u\psi_r\mathcal{A}\varphi dx\right| \le C\|\varphi\|_{W^{1,p'}(\mathbb{R}^N)}$$

for some positive constant, independent of φ.

Now, we consider a function $\eta \in C_c^\infty(\mathbb{R}^N)$ such that $\chi_{B_{2r}} \le \eta \le \chi_{B_{4r}}$ and the operator $\widetilde{\mathcal{A}} = \mathrm{Tr}(\widetilde{Q}D^2) + \langle b, \nabla\rangle$, where $\widetilde{q}_{ij} = \eta q_{ij} + (1-\eta)\delta_{ij}$ and $\widetilde{b}_i = \eta b_i$ for any $i,j = 1,\ldots,N$. Clearly,

$$\int_{\mathbb{R}^N} u\psi_r\mathcal{A}\varphi dx = \int_{\mathbb{R}^N} u\psi_r\widetilde{\mathcal{A}}\varphi dx.$$

The operator $\widetilde{\mathcal{A}}$ is uniformly elliptic in $\mathbb{R}^N$ and has the diffusion and the drift coefficients belonging, respectively, to $C_b^1(\mathbb{R}^N)$ and to $C_b(\mathbb{R}^N)$. Hence, we can apply property (i) to the function $u\psi_r$, obtaining that $u\psi_r \in W^{1,p}(\mathbb{R}^N)$ and, consequently, that $u \in W^{1,p}(B_r)$. The arbitrariness of $r > 0$ implies that $u \in W^{1,p}_{\mathrm{loc}}(\mathbb{R}^N)$.

(iii) We first assume that $\mathcal{A}$ is in nondivergence form. By property (i), u belongs to $W^{1,p}(\mathbb{R}^N)$. Therefore, taking (D.1.4) into account, we get

$$\int_{\mathbb{R}^N} uA_1\varphi dx = \int_{\mathbb{R}^N} f_1\varphi dx, \qquad \varphi \in C_c^\infty(\mathbb{R}^N), \tag{D.1.9}$$

where A_1 is the same operator as in the proof of (i) and $f_1 = f + u\,\mathrm{div}\, b + \langle b, \nabla u\rangle$ belongs to $L^p(\mathbb{R}^N)$. Hence,

$$\int_{\mathbb{R}^N} u(\lambda\varphi - A_1\varphi)dx = \int_{\mathbb{R}^N} g\varphi dx, \qquad \varphi \in C_c^\infty(\mathbb{R}^N), \tag{D.1.10}$$

where $g = \lambda u - f_1$ belongs to $L^p(\mathbb{R}^N)$. By density, we can extend (D.1.10) to any function $\varphi \in W^{2,p'}(\mathbb{R}^N)$. Now, we fix λ in the resolvent sets of both the operator A_1 and its adjoint A_1^* defined by $A_1^* u = \mathrm{Tr}(QD^2u) + 2\mathrm{div}(Q\nabla u) + \sum_{i,j=1}^N (D_{ij}q_{ij})u$ for any $u \in D(A_1^*) = W^{2,p}(\mathbb{R}^N)$ (see again the forthcoming Theorem D.3.5(i)). To prove that $u \in W^{2,p}(\mathbb{R}^N)$, we show that $u = R(\lambda, A_1^*)g$. For this purpose, it suffices to observe that (D.1.10) is satisfied if we insert $R(\lambda, A_1^*)g$ in place of u and this shows that $\int_{\mathbb{R}^N} z(\lambda\varphi - A_1\varphi)dx = 0$ for any $\varphi \in W^{2,p'}(\mathbb{R}^N)$, where $z = R(\lambda, A_1^*)g - u$. Since $\lambda I - A_1$ is surjective from $W^{2,p'}(\mathbb{R}^N)$ into $L^{p'}(\mathbb{R}^N)$, it follows that $z = 0$.

In the case when $\mathcal{A}$ is in divergence form, repeating the same computations as above (with the obvious changes), we can easily check that the same result holds true also in the case when q_{ij} $(i, j = 1, \ldots, N)$ belong to $C_b^1(\mathbb{R}^N)$. Indeed, in the above proof the second-order derivatives of the diffusion coefficients q_{ij} $(i, j = 1, \ldots, N)$ appeared only in the definition of the adjoint operator A_1^*. In the case when $\mathcal{A}$ is in divergence form, the operator A_1^* is given by $A_1^* u = \mathrm{div}(Q\nabla)$ for any $u \in D(A_1^*) = W^{2,p}(\mathbb{R}^N)$ and, therefore, there we do not need to require the existence of the second-order derivatives of the diffusion coefficients.

(iv) We first consider the case when $\mathcal{A}$ is not in divergence form. According to property (ii), u belongs to $W^{1,p}_{\rm loc}(\mathbb{R}^N)$. Therefore, also in this situation, u satisfies (D.1.9) for any $\varphi \in C_c^\infty(\mathbb{R}^N)$, where now $f_1 \in L^p_{\rm loc}(\mathbb{R}^N)$. Fix $r > 0$, let ψ_r be as in the proof of property (ii) and set $v = u\psi_r$. Taking (D.1.8) (written with A_1 instead of $\mathcal{A}$) into account, it can be shown that

$$\int_{\mathbb{R}^N} (\lambda\varphi - A_1\varphi)v dx = \int_{\mathbb{R}^N} g\varphi dx, \qquad \varphi \in C_c^\infty(\mathbb{R}^N),\ \lambda > 0,$$

where $g := \lambda v + uA_1\psi_r - \psi_r(f + u\,\mathrm{div}\, b + \langle b, \nabla u\rangle) - 2\mathrm{div}(uQ\nabla\psi_r)$ belongs to $L^p(\mathbb{R}^N)$. The function v satisfies the previous formula also if we replace A_1 with the operator $\widetilde{A}_1 = \mathrm{Tr}(\widetilde{Q}D^2)$ when the coefficients $\widetilde{q}_{ij}$ $(i, j = 1, \ldots, N)$ are as in the proof of property (ii). Since $\widetilde{q}_{ij} \in C_b^2(\mathbb{R}^N)$ $(i, j = 1, \ldots, N)$, according to property (iii) v belongs to $W^{2,p}(\mathbb{R}^N)$ so that $u \in W^{2,p}(B_r)$ and the arbitrariness of $r > 0$ allows us to conclude. The case when $\mathcal{A}$ is in divergence form can be addressed as in the last part of the proof of property (iii). □

We now consider the parabolic problems associated with the operator $\mathcal{A}$. For this purpose, we introduce the function $d : [0, +\infty) \times \overline{\Omega} \to \mathbb{R}$ defined by $d(t, x) = \mathrm{dist}(x, \partial\Omega) \wedge \sqrt{t}$ for any $t > 0$ and $x \in \overline{\Omega}$. Moreover, for any function $v : I \times \Omega \to \mathbb{R}$, $m \in \mathbb{N}$ and $\alpha \in (0, 1)$, we set

$$\begin{aligned}[[v]]_{m+\alpha} = & \sup_{(t,x)\in(0,T)\times\Omega} |(d(t, x))^m v(t, x)| \\ & + \sup_{\substack{(t,x),(s,y)\in(0,T)\times\Omega \\ (t,x)\neq(s,y)}} (d(t, x) \wedge d(s, y))^{m+\alpha} \frac{|v(t, x) - v(s, y)|}{(|x - y|^2 + |t - s|)^{\alpha/2}}.\end{aligned}$$

Theorem D.1.5 ([107], Theorems 3.5, 3.10, and [166]) *Let $a, b \in \mathbb{R}$ with $a < b$ and assume that the coefficients q_{ij}, b_i $(i, j = 1, \ldots, N)$ and c of the operator $\mathcal{A}$ belong to $C^{\alpha/2,\alpha}((a, b) \times \Omega)$ for some $\alpha \in (0, 1)$. Then, the following properties are satisfied.*

(i) *Let $u \in C^{1+\alpha/2,2+\alpha}_{\rm loc}((a, b) \times \Omega)$ be a bounded (with respect to the sup-norm) solution of the equation $D_t u = \mathcal{A}u$ in $(a, b) \times \Omega$. Then, there exists a positive constant C_1, depending only on the coefficients of the operator $\mathcal{A}$, such that*

$$[[u]]_\alpha + \sum_{i=1}^N [[D_i u]]_{1+\alpha} + \sum_{i,j=1}^N [[D_{ij} u]]_{2+\alpha} + [[D_t u]]_{2+\alpha} \le C_1 \sup_{(0,T)\times\Omega} |u|.$$

In particular, for any open set $\Omega' \Subset \Omega$ and any $a < s < T < b$ there exists a positive constant C_2, depending on s, T, the coefficients of the operator $\mathcal{A}$, Ω, Ω' and T, such that

$$\|u\|_{C^{1+\alpha/2,2+\alpha}([s,T)\times\Omega')} \leq C_2 \sup_{(0,T)\times\Omega} |u|. \tag{D.1.11}$$

If further the coefficients of the operator $\mathcal{A}$ are independent of t, $(a,b) = (0,T)$ and $\text{dist}(\Omega, \Omega') > \sqrt{T}$, then

$$\sup_{(t,x)\in(0,T)\times\Omega'} (\sqrt{t}|\nabla_x u(t,x)| + t|D_x^2 u(t,x)|) \leq C_3 \sup_{(t,x)\in(0,T)\times\Omega} |u(t,x)| \tag{D.1.12}$$

for some positive constant C_3, depending on the coefficients of the operator $\mathcal{A}$, Ω, Ω' and T.

(ii) *Let the coefficients q_{ij}, b_i $(i,j = 1,\ldots,N)$ and c be continuously differentiable with respect to the spatial variables up to the k-th order, for some $k \in \mathbb{N}$, with derivatives in $C^{\alpha/2,\alpha}((a,b)\times\Omega)$. Then, u is continuously differentiable in $(a,b)\times\Omega$ up to the $(k+2)$-th order, with respect to the spatial variables. Moreover, $D_x^k u \in C_{\text{loc}}^{1+\alpha/2,2+\alpha}((a,b)\times\Omega)$ and*

$$\|D_x^k u\|_{C^{1+\alpha/2,2+\alpha}((a+\varepsilon',b-\varepsilon')\times\Omega')} \leq C_4 \sup_{(t,x)\in(\varepsilon,b)\times\Omega} |u(t,x)|$$

for any $\varepsilon, \varepsilon' > 0$ $(2\varepsilon < 2\varepsilon' < b-a)$, any open set $\Omega' \Subset \Omega$ and some positive constant C_4, depending on $\varepsilon, \varepsilon'$, a, b, Ω, Ω', $\|D_x^h q_{ij}\|_{C^{\alpha/2,\alpha}((a,b)\times\Omega)}$, $\|D_x^h b_i\|_{C^{\alpha/2,\alpha}((a,b)\times\Omega)}$ $(i,j = 1,\ldots,N)$ and $\|D_x^h c\|_{C^{\alpha/2,\alpha}((a,b)\times\Omega)}$ for any $h \leq k$. In particular, if the coefficients are independent of t, $(a,b) = (0,T)$ and $\text{dist}(\Omega', \Omega) \geq \sqrt{T}$, then the estimate

$$\sup_{(t,x)\in(0,T)\times\Omega'} t^{\frac{3}{2}}|D_x^3 u(t,x)| \leq C_5 \sup_{(t,x)\in(0,T)\times\Omega} |u(t,x)|$$

holds true with some constant C_5 which depends on the coefficients of the operator $\mathcal{A}$, Ω, Ω' and T.

The next theorem provides us with Schauder estimates up to (a part of) the boundary of Ω.

Theorem D.1.6 ([159], Theorem 4.10.1) *Let Ω be an open subset of $\mathbb{R}^N$ with boundary of class $C^{2+\alpha}$ for some $\alpha \in (0,1)$ and let Ω' and Ω'' be two bounded subsets of Ω such that $\Omega' \subset \Omega'' \subset \Omega$ and $\text{dist}(\Omega', \Omega \setminus \Omega'') > 0$. Moreover, assume that the coefficients q_{ij}, b_i $(i,j = 1,\ldots,N)$ and c of the operator $\mathcal{A}$ belong to $C_{\text{loc}}^{\alpha/2,\alpha}(\overline{I\times\Omega})$. Finally, assume that $u \in C^{1+\alpha/2,2+\alpha}((T_1,T_2)\times\Omega'')$ solves the differential equation $D_t u = \mathcal{A}u$ in $(T_1,T_2)\times\Omega''$ for some $0 \leq T_1 < T_2$. Then, the following properties are satisfied:*

(i) *if u (resp. the normal derivative of u) vanishes on $(T_1,T_2)\times\partial\Omega''$, then, for any $T^* \in (T_1,T_2)$, there exists a positive constant $C = C_{T^*,T_1,T_2,\Omega',\Omega''}$ such that*

$$\|u\|_{C^{1+\alpha/2,2+\alpha}((T^*,T_2)\times\Omega')} \leq C\|u\|_{L^\infty((T_1,T_2)\times\Omega'')}; \tag{D.1.13}$$

(ii) *if $T_1 = T^* = 0$ and u (resp. the normal derivative of u) vanishes on $(0,T_2)\times\partial\Omega''$, then there exists a positive constant $C = C_{T_1,T_2,\Omega',\Omega''}$ such that*

$$\|u\|_{C^{1+\alpha/2,2+\alpha}((0,T_2)\times\Omega')} \leq C(\|u\|_{L^\infty((0,T_2)\times\Omega'')} + \|u(0,\cdot)\|_{C^{2+\alpha}(\Omega'')}).$$

D.2 Classical maximum principles

In this section we collect some classical maximum principles that we use in this book. For the proofs we refer the reader e.g. to [43], [146, Chapter 8] and [221].

Throughout this section we assume that Ω is either an open bounded set with boundary of class C^2 or $\mathbb{R}^N$.

Theorem D.2.1 *Suppose that the coefficients of the operator $\mathcal{A}$ do not depend on t and belong to $C_b(\overline{\Omega})$. Let $\lambda > c_0$ and suppose that $u \in W^{2,p}_{\rm loc}(\Omega)$ for any $p \in (1, +\infty)$ satisfies the differential inequality $\lambda u - \mathcal{A}u \geq 0$. Then, if u (resp. the normal derivative of u) is nonnegative on $\partial\Omega$, then $u \geq 0$ in Ω. Moreover, if $\lambda > c_0 := \sup_{x\in\Omega} c(x)$, $f \in C_b(\overline{\Omega})$ and $u \in \bigcap_{1\leq p<+\infty} W^{2,p}_{\rm loc}(\Omega)$ solves the problem*

$$\begin{cases} \lambda u(x) - \mathcal{A}u(x) = f(x), & x \in \Omega, \\ u(x) = 0, & x \in \partial\Omega, \end{cases} \tag{D.2.1}$$

or

$$\begin{cases} \lambda u(x) - \mathcal{A}u(x) = f(x), & x \in \Omega, \\ \dfrac{\partial u}{\partial \nu}(x) = 0, & x \in \partial\Omega, \end{cases} \tag{D.2.2}$$

then it holds that

$$\|u\|_\infty \leq (\lambda - c_0)^{-1}\|f\|_\infty. \tag{D.2.3}$$

Next proposition deals with solutions to the parabolic equation $D_t u = \mathcal{A}u$.

Proposition D.2.2 *Suppose that the coefficients of the operator $\mathcal{A}$ belong to $C_b(\overline{I \times \Omega})$. Fix $s, T \in I$, with $s < T$. Then, the following properties are satisfied:*

(i) *if $z \in C_b([s,T] \times \overline{\Omega}) \cap C^{1,2}((s,T] \times \Omega)$ (resp. $z \in C_b([s,T] \times \overline{\Omega}) \cap C^{0,1}((s,T] \times \overline{\Omega}) \cap C^{1,2}((s,T] \times \Omega)$) satisfies the inequalities $D_t z - \mathcal{A}z \leq 0$ in $(s,T] \times \Omega$, $z(s,\cdot) \leq 0$ in $\overline{\Omega}$ and z (resp. the normal derivative of z) is nonpositive in $(s,T] \times \partial\Omega$, then, $z \leq 0$ in $[0,T] \times \overline{\Omega}$;*

(ii) *if $z \in C_b([s,T]\times\overline{\Omega}) \cap C^{0,1}((s,T]\times\overline{\Omega}) \cap C^{1,2}((s,T]\times\Omega)$ satisfies the inequality $D_t z - \mathcal{A}z \leq 0$ in $(s,T] \times \Omega$ and attains its maximum value in $(s,T] \times \Omega$ at some point (t_0, x_0), then z is constant in $[s, t_0] \times \Omega$. Further, if z attains its maximum value at some point $(t,x) \in (s,T] \times \partial\Omega$, then the normal derivative $\partial z/\partial\nu$ is positive at (t,x).*

D.3 Existence of classical solutions to PDE's and analytic semigroups

As in all the book, for notational convenience, we set $c_0 = \sup_{x\in\overline{\Omega}} c(x)$.

For the proof of the following results, we refer the reader to [127, 128, 132, 133], [159, Chapter 4, Theorem 5.2] and [248, Chapter 5, Sections 3 and 5]. For simplicity, we confine ourselves to the particular case when the coefficients of the operator $\mathcal{A}$ are independent of t but most of the results that we present hold true, with obvious changes, also in the case when the coefficients depend both on t and x.

Proposition D.3.1 *Suppose that the coefficients of the operator $\mathcal{A}$ belong to $C_b^\alpha(\overline{\Omega})$ and Ω is an open set with a boundary which is uniformly of class $C^{2+\alpha}$ for some $\alpha \in (0,1)$ (or, possibly, $\Omega = \mathbb{R}^N$). Then, for any $f \in C_b(\Omega)$ the Cauchy-Dirichlet problem*

$$\begin{cases} D_t u(t,x) = \mathcal{A}u(t,x), & t > 0, \quad x \in \Omega, \\ u(t,x) = 0, & t > 0, \quad x \in \partial\Omega, \\ u(0,x) = f(x), & \qquad\quad x \in \Omega, \end{cases} \tag{D.3.1}$$

admits a unique solution $u \in C([0,+\infty) \times \overline{\Omega} \setminus (\{0\} \times \partial\Omega)) \cap C^{1,2}((0,+\infty) \times \Omega)$, which is bounded in $(0,T) \times \Omega$ for any $T > 0$. Moreover, it satisfies the estimate

$$\|u(t,\cdot)\|_\infty \le e^{c_0 t}\|f\|_\infty, \qquad t > 0, \tag{D.3.2}$$

and $t^{k/2}\|D_x^k u(t,\cdot)\|_\infty \le C\|f\|_\infty$ for any $t \in (0,T)$, any $T > 0$ and some positive constant $C = C(T)$, independent of f, and any $k = 1,2,3$. Moreover, there exists a unique function $g_\Omega \in C((0,+\infty) \times \Omega \times \Omega)$ such that

$$u(t,x) = \int_\Omega g_\Omega(t,x,y) f(y)dy, \qquad t > 0, \ x \in \Omega.$$

The function g_Ω is called the Green function of problem (D.3.1), *it is positive in $(0,+\infty) \times \Omega \times \Omega$ and*

$$g_\Omega(t+s,x,y) = \int_\Omega g_\Omega(s,x,z) g_\Omega(t,z,y)dz, \qquad s,t > 0, \ x,y \in \Omega. \tag{D.3.3}$$

Moreover, for any $t > 0$ the function $g_\Omega(t,\cdot,\cdot)$ is bounded in $\Omega \times \Omega$ and, for $y \in \Omega$, the function $g_\Omega(\cdot,\cdot,y)$ satisfies the equation $D_t g_\Omega(\cdot,\cdot,y) = \mathcal{A}g_\Omega(\cdot,\cdot,y)$ in $(0,+\infty) \times \Omega$.

Finally, if $f \in C_0(\Omega)$, then $u \in C([0,+\infty) \times \overline{\Omega}) \cap C_{\rm loc}^{1+\alpha/2,2+\alpha}((0,+\infty) \times \overline{\Omega})$; if $f \in C_c^{2+\alpha}(\mathbb{R}^N)$, then $u \in C^{1+\alpha/2,2+\alpha}((0,T) \times \Omega)$ for any $T > 0$, and, if $f \in C_c^{2+k+\alpha}(\Omega)$, $q_{ij}, b_j \in C_b^{k+\alpha}(\overline{\Omega})$ $(i,j = 1,\ldots,N)$ and Ω is of class $C^{2+k+\alpha}$ for some $k \in \mathbb{N}$, then all the spatial derivatives of u up to the k-th order belong to $C^{1+\alpha/2,2+\alpha}((0,T) \times \Omega)$ for any $T > 0$.

Similar results hold for the Cauchy-Neumann problem (D.3.4). Here, we limit ourselves to stating the results that we actually need in this book.

Proposition D.3.2 *Suppose that the coefficients of the operator $\mathcal{A}$ belong to $C_b^\alpha(\overline{\Omega})$ and Ω is an open set with a boundary which is uniformly of class $C^{2+\alpha}$ for some $\alpha \in (0,1)$. Then, for any $f \in C_b(\overline{\Omega})$ there exists a unique solution $u \in C([0,+\infty) \times \overline{\Omega}) \cap C_{\rm loc}^{1+\alpha/2,2+\alpha}((0,+\infty) \times \overline{\Omega})$ to the Cauchy-Neumann problem*

$$\begin{cases} D_t u(t,x) = \mathcal{A}u(t,x), & t > 0, \quad x \in \Omega, \\ \dfrac{\partial u}{\partial \nu}(t,x) = 0, & t > 0, \quad x \in \partial\Omega, \\ u(0,x) = f(x), & \qquad\quad x \in \Omega, \end{cases} \tag{D.3.4}$$

which is bounded in $[0,T] \times \Omega$ for any $T > 0$. Moreover, $\|u(t,\cdot)\|_\infty \le e^{c_0 t}\|f\|_\infty$ for any $t > 0$, and $t^{k/2}\|D_x^k u(t,\cdot)\|_\infty \le C\|f\|_\infty$ for any $t \in (0,T)$, any $T > 0$, any $k = 1,2,3$ and some positive constant $C = C(T)$.

Concerning the elliptic equation $\lambda u - \mathcal{A}u = f$, we state the following results.

Proposition D.3.3 *Suppose that the coefficients of the operator $\mathcal{A}$ belong to $C_b^\alpha(\overline{\Omega})$ and Ω is an open set with a boundary which is uniformly of class $C^{2+\alpha}$ for some $\alpha \in (0,1)$. Then, for any $f \in C_b(\overline{\Omega})$ and $\lambda > c_0$, there exists a unique solution $u \in \bigcap_{1\le p<+\infty} W^{2,p}(\Omega)$ to the Dirichlet problem* (D.2.1)*. The function u can be represented by*

$$u(x) = \int_\Omega K_\lambda^\Omega(x,y) f(y)dy, \qquad x \in \Omega, \tag{D.3.5}$$

where K_λ^Ω, the so-called Green function, is continuous in $\Omega \times \Omega$ and it is given by

$$K_\lambda^\Omega(x,y) = \int_0^{+\infty} e^{-\lambda t} g_\Omega(t,x,y)dt, \qquad x,y \in \Omega, \tag{D.3.6}$$

where g_Ω is the function in Proposition D.3.1.

Proposition D.3.4 *Suppose that the coefficients of the operator $\mathcal{A}$ belong to $C_b^\alpha(\Omega)$ and Ω is an open set with a boundary which is uniformly of class $C^{2+\alpha}$ for some $\alpha \in (0,1)$. Then, for any $f \in C_b(\overline{\Omega})$ and $\lambda > c_0$ there exists a unique solution $u \in \bigcap_{1\le p<+\infty} W^{2,p}(\Omega)$ to the Neumann problem* (D.2.2).

Most of the results of this section may be rephrased in terms of analytic semigroups. In the next theorem, we assume that the coefficients of the operator $\mathcal{A}$ belong to $C_b^{2\alpha}(\Omega)$ and Ω is an open set with a boundary which is uniformly of class $C^{2+2\alpha}$ for some $\alpha \in (0,1)$.

Theorem D.3.5 *The following properties are satisfied:*

(i) *for any $p \in (1,+\infty)$ the operator $A_p : D(A_p) \to L^p(\mathbb{R}^N)$, defined by $A_p u = \mathcal{A}u$ for any $u \in D(A_p) := W^{2,p}(\mathbb{R}^N)$, is sectorial in $L^p(\mathbb{R}^N)$ and $D(A_p)$ is dense in $L^p(\mathbb{R}^N)$. Moreover, there exist M_p and $C_p > 0$ such that $\|D^2u\|_{L^p(\mathbb{R}^N)} \le C_p\|A_p u - \lambda u\|_{L^p(\mathbb{R}^N)}$ for any $u \in W^{2,p}(\mathbb{R}^N)$ and $\lambda \in \mathbb{C}$, with $\mathrm{Re}\lambda \ge M_p$;*

(ii) *let A_p be the operator defined by $A_p u = \mathcal{A}u$ for any $u \in D(A_p) = W^{2,p}(\Omega) \cap W_0^{1,p}(\Omega)$. Then, A_p is sectorial in $L^p(\Omega)$, and $D(A_p)$ is dense in $L^p(\Omega)$;*

(iii) *for any $p \in (1,+\infty)$ let A_p be the operator defined by $A_p u = \mathcal{A}u$ for any $u \in D(A_p) = \{u \in W^{2,p}(\Omega) : \partial u/\partial\nu = 0$ on $\partial\Omega\}$. Then, A_p is sectorial in $L^p(\Omega)$ and $D(A_p)$ is dense in $L^p(\Omega)$. Moreover, $\rho(A) \supset (c_0,+\infty)$;*

(iv) *let $A : D(A) \to C_b(\overline{\Omega})$ be the operator defined by $Au = \mathcal{A}u$ for any $u \in D(A) = \left\{u \in \bigcap_{1\le p<+\infty} W^{2,p}_{\rm loc}(\Omega) : u_{|\partial\Omega} = 0,\ \mathcal{A}u \in C_b(\overline{\Omega})\right\}$. Then, A is sectorial in $C_b(\overline{\Omega})$ and $\overline{D(A)} = C_0(\Omega)$. Moreover, $\rho(A) \supset (c_0,+\infty)$ and*

$$D_A(\alpha,\infty) = \{u \in C_b^{2\alpha}(\overline{\Omega}) : u \in C_0(\Omega)\},$$
$$D_A(1+\alpha,\infty) = \{u \in C_b^{2+2\alpha}(\overline{\Omega}) : u, \mathcal{A}u \in C_0(\Omega)\}.$$

Finally, $C_b^1(\overline{\Omega}) \cap C_0(\Omega)$ belongs to the class $J_{1/2}$ between $C_b(\overline{\Omega})$ and $D(A)$. All the previous results hold also in the case when Ω is replaced with $\mathbb{R}^N$, provided we drop out any boundary condition.

(v) *Let $A : D(A) \to C_b(\overline{\Omega})$ be the operator defined by $Au = \mathcal{A}u$ for any $u \in D(A) = \left\{u \in \bigcap_{1\le p<+\infty} W^{2,p}_{\rm loc}(\Omega) : \frac{\partial u}{\partial \nu} = 0 \text{ on } \partial\Omega,\ u, \mathcal{A}u \in C_b(\overline{\Omega})\right\}$. Then, A is sectorial in*

$C_b(\overline{\Omega})$, its domain is dense in $C(\overline{\Omega})$ and the space $C^1_\nu(\overline{\Omega}) := \{u \in C^1_b(\overline{\Omega}) : \partial u/\partial\nu = 0$ on $\partial\Omega\}$ belongs to the class $J_{1/2}$ between $C_b(\overline{\Omega})$ and $D(A)$. Moreover,

$$D_A(\alpha,\infty) = \begin{cases} C^{2\alpha}_b(\overline{\Omega}), & \alpha \in (0,1/2), \\ C^{2\alpha}_b(\overline{\Omega}) \cap C^1_\nu(\overline{\Omega}), & \alpha \in (1/2,1); \end{cases}$$

$$D_A(1+\alpha,\infty) = \begin{cases} C^{2+2\alpha}_b(\overline{\Omega}) \cap C^1_\nu(\overline{\Omega}), & \alpha \in (0,1/2), \\ \{u \in C^{2+2\alpha}_b(\overline{\Omega}) : u, \mathcal{A}u \in C^1_\nu(\overline{\Omega})\}, & \alpha \in (1/2,1). \end{cases}$$

For the proof we refer the reader e.g. to [2, 5, 6, 7, 48, 49, 243, 244].

Remark D.3.6 Properties (ii) and (iii) in Theorem D.3.5, combined with the results in Section C.2, show that, for any $f \in C_b(\overline{\Omega})$, the classical solution to problem (D.3.1) (resp. (D.3.4)) in Proposition D.3.1 (resp. in Proposition D.3.2) is given by $u = T(\cdot)f$, where $\{T(t)\}$ is the analytic semigroup generated in $C_b(\overline{\Omega})$ by the realization of the operator $\mathcal{A}$ with homogeneous Dirichlet boundary conditions (resp. with homogeneous Neumann boundary conditions).

Similar results hold for the solutions to the elliptic equations considered in Propositions D.3.3 and D.3.4. Namely, the solution u to (D.2.1) (resp. to (D.2.2)) is given by $u = R(\lambda, A)f$, where $R(\lambda, A)$ is the resolvent operator associated with the operator A in Theorem D.3.5(iv) (resp. Theorem D.3.5(v)).

In Chapter 7, we use the following consequence of the previous estimates.

Proposition D.3.7 *Fix $p \in (1,+\infty)$ and let the coefficients of the operator $\mathcal{A}$ be bounded and continuous in $\mathbb{R}^N$, with the diffusion coefficients being α-Hölder continuous for some $\alpha \in (0,1)$. Then, there exists a positive constant ω such that, for any $r > 0$ and $\lambda \in \mathbb{C}$ with $\mathrm{Re}\lambda \geq \omega$,*

$$\|D^2u\|_{L^p(B_r)} \leq M_r(\|\lambda u - \mathcal{A}u\|_{L^p(B_{2r})} + \|u\|_{L^p(B_{2r})}), \tag{D.3.7}$$

for any $u \in W^{2,p}(\mathbb{R}^N)$ and some positive constant M_r, which depends on r, the sup-norm of the coefficients of the operator $\mathcal{A}$, the ellipticity constant, the Hölder seminorm of the diffusion coefficients, but it is independent of u and λ.

Proof Let us fix $r > 0$ and, for any $n \in \mathbb{N}$, set $r_n = 2r(1-2^{-n-1})$. Clearly, $r_0 = r$ and $\lim_{n\to+\infty} r_n = 2r$. Let $\{\varphi_n\}$ be a sequence of smooth functions such that $\chi_{B_{r_n}} \leq \varphi_n \leq \chi_{B_{r_{n+1}}}$, $|\nabla\varphi_n| \leq 2^n cr^{-1}$, $|D^2\varphi_n| \leq 4^nCr^{-2}$ in $\mathbb{R}^N$ for any $n \in \mathbb{N}$ and some positive constant C. Fix $u \in W^{2,p}(\mathbb{R}^N)$ and apply the estimate in Theorem D.3.5(i) to the function $v_n = u\varphi_n$. We get

$$\|D^2v_n\|_{L^p(\mathbb{R}^N)} \leq C_{p,1}(\|\lambda u - \mathcal{A}u\|_{L^p(B_{2r})} + 4^n\|u\|_{L^p(B_{2r})} + 2^n\|\nabla v_{n+1}\|_{L^p(\mathbb{R}^N)}) \tag{D.3.8}$$

for any $\lambda \in \mathbb{C}$ with $\mathrm{Re}\lambda \geq M_p$. The constant appearing in (D.3.8), as well as the constants appearing in all the forthcoming estimates, are all independent of λ, u and n. Using the estimate

$$\|\nabla\psi\|_{L^p(\mathbb{R}^N)} \leq C_N\|\psi\|^{\frac{1}{2}}_{L^p(\mathbb{R}^N)}\|D^2\psi\|^{\frac{1}{2}}_{L^p(\mathbb{R}^N)}, \qquad \psi \in W^{2,p}(\mathbb{R}^N),$$

and recalling that $\|v_{n+1}\|_{L^p(\mathbb{R}^N)} \leq \|u\|_{L^p(B_{2r})}$, we can estimate

$$\|\nabla v_{n+1}\|_{L^p(\mathbb{R}^N)} \leq C_N\varepsilon\|D^2v_{n+1}\|_{L^p(\mathbb{R}^N)} + \frac{C_N}{4\varepsilon}\|u\|_{L^p(B_{2r})}, \qquad \varepsilon > 0. \tag{D.3.9}$$

Plugging (D.3.9) into (D.3.8) and choosing $\varepsilon = 2^{-n-4}(C_{p,1}C_N)^{-1}$, we get

$$\|D^2 v_n\|_{L^p(\mathbb{R}^N)} \leq \frac{1}{16}\|D^2 v_{n+1}\|_{L^p(\mathbb{R}^N)} + C_{p,2}(\|\lambda u - \mathcal{A}u\|_{L^p(B_{2r})} + 4^n\|u\|_{L^p(B_{2r})}).$$

Let us multiply both sides of the previous inequality by 16^{-n} and then sum over $n = 0, \ldots, K$. We obtain

$$\|D^2 v_0\|_{L^p(\mathbb{R}^N)} - \frac{1}{16^{K+1}}\|D^2 v_{K+1}\|_{L^p(\mathbb{R}^N)} \leq C_{p,2}(\|\lambda u - \mathcal{A}u\|_{L^p(B_{2r})} + \|u\|_{L^p(B_{2r})}).$$

The assumptions on φ_n imply that $\|D^2 v_{K+1}\|_{L^p(\mathbb{R}^N)} \leq 4^{K+1}C_{p,3}\|u\|_{W^{2,p}(\mathbb{R}^N)}$ for some positive constant $C_{p,3}$, independent of M. Hence, we obtain (D.3.7) letting K tend to $+\infty$, recalling that $v_0 = u$ in B_r. □

Bibliography

[1] P. Acquistapace. Evolution operators and strong solutions of abstract linear parabolic equations. *Differential Integral Equations*, 1(4):433–457, 1988.

[2] P. Acquistapace and B. Terreni. Hölder classes with boundary conditions as interpolation spaces. *Math. Z.*, 195(4):451–471, 1987.

[3] R.A. Adams. *Sobolev spaces.* Pure and Applied Mathematics, Vol. 65. Academic Press, New York-London, 1975.

[4] S. Agmon. The L_p approach to the Dirichlet problem. I. Regularity theorems. *Ann. Scuola Norm. Sup. Pisa (3)*, 13:405–448, 1959.

[5] S. Agmon. On the eigenfunctions and on the eigenvalues of general elliptic boundary value problems. *Comm. Pure Appl. Math*, 15:119–147, 1962.

[6] S. Agmon, A. Douglis, and L. Nirenberg. Estimates near the boundary for solutions of elliptic partial differential equations satisfying general boundary conditions. I. *Comm. Pure Appl. Math.*, 12:623–727, 1959.

[7] S. Agmon, A. Douglis, and L. Nirenberg. Estimates near the boundary for solutions of elliptic partial differential equations satisfying general boundary conditions. II. *Comm. Pure Appl. Math.*, 17:35–92, 1964.

[8] A. Aibeche, K. Laidoune, and A. Rhandi. Time dependent Lyapunov functions for some Kolmogorov semigroups perturbed by unbounded potentials. *Arch. Math. (Basel)*, 94(6):565–577, 2010.

[9] A.A. Albanese, L. Lorenzi, and E. Mangino. L^p-uniqueness for elliptic operators with unbounded coefficients in $\mathbb{R}^N$. *J. Funct. Anal.*, 256(4):1238–1257, 2009.

[10] A.A. Albanese and E. Mangino. Cores for Feller semigroups with an invariant measure. *J. Differential Equations*, 225(1):361–377, 2006.

[11] A.A. Albanese and E. Mangino. Corrigendum to: "Cores for Feller semigroups with an invariant measure". *J. Differential Equations*, 244(11):2980–2982, 2008.

[12] L. Ambrosio, N. Fusco, and D. Pallara. *Functions of bounded variation and free discontinuity problems.* Oxford Mathematical Monographs. The Clarendon Press, Oxford University Press, New York, 2000.

[13] C. Ané, S. Blachère, D. Chafaï, P. Fougères, I. Gentil, F. Malrieu, C. Roberto, and G. Scheffer. *Sur les inégalités de Sobolev logarithmiques*, volume 10 of *Panoramas et Synthèses*. Société Mathématique de France, Paris, 2000.

[14] L. Angiuli. Pointwise gradient estimates for evolution operators associated with Kolmogorov operators. *Arch. Math. (Basel)*, 101(2):159–170, 2013.

[15] L. Angiuli and L. Lorenzi. Compactness and invariance properties of evolution operators associated with Kolmogorov operators with unbounded coefficients. *J. Math. Anal. Appl.*, 379(1):125–149, 2011.

[16] L. Angiuli, L. Lorenzi, and A. Lunardi. Hypercontractivity and asymptotic behaviour in nonautonomous Kolmogorov equations. *Comm. Partial Differential Equations*, 38(12):2049–2080, 2013.

[17] W. Arendt. Gaussian estimates and interpolation of the spectrum in L^p. *Differential Integral Equations*, 7(5-6):1153–1168, 1994.

[18] W. Arendt. Semigroups and evolution equations: functional calculus, regularity and kernel estimates. In *Evolutionary equations. Vol. I*, Handb. Differ. Equ., pages 1–85. North-Holland, Amsterdam, 2004.

[19] W. Arendt, A. Grabosch, G. Greiner, U. Groh, H. P. Lotz, U. Moustakas, R. Nagel, F. Neubrander, and U. Schlotterbeck. *One-parameter semigroups of positive operators*, volume 1184 of *Lecture Notes in Mathematics*. Springer-Verlag, Berlin, 1986.

[20] W. Arendt, G. Metafune, and D. Pallara. Gaussian estimates for elliptic operators with unbounded drift. *J. Math. Anal. Appl.*, 338(1):505–517, 2008.

[21] W. Arendt and A.F.M. ter Elst. Gaussian estimates for second order elliptic operators with boundary conditions. *J. Operator Theory*, 38(1):87–130, 1997.

[22] D.G. Aronson. Non-negative solutions of linear parabolic equations. *Ann. Scuola Norm. Sup. Pisa (3)*, 22:607–694, 1968.

[23] D.G. Aronson and P. Besala. Parabolic equations with unbounded coefficients. *J. Differential Equations*, 3:1–14, 1967.

[24] J. Assaad and J. van Neerven. L^2-theory for non-symmetric Ornstein-Uhlenbeck semigroups on domains. *J. Evol. Equ.*, 13(1):107–134, 2013.

[25] R. Azencott. Behavior of diffusion semi-groups at infinity. *Bull. Soc. Math. France*, 102:193–240, 1974.

[26] D. Bakry. Transformations de Riesz pour les semi-groupes symétriques. II. Étude sous la condition $\Gamma_2 \geq 0$. In *Séminaire de probabilités, XIX, 1983/84*, volume 1123 of *Lecture Notes in Math.*, pages 145–174. Springer, Berlin, 1985. (French).

[27] D. Bakry, F. Bolley, I. Gentil, and P. Maheux. Weighted Nash inequalities. *Rev. Mat. Iberoam.*, 28(3):879–906, 2012.

[28] D. Bakry and M. Émery. Diffusions hypercontractives. In *Séminaire de probabilités, XIX, 1983/84*, volume 1123 of *Lecture Notes in Math.*, pages 177–206. Springer, Berlin, 1985. (French).

[29] D. Bakry and M. Ledoux. Lévy-Gromov's isoperimetric inequality for an infinite dimensional diffusion generator. *Invent. Math.*, 123(2):259–281, 1996.

[30] S. Bernstein. Sur la généralisation du problème de Dirichlet. Première partie. *Math. Ann.*, 62(2):253–271, 1906. (French).

[31] M. Bertoldi and S. Fornaro. Gradient estimates in parabolic problems with unbounded coefficients. *Studia Math.*, 165(3):221–254, 2004.

[32] M. Bertoldi, S. Fornaro, and L. Lorenzi. Gradient estimates for parabolic problems with unbounded coefficients in non convex unbounded domains. *Forum Math.*, 19(4):603–632, 2007.

[33] M. Bertoldi, S. Fornaro, and L. Lorenzi. Pointwise gradient estimates in exterior domains. *Arch. Math. (Basel)*, 88(1):77–89, 2007.

[34] M. Bertoldi and L. Lorenzi. Estimates of the derivatives for parabolic problems with unbounded coefficients. *Trans. Amer. Math. Soc.*, 357(7):2627–2664, 2005.

[35] W. Bodanko. Sur le problème de Cauchy et les problèmes de Fourier pour les équations paraboliques dans un domaine non bornè. *Ann. Polon. Math.*, 18:79–94, 1966. (French).

[36] V.I. Bogachev, G. Da Prato, and M. Röckner. Existence of solutions to weak parabolic equations for measures. *Proc. London Math. Soc. (3)*, 88(3):753–774, 2004.

[37] V.I. Bogachev, G. Da Prato, and M. Röckner. On parabolic equations for measures. *Comm. Partial Differential equations*, 33(1-3):397–418, 2008.

[38] V.I. Bogachev, G. Da Prato, M. Röckner, and W. Stannat. Uniqueness of solutions to weak parabolic equations for measures. *Bull. Lond. Math. Soc.*, 39(4):631–640, 2007.

[39] V.I. Bogachev, N.V. Krylov, and M. Röckner. Differentiability of invariant measures and of transition probabilities of singular diffusions. *Dokl. Akad. Nauk*, 376(2):151–154, 2001. (Russian).

[40] V.I. Bogachev, N.V. Krylov, and M. Röckner. On regularity of transition probabilities and invariant measures of singular diffusions under minimal conditions. *Comm. Partial Differential Equations*, 26(11-12):2037–2080, 2001.

[41] V.I. Bogachev, M. Röckner, and S.V. Shaposhnikov. Global regularity and estimates for solutions of parabolic equations. *Teor. Veroyatn. Primen.*, 50(4):652–674, 2005. (Russian). Translation in *Theory Probab. Appl.* 50(4):561-581, 2006.

[42] V.I. Bogachev, M. Röckner, and S.V. Shaposhnikov. Estimates for the densities of stationary distributions and transition probabilities of diffusion processes. *Teor. Veroyatn. Primen.*, 52(2):240–270, 2007. (Russian). Translation in *Theory Probab. Appl.* 52(2):209-236, 2008.

[43] J.-M. Bony. Principe du maximum dans les espaces de Sobolev. *C. R. Acad. Sci. Paris Sér. A-B*, 265:A333–A336, 1967. (French).

[44] M. Bramanti, G. Cupini, E. Lanconelli, and E. Priola. Global L^p estimates for degenerate Ornstein-Uhlenbeck operators. *Math. Z.*, 266(4):789–816, 2010.

[45] M. Bramanti, G. Cupini, E. Lanconelli, and E. Priola. Global L^p estimates for degenerate Ornstein-Uhlenbeck operators with variable coefficients. *Math. Nachr.*, 286(11-12):1087–1101, 2013.

[46] D.L. Burkholder. Martingales and Fourier analysis in Banach spaces. In *Probability and analysis (Varenna, 1985)*, volume 1206 of *Lecture Notes in Math.*, pages 61–108. Springer, Berlin, 1986.

[47] O. Calin, D.-C. Chang, K. Furutani, and C. Iwasaki. *Heat kernels for elliptic and sub-elliptic operators. Methods and techniques.* Applied and Numerical Harmonic Analysis. Birkhäuser/Springer, New York, 2011.

[48] S. Campanato. Generation of analytic semigroups, in the Hölder topology, by elliptic operators of second order with Neumann boundary condition. *Matematiche (Catania)*, 35(1-2):61–72 (1983), 1980.

[49] S. Campanato. Generation of analytic semigroups by elliptic operators of second order in Hölder spaces. *Ann. Scuola Norm. Sup. Pisa Cl. Sci. (4)*, 8(3):495–512, 1981.

[50] P. Cannarsa and V. Vespri. Generation of analytic semigroups by elliptic operators with unbounded coefficients. *SIAM J. Math. Anal.*, 18(3):857–872, 1987.

[51] P. Cannarsa and V. Vespri. Generation of analytic semigroups in the L^p topology by elliptic operators in $\mathbb{R}^n$. *Israel J. Math.*, 61(3):235–255, 1988.

[52] E.A. Carlen, S. Kusuoka, and D.W. Stroock. Upper bounds for symmetric Markov transition functions. *Ann. Inst. H. Poincaré Probab. Statist.*, 23(2, suppl.):245–287, 1987.

[53] S. Cerrai. A Hille-Yosida theorem for weakly continuous semigroups. *Semigroup Forum*, 49(3):349–367, 1994.

[54] S. Cerrai. Some results for second order elliptic operators having unbounded coefficients. *Differential Integral Equations*, 11(4):561–588, 1998.

[55] S. Cerrai. *Second order PDE's in finite and infinite dimension. A probabilistic approach*, volume 1762 of *Lecture Notes in Mathematics*. Springer-Verlag, Berlin, 2001.

[56] C. Chicone and Y. Latushkin. *Evolution semigroups in dynamical systems and differential equations*, volume 70 of *Mathematical Surveys and Monographs*. American Mathematical Society, Providence, RI, 1999.

[57] R. Chill, , E. Fašangová, G. Metafune, and D. Pallara. The sector of analyticity of nonsymmetric submarkovian semigroups generated by elliptic operators. *C. R. Math. Acad. Sci. Paris*, 342(12):909–914, 2006.

[58] R. Chill, E. Fašangová, G. Metafune, and D. Pallara. The sector of analyticity of the Ornstein-Uhlenbeck semigroup on L^p spaces with respect to invariant measure. *J. London Math. Soc. (2)*, 71(3):703–722, 2005.

[59] A. Chojnowska-Michalik and B. Goldys. Nonsymmetric Ornstein-Uhlenbeck semigroups as second quantized operator. *J. Math. Kyoto Univ.*, 36(3):481–498, 1996.

[60] A. Chojnowska-Michalik and B. Goldys. On regularity properties of nonsymmetric Ornstein-Uhlenbeck semigroup in L^p spaces. *Stochastics Stochastics Rep.*, 59(3-4):183–209, 1996.

[61] A. Chojnowska-Michalik and B. Goldys. Symmetric Ornstein-Uhlenbeck semigroups and their generators. *Probab. Theory Related Fields*, 124(4):459–486, 2002.

[62] P. Clemént and J. Prüss. Completely positive measures and Feller semigroups. *Math. Ann.*, 287(1):73–105, 1990.

[63] R.R. Coifman and G. Weiss. *Transference methods in analysis*. American Mathematical Society, Providence, R.I., 1976. Conference Board of the Mathematical Sciences Regional Conference Series in Mathematics, No. 31.

[64] G. Cupini and S. Fornaro. Maximal regularity in $L^p(\mathbb{R}^N)$ for a class of elliptic operators with unbounded coefficients. *Differential Integral Equations*, 17(3-4):259–296, 2004.

[65] G. Da Prato. Perturbation of the Ornstein-Uhlenbeck semigroups. *Rend. Istit. Mat. Univ. Trieste*, 28(suppl.):101–126, 1997.

[66] G. Da Prato. Regularity results for some degenerate parabolic equations. *Riv. Mat. Univ. Parma*, 6(2*):245–257 (2000), 1999.

[67] G. Da Prato. A new regularity result for Ornstein-Uhlenbeck generators and applications. *J. Evol. Equ.*, 3(3):485–498, 2003.

[68] G. Da Prato. Optimal regularity results in spaces of Hölder continuous functions for some infinite dimensional Ornstein-Uhlenbeck semigroup. *Atti Accad. Naz. Lincei Cl. Sci. Fis. Mat. Natur. Rend. Lincei (9) Mat. Appl.*, 21(1):15–31, 2010.

[69] G. Da Prato and A. Debussche. 2D stochastic Navier-Stokes equations with a time-periodic forcing term. *J. Dynam. Differential Equations*, 20(2):301–335, 2008.

[70] G. Da Prato and B. Goldys. Elliptic operators on $\mathbb{R}^d$ with unbounded coefficients. *J. Differential Equations*, 172(2):333–358, 2001.

[71] G. Da Prato, B. Goldys, and J. Zabczyk. Ornstein-Uhlenbeck semigroups in open sets of Hilbert spaces. *C. R. Acad. Sci. Paris Sér. I Math.*, 325(4):433–438, 1997.

[72] G. Da Prato and P. Grisvard. Sommes d'opérateurs linéaires et équations différentielles opérationelles. *J. Math. Pures Appl. (9)*, 54(3):305–387, 1975. (French).

[73] G. Da Prato and A. Lunardi. On the Ornstein-Uhlenbeck operator in spaces of continuous functions. *J. Funct. Anal.*, 131(1):94–114, 1995.

[74] G. Da Prato and A. Lunardi. Elliptic operators with unbounded drift coefficients and Neumann boundary condition. *J. Differential Equations*, 198(1):35–52, 2004.

[75] G. Da Prato and A. Lunardi. On the Dirichlet semigroup for Ornstein-Uhlenbeck operators in subsets of Hilbert spaces. *J. Funct. Anal.*, 259(10):2642–2672, 2010.

[76] G. Da Prato and A. Lunardi. Ultraboundedness for parabolic equations in convex domains without boundary conditions. *Phys. D*, 239(15):1453–1457, 2010.

[77] G. Da Prato and M. Röckner. A note on evolution systems of measures for time-dependent stochastic differential equations. In *Seminar on Stochastic Analysis, Random Fields and Applications V*, volume 59 of *Progr. Probab.*, pages 115–122. Birkhäuser, Basel, 2008.

[78] G. Da Prato and V. Vespri. Maximal L^p regularity for elliptic equations with unbounded coefficients. *Nonlin. Anal., Ser. A: Theory Methods*, 49(6):747–755, 2002.

[79] G. Da Prato and J. Zabczyk. *Stochastic equations in infinite dimensions*, volume 44 of *Encyclopedia of Mathematics and its Applications*. Cambridge University Press, Cambridge, 1992.

[80] G. Da Prato and J. Zabczyk. *Ergodicity for infinite-dimensional systems*, volume 229 of *London Mathematical Society Lecture Note Series*. Cambridge University Press, Cambridge, 1996.

[81] D. Daners. Heat kernel estimates for operators with boundary conditions. *Math. Nachr.*, 217:13–41, 2000.

[82] E.B. Davies. L^1 properties of second order elliptic operators. *Bull. London Math. Soc.*, 17(5):417–436, 1985.

[83] E.B. Davies. Explicit constants for Gaussian upper bounds on heat kernels. *Amer. J. Math.*, 109(2):319–333, 1987.

[84] E.B. Davies. *Heat kernels and spectral theory*, volume 92 of *Cambridge Tracts in Mathematics*. Cambridge University Press, Cambridge, 1989.

[85] E.B. Davies and B. Simon. Ultracontractivity and the heat kernel for Schrödinger operators and Dirichlet Laplacians. *J. Funct. Anal.*, 59(2):335–395, 1984.

[86] J.-D. Deuschel and D.W. Stroock. *Large deviations*, volume 137 of *Pure and Applied Mathematics*. Academic Press, Inc., Boston, MA, 1989.

[87] G. Di Blasio. Linear parabolic evolution equations in L^p-spaces. *Ann. Mat. Pura Appl. (4)*, 138:55–104, 1984.

[88] J. Diestel and J.J. Uhl. *Vector measures*, volume 15 of *Amer. Math. Soc. Math. Surveys*. Amer. Math. Soc., Providence, R.I., 1977.

[89] G. Dore and A. Venni. On the closedness of the sum of two closed operators. *Math. Z.*, 196(2):189–201, 1987.

[90] R.M. Dudley. *Real analysis and probability*, volume 74 of *Cambridge Studies in Advanced Mathematics*. Cambridge University Press, Cambridge, 2002. Revised reprint of the 1989 original.

[91] N. Dunford and J.T. Schwartz. *Linear Operators. I. General Theory.* Pure and Applied Mathematics, Vol. 7. Interscience Publishers, Inc., New York; Interscience Publishers, Ltd., London, 1958.

[92] E.B. Dynkin. *Markov processes. Vols. I, II*, volume 122 of *Die Grundlehren der Mathematischen Wissenschaften, Bände 121*. Academic Press Inc., Publishers, New York; Springer-Verlag, Berlin-Göttingen-Heidelberg, 1965.

[93] E.B. Dynkin. Three classes of infinite-dimensional diffusions. *J. Funct. Anal.*, 86(1):75–110, 1989.

[94] S. Eidelman, S. Kamin, and F. Porper. Uniqueness of solutions of the Cauchy problem for parabolic equations degenerating at infinity. *Asymptot. Anal.*, 22(3-4):349–358, 2000.

[95] K.J. Engel and R. Nagel. *One-parameter semigroups for linear evolution equations*, volume 194 of *Graduate Texts in Mathematics*. Springer-Verlag, New York, 2000.

[96] J.B. Epperson. The hypercontractive approach to exactly bounding an operator with complex Gaussian kernel. *J. Funct. Anal.*, 87(1):1–30, 1989.

[97] S.N. Ethier and T.G. Kurtz. *Markov processes. Characterization and convergence.* Wiley Series in Probability and Mathematical Statistics: Probability and Mathematical Statistics. John Wiley & Sons, Inc., New York, 1986.

[98] L.C. Evans and R.F. Gariepy. *Measure theory and fine properties of functions.* CRC Press, Boca Raton, FL, 1992.

[99] B. Farkas and L. Lorenzi. On a class of hypoelliptic operators with unbounded coefficients in $\mathbb{R}^N$. *Commun. Pure Appl. Anal.*, 8(4):1159–1201, 2009.

[100] G.F. Feissner. *A Gaussian measure analogue to Sobolev's inequality.* PhD thesis, 1972. Thesis (Ph.D.)-Cornell University.

[101] G.F. Feissner. Hypercontractive semigroups and Sobolev's inequality. *Trans. Amer. Math. Soc.*, 210:51–62, 1975.

[102] W. Feller. Diffusion processes in one dimension. *Trans. Amer. Math. Soc.*, 77:1–31, 1954.

[103] S. Fornaro, N. Fusco, G. Metafune, and D. Pallara. Sharp upper bounds for the density of some invariant measures. *Proc. Roy. Soc. Edinburgh Sect. A*, 139(6):1145–1161, 2009.

[104] S. Fornaro and L. Lorenzi. Generation results for elliptic operators with unbounded diffusion coefficients in L^p- and C_b-spaces. *Discrete. Contin. Dyn. Syst.*, 18(4):747–772, 2007.

[105] S. Fornaro and V. Manco. On the domain of some ordinary differential operators in spaces of continuous functions. *Arch. Math. (Basel)*, 82(4):335–343, 2004.

[106] S. Fornaro, G. Metafune, and E. Priola. Gradient estimates for Dirichlet parabolic problems in unbounded domains. *J. Differential Equations*, 205(2):329–353, 2004.

[107] A. Friedman. *Partial differential equations of parabolic type.* Prentice-Hall, Inc., Englewood Cliffs, N.J., 1964.

[108] M. Fuhrman. Analyticity of transition semigroups and closability of bilinear forms in Hilbert spaces. *Studia Math.*, 115(1):53–71, 1995.

[109] M. Fuhrman. Hypercontractivity properties of nonsymmetric Ornstein-Uhlenbeck semigroups in Hilbert spaces. *Stochastic Anal. Appl.*, 16(2):241–260, 1998.

[110] M. Fukushima, Y. Ōshima, and M. Takeda. *Dirichlet forms and symmetric Markov processes*, volume 19 of *de Gruyter Studies in Mathematics.* Walter de Gruyter & Co., Berlin, 1994.

[111] M. Geissert, L. Lorenzi, and R. Schnaubelt. L^p-regularity for parabolic operators with unbounded time-dependent coefficients. *Annali Mat. Pura Appl. (4)*, 189(2):303–333, 2010.

[112] M. Geissert and A. Lunardi. Invariant measures and maximal L^2 regularity for nonautonomous Ornstein-Uhlenbeck equations. *J. Lond. Math. Soc. (2)*, 77(3):719–740, 2008.

[113] M. Geissert and A. Lunardi. Asymptotic behavior and hypercontractivity in nonautonomoous Ornstein-Uhlenbeck equations. *J. Lond. Math. Soc. (2)*, 79(1):85–106, 2009.

[114] M. Giaquinta. *Introduction to regularity theory for nonlinear elliptic systems.* Lectures in Mathematics ETH Zürich. Birkhäuser Verlag, Basel, 1993.

[115] I.I. Gihman and A.V. Skorokhod. *The theory of stochastic processes. II.* Die Grundlehren der Mathematischen Wissenschaften, Band 218. Springer-Verlag, New York-Heidelberg, 1975.

[116] D. Gilbarg and N.S. Trudinger. *Elliptic partial differential equations of second order.* Classics in Mathematics. Springer-Verlag, Berlin, 2001. Reprint of the 1998 edition.

[117] B. Goldys. On analyticity of Ornstein-Uhlenbeck semigroups. *Atti Accad. Naz. Lincei Cl. Sci. Fis. Mat. Natur. Rend. Lincei (9) Mat. Appl.*, 10(3):131–140, 1999.

[118] B. Goldys and M. Kocan. Diffusion semigroups in spaces of continuous functions with mixed topology. *J. Differential Equations*, 173(1):17–39, 2001.

[119] L. Gross. Logarithmic Sobolev inequalities. *Amer. J. Math.*, 97(4):1061–1083, 1975.

[120] L. Gross. Logarithmic Sobolev inequalities and contractivity properties of semigroups. In *Dirichlet forms (Varenna, 1992)*, volume 1563 of *Lecture Notes in Math.*, pages 54–88. Springer, Berlin, 1993.

[121] L. Gross. Hypercontractivity, logarithmic Sobolev inequalities, and applications: a survey of surveys. In *Diffusion, quantum theory, and radically elementary mathematics*, volume 47 of *Math. Notes*, pages 45–73. Princeton Univ. Press, Princeton, NJ, 2006.

[122] M. Haase. *The functional calculus for sectorial operators*, volume 169 of *Operator Theory: Advances and Applications.* Birkhäuser Verlag, Basel, 2006.

[123] M. Hieber, L. Lorenzi, and A. Rhandi. Second-order parabolic equations with unbounded coefficients in exterior domains. *Differential Integral Equations*, 20(11):1253–1284, 2007.

[124] E. Hille and R.S. Phillips. *Functional analysis and semi-groups.* American Mathematical Society, Providence, R. I., 1974. Third printing of the revised edition of 1957, American Mathematical Society Colloquium Publications, Vol. XXXI.

[125] L. Hörmander. Hypoelliptic second order differential equations. *Acta Math.*, 119:147–171, 1967.

[126] N. Ikeda and S. Watanabe. *Stochastic differential equations and diffusion processes*, volume 24 of *North-Holland Mathematical Library.* North-Holland Publishing Co., Amsterdam; Kodansha, Ltd., Tokyo, second edition, 1989.

[127] A.M. Il′in, A.S. Kalashnikov, and O.A. Oleinik. Second-order linear equations of parabolic type. *Uspehi Mat. Nauk*, 17(3 (105)):3–146, 1962. (Russian).

[128] A.M. Il′in, A.S. Kalashnikov, and O.A. Oleinik. Second-order linear equations of parabolic type. *Tr. Semin. im. I. G. Petrovskogo*, (21):9–193, 341, 2001. (Russian). Translation in *J. Math. Sci. (New York)* 108(4):435-542, 2006.

[129] K. Itô and H.P. McKean Jr. *Diffusion processes and their sample paths.* Die Grundlehren der Mathematischen Wissenschaften, Band 125. Academic Press, Inc., Publishers, New York; Springer-Verlag, Berlin-New York, New York, 1965.

[130] S. Itô. Fundamental solutions of parabolic differential equations and boundary value problems. *Jap. J. Math.*, 27:55–102, 1957.

[131] B. Jefferies. Weakly integrable semigroups on locally convex spaces. *J. Funct. Anal.*, 66(3):347–364, 1986.

[132] F. John. General properties of solutions of linear elliptic partial differential equations. In *Proceedings of the Symposium on Spectral Theory and Differential Problems*, pages 113–175. Oklahoma Agricultural and Mechanical College, Stillwater, Okla., 1951.

[133] F. John. *Plane waves and spherical means applied to partial differential equations.* Interscience Publishers, Inc., New York, London, 1955.

[134] I. Karatzas and S.E. Shreve. Brownian motion and stochastic calculus. volume 113 of *Graduate Texts in Mathematics.* Springer-Verlag, New York, second edition, 1991.

[135] T. Kato. *Perturbation theory for linear operators.* Grundlehren der Mathematischen Wissenschaften, Band 132. Springer-Verlag, Berlin-New York, second edition, 1976.

[136] T. Kato. L^p-theory of Schrödinger operators with a singular potential. In *Aspects of Positivity in Functional Analysis (Tübingen (1985))*, volume 122 of *North-Holland Math. Stud.*, pages 63–78. North-Holland, Amsterdam, 1986.

[137] O. Kavian, G. Kerkyacharian, and B. Roynette. Quelques remarques sur l'ultracontractivité. *J. Funct. Anal.*, 11(1):155–196, 1993. (French).

[138] R.Z. Khas'minskii. *Stochastic stability of differential equations*, volume 7 of *Monographs and Textbooks on Mechanics of Solids and Fluids: Mechanics and Analysis.* Sijthoff & Noordhoff, Alphen aan den Rijn-Germantown, Md., 1980.

[139] A. Kolmogorov. Zufällige Bewegungen (zur Theorie der Brownschen Bewegung). *Ann. of Math. (2)*, 35(1):116–117, 1934. (German).

[140] H. Komatsu. Fractional powers of operators. *Pacific J. Math.*, 19:285–346, 1966.

[141] S.G. Krein. *Linear differential equations in Banach space*, volume 29 of *Translations of Mathematical Monographs.* American Mathematical Society, Providence, R.I., 1971.

[142] S.N. Kružhkov, A. Castro, and M. López. Schauder type estimates, and theorems on the existence of the solution of fundamental problems for linear and nonlinear parabolic equations. *Dokl. Akad. Nauk SSSR*, 220:277–280, 1975.

[143] S.N. Kružhkov, A. Castro, and M. López. Schauder bounds and an existence theorem for the solutions of the Cauchy problem for linear and nonlinear parabolic equations. I. *Cienc. Mat. (Havana)*, 1(1):55–76, 1980.

[144] S.N. Kružhkov, A. Castro, and M. López. Schauder estimates and existence theorems for the solutions of the Cauchy problem for linear and nonlinear parabolic equations. II. *Cienc. Mat. (Havana)*, 3(1):37–56, 1982.

[145] N.V. Krylov. *Introduction to the theory of diffusion processes*, volume 142 of *Translations of Mathematical Monographs.* American Mathematical Society, Providence, RI, 1995.

[146] N.V. Krylov. *Lectures on elliptic and parabolic equations in Hölder spaces*, volume 12 of *Graduate Studies in Mathematics.* American Mathematical Society, Providence, R.I., 1996.

[147] N.V. Krylov. Some properties of traces for stochastic and deterministic parabolic weighted Sobolev spaces. *J. Funct. Anal.*, 183(1):1–41, 2001.

[148] N.V. Krylov. *Lectures on elliptic and parabolic equations in Sobolev spaces*, volume 96 of *Graduate Studies in Mathematics*. American Mathematical Society, Providence, RI, 2008.

[149] N.V. Krylov and E. Priola. Elliptic and parabolic second-order PDEs with growing coefficients. *Comm. Partial Differential Equations*, 35(1):1–22, 2010.

[150] M. Krzyżański. Construction et étude de la solution fondamentale de l'équation linéaire du type parabolique dont le dernier coefficient est non borné. *Atti Accad. Naz. Lincei. Rend. Cl. Sci. Fis. Mat. Nat. (8)*, 27:26–30, 113–117, 1959. (French).

[151] M. Krzyżański. Sur la solution fondamentale de l'équation linéaire normale du type parabolique dont le dernier coefficient est non borné. I. *Atti Accad. Naz. Lincei Rend. Cl. Sci. Fis. Mat. Nat. (8)*, 32:326–330, 1962. (French).

[152] M. Krzyżański. Sur la solution fondamentale de l'équation linéaire normale du type parabolique dont le dernier coefficient est non borné. II. *Atti Accad. Naz. Lincei Rend. Cl. Sci. Fis. Mat. Nat. (8)*, 32:471–476, 1962. (French).

[153] M. Krzyżański and A. Szybiak. Construction et étude de la solution fondamentale de l'équation lineaire parabolique aux coefficients non bornés. In *Symposium on the numerical treatment of partial differential equations with real characteristics: Proceedings of the Rome Symposium (28-29-30 January 1959) organized by the Provisional International Computation Centre*, pages 32–36. Libreria Eredi Virgilio Veschi, Rome, 1959. (French).

[154] H. Kunita. *Stochastic flows and stochastic differential equations*, volume 24 of *Cambridge Studies in Advanced Mathematics*. Cambridge University Press, Cambridge, 1997. Reprint of the 1990 original.

[155] M. Kunze, L. Lorenzi, and A. Lunardi. Nonautonomous Kolmogorov parabolic equations with unbounded coefficients. *Trans. Amer. Math. Soc.*, 362(1):169–198, 2010.

[156] M. Kunze, L. Lorenzi, and A. Rhandi. Kernel estimates for nonautonomous Kolmogorov equations with potential term. In *New prospects in direct, inverse and control problems for evolution equations*, volume 10 of *Springer INdAM Ser.*, pages 229–251. Springer, Cham, 2014.

[157] M. Kunze, L. Lorenzi, and A. Rhandi. Kernel estimates for nonautonomous Kolmogorov equations. *Adv. Math.*, 287:600–639, 2016.

[158] S. Kusuoka and D. Stroock. Applications of the Malliavin calculus. III. *J. Fac. Sci. Univ. Tokyo Sect. IA Math.*, 34(2):391–442, 1987.

[159] O.A. Ladyženskaja, V.A. Solonnikov, and N.N. Ural′ceva. *Linear and quasilinear equations of parabolic type*, volume 23 of *Translations of Mathematical Monographs*. American Mathematical Society, Providence, R.I., 1968.

[160] E. Lanconelli and S. Polidoro. On a class of hypoelliptic evolution operators. *Rend. Sem. Mat. Univ. Politec. Torino*, 52(1):29–63, 1994. Partial differential equations, II (Turin, 1993).

[161] E.M. Landis. *Second order equations of elliptic and parabolic type*, volume 171 of *Translations of Mathematical Monographs*. American Mathematical Society, Providence, RI, 1998.

[162] S. Lang. *Linear algebra.* Undergraduate Texts in Mathematics. Springer-Verlag, New York, third edition, 1987.

[163] Y. Latushkin and S. Montgomery-Smith. Evolutionary semigroups and Lyapunov theorems in Banach spaces. *J. Funct. Anal.*, 127(1):173–197, 1995.

[164] M. Ledoux. Remarks on logarithmic Sobolev constants, exponential integrability and bounds on the diameter. *J. Math. Kyoto Univ.*, 35(2):211–220, 1995.

[165] G. Leoni. *A first course in Sobolev spaces*, volume 105 of *Graduate Studies in Mathematics.* American Mathematical Society, Providence, RI, 2009.

[166] G. Lieberman. *Second order parabolic differential equations.* World Scientific Publishing Co., Inc., River Edge, NJ, 1996.

[167] V.A. Liskevich and M.A. Perel′muter. Analyticity of sub-Markovian semigroups. *Proc. Amer. Math. Soc.*, 123(4):1097–1104, 1995.

[168] V.A. Liskevich and Yu.A. Semenov. Some problems on Markov semigroups. In *Schrödinger operators, Markov semigroups, wavelet analysis, operator algebras*, volume 11 of *Math. Top.*, pages 163–217. Akademie Verlag, Berlin, 1996.

[169] V.A. Liskevich and Z. Sobol. Estimates of integral kernels for semigroups associated with second-order elliptic operators with singular coefficients. *Potential Anal.*, 18(4):359–390, 2003.

[170] L. Lorenzi. Estimates of the derivatives for a class of parabolic degenerate operators with unbounded coefficients in $\mathbb{R}^N$. *Ann. Scuola Norm. Sup. Pisa Cl. Sci. (5)*, 4(2):255–293, 2005.

[171] L. Lorenzi. Schauder estimates for degenerate elliptic and parabolic problems with unbounded coefficients in $\mathbb{R}^N$. *Differential Integral Equations*, 18(5):531–566, 2005.

[172] L. Lorenzi. Optimal Hölder regularity for nonautonomous Kolmogorov equations. *Discrete Contin. Dyn. Syst. Ser. S*, 4(1):169–191, 2011.

[173] L. Lorenzi and A. Lunardi. Elliptic operators with unbounded diffusion coefficients in L^2 spaces with respect to invariant measures. *J. Evol. Equ,*, 6(4):691–709, 2006.

[174] L. Lorenzi, A. Lunardi, and R. Schnaubelt. Strong convergence of solutions to nonautonomous Kolmogorov equations. *Proc. Amer. Math. Soc. (to appear).*

[175] L. Lorenzi, A. Lunardi, and A. Zamboni. Asymptotic behavior in time periodic parabolic problems with unbounded coefficients. *J. Differential Equations*, 249(12):3377–3418, 2010.

[176] L. Lorenzi and A. Rhandi. On Schrödinger type operators with unbounded coefficients: generation and heat kernel estimates. *J. Evol. Equ.*, 15(1):53–88, 2015.

[177] L. Lorenzi and A. Zamboni. Cores for parabolic operators with unbounded coefficients. *J. Differential Equations*, 246(7):2724–2761, 2009.

[178] A. Lunardi. *Analytic semigroups and optimal regularity in parabolic problems.* Progress in Nonlinear Differential Equations and their Applications, 16. Birkhäuser Verlag, Basel, 1995.

[179] A. Lunardi. An interpolation method to characterize domains of generators of semigroups. *Semigroup Forum*, 53(3):321–329, 1996.

[180] A. Lunardi. On the Ornstein-Uhlenbeck operator in L^2 spaces with respect to invariant measures. *Trans. Amer. Math. Soc.*, 349(1):155–169, 1997.

[181] A. Lunardi. Schauder estimates for a class of degenerate elliptic and parabolic operators with unbounded coefficients in $\mathbb{R}^N$. *Ann. Scuola Norm. Sup. Pisa Cl. Sci. (4)*, 24(1):133–164, 1997.

[182] A. Lunardi. Schauder theorems for linear elliptic and parabolic problems with unbounded coefficients in $\mathbb{R}^n$. *Studia Math.*, 128(2):171–198, 1998.

[183] A. Lunardi. Compactness and asymptotic behavior in nonautonomous linear parabolic equations with unbounded coefficients in $\mathbb{R}^d$. In *Parabolic problems*, volume 80 of *Progr. Nonlinear Differential Equations Appl.*, pages 447–461. Birkhäuser/Springer Basel AG, Basel, 2011.

[184] A. Lunardi, G. Metafune, and D. Pallara. Dirichlet boundary conditions for elliptic operators with unbounded drift. *Proc. Amer. Math. Soc.*, 133(9):2625–2635, 2005.

[185] J. Maas and J. van Neerven. On the domain of nonsymmetric Ornstein-Uhlenbeck operators in Banach spaces. *Infin. Dimens. Anal. Quantum Probab. Relat. Top.*, 11(4):603–626, 2008.

[186] J. Maas and J. van Neerven. Gradient estimates and domain identification for analytic Ornstein-Uhlenbeck operators. In *Parabolic problems*, volume 80 of *Progr. Nonlinear Differential Equations Appl.*, pages 463–477. Birkhäuser/Springer Basel AG, Basel, 2011.

[187] P. Maheux. New proofs of Davies-Simon's theorems about ultracontractivity and logarithmic Sobolev inequalities related to Nash type inequalities. Technical report, arXiv: https://protect-us.mimecast.com/s/1drmB5uK070DFK, 2006.

[188] P. Malliavin. *Stochastic analysis*, volume 313 of *Grundlehren der Mathematischen Wissenschaften.* Springer-Verlag, Berlin, 1997.

[189] V. Maz'ya. *Sobolev spaces with applications to elliptic partial differential equations*, volume 342 of *Grundlehren der Mathematischen Wissenschaften.* Springer, Heidelberg, augmented edition, 2011.

[190] G. Metafune. L^p-spectrum of Ornstein-Uhlenbeck operators. *Ann. Scuola Norm. Sup. Pisa Cl. Sci. (4)*, 30(1):97–124, 2001.

[191] G. Metafune, E.M. Ouhabaz, and D. Pallara. Long time behavior of heat kernels of operators with unbounded drift terms. *J. Math. Anal. Appl.*, 377(1):170–179, 2011.

[192] G. Metafune, D. Pallara, and E. Priola. Spectrum of Ornstein-Uhlenbeck operators in L^p spaces with respect to invariant measures. *J. Funct. Anal.*, 196(1):40–60, 2002.

[193] G. Metafune, D. Pallara, and A. Rhandi. Global properties of invariant measures. *J. Funct. Anal.*, 223(2):396–424, 2005.

[194] G. Metafune, D. Pallara, and A. Rhandi. Kernel estimates for Schrödinger operators. *J. Evol. Equ.*, 6(3):433–457, 2006.

[195] G. Metafune, D. Pallara, and A. Rhandi. Global properties of transition probabilities of singular diffusions. *Teor. Veroyatn. Primen.*, 54(1):116–148, 2009. Translation in *Theory Probab. Appl.* 54(1):68-96, 2010.

[196] G. Metafune, D. Pallara, and M. Wacker. Compactness properties of Feller semigroups. *Studia Math.*, 153(2):179–206, 2002.

[197] G. Metafune, D. Pallara, and M. Wacker. Feller semigroups on $\mathbb{R}^N$. *Semigroup Forum*, 65(2):159–205, 2002.

[198] G. Metafune and E. Priola. Some classes of non-analytic Markov semigroups. *J. Math. Anal. Appl.*, 294(2):596–613, 2004.

[199] G. Metafune, J. Prüss, A. Rhandi, and R. Schnaubelt. The domain of the Ornstein-Uhlenbeck operator on an L^p-space with invariant measure. *Ann. Sc. Norm. Super. Pisa Cl. Sci. (5)*, 1(2):471–485, 2002.

[200] G. Metafune, J. Prüss, A. Rhandi, and R. Schnaubelt. L^p-regularity for elliptic operators with unbounded coefficients. *Adv. Differential Equations*, 10(10):1131–1164, 2005.

[201] G. Metafune and C. Spina. Kernel estimates for a class of Schrödinger semigroups. *J. Evol. Equ.*, 7(4):719–742, 2007.

[202] G. Metafune and C. Spina. An integration by parts formula in Sobolev spaces. *Mediterr. J. Math.*, 5(3):357–369, 2008.

[203] G. Metafune and C. Spina. Elliptic operators with unbounded diffusion coefficients in L^p spaces. *Ann. Sc. Norm. Super. Pisa Cl. Sci. (5)*, 11(2):303–340, 2012.

[204] G. Metafune and C. Spina. Heat kernel estimates for some elliptic operators with unbounded diffusion coefficients. *Discrete Contin. Dyn. Syst.*, 32(6):2285–2299, 2012.

[205] G. Metafune, C. Spina, and C. Tacelli. Elliptic operators with unbounded diffusion and drift coefficients in L^p spaces. *Adv. Differential Equations*, 19(5-6):473–526, 2014.

[206] G. Metafune, C. Spina, and C. Tacelli. On a class of elliptic operators with unbounded diffusion coefficients. *Evol. Equ. Control Theory*, 3(4):671–680, 2014.

[207] S. Monniaux and J. Prüss. A theorem of Dore-Venni type for noncommuting operators. *Trans. Amer. Math. Soc.*, 349(12):4787–4814, 1997.

[208] C.B. Morrey Jr. *Multiple integrals in the calculus of variations.* Die Grundlehren der mathematischen Wissenschaften, Band 130. Springer-Verlag New York, Inc., New York, 1966.

[209] J. Nash. Continuity of solutions of parabolic and elliptic equations. *Amer. J. Math.*, 80:931–954, 1958.

[210] E. Nelson. The free Markoff field. *J. Functional Analysis*, 12:211–227, 1973.

[211] N. Okazawa. An L^p theory for Schrödinger operators with nonnegative potentials. *J. Math. Soc. Japan*, 36(4):675–688, 1984.

[212] E.M. Ouhabaz. *Analysis of heat equations on domains*, volume 31 of *London Mathematical Society Monographs Series.* Princeton University Press, Princeton, NJ, 2005.

[213] A. Pazy. *Semigroups of linear operators and applications to partial differential equations*, volume 44 of *Applied Mathematical Sciences*. Springer-Verlag, New York, 1983.

[214] K. Petersen. *Ergodic theory*, volume 2 of *Cambridge Studies in Advanced Mathematics*. Cambridge University Press, Cambridge, 1983.

[215] E. Priola. On a class of Markov type semigroups in spaces of uniformly continuous and bounded functions. *Studia Math.*, 136(3):271–295, 1999.

[216] E. Priola. The Cauchy problem for a class of Markov-type semigroups. *Commun. Appl. Anal.*, 5(1):49–75, 2001.

[217] E. Priola. On a Dirichlet problem involving an Ornstein-Uhlenbeck operator. *Potential Anal.*, 18(3):251–287, 2003.

[218] E. Priola. Formulae for the derivatives of degenerate diffusion semigroups. *J. Evol. Equ.*, 6(4):577–600, 2006.

[219] E. Priola. Global Schauder estimates for a class of degenerate Kolmogorov equations. *Studia Math.*, 194(2):117–153, 2009.

[220] E. Priola and J. Zabczyk. Liouville theorems for non-local operators. *J. Funct. Anal.*, 216(2):455–490, 2004.

[221] M.H. Protter and H.F. Weinberger. *Maximum principles in differential equations*. Prentice-Hall, Inc., Englewood Cliffs, N.J., 1967.

[222] J. Prüss. *Evolutionary integral equations and applications*, volume 87 of *Monographs in Mathematics*. Birkhäuser Verlag, Basel, 1993.

[223] J. Prüss, A. Rhandi, and R. Schnaubelt. The domain of elliptic operators on $L^p(\mathbb{R}^d)$ with unbounded drift coefficients. *Houstion J. Math*, 32:563–576, 2006.

[224] J. Prüss and H. Sohr. On operators with bounded imaginary powers in Banach spaces. *Math. Z.*, 203(3):429–452, 1990.

[225] J. Prüss and H. Sohr. Imaginary powers of elliptic second order differential operators in L^p-spaces. *Hiroshima Math. J.*, 23(1):161–192, 1993.

[226] M. Reed and B. Simon. *Methods of modern mathematical physics. I. Functional analysis*. Academic Press, New York-London, 1972.

[227] M. Reed and B. Simon. *Methods of modern mathematical physics. II. Fourier analysis, self adjontness*. Academic Press, New York, London, 1975.

[228] D.W. Robinson. *Elliptic operators and Lie groups*. Oxford Mathematical Monographs. The Clarendon Press, Oxford University Press, New York, 1991. Oxford Science Publications.

[229] M. Röckner and F.-Y. Wang. Supercontractivity and ultracontractivity for (non-symmetric) diffusion semigroups on manifolds. *Forum Math.*, 15(6):893–921, 2003.

[230] O.S. Rothaus. Logarithmic Sobolev inequalities and the spectrum of Schrödinger operators. *J. Funct. Anal.*, 42(1):110–120, 1981.

[231] W. Rudin. *Real and complex analysis*. McGraw-Hill Book Co., New York-Toronto, Ont.-London, 1966.

[232] N. Saintier. Schauder estimates for degenerate elliptic and parabolic equations in $\mathbb{R}^N$ with Lipschitz drift. *Differential Integral Equations*, 20(4):397–428, 2007.

[233] S.V. Shaposhnikov. Lower bounds for the densities of solutions of parabolic equations for measures. *Dokl. Akad. Nauk*, 429(5):600–604, 2009. (Russian). Translation in *Dokl. Math.* 80(3):877881, 2009.

[234] S.V. Shaposhnikov. Regularity and qualitative properties of solutions for parabolic equations for measures. *Teor. Veroyatn. Primen.*, 56(2):318–350, 2011. (Russian). Translation in *Theory Probab. Appl.* 56(2):252-279, 2012.

[235] S.V. Shaposhnikov. The Fokker-Planck-Kolmogorov equations with a potential and a non-uniformly elliptic diffusion matrix. *Trans. Moscow Math. Soc.*, pages 15–29, 2013.

[236] A. Sikora. On-diagonal estimates on Schrödinger semigroup kernels and reduced heat kernels. *Comm. Math. Phys.*, 188(1):233–249, 1997.

[237] B. Simon. *Functional integration and quantum physics*, volume 86 of *Pure and Applied Mathematics*. Academic Press, Inc., New York-London, 1979.

[238] M. Sobajima and C. Spina. Second order elliptic operators with diffusion coefficients growing as $|x|^\alpha$ at infinity. *Forum Math.*, 28(2):391–402, 2016.

[239] C. Spina. Kernel estimates for a class of Kolmogorov semigroups. *Arch. Math. (Basel)*, 91(3):265–279, 2008.

[240] C. Spina. *Kernel estimates for Markov semigroups and parabolic Schrödinger operators.* PhD thesis, Università del Salento, 2008. Available online at the address https://protect-us.mimecast.com/s/VAz4BRUrO3ODuY.

[241] W. Stannat. (Nonsymmetric) Dirichlet operators on L^1: existence, uniqueness and associated Markov processes. *Ann. Scuola Norm. Sup. Pisa Cl. Sci. (4)*, 28(1):99–140, 1999.

[242] W. Stannat. Time-dependent diffusion operators on L^1. *J. Evol. Equ.*, 4(4):463–495, 2004.

[243] H.B. Stewart. Generation of analytic semigroups by strongly elliptic operators. *Trans. Amer. Math. Soc.*, 199:141–162, 1974.

[244] H.B. Stewart. Generation of analytic semigroups by strongly elliptic operators under general boundary conditions. *Trans. Amer. Math. Soc.*, 259(1):299–310, 1980.

[245] D.W. Stroock and S.R.S. Varadhan. *Multidimensional diffusion processes*, volume 233 of *Grundlehren der Mathematischen Wissenschaften*. Springer-Verlag, Berlin-New York, 1979.

[246] K.T. Sturm. Harnack's inequality for parabolic operators with singular low order terms. *Math. Z.*, 216(4):593–611, 1994.

[247] H. Tanabe. *Equations of evolution*, volume 6 of *Monographs and Studies in Mathematics*. Pitman (Advanced Publishing Program), Boston, Mass.-London, 1979.

[248] H. Tanabe. *Functional analytic methods for partial differential equations*, volume 204 of *Monographs and Textbooks in Pure and Applied Mathematics*. Marcel Dekker, Inc., New York, 1997.

[249] A.F.M. ter Elst and D.W. Robinson. Second-order subelliptic operators on Lie groups. II. Real measurable principal coefficients. In *Semigroups of operators: theory and applications (Newport Beach, CA, 1998)*, volume 42 of *Progr. Nonlinear Differential Equations Appl.*, pages 103–124. Birkhäuser, Basel, 2000.

[250] H. Triebel. *Interpolation theory, function spaces, differential operators*, volume 18 of *North-Holland Mathematical Library*. North-Holland Publishing Co., Amsterdam-New York, 1978.

[251] N.S. Trudinger. Linear elliptic operators with measurable coefficients. *Ann. Scuola Norm. Sup. Pisa (3)*, 27:265–308, 1973.

[252] J.M.A.M. van Neerven. Nonsymmetric Ornstein-Uhlenbeck semigroups in Banach spaces. *J. Funct. Anal.*, 155(2):495–535, 1998.

[253] F.-Y. Wang. Logarithmic Sobolev inequalities on noncompact Riemannian manifolds. *Probab. Theory Related Fields*, 109(3):417–424, 1997.

[254] F.-Y. Wang. On estimation of the logarithmic Sobolev constant and gradient estimates of heat semigroups. *Probab. Theory Related Fields*, 108(1):87–101, 1997.

[255] F.-Y. Wang. Functional inequalities for empty essential spectrum. *J. Funct. Anal.*, 170(1):219–245, 2000.

[256] F.-Y. Wang. *Functional inequalities, Markov semigroups and spectral theory.* Science Press, Beijing, 2004.

[257] F.-Y. Wang. Gradient estimates of Dirichlet heat semigroups and application to isoperimetric inequalities. *Ann. Probab.*, 32(1A):424–440, 2004.

[258] A. Wiweger. A topologisation of Saks spaces. *Bull. Acad. Polon. Sci. Cl. III*, 5:773–777, 1957.

[259] J. Wloka. *Partial differential equations.* Cambridge University Press, Cambridge, 1987.

[260] K. Yosida. *Functional analysis.* Die Grundlehren der Mathematischen Wissenschaften, Band 123. Academic Press, Inc., New York; Springer-Verlag, Berlin, 1965.

[261] J. Zabczyk. Parabolic equations on Hilbert spaces. In *Stochastic PDE's and Kolmogorov equations in infinite dimensions (Cetraro, 1998)*, volume 1715 of *Lecture Notes in Math.*, pages 117–213. Springer, Berlin, 1999.

[262] Q.S. Zhang. Gaussian bounds for the fundamental solutions of $\nabla(A\nabla u)+B\nabla u-u_t=0$. *Manuscripta Math.*, 93(3):381–390, 1997.

Index

L

M

N

O

P

Q

R

S

An environmentally friendly book printed and bound in England by www.printondemand-worldwide.com

PEFC Certified

This product is from sustainably managed forests and controlled sources

PEFC/16-33-415

www.pefc.org

This book is made of chain-of-custody materials; FSC materials for the cover and PEFC materials for the text pages.